Mathematics, Informatics, Science and Education International Conference (MISEIC 2018)

Emerging Trends of Research in Mathematics, Informatics, Sciences, and Education

Journal of Physics: Conference Series
Volume 1108

Surabaya, Indonesia
21 July 2018

Editors:

Yuni Sri Rahayu
Rooselyna Ekawati
Nadi Suprapto
Asep Bayu Dani Nandiyanto
Setya Chendra Wibawa
Shofan Fiangga
M. Jakfar

ISBN: 978-1-5108-7714-6
ISSN: 1742-6588

Printed from e-media with permission by:

Curran Associates, Inc.
57 Morehouse Lane
Red Hook, NY 12571

Some format issues inherent in the e-media version may also appear in this print version.

Printed by Curran Associates, Inc. (2019)

For permission requests, please contact the Institute of Physics at the address below.

Institute of Physics
Dirac House, Temple Back
Bristol BS1 6BE UK

Phone: 44 1 17 929 7481
Fax: 44 1 17 920 0979

techtracking@iop.org

Additional copies of this publication are available from:

Curran Associates, Inc.
57 Morehouse Lane
Red Hook, NY 12571 USA
Phone: 845-758-0400
Fax: 845-758-2633
Email: curran@proceedings.com
Web: www.proceedings.com

Table of contents

Volume 1108

Mathematics, Informatics, Science and Education International Conference (MISEIC) 2018

21 July 2018, Surabaya, Indonesia

Accepted papers received: 10 October 2018
Published online: 4 December 2018

Preface

Papers

IOP Conf. Series: Journal of Physics: Conf. Series **1108** (2018) 011001 doi:10.1088/1742-6596/1108/1/011001

2nd Mathematics, Informatics, Science and Education International Conference (MISEIC).

PREFACE

The 2nd Mathematics, Informatics, Science and Education International Conference (MISEIC) is organized by Faculty of Mathematics and Natural Sciences Universitas Negeri Surabaya. This international conference aims to bridge the scientists, education experts and practicioners, and students in the scientific forum through sharing ideas and issues about theoretical and practical knowledge in mathematics, informatics, science and STEM education. The theme for this conference is **"Emerging Trends of Research in Mathematics, Informatics, Sciences, and Education"**.

This conference hosted by the Universitas Negeri Surabaya is considered to accommodate discussion among researchers in the field of mathematics, informatics, science and education as a scientific forum. Therefore, the invited speakers in this conference are experts in the field of mathematics, informatics, science and education.

The 2nd MISEIC took place in Surabaya, Indonesia, on 21 July 2018 involving more than 300 papers are participated from various topics including pure and applied mathematics, science and technology innovation, computer science, and innovation in mathematics, science, and computer science education. There are 129 selected papers that go through a strict peer reviewed process, and these papers will be published in the present conference proceedings.

We would like to thank the organizing committee and the members of reviewers for their kind assistance in reviewing the papers. We would also extend our best gratitude to keynote speakers for their invaluable contribution and worthwhile ideas shared in the conference.

The Editors
Dr. Yuni Sri Rahayu, M.Si. (Universitas Negeri Surabaya, Indonesia)
Rooselyna Ekawati, S.Si., M.Sc., Ph.D. (Universitas Negeri Surabaya. Indonesia)
Nadi Suprapto, S.Pd., M.Pd., Ph.D. (Universitas Negeri Surabaya. Indonesia)
Dr.Eng. Asep Bayu Dani Nandiyanto, S.T., M.Eng. (Universitas Pendidikan Indonesia, Indonesia)
Setya Chendra Wibawa, S.Pd., M.T. (Universitas Negeri Surabaya, Indonesia)
Shofan Fiangga, M.Sc. (Universitas Negeri Surabaya, Indonesia)
M. Jakfar, M.Si. (Universitas Negeri Surabaya, Indonesia)

IOP Conf. Series: Journal of Physics: Conf. Series **1108** (2018) 011001 doi:10.1088/1742-6596/1108/1/011001

CONFERENCE PHOTOGRAPH

IOP Conf. Series: Journal of Physics: Conf. Series **1108** (2018) 011001 doi:10.1088/1742-6596/1108/1/011001

IOP Conf. Series: Journal of Physics: Conf. Series **1108** (2018) 011001 doi:10.1088/1742-6596/1108/1/011001

The Problem

LIST OF COMMITTEE Chairman:
Rooselyna Ekawati, S.Si., M.Sc., Ph.D. (Universitas Negeri Surabaya)

Technical Chairman:
Nadi Suprapto, S.Pd., M.Pd., Ph.D. (Universitas Negeri Surabaya. Indonesia)

Scientific Commitee:
Dr. Yuni Sri Rahayu, M.Si. (Universitas Negeri Surabaya, Indonesia)
Husam Al-Najar, Ph.D. (The Islamic University of Gaza, Palestine)
Dr.ir. W.T. van Horssen* (TU Delft, The Netherlands)
Prof. Pitoyo Hartono* (Chukyo University, Japan)
Prof. Fou-Lai Lin* (National Taiwan Normal University, Taiwan)
Dr. Ade Gafar Abdullah (Universitas Pendidikan Indonesia, Indonesia)
Dr. Eng. Asep Bayu Dani Nandiyanto (Universitas Pendidikan Indonesia, Indonesia)
Dr. Dieky Adzkiya, M.Si (Institut Teknologi 10 Nopember, Indonesia)
Prof. Dr. Sari Edi Cahyaningrum, M.Si (Universitas Negeri Surabaya, Indonesia)
Prof. Dr. Dwi Juniati, M.Si (Universitas Negeri Surabaya, Indonesia)
Dr. Sven Schulte* (University of Bremen, Germany).
Prof. Dr. Muslimin Ibrahim (Universitas Negeri Surabaya, Indonesia)
Prof. George Spottl* (University of Bremen, Germany)
Dr. Yusuf Fuad, M.App.Sc. (Universitas Negeri Surabaya, Indonesia)
Dr. Elly Matul Imah, S.Si, M.Kom. (Universitas Negeri Surabaya, Indonesia)
Prof. Dr. Asep Kadarohman* (Universitas Pendidikan Indonesia, Indonesia)
Dr. I.G.P. Asto Buditjahjanto, S.T., M.T. (Universitas Negeri Surabaya, Indonesia)
Arie Wardhono, Ph.D. (Universitas Negeri Surabaya, Indonesia)
Dr. Abadi, M.Sc (Universitas Negeri Surabaya, Indonesia)
Dr. Manuharawati, M.Si (Universitas Negeri Surabaya, Indonesia)
Prof. Dr. Endang Susantini, M.Pd (Universitas Negeri Surabaya, Indonesia)

Peer review statement

All papers published in this volume of *Journal of Physics: Conference Series* have been peer reviewed through processes administered by the proceedings Editors. Reviews were conducted by expert referees to the professional and scientific standards expected of a proceedings journal published by IOP Publishing.

Cost-Efficient Orchestration of Containers in Clouds: A Vision, Architectural Elements, and Future Directions

Rajkumar Buyya[1], Maria A. Rodriguez[1], Adel Nadjaran Toosi[2], Jaeman Park[3]

[1]Cloud Computing and Distributed Systems (CLOUDS) Lab
School of Computing and Information Systems
The University of Melbourne, Australia

[2]Faculty of Information Technology
Monash University, Australia

[3]Data Cloud Lab, AI Center, Samsung Research
Samsung Electronics, South Korea

Abstract. This paper proposes an architectural framework for the efficient orchestration of containers in cloud environments. It centres around resource scheduling and rescheduling policies as well as autoscaling algorithms that enable the creation of elastic virtual clusters. In this way, the proposed framework enables the sharing of a computing environment between differing client applications packaged in containers, including web services, offline analytics jobs, and backend pre-processing tasks. The devised resource management algorithms and policies will improve utilization of the available virtual resources to reduce operational cost for the provider while satisfying the resource needs of various types of applications. The proposed algorithms will take factors that are previously omitted by other solutions into consideration, including 1) the pricing models of the acquired resources, 2) and the fault-tolerability of the applications, and 3) the QoS requirements of the running applications, such as the latencies and throughputs of the web services and the deadline of the analytical and pre-processing jobs. The proposed solutions will be evaluated by developing a prototype platform based on one of the existing container orchestration platforms.

1. Introduction

Containers, enabling lightweight environment and performance isolation, fast and flexible deployment, and fine-grained resource sharing, have gained popularity in better management and application deployment in addition to hardware virtualization [19][20]. Recently, various container solutions, such as Docker [1] and LXC [2], have gained acceptance by organizations to run various kinds of applications on either proprietary data centers or virtual clusters in Clouds. To efficiently utilize resources, instead of running separate clusters of homogeneous containers, organizations prefer to run containers of different applications on a shared cluster, which drives the creation of modern container orchestration platforms, such as Kubernetes [3], Docker Swarm [4], and Apache Mesos [5]. These platforms provide efficient bin-packing algorithms to schedule containers on the shared clusters. However, their solutions are static and cannot meet the dynamic resource needs of the applications by failing to consider the characteristics of the Cloud environments and applications.

In the context of Cloud, container orchestration platforms are usually deployed on virtual clusters acquired from a single or multiple Cloud data centers. To exploit the elasticity feature of Cloud, they support dynamically adding or removing resources from the cluster. Therefore, organizations can utilize the autoscaling systems provided by Cloud providers or their own solutions to automatically adjust the size of the shared clusters according to their real-time utilization. To save operational cost in Cloud, it is essential to consolidate containers onto as fewer virtual machines as possible because the initial placement may deteriorate along with dynamisms caused by workload fluctuation, application launches and terminations, and pricing variations. To realize this goal, the platform should be able to migrate the containers running on underutilized VMs to other VMs. Using the state-of-the-art techniques, container migration can be implemented in three steps: 1) checkpointing the target container (if necessary) [6], 2) killing it on the original host, and 3) then resuming it on the target host, which is only applicable to stateless and fault-tolerant applications. Therefore, the initial container placement strategy is as important as the dynamic rescheduling algorithm to the success of a container management platform in Cloud.

Currently, Cloud providers offer three types of pricing mechanisms, including reserved, on-demand, and rebated pre-emptive mode. In order to balance cost-efficiency, reliability, and QoS, organizations can provision their clusters dynamically using mixed instances of the three pricing models according to real-time workloads. This requires the orchestration platforms to be aware of the pricing models of the underlying resources, and the characteristics (e.g., ability to tolerate faults/failures) of the applications in order to correctly and wisely schedule and reschedule them on appropriate resources. For example, the orchestration algorithm should never place customer-facing applications on rebated preemptive resources since user experience will be disrupted by preemptions of the VMs. In addition, long-running jobs like web services should be avoided to be deployed on on-demand resources as it will prolong the running time of expensive on-demand resources. On the other hand, fault-tolerant applications are welcomed to run on rebated preemptive resources and on-demand resources as they can tolerate resource failures and can be terminated anytime to free the expensive on-demand resources.

Taking all these factors and requirements into account, the container orchestration problem becomes a challenging task. In summary, it involves three sub-problems: 1) efficiently schedule the containers onto suitable resources in their initial placement attempts so that later migrations are minimized and VMs can be freed easily; 2) dynamically reschedule movable containers to free up more resources while satisfying QoS requirements of the applications; 3) dynamically provision resources of different pricing models to the virtual cluster in order to satisfy growing needs of the applications with minimum cost.

1.1. Research Methodology

The methodology for solving the problem of scheduling, rescheduling and scaling policies for shared container-based virtual clusters in Clouds so that resource utilization is maximized, and resource cost is minimized while meeting QoS requirements of various applications running on the clusters is as follows:

- Define an architectural framework and principles for QoS-aware cluster management for containers.
- Investigate and develop efficient scheduling and rescheduling algorithms and scaling policies for container-based clusters in Clouds.
- Develop techniques that overcome the complexity caused by heterogeneities in pricing, application fault-tolerability, and QoS requirements, and exploit those heterogeneities to improve resource utilization and reduce cost.
- Develop a prototype system incorporating the techniques proposed above.

The rest of this paper is organized as follows: First, a brief survey on the existing literature in container orchestration systems is presented. Next, we present the proposed system architecture and explain its elements in detail, followed by the prototype framework implementation and some preliminary experiments and results to show the potential effectiveness of the proposed approaches. Finally, the paper concludes with a discussion on approaches for realizing future directions.

2. Related Work

There have been considerable attempts to build a reliable and scalable container management platform for clusters from the research and open-source communities. These systems target to ease the deployment and scaling of general-purpose applications packaged using container technologies such as Docker [1] and LXC [2] containers. Many of the solutions have been developed and deployed in very large-scale environments for production purposes by leading companies in various industries either on proprietary infrastructures or on public Clouds.

Kubernetes [3], derived from Google's in-house container management platform Borg [7], is an open-source system that enables automatic deployment and scaling of containerized applications. In addition, it also offers self-healing capacities that kill and replace unresponsive nodes and containers, and automated rollout and rollback features for applications and their configurations. The scheduler of Kubernetes is pluggable, and it provides a default scheduler that in turn schedules each container group (known as pods) on available resources filtered by user-defined requirements and ranked based on individually defined application affinities.

Docker Swarm [4] is the native clustering solution for Docker containers. It provides discovery services and schedules containers onto hosts using pre-defined strategies and filters. By default, it supports two simple strategies: (1) spread strategy, which favors running new containers on least loaded hosts, and (2) BinPack strategy, which selects the most loaded host that have enough resources to run the containers. Docker Swarm also employs a filtering mechanism to map containers to resources. Users can define filters regarding host statuses (e.g., based on available CPU and memory resources and health check results) and container configurations (e.g., resource and file affinities and dependencies to other containers).

Nomad [8] is an enterprise cluster management product from HashiCorp. It not only can deploy containers but also virtual machines and application runtimes like JVM. In addition, it is also capable of operating across multiple data centers and providing high-availability.

Apache Mesos [5] enables cluster sharing among various applications. Different from the previous platforms, Mesos can be viewed as a meta-platform that operates above them. It employs a two-stage scheduling approach. In the first stage, it divides the resources of the cluster and respectively offers available resources to each application, called framework. The framework once accepts the offer can then schedule its tasks on the obtained resources using its own framework scheduler. After that, Mesos actually launches the tasks for the framework on the corresponding hosts. Kubernetes [3] has been integrated into Mesos as a framework to run general-purpose containers. Apache Marathon [9] and Apache Aurora [10] are two other popular general-purpose frameworks built on Mesos. In addition, common data analytics platforms, such as Apache Hadoop [11], Apache Spark [12], and Apache Storm [13], and distributed job schedulers, such as Dkron [14] and Chronos [15], all support running on Mesos. The unique feature of its two-phase scheduling enables Mesos to run frameworks based on their urgency and priority while easing dynamic resource provisioning.

The scheduling and scaling modules of the existing platforms noted above only provide rudimentary strategies that overlook the heterogeneities in pricing models of Cloud resources, fault-tolerability of applications, and QoS requirements of applications. It is essential to take these factors into account to further improve resource utilization and reduce the cost of virtual clusters in public Clouds.

3. Significance

Along with the development and application of new technologies, including IoT, and artificial intelligence, the need for computing power will continue to grow significantly in the future. Cloud computing which has become the backbone of the IT infrastructure is expected to play a major role to satisfy this incoming computing demands. Therefore, minimizing the expenses of leasing Cloud resources becomes ever increasingly important for organizations relying on Clouds to host their applications and support their core business operation.

In addition, with the container management platforms enabling easy and flexible deployment and resource sharing on the same virtual cluster, organizations have been attracted to provision their computing infrastructure in this manner instead of managing separate virtual clusters for different applications.

Our research comes into the center of helping organizations to reduce their Cloud resource leasing expenses by utilizing this new resource management paradigm to improve their profitability and sustainability of businesses. This paper identifies novel resource scheduling/rescheduling and scaling policies that comprehensively consider the pricing mechanisms of Cloud resources, and fault-tolerability of applications to further reduce the cost of running a containerized cluster in Clouds while satisfying the individual QoS requirements of the applications sharing the platform.

4. System Architecture

The paper explores the problem of pricing, fault-tolerability, and QoS-aware scheduling/rescheduling algorithms and scaling policies for containerized clusters in Clouds. An architectural framework for the realization of these goals is shown in Figure 1. There are three main entities involved in the illustrated architecture; the users and applications, the containerized cluster management system, and the cloud resources.

Consumers and Applications: Cloud consumers acquire virtual or physical resources from Cloud data centers located anywhere in the world and compose a shared cluster using container management software. They submit their applications packaged in containers to the cluster to execute. The concept of consumers here refers to the organizations that rely on Clouds as the computing platform to deliver operations. They themselves may target to serve the end "users" submitting requests from terminal devices. An application refers to a set of one or more jobs packaged in one or more containers. Various applications are deployed on a shared platform or cluster. These can be user-facing web services, internal backend real-time services, data analytics jobs, data pre-processing tasks, cron jobs, etc. The cloud consumer should specify some meta-data for each application, such as the detailed QoS requirements in measurable metrics and whether the framework can tolerate failures or not.

Containerized Cluster Manager: Acts as the interface between the obtained virtual/physical resources and application frameworks and manager of the running applications. It requires the interaction of the following components to support cost-efficient and QoS-aware resource management:

Scheduler: Allocates jobs to the cluster resources. This is done while considering various factors, including but not limited to resource pricing, application fault-tolerability, QoS of the applications, resource utilization, and cost.

Autoscaler: If the scheduler finds that there are not enough available resources to allocate to critical jobs, it will interact with the auto-scaler to provide additional resources. The auto-scaler will then decide the type (on-demand, reserved, or rebated) and the amount of resources to be added to the virtual cluster and then provision the new resources through calling the API of the underlying Cloud provider. The autoscaler is also responsible for terminating or migrating containers running on a specific host in order to shut down the host at the end of the host's billing period if the cluster is underutilized. The killed containers will be rescheduled by the scheduler immediately after their termination or be queued and rescheduled later.

Task Launcher: This entity is responsible for launching containers on specific machines and specifying the amount of resources that should be allocated to each container.

Resource Monitor: Monitors the real-time resource consumption such as CPU and memory on each host in the cluster and provides the information to the scheduler to make resource allocation decisions.

Task Monitor: This component is responsible for auditing a running containerized task by recording its resource consumption and monitoring its QoS metrics. This information aides in detecting faults or QoS violations and enables the system to make better scheduling or relocation decisions.

Resource Estimator: This module is used to predict and estimate the amount of resources such as memory and CPU that a container consumes at different points in time. It aims to reduce the framework's reliance on the amount of resources requested by users when submitting their applications. The existence of this component is based on two main assumptions. Firstly, resource requests are usually misestimated, and overestimated, by users. Secondly, the resource consumption of a task is likely to vary over time, with the peak consumption spanning only over a fraction of its lifetime. Both scenarios lead to resources that are reserved but are idle most of the time and hence lead to the cluster being underutilized. By monitoring and estimating the resource consumption of containers, better oversubscription and opportunistic scheduling decisions can be made by the system.

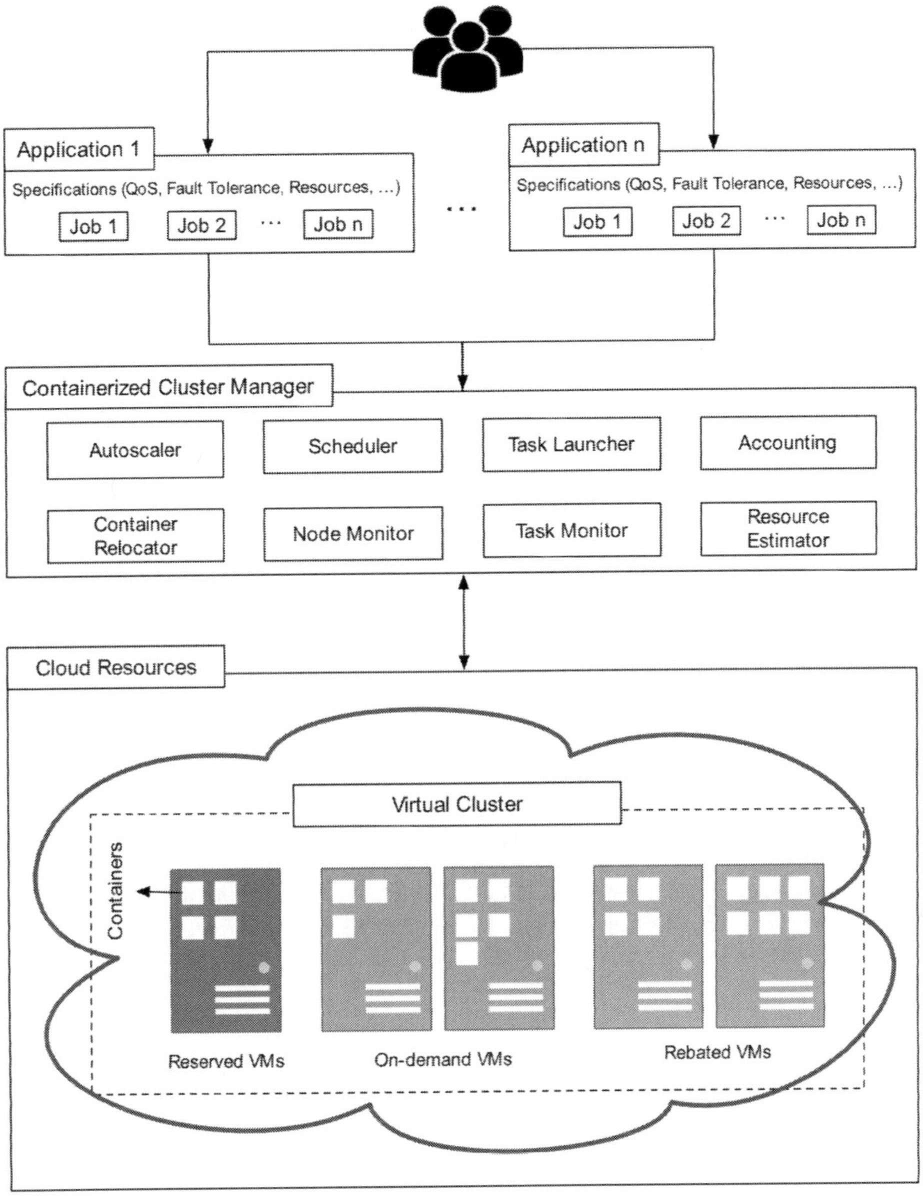

Figure 1: A high-level system architectural framework.

Accounting: Maintains the actual usage of resources by requests to calculate real-time usage costs. Historical usage information can also be used to improve service allocation decisions.

Cloud Resources: Infrastructure as a Service Clouds provide the main compute infrastructure for the deployment of the containerized applications. VMs can be leased on demand and their prices vary based on their capabilities (e.g., amount of CPU and RAM) and their pricing model (e.g., reserved vs. on demand).

5. Architectural Elements, Challenges, and Directions

In this section, we discuss various key elements identified in system architecture along with challenges and issues involved in their realization.

5.1 Initial Container Placement

Since container migration has to be conducted in a stop and resume manner, its feasibility depends on the nature of the applications. For example, a stateless web service container can be straightforwardly killed and restarted somewhere else; a fault-tolerant data analytics job running in a container can be check-pointed and resume its execution remotely. These containers can be flexibly migrated during runtime. On the other hand, an interactive web service with memory states cannot be migrated without affecting the end users; a time-critical map-reduce job cannot afford the time cost of being interrupted. For these containers, the management platform can only release the resources they use after they completed execution. Therefore, without live migration support, the initial placement algorithm needs to be aware of the application characteristics and as much as possible, avoid the co-location of these two types of containers on the same VM in order to maximize the opportunity in the future for migrating all the containers on it and shut it down. Currently, there is no such consideration in the existing container orchestration platforms as containers are not moveable in these systems. Their target is to simply bin-pack the containers onto as few VMs as possible.

Besides, we believe the placement algorithm in Cloud should also be aware of the heterogeneities of the underlying Cloud resources, including different pricing models, locations, and resource types and sizes. Taking the influence of the pricing to the resource provisioning as an example, a customer-facing application should not be placed on unreliable rebated resources, whose sudden termination will disrupt the end user experience, and should be avoided to be placed on on-demand resources, as it prolongs the running time of expensive on-demand resources. To realize these goals, it is required to filter unqualified resources and propose new resource affinity models to rank the resources when provisioning each framework. The newly proposed policies can be implemented as extensions of the existing filtering and affinity ranking mechanisms of the current platforms or as pluggable schedulers.

5.2 Dynamic Rescheduling

The quality of the placement of the containers may degrade as the time passes with workload fluctuations, launchings of containers, and terminations of containers. Therefore, it is necessary to optimize the placement of containers during runtime through migration.

As mentioned earlier, migration is not applicable to all the containers. Thus, the optimization is limited to the VMs whose hosted containers are all moveable. In addition, to fully utilize the acquired VMs, the optimization is conducted at the end of the billing period of each VM. Once the VM is close to the end of the current billing period, the platform judges whether it is possible to terminate all the containers currently running on that VM while not breaking any QoS constraints. If it is feasible, the containers are check-pointed if necessary and killed. The VM is then removed from the virtual cluster and is terminated. The scheduler immediately reacts to the killing of the corresponding containers. If there are free resources, they can be immediately assigned to the killed containers; otherwise, some tasks are queued until other tasks finish and resources are freed.

5.3 Dynamic Virtual Cluster Provisioning

Resource contention can happen when some applications ask for a large amount of resources to meet their QoS requirements or urgent tasks are submitted to the virtual cluster. To meet the dynamic resource needs, it is essential to timely provision new resources. On the other hand, the provisioned resources should be just enough and carefully composed regarding pricing models and sizes to minimize the resource cost. Furthermore, the provisioning policy should also be aware of the application nature in order to satisfy the affinity requirements of the applications and as much as possible separate moveable and non-moveable containers.

To integrate dynamic provisioning capabilities into the existing platforms, it is required to modify the architecture of the platform. The new architecture should be able to provide interfaces to monitor the QoS of each application and estimate the resource needs of the application using individually-defined metrics and models.

5.4 Application QoS Management

It is not unusual for applications to have specific QoS requirements, support for which is limited in existing systems. For instance, long-running services commonly have to serve a minimum amount of request per time unit or have stringent latency requirements. Batch jobs on the other hand can have a deadline as a time constraint for their execution or may need to be completed as fast as possible. For the first scenario, many systems offer a basic autoscaling mechanism. It monitors the CPU utilization of a service, and if a predefined threshold is exceeded, another instance of the service is launched. This however, is a baseline approach to application autoscaling and integrating more sophisticated approaches to container-based management systems is required. For batch jobs, orchestrating them and assigning them to resources so that their QoS are met is another open research area. For example, mapping tasks to resources so that their makespan is minimized is a useful feature lacking in current open source cluster management systems.

5.5 Resource Consumption Estimation

Especially useful when it comes to long-running services, estimating the amount of resources such as memory and CPU that they consume over time can aid in the efficient use of the cluster resources. This mechanism is absent in existing open-source systems, and, to the best of our knowledge, only Google's proprietary platform, Borg [7], has this feature in place. To deal with users overestimating their request for resources when submitting jobs, Borg monitors their resource consumption over time and based on the collected data, predicts their actual resource consumption. Based on this prediction, servers are oversubscribed to run lower-priority tasks that are able utilize the requested but unused resources. Another motivating factor for predicting the resource consumption of long-running services is the fact that it is likely to vary over time, with the peak consumption spanning only over a fraction of its lifetime. If not accounted for, this will also lead to underutilized resources that could be used opportunistically instead.

There are various challenges associated with achieving this goal. Firstly, accurately predicting the resource consumption of jobs is a difficult endeavor. A method capable of incrementally building a model based on data collected over time is essential. Furthermore, such method must be capable of processing streaming data in real time, as this is the nature of resource consumption measurements in a multi-tenant cluster environment. Finally, since the data corresponds to measurements taken in a cloud environment, being able to reflect the dynamicity and performance variability inherent to such platforms is essential. Hence, a method capable of handling concept drift that captures the change in the statistical properties of the collected data is essential. Aside from the analytics and machine learning point of view, successfully using the obtained knowledge to better utilize resources is another challenging area. Integrating the gained knowledge into schedulers and autoscalers must be done while ensuring that the overall system goals such as scalability and job throughput are met while respecting the applications' requirements and QoS goals.

6. Performance Evaluation – Sample Results

In this section, we present a prototype software platform that supports the orchestration of containers in clouds and show some of our preliminary experiment results.

6.1 System Prototype Architecture

We have implemented a system prototype including the components depicted in Figure 2 by extending the Kubernetes (K8s) platform. The Kubernetes framework is designed to manage containerized workloads on clusters with its basic building block being a pod. A pod encapsulates one or more tightly coupled containers that are co-located on the same machine and share the same set of resources; they also encapsulate storage resources, a network IP, and a set of options that govern how the pod container(s) should run. A pod is designed to run a single instance of an application; in this way multiple pods can be used to scale an application horizontally for example. The amount of CPU, memory, and ephemeral storage a container needs can be specified when creating a pod. This information can then be used by the scheduler to make decisions on pod placement.

Kubernetes is a highly mature system; it stemmed from ten years of experience at Google and is the leading container-based cluster management system with an extensive community-driven support and development base. It provides users with a wide range of options for managing their pods and the way in which they are scheduled, even allowing for pluggable customized schedulers to be easily integrated into the system. We take advantage of this feature to build our platform. Furthermore, as of version 1.10, Kubernetes is capable of supporting clusters of up to 5000 hundred nodes [17], which suits the needs of many organizations nowadays.

A custom scheduler interacts with the Kubernetes API server to continuously monitor the state of pods in the cluster. In particular, it focuses on processing pending pods (i.e., those that need to be scheduled). For each pending pod, a set of suitable resources (i.e., Kubernetes workers) is filtered from the entire cluster pool. One of these resources is then selected and a binding between the pod and the resources is created. This binding leads to Kubernetes running the pod on the chosen node. It is worthwhile noticing that different policies to select the set of suitable nodes and assign a task to one of them are easily pluggable into the system.

The custom scheduler interacts with the custom autoscaler when it determines that more resources are needed to place a pending pod. The autoscaler then decides the number and type of VMs to launch and instructs the Cloud Adapter to create the new instances via the specific IaaS cloud provider API. In the meantime, the unschedulable pod can be left in the scheduling queue in a pending state so that it can be scheduled in a later cycle when the newly provisioned (or recently freed) resources become available. The pod can also be removed from the general scheduling queue so that it can be directly assigned to the newly created node once it is available for use.

The custom scheduler can also make use of the Resource Consumption Estimator when making decisions. For instance, a batch job may be placed in a node in which no all the requested resources are predicted to be used. Different analytical techniques can be plugged into the Resource Consumption Estimator to make predictions on the amount of resources consumed by applications. Currently, a simple statistical technique has been implemented in which the median or average CPU and memory usage is estimated based on historical data. Kubernetes provides its own metric monitoring system as default. That consists of multiple components but the main three components are Heapster as a aggregator, InfluxDB as a time series database and Grafana as a visualizing and alerting solution. Our Resource Consumption Estimator can query to the InfluxDB directly.

6.2 Scheduling and Autoscaling Policies

Currently, we have implemented one simple scheduling algorithm and two autoscaling policies for validation purposes.

For each pending pod, a random scheduler first filters all available nodes based on their remaining resource capacity and the amount of resources requested by the pod. These resources are expressed in terms of memory and CPU. Once all nodes with sufficient capacity to execute the pod are identified, a random one is selected, and the pod assigned to it. If there are no nodes that can fulfil this condition, then the scheduler instructs the autoscaler to scale out. The unschedulable pod is left unchanged and an attempt to schedule it again will be made in the next cycle.

The scale out operation will depend on the autoscaler being used. The void autoscaler will simply ignore the request and hence simulates a system without autoscaling capabilities. The simple autoscaler on the other hand, will launch a new instance of a predefined type. The number of instances launched is capped to one every provisioning_interval. The motivation behind this limit is based on the following observation. Unschedulable pods are likely to be found in batches. That is, if there are insufficient resources to deploy one pod, there may be insufficient resources to deploy the next pending pod in the queue. Hence, scaling out requests are likely to be made several times during the same scheduling cycle.

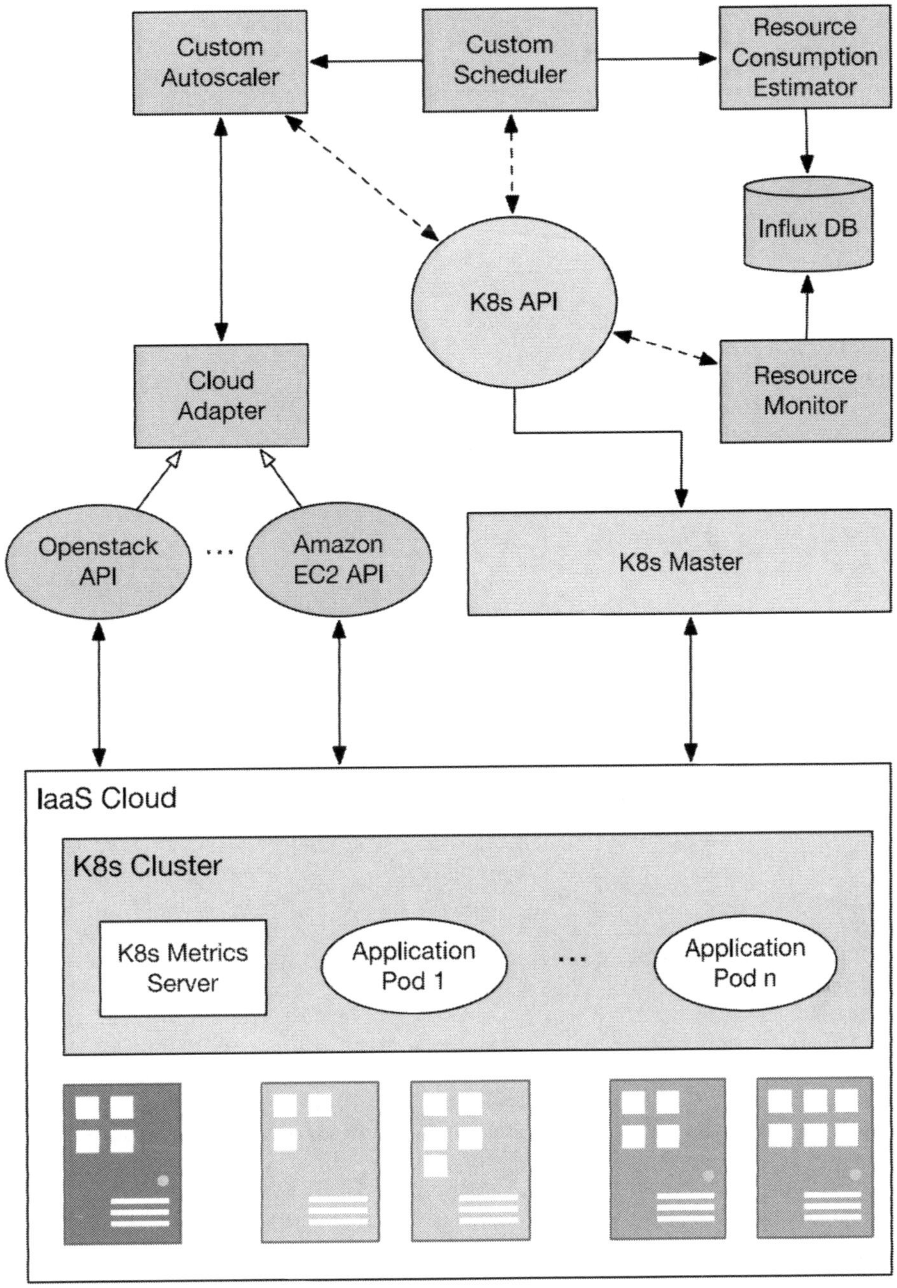

Figure 2: System Prototype Architecture.

This may lead to an excessive number of instances being launched which may end up being underutilized as a single one may have sufficed to execute the unschedulable pods. In fact, we set the provisioning_interval based on an estimate of the instance provisioning delay (i.e., the time it takes for the VM to boot and join the K8s cluster) plus a small contingency value. Notice however that this parameter is configurable by users.

Table 1. Master and worker nodes VM specifications.

	VM Type	# of vCPUs	RAM	Operating System
Master	m2.medium	2	6 GB	Ubuntu 17.01
Worker	m2.small	1	4 GB	Ubuntu 17.01

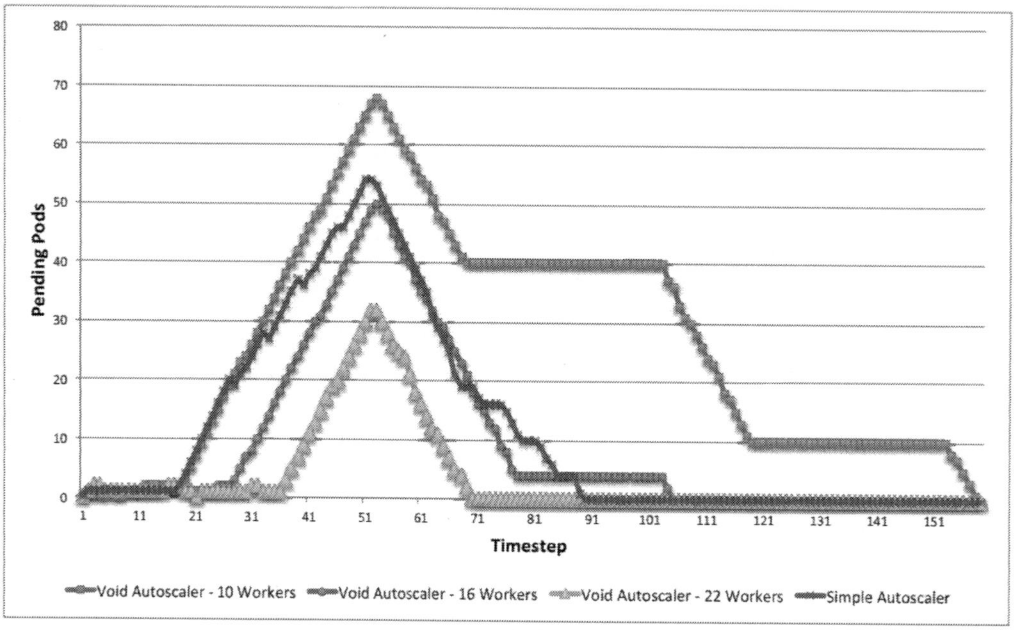

Figure 3. Number of pending pods throughout multiple scheduling cycles for different autoscaling scenarios.

6.3 Testbed
To demonstrate the potential capabilities of the proposed framework and the benefits of autoscaling, we performed a set of medium-scale validating experiments. For this purpose, we used a workload composed of 100 homogeneous batch jobs, each submitted every 10 seconds to Kubernetes. Each job requests 64 MB of memory and 250 millicores. The jobs do not actually use the requested resources as they only print out a statement and sleep for 1000 seconds (approximately 16 minutes). The purpose of the requests however is to validate the functionality of the scheduler.

The platform is deployed on Nectar [18], an Australian research cloud based on Openstack. The VM specifications used to deploy the Kubernetes master and worker nodes are depicted in Table 1. The custom architectural components were deployed outside Nectar on a MacBook Pro with a 2.9 GHz Intel Core i7 processor and 8 GB of RAM and the Kubernetes version used was 1.10.

6.4 Results
The evaluation was performed by comparing 4 different scenarios. The first one, referred to as Simple Autoscaler corresponds to an approach that uses the random scheduler along with the simple autoscaler with an initial cluster size of 10 worker nodes. The three other scenarios all refer to approaches using the random scheduler along with the void autoscaler. They differ from each other in the number of worker nodes in the cluster, which remains static throughout the execution of the workload and ranges from 10 to 22 worker nodes. We refer to these approaches as Void Autoscaler – N workers, where N corresponds to the size of the cluster.

To illustrate the benefits of autoscaling, we evaluate each solution from two perspectives, the scheduling performance and the total infrastructure cost. Figure 3 illustrates the total number of pending pods at different timesteps of the workload execution. Pending pods are those that require scheduling and each monitoring timestep is 20 seconds apart. Clearly, the void autoscaler with 10 nodes takes the longer to schedule all pending pods, followed by the 16-worker node approach. The void autoscaler with 22 nodes is the fastest out of all the approaches in scheduling the pending pods. Finally, the simple autoscaler gradually reduces the number of pending pods as more worker nodes are added to the cluster. Figure 4 shows the number of worker nodes used over time by the simple autoscaler. This approach is able to complete the scheduling the workload soon after the 22-worker approach and before the 16-worker one.

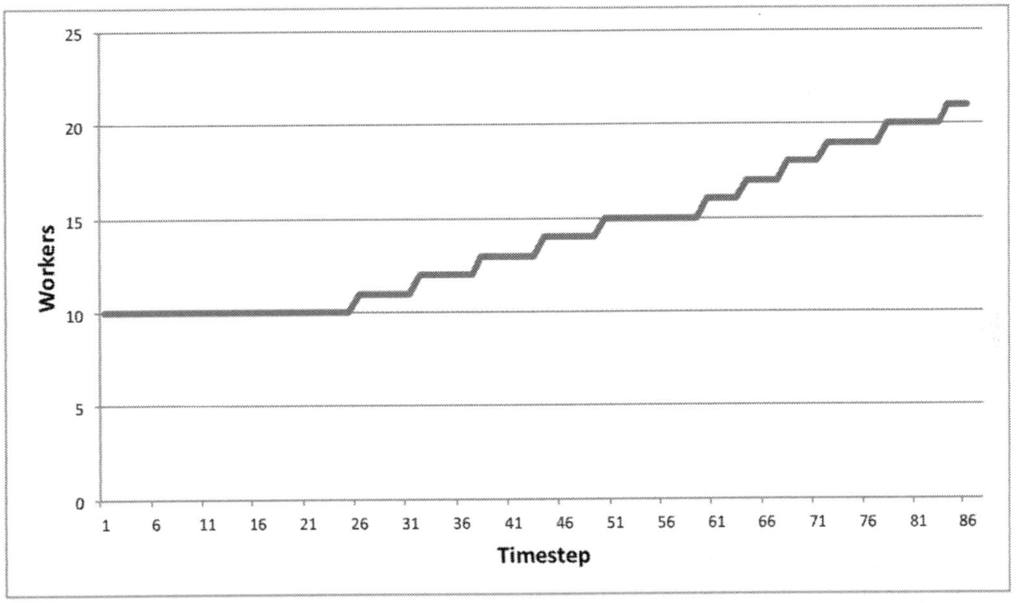

Figure 4. Number of worker nodes used by the simple autoscaler at different scheduling cycles.

To demonstrate that despite taking slightly longer to schedule all pods, the simple austoscaler has other benefits, primarily in terms cost, Table 2 depicts various scheduling performance metrics and the total infrastructure cost for each of the solutions. As already mentioned, the 22-worker void autoscaler approach presents the shortest scheduling duration, followed by the simple autoscaler and the 16 and 10-worker void autoscalers respectively. Based on this, the simple autoscaler achieves the second highest pod throughput, followed by the 16 and 10-worker approaches. It is clear that 10 nodes are not sufficient to execute the submitted workload efficiently, with significantly lower throughput measures.

Since Nectar does not charge for the use of resources, we estimate the cost of each approach based on the billing model of existing cloud providers. In particular, we assume a per-minute billing of $0.011 for each worker based on Microsoft Azure's general purpose *B2S* instance type, with any partial use being rounded up to the nearest minute. For the void autoscaler approaches, the number of minutes each worker is billed for is estimated based on the total scheduling duration. For the simple autoscaler, 10 workers are billed for the entire scheduling duration while the remaining 11 are billed from the moment they were launched until the moment when the scheduling of all pods completes. As expected, the simple autoscaler obtains the lowest cost as VMs are only launched when needed. The experiments demonstrate as well that having 22 worker nodes always available only improves the scheduling throughput by 1 pod per minute but leads to higher costs. Also, having a small number of nodes, such as 10 in this case, does not necessarily reduce the cost, as they must remain active for a longer period of time, inevitably incurring in more billing periods and higher costs.

Table 2. Scheduling performance metrics and cost for different autoscaling scenarios.

Autoscaler	Average Scheduling Delay (min)	Total Scheduling Duration (min)	Throughput (pods/min)	Cost
Void - 22 workers	1.95	23.08	4.33	$0.41
Void - 16 workers	4.85	34.58	2.89	$0.44
Void - 10 workers	14.03	52.71	1.89	$0.42
Simple	6.13	29.59	3.37	$0.32

7. Summary and Conclusions

This paper presented a vision, an architectural framework, elements, principles, and some preliminary experimental results for Pricing-, Fault-, and QoS-aware Containerized Cluster Management in Clouds. Many opportunities and open challenges exist in this context. Determining the initial placement of containers is one of them; this must be done while considering the characteristics of different applications, cloud providers and pricing models, resource types, geographical location, etc. Another challenge is optimizing the placement of containers at runtime by using rescheduling and migration techniques. The fault-tolerance and QoS requirements of applications must be considered to successfully achieve this objective. Autoscaling the cluster is another important requirement that needs attention; to meet the dynamic resource needs and to reduce cost it is essential to timely provision new resources when needed and shutdown resources when they are being underutilized. Orchestrating containers so that the applications' QoS constraints are met is another existing challenge. For example, mapping tasks to resources so that their makespan is minimized is a useful feature lacking in current open source cluster management systems. Finally, estimating the amount of resources such as memory and CPU that applications consume over time can aid in the efficient use of the cluster resources. This however is a challenging task that requires further research.

The evaluation of the new architecture and its elements along with the proposed or new approaches addressing the aforementioned challenges can be fostered using our ContainerCloudsim [16] simulator and the presented empirical platform. For a large-scale evaluation, our simulation toolkit can speed up the evaluation process with various measurements. For a practical proof-of-concept experiment, the empirical evaluation platform can be exploited to see the effectiveness of the proposed methodology in the real world.

Acknowledgments

We acknowledge Dr. Chenhao Qu for his contributions towards various ideas presented in this paper. This work supported through a collaborative research agreement between the University of Melbourne and Samsung Electronics (South Korea) as part of the Samsung GRO (Global Research Outreach) program. The work is carried out within the Melbourne CLOUDS Laboratory at the University of Melbourne.

References

[1] Docker https://www.docker.com
[2] LXC https://linuxcontainers.org
[3] Kubernetes https://kubernetes.io/
[4] Docker Swarm https://docs.docker.com/engine/swarm/
[5] Benjamin Hindman, Andy Konwinski, Matei Zaharia, Ali Ghodsi, Anthony D Joseph, Randy H Katz, Scott Shenker, and Ion Stoica. Mesos: A Platform for Fine-Grained Resource Sharing in

the Data Center. Proceedings of the 8th USENIX conference on Networked Systems Design and Implementation (NSDI), Boston, MA, 2011.
[6] Checkpoint Docker Container using CRIU https://criu.org/Docker
[7] Abhishek Verma, Luis Pedrosa, Madhukar Korupolu, David Oppenheimer, Eric Tune, and John Wilkes. Large-scale cluster management at Google with Borg. In Proceedings of the Tenth European Conference on Computer Systems (EuroSys '15). ACM, New York, USA.
[8] Nomad https://www.hashicorp.com/
[9] Apache Marathon https://mesosphere.github.io/marathon/
[10] Apache Aurora http://aurora.apache.org/
[11] Apache Hadoop on Apache Mesos https://github.com/mesos/hadoop
[12] Apache Spark on Apache Mesos https://spark.apache.org/docs/1.6.1/running-on-mesos.html
[13] Apache Storm on Apache Mesos https://github.com/mesos/storm
[14] Dkron http://dkron.io/
[15] Chronos https://mesos.github.io/chronos/
[16] Sareh Fotuhi Piraghaj, Amir Vahid Dastjerdi, Rodrigo N Calheiros, and Rajkumar Buyya. ContainerCloudSim: An environment for modeling and simulation of containers in cloud data centers. Software Practice and Experience. 47: 505–521, 2017.
[17] Kubernetes - Building Large Clusters on https://kubernetes.io/docs/admin/cluster-large/
[18] Nectar on https://nectar.org.au/
[19] Wes Felter, Alexandre Ferreira, Ram Rajamony, and Juan Rubio. 2015. An updated performance comparison of virtual machines and linux containers. In 2015 IEEE International Symposium on Performance Analysis of Systems and Software (ISPASS), 171–172, 2015.
[20] Cristian Ruiz, Emmanuel Jeanvoine, and Lucas Nussbaum.. Performance evaluation of containers for HPC. In Proceedings of the European Conference on Parallel Processing. Springer, 813–824, 2015.

Prospective teachers' difficulties in second order linear differential equation: a case of constructing methods in solving a non-homogeneous problem

B P Prawoto, S Hartono and D K Fardah
Department of Mathematics, Universitas Negeri Surabaya, Indonesia

Abstract. This study aims to explore students' difficulty in constructing a method used to solve non-homogeneous problems in 2nd order ordinary differential equations. The research method used is descriptive-explorative research. The data collected were in the form of responses of 46 prospective teachers of the class of ordinary differential equations semester 6 studying at Universitas Negeri Surabaya in Indonesia. Data obtained using a student work sheet in which consist of several steps in order to construct a method can be used to find the solution of non-homogeneous in 2nd order ordinary differential equations. The results show that prospective teachers have some difficulties in making relations between two or more equations they had found. We found that 69.5% of students had most difficulty in emerging an equation consists of $u_1'(t)$ and $u_2'(t)$ so that they didn't really understand the concept of solutions of equations.

1. Introduction

Ordinary differential equation is a course that often scores average less than of other courses in mathematics education courses. Especially, on the topic of 2nd order non-homogeneous ordinary differential equations (ODE), prospective teachers often experience difficult in using the method of variation of parameter. This is caused by several procedures that require students to recall the concepts that have been obtained before, mainly derivative and integral [1].

Learning differential equations is difficult [2]. According to Finizio & Ladaz [3] that differential equations are courses that describe equations that contain one or more unknown functions. Mallet [4] also explains that differential equations focus on algorithmic techniques in determining the solutions of some specific types of differential equations. There are two points of concern, namely to determine the type of differential equation and choose the right method to determine the solution of the differential equation. The tendency of students in constructing methods on differential equations often has difficulty. This is supported by data of student test results in 2017 which shows that 70% of students have difficulty determining the solution of the differential equation.

Based on some research results show that students who take the courses of differential equations can't apply the concept of derivative functions correctly [5]. In addition, Kristayulita & Nurhardiani [6] said that students in solving the problem of 1st order differential equation have difficulties, due to the lack of the concept of integral and logarithmic properties. In line with this study, Budiyono & Guspriati [1] also explained that students in following the courses of differential equations also often have difficulty in determining the results of roots m1 and m2 on the characteristic equations of 2nd order differential equations, determining of general solutions, derived functions and simplify form of equation on the integral form of function.

One way to know the difficulties of students in following this course is by identifying student errors in constructing the method when solving the problem of ordinary 2nd order differential equations. According to Dubinsky & McDonald [7] also said that to know a student concept has been obtained, even though the student had a failure or not, it can be done by looking at the results of construction methods undertaken by the student. Errors made by students need to be identified, in order to know what kind of difficulties made by students in this course. These difficulties can later be reduced when solving the problem with the same method. It shows that when solving a problem of ordinary differential equations, we can analyse the difficulty in the problem or the construct method in solving the problem.

2. Method

This research is a descriptive explorative research that to explore students' difficulty in constructing a method used to solve non-homogeneous problems in 2nd order ordinary differential equations.

The subjects of this research are 6th semester students, Mathematics Education Program, Surabaya State University who are taking a course of Differential Equal Equations. This research was conducted from February to May 2018. Research subjects consisted of 46 students, 23 students of class 2015U and 23 students of class 2015A.

The data of this study were collected using question about the non-homogeneous problems in 2nd order ordinary differential equations. The questions used are as follows:

Consider the following 2nd order ODE

$$y'' + p(t)y' + q(t)y = G(t) \quad (1)$$

where p, q, and G are given continuous functions.

The general solution of the corresponding homogeneous equation is

$$y_c(t) = c_1y_1(t) + c_2y_2(t) \quad (2)$$

Where y_1 and y_2 are the solutions of $y'' + p(t)y' + q(t)y = 0$.

The idea is to replace the constants c_1 and c_2 by functions $u_1(t)$ and $u_2(t)$, respectively; thus we have

$$y(t) = u_1(t)y_1(t) + u_2(t)y_2(t). \quad (3)$$

Equation (3) is the particular solution of equation (1).

Then we try to determine $u_1(t)$ and $u_2(t)$ so that the expression in equation (3) is a solution of the non-homogeneous equation (1).

In the worksheet is given several stages, so it can be known at what stage students have difficulty in constructing methods of parameter variation.

3. Results

Data obtained from research subjects are analysed and discussed then the results are presented in tabular form. The next will be presented in the form of a related table of prospective teachers' answers on the stages in constructing the method of parameter variation, as follows:

Tabel1. The result of prospective teachers' answer about the difficulty in constructing the parametric method

Type of Difficulty	Total	Percentage
Derivating $y(t)$ to obtain $y'(t)$	5	10.9%
Substituting of a derivative $y(t)$ to ODE	8	17.4%
Finding $u_1'(t)$ and $u_2'(t)$	32	69.5%
Integrating $u_1'(t)$ and $u_2'(t)$	1	2.2%

Based on the above table, the results of the answers from the prospective teachers show that 10.9% of students have difficulty in determining $y''(t)$ from $y'(t)$, the difficulty in substituting the $y(t)$ to differential equation is 17.4%. In addition, 69.5% of students experiencing the greatest difficulty lie in finding U_1 '(t) and U_2' (t). While, the student having the least difficulty in integrating $u_1'(t)$ and $u_2'(t)$ is 2.2%.

Then will be explained and shown sample of student work on type difficulty in constructing method, as follows:

a. Difficulty in determining the derivative of y

Figure 1 shows that students have difficulties in determining the derivative of y, y'. This is seen in the results of their work that concerned the result $y''(t)$ is wrong in deriving the multiplication form at function $y'(t)$. Based on the data obtained as much as 10.9% of 46 students who have difficulty at this stage.

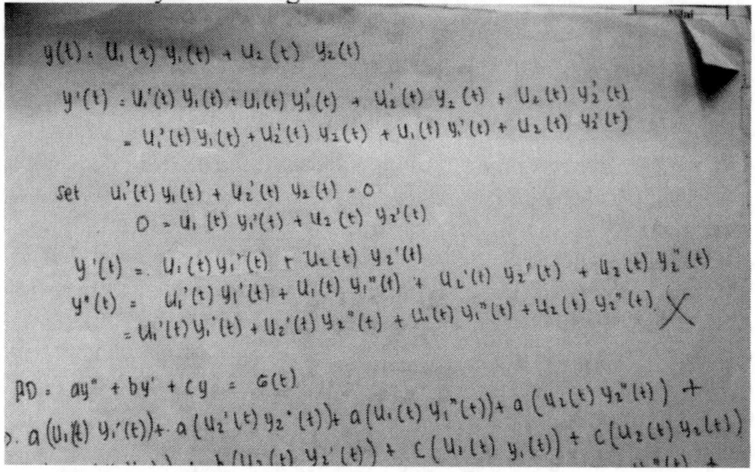

Figure 1. Difficulty in determining the derivative of y

b. Difficulty in substituting the derivatives from y(t) to ODE

Figure 2 shows that students have difficulty in substituting $y(t), y'(t)$, and $y''(t)$ into ODE. This is seen in the results of their work related to substitution $p(t)$ and $q(t)$. Based on the data obtained as much as 17.4% of 46 students who have difficulty at this stage.

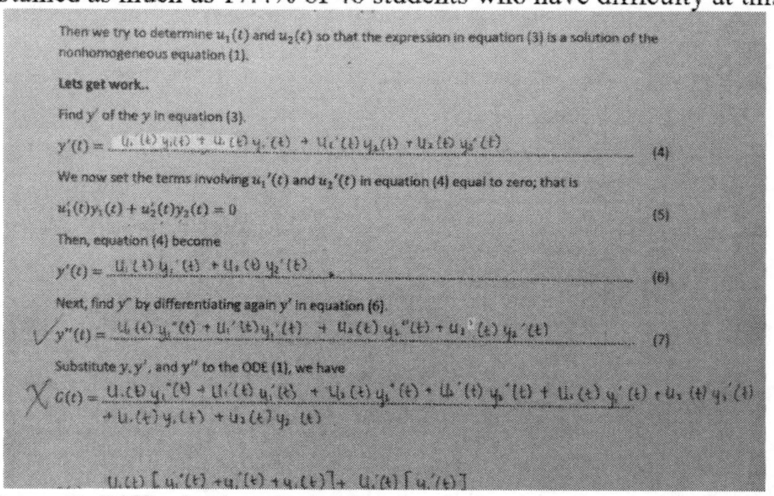

Figure 2. Difficulty in substituting the derivatives from $y(t)$ to ODE

c. Difficulty in finding $u_1'(t)$ and $u_2'(t)$

Figure 3 shows that the student difficulties in finding $u_1'(t)$ and $u_2'(t)$ after $y(t), y'(t), y''(t)$ is substituted to ODE. The student is unable to recall the equation $u_1'(t).y_1(t) + u_2'(t).y_2(t) = 0$ and has not been able to make the relationship between $y_1(t)$ and $y_1(t)$ which is the solution of equation (1) so they cannot find the final result $u_1'(t)$ and $u_2'(t)$. Based on the data obtained as much as 69.5% of 46 students who have difficulty at this stage.

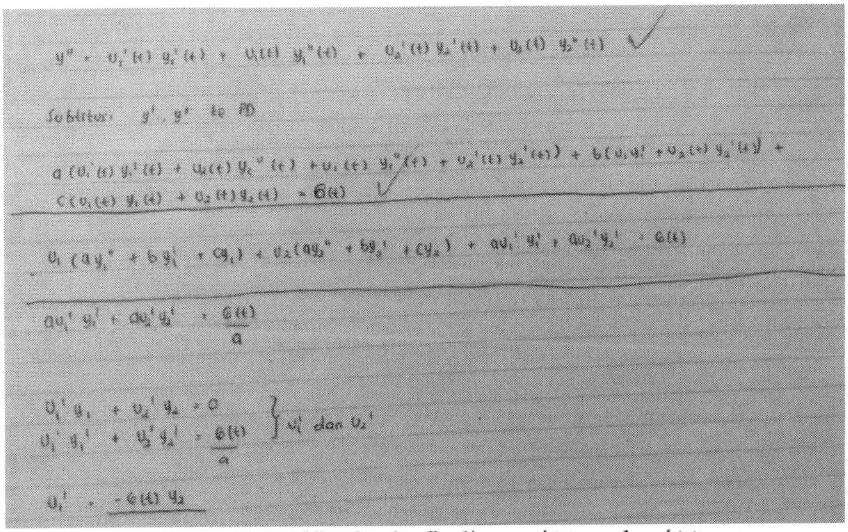

Figure 3. Difficulty in finding $u_1'(t)$ and $u_2'(t)$

d. Difficulty in integrating $u_1'(t)$ and $u_2'(t)$

Figure 4 shows that students have difficulty in integrating $u_1'(t)$ and $u_2'(t)$ after $y(t), y'(t)$, and $y''(t)$ are substituted to ODE. It is seen that the student does not continue the integral of $u_1'(t)$. Based on data obtained as much as 2.2% of 46 students who have difficulty at this stage.

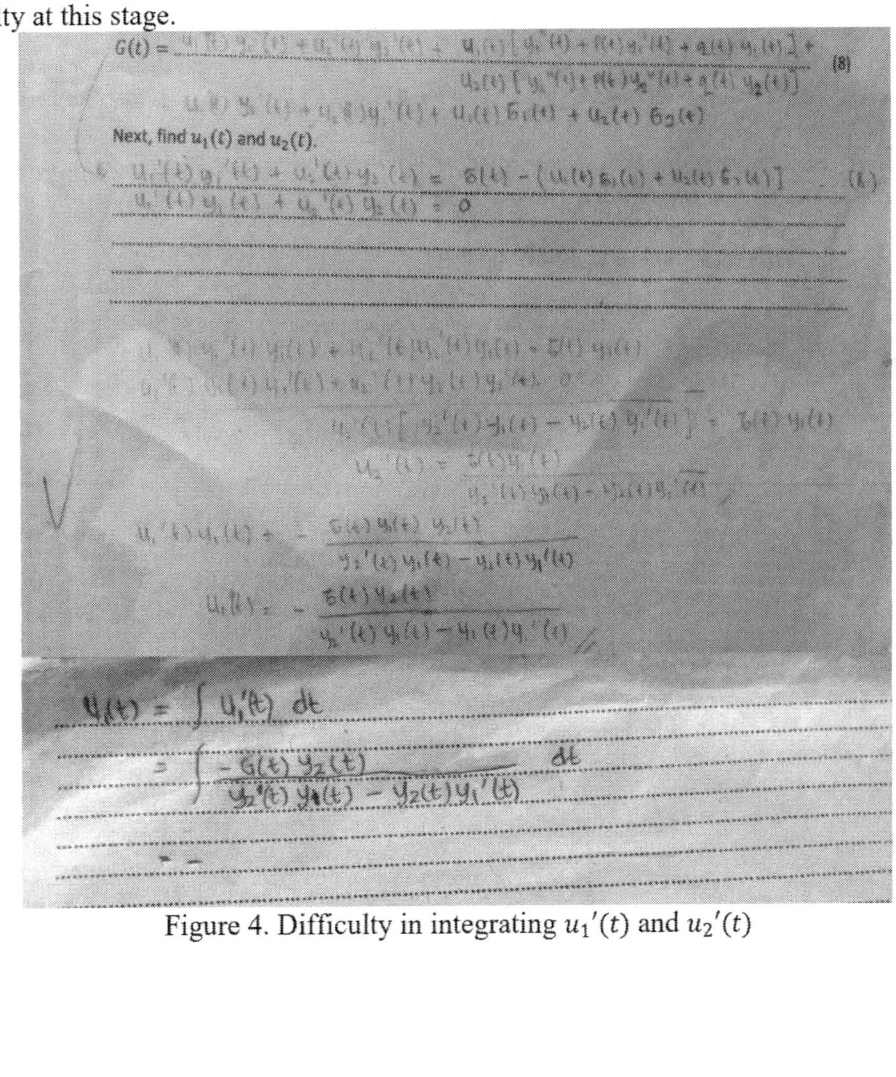

Figure 4. Difficulty in integrating $u_1'(t)$ and $u_2'(t)$

4. Discussion and Conclusion

The findings of the answers from the prospective teachers showed that 69.5% of students having the greatest difficulty lie in finding $u_1'(t)$ and $u_2'(t)$. The difficulties occur because students are less understood in understanding and bring up the functions known in the problem. The students cannot continue to search $u_1'(t)$ and $u_2'(t)$. This is shown in Figure 3, the student cannot substitute $y'' + p(t)y' + q(t)y = 0$ to the ODE. As a result many students cannot continue to solve the non-homogeneous problems in 2nd order ordinary differential equations.

In addition, the least difficulty lies in integrating $u_1'(t)$ and $u_2'(t)$ by 2.2%. Students are very difficult to change the form $u_1'(t)$ to the form $u_1(t)$ and $u_2'(t)$ to $u_2(t)$. This is seen in the student's answer Figure 4 which shows the student's difficulty to continue the integral process from $u_1'(t)$ to $u_1(t)$. In line with that, Kristayulita & Nurhardiani [6] said that the students are difficulty in integrating on the problem of ODE because students' concepts about integral are very lack when they have been obtained in the previous course.

Another difficulty is also seen that 10.9% of students have difficulty in derivating $y(t)$. Students are less conscientious in finding for the derivatives of the functions $f'(t)$, seen in Figure 1 that the student is still wrong in finding $y''(t)$ where there are some symbols still not appropriate. According to Budiyono & Guspriati [1], the most error in solving the problem of ordinary differential equations is in derivating function. While other difficulties, also seen in Figure2 that students are also less thorough in substituting the results of the derivative y (t) to ODE. As many as 17.4% of students have difficulty in substituting $y''(t)$ to PD, this is because students have difficulties from scratch in finding $y''(t)$ and if we substituting it is also inappropriate.

Based on the result of the research, it can be concluded that the students' understanding on the concept of non-homogeneous problems in 2nd order ordinary differential equations is less because students are still difficulty in derivating, substituting, integrating and calculating a solution function y(t). The results also show the greatest difficulty in finding $u_1'(t)$ and $u_2'(t)$ and the least difficulty lies in integrating $u_1'(t)$ and $u_2'(t)$ into $u_1(t)$ and $u_2(t)$. Further research, need to find a method of learning where students will easily understand the material non-homogeneous problems in 2nd order ordinary differential equations so that the difficulties that have been obtained in this study will be reduced.

5. Acknowledgement

We would like to thank Universitas Negeri Surabaya, Indonesia, which give us a grant and chance to disseminate an intellectual output of our research.

6. References

[1] Budiyono & Guspriati W 2009 Jenis-Jenis Kesalahan dalam Menyelesaikan Soal Persamaan Diferensial Biasa (PDB) Studi Kasus pada Mahasiswa Semester V Program Studi Pendidikan Matematika Universitas Muhammadiyah Purworejo Prosiding Seminar Nasional Matematika dan Pendidikan Matematika Jurusan Pendidikan Matematika *(UNY FMIPA)*

[2] Czocher J A Tague J & Baker G 2013 Coherence from Calculus to Diferential Equations In Proceedings of the 16th Annual Conference on Research in Undergraduate Mathematics Education (Vol. 2)

[3] Finizio N & Ladas G 1988 Persamaan Diferensial Biasa dengan Penerapan Modern (Jakarta : Erlangga)

[4] Mallet D G & McCue S W 2009 Journal of Mathematical Education in Science and Technology 40(5) 587-595

[5] Valcare M C & Díaz J P 2013 Initial Diagnostic Assessment to Support Learning Eighth *Congress of European Research in Mathematics Education (CERME 8)* Turkey

[6] Kristayulita K & Nurhardiani N 2011 Beta Jurnal Tadris Matematika 4(1) 30-52

[7] Dubinsky E & McDonald M A 2001 APOS: A constructivist theory of learning in undergraduate mathematics education research In The teaching and learning of mathematics at university level (Springer Dordrecht)

Design of Countenance Evaluation Model Based on *ANEKA-Tri Hita Karana* in Computer Learning for Vocational Students of Information Technology in Bali

D G H Divayana[1*], I P W Ariawan[2], A Adiarta[3], D P Parmiti[4], D B Sanjaya[5], I W Kertih[6], S C Wibawa[7]

[1]Department of Informatics Education, Universitas Pendidikan Ganesha
[2]Department of Mathematics Education, Universitas Pendidikan Ganesha
[3]Department of Electrical Education, Universitas Pendidikan Ganesha
[4]Department of Educational Technology, Universitas Pendidikan Ganesha
[5,6]Department of Civics Education, Universitas Pendidikan Ganesha
[7]Department of Educational of Information Technology, Universitas Negeri Surabaya
*Corresponding Author's Email : hendra.divayana@undiksha.ac.id

Abstract. This study was intended to provide an overview of the countenance evaluation model design based on *ANEKA* values integrated with *Tri Hita Karana* concept in computer learning, generally for vocational students of information technology and specifically for vocational students of information technology in Bali. The approach that was used in the design of this evaluation model was a research and development (RnD) of Borg and Gall model, which focuses on the stage of developing the preliminary form of a product. Parties who were involved in model testing were two educational experts and two informatics experts. The tools that were used in model testing were questionnaires. The analysis technique used in this study was quantitative descriptive manner to describe the data. The result showed that the initial form was ready to be used with a good level of effectiveness.

1. Introduction

Recent implication of industrial revolution 4.0 is now inevitable. This revolution occurs because the support of information technology development which rapidly grows up. The positive impact of the present industry revolution 4.0 is that all things can be done quickly and accurately with the help of information technology. Currently, information technology has been able to facilitate all needs in the areas of health, economy, political condition, education, and so forth. However, aside from various positive impacts, it cannot be denied the emergence of negative impacts including a lot of hoax news, sexual abuse, and violence. The negative impacts would then lead to decrease student characters. The decrease in the student characters, especially in senior high school or vocational high school now begins to appear. It can be seen from some examples such as many students not coming to school, students do violence, and students disobeying the rules.

The decrease in student character has also happened in several regencies in Bali province, such as Tabanan, Denpasar, and Badung. These problems occur because the rupture of understanding about Pancasila values and the concept of state defense [1]. In addition, character decrease is also found in students of the vocational high school and especially in ICT field. It occurs because students of ICT in vocational high school have been accustomed to operating information technology, so they tend to be more quickly affected by the advances in information technology and very risky to be affected by negative things. Based on the attention to that situation, it will greatly affect the learning process in school. Learning is an activity to change the mind, attitude and mental learners from unpleasant to pleasant condition.

The quality of learning can be said to be good if learners can improve insight/knowledge, and can show/maintain good manners and character. The quality of good learning can be obtained through a good learning process. However, in reality, today there are still many learning processes in schools do not run optimally, for example, the use of information technology in helping the learning process is not used in accordance with the allocation, the values of character has been decreased indicated by promiscuity, participants students have dared to oppose school rules, and others. The decrease in the student characters is caused by the low implementation and actualization of *ANEKA* values (Accountability, Nationalism, Public Ethics, Quality Commitment, and Anti-Corruption) in the learning process.

The decrease in accountability values, such as not having a sense of being responsible to finish the task/homework provided by the teacher and cheating the culture. The decrease in nationalism values, such as mutual mockery and disrespect the opinion of friends during group discussion, often noisy while the teacher is explaining the teaching materials. The decrease in public ethics values, such as: dare to violate school rules, dare against the teacher, and not discipline, disfigure the name of a school. The decrease in quality commitment values, such as: not creative, make the task careless and lack of focus in following the learning process. The decrease in anti-corruption values, such as: littering and being afraid of expressing opinion and action.

In anticipating the increasingly widespread problem in the learning process, it is necessary to evaluate the learning process that occurred in the vocational school (especially in ICT field) in Bali Province. Through the evaluation, then it can be found a recommendation to solve the problems encountered in the learning process. In general, evaluation is the activity of collecting and analyzing data to obtain recommendations that will be used in taking a decision.

Those statement according with the definition of evaluation expressed by Mahayukti, et al [2], Divayana, Adiarta and Abadi [3], Pratama, Setiawan and Ruslan [4]; Jampel, et al [5]; Divayana, et al [6],[7]; Arnyana, et al [8]; Mak [9]; Chu, Liu and Liu [10]; Kamaludin, et al [11]; Pauline, et al [12]; Sern, et al [13]. One suitable evaluation model used to evaluate the learning process in vocational high school (especially ICT field) in Bali Province is a model capable of performing evaluation ranging from components that measure the ability level of learners to ethical values and character of learners based on consideration of *ANEKA* value integrated with the *Tri Hita Karana* concept. Therefore, it is necessary to develop a countenance evaluation model based on *ANEKA* values (Accountability, Nationalism, Public Ethics, Quality Commitment and Anti-Corruption) integrated with *Tri Hita Karana* (*Parahyangan*, *Palemahan*, and *Pawongan*) concepts.

The research results about character education through school culture, that conducted by Furkan [14] have similarities with research conducted by researcher in terms of their research purpose was about improving the education quality or learning quality, and the research approach used in qualitative. The research difference lies in the model was used to achieve of expected objectives, where Furkan used the model of character values development of learners, while the researcher used countenance evaluation model based on *ANEKA* value integrated with *Tri Hita Karana* concept.

Research about the teachers' views on values education, conducted by Thornberg and Oğuz [15], has similarities with research conducted by the researcher in terms of his research purpose, that is, improving the education quality or learning quality, and also research approach used in qualitative. The difference lies in the use of appropriate strategies to achieve expected objectives, where Thornberg and Oğuz only used the internalization of values education in learning process, while researcher used the internalization of *ANEKA* values integrated with the *Tri Hita Karana* concept in the learning process.

Research about the implementation of school culture values in realizing quality school, that conducted by Yuniarsih, Sutarman, and Riyanto [16] has similarities with research conducted by the researcher regarding his research purpose was about improving the education quality or learning quality, and research approach used in qualitative. The difference lies in the efforts made to achieve the expected goal, where Yuniarsih, Sutarman, and Riyanto seeks to implement the values of school culture in realizing quality learning process, while researcher implements the of *ANEKA* values integrated with *Tri Hita Karana* concept in a learning process.

Based on the problems which was found and the study of several studies that have been done by some previous researchers, the researchers are interested in research the development of countenance evaluation model based on *ANEKA* values integrated with *Tri Hita Karana* concept to improve the character and quality of computer learning, especially at ICT vocational school in Bali Province. The primary objective of this paper is to introduce the design of countenance evaluation model based on *ANEKA* values integrated with *Tri Hita Karana* concept in computer learning for vocational students of information technology in Bali.

2. Method

This research was RnD based on Borg and Gall model, with ten stages, such as 1) research and information collecting, 2) planning, 3) developing preliminary form of product, 4) preliminary field testing, 5) main product revising, 6) main field testing, 7) operational product revising, 8) operational field testing, 9) final product revising, and 10) disseminating and implementating [17].

This paper study only focuses on the development stage design (develop the preliminary form of product). The participants involved in the design testing process are two education experts and two informatics experts. The tool was used in testing the evaluation model design that had been formed was a questionnaire. The number of instruments item to measure evaluation model design as much as ten items.

The analysis technique was used in this research was quantitative descriptive by using descriptive percentage calculation. The percentage descriptive calculation yields assessment percentage that subsequently was used to provide an interpretation or assessment of the test results of evaluation model that was developed. The formula that was used for descriptive percentage calculation was as follows [18].

$$\text{Percentage} = \frac{\sum(\text{Answer} * \text{Weight of Each Answer Choice})}{\text{n} * \text{Highest Weight}} * 100\% \qquad (1)$$

Notes: $\sum$ = total
n = The total number of questionnaires item

3. Results and Discussion

The results obtained in this research are design of countenance evaluation model based on *ANEKA* values integrated with *Tri Hita Karana* concept in computer learning, so later can be used as an overview of the stages that can be done by the education evaluator in evaluating the character and quality of learning computer on students in generally, and specifically for students in Bali Province. The design view of countenance evaluation model based on *ANEKA-Tri Hita Karana* in computer learning can be shown in Figure 1.

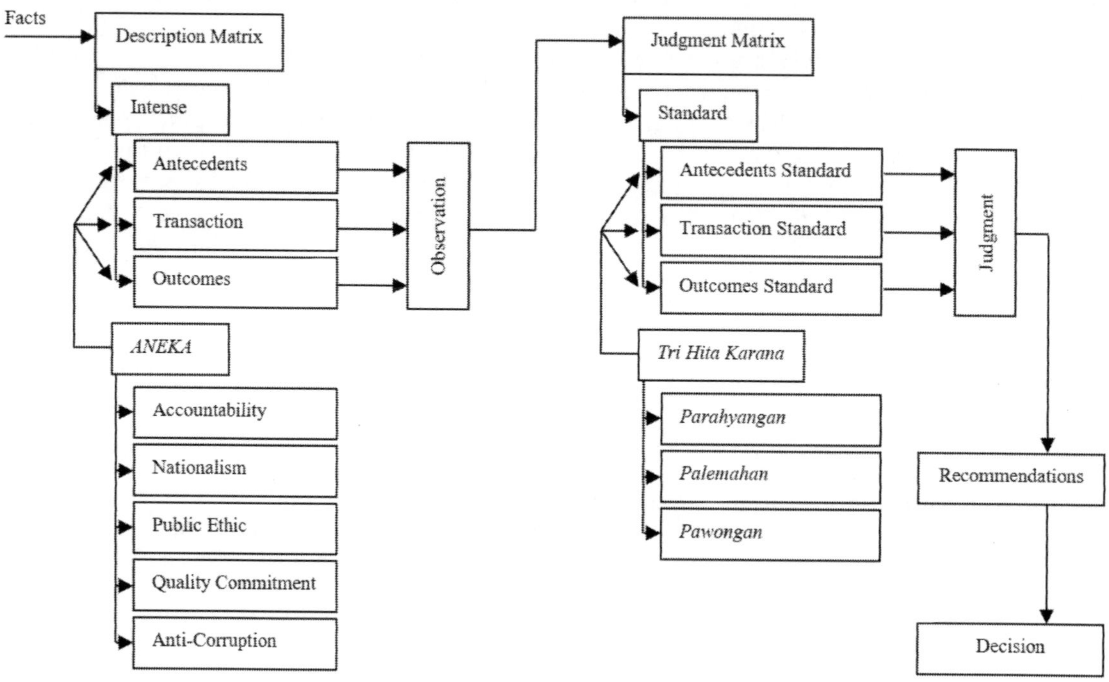

Figure 1. Design of Countenance Evaluation Model Based on *ANEKA-Tri Hita Karana* in Computer Learning

Figure 1 shows that the countenance evaluation model based on *ANEKA-Tri Hita Karana* in Computer Learning consists of two main parts, namely: description and judgment matrix. Rational facts that were inputted into the description matrix section are measured intensively with a review of three important things, such as 1) expected context (often called *Antecedents*), 2) expected processes (often called *transactions*), and 3) expected results and effects (often called *outcomes*). Those three important things based on the *ANEKA* concept, which consists of 5 components, such as 1) Accountability, 2) Nationalism, 3) Public Ethics, 4) quality commitment and 5) Anti-Corruption. In principle, *ANEKA* is the internalization concept in five positive values of person when performing a particular job, to manifest a professional person. The term of *ANEKA* was taken based on the initial letters of the five positive values that must be internalized, such as 1) *Akuntabilitas* (in Indonesian) or Accountability (in English), 2) *Nasionalisme* (in Indonesian) or Nationalism (in English), 3) *Etika Publik* (in Indonesian) or Public Ethics, 4) *Komitmen Mutu* (in Indonesian) or Quality Commitment (in English), and 5) *Anti-Korupsi* (in Indonesian) or Anti-Corruption (in English).

The facts that are measured by using three important things in the description matrix (antecedents, transactions, and outcomes) are examined and compared against the standard (antecedent standard, transaction standard, and outcomes standard) in the judgment matrix section. All those standards must be integrated with the *Tri Hita Karana* concept, which consists of 3 components: 1) *Parahyangan*, 2) *Palemahan*, and 3) *Pawongan*. In general, *Tri Hita Karana* is the three causes of happiness in human life. The happiness occurs because of the good relationship between peoples and God (usually called *Parahyangan*), the good relationship between peoples (usually called *Pawongan*), and the good relationship between peoples and their natural environment (usually called *Palemahan*). The observation results and comparative analysis between facts that was input and the existing standard were used as the base on the judgment section to give the right recommendations, so it easier in making a decision.

An evaluation model design can be said to be ready for use if it has been through a trial process and obtain results with a minimum value is a good category. The test results conducted by four experts on the design of countenance evaluation model based on *ANEKA* integrated with *Tri Hita Karana* concept can be seen in Table 1.

Table 1. Test Result of Design of Countenance Evaluation Model Based on *ANEKA-Tri Hita Karana*

Respondents	Item-										Σ	Effectiveness Percentage
	1	2	3	4	5	6	7	8	9	10		
Education Expert-1	5	4	4	5	5	5	4	4	4	5	45	90%
Education Expert-2	5	5	5	4	4	4	4	5	5	5	46	92%
Informatics Expert-1	5	4	4	5	5	4	4	4	4	4	43	86%
Informatics Expert-2	5	4	4	4	4	4	5	4	4	4	42	84%
											Average	88%

Determination of 10 items of instruments used to measure the effectiveness of Design of Countenance Evaluation Model Based on *ANEKA-Tri Hita Karana* refers to several criteria determining the effectiveness of a program, such as productivity, efficiency, stability, adaptation, quality, and satisfaction. Some of those criteria for determining program effectiveness have accordance with several effectiveness criteria expressed by Goodman and Pennings in 1977 [19], such as: quality, productivity, alertness, efficiency, stability, motivation, satisfaction, and forth.

Categorization of percentage effectiveness level of five scale, consist of the excellent category with percentage range 90-100%, the good category with percentage range 80-89%, the moderate category with percentage range 65-79%, the less category with percentage range 55-64%, and the poor category with percentage range of 0-54% [18]. Based on the average results of effectiveness percentage is shown in Table 1, so the design of countenance evaluation model based on *ANEKA-Tri Hita Karana* belongs to a good category in the categorization of percentage effectiveness level of five scale. Therefore, that evaluation model design not revised again and ready to be used by the evaluator to evaluate the quality and character of vocational students of ICT in computer learning. In addition to the advantages in the form of new evaluation model obtained from the results of this study, there are several weaknesses that are still found in this study, including 1) difficulty determining minimum standards in measuring the student characters viewed from *Parahyangan* aspects, and 2) there aren't applications can automatically perform the percentage calculation of effectiveness level each evaluation aspects in views from *ANEKA* components and *Tri Hita Karana* components.

4. Conclusions

This research has been able to make positive contribution in educational evaluation field with the invention of new breakthrough in the form of countenance evaluation model design based on *ANEKA* integrated with *Tri Hita Karana* concept that has been tested and categorized as good, and ready to use to evaluate the character and quality of computer learning on vocational students of ICT in Bali province.

The solutions that can be done to solve the obstacle found in this research, are: 1) to set the minimum standard in *Parahyangan* aspect based on the validity test against standard on judgment matrix involving education experts, informatics experts, and also at least two Hinduism experts, so being able to know the depth of student character values in following the computer learning process, 2) create a desktop-based application that can automatically perform the percentage calculation of effectiveness level each evaluation aspect.

5. Acknowledgment

The authors would like to thank Directorate of Research and Community Service, Ministry of Research, Technology and Higher Education of the Republic of Indonesia, who give financial support in this research. The author is grateful to Rector of Universitas Pendidikan Ganesha and Chairman of Research and Community Service Institute of Universitas Pendidikan Ganesha, who have granted permission to carry out this research.

6. References

[1] https://badungkab.go.id/index.php/baca-berita/1202/badungkab.go.id (accessed 20 March 2018)
[2] Mahayukti G A, Dantes N, Candiasa I M, Marhaeni A A I N, Gita I N and Divayana D G H 2018 Computer-based portfolio assessment to enhance students' self-regulated learning. *J. Theor. Appl. Inf. Technol.* **96** pp 2351–2360

[3] Divayana D G H, Adiarta A and Abadi I B G S 2018 Initial draft of cse-ucla evaluation model based on weighted product in order to optimize digital library services in computer college in Bali *The Consortium of Asia-Pacific Education Universities (CAPEU) 2017, Universitas Negeri Surabaya, IOP Conf. Ser.: Mater. Sci. Eng.* **296** pp 1–6

[4] Pratama H R A, Setiawan I H R and Ruslan 2018 Program evaluation coaching on abandoned children who drop out of school in PPSBR Makkareso in Maros *2nd International Conference on Statistics, Mathematics, Teaching, and Research 2017 J. Phys.: Conf. Ser.* **1028** pp 1-6

[5] Jampel I N, Lasmawan I W, Ardana I M, Ariawan I P W, Sugiarta I M and Divayana D G H 2017 Evaluation of learning programs and computer certification at course institute in bali using cse-ucla based on saw simulation model *J. Theor. Appl. Inf. Technol* **95** pp 6934–6949

[6] Divayana D G H, Sappaile B I, Pujawan I G N, Dibia I K, Artaningsih L, Sundayana I M and Sugiharni G A D 2017 An evaluation of instructional process of expert system course program by using mobile technology-based cse-ucla model, *J. Interact. Mob. Technol.* **11** pp 18–31

[7] Divayana D G H, Agung A A G, Sappaile B I, Simatupang W, Sastrawijaya Y, Sundayana I M and Sugiharni G A D 2017 Utilization of open source technology in determining of validity and reliability of evaluation model instruments based on ANEKA values in order to evaluate the quality of computer learning, *J. Theor. Appl. Inf. Technol.* **95** pp 5517–5534

[8] Arnyana I B P, Sadia I W, Suma I K and Divayana D G H 2017 Determination of Effectiveness of Evaluation Results on School Culture and Character of Junior High School Students Using Character Assessment Instruments with the Local Wisdom of Bali Based on Mobile Phone *J. Theor. Appl. Inf. Technol.* **95**, pp 5348–5359

[9] Mak W S 2014 Evaluation of a moral and character education group for primary school students *Discovery – SS Student E-journal* **3** pp 142–164

[10] Chu Z, Liu P and Liu C 2018 Evaluation of the education quality of innovation and entrepreneurship in applied colleges and universities based on AHP and BP neural network *International Conference on Manufacturing Technology, Materials and Chemical Engineering (MTMCE), IOP Conf. Ser.: Mater. Sci. Eng.* **392** pp 1–9

[11] Kamaludin M, Munawar W, Mahdan D, Simanjuntak M V and Wendi H F 2018 The evaluation of industry practical of mechanical engineering in vocational education: A CIPP model approach *2nd International Conference on Innovation in Engineering and Vocational Education, IOP Conf. Ser.: Mater. Sci. Eng.* **306** pp 1-4

[12] Pauline O, Taib H, Azlan M A, Nasir N F, Azmi M A, Salleh S M, Zain B A M, Mahzan S, Zainulabidin M H, Ahmad S, Rahman N A, Ngali Z and Rahman H A 2017 Assessment and evaluation for programme learning outcomes in faculty of mechanical and manufacturing engineering, Universiti Tun Hussein on Malaysia *Colloquium of Advanced Mechanics (CAMS2016), IOP Conf. Ser.: Mater. Sci. Eng.* **165** pp 1-11

[13] Sern L C, Nor N A H, Foong L M and Hassan R 2017 Students' perception on teaching practicum evaluation using video technology *International Research and Innovation Summit (IRIS2017), IOP Conf. Ser.: Mater. Sci. Eng.* **226** pp 1-8

[14] Furkan N 2014 The implentation of character education through the school culture in sma negeri 1 dompu and sma negeri kilo dompu regency *J. of Lit. Lang. & Linguist.* **3** pp 14–44

[15] Thornberg R and Oğuz E 2013 Teachers' views on values education: A qualitative study in Sweden and Turkey *Int. J. Educ. Res.* **59** pp 49–56

[16] Yuniarsih Y, Sutarman and Riyanto S 2017 The principal strategy in implementation of cultural character education on student participants (research on vocational secondary schools IT NU Saguling West Bandung) *IOSR J. Res. & Meth. in Educ.* **7** pp 65–71

[17] Borg W R and Gall M D 2008 *Educational research: an introduction, fifth edition* (New York: Longman)

[18] Aliah N 2010 *Using reading race in improving students' reading ability at islamic junior high school bahrul ulum, gowa regency* (Makassar: UIN Alauddin Makassar)

[19] Goodman P S and Pennings J M 1977 *New perspectives on organizational effectiveness* (San Francisco: Jossey-Bass)

Performance Analysis of the Windowing Technique on Elbow Joint Angle Estimation Using Electromyography Signal

Triwiyanto[1,3,*], O Wahyunggoro[1], H A Nugroho[1], Herianto[2]

[1] Department of Electrical Engineering & Information Technology, Universitas Gadjah Mada, Jl. Grafika No. 2, Yogyakarta, Indonesia;
[2] Department of Mechanical & Industrial Engineering, Universitas Gadjah Mada, Jl. Grafika No. 2, Yogyakarta, Indonesia;
[3] Department of Electromedical Engineering, POLTEKKES KEMENKES, Surabaya, Ministry of Health, Jl. Pucang Jajar Timur No.10, Surabaya, Indonesia
*Triwiyanto123@gmail.com

Abstract. In the digital signal processing, the windowing technique is essential in the feature extraction process because it influences the performance of the proposed models. The objective of this study is to evaluate the performance of the windowing technique in the elbow joint angle estimation based on electromyography (EMG) signal. In the elbow joint angle estimation, the results showed that the window length and percentage of overlap in windowing technique affected the performance of the estimation for all time-domain features. In the adjacent windowing technique, a window length of 100 milliseconds has the highest performance in estimation. In the overlap windowing technique, the percentage of overlap is 10% .It is the highest performance of the estimation. Features ZC, SSC, WAMP, and MYOP have better RMSE than the others. In the window length of 100 milliseconds, the mean and standard deviation of RMSE for the features ZC, SSC, WAMP, and MYOP are 14.41°±3.86°, 15.20°±4.59°, 14.50°±3.33° and 13.56°±3.53°, respectively.

1. Introduction

High performance in the elbow joint angle estimation using electromyography (EMG) is important in the devices that is based on EMG control [1]. EMG signal is bioelectric which is generated from muscle when the muscle performs a contraction. EMG signal can provide any information which relates to the kinematic parameters such as angle, force, and torque of the joint of human limbs. A digital signal processing (DSP) is required to obtain those parameters. Feature extraction is an important part of the DSP to obtain the characteristics of the information and to reduce the number of EMG samples. Choosing a correct windowing technique is important in the feature extraction process in order to obtain high performance in estimating the parameters. Furthermore, the failure in choosing the type and length of window will influence the performance of the estimation. Some efforts have been proposed to find the best type and window length in order to improve the performance.

Smith investigated numbers of windows length which ranged between 50 and 550 milliseconds for machine classifier to discriminate 7 motions. The results show that the optimal windows length covers from 150 to 250 milliseconds [2]. In the development of the force prediction, it was found that the length of the window determined the Pearson's correlation coefficient of the prediction [3]. St. Amant stated that window length is also influenced the signal to noise ratio (SNR) in determining the amplitude of the EMG signal [4]. The findings were that the SNR increased along with the window length. Some studies reported that the longer of the window length, the more increase the accuracy is

[5]–[7]. However, in the development of a real-time pattern recognition system, it was suggested that the window length should be less than 300 milliseconds [8].

Previous studies have revealed that window length will influence the performance of the classifier. Therefore, choosing a correct windowing technique and determining the length of the window are very important in the feature extraction process in order to obtain high performance in estimating the parameters such as angle, force, and torque of elbow joint. However, mostly, previous studies investigated the window length to evaluate the performance of the classifier. It was rare to be found in the literature which investigate the relationship between window length and the estimation performance. Therefore, the objective of this study is to evaluate the effect of the windowing technique to the performance of the elbow joint angle estimation using EMG signal and to determine the optimum of window length to estimate the elbow joint angle.

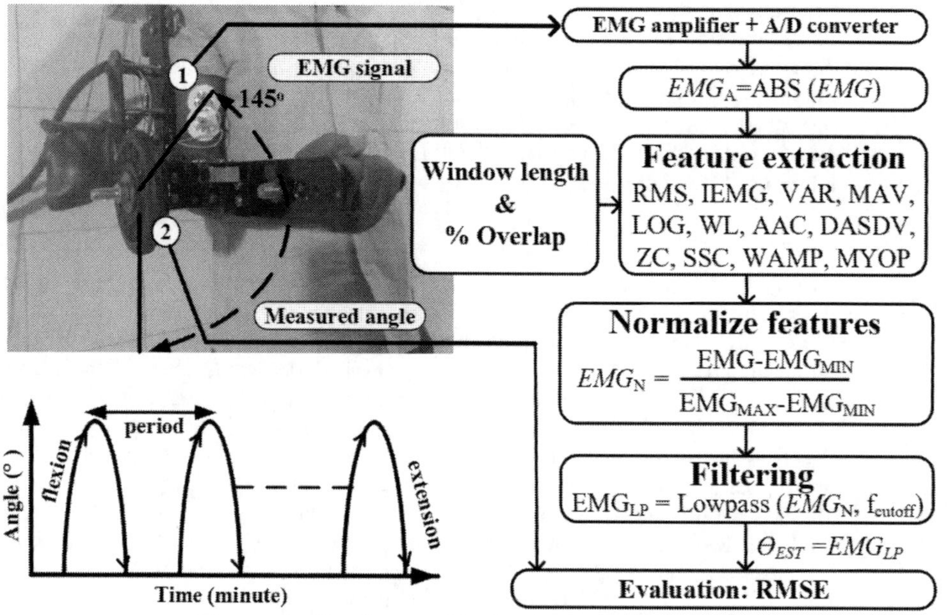

Figure 1. The diagram block of the proposed method.

2. Material and Methods

2.1. Data Collection

In order to perform the purpose of this study, four male subjects were involved. The EMG signal was collected only from bicep muscle using disposable electrode (Ag/AgCl, size: 57 x 48 mm, Ambu, Bluesensor R, Malaysia) while the elbow performed a flexion and extension motion. A metronome program was run to guide the elbow in performing the flexion-extension motion. In this work, the period of motion was set at 2 seconds. A linear potentiometer was installed in the exoskeleton frame in order to measure the real elbow joint angle. The measured angle was used as a gold standard in the evaluation. This experiment protocol had been approved by the Ethics Committee of Health Polytechnic of Surabaya, Ministry of Health.

2.2. Data Processing

In order to estimate the elbow joint angle, the DSP of the EMG signal was performed. The DSP consisted of rectifying, normalizing, windowing, feature extraction, and a lowpass filtering (**Figure 1**). The EMG signal is extracted using twelve of time-domain features. Those time domain features were the root mean square (RMS), integrated EMG (IEMG), variance (VAR), mean absolute value (MAV), logarithmic (LOG), waveform length (WL), average amplitude change (AAC), difference absolute standard deviation value (DASDV), zero crossing (ZC), sign slope change (SSC), Wilson amplitude (WAMP), and myopulse percentage rate (MYOP) [9]. Finally, the results of the estimation are evaluated using root mean square error (RMSE). The data processing of the elbow joint angle estimation is shown in **Figure 1**.

2.2.1. Windowing Technique. Windowing technique is an important part in the data processing which is used to find the optimum length of window for feature extraction process. Windowing technique is classified into adjacent and overlap windowing [7][10]. In the adjacent windowing, the signal is segmented into some windows with the end of the previous window is connected to the following window (**Figure 2(a)**). In the overlap windowing, some part of the previous and next window is overlapped (**Figure 2(b)**). In this study, various window lengths used to evaluate the performance of the windowing technique are 50, 100, 150, 200, 250, 300, 350, 400, 450, and 500 milliseconds. In the overlap windowing technique, the percentages of overlap are 10, 20, 30, 40, 50, 60, 70, 80, and 90%.

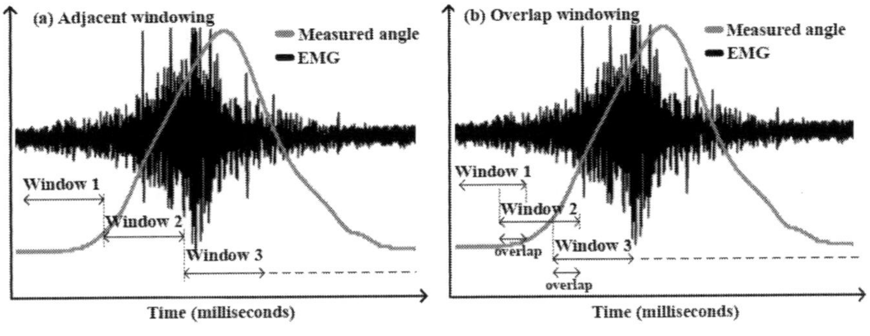

Figure 2. Windowing technique to extract the EMG signal. (a) Adjacent and (b) overlap windowing.

2.2.2. Time Domain Features. The feature extraction process is used to reduce the dimension and to obtain the characteristics of the EMG signal. The EMG signal was extracted using twelve of time-domain features. Those time domain features were the root mean square (RMS), integrate EMG (IEMG), variance (VAR), mean absolute value (MAV), logarithmic (LOG), waveform length (WL), average amplitude change (AAC), difference absolute standard deviation value (DASDV), zero crossing (ZC), sign slope change (SSC), Wilson amplitude (WAMP), and myopulse percentage rate (MYOP) [9]. Windowing techniques in feature extraction process used are adjacent and overlap window (**Figure 2**).

2.2.3. Normalize the Features. Each of 12 of the time domain features has a different unit. So that the unit of the features needs to be normalized. The equation (1) was used to normalize the features.

$$EMG_N = \frac{EMG_F - EMG_{F(MIN)}}{EMG_{F(MAX)} - EMG_{F(MIN)}} \quad (1)$$

When EMG_N is the normalized features, EMG_F is the EMG features, $EMG_{F(MIN)}$ is the minimum value of the features, and $EMG_{F(MAX)}$ the maximum value of the features. In this study, normalization was used to obtain the same range value for all the features which ranged between 0 and 1 [9].

2.2.4. IIR Butterworth low pass filter. The digital filter was applied to reduce the noise in the EMG features. The IIR filter, as well as the Kalman filter, have been used a lot in previous studies to decrease the noise [9][11]. Infinite impulse response (IIR) was selected because the number of the tap is minimal. In this study, the IIR was designed as a Butterworth low pass filter. The filter coefficients were calculated using MATLAB program. The basic equation of the IIR filter is shown in equation (2) as follows:

$$y[n] = b_0x[n] + b_1x[n-1] + .. + b_Mx[n-M] \\ - a_1y[n-1] - a_2y[n-2] - .. - a_Ny[n-N] \quad (2)$$

2.3. Data Analysis

The effect of window length and percentage of overlap to the performance of the elbow joint angle estimation was evaluated by calculating the root mean square error (RMSE). The variance of performance was represented in the box plot diagram so that some parameters can be read simultaneously such as minimum, Q1, median (Q2), mean, Q3, and maximum value.

3. Result and Discussion

In the evaluation, the performance was calculated using RMSE which was based on the estimated angle and the measured angle (as the gold standard). Figure 3 and Figure 4 show that the window length and percentage of overlap in windowing technique affected the performance of the elbow joint angle estimation for all time-domain features. In the adjacent windowing technique, a window length of 100 milliseconds has the highest performance in estimation. In the overlap windowing technique, the percentage overlap is 10% and it is the highest performance of the estimation. Features ZC, SSC, WAMP and MYOP have the best RMSE. In the window length of 100 milliseconds, the mean and standard deviation of RMSE for features ZC, SSC, WAMP and MYOP are 14.41°±3.86°, 15.20°±4.59°, 14.50°±3.33° and 13.56°±3.53°, respectively.

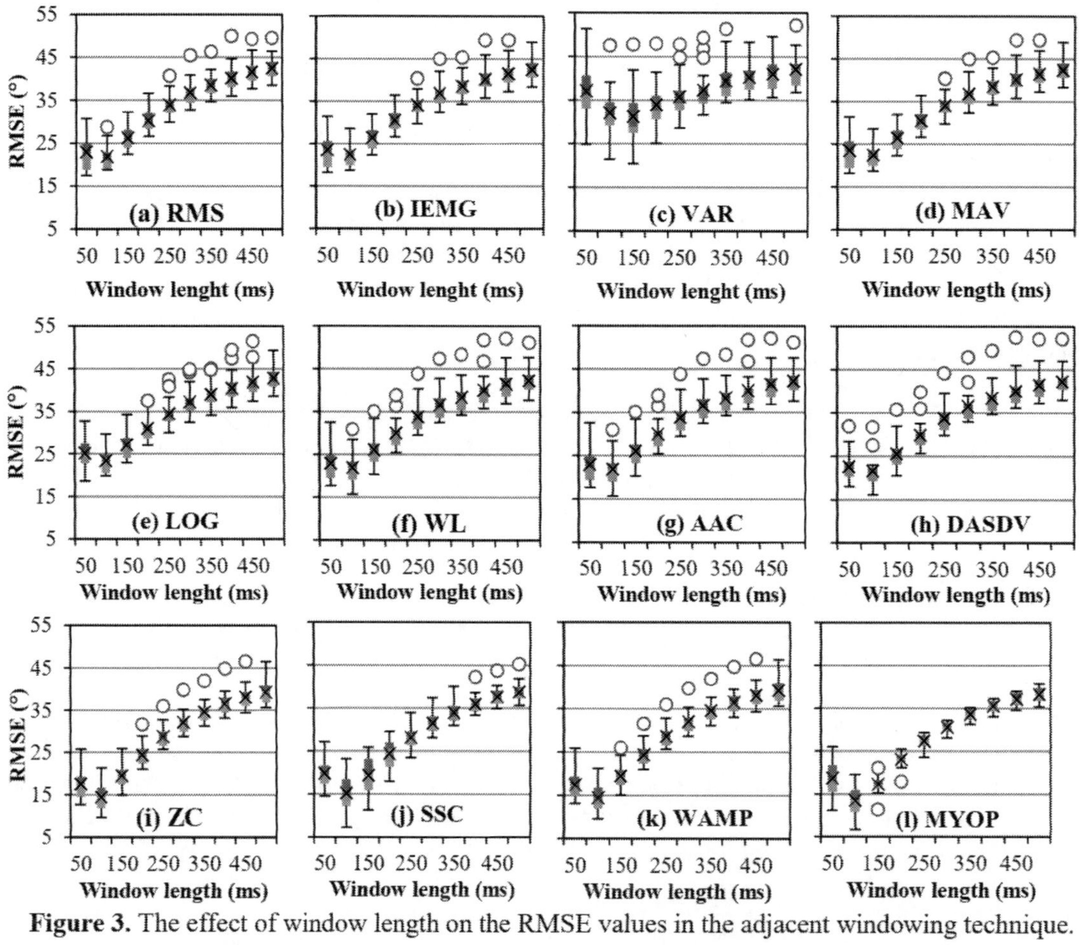

Figure 3. The effect of window length on the RMSE values in the adjacent windowing technique.

In the adjacent windowing technique, the mean RMSE for all features increased after window length of 100 milliseconds (Figure 3). The box plot diagrams (Figure 3) of RMSE show that the RMSE of MYOP feature has the lowest RMSE in the window length of 100 milliseconds (mean±SD.:13.56°±3.53°). The effect of the windows length on the performance of the estimation was also observed in the overlap windowing technique. The RMSE values were increased with length of window and the increase of the RMSE are close to quadratic function (Figure 4).

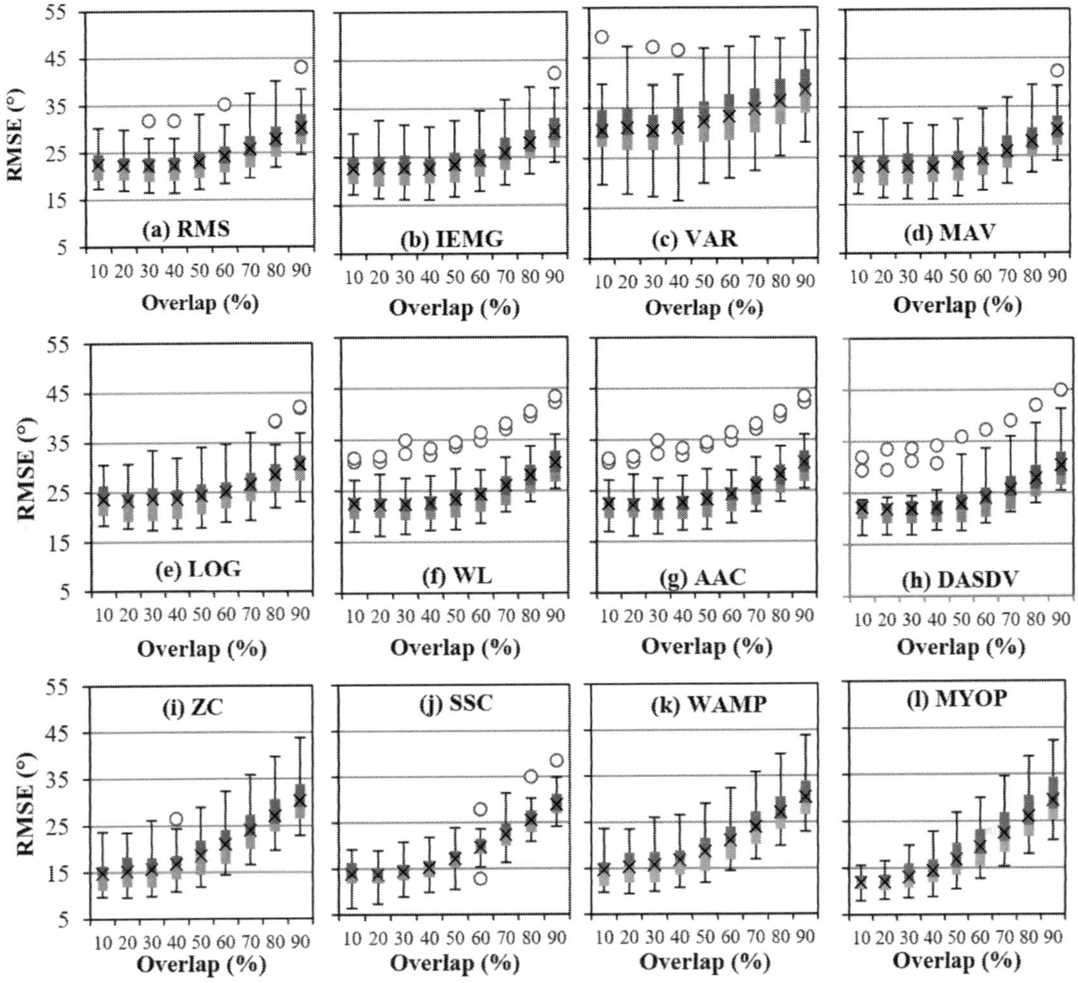

Figure 4. The effect of overlap percentage in RMSE values in the overlap windowing technique.

The impact of the window length to the performance was also showed by some previous studies [3], [5]–[7]. However, generally, this finding is contrary to the previous study which stated that the longer window length, the better the accuracy is. In this study, the performance increased for window length of 50 to 100 milliseconds, however, the performance will decrease after the window length more than 100 milliseconds. The reason is that in the classifier based on machine learning, has more input or longer window of the learning machine. It will eventually close to the real input objects.

The limitation of this study is the experiment that was performed in the continuous motion and in the non-fatigue condition. The evaluation should also be examined when the motion was conducted in the random motion due to the complexities of the EMG signal which will increase in the random motion. Muscle fatigue was an essential problem in the study which used the EMG signal [12][13]. Therefore, it should be considered in the next research when evaluating the effect of window length on the performance of the estimation. In the next research, the period of motion should be considered whether it is affected the estimation or not.

4. Conclusion.

This study evaluates the performance of the elbow joint angle estimation based on the windowing technique. Two windowing techniques, the adjacent and overlap window were applied in this experiment to find the best window length and the percentage of overlap which resulted to the best performance to estimate the elbow joint angle. The performance was indicated by the RMSE. The results of the study suggested that window length of 100 milliseconds and percentage of overlap 10% are effective in the feature extraction process for elbow joint angle estimation using EMG signal. Feature ZC, SSC, WAMP, and MYOP have higher performance than the other features. The performance of the estimation (RMSE) is affected by the length of the windows.

5. References

[1] M Asghari Oskoei and H Hu2007 Myoelectric control systems-A survey*Biomed Signal Process Control*vol 2 no 4 pp 275–294.

[2] L H Smith, L J Hargrove, B a Lock, and T a Kuiken2011 Determining the optimal window length for pattern recognition-based myoelectric control: Balancing the competing effects of classification error and controller delay*IEEE Trans. Neural Syst. Rehabil. Eng*, vol 19 no 2, pp 186–192.

[3] E N Kamavuako, D Farina, K Yoshida, and W Jensen 2009 Relationship between grasping force and features of single-channel intramuscular EMG signals*J. Neurosci. Methods*. Vol 185 no 1 pp 143–150.

[4] Y St-Amant D Rancourt and E a Clancy1998Influence of smoothing window length on electromyogram amplitude estimates*IEEE Trans. Biomed. Eng*. vol. 45 no 6 pp 795–800,.

[5] K Englehart, B Hudgins, and P a Parker2001 A wavelet-based continuous classification scheme for multifunction myoelectric control*IEEE Trans. Biomed. Eng*. vol 48 no 3 pp. 302–311.

[6] E Scheme and K Englehart2011Electromyogram pattern recognition for control of powered upper-limb prostheses: State of the art and challenges for clinical use *J. Rehabil. Res. Dev*. vol. 48 no 6 p 643.

[7] M a Oskoei and H Hu2008Support Vector Machine-Based Classification Scheme for Myoelectric Control Applied to Upper Limb*IEEE Trans. Biomed. Eng*. vol 55 no 8 pp 1956–1965.

[8] JU Chu I Moon and M S Mun,2006A real-time EMG pattern recognition system based on linear-nonlinear feature projection for a multifunction myoelectric hand*IEEE Trans. Biomed. Eng*. vol 53 no 11 pp 2232–2239.

[9] Triwiyanto O Wahyunggoro H A Nugroho and Herianto 2017An investigation into time domain features of surface electromyography to estimate the elbow joint angle*Adv. Electr. Electron. En,* vol 15 no 3 pp 448–458.

[10] M Hakonen H Piitulainen and AVisala2015Current state of digital signal processing in myoelectric interfaces and related applications*Biomed. Signal Process Control* vol 18 pp 334–359.

[11] T Triwiyanto O Wahyunggoro H A Nugroho and H Herianto2017Evaluating the performance of Kalman filter on elbow joint angle prediction based on electromyography*Int. J. Precis. Eng. Manuf*. vol 18 no 12 pp 1739–1748.

[12] T Triwiyanto O Wahyunggoro H A Nugroho and H Herianto 2018Muscle fatigue compensation of the electromyography signal for elbow joint angle estimation using adaptive feature*Comput. Electr. Eng*. vol 71 no 7 pp 284–293.

[13] T Triwiyanto O Wahyunggoro H A Nugroho and H Herianto2017Continuous wavelet transform analysis of surface electromyography for muscle fatigue assessment on the elbow joint motion *Adv. Electr. Electron. Eng*. vol 15 no 3 pp 424–434.

Development of Inquiry-Creative-Process Learning Model to Promote Critical Thinking Ability of Physics Prospective Teachers

Wahyudi[1], N N S P Verawati[1,*], S Ayub[1] and S Prayogi[2]

[1] Universitas Mataram, Jl. Majapahit No.62 Mataram 83115, Indonesia.
[2] Institut Keguruan dan Ilmu Pendidikan (IKIP) Mataram, Jl. Pemuda No. 59A, Mataram 83126, Indonesia.

*Corresponding Author: veyra@unram.ac.id

Abstract. Critical thinking (CT) is one of the skills of the 21st century and becoming one of the goals in the development of science learning. Teaching CT to the prospective teachers have garnered attention for a while, and the role of future teachers seems more crucial than ever for educational systems in terms of improving CT. This study aims to develop Inquiry-Creative-Process (ICP) learning model to promote CT ability of physics prospective teachers (PPT). This study was a development research that produce a product of learning models that meet the criteria of validity, practicality, and effectiveness. The ICP learning model was validated through the focus group discussion mechanism (*for validity aspect*), the implementation of the ICP learning model in the classroom was observed by a number of observers (*for practicality aspect*). The assessment of CT ability was done after the learning process using the ICP learning model (*for effectiveness aspects*) and then analysed. The findings of this research showed the ICP learning model that developed has been declared valid, practice, and effective to promote critical thinking ability of physics prospective teachers. The results of this study could be empirical evidence that the ICP learning model could promote CT ability of physics prospective teachers.

1. Introduction

One of the essential skills that the learners must have in the 21st century is Critical Thinking (CT) skill [1]. In some countries, CT has become a major focus and competency in learning at all levels of their education [2]. In Indonesia, CT has also become a very important part of the competence to be achieved at higher education level, as written in the Regulation of the Minister of Research, Technology and Higher Education of the Republic of Indonesia. Global Citizenship Education (GCE) recommends that universities have to facilitate students to analyse issues critically and identify the creative and innovative solutions. A function of higher education is to teach the students to think. University accreditation boards in some advanced countries, for example, the National Association of Industrial Technology (NAIT), the Accreditation Board of Engineering and Technology (ABET), and the International Technology Education Association (ITEA) recognise competences such as CT, problem solving, communication, and teamwork in their accreditation criteria [3].

CT is a component of high-level thinking skills that must be mastered and taught. CT is reflective and reasonable thinking that is focused on deciding what to believe or do [4], its purpose, and self-regulatory judgment which results in interpretation, analysis, evaluation, and inference, as well as explanation of the evidential, conceptual, methodological, or conceptual considerations upon which that judgment is based [5]. CT is a propensity and skill to engage in an activity with reflective skepticism [6]. CT is used to pass judgment on any information, explain the reasons, and able to solve the problem of the unknown [7], so that each individual are able to understand any information or content on a particular thing [8]. Critical Thinking is best understood as the ability of thinkers to take charge of their own thinking [9]. Despite differences among of thought and their approaches to defining critical thinking, there exist areas for agreement. The researchers of critical thinking typically agree on the specific abilities encompassed by the definition, which include; a) analysing arguments, claims, or evidence, b) making inferences using inductive or deductive reasoning, c) judging or evaluating, and d) decision making or problem solving.

Educational and professional success requires nurturing one's consistent internal willingness to think critically. To do this, the teacher must provide students with as many models, opportunities, exemplars, and explanations as possible in order to help them operationalise their skills. Educators have long been aware of the importance of CT skills as an outcome of students learning [10]. However, teaching CT remains confusing for many instructors [11]. This is partly due to the lack of clarity of various methods proposed to best teach of CT [11, 12]. At the higher education level, Bissell and Lemons [13] ascertained faculties who teach at universities consider critical thinking a primary objective. It is a sad truth that the average of college students does not think critically, and not all courses include CT. In learning CT requires a holistic approach and should involve a set of appropriate learning models. Therefore, it is necessary to develop a set of specific learning models to increase learners' CT ability.

Learning models based on inquiry activities have been widely developed for the purpose to increase students' CT ability, since inquiry is an instructional model that aims to guide about how learners think. The processes of scientific creativity in the inquiry activities need to be revealed as a way to increase the CT ability of physics prospective teachers. The development of learning models by integrating creativity processes with scientific inquiry activities needs to be explored and developed for that purpose. Creative processes or so-called scientific creativity potentially train the CT ability of learners [14]. Aspects of scientific creativity are in the form of problem finding, problem solving, hyphothesis creating, experiment designing, and product designing [15-19]. These aspects will be integrated with the inquiry model into a set of learning model, the learning model in this study is called Inquiry Creativity Process (ICP). The learning phases of the ICP model are; a) Establishing set and finding problem; b) Creating hyphothesis; c) Designing experiment creatively; d) Solving problem science creatively; and e) Designing product creatively.

This study aims to develop an Inquiry Creative Process (ICP) learning model to increase CT ability of physics prospective teachers. In this study, ICP is seen as a product of learning model. The criteria of good quality products according to Nieveen [20] are valid, practical, and effective.

2. Methods

This research was a development research that paired theory of Borg and Gall [21], and theory of Nieveen [20]. The validation method was used to collect the validity data of the ICP learning model. The validation aspects of the ICP model include content validity and construct validity. Content validity refer to the components of the model should be based on state-of-the-art knowledge, and construct validity refer to the all components of the model should be consistently linked to each other and logically. Validation is intended to obtain suggestions and feedback from validators. The validators were experts who also as user of learning model (lecturer). Technically, the ICP model validation was done through a focus group discussion (FGD) mechanism, its involving 5 (five) experts. Feedbacks from the validators in the FGD were then followed up to improve the ICP model.

The implementation subjects of the learning model are 21 physics prospective teachers (PPT) in the faculty of teacher training and education (FKIP), Mataram University, Indonesia. Practicality of the model evaluated from the learning feasibility (LF), the model declared to be practice if LF at least good criteria. The effectiveness of the model was evaluated from the improvement CT ability after the implementation of the model. CT ability was evaluated using the scoring technique adapted from Prayogi et al [22]. The indicators of CT ability in this study were indicators that have been used by previous researchers those are analysis, inference, evaluation, and decision making [1, 22]. While to know the score change of CT ability, it's analysed by using *n-gain* equation. The effectiveness criteria of the learning model is satisfied if there was an increase in critical thinking skills after the implementation of the model, and score of CT ability at least critically criteria in the posttest. Criteria of validity, learning feasibility and critical thinking ability shown as Table 1.

Table 1. Criteria of validity, learning feasibility and critical thinking ability

Validity		**Learning Feasibility**		**Critical Thinking Ability**	
Interval	***Criteria***	***Interval***	***Criteria***	***Interval***	***Criteria***
Va > 4,21	Very valid	Va > 4,21	Very good	X > 17,6	Very critically
3,40 < Va < 4,21	Valid	3,40 < Va < 4,21	Good	11,2 < X ≤ 17,6	Critically
2,60 < Va < 3,40	Quite valid	2,60 < Va < 3,40	Adequate	4,8 < X ≤ 11,2	Critically enough
1,79 < Va < 2,60	Less valid	1,79 < Va < 2,60	Less	-1,6 < X ≤ 4,8	Less critically
Va < 1,79	Invalid	Va < 1,79	Poor	X ≤ -1,6	Not critically

Note: *Va= validity level score; X=CT ability level score*

3. Results and Discussion

The validity results show that the ICP learning model which is developed was declared valid (Va = 4.12). The result of the validators assessment of the ICP learning model are provide in Table 2.

Table 2. Expert validation results on the ICP learning model

	Aspects of validation	**Average score**	**Category**
1. Content validity	The need for development of model.	4.20	Valid
	The model designed based on state-of-the-art of knowledge.	3.95	Valid
2. Construct validity	Consistency and logically of all arrangement components of model.	4.20	Valid
	Va	4.12	Valid

The content validity on aspects of the need for development of the model has a score of Va of 4.20 with valid criteria. This result cannot be separated from the purpose of ICP model development to increase the CT ability of prospective teachers as the need for the main competence of graduates of 21st century skills. The ICP model has also fulfilled the expectation of the higher education requirement in Indonesia. The aspect of needs in this study also becomes the answer of previous research findings that the model of ICP learning as an alternative model that can be used to increase CT ability of prospective teachers, as explained by Fahim and Masouleh [9] that educational success requires nurturing one's consistent internal willingness to think as well as developing one's thinking skills, and the educator should find ways in order to make learners willing and disposed to think critically. In this study, the experts as the validator have agreed that the ICP model has met several aspects of needs both globally and internally, and the most important that the ICP learning model has met the needs aspects in accordance with the regulation of competence to be achieved in learning in Indonesia, for the purpose of CT. Internally in the context of university learning, CT is a crucial skill that students need to develop while at university. It is important for a well-educated person to be able to make well-informed judgments, be able to explain their reasoning and be able to solve unknown problems. Therefore, CT should be developed to the learner from the early learning in university [7]. Content validity on the state-of-the-art aspects has a Va score of 3.95. The development of the ICP model was supported by empirical studies of scientific creativity from recent research references that true scientific creativity can serve as a bridge for CT purposes [14-19].

In construct validity component with main aspect was consistency and logically of all arrangement of component of model, the experts as validator give average score of Va equal to 4.20 with valid criteria. Experts have agreed that the ICP learning model has been consistent and logical in terms of the learning phase, sequence and interrelationships between components in the learning activities. According to Nieveen [20], if all components are consistently linked to each other (construct validity) then the product was considered to be valid.

At the implementation step in the classroom, the ICP learning model was considered practice, it was measured from the learning feasibility with the very good criteria (LF=4.25). Learning feasibility of each meeting shows in the Table 3.

Table 3. Learning feasibility of ICP learning model based on assessment of observers

Learning phases	*Learning feasibility*				*Mean*	*Mean of LF*	*Category*
	Meeting of learning						
	1st	*2nd*	*3rd*	*4th*			
I	3.4	4.0	4.5	4.0	3.97	4.25	Very good
II	4.0	4.0	4.0	4.5	4.13		
III	4.0	4.0	4.0	4.5	4.13		
IV	4.5	5.0	4.0	5.0	4.63		
V	4.5	4.5	4.5	4.0	4.37		

The ICP learning model consists of five learning phases, those are; establishing set and finding problem; creating hyphotheses; designing experiment creatively; solving problem science creatively; and designing product creatively. Observation result of learning feasibility using ICP learning model were done very good (LF = 4.25). It's because of the supports, especially the availability of the learning tools, including handbook (module) and worksheet. When learning tools were designed well, it can give information which help learner more effective to accomplish learning objectives. Good designed learning tools were functioned as communication tool, learning plan tool, learning plan for students, learning resources, and learning evaluation tool. The support from module was also very important in this study. The material in the book was arranged systematically so it can condition students to learn [23]. The worksheet in this study were designed as the guidelines of learning feasibility in inquiry activity according to ICP learning model to train students' CT ability.

Measurement results of CT ability showed that the mean score on pretest consist of 0.52 with the criteria of "less critically" and posttest consist of 17.5 with the criteria of "critically." N-gain value of 0.72 with the criteria of "high." The results are thoroughly provided on Table 4.

Table 4. Measurement results of CT ability

CTs interval	*Criteria*	*Pre test*		*Post test*		*N-gain*	*Criteria*
		Freq.	*Mean*	*Freq.*	*Mean*		
$X > 17{,}6$	VC	0	0.52	11	17.5	0.72	High
$11{,}2 < X \leq 17{,}6$	C	0	(Less critically)	10	(Critically)		
$4{,}8 < X \leq 11{,}2$	CE	1		0			
$-1{,}6 < X \leq 4{,}8$	LC	17		0			
$X \leq -1{,}6$	NC	3		0			
Amount		21		21			

Note: *VC (Very critically), C (Critically), CE (Critically enough), LC (Less critically), NC (Not critically)*

The effectiveness of the ICP learning model is measured by improving physic prospective teachers CT ability. The results indicate that there is an increasing ofCT ability after the implementation of the ICP learning model from "less critically" to "critically." This indicates that the developed ICP learning model was considered effective. The implementation of ICP learning model to the prospective teachers of physics is in line with the demand that a physics learning has to master CT ability to correlate and interrelate between two or more theories and concepts in learning physics. The use of ICP learning model obviously increased the PPT ability in developing their CT. The result of this current study was parallel with the previous relevant study which found that the intervention of scientific creativity to develop CT ability was significantly effective [24], in line with the result of this study, the implementation of exploration an creative ideas in the acquisition inquiry activity is highly enhance prospective teachers ability in developing their CT [1].

The enhancement of CT ability of PPT through ICP learning model cannot be separated from the intervention of each phases implemented during the learning process in ICP model in which the phases of learning were consistently training the CT ability. The problem finding and science creatively problem solving which are the dimension of scientific creativity, both have a correlation in the context to train critical thinking. The cognitive dimensions of creative thinking certainly correlated with some of the dimensions of CT, especially when the learners were thinking in the context of problem solving. When the learners were thinking in a given context (critical thinking), they used various thinking processes (creative thinking). The properties of CT were linked to the creative abilities during problem finding and problem solving [25]. Creative problem finding ability was defined as a kind of intellectual trait or ability that was demonstrated in the process of producing and expressing new-found questions in a unique, novel and useful and purposeful way, using existing contexts and experience. Furthermore, when problems were facing, hypothesis were needed to define the most appropriate way to solve those problems, those will automatically reinforce to think critically [26]. Creatively product design was a part of scientific creativity in which the learners are demanded to be able to design scientific product as the result of scientific creativity [19], those points were the important invention in terms of developing prospective teachers CT ability.

The ICP learning model has been effective in improving the CT ability of prospective teachers on aspects of analytical, inference, evaluation, and decision making. Its effectiveness is influenced by the validity of the model to improve CT, and also learning feasibility which are very good. When learning was well designed and properly planned it can help learners to achieve the expected learning objectives.

4. Conclusion

The ICP learning model was developed consist of five phases of learning, those are; establishing set and finding problem; creating hyphotheses; designing experiment creatively; solving problem science creatively; and designing product creatively. The results showed that the ICP learning model have been valid, practice, and effective to improve the CT ability of physics prospective teachers, including aspects of analysis, inference, evaluation and decision making.

5. References

[1] Prayogi S, Yuanita L & Wasis 2018 Critical Inquiry Based Learning: Model of Learning to Promote Critical Thinking Ability of Pre-service Teachers *J. Phys.: Conf. Ser.* **947** 1-6

[2] Schmaltz R M, Jansen E & Wenckowski N 2017 Redefining Critical Thinking: Teaching Students to Think like Scientists. *Front. in Psych* **8**(459) 1-4

[3] Scott S 2008 Perceptions of Students' Learning Critical Thinking through Debate in a Technology Classroom: A Case Study *Jour. of Tech. Stud* **34**(1) 39-44

[4] Ennis R H 1989 Critical thinking and subject specificity: Clarification and needed research *Edu. Resear* **18**(3) 4-10

[5] Facione P A 1990 *Critical thinking: A statement of expert consensus for purposes of educational assessment and instruction* (Millbrae, CA: The California Academic Press)

[6] McPeck J E 1990 Critical thinking and subject specificity: A reply to Ennis *Edu. Resear* **19**(4) 10-12

[7] Thomas T 2011 Developing First Year Students' Critical Thinking Skills *Asian Soc. Sci* **7**(4) 26-35

[8] Zane T 2013 *Implementing Critical Thinking with Signature Assignments* (Salt Lake Community College: Springer)

[9] Fahim M & Masouleh N S 2012 Critical Thinking in Higher Education: A Pedagogical Look *Theory and Prac. in Langu. Stud* **2**(7) 1370-1375

[10] Lai E R 2011 *Critical Thinking: A Literature Review* (Pearson)

[11] Bensley D A & Murtagh M P 2012 Guidelines for a scientific approach to critical thinking assessment *Teac. Psycho* **39**(1) 5-16

[12] Abrami P C, Bernard R M, Borokhovski E, Wade A, Surkes M A and Tamim R 2008 Instructional interventions affecting critical thinking skills and dispositions: a stage 1 meta-analysis *Rev. of Edu. Resear* **78**(4) 1102-1134
[13] Bissell A N and Lemons P P 2006 A new method for assessing critical thinking in the classroom *BioScience* **56**(1) 66-72
[14] Adams K 2006 *The Sources of Innovation and Creativity* (National Center on Education and Economy: USA)
[15] Turkmen H and Sertkahya M 2015 Creative Thinking Skills Analyzes of Vocational High School Students *Jour. of Edu. and Instruc. Stud. in the World* **5**(1) 74-84
[16] Ayas B and Sak U 2014 Objective Measure of Scientific Creativity: Psychometric Validity of the Creative Scientific Ability Test *Think. Skills and Creat* **13** 195-205
[17] Hu W, Shi Q Z, Han Q, Wang X and Adey P 2010 Creative Scientific Problem Finding and Its Developmental Trend. *Creat. Resear. Jour* **22**(1) 1-7
[18] Aktamis H & Ergin O 2008 The Effect of Scientific Proccess Skill Education on Student's Scientific Creativity, Science Attitudes, and Academic Achivements *Asia-Facific Forum on Sci. Lear. and Teac* **9**(1) 1-21
[19] Hu W & Adey P 2010 A Scientific Creativity Test for Secondary School Students *Inter. Jour. for Sci. Edu* **24**(4) 389-403
[20] Nieveen N 1999 *Prototyping to Reach Product Quality* (Kluwer: Academic Publisher)
[21] Borg W R and Gall M D 1983 *Educational Research: An Introduction* (London: Longman Inc)
[22] Prayogi S, Yuanita L and Wasis 2018 Critical Inquiry Based Learning: A Model of Learning to Promote Critical Thinking Among Prospective Teachers of Physic *Journal of Turkish Science Education* **15**(1) 43-56
[23] Levin B 2008 *How to Change 5000 Schools: A Practical and Positive Approach for Leading Change at Every Level* (Cambridge: Cambridge University Press)
[24] Koray Ö and Köksal M S 2009 The effect of creative and critical thinking based laboratory applications on creative and logical thinking abilities of prospective teachers *Asia-Pacific Forum on Sci. Lear. and Teac* **10**(1) 1-13
[25] Kousoulas F & Mega G 2008 *Creative and critical thinking in the context of problem finding and problem solving: A research among students in primary school.* (http://www.ep.liu.se/ecp/021/vol1/011/ecp2107011.pdf)
[26] Alberta Education 2010 *Inspiring Education: A Dialogue with Albertans* (Edmonton AB: Alberta Education)

Acknowledgement
This study is a part of decentralized research grant that financed fully by the Ministry of Research, Technology, and Higher Education of the Republic of Indonesia.

Indigenous Bacteria Community Fluctuation During Fermentation of Water Hyacinth (*Eichhornia crassipes*) and Corncob (*Zea mays*)

Isnawati[1], Ni'matuzahroh[2], and T Surtiningsih[2]

[1] Faculty of Mathematics and Natural Science, State University of Surabaya, Jalan Ketintang, 60231, Surabaya, Indonesia
[2] Faculty of Science and Technology, Airlangga University, Jalan Mulyorejo, 60114, Surabaya, Indonesia

Abstract. Water hyacinth nowadays is a disturbing water-weeds. In fact, water hyacinth is potential for cattle feed if it is mixed with corncob, and thus increase the nutritional value. In cattle feed production, there are some indigenous fungi and bacteria to be used for fermentation. This research was aimed to know the fluctuation of bacteria and fungi community for 15 days of fermentation process. The bacteria and fungi were first inoculated and isolated in both NBA and PDA agar plate. Then, they were purified so as to make pure-culture that will be easily identified. Bacteria identification was conducted using *Microbact Identification Kits (Microbact™ GNB12A and 12B)* and identification book *Bergey's Manual of Determinative Bacteriology*. Fungi identification were conducted based on macroscopic and microscopic characteristics. Ecological parameters such as diversity, evenness, and dominancy were analyzed on this research. There were eight bacteria species found, including *Bacillus laterosporus*, *B. badius*, *B. pantothenticus*, *B. brevis*, *Staphylococcus sciuri*, *B. stearothermophilus*, *Burkholderia pseudomallei*, and *Enterococcus duran*. The highest bacteria diversity were in the fourth day, the highest evenness occured at the first, tenth, twelveth, and fiftenth day. All of the bacteria species were dominant during fermentation, except *Bacillus badius* at the eleventh day that was a subdominant bacteria.

1. Introduction

Up to now, the feed stock especially the cheap and nutritious one has become a main problem in breeding. The cheap and nutritious feed can be produced through fermentation process by utilizing farm-waste and weeds materials. The efforts in producing cheap and nutritious fermented feed can be made of farm-waste, such as rice-straw, corn-straw (its leaves and stems), corncob, *Leguminoceae* waste, and other farm wastes. From all farm wastes, corncob has not been utilized optimally. Rice-straw is generally big-sized ruminants feed, such as cows so that in this research fermented feed is made from corncob as the main material.

Corncob is potential to be used as cattle feed because it contains 5,6% protein which is higher than the protein concentration in rice-straw (4,9%). The utilization of corncob is only for fuel (as substitutive component for wood), also for mushrooms growth media [1]. Some researchers had been utilized corncob as cattle feed, such as Sumaoang (corncob powder) [1]. Through fermentation process by utilizing microorganisms and molase addition, corncob not only becomes a nutritious feed but also becomes favourite food for pigs, goats, and poultries [1]. The corncob that has been blended also can be used to replace cassava slices as buffalo feeds [2]. Then, Rostika and Safitri [3] used corncob that has been blended and fermented with various mushrooms as fish feed. Lardy and Anderson [4] stated that corncob can be utilized as ruminant feed alternative.

However, there are some weaknesses in using corncob as feed materials that up to now can not be overcome. The weaknesses including of the high natural lignocellulose concentration (45-55% cellulose, 25-35% hemicellulose, 20-30% lignin) and also high fiber concentration (930 g netral detergent fibers/kg dry materials; 573 g acid detergent fibers/kg dry materials) that are difficult to digest in cattle [5]. The corncob processing before it is used in fermentation, such as rarefaction, fermentation itself and addition other materials will increase the overall nutrient value and the quality.

The nutrient enhancement of fermented feed can be conducted by adding other materials, so that the fermented feed does not contains a single material only. Water weeds, water hyacinth (*Eichhornia crassipes*) is a very suitable added material. It is because the water hyacinth can grow very fast [6], contains high protein around 11.87% and 14.28 % percentage and contains high calcium and phosphorous concentration, and therefore stimulate the milk production if it is combined with suitable concentration [7]. The nutrient concentration did not change because of its growth geographical difference [8]. However, the use of water hyacinth as cattle feed has some weaknesses. Water hyacinth contains a very high rough fiber concentration, so that the carbon source is less good. The other weakness is that the water concentration of water hyacinth is relatively very high, the texture is smooth, and the protein is difficult to digest [9]. Based on Umsakul, Dissara, and Srimuang [10] research the mixing process of water hyacinth in feed production can increase the nutrient value of the feed.

Saputro [9] had conducted fermentation process in water hyacinth by using *Aspergillus niger* obtained the duration of fermentation in 8 days to obtain fermented feed woth rough protein 18,84% percentage and 15,73% rough fiber. The other research related to the utilization of water hyacinth as fermented feed material has used tempe yeast as microorganism sources [11]. The feed produced can increase goat body weight. Research result conducted by Phioneer, Yurmiati, and Sinaga [12] on *New Zealand White* rabbit showed that the usage 20% water hyacinth provides body weight addition, but not influence the ratio usage efficiency.

There have been some efforts in increasing the feed quality by using fermentation technique. The fermented feed has some strengths such as high nutrient concentration, high digestibility, and high palatibility [13].

Fermentation process always involves microorganisms, including bacteria and fungi. The rate of fermentation process depends on some factors, mainly the compatibility of microbes related to fermentation process of materials and physical-chemical parameters. It means that fermentation process using certain materials will have specific and selected microorganisms [8]. This has relation with the enzymes of the microbes, whether it is suitable to degrade the chemical compounds in the materials that will be fermented or not. The diversity of indigenous microorganisms in every fermentation phase need to be known to give scientific reasons that are important to make a fit formula. The indigenous microorganisms diversity that have been known will give information about microorganisms population that have dominant role in fermentation, so that the suitable starter formulation will be easier.

The other factor determining the fermentation rate is physical condition and chemical condition that support fermentation process. Physical condition consists of temprature, pH, and humidity [14]. The chemical condition that influences fermentation process include of the presence of carbon and nitrogen sources, also respiratory gases (in aerobic and anaerobic condition) and suplement molecules to support microorganisms population growth. Research result of Belal [15] showed that different carbon sources will cause different fermentation result in rice-straw to produce bioethanol.

In the research that have been conducted, the nutrient value of water hyacinth and corncob mixture was high with fermentation duration for 5 days [16]. The goats that have been given fermented feed of water hyacinth and corncob mixture show higher biomass compared to the biomass when using conventional feed [16, 17]. The other strength of water hyacinth and corncob mixture fermented feed, based on Saputro [9], is that it can improve the goat carcass, signing by increasing protein concentration and lowering fat content. Those studies focus mainly on the use of tempe-yeast for fermenting agent which fullfil the standard needed 5 days.

The research about bacteria and fungi species involved in fermentation process, especially with water hyacinth and corncob mixture as materials has not been conducted. Knowledge about bacteria and fungi species related in fermentation process of certain materials is very essential to increase fermentation result quality and the rate of fermentation itself. This research will investigate the fluctuation of bacteria and fungi species incolving in fermentation process from day to day during fermentation process, then the diversity, evenness, and dominancy of bacteria and fungi species will be analyzed.

2. Method

Basically in this research, there were some stages, each of stages will be detailed as follow:

2.1 Indigenous microbe sources preparation and isolation

Water hyacinth as microbes sources was taken from a river that the water had been ensured unpolluted by doing water quality test and heavy metal concentration test. The corncob was taken from the rice mill that had been ensured uncontaminated from dangerous substances. As initial indigenous microbes sources, firstly the production of fermented feed with water hyacinth and corncob mixture was conducted with ratio 1:1. This feed was made by some stages icluding materials cutting, materials drying, steaming, and materials incubation so that fermentation will occur naturally [17]. Everyday during fermentation process, the indigenous microbes isolation by taking materials randomly for 30 gr, then the materials suspensed in sterile aquades, filtered, and cultured by pour plate method. Bacteria media culture used NA (Nutrient Agar for Microbiology Merck). Culture then incubated in 37°C temprature for 24-72 hours.

2.2 Purification, characterization, and identification of indigenous bacteria and fungi

The culture mixture was then purified. Characterization of pure bacteria isolates was conducted by macroscopic and microscopic observation. Identification of pure bacteria isolates was conducted using *Microbact Identification Kits (Microbact™GNB12A and 12B).*

2.3 Indigenous bacteria and fungi diversity observation during fermentation process

During fermentation process (for 15 days), observation and documentation including numbers and types of isolates that grow were conducted everyday. Then, counting of diversity parameter was conducted index of heterogeneity or index of diversity, index of evennes, and index of species richness. Index of diversity was counted using Shannon-Wiener formula as follow: $H' = - \sum (pi \log pi)$ (H' is index of diversity, pi adalah ratio of individual number in a species and total individual numbers). Index of evenness is counted by using formula $E = H'/\ln (S)$ (E is index of evenness of a species , H' adalah Shannon-Wiener index of diversity, S is numbers of species found, ln is natural logaritm). Index of species richness is caunted using formula $Di = ni/N$ (Di is species dominancy, ni adalah number of individuals in a species, N adalah number of individuals of all species).

3. Result and Discussions

The result obtained from daily observation from day to day in 15 observation days. The data obtained then was analyzed the community diversity, evenness, and daily species dominancy was conducted during fermentation process. The result obtained then presented in Table 1 as follow.

Table 1. Diversity, evenness, daily species dominancy during water hyacinth and corncob mixture fermentation process

Day of fermentation	Total Bacteria Numbers	Diversity	Evenness	Bacteria Dominancy
1	134,5	0,2981	0,4301	*Bacillus laterosporus (56,13%)* *Bacillus stearothermophilus (43,87%)*
2	242	0,4708	0,4286	*Bacillus laterosporus (39,67%)* *Bacillus stearothermophilus (26,03%)* *Enterococcus durans (34,30%)*
3	304,5	0,5346	0,3856	*Bacillus laterosporus (33,66%)* *Bacillus badius (5,42%)* *Bacillus stearothermophilus (23,15%)* *Enterococcus durans (37,77%)*
4	308	0,7146	0,3988	*Bacillus laterosporus (33,77%)* *Bacillus badius (7,63%)* *Bacillus pantothenticus (10,55%)* *Bacillus brevis (8,60%)* *Bacillus stearothermophilus (21,1%)* *Burkholderia pseudomallei (18, 34%)*
5	307,5	0,6216	0,3862	*Bacillus laterosporus (41,63%)* *Bacillus badius (18,21%)* *Bacillus pantothenticus (5,69%)* *Bacillus brevis (12,52%)* *Bacillus stearothermophilus (21,99%)*
6	284	0,4591	0,4179	*Bacillus laterosporus (45,78%)* *Bacillus badius (22,36%)* *Bacillus stearothermophilus (31,87%)*
7	280	0,4818	0,3475	*Bacillus laterosporus (49,46%)* *Bacillus badius (8,39%)* *Bacillus pantothenticus (6,96%)* *Bacillus stearothermophilus (35,18%)*
8	278	0,3872	0,3525	*Bacillus laterosporus (52,16%)* *Staphylococcus sciuri (7,01%)* *Bacillus stearothermophilus (40,83%)*
9	298	0,4687	0,4268	*Bacillus laterosporus (47,82%)* *Bacillus badius (5,20%)* *Bacillus brevis (8,22%)* *Bacillus stearothermophilus (38,76%)*
10	210	0,3004	0,4334	*Bacillus laterosporus (47,51%)* *Bacillus stearothermophilus (52,49%)*
11	391	0,6525	0,4054	*Bacillus laterosporus (24,78%)* *Bacillus badius (4,47%)* *Bacillus stearothermophilus (25,54%)* *Burkholderia pseudomallei (20,05%)* *Enterococcus durans (25,16%)*
12	289	0,4764	0,4336	*Bacillus laterosporus (30,62%)* *Bacillus stearothermophilus (34,08%)* *Enterococcus durans (35,29%)*
13	342	0.7502	0,3855	*Bacillus laterosporus (23,39%)*

Day of fermentation	Total Bacteria Numbers	Diversity	Evenness	Bacteria Dominancy
				Bacillus badius (6,58%) *Bacillus pantothenticus (5,70%)* *Bacillus brevis (5,99%)* *Staphylococcus sciuri (6,43%)* *Bacillus stearothermophilus (25,59%)* *Enterococcus durans (26,32%)*
14	298	0,7888	0,4054	*Bacillus laterosporus (25,67%)* *Bacillus badius (10,91%)* *Bacillus pantothenticus (7,72%)* *Bacillus brevis (10,74%)* *Staphylococcus sciuri (8,56%)* *Bacillus stearothermophilus (26,85%)* *Burkholderia pseudomallei (9,56%)*
15	306,5	0,6010	0,4335	*Bacillus laterosporus (22,35%)* *Bacillus badius (25,78%)* *Bacillus pantothenticus (27,24%)* *Bacillus stearothermophilus (24,63%)*

Based on Table 1, the diversity, evenness, and species dominancy were different in day-to-day observation. By using culture method in Nutrient Broth, eight bacteria species were successfully isolated, consisted of *Bacillus laterosporus*, *B. badius*, *B. pantothenticus*, *B. brevis*, *Staphylococcus sciuri*, *B. stearothermophilus*, *Burkholderia pseudomallei*, and *Enterococcus duran*. The highest bacteria diversity was in the fourth day, the highest evenness occured at the first, tenth, twelveth, and fifteenth day. All of the bacteria species were dominant during fermentation, except *Bacillus badius* at the eleventh day that was a subdominant bacteria.

Feed materials consisting of water hyacinth and corncob contain high cellulose. Because of that, the indigenous bacteria that previously have been isolated from the material are dominated by bacteria that have cellulolytic activity. From those, there are 2 bacteria that do not show cellulolytic activity, namely *Bacillus brevis* and *Burkholderia pseudomallei*. Even though the two bacteria cannot degrade cellulose, this two species can inhibit water hyacinth and corncob mixture because they can use cellulose degradation result by other bacteria as carbon and energy sources. Based on bacteria diversity data during fermentation process as in Table 1, the presence of *Bacillus brevis* and *Burkholderia pseudomallei* have been just detected in the fourth day of fermentation process. This gives signal that the two bacteria can utilize the cellulose degradation result by other bacteria to fulfill their needs. This is very different to *Bacillus laterosporus* which is present in the feed since the first day until fifteenth day, when the isolates taking stopped.

The indigenous bacteria diversity involved in fermentation process of water hyacinth and corncob mixture are different from day to day. The diversity of indigenous bacteria consisted of species variation and the numbers of bacteria that inhibit those materials. The overall bacteria diversity during fermentation process are classified low (Shannon-Wiener index of diversity value = 0,7065) ,which means that the diversity is low. This indicates that during fermentation process of water hyacinth and corncob mixture, there was not a big species diversity, only a few species involved in the fermentation process. In every species, individual numbers are not even with index of evenness 0,3398 which means that the evenness is low. This indicates that the species diversity are not distributed, only certain species involved in the fermentation process of water hyacinth and corncob mixture. The dominant species during fermentation process are *Bacillus laterosporus*, *B. badius*, and *Enterococcus durans*. Each different fermentation (mesophylic, termophylic, cooling and maturing) has different microorganisms variation. Fermentation process of a material involve some kinds of bacteria and fungi [18].

4. Conclusion

In fermentation process, the mixture of water hyacinth and corncob involves eight bacteria species including *Bacillus laterosporus, B. badius, B. pantothenticus, B. brevis, Staphylococcus sciuri, B. stearothermophilus, Burkholderia pseudomallei*, and *Enterococcus duran*. The diversity, evenness, and dominancy of related bacteria in those fermentation materials from day to day are various. The highest diversity is at the fourth day, while the highest evenness occured at the first, tenth, twelveth, and fiftenth days. All of bacteria species are dominant during fermentation process, except *Bacillus badius* at the eleventh day that was a subdominant bacteria.

Based on the research result obtained, there are bacteria diversity, evenness, and dominancy fluctuation from day to day during fermentation. Therefore, in the utilization of fermented feed as indigenous bacteria consorcium starter, all the catched isolates have to be sponsored everyday.

5. References

[1] Sarian Z B 2016(online) 'Corn Cobs Converted Into Nutritious Animal Feed' Version: 30 Oktober 2016 (http://www.zacsarian.com/category/agri-ideas)

[2] Wanapat M, Pilajun R, Kang S, Setyaningsih K, and Setyawan A R 2012 Effect of Ground Corn Cob Replacement for Cassava Chip on Fermentation and Urinary Derivatives in Swamp Buffaloes *Asian-Aust. J. Anim. Sci.* **25**(8) pp. 1124-1131

[3] Rostika R and Safitri R 2012 Influence of Fish Feed Corn-Cob Was Fermented By *Trichoderma sp.*, Aspegillus *sp., Rhizopus oligosporus* To The Rate of Growth of Java Barb *(Puntius Gonionitus) APCBEE Procedia* **2** pp. 148-152

[4] Lardy G and Anderson V 2009 *Alternative Feeds for Ruminant* (NDSU: Dakota)

[5] Kanengoni A T, Chimonyo M, Ndimba B K, and Dzama K 2015 Potential of Using Maize Cobs in Pig Diets — A Review *Asian Australas. J. Anim. Sci.* **28**(12) pp. 1669-1679

[6] Tham H T 2012 *Water Hyacinth (Eichhornia crassipes)-Biomass Production, Ensilability and Feeding Value to Growing Cattle* (Thailand: Disertasi Swedish University of Agricultural Sciences. Uppsala Thailand)

[7] Kumar A, Sharma P C, Kumar A, and Negi V 2011 A Study on Phenotypic Traits of *Candida* Species Isolated from Blood Stream Infections and Their *In Vitro* Susceptibility to Fluconazole *Al Am een J Med Sci.* **7**(1) pp. 83-91

[8] Boboescu I Z, Ilie M, Gherman V D, Mirel I, Pap B, Negra A, Kondorosi E, Biro T, and Maroti G 2014 Revealing the Factors Influencing A Fermentative Biohydrogen Production Process Using Industrial Wastewater as Fermentation Substrate *Biotechnology for Biofuels*. **7** pp. 139-154

[9] Saputro T 2016(online) Pakan untuk Ternak Domba. *Ilmu Ternak* (http://www.ilmuternak.com/2015/03/pakan-untuk-ternak-domba.html)

[10] Umsakul K, Dissara Y, and Srimuang N 2010 Chemical, Physical and Microbiological Changed during Composting of the Water Hyacinth. Pakistan *Journal of Biological Sciences* **13**(20) pp. 985-992

[11] Fitrihidajati H, Isnawati, and Suparno G 2013 Pemanfaatan Eceng Gondok (*Eichornia crassipes*) untuk Pakan Ternak Ruminansia sebagai Salah Satu Cara Mengatasi Gulma Perairan (Surabaya: Laporan Penelitian Hibah Bersaing Universitas Negeri Surabaya)

[12] Phioneer H R, Yurmiati H, and Sinaga S 2016(online) Tingkat Penggunaan Eceng Gondok (Eichhornia *crassipes*) dalam Silase Ransum Komplit terhadap Pertambahan Bobot Badan dan Efisiensi Ransum Kelinci Peranakan New Zealand White (http://download.portalgaruda.org/article.php)

[13] Seo J, Jung J K, and Seo S 2015 Evaluation of Nutritional and Economic Feed Values of Spent

Coffee Grounds and Artemisia Princeps Residues as a Ruminant Feed Using in Vitro Ruminal Fermentation. *Peer J.* **3** e1343

[14] Alrumman S A 2016 Enzymatic Saccharification and Fermentation of Cellulosic Date Palm Waste to Glukose and Lactic Acid *Brazilian Journal of Microbiology* **47** pp. 110-119

[15] Belal E B 2014 Bioethanol Production from Rice Straw Residu *Biotechnol Biofuels* **7** pp. 139-142

[16] Fitrihidajati H, Isnawati, and Suparno G 2014 *Pemanfaatan Eceng Gondok (Eichornia crassipes)* untuk *Pakan Ternak Ruminansia sebagai Salah Satu Cara Mengatasi Gulma Perairan* (Surabaya: Penelitian Hibah Bersaing. Universitas Negeri Surabaya)

[17] Fitrihidajati H, *Ratnasari E, Isnawati,* and *Soeparno G 2015* Kualitas Hasil Fermentasi Pada Pembuatan Pakan Ternak Ruminansia Berbahan Baku Eceng Gondok (*Eichhornia crassipes*) *Journal of Biosaintifika* **7**(1) pp. 62-67

[18] Ramos C L, de Almeida E G, Freire A L, and Schwan R F 2011 Diversity of Bacteria and Yeast in The Naturally Fermented Cotton Seedand Rice Beverage Produced Food Microbiology **28**(7) pp. 1380–1386

Utilising "Meriam Karbit" Indigenous Knowledge to Construct Alternative Physics Experiments Based Smartphone Camera

A Pramuda[1,*], Mundilarto[2], H Kuswanto[3], S Hadiati[4]

[1, 4] Physics Education Program, IKIP-PGRI Pontianak, INDONESIA
Student Postgraduate Program, Universitas Negeri Yogyakarta, INDONESIA
[2] Physics Education Program, Universitas Negeri Yogyakarta, INDONESIA
[3] Physics Study Program, Universitas Negeri Yogyakarta, INDONESIA

*Corresponding Author: adipramuda@ikippgriptk.ac.id

Abstract. Many research proved that indigenous knowledge is closely related to the context of today's concept of science and the recent smartphone technology can be used to reveal it. The motivation of this research is to suggest alternative physics experiments that could be conducted by the teacher using relevant traditional technology and smartphone sensor. In order to find alternative experiments and to explore the physics concepts behind the use of paper projectile in real "Meriam Karbit" attractions, we used a modified Meriam Karbit into a miniature form and its projectile with two mass variations. Recorded videos of Meriam Karbit miniature explosion from smartphone camera were analysed using VidAnalysis applications. The study succeeded in revealed that the higher the mass ratio of the Meriam Karbit system to the mass of the projectile, the velocity ratio of the two systems was even higher, and vice versa. The mass difference between paper used with the mass of real Meriam Karbit made from very large wood is enormous, the paper gets a huge portion of the burst velocity of the explosion while the Meriam Karbit is apparently unmoved. This was a more rational explanation of the use of paper projectile and could be used to study the momentum conservation concepts.

1. Introduction

The problems of physics education in West Borneo that arise was standardised experimental physics apparatus that the number has not been met, and many teachers didn't conduct alternative physics experiments. In many studies found that the real impact of learning physics without experiment activities was students become passive when learning physics [1]. On the other hand, many research at the global level has uncovered the benefits of smartphones and sensors integrated with it to overcome the limitations of conventional experimental apparatus and created an innovative and exciting atmosphere of physics learning [2-4]. However, the learning settings taken were not in accordance with Indonesian culture, this will lead to the phenomenon presented to be more imaginary and not touch the phenomenon that exists in the student's life, whereas many local wisdom that can be utilised for the development of physics learning, as one example is traditional technology. Technology in an indigenous tribe containing the correct knowledge and logic when associated with the scientific context at present [5]. Specifically reviewing the potential of traditional technology in West Borneo was discovered a technology locally named as Meriam Karbit. Meriam Karbit famous because it could produce canon-like sound and explosion. At present many people at the school age who held the celebration of incoming month Ramadhan using this Meriam Karbit didn't know what is the physics concept behind the functional system of Meriam Karbit [6]. Correlated with the results of the gun analysis, Meriam Karbit technology showed the knowledge in processing sound concept, pressure due to gas expansion, and explosion [7]. Despite the loud burst of explosions, this Meriam Karbit end is only covered with newsprint. The paper mass used as the cover is much lighter when compared to the mass of iron balls on the guns and the proportions are not as proportional to the size of the wood. In addition to this, the indigenous community apparently has knowledge of setting this Merriam Karbit on the banks of the Kapuas river with good stabilization, especially when reviewing its dimensions. The use of thin paper as a projectile and the installation of Meriam Karbit in such a way on the banks of the river is closely related to the concept of momentum and the law of conservation of momentum [8-11]. The problem is how to package and explore this indigenous knowledge in a class scale through experimentation activities, easy to do and understood by learners.

2. Literature Review

The explosive phenomenon of a weapon is an interaction that conceptually opposite of the collision phenomenon [8]. In Indonesia, it was very difficult to find the reality of events where archers or hunters were on the ice and other slippery places when firing shots and also the space shuttle releases its cargo charge [9-11]. Most schools have no experimentation tools similar to tubes with a massless spring that can be compressed between the two balls [8]. This is not in line with evidence-based strategies whereas this topic is generally placed as an introduction to lead to the law of conservation of momentum, so this simple explosion becomes difficult to understand [12]. Reviewing the law of the momentum, it can be stated that the total momentum ($\vec{P}$) of an isolated system is constant. Interaction within the system does not change its total momentum. The law of conservation of momentum directly has the consequence that the interaction between an isolated system is a pair of actions-reactions [8]. It could be written as in equation (1), with $\vec{P_i}$ is the total momentum before the interaction and $\vec{P_f}$ the total momentum after the interaction.

$$\vec{P_f} = \vec{P_i} \tag{1}$$

Since equation (1) is a vector equation, then when it described in its vector component for N the number of interacting particles, it can be expressed as in equation (2).

$$\begin{gathered}(p_{fx})_1 + (p_{fx})_2 + (p_{fx})_3 + \cdots = (p_{ix})_1 + (p_{ix})_2 + (p_{ix})_3 + \cdots \\ (p_{fy})_1 + (p_{fy})_2 + (p_{fy})_3 + \cdots = (p_{iy})_1 + (p_{iy})_2 + (p_{iy})_3 + \cdots\end{gathered} \tag{2}$$

If there are two objects A and B in a system. When these two objects interacting with each other can be regarded as an isolated form of a system, if the upward normal forces compensate for the force of gravity so that $\vec{F}_{netto} = 0$, for example as in the case of two spheres placed on a tube with a massless spring compressed between the second ball. The total momentum of this system is conserved. The total system momentum before it is detonated is zero because the two objects are initially in an equilibrium state, as a consequence, the total momentum will also be zero [8]. After being detonated when reviewing the direction of x, it can be expressed as in equation (3) and because the initial conditions of the velocities A and B are zero, the equation (4) can be obtained.

$$m_A(v_{fx})_A + m_B(v_{fx})_B = m_A(v_{ix})_A + m_B(v_{ix})_B = 0 \tag{3}$$

$$\frac{m_A}{m_B} = -\frac{(v_{fx})_B}{(v_{fx})_A} \tag{4}$$

Equation (3) and equation (4) show that the total momentum of the system is zero, but the momentum of each object is not zero [8]. Object A must have the same momentum as the magnitude of the momentum of object B, but with a negative sign indicating that the object will move in the opposite direction. Phenomena in mechanics with real-world condition settings can be easily displayed and analysed using computers, tablets, and smartphones [13-16]. Momentum and collisions conventionally used wireless cart experiments that the velocity can be visualised to a computer or PC. Cart equipped with this wireless device when used for very frequent experimentation intensity and rough handling were susceptible to damage and costs predicted to be much more expensive when compared to using video analysis especially using a device to mirror the smartphone screen to projectors [17]. Technology is now available, although there was a few research that uses it primarily to explore the concept of the explosion.

3. Material & Methodology

In the preliminary of this study, we reviewed various ethnographic documents that describe the traditional technologies in West Borneo. Information relating to physical form including the selection of materials and functional were matched with the structure of physics concepts based smartphone experimental activities. This study used projectors, Dongles and WiFi connections to clarify visualisation in class scale settings as shown in Figure 1. Explosive forces and smoke effects such as Meriam Karbit were obtained from gas expansion coming out of the butane fill cans which are often used for gas lighter fillers, associated with Meriam Karbit miniature used elastic capillary pipes.

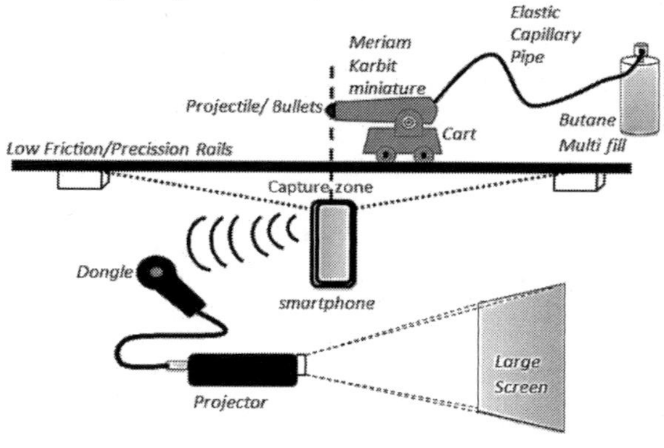

Figure 1. The experimental set up with Meriam Karbit Miniature and cameras is configured with the Dongle or AnyCast to be displayed in front of the class and real-time manner.

The projectile used two variations, in the first variation used a projectile that was not given additional mass so that was lighter and on the second variation used projectile was given thin metal as additional mass so that was heavier. The focus of the study was on the effect of the mass ratio of the system on the use of the projectile before and after adding the mass to the velocity ratio of Meriam Karbit and cart system after the explosion. From equation (4) it could be counting that higher the mass ratio of the Meriam Karbit system to the mass of the projectile, the velocity ratio of the two systems is even higher, this is in line with the first variation when used projectile without additional mass [8,10]. Conversely, lower the mass ratio of the Meriam Karbit systems to the mass of the projectile, the velocity ratio of these two systems is even lower, and this is in line with the second variation. The obtained video frame first was in trim or cut, so positioning from frame to frame becomes easier using VidAnalysis App.[15]. If the time and position of the moving object were known, then it can be determined its velocity with the equation (5).

$$\vec{v}_{average} = \frac{\Delta \vec{r}}{\Delta t} \tag{5}$$

In equation (5) $\left(\vec{v}_{average}\right)$ is the average velocity vector of an object during the time interval (Δt) in which the object undergoes a displacement $(\Delta \vec{r})$.

4. Results and Discussion

Figure 1 shows the experiments using the Meriam Karbit experimental system with the projectile without additional mass, Figure 2 and Figure 3 show the using of the VidAnalysis application on the Android smartphone that was generated the analytical output.

IOP Conf. Series: Journal of Physics: Conf. Series **1108** (2018) 012007 doi:10.1088/1742-6596/1108/1/012007

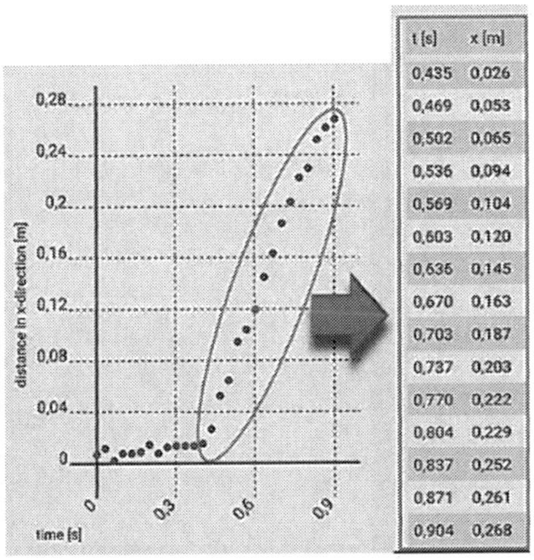

Figure 2. The output of VidAnalysis App. of the Meriam Karbit miniature motion without an additional mass on the projectile.

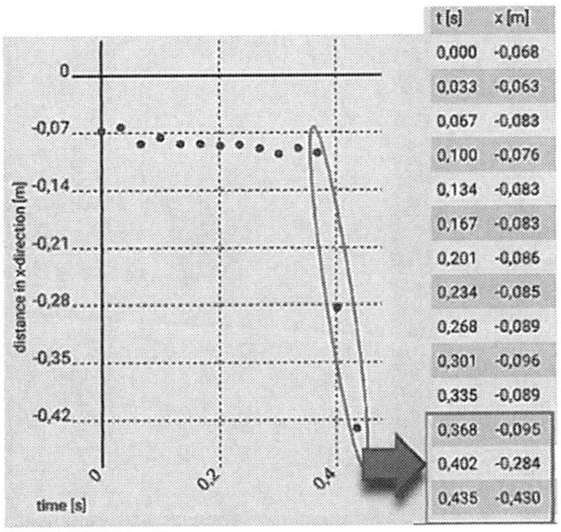

Figure 3. The output of VidAnalysis App. of the Meriam Karbit projectile motion without additional mass.

Both figures were produced from a recorded video at the same time, despite the focus of the frame to frame analysis was different. Figure 2 shows that the object to which the observation focuses was the Meriam Karbit on the dynamics cart and precision rails (on projectile without additional mass variations setting). While in Figure 3 shows the object that became the focus of observation is the Meriam Karbit projectile (on projectile without additional mass variations setting). The VidAnalysis application could analyse the position of the object according to the cartesian coordinates in the x and y directions, so to begin analysing after determining which video recording to analyse, would be asked to set the center of the motion coordinate axis and calibrate how many meters the actual distance from the two reference points calibration on video. The use of precision rails that have been equipped with a measuring scale facilitates the calibration process performed. The motion of the Meriam Karbit and the projectile were fixed in the x-axis direction. Figure 2 shows that the Meriam Karbit on cart motion toward the positive x-axis and its opposite with projectile motion toward the negative x-axis. This opposite direction of motion was clearly in line with the concepts presented earlier, which has consequences for the occurrence of action-reaction pairs. The graph of the projectile motion in Figure 3 was steeper than the Meriam Karbit motion graph in Figure 2, which indicates that the projectile moved faster than the Meriam Karbit on cart motion. The data points on the graph that were marked with the ellipse line correspond to the position and time values in the table marked with the square line on the right. The unmarked dot was a condition before the explosion, so close to constant indicated there was no movement. Non-zero value because the screen size of the smartphone is smaller than the tablet, we found the positions determined in the frame-by-frame on VidAnalysis become more difficult to precise. However, when the object velocity calculated using equation (5), this inaccuracy can be further eliminated. Figure 4 shows Meriam Karbit on the dynamics cart and the precision rail system with projectile weighted (used additional mass). Figure 5 shows that the focus of observation was Meriam Karbit projectile. The video trim process in Figure 4 and Figure 5 produces data that was very close to the event of the explosion, so the unimportant motion data of the object at rest before the explosion wouldn't be raised too much.

IOP Conf. Series: Journal of Physics: Conf. Series **1108** (2018) 012007 doi:10.1088/1742-6596/1108/1/012007

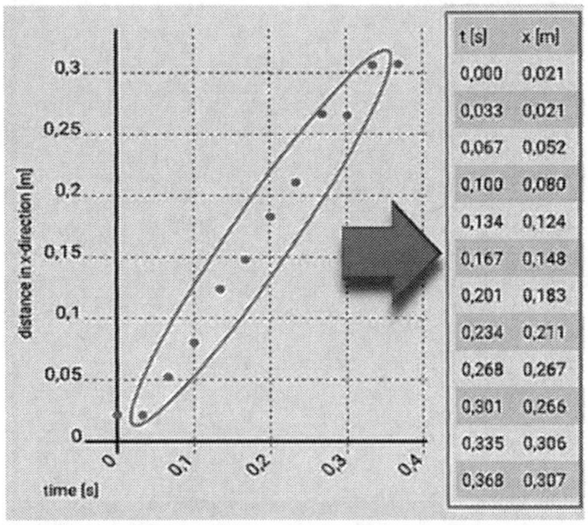

Figure 4. The output of the VidAnalysis App. of the Meriam Karbit miniature motion with the projectile was weighted.

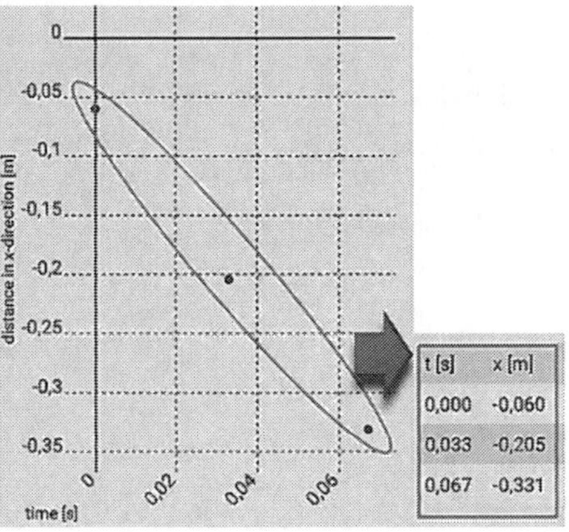

Figure 5. The output of the VidAnalysis App. of the Meriam Karbit projectile motion with the projectile was weighted.

As in the previous variations, the graphs in Figure 4 and Figure 5 show that the Meriam Karbit on cart motion toward the positive x-axis and its projectile motion was aimed toward the negative x-axis. The projectile still moves faster than the Meriam Karbit motion. Table 1 is presents the average velocity of each system that was calculated as the results in order to compare easily the difference in the results of the analysis between two variations of the projectile.

Table 1. The Meriam Karbit (MK) miniature and projectile velocity calculated from VidAnalysis.

Projectile Variation	Object of VidAnalysis	x_i (m)	x_f (m)	$\Delta\vec{r}$ (m)	t_i (s)	t_f (s)	Δt (s)	$\vec{v}_{average}$ (ms^{-1})
Light Projectile	MK	0,026	0,268	0,242	0,435	0,904	0,469	0,515
	Projectile	-0,095	-0,430	0,335	0,368	0,435	0,067	5,000
Heavy Projectile	MK	0,021	0,307	0,286	0,000	0,368	0,368	0,777
	Projectile	-0,060	-0,331	0,271	0,000	0,067	0,067	4,044

The calculation results that the average velocity of each projectile variation and the system in Table 1 shows that in the second variation the Meriam Karbit system moved faster when firing the projectile with additional mass, but the projectile with additional mass became slower than the first variation. The experimental results using Meriam Karbit miniatures and the underlying basis revealed by Indigenous knowledge show clearly and in accordance with the concepts presented, that the higher the ratio or mass ratio of the Meriam Karbit system to the mass of the projectile system, the velocity ratio of the two systems even higher, and vice versa for mass ratio of Meriam Karbit system with the mass of the projectile system getting lower. This is a more rationale of the use of paper projectiles at the attractions of Meriam Karbit. The mass difference between newsprint used with the mass of Meriam Karbit is enormous. The paper gets a huge portion of the burst velocity of the explosion and can be ripped instantly, while the Meriam Karbit is apparently unmoved. Meriam Karbit mooring in such way with Gertak on river banks more related to equilibrium concepts especially related to stability and balance because the size of the Meriam Karbit used. This, of course, will be much different if the projectile used has a larger mass, so the explosion will result in a larger Meriam Karbit velocity in the opposite direction from the direction of the projectile motion and it is definitely necessary to structure the Meriam Karbit more forcefully. In the context of everyday life in West Borneo the experimental system used was predicted to be more contextual and reasoned by students than by using two ball systems that is detonated using a spring compression in a tube, or an athlete with an arrow and a hunter with a gun that stands up on the skis above slippery ice floors, and space shuttles that release cargo in the deep space. In the broader context, the method used in this study is in line with the research paradigm related to traditional technological analysis. Indigenous knowledge can be introduced in physics learning using an advanced organizer because the cognitive load related to learning content increased. The advance organizer can also be presented using a smartphone and to clarify the visualization in front of the classroom can be used also Dongle and Anycast, so it can go hand in hand with real-time experimentation.

5. Conclusion

This study has succeeded in getting alternative physics experiment related to the reason of the use of thin paper as a projectile on the Meriam Karbit attractions. In the class scale to be easy to do and understood by learners, we used the modified Meriam Karbit into the form of a miniature. The projectile used two mass variations. Although in miniature form, the explosion event still occurs very fast when viewed with the eyes. The effect of this variation on the motion of the projectile and the miniature system can be recorded using a smartphone camera and VidAnalysis application used to get the position data versus time so that the average velocity can be calculated easily. Meriam Karbit miniature system moved slower when firing the lighter projectile or without additional mass, but the projectile velocity became faster. This results have a good agreement with momentum conservation and the real Meriam Karbit attractions, the paper gets a huge portion of the burst velocity of the explosion while the Meriam Karbit is apparently unmoved. This study has the potential to create an enjoyable learning and showed the contribution to the efforts of indigenous knowledge preservation.

6. References

[1] Mundilarto Ismoyo H 2017 *J. Baltic Sci. Educ.* **16**(5) 761–79
[2] Dilek U and Erol M 2018 *Phys. Educ.* **53** 1–4
[3] Vollmer M and Möllmann K P 2018 *Phys. Educ.* **53** 035019 1–11
[4] Pramuda A, Hadiati S and Sasono M 2016 *JPFK* **2**(2) 70–77
[5] Tapper R and Mclachlan K 2005 *Technology Tradition and Survival Aspects of Material Culture in the Middle East and Central Asia* (London: Frank Cass Publishers)
[6] Oxtora R 2017 (online) Meriam karbit resmi warisan budaya Pontianak ed Marboen A P (http://www.antaranews.com/berita/637447/meriam-karbit-resmi-warisan-budaya-pontianak)
[7] Štiavnický M and Lisý P 2011 *Sci. and Military* **1** 66–74
[8] Knight R D 2008 *Physics for Scientists and Engineers a Strategic Approach 2nd edition* (San Francisco: Pearson Education Inc.) chapter 9 p 240
[9] Halliday D, Resnick R and Walker J 2011 *Fundamentals of Physics 9th edition* (Hoboken NJ: John Wiley and Sons Inc.) chapter 9 p 210
[10] Serway R A and Jewett J W 2004 *Physics for Scientists and Engineers 6th edition* (Belmont, CA: Thomson Brooks/Cole) chapter 9 p 251
[11] Giancolli D C 2001 *Principles with Application 6th edition* ed Challice J (Upper Saddle River NJ: Pearson Education Inc.) chapter 7 p 167
[12] Streepey J W 2013 *The Phys. Teach.* **51** 54–55
[13] Klein P, Gröber S, Kuhn J and Müller A 2014 *Phys. Educ.* **49**(1) 37–40
[14] De Jesus V L B and Sasaki D G G 2016 *Phys. Educ.* **51** 055006 1–7
[15] tsaedek 2017 (online) VidAnalysis free (https://play.google.com/store/apps/details?id=com.vidanalysis.free)
[16] Pereira V, Ramos P M, da Silva P P and Silva M R 2017 *The Phys. Teach* **55** 312–13
[17] Vandermarlière J 2016 *The Phys. Teach.* **54** 566–7

Acknowledgement

This research is fully supported by The Ministry of Research Technonolgy and Higher Education Republic of Indonesia. Research Grant Number: 001/L.202.103/PDD/III/2018.

IOP Conf. Series: Journal of Physics: Conf. Series **1108** (2018) 012008 doi:10.1088/1742-6596/1108/1/012008

The oral mathematical communication profile of prospective mathematics teacher in mathematics proving

K V Pantaleon[1,2], D Juniati[3], and A Lukito[4]

[1] STKIP Santu Paulus Ruteng, Flores, NTT, Indonesia
[2,3,4] Faculty of Mathematics and Science, Universitas Negeri Surabaya, Indonesia
Corresponding author: christianvictor1979@gmail.com

Abstract. As reported by many recent researchers, communicating mathematical ideas verbally is not an easy task, especially in mathematics proving. A qualitative research was undertaken to describe the processes of oral mathematical communication emerged when the subject did the proving task. The subject was one prospective mathematics teacher, with the high mathematical skill, selected from 48 students who took the proving tasks of geometry and algebra. Data were collected through tests, assignments, and depth interviews. The results of this study showed that the oral mathematical communication in the proving of geometry task complied with the following stages: explaining what is understood, expressing the idea in the form of drawing/symbol, explaining the idea/argument, presenting the solution steps, and reinforcing the results obtained. Furthermore, the oral mathematical communication in the proving of algebra tasks followed the following stages: explaining what is understood, explaining the idea about the form of mathematical logic, putting forward the argument, re-explaining what is understood, explaining the idea/argument, presenting the solution steps, presenting ideas in a more understandable form, putting forward the argument, and reinforcing the results obtained. In general, the subject could communicate her ideas precisely, coherently, and clearly.

1. Introduction

Oral mathematical communication is the process of conveying mathematics ideas or understanding verbally by speaking it up, the art of transmitting mathematics knowledge directly to another. Speaking is a productive skill that challenges students' capability to perform a task [1] which can take place in the form of a conversation or student talk. The exploration of mathematical ideas from various perspectives through conversation or student talk can help students enhanced their thinking and make connections [2]. Student talk is the most effective way for students to demonstrate and improve their understanding [3]. Through oral communication, students are trained to express ideas and understanding about mathematics by speaking them directly in front of others. In this paper, oral mathematical communication is defined as the process of conveying ideas and understanding of mathematics done by the prospective mathematics teacher verbally by articulating it up directly in front of other students.

Like any other parts of mathematics, the proof needs to be communicated both in writing and speaking. Communicating proof in writing means step by step verbal explanation of the truth of a mathematical statement so that it can be understood and accepted with common sense. This truth will only be acceptable if it is supported by a variety of previously proven axioms, theories, definitions, or prior knowledge. Obviously this is not an easy task. Many studies at the high school level and above have shown that developing an understanding of mathematical proof and proving remains a challenge for many students [4]. Mathematical proof is known as the most difficult topic to learn [5, 6]. The problem of proof seems to be more complicated because the aim of a problem to prove is to show conclusively that a certain clearly stated assertions is true, or else to show it is false [7].

The research about oral mathematical communication has been investigated in many studies. Kabael [8] found out that junior high school mathematical teachers were less able to communicate using mathematical language. Though, mathematics is very important [9], and can be used as a tool to communicate effectively. Dewi [10] studied about mathematical communication of prospective teachers in solving mathematical problems. She also found out that prospective mathematics teacher were less able to communicate mathematical ideas and understanding. In fact, teachers are one of the key aspects of students success in learning [11]. Related to the mathematics proving, Yackel and Hanna [4] found out that developing an understanding of mathematics proving remains a challenge for many students. To communicate the proof requires a broader and more comprehensive understanding of knowledge such as conceptual understanding, procedural knowledge, proof methods, and reasoning abilities [12,13]. The various studies that have been done above emphasize the end result of communication rather than the communication process itself. In fact, communication is a process, that is the process of delivering ideas. Therefore, the process should be more emphasized.

2. Method

This research was a qualitative research. Data were collected by using tests, assignments, in-depth interviews, and questionnaire. A mathematical ability test consists of 10 questions was given to 48 students of mathematics education program of STKIP Santu Paulus Ruteng, Flores and used as a basis for determining the subject of research. Based on the results of the test, only six students have high mathematical ability. All of them are female students. One of them was chosen as a research subject. High-ability student was selected as research subject with consideration so that researchers can explore as many communication activities as possible in the proving process. In the next stage, the researcher gave the proving task to 48 students including the subject in it. The task consisted of two questions that were about geometry and algebra. After performing this task, the subject was asked to explain it verbally in front of the 11th grade students of the science program of SMAK Santu Fransiskus Xaverius Ruteng. During describing the task in front of the class, the researcher observed and recorded the performance of the research subject by using the handy-cam from beginning up to the end without interruption. The results of this recording were transcribed to serve as one of the guidelines in conducting interviews other than subject's writing on the board. The transcribed data and the results of the interview were analysed based on the theoretical framework of oral mathematical communication. Triangulation was conducted by giving Task II which was similar to Task I with the same step to ensure that the data was valid.

3. Result and Discussion

3.1. Data Description Based on Task I

Task I consists of two proving problems, each measuring the proving skill in the field of geometry and algebra, those are: 1) prove that if *ABCDEF* is a regular hexagon with the side length is *a* unit then the area of *BCEF* = $a^2\sqrt{3}$ units, and 2) prove that for every integer *n*, if n^2 is odd then *n* is odd.

For the first problem, the subject began the proof by writing down the problem that would be proven. Based on the problem, the subject asked questions about what is known and what will be proved. According to the subject, it was known that *ABCDEF* is a regular hexagon with the side length is *a* unit, and would be proved that the area of *BCEF* is $a^2\sqrt{3}$ unit. Based on the known elements, the subject drew a regular hexagon *ABCDEF* (See Figure 1a). The subject also affixed the letter *a* on each side of the hexagon to indicate that the length of the side is *a* unit. Then, the subject argued that the regular hexagon consisted of six congruent equilateral triangles. Based on this understanding, the subject determined the centre point O, and connected it to each vertex with a straight line to form six congruent triangles. The subject also affixed the letter *a* on each side of the triangle formed to express the length of the sides. In addition, the subject also created a dashed line connecting B to F and C to E so as to form the *BCEF* region. To demonstrate the area of *BCEF*, the subject drew an *ABCDEF* regular hexagon which contained *BCEF* rectangle (See Figure 1b). Then, the subject determined the length of *BE* and used it as one element to determine the length of *CE*. The subject drew again the *BCEF* rectangle and created diagonal lines of *BE* and *CE* that intersected at the center point O (See Figure 1c). Based on the picture, the subject explained that the length of $BE = 2a$. Furthermore, using the Phytagoras formula, the subject determined the length of *BF*. To apply this formula, first, the subject drew a rectangle *BCEF* and $\overline{BE}$ diagonal line that formed two right triangles (See Figure 1d). The subject affixed a little corner (∟) at the angle *EFB* to confirm that the triangle formed is a right triangle. After the length of *BE* was obtained, using the Phytagoras formula, the subject showed that $BF = a\sqrt{3}$. Finally, the subject showed that the area of the rectangle $BCEF = a^2\sqrt{3}$.

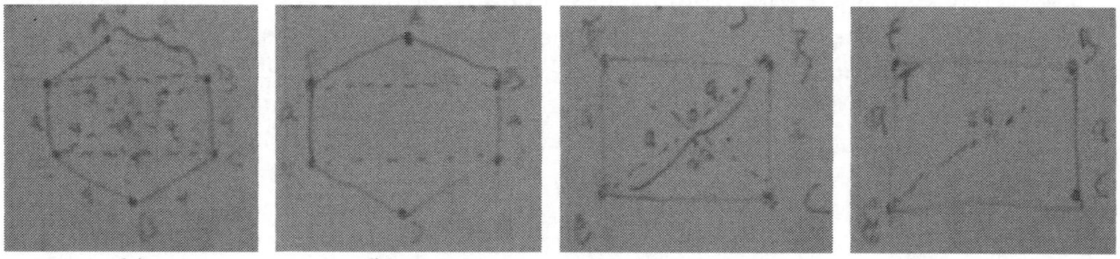

(a) (b) (c) (d)

Figure 1. (a) The process flow of completion for the first question in task I, starting for (a), (b), (c), and the last is (d)

As in the previous question, on the second question, the subject began the proof by writing the question and then asked the students to read it together. To ensure that students understand the meaning of the theorem, the subject gave a concrete example using integer 3. The subject explained that $3^2 = 9$ is an odd and it turned out that 3 is also an odd number. However, subject said that is just an example that could not be used as a basis to make general conclusion, therefore it needed to be proven. Furthermore, with the students, the subject identified the known elements and the elements to be proven based on the question. The subject found that, the known elements are $n \in Z$ and n^2 is odd, while the element to be proved is n is odd. Then the subject explained the logic model of the implications contained in the statement. The subject said that the implication model is $p \to q$ where p is "n^2 is odd" while q is "n is odd". Moreover, the subject argued that the form of the implication $p \to q$ is equivalent to its contraposition, $\neg q \to \neg p$, which brought about the statement of the implication could be proven by using contraposition. The subject said that the contraposition of "if n^2 is odd then n is odd" is "if n is even then n^2 is even". Based on this argument, the subject re-identified the elements known and the elements to be proven. According to the subject based on the contraposition statement, the known element, n, is even, and n^2, the element which was to be proved is also even. Since the known n is even, the subject explained the concept of even number. According to the subject, the even number is the integer that can be divided exactly by 2, therefore if n is even then $n = 2k$ where k is an integer, so $n^2 = (2k)^2 = 4k^2$. Then, the subject converted $4k^2$ to $2(2k^2)$ according to the pattern of even numbers. The subject argued that since k is an integer then by the closure property of the integer multiplication operation, $l = 2k^2$ is also an integer and $n^2 = 2l$. The subject argued that since $2l$ is divisible by 2 then n^2 is even. At the end of the proof, the subject explained that since the statement "if n is even then n^2 is even" is true then the statement "if n^2 is odd then n is odd" is true as well because both are equivalently related.

3.2. Data Description Based on Task II

Task II was given as an effort to check the consistency of the process of oral mathematical communication that emerged during the subject proved the statements of mathematics. In doing so, the two questions in Task II were similar to the two questions in Task I. The two given questions were: (1) prove that the regular hexagonal area with a side length is a is $\frac{3}{2}a^2\sqrt{3}$ units, and (2) prove that for every integer n, if n^2 is even then n is even.

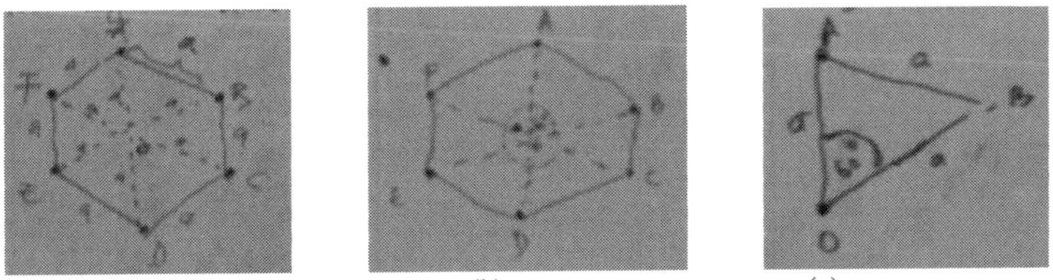

(a) (b) (c)

Figure 2. (a) The process flow of completion for the first question in task II, starting for (a), (b), (c), and the last is (d)

To deal with the first question, the subject began the proof by writing the question then asked the students to read it together. Based on the question, the subject asked questions about what is known and what will be proved. Through this way, the subject conveyed what she understood. According to the subject, it is known that there is a regular hexagon with the side length is a unit, and will be proved that its area is $\frac{3}{2}a^2\sqrt{3}$. Since the known is the hexagon, the subject drew a regular hexagon named *ABCDEF* by writing the constituent letters at each vertex (See Figure 2a). The subject also affixed a letter a on each side of the hexagon to indicate that the length of the sides is a unit. The subject explained that the sides of a regular hexagon have the same length. Then the subject determined the centre point O, and connected it to each hexagon angle with a straight line to form six triangles. The subject argued that the regular hexagon consisted of six congruent equilateral triangles so that the side lengths of the triangles is equal to the side lengths of the regular hexagon. Therefore, the subject affixed the letter a on each side of the formed triangle to express the side length. Furthermore, the subject determined the angle of the triangle right at the centre point O. The subject drew again the regular hexagon *ABCDEF*, determined its midpoint O and connected it to each vertex on the hexagon. In addition, the subject also created a circle centred at the point O (See Figure 2b). The subject explained that the degree measure of circle is 360°, so that the degree measure of every angle of the triangle coincided with the centre point O is $\frac{360^\circ}{6} = 60^0$. Furthermore, the subject determined the area of triangle *AOB*. For that reason, the subject drew the triangle *AOB* (See Figure 2c) and affixed the letter a on its sides. Then the subject argued that because the regular hexagon is made up of six equilateral triangles, the area of the regular hexagon is equal to six times the area of one of its constituent triangles. Using this concept, the subject then showed that the regular hexagonal area is = $\frac{3}{2}a^2\sqrt{3}$. At the end of the proof, the subject briefly confirmed the results obtained.

In the second question, the subject began the proof by writing the question then asked the students to read it together. Furthermore, together with the students, the subject identified the known elements and the elements to be proven based on the question. From the results of the identification it was found that, the known element is "n is an integer and n^2 is even", while the element to be proved is "n is even". Then the subject explained the logic model of the implications contained in the problem. The subject explained that the implication model is $p \rightarrow q$. Moreover, the subject argued that the form of implication $p \rightarrow q$ is equivalent to its contraposition, $\neg q \rightarrow \neg p$, which caused that the statement of implication can be proved by using contraposition. Based on this argument, the subject re-identified and re-explained the known elements and the elements to be proved based on its contraposition. According to the subject, the known element is "n is odd", whereas the element to be proved is "n^2 is odd". According to the subject, the odd number is an integer that is not divisible by 2, so that if n is odd then $n = 2k + 1$ where k is an integer, so $n^2 = (2k+1)^2 = 4k^2 + 4k + 1 = 4(k^2 + k) + 1 = 4l + 1$ with $l = k^2 + k$. The subject argued that since k is an integer then based on the closure property, $l = k^2 + k$ is an integer as well. To bring $4l + 1$ to an odd number form, the subject manipulated it to $2(2l) + 1 = 2m + 1$ with $m = 2l$. The subject argued that since l is an integer then $m = 2l$ is also an integer. Furthermore, the subject argued that since $2m + 1$ met the notation of the odd number pattern then $n^2 = 2m + 1$ is also a member of an odd number. At the end of the proof, the subject briefly explained: "it is proven that if n is odd, then n^2 is odd, so it is true that "if n^2 is even then n is even".

3.3. Discussion

The results of this study showed that the subject always started her oral explanation by rewriting the problem on the board. This is a good thing in communication. By rewriting the questions, the teacher ensures that the student knows clearly what he or she will learn. In addition, in this way, students have the opportunity to mobilize the various prerequisite knowledge needed to understand what the teacher will explain. Subsequently, the subject explained what she understands from a matter that includes known elements and elements to be proved. This step is very important in the process of proof. Polya asserts that in the early stages of proof we must identify the main points in the matter, such as the hypothesis (the known element), and the conclusion (the element to be shown) [7]. The hypothesis will be the assumption underlying the subsequent reasoning process to show that conclusions are acceptable. In this study, the subject could explain the main elements correctly. During the process of explanation, the subject also always put forward ideas to justify the reasoning process that she did. This is consistent with what the NCTM emphasizes that the subject should be able to explain the logical and systematic reasons for each reasoning process chosen [2]. In this study, using such logical reasoning, the subject could present the steps of completion systematically. In the process of completion, the subject often manipulate ideas and present them in other more easily understood forms. This is also one of the communication skills that needs to be developed. By re-envisioning ideas in a more suitable form, solving the problem would be much easier. At the end of her explanation, the subject always reaffirmed the results already obtained to confirm that the conclusion is proved. This also fits the rules of writing proof [7].

In addition to the results described above, it also appears that there is a difference in the communication strategy the subject chooses in explaining both types of questions given. For geometry problems, the subject tend to communicate her ideas by using images as symbolic representations of the concepts to be explained. This is in line with other findings which showed that in the geometric proof the subject tended to present the idea in the form of images [12,14]. The images are needed to form a mental model of the problem to be solved. The images make it easier for the subject to explain important elements and to find mathematical relationships that enable the subject to sequentially and systematically summarize the argument. Therefore, the completion steps made by the subject always refers to the image that has been made. As for communicating algebraic problems, subject rely on their mastery of various knowledge of mathematical logic. This is in line with previous findings which indicate that mastery of content knowledge is very important in solving mathematical problems [15]. In both given algebraic questions, the subject uses her knowledge of implication and contraposition. The subject knows that the contraposition of a statement is equivalent to its implications. Based on that knowledge, the subject presents the next steps of settlement by referring to the contraposition of the given statement.

4. Conclusion

Based on the results of the research described above, it can be concluded that the subject's mathematical communication in proving begins by writing the problem, explaining the main elements in the matter, constructing the argument and presenting the settlement steps, explaining ideas in a more understandable form, and ending by reaffirming the results obtained. Subject can communicate her authentication ideas correctly, systematically and logically. In addition, the subject also uses various communication strategies, such as by using images and other mathematical representations. The strategy chosen depends very much on the type of problem given.

Acknowledgments
Researchers would like to thank Ministry of Higher Education and Research on lecturer dissertation research grant in 2017, and the Head of Sekolah Tinggi Keguruan dan Ilmu Pendidikan St. Paulus Ruteng, Flores, NTT, Indonesia who have supported me in my lectures.

5. References

[1] Khatib F M M and Maarof N 20015 Self-efficacy perception of oral communication ability among english as a second language (ESL) technical students *Procedia - Social and Behavioral Sciences* **204** p 99
[2] NCTM 2000 *Principles and Standards for School Mathematics* (Reston: The National Council of Teacher of Mathematics, Inc) p 60
[3] Ministry of Education 2005 A Guide to Effective Instruction in Mathematics, Kindergarten to Grade 6 Vol 2 (Toronto: ON: Queen's Printer for Ontorio) p 66
[4] Yackel E and Hanna G 2003 Reasoning and proof in *A Research Companion to Principles and Standards for School Mathematics*, ed J Kilpatrick, W G Martin and D Schifter (The National Council of Teacher of Mathematics, Inc., USA) p 231
[5] Heinze A 2004 *Proc. of the 28th Conf. of the Int. Group for the Psychology of Mathematics Education* Vol 3 (Germany: University of Augsburg) p 41
[6] Mejia-Ramos J P and Inglis M 2009 *Proc.. of the ICMI Study 19 Conf.: Proof and Proving in Mathematics Education* Vol 2 (Taiwan: The Department of Mathematics, National Taiwan Normal University Taipei) p 88
[7] Polya G 1973 *How to Solve It; a New Aspect of Mathematical Method* (New Jersey: Princeton University Press) p 154
[8] Kabael T 2012 Graduate student middle school mathematics teachers' communication abilities in the language of mathematics *Procedia-Social and Behavioral Sciences* 55, 809 – 815
[9] Tamur M, Sennen E, and Pantaleon K V 2017 Etnomatematika daerah Manggarai Flores –NTT dalam tradisi belis dan pengukuran *Prosiding Seminar Nasional Matematika dan Pendidikan Matematika,vol.5 pp.15-21*
[10] Dewi I 2009 Profil komunikasi matematika mahasiswa calon guru ditinjau dari perbedaan jenis kelamin *Disertasi* (Surabaya: Unesa)
[11] Ma'rufi, Budayasa I K, and Juniati D 2017 The analysis of mathematics teachers' learning on algebra function limit material based on teaching experience difference *AIP Conference Proceedings* 1867, 020056 (2017); doi: 10.1063/1.4994459
[12] Pantaleon K V, Juniati D, Lukito A, and Mandur K 2018 The written mathematical communication profile of prospective math teacher in mathematical proving *Journal of Physics: Conf. Series,***947** 012070
[13] Lestari N D S, Juniati D, and Suwarsono St *Journal of Physics: Conference Series,***1008**(2018)012074
[14] Pantaleon K V, Kurnila V S, Tamur M, and Nendi F 2017 Profil komunikasi matematik tertulis calon guru matematika dengan tingkat kecemasan tinggi dalam pembuktian matematika *Prosiding Seminar Nasional Matematika dan Aplikasinya,* pp. 134 – 141
[15] Pattimukay N, Juniati D, and Budiarto M T 2018 Content knowledge of prospective elementary school teacher for fractional concepts *Journal of Physics: Conf. Series* **974** 012026 p 1

Characteristic Polynomial of Antiadjacency Matrix of Directed Cyclic Wheel Graph ($\overrightarrow{W_n}$)

L Widiastuti[1], S Utama[1,*] and S Aminah[1]
[1] Department of Mathematics Faculty of Mathematics and Natural Sciences Universitas Indonesia, Indonesia
[*]Corresponding author: suarsih.utama@sci.ui.ac.id

Abstract. A directed cyclic wheel graph with order n, where n ≥ 4 can be represented by an anti-adjacency matrix. The anti-adjacency matrix is a square matrix that has entries only 0 and 1. The number 0 denotes an edge that connects two vertices, whereas the number 1 denotes otherwise. The norm of every coefficient in characteristic polynomial of the anti-adjacency matrix of a directed cyclic wheel graph represents the number of Hamiltonian paths contained in the induced sub-graphs minus the number of the cyclic induced sub-graphs. In addition, the eigenvalues can be found through the anti-adjacency matrix of directed cyclic wheel graph. The result is, the anti-adjacency matrix of directed cyclic wheel graph has two real eigenvalues and some complex eigenvalues that conjugate to each other. The real eigenvalues are obtained by Horner method, while the complex eigenvalues are obtained by finding the complex roots from the factorization of the characteristic polynomial.

1. Introduction

Graph theory is a branch of mathematics that developes very rapidly. There are still many properties of a graph that have not been studied further, such as the antiadjacency matrices of directed graphs. Based on its direction, directed graphs are divided into directed acyclic graphs (a graph that does not contain a cycle graph) and directed cyclic graphs (a graph that contains a cycle graph). A directed graph can be represented in the form of an adjacency matrix or antiadjacency matrix. In a previous study, Fery had discussed about the antiadjacency matrix of some directed acyclic graphs [1]. Today's researches have not discussed widely about the antiadjacency matrix of directed cyclic graphs. Therefore, in this paper, the antiadjaceny matrix of a directed cyclic wheel graph $\overrightarrow{W_n}$ will be discussed. From the antiadjacency matrix of a directed cyclic wheel graph $\overrightarrow{W_n}$, the general form of its characteristic polynomial can be found. The characteristic polynomial of directed cyclic wheel graph $\overrightarrow{W_n}$ has n coefficients, those are $b_1, b_2, b_3, \ldots, b_n$. The coefficients represent structures that belong to the directed cyclic wheel graph $\overrightarrow{W_n}$. Moreover, we find the eigenvalues of the antiadjacency matrix of a directed cyclic wheel graph $\overrightarrow{W_n}$ which both real and complex eigenvalues. Eigenvalues have many applications in various fields. In graph theory, eigenvalues characterized the topological structures of a graph. In addition, we can use the eigenvalues to calculate the number of structures that belong to a graph, such as spanning trees, Hamiltonian paths, Hamiltonian circuits, induced sub-graphs and etc.

2. Basic Theory

The antiadjacency matrix of directed graph $G$is a matrix $B(G) = J - A(G)$, where $A(G)$is an adjacency matrix of graph G and $J$is a $n \times n$ matrix whose all entries are 1 [2]. Suppose A is a $n \times n$ matrix. The principal-submatrix of A of $(n-r) \times (n-r)$ matrix is the submatrix that obtained deleting r rows & r columns with the same index of A simultaneously. The principal-minor of A is the determinant of the principal submatrix of A [3]. A wheel graph W_n with order n is a graph that contains a cycle graph C_{n-1}, and every vertex in the cycle graph $C_{n-1}$is connected to the middle vertex [4]. Thus, a directed cyclic wheel graph $\overrightarrow{W_n}$ is a wheel graph that has a certain directed path with a cyclic direction.

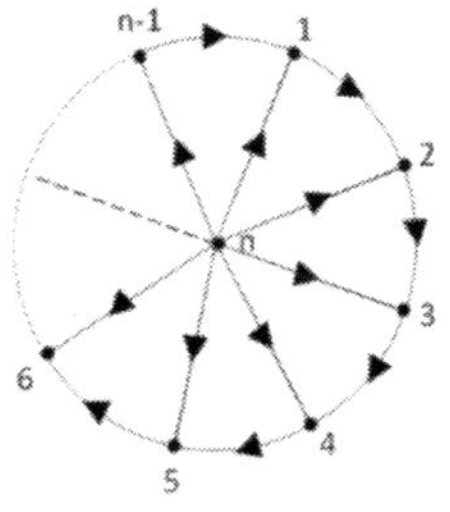

Figure 1. A directed cyclic wheel graph $\overrightarrow{W_n}$ with order n,($n \geq 4$), that has a certain named vertex

Here are some theorems, lemma and corollary of previous studies that support the research in this paper.

Theorem 2.1: Suppose B is the antiadjacency matrix of a directed cyclic cycle graph $\overrightarrow{C_n}$, then$det(B) = n - 1$ [5].

Corollary 2.2: Suppose G is an directed acyclic graph and B is the antiadjacency matrix of G. If Gcontains a Hamiltonian path then $\det(B) = 1$, and otherwise $\det(B) = 0$ [2].

Lemma 2.3: Suppose B is a $n \times n$ matrix that has entries $(0, 1)$ such that $b_{ij} = 1 \text{if} i \geq j$. If$b_{12} = b_{23} = b_{34} = b_{45} = \ldots\ldots = b_{(n-1)n} = 0$ then $det(B) = 1$, and otherwise $det(B) = 0$ [2].

Theorem 2.4: If$\lambda^n + c_1\lambda^{n-1} + c_2\lambda^{n-2} + \ldots + c_{n-1}\lambda + c_n = 0$ is the characteristic equation of a matrix $A_{n\times n}$, then $c_i = (-1)^i \sum_{j=1}^{w} |A_i^{(j)}|$, $i = 1,2,\ldots,n$, where $|A_i^{(j)}|$ is a principal minor of size $i \times i$ of A and $j = 1,2,3,\ldots,w$, and w is the number of the principal minor of size $i \times i$ of A [3].

Theorem 2.5: If A is a $n \times n$ triangular matrix, then $det(A) = a_{11}a_{22} \cdots a_{nn}$ [6].

Theorem 2.6: Let A be a $n \times n$ matrix, then (a) If B is the matrix that has results when one row or one column of A is multiplied by a scalar k, then $det(B) = k\, det(A)$; (b) If B is the matrix that has results when two rows or two columns of A are interchanged, then $det(B) = -det(A)$; (c) If B is the matrix that has results when a multiple of one row or one column of A is added to another, then $det(B) = det(A)$ [7].

3. Main Result

In this section we discuss about the theorems related to the main results. The following is a general form of an antiadjacency matrix of graph $\overrightarrow{W_n}$ of $n \times n$ matrix that denoted by $\left(B(\overrightarrow{W_n})\right) = [b_{ij}]$, where

$$b_{ij} = \begin{cases} 0, & i = n-1, j = 1\,; or\ i = 1,2,\ldots,n-2, j = i+1\,; or\ \ i = n, j = 1,2,\ldots,n-1 \\ 1, & \text{other} \end{cases}$$

Theorem 3.1

Suppose $B(\overrightarrow{W_n})$ is the antiadjacency matrix of a graph $\overrightarrow{W_n}$, then $det\left(B(\overrightarrow{W_n})\right) = n - 2$.

Proof

By using cofactor expansion, applying row reduction process and column reduction process, it was obtained the upper triangular matrix. By applying Lemma 2.3, Theorem 2.5 and Theorem 2.6, it followed that $det\left(B(\overrightarrow{W_n})\right) = n - 2$. ■

Theorem 3.2
Suppose $P(B(\overrightarrow{W_n})) = \lambda^n + b_1\lambda^{n-1} + \sum_{i=2}^{n-2} b_i \lambda^{n-i} + b_{n-1}\lambda + b_n$ is the characteristic polynomial of $B(\overrightarrow{W_n})$. Then $\|b_i\|$, $i = 1,\ldots,n$ represents the number of Hamiltonian paths, each of them has length equal to$(i-1)$, contained in the induced sub-graphs minus the number of the cyclic induced sub-graphs.

Proof
An induced sub-graph that contains Hamiltonian paths, each of them has length equal to $(i-1)$, of graph $\overrightarrow{W_n}$ can be obtained by deleting $(n-i)$ connected vertices, included all the edges that connected with the $(n-i)$vertices on graph $\overrightarrow{W_n}$, where $i = 1,\ldots,n$. The proof was divided into 4 cases:

i. $\|b_1\| = n$ was obtained from deleting $(n-1)$ vertices.
ii. $\|b_i\| = 2(n-1), i = 2,\ldots,n-2$ was obtained from deleting$(n-i)$ vertices.
iii. $\|b_{n-1}\| = 2n-3$ was obtained from deleting $n-(n-1) = 1$ vertex.
iv. $\|b_n\| = n-2$ was obtained from deleting $n-n = 0$ vertex

In every case, except the fourth case, there were two types of deletion. In the first type, the middle vertex was deleted first. In the second type, the vertices on the cycle path were deleted first. The deletion of $(n-i)$ vertices $(i = 1,\ldots,n)$ generated some induced sub-graphs which each of them could be represented in the form of a $i \times i$ principal-submatrix. Referring to Theorem 2.4 by expanding $\sum_{j=1}^{w}|A_i^{(j)}|$, the norm of every coefficient of the characteristic polynomial, $\|b_i\|$, $i = 1,\ldots,n$, was obtained.

i. In the first type, the middle vertex was deleted, then $(n-2)$ vertices on the cycle path were deleted, it remained one vertex on the cycle path. Because there were $(n-1)$ vertices on the cycle path, then there were $(n-1)$ ways of deleting $(n-2)$vertices on the cycle path. In the second type, after deleting $(n-1)$ vertices on the cycle path, it remained the middle vertex. From these two types of deletion, it followed that there were $n-1+1 = n$ induced sub-graphs which each of them contained one vertex and a Hamiltonian path with length equal to zero. Every induced sub-graph could be represented by the principal-submatrix of 1×1 matrix. Since the induced sub-graph was acyclic and contained a Hamiltonian path, then according to Corollary 2.2 the determinant of the principal-submatrix was equal to 1. As there were n induced sub-graphs, according to Theorem 2.4, it follows that $\|b_1\| = n$. Since there was no cyclic induced sub-graph, it could be concluded that $\|b_1\| = n$ denotes the number of Hamiltonian paths which each of them had length equal to zero and contained in the induced sub-graphs minus the number of the cyclic induced subgraphs.■

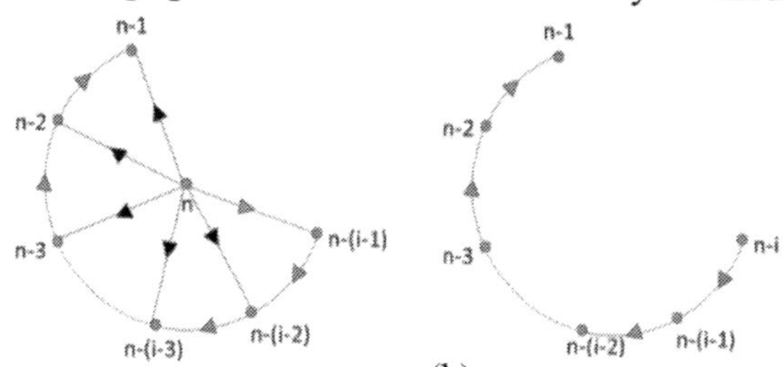

Figure 2. The induced sub-graphs that contained i vertices $(i = 2,\ldots,n-2)$ and Hamiltonian paths which each of them had length equal to $(i-1)$ as a result of deleting $(n-i)$ connected vertices. (a) The result of the first type. (b) The result of the second type.

ii. In the First type, the middle vertex was deleted, then $(n-i-1)$ vertices$(i = 2,\ldots,n-2)$ on the cycle path were deleted, it remains $i$vertices on the cycle path that connected to each other. Since there were $(n-1)$ vertices on the cycle path, there were $(n-1)$ ways of deleting $(n-i-1)$ vertices on the cycle path. In the second type, after deleting $(n-i)$ vertices$(i = 2,\ldots,n-2)$on the cycle path, it remained $(i-1)$vertices on the cycle path where every vertex was connected to the middle vertex. By taking the middle vertex as the first vertex, it was obtained a Hamiltonian path with length equal to $(i-1)$. As there were $(n-1)$ vertices on the cycle path, then there were $(n-1)$ ways of deleting $(n-i)$ vertices on the cycle path.

From those two following types, there were $n-1+n-1=2(n-1)$induced sub-graphs which each of them contains i vertices and contains a Hamiltonian path with length equal to $(i-1)$. Every induced sub-graph could be represented by the principal-submatrix of $i \times i$ matrix. Since the induced sub-graph was acyclic and contained a Hamiltonian path, then according to corollary 2.2 the determinant of the principal-submatrix was equal to 1. Because there were $2(n-1)$ induced sub-graphs, according to Theorem 2.4, it followed $\|b_i\| = 2(n-1), i = 2, \dots, n-2$. As there was no cyclic induced sub-graph, it could be concluded that $\|b_i\| = 2(n-1), i = 2,\dots,n-2$ denoted the number of Hamiltonian paths which each of them has length equals to $(i-1)$ and contained in the induced sub-graphs minus the number of the cyclic induced sub-graphs. ■

iii. In the first type, after deleting the middle vertex, it remained a cyclic induced sub-graph $\overrightarrow{C_{n-1}}$. Moreover, by deleting one edge of $\overrightarrow{C_{n-1}}$, it was obtained a Hamiltonian path with length equal to $(n-2)$, and those deletions were related to the choice of a vertex as the first vertex of any Hamiltonian path. Since $\overrightarrow{C_{n-1}}$ contained $(n-1)$ vertices, there were $(n-1)$ways to choose a vertex as the first vertex. In the second type, after deleting one vertex on the cycle path, it remained $(n-2)$ vertices on the cycle path where every vertex was connected to the middle vertex. By taking the middle vertex as the first vertex, it was obtained a Hamiltonian path with length equal to $(n-2)$. Since there were $(n-1)$vertices on the cycle path, there were $(n-1)$ ways of deleting one vertex on the cycle path.

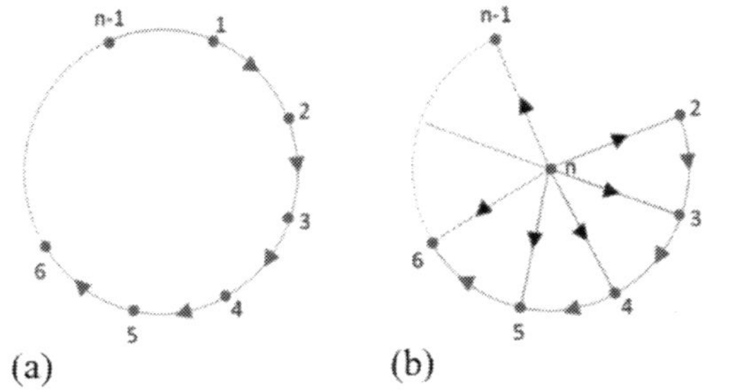

Figure 3. The induced sub-graphs that contained $n-1$ vertices and Hamiltonian paths, each of them had length equal to $(n-2)$, as a result of deleting one vertex. (a) The result of the first type. (b) The result of the second type.

According to theorem 2.1, the determinant of the antiadjacency matrix of $(\overrightarrow{C_{n-1}})$ was $(n-2)$. Secondly, it was obtained $(n-1)$ induced sub-graphs which each of them contains $(n-1)$ vertices and could be represented the principal-submatrix of $(n-1) \times (n-1)$ matrix. Since every induced sub-graph was acyclic and contained a Hamiltonian path, according to Corollary 2.2, the determinant of the principal submatrix was equal to 1. As there were $(n-1)$ induced subgraphs, then the sum of the determinants of the principal submatrices were $(n-1)$. Thus, by combining the result of the first and the second steps and also referring to Theorem 2.4, it followed that $\|b_{n-1}\| = 2n-3$. For there was only one cyclic induced sub-graph, it could be concluded that $\|b_{n-1}\| = 2n-3$ denoted the number of Hamiltonian paths which each of them had length equal to $(n-2)$ and contained in the induced sub-graphs minus the number of the cyclic induced sub-graphs. ■

iv. After deleting the zero vertexes, it was obtained $\overrightarrow{W_n}$as the cyclic induced sub-graph. According to Theorem 3.1, the determinant of $B(\overrightarrow{W_n})$ was $(n-2)$. Since there was only one induced sub-graph, according to Theorem 2.4, it follows that $\|b_n\| = n-2$. Moreover, the graph $\overrightarrow{W_n}$ contained Hamiltonian paths which each of them had length equal to $(n-1)$ and had the middle vertex as the starting point. For the middle vertex should be the starting point, the end point should be one of the vertices on the cycle path. As there were $(n-1)$ vertices on the cycle path, it followed that there were $(n-1)$ Hamiltonian paths. Furthermore, there was only one cyclic induced sub-graph. Thus, it could be concluded that $\|b_n\| = n-2$

denoted the number of Hamiltonian paths which each of them has length equal to$(n-1)$ and contained in the induced sub-graphs minus the number of the cyclic induced sub-graphs. ■

Corollary 3.3
Let $P(B(\overrightarrow{W_n})) = \lambda^n + b_1\lambda^{n-1} + \sum_{i=2}^{n-2} b_i\,\lambda^{n-i} + b_{n-1}\lambda + b_n$ is the characteristic polynomial of $B(\overrightarrow{W_n})$.Then$b_1 = -n; b_i = (-1)^i 2(n-1), i = 2, \dots, n-2; b_{n-1} = (-1)^{n-1}(2n-3)$&$b_n = (-1)^n(n-2)$.
Proof
From Theorem 3.2 it followed that $\|b_1\| = n$; $\|b_i\| = 2(n-1), i = 2, \dots, n-2$; $\|b_{n-1}\| = 2n-3$ and $\|b_n\| = n-2$. By using the equation $b_i = (-1)^i \sum_{j=1}^{w} |A_i{}^{(j)}|$ in Theorem 2.4 it follows that;

- For $i = 1$, $b_1 = (-1)^1 \sum_{j=1}^{w} |A_1{}^{(j)}| = (-1)^1\|b_1\| = -n$.
- For $i = 2, \dots, n-2$, $b_i = (-1)^i \sum_{j=1}^{w} |A_i{}^{(j)}| = (-1)^i\|b_i\| = (-1)^i 2(n-1)$.
- For $i = n-1$, $b_{n-1} = (-1)^{n-1} \sum_{j=1}^{w} |A_{n-1}{}^{(j)}| = (-1)^{n-1}\|b_{n-1}\| = (-1)^{n-1}(2n-3)$.
- For $i = n$, $b_n = (-1)^n \sum_{j=1}^{w} |A_n{}^{(j)}| = (-1)^n\|b_n\| = (-1)^n(n-2)$.

Therefore, this corollary was proved. ■

3.4. Theorem 3.4
The real eigenvalues of $B(\overrightarrow{W_n})$ are 1 and$n-2$ for every n.
Proof
By substituting $\lambda = 1$ in $P\left(B(\overrightarrow{W_n})\right)$, it followed that $P\left(B(\overrightarrow{W_n})\right) = 0$ for every n. It means that $\lambda = 1$ was one of the real eigenvalues. By using Horner method, it was obtained that the other real eigenvalues were $(2k-2)$for $n = 2k$(even) & $(2k-1)$for $n = 2k+1$(odd), $k \geq 2$.So it can be concluded that $B(\overrightarrow{W_{2k}})$ has two real eigenvalues namely $\lambda = 1$ and $\lambda = 2k-2$. Furthermore, $B(\overrightarrow{W_{2k+1}})$ had $\lambda = 1$ whose multiplicity equals 2 and $\lambda = 2k-1$ as its eigenvalues. Figure 4 represented the Horner method which used to find the real roots of $P\left(B(\overrightarrow{W_{2k}})\right)$ (the real eigenvalues of $B(\overrightarrow{W_{2k}})$).

1	1	-2k	4k-2	-(4k-2)	4k-2	-(4k-2)	...	4k-2	-(4k-3)	2k-2
		1	-(2k-1)	2k-1	-(2k-1)	2k-1	...	-(2k-1)	2k-1	-(2k-2)
2k-2	1	-(2k-1)	2k-1	-(2k-1)	2k-1	-(2k-1)	...	2k-1	-(2k-2)	0
		2k-2	-(2k-2)	2k-2	-(2k-2)	2k-2	...	-(2k-2)	2k-2	
	1	-1	1	-1	1	-1	...	1	0	

Figure 4. The result from Horner method for $P\left(B(\overrightarrow{W_{2k}})\right)$

In fact, by using Horner method, the coefficients of λ^i of $P\left(B(\overrightarrow{W_n})\right)$ were also found (see Figure 4). The coefficients of $\lambda^i, i = 1, \dots, 2k-2$in $P\left(B(\overrightarrow{W_{2k}})\right)$ is $(-1)^i$ while the coefficients of $\lambda^{2i}, i = 1, \dots, k-1$ in $P\left(B(\overrightarrow{W_{2k+1}})\right)$ is $(-1)^{2i}$. Hence, $P\left(B(\overrightarrow{W_n})\right)$ could be written as the following $P(B(W_{2k})) = (\lambda-1)(\lambda-(2k-2))(1+\sum_{i=1}^{2k-2}(-1)^i\lambda^i)$& $P(B(W_{2k+1})) = (\lambda-1)^2(\lambda-(2k-1))(1+\sum_{i=1}^{k-1}(-1)^{2i}\lambda^{2i})$.■

Theorem3.5
The number of the complex eigenvalues of $B(\overrightarrow{W_n})$ is always even and it conjugate to each other for every n.
Proof
The complex eigenvalues of $B(\overrightarrow{W_{2k}})$ could be obtained from the equation $(1+\sum_{i=1}^{2k-2}(-1)^i\lambda^i) = 0$ which was similar to a finite geometric series form $(a = 1, r = (-\lambda), n = 2k-1)$.Then, the sum of the series was

$$\frac{1((-\lambda)^{2k-1}-1)}{(-\lambda)-1} = 0,\ \lambda \neq -1 \qquad (1)$$

From the equation (1), it followed that the complex eigenvalues were

$$\lambda = e^{i(\frac{\pi}{2k-1}+\frac{2m\pi}{2k-1})}\ ; m = 0, \dots , 2k - 2 \tag{2}$$

Since $\lambda = e^{i\pi}$ was achieved when $m = k - 1$ in the Equation (2) and since $\lambda \neq -1$ in the Equation (1), it followed that there were $(2k - 2)$ complex eigenvalues of $B(\overrightarrow{W_{2k}})$. Hence, the number of complex eigenvalues of $B(\overrightarrow{W_{2k}})$ were even. It could be proved that the eigenvalues $\lambda_{k-1+j} = e^{i(\frac{\pi}{2k-1}+\frac{(2k-2+2j)\pi}{2k-1})}$ was the conjugate of the eigenvalue $\lambda_{k-1-j} = e^{i(\frac{\pi}{2k-1}+\frac{(2k-2-2j)\pi}{2k-1})}$,$j = 1, \dots , k - 1$. ■

The complex eigenvalues of $B(\overrightarrow{W_{2k+1}})$could be obtained from the equation $(1 + \sum_{i=1}^{k-1}(-1)^{2i}\lambda^{2i}) = 0$ which was similar to a finite geometric series form $(a = 1, r = (\lambda^2), n = k)$. Then, the sum of the series was

$$\frac{1((\lambda^2)^k-1)}{\lambda^2-1} = 0, \lambda \neq 1\ and\ \lambda \neq -1 \tag{3}$$

From the Equation(3), it follows that the complex eigenvalues were

$$\lambda = e^{i(\frac{m\pi}{k})}\ ; m = 0, \dots , 2k - 1 \tag{4}$$

Since$\lambda = e^0$&$\lambda = e^{i\pi}$ were achieved when $m = 0$&$m = k$ respectively in the Equation (4) and since $\lambda \neq 1$&$\lambda \neq -1$in the Equation (3), it follows that there are $(2k - 2)$ complex eigenvalues of $B(\overrightarrow{W_{2k+1}})$. Hence, the number of complex eigenvalues of $B(\overrightarrow{W_{2k+1}})$ were even. It could be proved that the eigenvalues $\lambda_{k+j} = e^{i(\frac{(k+j)\pi}{k})}$ was the conjugate of the eigenvalue $\lambda_{k-j} = e^{i(\frac{(2k-j)\pi}{k})}$,$j = 1, \dots , k - 1$.■

4. Conclusion

Based on the main results, $\lambda^n + b_1\lambda^{n-1} + \cdots + b_{n-1}\lambda + b_n$ was considered as the characteristic polynomial of $B(\overrightarrow{W_n})$. $\|b_i\|, i = 1, \dots , n$ represented the number of Hamiltonian paths which each of them had length equal to $(i - 1)$and contained in the induced sub-graphs minus the number of the cyclic induced sub-graphs. The values of the coefficients of the characteristic polynomial of $B(\overrightarrow{W_n})$ were given in Table 1.

Table 1. The coefficientsof the characteristic polynomial of$B(\overrightarrow{W_n})$

Coefficient	Coefficient value	
b_1	$-n$	
$b_i, i = 2, \dots , n - 2$	$(-1)^i\, 2(n - 1)$,	$i = 2, \dots , n - 2$
b_{n-1}	$(-1)^{n-1}(2n - 3)$	
b_n	$(-1)^{n-1}(n - 2)$	

It was also obtained that $B(\overrightarrow{W_n})$had two real eigenvalues namely 1 and $(n - 2)$for every n. While thenumber of the complex eigenvalues were always even and it conjugate to each other for every n. The eigenvalues of $B(\overrightarrow{W_n})$ were given in Table 2.

Table 2. The real and complex eigenvalues of $B(\overrightarrow{W_n})$

The number of$n$	Eigenvalues	
	Real	Complex
$n = 2k;\ k \geq 2$	$\lambda = 1$&$\lambda = 2k - 2$	$\lambda = e^{i(\frac{\pi+2m\pi}{2k-1})}\ ; m = 0, \dots , 2k - 2$. Except$\lambda = e^{i\pi}$
$n = 2k + 1;\ k \geq 2$	$\lambda = 1$&$\lambda = 2k - 1$	$\lambda = e^{i(\frac{m\pi}{k})}\ ; m = 0, \dots , 2k - 1$. Except $\lambda = e^0$&$\lambda = e^{i\pi}$

5. Reference

[1] Firmansyah F 2014 *Polinomial Karakteristik Matriks Anti-adjacency dari Graf Berarah yang Acyclic (*Thesis: Universitas Indonesia)
[2] Bapat R 2010 Graphs and Matrices (India: Hindustan Book Agency)
[3] Meyer C D 2000 Matrix Analysis and Applied Linier Algebra (New Jersey: SIAM)
[4] Skiena S 1990 Implementing Discrete Mathematics: Combinatorics and Graph Theory with Mathematica Reading (MA: Addison-Wesley)
[5] Putra A P 2015 Polinomial Karakteristik Matriks Antiadjacency dari Graf Lingkaran Berarah
[6] $\overrightarrow{C_n}$ dan Graf Lingkaran Berarah dengan Penambahan Satu Chord $\overrightarrow{C_n^t}$ (Thesis: Universitas Indonesia)
[7] Anton H and Chris R 2010 Elementary of Linear Algebra Application Version (10th ed.) (New Jersey: John Wiley and Sons)
[8] Anton H and Chris R 2011 Elementary of Linear Algebra (11th ed.) (New Jersey: John Wiley and Sons)

Acknowledgments
This research is funded by Hibah PITTA UI [2307/UN2.R3.1/HKP.05.00/2018].

An Inventory-Dependent Demand Model with Deterioration, All-Units Discount, and Return

N Loedy[1], D Lesmono[1], and T Limansyah[1]

[1]Department of Mathematics, Universitas Katolik Parahyangan, Jalan Ciumbuleuit 94 Bandung 40141

Abstract. In this paper, we propose an inventory model with inventory-dependent demand, deterioration, all-units discount and return. Conditions considered in the model are reflected in the real life when a retailer deal with products with deterministic demand having constant deterioration rate. Also, occasionally the supplier offers an all-units discount to the retailer and on the other occasion retailer can return some products to the supplier at some costs. We develop the model that will determine the optimal return time and optimal order quantity that minimize the total cost. We also assume that all shortages are backordered and one tenth of the order quantity is returned to the supplier. We provide some numerical examples and find that as the deterioration rate, and demand increase, the optimal return time decreases making the optimal order quantity and the total cost increase. Sensitivity analysis is also conducted to analyse the effect of changes in parameter values to the optimal return time, optimal order quantity and the total cost. It is found that changes in return cost do not significantly change the optimal return time, the optimal order quantity and the total cost, compared with changes in deterioration rate, inventory-dependent demand rate and time between replenishment.

1. Introduction

Inventory for retailers is a number of products that are stored to be used or sold within a certain period of time. In general, American Production Inventory Control Society (APICS) Dictionary defines inventory as items used to support production, supporting activities and customer service [1]. There are some characteristics owned by inventory such as demand, replenishment lead time, inventory level and review times, lifetime and reparability [1]. Managing inventory related with determination of order quantity and ordering time. This task becomes more complicated when we are dealing with other factors such as deterioration, discount, inventory-dependent demand and return. In real world, deterioration almost occur in every product, but especially in food industry, chemical material and dairy products. Inventory-dependent demand relates to the situation where demands are increasing when customers see a lot of product in the shelf such as in department store or supermarket ([2], [3]). This situation will make retailer to provide more products in its shelf but in reality the shelf capacity is limited. Another factor that make retailer tend to provide more product is the amount of discount offered by the supplier and the possibility to return some deteriorated products to the supplier. In this paper, we develop an inventory-dependent demand model with deterioration, all-units discount, and return. The model we develop is an extension and combination of the model in [4], [5] and [6]. We also propose an algorithm to find the optimal solution and provide numerical experiments and sensitivity analysis to study the effect of parameter changes on the optimal solution. The aim of our model is to determine the optimal return time and order quantity that minimize the total cost.

In the last few decades, there are a lot of research regarding deteriorating goods, making this topic as one of important factors that should be considered by retailers. Managing goods with deteriorating factor or perishable goods needs a good strategy, especially in determining the optimal order quantity. Determining the optimal order quantity is crucial, because it is related with many factors faced by retailers including some costs such as holding, ordering and shortage. Shortage cost is sometimes intangible if it is related to the customer satisfaction. It is also related to the potential profit loss. Several strategies in managing perishable goods include giving discount and the possibility for the retailer to return deteriorated goods to the supplier. Some mathematical models have also been developed and implemented regarding these issues. Nafisah et al. [4] have developed an inventory model for perishable goods by considering return and discount, while Nagare and Dutta [5] developed a continuous review inventory model with inventory-dependent demand and deterioration. Setiawan, et.al [6] proposed an inventory-dependent demand model with deterioration rate and return. Rikardo et al. [7] proposed a continuous review model with deterioration and all-units discount. We use the same model of differential equation to describe the inventory level as in Nagare and Dutta [5], but we add all-unit discount and return in the model to describe more realistic situation. The inventory-dependent demand functions used in Nafisah et.al. are constant in each interval between $0 \leq t \leq T$, while we use a continuous function that depends on the number of inventory at time *t*. Comparing with Setiawan, et.al [6], our model has considered all-unit discount and has sensitivity analysis on the optimal solution based on some changes in the parameters' values. The aim of our model is to determine the optimal return time and optimal order quantity that give the minimum total cost. We extend the model of [4], [5] and [6] by incorporating inventory-dependent demand, all-units discount and constant deterioration rate. The organization of this paper is as follows. In Section 2, we introduce the notations, assumptions, our model and propose an algorithm to find the optimal solution. Section 3 deals with numerical experiments and sensitivity analysis. Conclusions and further research are provided in Section 4.

2. The Model

2.1 Notations and Assumptions

The notations we use in this paper are as follows:

TC : total inventory cost
PC : purchasing cost
OC : ordering cost
HC : holding cost
SC : shortage cost
RC : return cost
T : time between order (months)
$O(t)$: demand at time t $[O(t) = a + bI(t)]$
b : inventory-dependent demand factor
Q : order quantity
T_l : return time (months)
T_l^* : optimum return time
$I(t)$: inventory level at time t.
n : number of inventory during period $[0, T_l]$
S : number of shortage during period $[T_l, T]$
P : purchasing cost per unit
θ : deteriorating rate
h : holding cost per unit per year
A : ordering cost per order
A_l : return cost per return
R : return cost per unit
ψ : backorder cost per unit
Q_l : inventory level at returns time

We have several assumptions in developing our model: (1) The inventory-planning horizon is infinite and the inventory system only deals with one product; (2) The amount of inventory returned (Q_l) is 10% of order quantity (Q); (3) Replenishments are instantaneous with zero lead-time and the entire lot is delivered in one batch (fresh and new); (4) When $t = T_r$, returns will occur for Q_l instantaneously; (5) Demand at time t $(O(t))$ is a linear function of the inventory level at time t (I(t)), (6) Supplier offers discount in terms of all-units discount.

2.2 Model and Algorithm

Based on the fifth assumption, the inventory-dependent demand is given by the following equation.

$$O(t) = \begin{cases} a + bI(t) & ,0 \le t \le T_l \\ \delta & ,\ T_l \le t \le T \end{cases}$$

The inventory level decreases in time due to demand and deteriorating rate in the interval [0,T] until it reaches its minimum at T_l . At time $t = T_l$ return will be made for all deteriorated item at the amount of Q_l. In the interval $[T_l, T]$, shortage occurs and will be handled by full back order since all returned product will only be replaced on the next shipment. Based on the condition above, in general the inventory level during interval $0 \le t \le T$ can be described using the following differential equation

$$\frac{dI(t)}{dt} + \theta I(t) = -O(t)$$

Substituting $O(t) = a + bI(t)$, we have

$$\frac{dI(t)}{dt} = -\theta I(t) - \big(a + bI(t)\big) \quad ,0 \le t \le T_l$$

$$\frac{dI(t)}{dt} = -\theta I(t) - \delta \qquad ,T_l \le t \le T$$

Using the boundary condition $I(T_l) = 0$, we have

$$\frac{dI(t)}{dt} = -\delta;\ I(t) = \delta(T_l - t)$$

In the interval $0 \le t \le T_l$, using the boundary condition $I(T_l) = Q_l$, we have

$$I(t) = -\frac{a}{b+\theta} + \frac{e^{-(b+\theta)t}\left(Q_l + \frac{a}{b+\theta}\right)}{e^{-(b+\theta)T_r}}$$

When $t = 0$, $I(t = 0) = Q - \delta(T_l - T)$ the following expression for Q is obtained.

$$Q = -\frac{a}{b+\theta} + \frac{e^{-(b+\theta)t}\left(Q_l + \frac{a}{b+\theta}\right)}{e^{-(b+\theta)T_l}} + \delta(T - T_l)$$

By employing the second assumption, after some calculations we have the following

$$Q_l = \frac{-\delta T_l e^{-(b+\theta)T_l}\beta - \delta T_l e^{-(b+\theta)T_l}\theta + \delta T e^{-(b+\theta)T_l}\beta + \delta T e^{-(b+\theta)T_l}\theta - ae^{-(b+\theta)T_l} + a}{10e^{-(b+\theta)T_l}b + 10e^{-(b+\theta)T_l}\theta - b - \theta}$$

The number of inventory during interval $[0, T_l]$, n and the number of shortages during interval $[T_l, T]$, S are provided by

$$n = \int_0^{T_l} I(t)dt\,;\ S = \delta T_l(T - T_l) - \frac{B\left(T^2 - T_l^2\right)}{2}$$

The objective of our model is to find the return time that minimizes the total cost. The total cost (TC) consists of purchasing cost (PC), ordering cost (OC), annual holding cost (HC), shortage cost (SC) and return cost (RC). The formulation of the total cost can be written as:

$$TC = P\frac{Q}{T} + \frac{A}{T} + \frac{h.n}{T} + \frac{-\psi\left(BT_l(T - T_l) - \frac{\delta\left(T^2 - T_l^2\right)}{2}\right)}{T} + \frac{A_l + Q_l R}{T}$$

The optimal return time is found by using the conditions of $\frac{dTC}{dT_l} = 0$ and $\frac{d^2TC}{dT_l^2} > 0$. In order to find the optimal solution, we develop an algorithm in Figure 1 below.

3. Numerical Experiments and Sensitivity Analysis

In this section, we will give numerical experiments to illustrate and give a better understanding of our model. Next, sensitivity analysis in order to study the effect of parameters changes to the optimal solution is also performed. First, we consider the following parameters in order to find the optimal return time, the optimal order quantity and the minimum total cost.

T = 6 months, a = 250, A =\$10, h = \$5, ψ = \$2, R = \$2, A_l = \$10, b = 0.1, θ = 0.25 and δ = 5000

The purchasing cost with all-units discount offer is as follows

$$P = \begin{cases} \$5 \ , Q \leq 25{,}000 \\ \$3{,}5 \, , Q > 25{,}000 \end{cases}$$

Based on our algorithm, using Maple, we found the following optimal return time and order quantity.

i	Price/unit	T_i^*	Q	Status
0	\$5	3.51	21,519	Valid
1	\$3.5	3.37	21,867	Not valid

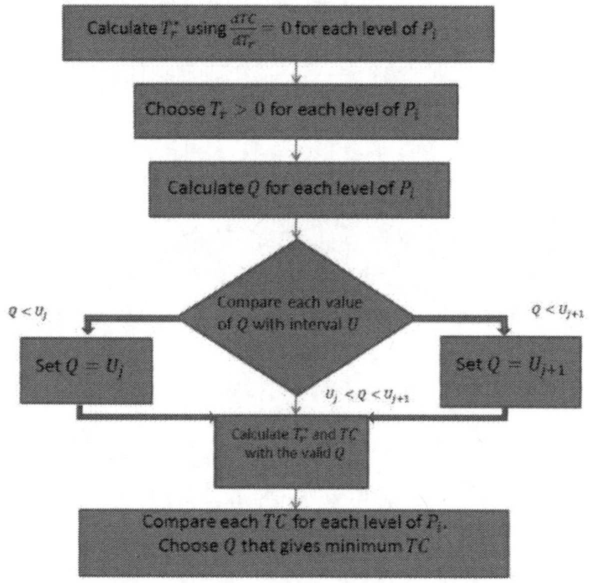

Figure 1. Algorithm in finding the optimal solution

Figure 1 showed that Algorithm in finding the optimal solution. The optimal order quantity Q is 21,519 units at the purchasing cost of \$5. In the all-units discount condition, this number is valid because it is in the interval of discount scheme that is less than 25,000 units. Therefore, this Q is valid. However, for the purchasing cost of \$3.5, the obtained Q of 21,867 units is not valid since it must be higher than 25,000 units must. This Q is not valid; therefore, we choose $Q = 25{,}001$ units. The optimal return time, order quantity and minimum total cost are given below. The minimum cost of \$38,235.90 occurs for $Q = 21{,}519$ units and T_i^*=3.51 months.

i	Q (valid)	T_i^*	TC
0	21,519	3.51	\$38,235.90
1	25,001	5.87	\$63,914.84

We also perform sensitivity analysis to study the effect of parameters changes in the optimal solution. We consider the effect of parameters' changes of R,A_l, b, T and θ on the optimal solution. In this sensitivity analysis, we increase parameters' values by 10% and 15% and decrease them with the same percentage. The results are given in Table 1. Based on Table 1, we can see that increasing the deterioration rate θ, will shorten the optimal return time T_i^*, increase the optimal order quantity Q and total cost TC. When θ increases, the number of deteriorated goods increases, therefore Q increases to fulfil demand and T_i^* decreases making then total cost TC increases. Increasing the inventory-dependent demand factor b, will have the same effect as increasing the deterioration rate θ. This is because as b increases, demands are increasing and hence the total cost are also increasing. We can also see from

Table 1 that A_l does not significantly affect the total cost. This is because A_l is relatively small (between \$0.85 and \$1.15) compared with other costs. Therefore its changes does not substantially affect Q, T_l^* and TC. For return cost per unit R, as we can see in Table 1, decreasing R will make optimal return time T_l^* shorter, increasing the optimal order quantity Q and decreasing the total cost TC although the increase is not too significant. Also, when time between replenishment T becomes longer, the optimal return time T_l^* becomes shorter, making Q and TC increase. We can also infer from Table 1 that all-units discount does not necessarily make the total cost decreases.

4. Conclusions

We have developed an inventory-dependent demand model by considering deterioration rate, all-units discount and return and introduced the algorithm to find the optimal solution. Based on our numerical experiments and sensitivity analysis, we found that the minimum total cost is obtained when the deterioration rate θ, inventory-dependent demand factor b, time between replenishment T and return cost per unit R are lower. Furthermore, we also found that the total cost is not sensitive to the changes of return cost per each return (A_l). All-units discount does not guarantee minimum total cost since there are other factors affecting the total cost such as deterioration rate, return time, time between replenishment, the optimal order quantity and inventory-dependent demand factor. In our model, we assume that the model is deterministic where demands, the rate of inventory-dependent demand and deterioration rate are known. Incorporating probabilistic demands and deterioration rate in the model are possible further research directions in this area. Another avenue for further research direction is by considering multi-item inventory problem, where the replenishment decision can be in terms of individual replenishment policy or joint replenishment policy. Using another discount scheme such as incremental discount is also another interesting topic to pursue. The Sensitivity Analysis shown in Table 1 below.

Table 1. Sensitivity Analysis

Parameter	% Difference	P (\$)	T_l^* (months)	Q (unit)	TC (\$/year)	% Difference of TC
	+15	5	2.960	24,209	41,263	0.0792
	+10	5	3.122	22,320	40,350	0.0553
	-10	5	4.069	18,843	35,562	-0.0700
	-15	5	4.482	16,890	33,868	-0.1142
θ						
	+15	3.5	4.58	25,001	35,399.12	0.0913
	+10	3.5	5.01	25,001	35,162.28	0.0750
	-10	3.5	6.80	25,001	Not valid	-
	-15	3.5	7.3	25,001	Not valid	-
	+15	5	3.267	22,727	39,556	0.0345
	+10	5	3.345	22,348	39,135	0.0235
	-10	5	3.710	20,574	37,251	-0.0258
	-15	5	3.819	20,047	36,720	-0.0397
b						
	+15	3.5	5.346	25,001	56,633.20	0.4812
	+10	3.5	5.517	25,001	58,920.91	0.5410
	-10	3.5	6.225	25,001	69,557.87	0.8192
	-15	3.5	6.410	25,001	72,650.36	0.9001
	+15	5	3.515	21,519	38,235.74	0.0000
	+10	5	3.515	21,519	38,235.74	0.0000

	-10	5	3.515	21,519	38,235.74	0.0000
	-15	5	3.515	21,519	38,235.74	0.0000
A_l	+15	3.5	5.866	25,001	63,918.76	0.6717
	+10	3.5	5.866	25,001	63,918.76	0.6717
	-10	3.5	5.866	25,001	63,918.76	0.6717
	-15	3.5	5.866	25,001	63,918.76	0.6717
	+15	5	3.518	21,512	38,343.48	0.0028
	+10	5	3.517	21,514	38,307.62	0.0019
	-10	5	3.514	21,523	38,164.17	-0.0019
	-15	5	3.513	21,525	38,128.30	-0.0028
R						
	+15	3.5	5.866	25,001	64,043.58	0.6750
	+10	3.5	5.866	25,001	64,001.84	0.6739
	-10	3.5	5.866	25,001	63,835.17	0.6695
	-15	3.5	5.866	25,001	63,793.50	0.6684
	+15	5	3.17	28,852	45,337.56	0.1857
	+10	5	3.26	26,521	43,107.20	0.1274
	-10	5	4.03	15,341	32,302.27	-0.1552
	-15	5	7.55	-	-	-
T						
	+15	3.5	3.09	28,859	39,064.00	0.0217
	+10	3.5	3.17	27,707	38,561.37	0.085
	-10	3.5	7,8	25,001	Not valid	-
	-15	3.5	8,01	25,001	Not valid	-

5. References

[1] Shenoy D and Rosas R 2018 Problems & Solutions in Inventory Management Switzerland: Springer

[2] Levin R I, McLaughlin C P, Lamone R P, and Kottas J F 1972 *Productions/Operations Management: Contemporary Policy for Managing Operating Systems* New York: McGraw-Hill

[3] Silver E A and Peterson R 1985 *Decision Systems for Inventory Management and Production Planning*, 2nd ed (New York: Wiley)

[4] Nafisah L, Sally W and Puryani 2016 Model persediaan pada produk yang mendekati masa kadaluwarsa: mempertimbangkan diskon penjualan dan retur *Jurnal Teknik Industri* **18** 63-72

[5] Nagare M and Dutta P 2012 Continuous review model for perishable products with inventory dependent demand. *Proc. of the International MultiConference of Engineers and Computer Scientists March 24-16 2012* pp. 1513-17

[6] Setiawan, S W, Lesmono D and Limansyah T 2018 A perishable inventory model with return. *IOP Conf. Ser: Mater. Sci. Eng* **335** 012049

[7] Rikardo C, Lesmono D and Limansyah T 2017 Pengembangan model continuous review dengan all-unit discount dan faktor kadaluwarsa *Jurnal Teknik Industri* **19** 29-38

Acknowledgement

Funding of this research from The Institute of Research and Community Service Universitas Katolik Parahyangan are gratefully acknowledged.

The Development Test of Practical Work Module of Animal Structure Using Modified Free Inquiry

Nurhadi*[1], M M Zural[2], Y Zubir[3] and E Rosba[4]

[1, 2, 4]Program Studi Pendidikan Biologi, STKIP PGRI Sumatera Barat, Padang, West Sumatera (25137), Indonesia
(E-mail: nurhadisadira@gmail.com,miminzural@gmail.com, rosba.evrialiani@gmail.com)
[3]Bimbingan Konseling, STKIP PGRI Sumatera Barat, Padang, West Sumatera (25137), Indonesia
(E-mail: yuzarionzubir11@gmail.com)

Abstract. This was the continuation research of validity test at a development stage in the form of practicability test for practical work module of animal structure by using modified free inquiry approach. This research used Four-D model that consisted of four stages (*Define, Design, Develop* and *Disseminate*). In this development test stage, the product practicality test was carried out to the students who were taking the animal structure course, to the lecturers of the animal structure course (expert lecturers) and to other lecturers of other courses (as non-expert lecturers). The evaluation of practicality used the instrument of practicality in the form of questionnaires. The instruments used Likert Scale consisting of four options of answers and the results of the instrument were processed using the percentage formula. From the instrument result, it was found that the practicality of animal structure module that used modified free inquiry approach for the students and the lecturer were very practical.

1. Introduction

Practical work module of animal structure that uses modified free inquiry is the development of animal structure textbooks for practical work activity in laboratories. The use of modified free inquiry in the module is intended for students to realize their learning outcomes by finding their own knowledge. It is also intended that the learning of animal structure will be more meaningful and able to improve their scientific skills such as recognizing the problems given by the practical work module, formulating hypotheses and collecting data on the practical work activities of animal structure in the laboratory. Inquiry is an investigation of ideas, questions, or problems, in which the investigation may be a laboratory activity or other activity that can be used to gather information [1]. Furthermore, the type of technology integration from the application of inquiry learning is printed media such as books, journals, modules and the like [2].

The practical work module of animal structure that uses modified free inquiry has been assessed by experts. After the expert assessed, the learning materials on this module are tested to the lecturers and students during animal structure course. The test is conducted on practical work groups. The development test involves testing of the instructional materials on the target group members of the trainee [3]. In this case the group members of trainee refer to the lecturers and students. The purpose of this development test is to collect feedback to make the instructional and motivational material to be more effective. Frequently, the large gap is found between the judgments of validator and predictions of students practicing in using the developed products. Thus, the data from students' testing remains the most valid predictor of potential problems.

A development test at the initial testing stage is the practicality stage by conducting a suitability test (the practicality of the practical work module of animal structure using modified free inquiry) to the user, tested individually to the lectures or students on face-to-face practical work activity with instructional developer (researchers) in the laboratory. Practically, it means the ease of utilisation [4]. Thus, this development test aims to assess the degree of fondness and ease in the utilisation of practical work module about animal structure by using modified free inquiry. Practicality assessment evaluates whether target users can work with intervention (practicality) and are willing to apply it in their teaching (relevance & sustainability) [5].

Tested textbooks are used by their authors or lecturers (users) during learning practice. The level of usability can be known and felt by the users. It can be recognized by finding the strength and weakness of the textbook in terms its relevance, accuracy, readability, and compatibility with student-centered learning. Based on these criterions, users can give feedbacks about the developed textbook [6].This research founding is practical work module of animal structure using modified free inquiry which is very practical based on whether users (students, lecturers or other experts) consider the intervention can be used and preferable under normal conditions. In development research, developed products can be considered as a practical product if the experts and practitioners state that the developed theory can be implemented and the indicators of practicality were satisfied.

2. Method

This research was a Four-D development (Define, Design, Develop and Disseminate). This research was in the develop stage of testing material development involving actual trainees to find sections for revisions on the basis of responses, reactions, comments from trainees, and modified material.

The testing was a practical test toward practical work module of animal structure using modified free inquiry. In the practical test stage, the usability of the practical work module was tested to the students of Biology Education Program STKIP PGRI West Sumatera taking the course of animal structure. The testing was carried out on practical work activities of: 1) digestion system, 2) respiratory system and circulation system, 3) excretion and reproduction system, and 4) central nervous system.

The instrument that used to assess the practicality of practical work module of animal structure using modified free inquiry was a questionnaire of practicality given at the end of practical work activity. The questionnaire was used as a data-gathering tool because the collected could be more objective than using interviews. Moreover, the correspondents could answer more freely without being influenced by the mental attitude of the researcher's relationship with the research subject, or the time available in thinking of the answer [7]. Questionnaires were arranged in several aspects and in one aspect arranged in several indicators (Table 1).

Table 1. Practicality aspects assessed

No	Practicality Aspects		Indicators	Statement No.	Total
1	Ease of use	a.	The practical work module of animal structure using modified free inquiry has guidance in using it.	1	12
		b.	The practical work module of animal structure using modified free inquiry which has a goal adjusted toward student learning achievement.	2	
		c.	The practical work module of animal structure using modified free inquiry has clear and systematic scientific steps.	3	
		d.	The practical work module of animal structure using modified free inquiry leads the students in doing scientific steps.	4	
		e.	The practical work module of animal structure using modified free inquiry has a theoretical review with the drawing which	5,6	

Table 1. Practicality aspects assessed

No	Practicality Aspects		Indicators	Statement No.	Total
			makes it easier for students to understand practical work activities.		
		f.	The practical work module of animal structure using modified free inquiry has an easy to follow steps.	7	
		g.	The practical work module of animal structure using modified free inquiry has helpful tips and tricks for the practical work.	8	
		h.	The practical work module of animal structure using modified free inquiry has a place / column / table for the data / results obtained to be used easily and in accordance with the needs.	9,10,11	
		i.	The practical work module of animal structure using modified free inquiry has evaluation questions appropriate to practical work activity.	12	
2	Time required for implementatio n	a.	In practical work activity in the laboratory, the time allocated to carry out the experiment / practical work activities is sufficient	13	6
		b.	In practical work activity in the laboratory, time to draw the observation result is sufficient	14	
		c.	In practical work activity in the laboratory, time to draw the observation result is sufficient	15	
		d.	In practical work activity in the laboratory, the time to answer the hypothesis is sufficient	16	
		e.	Using the practical work module of animal structure using modified free inquiry is not boring, practical and easy.	17,18	
3	Ease to interpret		Easily interpreted by expert lecturers as well as by other lecturers.	19,20	2
4	equivalency		It has the same equivalence.	21	1
5	Motivation		Provide motivation to have, do practical work activities and process the result of practical work.	22,23,24	3
				Total	24

The practicality was assessed by using the percentage formula [8]

$$V = \frac{item\ score\ obtained}{maximum\ score} \times 100\%$$

Based on the "V" score obtained, criteria for appraisal of practicality [8] of the practical work module of animal structure using modified free inquiry were as set out in Table 2.

Table 2. Categories of practicality of the practical work module of animal structure using modified free inquiry

Practicality score (%)	Categories
0 - 20	Not practical
21 – 40	Less practical
41 – 60	Quite practical
61 – 80	Practical
81 – 100	Very practical

3. Result and Discussion

From the questionnaire recapitulation result, the practicality of practical work module of animal structure using modified free inquiry was shown in Table 3 and 4.

Table 3. The practicality result of practical work module of animal structure using modified free inquiry by the students

No.	Aspect of Assessment	Total Score	Score maximum	Result	Criteria
1	Ease of use	1799	2112	85,18	Very Practical
2	Time required for implementation	868	1056	82,20	Very Practical
3	Equivalency	146	176	82,95	Very Practical
4	Motivation	441	528	83,52	Very Practical
	Mean			83,46	Very Practical

Table 4. The practicality result of practical work module of animal structure using modified free inquiry by the lecturers

No.	Aspect of Assessment	Total Score	Score maximum	Result	Criteria
1	Ease of use	84	96	87,50	Very Practical
2	Time required for implementation	42	48	87,50	Very Practical
3	Easy to interpret (only for lecturer)	15	16	93,75	Very Practical
4	Equivalency	7	8	87,50	Very Practical
5	Motivation	20	24	83,33	Very Practical
	Mean			87,92	Very Practical

Based on Table 3 and Table 4, it could be shown that the practical work module of animal structure using modified free inquiry was very practical to be used by students and lecturers during the process of practical work in laboratory.

Practicality in terms of ease of use assessed the ease of using and understanding of the contents in the module. [9] Learning media was said to be practical related to whether or not the media was easy to be used. [10] Practicality referred to the condition of developed learning modules that could be easily used by learners so that learning was meaningful, interesting, fun, and useful for the learner's life. It was also able to improve their creativity in learning and having a degree of effectiveness to the participants' learning outcomes. The practical work module of animal structure using modified free inquiry was completed with guidance on the use of modules, practical purposes tailored to student learning achievements, clear and systematic scientific steps, guidance for students in doing scientific steps, theoretical review with picture shows that make it easier for students to understand practical activities, easy-to-follow steps, tips and tricks that help in the practical work, place/column/table for the data/results obtained that can be used easily and in accordance with the needs, and the questions of evaluation in accordance with practical work activities. A guidance to use the module was provided at the beginning to guide students and lecturers in filling out the module assigned tasks, introducing practical work activities using modified free inquiry and listing criteria of practical work report. Thus, in terms of ease of use, the practical work module of animal structure using modified free inquiry as practical work learning media was considered as very practical by both students (85,18%) and lecturer (87,50%).

The practicality in terms of time required for implementation assesses the use of time in practical work activity using the module, such as the time allocated to carry out experimental / practical work activities, drawing observations result, filling out the observation table, answering the hypothesis, and using the practical work module of animal structure using modified free inquiry was not boring, practical and easy. Although students within their group had different abilities in understanding learning, the time used to collect data, such as dissecting and decomposing the observed organs and arranging them in an integral set of organ systems, was still sufficient. Then, to draw the observations, the students individually could finish drawing the organ systems of each species according to the time they agreed with. To conduct the discussion, fill out the table of observations and conclusions, students could complete the practical work activities accordance with the time needed. The developed module should provide an opportunity for individual differences, meaning that learners should be the focus of attention in designing objectives, selecting materials, and using methods and media as learning resources [2].

In addition, [11] learning with modified free inquiry was very effective to train students' critical thinking skills. Students' interest in this part was very positive because students were motivated to learn even outside of the learning time. For example, after the practical work time was over, some students were still stay in the practice room to ask questions with their or other group members as well as with the lectures, to reinforce the results of the lab work they get. Thus, it could be seen from the statement points in the questionnaire "practical work by using module of animal structure using modified free inquiry was not boring even for a long time" received positive responses from the students. Therefore, as a result of the assessment for practicality in terms of time required for implementing was considered very practical both by students (82.20%) and lecturers (87.50%).

Practicality in terms of ease to interpret, the practical work module of animal structure using modified free inquiry was considered very practical (93.75%) by the expert lecturers as well as by other lecturers. Thus, this module was very easy to be interpreted, either by the lecturer of animal structure or lecturer who is not teaching animal structure course. Interpretation of the same object could be different for different people. However, judging from the assessment, this module had a message that can give the same interpretation, explanation, meaning, impression, opinion or view to every lecturer who read it. The instruction sentences, commands, learning materials, practical work procedures, and modified free inquiry steps could give the readers the same message. In other words, it was communicative. [6] Communicative meant that the content of the book was easy to be understood by the reader, systematic, clear, and contained no language errors.

The practicality in terms of equivalency of the practical work module of animal structure using modified free inquiry was rated very practical by the students (82.95%) and by the lecturers (87.50%). This module has a language quality that contains learning messages, in which [12] language is a tool for thinking. The messages have been arranged in such a way that the responses of all readers are the same. This means that the practical work module of animal structure using modified free inquiry has its own features in order to be easily understood by students and lecturers who read it. [6] Textbooks with high legibility contain sentences length and structure according to the reader's comprehension, and the length of the paragraph corresponds to the reader's understanding.

The practicality in terms of giving motivation to do practical work activity and to process the result data of practical work were rated very practical by student (83.52%) and lecturer (83.33%). Learning by using the practical work module of animal structure using modified free inquiry could awaken students' curiosity and interest so that students were motivated to do practical work and process the data obtained. [13] Students' learning motivation of student groups taught through inquiry learning was better than that of students taught through direct learning methods in both high-score student groups and low-score student groups.

In practical work activities, students were arranged to work in groups in order to make a cooperative learning. Using this inquiry learning, students could share and discuss with their group mates. [14] When students work together to achieve common goals, as they do when the structure of cooperative awards was applied, they learned about the work that could help the success of one's group. Therefore, the learners could encourage each other's learning, academic efforts, and express norms that were appropriate to academic achievement. Furthermore, [15] the impetus of a person to do something is called motivation. In which, motivated behaviour was selective, goal-oriented, and persistent behaviour.

IOP Conf. Series: Journal of Physics: Conf. Series **1108** (2018) 012011 doi:10.1088/1742-6596/1108/1/012011

4. Conclusion

From the result of the research, it could be concluded that the practical work module of animal structure using modified free inquiry that had been developed was practical in terms of ease of use, time required for implementation, easy to interpret, equivalency, and motivation to be used in the learning process of practical work of animal structure in college.

Acknowledgments

Special thanks to Kemenristekdikti which has supported as sponsors of Penelitian Strategi Nasional Institusi (PSNI) [Project No: SP DIPA-042.06.1.401516/2018, tanggal 5 Desember 2017]

References

[1] Sani R 2015 *Pembelajaran Saintitik untuk Implementasi Kurikulum* 2013 (Jakarta: Bumi Aksara) p. 89.
[2] Yaumi M 2018 *Media dan Teknologi Pembelajaran* (Jakarta: Prenada Media Group)
[3] Thiagarajan S, Semmel D and Semmel M 1975 Media Review: Books: Instructional Alternatives for Exceptional ChildrenInstructional Alternatives for Exceptional Children.DenoEvelyn N. (Ed.). Minneapolis: Leadership Training Institute/Special Education, University of Minnesota, 1973. *Exceptional Children* **41**(6) pp. 441-443
[4] Setiawan E 2018 Arti kata praktis- Kamus Besar Bahasa Indonesia (KBBI) [online] (https://www.kbbi.web.id/praktis)
[5] Plomp T and N Nieveen N 201007 *An Introduction to Educational Design Research* (Enschede: SLO)
[6] Akbar 2013 *Instrumen Perangkat Pembelajaran* (Bandung: Remaja Rosdakarya)
[7] Ali M and Asrori M 2014 *Metodologi & Aplikasi Riset Pendidikan* (Jakarta: Bumi Aksara)
[8] Riduwan, 2012 *Dasar-dasar Statistika* (Bandung: Alfabeta)
[9] Hestari S, Susantini E dan and Lisdiana L 2016 Validitas, Kepraktisan dan Efektivitas Media Pembelajaran Papan Magnetik pada Materi Mutasi Gen *Jurnal BioEdu Berkala Ilmiah Pendidikan Biologi* **5**(1) 7-13
[10] Alfiriani A and Hutabri E 2017 Kepraktisan dan Keefektifan Modul Pembelajaran Bilingual Berbasis Komputer *Jurnal Kependidikan* **1**(1) pp. 12-23
[11] Hadi S, Susantini E and Agustini R 2018 Training of Students' Critical Thinking Skills through the implementation of a Modified Free Inquiry Model *Journal of Physics Conference Series* **947** 012063.
[12] Sukmadinata N S 2016 *Landasan Psikologi Proses Pendidikan* (Bandung: PT. Remaja Rosda Karya)
[13] Ginanjar A 2015 Pengaruh Metode Inkuiri Terhadap Motivasi Belajar Siswa SMP *Jurnal Kependidikan* **45**(2) pp. 123-129
[14] Slavin R E 2015 *Cooperative Learning Teori, Riset dan Praktik* (Bandung: Nusa Media)
[15] Latipah E 2017 *Psikologi Dasar Bagi Guru* (Bandung: Remaja Rosda Karya)

IOP Conf. Series: Journal of Physics: Conf. Series **1108** (2018) 012012 doi:10.1088/1742-6596/1108/1/012012

Support Vector Machine with Fisher Score Feature Selection to Predict Disease-Resistant Gene in Rice

G S Saragih[1*], Z Rustam[1]

[1]Department of Mathematics, FMIPA Universitas Indonesia, Kampus UI Depok, Depok 16424, Indonesia

*Corresponding author : glori.stephani@sci.ui.ac.id

Abstract. Indonesia is one of the leading rice producers and biggest rice consumers in the world. Hence, the sustainability of rice production systems needs to be considered. Bacterial leaf blight is one of the diseases that distract the growth of rice. But, BLB can be controlled through the development of BLB resistant. The experimental method (in vitro) is one of common ways to prevent it, but this method is not efficient and having a big error. Thus, a new method called computational method required to overcome these shortcomings. There are three steps of that method in predicting genes resistance disease which are extraction, feature selection, and machine learning. The global encoding was used in the feature extraction process, then fisher score was used to select the feature. After that, a model that represents the data to analyze disease-resistant gene in rice was built. The Support Vector Machine (SVM) was applied. The result was obtained that with only ten features and the performance of purpose method, the model could represent protein information 90.91% with training data used 90%. It indicated that it was very useful. It could predict disease-resistant gene in rice with high accuracy short running time and low dimension data.

1. Introduction

Rice is one of the most important and major food-grain of the world, especially for Indonesia. Rice is the staple food for Indonesian, but unfortunately there are many diseases that distract the growth of rice. In this paper we focus for the Bacterial Leaf Blight (BLB) disease caused by Xanthomonas oryzae pv. oryzae (Xoo). This disease causes wilting of seeds and yellowing and drying of leaves. It can significantly decrease rice production and causing the loss of yields. Currently, BLB is reported to not only damage wetland rice, but also upland rice in Indonesia. BLB disease control through the development of BLB resistant varieties is one of the effective and easiest way to be applied by farmers. Nowadays, six gene Xa1, xa5, xa13, Xa21, Xa3/Xa26 and Xa27 have been reported to be isolated for bacterial blight resistance [1]. The experimental method is one of the usual ways to find disease resistance gene. This method is costly, requires a lot of time, and known to have a big error. Thus, a new method is required to overcome these shortcomings [2].

Computational method is one solution to solve this problem. There are three steps from computation methods in predicting genes resistance disease which are extraction, feature selection, and machine learning. Feature extraction is digging up useful information in proteins and describe it as a normalized vector feature. In this research, we used global encoding in the feature extraction process. This method will, firstly, classify the 20 types of amino acids into 6. We then get 10 combinations where each combination contains 3 different classes. At the end of this method, there are two types of descriptor, composition and transition, which will be extracted from the obtained subsequence to illustrate the global composition from each sequence. Then, it will be used as a final vector feature for the classification method [2]. The new dataset from extraction feature that used in prediction of disease-resistant dataset consists of hundreds of features. However, high dimensional will make over fitting in classification and high computational cost. Therefore, it takes feature selection by reducing the number of features that will be used in the classification and generate a new data set containing the best and most relevant features. Feature selection is one of pre-processing data method in machine learning. In addition, to reducing the amount of cost, time, and memory capacities, this process can also improve the accuracy of the classification [2].

Then, a model representing the data to analyze disease-resistant gene in rice was built. Using the development of machine learning research, there are multiple ways to analyze disease-resistant gene data using machine learning methods. Jingbo, Xia, et al. have predict disease resistant gene in rice with Artificial Neural Network [1]. However, It will be tried another purpose method applying the popular and effective methods with a high accuracy called Support Vector Machine (SVM). SVM has demonstrated high classification capabilities in the field of protein prediction, functional classification of proteins, multiplication of protein folds, and sub-cellular location prediction. It has been applied by my colleagues in cancer classification [4], predictive protein interactions in HIV [5], and intrusion data systems [6], etc.

2. Data
The data was classified into disease-resistant and non-disease-resistant. Disease-resistant amino acidacid sequence in this paper were chosen based on literature reported on Xanthomonas resistance and the sequences were collected from National Center for Biotechnology Information (NCBI, http://www.ncbi.nlm.nih.gov/) by using the gene names as the key words [1]. While non-disease-resistant amino acid sequences in rice we took through the UniPort (https://uniport.org) Protein database. The dataset was divided into training and testing dataset.

3. Method
3.1. Global Encoding(GE) of Amino Acid Sequence
There were three steps to convert each amino acid sequence into a vector input to be used in the classification method.

Step 1. Transformation of protein sequence: Researches [2] classified 20 kinds of amino acids into 6 classes based on physicochemical characteristic (see Table 1). Every protein sequence would be represented by six symbols: A1, A2, A3, …, A6. After that, it was obtained 10 combinations which contained three different classes. Then, ten modes could be obtained as follows: {A1, A2, A3} vs {A4, A5, A6}, {A1, A2, A4} vs {A3, A5, A6}, {A1, A2, A5} vs {A3, A4, A6}, {A1, A2, A6} vs {A3, A4, A5}, {A1, A3, A4} vs {A2, A5, A6}, {A1, A3, A5} vs {A2, A4, A6}, {A1, A3, A6} vs {A2, A4, A5}, {A1, A4, A5} vs {A2, A3, A6}, {A1, A4, A6} vs {A2, A3, A5} and {A1, A5, A6} vs {A2, A3, A4} [7]. Furthermore, it continued by transforming those ten combinations into ten binary sequences. Let's symbolize protein sequence $S = s_1, s_2, \dots, s_{10}$and ten transformed sequences of S as $T_1, T_2, \dots, T_{10}$. Then, the first two numerical sequences could be calculated, $T_1(s_1)$and $T_2(s_2)$ [8], they are defined as follows :

$$T_1(s_i) = \begin{cases} 1, & s_i \in \{A_1, A_2, A_3\} \\ 0, & s_i \in \{A_4, A_5, A_6\} \end{cases} i = 1, 2, \dots, n \quad (1)$$

$$T_2(s_i) = \begin{cases} 1, & s_i \in \{A_1, A_2, A_4\} \\ 0, & s_i \in \{A_3, A_5, A_6\} \end{cases} i = 1, 2, \dots, n \quad (2)$$

T_i referred as the i-th characteristic sequence of amino acid sequence and s_i referred as the i-th amino acid of the given protein sequence.

Table 1. Amino Acid Classification

Amino Acid Classification	
Aliphatic amino acid :	A1 = { A, V, L, I, M, C }
Aromatic amino acid :	A2 = { F, W, Y, H }
Polar amino acid :	A3 = { S, T, N, Q }
Positive amino acid :	A4 = { K, R }
Negative amino acid :	A5 = { D, E }
Special conformations :	A6 = { G, P }

Step 2. Partition from characteristic sequences: In this step, each characteristics sequence would be divided into some of sub sequences with different length and specific strategy. Each characteristics sequence would be divided into specific numbers of sub sequences according to a partition method [6]. For any characteristic sequence $T_n = t_1, t_2, \dots, t_n$, was given a positive integer *N*, T_n that would be divided into *N* sub sequences. $SubT_k$ was called as *k*-th subsequence, where *k* = *1,2,…,N* and $SubT_k$ was composed of the first $\lfloor k_n/N \rfloor$ numbers of T_n [2].
Step 3. Extraction of feature vectors: In the last step, feature vectors which consisted of two descriptors namely composition and transition would be extracted from subsequence. The composition descriptor was the frequencies of '0' and '1' in each subsequence. From composition descriptor, one of subsequence contained two frequency values, so each characteristic would be represented by a *2*N* dimensional feature vector. Then, transition was second descriptor that calculate the switch frequency between '0' and '1' in every subsequence. The times where '0' changes '1' and '1' changes '0' [2].

3.2. Fisher Score Feature Selection
Nowadays, there are many high dimensional datasets like gene expression microarrays, text documents, digital images, clinical data, and SNP data. But, learning machine is very hard in high dimensional data; it can make classification over fitting and high computational cost. Hence, a further challenge for modelling in high dimensional space could avoid over fitting of the training data [3]. Too many features which made by the model could identify single data points by single features and build a special case just for a single data point. Therefore, it was used feature selection to reduce the dimensional of dataset after extraction feature with global encoding. The score of each feature that was obtained after feature selection examined the level of contribution of each gene to cause discrimination between disease-resistant (class I) and non-disease-resistant (class II). In this research, the distributions of class I and class II were known. Fisher score method was used to choose the most informative features. The formula was defined as follows:

$$V(i) = \frac{(\mu_1(i)-\mu_2(i))^2}{\sigma_1^2(i)-\sigma_2^2(i)} \tag{3}$$

Where,
$\mu_1(i) = mean\ for\ the\ first\ class$
$\mu_2(i) = \ mean\ for\ the\ second\ class$
$\sigma_1^2(i) = variance\ for\ the\ first\ class$
$\sigma_2^2(i) = \ variance\ for\ the\ second\ class$
The lowest scores from fisher score calculation was the best feature for our data.

3.3. Support Vector Machine (SVM)
SVM was supervised learning model and binary classification algorithm. Supervised learning model, it meant that a dataset which has been label was needed, so SVM was suitable for the task of distinguishing between disease-resistant and non-disease-resistant gene in rice. Data training used to build a model and we must suppose that data which could be used to classify and could be separated by a line. The line could be represented by the equation:

$$y_i(< w.x_i > +b) \geq 1 - \xi_i,\quad \xi_i \geq 0\ (i = 1,2,3,\dots,N) \tag{4}$$

where w was the normal vector of hyperplane; b was the bias hyperplane; ξ_i was the slack variables. It was known that there was an infinite possible line obtained by changing the value of w and b. The goal was to design a hyperplane that classify all training vectors in two classes. The optimal hyperplane was the one which maximizes the margin of the data training. The biggest margin might be found to get an optimal hyperplane by selecting two hyperplanes separated the data with no points between them and maximize their distance (the margin). Then, maximizing margin was the same thing as minimizing the norm w.

3.4. Algorithm
The following stepwise procedure was employed so as to implement the algorithm:

1. Download and prepare the amino acid sequence data.
2. Assign labels 1 or 0 to disease-resistant or non disease-resistant gene separately.
3. Extract the amino acid sequence use global encoding to convert amino acid sequence into a vector input to be used in the classification method.
4. Divide the data into training and testing dataset.
5. Run fisher score to get an order number of features
6. Run SVM and obtained the trained model which have derived the classification rules.
7. Run classifier on the test data to assess the prediction.

4. Experiment and Results
The data which used was a sequence of amino acids that have been extracted to be numeric using global encoding. Then, we would select the features with fisher score. In this experiment, it was used parameter $N = 5$ in feature extraction model. Feature selection with fisher score gave us the following results.

Table 2. SVM Results with Fisher Score in Predicting Disease-Resistant Gene in Rice

Number of Features	% Training Data	%Accuracy[a]	Running Time
10	70	85.29	0.59
10	80	82.61	0.66
10	90	90.91	0.69
20	70	88.24	0.59
20	80	86.96	0.63
20	90	90.91	0.66
60	70	82.35	0.63
60	80	86.96	0.64
60	90	90.91	0.67
80	70	88.24	0.59
80	80	86.96	0.59
80	90	90.91	0.69
90	70	88.24	0.55
90	80	86.96	0.59
90	90	90.91	0.67
ALL	70	88.24	0.55
ALL	80	86.96	0.59
ALL	90	90.91	0.67

[a]Accuracy = (number of prediction/(training + testing)) * 100%

From Table 2, it could be seen that only ten features were used. However, it could represent the model that could reduce the dimensional of data set to be a new data set that will be classified in Support Vector Machine by using the feature of sequence number as follows: 7, 8, 3, 4, 6, 5, 10, 9, 1, 2. It meant that the used of fisher score feature selection gave advantages to reduce dataset into the optimal one. Data on table 2 shown us data related the running time. It was about how long model could predict data, in the sense of outputting. It shown that it just needed less than 1 second to get output of predicting disease-resistant gene in rice. It was used SVM non-linear with kernel Radial Basis Function with parameter 0.05. Then, the calculation of performance model with number of features with only 10 different composition of data training was selected randomly and shown in the following table.

Table 3. SVM Results with Fisher Score (10 Features)

Number of Features	% Training Data	%Accuracy[a]	Running Time
10	10	44.66	0.13
10	20	79.35	0.20
10	30	81.25	0.30
10	40	84.06	0.45
10	50	85.96	0.53
10	60	86.96	0.52
10	70	85.29	0.59
10	80	82.61	0.66
10	90	90.91	0.69

[a]Accuracy = (number of prediction/(training + testing)) * 100%

From Table 3, it was known that the performance of purposed method indicated that model can represent protein information 90.91% with training data used 90%.

5. Conclusion

In this research, it proposed an accurate and robust computational method for predicting a disease-resistant gene in rice. This method combined the global encoding as features extraction method then selection the feature new dataset with fisher score and Support Vector Machine as a machine learning model. The accuracy result shown that feature selection could reduce the number of features with the performance of the proposed method. It indicated that, using only ten features model, it could represent protein information 90.91%.This result was very encouraging and shown that our purpose method was very useful. In short running time and low dimension data, it could predict disease-resistant gene in rice with high accuracy.

References

[1] Jingbo Xia., et. al. 2009 *Prediction of Disease-Resistant Gene by Using Artificial Neural Network* (Wuhan:International Conference on Research Challenges in Computer Science, College of Science, Huazhong Agricultural University)

[2] Huang Y, You Z, Chen X, Chan K and Luo X 2016 Sequence-Based Prediction of Protein-Protein Interaction using Weighted Sparse Representation Model Combined with Global Encoding *BMC Bioinformatics* 74-184

[3] Clarke R, Resom HW, Wang A, Xuan J, Liu MC, Gehan EA and Wang Y 2008 The Properties of High-Dimensional Data Spaces: Implications for Exploring Gene and Protein Data *Nature Reviews Cancer* **8** pp 37-49

[4] Rustam Z and Maghfirah N 2017. *Correlated based SVM-RFE and Feature Selection for Cancer Classification using Microarray Databases* (Depok : The 3rd International

Symposium on Current Progress in Mathematics and Sciences, Universitas Indonesia)

[5] Lestari D, Musti M I S and Bustamam A 2017 *Sequence-Based Prediction of Protein-Protein Interactions Using Ensemble Based Classifier Combined with Global Encoding in HIV (Human Immunodeficiency Virus)* (Depok : The 3rd International Symposium on Current Progress in Mathematics and Sciences, Universitas Indonesia)'

[6] Rustam Z and Audia N P A 2017 *Comparison between Support Vector Machine and Fuzzy Kernel C-Means as Classifiers for Intrusion Detection System using Chi-Square Feature Selection* (Depok : The 3rd International Symposium on Current Progress in Mathematics and Sciences, Universitas Indonesia)

[7] Li X, Liao B, Shu Y, Zeng Q, and Luo J 2009 Protein functional class prediction using global encoding of amino acid sequence *Journal of theoretical biology* **261**(2) 290-293

[8] He P A and Wang J 2002 Numerical characterization of DNA primary sequence *Internet Elec J Mol Des* **1** pp 668-674

[9] Ren Y, Wang D, Wang Y, Zhou J, Zhang H, Zhou Y and Liang Y 2010 Prediction of disease-resistant gene in rice based on SVM-RFE *Biomedical Engineering and Informatics (BMEI), 2010 3rd International Conference **6** pp 2343-2346 IEEE.V. Panca and Z. Rustam 2017 Application of Machine Learning on Brain Cancer Multiclass Classification (The American Institute of Physics (AIP) Conference)*

[10] Suryadi Y, Susilowati D N, Lestari P, Sutoro S, Manzila I. Kadir T S, ... & Artika I M 2014 Analisis Keragaman Genetik Isolat Bakteri Xanthomonas oryzae pv. oryzae dari Jawa Barat dan Jawa Tengah Berdasarkan Analisis ARDRA Gen 16SrRNA *Jurnal Fitopatologi Indonesia **10***(2) 53

[11] Suryadi Y, Samudra I, Priyatno T P, Susilowati D N, Lestari P, Fatimah F and Kadir T S 2016 Determination of Pathotypes from Indonesian Xanthomonas oryzae Pv. Oryzae Population causing Bacterial Leaf Blight and their Reactions on Differential Rice. *Makara Journal of Science* 109-118

Contextual Learning On The Basis of Coastal Culture to Enhance Students' Competency in Mathematical Problems Solving

M A Kleden[1], Y Sugi[2] and U Gerardus[3,*]

[1, 2]Mathematic Department, Universitas Nusa Cendana Kupang, Indonesia.
(E-mail: maria_kleden@staf.undana.ac.id, yoseph.sugi@staf.undana.ac.id)
[3] PGSD, Universitas Nusa CendanaKupang, Indonesia

*Corresponding author: udagerardus@gmail.com

Abstract. Problem solving is one of the competencies in learning mathematics. It is expected that students have this competency to be able to solve problems in their life by using math. The aim of this research is to improve problem solving competence through contextual learning based on coastal culture. The research design used was pretest-posttest control group design. The population is junior high school students in East Flores District, East Nusa Tenggara Province. 119 students were chosen by random purposive sampling technique. Four classes from grade 8 from a Junior High School at Larantuka were randomly selected as experimental and control class. The experimental class students employ contextual learning based on coastal culture, whereas students in the control class use conventional learning. The instrument of this research is test of mathematical problem solving competence. The results showed that there was an enhancement in mathematical problem solving competence in students who received contextual learning based on coastal culture (CLBCC) and conventional learning approach (CLA). Improvement competence of mathematical problem solving on CLBCC group is higher than CLA group. This also happened to the achievement of mathematical problem-solving competency. There are differences in the improvement and achievement of mathematical problem-solving competency between CLCBB and CLA students.

1. Introduction

East Flores regency is one of the regencies in NTT province located on the eastern edge of Flores Island. The total area of East Flores Regency covers 17 islands, namely 3 inhabited islands and 14 other uninhabited islands with 1,812.85 km2 (31%) and ocean area which covers 4,170,53 km2 (69%). This regency consists of 19 districts living in coastal areas. As many as 90 percent of the people of East Flores are coastal communities so that the livelihoods of most people of East Flores are fishermen and farmers Profile of East Flores Regency, 2016

The enrollment rate for coastal high schools is only 39%, which is below the Indonesian average rate of 80.49% (UNESCO, 2013). In addition, coastal communities are still having low awareness of culture safe guarding and preserving coastal environments. UNESCO further confirms that education for sustainable development should continue to focus on the importance of natural resource issues. Through education, it is expected that people will aware of human resources who realize the need for harmonization between human resources and natural resources to improve community's' welfare.

Establishing qualified human resources with good mathematical skills requires meaningful mathematics learning. Learning that utilizes the surrounding environment contextually will build a good understanding of the material learned as well as the behavior of maintaining the surrounding environment. Meaningful learning is needed to train ways of reasoning, communicating, connecting, and solving problems they face in their environment. In this learning, students were given opportunity to directly solve problems in everyday life using mathematical concepts.

This shows that learning mathematics is a very important activity. It cannot be denied that students' mathematical ability is still quite inadequate. Low mathematical ability is not only found in elementary and high school students but also among all students. To the best of our knowledge, no research has been conducted to look at the student's mathematical abilities in this area. Therefore, this work is the first study to assess the student's mathematical abilities focusing on the coastal areas of district Flores Timur (East Flores). The lack of mathematical abilities is found in the students of the Mathematics Study Program [14]. This research reveals that after getting metacognitive learning there is improvement of mathematical communication ability and logical thinking mathematically, but the improvement is still classified in medium category.

Contextual learning based on coastal culture not only reinforces students' understanding of the material learned but also leads to cultural awareness. Such learning integrates cultural and environmental contextual issues in mathematical concepts. The integration of culture and environment in mathematics learning is expected to improve students' mathematical competency.

The above exposure indicates the importance of context in every learning implementation. Contextual learning is referred to contextual teaching and learning (CTL). In this learning, all student activities are related to relating the subject matter to the real-life context they find. This means that students seek the meaning of tasks in the form of finding interesting mathematical problems, seeking information and drawing conclusions, actively choosing the right strategy, composing, planning, investigating, questioning, and making conclusions about solving a mathematical problem, and linking mathematical concepts they studied with context in life situations[5]. Therefore, a research that analyze student problem solving ability is needed. This is in line with [2] who say that through problem solving, students can find mathematical concepts. The process of learning mathematics is not just transferring concepts or ideas from the teacher to students, but it is a process in which the teacher gives the opportunity for students to understand and construct ideas given for later use in solving various problems faced in accordance with the level of development. Based on this view, the problem solving ability is one of the essential and fundamental abilities in mathematics learning that must be possessed by every student (Rahayu & Afriansyah, 2015). Having a problem solving competency not only train students to solve mathematical problems but also make them able solve problem in everyday life.

Based on the above background, the research problems are formulated as follow: (1) is there any difference of problem solving competencies between groups of contextual learning students based on the coastal culture approach and group of students from conventional learning approach?. (2) Is there any difference of problem-solving competencies between groups of contextual learning students based on coastal culture approaches and groups of students with conventional learning approaches?

The purpose of this study were to identify: (1) differences of problem-solving competencies between contextual learning group of students based on coastal culture approach and group of students who acquired conventional learning approach; (2) the differences of problem-solving competencies between contextual learning group students based on coastal culture approach and group of students from conventional learning approach;

2. Literature Review

2.1 Problem Solving Competency

Solving a problem is part of human life. Humans always get in touch by both social and personal issues. For that reason, human should have ability and skill to solve every problem. Dahar [3] says that problem solving is a human activity to combine concepts and rules that have been obtained before and not as generic. Further [7] discloses that problem solving is an attempt to find a solution or solution of a situation to achieve certain goals. To measure problem solving competency some mathematics indicators are required. These indicators are used as a reference to assess student' ability in problem solving. The steps used refer to Polya's troubleshooting steps. The problem-solving indicators in this study refers to the indicators of problem-solving capabilities summarized by[10], they can be stated as follows: 1) identifying what is known, questioned, and elemental elements; 2) create a mathematical model; 3) implementing problem solving strategies in/outside mathematics; 4) explain or interpret the results; 5) solve mathematical models and real problems; 6) using mathematics meaningfully.

Based on the description of problem solving competency above, it can be concluded that in solving problem, students is not only required to have the knowledge of mathematical procedures but students should have confidence in solving problems encountered. Thus it will generate motivation and willingness to find solutions to problems with strategies used. Polya [7] defines problem solving as an attempt to find a way out of a difficulty to achieve a goal that is not so immediately achieved. In other hand, Sujono (1988) describes a mathematical problem as a challenge when the solution requires creativity, understanding and original thought or imagination. In sum, the problem solving ability is the ability to understand problems and find solutions to mathematical problems related to coastal culture in junior high school students. Indicators of problem solving abilities include identifying problems, planning solutions, solving problems, and interpreting results.

2.2 Contextual Learning Based on Coastal Culture

Contextual learning is learning using a contextual approach that occurs naturally in the form of work and experience. In this learning, students are trained to experience a process of learning-related materials with the daily real-world context. In contextual learning there are seven main components: (1) constructivism; (2) questions; (3) inquiry; (4) learning community; (5) modeling; (6) reflection, and (7) authentic assessment. In this case, constructivism means building new knowledge based on initial knowledge through a process of social interaction and assimilation accommodation. The first stage in learning is learning community. Students are divided into groups of three to five. The division of this group aims to motivate students to share information, discuss, and work together in groups.

Asking questions is one of the activities undertaken by teachers and students in CTL. Questions posed by teachers aim to encourage, guide, and assess students' thinking abilities. While the question posed by students is related to the lack of understanding of concepts being studied. Blakey and Spence [11] states that learners should ask themselves what they know and what they do not know at the beginning of the learning activities. Question is one indicator to know how far students understand the concept being studied.

In mathematics learning, modeling is an important aspect. Modeling is helpful in solving a problem. Various phenomena of everyday life can be expressed in mathematical models. However, it is realized that the ability of learners in translating the everyday phenomena in mathematical models is still low. Muijs & Reynolds, Killen, [11]) states that modeling occurs when teachers show the processes involved in performing difficult tasks, or when teachers tell students about their thinking and are motivated to choose specific strategies when solving problems. Modeling and discussion enhance learners' thinking and talk about their own thinking [11].

Inquiry is a process of discovering concepts, principles, knowledge that begins with the giving of contextual problems. Inquiry process can be done in group discussion or after discussion. Ruseffendi [9] asserts that one of the learning objectives with the question is that students learn scientific method through discovery and its application in other situations. Self-discovery gives students a great deal of satisfaction and can encourage them to get new findings. This process has an impact on increasing motivation to learn. After that, students are directed to reflect through activities of recording what has been studied, asked, reported.

Reflection aims to check how students understand the mathematical concepts learned. The ability to reflect is a prerequisite for articulation and articulation itself requires identification of the essentials of an action. Students can reflect on what they have done based on important aspects of their thinking and actions.

3. METHODS

This research undertaken a pretest-posttest control group design [1] Group of experiment are treated with a contextual learning based on coastal culture approach, whilst group of control are treated with conventional learning approach.

The population of this study is all 8th grade students in East Flores District. The sample of research was chosen from two classes from four classes through purposive sampling technique. 119 students became the object of this study. The initial competence of both groups is homogeneous. The research instrument was a test of competence of mathematical problem solving that has been tested on 9th grade students. The test consists of five valid questions with a high reliability category that is 0.77.

Quantitative data in the form of improvement and achievement of problem solving ability are analyzed through descriptive and inferential approach. Hypothesis testing performs parametric and non-parametric statistical analysis. The t-test used for data was normally distributed, whereas the Mann Whitney test was used in data that are not normally distributed.

4. Result And Discussion

4.1 Achievement of Mathematical Problem Solving Competency

The data of Mathematical problem solving achievement was not normally distributed. To test the difference in achievement of mathematical problem-solving ability among students who received CLBCC and who received conventional learning, it was used Mann Whitney U-test.

Hypothesis of difference tests upon achievement of mathematical problem solving competency outcomes are as follows:

H_0: There is no different achievement of mathematical problem solving competency amongst groups of different learning approaches.

H_1:There is different achievement of student mathematical problem solving competency amongst groups of different learning approaches.

Table 1 presents the results of different test data achievement of mathematical problem solving abilities between CLCB students and conventional learning students.

Table 1. Difference Test of Mathematical Problem Solving Competency Outcomes Between 2 Test Groups

Learning Method	N	Avg.	Mann-	Wilcoxon	Z	*Sig.*	Declaration

			Whitney U	W			
CLBCC	59	17.525	744,500	2574,500	-5,461	0,000	H_0 rejected
CLA	60	12.083					

Based on the t-test results presented in Table 1, value of probability (sig.) is smaller than $\alpha = 0.05$. Therefore H_0 is rejected. This signifies that group of students being taught through contextual learning based on coastal culture approach yield higher outcomes in terms of mathematical problem solving competency than students of conventional learning approach group.

4.2 Mathematical Problem Solving Competency Enhancement

In contrast to achievement data, mathematical problem-solving data are normally distributed and homogeneous. For that reason, test difference of problem solving competency used *t* test. The analysis results are presented in Table 2. The hypothesis used in this test is:

H_0 :There is no different enhancement upon mathematical problem solving competency between two groups of different learning methods.

H_1 :There is different enhancement upon mathematical problem solving competency between two groups of different learning methods.

Table 2. Difference Test of Mathematical Problem Solving Competency Enhancement Between 2 Test Groups

Learning Method	N	Avg.	Avg. Dif.	*t*	dof	*Sig.*	Declaration
CLBCC	59	0,591	0,215	6,517	111	0,000	H_0 is rejected
CLA	60	0,376					

Table 2 shows significant smaller values than α, so H0 is rejected. This shows that there are differences in the ability to solve mathematical problems between CLBCC students and conventional learning students. Based on the average value, it appears that the average value of CLBCC class problem-solving enhancement is higher than CLA students. This means that CLBCC is better to improve mathematical problem solving compared to CLA.

Enhancement in mathematical problem solving competency for both test groups generally relies on the average scale, however enhancement for contextual learning based on coastal culture group were higher than the conventional learning group. In a contextual approach, students are habituated to understand the concept of mathematics through problems in everyday life. Students explore and study of various learning resources, identify constraints, build questions, answer questions and express ideas. These special activities trigger students to solve their daily problems using mathematical concepts. In addition, students can explore various strategies in solving problems.

Contextual learning based on a coastal culture approach encourages students to not only understanding mathematical concepts being studied but it also aims to understand coastal culture. Understanding the coastal culture is needed to preserve environment. In this class, the students are given various problems related to the lives of coastal communities such as the number of fish caught, the fishing process, the use of the beach for drying fish, wind direction and boat speed, planting and mangrove functions, ocean currents and marine biota. Students discussed in small groups to solve problems in student worksheets and presented the solved problem. Students are given the opportunity to link their understanding of coastal culture and the concept of two-variable linear equations. This activity encourages students to use their current knowledge in solving mathematical problems as well as other problems and expressing situations. In contextual learning based on coastal culture, students are guided to identify math concepts through coastal community habits. After identifying concept, students further understand the concepts that have been identified through questions being asked. This method encourages students to understand mathematical concepts and to understand and preserve the culture of coastal communities.

There is no doubt that there are still difficulties encountered. Some students have difficulty translating the story into a mathematical sentence. There are even some students who have difficulties writing what concepts are known and what funds are being asked. In addition, they obtain a reference in one of the obstacles experienced by students. They only rely on books shared from schools which are also limited. Reference access from internet is unlikely to be experienced by students due to inadequate internet connection.

Contextual learning based on coastal culture encourages students to be actively involved in discussions. Students are free to ask, answer and refute from friends who different point of view. Interviews showed that in this kind of learning, students feel very close to the problems they face in their daily lives. This helps them understand the concept of math. The classroom' atmosphere is created to find their own answers in the discussion and it makes the students feel the usefulness of the mathematic. Students have a higher curiosity about the next concept because they feel that the concept is meaningful to their lives.

5. Conclusion And Suggestion

5.1 Conclusion

Referring to research results and discussions in the previous section, a conclusion is drawn as follows: (a) There are differences in achievement and enhancements of students mathematical problem solving competency in general; (b) The enhancements of mathematical problem solving competency for both groups lie on an average scale, however, enhancements found upon the BLBCC group was higher than those of the CLA group; (c) Contextual learning based on coastal culture encourages students to more actively solve the problems they face in their daily lives.

5.2 Suggestion

Based on the results of the research and the conclusion above, some suggestions are stated as follows: Contextual learning based on a coastal culture approach in mathematics improves students' competence in math problem solving. Therefore contextual learning based on coastal culture approach should be considered as an alternative learning approach for coastal area students.

6. References

[1] Cresswell, J. W. (2005). Educational research: Planning, conducting and evaluating qualitative and quantitative research. *Upper Saddle River, NJ: Merrill & Prentice Hall.*

[2] Hendriana, H., & Soemarmo, U. (2014). Penilaian pembelajaran matematika. *Bandung: Refika Aditama.*

[3] Herlambang 2013 *Analisis Kemampuan Pemecahan Masalah Matematika Siswa Kelas VII-A SMP Negeri 1 Kepahiang Tentang Bangun Datas Ditinjau Dari Teori Van Hiele* (Bengkulu: Pascasarjana Universitas Bengkulu)

[4] Kleden M A 2015 *Meningkatkan Kemampuan Berpikir Logis Matematis, Komunikasi Matematis dan Self Directed Learning Mahasiswa Melalui Pembelajaran Metakognitif* (Bandung: Disertasi Sekolah Pascasarjana UPI).

[5] Kleden M A and Geradus U 2018 Enhancement of Mathematical Communication Competency Upon Students of Junior High School Through Contextual Learning Based on Coastal Culture. *Education QuarterlyReviews* 1(1) 9-17.

[6] National Council of Teachers of Mathematics (NCTM) 2000 *Professional Standars for Teaching Mathematics.* (Reston. VA : NCTM)

[7] Polya G 1985 *How To Solve It* 2nd ed (New Jersey: Princeton University Press)

[8] Rahayu D V and Afriansyah E A 2015 Meningkatkan Kemampuan Pemecahan Masalah Matematik Siswa Melalui Model Pembelajaran Pelangi Matematika *Jurnal Pendidikan Matematika* **5** (1) 2086-4299

[9] Russeffendi E T 2006 *Pengantar kepada Membantu Guru Mengembangkan Kompetensinya dalam Pengajaran Matematika untuk Meningkatkan CBSA* (Bandung: Tarsito)

[10] Sumarmo U 2014 *Berpikir dan Disposisi Matematik serta Pembelajarannya* (Bandung: UPI)

[11] Toit S and Kotze G 2009 [online] *Metacognitive Strategies in the Teaching and Learning of Mathematics* http://ww.google.com/url?sa=t&rct=j&q=metacognitive+strategies+in+math&source=web&cd=6&cad=rja&ved=0CFEQFjAF&url=http%3A%2F%2Fwww.pythagoras.org.za%2Findex.php%2Fpythagoras%2Farticle%2Fdownload%2F39%2F30&ei=4LImUbzEouMrgf4t4DYAQ&usg=AFQjCNHVj8Pc5sv 12XID57q2yXObEkvikA&bvm=bv.42768644,d.bmk

Students' activities for understanding function shifting by using GeoGebra

Abadi[1,*], D K Fardah[1]

[1] Mathematics Department, Faculty of Mathematic and Natural Science, Universitas Negeri Surabaya, Indonesia

*Corresponding Author: abadi@unesa.ac.id

Abstract. The needs of worker for STEM-related jobs caused STEM education becomes an important focus. Designing activities by combining Science, Technology, Engineering and Mathematics itself (STEM) is possible to do in mathematics learning. This paper aimed to describe students' understanding on function shifting rules in calculus course through activities using GeoGebra as a technology application. A contextual problem was given to students and required them to solve it by using GeoGebra and using their prior knowledge in mathematics and physics. Teaching experiment was conducted to first-year mathematics education students. The data were obtained from the students' worksheet and depth interview. The results showed that the students were interested with the learning approach and they could understand the rules of function shifting.

1. Introduction

The needs of worker for STEM-related jobs these days caused STEM education becomes an important focus [1]. However, the fact revealed that the interest of Students in learning STEM subjects declined this last decade [2]. STEM education concerned on improving science and mathematics with little integration and attention given to technology or engineering several decades recently [2]. Integrated STEM education defined as approaches that explore teaching and learning between/among any two or more of the STEM subject areas, and/or between a STEM subject and one or more other school subjects [3]. The outcomes for learning at least one of the other STEM subjects should be purposely designed in a course—such as a math or science learning outcome in a technology or engineering class. Integrated STEM education is also described as an effort to combine some or all of the four disciplines of STEM subjects into one class, unit, or lesson that is based on connections between the subjects and real-world problems [4]. In summary, STEM Education is an approach that combining some or all of the four STEM subjects into one lesson that is designed intentionally.

STEM education programs of high quality should include (a) integration of technology and engineering into science and math curriculum at a minimum; (b) promote scientific inquiry and engineering design, include rigorous mathematics and science instruction; (c) collaborative approaches to learning, connect students and educators with STEM fields and professionals; (d) provide global and multiperspective viewpoints; (e) incorporate strategies such as project-based learning, provide formal and informal learning experiences; and (f) incorporate appropriate technologies to enhance learning[5]. Based on those characteristics of STEM education program, students' activity using GeoGebra for understanding the rules of function shifting, a subtopic in Calculus Differential course was designed and implemented in a class of first-year mathematics students.

Educators are as the center of the learning activity in most of higher mathematics and science learning processes [6]. Recent study by Căprioară (2013) showed that there are some didactic practices of mathematics teacher that negatively impact to the learning process. Mathematics teachers were more subject oriented rather than to the learning process itself. In other words, most of mathematics teachers still involved the students into the learning process at a minimum level of participation. Considering those research results, mathematics learning should involve the students actively and should be students oriented. Context in science and engineering are possible to be engaged in mathematics. Technology is also provided in vary to support mathematics learning. GeoGebra is one of them. GeoGebra has helped the students to enhance their performance in learning mathematics, one of them is in coordinate system topic [8]. Combining real world problem and the use of technology have potential to increase students' engagement, interest and sense of ownership and autonomy over their learning process [9]. Function shifting rules usually only memorized by some students. When they failed to memorize the rules, they could not determine the graph of the new function when it was shifted or they could not justify what rules has applied so that we can obtain the new graph from the old one. Function shifting is flexible to be learnt by using both of real life context and involving technology application. Accordingly, to help the students understand the concept of function shifting, the activity involving real life context and integrated with GeoGebra was design and implemented.

2. Research Method

This study was qualitative research. The method started with content analysis. Function shifting as subtopic of Graph Transformation in calculus course was selected to be developed because it is possible to use GeoGebra actively and effectively in the learning process. This topic is also can be generated by using real life context to be modelized mathematically. In this research, some activities were designed in a form of students' worksheet. The learning resources designed including the worksheet was reviewed by an expert to state the validity. The next step was implementing the activities designed in an experiment class and observing the teaching and learning process in the classroom in order to look for students' response toward the tasks being given. Finally, interviews were undertaken to find information about students' perception and response toward the learning approach as well as their understanding to the topic.

The learning activity including some tasks was implemented to a class mathematics education program as the experiment class. During the implementation, video recorder and field note was employed to record the teaching and learning process including students' activities when doing the tasks given or when specific related responses of the teachers appear. Data from the video record and the field note were coded and analyzed after the data reduction.

3. Results

The results of this study weredivided into two main parts. The first result the implementation of the learning activity and the students understanding toward the topic, and the second is the students respond towards the learning activity.

3.1. *The implementation of the learning activity and the students understanding toward function shifting*

There is a contextual problem formulated in several tasks that were given in the worksheets. The problem is about finding a basketball trajectory equation when it is thrown upward. The first task of the problem is asking the students to sketch the graph of the function based on given data of the height of the ball at some specific times and then confirm their work by using GeoGebra, a dynamics and interactive mathematics software for learning and teaching mathematics in science.

Table 1. Data of the ball height at some instantaneous times

t (time)	$f(t)$ (height)

0	2
0.5	5.25
1	6
1.5	4.25

The second task of the problem is collecting similar data of the height of the ball when it is thrown by the player jumping upright 0.5 m above ground level. Based on the data obtained, students were asked to draw the graph of the function by using GeoGebra. The final task is asking the students to conclude from the two graphs how to obtain the second graph from the initial graph and give the reason. As an exercise, students were asked to do ball throwing 0.5 m lower by following similar lines as the previous case.

The class was divided into small groups that consist of 3 students. Each group was supplied with worksheets which started with finding the formula of basketball trajectory. Students did some analysis to obtain the rules of shifting by following the guidance in the worksheet and utilizing GeoGebra to do the experiment. When technological tools are available, students can focus on higher level of thinking such as decision making, reflection, reasoning, and problem solving [8].

In this research, there were some tasks that should be done by the students separated into two worksheets. The first worksheet consist of activities that require the students to: 1) modelling the problems into mathematical function; 2) determining the domain and the range of the function formed; 3) sketching the graph manually; 4) sketching the graph using GeoGebra and comparing with the graph that was sketched manually. The second worksheet was required students to: 1) do manipulation to the function that was obtained in the activity 1 by adding or subtracting several arbitrary constants; 2) comparing the graph function that was obtained by using GeoGebra before and after manipulation; 3) draw the conclusion of shifting rule; 4) solve the problems in the figure 1. Table 2 below described the implementation of the activities and divided into three parts i.e. the activities being implemented, the problems occurred, and the intervention of the educator to solve the problems.

Table 2. Summary of The Learning Activity

Worksheet	Activity	Problems	Intervention
	Modelling the problems into mathematical function	Students do not know how to start solving the problems.	Questioning the students. The educator tried to remind the students by asking, "did you remember the shape of trajectory of a ball throwing? All of you had already learned it in physics, hadn't you?"
1	Determining the domain and the range of the function formed	Give the incorrect domain and the range. The domain and the range are $(-\infty,\infty)(-\infty,\infty)$ and $(-\infty,-\frac{D}{4a})(-\infty,-\frac{D}{4a})$, respectively because they forgot that time which corresponds to the domain and height of the ball which corresponds to the range cannot be negative.	Remind the context of the problem by questioning.
	Sketching the graph manually	Initially only drew the points of the given data without connecting them and without continuing it for the possible	Remind the context of the problem questioning.

		domain	
	Sketching the graph using GeoGebra and comparing with the graph that was sketched manually	Students inputted 4 points to Polynomial[<ListofPoints>] and obtained a cubic equation	Reminded the students that quadratic equation used 3 points.
		Find that the manual graph was different with the graph sketched by using GeoGebra	Asked the students to check both of graph.
2	Doing manipulation to the function that was obtained in the activity 1 by adding or subtracting several arbitrary constants into the function $f(t)$	-	-
	Doing manipulation to the function that was obtained in the activity 1 by adding or subtracting several arbitrary constants into absis of the function $f(t)$	Students confused when determining the sample points for graphing the function $y = f(t + c)$, where c is an arbitrary real constant.	The educator asked the students to check the data table they constructed and to inspect each ordered pair.
	Comparing the graph function that was obtained by using GeoGebra before and after manipulation	-	-
	Drawing the conclusion of function shifting rules	Students find difficulties in drawing conclusion of the second activities, that by adding or substracting several arbitrary constants into absis of the function $f(t)$ which one is move to the right, which one is to the left.	Ask the other group of student to demonstrate classically one example by using GeoGebra.

We can see from the Table 2 above that each step of activity students had their own difficulties and to help the students, the educator gave interventions such as questioning, giving a hint, or asking other students to share their opinion.

There are few problems experienced by the students in doing the activities. One of them was determining the data for the shifted graph. For example, students confused when determining the sample points for graphing the function $y = f(t + c)$, where c is an arbitrary real constant. This was because they wrongly paired each value t taken from the domain with its corresponding range. For example, for $f(t) = -5t^2 + 9t + 2$ and taking $c = 1$, they inputted Polynomial[{(1,6),(1.5,4.25),(2,0)}]. As a result, they obtained exactly the same function and the same graph as the initial one as seen in figure 1. The educator asked the students to check the data table they constructed and to inspect each ordered pair. He reminded that the students should not put the sum $t + 1$ as the absisca of the points. Instead, they should put each value of t which were taken as the absisca and pairing it with the map of $t + 1$ under the function. So, they would have inputted the following command: Polynomial[{(0,6),(0.5,4.25),(1,0)}] to obtain a new graph that was actually obtained by shifting the initial graph to the left 1 unit as far as seen in figure 2.

IOP Conf. Series: Journal of Physics: Conf. Series **1108** (2018) 012014 doi:10.1088/1742-6596/1108/1/012014

The other problem that students met was drawing the conclusion for $f(x \pm c)$. They got confuse to decide which one is move to the right and which is move to the left. It was because they were still influenced by the first conclusion for $f(x) \pm c$. The conclusion of $f(x) \pm c$ was easier to understand. The movement of of $f(x)$ is depended to the value of c. If c is positive, $f(x)$ move upward as far as c units and vice versa.

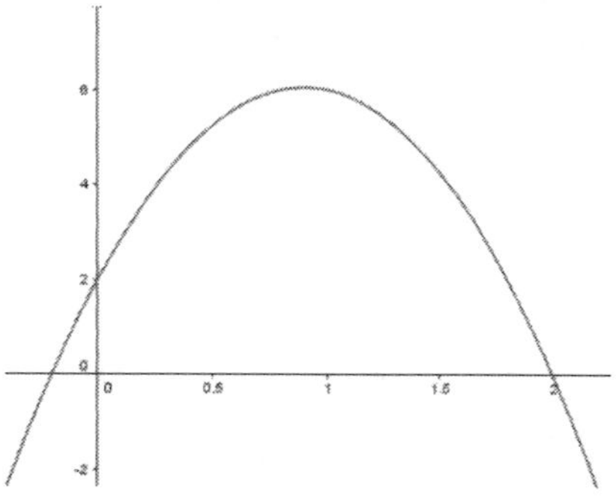

Figure 1. Initial graph function

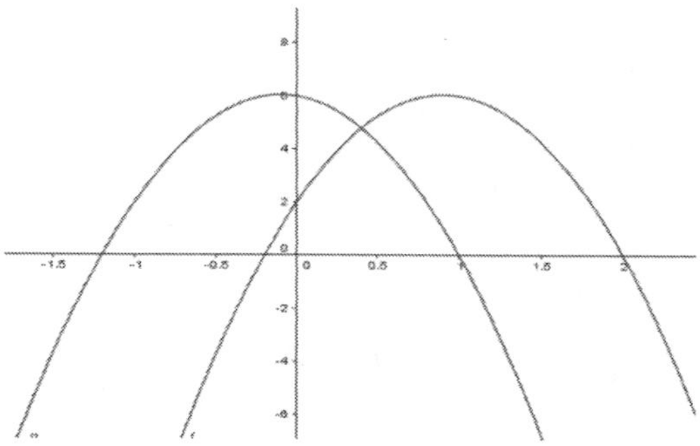

Figure 2. Initial graph (green) and shifted graph graph

It was easier because the movement did not counter the y-axis. But the movement of $f(x \pm c)$ countered the x-axis. When the value of c is positive, the graph of the function moved to left. At this stage, educator asked a group to explain and demonstrate the problem using GeoGebra in front of the class for several arbitrary c. Then the educator and students made the conclusion. Many groups could draw a correct conclusion although not in a complete sentence as the following.

"The second graph was obtained by shifting the first graph 0.5 unit upward".

This is interesting since they understood the idea but could not explain using a clear and complete sentence. At this stage, the educator checked each group by asking them to repeat the conclusion for the purpose to assure that they draw a correct conclusion.

3.2. Students response toward the learning process

The other interesting result is the information about students' response toward the teaching approach. From the interview, these data was obtained. Some students confessed that the approach was interesting (Vignette 1, line 16 - 18). Although, they were confused at the beginning, they found that the approach helped them to understand the topic better because they experienced themselves how the formulas come from (Vignette 1, line 3 – 5). They felt that from the learning process, they received some advantages (Vignette 1, line 9-12). The transcript of vignette 1 below reveals the facts.

Vignette 1. Transcript of students response toward the learning process

Interviewer	:	"What do you think about the teaching and learning	[1]
		approach?"	[2]
Student	:	"At first, it was confusing because we don't know where	[3]
		we are up to. But, after the educator brought the results of	[4]
		each group classically, we realized our mistakes.	[5]
		This made us understood better."	[6]
Interviewer	:	"What benefit do you get from such an approach of	[7]

learning?"

Student : "We could learn to make reflection of our work. Firstly, we work separately in groups and then when the educator discuss each group work in class we realized our mistakes and value other group's work."

Interviewer : "What do you prefer? Using this learning approach or the educator explains straight away?"

Student : "I prefer this approach because, for me, it is really well-suited. The learning process was more interesting beacause I experienced myself constructing my understanding together with my group's member about the concept. If it is explained by the educator straight away usually I will easily forget what I understood from the educator's explanation."

The transcript in the vignette 1 shows positive response from the students toward the approach being implemented. They enjoyed the learning process which challenge and help them at the same time to understand the concept of function shifting.

4. Conclusion

The learning activities using GeoGebra had been implemented in this study could facilitate the students to derive and use the formula of shifting function. In the learning process, the educator could implement approaches of teaching in which the educator must be perceptive toward all students' responses. The students found few problems in the learning activities such as give the incorrect domain and the range, found difficulties in sketching the graph, and drawing a conclusion. The educator may questioning, give a hint, invite others' opinion or else such that students will think deeper, explore, or share opinion.

The implementation of the activity using GeoGebra supported students' to have a good understanding about the topic. GeoGebra helped the students in to do trial and error until the find the hypothesis of the general form of function shifting. Students are not only able to answer the questions related to the shifting rules but also able to explain their answers by providing examples to support their answers. In addition, being experienced with the learning activity, the students were more comfortable in learning mathematics. They enjoyed the learning process such as: they could discuss with teammates, could reflect their works with other teams' result, and on the other hand, they felt that the educator's role really helped them when they stuck or had a deadlock in the discussion. Students satisfied with the learning process where they could construct their own concept of function shifting.

5. References

[1] Hom E J 2014 (online) *What is STEM Education?* (http://www.livescience.com/43296-what-is-stem-education.html)

[2] Kelley T R and Knowles J G 2016 A conceptual framework for integrated STEM education *Int. J. STEM Educ.* **3**(11) pp. 1-11

[3] Sanders B M 2009 STEM STEM Education STEM mania *Technol Teach* pp. 20–7

[4] Moore T J, Stohlmann M S, Wang H-H, Tank K M, Glancy A W, and Roehrig G H 2014 (online) Implementation and Integration of Engineering in K–12 Stem Education In:

Engineering in Pre-College Settings: Synthesizing Research, Policy, and Practices [Internet]. Purdue University Press p. 35–60 (http://www.jstor.org/stable/j.ctt6wq7bh.7)

[5] Kennedy T J 2014 *Engaging Students In STEM Education* **25**(3) pp. 246–58

[6] Timmerman B E, Strickland D C, and Carstensen S M 2008 Curricular reform and inquiry teaching in biology: where are our efforts most fruitfully invested? *Integr. Comp. Biol.* 48(2) pp. 226–40

[7] Căprioară D 2013 Some Specific Features of the Educational Learning Context – an Outlook on Secondary Mathematics *Procedia - Soc Behav Sci* **93** pp. 2068–72

[8] Saha R A, Ayub A F M, and Tarmizi R A 2010 The effects of GeoGebra on mathematics achievement: Enlightening Coordinate Geometry learning *Procedia - Soc Behav Sci* **8**(5) pp. 686–93

[9] Bray A and Tangney B 2015 Enhancing student engagement through the affordances of mobile technology : a 21st century learning perspective on Realistic Mathematics Education *Math Educ Res J.* **28**(1) pp. 173–97

Acknowledgments

This research is funded by DP2M Dirjen Dikti, The Ministry of Research, Technology, and Higher Education (Kemenristekdikti), Republic of Indonesia Under the scheme of Penelitian Unggulan Perguruan Tinggi, year 2016, DIPA Universitas Negeri Surabaya, contract number: 004.12/UN38.11-P/LT/2016. This research also got full support by the Department of Mathematics, Universitas Negeri Surabaya, Indonesia.

Eigenvalues of Antiadjacency Matrix of Directed Cyclic Dumbbell Graph

S Budiyanto[1], S Utama[1,*] and S Aminah[1]

[1] Department of Mathematics, Faculty of Mathematics and Natural Sciences, Universitas Indonesia, Indonesia

[*] Corresponding author: suarsih.utama@sci.ui.ac.id

Abstract. This paper explains the steps used to find the characteristic polynomial of the antiadjacency matrix of a cyclic dumbbell graph. The antiadjacency matrix of a graph is a matrix whose entries represent whether there is an edge that connects two vertices or not. The general form of the characteristic polynomial includes its eigenvalues. The antiadjacency matrix is obtained by using some theorems, the number of solutions of integer equations, quadratic formula, and polynomials factorization. Finally, results showed that the coefficients of the characteristic polynomial and its eigenvalues were dependent on the number of vertices of the cyclic dumbbell graph.

1. Introduction

There are many applications of eigenvalues of adjacency matrices of graphs such as representation of linear codes in coding theory. The minimum Hamming distance of linear codes is represented by the second largest eigenvalue of adjacency matrix of a certain regular graph [1]. Regular graph is a graph whose every vertex has an equal degree. Hamming distance is the number of entries in which two codewords differ. The adjacency matrices are more commonly used to represent graphs in the form of matrices. There are only a few of researches about antiadjacency matrices and their applications. One of this few researches is carried out by Ferry who wrote about the eigenvalues of antiadjacency matrix of directed acyclic graphs [2]. This paper discusses about the eigenvalues of the antiadjacency matrix of directed cyclic dumbbell graph. This paper also discusses about dumbbell graph because it can be represented as cyclic graph and the number of researches about dumbbell graph is not many.

2. Theoretical basis

A graph G is a pair of sets $G=(V(G),E(G))$ where $V(G)$ is finite non-empty set called the set of vertices, and $E(G)$ is a set of ordered pair of two elements of $V(G)$, called the set of edges. If the edges of a graph are in the form of (u,v) where u is the starting vertex and v is the end vertex, then the edges are called directed edges. Directed graph is a graph whose edges are directed edges [3]. Dumbbell graph, denoted by G_n, is a graph that consists of two cycle graphs, V_1 and V_2, $V_1=\{v_1,v_2,\ldots,v_n\}$ and $V_2=\{v_{n+1},v_{n+2},\ldots,v_{2n}\}$, such that $V(G_n)=V_1\cup V_2$ and $V_1\cap V_2=\varnothing$, and there exist one $v_i\in V_1$ and one $v_j\in V_2$ such as $v_iv_j\in E(G_n)$. A dumbbell graph G_n is said to directed cyclic graph if each of V_1 and

V_2 is directed cyclic cycle graphs with $v_n v_{n+1} \in E(G_n)$. In this paper, 'G_n' will be referred as directed cyclic dumbbell graph as illustrated in figure 1.

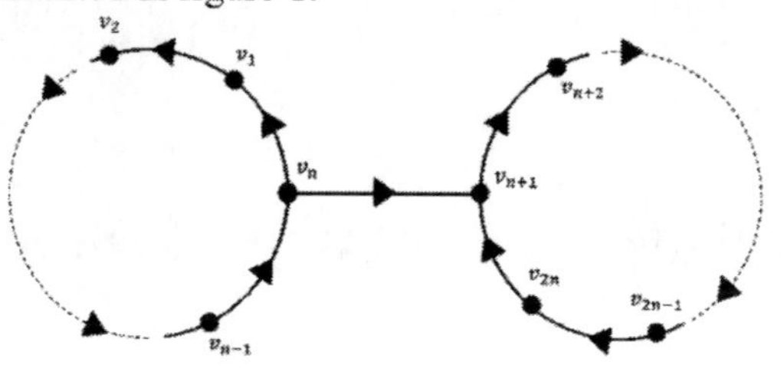

Figure 1. Directed cyclic dumbbell graph G_n

The antiadjacency matrix of a directed graph G which has n vertices, is an $n \times n$ matrix, $\mathrm{B}(G) = \left[b_{ij}\right]$, such as b_{ij} is equal to 0 if there exists a directed edge from v_i to v_j, equal to 1 if there is no directed edge from v_i to v_j [4]. The following theorems are further needed to get the main results in section 3.

Theorem 1. Let $\lambda^n + b_1\lambda^{n-1} + b_2\lambda^{n-2} + \ldots + b_{n-1}\lambda + b_n$ be the characteristic polynomial of the antiadjacency matrix $\mathrm{B}(G)$ and S_i is the number of induced subgraphs of G in the form of Hamiltonian paths, each of them has i vertices. Let $\left|\mathrm{B}\left(\langle U\rangle_{cyclic}\right)_i^{j_2}\right|$ be the determinant of the antiadjacency matrix of the cyclic induced subgraph of G that has i vertices and $j_2 = 1, 2, \ldots, w_2$, with w_2 is equal to the number of directed cyclic induced subgraphs $\langle U\rangle_{cyclic}$ that has i vertices from a directed cyclic graph G, then

$$b_i = (-1)^i \left(S_i + \sum_{j_2=1}^{w_2} \left|\mathrm{B}\left(\langle U\rangle_{cyclic}\right)_i^{j_2}\right| \right), \text{ for } i = 1, 2, \ldots, n \text{ [4] [5].}$$

Theorem 2. If C_n is directed cyclic cycle graph that has n vertices and $\mathrm{B}(C_n)$ is the antiadjacency matrix of C_n, then $\det\left(\mathrm{B}\left(\mathrm{C}_n\right)\right) = n-1$ [4].

Theorem 3. If A is a square matrix that has 2 proportional rows or 2 proportional columns, then $\det(A) = 0$ [6].

3. Main Results

The main results consisted of three outcomes. The first finding was the induced subgraphs of G_n in the form of Hamiltonian paths as stated in Equation (4). The second finding was the sum of determinants of directed cyclic induced subgraphs of G_n as stated in Equation (5). The last was a finding the characteristic polynomial and the eigenvalues of the antiadjacency matrix of G_n as stated in the Equations (6), (7), (8), (9).

3.1. The number of induced subgraphs of G_n in the form of Hamiltonian paths

The number of directed acyclic induced subgraphs in the form of Hamiltonian paths S_i can be found by considering 2 cases, which were:

- Case 1: acyclic induced subgraphs in the form of Hamiltonian paths that passed through $v_n v_{n+1}$
- Case 2: acyclic induced subgraphs in the form of Hamiltonian paths that did not pass through $v_n v_{n+1}$.

IOP Conf. Series: Journal of Physics: Conf. Series **1108** (2018) 012015 doi:10.1088/1742-6596/1108/1/012015

In each case, theorem would be constructed and proved to obtain the results.

Theorem 4. if $S_i^{(1)}$ was the number of directed acyclic induced subgraphs of G_n in the form of Hamiltonian paths that has i vertices and pass through $v_n v_{n+1}$, then

$$S_i^{(1)} = \begin{cases} i-1, & i=1,2,\ldots,n \\ 2n-i-1, & i=n+1,n+2,\ldots,2n-2 \end{cases}.$$

Proof: For i = 1, there was no Hamiltonian path that passed through $v_n v_{n+1}$, since it required at least 2 vertices to form a path. So, $S_1^{(1)} = 0 = 1-1 = i-1$. ...(1)
For $i = 2,3,\ldots,2n-2$, Let X be the number of vertices in V_1 beside v_n, and Y be the number of vertices in V_2 beside v_{n+1}. Since Hamiltonian paths passed through $v_n v_{n+1}$, then $i = X+Y+2$. Since V_1 and V_2 were cycle graphs that had n vertices respectively, then the vertices which could be used in V_1 and V_2 were at most $n-1$ vertices respectively, as a result $X = 0,1,2,\ldots,n-2$ and $Y = 0,1,2,\ldots,n-2$. From this information, $S_i^{(1)}$ could be obtained by counting the number of solutions to the following equations

$$\begin{array}{l} X+Y+2=i, \quad i=2,3,\ldots,2n-2;\ X=0,1,\ldots,n-2;\ Y=0,1,\ldots,n-2 \\ E_i = \{(X,Y) \mid X+Y+2=i\}, \quad i=2,3,\ldots,2n-2 \end{array}.$$

$S_i^{(1)} = |E_i|$, and $|E_i|$ was counted by finding the possible minimum and maximum values of X or Y, so
$S_i^{(1)} = i-1, \quad i=2,3,\ldots,n$. ...(2)
And, for $i = n+1, n+2,\ldots,2n-2$, the possible values of X were $X = i-n, i-n+1,\ldots,n-2$, such that
$S_i^{(1)} = |E_i| = (n-2)-(i-n)+1 = 2n-i-1, \quad i=n+1,n+2,\ldots,2n-2$ ...(3)

From the Equations (1), (2), (3), it followed that,

$$S_i^{(1)} = \begin{cases} i-1, & i=1,2,\ldots,n \\ 2n-i-1, & i=n+1,n+2,\ldots,2n-2 \end{cases}. \qquad \blacksquare$$

Theorem 5. If $S_i^{(2)}$ was the number of directed acyclic induced subgraphs of G_n in the form of Hamiltonian paths that had i vertices and did not pass through $v_n v_{n+1}$, then

$$S_i^{(2)} = \begin{cases} 2n, & i=1,2,\ldots,n-1 \\ 0, & i=n,n+1,\ldots,2n \end{cases}.$$

Proof: The value of $S_i^{(2)}$ could be determined by counting the number of ways to choose a vertex in a cycle graph (V_1 or V_2) as the start of the Hamiltonian path. So,

$$S_i^{(2)} = \begin{cases} 2n, & i=1,2,\ldots,n-1 \\ 0, & i=n,n+1,\ldots,2n \end{cases}. \qquad \blacksquare$$

S_i could be obtained by adding $S_i^{(1)}$ and $S_i^{(2)}$. The result was

$$S_i = \begin{cases} 2n+i-1, & i=1,2,\ldots,n-1 \\ 2n-i-1, & i=n,n+1,n+2,\ldots,2n-2 \\ 0, & i=2n-1,2n \end{cases}. \qquad \ldots(4)$$

3.2. Directed cyclic induced subgraphs of G_n

To form directed cyclic induced subgraphs of G_n, the number of vertices used had to be greater than or was equal to n, and all vertices in V_1 or V_2 (directed cyclic cycle graphs that has n vertices) had to be used. The following subsections will list every possible form of directed cyclic induced subgraph of G_n.

3.2.1. Fundamental forms of directed cyclic induced subgraphs of G_n. The following graphs, as in Table 1, were said to be the fundamental forms of directed cyclic induced subgraphs of G_n because every directed cyclic induced subgraph of G_n could be obtained by combining those graphs. The names of the vertices used in Table 1 was not related to the names of the vertices in the definition of G_n in Figure 1.

Table 1. List of fundamental forms of directed cyclic induced subgraphs of G_n

Graph illustration	**Notation**
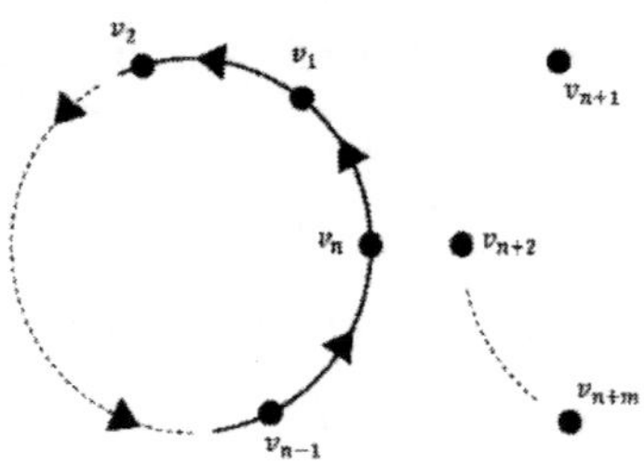	$T_1(n,m)$ n = the number of vertices in cycle graph m = the number of vertices that did not have directed edge.
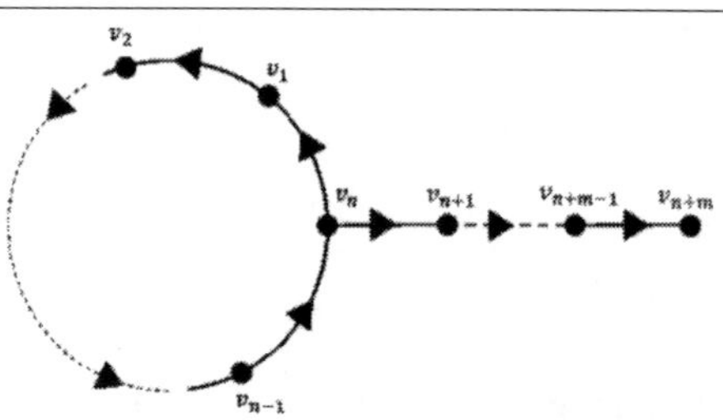	$T_3(n,m)$ n = the number of vertices in cycle graph m = the number of vertices on the path that was directed out of the cycle graph
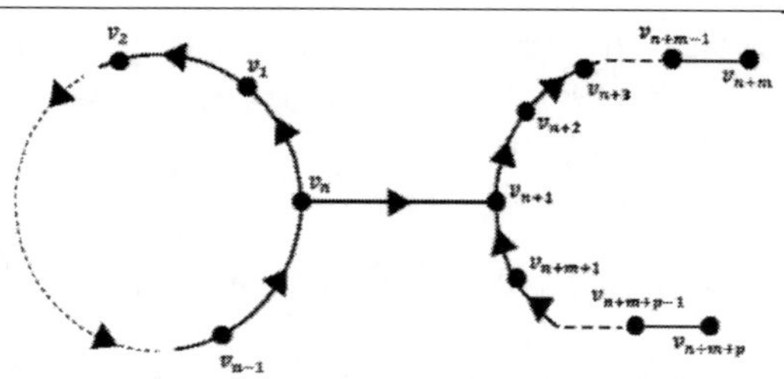	$T_5(n,m,p)$ n = the number of vertices in cycle graph m = the number of vertices on the path that is directed out of the cycle graph p = the number of vertices on the path that is directed to v_{n+1}

It should be noted that the graphs T_2, T_4, T_6 and C_n were members of Table 1 but were not listed. T_2 was similar to T_1, except there was a directed path from v_{n+1} to v_{n+m}. T_4 that was similar to T_3, except the direction of the path from v_n to v_{n+m} was inverted. T_6 was similar to T_5, except the direction of the path from v_n to v_{n+m} and the path from v_{n+1} to v_{n+m+p} were inverted. C_n was directed cyclic cycle graph.

3.2.2. Combinations of the fundamental forms of directed cyclic induced subgraphs of G_n. The notation $T_{i.j.k}$ was a graph obtained by combining the graphs T_i, T_j, T_k which were listed in Table 1. So, $T_{1.4}$ was a graph that had a cycle graph, vertices that did not have directed edge and path that was directed into the cycle graph.

IOP Conf. Series: Journal of Physics: Conf. Series **1108** (2018) 012015 doi:10.1088/1742-6596/1108/1/012015

3.3. *Determinant of antiadjacency matrix of the directed cyclic induced subgraph of G_n*

Determinant of the antiadjacency matrix of the directed cyclic induced subgraph of G_n was obtained by applying elementary row reductions to the determinant of the matrix. The results were shown as follows:

Table 2. Determinants of the directed cyclic induced subgraphs of G_n

Subgraph Notation	Determinant	Subgraph Notation	Determinant
$T_1(n,1)$	-1	$T_5(n,m,p)$	-1
$T_1(n,m>1)$	0	$T_6(n,m,p)$	-1
$T_2(n,m)$	-1	C_n	$n-1$
$T_3(n,m)$	0	G_n	$-2(n-1)$
$T_4(n,m)$	0	$T_{1,4}(n,1,m_4)$	0

There were some other directed cyclic induced subgraphs from Table 2 which were not listed in Table 3. Every subgraph which was not listed in Table 2 had at least 2 vertices that did not have any directed edge to all the other vertices. As a result, the antiadjacency matrix of the subgraph had at least 2 proportional rows and as a result, the determinant was 0.

3.4. *The sum of determinants of the directed cyclic induced subgraphs of G_n*

Let *i* be the number of vertices. For $i=n$, the directed cyclic induced subgraphs of G_n were in the form of directed cyclic cycle graph C_n. By the definition of G_n, the number of such subgraph in G_n was 2. For $i=n+1$, the directed cyclic induced subgraphs of G_n were in the form of $T_1(n,1)$. If $T_1^{(n)}$ represented the number of such subgraphs in G_n, its value was equal to the number of ways of choosing 1 vertex outside cycle graph (V_1 or V_2) from the available $n-1$ vertices, so $T_1^{(n)}=2(n-1)$.

Theorem 6. For $i>n+1$, the directed cyclic induced subgraphs of G_n were in the form of $T_2(n,m)$, $T_5(n,m,p)$ or $T_6(n,m,p)$. Let $T_2^{(i)}$, $T_5^{(i)}$ and $T_6^{(i)}$ be the number of directed cyclic induced subgraphs of G_n in the forms of $T_2(n,m)$, $T_5(n,m,p)$ and $T_6(n,m,p)$ respectively, then $T_2^{(i)}=2(2n-i)$, and $T_5^{(i)}=T_6^{(i)}=i-n-1$.

Proof: The value of $T_2^{(i)}$ could be determined by counting the number of ways to choose a vertex in a path that had $n-1$ vertices as the starting point of the path that has *m* vertices, so $T_2^{(i)}=2(2n-i)$. For $T_5^{(i)}$, choose all vertices of V_1 and *m* and *p* vertices of V_2 to form $T_5(n,m,p)$. The number of $T_5(n,m,p)$ which consisted of *i* vertices was the number of solutions that satisfied the equations

$$m+p=i-n;\ m=1,2,\dots,n-2;\ p=1,2,\dots,n-2,$$
$$E_i=\{(m,p)|m+p=i-n\},$$

$T_5^{(i)}=|E_i|$ for $i=n+2,n+3,\dots,2n-1$. $|E_i|$ could be obtained by determining the possible minimum and maximum values of *m*, so $T_5^{(i)}=i-n-1$. By applying the same method to $T_6^{(i)}$ (related to the graph $T_6(n,m,p)$), it followed that $T_6^{(i)}=i-n-1$. ■

Therefore, $\sum_{j_2=1}^{w_2}\left|B\left(\langle U\rangle_{cyclic}\right)_i^{j_2}\right|$ could be obtained by multiplying the number of each subgraph in Table 3 with its corresponding determinants, then summing the results of the multiplications. Therefore,

$$\sum_{j_2=1}^{w_2}\left|B\left(\langle U\rangle_{cyclic}\right)_i^{j_2}\right| = \begin{cases} 0, & \text{for } i=1,2,\dots,n-1, \\ 2(n-1), & \text{for } i=n, \\ -2(n-1), & \text{for } i=n+1,n+2,\dots,2n \end{cases} \qquad \dots(5)$$

3.5. *Characteristic polynomial and eigenvalues of the antiadjacency matrix of G_n*

Let $P(G)=\lambda^{2n}+b_1\lambda^{2n-1}+b_2\lambda^{2n-2}+\dots+b_{2n-1}\lambda+b_{2n}$ was the characteristic polynomial of the antiadjacency matrix $\mathrm{B}(G_n)$. From the Equations (4) and (5), and by using Theorem 1, it followed that

$$b_i = \begin{cases} (-1)^i(2n+i-1), & \text{for } i=1,2,\dots,n-1, \\ (-1)^n(3(n-1)), & \text{for } i=n, \\ (-1)^i(-i+1), & \text{for } i=n+1,n+2,\dots,2n-1 \\ -2(n-1), & \text{for } i=2n \end{cases} \qquad \dots(6)$$

Therefore, the characteristic polynomial of the antiadjacency matrix $\mathrm{B}(G_n)$ could be written as

$$P(\lambda)=\lambda^{2n}+\left(\sum_{i=1}^{n-1}(2n+i-1)(-\lambda)^{2n-i}\right)+3(n-1)(-\lambda)^n+\left(\sum_{i=n+1}^{2n-1}(-i+1)(-\lambda)^{2n-i}\right)-2(n-1)$$

$$P(\lambda)=\frac{\left((-\lambda)^n-1\right)^2\left(\lambda^2-(2n-2)\lambda-(2n-2)\right)}{(\lambda+1)^2}, \quad \lambda\neq -1,$$

The eigenvalues could be obtained by solving the equation $P(\lambda)=0$.

$$\left((-\lambda)^n-1=0 \text{ or } \lambda^2-(2n-2)\lambda-(2n-2)=0\right) \text{ and } (\lambda\neq -1)$$

For $\lambda^2-(2n-2)\lambda-(2n-2)=0$, by applying quadratic formula, $\lambda_{1,2}=(n-1)\pm\left(n^2-1\right)^{1/2}$ $\dots(7)$

For $(-\lambda)^n-1=0$, $(-\lambda)^n=1 \Leftrightarrow \lambda^n=(-1)^n$, 2 cases were considered. For n even, $\lambda^n=1=\mathrm{e}^{0+2k\pi i}$

$$\lambda_k=\mathrm{e}^{\frac{2k\pi i}{n}}=\cos\left(\frac{2k\pi}{n}\right)+i\sin\left(\frac{2k\pi}{n}\right), \quad k=0,1,\dots n-1, \qquad \dots(8)$$

Since $\lambda\neq -1$, then $k\neq\frac{n}{2}$. For n odd, $\lambda^n=-1=\mathrm{e}^{(\pi+2k\pi)i}$

$$\lambda_k=\mathrm{e}^{\frac{(\pi+2k\pi)i}{n}}=\cos\left(\frac{(\pi+2k\pi)}{n}\right)+i\sin\left(\frac{(\pi+2k\pi)}{n}\right), \quad k=0,1,\dots n-1, \ k\neq\frac{n-1}{2} \qquad \dots(9)$$

4. Conclusion

The eigenvalues and the coefficients of the characteristic polynomial of the antiadjacency matrix of directed cyclic dumbbell graph can be written as functions of n which can be seen in Equations (6), (7), (8) and (9), the number of vertices of G_n. Also, the eigenvalues in Equations (8) and (9) all have multiplicity of 2.

IOP Conf. Series: Journal of Physics: Conf. Series **1108** (2018) 012015 doi:10.1088/1742-6596/1108/1/012015

5. References

[1] Spielman D A 1999 *IMA Vol. Math. Appl.* **109** 591
[2] Firmansah F 2014 *Characteristic polynomials of* antiadjacency *matrix of directed* directed acyclic *graph* thesis: Universitas Indonesia
[3] Harary F 1996 *Graph Theory* (Massachusetts: Addison Wesley) p 9
[4] Bapat R 2010 *Graphs and Matrices* (New Delhi: Hindustan Book Agency)
[5] Wildan 2015 *Characteristic polynomials of antiadjacency and adjacency matrix of simple oriented graphs* (Thesis: Universitas Indonesia) p 32
[6] Anton H and Chris R 2010 *Elementary Linear Algebra* ed 10 (New York: Wiley) p 145

Acknowledgement
This research is funded by Hibah PITTA UI [2307/UN2.R3.1/HKP.05.00/2018].

IOP Conf. Series: Journal of Physics: Conf. Series **1108** (2018) 012016 doi:10.1088/1742-6596/1108/1/012016

Common False of Student's Scientific Reasoning in Physics Problems

N Erlina[1], E Susantini[2,*], Wasis[3]

[1] Science Education, Postgraduate Program Universitas Negeri Surabaya, Indonesia
[2] Biology Education, Postgraduate Program Universitas Negeri Surabaya, Indonesia
[3] Physics Education, Postgraduate Program Universitas Negeri Surabaya, Indonesia

*Corresponding Author: endangsusantini@unesa.ac.id

Abstract. Scientific reasoning is the ability to think about the characteristics of formal development to master concepts and solve physics problems. Physical problem solving requires a high level of thinking that is a combination of reasoning namely conceptual and general mathematical equations. The student's thinking ability that is inconsistent with their formal developmental stage is the cause of the student's scientific reasoning lack. The purpose of this study is to analyze the common false made by students during their scientific reasoning process. This research is a qualitative descriptive with the aim at describing common error of scientific reasoning with the physics problem. Scientific reasoning tests were given to 108 students who had studied physics using mixed methods of lectures and assignments. Results showed the false of each component of the student's scientific reasoning. Student's scientific reasoning was dominated by competence level. The quality of claims, arguments, and levels of student's confidence in giving answers was still low. Students experienced false scientific reasoning and weak concepts when confronted with physics problems. The important recommendation of this study is that the learning process needs an innovation to overcome false and inconfident feeling in the process of scientific reasoning and mastery of student concepts.

1. Introduction

Scientific reasoning is a cognitive ability in formal operations that include activity of minds-on and hands-on to support the process of physics learning [1]. Physics-based learning of scientific reasoning stimulates student's mental and physical activity to achieve scientific knowledge [2]. Students are not only required to know the content of physics in the form of facts, concepts, or principles, but also must be able to investigate, collect evidence, analyse and evaluate the understanding of physics [3]. [4] conveyed that an understanding of physics concepts would be better when they could touch, feel, measure, manipulate, draw, graph, compile data, and when they find and able to explain answers using their understanding. The transfer process in physics learning requires high-order thinking skills that form a combination between reasoning and conceptualization of general mathematical equations [5].

Scientific reasoning is required as a billing of high-level cognitive abilities but this skill has not yet been trained to activate the students' mindset [6]. Evaluation bills in learning are still simple fact-recognition as the lowest cognitive level although scientific methods and scientific reasoning efforts become the provision of learning competencies [7]. Relevant research results show that teachers rarely use questions (e.g. analysis and evaluation) and help students develop their scientific thinking skills [8]. The student's thinking ability that is inconsistent with their formal developmental level causes a person not to reach the level of formal thinking during his lifetime [9].

The fact of scientific reasoning ability in physics learning has not been satisfactory. Results showed student's scientific reasoning in 11th grade of high school only reached the average score of 38.7 (low category) [12]. Students have difficulties in defining experimental variables, interpreting, and analyzing graphs or data. The result of the preliminary study also showed that the high school students' scientific reasoning is still low amounting the average score of 27.22. The questionnaire data included physics learning process conducted by physics teacher in the high school that had not trained the skill yet [13]. The involvement of students in mastering physics is limited to applying the principles, methods, requirements, and theories of physics [14].

The involvement of scholarly reasoning requires students in connecting evidence on investigative activities to support scientific arguments. The ability of scientific reasoning is increasingly structured and evident from the quality of scientific arguments. Reasoning results in conclusion of thought, clarity, and firmness to explain why something happened or what will happen. Reasoning refers to the mental processes involved in making and evaluating logical arguments [16]. The preparation and evaluation of scientific arguments related to evidence and reasoning [17]. The student's ability to explain information is based on belief and desire in the process of reasoning [18].

This study focus on enabling students to do scientific reasoning with arguments and students' metacognitive skills in the form of three-tier test. Three-tier test is a better development of two-tier test applied in scientific reasoning component in which the three-tier test of scientific reasoning requires students to choose the alternative choice of answers. The test comes with a level of confidence in producing answers and reasons that show the process of realizing scientific reasoning. The research provides easy analysis and description of the common false about the process of scientific reasoning.

2. Research method

This research used qualitative and descriptive approach by the aim of describing common false of scientific reasoning in the physics problem. The subjects of this study were 108 high school students in Jember, East Java, Indonesia, with mechanical topics in the odd semester of 2017. The test was used as a data collection technique. The study was conducted by giving the test to the students after conducting the learning process with lecturing mode, task and question, and answer. The test used was a three-tier test based on scientific reasoning components covering: (a) Proportional Thinking (PPT) to determine relationships and to compare between variables using numbers, mathematical equations, tables, and graphs, (b) Control of Variables (CV) to identify independent variables and bound variables, (c) Probabilistic Thinking (PBT) to predict the probability of results obtained during repetition, (d) Correlational Thinking (CT) to create a reciprocal relationship (related or unrelated) between variables, and (e) Hypothetical-Deductive Reasoning (HDR) to construct hypotheses based on general concepts to specific concepts [10]. Data obtained from the three-tier test were analyzed based on scientific reasoning component. Expert level was achieved when the students were able to choose the first choice correctly, the reason was perfect for the second level, and was confident with the given answer (score 4). In coping with the level of experience, students were able to choose the first option correctly, the reason was less perfect on the second level matter and confident with the answer (score 3). The competent level was shown by the students who were able to choose the first option correctly, the wrong reason on the second level and confident with the answer (score 2). At the novice level, students were convinced by the wrong choice from the first answer (score 1) and the students were convinced by the wrong answer or was not sure of the answer (score 0) [13].

3. Results and Discussion

This research produced the level and description of common false of each component of student's scientific reasoning in physics learning. Figure 1 shows the frequency of scientific reasoning ability in each component determined by using scientific reasoning tests.

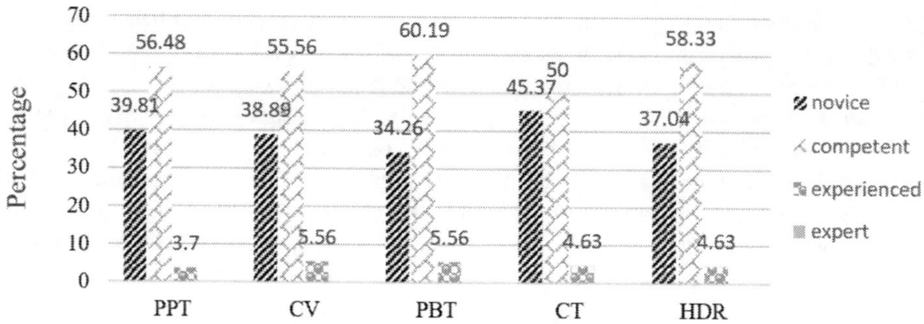

Figure 1. Level of scientific reasoning

Figure 1 shows the percentage of levels based on student's scientific reasoning components. The largest percentage at this level was achieved on the PBT component. The competent level indicated the ability of the students in choosing the right and convinced choice of answers even though the argument against the answer was wrong. Students gave answers only based on their memory. This result explained that learning process using lecturing mode, duty and question, and answer had not been effective in delivering the students' ability at expert level. Theoretical learning tent to ignore evidence to determine consistent answers [19]. Learning that had become students' routine had not been able to build arguments and to summarize the knowledge correctly and logically based on the process and the results of scientific reasoning. Scientific reasoning can be trained and transferred, also giving a long-term impact on student's achievement [11].

Figure 1 presents the percentage of novice levels occurring in each component of scientific reasoning of more than 30 percent. Students at the novice level had a weak belief in the outcome of the answer. The student's knowledge still lacked and there was a misconception in doing scientific reasoning. Teachers usually did not emphasize student's reasoning about the predictions they made [20]. Reasoning referred to the mental processes involved in making and evaluating logical arguments and problem-solving. Problem-solving could develop student's reasoning because it required students to explain why something happened or what would happen [16]. The development of scientific reasoning was influenced by conceptual change as a constructivist approach. It related to the basic reasoning process involved in the investigation, experimentation, evaluation of evidence, inference and argument made to achieve conceptual change or scientific understanding [21].

The level of scientific reasoning achieved in Figure 1 was dominated by the competent level. Students 'answers were based on the results of memory that had not been yet clarified and the results tent to lead to weak support for the development of student's thinking. Table 1 describes the test results illustrating the false of student's thinking patterns.

Table 1. Causes of common false of scientific reasoning

Components	Causes of False
Proportional Thinking	1.Not applying the concept of physics in comparison operations 2.Unable to use mathematical operations in comparison relationships
Control of Variable	1. Unable to describe them operational definition of the variable 2. Unable to determine free and bound variable
Probabilistic Thinking	1. Fail to predict though to give the perfect argument 2. Provide the same assumption on different circumstances 3. Incompleted arguments 4. Weak concept
Correlational Thinking	1. The truth about the concept in the argument did not have a logical relationship to the problem 2. Weak concept

Components	Causes of False
Hypothetical-Deductive Reasoning	1. The students' logical thinking ability was not used to relate concepts to the solutions needed in the problem 2. Application of the wrong concept

The common false of the components of scientific reasoning were described below with the accompanying examples. Proportional thinking could be conceptualized by identifying two variables, i.e. extensive variables and intensive variables. Extensive variables (depending on the amount of material measured) applied to the problem as a linear function and comparison where intensive variables remained constant [10]. The first and second common false of alternatives selected by the students indicated an error when doing the proportional thinking. Students gave an incomplete reason to support the answer. The first error occured when faced with a proportional problem in a graph showing a circular moving object. Students were able to provide a logical argument using a comparison of values of $v_{0A}/v_{0B}=R_A/R_B$. Students showed correct calculation results based on the value comparison relationship. The student's strategy error was that they did not include the centripetal force equation into the mathematical relationship including the square of a magnitude so that the results obtained by the student were false. A common false that also occured in other mechanical topics was that students did not give mathematical arguments. The first and second common false were relevant to the results of the study showing student's false when doing the proportional thinking. Students more struggled when swapping the ratios or units around and failed to use proportions [15].

Control variable was the ability to evaluate the evidence generated from controlled experiments in scientific investigations and experiments. Constant variables were required during experimental activities [22]. The first common false was that the student was able to answer the alternatives correctly but the reason for supporting the answer was wrong. Students were able to determine the distance between points as a result of the speed of an object. Students gave false arguments about the description of the variables involved, for example, the times required from the starting to the end points was described as the distance. The second false was that the students was unable to determine the free and bound variable. This error was supported by research results which showed that students generally failed to determine constantly the maintained variables [15]. For example, students were unable to determine the magnitudes that affected the linear velocity of an object placed on a rotating record.

Probabilistic thinking skills included high-level thinking processes. Probabilistic thinking was a situation that generated a certain amount when repeated in the same state in a larger context [10]. The first and second false when students did probabilistic thinking was that students failed to predict so the answer was wrong. For example, a student was expected to predict the time required for 1, 2, and 3 cars with some types and directions of the passage. Students were able to analyze the length of the passage through the case of car 1, but they failed to determine the type and direction of the track. Students assumed that car 1, 2 and 3 went through the same type and direction (students did not support the type of track used by car 1). Students were able to provide perfect arguments but students' predictions were wrong. Common false that occured in the third and fourth false were those that made students unable to present a complete and correct argument. For example, how did the student's ability to predict the speed, the influence of gravity, and the acceleration of the ball at the highest point if the player repeated the ball with a trajectory much farther than before? Students did not understand the concept of speed at the highest point and the influence of gravity on the vertical and horizontal planes. The weaknesses experienced by students who tried to achieve reasoning skills were still not able to reach the expert level. Students did not consider any factors that might have an effect on probability [15].

Correlational thinking determined the strength of the interrelationships between variables that made possibility to manufacture predictions during scientific exploration [10]. Correlations were used to describe the degree of dependence between two or more variables [11]. The first common false could be identified based on the wrong student's choice of answers with the correct and complete reason. The reasons given by the students were not related to the student's choice of alternatives. For example, students were able to provide arguments in the form of trueness but illogical concepts with problems.

Students' argument stating that every object connected would be related to gravity, time, acceleration and speed. Things that fell to the ground were determined by gravity, time, acceleration and speed. Students showed a wrong answer that the large speed of the horizontal object was related to the gravity. Student answering to other mechanical topics showed weak concepts during the correlational thinking process. Students' correlational thinking was based on their original knowledge [15].

Hypothetical-deductive reasoning as a characteristic of the reasoning process resulted in the development and organization of possible solutions to deal with problems in every step and domain of life [10]. Hypothetical reasoning was the process of creating and making conclusions about the possibilities so that what was believed (reality) could be compared to some alternatives [23]. Table 5 shows two common false in the deductive-hypothesis reasoning component. The first false was that the student gave an incomplete reason even though the chosen alternative was correct. Students were able to provide a logical argument in which the path taken by an increasingly long boat caused the boat longer to cross. The students' logical thinking ability was not adapted to the topic in question so that the students were not able to formulate the argument correctly. The second error was a special error that could lead to failure to construct a deductive hypothesis. The hypothesis was based on the theoretical basis so that the application of concepts to solve the problems affect the success. For example, the concept of horizontal axis velocity was not appropriately applied to motion problems in the vertical axis direction. Common false that supported false in reasoning deductive hypotheses was predicted insufficient to support and, moreover, did not match the hypothesis [15].

4. Conclussion

The conclusions of this research indicate the emergence of common false made by students during the scientific reasoning process. Student's scientific reasoning was dominated by competent level. Student's answers were based on the results of memory that had not been clarified yet and the results tent to lead to weak support for the development of student's thinking. The quality of claims, arguments, and levels of student confidence in the answers was still low. Students experienced false scientific reasoning and weak concepts when confronted with physics problems. Types of common false in every component of scientific reasoning in learning targets were in the forms of innovations to overcome false and confidence in the process of scientific reasoning and the mastery of student concepts.

Acknowledgment

The authors would like to thank the Government of the Republic of Indonesia through the Ministry of Finance, especially LPDP for the added funds to complete this research. [Project Number = 20170961080036]

Reference

[1] Zimmerman C 2005 *The Development of Scientific Reasoning Skills: What Psychologists Contribute to an Understanding of Elementary Science Learning* (New York: Normal Illinois State University)

[2] Ates Ö and Eryilmaz A 2011 Effectiveness of Hands-on and Minds-on Activities on Student's Achievement and Attitudes Towards Physics *Asia-Pacific Forum on Science Learning and Teaching* **12**(1)

[3] Frost J 2010 *Learning to Teach Science in the Secondary School* (New York: Routledge)

[4] Lebuffe J R 1994 *Hands-On Science in The Elementary School* East Lansing MI National Center for Research on Teacher Learning ERIC Document Reproduction Service No.ED 375003

[5] Tajudin N M and Chinnappan M 2016 The Link between Higher Order Thinking Skills Representation and Concepts in Enhancing TIMSS Tasks *International Journal of Instruction* **9**(2) 199-214

[6] Thuneberg H, Hautamäki J and Hotulainen R 2015 Scientific reasoning, school achievement and gender: A multilevel study of between and within school effects in Finland Scandinavian *Journal of Educational Research* **59**(3) 337-356

[7] Susantini E, Faizah U, Prastiwi M S and Suryanti 2016 Developing Educational Video to Improve the Use of Scientific Approach In Cooperative Learning *Journal of Baltic Science Education* **15**(6) 725-737

[8] Azar A 2005 Analysis of turkish high-school physics-examination questions and university entarence exams questions according to Bloom's taxonomy *Türk Fen Eğitimi Dergisi* **2**(2) 144-150

[9] Erman E 2010 Intervensi Kognitif Dalam Pembelajaran IPA Untuk Memacu Perkembangan Kemampuan Berpikir Abstrak Siswa SD *Jurnal Pendidikan dan Pembelajaran (JPP)* **15**(1) 270-277

[10] Piraksa C, Srisawasdi N, and Koul R 2014 Effect of Gender on Student's Scientific Reasoning Ability: A Case Study in Thailand *Procedia- Social and Behvioral Sciences* **116** 486–491

[11] Han J 2013 *Scientific Reasoning: Research, Development, And Assessment* (Columbus: The Ohio State University)

[12] Shofiyah N, Supardi Z A, and Jatmiko B 2013 Mengembangkan Penalaran Ilmiah (Scientific Reasoning) Siswa Melalui Model Pembelajran 5E Pada Siswa Kelas X Sman 15 Surabaya *Jurnal Pendidikan IPA Indonesia* 83-87

[13] Erlina N, Wasis, & Rosyid 2016 Pengembangan dan Penerapan Three-Tier Test Untuk Mengukur Keterampilan Penalaran Ilmiah Siswa SMA *Prosiding Seminar Nasional Pendidikan 2016* **1** pp 214-222

[14] Motlhabane A 2017 Unpacking The South African Physics-Examination Questions According To Blooms'revised Taxonomy *Journal of Baltic Science Education* **16**(6) 919-931

[15] Woolley J S, Deal A M, Green J, Hathenbruck F, Kurtz S A, Park T K, Jensen J L 2018 Undergraduate students demonstrate common false scientific reasoning strategies *Thinking Skills and Creativity* **27** 101-113

[16] Schunk D H 2012 Learning Theories: an educational perspective 6th ed (Carolina: The University of North Carolina at Greensboro)

[17] Hsu C-C, Chiu C-H, Lin C-H, and Wang T-I 2015 Enhancing Skill in Constructing Scientific Explanations Using A Structured Argumentation Scaffold in Scientific Inquiry *Computers & Education* **91** 46-59

[18] Annie E Wertz T C 2007 Belief–Desire Reasoning in the Explanation of Behavior: Do Actions Speak Louder Than Words? *ScienceDirect Cognition* **105** 184–194

[19] Croker S and Buchanan h 2016 Scientific Reasoning in A Real-World Context: The Effect of Prior Belief and Outcome on Children's Hypothesis-Testing Strategies *British Journal of Developmental Psychology* 409–424

[20] Biggers M, Forbes C T and Zangori L 2013 Elementary Teachers' Curriculum Design and Pedagogical Reasoning for Supporting Student's Comparison and Evaluation of Evidence-Based Explanations. *The Elementary School Journal* **114**(1) 48-72

[21] Tajudin N a M and Chinnappan M 2016 The Link between Higher Order Thinking Skills Representation and Concepts in Enhancing TIMSS Tasks *International Journal of Instruction* **9**(2) 199-214

[22] Schwichow M, Croker S, Zimmerman C, Höffler T and Härtig H 2016 Teaching the control-of-variables strategy: A meta-analysis *Developmental Review* **39** 37-63

[23] Amsel E and Moshman D 2015 Logical and Hypothetical Reasoning in Adolescence *Development of International Encyclopedia of the Social & Behavioral Sciences 2nd edition* 306–312

Student's Strategic Competence toward Open-Ended Problems Before and After the Transition to Junior High School

I Sabilah[1,*], T Y E Siswono[2], and Masriyah[2]

[1]Postgradute Programme in Mathematics Education, Universitas Negeri Surabaya, Ketintang, Surabaya 60231, Indonesia
[2]Departement of Mathematics, Universitas Negeri Surabaya, Ketintang, Surabaya 60231, Indonesia

[*]Corresponding Author's email : imamasabilah16070785050@mhs.unesa.ac.id

Abstract. Strategic competence refers to mental activities that apply strategies to formulate, to represent, and to solve mathematical problems. The aim of this qualitative research is to explore strategic competence of the sixth, seventh, and eighth grade students in solving open-ended problems. There were 25 students with different mathematical competencies involved to take part in this research. Among them, three selected students with higher mathematical competency represented each grade. This research used interview-based task to explore students' strategic competence. Results showed that students understood the problems by reading and drawing/graphing them, formulated problem solving by recalling their prior knowledge about quadrilateral topic, and conveyed numerical and visual strategy to solve the open-ended problems. At last, they could determine effective solutions using arithmetical method, find all solutions correctly, and choose the cheaper ones. As an implication, we suggest further research to explore students' strategic competence with other types of mathematical competences.

1. Introduction

One major purpose for students learning mathematics is to accelerate mathematical proficiency including conceptual understanding, procedural fluency, strategic competence, adaptive reasoning, and productive disposition [1]. Eventhough the five strands are interrelated, students essentially need to develop and focus on strategic competence. Those with good strategic competence indirectly master the other four competences [2]. Strategic competence is a mental activity using formulating-problem strategy, representing problems mathematically, and solving as well as choosing the most effective solution for the problems [1, 3, 4]. Students must have excellent thinking skills and do mathematics creatively, productively, critically, independently, collaboratively and communicatively through appropriate scientific approach learned in education units and other sources [5]. Creative students with good communication skills will be easier to solve any problems.

Open-ended question is a problem formulated by using multiple correct answers [6]. Learning with open-ended problem also refers to implementing an open learning that presents problems with a variety of problem solving ways *(flexibility)* and the solution can be varied *(fluency)* [7]. The main purpose of the provision of open-ended problem is not to get an answer, but to emphasis on how students can arrive at an answer, which is not fixated on the answers that must be collected to the teacher [8]. Moreover, the "openness" of the problem is meant to be lost if the teacher only gives alternatives to answer the problems. Henceforth, there must be many approaches to get the answers of the given problems. There were three types of open-ended problems. The first is *open the process of completion* if the question has various ways of settlement. The second type is *open the result* if the question has many correct answers. The last type is *open development of the sequel* meaning that when students have completed something, then they can develop new problems by changing the terms or conditions on an already solved matter [9]. Open-ended problem is a mathematics question that has one or more answers to a various strategies of completion [10].

So far, students who do not learn the material face difficulty in solving the addressed problems [11, 12, 13]. Those who cannot alleviate such difficulty tend to solve problems by guessing strategy. On contrasts, those who know the materials tend to determine appropriate solvencies to troubleshoot the problems. These typical students then have stronger strategic competences and are able to encounter various problems [14, 15, 16].

The aim of this qualitative research is to explore strategic competence of the sixth, seventh, and eighth grade students in solving open-ended problems. Results depicted that student understood the problems by reading and drawing/graphing them. They formulate problem solving by recalling their prior knowledge about quadrilateral topic. They conveyed numerical and visual strategy to solve open-ended problems. Finally, they could determine an effective solution using arithmetical method, find all solutions correctly, and choose the cheaper ones. As an implication, we suggest further researchers to explore students' strategic competence with other types of mathematical competences.

2. Method

This qualitative study involved three of 25 students in an interview-based task. Twenty-five students were included into quadrilateral learning class and given (1) mathematical competence test to categorise students' mathematical competencies and (2) task in a form open-ended problems. The subjects selected for interview were those who met the following criteria: had studied the quadrilateral material, able to communicate their ideas clearly, and high students' mathematical competency. Interview data were analysed in three stages: data condensation, data presentation, and drawing conclusions [17]. Data condensation referred to the process of selecting, focusing, simplifying, abstracting or transforming the obtained data that appeared in the full corpus of field notes, interview transcripts, documents, and other empirical materials. Data presentation was in the form of a set of narrative text information/ data arranged orderly and systematically, which further eased to draw conclusion. Based on the data obtained through tests and interviews, researchers then made indicators for each aspect of strategic competence (see Table 1).

Table 1. Indicators of Strategic Competence.

Strategic Competence Aspects	Indicators
Formulating Problem	• Students could write and explain the given information and question. • Students could explain the relationship between the given information and question and relevant prior knowledge.
Representing Problem	• Students could represent problems using diagram/ table/ picture/ mathematics model/ equation/ numeric/ symbolic/ verbal/ graphic/ etc.
Solving Problem	• Students could write and explain a plan to solve. • Students solved problems correctly. • Students could explain problem-solving steps correctly.

3. Results and Discussion

3.1. Formulating

Students formulated the solution of the given problems to understand comprehensively where a mathematical formula could be used. Sixth grade students directly drew and estimated the area by considering her calculation, comparing prices on different size, and choosing the cheaper solutions. The seventh grade student used multiplication and division operations and got the total of all the area, calculated and compared cost reduction and operating expenses using the formula for a rectangular area, and determined the effective resolution on a more economical choice. The eighth grade student wrote down given information and answers addressed on problems with enclosed conclusion. First interview was for the sixth grade student (S1), the second interview was for the seventh grade student (S2), and the third interview was for the eighth grade student (S3), of which all used different strategy to formulate completion of financial problems in a form of open-ended problems given by researcher (R). All of them were compared the different size of the ceramics option and they chose the cheaper prices for the solution. The difference was their perspectives about systematic formula addressed to the problems.

R : *Please explain the steps of completion that you will use!*

S1 : *First, I drew and estimated the size of ceramics. Then, I counted the price for each ceramics' size before deciding the cheaper ceramics' price, I compared the prices of the cost with different sizes.*

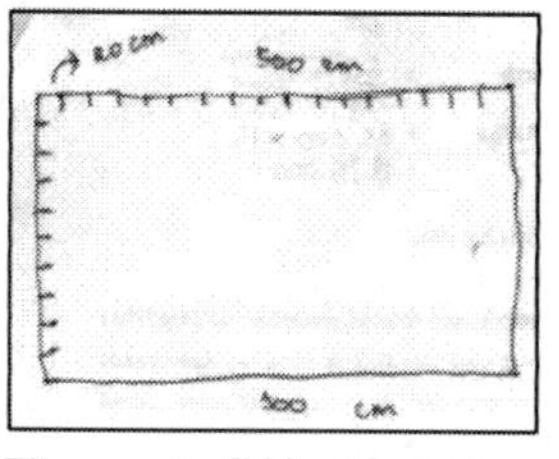

Figure 1. S1's First Step

S2 : *First, I found that the key of the problems was "minimal price to buy the ceramics until the area closed"*
Then, I wrote the given information and question from the problems. Finally, I solved the problem by comparing the size of the ceramics and their prices and choosing the cheaper ceramics price.

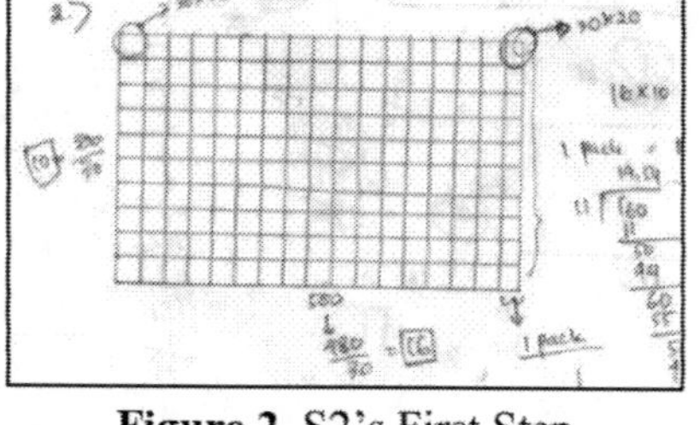

Figure 2. S2's First Step

S3 : *I calculated all prices of each ceramics' size and then compared all of them with a same quantity of living room area (multiplying 3 meters and 5 meters) to show the cheaper ceramics' price.*

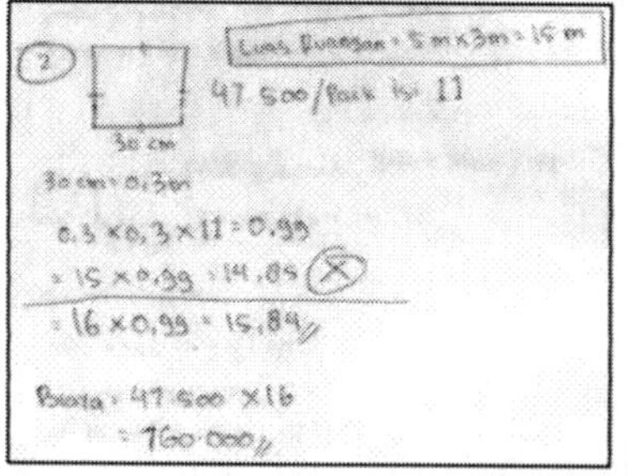

Figure 3. S3's First Step

3.2. Representing

Student's representation strategy in open-ended problems used numerical and visual strategies. The sixth and seventh grade students drew a picture to represent the given information and idea regarding the problem solving statements. Meanwhile, the eighth grade student used a simple strategy such as numerical calculations and numbers in representation.

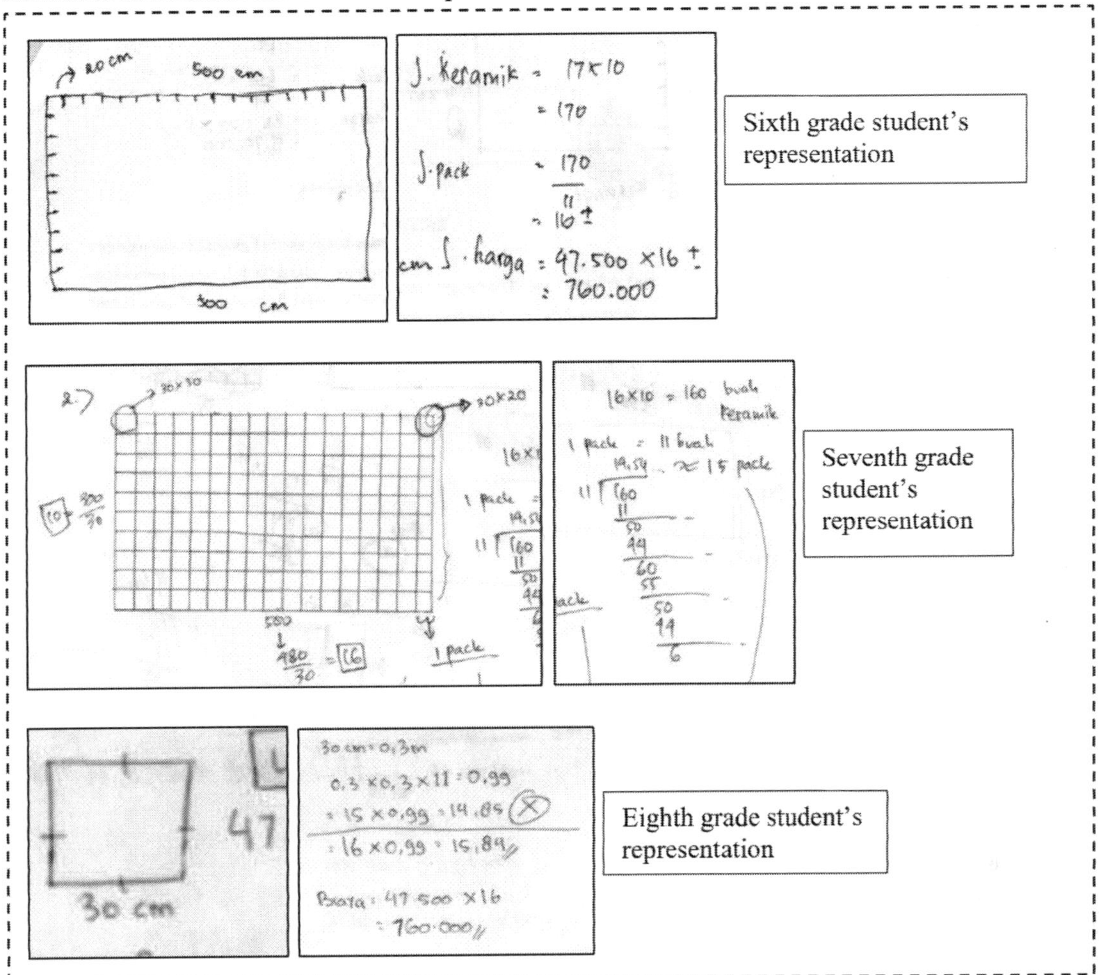

Figure 4. Students' Representations

Figure 1 shows that the sixth, seventh, and eighth grade students used numerical and visual strategies to represent problems and problem-solving ideas.

3.3. Solving

The sixth grade student directly drew and determined the cost by estimating the calculation. The seventh grade student performed various arithmetic operations such as multiplication and division and used a rectangle area formula to resolve the cost. The eighth grade student chose to use a sequence of completion by writing down the information about the problems and then conducted the calculations including drawing the conclusion for each ceramics' size.

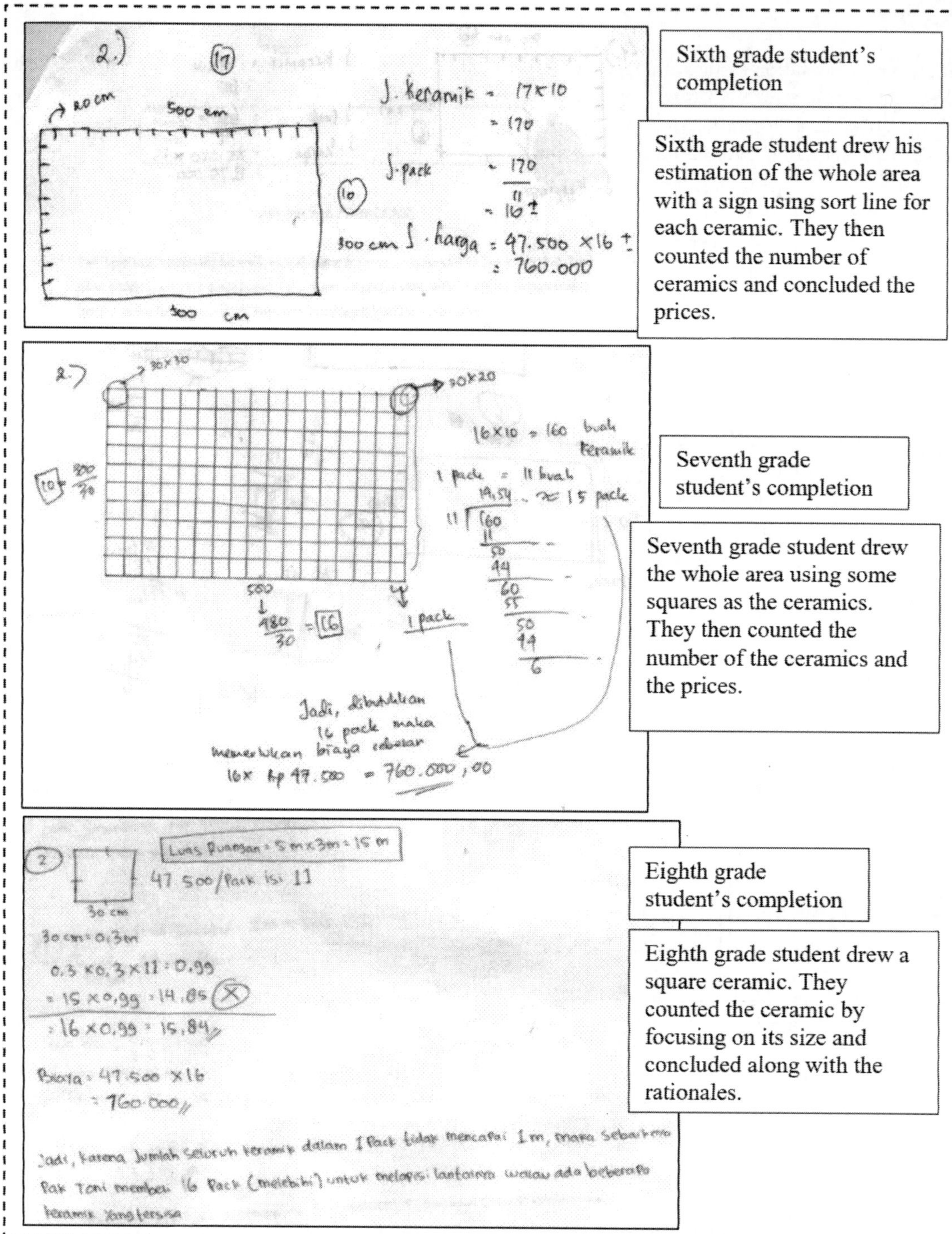

Figure 5. Student's Answer to Solving Open-ended Problems

Figure 2 showed that the sixth, seventh, and eighth grade students determined the effective solution using arithmetic calculation and choose the cheapest solution.

That is, students who did not learn the materials found difficulty in solving the problems [11, 12, 13]. Consequently, they used guessing strategy to answer or solve the problems. However, those who got the basic materials might have high strategic competences and be able to solve various problems [14,15,16].

4. Conclusion
Results showed that students understood the problems by reading and drawing/graphing them. They formulated problem solving by recalling their prior knowledge in quadrilateral topics. They solved the open-ended problems by numerical and visual strategy. Finally, they determined an effective solution using arithmetical method, found all solutions correctly, and chose the cheaper ones.

5. Implication
We suggest the next research to explore students' strategic competence with other types of mathematical competences.

6. Appendices
The following is the open-ended problems used in this research constructed by the researchers

Mr. Toni will set new ceramics in his living room measuring 5m × 3m. Mr. Toni will buy a ceramic in one of the shops near his house. The ceramic price list at the store of the same quality is shown below.

Ceramics Size	Prices	Contents /pack
20 × 20	44.500,00 IDR/pack	25 pcs
30 × 30	47.500,00 IDR/pack	11 pcs
40 × 40	76.000,00 IDR/pack	6 pcs
50 × 50	85.000,00 IDR/pack	4 pcs

Please help Mr. Toni to count minimal cost to buy the ceramics until the area closed. Write and explain your answer and this reason.

7. References

[1] Kilpatrick J, Swafford J, and Findell B 2001 *Adding It Up: Helping Children LearningMathematics* (Washington DC: National Academy Press)

[2] Garderen D V, Scheurmann A and Poch A 2013 *ZDM Math.Educ.* **46(1)** 135-149

[3] Syukriyani A 2017 *AIP Conf. Proceed.* **1868** 05009

[4] Ozdemir E Y and Pape S J 2012 *Mathematics Education Research Journals* **24(2)** pp. 153-168

[5] Kemendikbud 2016 *Permendikbud Nomor 20 Tahun 2016 Tentang Standar Kompetensi Lulusan Pendidikan Dasar dan Menengah* (Jakarta: Kementerian Pendidikan dan Kebudayaan)

[6] E Suherman et all 2003 *Soal Terbuka* dalam *Strategi Pembelajaran Matematika Kontemporer* (Bandung: JICA-Universitas Pendidikan Indonesia) p 123

[7] A R Luthfiana 2015 *Kemampuan Berfikir Kreatif Siswa dalam Menyelesaikan Soal Open-Ended ditinjau dari Gaya Belajar VARK* (Surabaya: Universitas Negeri Surabaya)

[8] N Saádah 2013 *Profil Pemecahan Masalah Matematika Open-Ended Siswa SMP Berdasarkan Tingkat Kecerdasan Emosional* (Surabaya: Universitas Negeri Surabaya)

[9] A Mahmudi 2008 *Mengembangkan Soal Terbuka (Open-Ended Problem) dalam Pembelajaran Matematika* [Online]. Available: http://staff.uny.ac.id/sites/default/files/penelitian/Ali%20 Mahmudi,%20S.Pd,%20M.Pd,%20Dr./Makalah%2002%20PIPM%202008%20_Mengembangk an%20Soal%20Terbuka_.pdf. [Accessed 2 October 2015]

[10] I Sabilah and J T Manoy 2018 The Use of Open-Ended Questions with Giving Feedback (OEQGF) for Effective Mathematic Learning *J. of Phys.: Conf. Series* **947** 012032

[11] Awofala A O A 2017 Assesing Senior Secondary School Students'Mathematical Proficiency as Related to Gender and Performance in Mathematics in Nigeria *Int. J. of Res. Educ. and Sci.* **3(2)** 488-502

[12] Groves S 2012 Developing Mathematical Proficiency *Journal of Science and Mathematics Education in Southest Asia* **35(2)** 119-145
[13] Ostler E 2011 Teaching Adaptive AND Strategic Reasoning through Formula Derivation: Beyond Formal Semiotics *Technomath. Res. Found.* **4(2)** 16-26
[14] Khairani A Z and Nordin M S 2011 The Development and Construct Validation of The Mathematics Proficiency Test For 14-year-old *Asia Pasific J. of Educ. and Educ.* **26(1)** 33-50
[15] Rababah G 2016 *The Effect of Communication Strategy Training on The Development of EFL Learners' Strategic Competence and Oral Communicative Ability* (Jordan: Springer)
[16] Nugraheni Z, Budiyono and Slamet I 2017 *AIP Conf. Proceed.* **1913** 020029
[17] Miles, Huberman, and Saldana 2014 *Qualitative Data Analysis* (Los Angeles: SAGE Publications)

Analysis of Churn Rate Significantly Factors in Telecommunication Industry Using Support Vector Machines Method

S Arifin[1], **F Samopa**[1,2]

[1] Magister Management of Technology Department, InstitutTeknologi SepuluhNopember, Surabaya, Jawa Timur 60264, Indonesia.
[2] Research Center Building, InstitutTeknologiSepuluhNopember, Teknik Kimia Street, Surabaya, Jawa Timur 60117, Indonesia.

Abstract. This research was intended to know factors that influenced the churn rate significantly in telecommunication company through research of historical billing and profiling data of customers. This study consisted of seven variables which are four billing historical variables and three profiling variables. The data was taken from a telecommunication company in Indonesia by taking active customer with minimum 6 months and historical data form January until March 2018. The data were tested using Support Vector Machine method by taking the result of classification performance either on performance of total variable or performance in each variable. The significance threshold was defined 5%. The results show that there were three variables that influenced the churn rate significantly namely voice usage, data usage and reload with performance percentage less than 5% of total performance. Those three variables were historical billing data.

1. Introduction

In the telecommunications industry, customer is the most valuable asset for the company to sustain the business process. Reducing number of customer is certainly unexpected thing for the companies. Thus, companies need to analyze the customers profile to facilitate the conduct of business segmentation in support of a proper decision making. The right decision is expected to generate a value that can improve the performance and profit of the company. In analysis customer data, the data history is absolutely necessary. It can describe the customer's value in the future whether the customer is potentially churn or non churn. In this case, Support Vector Machine method can be used for classification [1].

Several methods of classification lately have been used for main purpose especially in the data processing. The popular method that used by researchers was data mining based; either using classification or clustering such as Decision Tree and K-Mean. However, most papers only discuss churn predict ions without analyzing historical accounts of customer billing. In this study, in order to predict churn and also analyzed historical billing that included the use of voice in minutes, short message service in the event, and the internet data in kb. During the last 3 months in 2017, Support Vector Machine method was applied by taking voice, short message service (sms) and the data service as the main variable in the test. Based on the test results, it could be used as a reference in analyzing the factors that affect customer churn [2].

Table 1. Churn percentage in one of Telecommunication company in Indonesia 2017

Region	Q1-Q2	Q2-Q3	Q3-Q4	Q4-Q1
Central	15%	7%	16%	17%
East	39%	-19%	1%	23%
Capital	2%	7%	9%	6%
North	19%	11%	32%	30%
West	3%	5%	22%	36%
National	15%	0%	13%	20%

Table 1 illustrated that one of the internal sources of telecommunication companies from 1st quarter, 2nd quarter, 3rdquarter and 4thquarter explained significantly customers churn numbers which each region also quite large numbers. If it was viewed more detail as a whole data during 2017, there was only a region that experienced decreasing in churn rate, namely the East region in the second quarter. Nationally, the churn rate increase in each quarter grew by 20%. This was a corporate homework that requires evaluation and analysis of factors that can potentially be affectedto churn rate.

Using the prediction of customer churn and customer behavior analysis, it were expected to help the marketing and CRM division in segmenting customers who could simplify the product campaign to minimize the existence of customer churn and increase the profit of the company [3]. The analysis conducted to produce variables which have great potential to make the customer into churn. Based on these variables, it can be used as a benchmark in determining a particular focus product in accordance with the profile of the customer.

Based on the background above, the problem in this study was unknown factors that affect the churn rate significantly. The existing services in telecommunication companies have not reflected their effectiveness in influencing the high churn rate. So, it takes effort in analyzing the billing behavior of consumers in determining the potential of churn rate [4]. From this case, the formulation of the problem in this study is "analysis of churn rate significantly factors in telecommunication industry using Support Vector Model ".

The purpose of this study was to determine what factors have a significant effect on the occurrence of churn rate using customer billing services and customer profilingwithin focus on the correlation between independent variables with dependent variable. In addition, this study aimed to classify and predict future churn rates based on patterns in the data history.

2. Support Vector Machine Principle and Algorithm

Support Vector Machine (SVM) is defined as a set of related learning methods that analyze data and recognize patterns, which are used for classification and regression analysis [5]. SVM takes a set of input data and predicts for each given input, derived from two classes that are classified by finding the best hyperplane value. Hyperplane is separator between the two classes that can be found by measuring the hyperplane's margins and searching for the maximum point[6].

For more detail about SVM, here given anexample using dummy data test in the table 2.

Table 2. Data set

Voices	Data Usage	y
2.947814	6.626878	1
2.530388	7.785050	1
3.566991	5.651046	1
3.156983	5.467077	1
2.582346	4.457777	-1

2.155826	6.222343	-1
3.273418	3.520687	-1

Table 2 describe that the dummy data consists of two attributes namely Voices and Data usage and there are two classes 1 as non-churn and -1 as churn. The next step is to determine the constraint for both attributes. Constraint is used for quadratic programming to looking for the value of w1, w2 and b value by using the formula:

$$y_i(\mathbf{w}^{\mathrm{T}}\mathbf{x}_i + b) \tag{1}$$

The constraint results can be described in the table 3 below:

Table 3. Finding the constraint.

W1.Voices	W2.Data Usage	**Constraint**
8.560183	7.960593	1.785812
7.348017	9.351858	1.96491
10.35822	6.788367	2.41162
9.167592	6.567373	1
7.498897	5.354943	1.881125
6.260322	7.474642	1
9.50571	4.229255	1

Table 3 mention how the constraint obtained by multiple w1 with voice variable and w2 with data usage variable as in the formula the plus b. After the constraint have been obtained, then find the new value of w1, w2 and b value by using quadratic programming that subject to constraint value. The quadratic programming can be done by using the below formula:

$$\min_{w,b} \frac{1}{2}\mathbf{W}^{\mathrm{T}}\mathbf{W} \tag{2}$$

$$\textit{subject to}$$
$$y_i(\boldsymbol{w}^T\boldsymbol{x}_i + b) \geq 1 \ \textit{for} \ \forall i$$

After compute the quadratic programming formula, the value of w1,w2 and b can be found in the table 4 below.

Table 4.new w1, w2 and b value

w1	w2	b
2.903909	1.201258	-14.735

Table 4, explain the result of quadratic programming value to looking for the new value of w1, w2 and b. Those value are used to determine the margin score. The margin score of each data are used to find the classification classes. Margin score indicates the not-normalised distance between each data point to the decision hyperplane.Margin scoring data can be found by the formula below:

$$\Delta^i = W^T X_i + b \tag{3}$$

Where b is the value of the new b after quadratic programming. The margin score value can be found in the table 5 below.

Table 5. Margin score

W1.Voice	W2.Data Usage	Margin Score
8.560183	7.960593	1.7858
7.348017	9.351858	1.9649
10.35822	6.788367	2.4116
9.167592	6.567373	1
7.498897	5.354943	-1.8811
6.260322	7.474642	-1
9.50571	4.229255	-1

Table 5 explain how margin score can be found. The margin score are used to determine classification classes by simply take the sign of margin score in each data using function below.

$$sign(x) = \begin{cases} +1 \; if \; x \geq 0 \\ -1 \; if \; x < 0 \end{cases} \quad (4)$$

Sign function will return 1 if the input argument is positive or zero and return -1 if the input argument is negative. Finally classification result as shown in table 6 below.

Table 6. Classification result.

Score	Classification
1.785812	1
1.96491	1
2.41162	1
1	1
-1.88113	-1
-1	-1
-1	-1

In the table 6 show that classification data are same with original data set. There are four data for 1 class and three data for -1 class. It can be concluded that SVM 100% classified correctly.

3. Research Methodology

In this research, the data were tested amounted to 819 MSISDN with 7 variables. The data taken from a telecommunication company in Indonesia with MSISDN active at least 6 months and there were usage for 3 months from January until March 2018. In this test, it was conducted and supervised a learning technique, that in each variable there are churn and non churn data that become data reference learning.

Table 7. Data set and the index.

Variables	Index
POC	Consist of 22 POC Area
Handset	2.5G, 2.75G, 2G, 3.5G, 3.75G, 3G, 4G

Voice in Minutes	0>1000, 1001-2000, 2001-3000, 3001-4000, >4000 in Minutes
SMS in Event	0>1000, 1001-2000, 2001-3000, 3001-4000, >4000 in Event
Data in Kb	<=20GB, >20GB - <=40GB, >40GB - <=60GB, >60GB
Package Name	More than 12 package
Reload	19 Reload categories from 5000 to 200000

Table 7 described that there were seventh variables, namely POC (point of control), Handset in devices, Voice usage in Minutes, Short Message Services in event, data usage in kb, packet name that attached in the MSISDN and reload which each variable has an index. The index means the categorical data in every single rows. However, the data obtained was not in single data but in the form of categorical. After variables data was defined, and then it was continued to the research flow. The research flow explains how data is processed and produce a result for getting knowledge analysis as shown in Figure 1 and Figure 2.

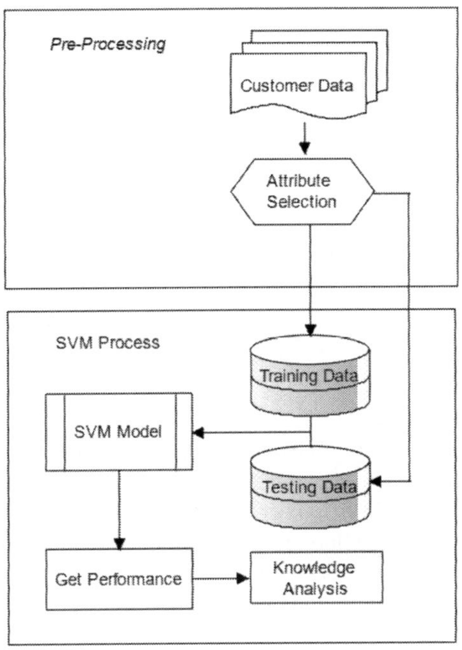

Figure 1. Data Processing.

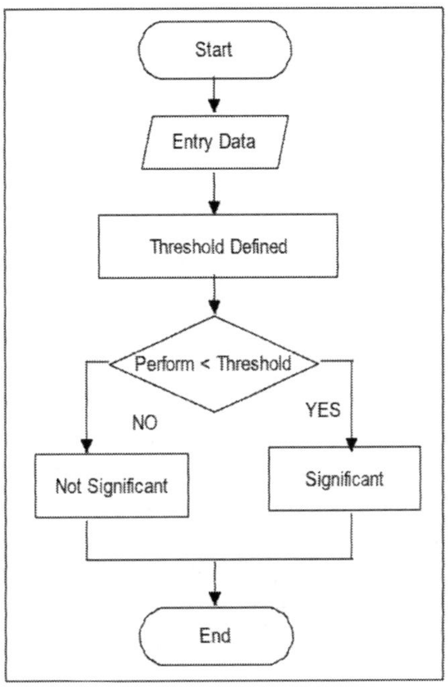

Figure 2. SVM Model.

In Figure 1, it was explained that from each attribute in the tested data, consist of training data and testing data. Both data would be tested with SVM modelling in Figure 2 for getting performance that would be used for knowledge analysis.

In figure 2, the threshold was determined at 5% [7] of the total performance, if the attribute performance tested was smaller than the threshold, then the attribute had a significant effect, otherwise if greater than 5%, then the attribute did not significantly influence.

4. Finding and Discussion

In this section, it would be explained how to analysis the data and find the churn rate significantly variables based on hypothesis that had been decided. Firstly, the zero hypothesis was determined as shown in figure 3.

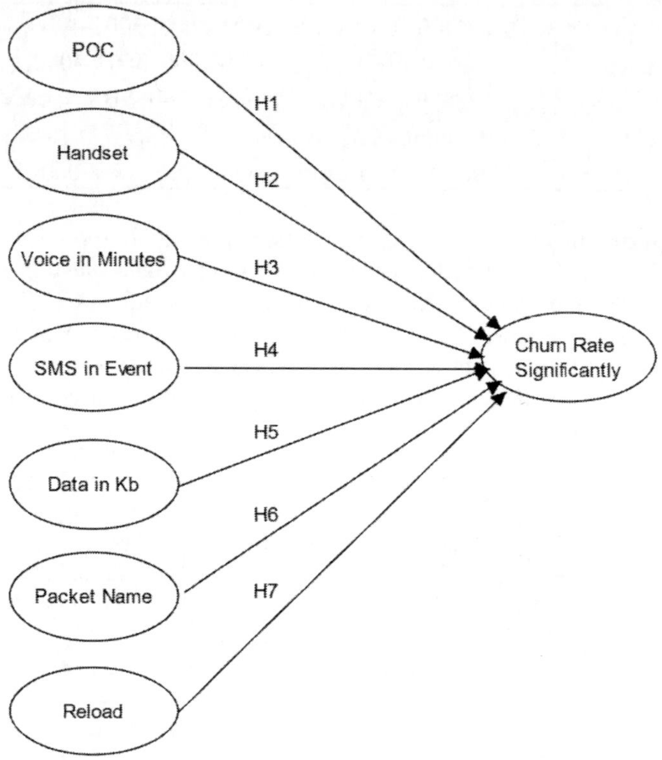

Figure 3. Research Model

In Figure 3, there were seven hypotheses as zero hypothesis[8] with POC as hypothesis 1, Handset as hypothesis 2, Voice in Minutes as hypothesis 3, SMS in Event as hypothesis 4, Data in Kb as hypothesis 5, Packet Name as hypothesis 6 and Reload as hypothesis 7 with respectively H0 as the comparative hypothesis. The initial hypothesis was all variables which not significantly affected to churn rate.

To prove the hypothesis, then it was conducted a testing influence of each variable on the overall result of the variable. It would determine the significant value based on the predetermined threshold that was 0.05 or 5% [9].

Table 7. Research Result.

Hypothesis	Performance Classification	Distance	Sig (5%)	ROC Area	Mean absolute error
H1	60.20%	22.80%	FALSE	0.40	0.53
H2	58.90%	24.00%	FALSE	0.41	0.51
H3	79.30%	3.70%	**TRUE**	0.78	0.20
H4	74.00%	8.90%	FALSE	0.71	0.26
H5	82.50%	0.40%	**TRUE**	0.82	0.17
H6	64.60%	18.30%	FALSE	0.61	0.35
H7	80.60%	2.30%	**TRUE**	0.75	0.19

Based on table 7, it could be concluded that there were three attributes that significantly influence the churn rate of total 7 attributes which have been tested. The usage data in kb (H5), voice in minutes (H3) and reload (H7) had a performance smaller value than 5% [10] of the total performance of the overall variables. Meanwhile, the variable was approaching the significant in SMS (H4) that was equal to 9%. Based on these results, the telecommunications companies XYZ should maintain Data and Voice services to minimise the churn rate.

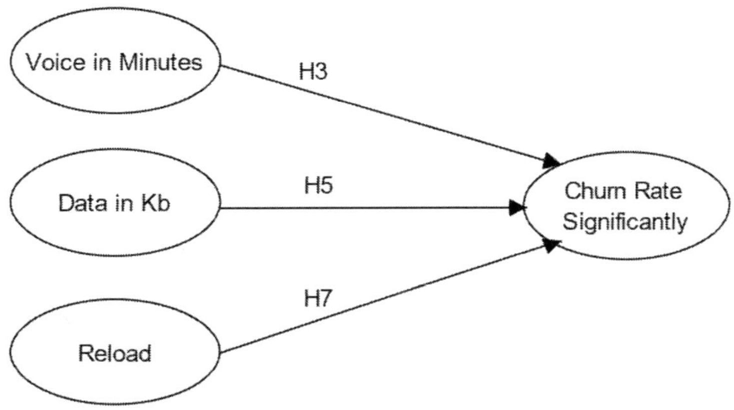

Figure 4. NewResearch Model

After the test, it was obtained a new research model that included three variables as the result namely of voice in minutes, data in Kb and reload. Those variables showed significant effect to the churn rate.

5. Conclusion

Based on the result, it could be concluded that variable voice in minutes, data in Kb and reload significantly influenced the occurrence of churn rate. The telecommunications company especially in Indonesia needed to maintain three main services, namely voice, data and reload service.

6. Reference

[1] Guo-enXIA 2008Model of Customer Churn Prediction on Support Vector Machine*SETP***28**(1)pp. 71-77.
[2] Shaaban E 2012 A Proposed Churn Prediction Model*MUST University***24**
[3] Brandusoiu I 2013 Churn Prediction In The Telecommunications Sector Using Support Vector Machines *Fascicle of Management and Technological Engineering***1** 3
[4] Yudhistira A 2017 (online) Analisa Customer Churn pada perusahaan Internet Service Provider XYZ menggunakan Backpropagation Neural Network (http://repository.its.ac.id/2564/)
[5] Srivastava D 2009 Data Classification using Support Vector Machine *Journal of Theoretical and Applied Information Technology JATIT***2**(1) pp. 1-7.
[6] Yin Y, Han D, and Cai Z 2011 Explore Data Classification Algorithm Based on SVM and PSO for Education Decision*Journal of Convergence Information Technology***6**(10) 122–128
[7] HerawatiM 2016 Prediksi Customer Churn MenggunakanAlgoritma Fuzzy Iterative Dichotomiser 3 J. Math. and Its Appl. **13**(1)
[8] Tanudjaja H 2007 *AnalisisHubungan Dan PengaruhVariabelMakroekonomiTerhadapKreditBermasalahPerbankan Indonesia* (Jakarta: Universitas Indonesia)
[9] Ardiansya, IH, Lubis D 2017 PengaruhVariabelMakroekonomiterhadapPertumbuhan Sukuk Korporasi di Indonesia IAEI **5**(1)
[10] Fraticasari S 2018 OptimasiPemodelanRegresi Linier Berganda Pada PrediksiJumlahKecelakaanSepeda Motor DenganAlgoritmeGenetika*JurnalPengembanganTeknologiInformasi dan IlmuKomputer***2**(5)
[11] Teknomo K 2012 *Support Vector Machine Tutorial* (Columbia: Revoledu.com publishing)

Acknowledgments

This research has been acknowledged by advisors of Magister Management of Technology Department in Institut Teknologi Sepuluh Nopember (ITS), Surabaya and Revoledu.com by TeknomoKardi [11].

IOP Conf. Series: Journal of Physics: Conf. Series **1108** (2018) 012019 doi:10.1088/1742-6596/1108/1/012019

Multinomial Logistic Regression and Spline Regression for Credit Risk Modelling

M R Adha[1,*], S Nurrohmah[1] and S Abdullah[1]

[1]Department of Mathematics, Universitas Indonesia, Kampus Baru UI, Depok, 16424, Indonesia

*Corresponding Author: muhammad.rizky41@sci.ui.ac.id

Abstract. Regression modelling has been adapted in retail banking because of its capability to analyze the continuous and discrete data. It is an important tool for credit risk scoring, stress testing and credit asset evaluation. In this paper, the approach used is multinomial logistic regression model to gain the information regarding the factors that affect the occurrence of default and attrition events on credits. In addition, this paper will also introduce spline regression approach using truncated power basis to model the hazard functions of default and attrition events. The flexibility of spline function allows us to model the nonlinear and irregular shapes of the hazard functions. Then, by using spline regression and multinomial logistic regression model, there will be a better result and interpretation. There are several advantages by using those both models. First, by using the flexible spline regression function, it can model nonlinear and irregular shapes of the hazard functions. Second, it is easy to understand and implement, and its simple parametric form from multinomial logistic regression model can make it easy in model interpretation. Third, the multinomial logistic regression model has the ability to do prediction. Furthermore, by using a credit card dataset, we will demonstrate how to build these models, and we also provide statistical explanatory and the prediction accuracy of multinomial logistic regression model in classifying customers based on the prediction of default and attrition is 95.3%.

1. Introduction

In the last 50 years, credit users have been the driving force of the economy in almost all developed countries [1]. By 2018, Bank Indonesia expects credit growth to increase to 10-12%. In line with these figures, Indonesia's economic growth is in the range of 5.1-5.5%. Therefore, the magnitude of credit growth will affect the magnitude of economic growth in a country.

The rapid growth of consumer credit brings the demand for better credit scoring system to managing credit lending business. The credit scoring has been successfully applied in credit risk management [2]. However, with the fierce competition, the credit scoring has also been expanded from only controlling the credit default risk to maximizing the profit [1]. The profitability has become the most significant business consideration in credit decision making. To better estimate customer's profitability, the lenders need more sophisticated models which can estimate the dynamics of risk and revenue over time.

Based on the consideration of profitability, required a model that can be used as the basis of decision making related to the provision of credit, interest rates and credit line. For this purpose, in this paper will be used approach by using multinomial logistic regression model to know the factors that influence the default and attrition events on a credit. We use multinomial logistic regression model because this model also can handle time-dependent and time-independent covariates and deal with competing risks. In this paper, the object of competing risk is credit because it can risk default, attrition (i.e.,customers pay off their balance and close their accounts) or not both (open status).

In addition, we also introduce spline regression approach using truncated power basis to model hazard functions of default and attrition events. In this paper, the dependent variable is hazard (the number of individuals who experience the target event in certain time period by the number of individuals at risk during that time period) and the independent variable is time (in months). The flexibility of spline function allows us to model the nonlinear and irregular shape of the hazard functions.

In the end of the discussion, by using a credit card dataset, we will demonstrate how to build these models, and we also provide statistical explanatory and prediction accuracy.

2. Multinomial Logistic Regression Model

In this paper, there are 3 types of event in credit dataset. Define $j = 1$ as the default event, $j = 2$ as the attrition event, and $j = 0$ as the open status. Thus, in other words J states competing risk in the data.

In the multinomial logistic regression model with the dependent variable having three categories, model formation requires two logit functions. Assume the categories of the dependent variable, y_i, are coded with 0, 1, and 2. In this paper, $y_i = 0$ will be used as a reference or basis. To build the model, assume the loan portfolio has p covariates, denote by vector, $\mathbf{x}_i, i = 1,2,...,n$. Thus, obtained a logit model as follows.

$$g_j(x_i) = \text{In}\left[\frac{\Pr(y_i = j \mid \mathbf{x}_i)}{\Pr(y_i = 0 \mid \mathbf{x}_i)}\right] = \text{In}\left[\frac{\pi_j(\mathbf{x}_i)}{\pi_0(\mathbf{x}_i)}\right] = \text{In}\left[\frac{\pi_j(\mathbf{x}_i)}{1-\sum_{j=1}^{2}\pi_j(\mathbf{x}_i)}\right] = \beta_{0j} + \beta_{1j}x_{1i} + ... + \beta_{pj}x_{pi} \qquad \begin{matrix} j = 1,2 \\ i = 1,2,...,n \end{matrix}$$

Then, the probability of each event is as follows.

$$\pi_0(\mathbf{x}_i) = \frac{1}{1+\exp(g_1(\mathbf{x}_i))+\exp(g_2(\mathbf{x}_i))} \qquad i = 1,2,...,n$$

$$\pi_1(\mathbf{x}_i) = \frac{\exp(g_1(\mathbf{x}_i))}{1+\exp(g_1(\mathbf{x}_i))+\exp(g_2(\mathbf{x}_i))} \qquad i = 1,2,...,n$$

$$\pi_2(\mathbf{x}_i) = \frac{\exp(g_2(\mathbf{x}_i))}{1+\exp(g_1(\mathbf{x}_i))+\exp(g_2(\mathbf{x}_i))} \qquad i = 1,2,...,n$$

2.1 Parameter Estimation

The method used to estimate the parameters in multinomial logistic regression is the maximum likelihood method. In this method, we will find the parameter value of $\boldsymbol{\beta} = (\boldsymbol{\beta}_1, \boldsymbol{\beta}_2)$ which maximizes likelihood function. To simplify the notation, consider $\pi_{ji} = \pi_j(\mathbf{x}_i)$. The $\boldsymbol{\beta}$ value that maximizes $\text{In}\, L(\boldsymbol{\beta})$ is obtained by differentiating $\text{In}\, L(\boldsymbol{\beta})$ against parameters $\beta_{0j}, \beta_{1j}, ..., \beta_{pj}$, which is then equated with 0 as follows.

$$\frac{\partial\, \text{In}\, L(\boldsymbol{\beta})}{\partial \beta_{kj}} = \sum_{i=1}^{n} x_{ki}(y_{ji} - \pi_{ji}) = 0$$

for $j = 1,2$ and $k = 0,1,2,..., p$, with $x_{0i} = 1$ for each observation.

2.2 Parameter Testing

Partial test is used to test the effect of each independent variable individually. The test statistic used is Wald test.

3. Spline Regression Model

Spline regression is a regression analysis method that is piecewise polynomial, i.e., a polynomial that has a segmented property on r intervals formed by knot points [3]. Knot point is a common fusion point that occurs because there are changes in pattern behavior at different intervals. Spline has an advantage in overcoming the pattern of data that shows a sharp rise or fall with the assist of knot points, and the resulting curve is relatively smooth [4].

The general form of spline regression using the truncated power basis is as follows.

$$y=\sum_{j=0}^{d}\alpha_j t^j+\sum_{p=1}^{r}\alpha_{p+d}\left(t-K_p\right)_+^d+\varepsilon$$

with

$$\left(t-K_p\right)_+^d=\begin{cases}\left(t-K_p\right)^d, & t>K_p\\ 0, & t\le K_p\end{cases}$$

in which $\boldsymbol{\alpha}=\left(\alpha_0,\alpha_1,\ldots,\alpha_p,\alpha_{p+d},\ldots,\alpha_{r+d}\right)^T$ is the parameter or regression coefficient for the truncated power basis function, d is degree of polynomial on truncated power basis, r is number of knots used, and $K_1,K_2,\ldots,K_r$ are location of knots.

3.1 *Parameter Estimation*

Eubank [5] states that parameter estimation on spline regression using OLS (Ordinary Least Square). The parameter estimator for spline regression using truncated power basis is

$$\boldsymbol{\alpha}=\left(T^TT\right)^{-1}T^T\mathbf{y}$$

with

$$\boldsymbol{\alpha}=\begin{pmatrix}\alpha_0\\ \vdots\\ \alpha_d\\ \alpha_{d+p}\\ \vdots\\ \alpha_{d+r}\end{pmatrix},\ T=\begin{pmatrix}1 & x_1^1 & \cdots & x_1^d & (x_1-K_1)_+^d & \cdots & (x_1-K_r)_+^d\\ 1 & x_2^1 & \cdots & x_2^d & (x_2-K_1)_+^d & \cdots & (x_2-K_r)_+^d\\ 1 & x_3^1 & \cdots & x_3^d & (x_3-K_1)_+^d & \cdots & (x_3-K_r)_+^d\\ \vdots & \vdots & \ddots & \vdots & \vdots & \ddots & \vdots\\ 1 & x_n^1 & \cdots & x_n^d & (x_n-K_1)_+^d & \cdots & (x_n-K_r)_+^d\end{pmatrix},\ \mathbf{y}=\begin{pmatrix}y_1\\ y_2\\ y_3\\ y_4\\ \vdots\\ y_n\end{pmatrix}$$

3.2 *Location and Number of Knots*

Ngo & Wand [6] uses a rule of determining the optimal number of knots on spline regression as follows.

$$r=\max\left(5,\min\left(\frac{1}{4}N,35\right)\right)$$

with N is number of unique x_i 's, and rule for the knot locations is $K_s=\dfrac{s+1}{r+2}$ th sample quantile of unique x_i 's for $s=1,2,\ldots,r$.

3.3 *Error Variance Estimation*

To measure the quality of regression parameter estimates by using OLS (Ordinary Least Square), it is necessary to assess the unknown variance of errors. The estimator for error variance is MSE (Mean Square Error). MSE obtained from Sum Square Error divided by error's degrees of freedom.

3.4 *Coefficient of Determination*

The coefficient of determination is a measure that states how considerable the contribution of independent variables in predicting the value of the dependent variable.

4. Numerical Example

The dataset used is credit card data of 400 customers with 24 months observation period, in which the data contains information about the occurance time of the event, customer status (default, attrition, or open status), credit line, monthly income, cedit score, average monthly debt, number of credit cards, gender, type of credit card (additional or major), relationship status (married or not).

4.1 Multinomial Logistic Regression Model

In the model, the event that will be used as a reference/basis is the open status event ($j = 0$), where x_{1i} represents credit line, x_{2i} represents income, x_{3i} represents credit score, x_{4i} represents average debt, x_{5i} represents the number of credit card, x_{6i} represents gender, x_{7i} represents the type of credit card, and x_{8i} represents the relationship status.

The parameters in multinomial logistic regression are estimated using the maximum likelihood method. The results of parameter estimation are shown in Table 1 as follows.

Table 1. Results of parameter estimation for default event and attrition event

Variable	Default Event			Attrition Event		
	Parameter	Coefficient	p-value	Parameter	Coefficient	p-value
Intercept	β_{01}	15,420	0,000	β_{02}	26,057	0,000
Credit line	β_{11}	0,135	0,002	β_{12}	-0,015	0,789
Income	β_{21}	-0,516	0,000	β_{22}	-1,092	0,000
Credit score	β_{31}	-0,029	0,000	β_{32}	-0,022	0,000
Average debt	β_{41}	0,122	0,397	β_{42}	0,076	0,790
Number of credit cards	β_{51}	0,791	0,057	β_{52}	1,677	0,000
Gender	β_{61}	0,440	0,502	β_{62}	-0,392	0,672
Type of credit card	β_{71}	1,964	0.121	β_{72}	0,257	0,813
Relationship status	β_{81}	0,661	0,406	β_{82}	0,157	0,845

For partial test can be seen in Table 1. at $\alpha = 0.05$, it can be seen that the independent variable has significant effect, that is, p-value $< \alpha = 0.05$. The independent variables that significantly influence the default event are credit line, income, and credit score. Then, the independent variables that significantly influence the attrition event are income, credit score, and number of credit cards.

The logistic regression model obtained is used to classify customers based on predictions of default events and attrition events. The degree of accuracy of the logistic regression model in the classification is calculated using the classification table shown in Table 2.

Table 2. The classification table

Observed	Predicted			
	Open Status	Default	Attrition	Percent Correct
Open Status	290	2	6	97.3%
Default	2	60	0	96.8%
Attrition	7	2	31	77.5%
Overall Percentage	74.8%	16.0%	9.3%	95.3%

The percentage of classification accuracy is calculated as the ratio between the number of observations classified precisely by the model to the total number of observations. Thus, the percentage accuracy of multinomial logistic regression model obtained in classifying customers based on the prediction of default and attrition is 95.3%.

4.2 Spline Regression Model

In this paper, spline regression is used to model the hazard functions for default and attrition events. In this numerical example, y is the dependent variable which represents hazard and t is the independent variable which represents time (in months). The degree of the polynomial on the truncated power basis used is d = 3 (cubic spline), and we will use the rules of location and number of knots that introduced by [7], then the required number of knots is $r = 6$.

Next, we will find the model of spline regression using the cubic truncated power basis with $r = 6$. For comparison, we also use the spline regression with $r = 5$ and $r = 7$. In order to know which regression model is better, it can be known through MSE and R^2 of each regression model as follows.

Table 3. MSE and R^2 of spline regression for default event

Type of Event	Number of Knots	MSE	R^2
	$r = 5$	0,0000128675	88,99%
Default	$r = 6$	0,0000112607	91,01%
	$r = 7$	0,0000143916	89,33%

From the three models above, the best model will be chosen based on the criteria of MSE and R^2, i.e., the smallest MSE and the largest R^2. Thus, the best model chosen is a spline regression model using cubic truncated power basis with 6 knots, with estimated models as follows.

$$\hat{y} = 0.0145062 - 0.0129881t + 0.0041186t^2 - 0.0003220t^3 + 0.0008372(t - K_1)_+^3$$
$$-0.0008378(t - K_2)_+^3 + 0.0004749(t - K_3)_+^3 - 0.0003707(t - K_4)_+^3$$
$$+0.0006080(t - K_5)_+^3 - 0.0009193(t - K_6)_+^3$$

The plots for spline regression using the cubic truncated power basis with $r = 6$ is shown on Figure 1 as follows.

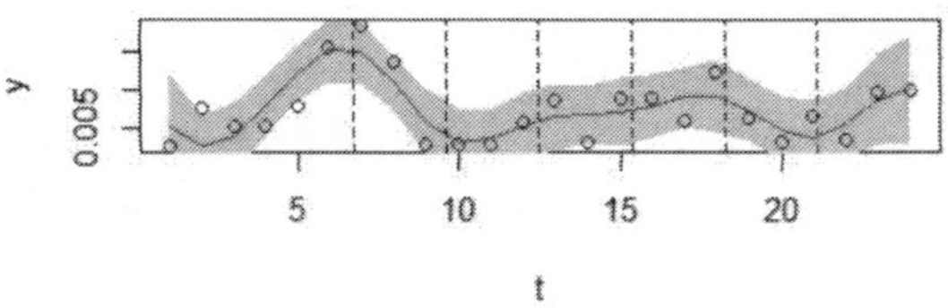

Figure 1. Cubic spline with $r = 6$ for default event

The interpretation of the best model chosen or in accordance with Figure 1, namely that the risk of default increased significantly until the 7th month, after which, there was a significant decrease as well until the 10th month. The gradual increase in risk occurs after the 10th month, followed by a decrease in the 16th month until the 19th month, and after that the risk increases back until the end of observation period.

For the attrition event, we get the models with results as follows.

Table 4. MSE and R^2 of spline regression for attrition event

Type of Event	Number of Knots	MSE	R^2
	$r = 5$	0,00000695429	86,72%
Attrition	$r = 6$	0,00000673517	87,99%
	$r = 7$	0,00000716070	88,15%

From the three models above, the best model will be chosen based on the criteria of MSE and R^2, i.e., the smallest MSE and the largest R^2. Thus, the best model chosen is a spline regression model using a cubic truncated power basis with 6 knots, with estimated models as follows.

$$\hat{y} = 0.01278 - 0.008524t + 0.002081t^2 - 0.0001407t^3 + 0.0002279(t-K_1)_+^3$$
$$+0.00008648(t-K_2)_+^3 - 0.0006001(t-K_3)_+^3 + 0.0008499(t-K_4)_+^3$$
$$-0.0006539(t-K_5)_+^3 + 0.0004447(t-K_6)_+^3$$

The plots for spline regression using the cubic truncated power basis with $r = 6$ is shown on Figure 2 as follows.

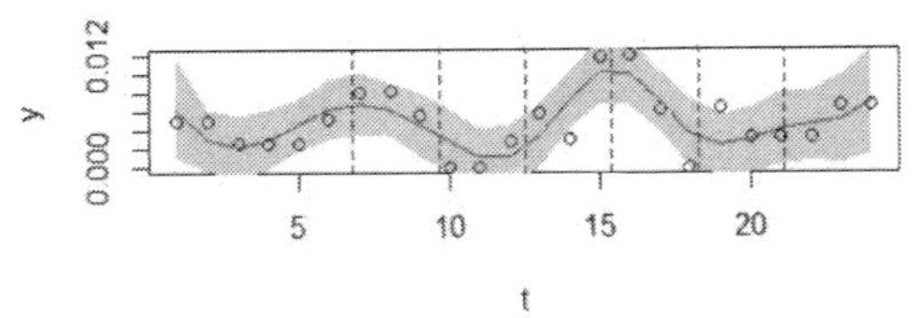

Figure 2. Cubic spline with $r = 6$ for attrition event

Interpretation of the best model chosen or in accordance with Figure 4.5, namely that the risk of attrition begins to increase after the 3rd month until the 8th month. After that, the risk of attrition always decreases until the 12th month, then experienced a significant increase risk until the 15th month. After the 15th month, the risk decreases again until the 19th month, and increases slowly until the end of observation period.

5. Conclusion

In the formation of a multinomial logistic model with categories which mutually exclusive and exhaustive, the parameters are represented in the logit function of category j compared to one of the J categories that serve as the reference/basis, so that in the formation of the model $J-1$ is required for the logit function. Based on numerical example in this paper, factors that significantly affect the default event are the credit line, income, and credit score. Then, the factors that significantly affect the attrition event are income, credit score, and number of credit cards. Presentation of accuracy of multinomial logistic regression model obtained in classifying customers based on the prediction of default and attrition is 95.3%.

The relationship between the dependent variable and the independent variable is represented by a curve called the regression curve. If the form of the regression curve is unknown, then the regression curve form can be estimated through the spline regression approach using the truncated power basis. Based on numerical example in this paper, best model chosen for default event and attrition event is spline regression model using cubic truncated power basis with 6 knots.

6. References

[1] Sirong L, Xiao K, and Tingting N 2016 *Spline Based Survival Model for Credit Risk Modelling* (Shanghai: Shanghai University of Finance and Economics)
[2] Baklouti I and Bouri A 2014 Can Credit-Scoring Models Effectively Predict Microloans Default? Statistical Evidence from the Tunisian Microfinance Bank pp. 57-69
[3] Wang J and Yang L 2009 Polynomial Spline Confidence Bands for Regression Curves *Statistica Sinica* **19**, pp. 325-342
[4] Härdle W 1990 *Applied Nonparametric Regression.* (New York: Cambridge University Press)
[5] Eubank R L 1999 *Nonparametric Regression and Spline Smoothing Second Edition* (New York: Marcel Dekker)
[6] Ngo L and Wand M P 2004 *Smoothing with Mixed Model Software* (Massachusetts: Harvard University)
[7] Nielsen D Jason 2007 *Splines Models for The Analysis of Recurrent Event Panel Data* (Columbia: Simon Fraser University)

Acknowledgements
This work is supported by Hibah PITTA 2018 funded by Directorate of Research and Community Service of Universitas Indonesia (DRPM UI) No.2308/UN2.R3.1/HKP/05.00/2018.

IOP Conf. Series: Journal of Physics: Conf. Series **1108** (2018) 012020 doi:10.1088/1742-6596/1108/1/012020

The Factors Affecting Maternal Mortality in Java: Spatial Modeling Application on Demographic Data

C H P Tatipatta[1], Y Widyaningsih[2] and S Nurrohmah[3]

[1,2,3]Departement of Mathematics, Universitas Indonesia, KampusBaru UI, Depok, 16424, Jawa Barat, Indonesia.

Abstract. Maternal deaths are considered very serious issue worldwide, according to World Health Organization (WHO) everyday approximately 830 women die from preventable causes related to pregnancy and childbirth. According to Indonesia's health profile in 2014, West Java, Central Java and East Java provinces contributed 26 percent of the total number of maternal deaths in Indonesia. The objective of this study is to model maternal deaths in West Java, Central Java, and East Java using Spatial Autoregressive Model (SAM) and Spatial Error Model (SEM), inspect the factors affecting maternal deaths, and compare the models. The results showed that number of health personnel, the number of residents received Raskin (BerasMiskin), and number of illiterate residents were significant factors that affect maternal deaths. Based on the result of the Moran's I test, spatial autocorrelation in the residual of linear regression model, and the response variable (maternal deaths) are significant. The likelihood ratio test showed that the spatial regression models were better than multiple linear regression model in predicting the maternal deaths. The AIC value of the OLS model is 243.1, SAM is 240.3, and SEM is 234.4. Based on the AIC values, SEM is better than the SAM in predicting maternal deaths.

1. Introduction

Maternal health problems occurring in a country are very serious, as maternal health is an important indicator in the development of the health sector to describe the state of health in a region. One of most important maternal health indicators are maternal mortality ratio. One of the Millennium Development Goals (MDGs) of Indonesia is to reduce maternal mortality ratio from 446 deaths per one hundred thousand live births to 110 deaths per one hundred thousand live births during the period 1990 to 2015.

The maternal mortality ratio of Indonesia has not reached the target of the established MGDs, where the ratio is 133 deaths per one hundred thousand live births and the maternal mortality ratio is higher compared to the Philippines, 117 deaths per one hundred thousand live births, Vietnam is at 54 deaths per one hundred thousand births life, Malaysia is at 41 deaths per one hundred thousand live births, Brunei Darussalam is at 23 deaths per hundred thousand live births, Thailand is at 21 deaths per one hundred thousand live births, and Singapore is at the ratio of 10 maternal deaths per one hundred thousand live births. That puts Indonesia as the country with the fourth highest maternal mortality ratio in Southeast Asia.

Maternal mortality ratio depends on the number of maternal death. Maternal death is the death of a woman while pregnant or within 42 days of termination of pregnancy, irrespective of the duration and site of the pregnancy, from any cause related to or aggravated by the pregnancy or its management but not from accidental or incidental causes. To facilitate the identification of maternal deaths in circumstances in which cause of death attribution is inadequate, a new category has been introduced:

Pregnancy-related death is defined as the death of a woman while pregnant or within 42 days of termination of pregnancy, irrespective of the cause of death. According Indonesia's health profile in 2014, West Java, Central Java, and East Java contribute 26 percent of maternal deaths in Indonesia.

Based on literature study, there are 5 variables suspected to affect maternal deaths. The first variable is the number of residents received Raskin (Beras Miskin) which is involved based on [5], that nutritional support provided by the government for the poor affects the number of maternal deaths occurring in an area. The second variable is the proportion of people that have access to clean water [4], that the availability of clean water affects the number of maternal deaths in an area. The third and fourth variables are number of health personnel and the number of births without health personnel [9], that the number of health personnel and the number of births without health personnel affect the number of maternal deaths in an area. The fifth variable is the number of illiterate residents [8], that the number of illiterate residents affects the number of maternal deaths in an area. Therefore, the objective of this study is to modelling maternal deaths using spatial autoregressive model and spatial error model, inspect the factors affecting maternal death, and determining the better model based on AIC values.

2. Study Area and Methods

2.1 Study Area

The study area is located in West Java, Central Java, and East Java, Indonesia with approximately 117,775km^2 (Figure 1). Total number of pregnant woman in the study area is approximately 2,225,662.

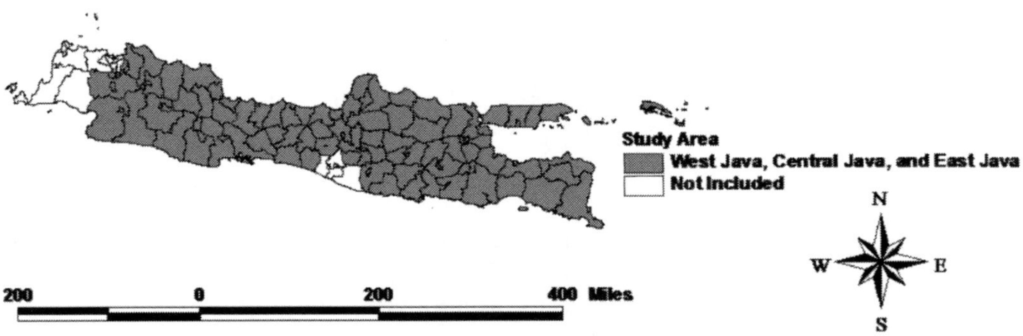

Figure 1. Study Area West Java, Central Java, and East Java

2.2 Methods

In this study, multiple linear regression is used to describe the relationship between the dependent and explanatory variable. There are four major assumption for regression analysis, the error term has zero mean, the error term has constant variance σ^2, the error are uncorrelated and the errors are normally distributed. When one or more of those major assumption are violated, we need to transform the dependent variables or explanatory variables. If the error is still correlated after the transformation, it is suspected that spatial relationship is exist. The spatial relationships between the dependent and the error terms were analyzed (modelled) and spdep library in *R* Software v.3.3.3 and *Geoda* v1.12.1.59 are used for computation. The output of the analysis is a single regression equation describing the relationship between the dependent and explanatory variables across the whole study area Montgomery [7]. The general model of multiple linear regression is:

$$\mathbf{y} = \mathbf{X}\boldsymbol{\beta} + \boldsymbol{\varepsilon}$$

where $\mathbf{y}$ is an $n \times 1$ vector of dependent (response) variable, $\mathbf{X}$ is an $n \times (k+1)$ matrix of explanatory variables, $\boldsymbol{\beta}$ is an $(k+1) \times 1$ vector of regression parameter, $\boldsymbol{\varepsilon}$ is an $n \times 1$ vector of error term, k is the number of explanatory variable, and n is the number of observations. The regression model in a matrix form is:

$$\begin{bmatrix} y_1 \\ y_2 \\ \vdots \\ y_n \end{bmatrix} = \begin{bmatrix} 1 & x_{11} & \cdots & x_{1k} \\ 1 & x_{21} & \cdots & x_{2k} \\ \vdots & \vdots & & \vdots \\ 1 & x_{n1} & \cdots & x_{nk} \end{bmatrix} \begin{bmatrix} \beta_0 \\ \beta_1 \\ \vdots \\ \beta_k \end{bmatrix} + \begin{bmatrix} \varepsilon_1 \\ \varepsilon_2 \\ \vdots \\ \varepsilon_n \end{bmatrix}$$

The parameter of $\boldsymbol{\beta}$ were estimated using *Ordinary Least Square* (OLS) method.OLS is a method to find the estimator of parameters of the linear regression by minimizing the sum of square error (SSE) of the linear regression model. The least square function is written as:

$$S(\boldsymbol{\beta}) = \boldsymbol{\varepsilon}'\boldsymbol{\varepsilon} \qquad (1)$$

by substituting $\varepsilon = (y - X\beta)$ to equation (1), obtained

$$S(\beta) = (y - X\beta)'(y - X\beta)$$
$$= \mathbf{y}'\mathbf{y} - 2\boldsymbol{\beta}'\mathbf{X}'\mathbf{y} + \boldsymbol{\beta}'\mathbf{X}'\mathbf{X}\boldsymbol{\beta}$$

and minimizing the least square function by using first partial derivative of $\boldsymbol{\beta}$,

$$\left.\frac{\partial S(\beta)}{\partial \beta}\right|_{\hat{\beta}} = 0$$

$$-2X'y + 2X'X\hat{\beta} = 0$$

$$X'X\hat{\beta} = X'y$$

$$\hat{\beta} = (X'X)^{-1} X'y$$

so, the estimator of $\boldsymbol{\beta}$ is :

$$\hat{\boldsymbol{\beta}} = (\mathbf{X}'\mathbf{X})^{-1}\mathbf{X}'\mathbf{y}$$

The response variable have a tendency to spatially auto correlated where the value of a variable in a location correlated to the value of the variable in the location around it, and the problem that usually arise from this condition is the error term of the linear regression is auto correlated. Therefore, spatial regression models were performed to improve the prediction of maternal death. Starting with a standard linear regression model, two spatial models can be distinguished, which is Spatial Autoregressive Model (SAR) and Spatial Error Model (SEM).

The general model for Spatial Autoregressive is

$$y = \rho W y + X\beta + \varepsilon$$
$$\varepsilon \sim NIID(0, \sigma^2 I_n)$$

and the general model for Spatial Error Model is

$$y = X\beta + u$$
$$u = \lambda W u + \varepsilon$$
$$\varepsilon \sim NIID(0, \sigma^2 I_n)$$

Where $\mathbf{y}$ denotes an $n \times 1$ vector of dependent variable, $\mathbf{X}$ is an $n \times p$ matrix of independent variable, $\boldsymbol{\varepsilon}$ is a normally distributed error, $\boldsymbol{\beta}$ is a $p \times 1$ vector of regression parameter, ρ is the coefficient of spatial lag, λ is the coefficient of spatial autocorrelation in error term, $\boldsymbol{WY}$ denotes the endogenous interaction effects among the response variable, and $\boldsymbol{Wu}$ denotes interaction effects among the error term.

3. Results and Discussion

The observed maternal death, proportion of people who have access to clean water, the number of health personnel, the number of resident who received Raskin, the number of illiterate residents and the number of births without health personnel in West Java, Central Java, and East Java are shown in Figure 2.

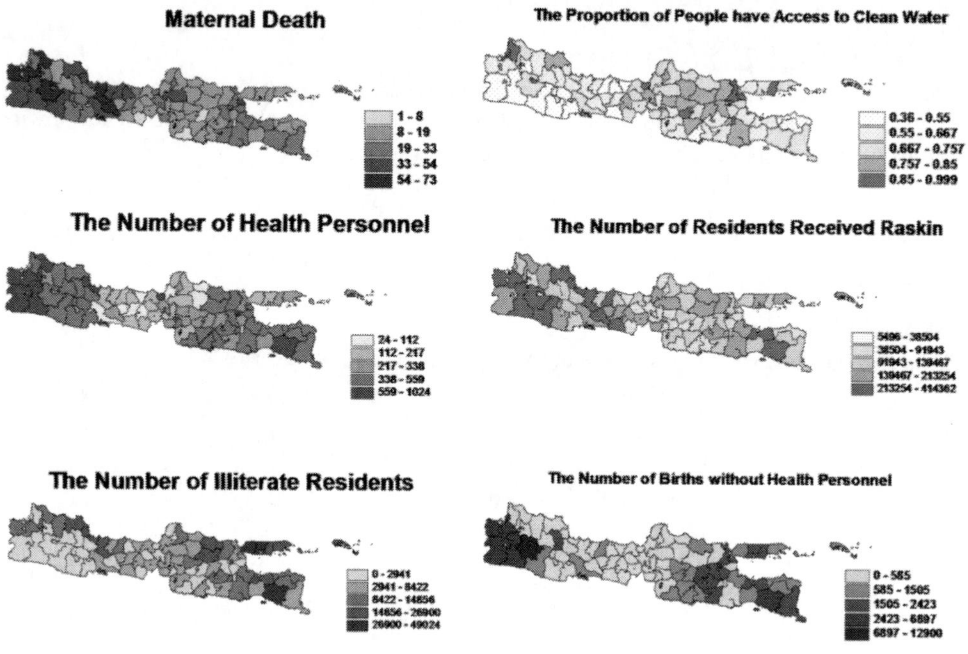

Figure 2. Map of Maternal Death and Explanatory Variables

The observed value of maternal death in West Java, Central Java, and East Java are vary from 1 to 73 with a mean of 21. Most of the area in the west and central area of Java have maternal death above the mean. The number of births without health personnel varies from 0 to 12900 with a mean of 982. Based on the map, the areas which have more number of births without health personnel have high maternal death. There are 14Districts/City that have only 36% – 55% people have access to clean water, while only 25Districts/City that have 55% – 66.7% people have access to clean water, 20Districts/City that have 66.7% –75.7% people have access to clean water, 20Districts/City have75.7% – 85% people have access to clean water, and 16 Districts/City have75.7% – 85% people have access to clean water (Figure 2). The map of the other land use variable can be seen in Figure 2.To determine the explanatory variables that have significant effect to the maternal death (*Y*), OLS model were used. The explanatory the number of health personnel (*HP*), the number of residents received RASKIN (*R*), proportion of people have access to clean water(*CW*), the number of births without health personnel (*NHP*), and the number of illiterate residents (*IR*). The proposed model is:

$$Y_i = \beta_0 + \beta_1 HP_i + \beta_2 R_i + \beta_3 CW_i + \beta_4 NHP_i + \beta_5 IR_i + \varepsilon_i$$

Table 1. Estimated Values and the Result of Residual Analysis

Variable	Estimation	SD	*t-value*	*p-value*	VIF
(Intercept)	1.628	7.977	0.204	0.83885	
HP	1.387×10^{-2}	6.282×10^{-4}	2.207	0.03038^{*}	1.277
R	1.555×10^{-4}	2.119×10^{-6}	7.341	$2.27 \times 10^{-10***}$	2.049
CW	-1.041	9.605	-0.108	0.91402	1.206

NHP	1.671×10^{-4}	7.615×10^{-4}	0.219	0.82689	1.360
IR	-4.115×10^{-4}	1.418×10^{-4}	−2.902	0.00489***	1.293
F-*statistics*	23.14			6.477×10^{-14}	
Adjusted R-Squared	0.5836				
Breusch-Pagan Test	20.443			0.001032	
Shapiro-Wilk	0.97051			0.0584	
Durbin-Watson	1.5173			0.007522	

The *F-statistic* on Table 1 show that the model is sufficient to predict maternal death (p-value = $6.477 \times 10^{-14} < 0.05$). The explanatory variable that significantly affect the maternal death at level 0.05 are the number of health personnel, the number of residents received Raskin, and the number of illiterate people. VIF test were performed to analyze the multicollinearity, and the result is all of the explanatory variables chosen have no multicollinearity problem (VIFs value < 10). Shapiro-Wilk, Breusch-Pagan, and Durbin-Watson test was performed to analyze the normality, homogeneity, and autocorrelation of the residual of the regression model. It is found that the residual have a heteroskedasticity problem (p-value = 0.001< 0.05) and auto correlated (p-value = 0.007522< 0.005).

Since the residual model is not normally distributed, one way to handle the misspecification of the model is to transform one or more explanatory variables or the dependent (response) variable using square root, log, arcsin, or inverse transformation. The best possible transformation is to transform the dependent variable (*Y*) using root transformation. The proposed model is:

$$\sqrt{Y} = \beta_0 + \beta_1 HP_i + \beta_2 R_i + \beta_3 CW_i + \beta_4 NHP_i + \beta_5 IR_i + \varepsilon_i$$

Table 2. Estimated Values and The Result of Residual Analysis After Root Transformation.

Variable	Estimation	SD	*t-value*	*p-value*	VIF
(Intercept)	2.815	8.028×10^{-1}	3.507	0.000775	
HP	1.639×10^{-3}	6.322×10^{-4}	2.593	0.011474*	1.26
R	1.427×10^{-5}	2.133×10^{-6}	6.691	3.68×10^{-9}***	1.96
CW	-6.666×10^{-1}	9.667×10^{-1}	−0.690	0.492653	1.20
NHP	-8.494×10^{-6}	7.664×10^{-5}	−0.111	0.912043	1.36
IR	-3.331×10^{-5}	1.427×10^{-5}	−2.334	0.022307*	1.26
F-*statistics*	21.47			3.287×10^{-13}	
AIC	243.135				
Adjusted R-Squared	0.5644				
Log Likelihood	−116.9259				
Breusch-Pagan Test	8.4468			0.1333	
Shapiro-Wilk	0.9831			0.3736	
Durbin-Watson	1.6037			0.02184	

After transformed the maternal death (*Y*), the residual of the regression model have no heteroskedasticity problem (BP = 8.44, p-value = 0.133> 0.05). However, the Durbin-Watson test result showed that the residual auto correlated even after transformed using root transformation (p-value = 0.0021< 0.05). It might caused by the existence of spatial autocorrelation in the data, as we can see in the Figure 2 that the area with a high value of maternal death have a tendency to adjacent to the area that have high value. The alternative solution is to modelling the data using spatial regression model.

The spatial auto correlated in the data was analyzed by using Moran's I test. The spatial weight matrix that was used in this study is queen contiguity where two areas are considered to have spatial relation one another if they share a side or a vertex. The weight matrix has 80×80 elements.

The Moran's I test was performed at the dependent variable, explanatory variables, and the residual of the regression model, and the result written in Table 3.

Table 3. Moran's I test result.

	Moran's I	$E[I]$	Mean	Sd	*z-value*	*p-value*
$\sqrt{Y}$	0.2643	−0.0128	−0.0142	0.0797	3.4941	0.001
Residual	0.2400	−0.0128	−0.0142	0.0853	2.9794	0.003

Based on Moran's I test result in Table 3, it is found that dependent variable as well as the residual of multiple linear regression model are spatially auto correlated (p-values < 0.05). We tried to model the data using spatial regression models that have been explained in Figure 2. The result is given in Table 4.

Tabel 4. Spatial Autoregressive Model and Spatial Error Model Test Result

Variable	Spatial Autoregressive (SAM)	Spatial Error (SEM)
(Intercept)	1.59871*	2.51665*
HP	0.0016*	0.0018*
R	1.387×10^{-5}*	1.37031×10^{-5}*
CW	−0.3475	−0.353
NHP	-3.1726×10^{-5}	9.18513×10^{-6}
IR	-2.6414×10^{-5}	-3.182×10^{-5}*
ρ	0.20458*	-
λ	-	0.4212*
LR-Test	0.02943*	00032*
Shapiro-Wilk Test	0.7654	0.2251
BP-Test	0.44623	0.30661
Moran's I	0.1163	0.4098
AIC	240.393	234.486
*: *p-value* < 0.05		

Likelihood ratio test was conducted to see if the spatial regression models is better than OLS to model the maternal death in West Java, Central Java, and East Java. Based on AIP value, the SEM model is slightly better than SAM. And the significant factors affecting maternal death in West Java, Central Java, and East Java are the number of health personnel, the numbers of residents received Raskin, and the number of illiterate residents.

The Spatial Error Model for the maternal death at i-th location is as the following.

$$\sqrt{Y} = 2.51665 + 0.0018{*}HP + 1.37031\times10^{-5}{*}R - 0.353{*}CW + 9.18513\times10^{-5}{*}NHP - 3.182\times10^{-5}{*}IR + 0.4212\sum_{j=1, i\neq j}^{80} w_{ij}u_j + \varepsilon_i$$

4. Conclusion

Maternal death in West Java, Central Java, and East Java is spatially auto correlated, make it difficult to construct the model using the ordinary least square (OLS). Since, the residual of the OLS is auto correlated even after the data was transformed, the alternative is spatial regression model. Based on the result SEM is better than SAM to model maternal death in West Java, Central Java, and East Java.

Based on SEM the explanatory variables that significantly affect maternal death are the number of health personnel, the number of residents who received Raskin, and the number of illiterate residents.

5. References

[1] Anselin, L 1988 *Spatial Econometrics: Methods and Models* (Dordrecht: Kluwer Academics Publisher)
[2] Cressie N A C 2015 *Statistics for Spatial Data* (New Jersey: John Wiley & Sons Inc)
[3] Hogg R V and Craig A T 1995 *Introduction to Mathematical Statistics: Fifth Edition* (Upper Saddle River, New Jersey: Prentice-Hall, Inc.)
[4] Benova L, Cumming O and Campbell O M 2014 Systematic Review and Meta-analysis: Association Between Water and Sanitation Environtment and Maternal Mortality *Tropical Medicine & International Health* **19** p 388
[5] Lanree-Abass B A 2008 Poverty and Maternal Mortality in Nigeria: Towards a More Viabel Ethics of Modern Medical Practice *International Journal for Equity in Health* **7** p 11
[6] Lee J and Wong D W S 2001 *Statistical Analysis with Arcview GIS* (New York: John Wiley & Sons, Inc)
[7] Montgomery D C, Peck E A and Vinning G G 2012 *Introduction to Linear Regression Analysis* (New York: John Wiley & Sons, Inc)
[8] Pillai V K, Maleku A and Fangshun W 2013 Maternal Mortality and Female Literacy Rates in Developing Countries during 1970-2000: A Latent Growth Curve Analysis *International Journal of Population Research* 163292
[9] Yego F, D'Este C, Byles J, Williams J S and Nyongesa P 2014 Risk Factors for Maternal Mortality in a Tertiary Hospital in Kenya: a Case Control Study *BMC Pregnancy Childbirth* **14** p 38
[10] Astuti W D, Khaqiqi Z and Lestari, D 2011 Perilaku Hidup Bersih dan Sehat (PHBS) dalam Rumah Tangga Ibu Hamil dan Ibu Pernah Hamil di Indonesia *Buletin Penelitian Sistem Kesehatan* **14** p 4

Acknowledgements
This research was funded by Directorate of Research and Community Service of Universitas Indonesia (DRPM UI) as a grant of PITTA (Publikasi Internasional Terindeks untuk Tugas Akhir) 2018.

IOP Conf. Series: Journal of Physics: Conf. Series **1108** (2018) 012021 doi:10.1088/1742-6596/1108/1/012021

Determination of factors associated with motor complications frequency in people with early parkinson's disease: bayesian method for zero-inflated poisson regression

A Awaliah[1], S Abdullah[2] and A Bustamam[3]

[1,2,3] Department of Mathematics, Universitas Indonesia, Depok, 16424, Indonesia

Abstract. Parkinson's disease (PD) is the second most common neurodegenerative disease worldwide that mainly affects motor system. Treatment given to PD patients may have further complications effect such as dyskinesias. People with PD receiving medication often experience complications. It is interest to identify factors associated with the complications. Data on 215 people with PD obtained from the Parkinson's Progression Markers Initiative (PPMI) database were analysed. Total scores for the Movement Disorder Society-Unified Parkinson Disease Rating Scale (MDS-UPDRS) Part 1, 2 , and 3 were used as the explanatory variables. We proposed Zero-Inflated Poisson (ZIP) to model the frequency of motor complications in people with PD. Therefore, the parameters in ZIP regression were estimated using Bayesian approach. Sampling from the posterior distribution of the parameters is conducted using Monte Carlo Markov Chain-Gibbs Sampling (MCMC-GS). The result shows that total score of MDS-UPDRS Part 1 and 2 are negatively associated with people for no need medication while the opposite is observed for total score of Part 3. Furthermore, in the second stage of the model, total score of Part 3 is negatively associated with frequency of complications, while the opposite trend is observed for total score of Part 1 and 2.

1. Introduction

Parkinson's disease (PD) is known as the second neurodegenerative disease in the world after Alzheimer's disease and affects 1% of the population aged 65 years and over. PD causes disruption of human motor system. In pathology, PD categorized as a disease caused by a dopamine reduction in the brain substance. These substances serve to organize and coordinate movements by the muscles [1].

According to the Parkinson's Foundation, the early symptoms of PD are the stiffness of the face, tremors in parts of fingers causing difficulty in writing, moving slowly with small steps, loss the ability of smelling, and often experience falls when standing or walking. The standard treatments to cure this disease have not found yet. Every Parkinson's sufferers have different treatment depending on the symptoms he felt. The treatment can be treated with medication or physical therapy.

Therapy given to patients with PD can cause more motor complications such as dyskinesias. This happens due to the influence of excessive medicines effects on patients with PD [7]. Data 215 patients with Parkinson's obtained from the Parkinson's Progression Markers Initiative (PPMI) database were analyzed.

The result showed that 144, or about 67% of patients do not experience motor complications and the rest suffer complications with different severity levels. The result is obtained based on the total score of count data from the test motor complications Movement Disorder Society-Unified Parkinson's Disease Rating Scales (MDS-UPDRS) Part 4. There were no complications demonstrated through zero response while complications are characterized by a response greater than zero. In other words, data has excess zero condition. Therefore, this aim research was to determined the factors are associated with the frequency of motor complications in PD using Zero-Inflated Poisson regression (ZIP). The parameters in ZIP regression were estimated using Bayesian approach.

2. Data and variables

The selected respondents for this study are Parkinson's patients with the following criterias.

- Early Parkinson's or Parkinson's patient is diagnosed with not more than 6 months.
- Level Hoehn and Yahr scale (HY) not more than 2.

The variables used in this research are following.

1. The response variable (Y)
   Y is total score of MDS-UPDRS Part 4 in the second year of treatment.
2. The predictor variables (X)
   - X_1 is total score of MDS-UPDRS Part 1. That is a test of non-motor experiences daily living such as dizziness, urinating problems, and so on.
   - X_2 is total score of MDS-UPDRS Part 2. That is a test of motor experiences daily living such as talking, eating, writing, and etc.
   - X_3 is total score of MDS-UPDRS Part 3. It assesses the motor sign of PD such as facial expression, rigidity, hand movements and etc.

3. Zero-inflated poisson (ZIP) regression model

ZIP regression splits the responses into two stages by identifying structural zeros at the first stage and the Poisson counts at the second stage. Zero response in Poisson counts is known as sampling zeros. Structural zeros are the response obtained from an object that the response prediction will always be zero, while sampling zeros are obtained as a response to sampling variability [8]. The first stage is done by assuming the probability of structural zeros as success (1) is $\boldsymbol{\pi}$ and probability the Poisson counts as fail (0) is $\mathbf{1}-\boldsymbol{\pi}$. The second stage is done by assuming the mean for Poisson counts is $\boldsymbol{\lambda}$. If $\boldsymbol{Y}$ is a random variable distributed ZIP $(\boldsymbol{\pi},\boldsymbol{\lambda})$ then the pdf of $\boldsymbol{Y}$ is

$$Pr(Y=y)=\begin{cases}\pi+(1-\pi)e^{-\lambda}, & y=0\\ (1-\pi)\dfrac{e^{-\lambda}\lambda^{y}}{y!}, & y>0.\end{cases} \qquad (1)$$

ZIP regression models have two sets of parameters corresponding to π and λ. There are α and β. The regression model with covariates associated with π is called the logistic regression function while dealing with λ is the Poisson regression function [6].

In many cases, researchers often can not determine the number of structural zeros or sampling zeros. If he knows the predictor variables that affect the zero response, he could define the logistic regression function and Poisson regression function by adjusting the independent variable information he obtained. Conversely, if the information about the variables that determine the response of structural zeros or Poisson counts, the researchers were able to define the function of a logistic regression equation and Poisson regression function using the same variables.

Let Y is variable random distributed of ZIP. $Y = (y_1, y_2, \dots, y_p) \sim ZIP(\pi, \lambda)$ by assuming y_i for $i = 1, 2, \dots p$ independent. Suppose that in the data, there are k variable predictors in first stage and l variables in second stage (Poisson counts). The regression model function for $Y \sim ZIP(\pi, \lambda)$ is

$$logit\ (\pi) = \ln\left(\frac{\pi}{1-\pi}\right) = \boldsymbol{X_1\alpha} = \alpha_0 + \alpha_1 X_{11} + \alpha_2 X_{12} + \dots + \alpha_m X_{1k} \quad (2)$$

$$\ln(\lambda) = \boldsymbol{X_2\beta} = \beta_0 + \beta_1 X_{21} + \beta_2 X_{22} + \dots + \beta_n X_{2l}, \quad (3)$$

where
π : probability for the structural zeros
λ : mean for the Poisson counts
$\boldsymbol{\alpha}$: regresion parameters in stage 1 (structural zeros)
$\boldsymbol{\beta}$: regresion parameters in stage 2 (Poisson counts)
X_{kj} : the j variable predictor at the stage k

4. Bayesian method
The main foundation of Bayesian method is Bayes theorem [2].

$$p(\theta|y) = \frac{p(y|\theta)p(\theta)}{p(y)} \quad (4)$$

Explanation:
- $p(\theta|y)$ is the posterior distribution which represents the probability of the parameter θ after given the data.
- $p(\theta)$ is a prior distribution which represents the probability of initial parameters θ before given the data.
- $p(y|\theta)$ is sampling distribution or likelihood. It represents the probability of the data y by assuming the parameter θ is true
- $p(y)$ is marginal likelihood that represents as summation or integral between prior and likelihod.

Equation (4) can be expressed in other forms as

$$p(\theta|y) \propto p(\theta)p(y|\theta) \quad (5)$$

or

$$posterior \propto prior \times likelihood. \quad (6)$$

The main challenge in Bayesian method is determining the prior distribution. Prior determination is influenced by the parameters of interest. Selecting the prior types are characterized by the value of the used variance. The more information about the parameters, the smaller the variance of the prior distribution. Conversely, the less information, the greater variance.

5. Modelling data with bayesian method for zero-inflated poisson (ZIP) regression
Based on information from the data we obtained, the first stage of a structural model of zeros can be defined as patients who are not taking medications yet so they will not experience motor complications. The model of the second stage of Poisson counts refer to patients who are taking medication. Zero responses on Poisson counts indicate that patients do not experience the

complications. The responses greater than zero indicate the frequency of motor complications. The greater the value of Poisson counts, the higher rate of complications suffered by patients.

5.1. Likelihood for counts data in ZIP model

Suppose Y is frequency motor complications variable random for ZIP distribution of 215 patients with early PD. $Y = (y_1, y_2, \dots, y_{215}) \sim ZIP(\pi, \lambda)$. Assuming independent and identically y_i for $i = 1, 2, \dots, 215$. Suppose that the observations contained m structural zeros responses and n Poisson counts, $m + n = 215$. Likelihood for $Y \sim ZIP(\pi, \lambda)$ as follow.

$$p(Y|\alpha, B) = \prod_{i=1}^{m} \left[\frac{e^{X_{1i}\alpha}}{1 + e^{X_{1i}\alpha}} + \left(1 - \frac{e^{X_{1i}\alpha}}{1 + e^{X_{1i}\alpha}}\right) e^{-e^{X_{2i}\beta}} \right] \times \prod_{i=1}^{n} \left[\left(1 - \frac{e^{X_{1i}\alpha}}{1 + e^{X_{1i}\alpha}}\right) e^{-e^{X_{2i}\beta}} \frac{\left(e^{X_{2i}\beta}\right)^{y_i}}{y_i!} \right]. \quad (7)$$

5.2. Prior distribusion for parameters α dan β in ZIP model

Without prior knowledge about the distribution of the parameter, we choose non-informative priors. In doing so, the role of the prior distribution is minimized and more weight is given to the data in determining the posterior distribution. Therefore, we determine normal distribution by assuming $\mu = 0$ and $\sigma^2 = 1000$. Prior distribution for $\alpha_j, \beta_j \sim \text{Normal}(0,1000)$ with $\pi = 22/7$ and $j = 0,1,2,3$ is

$$p(\boldsymbol{\alpha}, \boldsymbol{\beta}) = \prod_{j=0}^{3} \left[\frac{1}{\sqrt{2\pi}\sigma_{\alpha j}} e^{\frac{-(\alpha_j - \mu_{\alpha j})^2}{2\sigma_{\alpha j}^2}} \right] \times \prod_{j=0}^{3} \left[\frac{1}{\sqrt{2\pi}\sigma_{\beta j}} e^{\frac{-(\beta_j - \mu_{\beta j})^2}{2\sigma_{\beta j}^2}} \right]. \quad (8)$$

5.3. Posterior distribution for sampling parameters in ZIP model

Based on the likelihood and prior distribution we obtained in equation (7) and (8), the posterior distribution as follow.

$$posterior \propto prior \times likelihood$$

$$\propto \prod_{j=0}^{3} \left[\frac{1}{\sqrt{2\pi}\sigma_{\alpha j}} e^{\frac{-(\alpha_j - \mu_{\alpha j})^2}{2\sigma_{\alpha j}^2}} \right] \prod_{j=0}^{3} \left[\frac{1}{\sqrt{2\pi}\sigma_{\beta j}} e^{\frac{-(\beta_j - \mu_{\beta j})^2}{2\sigma_{\beta j}^2}} \right] \times \prod_{i=1}^{m} \left[\frac{e^{X_{1i}\alpha}}{1 + e^{X_{1i}\alpha}} + \left(1 - \frac{e^{X_{1i}\alpha}}{1 + e^{X_{1i}\alpha}}\right) e^{-e^{X_{2i}\beta}} \right] \times \prod_{i=1}^{n} \left[\left(1 - \frac{e^{X_{1i}\alpha}}{1 + e^{X_{1i}\alpha}}\right) e^{-e^{X_{2i}\beta}} \frac{\left(e^{X_{2i}\beta}\right)^{y_i}}{y_i!} \right]. \quad (9)$$

6. The results for modelling data by bayesian method for ZIP regression model

We apply algorithm of Markov Chain Monto Carlo-Gibbs sampling (MCMC-GS) with 200,000 iterations for sampling the parameters from the posterior distribution. The results shows in Table 1. The bolds typing show the variables associated with parameters affect the model significantly. The significant parameters is showed by no zero between 2.5 percentil and 97.5 percentil [5].

Table 1. The results of parameter estimates for 215 patients with early PD using algorithm MCMC-GS for ZIP regession model

Parameters	Mean	Std. Dev	2.5 Percentil	Median	97.5 Percentil
α_0	**0.648**	0.4283	0.1805	0.6457	1.496
α_1	-0.03436	0.03552	-0.1045	-0.0343	0.03457
α_2	**-0.1053**	0.03802	-0.1811	-0.1048	-0.013182
α_3	**0.0643**	0.01904	0.02789	0.06394	0.1027
β_0	**1.585**	0.139	1.308	1.585	1.856
β_1	**0.02077**	0.009477	0.002021	0.02073	0.03948
β_2	0.009587	0.01038	-0.0106	0.009499	0.03005
β_3	-0.005303	0.006328	-0.01766	-0.005336	0.007177

Based on Table 1, it can be seen that the total score of the MDS-UPDRS Part 2 and 3 associated with the frequency of motor complications at the first stage regression. It means, whether or not the consumption of medicines by patients is associated with these variables. While the total score of the MDS-UPDRS Part 1 associated with the second stage regression. This suggests that for patients taking the medicines, the variables associated with the frequency of experiencing motor complications. Thus, the logistic regression function and Poisson regression function for ZIP regression model is

$$logit\ (\pi) = \boldsymbol{X_1\alpha} = 0.648 - 0.03436X_{11} - 0.1053X_{12} + 0.0643X_{13}, \quad (10)$$

$$\ln(\lambda) = \boldsymbol{X_2\beta} = 1.585 + 0.02077X_{21} + 0.009587X_{22} - 0.005303X_{23}. \quad (11)$$

Logistic regression function in equation (10) indicates that the relation between the total score of MDS-UPDRS Part 1 and 2 inversely related with $logit\ (\pi)$. It suggest, the smaller the variable is, the greater probability for structural zeros. Thus the probability of patients do not experience the complications are increase. While the total score of the MDS-UPDRS Part 3 relates positively to $logit\ (\pi)$. That is, the greater the variable, the bigger the possibility of PD patients not suffering the motor complications. In equation (11), the correlation of total score MDS-UPDRS Part 1 and 2 are inversely related to $ln\ (\lambda)$. It shows that the smaller the value of the variables, the greater the probability of expriencing the frequency of complications. While the total score of the MDS-UPDRS Part 3 relates positively to $ln\ (\lambda)$. Thus, the greater the value of the variable then the frequency of experiencing complications are also getting bigger.

The full posterior distributions of the estimated regression coefficients from the ZIP regression model with covariates are displayed in Figure 1.

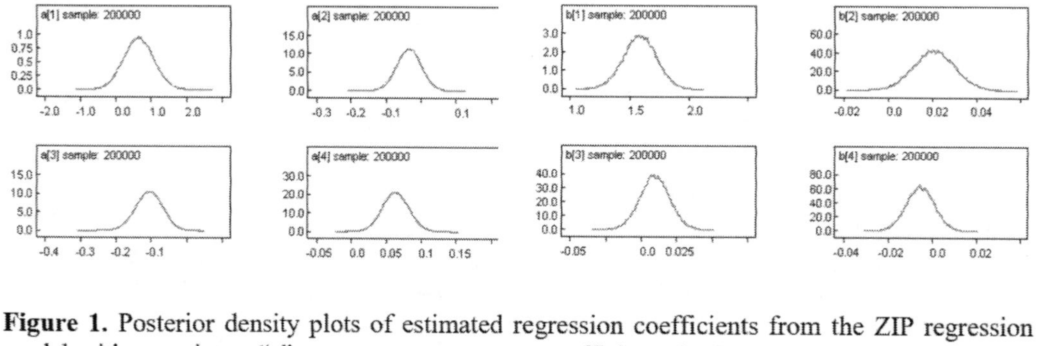

Figure 1. Posterior density plots of estimated regression coefficients from the ZIP regression model with covariates. "a" represents parameters coefficients in the structuran zeros stage and "b" represents parameters coefficients in the Poisson counts stage.

7. Conclusion

215 data with early PD Patients were analysed to determine factors that associated with frequency of motor complications. The conclusion of this reasearch is the patients who consume or not medicines are associated with total score of MDS-UPDRS Part 2 and 3. As for the patients consumethe medicines, the frequency of motor complications associated with total score of MDS-UPDRS Part 1.

8. References

[1] Gibrat C et al 2009 *Differences between Subacute and Chronic MPTP Models: Investigation of Dopaminergic Neural Degeneration and α-synuclein Inclusions* (Journal of Neurochemistry: no 109) p 1469-1482
[2] Hoff P D 2009 *A First Course in Bayesian Statistical Method* (New York: Springer)
[3] Hogg R V and Craig A T 1995 *Introduction to Mathematical Statistics* (New Jersey: Pearson Prentice Hall)
[4] Lambert D 1992 *Zero-Inflates Poisson Regression with an Application to Defects in Manufacturing (*Technometrics vol 34 no 1*)* p 1-14
[5] Liu H and Powers D A 2012 *Bayesian Inferences for Zero-Inflates Poisson Regression Model* vol 7 no 2 (Journal of Statistics: Advances in Theory and Applications) p 1-34
[6] Montgomery D C et al 2012 Introduction to Linear Regression Analysis (New Jersey: John Wiley and Sons)
[7] Parkinson's Progressive Markers Initiative https://www.ppmi.org
[8] Tang W et al 2014 *Structural Zeros and Zero-Inflated Models (*Journal of Mathematics vol 3*)*

9. Acknowledgments

This research is funded by Universitas Indonesia via PITTA 2018.

IOP Conf. Series: Journal of Physics: Conf. Series **1108** (2018) 012022 doi:10.1088/1742-6596/1108/1/012022

Analysis of Student's Understanding of Exponential Concept: A Perspective of Pirie-Kieren Theory

N Syafiqoh[1*], S M Amin[2], and T Y E Siswono[2]

[1] Mathematics Education Program, Postgraduate Program, Universitas Negeri Surabaya, Jl. Ketintang, Surabaya 60231, Indonesia
[2] Mathematics Education Program, Postgraduate Program, Universitas Negeri Surabaya, Jl. Ketintang Surabaya 60231, Indonesia

*Corresponding Author: nursyafiqoh16070785028@mhs.unesa.ac.id

Abstract. Solving mathematical problem is frequently used by employing exponential concept, such as logarithm. The aim of the research is to analyze students' understanding of exponential concept based on Pirie-Kieren theory. The participant of the research is one female student and one male student at the 11th grade. The data collection was conducted by giving two tasks and interviewing with semi-structure interview. The result showed that both students can solve the first task systematically with different process. The processes are correct although the last expression is not simple. For primitive understanding level, they had exponential basic knowledge. They could explain exponential definition by their own statements. At image making and image having level, they had different ideas to solve problem and they were able to deliver their solutions. At property noticing level, they had competences to identify and differentiate the exponential properties for multiplication and division. At formalizing level, they are able to express these properties to mathematics symbol. However they had no algebraic ability for proving exponential properties. At the second task, they couldn't reach observing level. They answered the problem incorrectly. Therefore, it can be concluded that the female student and male student's understanding have already reached formalizing level in the exponential concept.

1. Introduction

Mathematics is one of science which has important role in daily life. Therefore, mathematics has been taught since elementary school. In learning mathematics, students need understanding. National Council of Teachers of Mathematics (NCTM) stated that the students must learn mathematics with understanding, actively building new knowledge from the experience and the prior knowledge [1].According to Hiebert and Carpenter's explanation, the goal of mathematics education is to ensure learning with understanding [2]. Similarly, Kastberg explained that there are four important points about understanding. First, the goal of mathematics teaching is students' understanding. Second, students' understanding of mathematical concept exists in their mind. Third, what is in the mind of students cannot be known certainly but the workings of their minds can be concluded through working that is showed by evidence. Fourth, when students try to solve mathematical problems, they try to understand it by their own way [3]. In fact, there are still found many difficulties of students in solving mathematics problems, such as geometric problems [4], fraction problems [5], exponential problem [6], and so on. In addition, Confrey and Smith's research showed that student's understanding of exponential concept is quite limited [6]. It is caused by a lack of student's understanding about exponential concept. A common mistake in this concept is that the student thinks that the exponent is two values of the factors, for example $5^3 = 5 \times 3$ [7]. Exponents can provide a convenient shorthand notation. The symbol a^m represents the product of m factors of a [8]. Moreover, the students also often forget the exponential properties shortly after they have studied and rarely be able to explain why these properties are true [6]. How such understanding occurs can be identified to know whether students understand or not [9]. A growth understanding theory from Pirie & Kieren explains that mathematical understanding process is illustrated like embedded rings (figure 1) [10]. There are eight levels in this theory, namely primitive knowing, image making, image having, property noticing, formalizing, observing, structuring, and inventising.

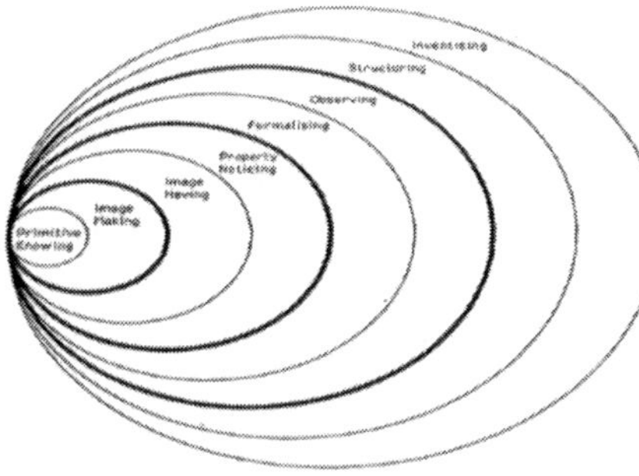

Figure 1. Growth of mathematical understanding on Pirie-Kieren model

1.1. Pirie Kieren's Understanding Model
Firstly, Pirie &Kieren defined mathematical understanding by developing a definition of understanding by Von Glasersfeld which stated that the construction of individual knowledge is needed during understanding process [11]. Pirie & Kieren revealed that the growth of mathematical understanding is as a whole, dynamic, levelled but non-linear, transcendently recursive process [10]. There are eight understanding levels within the growth of understanding. The first is primitive knowing. Primitive knowing is a level of students' understanding in using their preliminary knowledge to understand a new concept. The second level is image making. In this level, students try to make an image of the concept by using their prior knowledge with physical activities. Wright explained that the images can be pictures, symbols, words, or memories of events [12]. The next level is image having. In this level, students use a mental construction about a topic without having to perform certain activities that lead to the topic. It means that students have had ideas or images in their mind. The fourth level is property noticing. In this level, students use the ideas that had been obtained to establish the relevant properties about a topic. Formalizing is the next level. Students generalize what he has gained in the previous level into formal definition or algorithm about the topic. The sixth level is observing. Observing is a level where students coordinate formal activities on the formalizing level so they will be able to use it on the related problems faced. The next level is known as structuring. Structuring refers to the student's ability in proving theorem with a logical argument. The last level is inventising. Inventising is a level of students understanding which had gained a full understanding of a concept and they can ask a question which can generate a new concept. The four inner levels are described as informal levels, whereas the four outer levels are described as formal understanding of student [11]. Based on this theory, the growth of understanding does not always work in the same direction. A person understands often returns to the previous level to advance to the next level of understanding. The process is called by folding back.

1.2. Purpose of the study
The study was carried out to address the following question: how is the student's mathematical understanding level based on Pirie-Kieren theory in the exponential concept? Thus, based on the research question, the aim of this research is to analyze 11th grade students' understanding of

exponential concept in female and male students based on Pirie-Kieren theory. The researcher aimed to observe the students' constructional development of the exponential concept restricted in exponential properties for multiplication and division.

2. Method

The descriptive method was used in the study. This method aims to describe students' understanding level based on Pirie-Kieren Theory about exponential concept. The participant of the research is one female student and one male student at 11th grade who had middle mathematical ability. The participants were selected based on the mathematical score of the report and the score should be less than five points. The researcher selected these students because they had learnt about exponential in 9th grade and 10th grade. The students were given a test focused on exponential concept especially exponential properties for multiplication and division and then they were given semi-structured task based interviews. As a result, this research is delivered descriptively based on test result and interviews of participants.

In this study, the research data were collected by giving a test about exponential concept to participants. The test consists of two tasks about exponential properties for multiplication and division. They were asked to simplify the exponential form $\frac{3^2x^4y^{-2}}{6^3x^2y^{-3}}$ and were asked to determine the values a and b in the equation of $a^b = 2^{22} - 2^{21}$. After they completed the test, they were given semi-structured task based interviews. The interviews were constructed to identify student's mathematical understanding based on Pirie-Kieren theory. The process of the observation was equipped with the digital audio recorder. The recorder captures all of information which is given by participants during interviews, so that there is no left information.

The test of students' results and the interview results were used as data sources. All data collected were analysed using descriptive method of Miles and Huberman. The stages for analyzing qualitative research data are data condensation, data display and conclusion drawing/verification. In the condensation stage, the data has been obtained from interview result which then selected, focused and simplified so that they will be appropriate with the purpose of research. After that, the next step is presenting the data in the form of narrative text and concerning understanding levels indicators described previously. Then, the data were interpreted for making conclusion.

3. Result and Discussion

Firstly, the participants were given a test consist of two tasks about exponential concept. The researcher supervised the participants during completing the test. After completing, they were given interviews based on test result. Students' understanding level was identified based on test result and task based interviews and then analyzed according to the Pirie-Kieren Theory. Participants of this research are Kevin (K) as a male student and Nina (N) as female student. Both participants are pseudonym.

3.1. Female student's understanding level

Student' primitive knowing of exponential concept was identified based on her test result and task based interviews. Nina (N) had basic knowledge about exponential such as multiplication and she could express what the meaning of exponential. The following dialogue which occurred between Nina (N) and Researcher (R) supports this finding:

R: Nina, can you tell me what is asked in the question number 1?
N: Simplify the exponential form (Reading the question)
R: What is your idea to solve this?
N: I solve this by using exponential properties (Explained her answer step by step systematically) based on exponential properties.
R: Okay, before we continue to the next problem, I want to know, what do you know about exponential definition?

N: The number which is multiplied by the number itself, for example $3^2 = 3\times3$
R: If a^n, what it means?
N: a is multiplied as much as n.

Based on the short conversation above, she understood about what is the meaning of exponential by using her own statement. Nina could solve the given test systematically. The final result is correct; however it is not simple expression. The figure below depicts Nina's answers. Image making level was showed based on students' test result are as in the following:

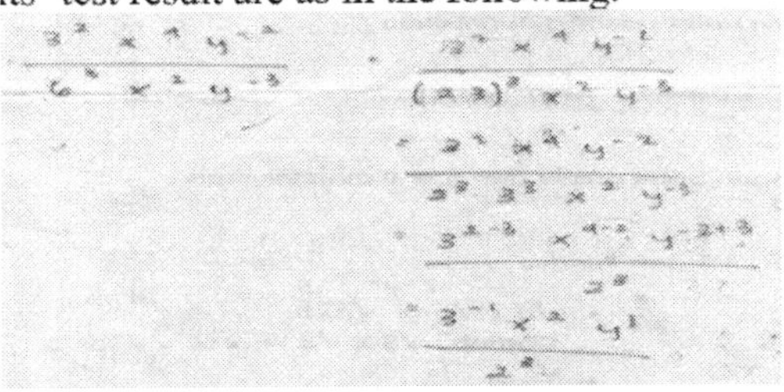

Figure 2. Nina's answer for the first task

In the figure 2, Nina spell out 6 became 2 times 3. She did it because there are numerators 3^2 so that she could use the exponential property of the division to simplify the exponential form. In this level, student is able to acquire an idea to solve the given problem. It is important because the idea helps them early to solve the problem and make them to think continually during the working process. So, this level is very important because it is the key for mathematical understanding [13].Similar with this, Borgen revealed that image making is an important and active process for the growth of understanding [14]. Image having level can be showed based on conversation between student and researcher above. The dialogue above indicated that Nina knew what the meaning of the question is. She knew what she should do for solving the problem. She also could explain her answer step by step systematically and clearly. She solved the given problem by using exponential property of division. She had has images in her mind for solving the problem by using exponential properties. In the property noticing level, she could distinguish about multiplication and division properties of exponential form. She told that the multiplication property of exponential form is by adding the exponents with notes of both bases which are actually the same, whereas for division of exponential form property is by subtracting the exponents with notes of both bases which are the same. For the formalizing level, she could generalize the exponential properties of multiplication and division into the mathematical symbols. She expressed the exponential properties for multiplication as $a^m \times a^n = a^{m+n}$ and for division as $a^m : a^n = a^{m-n}$. According to Skemp, an ability to connect mathematical symbolism and notation with relevant mathematical ideas and to combine these ideas into chains of logical reasoning is called by formal understanding or logical understanding [15]. For checking the ability for proving the properties, the researcher asked Nina to prove it. She didn't prove it algebraically, but she just showed that the value of equation in the left side is equal to the right side. The second problem was given to see students' understanding on the observing. Based on her answer below, it showed that she couldn't solve it correctly. She also didn't reach the highest level because she was incapable to create new questions that lead to new concept.

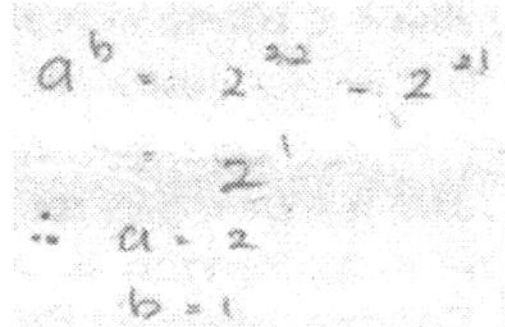

Figure 3. Nina's answer for second task

3.2. Male student's unders

Like Nina, Kevin had basic knowledge in completing the test. He could state the meaning of exponential in different statement compared to Nina. He gave an example about the meaning of

IOP Conf. Series: Journal of Physics: Conf. Series **1108** (2018) 012022 doi:10.1088/1742-6596/1108/1/012022

exponential. He expressed that $2^3 = 2\times2\times2$. He said that if a^n, it means a times a times a until n. He solved the problem in different way than Nina. He solved the first task systematically but the final result is not simplest expression. The answer can be seen as follows.

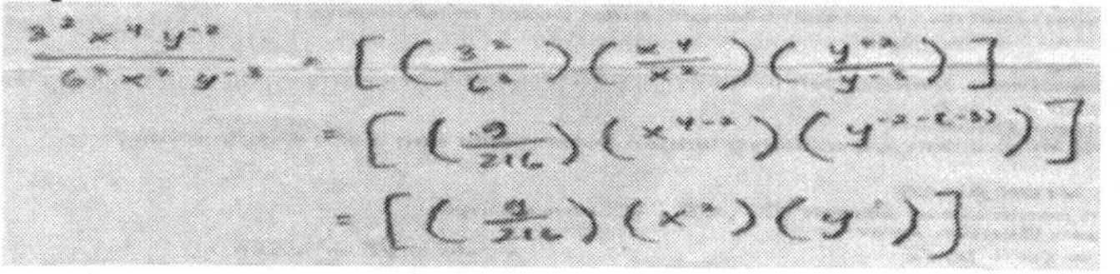

Figure 4. Kevin's answer for first task

In the figure 4, Kevin separated the coefficients, variables of x and variables of y in each bracket. He used exponential definition for outlining the coefficient and used exponential properties for simplified x and y variables. But, the final result is not the simplest expression. His final answer is $\frac{9}{216}x^2y$. He should simplify $\frac{9}{216}$, but he did not. For image having, it is showed in the excerpt of dialogue between Researcher (R) and Kevin (K) as follows:

R: What is asked in the question?
K: Simplify the exponent
R: What is your idea to solve this?
K: I think it is must be separated between coefficient, x variables and y variables. If the denominator is changed its position upwards, so the sign of the power is also changes. Positive becomes negative. Negative becomes positive.

If we consider Kevin's answer, he solved the problem by using exponential property of division. Actually, according to the conversation between Researcher and Kevin above, the image in his mind is not about exponential properties in solving the problem. But, rather than sign the power. If denominator is changed become numerator, the sign of power is also converted into the opponent's sign, positive becomes negative and vice versa. However, his answer was correct and he could explain his answer step by step systematically and clearly. When he was asked about exponential properties of multiplication and division, he could explain the difference between multiplication and division of exponential form properties. Like Nina, Kevin also said that the exponential properties for multiplication and division in the same expression. However, he couldn't prove why these properties are true, but rather applying the formula. So, he had no ability to prove them by providing a logical argument. For the second problem, he couldn't solve it correctly like the figure below.

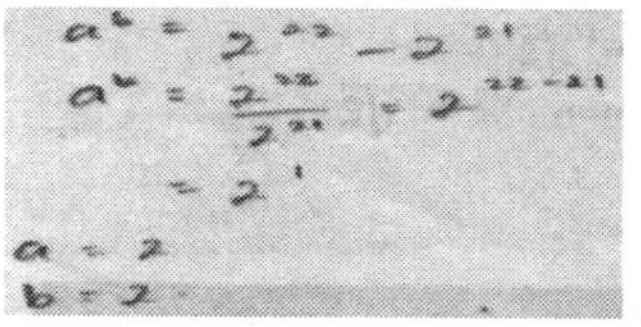

Figure 5. Kevin's answer for second task

4. Conclusion

Based on data analysis, the female student and male student could solve the first task systematically. The procedure of the solution is correct but the final expression is not simple. They had basic knowledge about exponential concept by explaining exponential definition in their own statements. Although they explained in different ways, their intentions were the same. They understood what the meaning of exponent is. For image making level, both students were able to acquire an idea to solve the given problem with different process and they could explain their solutions clearly. So they had indicator for image having. For the property noticing, they could identify exponential properties for multiplication and division. They also could distinguish the properties of multiplication and division in the exponential form. For the formalizing level, they could generalize the exponential properties of multiplication and division into the mathematical symbols. When the researcher asked both students for proving the exponential multiplication and division properties, they had no ability to prove it by providing a logical argument. For the second task, both students couldn't solve it. Therefore, it can be concluded that the female student and male student's understanding have reached formalizing level in the exponential concept. This study is useful for researchers who are interested on student's mathematical understanding based on Pirie-Kieren theory. In the next study, it is suggested that the next research should more communicative and more detail in exploring students' understanding level. It is caused by student's understanding which is not written. In addition, this study result is useful for teachers for planning their exponential lesson.

5. References

[1] National Council of Teachers of Mathematics (NCTM) 2000 *Principles and Standards for School Mathematics* (Reston, VA: NCTM)

[2] Hiebert J and Carpenter T P 1992 *Learning and teaching with undesrtanding* (Virginia: National Council of Teachers of Mathematics)

[3] Kastberg S E 2002 *Understanding Mathematical Concepts: The Case of the Logarithmic Function* (Dissertation: The University of Georgia)

[4] Shoimah R N, Lukito A, and Siswono T Y E 2018 The creativity of reflective and impulsive selected students in solving geometric problems *J. of Phys: Conf. Ser.* **947** 1-6

[5] Mamonto K, Juniati D, and Siswono T Y E 2018 Understanding fraction concepts of Indonesian junior high school students: A case of field independent and field depenent students, *J. Phys:Conf. Ser.* **947** 1-7

[6] Weber K 2002 Developing students' understanding of exponents and logarithms *Proc. annual meeting* **1**(4) 2

[7] Walle J A, Karp S K, and Williams J M B 2013 *Elementary and middle school mathematics teaching developmentally* (USA: Pearson)

[8] Musser G L, Peterson B E, and Burger W F 2014 *Matheamtics for elemantary teachers* (USA: Wiley)

[9] Gulkilik H, Ugurlu H, and Yuruk N 2015 Examining student's mathematical understanding of geomteric transformations using the pirie-kieren model *Edu. Sci.: Theo. & Pract.* **15** 1531

[10] Pirie S and Kieren, T 1994 Growth in mathematical understanding: how can we characterise it and how can we represent it? *Edu. Studiesin Math* **26** 165-190

[11] Sengul S and Yildiz, S G 2016 An examination of the domain of multivariable function using the pirie-kieren model*Soc. and Behav. Sci.* **197** 1263-1270

[12] Wright V 2014 Frequencies as proportions: using a teaching model based on pirie and kieren's model of mathematical understanding*Math. Ed. Res. J.* **26** 104

[13] Mokwebu D J 2013 *An Exploration of the Growth in Mathematical Understanding of Grade 10 Learners* (Polokwane: University of Limpopo)

[14] Simon M A 2006 Key developmental understandings in mathematics: a direction for investigating and establishing learning goals *Mathematical thinking and learning* **8**(4) 359-371

[15] Skemp R R 1987 *Psychology of learning mathematics* (New Jersey: Lawrence Erlbaum Associates)

Peat Endogenous Lignocellulolytic Bacteria for Humic Waste Decomposition

F Solikhah[1,*], W Assavalapsakul[2], and E Zulaika[1]

[1] Department of Biology, Institut Teknologi Sepuluh Nopember, Surabaya 60111, Indonesia
[2] Department of Microbiology, Chulalongkorn University, Bangkok, Thailand

*Corresponding author: enny@bio.its.ac.id

Abstract. The low carrying capacity of the peat soil makes a consideration to accelerate peatland decomposition through the addition of endogenous lignocellulolytic bacteria. These bacteria can use nutrients found in the peat soil and secrete enzymes to accelerate peat decomposition. In this study, endogenous lignocellulolytic bacteria from Palangkaraya peat soil, Kalimantan was used in lab scale decomposition experiments. These bacteria were added into the peat soil and incubated for six weeks. Analysis of peat fiber decomposition was performed weekly and at the end of the incubation period, bacteria colony was counted using total plate count method. The results showed that all bacteria can decomposed peat soil with an average value of 86%. The number of bacterial colonies at the end of incubation ranges from 3.0×10^3 to 6.7×10^4 CFU/ml. This study can be used as a preliminary study of peat soil decomposition that subsequently can be developed through analyzing several factors to stabilize peats oil area.

1. Introduction

The increasing development causes many buildings to be built on poor soil conditions, such as on peatlands. One of the largest peatlands in Indonesia is Kalimantan which has 4.78 million hectares of peatland [1]. Peat soil is known as a soil that has high pores and water content so that the carrying capacity and compression ability are very low [2]. Most of the peat soils are formed from the incomplete decomposition of various plant materials in the marsh environment. In general, the constituent of plant cell walls on undecomposed peatland contains lignocellulosic complex, comprising cellulose, hemicelluloses, and lignin [3]. Various methods of peat decomposition to improve soil stability have been developed, including mechanical and stabilization methods through the addition of chemical substances [4]. However, the improvement of peat soils by these methods can have long-term environmental damage. The alternative way is using the environmental-friendly biological method, through the addition of endogenous lignocellulolytic bacteria. The capability of lignocellulolytic bacteria to decompose peat fiber can reduce the fiber content. Decomposition leads to changes in the structure, number and size of peat fiber [5]. Several lignocellulolytic bacteria have been identified are genus*Cellulomonas, Pseudomonas, Bacillus, Clostridium,* and *Fibrobacter* [6,7]. This study aims to determine the percentage of peat soil decomposition by lignocellulolytic bacteria and the viability of bacteria after decomposition.

2. Material and Methods

2.1 Experimental bacteria

The lignocellulolytic screening had been performed and the screened isolates were isolated D1, D2, D3, U2, U3, and U4. Furthermore, each isolate was applied to decompose peat soil.

2.2 Inoculating the bacteria culture

A loop of each lignocellulolytic bacteria respectively inoculated into 25 ml Nutrient Broth and incubated on a rotary shaker (100 rpm, 24 h) at room temperature. Each of 10 ml bacterial culture then inoculated into 90 ml of NB medium and incubated on a rotary shaker (100 rpm,7 h) at room temperature until the bacterial density reaches 10^6-10^8 cells/ml. Two ml of lignocellulolytic starter bacteria were inoculated into 20 grams of peat soil homogeneously in a petri dish, closed and incubated at 30^{0}C for 6 weeks. This experiment was performed on 4 times replication.

2.3 Peat soil decomposition test

Twenty grams of treated peat soil were dried in the oven (105^{0}C, 5 h) and weighed as initial dry biomass. A total of 20 grams of treated peat soil was added by 100 ml of 5% sodium hexametaphosphate $(NaPO_3)_6$ solution as dispersing agent, stirred evenly, incubated for 3 hours and filtered (mesh size 0.15 mm). The soil was washed with tap water until the water passing through the sieve was clear. Then, the filtered soil was immersed in 2% HCl solution for 2 minutes to remove the remaining carbonate ions, washed again with tap water for 5 minutes to remove the remaining HCl. The remaining filtrate was placed on Whatman paper no. 4, washed through the funnel and dried by the oven at 105^{0}C for 5 h. The peat fiber biomass minus the mass of filter paper is recorded as the fiber dry biomass.The percentage of soil decomposition was calculated on the basis of the ratio of fiberdry biomass to the initial dry biomass of the soil, can be written as the following equation (1):

$$\text{Percentage of decomposed soil and fiber} = \frac{\text{(initial dry biomass-fiber dry biomass)}}{\text{initial dry biomass}} \times 100\% \quad (1)$$

[8].

2.4 Viability test

The viability test was used to determine the bacterial survival during the incubation period. The viability test was performed during the sixth week of the incubation period. A total of 1 gram of soil was added into 9 ml of physiological saline solution (NaCl 0.85%) and performed serial dilutions up to 10^{-5}. Each dilution was taken 0.1 ml and inoculated on Nutrient Agar by spread plate method, incubated for 24 hat 37^{0}C and counted the growing colonies [9].

3. Results and Discussions

3.1 Decomposition of peat soil by lignocellulolytic bacteria

The ability of each lignocellulolytic bacteria to decompose peat soil is shown in Table 1 and Figure 1.

Table 1. Percentage of peat soil decomposition by lignocellulolytic bacteria

Isolate code	Percentage of decomposed soil and fiber (%)					
	Week 1	Week 2	Week 3	Week 4	Week 5	Week 6
D1	83.24	81.07	85.37	86.87	86.94	89.25
D2	86.46	84.04	84.66	88.64	89.54	88.90
D3	86.4	82.2	88.81	85.47	88.86	88.65
U2	79.11	88.88	85.19	84.38	85.91	89.56
U3	84.48	86.78	87.25	88.41	88.71	87.97
U4	81.43	88.86	87.03	87.30	87.82	88.23
Control	73.04	74.29	83.21	80.23	83.41	84.65

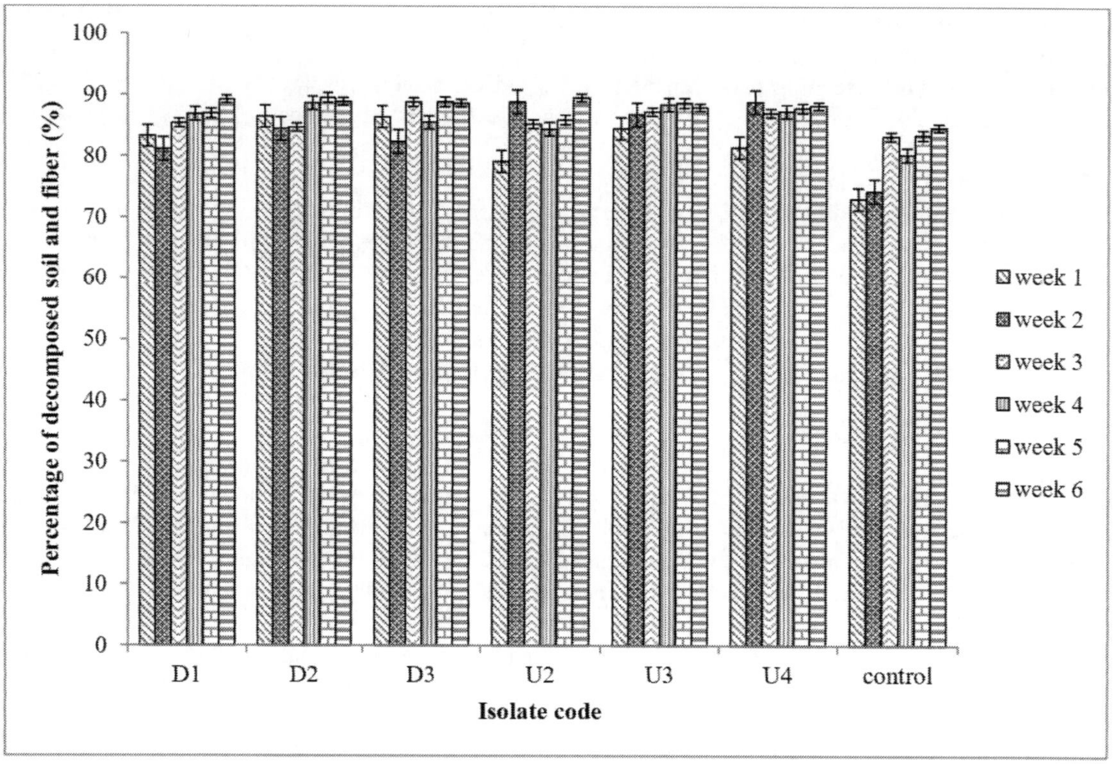

Figure 1. Percentage of Decomposed Peat Soil And Fiber on Each Lignocellulolytic Bacteria for 6 Weeks Incubation Period

The result showed that all the bacteria are able to decompose peat soil faster than the control. However, the soil decomposition charts tend to fluctuate for six weeks of incubation. During six weeks of incubation, the percentage of soil and fiber degradation increased, ranging from 2-10%, depending on each bacteria. The highest decomposition percentage was achieved by isolate U2 that can decomposed89.56% of peat soil on the sixth week, followed by isolate D1 that showed 89.25% decomposition rate. Control indicated a slight decomposition activity caused by endogenous bacteria in the peat soil without any addition.

Decomposition of plant material is an on-going process, driven by microorganisms that use the decaying organic matter as an energy source and building material [10,11]. The degree of peat decomposition showed the quantitative ratio of dark amorphous matter, consisting of humic compounds and the other products of the plant. However, in general, vegetative plant organs, especially leaves and stems, decompose rapidly, forming an amorphous peat mass [12]. In this study, high levels of decomposition were affected by experimental peat soil conditions that comprise of undecomposed plant material. The rate of decomposition of dry peat soils is relatively higher than wet peat[13]. The decomposition rate of fibrous peat can be accelerated by altering the carbon: nitrogen (C:N) ratio, pH, oxygen supply and temperature for optimum conditions [5].

3.2 Viability of lignocellulolytic bacteria

On the sixth week of the incubation period, the bacteria still remain in the peat soil, although the bacteria density decreased from the initial number (Table 2 and Figure 2).

Table 2. Number of Bacteria on The Early and End of Incubation Period

Decomposition medium	**Number of bacteria colony (CFU/ml)**	
	Early of incubation	End of incubation
1	61.8×10^7	2.9×10^4
2	58.1×10^7	1.63×10^4
3	64.7×10^7	2.87×10^4
4	46.5×10^7	6.7×10^4
5	46.8×10^7	6.2×10^4
6	58.4×10^7	4.24×10^4
Control	N/A	0.3×10^4

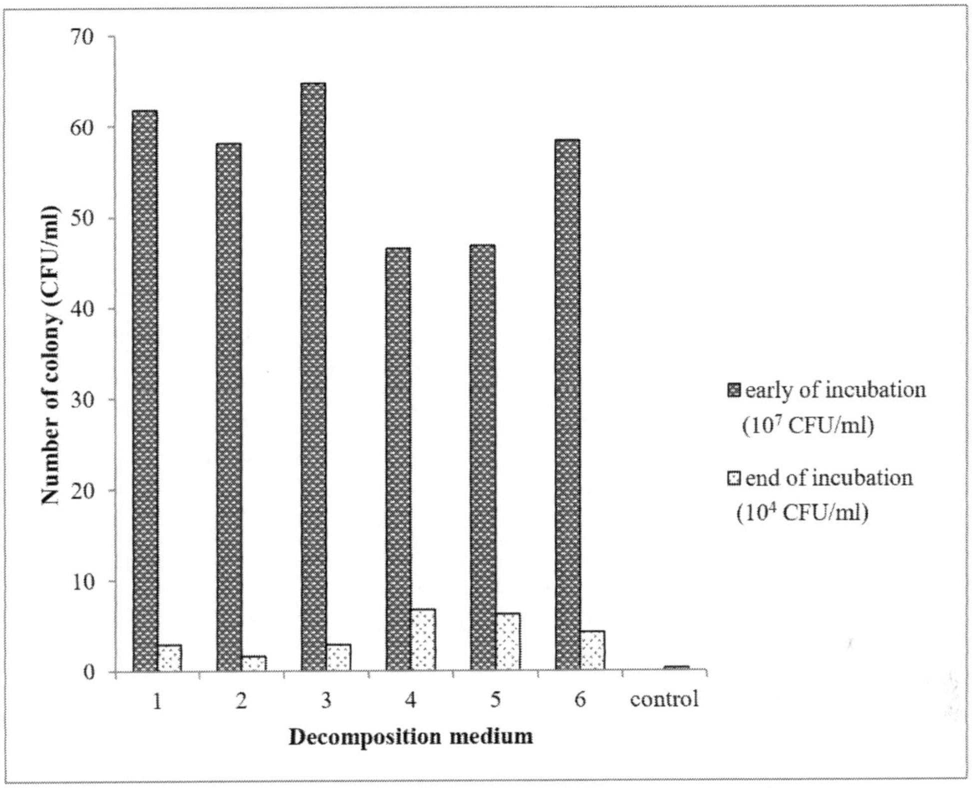

Figure 2. The viability of Each Bacteria on The Early and End of Incubation Period

The number of bacteria on the treated soil was higher than the control. The number of peat soil bacteria on the end of the incubation period ranging from 3.0×10^3 to 6.7×10^4 CFU/ml. At the end of incubation, the highest number of bacteria was found in medium 4 i.e, peat soil added by isolate culture U2, having 6.7×10^4 CFU/ml, followed by medium 5 i.e, peat soil added by isolate culture U3, showed 6.2×10^4 CFU/ml. The endogenous bacteria of control also showed the viability until the end of incubation, although the number of bacteria at the early of incubation was not calculated. Another research reported the mean microorganism counted for moderately decomposed *Sphagnum* peat by dry mass basis was 10^5 CFU/g after 10 days of incubation [14]. Similar findings have reported that the mean of microorganism measured for the natural peat are consistent with the value of 2.6×10^5 CFU/g dry peat [15]. The fewer number of bacteria on this research presumably because the available nutrients on the treated soil had already been exhausted during the incubation.

4. Conclusion
Peat undergoes continuous biological decomposition. This decomposition process is performed by lignocellulolytic bacteria that have been screened. All bacteria have the ability to decompose peat soils with an average decomposition value of 86%. After six weeks of the incubation period, the bacteria still viable, ranging from3.0 × 10^3 to 6.7 × 10^4 CFU/ml, includes the control. However, the application of these experimental findings in the field may not be straightforward and will require further studies in the field.

5. References
[1] Wahyunto, Nugroho K, Ritung S, and Sulaeman Y 2014 *Indonesia peatland map: method, certainty and uses* Conf. on Sustainable Management of Degraded Peatlands (Jakarta: Indonesian Center for Agricultural Land Resources Research and Development) pp 81-96
[2] Mochtar N E, Yulianto F E, and Satria T R 2014 The effect of stabilization age on fibered peatlands stabilized with a $CaCO_3$ and pozolan mixture *J. Civil Eng.***21(**1) pp. 57-64
[3] Souza W R D 2013 *Sustainable Degradation of Lignocellulosic Biomass – Techniques, Applications and Commercialization: Microbial Degradation of Lignocellulosic Biomass*, ed. A Chandel (InTech) chapter 9 pp. 208-247
[4] Yulianto F E and Harwadi F 2013(online)*Improvement Methods for Peatlands* (retrieved from http://ft.unira.ac.id)
[5] Pichan S P and O'Kelly B C 2012 Effect of decomposition on the compressibility of fibrous peat – a review *Int. J. of Geomechanics and Geoengineering***9**(4) pp 286-296
[6] Lynd L R, Weimer P J, Van Zyl W H and Pretorius I S 2002 Microbial cellulose utilization: fundamentals and biotechnology *Microbiology and Molecular Biology Reviews***66**(3) pp. 506-577
[7] Motta F L, Andrade C C P and Santana M H A 2013 *Sustainable Degradation of Lignocellulosic Biomass – Techniques, Application and CommercializationA Review of Xylanase Production by Fermentation of Xylan : Classification, Characterization and Application,*ed. Chandel A and da Silva S S (Shangai: InTech) pp. 251-275
[8] ASTM 1996 *Standard Test Method for Laboratory Determination of the Fiber Content of Peat Samples by Dry Mass* (United States: ASTM International) pp. 1-2
[9] Harley J P and Prescott L M 2002 *Laboratory Exercise in Microbiology 5th Edition* (USA: Mc Graw Hill)
[10] Kazemian S, Huat B B K, Prasad A and Barghchi M 2011 A state of art review of peat: geotechnical engineering perspective *International Journal of the Physical Sciences***6**(8) pp. 1974–1981
[11] Pankratov T A, Ivanova A O, Dedysh S N and Liesack W 2011 Bacterial populations and environmental factors controlling cellulose degradation in an acidic *Sphagnum* peat *Environmental Microbiology***13**(7) pp. 1800–1814
[12] Drzymulska D 2016 Peat decomposition – shaping factors, significance in environmental studies and methods of determination; a literature review *Geologos***22** (1) pp. 61-69
[13] Van der Linden M and van Geel B 2006 Late Holocene climate change and human impact recorded in a south Swedish ombrotrophic peat bog *Palaeogeography, Palaeoclimatology, Palaeoecology***240** pp. 649–667
[14] Pichan S P and O'Kelly B C 2013 Stimulated decomposition in peat for engineering applications *Proceedings of the Institution of Civil Engineers: Ground Improvement***166**(GI3) pp. 168-177
[15] Hunter P J, Petch G M, Calvo-Bado L A, Pettitt T R, Parsons N R, Morgan J A W and Whipps J M 2006 Differences in microbial activity and microbial populations of peat associated with suppression of damping-off disease caused by *PhythiumsylvaticumApplied and Environmental Microbiology***72**(10) pp. 6452–6460

Acknowledgements
The authors would like to express appreciation for the support of Department of Microbiology, Chulalongkorn University that has facilitated this research.

IOP Conf. Series: Journal of Physics: Conf. Series **1108** (2018) 012024 doi:10.1088/1742-6596/1108/1/012024

The improvement of students physics problem solving skills through the implementation of PO2E2W learning model assisted PhET media

A S D Sari*[1], B K Prahani[2], Munasir[3] and B Jatmiko[3]

[1] Department of Postgraduate Science Education, State University of Surabaya, Indonesia
[2] Department of Islamic Primary Teacher Education, Faculty of Tarbiyah and Teacher Education, State Islamic University of Sunan Ampel, Indonesia
[3] Department of Physics, Faculty of Matemathics and Natural Science, State University of Surabaya, Indonesia

*Corresponding Author: agungsucidiansari@gmail.com

Abstract. The purpose of this research was to improve students' physics problem solving skills through the implementation of PO2E2W learning model assisted PhET media on physics learning. This research used one group pre-test and post-test design at 64 students academic year 2017/2018. Students physics problem solving skills were measured by using Problem Solving Skills Assessment Sheet (PSSAS). Data analysis technique used Paired t-test, n-gain, and Independent t-test. The result of this research proved that: (1) there was improvement of student physics problem solving skill at $\alpha = 5\%$, (2) average score of n-gain physics problem solving skill 0.81 (high category), (3) there was no difference (consistency) n-gain students physics problem solving skills in all groups, and (4) students response were very positive toward the implementation of PO2E2W learning model assisted PhET media. Therefore the PO2E2W learning model assisted PhET media had been proven to be effective in improving the physics problem solving skills of senior high school students. The implication of PO2E2W learning model with PhET media study can be used as a solution to improve physics problem solving skills of senior high school students.

1. Introduction

The 21st century has just been running for a decade, but in the world of education, there has been a shift, and even a fundamental change in philosophy, its direction and its goals. It is no exaggeration to say that the progress of science is triggered by the advancement of science and computer technology. In the context of the utilization of information and communication technology, education world has proved to be more narrow and merging factor of space and time which has been the determinant aspect of speed and success of mastery of science by mankind. The problems which has been facing human in the 21st century are increasingly complex, interlocking hooks, rapidly changing and full of paradoxes. Generally, the futurists attribute the rapidly growing population of the world as a trigger. If in 2010 the world population is 6.9 billion, then, in 2050, the United Nations Population Division estimate at 9.2 billion people on world population. This means that, within forty years, there will be an increase of 2.5 billion inhabitants. The impacts of this growth on all human life are remarkable, ranging from survival, food, health, welfare, security and education issues [1]. In facing the challenges of the 21st century, one of the skills needed is problem solving skills. Someone who has problem solving skills can easily solve the problems that arise in his life. In addition, problem solving skills are needed in all professions and determine one's success in the future. A person who has problem solving skills will be more creative in thinking, having critical thinking skills in analyzing data, facts and information, and more independently in acting and working [2-3]. According to [4], the problem is a situation, quantitative or otherwise, faced by individuals or groups of individuals in need of solution.

Based on the results of preliminary study of problem solving skills on the material of kinetic theory of gas, especially the ideal gas law in class XII IPA 5 which amounted to 15 students at SMA Kemala Bhayangkari 1 Surabaya, the results obtained included: (1) improper understanding of problems related to ideal gas law (54.35%), (2) inappropriate in planning problem solving related to ideal gas law (56.52%), (3) improper in conducting problem solving plan related to ideal gas law (67.39%) and (4) incorrect in re-checking from problem solving related with ideal gas law (32.61%). Based on the results of these preliminary studies, it was necessary to have an effective learning model from which the student may improve their problem solving skills. Previous research [5] suggest that the use of developing a self-regulated learning PO2E2W learning model can improve students' problem solving skills on caloric materials in junior high. This was because the PO2E2W learning model has the following characteristics: (1) PO2E2W learning model is oriented to self-regulated learning from which, in each phase of activity, the students is guided in problem solving [5], (2) PO2E2W learning model has a characteristic of cooperative and collaborative works reflected in several phases in the PO2E2W learning model, although the learning model is oriented to self-regulated learning, but according to research results [6,18] suggests that collaborative work and small groups may develop self-regulated learning in various activities such as playing or problem solving, (3) The PO2E2W learning model has a phase of elaboration and write in science which is a phase of repetition of the problem orientation, observation and explanation phases that aim to enable the learning process to access the long term memory [5]. This stage is supported by the theory of information processing that explains processing, storage and retrieval of knowledge from the brain [8].

Based on the results of the study [5], it requires the improvement of the PO2E2W learning model for example from the aspect of a more varied learning media such as interactive multimedia [5]. So in this research want to use PO2E2W learning model which media assisted that is PhET media, such that it hopefully can improve effectively the students problem solving skill and challenge in facing 21st century about technological development. PhET media is a software simulation medium that can be used in physics learning both in online and offline state. PhET media is perfect for overcoming some abstract physics material. The use of PhET media in the process of learning activities is very suitable for senior high school students because senior high school students of category 17-19 years old students can already think abstractly [9]. This is supported based on the results of the study [10] which states that learning with the help of virtual media influences the ability of physics problem solving learners. In addition, based on the results of research [11], the PhET media assisted learning can improve the ability of high level thinking and responsibility students of class XI IPA on the material of kinetic theory of gas. The focus of this study is a follow-up study based on recommendations [5] to see the effectiveness of the PO2E2W learning model with PhET assisted media to improve the problem solving skills of high school students. The results of this study are expected to be an empirical evidence that the learning model of PO2E2W with PhET assisted media can improve the problem solving skills of senior high school students in physics learning.

2. Method

2.1 General Background of Research

The research was conducted at SMA Kemala Bhayangkari 1 (Surabaya, Indonesia). Scope of this research is senior high school students of class XI IPA 2 and XI IPA 3 who take physics year 2017/2018 academic year. The objective was to analyze the effectiveness of PhET media which was used for helping the implementation of the PO2E2W learning model. This study was analyzing the improvement of physics problem solving skills of senior high school students before and after using PO2E2W learning model with using PhET media as the assistant. The effectiveness of PO2E2W learning model using PhET media was determined based on: (1) there was improvement of students physics problem solving skills at $\alpha = 5\%$, (2) average score of n-gain students physics problem solving skills 0.81 (high category), (3) there was no difference (consistency) n-gain students physics problem solving skills in all groups, and (4) students response were very positive toward the implementation of PO2E2W learning model assisted PhET media.

2.2 Sample of Research
The samples were 64 students at SMA Kemala Bhayangkari 1 (Surabaya, Indonesia) using purposive sampling technique in two groups: group 1 (class XI IPA 2) and group 2 (class XI IPA 3) had homogeneity of physics problem solving skills. Each group consisted of students on physics subjects in the academic year 2017/2018.

2.3 Instrument and Procedures
Problem solving skills in Physics of senior high school students was measured using the Science Problem Solving Skills Assessment Sheet (SPSSAS), which had been declared valid and reliable. SPSSAS was prepared based on the measured physics problem solving indicator that was: (1) understand the problem, (2) plan problem solving, (3) do problem solving plan, (4) do check back from troubleshooting result [12]. Physics lesson materials used in this study were selected in accordance with the characteristics of PO2E2W learning model assisted PhET media, namely the kinetic theory of gas. This study used one group pre-test and post-test design, O1 X O2 [13]. The learning process was begun by giving pre-test (O1). Every senior high school student subjects was required to complete SPSSAS. After the pre-test, the teacher applied PO2E2W learning model and Physics learning device in each group (X). The implementation of the PO2E2W learning model had been conducted for five meetings on Physics subjects. Physics Learning used the PO2E2W learning model assisted PhET media with five syntaxes: (1) problem orientation, (2) observation, (3) explanation, (4) elaboration, and (5) Write in science. Physics learning tools consisted of: syllabus, learning implementation plan, student activity sheet, textbook, SPSSAS and students' response questionnaire (valid and reliable) [14]. After the implementation of the PO2E2W learning model assisted PhET media was ended with post-test (O2) using SPSSAS. Every senior high school student subjects was required to complete SPSSAS on post-test.

2.4 Data Analysis
Physics problem solving skills of senior high school students were analyzed based on the assessment determined before and after using the PhET media to assist the implementation of the PO2E2W learning model of learning. Pre-test, post-test, and n-gain problem solving skills of senior high school physics physicists were analyzed using inferential statistics with the help of SPSS software. n-gain is determined using the equation: n-gain = (maximum score-pre-test score) [15], with criteria: (1) if n-gain ≥ 0.7 (high), (2) if $0.3 <$n-gain < 0.7 (moderate) and (3) if n-gain ≤ 0.3 (low) [15]. The choice of statistical testing methods depends on fulfilling the assumptions of normality and homogeneity of variants for pre-test, post-test, and n-gain problem solving skills of senior high school students physics. Inferential statistical tests with Paired t-test (analysis of statistical improvement) and n-gain consistency analysis of all senior high school students groups was done using Independent t-test. The students responses were analyzed using quantitative descriptive technique with the percentage of students response = ((total number of students who voted/total students)$\times$100%), with criteria: (1) 76%-100% (very positive), (2) 51% -75% (positive), (3) 26%-50% (less positive), and (4) 0%-25% (not positive) [16].

3. Result and Discussion
The results are presented in Table 1, Table 2, Table 3, Table 4 and Figure 1 which will be explained as follows.

Table 1. The average value of pre-test, post-test and n-gain physics problem solving skills of senior high school students.

Group	Physics problem solving skills of senior high school students					
	Pre-test		Post-test		N-gain	
1 (Class XI IPA 2)	31.70	Low	86.44	High	0.80	High
2 (Class XI IPA 3)	28.24	Low	86.94	High	0.82	High

Table 1 described the mean of pre-test, post-test and n-gain physics problem solving skills of senior high school students. The low pre-test value was due to the students not yet having problem solving skills. According to Piaget [6], the process of adaptation can be through the process of assimilation and accommodation. The process of assimilation means that if students gain new knowledge or experience, then the new experience fits the scheme. While the accommodation process is the process of adaptation that occurs if students gain new knowledge, then the students change the schemes to be in accordance with the new knowledge. During the adaptation process, students interest in learning using PO2E2W model used PhET media, so that students' physics problem solving skills can be improved. This claim is supported by the n-gain class IX IPA 2 and class XI IPA 3 which is considered as high category, which meant pre-test and post-test values were increasing in both groups. The results of this implementation of learning model PO2E2W was assisted PhET media study was proved to be effective in improving the physics problem solving skills of high school students. This was because the developed PO2E2W learning model has met the validity (content and construct), the practicality, and the effectiveness of improving the physics problem solving skills of senior high school students [14]. This was supported by the results of the study [5] that the PO2E2W learning model is feasible to meet the validity (content and construct), practicality, and effectiveness will be able to improve and achieve the learning objectives of improving students' physics problem solving skills. Normality test results and homogeneity of variance suggested that the pre-test, post-test, and n-gain students physics problem solving skills were homogeneous and normally distributed for the whole group. Therefore, the impact of implementing the PO2E2W learning model on improving students' physics problem solving skills for all groups can be elicit by using Paired t-test and consistency test using Independent t-test. Paired t-test and Independent t-test results are presented in Table 2 and Table 3.

Table 2. Result of Paired t-test physics problem solving skills for all groups.

Group	N	Paired t-test, α = 5%			
		Mean	t	df	P
Group-1	32	-5.474	-66.135	31	.00
Group-2	32	-5.870	-93.903	31	.00

Table 2 showed the values of t = -66.135 and -93.903 for degrees of freedom (df) = 31 (group-1 and group-2). Each score was considered significant because p < 5%. Therefore, the results of the calculations were negative, thus indicating there was an improvement in students' physics problem solving skills after the implementation PO2E2W learning model assisted PhET media for all groups.

Table 3. Result of Independent t-test physics problem solving skills for all groups.

Independent t-test, α = 5%	t	df	Sig. (2-tailed)
N-gain_Group-1_Group-2	-1.473	62	0.146

Table 3 showed that t arithmetic gives t < t table with significance level P > 5%. This showed clearly that there was no difference in the improvement of students physics Problem solving skills after the implementation PO2E2W learning model assisted PhET media for all groups. Based on the results in Table 2 and Table 3, it had been shown that the results of improving students physics problem solving skills and outcomes for all groups had no significant differences (the existence of consistency) of students physics problem solving problem skills from the impact of the implementation of PO2E2W learning model assisted PhET media for all groups had been developed to improve students problem solving skills through syntax of problem orientation, observation, explanation, elaboration, and write in science. The results were reinforced with theoretical and empirical support that the PO2E2W learning model is an effective self-regulated learning oriented for improving students' problem solving skills [5] supported by learning theories, namely behavioral theory, social cognitive theory, constructivism theory and information processing theory [18-20]. This was supported by the results of the study [2-21] that the models, media, methods, instruments that are eligible for their validity, practicality and effectiveness will be able to improve and achieve learning objectives. Therefore, the PO2E2W learning model assisted PhET media is effective for improving the physics problem solving skills of senior high school students. The improvement of students physics problem solving skills reinforced qualitative data presented in Figure 1.

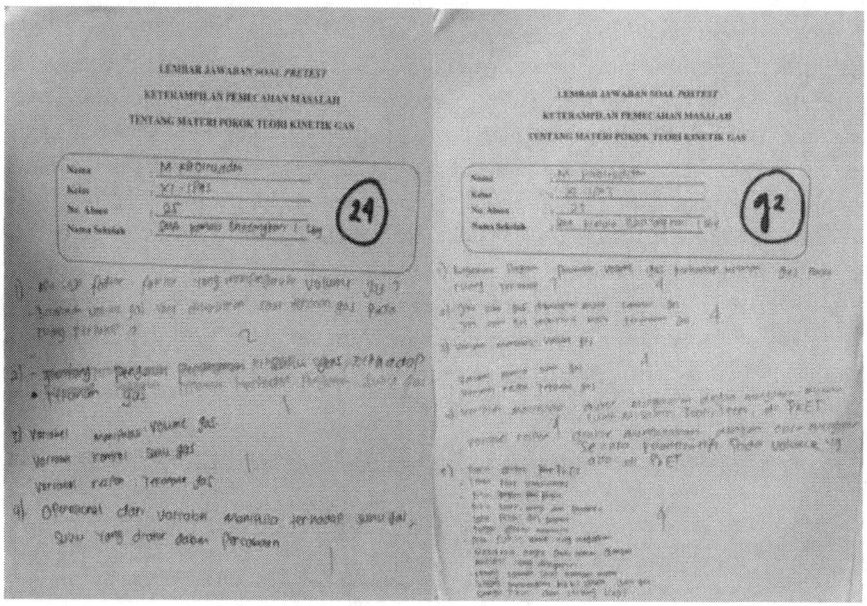

Figure 1. Differences of students physics problem solving skills before and after implementation of PO2E2W learning model assisted PhET media.

Based on Figure 1 explained that the students pre-test value is 24. This value was low. This showed that students problem solving skills were still very low. After the students got the learning by using PO2E2W learning model assisted PhET media, students' problem solving skills had been improved, marked with high post-test value that was 92. In the pre-test sheet, it can be seen that the students had not understood in answering the problem related to the indicators of problem solving skills on kinetic theory of gas material. For example, in question number 1 pre-test, students could not understand the problem to be able to formulate the problem by answering what are the factors affecting the volume of gas. The students also answered what is the volume of gas produced when gas pressure in the closed room so that the students instead and only got score 2 points from question number 1 that was the question answered and the answer was in question but still wrong in formulating the problem. After the students were given the treatment of implementing learning by using the PO2E2W learning model assisted PhET media, the students understood better the problem in post-test about the number 1 by being able to formulate the problem correctly. Based on the pre-test and post-test results, it can be seen that the students had reached the indicator problem solving skills. In this process, the students were given learning with the PO2E2W learning model assisted PhET media. The students adapted toward the learning model at several learning meetings until students learning interest emerging and can improve student physics problem solving skills. Improvement of students' problem solving skills were supported by the PO2E2W learning model assisted PhET media model instrument in each phase of self-regulated learning oriented activities [14]. Besides, it was also based on the results of the study [22,23] suggesting that problem solving relates to self-regulated learning and requires scaffolding. Other research results [24, 25] suggest that groups who were given self-regulated learning training achieved higher academic achievement than the group who do not have self-regulated learning training. The improvement of the students physics problem solving skills after implementation PO2E2W learning model assisted PhET media supported by student response data presented in Table 4.

Table 4. Senior high school students response to PO2E2W learning model assisted PhET media

Group I (Class XI IPA 2)			**Group II (Class XI IPA 3)**		
Sample	**Response**	**Category**	**Sample**	**Response**	**Category**
32	90.11%	very positive	32	91.14%	very positive

Table 4 showed that in general the students response was very positive toward the learning of the PO2E2W learning model assisted PhET media. Interview results showed that students feel the problem solving skills of physics increases. The results of this study became an empirical evidence of the effectiveness of PO2E2W learning model assisted PhET media to improve the skills of senior high school students problem solving.

4. Conclusion

The learning model of PO2E2W (problem orientation, observation, explanation, elaboration, and write in science) assisted PhET media was a self-regulated learning model in which its process of physics learning activities is carried out with the help of PhET media. The result of this study proved that: (1) There was improvement of students physics problem solving skills at $\alpha = 5\%$, (2) Average score of n-gain of students physics problem solving skills was 0.81 (high category), (3) there was no difference (consistency) n-gain significant physics problem solving skills in all groups and (4) students' responses were very positive toward the implementation of PO2E2W learning model assisted PhET media. Therefore, the PO2E2W learning model assisted PhET media had been proven to be effective in improving the physics problem solving skills of senior high school students. The implications of this research were as an alternative in improving the physics problem solving skills of senior high school students in Indonesia. Further research needs to replicate PO2E2W learning model assisted PhET media in improving students' physics problem solving skills at various levels.

References

[1] BSNP 2010 *Paradigma pendidikan nasional abad XXI* (Jakarta: Badan Standar Nasional Pendidikan)
[2] Sunarti T, Wasis, Madlazim, Suyidnoand, and Prahani B K 2018 *J. Phys: Conf. Ser.* **997** 012013
[3] Limatahu I, Suyatno, Wasis and Prahani B K 2018 *J. Phys: Conf. Ser.* **997** 012022

IOP Conf. Series: Journal of Physics: Conf. Series **1108** (2018) 012024 doi:10.1088/1742-6596/1108/1/012024

[4] Krulik Sand Rudnick J A 1996 *The new sourcebook for teaching reasoning and problem solving in junior and senior high school* (Boston: Allyn and Bacon)
[5] Odja A H 2017 *Disertasi* (Surabaya: Pascasarjana Unesa)
[6] Iiskana, Tuike, Vaurus, and Marjaand L E 2004 *Journal of Psychology* **1** 147
[7] Whitebread, David, Bingham, Sue, Grau, Valeska, Pasternak, Deborah P, and Sangster, C 2017 *Journal of Cognitive Education and Psychology* **6** 433
[8] Trianto 2007 *Model pembelajaran terpadu dalam teori dan praktik* (Jakarta: Prestasi Pustaka)
[9] Bulu S T and Pedersen S 2010 *Education Teaching Research* 507
[10] Hastuti A, Sahidu H, and Gunawan 2016 *Jurnal Pendidikan Fisika dan Teknologi* **23**
[11] Afifah R, Masjikur K, and Sutarman2014 *Jurnal Universitas Negeri Malang* **2** 1
[12] Polya G 1957 *How to solve it: An new aspect of mathematical method* (New York: Doubleday & Company, Inc)
[13] Fraenkel J R, Wallen N E, and Hyun H H 2012 *How to design and evaluate research in education* (8^{th}ed.) (New York: McGraw-Hill)
[14] Sari A S D 2018 *Makalah Komprehensif (*Surabaya:Pascasarjana Unesa)
[15] Hake R R 1998 *Am. J. Phys.***66** 64
[16] Riduwan 2012 *Skala Pengukuran Variabel-Variabel Penelitian* (Bandung: Alfabeta)
[17] Lefrancois G R 2000 *Theories of human learning* (Belmont: Cengage Learning)
[18] Moreno R 2010 *Educational psychology* (New York: John Wiley and Sons Inc)
[19] Arends R I 2012 *Learning to teach* (New York: Mc. Graw-Hill Companies)
[20] Slavin ER 2011 *Educational psychology: Theory and practice* (Boston: Pearson)
[21] Suyidno, Nur M, Yuanita L and Prahani B K 2017 *J. Res. Method. Educ.* **7** 56
[22] Bulu S T and Pedersen S 2010 *Education Teaching Research* 507
[23] GeX, and Land S M 2004 *Educational Technology Research & Development* **5** 22
[24] Fasikhah S M and Fatimah S 2013 *Sci. J. App. Psych.* **1** 145
[25] Al-Rawahi N M and Al-Balushi S M 2015 *I. J. Environmental Sci. Educ.* **10** 367

Acknowledgements
The authors would like to thank SMA Kemala Bhayangkari 1 (Surabaya, Indonesia) for allowing us to conduct research, as well as teachers and XI class students who have many roles in this research.

IOP Conf. Series: Journal of Physics: Conf. Series **1108** (2018) 012025 doi:10.1088/1742-6596/1108/1/012025

Mathematical problem solving skills analysis about word problems of linear program using IDEAL problem solver

L D Permata[1] *, T A Kusmayadi[1] and L Fitriana[1]

[1]Postgraduate School of Mathematics Education Universitas Sebelas Maret, Indonesia.
*E-mail: laelatul_dhian@yahoo.co.id

Abstract. IDEAL problem solver is one of mathematics problem-solving strategies, which function to investigate and describe student's way of thinking in solving problems process. This research aims to analyze mathematical problem-solving skills using modification IDEAL problem solver. This research aimed to explore, categorize, and analyze student's problem-solving skills about word problem of the linear program, which can be used as a basis to consider the improvement of learning. This research is a descriptive study including fifteen senior high school students in Grobogan as its subjects. The research variables are indicators of IDEAL problem solver that students performed including *identify problems*, *define the problem*, *explore solution*, *act on the strategies*, and *look back and evaluate*. The data were taken in a form of test and interview. The results showed that the indicator identifying the problems are categorized as very poor (8%), indicator defining the problem is categorized as very poor (8%), indicator exploring the solution is categorized as poor (49%), indicator acting on the strategies is categorized as enough (67%) and indicator looking back and learning is categorized as very poor (31%). Most students were not able to identify the problem and define the goal that resulted to lead the interpretation result.

1. Introduction

Mathematics can be used as a tool to make the job easier, more effective, and efficient. This is similar with the statement of [1] which states that mathematics should be taught because (1) it is very essential and useful in various fields, (2) as a powerful, concise, and unambiguous medium of communication, (3) can present the information in multiple ways, and (4) provide satisfaction in solving challenging problems. Mathematics regarding of both technical and content can help students deal with complex issues in the digital era nowadays [2]. Problem-solving is one of the skills that students must develop to meet the digital era in addition to critical thinking skills, creativity, innovation, communication and collaboration [3].

Problem-solving is the focus of the school's mathematics curriculum which in the learning process is expected to gain the experience to transfer and apply learned knowledge and skills to solve problems [4-6]. Regarding to that problem, skilled people solve problems to survive and to face the various issues in his life. In fact, the Government of Indonesia paid particular attention to the mathematical problem-solving skills shown in the 2013 curriculum where the ability of mathematical problem solving is one aspect of mathematical competence that is expected to be gained in the mathematics learning.

Bransford and Stein introduced an IDEAL problem solver consisting of five indictors including (1) identifying problems; identifying information, questioning, visualizing situation and making them creative thinking opportunities to determine the next stages. (2) Defining the problem; organizing information and question, finding and selecting the necessary/essential information to answer the question provided. (3) Exploring solution; finding/creating possible strategies: pattern, table, or models to solve the problem. (4) Acting on the strategies; using numeracy, algebra or geometric capabilities to solve the problem given, and (5) looking back and evaluate; rechecking answer, determining alternative solutions, discussing and developing answer to other situations [7]. IDEAL is a strategy that can be used to describe the ability of thinking skills in the process of problem-solving [8].

This study aims to analyze the mathematical problem-solving skills about word problem of the linear program. The analysis of mathematical problem-solving skills in this research is based on the modification of IDEAL problem solver indicator. This research is used to investigate, categorize, and analyze student's problem-solving skills about word problem of the linear program using IDEAL problem solver, which can be used as a basis tool to consider the learning improvement.

2. Research method

The research design is descriptive. The subjects of this research were fifteen 12th grade senior high school students in Grobogan who were selected by random sampling technique. These schools are chosen because they are considered to represent the average ability of high school student in Grobogan, which is shown from the absorption capacity of linear program material in the UN SMA 2016/2017. The average absorption capacity of linear program material has reached 39,32 %. The students who are as research subjects are students with heterogeneous learning abilities. The variables of this research used the indicators which is adopted from IDEAL problem solver including the ability of students to identify problems, determine the purpose of the problem given, choose most appropriate solving strategies, complete the procedure, and recheck the obtained solution. Furthermore, interviews with students about problem-solving strategies have been done to answer questions in the form of a given story. A description of the modification of IDEAL problem solver indicators used in this research is shown in Table 1.

Table 1. Description of IDEAL Problem Solver.

No	IDEAL's indicators	Description
1	Identify Problem	(a) Students write down known and needed information appropriately
		(b) Students write down known and needed information but not quite right
		(c) Students do not write down known and required information
2	Define Goal	(a) Students write down correctly what is asked in the problems
		(b) Students write what is asked in the problems but not quite right
		(c) Students do not write down what is asked in the problems they ask
3	Explore Possible Strategies	(a) Students write down problem-solving plan (table/model/graph) appropriately
		(b) Students write down problem-solving plan (table/model/graph), but not exact
		(c) Students do not write down a problem-solving plan (table, model or graph)
4	Anticipate Outcome and Act	(a) Students implement the plan, write the answers and computations appropriately
		(b) Students implement the plan, write down the answers and computations but not quite right
		(c) Students do not implement plans, write answers and computations.
5	Look Back and Learn	(a) Student interprets the obtained final answer by making a correct conclusion
		(b) Students interpret the obtained final answers by making conclusions but are inadequate
		(c) Students do not interpret the obtained final answer by making conclusions

Triangulation is used to confirm the validity of the data. Confirmation process of data validity is done by comparing information or test result data and interview. The stages of data analysis in this study include: (1) giving test about word problems of the linear program to students; (2) analyzing of the test results; (3) conducting interviews with students of various skills; and (4) analyzing the interview result.

3. Results and Discussion

3.1. Results

The result of problem solving test on first and third problem by using IDEAL indicators show that 15 subjects can be categorized into three categories. They are good categorized which is 13%, poor categorized which is 47% and very poor categorized which is 40%, meanwhile, most subject are not able to analyze the second problem. Furthermore, a summary of each IDEAL problem solver indicator with a maximum score of 2 can be seen in Table 2.

Table 2. Percentage of mathematical problem solving ability based on IDEAL.

IDEAL problem solver indicators	Analysis				
	The number of questions	Student's score	Total score	Percentage	Qualification
Identify Problem	3	7	90	8%	Very Poor
Define Goal	3	7	90	8%	Very Poor
Explore Possible Strategies	3	44	90	49%	Poor
Anticipate Outcome and Act	3	60	90	67%	Enough
Look Back and Learn	3	28	90	31%	Very Poor

Table 2 shows a summary of student's mathematical problem-solving skills from three given problems. In the first and third problems, the problems are found in indicators identifying problems; define goals item only reach 12%, look back and learn are categorized very poor by 28%. Meanwhile, indicators on exploring possible strategies and anticipate outcome and act were good categorized as either 72% or 83%. Figure 1 below presents the result of very poor category subject (LDF) on the first problem and in Figure 2 represents the result of poor category subject (BKJ) on the second problem.

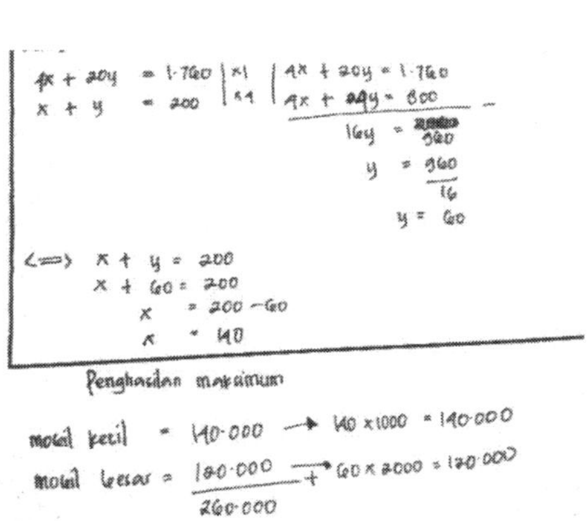

Figure1. Test's result of LDF on problem 1

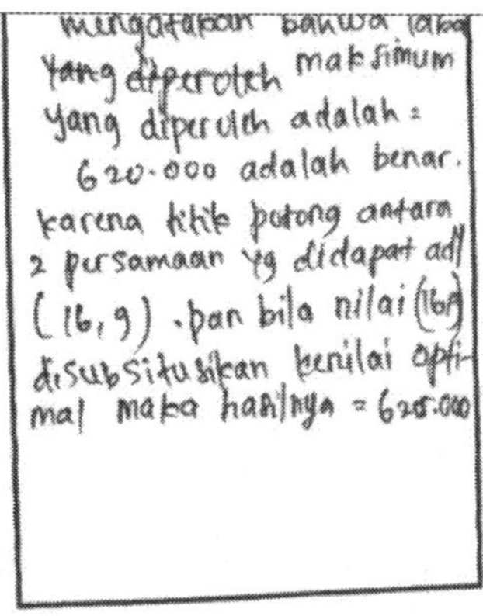

Figure2. Test's result of BKJ on problem 2

In the first problem, the LDF subject does not write down the known information and questions asked. The subject does not write a mathematical model that can be formed to the problem. The subject assumes the maximum value that can be searched from the result of the intersection of two

IOP Conf. Series: Journal of Physics: Conf. Series **1108** (2018) 012025 doi:10.1088/1742-6596/1108/1/012025

constraint function equations. Students do not write down the results of interpretation of final answers or conclusions about the last answer given. Subsequently, the interview was conducted on subjects with very poor categories (LDF) to confirm the subject's responses.

Question 1a	:	Of all the problems given, have you identified and written down the information provided? Is all the information is needed to answer the story?
Answer (Very Poor)	:	I did not write down the information provided, it might be useful to answer the question but I try to answer all the questions given
Question 1b	:	Have you written down the goals or questions asked on the problems?
Answer (Very Poor)	:	I did not write down all the questions asked
Question 1c	:	What strategy do you do to solve first problem? Have you worked on a chosen strategy?
Answer (Very Poor)	:	I use elimination and substitution, because to find the maximum value I have to find the value of x and y. The value of x and y that I get is $(60,140)$
Question 1d	:	Is that just the x and y values you get? What conclusion do you get? Have you written the conclusion of the final answer to all the questions you got?
Answer (Very Poor)	:	Yes, only one value of x and y. I did not write any conclusion but I know what to look for; the maximum income. I also did not write the final conclusions of the questions given

The Interviews were also conducted on subjects with good and poor categories. Written interview transcripts include indicators identifying problems, defining goals, and looking back and learning. These indicators are presented in three questions.

Question 1a	:	Of all the problems given, have you identified and written down the information provided? Is all the information is needed to answer the story?
Answer (Good)	:	I write down the information given in the problems, and the information provided is needed to answer the problems
Answer (Poor)	:	I sometimes write, but sometimes not, maybe all the information is needed to answer the problems
Question 1b	:	Have you written down the goals or questions asked on the problems?
Answer (Good)	:	I wrote down what is asked on the problems
Answer (Poor)	:	I did not write what is asked, but I knew what was being asked
Question 1d	:	Have you written the conclusion of the final answer to all the questions you got?
Answer (Good)	:	Yes, I wrote down the conclusion of the final answer I have got, but not all the problems
Answer (Poor)	:	Sometimes I write the conclusion of the final answer, but do not know whether they are right or not

In Figure 2, subject BKJ is able to analyze the location of errors of the presented work. The subject only analyzed the final answer of the task given without looking at the process of the given problem. Most of the subjects do the same analysis that they only see the final results that have been presented without looking back at the operation of the problem given.

3.2. Discussion

In general, the results show that the students are trying to solve the problems based on IDEAL problem solver indicators. Students have been attempting to explain the given word problem, although not all work is done correctly.

Students with a good category can often solve the questions correctly; even the obtained results are not optimal. Students with good category have been able to identify and write down the information provided accurately, although they are incomplete. Besides, good category students have also written down what is asked on the problems, writing down the problem-solving plan (mathematical models, graphs, tables) and their computations, and providing an interpretation of the obtained final answers.

Students with poor categories have tried to solve the word problems given, although they are less precise. Students with poor categories try to identify and write down the information provided and what is asked correctly, although they are incomplete. Students with poor categories have written problem-solving plan (mathematical models, graphics, tables) and their computations, but do not write the final interpretation of the obtained final answer.

Students with very poor categories have tried to solve the story problems given, although the results of the work presented are less precise. Other finding showed that there are students with very poor category who made conceptual error in the first problems. Students with very poor categories do not identify and write down the information provided and what is asked, they tried to write down problem solving plan (mathematical models, graphs, tables) and their computations, but do not write the final interpretation of the obtained final answer.

The phenomenon above is called by various factors. One of the factors in the problem-solving process is students' way of thinking. The things that are shown from the results of the research are the facts that most of the students are not able to identify the problem and define the goal, which influence the interpretation result. One way that can be taken to overcome the student's inability to identify problems is by doing a mathematical representation [9]. Through mathematical representation, students will develop their way of thinking in communicating mathematical ideas in various representations, making them easier to understand. Other than that, continuous training is required not only through routine exercise, but through learning process designed by the teacher that can help to find the right solution [10].

Learning process which become the pillar is used to construct student's knowledge. Therefore, teachers should provide learning opportunities to the students to solve the problems [11].In the learning process, choosing the right learning strategy is one of the important tasks for a teacher. Problem solving skills can be accessed through learning activities with appropriate learning strategies [12]. Learning strategy using contextual context is one of learning strategies that can be used to improve students' understanding concept in problem solving process [13, 14].

The process of choosing the right strategy of learning aims to develop a meaningful learning process, allows students to make flexible decisions in solving mathematical problems. Students with flexibility in problem solving will solve the problem with various strategies [15].In addition to paying attention to the learning strategy, the teacher should provide deep reinforcement of the prerequisite material before starting the new material [16]. Mathematics teachers are expected to be able to create learning which is not only focuses on result, but also the process, so the teacher can give experience to the students to develop their knowledge and aims to improve their problem-solving ability.

4. Conclusion

The results of the analysis provide an illustration that student's mathematical problem solving ability at high school level is not expected. The description of student's ability to solve problem, can be seen that at high school level percentage, the student's ability to identify problem reached 8%. It is classified in the category of very poor. The percentage of student's ability to define the goal reached 8%. It is classified in the category of very poor. The percentage of student's ability to explore possible strategies reached 49%, and it is classified in the category of poor. The percentage of student's ability to anticipate outcome and act reached 67%, and it is classified in the category of enough. The percentage of student's ability to look back and learn reached 31%

and it is classified in the category of very poor. Most of the students are unable to identify the problem and define the goal which influence the interpretation result. Mathematics teachers are expected to create learning, which is not only focuses on the result, but also the process, so the teacher can give experience to the students to develop their knowledge and aims to improve their problem solving ability. In addition, teacher should also pay attention to the reinforcement of prerequisite materials before starting new material. Through learning that focuses on the process, students will be able to make best planning to solve various problems.

Acknowledgments
The author would like to thank the researchers involved in the following references for improving the quality of this article and Senior High School in Grobogan region for the opportunity to do research.

5. References

[1] Cockcroft W H 1982 *Mathematics Counts: report of the Committee of Inquiry into the Teaching of Mathematics in Schools under the chairmanship of W.H Cockcroft* (London: H.M.S.O)
[2] Gravemaijer K, Stephan M, Julie C, Lin F and Ohtani M 2017 what mathematics education may prepare students for the society of the future *Int. J. of Sci. and Math. Edu.* **15**p S107
[3] Voogt J and Roblin N P 2010 *21st Century Skills: Discussion Paper* (Netherlands: Universiteit Twente)
[4] Dwiyogo W D 2016 Teaching and Learning Process: Thinking and Problem Solving *The Online Journal of New Horizons in Education* **6** 121-9
[5] Krulik S and Rudnick J A 1988 *Problem Solving: A handbook for Elementary School Teachers* (USA: ERIC)
[6] T Y E Siswono, A W Kohar and S Hartono 2017*J. Phys.: Conf. Ser* **812** 012046
[7] Bransford J D and Stein B 1993 *The Ideal Problem Solver: Second Edition* (New York: W H Freeman and Company) 20-37
[8] Rahayu R and Kartono 2012 the effect of mathematicaldisposition toward problem solving ability based on ideal problem solver*International Journal of Science and Research* **3** 1315-18
[9] Sahendra A, Budiarto M T and Fuad Y 2018 *J. Phys.: Conf. Ser* **947**012059
[10] Ulya H 2016 profil kemampuan pemecahan masalah siswa bermotivasi belajar tinggi berdasarkan ideal problem solving *Jurnal Konseling GUSJIGANG* **2** 96
[11] Riswan, Sudia M and Kadir 2018 profile of mathematical problem solving ability of grade vii students reviewed from students' cognitive style *International Journal of Science and Research* **6** 45-50
[12] Pujiadi, Kartono and Asikin M 2015 influence of creative problem solving aided with interactive compact disk towards mathematics learning achievement of grade x students *International Journal of Education and Research* **3** 617
[13] Jazuli A, Setyosari P, Sulthon and Kuswandi D 2017 improving conceptual understanding and problem solving in mathematics through a contextual learning strategy *Global Journal of Engineering Education* **19** 49-53
[14] Yee S P and Bostic J D 2014 developing a contextualization of students' mathematical problem solving *Journal of Mathematical Behavior* **36** 1-19
[15] Dina N A, Amin SM and Masriyah 2018 *J. Phys.: Conf. Ser* **947**012025
[16] Utomo, Kusmayadi T A and Pramudya I 2018 *J. Phys.: Conf. Ser***1008** 012070

Building Scientific Literacy and Physics Problem Solving Skills through Inquiry-Based Learning for STEM Education

L Yuliati[1,*], Parno[1], A A Hapsari[1], F Nurhidayah[1], L Halim[2]
[1]Physics Department, Universitas Negeri Malang, Indonesia
[2]Universiti Kebangsaan Malaysia
*Corresponding Author: lia.yuliati.fmipa@um.ac.id

Abstract. This study aims to determine the influence of Inquiry-Based Learning for STEM Education towards scientific literacy and physics problem-solving skills. This study used a quasi-experimental with pretest-posttest control group design. The population in this study is X grader students in SMA Negeri 1 Gondanglegi and the samples were chosen by cluster random sampling technique, resulted X MIA 1 as an experimental group and X MIA 2 as the control group. Each group consists of 26 students. Data collection was done by pretest and posttest for physics problem-solving skills. The result of t-test showed that p =0,013<0,05 and the average of the experiment group is 64,99 while the control group is 54,67. The experimental group was also measured in term of scientific literacy by using descriptive statistics and analyzed by using coding based on scientific literacy's rubric. The descriptive statistic showed that the average result of scientific literacy was 50,92. The result in this study by using inquiry-based learning for STEM Education showed that the physics problem-solving skills on the experimental group is different by the control group, and there was an effect on student's scientific literacy on experimental group.

1. Introduction

One of the materials in physics learning is work and energy. The concept of energy is an important concept in explaining the abstract phenomena of physics and often become the starting point in physics problems solving [1,2]. The scope of work and energy is broad enough for students because it discusses not only one principle but several principles [3]. Students difficulties are sometimes related to work and energy are commonly stated as misconception in applying the concept of negative and positive work without seeing the movement direction of the object [4]; students also don't recognize the relationship of cause and effect in work and energy theorem, and they don't recognize the difficulty in determining the magnitude of the work made by some components of particular effort [5]. If the students' understanding is inadequate, students will be unable to solve the physics problem, because the physics learning can never be separated from solving a problem [6].

Physics problem-solving skills is the expertise which is gained from an investigative activity, where the solver develops a solution to solve the problem [7]. The ability of students in solving the physics problem can be seen from understanding the concept and through the problem-solving procedures [6,7,8]. The four-stage problem-solving processes include the understanding of the problem, devising a plan, carrying out the plan, and looking back or evaluating [7].

Physics problem-solving skills is one of the basic concepts of scientific literacy that encompasses the elements of scientific inquiry and scientific knowledge [9,10]. Scientific literacy has been defined as a capability of students to be involved in scientific issues related to science manner[11]. The scientific literacy assessment does not only measure the level of understanding related to scientific knowledge but also measure the scientific process and student's abilities to apply in the real-life context [12]. Scientific literacy can be measured through 4 domains: scientific context, scientific competence, scientific knowledge and attitude [13].

To build the physics problem-solving skills, the application of learning that focuses on the activity of students can be performed. It aims to construct students' knowledge, through the inquiry-based learning model [14,15]. Inquiry-based learning is a learning that involves the process of inquiry, asking the question, and making an investigation to gain the new insights [16]. Currently, inquiry-based learning can be developed according to the 21stcentury demands by connecting real-world problems by involving STEM Education [17]. STEM Education is an integrative learning, that enables various learning methods to be used to support its application, one of them is the inquiry-based learning model [18].STEM Education is an integration of Science, Technology, Engineering, and Mathematics. It functions to prepare students in facing the challenges of the 21st century and makes the student as a center of learning activities [19]. Some research show the advantages of STEM Education; from the Engineering process in STEM. It is able to motivate the students in developing real-world problem-solving skills [20]; by submitting the scientific questions in the inquiry phase, the students will be able to generate curiosity. They can generate their curiosity to know more about STEM products [21]. Inquiry-based learning in STEM Education also helps students to achieve high achievement and be an active learners [22,23]. The purpose of this study is to know the influence of inquiry-based learning in STEM Education on students physics problem-solving skills and students' scientific literacy, especially on work and energy physics topic.

2. Method

This study used a quasi-experimental with pre-test and post-test control group design. The experimental group was taught by using inquiry learning in STEM Education and control group was taught by inquiry learning only. The population consists of all students of X Grader in SMA Negeri 1 Gondanglegi and the sample consists of 52 students; 26 students in the experimental group and 26 students in the control group. The sample was chosen by cluster random sampling technique. The students' problem-solving skills were measured by using an essay test. The instrument consists of using learning implementation plan with the worksheet. Pre-test and post-test results are analyzed by a prerequisite test, initial equality test, and hypothesis test. The prerequisite test results showed that the data is normally distributed and homogeneous so it can be used as the Independent Simple t-Test. The post-test result of the experimental group was analyzed by using descriptive statistics and coding based on problem-solving rubric and scientific literacy.

STEM Education was applied in inquiry-based learning in experimental groups where the components of science and mathematics were implemented through the presentation of problems of scientific phenomena to the preparation of investigation hypotheses. During the process of data collection and data analysis, the students utilized the existence of technology to design (engineering) STEM products as a solution of the topic of problems using theories and approaches of physics. Students were trained in their mathematical skills during the process of data analysis to take into account the accuracy of STEM products they create based on theories of science.

3. Result and Discussion

Based on the result of the hypothesis test, students' physics problem-solving skills of experimental group and control group is different ($p = 0.013 < 0,05$). It was shown by the higher score of experiment group which reaches average score of 64.99 compared to the control group which reaches score of 54.67. The average percentage of the procedures of students' physics problem-solving skills from the posttest data is shown in **Figure 1.**

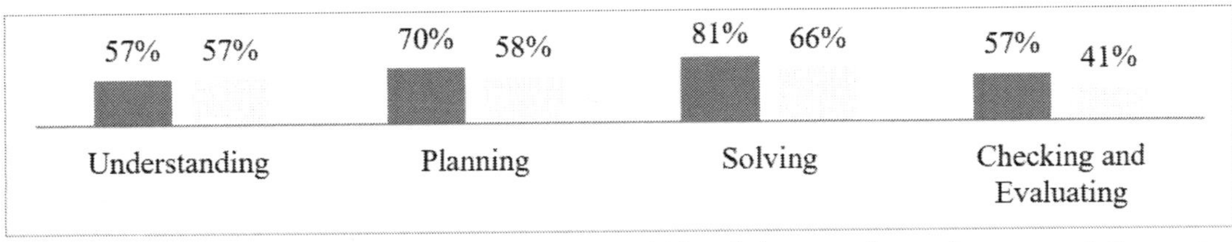

Figure 1. The Result of physics problem-solving skills of the experimental group and the control group. Figure 1 shows that the average physics problem-solving ability of students in the early stages of understanding the problem of both the experimental group and the control group is equal to 57%. In the second stage of planning problem solving, the results showed that in the experimental group reaches higher percentages of 70% and the control group as many as 58%. Furthermore, in the third stage it is showed that the average of students' ability in carrying out problem-solving of the experimental group is higher by 81% than control group which is 66%. The conclusion of the problem-solving result shows that the experiment group average is higher than 57% and the control group is 41%.

After measuring the students' physics problem-solving skills, the experimental group of posttest results was used to measure the scientific literacy of the students based on the scientific literacy indicator which is presented in **Figure 2** and completed with an example of problem-solving skills in **Figure 3**.

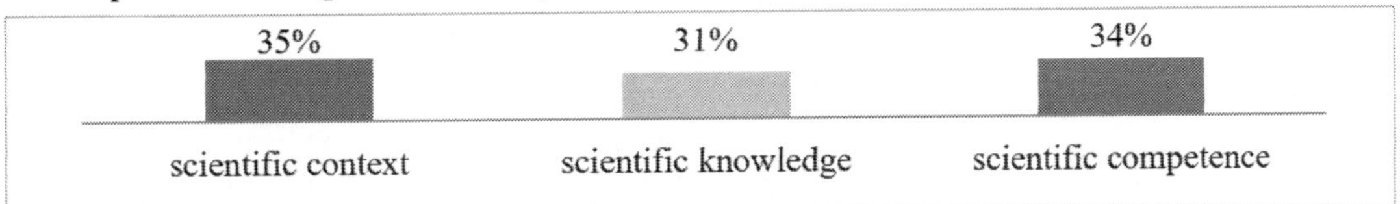

Figure 2. The result of the experimental group's scientific literacy. Figure 2 shows that although the scientific literacy of the students appears to be evenly distributed in all three aspects, the scientific context has the highest percentage than the other that is 35%. This is because the issue raised as the scientific phenomenon in the learning process was chosen based on the students' personal experience so that students can recognize scientific situation related to scientific knowledge and scientific competence to solve the problems. The implementation of scientific competence refers to the mental process in building scientific literacy capability. It is influenced by the understanding of scientific knowledge which refers to the basic concepts of science (physics) that needed to understand scientific issues. However, the scientific knowledge of students has the lowest percentage that is 31%, so that the scientific competence will be low too that is 34%.

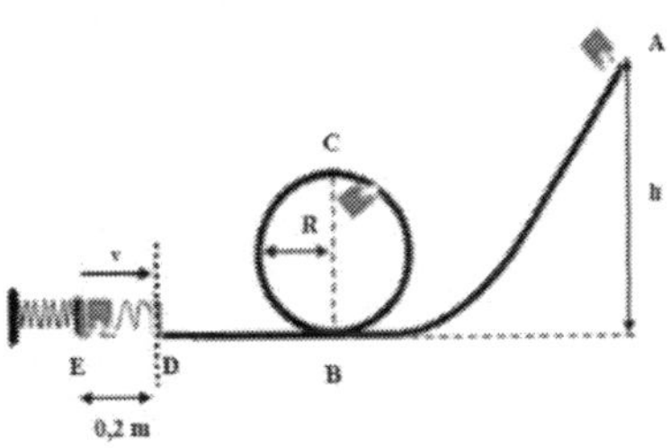

Figure 3. A roller coaster is designed to have a looping trajectory, with radius R = 4 m, then it is connected to a spring which the mass of the spring is ignored. Assume there is no friction between the train and the track (conservative). The train was launched from A and passed through the loop without bouncing off the track. Give your opinion, how to get the train through the loop path without going back and without bouncing out? What is the minimum height (h) required for the train to pass through the loop, without falling on point C? (include the free body diagram)

Tabel 1. The students' post-test answers are complemented by an assessment of problem-solving skills and scientific literacy.

Student Answers	Description of problem-solving skills	Description of Scientific Literacy
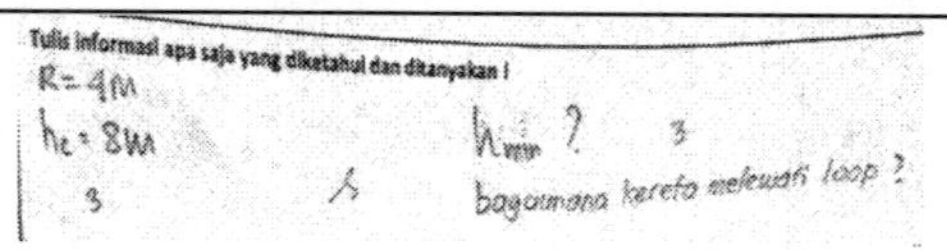	Superior understanding of the problem	Scientific literacy context
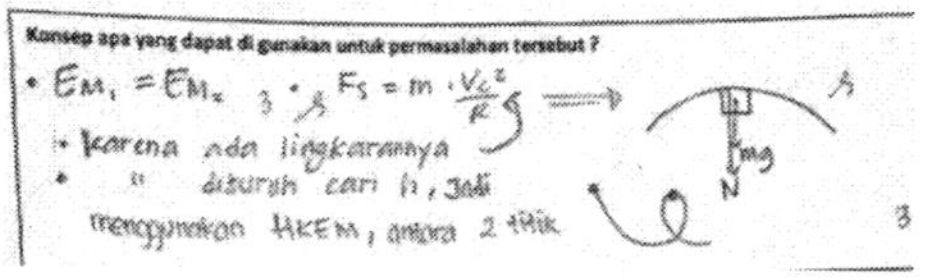	Superior problem-solving pathway	Scientific literacy knowledge
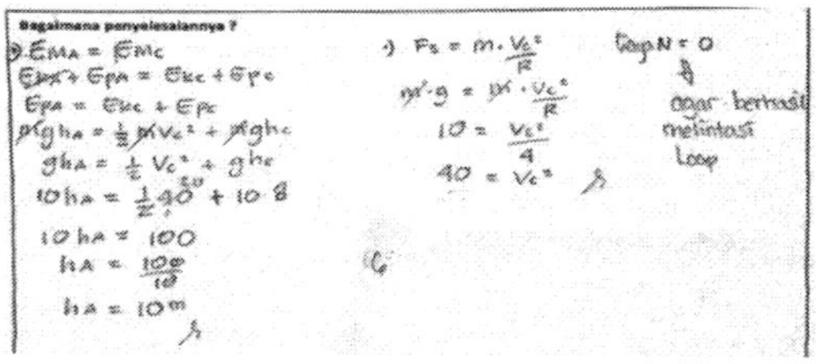	Calculation are complete and right	Scientific literacy knowledge & competence
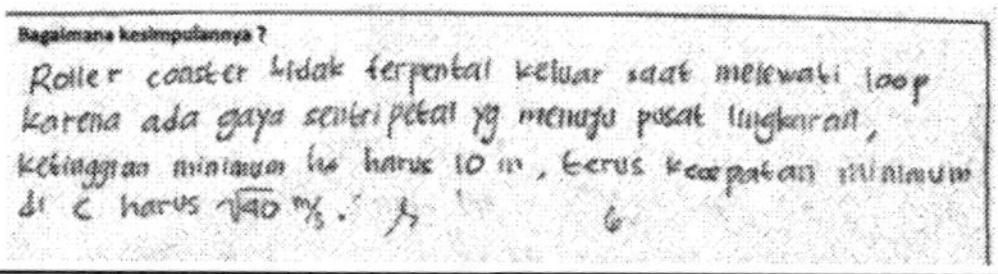	Superior reporting	Scientific literacy competence

Based on **Table 1**, the process of understanding the problem and planning the problem can be identified through the experimental group, so the assessment of understanding the problem and planning the problem is *superior understanding of the problem* and *superior problem-solving pathway* categories. On the planning phase, the students draw free body diagrams on the top of the loop. It is used to facilitate the analysis of problem-solving planning and to identify the loop radius and peak height of the loop using mechanical energy laws and the concept of circular motion. By carrying out the problem-solving phase, students can do the calculation based on the formulation that has been determined from the previous stage so that the assessment is *calculation are complete and right*. In the final step, students are able to make conclusions based on the problem-solving path and able to relate the results of answers with the concept of physics, so that the category is *superior reporting* assessment. **Table 1** also shows that the student recognizes the scientific situation well so they can write the issues that should be solved and identify the most important information that needed to solve the problem. It makes their scientific contexts are multidimensional. Scientific knowledge can be seen from how students plan problem solution. **Figure 3** use one of the scientific knowledge that is epistemic, which this knowledge helps students to provide scientific arguments based on scientific concepts and evidence correctly. The element of scientific competence can be recognized from how the student gave a conclusion at the end of the problem-solving. In this case, the expected element of competence is that students can interpret the problem situation with scientific argument and accompanied by systematic problem solution planning. In this case, students can answer the scientific knowledge correctly and scientific competence completely, so that students are at a multidimensional level.

Students are able to solve problems if they are able to recognize the concept of the problem [24]. The difference between students who have low and high problem-solving skills in solving physics problems can be seen from how students organize and use knowledge, as well as relating how the concepts they already mastered when solving the problem [25].Inquiry-based learning in STEM

Education, do not only involve students to investigate the questions but it is able to solve the real-world problems, one of them is through the design of a roller coaster trajectory.Through inquiry-based learning in STEM Education, students will carry out the problem-solving tasks from teachers and proven in previous research. There will be a significant relationship between learning, teaching and problem-solving tasks that teachers have given to students [26]; through the inquiry phase, it is expected to effect on students' math skills. Students were not only able to design, but they will learn how to calculate and how to design the balance and the successful operation of a STEM product while it is being operated [21].

Furthermore, the inquiry-based learning context for STEM Education was chosen based on the students' personal experiences related to the roller coaster game. This is in accordance with the statements submitted by the OECD (2014) that problems and solutions may arise from different situations or contexts based on individual experience. Context becomes a stimulus for students to gain knowledge and competence through scientific phenomenon. The experimental group is also trained to conduct a scientific investigation so it influences the scientific literacy of students [27]. Scientific literacy will be better if it is solved through scientific inquiry. Scientific inquiry can help students explaining what are students observing and remember what happened so that the knowledge and competence of the scientific literacy of students are growing [28]. Based on **Figure 2**, students in the superior category will be placed as in the multidimensional level. It can be interpreted that scientific literacy is based on problem-solving skills. In order for students to be scientifically literate, students then should have the basics concept of good problem-solving skills [29].

4. Conclusions

The results showed that students' physics problem-solving skills with inquiry-based learning in STEM Education were different from those with inquiry-based learning only. The difference is found between the average obtained in each group and the learning activities that have been done. The students who have high problem-solving skills are able to scientifically literate and make correct problem-solving procedures that affect the scientific literacy of students. It is resulted that students are able to make scientific decisions to solve the problems appropriately. This study is recommended for further researchers who wish to explore more in-depth problem-solving skills and scientific literacy of students on inquiry-based learning in STEM Education for the work and energy topic.

5. References

[1] Serway R A and Jewett J W 2014 *Physics for Scientists and Engineers with Modern Physics 9th Edition* (Boston: Cengage Learning)

[2] Constantiou C P and Papadouris N 2012 Teaching and Learning About Energy in Middle School: An Argument for An Epistemic Approach. *Studies in Science Education* **48**(2)161-186.

[3] Rahman A 2015 Dampak Work and Energy Macromedia Flash Program (WEM) Terhadap Kemampuan Pemecahan Masalah dan Berpikir Kritis Mahasiswa S1 Pendidikan Fisika Topik Usaha dan Energi. *Prosiding Seminar Nasional Fisika dan Pembelajarannya 2015 Pascasarjana UM* 33-35

[4] Nugraha H A, Kaniawati I, and Suhendi E 2014 Analisis Miskonsepsi Topik Usaha dan Energi Siswa Kelas XI Setelah Pembelajaran Kooperatif Menggunakan Simulasi Komputer. *Prosiding Seminar Kontribusi Fisika Bandung* 17-18

[5] Dalaklioğlu S, Mooc M, Şekercioğlu A 2015 Eleventh Grade Students' Difficulties And Misconceptions About Energy and momentum concepts.*International Journal on New Trends in Education an Their Implications* **6**(1)13-21

[6] Docktor JL, Dornfeld J, Frodermann E, Heller K, Hsu L and Jackson KA, Yang J 2016 Assessing student written problem solutions: A problem-solving rubric with application to introductory physics. *Physical Review Physics Education Research* **12**(1), 10130.

[7] Selçuk GS, Çal S and Erol M 2008 The Effects of Problem Solving Instruction on Physics Achievement, Problem Solving Performance, and Strategy Use.*Journal Physics Education,* **2**(3) 151-166

[8] Pradugawati D, Diantoro M and Sutopo 2016 Kemampuan Penyelesaian Masalah Siswa SMA pada Materi Usaha dan Energi. *Prosiding Seminar Nasional Pendidikan IPA. Universitas Negeri Malang*,**1**146-153

[9] Abd-El-Khalick F, Boujaoude S, Duschl R, Lederman NG, Mamlok-Naaman R, Hofstein A, Niaz M and Treagust D 2004 Inquiry in Science Education: International Perspectives. *Science Education* **88** 397-419

[10] NentwigP, Rönnebeck S, Schöps K, Rumann S and Carstensen C 2009 Performance and levels of contextualization in a selection of OECD countries in PISA 2006. *Journal of Research in Science Teaching*, **8**, 897-908

[11] PISA 2015 *Draft Science Framework 2015*, 1–54.

[12] Sukowati D, Ani R and Sugianto 2017. Analisis Kemampuan Literasi Sains dan Metakognisi Peserta Didik. *Phys. Comm***1**(1)

[13] OECD 2004 *Problem Solving for Tomorrow's World* (Paris: OECD)

[14] Wang J, Guo D and Minjou 2015 A Study on the Effect on Model-Based Inquiry Pedagogy on Student's Inquiry Skills in a Virtual Physics Lab. *Computers in Human Behavior*. (Online) (www.elseiver.com).

[15] Wenning C J 2011 Experimental Inquiry in Introductory Physics Courses. *Journal of Physics Teacher Education Online*. **6**(2), 9-16.

[16] NAP 2000 *Inquiry and the National Science Education Standards: A Guide for Teaching and Learning* (Washington DC: National Academic Press)

[17] Rustaman N and Lufri 2016 Pembelajaran Sains Masa Depan Berbasis STEM Education. *Prosiding Seminar Nasional Bio-EduPadang: PGRI Sumatera Barat*, ISBN: 978-602-74224-1-4

[18] Permanasari A 2016 STEM Education: Inovasi dalam Pembelajaran Sains. *Seminar Nasional Pendidikan SainsUPI* 23-34.

[19] Bybee R W 2013 *The Case For STEM Education: Challenges and Opportunity*(Arlington: National Science and Engineering Teacher)

[20] Hafizan E, Halim L and Shahali M 2016 STEM Learning through Engineering Design : Impact on Middle Secondary Students ' Interest towards STEM. *EURASIA Journal of Mathematics Science and Technology Education* **13**(5) 1189–1211.

[21] Adnan M, and C H 2018Integrating STEM education through Project-Based Inquiry Learning (PIL) in topic space among year one pupils *IOP Conference Series: Materials Science and Engineering* **296** 1-6

[22] Suwarma IR, Astuti P and Endah NE 2015 "Balloon Powered Car" Sebagai Media Pembelajaran IPA Berbasis STEM (*Science, Technology, Engineering, and Mathematics*). *Prosiding Simposium Nasional Inovasi dan Pembelajaran Sains* 373-376

[23] Nuangchalerm PNM 2017 Implementing Inquiry-based STEM Learning in Tenth Grade Students. *The International Academic Forum. Mahasarakham University*.

[24] Mason A and Singh C 2011 Assessing Expertise in Introductory Physics Using Categorization Task. *Physical Review Special Topics-Physics Education Research*, **7**

[25] Shih YL and Singh C 2013 Using an isomorphic problem pair to learn introductory physics: Transferring from a two-step problem to a three-step problem *Physical Review Special Topics-Physics Education Research* **9**

[26] Adeyemo S 2010 Students' Ability Level and Their Competence in Problem-Solving Task in Physics. *International Journal of Educational Research and Technology* **1**(2) 35–47

[27] OECD 2014 *PISA 2012 Results: Creative Problem Solving – Students' Skills in Tackling Real-Life Problems Vol. V* (Paris: OECD)

[28] Martin, Michael O, Mullis, Ina V, Chrostowski and Steven J 2008 *TIMSS 2007: International science report* (Chestnut Hill, MA: Boston College)

[29] Toharudin U, Hendrawati S, and Rustaman A 2011 *Membangun Literasi Sains Peserta Didik* (Bandung: Humaniora)

Mathematical communication skill of senior high school students based on their personality types

A C Septiana[1], T A Kusmayati[1] and L Fitriana[1]

[1]Postgraduate School of Mathematics Education, Universitas Sebelas Maret, Indonesia.
Corresponding author: anggrahinyclara14@gmail.com

Abstract. Every student with different type of personality, extrovert or introvert, has his or her own characteristic. The difference of the personality type can influence their way of having confident attitude, communication, and other activities including learning. This research aims at analysing student's ability of mathematical communication with extrovert and introvert personality type. This research is descriptive qualitative research. The data of this research was obtained from 34 students of senior high school in Sukoharjo region with heterogeneous abilities, collected using written test and interview. The result of the research gave an overview that mathematical communication skill of students having introvert personality can arrange conjecture, make an argument, and formulate generalization definition. However, they had difficulty in understanding a mathematical presentation. Meanwhile, the students having extrovert personality can explore their ideas, but they have difficulty in revealing the idea or mathematical paragraph in their language. In order to improve mathematical communication, students should be given mathematical problem related to mathematical communication and the teacher is expected to know the characteristics of the students so that the teacher can provide appropriate actions.

1. Introduction

Good education quality will result good academic achievement as well. Intelligence and personality are important factors to study in relation to academic achievement [1]. People have different characteristics that influence their life, these personal characteristics influence even their way of learning. These different characteristics are called person's personality type [2]. Jung (in [3]) distinguished personality types into two kinds namely extroversion and introversion, the two of which refer to how far someone's basic orientation directed outside (outside world) or inside himself.

According to Tieger & Barron-Tieger, introvert personality is a nature of imaginative people which is self and idealism-centered, often or characterized as thinking hard before taking action or doing talk, more easily stimulated by new ideas and unusual situation, thorough, honest, and consistent. Meanwhile, extrovert personality is a nature of imaginative people which is outside world and idealism-centered, often or characterized as an act without thinking too hard. It tends to show more the person's openly emotional situation, tends to be like acting directly which is better than fantasizing, and tend not to be consistent [4].

There is a meaningful connection between personality type and learning motivation, as well as the ability of solving problem faced by student [5]. A person with extrovert personality type more often uses movement and writing for communication compared to the introvert one [6]. Communication skill is strongly needed to sort and describe solution construction of analysis result, or logical description of mathematical problem appeared [7]. Communication is a transmission process of information, idea, emotion, ability, and so on through symbols, words, pictures, numbers, and others. Meanwhile, mathematical communication skill is defined as a person's ability to write a mathematical statement, write reason or provide explanation of each mathematical arguments he uses to solve the mathematical problem; use terms, tables, diagrams, notations, or mathematics formula properly; and check or evaluate another mathematical thought [8].

The aim of mathematics learning deals five main standards of mathematics learning covering problem-solving, reasoning and verification, communication, connection, and representation [9]. In curriculum process standards, it is stated that the aim of mathematics learning are followings: (1) to solve problems consisting of ability to understand the problems, design mathematical models, solve the models, and interpret solution; (2) to communicate ideas with symbols, tables, diagrams, and other media to clarify situation or specific problem; and (3) to have attitude in respecting mathematical practice in daily life, consisting of having curiosity, concerns and interest towards learning mathematics, along with the tenacity and confidence in solving mathematical problem [10]. However, the student's mathematical communication skill today has not yet fulfilled the expectation.

Some researchers have found one same problem that is low mathematical communication skill, seen from Trends International Mathematics and Science Study (TIMSS) report. It mentions that Indonesian students' ability in mathematical communication is still very far below the other countries [11]. Meanwhile, in TIMSS 2015 report, Indonesian students are in the 45th position from 50 countries. Indonesian score is 397. Related to the reasoning, the ability of Indonesian students is still very low. The ability to reason is closely related to communication skills. Another research states that 58.54% of Indonesian students still have difficulty in student's worksheet which has relation to difficulty in stating mathematical sentence and lack of student's bravery in asking questions and responding to problems [12]. Remembering the importance of communication in mathematics learning and low mathematical skill in Indonesia, the necessity of student's mathematical communication skill is fundamental so that in the following days the educator can determine appropriate methods and models to overcome various problems. Then, at the end, it will have an impact to the improvement of learning results [13].

The results [19] show that junior high school students had weak difficulties in visualizing mathematical ideas visually, making mistakes in demonstrating mathematical ideas and having errors when using mathematical terms. They also had problems in expressing mathematical ideas and in using structures of mathematics to present mathematical ideas. Based on those studies, researchers are interested in doing research on the communication skills of high school students. Because of the importance of mathematical communication in learning, then researcher was interested in having research about the student's mathematical communication skill from the point of view of introvert and extrovert personalities, as other researcher had studied about the influence of mathematical communication to mathematics learning achievement [14]. Besides, there is also another researcher who did research about the influence of prospective mathematics teacher's mathematics communication when teaching geometry [15]. Meanwhile, in this research, the questions used deals with linear program. There is also a researcher who compares mathematical communication between teacher and student [16]. Different from it, this research focuses on student's mathematical communication. Moreover, even though there is also another researcher who had researched about communication through social media by seeing from introvert and extrovert personalities point of view [17], this research focuses on mathematics communication.

In this research, the indicators of mathematics communication skill covers analysing and writing information obtained into mathematical symbols; reading with understanding a mathematical equation; expressing ideas, mathematical situation oral or with picture, graph and algebra form; arranging conjecture, argument or formulating generalization definition; and revealing a mathematical sentence into its own language.

2. Method

This research was done through descriptive qualitative method. The method in this research dealt with analyzing student's mathematics communication skill based on the introvert-extrovert personality he or has. The subject consists of 34 students of senior high school in Sukoharjo region with heterogeneous abilities. The students were selected through purposive sampling. The data used was mathematical questions and interview. Mathematical communication data were submitted through essay test carried out in 45 minutes. The essay topic for the test was about linear program. Next, the data was descriptively analysed. The data analysis techniques in this study included: (1) reducing the data, (2) the presentation of the data, and (3) conclusion.

3. Result and Discussion

In this study, indicators of mathematical communication skills were synthesized by Hendriana and Soemarmo and Yang, Chang, Cheng and Chan. The indicators are analyzing and writing information obtained into mathematical symbols; reading by understanding mathematical equations; express ideas, oral mathematical situations or with pictures, graphics and algebraic forms; regulate allegations, arguments or formulating definitions of generalizations; and expressing mathematical sentences into the students's own language.

The result of this research shows the score and percentage of students' mathematical communication indicators as provided in Table 1.

Table 1. Mathematical Communication Skill of Introvert-Extrovert Personality

No	Mathematical Communication Skill Indicators	Personality			
		Introvert		Extrovert	
		Score	%	Score	%
1	Analysing and writing information obtained into mathematical symbouls.	42	30,88%	107	78,68%
2	Reading with understanding a mathematical equation.	24	17,65 %	79	58,09%
3	Expressing idea, mathematical situation oral or with picture, graph and algebra form.	58	42,65%	110	80,88%
4	Arranging conjecture, argument or formulating generalization definition.	108	79,41%	24	17,65%
5	Revealing a mathematical sentence into its own language	10	7,36 %	41	30,15%

Based on Table 1, the students having introvert-extrovert personality could be classified into high, medium and low categories based on the percentage obtained. It was included in high category if the percentage was between 67%-100%, while it was included in medium category if the percentage was between 34%-66% and low category if the percentage was between 0%-33%. The following description is the research result consisting of general description about students' mathematical communication skill dealing with introvert and extrovert personality. Some of the students' answers can be analyzed as follows:

Figure 1 shows that the student analysed and wrote the information obtained into mathematical symbols by the permissibility variable x = clothes type A, and y = clothes type B. That student could express his idea shown by solving the problem with elimination way, but the student was not able to express his idea into the graph. The student explained that he still got confused and didn't understand yet when drawing graph to find the intersection of the line. They should express their ideas in the graph so that the truth of the answer they got was correct. The student was also able to arrange the steps of the execution or conjecture correctly and arrange the arguments and formulate the generalization definition shown by the finding of the maximum profit. Figure 1 also shows that the student was not able to reveal a mathematical sentence into his language yet, shown by his work in which the student did not write the final result or the conclusion of his work to answer the problem. The aforementioned statement was strengthened by the result of the student interview in which he stated: "...because I was confused to make a conclusion sentence. I think the important thing is that I have my answer already if the requested question is the value of x and y along with the maximum profit."

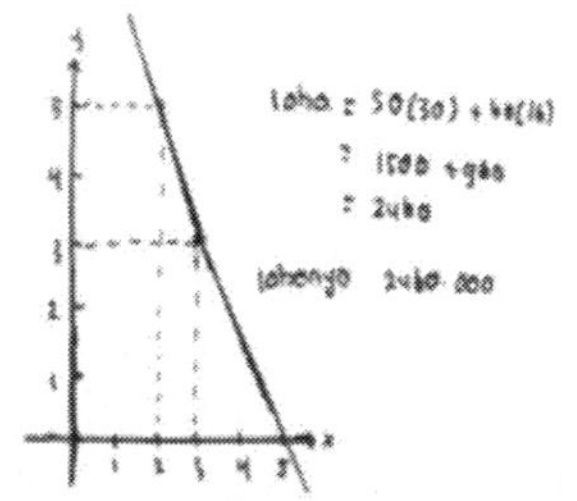

Figure 1. Student's answer with introvert personality

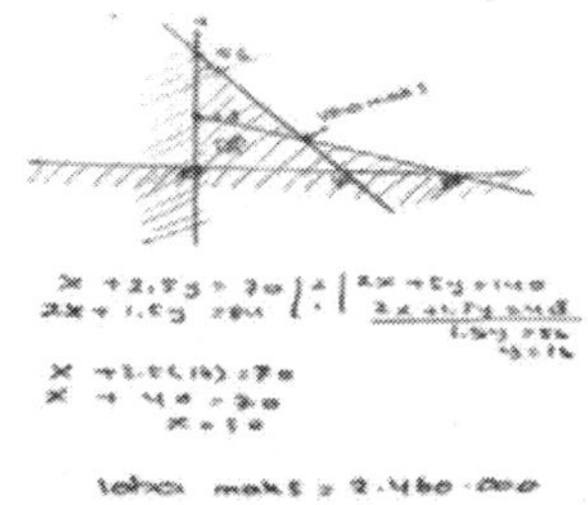

Figure 2. Student's answer with extrovert personality

Figure 2 shows that the student analysed and wrote the information obtained into mathematical symbols accurately. Although the student was not able to read the mathematical equation correctly yet, he could express the ideas, the mathematical situation with oral or graph and the algebra form. Figure 2 also shows that the student could solve the problem by expressing the ideas into the graph and elimination way. The student was also able to arrange the steps of the execution or conjecture correctly, even though he did not arrange the arguments and formulate the generalization definition shown by the absence of critical point. The result was according to the student's statement saying: "I am still confused to determine which the critical point is and how to draw the line." The student also did not write the final result or the conclusion of his work to answer the problem, which meant that he was not able to reveal a mathematical sentence in his language. It was based on the student's statement during the interview in which he said: "I was confused what sentence I should write for the conclusion, well, finally I just wrote the maximum profit."

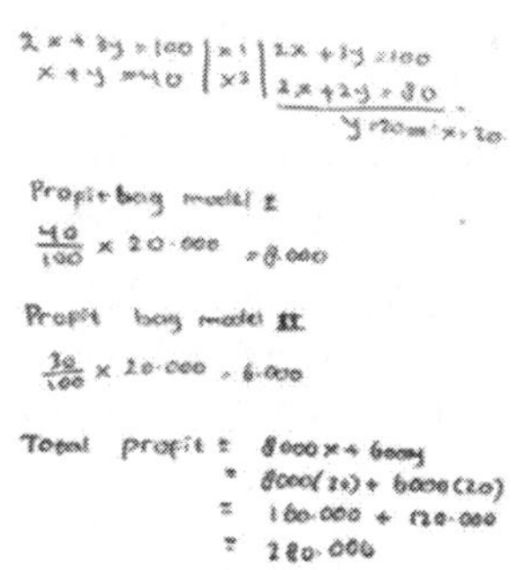

Figure 3. Student's answer with introvert personality

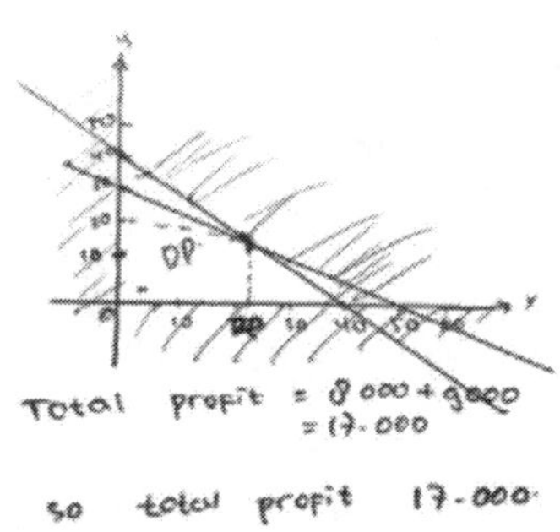

Figure 4. Student's answer with extrovert personality

Figure 3 shows that the student analysed and write the information obtained into mathematical symbols by the permissibility variable x = bag model I and y = bag model II. The student could express his idea, shown by his problem-solving of the elimination. However, the student was not able to express his idea yet into the graph. They should express their idea into the graph as well so that the truth of the answer they got was no doubt. Such result was based on the student's statement in the interview: "I was confused which points used in order to make the graph, hence I just did the elimination although I was, sometimes, not sure with my answer because it was not so clear whether it was correct or not if it was done only through elimination." The student was also able to arrange the steps of the execution or conjecture correctly and arrange the arguments and formulate the generalization definition shown by the finding of the profit of bag model I and II, although he was less accurate in substituting the value of the capital. Such result was according to the student when he said: "I was confused how to multiply it to find the profit. Then, I had a thought to use the value of x and y I had found in the beginning." Figure 3 also shows that the student was not able to reveal a mathematical sentence into his language yet, shown by his work where the student did not write the final result or the conclusion of his work, he wrote only the total profit. Such result was based on the student's statement in which he stated: "I was confused about arranging the sentence and I thought if the maximum profit is equal to the total profit, so I just wrote like that to answer it."

Figure 4 shows that the student was able to analyse and write the information obtained into mathematical symbols by the permissibility variable x =bag model I and y = bag model II. The student could express his idea shown with his problem-solving dealing with the elimination and graph as well as the ones in the form of algebra, which were shown by the graph and by the finding of the value of x and y done through the elimination. The student was also able to arrange the steps of the execution or conjecture correctly and arrange the arguments and formulate the generalization definition shown by the finding of the profit of bag model I and II, without writing the critical point obtained from the graph. Such finding was based on the student's statement in which he said: "I did not understand which number it was supposed to be substituted and how to complete the critical point." Figure 4 also shows that the student could reveal a mathematical sentence into his language, shown by his work where the student wrote the final result or the conclusion of his work. The total profit was 17.000 and it was the result when he was in a hurry to take the conclusion which made him solve the problem incorrectly. Such finding was based on the student's statement in which he stated: "I am sure if the total profit is the sum of profit I and profit II, and since I wished to finish it soon, therefore I just answered in that way so that I was able to finish it soon."

From some of the students' answers analyses above, it can be concluded that students with extrovert personality had better mathematical communication than those who had introvert personality. This is in line with a research which states that "professional commitment between teacher and students having extrovert personality is significantly different from those who have introvert personality [4]. It also supports that different personality type influenced their attitude a lot toward language learning through internet usage, where the students having more extrovert personality experienced high integration in ITC compared to those who have introvert personality [18]. Moreover, there was a difference in communication intensity through social media in which extrovert personality type had higher communication intensity in comparison to the introvert personality type [17]. However, if we see in Table 1, every person had high and low presentation. The students who had introvert personality could arrange conjectures, arguments, to formulate generalization definition, but had difficulty in reading with understanding a mathematical presentation. Meanwhile, the students who had extrovert personality could explore ideas, but had difficulty in uncovering an idea or a mathematical paragraph in their language.

4. Conclusion

The analysis result above gives an overview that mathematical communication skill of the students who had extrovert personality was better than those who had introvert personality. However, every person had advantages and weakness, as well as the mathematical communication skill of the student who had introvert personality, and he could arrange conjectures, arguments, formulate a generalization definition, but had difficulty in reading by understanding a mathematical presentation. Meanwhile, the students who had extrovert personality could explore his ideas, but also had difficulty in revealing an idea or mathematical paragraph in their language. Therefore, the students should be given mathematical problems exercises so that the mathematical communication for each student can be stimulated and be better.

Acknowledgment
The authors would like to thank the researchers involved in the following references for helping improving the quality of this paper and also from senior high school in Sukoharjo region with heterogeneous abilities for the opportunity to do this research.

5. References

[1] Ciorbea I and Pasarica F 2012 the study of the relationship betweenpersonality and academic performance *J.Elsevier Procedia-Socialand Behavioral Sciences* **3** 52-67
[2] Boroujeni, Roohani and Hasanimanesh 2015 the impact of extroversion and introversion personality types on efl learners' writing ability *J.Theory and Practice in Language Studies* **5** 212-29
[3] Suryabrata S 2002 *Psikologi kepribadian* (Jakarta: PT Rajawali Pers)
[4] Prakash S, Singh A and Yadav S K 2016 personality (introvert, and extrovert) and professional commitment effect among bed teacher educator students *J. Indian psychology* **3** 43-9
[5] Sari H and Shabari 2016 *Idea nursing journal* **7** 1-10
[6] O'Carroll S 2015 the effect of extroversion on communication: evidence from an interlocutor visibility manipulation *J. Elsevier Speech Communication* **69** 1–8
[7] Hariwijaya 2009 *Meningkatkan Kecerdasan Matematika* (Yogyakarta: Tugupublisher)
[8] Sari D S, Kusnandi K and Suhendra S 2017 a cognitive analysis of communication ability on geometry *Journal of Physics* Conf. Series **895** 012083 1-7
[9] NCTM 2000 *Principles and Standards for School Mathematics* (Reston Virginia: NCTM).
[10] Kemendiknas 2006 *Permendiknas No 22 Tahun 2006 Tentang Standart Isi Pendidikan Dasar Dan Menengah* (Jakarta: Kementrian Pendidikan Nasional)
[11] Fachrurozi 2011 *J. Penelitian* **1** 76-89
[12] Ranti M 2015 *J. Pendidikan Matematika* **1** 96-102
[13] Rohman A A 2017 *J. Ilmiah Pendidikan Matematika* **5** 7-20
[14] Lomibao L S, Luna C A and Namoco R A 2016 *J. of education research* **4** 378-82
[15] Dimmel J K and Herbst P G 2017 Secondary Mathematics Teachers' Attitudes Toward Alternative Communication Practices When Doing Proofs In Geometry *J. Elsevier* **68** 151-160
[16] Huang J, Normandia B and Greer S 2005 *J. communication educarion* **54** 34-51
[17] Widiantari K S and Herdiyanto Y K 2013 *J. Psikologi Udayana* **1** 106-15
[18] Mirzaee M. and Gharibeh G S 2016 *J. Teaching English Technology* **16** 57-70
[19] Firdaus H P E 2016 *International Conference Education (IECO) Proceding* **1** 344-351

The Development of Biology Comic Education with Compare and Contrast Learning (CaC) Method to Improve 10th Grade Students' Literacy

R A Putri[1,*], R Fida[2], and Wasis[3]

[1, 2, 3] Postgraduate Science Education Program Universitas Negeri Surabaya, Indonesia

*Corresponding Author: rimasaldillaputri@gmail.com

Abstract. Literacy in the biology does not only deal with reading activity but also deal with several other activities like identification, determinant, and communication of information to overcome life problems. Furthermore, in biology subject, the chapter related to virus is the one of the most difficult chapters due to the abstract cause and the complexity of the content. This research aims at producing a feasible biology comic education to increase student's literacy in Senior high school with compare and contrast (CAC) learning method. The development of this biology comic education refers to the 4D Model. It used one-group pretest-posttest design. The comic education was applied in normal class with CAC learning method. The data of the research were collected through validation, questionnaire, and pretest-posttest instruments. The data were analyzed through descriptive quantitative statistic technique. The result of this research revealed that: (1) the biology comic education was developed in valid categorizes, (2) the literacy learning was categorized improved; and (3) the students responded positively towards the learning. These results indicated that the biology comic education which had been developed was feasible to be used to improve the students' literacy. In addition, the biology comic education can be used as a reference or alternative learning resources in the learning in the school.

1. Introduction

Based on the results of PISA (Programme for International Student Assessment) 2012 study on literacy, Indonesia ranks in the 64th position out of 65 countries with a specific score on literacy reading of 396 or equivalent to rank 61 [1]. In response to this, the Ministry of Education and Culture Decree (Permendikbud) Number 23 2015 about building students' behaviors and character which requires students to read books 10-15 minutes every day before starting the lesson.

Reading is considered as the most important component in the 21st century in order to survive in the current era of globalization. The success of learners in participating in the teaching and learning activities in schools is much determined by the ability to read.

Students can read books in the classroom through a collection of class libraries with other types of books other than subject textbooks, which are categorized popular books but still within the appropriate levels and preferably for the students according to their level. Furthermore, it is mentioned also that popular books such as domestic and foreign literature, newspapers, magazines and educational comics are some examples of books to develop student's interest in reading. Such perception is supported by the international journal namely Journal of Science Communication [2] entitled Science comics as tools for science education and communication: a brief, exploratory study states that comics are used in promoting science literacy through education and communication.

A low reading interest causes low activity and low learning outcomes. The complexity of learning materials like the chapter of virus makes students less interested in learning and in reading textbooks including Biology textbooks. Students tend to be interested in observing picture books rather than textbooks, because they are packed with interesting attributes as well as completed with a coherent storyline and regularly which make it easy to remember. Therefore, educational comics can illustrate and explain the structure of the virus more clearly.

Hence, this study aims at fostering students' interest in reading through reading stuffs that contain learning materials that are packed with pictures and language that are easily understood. The biology comic education would be applied in the normal class with compare and contrast (CAC) leaning method.

Through this research, it is expected that such way will be used as an alternative mean for helping teachers in the learning process, especially developing educational comic learning materials on other chapters of Biology subject as well as becomes an effort to develop interest in reading (student literacy) and improving the quality of learning.

2. Theoretical framework

The image media is a medium that combines facts and ideas in a clear and powerful way through a combination of expressions of words and images; this medium consists of drawings, charts, diagrams, graphics, posters, cartoons and comics [3]. This study developed comic education to enhance students reading interest. Comics have many functions in advertising, entertainment and education. Educational comic is a comic that tells about the subject matter or matters relating to education. Comic education has two main functions as a medium of entertainment as well as a mean for conveying the purpose of education to the reader.

Interest in reading is a tendency and a high passion or a strong desire to read. Interest in reading would also determine the high quality of the learning achievement [4] Interest is not innate, but something that can be learned and developed. It means that an interest can be developed because of certain inputs or new insights and new patterns of thinking in the form of light reading stuffs such as educational comics.

Compare and contrast (CAC) learning method would help students in doing comic education literacy in the normal class. This learning model aims at assisting students in developing critical thinking skills through image descriptions, objects, and readings as well as providing a mean for organizing new information [5]. CAC learning method consists of four main steps: 1) Reading and learning; 2) reasoning and analysis; 3) communication; and 4) reflection and connect.

3. Method

The development of this educational comic with virus material used the development learning materials of Four-D (4-D) model because the researchers think that the 4-D development model is easier to understand and apply. The 4-D device development model was developed by [6], composed of two main steps namely planning or development and implementation.

At the planning or development stage, 4D model was carried out through four stages: Define, Design, Develop, and Disseminate [7]

The defining stage (Define) is a stage to define the materials and components that exist in educational comics. Meanwhile, the Design stage aims at defining the terms of the story and the material in the comic strip. The development stage (Develop) aims at producing a revised comic based on the input of experts. The final stage is the dissemination (Disseminate) which is the use of devices that have been developed with a wider scale [7]. After the main step of the planning or development, the next major step is the implementation. The implementation stage is the phase to implement the biology educational comics that have been developed.

Students literacy/reading interest in this study would be measured by using paper and pencil test method that used Likert scale questionnaire instruments including statements that were tailored to the indicator in order to measure students' reading interests. The indicators to be measured in this study embraced feelings during reading, frequency, interest, and reading awareness. The statements in the questionnaire consisted of 30 questions with optional answers that would be filled by 35 students of SMAN 3 Surabaya.

The questionnaire dealing with the student literacy was filled by the students in order to know the stage of students' reading interest in biology both before and after biology learning activities with the use of comic education and CAC learning method. In addition, this interest reading the questionnaire was also used to measure the ease of students in understanding the material contained in comic education.

The student literacy questionnaire used was closed-ended questionnaire so that students could choose an answer that matches their opinion on the option by checking (√) in the appropriate answer option (1=Disagree, 2=Less Agree, 3=Agree, and 4=Strongly Agree).

This research method used one group pretest-posttest design with formula: (O1 X O2) that gave the initial student literacy questionnaire before the learning activity progress in order to know their reading interest and their responses at the beginning of the lesson. After the 3rd meeting of the study (the last meeting), the students got questionnaires once again to know the development of students' literacy and students' responses about the learning activities through the use of educational comics.

The data of the student literacy was analyzed by using qualitative descriptive and entering questionnaires data in the following formula:

$$\mathbf{P} = \frac{\Sigma R}{\Sigma N} \mathbf{X\ 100\%.}$$

4. Analysis and Discussion

The data validation of the comic education coming from the two validators, obtained results that generally educational comic developed by researcher had *valid category* and *very valid category* with the score of 3 to 4 and reliability percentage of 86% until 100%. Based on those assessments, the comic education developed could be used with little revision.

The implementation of the research began with the collection of early data of the reading interest (preliminary research) conducted in SMAN 3 Surabaya in class X. The respondents were Class X MIA 3 which consisted of 35 students. The research was done through the student literacy questionnaire *in preliminary research, the recapitulation of which resulted 24%* and the student literacy questionnaire *in post research, the recapitulation of which resulted 56%.* The student literacy indicators used included the feeling of pleasure during reading without coercion, the frequency of reading, the interest of students to read books, and the students' awareness of the important and purpose of reading; which were presented in the following table.

Table 1. Results of questionnaires on students' reading interests

No	Indicator	Pre-Research (%)	Post-Research (%)
1	Feeling of pleasure during reading without any coercion	15	38.63
2	The frequency of reading	25	44.25
3	The students' interests to read books	25	56.13

4	The awareness of the purpose of reading	31.25	87.88
	Average	**24.06**	**56.72**

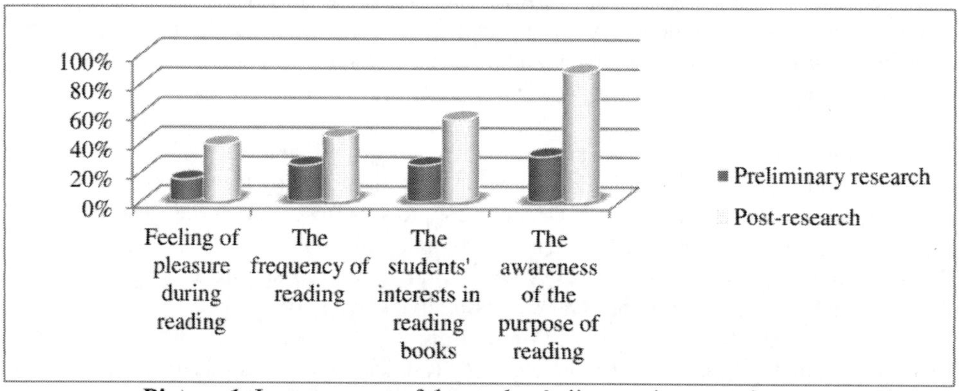

Picture 1. Improvement of the student's literacy in pre and post research

From the graph 1, it is known that the student's literacy on each indicator shows different results, but all the indicators showing the improvement between pre- and post-lessons using biological educational comics. The interest was not innate, but something that could be learned and developed. It meant that an interest can develop because of certain inputs or new insights and new patterns of thinking. Similar to the reading interest, it could be developed in several ways. According to [8] there are some tips to develop reading interest in students by growing motivation to be aware of the purpose of reading activity such as adding insight, being able to know the solutions to various problems, and reinforcing the knowledge of something.

The data of the students' responses towards the educational comics developed were obtained from the data of the students' questionnaires. All students filled the questionnaires, then the results were recapitulated. The recapitulation of the response of 35 students to the learning activities using educational comics got a strong and positive response with the percentage of response 70% to 100%. It was because the educational comics made the learning become more interesting, especially in understanding the material that was long and difficult to be understood by students.

The existence of an interesting storyline with colors in comic education that are not monotonous can attract the attention of students to learn it. According to [3], the use of appropriate colors in a learning medium can generate motivation, feelings, attention, and willingness of students in learning.

5. Conclusion

From the results of the development and application of biological education comics dealing with virus material at SMAN 3 Surabaya, it could be concluded that educational comics could be used to cultivate a positive attitude of the students since it improved the students' literacy or reading interests as well as it could be used as an alternative guide book for learning biology in another chapter. Some suggestions from the results of this study point out that comics can be used as one of the learning media in class, not limited to the biology course. However, it can be applied for other subjects or other learning methods to develop the aspects of the students' reading interests, students' activities, and students' learning outcomes.

IOP Conf. Series: Journal of Physics: Conf. Series **1108** (2018) 012028 doi:10.1088/1742-6596/1108/1/012028

6. References

[1] OECD 2012 *Measuring Student Knowledge and Skills, A New Frame Work for Assessment* (Paris: OECD)

[2] Vuleta, Sporis, Talovic and Jeleskovic 2010 Reliability and Factorial Validity of Power Tests For Handball Players *Journal Sport Science* **3**(1) 42-46.

[3] Sudjana 2015 *Media Pembelajaran* (Bandung: Sinar Baru Algensindo)

[4] Singer F M and Voica C 2015 *Is Problem Posing a Tool for Identifying and Developing Mathematical Creativity? In Florence Mihaela Singer, Nerida F. Ellerton and Jinfa Cai (Ed). Mathematical Problem Posing: From Research to Effective Practice* (New York: Springer)

[5] Silver H F 2010 *Compare and Contrast – Teaching comparative thinking to strengthen student learning* (US: ASCD Publication)

[6] Thiagarajan and Sivasailam 1974 *Instructional Development for Training Teachers of Exceptional Children* (Washinton DC: National Center for Improvement Educational System)

[7] Ibrahim M 2005 *Asesmen Berkelanjutan Konsep Dasar, Tahapan Pengembangan dan Contoh* (Surabaya: Unesa University Press)

[8] Manis H 2010 *Learning is Easy – Tip dan Panduan Praktis Agar Belajar Jadi Asyik, Efekktif, dan Menyenangkan* (Jakarta: PT Elex Media Komputindo)

Developing A Media 'Visual Design of Pop Up Mathematics Book' as a Supporting Tool in Inquiry-Based Learning for Learning Three-Dimensional Figures

T Susilo[1,*], I Sujadi[2] and D Indriati[3]

[1] Postgraduate Mathematics Education, Sebelas Maret University, Indonesia.
[2,3] Mathematics Education Department, Sebelas Maret University, Indonesia.

*Corresponding Author: trisusilo.math@gmail.com

Abstract. Inquiry-based learning is one of the learning models that involves more students with the invention of concepts or formulas as its main goal. Inquiry learning can improve the students' learning outcomes and learning quality. Inquiry learning steps include investigating problems, formulating hypotheses, designing experiments, analyzing data, and concluding. The main problem faced by the students when dealing with inquiry learning is to formulate the hypotheses and find the formulas. The objective of this research is to develop a media called *Visual Design of Pop Up Mathematics Book* to overcome the difficulty of inquiry learning on three-dimensional figures. This research is a media development research which is effective to support inquiry-based learning. The development processes carried out by the researcher through the selected development model include product analysis, the first product development, expert validation and revision, small-scale field test and product revision, large-scale field test and the final product. Subjects were 8th grade students of a state secondary school in Wonogiri district. Based on the small-scale test, students who were given the media of visual design of pop up mathematics book obtained better results. They were more able to guess the possible steps and find the right way or formula in solving the problem than the students who did not use the media.

1. Introduction

Geometry is an important branch of mathematics and has been identified as a basic mathematical skill [1]. Geometry is important for students as it is also applied in other branches of mathematics. For example, geometry is applied in other subjects such as engineering drawings, drawing geometry, and so on. Basically, there are two objectives of learning geometry, namely to develop the logic of thinking skills and to develop spatial intuition refering to how someone views the space and area in the real world [2, 3]. However, there are some difficulties encountered in studying geometry especially in three-dimensional figures, which include the lack of students' evidence, background knowledge, reasoning skills in geometry, geometric language understanding, visualization skills, teaching methods, unavailability of teaching materials, gender differences, and etc [4].

To prepare the learners who are able to face the education problem of 21st century, one of the learning models supporting the education has been modeled. It is the model of inquiry-based learning.

To improve the process quality and the learning outcomes, some experts have suggested the use of structured inquiry learning paradigms for teaching and learning activities in the classroom [5]. Inqury can also increase student involvement in the learning process [6]. Inquiry is the process of defining and investigating problems, formulating hypotheses, designing experiments, finding data, and describing the conclusions of those problems [7].

The major obstacle in inquiry learning is solving problems based on investigations and assumptions made by the students [8]. Then, the need for the existence of a technology-based learning is needed to face the education of the 21st century [9]. Therefore, the development of a teaching material in the form of technology-based media is appropriate in the current learning process. The use of learning media enables the teachers to explain, describe, distribute, and deliver lectures more easily and effectively than depending only on words [10].

In an effort to overcome the problems in inquiry-based learning on the study of three dimensional figures and to answer the challenges of 21st century education, it is offered a media called a *Visual Design of Pop Up Mathematics Book.* The pop up book can help students in learning geometry, particularly on the solid figures [11]. The book combined with ICT (Information and Communication technology) provides a good understanding in learning [12]. Thus, a visual design of pop up mathematics book is made with the hope that students are able to construct the concept of three-dimensional figures independently and improve the quality of their learning.

2. Method

This research is a development research which is a method to produce a certain product [13]. The product made is in the form of a media, visual design of pop up mathematics book. The procedures in this development conform to Borg and Gall that have been simplified. The procedures include conducting: product analysis to be developed, the first product development, expert validation and revision, small-scale field test and product revision, large-scale field test and the final product. Data collecting techniques used were question and questionnaire. The subjects of this study were two students of 8th grade of SMP Negeri 2 Wonogiri (public school). Both students were the subjects for small-scale test in an effort to know the effectiveness of visual design of pop up mathematics book. Of the subjects, one student conducted a learning and worked on media-aided question while the other student did not use the media or just used a regular handbook. The students' acknowledgment can be seen in table 1 below.

Table 1. The students' acknowledgment

Name	The use of 'Visual Design of Pop Up Mathematics Book'	Test
Student A (Lathifah Ninda Prilia)	Not using the media	Yes
Student B (Renata Amara Ginesti)	Using the media	Yes

After the test, students then filled out a questionnaire of response towards the learning using the Visual Design *Pop*-Up Book of Mathematics.

3. Results and Discussion

3.1. Product Analysis

This activity was carried out with literature review and field study on junior high school students in Wonogiri regency who have used the latest curriculum. The literature review is related to the importance of learning geometry, especially the material on the solid figures [1,2,3]; the difficulties in learning geometry especially the material on the solid figures [4]; the inquiry learning model which is seen as an appropriate learning model according to some experts [5]; the difficulties in running the inquiry learning model [7]; 21st century learning that puts technology ahead [9]; the need to develop a media that can be applied to the learning process [10]; and the effectiveness of the existing pop up book [11,12]. Meanwhile, the field study related to the selection of the selected research sites based

IOP Conf. Series: Journal of Physics: Conf. Series **1108** (2018) 012029 doi:10.1088/1742-6596/1108/1/012029

on the average score of the national school exams representing the medium-categorized school i.e. SMP Negeri 2 Wonogiri.

3.2. First Product Development

The product, a visual design of pop up mathematics book, is a combination form of media as well as teaching materials, especially on the solid figures. This media were created to support the inquiry learning model to overcome the existing difficulties. The beginning pages consist of the cover, the instructions of media usage, and the competencies to be achieved. Figure 1 shows the front page that contains a cover with the title while Figure 2 is about the media usage instruction and explanation of the navigation buttons. Figure 3 contains geometrical material that is adapted to inductive reasoning for students to support learning, and Figure 4 is the image of all solid figures to be used in the media of visual design of pop up mathematics book. The display of the product made are as follows.

Figure 1. The front page that contains a cover with the title

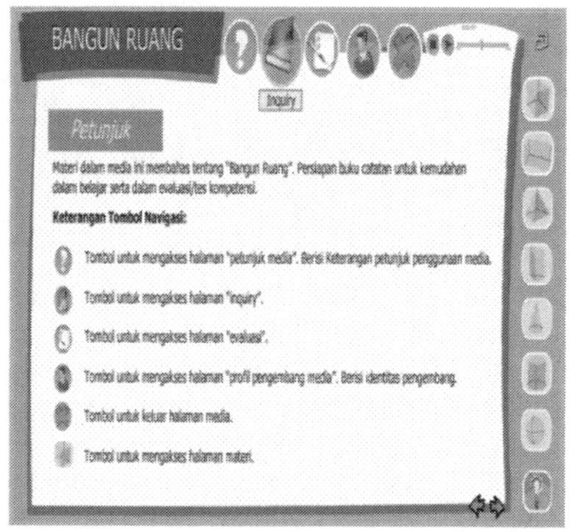

Figure 2. The media usage instruction and explanation of the navigation buttons

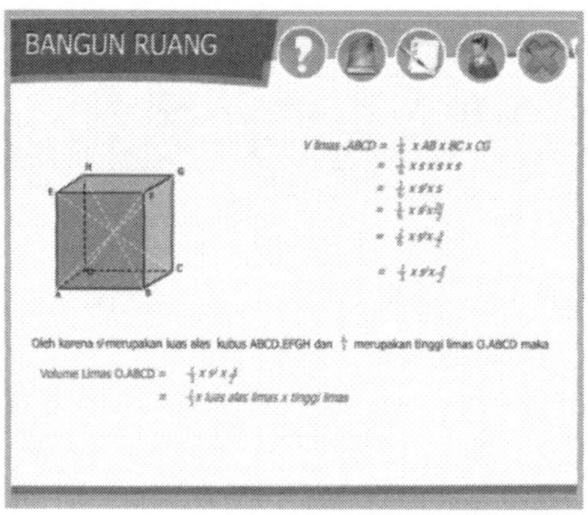

Figure 3. Geometrical material that is adapted to inductive reasoning for students to support inqury learning

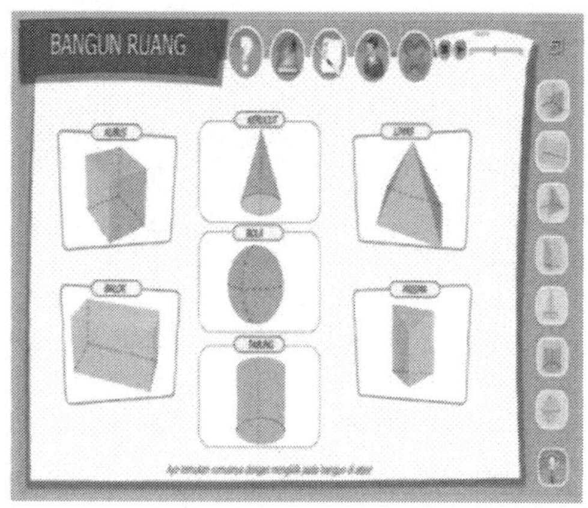

Figure 4. The image of all solid figures to be used in the media

3.3 Small Scale Test

IOP Conf. Series: Journal of Physics: Conf. Series **1108** (2018) 012029 doi:10.1088/1742-6596/1108/1/012029

In this small-scale test, the subjects are in accordance with table 1.1 mentioned above. To know the effectiveness of visual design of pop up mathematics book, two things were done to measure it. First, the student using the media and the other without media were given a similar test, the same time, and the initial ability assumption. Second, the students' response to the media used were taken. After a small-scale test, the results were different from the students before conducting the test. The use of visual design of pop up mathematics book influences the development of students' inductive reasoning, which can support inquiry learning.

3.3.1. A test to measure learning ability.

Both students were given some questions to determine whether there is a difference between student A and Student B after learning with different learning media. The question to solve is shown in Figure 5 below.

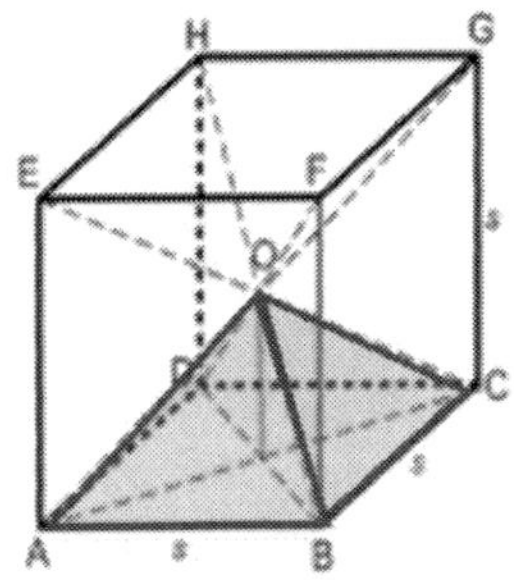

Problem:

A cube has sides (S) with the length of 20 cm. There is a shaded shape inside in which the height is the intersection of each space diagonal to the base center. Determine the volume of the shaded shape!

Figure 5. Problem to measure the volume of shaded shape

The problem has a single answer, but the result from students have several steps. It can be seen below.

The Results of Student A (without the media)

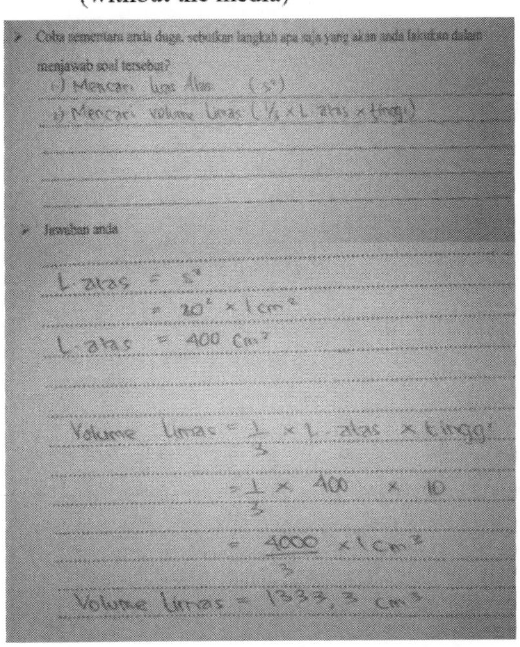

The Result of Student B (with the media)

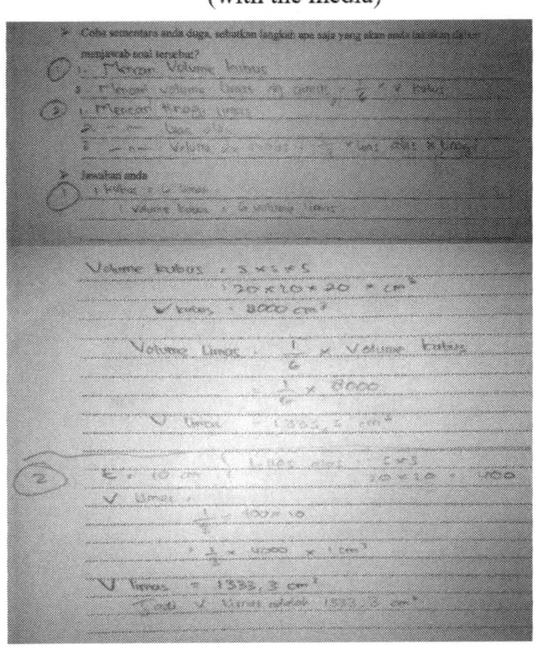

Figure 6. The result of the students

Based on the results of the two students above, it can be seen the differences as described in table 2 below:

Table 2. The differences between two students abilities

Working Indicators	Student A	Student B
1. Ability to estimate before answering questions	Student A estimated to: - find the base area - find the volume of pyramid	Student B estimated using formula I to: - find the volume of cube - find the volume of pyramid with a formula (1/6 x the volume of the cube) Student B estimated using the formula II to: - find the height of pyramid - find the base area - find the volume of pyramid
2. Ability to correlate to the previous materials	Not able to correlate the volume of pyramid with the volume of cube yet	Able to correlate the volume of pyramid with the volume of cube seen in the estimation using formula I
3. Ability to solve the problem	Able to answer the question correctly but not yet able to explore the other steps.	Able to answer the question correctly and able to provide other steps with the same final results.

Table 2 on the difference of the results of the two students who received different treatments shows that the student who learned using the visual design of pop up mathematics book was more able to explore the estimation in determining the steps to solve the problem. Meanwhile, the student who learned without any media or just with a handbook was able to answer the question but had not been able to explore more deeply the other steps in solving the problem in Figure 4. Thus, it can be said that the use of visual design of pop up mathematics book is more effective in supporting the inquiry learning in terms of students' ability to find estimation and relate it to the previous materials.

3.3.2. Student's response to media.

From questionnaires given to students related to the media, an assessment was obtained as seen in the table 1.3 below: (with category 1%-25%: Very Bad ; 26%-50%: Bad ; 51%-75%: Good ; 76%-100%: Very good).

Tabel 3. Student response to the use of media

Statements	**Score**	**Total**	**%**	**Category**
Response to use the media	100	125	80 %	Very good

From the student' assessment which showed "very good" judgment can be interpreted that the media is appropriate and effective to use.

4. Conclusion

Based on the research conducted, it shows that the difficulty of inquiry-based learning can be minimized by using a media, *Visual Design of Pop Up Mathematics Book*, especially on the material of the solid figures (three-dimension). The success is measured through two indicators of the effectiveness of the media, namely question and students' response questionnaire. The student who used the media obtained more learning experience than the student who did not use any media. Student B's ability to estimate and to explore was better than Student A's. Thus, it can be concluded that the media, *Visual Design of Pop Up Mathematics Book*, is more effective than the previous ones applied in the school. This is also the case with other studies, that the Pop Up Book that is visually designed is considered effective to use in today's learning [12], and it is very appropriate to overcome learning problems especially on three-dimensional figures [11].

5. References

[1] Abdulah A H 2012 *The Effects of Van Hiele's Phases of Learning Geometry on Students' Degree of Acquisition of Van Hiele's level* (Malaysia: Procedia-Social and Behaviorial Sciences) p 251-266
[2] Sherard W H 1981 Why is Geometry a Basic Skill? *Mathematics Teacher* **74**(1), 19-21
[3] Hong L T 2005 *Van Hiele Levels and Achievement in Writing Geometry Proofs Among Form 6 Students* (Kuala Lumpur : Universiti of Malaya)
[4] Fabiyi T 2017 Geometry Concepts in Mathematics Perceived Difficult To Learn By Senior Secondary School in Ekiti State, Nigeria *Journal of Research & Method in Education* **7**(1) pp 83-90
[5] Paivio A 2013 *Imagery and verbal processes* (Ontario: Psychology Press)
[6] Artigue M and Blomhøj M 2013 *Conceptualising inquiry based education in mathematics* (Prancis: Springer)
[7] Andrew S 2015 *Guided science inquiry instruction with students with special education needs* (America: National Center for Research on Rural Education)
[8] Fielding M 2012 *Beyond student voice: Patterns of partnership and the demands of deep democracy* (London: Revista de Educación)
[9] Anna H Lint 2013 *E-Learning Student Perceptions on Scholarly Persistence in the 21st Century with Social Media in Higher Education* (USA: Scientific Research Publishing)
[10] Fakomogbon M A 2012 Instructional media in teaching and learning: a nigerian prespective Global Media Journal African Edition 6(2) pp. 218-228.
[11] Mariani S 2014 The Effectiveness of Learning by PBL Assisted Mathematics Pop Up Book Againts The Spatial Ability in Grade VIII on Geometry Subject Matter *International Journal of Education and Research* **2** pp. 531-548
[12] Ulya S A 2017 The Development of Pop-up Book on the Role of Buffer in the Living Body *European Journal of Social Sciences* **4**(4) pp 213-221
[13] Sukmadinata N S 2007 *Metode penelitian pendidikan* (Bandung: PT Remaja Rosda Karya)

Acknowledgement

On this opportunity, I would like to give my special thank to all of those who have given me valuable contributions in accomplishing this research.

IOP Conf. Series: Journal of Physics: Conf. Series **1108** (2018) 012030 doi:10.1088/1742-6596/1108/1/012030

Steady Infiltration in Heterogeneous Soil

I Solekhudin

Department of Mathematics, Faculty of Mathematics and Natural Sciences, UniversitasGadjahMada

E-mail: imams@ugm.ac.id

Abstract. In this study, problems involving steady infiltration from periodic channels in heterogeneous soils are considered. The problems are governed by a Richards's equation. To study the problems, the equation is transformed into a modified Helmholtz equation. The modified Helmholtz equation is then solved numerically by employing a Dual Reciprocity Method (DRM). The method is applied to solve infiltration in different types of heterogeneous soils. Numerical solutions in terms of suction potential are compared. The solutions indicate that soil layers produce jumps in values of the suction potential. The results also indicate that the soil type influence to the water content in the soil.

1. Introduction

Problems involving infiltration in homogeneous soils has been considered by numerous researchers, such as Waechter and Mandal [8], Pullan and Collins [4], Clements and Lobo [3], and Solekhudin [6, 7]. Waechter and Mandal [8] have studied infiltration from a semicircular cylindrical trench and hemispherical pond. Pullan and Collins considered steady infiltration from buried and surface cavities [4]. In other hand, Solekhudin [6, 7] has considered the water flow from periodic channels in homogeneous soils. Clements and Lobo examined water ow from single channel in a homogeneous soil [3]. The works considered by such researchers were ow in homogeneous soil. Hence, a possible direction to continue their study is to study the infiltration in heterogeneous soils.

In this paper, a steady infiltration problem from periodic trapezoidal channels in three-layered soils is considered. The problem is governed by a Richards's equation. Since the governing equation may not be solved analytically, a numerical method called Dual Reciprocity Method (DRM) is employed. In order to employ the method, the governing equation is transformed into a modified Helmholtz equation using a set of transformation. The modified Helmholtz equation is then solved using the method. The results obtained are then used to compute values of suction potential.

2. Formulation of the problem

In this study, the researchers consider three-layered soils. The three-layered soils consist of Silt Loam (SL), Pima Clay Loam (PCL) and Touchet Silt Loam (TSL). The order of the layers, from the surface of soil, is SL-PCL-TSL. The depth of the first layer is 100 cm. The second layer has the same depth as the first layer. On the surface of the first layer, periodic trapezoidal channels are created. For every centimeter length, the channel has a sunken surface area of 100 cm^2. The distance between the centers of two adjacent channels is 200 cm. It is assumed that the channels are identical and sufficiently long. The channels are filled with water. It is also assumed that water fluxes are only from the surface of the channels, and such fluxes are constant. Given this situation, it is expected to determine the values of suction potential in the soils. To solve the problem, it was used a Cartesian coordinate system *OXYZ* with *OZ* vertically positive downwards. The problem described is illustrated in Figure 1.

Since the length of the channels are sufficiently long compared to the width of the channels,water flow may be assumed two-dimensional flow. From the assumptions, the problem is symmetrical about the plane $X = \pm 100k$, $k = 0, 1,2,\ldots$. Hence, it is sufficient to consider a semi-infinite region bounded by $0 \le X \le 100$ cm and $Z \ge 0$. R denotes such region where it is bounded by a curve C, as shown in Figure 2.

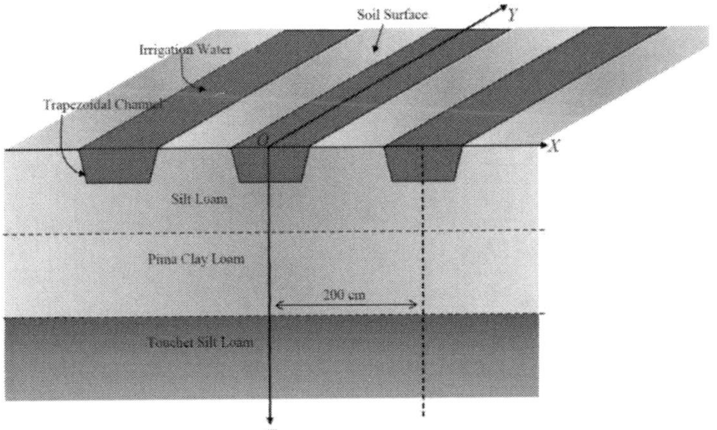

Figure 1. Geometry of periodic trapezoidal channels.

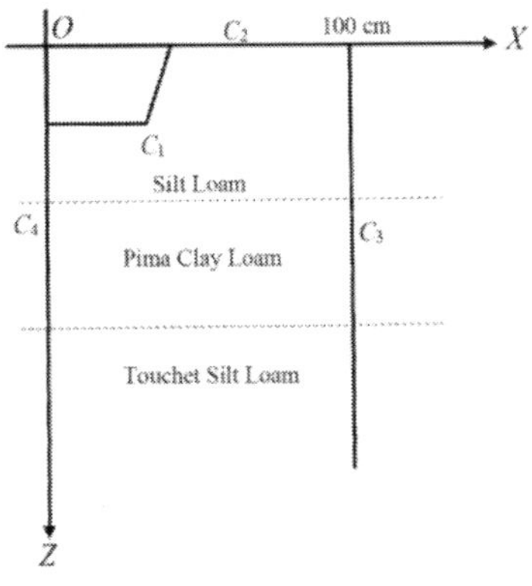

Figure 2. Region of the problem

From the assumption above, flux across the surface of channel is constant, v0, and flux across the surface of soil outside the channel is 0. Fluxes across $X = 0$ and $X= 100$ cm are 0, as the problem symmetrical about these two lines. The derivatives $\frac{\partial \Theta}{\partial X} \to 0$ and $\frac{\partial \Theta}{\partial Z} \to 0$ as $X^2 + Z^2 \to \infty$ [2].

3. Basic equations

The governing equation of steady infiltration problems from periodic channels in homogeneous soil has been presented in [5]. Set of transformations needed to transform the governing equation into a modified Helmholtz equation are also presented and discussed in [5]. In this study, the governing equation and the method used are similar to those in [5]. However, for the completeness and the convenience of readers, some details of the equation and the method are presented. The governing equation of steady infiltration in three-layered soils is in the form of following Richard's equation.

$$\frac{\partial}{\partial X}\left(K\frac{\partial \psi}{\partial X}\right) + \frac{\partial}{\partial Z}\left(K\frac{\partial \psi}{\partial Z}\right) - \frac{\partial K}{\partial Z} = 0, \qquad (1)$$

Where K is the hydraulic conductivity and ψ is the suction potential.

The Kirchhoff transformation

$$\Theta = \int_{-\infty}^{\psi} K(s)ds, \qquad (2)$$

WhereΘ is the Matric Flux Potential (MFP), and an exponential relationship between K andψ

$$K(\psi) = K_0 e^{\alpha\psi}, \qquad (3)$$

Where Ks is the saturated hydraulic conductivity and $\alpha$is a constant soil parameter corresponding to the roughness of the soil, transform Equation (1) to

$$\frac{\partial^2 \Theta}{\partial X^2} + \frac{\partial^2 \Theta}{\partial Z^2} = \alpha\frac{\partial \Theta}{\partial Z}, \qquad (4)$$

In the present study, there are distinct values of αin three different layers. Assume α_1, α_2 andα_3 are the values of αin the first, second, and third layer, respectively. Consequently, Equation (4) cannot be used to solve the problem described. Hence, it needs to overcome this issue. A method that may be employed to solve the problem is as that has been used by [9], which is presented as follows:

$$\alpha^* = \frac{\alpha_1 + \alpha_2 + \alpha_3}{3}.$$

Usingα^*, Equation (4) now is written as

$$\frac{\partial^2 \Theta}{\partial X^2} + \frac{\partial^2 \Theta}{\partial Z^2} = \alpha^* \frac{\partial \Theta}{\partial Z}, \quad (5)$$

The flux normal to the surface with outward pointing normal $\mathbf{n} = (n_1, n_2)$ is

$$F = U\, n_1 + V\, n_2, \quad (6)$$

Where

$$U = -\frac{\partial \Theta}{\partial X} \text{ and } V = \alpha\Theta - \frac{\partial \Theta}{\partial Z} \quad (7)$$

Now, substituting the dimensionless variables

$$\Phi = \frac{\pi}{50\alpha^*}\Theta, \qquad x = \frac{\alpha^*}{2}X, \qquad z = \frac{\alpha^*}{2}Z,$$
$$u = \frac{\pi}{25 v_0 \alpha^*}U, \qquad v = \frac{\pi}{25 v_0 \alpha^*}V, \; f = \frac{\pi}{25 v_0 \alpha^*}F, \quad (8)$$

Into Equation (5), it is obtained

$$\frac{\partial^2 \Phi}{\partial x^2} + \frac{\partial^2 \Phi}{\partial z^2} - 2\frac{\partial \Phi}{\partial z} = 0, \quad (9)$$

The dimensionless flux is

$$f = -\frac{\partial \Phi}{\partial x} n_1 + \left(2\Phi - \frac{\partial \Phi}{\partial z}\right) n_2. \quad (10)$$

Transformation

$$\Phi = \phi e^z, \quad (11)$$

Transforms Equation (9) into

$$\frac{\partial^2 \phi}{\partial x^2} + \frac{\partial^2 \phi}{\partial z^2} - \phi = 0, \quad (12)$$

and

$$f = -e^z\left(\frac{\partial \phi}{\partial n} - \phi\, n_2\right), \quad (13)$$

where $\frac{\partial\phi}{\partial n} = \frac{\partial\phi}{\partial x}n_1 + \frac{\partial\phi}{\partial y}n_2$.

Boundary conditions, which are described in the preceding section, are written in terms of $\partial\phi/\partial n$ is presented as follows.

$$\frac{\partial\phi}{\partial n} = \phi n_2 + \frac{\pi}{25\alpha}e^{-z}, \text{ on the surface of the channels,} \tag{14}$$

$$\frac{\partial\phi}{\partial n} = -\phi, \text{ on } z = 0 \text{ outside the channels,} \tag{15}$$

$$\frac{\partial\phi}{\partial n} = 0, \text{ on } x = 0 \text{ and } z \geq 0, \tag{16}$$

$$\frac{\partial\phi}{\partial n} = 0, \text{ on } x = 50\alpha \text{ and } z \geq 0, \tag{17}$$

$$\frac{\partial\phi}{\partial n} = -\phi, \text{ for } 0 \leq x \leq 50\alpha \text{ and } z = \infty. \tag{18}$$

Equation (12), which is a modified Helmholtz equation, subject to Boundary conditions (14) - (18) that may be solved numerically by using a DRM to obtain numerical values of ϕ. Using the values of ϕ obtained, form Equations (2), (3), (11) and Dimension less variables (8), values of ψ may be computed by the formula of:

$$\psi = \frac{1}{\alpha}\ln\left(\frac{50v_0\alpha\phi e^z}{\pi K_s}\right). \tag{19}$$

Readers may refer to [5] for the details of the DRM.

4. Results and discussion

In this section, the DRM is employed to solve a problem involving steady infiltration in three-layered soil. The first layer, which is Silt Loam, has a depth of 100 cm. Pima Clay Loam, the second layer, also has a depth of 100 cm. Touchet Silt Loam, which is the third layer, has an infinitely depth. It was set the width and the depth of the channels, which are $200/\pi$ cm and $150/2\pi$cm, respectively. As discussed in the preceding section, the value of α^*must be computed by taking average of the values of αin each layer. Moreover, to computeψ, it needs the values of K_s. The values of α and K_s of the soils are shown in Table 1. These values are as reported by Amoozegar-Fard et al [1].

Table 1. Values of α and K_s.

	α (cm^{-1})	K_s (cm/day)
Silt Loam	1.39×10^{-2}	4.9594
Pima Clay Loam	1.40×10^{-2}	9.9360
Touchet Silt Loam	1.56×10^{-2}	41.9904

To implement the method, the region R must be bounded by a simple closed curve. Hence, an imposed boundary is needed. After several computational experiments, it is obtained that the suitable imposed boundary is $z = 4$. The DRM needs discretization of the boundary and a number of interior collocation points in its implementation. To do so, the boundary is discredited into 197 elements, and 400 interior collocation points are chosen. The results obtained are presented in Table 2, Figure 3, and Figure 4.

Table 2. Values of ψat selected points.

Soil Type	Suction potential (cm)

(10 cm, 25 cm)	Silt Loam	-51.84549382
(10 cm, 75 cm)	Silt Loam	-66.91679566
(10 cm, 125 cm)	Pima Clay Loam	-119.99130522
(10 cm, 175 cm)	Pima Clay Loam	-121.26207966
(10 cm, 225 cm)	Touchet Silt Loam	-194.53090712
(10 cm, 275 cm)	Touchet Silt Loam	-194.48987849
(90 cm, 25 cm)	Silt Loam	-97.66829776
(90 cm, 75 cm)	Silt Loam	-80.29745086
(90 cm, 125 cm)	Pima Clay Loam	-123.80530011
(90 cm, 175 cm)	Pima Clay Loam	-122.36097650
(90 cm, 225 cm)	Touchet Silt Loam	-194.82079719
(90 cm, 275 cm)	Touchet Silt Loam	-194.58348375

Table 2 shows values of ψ at selected points or locations. In this table, values of locations under the channels, X = 10 cm, and further from the channels, X= 90 cm are presented. It can be seen that under the channels, at the first and the second layer, values of ψ decreases as soil goes deeper. In the first layer, which is Silt Loam, from Z = 25 cm to Z= 75 cm values of declines about 29.07 %. In Pima Clay Loam, ψ drops about 1.06 %, from Z= 125 cm to Z = 175 cm. In the third layer, values of ψ are about the same. However, at locations further from the channels, X = 90 cm, ψ increases as Z increases. For the same values of Z, increasing in Silt Loam is about 17.79 %. The rise in ψ in Pima Clay Loam is about 1.17 %. In Touchet Silt Loam, the values of $\psi$are also about the same.

The results indicate that at location under the channels, in the first and second layer, suction potential drops as soil goes deeper. The drops in at locations near the surface of the soil are higher than those deeper. The drops get lower as Z increases, and eventually after reaching some value of Z, there are no drops inψ. Conversely, at locations further from the channels, suction potential in the first and second layer increases as Z increases. The increase falls as soil goes deeper, and after reaching a level of soil depth, there are no more increase in ψ.

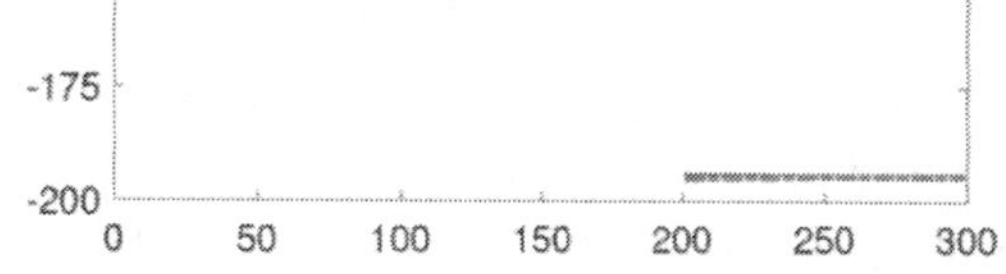

Figure 3. Graphs of ψ along Z-axis at selected values of X.

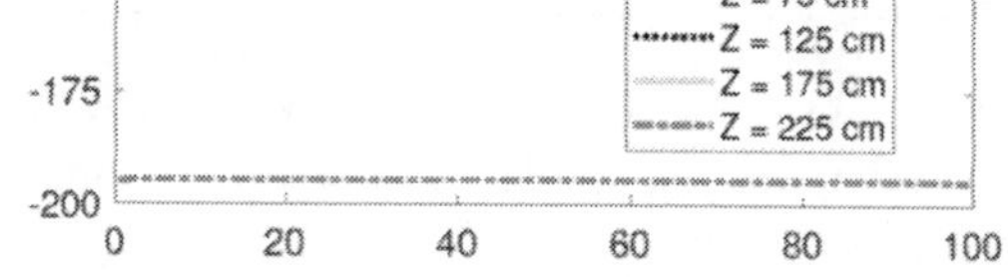

Figure 4. Graphs of ψ along X-axis at selected values of Z.

Graphs of ψ at selected values of X along Z-axis are presented in Figure 3. In the first layer,values of ψ at X = 10 cm, especially at the surface of channels, the highest compared to those at any locations. In other hand, values of ψ at X = 90 cm, especially at the surface of the soil, the lowest. These results indicate that the highest value of suction potential occurs at the location nearest the center of the channels, and the lowest value of ψ is at the furthest away from the channels at surface of the soil.

As those in the first layer, in the second layer ψ reaches its highest value at X= 10 cm, and the lowest value of ψ is attained at X = 90 cm. In the third layer, it can be observed that values of ψ have reached their convergent value. These results are expected as $\partial\Theta/\partial X \rightarrow 0$ and $\partial\Theta/\partial Z \rightarrow 0$as $Z \rightarrow \infty$.

It can also be observed that there are jumps in values of ψat Z = 100 cm and Z = 200 cm. These due to the situation that soil type at Z<100 cm is different from that Z>100 cm. Similarly, the situation occurs at Z = 200 cm. Among the three layers, the first layer results in higher values of ψ. In other hand, the third layer cause in lower values of ψ. These imply that soil with higher values of $\alpha$and K_s has lower values of ψ.

Figure 4 shows graphs of at selected values of Z along X-axis. It can be seen that there are variation in values ofψ at Z = 25 cm. At Z = 75 cm, there are less variation in the values of ψ. As Z goes deeper, the variation of ψ along X-axis is lesser, and eventually after reaching a deep level there are no variation in values of ψ.

5. Concluding remarks

Problems involving steady infiltration from periodic trapezoidal channels have been solved numerically by transforming the governing equation and using a Dual Reciprocity Method (DRM) to solve the resulting partial differential equation. The method is applied to obtain numerical values of suction potential.

The results obtained indicate the effect of soil type to the suction potential. Soil with higher hydraulic conductivity and more porous, produce such higher suction potential than less porous medium and lower hydraulic conductivity. More decrease or increase in the values of suction potential occurs at shallower level of soil. There are no decrease or increase in values of suction potential after a level of soil depth. In shallower level of soil, there are more variation in values of the suction potential than those in deeper level of soil. Moreover, there are jumps in the values of suction potential. The jumps occur at locations where the soil type changes.

References

[1] Amoozegar-Fard A, Warrick AW and Lomen DO 1984*J. of Irrig. and Drain. Eng.***110** 107
[2] Batu V 1978 *Soil Sci. Soc. Am. J.***42** 544
[3] Clements DL and Lobo M 2010 *Eng. Anal. with Bound. Elem.***34** 1100
[4] Pullan AJ and Collins IF 1987 *Water Resour. Res.***23** 1633
[5] SolekhudinI and AngKC 2013 *ANZIAM J.***54** 171
[6] SolekhudinI 2016 *Far East Journal of Mathematical Sciences***100** 2029
[7] SolekhudinI 2017 *Advances and Applications in Fluid Mechanics***20** 127
[8] Waechter RT and Mandal AC 1993 *Water Resour. Res.***29** 457
[9] Watson KK and WhislerFD 1972 *Soil Sci. Soc. Amer. Proc.***36** 251

A Comparative Study: Concept Number Skills of Children Aged 4-5 Years after Playing Two Different Games

E Oktavianingsih[1] and K Fahuzan[2]

[1]Department of Early Childhood Education, Graduated School, Yogyakarta State University, Indonesia.
[2]Department of Mathematics Education, Graduated School, Yogyakarta State University, Indonesia.
Corresponding Author: eka.oktavianingsih2016@student.uny.ac.id

Abstract. Mastering mathematics for preschoolers is essential because it will make children become more skilled in solving problem. Currently, learning number concepts becomes easier because the smartphones have provided the games, including "Belajar Angka" and "Marbel Belajar Berhitung". Games used to introduce the concept of numbers to children mostly in two kinds, text games, and illustrative games. A text game used many symbols, while an illustrative game used images of concrete objects. This study aimed to determine the understanding of the concept number skills of children between a text game and an illustrative game in learning through mobile games. This study used a quantitative approach. 20 children who are 4-5 years old were participated and they were divided into two groups. A test was involved to obtain information about the concept number skills of children. The results showed that the illustrative game was significantly better in improving the understanding of the concept of numbers through pre-test and post-test comprehension, $t(18) = 2.855$, $p = 0.011$. This paper also discussed the design of each game.

1. Introduction

Early age is an appropriate time for children to open their horizons about the surrounding phenomenon leading to the concept of mathematics [1]. Mathematics for early childhood is a way of looking at the children in seeing, experiencing, and understanding the environments around them. Therefore, learning mathematics for children is not only learning numbers, but it can include counting, sorting, measuring, building and describing shapes, finding patterns, estimating, among other experiences. Learning mathematics for children is important because it will make children become more skilled in solving problems in the future.

Mathematics mastery at the early stage is really crucial and becomes the indicator of the advanced mathematics achievement and mastery [2]. Experiences and knowledges obtained by children on mathematics during their preschool will create important foundations for their experiences in primary education [3]. The child's skill in understanding the concept of numbers is the ability to understand the value of small quantities rapidly, making decisions about the numbers and values of numbers (i.e. 4 close to 3 compared to 6) [4]; understanding the count principle (i.e. the symbol of number indicates number of objects); and combining and separating sets (i.e. 3 and 2 will be 5, and 5 can be split into 2 and 3). The Ministry of Education and Culture Law Number 137 in 2014 on the National Standard of Early Childhood described that the children's ability aged 4-5 years in recognizing the concept of numbers at least includes the ability in: (a) spelled out many objects, (b) recognizes the concept of numbers, and (c) recognizes the number symbols.

The introductions of the concept of numbers in preschool are still done by conventional methods such as a story telling, question and answer, assignment, and direct practice. In fact, today's technology is growing rapidly on all fronts, one of them is the smartphone. Most children are adept at playing smartphones for recreational purposes rather than educative purposes. Currently, many smartphone users in the world reached 1.86 billion or about 25.7% [5]. From the population of children around the world, early childhood was already starting to use the smartphone that is equal to 63% [6]. When viewed from the side of its usage, the most widely used smartphone is to play game, which is equal to 39% [7]. Viewed from groups of children under 8 years old, 7% use handled video game, and 6% use video game console [8].

Game is a physical or mental contest based on special rules with the aim of entertaining and rewarding the players [9]. Game based learning is the use of game to support learning process [11]. It is not an enemy to be avoided, as it provides the best chance of involving the child in the actual learning process [13]. As a learning media, the use of games is very useful, because games can enliven learning topics and are very effective in handling problem solving [14].

Currently, the development of game for children has been faster than before [15]. In quantity, the number of games for early childhood is more easily accessed by smartphone. The number of educational games has increased significantly but limited studies in assessing its effectiveness and validity [16]. Early childhood has been introduced to the concept of numbers [17]. To teach the concept of numbers in children aged 4-5 years, researchers found two types of games developed by application providers, namely games supported with the illustration of objects and games supported with using the symbol. Through this research, researchers analyses two types of games and compare their effectiveness in improving the skills of children to know the concept of numbers. The research questions are as follows:

a. Which game that is more effective for improving concept number skills of children aged 4-5 years old?
b. Why is that game effective in improving the child's skill to recognize the concept of numbers?

2. Method

2.1 Design

This study used an experimental method with two groups. Each group got the different maipulation. Before the manipulation was done, the measurements were done in both groups (pre-test), and the manipulation effect was measured in both groups (post-test). The selection of subjects in both groups used randomization so that the subject conditions in both groups were equivalent [18]. The randomization criteria were children aged 4-5 years.

2.2 Instrument

Measurements aimed to assess the skill of children to understand the concept of numbers. Measurements were made using a test that have been validated by the expert with the Cronbach alpha score, which shows the number 0.82 indicating the instrument is included in either category [19].

The skills to recognize the concept of numbers measured consisted of four indicators i.e: (1) knowing the concept of *many* and *little*, (2) calculating coherently, (3) calling the symbol of the number, and (4) calling the number symbol corresponding to the number of objects. The range of scores for each indicator was: (1 = Not Developed), (2 = Beginning), (3 = Expanding According to Expectations), and (4 = Very Good Develop). The overall maximum score of the indicator is 16.

Text-based games are represented by "Belajar Angka", a game developed by Solite Kids, while illustration-based games are represented by "Marbel Belajar Berhitung", a game developed by Educa Studio. "Belajar Angka" game was a series of educational applications for children under the age of 8 that aimed to help children learn about basic numbers. There were two features in this apps, learning and playing. Learning features consisted of learning basic numbers (0-9), learning teens and dozens (10-99), studying numbers hundreds (100-900), learning the numbers of tens thousands and hundreds thousands, and learning millions. The play features consisted of playing magic number lamps, playing number trucks, playing bianglala, playing figure dolls, playing fish ponds, playing exciting numbers, and playing shooting meteors. "Marbel Belajar Berhitung" game has characteristics using illustrations in the form of pictures of objects in the delivery of material, such as flowers, balloons, and others used to guide children in understanding the concept of numbers. Because of the limitations of space in displaying the objects that exist, then the calculation of objects only until the number 9. In addition, the assessment parts of the game also used the illustration of objects, by calculating the existing objects there and choosing the right answer. The differences of games' display were shown in Table 1.

Table 1. The differences of games' display

No.	Component	Text Game	Illustrative Game
1	Opening		
2	Introducing the numbers		

2.3 Procedure
Subjects in this study were 20 children aged 4-5 years in kindergarten ABA Kalimundu, Bantul Regency, which was divided into two groups. The subject selection requirements should never use a smartphone before. They were divided into 2 groups, an illustrative game and a text game groups. The flow of this research included: (a) giving the pretest, (b) introducing how to play the games, (c) learning with the games (the concept of many and little, number of objects, symbols of numbers), and (4) giving a postest.

The skills to recognize the concept of numbers were given before and after learning by games. The test involved four main indicators, i.e., knowing the concept of *many* and *little*, calculating coherently, calling the number symbol, and calling the number symbol corresponding to the number of objects. It divided into 32 question items (8 questions for each indicator). The total value will then be categorized according to four categories, namely Very Good Developed (12-16), Expanding According to Expectations (8-11), Beginning (4-7), and Not Developed (1-3).

2.4 Data Analysis
This research used two kinds of tests including a normality test and a hypothesis test through SPSS.21. The normality test aimed to decide the distribution of data and establish the statistical techniques that should be used in the next step. The next step was using a One Kolmogorov-Smirnov Test method with a significance level (α) 5% or 0.05. Hypothesis test used to obtain the differences of concept number skill level before and after using a game in each group. Independent Sample t-Test was used to test the hypothesis.

3. Result and Discussion

3.1 Result
Most of the participants in this study were boys (n = 12, 60%), and the rest of the participants were girls (n = 8, 40%). Normality and homogeneity tests were performed for pre-test results from both groups. Based

on the result of pre-test value, both groups showed there was no significant difference of value. This suggests that the pre-test data in both groups is normal and homogeneous. Furthermore, based on post-test results shows that games with the illustrations were better than games with the text. Full pre-test and post-test results were shown in Table 2.

Table 2. Data analysis of *pre-test* and *post-test* in two groups.

		Mean	N	Std. Deviation
Text Game	*Pre-test*	8.5	10	1.95789
	Post-test	10.1	10	1.66333
Illustration Game	*Pre-test*	8.4	10	2.54733
	Post-test	12.8	10	2.48551

Based on the result of pretest, the average score of the group with the text game was 8.50 with the standard deviation of 1.95789, and group with the illustrative game was 8.4 with the standard deviation 2.54733. The postest results obtained the average value of the text game group of 10.1 with the standard deviation 1.66333, while the illustrative game group of 12.8 with a standard deviation of 2.48551.

Based on Table 2, it shows that post-test score in the group of illustrative game is higher than the text game group. The results of pre-test and post-test in both groups then analysed by using One Kolmogorov-Smirnov Test, and data is normally distributed and homogeneous. The next analysis is t-test by using Independent Sample Test. From Table 3, it appears that the indicators of knowing the concept of *many* and *little*, calculating coherently, calling the symbol of number is less significant in the both groups. But on the indicator, the number symbol corresponding to many objects showed a significant difference (d = 1.3).

Table. 3 The difference of score in each indicator of concept number skills

Indicator / Score	knowing the concept of *many* and *little*	calculating coherently	calling the number symbol	calling the number symbol corresponding to the number of objects
Text Game Group	2.7	2.4	2.7	2.3
Illustrative Game Group	3.2	3	3	3.6
Deviation	0.5	0.6	0.3	1.3

3.2 Discussion

The results of this study indicate that, based on post-test results, the skills to comprehend the concept of numbers in the illustrative games was better than the text game with a balanced pre-test score. Of course, illustrative games could be said to be more effective in improving the ability of understanding the concept of numbers. The quality of educational games for early childhood was certainly influenced by the suitability of games on the principles of educational games. Miller & Kocurek argued that the principles of developing educational games for early childhood including: (a) game compatibility with child development, (b) fitness of games with learning objectives, (c) facilitating diversity, and (d) game compatibility with the principles of play [16].

Table 4. The suitability of the game "Belajar Angka" and game "Marbel Belajar Berhitung" against the principles of game development by Miller & Kocurek

No	**Principles**	**Component**	**The type of game**

			Belajar Angka	Marbel
1.	Compatibility with child development	Fine motor	S	S
		Cognitive	LS	S
		Language	S	S
2.	Compatibility with the learning's goals	-	LS	S
3.	Facilitating diversities	-	S	S
4.	Compatibility with the play's principles	Spontaneous	S	S
		Volunteer	S	S
		Interesting	LS	S
		Get rewards	S	S

Note: VS: very suitable, S: suitable, QS: quite suitable, LS: less suitable, NT: not suitable.

Based on Table 4., it showed that an illustrative game has fulfilled the principles in developing games compared with text game. First, the illustrative game was appropriate better than a text game in the cognitive development of children aged 4 to 5 years old because it was using concrete objects as a material to convey the concept of numbers. According to Jean Piaget, the children aged 4-5 years old was still at the pre-operational phase. Children at this stage requires concrete materials to help them in the learning process [20]. Furthermore, in teaching the concept of number to children has to consider the three phases of the process including the use of concrete objects, then gradually replace them into images representing objects, and finally to an abstract level using numbers and symbols [21]. An illustrative game has involved images representing concrete objects (flowers and balloons) to count, before displaying the numbers. In contrast, a text game directly displays the symbol of the numbers.

Both games have also considered the fine motor skills of the children, and also stimulate the "tap" gesture. The illustrative game has involved slide, drag, and drop gestures for children. Basically, children aged 4-5 years have mastered 7 basic movements (seven fundamental gestures), namely: tap, flick, slide, drag and drop, rotate, pinch and spread [22]. Educational games should pay special attention to the intended audience, which should dictate the level of communication and in particular, the language used in the game [16]. Both games have a child-friendly language, such as passing one command at a time.

The games have facilitated children's diversities from the gender perspective. They involved objects that were liked both of boys and girls, i.e. fishes, flowers, lamps, and trucks. The games also made children felt interested and introduced to the modern technology, so the children would explore by themselves [23]. The learning purposes of illustrative game were also more relevant than the text game. It focused on the introduction of the concept of numbers in more depth by giving meanings to many objects and the concept of numbers, while the text game focused on the introduction of the symbol of numbers. Actually, to recognize the numbers well, firstly children had to understand the numbers 1 to 10, then the children could be developed in the higher level. The use of illustrations in the form of concrete objects would be more interesting to the children than using number symbols alone without knowing the meaning. An illustrative game group was more excellent in calling the number symbol corresponding to the number of objects. It was influenced by the characteristics of the games, and if the illustrations involved in teaching concept number skills, it would give the children the meaningful senses [23].

4. Conclusion

The concept number skills of children aged 4-5 years old in the group of an illustrative game are higher than in the group of a text game. One of the indicators, the symbol of number corresponding to the number of objects, has a significant difference in both groups. However, the other three indicators are a less significant of both. Overall, the illustrative game had occupied the principles of developing game, and the text game was less meet the principles of developing game.

Parents or teachers could be used these games in learning number concepts if they have complied the principles of developing games for children. For the game developers, they should consider the principles of developing games in making games for children.

5. Acknowledgement

Acknowledgment is conveyed to all participants of this study.

References

[1] J A Brewer 2007 *Introduction early childhood education, preschool through Primary Grades.* (Boston: Pearson Education, Inc.) *Sixth Edition*
[2] J. Sarama, A. A. Lange, D. H. Clements, dan C. B. Wolfe 2012 *Early Child. Res. Q.* **27** 489
[3] T. Sezer, M. Güral, G. Güven, dan K. E. Azkeskin 2013 *J. Educ. Instr. Stud. World* **15**
[4] N. C. Jordan, D. Kaplan, C. Ramineni, dan M. N. Locuniak 2009 *Dev. Psychol.* **45** 850
[5] C. Liu 2015 *Worldwide internet and mobile users* (August. New York: eMarketer).
[6] A. Nanji 2013 *75% of American Children Under 8 Have Access to a Smartphone or Tablet*" (New York)
[7] A. Moscaritolo 2012 *Tablet Users Older Than Smartphone Users* (PC Mag UK), 2012. [Daring].
[8] F. Richter 2013 *Nearly 1 in 5 Kids Use Mobile Devices Every Day* (Hamburg)
[9] P. Noemí dan S. H. Máximo 2014 *Univers. J. Educ. Res.* **2** 230
[10] C. Perrotta, G. Featherstone, H. Aston, dan E. Houghton 2013 *Game-based learning: latest evidence and future directions* (Berkshire: NFER)
[11] R. Al-azawi, F. Al-faliti, dan M. Al-blushi 2016 *Int. J. Innov. Manag. Technol.* **7** 132
[12] B. Harriehausen-mühlbauer 2013 *J. Syst. Cybern. Inf.* **11** 82
[13] M. Prensk 2003 ACM Comput. Entertain. **1**
[14] S. Boyle 2011 UCD Teach. Learn., **5** 254
[15] J. Blow 2004 *Queue* **1** 28
[16] J. L. Miller dan C. A. Kocurek 2017 *J. Child. Media* **11** 314
[17] K. A. Gafoor dan A. Kurukkan 2015 *UGC Spons. Natl. Semin. Pedagog. Teach. Educ. Trends Challenges* **1**
[18] E. Nahartyo 2013 *Desain dan Implementasi Riset Eksperimen* (Yogyakarta: UPP STIM YKPN)
[19] N. A. Thompson 2013 *Reliability & Validity* (USA: whitepaper)
[20] L. C. Chin dan E. Zakaria 2015 *Creat. Educ.* **6** 1306
[21] R. Mononen 2014 *Early mathematics interventions: Supporting young children with low performance in mathematics.* (research report)
[22] N. Azah dan A. Aziz 2013 *Int. J. Comput. Inf. Technol.* **2** 447
[23] S. Petrovska, D. Sivevska, dan O. Cackov 2013 *Procedia - Soc. Behav. Sci.* **92** 880

IOP Conf. Series: Journal of Physics: Conf. Series **1108** (2018) 012032 doi:10.1088/1742-6596/1108/1/012032

Metacognitive Knowledge of a Student in Planning the Solution of Limit Problems

Pathuddin[1,2,*], I K Budayasa[1] , A Lukito[1]

[1]State university of Surabaya, Jl. Ketintang Surabaya, Indonesia 60231
[2]University of Tadulako, Jl. Soekarno Hatta KM 9, Indonesia

*Corresponding Author: pathuddin@yahoo.com

Abstract. This study aims to describe metacognitive knowledge of a student in choosing an approach or strategy in solving the limit problems. The type of this research is explorative with qualitative approach, that is from written works and depth interviews. These collected data will be analysed in depth how the subject used metacognitive knowledge. The results show that the student in choosing approach or strategy in solving the limit problems uses metacognitive knowledge, either declarative, procedural, or conditional ones.

1. Introduction

As a basic skill of calculus, an understanding of limit is needed in order to build an understanding of the basic concepts of calculus. This suggests that a good students' understanding of the limit is needed so that they understand the concepts that are based on the concept of limit. Metacognition is part of the ability of self–knowledge-monitoring. Furthermore, Solso explains that metacognition has an impact on the supervision and control of information retrieval processes and inference processes that occur in the memory system. Monitoring refers to the way we evaluate what we know or what we do not know. Metacognition as an advanced thought involves active control over cognitive processes [1, 2].

Metacognition is widely used, among others related to efforts in optimizing the ability of students in solving problems [3-6], optimizing learning outcomes that can be achieved by students [5,7,8], understanding the contents of a reading topic [9, 11], or improving someone's abilities to become a successful learner [12,13]. Therefore, the efforts that involve metacognition in various learning activities are expected to provide benefits to improve the quality of learning being implemented.

Metacognition as a thought is about thinking itself which is the interaction between three important aspects they are: 1). knowledge of self-thinking process, 2) control or self-regulation, and 3). beliefs and intuition. Knowledge of thinking processes concerns how accurate a person is in expressing their thinking processes, self-awareness or self-regulation, the accuracy of a person in maintaining and managing what they must do when it comes to solving mathematical problems, and how accurately someone in using input from his/her observations to direct activities to solve problems, whereas beliefs and intuitions concern any mathematical ideas that are prepared to solve mathematical problems and how those ideas form ways to solve math problems [3,6,14].

This interaction is very important because our knowledge of our cognitive process can help in regulating the things around us and selecting strategies to improve our cognitive abilities in the future

[14]. The metacognitive process includes the ability to ask and answer questions about (1) What do I know about this subject, topic and subject matter? (2) Do I know what I should know? (3) Do I know where I can get some knowledge information? (4) How long will it take me to learn this? (5) What strategies and tactics can I use to learn this? (6) Do I understand what I hear, read or see? (7) How do I know if I am studying at the appropriate level? (8) How can I see if I make a mistake? (9) How should I revise my plan if it does not fit my expectations and satisfaction? [6, 14]

Metacognitive knowledge is, part of someone's knowledge that acquired through interaction with others as a cognitive being and with different tasks, goals, actions and experiences of cognition. Metacognitive knowledge consists primarily of knowledge or belief about what factors or variables play a role and interact and influence the direction and outcome of cognitive activity. This kind of knowledge is also known as awareness of cognition. There are three main factors or variables in metacognitive knowledge which are: (1) individual, (2) task, and (3) strategy. The "individual" category includes everything someone believes about their nature and others as a cognitive processor [12, 15, 16]. It relates to the kind of knowledge gained and the beliefs about human beings as cognitive beings. The "task" category relates to information available to a person during cognitive activity. The "strategy" category deals with an important issue of knowledge gained through a strategy that is likely to be effective for achieving goals in a cognitive endeavour [11].

Metacognitive knowledge is distinguished by three types of metacognition, that are: (1) person metacognition that is knowledge of the ability of memory and limitations of a person, including the limitations of self, and ability to monitor the experience of his/her memory directly, (2) task metacognition is knowledge of how difficult a memory problem related to task specificity, and (3) strategy metacognition is the knowledge of something that can be done to influence the effectiveness of memory [5, 17].

Brown classifies metacognitive knowledge based on someone's consciousness of their metacognitive knowledge which are: declarative, procedural and conditional [16,18,19]. And then Pierce suggests that: "To increase their metacognitive abilities, students need to possess and be aware of three kinds of content knowledge: declarative, procedural, and conditional. Declarative knowledge is the factual information that one knows; it can be declared—spoken or written. Procedural knowledge is knowledge of how to do something, of how to perform the steps in a process. Conditional knowledge is knowledge about when to use a procedure, skill, or strategy and when not to use it [11, 20].

Metacognitive components in solving mathematical problems in learning are: (a) metacognitive knowledge, (b) metacognitive skills. Metacognitive knowledge of problem solving refers to declarative knowledge, procedural knowledge, and a person's conditional knowledge on problem solving. While metacognitive skills are the controls of their own thinking processes that are differentiated in prediction skills, such as "how difficult is the task", planning skills, such as "what will I do to execute this task?" , monitoring skills, such as "What do I not know I've been getting to achieve goals?" and evaluation skills, such as "are all the tables I created are complete for problem solving"? [13, 21- 23].

Brown further argues that metacognitive knowledge is related to resolute or state, the ability to reflect on complex cognitive processes and discuss them with others (stateable), the ability to know certain facts about cognition that are not true (fallible), and its development is affected by age [7, 16]. This Brown's view indicates that one's metacognitive knowledge will develop as the age and learning experience progresses.

Considering the above opinion, it can be said that metacognitive knowledge is a knowledge that is possessed by a person about his/her cognitive ability, tasks to be performed and cognitive strategies that must be selected. This metacognitive knowledge is the knowledge of a person acquired through interaction with others, through the execution of a task, or through an attempt to achieve a goal. Metacognitive knowledge consists of declarative knowledge, procedural knowledge, and conditional knowledge.

Problem solving is one of the proper methods for learning and teaching mathematics. A person with problem-solving skills has several advantages, including developing critical thinking skills and strengthening math skills. There are four categories in teaching problem solving, namely: solving the problem develops cognitive skills in general, problem solving encourages creativity, problem solving is part of the mathematical application process, and problem solving motivates learners to learn math [6, 22]. With the ability to solve problems obtained from mathematics lessons, students are expected to use to solve problems in everyday life, as revealed by Cooney that is, teaching learners to solve problems, allows learners to be more analytical in making decisions in his/her life [24, 25].

2. Method

This research is explorative type with qualitative approach. The subject is one-student from Mathematics Education department, Faculty of Teacher Training and Education, Tadulako University. He is the student who have followed the calculus course, we assumed that he has adequate knowledge about the concept of limit of a function in one variable.

The data collection includes written assignment and in-depth, unstructured interview for assuring credible data we utilized a time triangulation. To analyse the credible data, we followed the Miles and Huberman model with stages: 1) data reduction, 2) data display, 3) conclusion drawing [10].

3. Results and Discussion

In the following we label the subject of the research with S and the researcher with P. The attached number string indicates the sequence of the interview stage (e.g. (005)S means that the subject respond is in fifth stage in the interview sequence).

The second stage in solving the problems proposed by Polya is the stage of preparing a settlement plan, that is choosing an approach or problem-solving strategy. After understanding the problem, S is to arrange a plan (devising a plan), that is choosing an approach or problem-solving strategy [26]. At this stage S is convinced to draw up a plan of completing the task using a certain strategy to achieve his cognitive goals ((035)S, (036)S). Therefore, S predicts that he is able and sure to retrieve information that has been obtained and contained in long-term memory. Explanation given by S about it can be known from the following interview transcripts:

(035)P : Because you already understand the problem, then what will you do?
(035)S : *Prepare a settlement plan, Sir.*
(036)P : Are you sure you can make that plan?
(036)S : *Yes, I am, Sir.*
(037)P : Why are you so convinced?
(037)S : *I've been taught about this kind of thing, and the limit material has also been studied in Calculus I, Sir.*

S predicts the time to be used in developing an approach or problem-solving strategy that is about nine minutes. S can devise a problem-solving plan because he has already worked on the problem of function limit ((038)S, (039)S). In solving the problem S plans to use the definitions of left and right limits, as well as limit theorems ((040)S). The cognitive strategy was chosen to make it easier in solving the problem ((041) S). The explanation given by S can be known through the following interview transcripts.

(038)P : How much time do you need to create a settlement plan in this question?
(038)S : *Approximately nine minutes, Sir.*
(039)P : Roughly, do you able to do with that time?
(039)S : (S bows the head)
Yes I am, Sir, because I've learned limit of a function.
(040)P : What your plan is like?/What is your plan?
(040)S : *My plan is, I will solve it by using the definition and limit theorem, Sir.*
(041)P : Why do you use that?
(041)S : *To make it easier and normally, a question like that requires the definition and limit theorem.*

Based on the result of the data presentation above, then it is concluded that S used a metacognitive knowledge at the stage of preparing problem-solving plan. To plan, S involves his metacognitive activity with planning the preparation of problem-solving strategy consciously using certain strategy to reach his cognitive goals and by considering the initial knowledge as his declarative knowledge. At this moment S can mention the symbols related to function limit, write the limit definition, and this knowledge helps to prepare a settlement plan. S predicts the time to be used in preparing a problem-solving plan by considering his initial knowledge. This is in line with some expert's opinion that declarative knowledge is factual information or knowledge of known facts [21, 22, 27].

In order to achieve his cognitive goals, firstly S planned to determine the value of the left and right limits ((045)S), so that, from the result he will be able to determine the value of the limit ((046)S). S realized that the limit value of a function at a point does not always exist and the value of a limit of a function at a point can either exist or not. So it depends on the value of the left and right limits ((047)S, (048)S, (049)S, (050)S). Explanation given by S about it can be known from his work and the following interview transcripts:

(042)S : 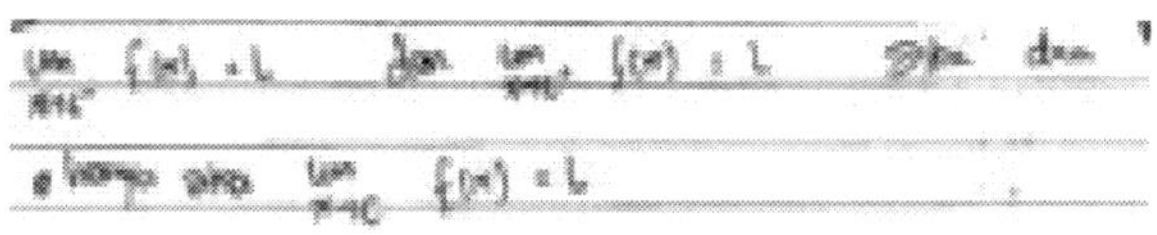

(handwriting S write down the theorem to be used)

....

(045)P : Continue.
(045)S : *Firstly, determine left and right limits.*
(046)P : Continue.
(046)S : *From that results I will determine the value of the limit, Sir.*
(047)P : What do you mean?
(047)S : (S bows the head)
I mean, the value of the function limit at a point does not always exist, Sir.
(048)P : And then?
(048)S : *The limit value of a function at a point can exist and can't exist.*
(049)P : How to make it sure?
(049)S : *Pardon, Sir?*
(050)P : The value of the limit that you said before.
(050)S : *The value of the limit can be known after calculating the left and right limits.*

Based on the results of the data presentation above, it can be concluded that the used his metacognitive knowledge at the stage of preparing a problem-solving plan. To be able to plan, S involves his metacognitive activity, that is knowing the procedure that will be used in solving the problem so that the troubleshooting plan can be achieved. Accordingly, this is in line with the notion of procedural knowledge, that is knowledge of how to do something, how to do every step in the process [8, 21, 22, 27, 28]. In this case, S shows its awareness of its procedural knowledge, where S can perform a procedure or stages by using the limit theorem, so the value of the limit can be obtained.

In evaluating the solution plan that has been settled, S is rechecking whether the plans that have been made are in accordance with the desired results and evaluating the appropriateness of the procedures that had been selected as a basis for concluding that the purpose of the planned cognition has been reached ((052) S (053) F). So that, S can state that the arranged plan can be used to solve the problem and suggest that there may be other ways, but that is all that he thought of ((054) S, (055) S and (056) S). The explanation by S about it can be known from the following interview transcripts:
(052)P: You said that you will make a plan. Is that enough?
(052)S : (S is silent)
I think that is enough, Sir.

(053)P : Do you think this plan can be used to solve the problem?
(053)S : (S looks back at the theorem that he wrote)
Yes, Sir.
(054)P : Are you sure?
(054)S : *Sure, Sir.*
(055)P : Do you have any other plans that can be used to solve that problem?
(055)S : *No, Sir.*
(056)P : Why?
(056)S : *Maybe there is, but that's just what I think, Sir.*

Based on the results of the data presentation above, it can be concluded that he used his metacognitive knowledge S at the stage of preparing a problem-solving plan. To be able to plan, S involves his metacognitive activity, that is by monitoring the implementation of a problem-solving plan with his consciousness planning to use the limit theorem for solving the problems by considering an easy completion. In this case, his awareness of utilizing conditional knowledge, can be seen when setting out to use the limit theorem in solving the problem. S consciously evaluates the steps of the solution plan for the problem in how to re-examine the steps that have been made, whether these have led to the achievement of goals or not, and checks the adequacy of information needed to solve the problem. So, S does not need to plan another way to develop a problem-solving plan [7, 20, 27].

4. Conclusion

Based on a description of metacognitive activity, when choosing the strategy in solving the limit problem he has shown the metacognitive activities, that is by utilizing his metacognitive knowledge, whether in the form of declarative knowledge, procedural knowledge, or conditional knowledge.

Involving metacognitive knowledge is very useful in building the subject's awareness of existing knowledge when solving problems, in the proper context. Therefore, it is suggested to the researcher having a relevant problem to examine more deeply the involvement of metacognitive knowledge in solving the function limit problems in particular and for other problems in everyday life.

5. Acknowlegments

The authors are grateful for financial support of the scholarship *Beasiswa Pendidikan Pascasarjana Dalam Negeri* (BPP-DN), Ministry of Research, Technology and Higher Education, through Doctoral Program of Graduate School of Surabaya State University, Indonesia.

References

[1] Solso R L, Maclin O H and Maclin M K 2008 *Cognitive Psychology* (Jakarta: Penerbit Erlangga)
[2] De Bruin A B and van Gog T 2012 Improving self-monitoring and self-regulation: From cognitive psychology to the classroom *Learning and Instruction* **22** 245–252
[3] De Corte E 2003 Intervention Research: A Tool for Bridging the Theory Practice Gap in Mathematics Education *Proceedings of the International Conference, The Mathematics Education into 21st Century Project* Brno Czech Republic
[4] Gartman S and Freiberg M 1993 Metacognition and Mathematical Problem Solving: Helping Students to Ask the Right Questions *The Mathematics Educator* **6**(1) 9-13
[5] Kirbulut Z D and Gokalp M S 2014 The relationship between pre-service elementary school teachers' metacognitive science learning orientations and their use of constructivist learning environment *International Journal of Innovation in Science and Mathematics Education* **22**(6) 1-10
[6] Schoenfeld E A H 2016 Learning to think mathematically: problem solving, metacognition, and sense-making in mathematics learning to think mathematically (Reprint) **196**(2) 1-38.
[7] Gama C A 2004 *Integrating Metacognition and Mathematical Problem Learning Environment*, (New York: D Phil Dissertation University of Sussex)

[8] Abdellah R 2015 Metacognitive awareness and its relation to academic achievement and teaching performance of pre-service female teachers in Ajman University in UAE *Procedia-Social and Behavioral Sciences* **174** 560-567
[9] Wellman L L 2008 (online) *Metacognition and Reading Comprehension* www.speechpathology.com
[10] Miles M B, Huberman A M and Saldana J 2014 *Qualitative Data Analysis: A Methods Soucebook* (New York: Arizona State University)
[11] Kompal W 2015 Metacognition and Motivation: In context with the academic success of students International *Journal of Multidisciplinary Research and Development* **2**(12) 66-68
[12] Lai E R 2011 (online) *Metacognition: A literature review*. Pearson Research Reports. (http://www.pearsonassessments.com)
[13] Pantiwati Y 2015 Learning strategies, self assessment, and Metacognition in learning science *Proceedings of the National Seminar on Education Biology*
[14] Schoenfeld AH 1992 *Learning to Think Mathematicaly, Problem Solving, Metacognition, and Sense Making in Mathematics The University of California, Berkley. In Grouws. A. D Handbook of Research on Mathematics Teaching and Learning. A project of the NCTM* (New York: Macmillan Publishing Company)
[15] Flavell J H 1979 *Metacognition and Cognitive Monitoring, A New Area of Cognitive-* Developmental *Inquiry, in Nelson, T. O. (Ed)* (Boston: Allyn and Bacon)
[16] Quiles C, Prouteau A, and Verdoux H 2015 Associations between self-esteem, anxiety and depression and metacognitive awareness or metacognitive knowledge *Psychiatry Research* **230** 738–741
[17] Jayapraba G 2013 Metacognitive instruction and cooperative learning- strategies for promoting insightful learning in science *International Journal on New Trends in Education and Their Implications* **4**(1) 165-172
[18] Pantiwati Y and Husamah 2017 Self and Peer Assessments in Active Learning Model to Increase Metacognitive Awareness and Cognitive Abilities *International Journal of Instruction* **10**(104) 185–202
[19] Gassner L 2009 *Developing metacognitive awareness: A modified model of a PBL tutorial* (Bachelor thesis: Malmo University)
[20] Pierce W 2003 (online) *Metacognition: Study Strategies, Monitoring, and Motivation*, A greatly expanded text version of a workshop presented November 17, 2004, at Prince George's Community Colleg. http://academic.pg.cc.md.us/~wpeirce/MCCCTR/metacognition.htm,
[21] Iiskala T, Vauras M, Lehtinen E, and Salonen P 2011 Socially shared metacognition of dyads of pupils in collaborative mathematical problem-solving processes *Learning and Instruction* **21** 379-393.
[22] Hutkemri E Z 2014 Impact of using Geogebra on Students' Conceptual and Procedural Knowledge of Limit Function *Mediterranean Journal of Social Sciences* **5**(23) 873-881
[23] Desoete A 2007) Evaluating and Improving the Mathematics Teaching – Learning process Through Metacognition Departemen of Experimental Clincal and health Psychology, Ghent University & Arteveldehogeschool Spain *Electronic Journal of Research in Educational Psychology* **5**(3) 705-730
[24] Hudojo H 2003 *Pengembangan Kurikulum dan Pembelajaran Matematika* (Malang: Universitas Negeri Malang)
[25] Engelke-Infante N 2007 Students' understanding of related rates problems in calculus (Tempe, AZ: Unpublished Doctoral Dissertation, Arizona State University)
[26] Polya G 1973 *How to Solve it* (Washington DC: Princeton University Press)
[27] Lee M and Baylor A L 2006 Designing Metacognitive Maps for Web-Based Learning. *Educational Technology & Society* **9**(1) 344-348
[28] Cihanoglu M O 2012 Metacognitive Awareness of Teacher Candidates *Procedia – Social and Behavioral* Sciences **46** 4529 – 453

The Way of Students' Reasoning of Elementary Students in Solving Integer Problem

R Febriyanti[1]*, Y Fuad[1], and Masriyah[1]

[1, 2] Program Pascasarjana, Universitas Negeri Surabaya, Surabaya, East Java, INDONESIA

*Corresponding author: rahmafebriyanti16070785041@mhs.unesa.ac.id

Abstract. Students' reasoning may implement to investigate how students learn to think mathematically in solving mathematical problems. This study investigates the way of students' reasoning in solving some integer comparisons. The sample were the 5th grade students of Labschool Unesa, consisting 11 girls and 8 boys. Considering the existing students' portfolio, all students completed both maths ability and integer comparisons tests, and were ranked the result based on the results. Three students were voluntarily chosen from each mathematic ability level to assess their reasoning way in solving integer comparison problem. The data was obtained from the reasoning test about integer comparison and a semi-structured interview based on reasoning test. It is underlined that students used a variety of ways to show their conclusions. Students presented appropriately ways and reasons in solving the problem. The student with high mathematics ability often used the number line to illustrate the problem, its mean used order. Meanwhile the student with medium ability also utilized the number line. The student with low mathematics ability applied commonly the other context to solve the problem. its mean used magnitude reasoning. This result suggest that teachers should exercises the students to more explore their reasoning way in any problem.

1. Introduction

One of material in elementary school is integer. Integers are an important topic in arithmetic. Children in elementary school learn about integer, because they will use it to counting of items in the world[1]. In many everyday contexts or activities, such as the numbers of balls, pencils, books, and so on. But sometimes, students feel confused to ilustrate negative number. When students are asked to compare two integers with negative signs, they are often answered incorrectly, for example $-5 > 2$[2]. In comparing integers, 3 students of the 5th grade gave differently ways of comparing between five pencils and the number of pencils. While another student made a sequence of numbers 1 through 5 to show that the number five is greater than the number one. There are various ways of students in comparing integers[3]. The way of students' reasoning is very rich and interesting, especially the solving of integers problems. Reasoning is an important element that is not only used to solve problems but is also used during classroom learning activities. It has long been agreed that reasoning is important in mathematics, that is in learning and its use[4].

In reasoning, students have different way. Similarly, the use of reasoning in comparing integer. First-grade students who play with the number line argue that the location of numbers on the number lines gives the possibility that the numbers on the left of zero can be compared with debt. Thus, reasoning by sequence allows the number to be on the left of zero[5]. In the other hand, the students' reasoning of the 2nd and 4th grade tend to use the sign of integer as operation in non-negative integers. Meanwhile when involving negative integers, students are more likely to use the number line to show the position of number because it make easier to compare integer[6]. When the first grade students had to order integers and compare their values, there were 71% of students removed automatically negative signs, 23% knew the concept of negative numbers so that they can compare negative numbers with positive numbers, and only 11% can sort and compare numbers consistently[7]. Students are said to reason when doing a logical thinking process related to the withdrawal of conclusions. Logical thinking here means mental activity derived from behaviors that arise based on statements and problem-solving results in accordance with previously known knowledge. The students' mental behavior that appears will show how students are reasoning[8]. The ability of one's math is one aspect that influences one's reasoning in solving problems related to mathematics[9]. Basically every student has different abilities, including mathemathics ability. Many factors that affect it. The differences in students' ability will affect the differences in behavior, development, and cognitive processing of each student [10]. Based on the explanation above, we may answer the following questions how do elementary students reason in solving integer problem, especially integer comparison.

2. Method

2.1 Participant

This study investigates the way of students' reasoning in solving integer problem especially integer comparison. The sample were the 5th grade of elementary students in Labschool Unesa, consists of 11 girls and 8 boys. All students were given mathematics test to know students' mathematics ability. Based on both the tests and students' portfolio were ranked. Three students were chosen as participants from each group to assess their way of reasoning in solving integers comparison problem and will be analyzed based on the way of reasoning indicator in Table 1. The given problem in reasoning test about integer comparison : *In February 2018, there was an uncertain weather change in Indonesia. The last temperature change was recorded in Batu –4° C and Semarang 1° C. Will the temperature change in Semarang be the same as Batu if there is an increase of 3° C?*

2.2 Analysis

The data was obtained from the reasoning test about comparison temperature and a semi-structured interview based on reasoning's test. In this paper, we focus on three students' reasoning way to the integer comparison problem. The problem was printed on a sheet and posed to students one at time. First, the students were asked to give justification then give the reason for their justification. After the data were identified their justification, the justification within each way of reasoning were combined to determine how the ways of reasoning were used. The interviewer posed some questions to know the way of students' reasoning, such as "Why do you agree with that statement? How did you think about that?. Then posed additional question if necessary to know deeply about the way of students' reasoning in comparison integer. Each student way was based on their justification. In particular, we present the way of reasoning [6] below.

2.2.1 Ordering-based reasoning

This way of reasoning is based on the sequence of integers in the count or its position on the number line. This method focuses on numbers that lie before or after other numbers when counting, or that are located more to the right, more to the left, or near zero on the number line.

2.2.2 Magitude-based reasoning

Students who use reason-based reasoning tend to compare the magnitude or the cardinality of a number. In accordance with this reasoning, students argue that zero and negative numbers are meaningless, as well as some integers associated with a specific context.

2.2.3 Developmental-based reasoning

In this reasoning, students are not familiar with negative numbers. This indicates that the students are more knowledgeable about the number of counts and their operations. As a result of that, students prefer to remove the minus sign, such as replacing –7 as 7. There is also a sign indicating minus as a reduction and there are also assume that there is no negative number.

Table 1. The way of reasoning

The way of reasoning categories	Students' activities
Ordering-based reasoning	• Comparing two integers, a negative integer is always less than a positive integer or zero, or a positive integer is always greater than a negative or zero integer. • Comparing two integers, integers close to zero or close to positive integers will be larger. • Comparing two integers, more to right integers are greater than, or more to left integers are less than. • Comparing two integers, the more left integer is greater (assuming a negative integer of the positive integer)
Magitude-based reasoning	• Comparing two integers, negative integers will be more or less negative than other negative numbers. • Compare non-zero integers to zeros, on the ground that zero has no meaning. • Determine that all positive integers are larger than negative integers, since negative integers do not have any magnitudes. • Describes the magnitude of an integer in a context.
Developmental-based reasoning	• Removes the minus sign so that it becomes a natural number. • Treats a minus sign as an operation.

3. Result

About 80% of all 19 students in the sample answered correctly, but only few students give reasonable argument about their answer. At the way of reasoning, After the sample is clasified based on the both test (mathematics test and reasoning test) and students' portfolio, there were 2 high students' ability, 4 medium students' ability, and 2 low students' ability. Three students were voluntarily choosen from each group to assess their reasoning in solving integers comparison problem deeply. The selection of three volunteer based on consistency and uniqueness in solving integer comparison problem, fluently communication, and willingness. Here, we present three description of students' reasoning way based on different ability as follows:

3.1. Reasoning Way of High Mathematics Ability Student (HMAs)

When the subject was asked to compare the temperature changes of Semarang and Batu, the subject stated that the given statement was wrong because "Semarang is 1° C, while Batu is –4° C. If Semarang increased 3° C equal to Semarang 4° C, while Batu is –4° C". From the subject's worksheet, she also describes the number line by stating that -4°C for Batu. Meanwhile 1° C for Semarang. The subject also added a sum operation with number 3 to obtain a new temperature change of 4° C.

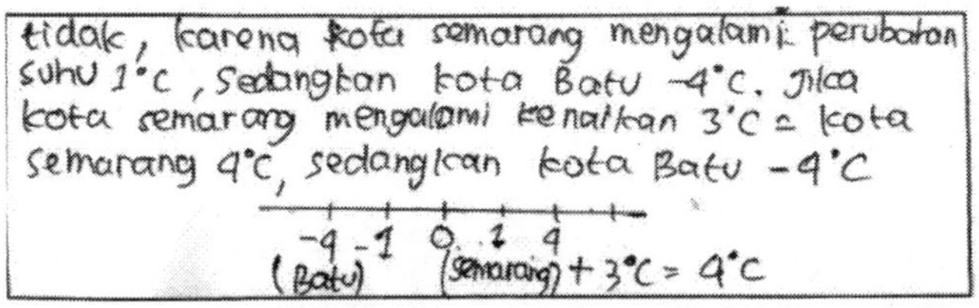

Figure 1. The way of students reasoning of high mathematics ability

In the interview, the subject also said

R: Is the temperature changes in Semarang equal to Batu?
S: No. Batu has a temperature changes negative 4, while Semarang positive 4 so they are not same.
R: Whereas the two cities have the same value of 4, how can they not be the same?
S: Different, because Semarang is to the right of zero, if Batu is to the left of zero. So if the number is getting to the left will be the smaller. If the number is more to the right will be the greater.
R: When viewed from the number line, both of Semarang and Batu jump as much as 4 away from zero. Are they not the same?
S: Not the same, the temperature change in Batu is smaller because of the temperature decreasing. While Semarang is greater because of the temperature rising.

Based on interview data, the subject explains that the temperature change for Batu is on the left side of zero so it is marked with a negative sign. While the temperature change for Semarang is on the right zero so it will be positive sign. If the temperature changes are located increasingly to the left then the smaller. So the subject states the temperature changes Batu is smaller than Semarang.

3.2. Reasoning Way of Medium Mathematics Ability Student (MMAs)

When subject was asked to compare temperature changes from two cities, the subject stated disagree that the temperature change in Semarang and Batu is same. The subject gave the reason "if Semarang added by 3° C it will be +4° C". The subject also made a number line. The subject gave a description of "Bat" on the number –4, "Se" on +1 and +4. The description written on number line indicates that –4 for Batu, while +1 for Semarang, and +4 is also for Semarang after experiencing temperature rise 3° C.

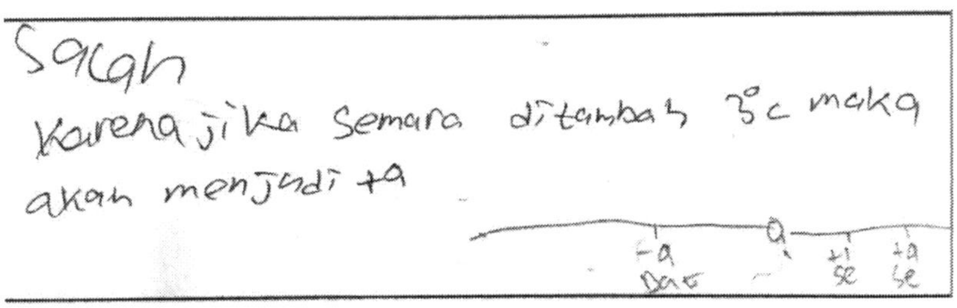

Figure 2. The way of students reasoning of medium mathematics ability

In the interview, the subject also said

R: Is it true that the temperature change in Semarang will be the same as Batu if it increases 3° C?
S: Em .. wrong
R: Why is it wrong?
S: Because at first Semarang is 1° C, then up 3° C become 4° C. Differences with Batu –4° C.
R: How can that be?
S: If we drawn a line of numbers like this (make number line), –4 is for Bat (its mean Batu) and this is +4 for Se (its mean Semarang)
R: Are these two cities not the same when both are worth 4?
S: Not the same, Batu has decreased means on the left. If Semarang increased means on the right.
R: Which cities have smaller or greater temperature changes?
S: The city that experienced a bigger temperature change is Semarang, because Semarang is located in the right of Batu. The sign of Batu is negative and Semarang is positive. Positive definitely greater than negative.

Based on interview data, the subject explained that the temperature change in Semarang will not be equal to Batu because Semarang has positive sign while Batu has negative sign. Which means positive definitely greater than negative. It means Semarang experienced a greater temperature change compared to Batu.

3.3. Reasoning Way of Low Mathematics Ability Student (LMAs)

When the subject was asked to compare the temperature changes of Semarang and Batu, the subject stated that the statement given was wrong with the reason of "far difference". Subject does not give any information or reason on the worksheet provided.

Figure 3. The way of students reasoning of low mathematics ability

In the interview, the subject also said

R: Is the temperature change in Semarang the same as Batu after increased 3°C?
S: Wrong, because it is much different.
R: How can it be different?
S: If Semarang 4, while the city of Batu –4. Means Batu is below and Semarang is above.
R: What does it mean below and above?
S: So it is like a stone that is arranged. Semarang means 4 pieces of stone arranged on the ground for positive, if Batu, the stone is arranged underground for negative.
R: So which temperature change is greater or smaller between Batu and Semarang?
S: The smaller one is Batu because it is underground in its stone. If Semarang is on the ground so it will be greater.

Based on interview data, the subject explained that Semarang experiencing temperature rise is illustrated by arranging stone on the ground as many as 4 pieces of stone because it is positive, while Batu is illustrated with 4 pieces of stone arranged in underground because of negative value.

4. Discussion

Our aim in this study was to investigate the way of students' reasoning with various mathematics ability about integer problem, especially integer comparison. In the results reported here, the students compared integers : positive versus negative. We discuss findings in students' ways of reasoning.

4.1. Reasoning Way of High Mathematics Ability Student (HMAs)

In comparing integers, the way in which the subject is used is to draw the number line but not to write the numbers in sequence on the number line, only on the number required. Next, the subject gives information on the number to indicate the location of the city in question. The subject also states that the number located on the left of zero is smaller than the number located to the right of zero. As in the case, Batu is located on the left of zero and Semarang is located on the right side of zero. This means that the temperature change in Batu is smaller than Semarang. In addition, the subject also states the numbers are getting to the left will be smaller [7]. This is shown by the subject that the location of Batu is more to the left than Semarang. So it can be concluded that the temperature changes in the city of Batu is smaller than the city of Semarang. The way of reasoning that subject used include in order-based.

4.2. Reasoning Way of Medium Mathematics Ability Student (MMAs)

In comparing the integers, almost the same as the subject of high-ability that is making a number line to show the temperature changes experienced by the two cities of Batu and Semarang. However, the subject only gives an explanation by writing down the initials of the city. As in pointing out of Batu, the subject only wrote "Bat" just below the number –4. As for showing the city of Semarang, the subject wrote "Se" under +1 and +4. The reason the subject wrote under +1 and +4 because Semarang experience temperature rise 3°C. In addition, the subject also states that the city located on the left of the other city will be smaller. So if it is seen on the number line the subject has made, he explains that "Batu is located on the left side of Semarang, meaning that the temperature of Batu city is smaller than Semarang". It shows that numbers decrease in value further to the left on the mental number line and increase in value further to the right [7]. Based on the explanation above, this subject used order-based way of reasoning.

4.3. Reasoning Way of Low Mathematics Ability Student (LMAs)
In comparing integers, subjects with low mathematics ability provide true justification. But the reason given in writing is very brief and does not provide any information about the way the subject is used. When the subject explains the idea used, he says that the two comparable cities are illustrated by arranging several stones. Semarang experiences positive temperature changes is illustrated with 4 pieces of stone arranged on the ground. While Batu with a negative value is illustrated with 4 pieces of stone arranged underground. Then the subject explains that the stone underground will be worth less than the stone on the ground. So it can be concluded that the temperature change in Batu is smaller than Semarang. This indicates that the subject uses another context contained in his environment in comparing integers. It means that integer reasoning can be referred to contexts such as lending and owing money, traveling forward and backward, elevation (above and below some relative zero), temperature (above and below zero degrees), happy-sad day, and so on [11]. Its mean that the subject used magnitude-based way of reasoning.

We relate our research to the literature bases that informed this study. We focus on the way of students' reasoning in solving integer problem. In the relevant mathematics literature order and magnitude reasoning have been particular ways that students may use when reasoning about integer comparison[6].

5. Conclusion
This study had demonstrated the way of students reasoning in solving the integer problem. It is underlined that students used a variety of way to draw their conclusions. Students presented appropriately ways and reasons in solving the problem. Subjects with high mathematics ability use a number line to show where the temperature changes experienced by both cities. Besides the subject gives the reason that the location of the number that is on the left of the zero will be smaller than the number located on the right of zero. The subject also says that on the number line, the numbers that are located increasingly to the left will be worth less. The way of reasoning that subject used include in order-based reasoning.Subjects with medium mathematics ability also use a number line to indicate the temperature changes experienced by both cities. The subject explains that on the number line, the number is located more to the left of the other number then the number will be smaller, and vice versa. This subject used order-based way of reasoning. Meanwhile, subjects with low mathematics ability, using a different context from those given. He uses an illustration of the arrangement of stones on the ground. Stones that are arranged on the ground (positive value) will be greater in value than stones arranged underground (negative value). Its mean that the subject used magnitude-based way of reasoning. We offer this result as a tool for mathematics teachers to inform students' reasoning about integer comparison more deeply [2] and explore the way of students' reasoning in integer problem.

6. Acknowledgments
The authors thank to the cooperating teacher and students in Labschool Unesa for helping and supporting this study.

7. References

[1] Stephan M and Akyuz D 2012 A Proposed Instructional Theory for Integer Addition and Subtraction, *National Council of Teachers of Mathematics Inc.* **43**(4) 428-464.

[2] Bishop J P, Lamb L L, Whitacre I, Schappelle B P and Lewis M L 2014 Obstacles and Affordances for Integer Reasoning: An Analysis of Children's Thinking and the History of Mathematics *National Council of Teachers of Mathematics Inc.* **45**(1) 19–61.

[3] Charles R I 2005 Big Ideas and Understandings as the Foundation for Elementary and Middle School Mathematics, *NCSM*, **7**(3), 9–24.

[4] Conner A M, Singletary L M, Smith R C, Wagner P A and Fransisco R T 2014 Identifying Kinds of Reasoning in Collective Argumentation, *Taylor and Francis Group LLC* **16**(3), 181-200.

[5] Wilcox V B 2008 Questioning Zero and Negative Numbers, *Teaching Children Mathematics*, **15**(4), 202–206.

[6] Whitacre I, Azuz B, Lamb L L, Bishop J P, Schappelle B P and Phillipp R A 2017 Integer comparisons across the grades: Students' justifications and ways of reasoning, *Journal of Mathematical Behavior* **45** 47–62.

[7] Bofferding L 2014 Negative Integer Understanding : Characterizing First Graders ' Mental Models, *J. Res. Math. Educ.* **45**(2) 194–245.

[8] Holisin I, Budayasa I K and Suwarsono 2017 Comparison of Male and Female Primary School Student Reasoning Profiles in Solving Fractional Problems, *International Journal of Environmental & Science Education*. **12**(6) 1553–1565.

[9] Lohman D F and Lakin J M 2009 *Reasoning and Intelligence* (New York: Cambridge University Press) p 1–47.

[10] Jensen E 2011 *Pembelajaran Berbasis Otak Paradigma Pengajaran Baru. Edisi Kedua* (Jakarta: PT Indeks)

[11] Whitacre I, Bishop J P, Lamb L C, Philipp R A, Schappelle B P and Lewis M L 2012 Happy and sad thoughts: An exploration of children's integer reasoning. *Journal of Mathematical Behavior* **31**(3) 356–365.

Validity of Teaching Materials Based on Socio-Scientific Issues Approach on The Topic of Vibration, Waves, and Sound

K W Hariapsari[1,*], Tukiran[2] and E Sudibyo[3]

[1]Student of Science Education Program, Postgraduate of Universitas Negeri Surabaya, Indonesia.
[2,3]Lecturer of Science Education Program, Postgraduate of Universitas Negeri Surabaya, Indonesia.

*Corresponding Author: kiranahariapsari16070795024@mhs.unesa.ac.id

Abstract. Participation in TIMSS and PISA showed that the achievements of Indonesian children are still under the average and low in the scientific literacy. Scientific literacy is the ability to make decisions about scientific issues and scientific phenomena that can be applied in daily life. This ability can be developed using innovative learning approaches such as socio-scientific issues (SSI). SSI is an approach that represents issues or problem in social life that are conceptually related to science. The material discusses on the topic of vibration, waves, and sound because it related to scientific issues and phenomena in daily life. This study aims to describe validity of teaching materials based on socio-scientific issues approach including Syllabus, Lesson Plan, Students' Book, Students' Worksheet, and Scientific Literacy Test. The data are then collected through validation questionnaire to assess validity of teaching materials. Data collection instruments used validation sheets. The data are analysed by using both descriptive quantitative and qualitative methods. The result of teaching materials validation showed that teaching materials is very valid and can be used for learning process.

1. Introduction

Science lessons in junior high schools is implemented 2013 Curriculum based on internal challenges and external challenges. One of the internal challenges is the readiness of education personnel in developing teaching materials as an innovation in the learning process, while external challenges are related to the development of education at the international level. Indonesia faced these external challenges by participating *Trends in International Mathematics and Science Study* (TIMSS) in 1999 and *the International Student Assessment Program* (PISA) in 2000 [1]. Indonesia's participation in TIMSS aims to gain information on Indonesian students' competence in Mathematics and Natural Sciences compared to other countries in the world. Participation in PISA aims to examine and compare the achievements of students around the world with a view to improve educational methods and results. The participation indicates that the achievements of Indonesian students are not encouraging in the results of TIMSS and PISA are presented [2].

The results of TIMSS data for 2011 for the field of science class VIII, Indonesia ranked at number 39th with a score of 406 points from 42 participating countries. In 2015 for class IV, the results are ranked 45th with a score of 397 points from 48 participating countries. Both of these results indicate that Indonesia is below the average score of the TIMSS scale of 500 points. PISA results are also not encouraging in 2012 and 2015, Indonesia ranked 64th with a score of 382 points from 65 countries and ranked 62th with a score of 403 points from 70 participating countries [2]. Indonesia have under-average scores in TIMSS and PISA scores that show Indonesian students are still low in scientific literacy such as identifying scientific problems, using scientific facts, understanding living systems and understanding to use of science equipment [3]. Scientific literacy is the ability to understand the science of content, competence, processes and attitudes to make decisions regarding evolving scientific issues and scientific phenomena that can be applied in daily life [4-6].

Understanding of science learning that leads to the formation of scientific literacy of students is still not fully understood by science teachers. It results the learning process is still conventional and based on the conceptual mastery of students so that the achievement of scientific literacy of students is low. Other causes were due to lack of learning that involves the process of science related to the students' daily life [7].

Learning process that can be taught or increase scientific literacy is using approach of socio-scientific issues [8]. Socio-Scientific Issues (SSI) approach is a learning approach that involves socio-scientific issues by orienting learning to the context of science and its relation to social life in society [9-10]. This approach aims to improve the ability of making decisions on discussed or developed issues [11-13].

The topic of vibration, wave and sound is the science topic of Junior High School grade 8th in basic competence such as analysing the concept of vibration, waves, and sound in daily life including auditory system in human and sonar system in animals and presents experimental results about vibration, waves, and sound. The sub topic are vibrations, waves, sound, applications of vibration and waves in technology [14]. In this topic, students are expected to identify, analyse, evaluate, and criticize the utilization of sound waves in daily life. This topic is chosen because it is closely related in the students' daily life application. In addition to its knowledge of science content in PISA 2015, the physical system requires knowledge of "Interactions between energy and the topic" e.g. light and radio waves, sound waves and seismic [15]. Thus, this material has authentic issues that can be used as learning materials that are linked to the scope of context, knowledge, competence and attitudes in scientific literacy.

The mastery of scientific literacy can be enhanced by teaching materials with the scientific issues discussed. The development of teaching materials is a process of designing a logical and systematic learning of a set of media used by teachers and students in the learning process in the classroom. It becomes a reference for learning that teachers should prepare in teaching and learning activities [16]. Teaching materials usually contain syllabus, learning plan, students' book, students' worksheet, and scientific literacy test. The purpose of this study is to describe the validity of teaching materials based on socio-scientific issues approach on the topic of vibration, wave, and sound that can improve the students' scientific literacy.

2. Research Method

The type of this research is a development research. It refers to a process of developing teaching materials based on socio-scientific issues approach on the topic of vibration, waves, and sound associated with scientific literacy. This research uses Dick and Carey development model [17]. This development model has 10 procedural steps, they are: identify instructional goals; conduct instructional analysis; identify entry behaviours, characteristics; write performance objectives; develop criterion referenced test items; develop instructional strategy; develop and select instructional materials; design and conduct formative evaluation; revise instruction; design and conduct summative evaluation [17].

The variables observed in this study is the feasibility of instructional devices in terms of validity. Validity in the form of validation result of instructional device. The developed teaching materials is validated by 3 validators to provide assessment in accordance with validation instrument.

Validation instrument sheet of teaching materials in this study consist of (1) Validation Sheet of Syllabus, (2) Validation Sheet of Lesson Plan, (3) Validation Sheet of Students' Book, (4) Validation Sheet of Students' Worksheet, and (5) Validation Sheet of Scientific Literacy Test. It contains instructions and requests to validator that provide an assessment of the teaching materials. The validation instrument of teaching material is in the following Table 1 [18].

Table 1. Rubric on Validity of Teaching Materials.

Score	Criteria
1	Low
2	Enough
3	Good
4	Very Good

Validation data were analysed quantitatively with the following steps: (1) tabulating all obtained data from validators for each item of assessment available in the validation instrument, (2) calculating the average score of every component aspects of teaching materials.

$$Average\ Score = \frac{Score\ Counts\ Obtained}{Observers} \quad (1)$$

and (3) giving the average score criterion from interpretation score of validation as seen in Table 2 [19].

Table 2. Score Interpretation of Validation.

Average Score	Criteria	Interpretation
0.00 – 1.49	Invalid	Not applicable
1.50 – 2.49	Less Valid	Can be used with larger revisions
2.50 – 3.49	Valid	Can be used with smaller revisions
3.50 – 4.00	Very Valid	Can be used without revisions

Teaching materials are said to be valid when teaching materials reach valid or very valid criteria. Reliability is calculated using the following percentage agreement formula.

$$R\ (\%) = 100\left[1 - \frac{A-B}{A+B}\right] \quad (2)$$

Where R is the percentage agreement (%), A is the frequency counts of observers who gives a larger frequency, B is the frequency counts of observers who gives a smaller frequency. Teaching materials are said to be reliable if the percentage agreement is more than 75% [20].

3. Result and Discussion

The validity of teaching materials can be seen from the validation results. Teaching materials are validated by 2 expert lecturers and 1 science teacher before the research is conducted. Validity of teaching materials are then revised based on the suggestions given by the validator. Recapitulation of teaching materials validation results can be seen in Table 3.

Table 3. Recapitulation of Teaching Materials Validation Results.

No.	Components of Teaching Materials	Score			Average Score	Criteria	Reliability (%)
		V-1	V-2	V-3			
1.	Syllabus	3.62	4.00	3.12	3.58	Very Valid	87.64
2.	Lesson Plan	3.86	3.95	3.64	3.82	Very Valid	95.91
3.	Students' Book	3.82	3.69	3.81	3.77	Very Valid	98.27
4.	Students' Worksheet	3.60	3.90	3.80	3.77	Very Valid	96.00
5.	Scientific Literacy Test	3.69	4.00	3.77	3.82	Very Valid	95.97

All of components of teaching materials get more than 3.50 with very valid criteria. Otherwise, the reliability of all components also represented more than 75%, so we can state that components of teaching materials are reliable.

As mentioned above that the aim of this research was to describe the validity of teaching materials based on socio-scientific issues approach including Syllabus, Lesson Plan, Students' Book, Students' Worksheet, and Scientific Literacy Test. The teaching materials were developed using Dick and Carey development model. After the materials were developed, the next step was to describe their validity.

3.1 Syllabus

Principles of syllabus development include scientific, relevant, systematic, consistent, adequate, actual and contextual, flexible, and comprehensive. The steps in developing syllabus are the study of core competence and basic competence, identify the topic subject, develop the learning activities, formulate the indicators of achievement of competence, determine the type of assessment, determine the allocation of time, and determine the source of learning.

The validity of the syllabus had been stated to be very valid. This is supported by the average score to be more than 3.50. The validator's suggestions are emphasizing literacy activities by reading socio-scientific issues and informal reasoning activities, adding experimental or observational activities then reasoning and discussing observations. The results of revision are giving motivation by reading introductory issues such as "BMKG: News that there will be a secondary earthquake with 7.5 SR: Hoax", and adding activities to collecting data by working on students' worksheet 1 and reasoning/associating the results of the discussion to interpret the data and scientific evidence of the occurrence of waves and types of waves.

3.2 Lesson Plan

The learning process is implemented based on syllabus and the teacher must prepare the lesson plan. Lesson plan is a grip for teachers in implementing learning to achieve one or more basic competencies set out in the standard competencies. There are several principles that must be considered by a teacher in the development of lesson plan, such as the competencies formulated in the lesson plan must be clear or easily observed; the developed activities in the lesson plan should be supportive and in conformity with the basic competencies; lesson plan should be simple and flexible so that it can be implemented in the learning activities and the formation of competence of learners; must be complete, comprehensive, clear stages and steps of achievement.

The developed lesson plan refers to the analysis of the characteristics of learners and the format of the Curriculum 2013. It was made in 3 meetings with time allocation of 3 x 40 minutes at the first and third meetings, and 2 x 40 minutes at the second meeting. It was structured based on the stages in socio-scientific issues developed to improve the scientific literacy.

The validity of the lesson plan had been stated to be very valid. This is supported by the average score that is more than 3.50. The validator's suggestions are clarifying the learning activities by emphasizing the SSI approach that it is the provision of reading scientific or socio-scientific issues. In each learning activity, there are issues discussed such as: "BMKG: News that there will be a secondary earthquake with 7.5 SR", "'Ship is safer in the middle of the sea in the event of a tsunami", "Lightning", "Uncomfortable ear when riding an airplane", "Myth: Blind Bats", "Animals Can Predict Earthquakes, Myths or Facts?", and "Ultrasound is Harmful to Pregnancy, Myths or Facts".

3.3 Students' Book

Students' book is a set of teaching materials that it is designed in a systematic and interesting in achieving the learning objectives and competencies. It is good that it should be designed according to the instructional rules. It can help students to gain new knowledge.

Students' book should be developed in accordance with the rules of which must be tailored to the students, able to change the students' behaviour, included specific learning objectives, and contained more detail learning materials. The steps in the preparation of students' book is setting the title of teaching materials, setting the learning objectives, establishing the outline or concept of learning materials, developing the material based on basic concepts, and re-examining the draft of teaching materials produced [21]

Students' book-oriented science literacy with the SSI approach that is developed contain readings materials to the concept of vibration, waves, and sound. Scientific literature in students' book can improve students' interest and motivation to read and relate them with science concepts so that science literacy can also be improved. The students' book is also equipped with a Mini Lab to facilitate students in initial observation.

The validity of the students' book had been stated to be very valid. This is supported by the average score that is more than 3.50. The validator's suggestions are replacing the term module with students' book, source writing needs to be improved, and the picture needs to be clarified.

3.4 Students' Worksheet

Students' worksheet is a learner's guide that is used to conduct investigation or problem solving activities. It contains a set of basic activities that must be done by the learners to maximize the understanding in the effort of forming basic skills according to indicators of achievement of learning outcomes pursued [6].

It contains socio-scientific issues in the form of reading related to the purpose of observation, tools and materials, work steps, observations, discussions, and conclusions. Students are trained to read literature by finding concepts related to the material and analysing the results of the facilitation in the discussion and questions. Students are also trained in making decisions from the answers of discussion and questions and making conclusions from observations and discussions.

The validity of the student worksheet had been stated to be very valid. This is supported by the average score that is more than 3.50. The validator's suggestions are the purpose of students' worksheets such as "evaluating and designing scientific investigations about types of waves" should replaced by "collecting data and analysing types of waves"; "interpreting the data and scientific evidence of the wave" should replaced by "presenting the results of discussions about waves"; and the sentence in the discussion quest is "analyse the graph of the relationship between the direction of vibration and direction of wave travel" should replaced by "Draw a wave pattern and complete it with an amplitude position (A), wavelength (λ), hill, and valley!".

3.5 Scientific Literacy Test

Science literacy test is an assessment done in the process of collecting and processing information to measure the achievement of scientific literacy who refers to the instrument of scientific literacy test in PISA 2015 with the assessment using the score and level. Indicators of the scientific literacy test include the suitability of the item with the purpose of learning/ specification and the item on the reading is given. Scientific literature assessment includes aspects of context, competence, knowledge, and attitude of science.

The validity of the scientific literacy test had been stated to be very valid. This is supported by the average score that is more than 3.50. The validator's suggestions are improving some questions that have the same answer and have different alternate lengths, replacing the socio-scientific issues, and creating questions that can increase students' scientific literacy to make informed decisions about what to do.

The novelty or research findings of this paper are validity of teaching materials based on socio-scientific issues approach on the topic of vibration, waves, and sound; this research can be implemented in the scientific learning process and be able to increase the students' scientific literacy.

4. Conclusion

Based on the results of the research, the teaching materials were very valid. It can be concluded that teaching materials based on socio-scientific issues on vibration, waves, and sound subject can be used for learning process. The teaching materials are related with the increased of students' scientific literacy.

There are several suggestions that can be given as follows: (1) content of scientific literacy test is more increased, (2) literature of socio-scientific issues is more increased, and (3) time allocation must be considered and regulated appropriately to achieve maximum results.

5. References

[1] Kemdikbud 2014 *Lampiran I Peraturan Menteri Pendidikan dan Kebudayaan No. 58 Tahun 2014 tentang Kurikulum 2013 di Sekolah Menengah Pertama/ Madrasah Tsanawiyah* (Jakarta: Kementrian Pendidikan dan Kebudayaan)

[2] NCES 2015(online) *PISA Profiency Levels* https://nces.ed.gov/surveys/pisa/pisa2015/pisa2015highlights_8f.asp

[3] OECD 2016 (Online) *Programme for International Student Assessment (PISA) Results from PISA 2015 – Indonesia* https://www.oecd.org/pisa/PISA-2015-Indonesia.pdf)

[4] Holbrook J 2011 *Enhancing Scientific and Technological Literacy (STL): A Major Focusfor Science Teaching at School* http://eric.ed.gov/?id=EJ884397 accessed 2 December 2017.

[5] OECD (*Organization for Economic Co-operation and Development)* 2016 (Online) *PISA 2015 Assesment and Analytical Framework: Science, Reading, Mathematics and Financial Literacy.* Online. Diunduh dari (http://www.mecd.gob.es/dctm/inee/internacional/pisa-2015-frameworks.pdf?documentId=0901e72b820fee48)
[6] Toharudin U 2011 *Membangun Literasi Sains Peserta Didik* (Bandung: Humaniora)
[7] Zeidler D, Nichols B 2009 (Online) *Socioscientific Issues: Theory and Practice* https://files.eric.ed.gov/fulltext/EJ849716.pdf)
[8] Gutierez S 2014 (Online) *Integrating Socio-Scientific Issues to Enhance the Bioethical Decicsion Making Skills of High School Students* https://files.eric.ed.gov/fulltext/EJ1060853.pdf
[9] Karisan D and Zeidler D 2011 Contextualization of nature of science within the socioscientific issues framework: A review of research *International Journal of Education in Mathematics Science and Technology* **5**(2) 139-152
[10] Siribunnam, S, Nuangchalerm P, and Jansawang N 2014 Socio-scientific Decision Making in the Science Classroom *International Journal for Cross-Discilinary Subjects in Education* **5**(4)
[11] Belova, N and Eilks I 2014 (Online) *Promoting Societal-Oriented Communication and Decision Making Skills by Learning about Advertising in Science Education* https://files.eric.ed.gov/fulltext/EJ1129500.pdf
[12] Stuckey M, Lippel M, Eilks I 2014 (Online) *Teaching Chemistry about 'Stevia' – A Case of Cooperative Curriculum Innovation within PROFILES in Germany* https://files.eric.ed.gov/fulltext/EJ1129489.pdf
[13] Zeidler D, Sadler, Applebaum S, Callahan B 2009a (Online) *Advancing Reflective Judgment through Socioscientific Issues* https://onlinelibrary.wiley.com/doi/epdf/10.1002/tea.20281
[14] OECD (*Organization for Economic Co-operation and Development)* 2016 (Online) *PISA 2015 Assesment and Analytical Framework: Science, Reading, Mathematics and Financial Literacy* http://www.mecd.gob.es/dctm/inee/internacional/pisa-2015-frameworks.pdf?documentId=0901e72b820fee48)
[15] Kemdikbud 2016 *Peraturan Menteri Pendidikan dan Kebudayaan No. 24 Tahun 2016 tentang Kompetensi Inti dan Kompetensi Dasar Pelajaran pada Kurikulum 2013 pada Pendidikan dasar dan Pendidikan Menengah* (Jakarta: Kementrian Pendidikan dan Kebudayaan)
[16] Ibrahim M 2010 *Dasar – Dasar Proses Belajar Mengajar* (Surabaya: Unesa *University Press*)
[17] Dick W, Carey L, Carey J 2009 *The Systematic Design of Instruction: seventh edition* (New York: Pearson)
[18] Riduwan 2012 *Skala Pengukuran Variabel-Variabel Penelitian* Cetakan ke-VII (Bandung: Alfabeta)
[19] Ratumanan T and Laurens T 2011 *Penilaian Hasil Belajar pada Tingkat Satuan Pendidikan Edisi 2* (Surabaya: Unesa University Press)
[20] Borich, G 1994 *Observation Skills for Effective Teachin* (USA: Macmilan Publishing Company)
[21] Widodo, Chomsin S, Jasmadi 2007 (Online) *Panduan Menyusun Bahan Ajar Berbasis Kompetensi* https://books.google.co.id/books?id=Gv3ZYRgsPkAC &pg=PA40&dq=bahan+ajar+adalah&hl=en&sa=X&ved=0ahUKEwiA37uy9fzaAhXGtI8 KHeq2ACIQ6AEIJzAA#v=onepage&q=bahan%20ajar%20adalah&f=false

Disturbance compensating model predictive control for warship heading control in missile firing mission

Heri Purnawan[1], Tahiyatul Asfihani[2], Dieky Adzkiya[3] and Subchan[4]

Modeling and Simulation Systems Laboratory, Department of Mathematics
Faculty of Mathematics, Computing and Data Science, Institut Teknologi Sepuluh Nopember, Surabaya 60111, Indonesia

E-mail: heripurnawan93@gmail.com[1], t_asfihani@matematika.its.ac.id[2], dieky@matematika.its.ac.id[3] and subchan@matematika.its.ac.id[4]

Abstract. One of warship missions is shooting and avoiding enemy attacks, which can be done by missile firing strategy. Due to missile firing, the movement of ship becomes unstable. To overcome this problem, we need a control system that can overcome the disturbance caused by the impact of the missile firing. So, in this paper, we propose a method called disturbance compensating MPC (DC-MPC) control system. This control is chosen, because DC-MPC is a robust method for measurable disturbance. Based on the result of simulation, DC-MPC can overcome disturbance on the system and make ship's actual heading angle to follow the desired reference quickly which is about 10 s. After the system gets disturbance, controller is also able to adjust the value of the ship heading angle to original position.

1. Introduction

One of warship missions is shooting and avoiding enemy attacks, which can be done by missile firing strategy [1]. Based on the mission, impact disturbance modelling is also needed to find out how much force a ship receives when firing a missile. In this study, the model of impact disturbance when firing missile is based on the laws of physics. The magnitude of the impact force of the missile firing on the vessel is calculated using the Newton's second law and Newton's third law on translational motion, whereas to know the magnitude of its impact force moment is calculated using the Newton's second law on the rotational motion.

The impact disturbance caused by missile firing is an external disturbance that will affect the stability on a ship when maneuvering in the ocean. To overcome these problems, it is necessary to use a robust control system in overcoming the impact disturbance of missile firing. The control method developed in the field of ship system control is Model Predictive Control (MPC) [2, 3, 4, 5, 6]. Basically, MPC works based on mathematical model of plant to predict the output process in the horizon time. The purpose of the control calculation is to minimize the objective function [7]. The following four main aspects of MPC make the design of this method attractive to practitioners and academics [8]. The first aspect is the design formulation using multivariable system [9] (multi input multi output). The second aspect is the ability of the method to handle the constraints on the system. The third aspect is the ability to perform an online optimization process. The fourth aspect is the simplicity of the design of control in dealing with complex problems.

Disturbance Compensating Model Predictive Control (DC-MPC) is a development of the

MPC method. The idea of developing the MPC method for measurable disturbances was first introduced by Li and Sun in a study of ship heading control [4]. In their research, it was shown that DC-MPC method is more robust than MPC in overcoming the measured disturbance of sea wave disturbance which is modeled through constant and sinusoidal wave equation. Based on the research, we implement the design of DC-MPC method to overcome the impact of missile firing on the ship, with aims to make the ship's actual heading angle can follow the desired reference after the missile is fired.

2. 2 DOF Ship Model

Model used in this paper for designing ship heading control is Davidson and Schiff model [10]. This model considered 2 DOF, namely, sway velocity v, yaw rate r dan one control input, namely, rudder angle δ. Mathematical model for 2 DOF Davidson and Schiff is showed by equation (1):

$$\boldsymbol{M}\dot{\boldsymbol{\nu}} + \boldsymbol{N}(u_0)\boldsymbol{\nu} = \boldsymbol{b}\delta \tag{1}$$

where $\boldsymbol{\nu} = \begin{bmatrix} v & r \end{bmatrix}^T$ and matrices $\boldsymbol{M}, \boldsymbol{N}(u_0)$, and $\boldsymbol{b}$ as in [10]. From (1) and by adding kinematics equation [10] $\dot{\psi} = r$, we obtain the following state space equation:

$$\dot{\boldsymbol{x}} = \boldsymbol{A}_c\boldsymbol{x} + \boldsymbol{B}_c\boldsymbol{u} \tag{2}$$

where

$$\boldsymbol{x} = \begin{bmatrix} \boldsymbol{\nu} \\ \psi \end{bmatrix}, \boldsymbol{A}_c = \begin{bmatrix} -\boldsymbol{M}^{-1}\boldsymbol{N}(u_0) & \boldsymbol{0} \\ 0 \quad 1 & 0 \end{bmatrix}, \boldsymbol{B}_c = \begin{bmatrix} \boldsymbol{M}^{-1}\boldsymbol{b} \\ 0 \end{bmatrix} \text{ and } \boldsymbol{u} = \delta$$

3. Modeling of Impact Force and Moment

The force caused by missile firing on a ship, will cause the ship to change motion. The magnitude of the missile impact force is equal to the magnitude of the missile force, but with different direction (counterclockwise) [11]. Based on the Newton's second law, the magnitude of the F_t fire force depends on the mass and acceleration of the missile. Mathematically, it can be formulated as: $F_t = m_M \cdot a_M$, where m_M is mass of a missile and a_M is acceleration of a missile. The illustration of the direction of the impact force from missile firing on the ship is shown in Figure 1.

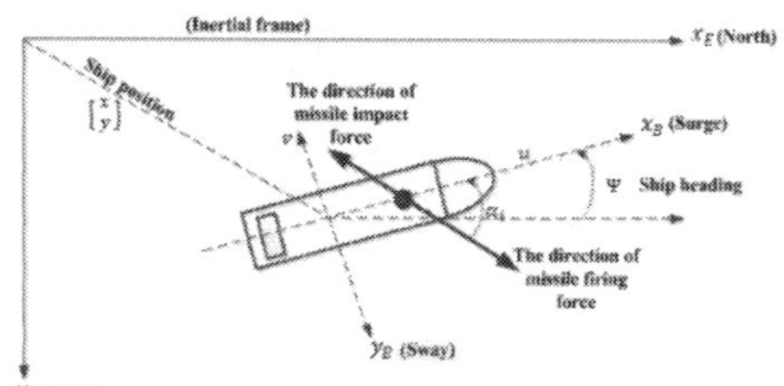

Figure 1. The illustration of the direction of the impact force from missile firing

Figure 1 showed a condition when the ship's current heading angle is ψ and a missile is fired with variation of the missile launcher heading angle relative to the ship's surge α_i. The process to obtain impact force equation from missile firing on the ship is given by:

1. Determining the horizontal direction force (F_{t_x}) from F_t

$$F_{t_x} = F_t \cos\theta_M$$

2. Determining sway force $(F_{t_{sway}})$

$$F_{t_{sway}} = F_{t_x} \sin \alpha_i = m_M a_M \cos \theta_M \sin \alpha_i$$

Based on the Newton's second law, impact force equation of missile firing can be written as follows: $F_{i_{sway}} = -F_{t_{sway}}$. Because $Y_i = F_{i_{sway}}$ is external force of missile impact, so we obtained

$$Y_i = -F_{t_{sway}} = -m_M a_M \cos \theta_M \sin \alpha_i, \quad -\frac{\pi}{2} < \alpha_i < \frac{\pi}{2} \tag{3}$$

The impact force moment equation of missile firing is derived from the second Newton law on rotational motion, i.e, $\tau = I \cdot \alpha$, where I is inertia moment and α is angle acceleration. Because $\alpha = \frac{a}{d}$ and $I = md^2$, so torsion equation can be written as:

$$\tau = F \cdot d \tag{4}$$

where d is arm of the moment or the distance of force F to centre of gravity. Based on equation (4), the magnitude of yaw impact force moment of missile firing can be formulated as follows:

$$N_i = -m_M a_M d_L \cos \theta_M \sin \alpha_i, \quad -\frac{\pi}{2} < \alpha_i < \frac{\pi}{2} \tag{5}$$

where d_L is the distance of missile launcher to center of gravity on the ship. In equation (3) and (5), model still has a dimension, so it must be changed into nondimension.

4. Design Control System

The purpose of DC-MPC controller is ship's actual heading angle can follow the desired heading angle when there is impact disturbance of missile firing. The objective function of DC-MPC is defined by [7]:

$$J = \min_{\boldsymbol{u}(\cdot|k)} \sum_{j=1}^{N_p} \left[\|\boldsymbol{y}_d(k+j|k) - \boldsymbol{y}(k+j|k)\|_{\boldsymbol{Q}}^2 + \|\boldsymbol{u}(k+j-1|k)\|_{\boldsymbol{R}}^2 \right] \tag{6}$$

subject to [4]

$$\begin{aligned}
&\boldsymbol{x}(k+j+1|k) = \boldsymbol{A}\boldsymbol{x}(k+j|k) + \boldsymbol{B}\boldsymbol{u}(k+j|k) \\
&\boldsymbol{x}(k|k) = \boldsymbol{x}(k) \\
&\boldsymbol{C}_1\boldsymbol{x}(k+j+1|k) \leq \boldsymbol{D}_1, \qquad j = 0, 2, \cdots, N_p - 1 \\
&\boldsymbol{S}\boldsymbol{u}(k|k) \leq \boldsymbol{T} - \boldsymbol{S}\Delta\boldsymbol{u}^* \\
&\boldsymbol{S}\boldsymbol{u}(k+j|k) \leq \boldsymbol{T}, \qquad j = 1, 2, \cdots, N_p - 1
\end{aligned}$$

The disturbance at time step $k-1$, i.e., $\hat{w}(k-1)$, can be estimated by the following equation if the state and control are measurable [12]. The disturbance compensation $\Delta\boldsymbol{u}^*$ is calculated as in [4]. We define vector [8]: $\boldsymbol{Y} = \left[\ \boldsymbol{y}(k+1|k), \cdots, \boldsymbol{y}(k+N_p|k)\ \right]^T, \boldsymbol{U} = \left[\ \boldsymbol{u}(k|k), \cdots, \boldsymbol{u}(k+N_p-1|k)\ \right]^T$, so the output of system $\boldsymbol{Y}$ can be written into matrix form as follows [8]:

$$\boldsymbol{Y} = \boldsymbol{F}\boldsymbol{x}(k|k) + \boldsymbol{\Phi}\boldsymbol{U} \tag{7}$$

where

$$\boldsymbol{F} = \begin{bmatrix} CA \\ CA^2 \\ CA^3 \\ \vdots \\ CA^{N_p} \end{bmatrix}; \quad \boldsymbol{\Phi} = \begin{bmatrix} CB & 0 & 0 & \cdots & 0 \\ CAB & CB & 0 & \cdots & 0 \\ CA^2B & CAB & CB & \cdots & 0 \\ \vdots & & & & \\ CA^{N_p-1}B & CA^{N_p-2}B & CA^{N_p-3}B & \cdots & CB \end{bmatrix}$$

IOP Conf. Series: Journal of Physics: Conf. Series **1108** (2018) 012035 doi:10.1088/1742-6596/1108/1/012035

so by using equation (7), the objective function in equation (6) can be written into:

$$\begin{aligned} J &= (\boldsymbol{R}_s - \boldsymbol{Y})^T\boldsymbol{Q}_c(\boldsymbol{R}_s - \boldsymbol{Y}) + \boldsymbol{U}^T\boldsymbol{R}_c\boldsymbol{U} \\ &= \boldsymbol{U}^T\left(2\boldsymbol{\Phi}^T\boldsymbol{Q}_c\left[\boldsymbol{F}\boldsymbol{x}(k|k) - \boldsymbol{R}_s\right]\right) + \frac{1}{2}\boldsymbol{U}^T\left(2\left[\boldsymbol{\Phi}^T\boldsymbol{Q}_c\boldsymbol{\Phi} + \boldsymbol{R}_c\right]\right)\boldsymbol{U} \end{aligned} \tag{8}$$

where $\boldsymbol{R}_s = \left[\begin{array}{cccc} \boldsymbol{y}_d(k+1|k), & \boldsymbol{y}_d(k+2|k), & \cdots, & \boldsymbol{y}_d(k+N_p|k) \end{array}\right]^T$, $\boldsymbol{Q}_c = \text{diag}(\boldsymbol{Q})$, $\boldsymbol{R}_c = \text{diag}(\boldsymbol{R})$ and the constraints for the objective function (8) are as follow:

$$\boldsymbol{S}_1\boldsymbol{U} \leq \boldsymbol{T}_1 \quad \text{and} \quad \boldsymbol{\Phi}_1\boldsymbol{U} \leq \boldsymbol{D} - \boldsymbol{K}\boldsymbol{x}(k|k)$$

where

$$\boldsymbol{S}_1 = \begin{bmatrix} \boldsymbol{S} & \boldsymbol{0} & \boldsymbol{0} & \cdots & \boldsymbol{0} \\ \boldsymbol{0} & \boldsymbol{S} & \boldsymbol{0} & \cdots & \boldsymbol{0} \\ \boldsymbol{0} & \boldsymbol{0} & \boldsymbol{S} & \cdots & \boldsymbol{0} \\ \vdots & & & \ddots & \\ \boldsymbol{0} & \boldsymbol{0} & \boldsymbol{0} & \cdots & \boldsymbol{S} \end{bmatrix}; \boldsymbol{T}_1 = \begin{bmatrix} \boldsymbol{T} - \boldsymbol{S}\Delta\boldsymbol{u}^* \\ \boldsymbol{T} \\ \boldsymbol{T} \\ \vdots \\ \boldsymbol{T} \end{bmatrix}; \boldsymbol{K} = \begin{bmatrix} \boldsymbol{C}_1\boldsymbol{A} \\ \boldsymbol{C}_1\boldsymbol{A}^2 \\ \boldsymbol{C}_1\boldsymbol{A}^3 \\ \vdots \\ \boldsymbol{C}_1\boldsymbol{A}^{N_p} \end{bmatrix}$$

$$\boldsymbol{D} = \begin{bmatrix} \boldsymbol{D}_1 \\ \boldsymbol{D}_1 \\ \boldsymbol{D}_1 \\ \vdots \\ \boldsymbol{D}_1 \end{bmatrix}; \boldsymbol{\Phi}_1 = \begin{bmatrix} \boldsymbol{C}_1\boldsymbol{B} & \boldsymbol{0} & \boldsymbol{0} & \cdots & \boldsymbol{0} \\ \boldsymbol{C}_1\boldsymbol{A}\boldsymbol{B} & \boldsymbol{C}_1\boldsymbol{B} & \boldsymbol{0} & \cdots & \boldsymbol{0} \\ \boldsymbol{C}_1\boldsymbol{A}^2\boldsymbol{B} & \boldsymbol{C}_1\boldsymbol{A}\boldsymbol{B} & \boldsymbol{C}_1\boldsymbol{B} & \cdots & \boldsymbol{0} \\ \vdots & & & & \\ \boldsymbol{C}_1\boldsymbol{A}^{N_p-1}\boldsymbol{B} & \boldsymbol{C}_1\boldsymbol{A}^{N_p-2}\boldsymbol{B} & \boldsymbol{C}_1\boldsymbol{A}^{N_p-3}\boldsymbol{B} & \cdots & \boldsymbol{C}_1\boldsymbol{B} \end{bmatrix}$$

5. Simulation

In this simulation, we show performance of DC-MPC for the ship motion control when the ship gets disturbance in form impact of missile firing. The system used in designing DC-MPC is discrete time system [4], so model in equation (2) must be changed into discrete time system. By taking sampling time $T_s = 0.5$ s, then the discrete time system matrices are given by

$$\boldsymbol{A} = \begin{bmatrix} 0.7676 & -0.0206 & 0 \\ -1.0313 & 0.2020 & 0 \\ -0.3507 & 0.2446 & 1 \end{bmatrix}; \quad \boldsymbol{B} = \begin{bmatrix} -0.0070 \\ 0.2446 \\ 0.0770 \end{bmatrix}$$

In this simulation, the rudder constraints are $|\delta| \leq 35°$ and the yaw rate constraints are $|r| \leq 0.0932$ rad/s, so, the corresponding matrices $\boldsymbol{C}_1, \boldsymbol{D}_1, \boldsymbol{S}$ and $\boldsymbol{T}$ are given by

$$\boldsymbol{C}_1 = \begin{bmatrix} 0 & 1 & 0 \\ 0 & -1 & 0 \end{bmatrix}; \boldsymbol{D}_1 = \begin{bmatrix} 0.0932 \\ 0.0932 \end{bmatrix}; \boldsymbol{S} = \begin{bmatrix} 1 \\ -1 \end{bmatrix}; \boldsymbol{T} = \begin{bmatrix} 35\pi/180 \\ 35\pi/180 \end{bmatrix}$$

whereas the disturbance value $\boldsymbol{w}$ is calculated using equation (3) and (5) multiplied by matrix $\boldsymbol{M}^{-1}$. The parameters used to obtain the value of w are $a_M = -g = -10\ m/s^2$ as opposed to the gravitational acceleration of the earth, $\theta_M = 45°$, $\alpha_i = 30°$, $d_L = 15$ m from center of gravity (assumed). The kind of missile used in this paper is Torpedo A244/S with mass $m_M = 244$ kg.

The simulation is divided into 2 scenarios i.e (i) missile is fired when the ship has followed the specified heading angle and (ii) the missile is fired when the ship is adjusting the heading angle. The value of the control parameters is $\boldsymbol{Q} = \text{diag}\{100, 100\}$ and $\boldsymbol{R} = 1$. The simulation results from scenario 1 with different predictive horizons can be seen in Figure 2.

In Figure 2, the simulation is done with different prediction horizons to measure the performance of each prediction horizon value when the system gets the impact of missile firing.

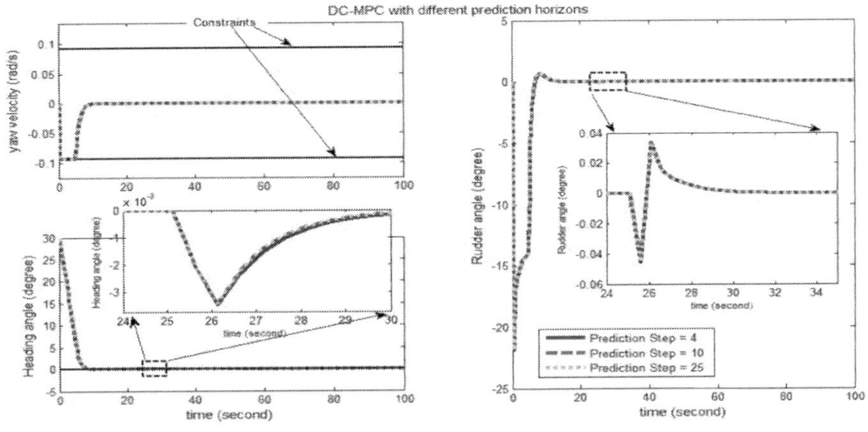

Figure 2. Simulation of the DC-MPC ship heading controller with disturbance for different prediction horizons in scenario 1

In this simulation, the initial value for the ship heading angle is 30°, then we want the ship's actual heading angle to follow desired heading angle $\psi_d = 0°$. In this simulation the missile is fired at the 25th seconds, so at that time, the system is added with disturbance $\boldsymbol{w}$. The simulation results show that DC-MPC has the ability to handle state constraints (yaw rate) for all prediction horizons used with the impact disturbance of missile firing. To see the comparison of the prediction horizons value, then we use the RMSE value between the ship's actual heading angle and the desired heading angle.

Table 1. RMSE for the different prediction horizons

Prediction horizon value	RMSE value
4	4.5693
10	4.5682
25	4.5682

Based on Table 1, when the prediction horizon is longer, the performance is better, because RMSE value of longer prediction horizons is smaller than RMSE value of short prediction horizons. For the prediction horizon values greater than 10, they have the same RMSE value.

The simulation for scenario 2 is done with the same initial value as in scenario 1, but the missile is fired when the ship is adjusting the heading angle from 30° to 0°. This simulation is done with the same prediction horizon. The simulation results can be seen in Figure 3.

Figure 3 shows the standard MPC and DC-MPC simulations without and with impact disturbance of missile firing at the 1st seconds. For the disturbance-free system, the DC-MPC scheme is the same as the standard MPC. In Figure 3, after the system has a disturbance, the ship's actual heading angle is slower toward the desired reference than system without disturbance. With this condition, the rudder angle oscillates at the 1st to 2nd seconds. Its purpose is to adjust the ship heading angle as quickly as possible to the desired reference.

6. Conclusion

In this paper, the DC-MPC scheme was developed to compensate impact disturbance of missile firing. Based on the result of simulation in previous section, DC-MPC can overcome disturbance on the system although the value of its disturbance is very small. For both scenario 1 and scenario

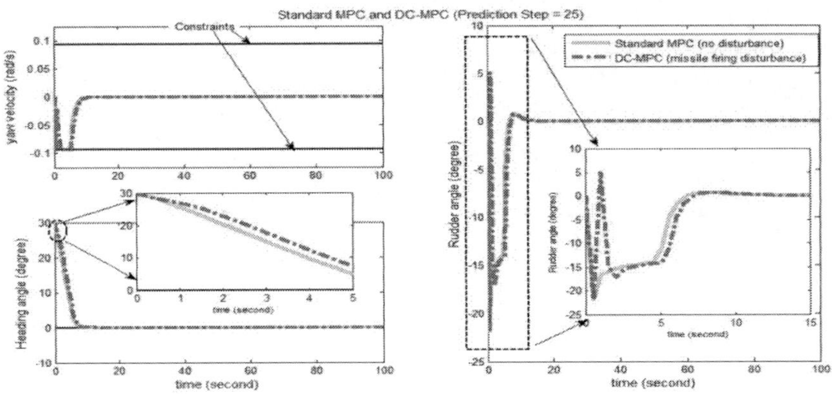

Figure 3. Simulation of Standard MPC ship heading controller without missile firing and DC-MPC ship heading controller with missile firing

2, the DC-MPC controller can make actual heading angle to follow the given reference quickly which is about 10 s. DC-MPC can also adjust the value of the ship heading angle to 0° after the disturbance occurs at the 25th seconds. The performance of the system for different prediction horizons is almost the same, because the error between actual ship heading angle and desired heading angle is very small. In the future work, the disturbance of missile firing can be applied on the ship which has the smaller size. So, the DC-MPC can perform well when the ship gets the disturbance caused by missile firing.

Acknowledgment
The authors wish to thank DRPM RISTEKDIKTI for funding this research (No. 928/PKS/ITS/2018)

References

[1] Asfihani T, Subchan S, Adzkiya D, Rosyid D M, Purnawan H and Kamilah R 2017 Estimation of the corvette sigma motion in missile firing mission *Instrumentation, Control, and Automation (ICA), 2017 5th International Conference on* (IEEE) pp 203–207
[2] Cahyaningtias S 2014 *Penerapan Disturbance Compensating Model Predictive Control (DC-MPC) pada Kendali Gerak Kapal* Master's thesis Institut Teknologi Sepuluh Nopember
[3] Ghaemi R, Oh S and Sun J 2010 Path following of a model ship using model predictive control with experimental verification *American Control Conference (ACC)* (IEEE) pp 5236–5241
[4] Li Z and Sun J 2012 *IEEE Transactions on Control Systems Technology* **20** 257–265
[5] Subchan S, Syaifudin W and Asfihani T 2014 *Far East Journal of Applied Mathematics* **87** 245
[6] Zheng H, Negenborn R R and Lodewijks G 2014 *IFAC Proceedings Volumes* **47** 8812–8818
[7] Putri D K R, Asfihani T and Subchan S 2018 Steering angle control of car for dubins path-tracking using model predictive control *Journal of Physics: Conference Series* vol 974 (IOP Publishing) p 012066
[8] Wang L 2009 *Model predictive control system design and implementation using MATLAB®* (Springer Science & Business Media)
[9] Camacho E F and Bordons C A 2012 *Model predictive control in the process industry* (Springer Science & Business Media)
[10] Fossen T I 1994 *Guidance and control of ocean vehicles* vol 199 (Wiley New York)
[11] Munadhif I, Aisjah A S and Masroeri A 2016 *IPTEK Journal of Proceedings Series* **2**
[12] Ghaemi R, Sun J and Kolmanovsky I 2006 Computationally efficient model predictive control with explicit disturbance mitigation and constraint enforcement *45th IEEE Conference on Decision and Control* (IEEE) pp 4842–4847

The Influence of Inquiry Method on Student Learning Result With Different Class Learning Style on Plantae Material.

S Ariani[1,*], Y S Rahayu[2], and E Susantini[2]

[1] SMA Yadika Bangil, Indonesia
[2] Universitas Negeri Surabaya, Indonesia

*Corresponding author: silfiariani16070795031@mhs.unesa.ac.id

Abstract. The fundamental issue that rarely gets the attention of schools in the learning process is how teachers teach the internal environment of learning is prepared. The Learning process done so far rarely touches how the internal environment is formed so that the learning process can provide maximum results. The strategy with inquiry methodology with class division based on learning style was developed as an effort to improve the learning outcomes also look for alternative learning environment more conducive so that the teacher more easily choose the right learning method. This research was an experimental research with aims to: 1) Describe the influence of Inquiry learning methods and different learning styles on student learning outcomes, 2) To explain which learning styles are the most influential in the implementation of guided Inquiry learning methods to improve student learning outcomes. This research used experimental squasize (quasi experiment) with 1x3 factorial design. Student data analysis technique result obtained in test with ANACOVA. Based on ANACOVA statistic test, the value of A.Sign learning result was 0,019 and smaller than the level of research significance (α=0,05). Post Hoc follow-up Tests on learning outcomes showed that kinesthetic learning styles provided the best results when applied with inquiry methods.

1. Introduction

The learning process has the goal to change a person's behavior based on his experience and interaction with the environment. This change should be settled and applicable in life. Learners in the learning process is a unity of individuals who have their own foreign characteristics. Therefore the learning process must be able to accommodate the diversity of these individuals so that each of these individuals have the ability to actively seek, process, construct and use of knowledge [1]. For that learning should be concerned with the opportunities given to learners to construct knowledge in their cognitive processes. In its implementation, the application of learning methods according to the curriculum of 2013 should be able to make students involved in process skills such as observing, classifying, measuring, predicting, explaining, and concluding as well as finding the right form of interaction that is able to accommodate the diversity of students in absorbing information [2].

To achieve that goal, guided inquiry method applied to the class with different modalities will give maximum result to the achievement of student learning result because the mercury in its execution there are still problems especially related to the spirit of students who are not yet accustomed to do scientific method and the provision of learning environment according to the way easiest for students in learning [3].

Based on the results of the analysis of the student's daily test score shows that the average biological value obtained by students is 5.5 This value is still below the Minimum Mastery Learning Criteria (KKM) SMA Yadika Bangil namely: 70.00. Level achievement of low learning outcomes by learners in class X SMA Yadika Bangil Kab. Pasuruan is caused by several factors including low student input, inadequate learning facilities, less learning time, teacher dominance in learning, teachers have not found the right method of learning and has not found the proper form of interaction that is able to motivate students in teaching and learning process. One of the factors that arise from within the learners that enough to give influence in the learning process and rarely used as the basis of initial ability is the learning style of students. Learning style becomes very important because it is an expression of education that is closely related to the uniqueness of each individual, in absorbing or understanding the information received [3].

Inkuiry learning strategy has a significant effect on the biology study of cognitive domain [4]. But the reality in the class shows, in Inquiry learning there are still some students who are less excited especially the students who are not yet accustomed to perform there should be an explanation about inquiry . Psychological condition of students who are not good makes the learning process is not running optimally. For that we need to find a special strategy for Inkuiry method can be run optimally for all students, one of them grouping the process of learning Inquiry in class with different learning styles. The application of inquiry methods in the classroom with different modalities is expected to improve students' learning outcomes and provide their own motivation in teaching and learning activities and ultimately students have the spirit, discipline, responsibility and seriousness in the learning process.

2. Experimental Method

2.1 General Background of Research

This research was conducted in Senior High School (SMA) of Yadika Bangil (Pasuruan, Indonesia) with the senior high school students of 10th grade who took bioogy subject in school year 2017/2018. This research was an experimental research with aims to: 1) describe the influence of Inquiry learning methods and different learning styles on student learning outcomes, 2) know which learning styles are the most influential in the implementation of guided Inquiry learning methods to improve student learning outcomes

2.2 Sample of Research

The participants were the students of grade X SMA Yadika Bangil Kab. Pasuruan 2017/2018 academic year, amounting to 133 students. The classes that are the subjects of the study are Class X IPA 1 (visual class), X IPA 2 (auditory class) and X IPA 3 (kinesthetic grade). The students' learning modalities are obtained from school data tested on each new student. The research design as follows:
Research design.

Table 1. Research Design

Treatment / learning Sthyle	Guided Inquiry Method
Visual	O_1
Audiori	O_2
Kinetetik	O_3

O1 = The value of visual learning result test
O2 = The value of the Auditory learning result test
O3 = The value of kinesthetic learning test results

2.3 Instrument and Procedures

This research used experimental squasize (Squasi experiment) with 1x3 factorial design. The exsperiment squares carried out for research residing in situations where controlling or managing the study are perfectly difficult, for example in a school system that is intake and indivisible to produce a suitable or random sample [5]. The design was designed to determine the causal relationships of the learning method by involving the experimental class (using guided Inquiry method) on learning outcomes with different learning styles.The factorial design model 3 x 1 is used to determine the effect of a treatment in research involving two variables with different levels. One variable with two levels, the learning method, and other variables consisting of three levels of learning styles of students, namely visual, auditory and kinestetik.The students' cognitive learning outcomes were measured by means of an objective test consisting of 25 items. The data of the research were analyzed by using N-Gain analysis to know the increase of pre-test and post-test value. Further data is tested by ANACOVA satellites test and acting as covariate is student's learning modality. Post Hoc tests are performed, to find out the best learning styles that can be used to apply with inquiry methods.

2.4 Data Analysis

Data analysis conducted in this study include analysis of learning material validity, lesson plan implementation, observation of student activities during the learning process, test result learning after the learning process takes place Analysis of learning outcomes by appendix and assessment guidance on aspects of knowledge using a scale of 0-100 with the provisions of predicate as follows:Very good (A) 86 -100, Good (B) 71 – 85, Quite 56 – 70 Less <55.Basic competence (KD) on core competence (KI) 3 and KI (4) learning outcomes, is said to achieve completeness individually if it meets the minimum completeness criteria established by the school of 75 and has a minimum prediction of B (good). Data in the form of student learning outcomes from the post-test is processed in three main groups namely groups of students who have auditory, visual and kinesthetic learning styles. To know the improvement of learning result before and after treatment of the method of mercury was done statistical analysis in the form of N-Gain analysis. N-gain was determined using this equation: n-gain = (score post-test - score pre-test) / (maximum score - pre-test score), with the criteria: (1) if n-gain $\geq .7$ (high), (2) if $.3 <$n-gain $< .7$ (moderate), and (3) if n-gain $\leq .3$ (low). Statistical analysis is used to test the research hypothesis. This analysis is applied to data of learning result that is pre-test and post-test data. Hypothesis testing used to know differences in student learning outcomes after the application of inquiry methods with different learning styles of students. To see any difference to these variables ANCOVA (Analysis of Covariance) method is used in SPSS 20.0 for windows program.

3. Result and Discussion

The results of the study were based on the descriptive analysis of the learning outcomes using the average N-Gain ratio of pre-test and post-test for the visual class were 0.74%, auditory class 0.73 %% and kinesthetic grade 0.80%. Based on learning comprehensiveness obtained information that the pre test of all students with learning styles in different classes no one completed, but after inquiry conducted post-test results show the changes in completeness as expected. Visual class completeness of post-test result that is 91,66%, class of auditory mastery of post-test result that is 0,89% and kinesthetic grade of completeness of learning result of post-test that is 100%. ANACOVA statistic test obtained by value at A.sign learning result that is 0,019 and smaller than level of significance of research (α), that is equal to 0,05. Hence Ho is rejected and H1 accepted or can be concluded that there is influence of learning style of student in different class to student learning outcomes. Post Hoc follow-up Tests on learning outcomes suggest that kinesthetic learning styles (0,013) deliver the best results compared to visual learning styles and auidtori when applied with inquiry methods. The following bar chart shows the improvement of learning outcomes in pre-test and post-test by inquiry methods on different classroom learning styles on the N-Gain and ANCOVA statistical test.

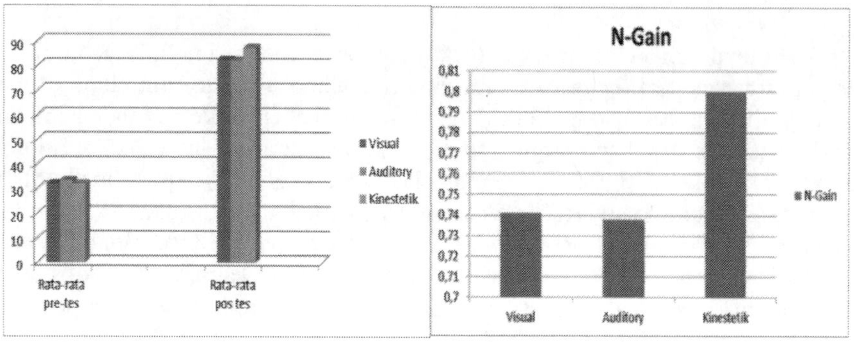

Figure 1. Diagrams of learning outcomes before and after inquiry are applied with different modes of class.

Based on the research results obtained information that the use of methods of inquiry with different learning styles in a descriptive analysis showed significant changes. The results of this study are in accordance with previous research conducted among others [6] which proves that students who are taught by inquiry based learning achieves higher academic assessment on the value of post-test compared with conventional learning. Learning with guided inquiry method has an increase in the value of post-test when compared with pre-test value, it is charged because with guided inquiry students will have more confidence than at the end of learning and students tend to favor laboratory experiments that have a method that combines pure mercury-based laboratory processes with a teacher-based process-based lab-laboratory [7]. Guided inquiry with the process that makes students more experienced in the learning process so that learning can be more meaningful, and make students more master the learning materials [8]. Guided inquiry learning method is a learning method with inquiri syntax that gets attention by explaining inquiry process, identification and problem formulation, hypothesis formulation, data collection, data interpretation, formulating conclusion, reflecting problem situation and thinking process used to investigate according to learning style requirement students [9]. In guided inquiry activities, the teacher asks questions for investigation, then gives the learner time to deliver inquiry solutions in the delivery process capable of providing experience elements through direct observation, conversation or student discussions. This inquiry activity is able to emphasize feelings, sensations and imagination as a complement in improving analytical, reasoning, and problem solving skills in a sequential and logical order. Learning by inquiry method in accordance with visual student learning style because with the inquiry syntax students learn by seeing, observing by demonstration, or viewing the lesson material presented in the form of video. Learning with guided inquiry method in accordance with auditory learning style because in inquiry is guided by learning activities by listening, discussing, questioning and answering. While students with kinesthetic learning styles correspond to guided inquiri because in the process of inquiry there is physical activity, direct involvement, responding to physical attention, moving, learning through practice, memorizing by seeing, using lots of body cues and often talking with gratification. Students with kinesthetic styles have a tendency to be silent and sit for hours because the desire for self-exploration and activity is facilitated by the implementation of guided inquiry method [10]. With the high post-test results achieved in the three different learning styles, it indicates that guided enchanting application is very effective when applied in classes with different student learning styles.

Based on the statistical analysis, it is found that the use of inquiry method with different learning styles has an effect on the students' learning outcomes. This is indicated by the value in A.sign learning outcome is 0.019 and smaller than the level of significance of research (α), which is 0.05.so Ho is rejected and H1 accepted or can be concluded that there is influence of learning styles of students in different classes against student learning outcomes. Guided inkuiry based on classroom learning styles with different shapes and characters requires teachers to pay more attention to the needs of students in learning. Individual differences will affect the speed, methods and activities of students in learning. Learning will give good results if the teaching and learning process takes into account individual needs.

Style in accordance with the characteristics of learning to make students more motivated and further improved academic achievement. Conversely, if the adjustment between teaching and learning is not appropriate, then academic and attitudinal attitudes will degenerate [3]. Teachers as facilitators in the inquiry process have skills in managing the learning process, providing conditions that make students interested and always pay attention in the learning process, responding to the needs of the students by observing the way students learn and provide opportunities for students to learn by building a more productive environment, fun, so students can be motivated to do teaching and learning activities [11].

The learning style of learning which is the intrinsic motivation of the teacher with the inquiry syntax makes the teacher always able to nurture and nurture and provide positive attitude encouragement, providing comfortable environmental conditions for more meaningful learning outcomes. Post Hoc advanced tests The results of the study show that Ho is accepted or rejected on a visual-auditory, visual auditory comparison. Comparison between visual-kinesthetic class, auditory-kinesthetic, Ho is rejected and H1 accepted, so it can be concluded that inquiry method with kinesthetic learning style has the best effect on learners' learning outcomes. Guided inkuri based on classroom learning styles with different shapes and characters requires teachers to pay more attention to the needs of students in learning. Individual differences will affect the speed, methods and activities of students in learning. Learning will give good results when the teaching and learning process takes into account individual needs. Style in accordance with the characteristics of learning to make students more motivated and further improve academic achievement. Conversely, if the adjustment between teaching and learning is not appropriate, then academic and attitudinal attitudes will degenerate [3]. Teachers as facilitators in the learning process should have skills in managing the learning process, providing conditions that make students interested and always pay attention in the learning process, responding to the needs of students by observing the way students learn and provide opportunities for students to learn by building a more productive environment, fun, so that students are expected to be motivated to do teaching and learning activities [11]. Learning styles of learning are intrinsic motivation, teachers and schools are required to nurture, safeguard, always give positive attitude encouragement, provide comfortable environmental conditions for more meaningful learning outcomes. The kinesthetic learning style is better when applied with the inquiry method.

4. Conclusion

Based on the research that has been done we can conclude that the precise inquiry method is used as an alternative to effective learning methods for the application of the scientific approach to the 2013 curriculum at the level of Senior High School with various students' learning modalities. In addition, Inquiry with kinesthetic modalities has the best effect in tackling student learning outcomes.

5. References

[1] Wenning C J 2012 Levels of Inquiry: Hierarchies of Pedagogical Practices and Inquiry Processes *J. of Physic Teacher Education* **2**(3) pp 3-12

[2] Kemendikbud 2015 *Materi Pelatihan Implementasi Kurikulum 2013* (Jakarta: Kementerian Pendidikan dan Kebudayaan)

[3] DePorter B and Hernacki M 2006 *Quantum Learning: Membiaasakan Belajar Nyaman & MEnyenangkan* (Bandung: PT.Mizah Pustaka)

[4] Agustini R 2013 *Efektifitas Pembelajaran Biologi dengan Pendekatan Inkuiri Terbimbing Berbantuan Media Animasi Digital terhadap motivasi dan Hasil Belajar Siswa* (Malang: PPS Pendidikan Biologi -Universitas Negeri Malang)

[5] Tuckman B W 2012 *Conducting Educational Research (Sixtth Edition)* (New York: Harcourt Brace Jovanovich)

[6] Maxwell D O 2015 Effects of Using Inquiry-Based Learning on Science Achievement for Fifth-Grade Students *J. Asia-Pacific Forum on Science Learning and Teaching* **16**(1)

[7] Banerjee A 2010 Teaching Science Using Guided Inquiry as the Central Theme: A Professional Development Model for High School Science Teachers *J. of Science Educator* **19**(2) 1-9
[8] Pentecost T C and Scott P 2013 From Verification to Guided Inquiry: What Happens When a Chemistry Laboratory Curriculum Changes? *J. of College Science Teaching* **42**(3) 8388
[9] Kulthau C C and Maniothes 2007 *Guided Inquiry:Learning in the 21 St Century School* (London: Libraries Unlimited)
[10] Suyono and Harianto 2011 *Belajar dan Pembelajaran* (Bandung: Remaja Rosdakarya)
[11] Arends R I 2007 *Learning to Teach Seventh Edition* (New York: Mc. Graw Hill)

Acknowledgments
A class division in the same department at a SMA Yadika Bangil should pay attention to the student's learning modality, in an effort to create a comfortable and enjoyable environment in the learning process.

Improving Students' Scientific Asking Skills through the Implementation of Question Webs Based Learning Model

Evendi[1,]*, E Susantini[2,] Wasis[3,] and B K Prahani[4]

[1]Physics Education, University of Syiah Kuala, Jl. T. Hasan Kreung Kalee, Darussalam, Banda Aceh 24415, Indonesia.
[2]Biology Department, Surabaya State University, Jl. Ketintang, Surabaya 60231, Indonesia.
[3]Physics Department, Surabaya State University, Jl. Ketintang, Surabaya 60231, Indonesia.
[4]Teacher Education of Madrasah Ibtidaiyah, State Islamic University Sunan Ampel, Jl. A.YaniNo.117, Surabaya 60237, Indonesia.

*Corresponding Author's Email: evendi@unsyiah.ac.id

Abstract. This research aims to improve students' scientific asking skills through the implementation of Question Webs Based Learning (QWBL) model. This research used one group pre-test and post-test design implemented for senior high school students enrolled in 2017/2018 academic period. The Students' scientific asking skills were measured by using Scientific Asking Skills Assessment Sheet (SASAS). The obtained data were analysed using paired t-test and n-gain. Results showed that: (1) The average post-test score was 3.42 (high category), (2) the average score of n-gain students' scientific asking skills was 0.67 (medium category), and (3) the students responded very positively (95%) to the implementation of QWBL Model. Another result found that more than 75% of the students felt courageous to ask scientific questions through writing. This research implies that QWBL Model can be used to improve students' scientific asking skills in various education levels in Indonesia.

1. Introduction

The process of teaching physics is done by the application of scientific method. One of the components students have to master in applying the method is *scientific asking skills*, student's skill in asking scientific questions. Scientific question is a typical question that can be answered through experiment and based on factual reasons or evidence [1-5]. Scientific questions play an important role in the process of education in schools because such questions will have a positive impact on students, such as (1) increasing their active learning participation, (2) developing their thinking processes to help maintain their attention, (3) assessing their learning to arouse curiosity, (4) stimulating interest and making them more confident, (5) and motivating them to seek new information [6-9].

Early research results [9] on scientific questions showed that students' ability to ask scientific questions in oral or in writing in physics learning was still low. The number of students who were able to ask questions verbally were only 20% of the total students and 80% of them all are unable to ask questions verbally. Most students did not want to ask verbally because they (1) did not know what to ask, (2) did not dare to ask, (3) felt fear in conveying wrong question or not in line with the taught concept, (4) felt unsatisfied with the teacher's responses, (5) preferred to hear rather than to ask questions. Similarly, previous research resulted that students' ability in formulating questions based on Blooms' taxonomy showed low levels; it was 86% recall, 7% application, and 7% not yet able to formulate questions [9-10]. The results of this study are in accordance with the opinion initiated by Kaberman & Dori that along with the increasing level of education, the phenomenon of many students who rarely ask questions and do not even want to ask during the learning process often happens [9,11].

One of the innovative learning models to improve scientific asking skills of senior high school students is Question Webs Based Learning (QWBL) model. The web question is a chart of questions designed to help students inquire about the predictions, theories, and the results of an experiment or concept studied [3]. The model has a syntax that consists of 5 phases, they are: (1) Identifying problems and conveying goals, (2) Delivering information and compiling questionnaires, (3) Discussing scientific questions, (4) Reflecting on scientific questions and (5) Evaluation [9]. Each phase or step in the model is supported by the latest theoretical and empirical foundations. In addition, the model and its instruments have been validated by experts or validators in a Focus Group Discussion (FGD) and stated to be very valid (on content validity and construct validity) and reliable, so that the QWBL model is declared eligible to be applied in school [9].

2. Experimental Method

2.1 General Background of Research

This research belongs to Educational Design Research (EDR) [10]. The purpose of this research was to analyse the improvement of scientific asking skills of senior high school students through the implementation of QWBL model on physics learning. This research used one-group pre-test and post-test design. The improvement of students' scientific asking skills through the implementation of the QWBL model was determined based on: (1) The average post-test scores involving students' scientific skills at least in scientific category, (2) a significant increase (statistically) on the pre-test and post-test scores, (3) average n-gain determined at least on the middle improvement criteria, and (4) the students' response that must be at least positive.

2.2 Sample of Research

The samples of this research were 20 students of senior high school in Aceh, Indonesia at academic year 2017/2018 who took physics course especially in static fluid materials. The samples determination used purposive sampling technique [13].

2.3 Instrument and Procedures

Scientific asking skills of senior high school students were measured using the Scientific Asking Skills Assessment Sheet (SASAS). SASAS was prepared based on the scientific indicators of the measured skills like shown in Table 1. The high school students' resposes to the implementation of the QWBL model was measured by using a student response questionnaire [5,9, 13-14].

This study was classified as a pre-experimental study that used one group pre-test and post-test design, which was O1 X O2 (13, 16-17). The learning started by giving pre-test (O1). Then, the teacher applied QWBL model and physics learning instruments in the experimental class (X) for four meetings. In each meeting, students were given a questionnaire webs sheet to train them ably composing questions in accordance with the taught concept. The teacher gave a fundamental question written on the middle circle of the questionnaire webs sheet. Students were asked to write sub-questions in general based on the given margin of the first page of the questionnaire (Figure 1a) by using the question word "H&5W Questions" (what, why, who, whom, where, which, how). The teacher asked students to write scientific questions on the second page (Figure 1b) based on the sub-questions on the first page by using one of the scientific questions formats: a) is there a relationship between..... With.....? b) What factors are causing......? c) Is there any influence.....to .....? d) What is the relationship between ...... to......? [2,9]. The preparation of questionnaire webs sheets pages was done by students in groups to produce qualified scientific questions. The implementation of the QWBL model on physics learning was ended by post-test (O2) using SASAS. Each senior high school student was required to complete SASAS and filled out a response questionnaire on the post-test.

2.4 Data Analysis

Table 1 and Table 2 show the data analysis of students' scientific asking skills before and after the implementation the QWBL Model based rubric.

Tabel 1. The rubric to assess indicator of the scientific asking skills.

Indicator	Assessment aspect	Score
1. Using one of the scientific question formats	Formulated by using one of the scientific question formats.	1.00
2. Shows two or more variables.	The scientific question shows the relationship between two or more variables that can be tested.	1.00
3. Contains facts or evidence	The scientific question contains facts or evidence that can be investigated or experimentation.	1.00
4. Not based on values and opinions	The scientific question not based on values and opinions.	1.00
	Total	**4.00**

Table 2. The assessment category of scientific asking skills.

Score	Scientific asking skills category	Notes
$3.20 < x \leq 4.00$	Very scientific	It contains all aspects of the assessment
$2.40 < x \leq 3.20$	Scientific	It contains three aspects of assessment
$1.60 < x \leq 2.40$	Less scientific	It contains two aspects of assessment
$0.80 < x \leq 1.60$	Very less scientific	It contains only one aspect of the assessment
$0.00 < x \leq 0.80$	Not scientific	It does not contain any aspect of the assessment

The degree of increased score of students' scientific asking skills was analysed by using the n-gain equation [9,18]. The n-gain score was determined by the following criteria: High> 0.70; medium 0.30 - 0.70; Low <0.30. The pre-test and post-test data of scientific asking skills of high school students were analysed by using inferential statistic with the help of SPSS software.

3. Result and Discussion

Table 3 presents the results of the QWBL model implementation on physics learning. The scientific asking skills of senior high school students at pre-test in 4 students (20%) were in very less scientific category, 11 students (55%) were less scientific and 5 students (25%) were in the scientific category. In general, students' scientific asking skills before the implementation of the QWBL Model were only 2.17 (less scientific category). These findings were supported by the research [9] that the scientific asking skills of high school students are still low because they are not trained by teachers and students are still afraid to ask even if they do not understand the concept of physics that is being described by the teacher.

Conversely, after the implementation of QWBL model, senior high school students' scientific asking skills increased. At post-test, 16 students (80%) were in very scientific category while the other 4 students (20%) were in scientific category. The application of the QWBL model could improve students' scientific asking skills from 75% less scientific and very less scientific to 80% high scientific. The increase of n-gain score was from 0.38 to 0.94 which meant in moderate and high category. This increase in n-gain indicated that the QWBL model was effective to improve scientific asking skills in physics learning especially on static fluid materials

Table 3. The average score of pre-test, post-test and n-gain on scientific asking skills of high school students in physics learning.

Students	Pre-test	Category	Post-test	Category	N-gain	Category
S1	1.60	Very Less Scientific	3.87	Very Scientific	0.94	High
S2	2.20	Less Scientific	3.47	Very Scientific	0.70	Medium
S3	2.80	Scientific	3.33	Very Scientific	0.44	Medium
S4	1.60	Very Less Scientific	3.40	Very Scientific	0.75	High
S5	2.13	Less Scientific	3.60	Very Scientific	0.79	High
S6	1.87	Less Scientific	3.27	Very Scientific	0.66	Medium
S7	3.13	Scientific	3.73	Very Scientific	0.69	Medium
S8	2.40	Scientific	2.93	Scientific	0.33	Medium
S9	2.13	Less Scientific	3.33	Very Scientific	0.64	Medium
S10	2.67	Scientific	3.60	Very Scientific	0.70	Medium
S11	1.87	Less Scientific	3.87	Very Scientific	0.94	High
S12	2.00	Less Scientific	3.53	Very Scientific	0.77	High
S13	3.00	Scientific	3.60	Very Scientific	0.60	Medium
S14	1.60	Very Less Scientific	3.53	Very Scientific	0.81	High
S15	2.20	Less Scientific	3.67	Very Scientific	0.81	High
S16	2.20	Less Scientific	3.33	Very Scientific	0.63	Medium
S17	1.33	Very Less Scientific	2.47	Scientific	0.43	Medium
S18	2.20	Less Scientific	3.20	Scientific	0.56	Medium
S19	2.27	Less Scientific	3.47	Very Scientific	0.69	Medium
S20	2.27	Less Scientific	3.13	Scientific	0.50	Medium
Average	**2.17**	**Less Scientific**	**3.42**	**Very Scientific**	**0.67**	**Medium**

In general, scientific asking skills of high school students during post-test is 3.42 (very scientific category) and n-gain was in moderate categories. The results of this study are supported by the results of previous studies [19-31] that learning, teaching, models, methods, instruments, and media that are eligible, qualified, valid, practical, and effective will enhance and achieve learning objectives. That is, different test of pre-test score and post-test of scientific asking skills on the implementation of QWBL Model used Paired t-test (see Table 4).

Table 4. The results of Paired t-test of students' scientific asking skills.

Paired Samples Test, α = 5%						
	Mean	**Std. Deviation**	**Std. Error Mean**	**T**	**df**	**Sig. (2-tailed)**
Pre-test – Post-test	-1.24	0.50	0.11	-11.07	19	0.00

Table 4 shows that the mean score of students' scientific asking skills was -1.24 and had degrees of freedom (df) of 19, t arithmetic of -11.07. This result was significant because $p < 5\%$. The result of the calculation had a negative sign, which meant that the pre-test was different from the post-test (scientific) perceived by the students after using QWBL model in physics learning. The students' positive response to the QWBL model was 95% of students who said that the QWBL Model could motivate students to ask questions. The students' learning outcomes increased as they had a good response to the learning [33]. Increasing student involvement can improve learning outcomes in terms of the effectiveness, measurable learning, high students' achievement, and their learning responses [34]. A teaching is said to be effective when a teacher has the right strategy for conveying his knowledge to students structurally and can integrate theory and practice into the learning process [35]. The active participation of students in learning by using the QWBL Model can be seen from the work of the student groups in preparing the question webs sheet as in Figure 1.

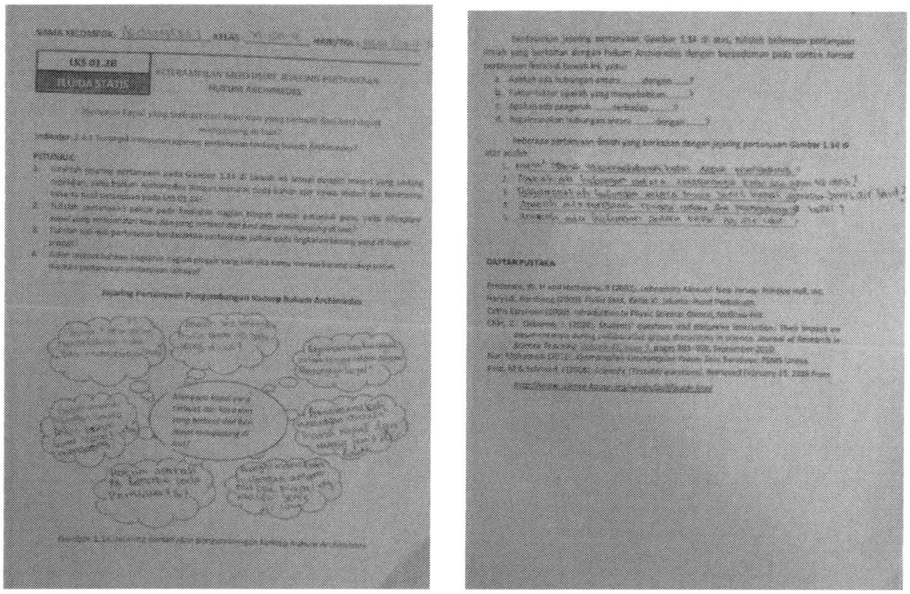

(a) (b)

Figure 1. Question webs sheet.

The question webs sheet were arranged in groups to produce qualified questions, although there were some general questions that had not led to scientific questions on the first sheet such as "What is the relationship between floating materials for ships?" and "on the event?" but based on the agreement in the students group discussion, the general questions on the first page were rearranged into several scientific questions on the second page such as" Is there a relationship between the density of the ship and the sea water mass?". These scientific questions were used by students to conduct class discussions. If students did not dare to ask questions verbally or were afraid making wrong questions, then students could ask questions in writing by reading the questions on the questionnaire.

Principle finding based on interview questions with several students was that more than 75% of students at first were not courageous to ask question because they were afraid the questions were wrong and fear of being laughed at by friends. However, after being taught by implementation of the QWBL Model, the students feel more courageous in scientific writing and can ask scientific questions through writing and oral with the question is very scientific. The results of this study proves that the implementation of the QWBL Model in physics learning proved to be effective in improving the scientific asking skills of senior high school students.

4. Conclusion

Question webs Based Learning (QWBL) model is specifically designed to improve students' scientific asking skills. The results of the research prove that: (1) The average post-test score was 3.42 (high scientific category); (2) Average n-gain score of students' scientific asking skills was 0.67 (medium category); and (3) Students responded very positively (95%) to the implementation of QWBL model. Principle finding based on interview results was that students feel more courageous in scientific writing and can ask scientific questions through writing and oral. Therefore, the QWBL Model has proven to be effective for improving the scientific asking skills of high school students. The implication of this research is that QWBL Model is expected to be implemented to improve students' scientific inquiring skills in various education levels in Indonesia. Further research is expected to be done at different levels of education and country.

5. References

[1] Research Council 2000 *Inquiry and the National Science Education Standards: A guide for Teaching and learning* (Washington D.C: National Academy Press)
[2] Nur M 2011 *Modul Keterampilan Proses Sains* (Surabaya: Universitas Negeri Surabaya)
[3] Chin C and Osborne J 2010 *J. Res. Sci. Teach.* **47** 883
[4] Nussbaum E and Edwards O V 2011 *J. Learn. Sci.* **20** 443
[5] Chiron G and Sandra R 20122012 *J. College Sci. Teach.* **14**46
[6] Wragg E and Brown G 2001 *Questioning in the Secondary School* (London: Routledge Falmer)
[7] Eggen Kauchak 2012 *Strategy and Model for Teacher: Teaching Content and Thingking Skills* (Boston: Pearson Education Inc)
[8] Harrold P 2013 *Any Questions? Designing An Observation Instrument To Record Teacher Questioning* (Japan: Kanda University of International Studies)
[9] Evendi 2018 *Makalah Seminar Hasil* (Surabaya: Pascasarjana Unesa)
[10] Nieveen N 2010 *Formative Evaluation in Educational Design Research* Enschede (the Netherlands: SLO)
[11] Liliasari 2003 *J. Pendidik. Matematika dan Sains* **3** 174
[12] Kaberman Z and Dori Y J 2009 *I. J. Sci. Math. Educ.* **7** 597
[13] Fraenkel J R, Wallen N E, and Hyun H H 2012 *How to Design and Evaluate Research in Education* (New York: McGraw-Hill)
[14] Erika F, Prahani B K, Supardi Z A I and Tukiran 2018 *WTE & TE* **17** 1
[15] Lan Y F and Lin P C 2011 Australasian J. Educ. Tech. **27** 581
[16] Prahani B K, Winata S W and Yuanita L 2015 *J. Penelit. Pend. Sains* **4** 503
[17] Prahani B K, Limatahu I, Nur M and Yuanita L 2016 *Int. J. Educ. Res.* **4** 231
[18] Hake R R 1998 *Am. J. Phys.* **66** 64
[19] Hadi S A, Susantini E and Agustini R 2018 *J. Phys: Conf. Ser.* **947** 012063
[20] Jatmiko B, Prahani B K, Munasir, Supardi Z A I, Wicaksono I, Erlina N, Pandiangan P, Althaf R and Zainuddin 2018 *J. Baltic Sci. Educ.* **17** 1
[21] Limatahu I, Suyatno S, Wasis and Prahani B K *J. Phys.: Conf. Ser.* **997** 0120022
[22] Madeali H and Prahani B K 2018 *J. Phys: Conf. Ser.* **997** 0120029
[23] Prahani B K, Suprapto N, Suliyanah, Lestari N A, Jauhariyah M N R, Admoko S and Wahyuni S 2018 *J. Phys.: Conf. Ser.* **997** 0120008
[24] Purwaningsih E, Suyatno S, Wasis and Prahani B K 2018 *J. Phys: Conf. Ser.* **997** 0120032
[25] Sudiarman W, Soegimin W and Susantini E 2015 *J. Penelit. Pendidik. Sains* **4** 636
[26] Sunarti T, Wasis, Madlazim, Suyidno and Prahani B K 2018 *J. Phys: Conf. Ser.* **997** 0120013
[27] Susantini E, Isnawati and Lisdiana L 2016 *J. Sci. Educ.* **2** 74
[28] Susantini E, Lisdiana L, Isnawati, Al Haq A T and Trimulyono G 2017 *Biochemistry and Molecular Bio. Educ.* **45** 216
[29] Susantini E and Qomariyah N 2013 *J. Pendidik. IPA Indonesia* **2** 142
[30] Susantini E, Thamrin M, Isnawati H and Lisdiana L 2012 *J. Pendidik. IPA Indonesia* **1** 102
[31] Suyidno, Nur M, Yuanita L, Prahani B K and Jatmiko B2018 *J. Baltic Sci. Educ.* **17** 136
[32] Wicaksono I, Wasis and Madlazim 2017 *J. Baltic Sci. Educ.* **16** 54
[33] Zimmerman B J and Schunk D H 2012 *Self-Regulated Learning and Academic Achievement: Theory, Research, and Practice* (New York: Springer Science & Business Media)
[34] Jatmiko B,Widodo W, Budiyanto M, and Wicaksono I, 2016 *J. Baltic Sci. Educ.* **15** 441
[35] Hughes J 2005 *J. Tech. Teach. Educ.* **13** 277

Acknowledgment
The author's gratitude goes to the Ministry of Research, Technology and Higher Education for funding the Research (SK Number: 15/UN11.2/PP/SP3/2018). Likewise, the author's gratitude goes to the University of Syiah Kuala, Surabaya State University, State Islamic University of Sunan Ampel Surabaya and Senior High School Students (Aceh, Indonesia) that have provided research opportunities.

Application of Support Vector Machines in Evaluating the Internationalization Success of Companies

Z Rustam*[1], F Yaurita*[1] and M J Segovia-Vergas[2]

[1] Department of Mathematics, Universitas Indonesia, Depok, 16424, Indonesia
(E-mail: rustam@ui.ac.id, frederica.yaurita@ui.ac.id)
[2] Departement of Financial Economy and Accounting I, Facultad de Ciencias Económicas y Empresariales, Universidad Complutense de Madrid 28223, Spain.

Abstract. The internationalization started to be seen as an opportunity for many companies. This is one of the most crucial growth strategies for companies. Internationalization can be defined as a corporative strategy for growing through foreign markets. It can enhance the product lifetime and improve productivity and business efficiency. However, there is no general model for a successful international company. Therefore, the success of an internationalization procedure must be estimated based on different variables such as the status, strategy, and market characteristics of the company. In this paper, we try to build a model in evaluating the internationalization success of a company based on existing past data by using Support Vector Machines. The results are very encouraging and show that Support Vector Machines can be a useful tool in this sector. We found that Support Vector Machines achieved 81.36% accuracy rate with RBF Kernel, 80% training set, and $\sigma = 0.05$

1. Introduction

Internationalization started to be seen as an opportunity for many companies. It is an important strategic choice to face economic globalization challenge [1]. Internationalization is defined as the process of involvement increase in international operations [2, 3]. It refers to the degree of sale incomes or operations which firms direct in foreign market [4]. Thomson et al in 2015 said there are five reasons why a company may opt to enter the foreign market. There are: to gain access to new customers, to gain access to resources in foreign market, to gain access to low-cost inputs of production, to spread the business risk, and the last is to achieve lower costs through the increase of purchasing power [5]. However, internationalization is a high risky business strategy. The firms should prepare for political and economic risks, they will face international market change, technology, market and financial risks.

There is no general model for a successful international company. The success of an internationalization procedure must be estimated based on different variables such as the status, strategy, and market characteristics of the company [6]. In this study, we will build a model by using existing past data that have faced an internationalization process before. This model will be able to classify internationalization success or failure of a company. It will help a company to predict the internationalization success of the company base on the current company's condition.

We used machine learning techniques to build the model. Machine learning is an application of artificial intelligence that provides systems the ability to automatically learn and improve from experience without being explicitly programmed (Arthur Samuel, 1959). Support Vector Machines (SVM) is known as the powerful machine learning tool for classification. Application of SVM has been used by Rustam Z and Yaurita F [7], Rustam Z and Zahras D [8], Rustam Z and Ariantari N P A [9], Wardana I P A and Rustam Z [10], Puspitasari D A and Rustam Z [11], Rustam Z, Vibranti D F,and

Widya D [12], respectively for insolvency prediction, classifier for intrusion detection system, classifying policyholders satisfactorily, decision-making in stock investment, stock analysis of Indonesian stock exchange, and predicting the direction of Indonesian stock price movement.

Since a lot of financial problems could be solved by SVM and the results are trustworthy, in this study we try to conduct the insolvency prediction model used SVM algorithms. We built the model by using 595 companies which have faced an internationalization process before. The results are very encouraging and show that SVM can be a useful tool to predict company's internationalization success. The rest of paper is structured as follows. In the next section, there are brief introduction to SVM algorithm and kernel function. The third section describes the data set that we used and steps of the experiment. And finally, the last section contains some concluding remarks.

2. Support Vector Machines and Kernel Function

2.1. Support Vector Machines (SVM)

SVM has received much attention in classification problems. This method first proposed by Vapnik in 1998. In a few years later, Nello Cristianini (Professor of Artificial Intelligence at the University of Bristol) did a research about SVM based on Vapnik's theory before and have written the result in [13]. After that in 2002, Bernhard Scholkopf developed SVM theory and published it in a book which is about kernels in SVM as we can see in [14]. We used [15], [16], [17] and [18] as our main reference in this SVM section.

Let $D = \{(\boldsymbol{x}_i, y_i)\}_{i=1}^{N}$ be a classification data set, with N points in d-dimensional space. Further, let us assume that there are two class labels, $y_i \in \{+1, -1\}$ denoting the positive and negative classes. The main purpose of SVM is to find the best hyperplane that is defined as follows:

$$f(x) = \boldsymbol{w}^T \cdot \boldsymbol{x} + b \quad (1)$$

Here, **w** is a d dimensional weight vector, and b is a scalar, called bias. For points that lie on the hyperplane, we have:

$$\boldsymbol{w}^T \cdot \boldsymbol{x} + b = 0 \quad (2)$$

Figure 1 shows the illustration of SVM. Let the red point belongs to the positive class and the blue point belongs to the negative class. The H line is called hyperplane, H_- and H_+ are called canonical hyperplane. Support vectors are the data points that lie closest to the hyperplane. There are many hyperplanes which able to separate two classes, but there is only one the optimal hyperplane [15]. It should be the hyperplane that maximizes the margin between to class, so that the distance from the hyperplane to the nearest data point on each side is maximized.

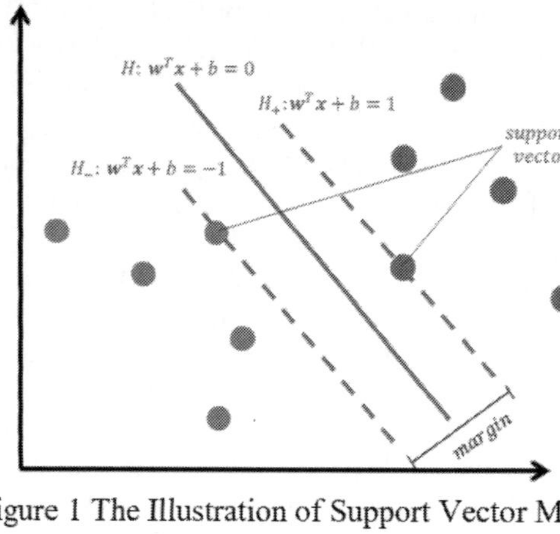

Figure 1 The Illustration of Support Vector Machines

The formula of margin is $\frac{2}{||\boldsymbol{w}||}$ where $\boldsymbol{w}$ is a vector that perpendicular to the hyperplane. Consider that instead of maximizing the margin $\frac{2}{||\boldsymbol{w}||}$, we get an equivalent formula if we minimize $||\boldsymbol{w}||$. Then we can obtain an equivalent minimization formula given as follows:

$$minimize \quad \frac{1}{2}||w||^2 \tag{3}$$

$$subject\ to \quad y_i(\boldsymbol{w}^T \cdot x_i + b) \geq 1,\ \forall i = 1, \dots, N \tag{4}$$

The **Eq. (3)** can be solved directly by using standard optimization algorithms. However, it is more common to solve the dual problem which is obtained by the use of Lagrange function. We introduce Lagrange multipliers α_i for each of the constraints in (4), thus we get:

$$L(w, b, \alpha) = \frac{1}{2}||w||^2 - \sum_{i=1}^{N} a_i\{y_i(w \cdot x_i + b) - 1\} \tag{5}$$

where $a = (a_1, a_2, \dots, a_n)^T$. Setting the derivatives of $L(w, b, \alpha)$ with respect to $\boldsymbol{w}$ and b equal to zero, the following two conditions were obtained:

$$\frac{\partial L}{\partial w} = \boldsymbol{w} - \sum_{i=1}^{N} a_i y_i x_i = 0 \rightarrow \boldsymbol{w} = \sum_{i=1}^{N} a_i y_i x_i \tag{6}$$

$$\frac{\partial L}{\partial b} = \sum_{i=1}^{N} a_i y_i = 0 \rightarrow \sum_{i=1}^{N} a_i y_i = 0 \tag{7}$$

The **Eq, (3)** can be transformed into a dual problem by substituting the derivatives of Lagrangian **(Eq. (6), (7))** back into the Lagrangian function [12], therefore we obtained the dual form as follows:

$$L(\alpha) = \max\left\{-\frac{1}{2}\sum_{i=1}^{N}\sum_{j=1}^{N} y_i y_j a_i a_j (x_i, x_j) + \sum_{i=1}^{N} a_i\right\} \tag{8}$$

$$subject\ to\left\{\sum_{i=1}^{N} y_i \alpha_i = 0\,, \alpha_i \geq 0\right\} \tag{9}$$

There is a case that the data set is not linearly separable, therefore the data set should transform into a higher dimension using kernel function $\kappa(x_i, x_j) = \phi(x_i)^T \phi(x_j)$. In this paper, we used RBF and polynomial kernel as described in **Sec 2.2**, and by changing the inner product of two vectors with $\kappa(x_i, x_j)$. **The Eq. (8)** can be rewritten as:

$$L(\alpha) = \max\left\{-\frac{1}{2}\sum_{i=1}^{N}\sum_{j=1}^{N} y_i y_j a_i a_j K(x_i, x_j) + \sum_{i=1}^{N} a_i\right\} \tag{10}$$

Solving **Eq. (10)** with constraints in **Eq. (9)** will determine the Lagrange multipliers $(a_i a_j)$. By using KKT condition the formula of **w** and b are finally obtained as follows:

$$\boldsymbol{w} = \sum_{i=1}^{N} a_i y_i \boldsymbol{x}_i \tag{11}$$

and

$$b = \frac{1}{N_s}\sum_{i \in S}\left(y_i - \sum_{m \in S} a_m y_m \boldsymbol{x}_m\right) \tag{12}$$

where N_s is the number of support vectors.

2.2. Kernel Function

Based on the prior experiments, most of training set are not linearly separable. To solve this problem, each vector $\mathbf{x}_i$ is mapped to a higher dimensional space which is called feature space. It is expected that the transformed data can approach the linearly separable data, so that classification accuracy can be improved. A kernel is a function κ that for all $\boldsymbol{x}_i, \boldsymbol{x}_j \in X$ satisfies [14]:

$$\kappa(\mathbf{x}_i, \mathbf{x}_j) = < \phi(\mathbf{x}_i), \phi(\mathbf{x}_j) > \tag{13}$$

where ϕ is a mapping from X to an (inner product) feature space F.

$$\phi: x_i \to \phi(x_i) \in F \tag{14}$$

In our research we used RBF and polynomial kernel that defined as follows:

$$RBF \quad \kappa(\mathbf{x}_i, \mathbf{x}_j) = \exp(-\frac{\|\boldsymbol{x}_i - \boldsymbol{x}_j\|^2}{2\sigma^2}) \tag{15}$$

$$Polynomial \quad \kappa(\mathbf{x}_i, \mathbf{x}_j) = (< \mathrm{x_i}, \mathrm{x_j} > +1)^{\mathrm{d}} \tag{16}$$

where d and σ are the adjustable parameters.

3. Experiment and Results

3.1. Data set

In this study, we used SVM algorithm to build the model that can predict internationalization's success of companies. The data set has got from Prof. M J Segovia [19]. This data set used is yearly elaborated by the Spanish Foundation of Statal Industrial Participation (SEPI), by investigated more than ten employees in manufacturing companies. This database is elaborated for the whole country and all variables involved are annual. In our study, there are 595 Spanish companies data including the success or failure of the company in its internationalization process. Each firm is described by 30 variables (V1, V2, ..., V30). All the respective variables describing the companies themselves and their internationalization strategy.

3.2. Results

The aim of this research is to build the model that able to predict the internationalization success of a company. We solved the problem by using SVM algorithm. In machine learning, before using the algorithm, we should split the data set into training set and test set. Training set is implemented to build up the model while the test set use to validate the model built (to check the accuracy). Therefore, the first step in our research is, we randomized all the data and split it into training set and test set. Let the training set is p% data from each class, then it is obvious the test data will be (100 - p)% data from each class.

Assume that $D = \{(\boldsymbol{x}_i, y_i)\}_{i=1}^{N}$ are the given training set, where $\{\boldsymbol{x}_i\}_{i=1}^{N}$ are the set of companies that have faced internationalization process with N points in d-dimensional space, and $y_i \in \{+1, -1\}$ denotes label of each company. Label +1 belongs to companies that succeeded in internationalization process, and -1 belongs to companies that failed in internationalization process. Therefore, we had two class in the experiment, the first class is the set of successful companies and the second class is the set of unsuccessful companies.

After randomized the data set, and determined several data to be training set, SVM algorithm worked by solving the quadratic problem in **Eq. (10)**. In short, the algorithm got the weight vector (**w**) and bias (b). Substituting **w** and b into **Eq. (1)** brought the algorithm to get the best function of hyperplane. The next step was to see how accurate the hyperplane for classifying the successful and unsuccessful companies. In this process, the algorithm used (100 - p)% data from each class. Every test set is substituted into function $f(x)$. $f(x) > 1$ belongs to the success class, and $f(x) < -1$ belongs to unsuccess class. If the input data is a successful company and $f(x) > 0$, this indicates that the model does a true prediction. Same as if the input data is an unsuccessful company and $f(x) < 0$, this indicates that the model does a true prediction. Other than these means that our prediction is incorrect. Then the accuracy calculated by using this form:

$$Accuracy = \frac{Number\ of\ true\ prediction}{Number\ of\ test\ set} \times 100\% \tag{17}$$

4. Results

Table 1. Accuracy and running time of SVM algorithm with polynomial kernel (degree=2) and RBF kernel (σ=0.5)

Training Set	Polynomial		RBF	
	% Accuracy	Running Time (s)	% Accuracy	Running Time (s)
20%	60.92%	0.0051	50.09%	0.0051
40%	70.23%	0.0137	60.96%	0.0091
60%	73.53%	0.0218	71.85%	0.0133
80%	**79.66%**	0.0357	**81.36%**	0.0174

Table 1 shows the summary of our experiments. In the first column, there is % training set, it is the percent data set (from each class) that implemented to build up a model. The second and third column is the accuracy and running time of experiments that used SVM algorithm with polynomial kernel while the fourth and fifth column is the accuracy and running time of experiments that used SVM algorithm with RBF kernel. Accuracy shows the percentage of true prediction that the model did, and the running time shows how long is the program can provide the result.

In the first experiment, we used polynomial kernel with degree = 2. It can be seen from **Table 1** that the best accuracy achieved when we used 80% data training, it is 79.66%, and the worst accuracy achieved when we used 20% training set. In the second experiment, we used RBF kernel with $\sigma = 0.05$. From **Table 1,** we know that the best accuracy achieved when we used 80% training set, it is 81.36%, and the worst accuracy achieved when we used 20% training set.

5. Conclusions

In this research, we have made a model for evaluating the internationalization success of companies by using 595 companies which have faced an internationalization process before. This new model can help a company to predict its internationalization success. The highest accuracy is obtained by the experiment that use RBF kernel with $\sigma = 0.05$, it is 81.36%. In future studies, we interested to try the other machine learning algorithms such as AdaBoost, kNN, Naïve Bayes, etc.

6. References

[1] Jiang M and Xu M 2011 Chinese Firms *2011 International Conference of Information Technology, Computer Engineering and Management Sciences* pp 186-189 IEEE.Behyan M 2016 *Investigating the Influence of External Networks on Intenationalization Success of Firms* IEEE 978-1-5090-2172-7/16, 2016.

[2] Welch, L S and Luostarinen R 1988 *Internationalisation: evolution of a concept* Journal of General Management, Vol. 14 No. 2, pp. 34-55.

[3] Holmlund M, Kock S and Vanyushyn V 2007 Small and medium-sized enterprises' internationalisation and the influence of importing on exporting *International Small Business Journal* **25** pp 459-477.

[4] Thompson A A, Strickland A J, Gamble J E, and Zeng'an Gao 2008 Crafting and executing strategy: The quest for competitive advantage: Concepts and cases.Elango B and Pattnaik C *Building capabilities for international operations through networks: a study of Indian firms Journal of International Business Studies* **38** pp 541-555

[5] Rustam Z and Yaurita F 2018 *Insolvency Prediction in Insurance Companies Using Support Vector Machines and Fuzzy Kernel C-Means* Journal of Physics: Conference Series J. Phys. 1028 012118.

[6] Rustam Z and Durrabida Z 2018 *Comparison between Support Vector Machine and Fuzzy C-Means as Classifier for Intrusion Detection System* Journal of Physics: Conference Series J. Phys. 1028 012118 1028 012227.

[7] Rustam Z and Ariantari NPA 2018 *Comparison between Support Vector Machine and Fuzzy C-Means as Classifier for Intrusion Detection System* Journal of Physics: Conference Series J. Phys. 1028 012005.

[8] Wardana I P A and Rustam Z 2017 Application of Support Vector Machines on Decision-Making in Stock Investment *The 3rd ISCPMS AIP Proceeding Conference.*

[9] Rustam, Z, Vibranti D F and Widya D 2017 Predicting The Direction of Indonesian Stock Price Movement using Support Vector Machines and Fuzzy Kernel C-Means *The 3rd ISCPMS AIP Proceeding Conference.*

[10] Rustam Z and Yaurita 2017 Application of Support Vector Machines for Reject Inference in Credit Scoring *The 3rd ISCPMS AIP Proceeding Conference.*

[11] Cristianini, N and Taylor J S 2000 *Support Vector Machines and other kernel-based learning methods* (New York: Cambridge Univercity Press)

[12] Taylor J S and Cristianini N 2004 *Kernel Methods for Pattern Analysis* (New York: Cambridge Univercity Press)

[13] Vapnik V N 1998 *Statistical Learning Theory* (USA: Wiley)

[14] Scholkopf and Smola 2002 *Learning with Kernels: Support Vector Machines, Regularization, Optimization, and Beyond* (USA: MIT)

[15] Campbell C and Ying Y 2011 Learning with support vector machines *Synthesis lectures on artificial intelligence and machine learning* **5**(1) pp1-95

[16] Zaki M J and Meira W 2013 *Data Mining and Analysis: Fundamental Concepts and Algorithms* (New York: Cambridge University Press)

[17] Landa-Torres I, Ortiz-Garcia E G, Salcedo-Sanz S, Segovia-Vargas M J, Gil-Lopez S, Miranda M, ... & Del Ser J 2012 Evaluating the internationalization success of companies through a hybrid grouping harmony search—extreme learning machine approach. *IEEE Journal of Selected Topics in Signal Processing* **6**(4) pp 388-398

[18] Stoean, C. & Stoean R. 2014 *Support Vector Machines and Evolutionary Algorithms for Classification* (London: Springer)

IOP Conf. Series: Journal of Physics: Conf. Series **1108** (2018) 012039 doi:10.1088/1742-6596/1108/1/012039

Sequence Variation of Metallothionein (MT) Gene in Poaceae Family

P Peresia[1,*] and D S Rini[2,*]

[1]Biology Department, Surya University, Indonesia
[2]Research Center for Biology, Indonesian Institute of Sciences (LIPI), Indonesia
*Corresponding Author. Email: pebepperesia@yahoo.co.id; dw.setyo19@gmail.com

Abstract. Several plant in Poaceae family can be used as a accumulator plant in phytoremediation technique to clean the contaminated area. Metallothionien (MT) is one of metal binding protein groups that regulate heavy metal absorption. MT has low molecular weight and has cysteine-rich residues. MT plays a role in essential metals homeostasis and also in toxic metals detoxification. Meanwhile, MT also functions as a chelator that can bind heavy metals. Due to no information about the functional domains of MT gene in plants, study the sequence variation of MT gene is required to find the conserved region, especially in Poaceae family, related to MT gene function in phytoremediation. The MT gene homology shows the gene genetic relatedness in interspecies. In this study, the MT gene sequences in Poaceae plants (Hordeum vulgare, Sorghum bicolor, and Setaria itálica) were obtained from NCBI. T-cofee tool was used to align sequences for identifying regions similarity. BLAST was used to compare the sequence database and calculate significant of matches. The results indicated that MT gene in Poaceae plants have high homology in several regions. MT sequence of Chloris virgata had the sequence identity around 92% with Allium sativum which was grouped in Alliaceae family.

1. Introduction

Poaceae is one of the largest and nearly most ubiquitous monocotyledonous angiosperm, comprising of proximately 650 until 765 genera and more than 6000 species. These family members are also known as grasses. Grasses can be used in multiple ways for human's life, for instance as a food (rice, wheat, and maize), as a building material, for making paper, and for making perfume (lemon grass) [1]. Grasses also has been known for their phytoremediation ability. Phytoremediation is defined as cleaning of pollutants by using plants including trees, grasses and aquatic plants [2].

Grasses are distinguished by a fast growth rate, well-developed root system, large biomass and a long-term growth cycle. In addition, they show a high tolerance to soil contamination with heavy metals and can accumulate large quantities of heavy metals in both roots and shoots [3]. Some of them have been recognized as potential accumulator plants, such as *Hordeum vulgare* [4], *Sorghum bicolor* [5], and *Setaria italica* [6].

The function in absorbing the heavy metals by the accumulator plant is regulated by many genes, by which, one is Metallothionein (MT) gene. Metallothionein is one of metal binding protein groups which is rich in amino acids cysteine (30% of total residues) and has a low molecular mass character [7]. MT was first discovered in 1957 by Margoshes & Vallee who isolated the MT from the horse's renal cortex and found that MT has a high affinity for Cd [8]. Since then, MT has been studied in fungi, cyanobacteria, mammals, and also plants. Plant MTs can be divided into four types based on their amino acid residues, namely MT1, MT2, MT3, and MT4. Type 1 MTs comprise of six C-Xaa-C motifs (Xaa symbolize another amino acid) that are distributed equally among two domains. Type 2 MTs comprises of two cysteine-rich domains separated by a spacer of approximately 40 amino acid residues. The first pair of cysteines is existing as a C-C motif in amino acid positions 3 and 4 of these proteins. C-G-G-C motif is present at the end of the N-terminal cysteine-rich domain. Type 3 MTs comprise of only four Cys residues in the N-terminal domain. The consensus sequence for the first three is C-G-N-C-D-C. The fourth cysteine is not part of a pair of cysteines but is contained within a highly conserved motif, G-C-Xaa-K-K-G. The six Cys residues in the C-terminal cysteine-rich domain are arranged in C-Xaa-C motifs. Type 4 MTs have three cysteine-rich domains, each containing 5 or 6 conserved C residues, which are separated by 10 to 15 amino acids. Most of the cysteines are present as C-Xaa-C motifs [9].

MT can works against free radicals, such as hydrogen peroxide, superoxide, nitric oxide and hydroxyl radicals. It also functions for metal detoxification, and homeostasis [10]. Meanwhile, MT also functions as a chelator that can bind wide range of heavy metals. Its cystein-rich residue leads the chelating process by its group which has high affinity for heavy metal ions. Also, it has been known that metallothioneins play a role in absorbing excess metal because an increased concentration of metal ions will induce its response [11].

MT gene is a heavy-metal inducible gene. Its protein encoded regulates the plants to be easier to absorb and clean up the metal pollution from the environment [10, 12]. Therefore, metallothionein has been used as a biomarker of metal exposure [13, 14]. Certainly, study the character of this gene in metal accumulator plants is absolutely necessary to understand its function and regulation related to the resistant mechanism of accumulator plant under heavy metal rich conditions. However, there is no information about the functional domains of MT gene in plants. It should be noticed that there will be variation in specific gene or genomic region in a natural population. The level of this variation is not constant, but varies in different genes or gene region reflecting different functional characters or gene products. Thereby, study the sequence variation of MT gene is required to find out the gene conserved region, especially in Poaceae family, related to MT gene function for phytoremediation. The homology of MT gene can be utilized for showing the information about the genetic relatedness of the gene in interspecies.

2. Materials and Method

2.1. Sequence materials

DNA sequences used in this study were obtained from National Centre for Biotechnology Information (NCBI) GenBank database (www.ncbi.nlm.nih.gov). The sequences then were converted into FASTA format and grouped according to type of Metallothionein gene in Poaceae family.

2.2. Data analysis

Phylogenetic trees of MT sequences of Poaceae family were constructed based on the distance matrix with the maximum-likelihood (ML) method (www.phylogeny.fr).In order to analyze the ML tree, the bootstrap test was performed based on 1000 replications. To confirm the identity of MT sequence in Poaceae family, the sequences were subjected to BLAST algorithms search against the nucleotide sequences database (National Center for Biotechnology Information, NCBI), under an E-value threshold of $<10^{-5}$.T-cofee tool (tcoffee.crg.cat/) was utilized to align the sequences, therefore this tool identified the homology region between the sequences.

3. Results and Discussion

DNA sequence variation of MT gene, both within and between species in Poaceae family, provides the powerful information in determining interspecific phylogenetic distances. Also, the phylogenetic tree aims to identify the pattern of individual Poaceae plant in every MT gene type. In this study, the

IOP Conf. Series: Journal of Physics: Conf. Series **1108** (2018) 012039 doi:10.1088/1742-6596/1108/1/012039

phylogenetic tree was performed using the maximum-likelihood (ML) method. The ML test is generally used to figure out the evolutionary tree as the best tool to observe the variation in a group of sequences. This ML method was already used to construct the phylogenetic relationship of endosulfan-degrading bacteria [15], *Hibiscus syriacus* and other plant species based on the chloroplast genome sequence [16], plant annexins based on cDNA sequences contained within the 933- nt homologous tetrad core region [17], plant sulfotransferase enzymes[18], 16S rRNA gene sequences of Micromonospora species [19],the LYK group of LysM-receptor like kinases in Legumes [20].

Figure 1 shows that MT1A of *Chloris virgata* is grouped in the same cluster with MT1 of *Eleusine indica*, whereas MT1A *of Sorghum bicolor* and *Setaria italica* are gathered in the same cluster with *Zea mays*. Metallothionein 1 type Bof*Hordeum vulgare*is separately clustered with type MT1 of *Aegilops tauschii subsp. tauschii*. MT1B of *Oryza sativa subsp. japonica* is located in the same cluster with *Oryza brachyantha*,by which both are in the separate cluster with MT1A of *Oryza sativa subsp. japonica*. Overall, phylogenetic tree of MT1 in Poaceae plants presents a different cluster between MT1A and MT1B. BLAST data of MT1A sequence of *Chloris virgata* resulted in a high identity (90%) with *Allium sativum* which is being a member of Amaryllidaceae family, not Poaceae (Fig. 2).

Phylogenetic tree by using the maximum-likelihood (ML) method in MT2 gene sequence of Poaceae plants (Fig 3) indicates that MT2 is noticeably grouped in three clusters of MT2A, MT2B, and MT2C. Metallothionein gene type 2C of *Sorghum bicolor*, *Setaria italica*, *Brachypodium distachyon*, *Hordeum vulgare*, and *Aegilops tauschii subsp. tauschii* are separeated in a different cluster with MT2A gene of *Oryza brachyantha*, *Setaria italica*, *Sorghum bicolor*, and *Zea mays*. The results of multiple sequence alignment (MSA) presents that the homology region of MT2 gene sequence in Poaceae plants is identified in the area close to N-terminal domain (Fig. 4). This homologous region has around 70 base nucleotides in length which mostly contains of repeated codon for Cystein amino acids, namely TGT or TGC.

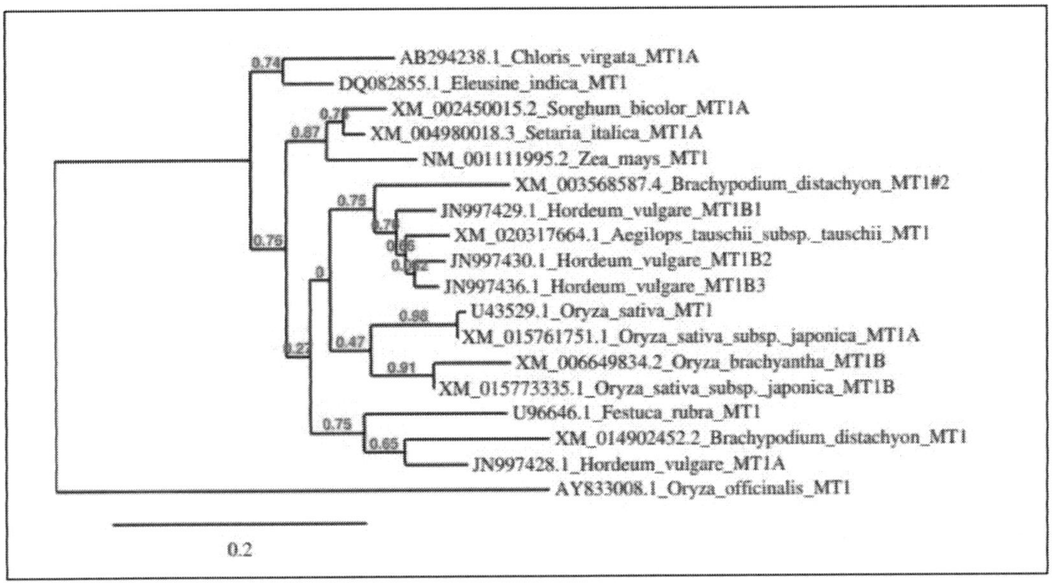

Figure 1. Phylogenetic analysis of 18 Poaceae plants based on nucleotide sequence variation of MT1 gene. Phylogenetic tree was performed using the maximum likelihood (PhyML) method. Branch support values are in red colour in above lines denoting branches.

```
Allium sativum type 1 metallothionein

Score           Expect    Identities       Gaps        Strand
340 bits(184)   2e-89     241/268(90%)     5/268(1%)   Plus/Plus

Query  12   CATCCACAGCTCTTCCGTTTAGCTTCGTCTCGCAGAT----CAAGCAAGATGTCGTGCAA  67
Sbjct  32   CATCCACA-CAGTTGCGTTTAGCTTCGTCTCTCAGATCAAGCAAGCAAGATGTCGTGCAG  90

Query  68   CTGCGGATCAGGCTGCAGCTGCGGCTCCAACTGCAAGTGCGGTAAGATGTACCCCGACCT  127
Sbjct  91   CTGCGGATCAAGCTGCAACTGCGGCTCGAGCTGCAACTGCGGCAAGATGTACCCTGACCT  150

Query  128  TGAGGAGAAGAGCACCGGCGCCCAGGCCACCGTCGTCCTCGGCGTTGCACCGGAGCACAA  187
Sbjct  151  TGAGGAGAAGAGCACCGGCGCTCAGGCCACCGTTGTCCTCGGCGTTGCACCCGAGCAGAA  210

Query  188  GGTCCAGCTGGAGGCGGACACCGAGTCCGGCGAGACCGCCCACGCCTGCGGCTGCGGCTC  247
Sbjct  211  GGTTCAGTTGGAGGCGGCCACCGAGTCCGGCGAGACCGCCCACGCCTGCGGCTGCGGCGC  270

Query  248  CAACTGCAAGTGCAACCCCTGCAACTGC  275
Sbjct  271  CAACTGCAAGTGCGACCCTTGCAACTGC  298
```

Figure 2. BLAST of Chloris *virgata* MT1 sequence. The result shows a high similarity (90%) of *Chloris virgata* MT1 sequence with Allium sativum as a member of Amaryllidaceae.

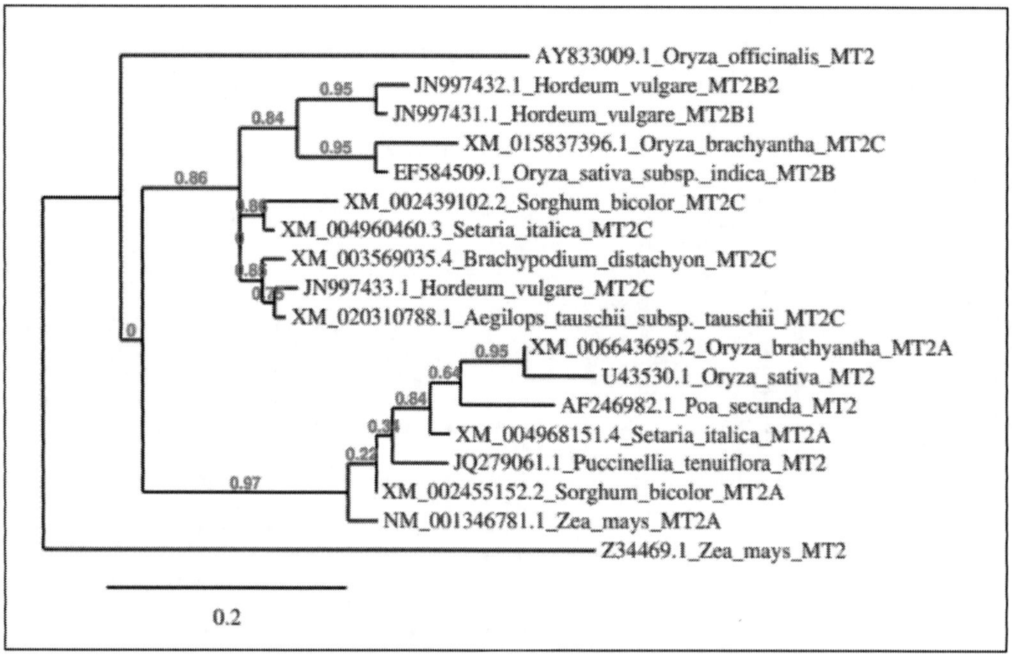

Figure 3. Phylogenetic analysis of Poaceae plants based on nucleotide sequence variation of MT2 gene. Phylogenetic tree was performed using the maximum likelihood (PhyML) method. Branch support values are in red colour in above lines denoting branches.

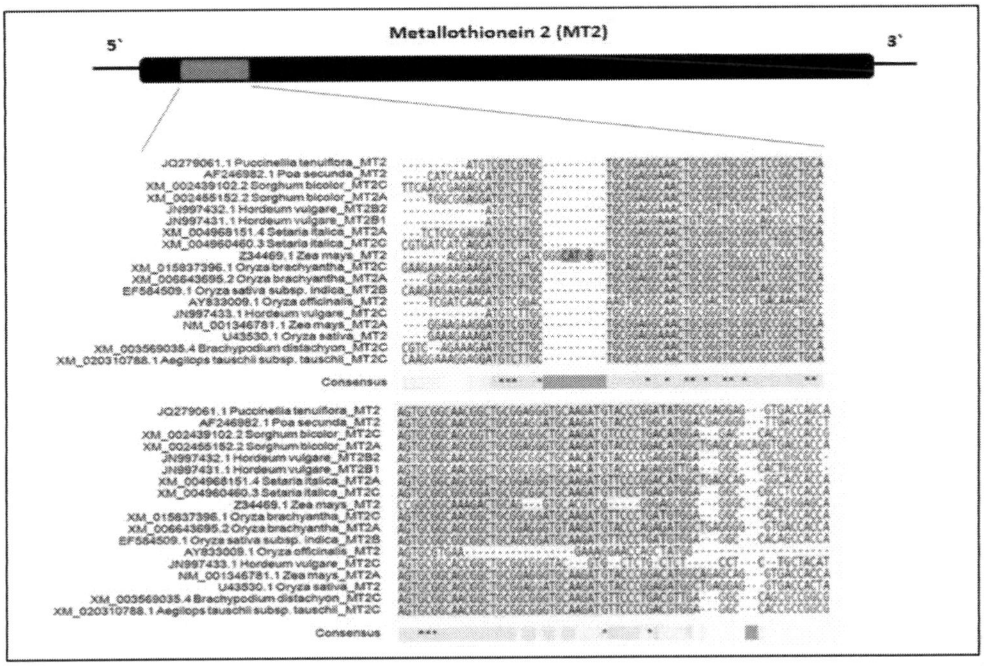

Figure 4. Homology sequence of MT2 gene in Poaceae plants. The homologous region is located close to N-terminal domain of MT2 gene.

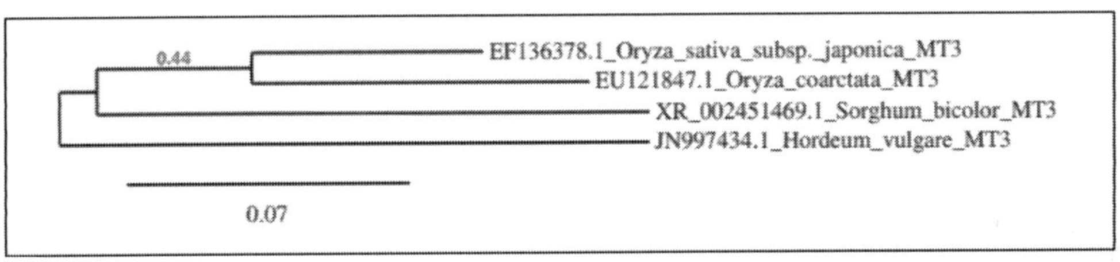

Figure 5. Phylogenetic analysis of Poaceae plants based on nucleotide sequence variation of MT3 gene. Phylogenetic tree was performed using the maximum likelihood (PhyML) method. Branch support values are in red colour in above lines denoting branches.

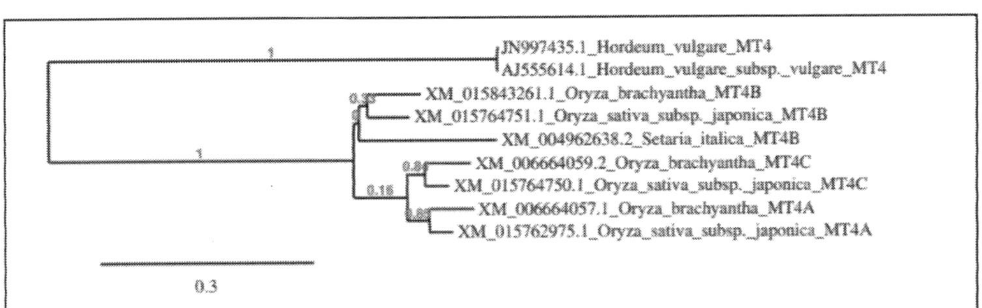

Figure 6. Phylogenetic analysis of Poaceae plants based on nucleotide sequence variation of MT4 gene. Phylogenetic tree was performed using the maximum likelihood (PhyML) method. Branch support values are in red colour in above lines denoting branches.

Only a few Poaceae plants used to construct phylogenetic relationship of MT3 gene (Fig.5). Metallothionein type 3 gene of *Oryza sativa subsp japonica* and *Oryza coarctata* bicolor are in-groups, whereas *Hordeum vulgare*and *Sorghum bicolor* are out-groups of *Oryza* plants. Figure 6 shows that separated clustersare presented in Metallothionein type 4A, 4B, and 4C between Poaceae plants.Interestingly, MT4 of *Hordeum vulgare* plants are outgroup of MT4A, MT4B, and MT4C in PhyML of Metallothionein genes in Poaceae plants.

When genes determining the genetic basis of a specific trait are known, there is a way to develop molecular marker for a particular purpose. Genetic markers are ultimately based on the variation of gene sequence [21]. The knowledge of an adaptive genetic variation in plants is a basis to evolve a strategy to overcome heterogeneous conditions. Hence, the presence of MT genes variation in Poaceae plants is very important for the long-term persistence under metal contaminate area.

4. References

[1] Anjum N A, Gill S S and Tuteja N 2017 *Enhancing Cleanup of Environmental Pollutants* (Switzerland: Springer International Publishing)
[2] Hidayati N 2005 *Hayati* **12** 35
[3] Golda S and Korzeniowska J 2016 *Env. Prot. And Nat. Resources* **27** 8
[4] Dahbia Z, Mhamed M, Benchaban H and Mohamed AH 2013 *IJER* **2** 49
[5] Wang X, Chen C and Wang J Cs 2017 *Int. J. Phytoremediation* **19** 402
[6] Chiang PN, Chiu CY, Wang MK and Chen BT 2011 *Soil Sci.* **176** 33
[7] Wong DL, Merrifield-MacRae ME and Stillman MJ 2017 *Metal Ions in Life Sciences* **17** 241
[8] Margoshes M and Vallee B 1957 *JACS* **79** 4813
[9] Cobbett and Goldsbrough P 2002 *Annu. Rev. Plant Biol* **53** 159
[10] Carpene E, Andreani G and Isani G *J. Trace Elem. Med. Biol.* **21** 35
[11] Robinson NJ, Tommey AM, Kuske C and Jackson PJ 1993 *Biochem. J.* **295** 1
[12] Ghoshal K and Jacob ST 2001 *Prog. Nucleic Acid RE* **66** 357
[13] Atli G and Canli M 2008 *Environ. Toxicol. Pharmacol.* **25** 33
[14] Espinoza H, Williams C and Gallagher E 2012 *Aquat Toxicol* **10-111** 37
[15] Mir Z, Ali S, Tyagi A, Ali A, Bhat J, Jaiswal P, Qari H and Oves M 2017 *3 Biotech.* **7** 211
[16] Kwon H, Kim J, Kim S, Park J and Lee H 2015 *Mitochondr. DNA* **27** 1
[17] Morgan R and Fernandez M 1997 *J. Mol. Evol.* **44** 178
[18] Piotrowski M, Schemenewitz A, Lopukhina A, Muller A, Janowitz T, Weiler E and Oecking C 2005 *J. Biol. Chem.* **279** 50717
[19] Trujillo M, Riesco E, Benito P and Carro L 2015 *Front. Microbiol.* **6** 1341
[20] Leppyanen I, Shakhnazorova V, Shtark O, Vishnevskaya N, Tikhonovich I and Dolgikh E 2017 *Int. J. Mol. Sci.* **19** 8
[21] Seifert S, Vornam B and Finkeldey R 2012 *Eur. J. Forest Res.* **131** 1761

Valuation of Survivorship Life Insurance with Stochastic Rates of Return and Dependent Mortality

M Y Tobing*, S Nurrohmah and B D Handari

Department of Mathematics, Universitas Indonesia, Kampus UI Depok, Depok 16424, INDONESIA

E-mail: *michelle.yeremia@ui.ac.id, snurrohmah@ui.ac.id, bevina@ui.ac.id

Abstract. Life insurance has been one of the many options for people with concerns about the uncertainty of their future, because it is designed to protect against the serious financial impact that results from an individual's death. An important variation of the single life insurance is the survivorship life insurance which covers two or more lives. Under such contract, a death benefit is paid out only on the last death. Insurance valuation is a very important tool for many insurance companies to be able to know its financial status in order to meet its future obligations. The valuation of an insurance policy comprises of 2 main components, they are: rates of return and mortality assumption. Traditionally, actuaries assume a constant interest rates and an independent mortality assumption in valuing joint-life insurance for the sake of simplicity. However, there has been considerable interest in the actuarial literature in studying the use of stochastic interest rate models and dependent mortality assumptions for insurance valuations. In this study, the mathematical expressions to value survivorship life insurance with stochastic rates of return and dependent mortality are presented. The valuation is conducted by the means of calculation of the expected value of the prospective loss random variable, by assuming an AR(1) process and Frank's Copula to model the rates of return and the dependent mortality of the lifetimes of the policy holders, respectively.

1. Introduction

In the world of insurance, valuation is an important tool for an insurance company to estimate a future liability or asset of the company at different points in time. Therefore, survivorship life insurance valuation is a necessity for the company issuing the policy. In this study, the valuation is done by calculating reserves the company needs to prepare to fulfill its future obligations. In Cunningham [1], reserve is expressed as the expected value of the loss random variable. In this study, we focus on the prospective loss random variable described in Chen *et al.* [2].

Suppose there is a portfolio consisting of m fully discrete survivorship life insurance policies that are still in force at current time ($t = 0$). The ith policy, with policy term n_i, is issued at time r_i, to a married couple aged $(x_i : y_i)$. Suppose the policy is valued at time t in the future, $0 < t < n_i - r_i$, with the following conditions:

(A1) The death benefit, b_i, is payable at the end of the year of the last death (second death) if the death of the last person occurs before the maturity time $n_i - r_i$.

(A2) The pure endowment benefit, c_i, is payable at maturity if at least one of the two of the policyholders is still alive at the maturity time $n_i - r_i$.

(A3) The level premium, π_i, payable by the policyholder at the beginning of each year as long as at least one of the policyholders is alive.

Unlike single-life policies, a survivorship policy that is in force could be in one of three states. At current valuation time, $t = 0$, at least one life insured in the ith contract ($i = 1, 2, ..., m$) has survived r_i years ($r_i < n_i$), and therefore its survivorship status, denoted as u_i, becomes:

$$u_i = \begin{cases} x_i + r_i : y_i + r_i, & \text{if both insureds are alive.} \\ x_i + r_i, & \text{if only the husband is alive.} \\ y_i + r_i, & \text{if only the wife is alive.} \end{cases} \tag{1}$$

2. The Prospective Loss Random Variable

As described earlier, the ith contract in the portfolio with current status u_i was issued r_i years ago. Let ${}_tL_{r_i}{}^{(u_i)}$ be the prospective loss random variable at time t for the ith contract with conditions (A1)-(A3), given that the policy is still in force at time $t < n_i - r_i$. For $t < n_i - r_i$, the prospective loss random variable for the ith policy, ${}_tL_{r_i}{}^{(u_i)}$, is expressed as:

$${}_tL_{r_i}{}^{(u_i)} = \begin{cases} b_i v(t, t + K_{\overline{u_i+t}}) - \pi_i \ddot{a}(t, t + K_{\overline{u_i+t}}), & 1 \le K_{\overline{u_i+t}} \le n_i - r_i - t. \\ c_i v(t, n_i - r_i) - \pi_i \ddot{a}(t, n_i - r_i), & K_{\overline{u_i+t}} > n_i - r_i - t. \end{cases} \tag{2}$$

where $K_{\overline{u_i+t}}$ is the discrete time-to-death random variable for the last death of the ith policy with status $\overline{u_i + t}$, and then $v(.)$ and $\ddot{a}(.)$ denote the discount factor and annuity factor for present value calculations. Equation (2) illustrates that in the occurrence of ${}_tL_{r_i}{}^{(u_i)}$, there are two possible events. The first event, denoted by Event I, occurs when $1 \le K_{\overline{u_i+t}} \le n_i - r_i - t$, and the second event, denoted by Event II, occurs when $K_{\overline{u_i+t}} > n_i - r_i$. Further explanations about these two events are described in figure 1.

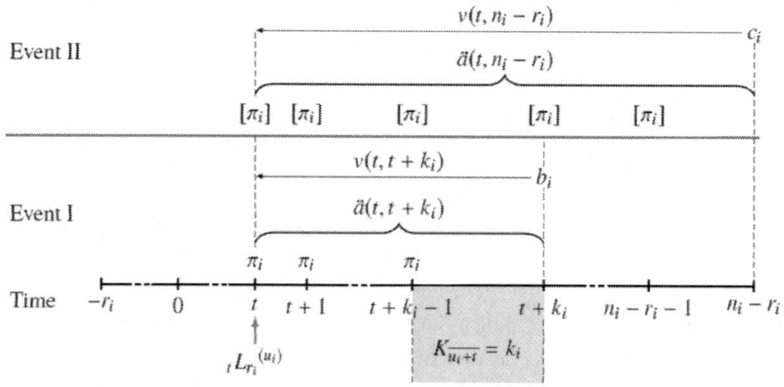

Figure 1. Timeline of loss of the ith policy ${}_tL_{r_i}{}^{(u_i)}$.

To simplify the notation, let $f_{i,k_i}(t)$ be the random variable that represents the occurrence of Event I and $s_i(t)$ be the random variable that represents the occurrence of Event II. $f_{i,k_i}(t)$ and $s_i(t)$ are defined as follows for $0 < t < n_i - r_i$:

$$f_{i,k_i}(t) = b_i v(t, t + k_i) - \pi_i \ddot{a}(t, t + k_i), \quad 1 \le k_i \le n_i - r_i - t, \tag{3}$$

$$s_i(t) = c_i v(t, n_i - r_i) - \pi_i \ddot{a}(t, n_i - r_i), \quad k_i > n_i - r_i - t, i = 1, 2, ..., m. \tag{4}$$

Then, ${}_tL_{r_i}{}^{(u_i)}$ for $i = 1, 2, ..., m$, can be rewritten in the following form:

$$ {}_tL_{r_i}{}^{(u_i)} = f_{i,k_i}(t)1_{\{K_{\overline{u_i+t}}=k_i, 1\leq k_i \leq n_i-r_i-t\}} + s_i(t)1_{\{K_{\overline{u_i+t}}=k_i, k_i>n_i-r_i-t\}}, \quad 0 < t < n_i - r_i. \quad (5) $$

where $1_{\{K_{\overline{u_i+t}}\}}$ is an indicator variable. Therefore, it follows from equation (5) that

$$ \mathbb{E}\left[{}_tL_{r_i}{}^{(u_i)}\right] = \sum_{k_i=1}^{n_i-r_i-t} \mathbb{E}\left[f_{i,k_i}(t)\right]{}_{k_i-1|}q_{\overline{u_i+t}} + \mathbb{E}\left[s_i(t)\right]{}_{n_i-r_i-t}p_{\overline{u_i+t}}, \qquad 0 < t < n_i - r_i. \quad (6) $$

where $\mathbb{E}\left[f_{i,k_i}(t)\right] = b_i\mathbb{E}[v(t, t+k_i)] - \pi_i\mathbb{E}[\ddot{a}(t, t+k_i)]$, $\mathbb{E}\left[s_i(t)\right] = c_i\mathbb{E}[v(t, n_i-r_i)] - \pi_i\mathbb{E}[\ddot{a}(t, n_i-r_i)]$.

Now let ${}_tL$ denote the prospective loss random variable of the portfolio evaluated at future time t, that is: ${}_tL = \sum_{i=1}^{m} {}_tL_{r_i}{}^{(u_i)} \cdot 1_{\{t<n_i-r_i\}}$, which sums the random variables ${}_tL_{r_i}{}^{(u_i)}$ defined in equation (5) over all policies in force, for $i = 1, 2, ..., m$. Therefore it follows that $\mathbb{E}\left[{}_tL\right] = \sum_{i=1}^{m} \mathbb{E}\left[{}_tL_{r_i}{}^{(u_i)}\right]$ which can be calculated based on equation (6).

3. AR(1) Process for the Rates of Return

In this section, we use the assumption in Chen *et al.* [2] to illustrate the valuation approach. Suppose the rate of return is constant throughout the year of each policy year, and let $\delta(k)$ denote the future rate of return earned during the kth year, that is $(k-1, k]$, for $k \in \mathbb{N}^+$. A stochastic process AR (1) for the sequence $\{\delta(k); k = 1, 2, ...\}$ is defined as follows:

$$ \delta(k) = \delta + \phi[\delta(k-1) - \delta] + e_k, \quad k \in \mathbb{N}^+, \quad (7) $$

where $\delta(k)$ is the rate of return in year k, $\delta(k-1)$ is the rate of return observed for the previous year, $\{e_k; k \in \mathbb{N}^+\}$ is a sequence of i.i.d. random variables following a normal distribution with mean zero and finite variance σ^2, i.e. $e_k \sim N(0, \sigma^2)$, δ is the long-term mean force of the process, and ϕ is the autoregressive coefficient of the process, with $|\phi| < 1$.

The alternative form of equation (7) is as follows:

$$ \delta(k) = \delta + \phi^k[\delta(0) - \delta] + \sum_{i=0}^{k-1} \phi^i e_{k-i}, \quad k \in \mathbb{N}^+, \quad (8) $$

so if it is known that $\delta(0) = \delta_0$, the expectation and variance of $\delta(k)$, given $\delta(0)$, are as follows:

$$ \mathbb{E}[\delta(k)] = \delta + \phi^k[\delta_0 - \delta], \quad Var[\delta(k)] = \frac{\sigma^2\left(1-\phi^{2k}\right)}{1-\phi^2}, \quad Cov[\delta(s), \delta(k)] = \frac{\sigma^2}{1-\phi^2}\phi^{k-s}\left(1-\phi^{2s}\right), $$

where $s < k$. Furthermore, the accumulated function of the rate of return from time t to time j is expressed as follows:

$$ I(t, j) = \begin{cases} \sum_{k=t+1}^{j} \delta(k), & t < j, \\ 0, & t = j. \end{cases} $$

By assuming an AR (1) process, Chen [2] proposes a proposition stating that if the rate of return in year k, $\delta(k)$, follows the AR (1) process, then according to $\mathbb{E}[\delta(k)]$, $Var[\delta(k)]$, and $Cov[\delta(s), \delta(k)]$, the mean and variance of $I(t, j)$, given $\delta(0) = \delta_0$, are as follows:

$$ \mathbb{E}[I(t, j)] = (j-t)\delta + (\delta_0 - \delta)\frac{\phi(\phi^t - \phi^j)}{1-\phi}, \quad (9) $$

$$Var[I(t,j)] = \frac{\sigma^2}{1-\phi^2}\left[(j-t)\frac{1+\phi}{1-\phi} - \frac{2\phi}{(1-\phi)^2}(1-\phi^{j-t}) - \frac{\phi^2}{(1-\phi)^2}(\phi^t - \phi^j)^2\right] \tag{10}$$

Suppose $v(t,j)$ denotes the discount factor which discounts a payment made at time j to time t, for $t < j$. $v(t,j)$ is defined as follows:

$$v(t,j) = e^{-I(t,j)}, \quad 0 \le t \le j, \quad t,j \in \mathbb{N}^+. \tag{11}$$

From equation (7), it is known that given $\delta(0)$, each $\delta(k)$, for $k \in \mathbb{N}^+$, is normally distributed, which implies that $v(t,j) = e^{-I(t,j)}$ is lognormally distributed. Therefore, the expectation of $v(t,j)$ based on the property of the lognormal distribution is as follows:

$$\mathbb{E}[v(t,j)] = e^{-\mathbb{E}[I(t,j)]+0.5Var[I(t,j)]}, \quad t \le j, \quad t,j \in \mathbb{N}^+. \tag{12}$$

Suppose $\ddot{a}(t,j)$ denotes the present value at time t of an annuity that pays 1 at the beginning of each year starting from time t to $j-1$. Then from equation (12) it follows that:

$$\mathbb{E}[\ddot{a}(t,j)] = \mathbb{E}\left[\sum_{k=t}^{j-1} v(t,k)\right] = \sum_{k=t}^{j-1}\mathbb{E}[v(t,k)], \quad t < j, \text{ and } t,j \in \mathbb{N}^+. \tag{13}$$

$\mathbb{E}[\ddot{a}(t,j)]$ and $\mathbb{E}[v(t,j)]$ is the annuity factor and the discount factor needed to calculate $\mathbb{E}[f_{i,k_i}(t)]$ and $\mathbb{E}[s_i(t)]$, which is then required to calculate $\mathbb{E}\left[{}_tL_{r_i}{}^{(u_i)}\right]$ in equation (6).

4. Frank's Copula Model for Dependent Mortality

In this section, we describe the mortality assumption for illustration, a Copula model with the marginal distribution function of Gompertz, to model the random variable $K_{\overline{u_i+t}}$. In this study, as in Frees *et al.* [3], we use Frank's Copula by Frank (1979) as follows:

$$C(u,v) = \frac{1}{\alpha}\ln\left(1 + \frac{(e^{\alpha u}-1)(e^{\alpha v}-1)}{e^{\alpha}-1}\right), \quad 0 \le u \le 1, \quad 0 \le v \le 1,$$

where U and V are random variables with marginal cumulative distributions u and v, $\alpha \neq 0$ is a parameter measuring the dependence between the two underlying random variables, and the resulting Frank's Copula, $C(u,v)$, is the joint distribution for U and V.

Suppose the lifetimes of the policyholders follows the Gompertz model. Suppose X and Y is Gompertz distributed with distribution functions $G_1(x)$ dan $G_2(y)$, that is:

$$G_1(x) = 1 - e^{e^{-v_1/\tau_1}(1-e^{x/\tau_1})}, \quad x > 0, \quad \text{and} \quad G_2(y) = 1 - e^{e^{-v_2/\tau_2}(1-e^{y/\tau_2})}, \quad y > 0, \tag{14}$$

where parameters v_i and τ_i, $i = 1,2$, represents the mode and the scale parameter of the distribution. Under the assumption of Frank's Copula, the joint distribution of X and Y for $x,y > 0$ is as follows:

$$G(x,y) = \frac{1}{\alpha}\ln\left(1 + \frac{\left(e^{\alpha G_1(x)}-1\right)\left(e^{\alpha G_2(y)}-1\right)}{e^{\alpha}-1}\right). \tag{15}$$

Thus, by using Frank's Copula, ${}_kp_{xy}$, which is the probability of status (xy) will survive up to time k is expressed in the form of G, G_1, and G_2 as follows:

$${}_kp_{xy} = \frac{1 - G_1(x+k) - G_2(y+k) + G(x+k,y+k)}{1 - G_1(x) - G_2(y) + G(x,y)}, \quad k \in \mathbb{N}^+, \tag{16}$$

and ${}_kq_{\overline{xy}}$, which is the probability of status $(\overline{xy})$ will fail or die at time k is as follows:

$$ {}_kq_{\overline{xy}} = \frac{G(x+k, y+k) - G(x, y+k) - G(x+k, y) + G(x, y)}{1 - G_1(x) - G_2(y) + G(x, y)}, \quad k \in \mathbb{N}^+. \tag{17}$$

In Cunningham [1], it is known that ${}_tp_x = \frac{S_X(x+t)}{S_X(x)} = \frac{1-F_X(x+t)}{1-F_X(x)}$, therefore we have

$$ {}_kp_x = 1 - {}_kq_x = \frac{1 - G_1(x+k)}{1 - G_1(x)}, \quad \text{and} \quad {}_kp_y = 1 - {}_kq_y = \frac{1 - G_2(y+k)}{1 - G_2(y)}. \tag{18}$$

The probabilities ${}_kp_{xy}$, ${}_kq_{\overline{xy}}$, ${}_kp_x$, and ${}_kp_y$ are useful for the calculations of ${}_{k_i-1|}q_{\overline{u_i+t}}$ and ${}_{n_i-r_i-t}p_{\overline{u_i+t}}$, which is required for the calculation of $\mathbb{E}\left[{}_tL_{r_i}{}^{(u_i)}\right]$ in equation (6).

5. The Calculation of the Expected Loss of the Portfolio $\mathbb{E}\left[{}_tL\right]$

In this section, we will show the calculation of an illustrative non-homogeneous portfolio that consists of 6 survivorship life insurance policies as described in table 1. Since the focus of this study is to conduct a valuation, then we refer the parameters for the Copula mortality model in Frees *et al.* [3], they are $v_1 = 85.82$, $\tau_1 = 9.98$, $v_2 = 89.40$, $\tau_2 = 8.12$, and $\alpha = -3.367$. The rate of return is modeled by the AR(1) process in equation (7), with the parameters $\delta = 0.06$, $\delta_0 = 0.04$, $\phi = 0.9$, and $\sigma = 0.01$. The calculation results are presented in table 2.

Table 1. Description of the illustrative non-homogeneous portfolio.

Policy i	Age at issue x_i	Age at issue y_i	Death benefit b_i	Pure endowment benefit c_i	Policy term n_i
1	35	30	3000	0	20
2	35	30	3000	3000	10
3	45	40	1000	0	20
4	45	40	1000	1000	30
5	55	50	2000	0	30
6	55	50	2000	2000	15

Table 2. Calculation results for the non-homogeneous portfolio.

Policy year elapsed r_i	$\mathbb{E}\left[{}_0L_{r_i}{}^{(u_i)}\right]$	Future valuation time t	$\mathbb{E}\left[{}_tL_0{}^{(u_i)}\right]$
1	360.9397	1	348.4374
2	739.5166	2	712.4657
3	1136.205	3	1093.988
4	1551.404	4	1494.927
5	1985.414	5	1917.216
6	2438.407	6	2362.797
7	2910.387	7	2833.575
8	3401.151	8	3331.627

The calculation results in table 2 will be explained. On the left side of the table, suppose an insurance company issues the non-homogeneous portfolio r_i years ago, for $r_i = 1, 2, ..., 8$. At the present time, namely $t = 0$, the insurance company wants to know how much fund it needs to prepare to cover the loss that might occur from the portfolio. Therefore, the company needs to prepare a fund of $360,9397$ for a portfolio issued 1 year ago, a fund of $739,5166$ for a portfolio issued 2 years ago, and so on. Furthermore, on the right side of the table, suppose an insurance company issues the same non-homogeneous portfolio at the present time, i.e. $r_i = 0$, and wants to know how much fund it needs to prepare t years from now, for $t = 1, 2, ..., 8$. Therefore, within 1 year from now, the company needs to set aside a fund of $348,4374$, within 2 years from now, a fund of $712,4657$, and so on.

In the end, information from this calculation can be useful for insurance companies to determine the amount of reserves they need to prepare at present time or in the future.

6. Conclusion

This study incorporates the valuation approach presented by Chen [4] and Chen *et al.* [2], which explored the valuation of a general (non-homogeneous) survivorship insurance portfolio in a stochastic interest rate environment and dependent mortality assumptions. Since it is still a new topic in its infancy, the idea of this study is to mainly show the formulas of the expected value of prospective loss random variable to value a survivorship life insurance with stochastic rates of return and dependent mortality assumption, by assuming an AR (1) process and Frank's Copula to illustrate the valuation approach. The models used in this study are in no way the best models and the authors recommend further research regarding this subject.

References

[1] Cunningham R J, Herzog T N and London R L 2006 *Models for Quantifying Risk 3rd Ed* (Connecticut: ACTEX Publications, Inc.)
[2] Chen L, Lin L, Lu Y and Parker G 2017 Analysis of Survivorship Life Insurance Portfolios with Stochastic Rates of Return. *Insurance: Mathematics and Economics.* **75** 16–31
[3] Frees E W, Carriere J and Valdez E 1996 Annuity valuation with dependent mortality. *J. Risk Insurance.* **63** 229–261
[4] Chen L 2010 *Analysis of Joint Life Insurance with Stochastic Interest Rates* (Canada: Simon Fraser University)
[5] Bowers N L, Gerber H U, Hickman J C, Jones D A and Nesbitt C J 1997 *Actuarial Mathematics, second ed* (Schaumburg: The Society of Actuaries)
[6] Cryer J D, Chan K-S 2008 *Time Series Analysis: With Applications in R 2nd Ed* (New York: Springer)
[7] Parker G 1997 Stochastic analysis of the interaction between investment and insurance risks. *N. Am. Actuar. J.* **1** 55–84

Acknowledgments

The authors would like to express appreciation to the DRPM Universitas Indonesia for supporting our study with Hibah PITTA UI 2018.

Formation Control of Multiple Dubins Airplane System with Geometric Approach

Asmianto[1], Hariyanto[1] and Miswanto[2]

[1] Department of mathematics , Faculty of Mathematics, Computing and Data Science, Institut Teknologi Sepuluh Nopember (ITS), Surabaya, Indonesia
[2] Department of mathematics, Airlangga University (UNAIR), Surabaya, Indonesia

Abstract. Formation Control is very important for multi agent systems (Swarm). This paper analyses control formation on the Dubins airplane that consists of three airplanes, where an airplane as the leader and two other airplanes move to follow leader airplane's track. First, Doing design of leader airplane's controls by using the method of tracking error dynamics. Controls on the leader follow the trajectory tracking that desired. Desired path is in the form of parametric functions. Second, geometric approaches method is used to design two other airplanes, so the move follows path of the airplane as the leader with a certain distance. Simulation result shows that airplane leader can be controlled to the desired path crosses with good results. Therefore, error has been defined as the distance between leader airplane's paths with the desired path while tracking also depends on the shape of path. The more difficult trajectory tracking of airplane also serves as a leader in passing through it. The result of the simulations also indicates that the value of the mean percentage error (MPE) of 26.93 percent. This suggests that the methods used in this study quite effective either used to control design leader or follower.

1. Introduction

Natural phenomenons are very interesting to analyse mathematically. One of the examples is the phenomenon called swarming which occurs in various groups of organism. Swarming's behaviour or aggregation of organism in groups is abundant in nature. For example, the behaviour can be seen in animal aggregation, such as flocks of birds (geese), schools of fish, herds of mammals etc. Flocks of geese often fly along in the inverted "**V**" formation. By flying in the inverted "**V**" formation they obtain some adventages. The motion of swarm can be used in engineering for cooperative control (multi robots) and formation control (aircraft and ship). The formation preservation of the swarm has received considerable attentions. Gazi, et al. [1],[2] and Miswanto, et.al[3] study the coordination and tracking control problem of the motion of the swarm. In the literature, some researchers have discussed the formation control of mobile robots. In [4], the authors study a stable and decentralized control strategy for multi agent system to capture a moving target in a specific formation. In [5] the authors study control and coordination for many robots moving in formation using decentralized controllers. They investigate feedback law used to

IOP Conf. Series: Journal of Physics: Conf. Series **1108** (2018) 012041 doi:10.1088/1742-6596/1108/1/012041

control multiple robots moving in formation. They proposed a method for controlling formations that uses only local sensor based information, in a leader – follower motion. They use methods of feedback linearization to exponentially stabilize the relative distance and orientation of follower and show that the zero dynamics of the system are also stable. Another researchers have discussed the control design of a mobile robot to track a desired path. In [6], the authors discuss the tracking control of mobile robot using integrator back stepping. Many mechanical systems with non-holonomic constrains can be locally or globally converted to the chained form under coordinate change. In [7], the authors study the tracking control problem of non-holonomic system in chained form. They derive semi global tracking controllers for general chained form systems by means of back stepping and they achieve global tracking result for some special cases. In [8] an adaptive tracking control problem is studied for a four wheel mobile robot. The authors propose a formulation for the adaptive tracking problem that meets the natural prerequisite such that it reduces to the state feedback tracking problem if the parameters are known. They derive a general methodology for solving their problem. In [9] the authors study the formation control of swarm whose agents are Dubins cars. The agents of swarm are moving to track a desired path. They consider the swarm model with presence of a leader. First, they design the control of the leader with tracking error dynamics. The control of the leader is designed for tracking the desired path. After that, geometry approach is used to design the control of the others.

In this paper, Formation control of the swarm model whose agents are airplane moving together in a specific geometry formation. In this model, an airplane has a role as a leader. The control of the leader is designed for tracking the desired path. Furthermore, this paper also carried out computation the tracking error of the path of the leader tracking. A desired path is sufficiently small and the distance between the leader path and the desired path is preserved. In the next section, the formal problem statement is described. In section 3, Doing design control of the leader using tracking error dynamics. In section 4, Design the control of each agent follower by using geometric approach and the last section will show the results of numerical simulation trajectory of the leader airplane and the desired path also error between the trajectory of the leader airplane and desired path.

2. Problem Formulation

In this paper, the dynamics system of airplane is taken from Fozen (2011) and three airplanes which described as:

$$\begin{aligned}
\dot{x}_i &= u_i \cos\theta_i + v_i \sin\theta_i \\
\dot{y}_i &= -u_i \sin\theta_i + v_i \cos\theta_i \\
\dot{\theta}_i &= w_i \\
\dot{u}_i &= a_i \qquad\qquad a_i, b_i, c_i > 0 \\
\dot{v}_i &= b_i \\
\dot{w}_i &= c_i
\end{aligned} \tag{1}$$

Where $(x_i, y_i) \in R^2$ represents the position of the $i-th$ airplane, and $\theta_i \in [0,2\pi]$ represents the orientation of the i – *th* airplane. The symbol u_i is the linear velocities for moving forward and back, v_i is the linear velocities to move to the side and w_i represents airplane maneuvering motion. In this paper, the desired path (γ) that would be tracked by the leader airplane is obtained by using calculus variation method. The path is denoted by $\gamma(t) = [\gamma_x(t), \gamma_y(t)]^T$, There are two problems that will be completed. First, Design the control of the leader airplane for tracking the desired path by tracking error dynamic. Second, Design the control of the other agents by geometric approach to follow the leader's path with a certain distance.

3. The Control Design of The Leader Airplane

Design control of the leader by tracking error dynamics for minimizing the tracking error in order to keep the position of the leader close to the desired path. Tracking error *e(t)* as the difference between the actual leader path and desired path.

$$e(t) = [x_1(t) - \gamma_x(t), y_1(t) - \gamma_y(t)]^T \quad (2)$$

Differentiating the error equation (2) with respect to time yields.

$$\dot{e}(t) = [\dot{x}_1(t) - \dot{\gamma}_x(t), \dot{y}_1(t) - \dot{\gamma}_y(t)]^T = [u_1 \cos\theta_1 + v_1 \sin\theta_1 - \dot{\gamma}_x(t), -u_1 \sin\theta_1 + v_1 \cos\theta_1 - \dot{\gamma}_y(t)]^T. \quad (3)$$

And,

$$\ddot{e}(t) = [\ddot{x}_1(t) - \ddot{\gamma}_x(t), \ddot{y}_1(t) - \ddot{\gamma}_y(t)]^T = [(\dot{u}_1 \cos\theta_1 - u_1 w_1 \sin\theta_1) + (\dot{v}_1 \sin\theta_1 + v_1 w_1 \cos\theta_1) - \ddot{\gamma}_x(t), (-\dot{u}_1 \sin\theta_1 - u_1 w_1 \cos\theta_1) + (\dot{v}_1 \cos\theta_1 + v_1 w_1 \sin\theta_1) - \ddot{\gamma}_y(t)]^T. \quad (4)$$

Now, we define the tracking error dynamics F where $F = [f_1, f_2]^T$ and $f_i = (e_i, \dot{e}_i) = 0, i = 1,2$

$$f_1(t) = \dot{e}_1(t) + k_1 e_1(t)$$
$$f_2(t) = \dot{e}_2(t) + k_2 e_2(t) \quad (5)$$

Where k_1 and k_2 are positive contants. Differenting the system (5) with respect to time, one obtains

$$\dot{f}_1(t) = \ddot{e}_1(t) + k_1 \dot{e}_1(t)$$
$$\dot{f}_2(t) = \ddot{e}_2(t) + k_2 \dot{e}_2(t) \quad (6)$$

The control of the leader can be determined from equation (1), (2), (3), (4), (5) and (6).

To get of the u_2, v_2 and w_2 done by way of elimination and multiplying with $\cos\theta_2$ and $\sin\theta_2$ so obtained;

$$w_1 = \frac{\begin{array}{c}(u_1 \sin\theta_1 + v_1 \cos\theta_1)(\dot{u}_1 \cos\theta_1 - \dot{v}_1 \sin\theta_1 + \ddot{\gamma}_x(t) + k_1(u_1 \cos\theta_1 + v_1 \sin\theta_1 - \dot{\gamma}_x(t))) + \\ (u_1 \cos\theta_1 + v_1 \sin\theta_1)(-\dot{u}_1 \sin\theta_1 + \dot{v}_1 \cos\theta_1 - \ddot{\gamma}_y(t) + k_1(-u_1 \sin\theta_1 + v_1 \cos\theta_1 - \dot{\gamma}_y(t)))\end{array}}{(u_1^2 - v_1^2)}$$

Where,

$$u_1 = \dot{\gamma}_x(t)\cos\theta_1 - \dot{\gamma}_y(t)\sin\theta_1 - k_1 x_1(t)\cos\theta_1 + k_1 y_1(t)\sin\theta_1 + k_1\gamma_x(t)\cos\theta_1 - k_1\gamma_y(t)\sin\theta_1$$

And,

$$v_1 = \dot{\gamma}_x(t)\sin\theta_1 + \dot{\gamma}_y(t)\cos\theta_1 - k_1 x_1(t)\sin\theta_1 - k_1 y_1(t)\cos\theta_1 + k_1\gamma_x(t)\sin\theta_1 + k_1\gamma_y(t)\cos\theta_1 \quad (7)$$

Then this control ($u_1, v_1, and\ w_1$) is substituted to the system (1). Thus, if one uses (7), one obtains differential equations:

$$\dot{x}_1 = (\dot{\gamma}_x(t)\cos\theta_1 - \dot{\gamma}_y(t)\sin\theta_1 - k_1 x_1(t)\cos\theta_1 + k_1 y_1(t)\sin\theta_1 + k_1\gamma_x(t)\cos\theta_1 - k_1\gamma_y(t)\sin\theta_1)\cos\theta_1 + (\dot{\gamma}_x(t)\sin\theta_1 + \dot{\gamma}_y(t)\cos\theta_1 - k_1 x_1(t)\sin\theta_1 - k_1 y_1(t)\cos\theta_1 + k_1\gamma_x(t)\sin\theta_1 + k_1\gamma_y(t)\cos\theta_1)\sin\theta_1$$

$$\dot{y}_1 = -(\dot{\gamma}_x(t)\cos\theta_1 - \dot{\gamma}_y(t)\sin\theta_1 - k_1 x_1(t)\cos\theta_1 + k_1 y_1(t)\sin\theta_1 + k_1\gamma_x(t)\cos\theta_1 - k_1\gamma_y(t)\sin\theta_1)\sin\theta_1 + (\dot{\gamma}_x(t)\sin\theta_1 + \dot{\gamma}_y(t)\cos\theta_1 - k_1 x_1(t)\sin\theta_1 - k_1 y_1(t)\cos\theta_1 + k_1\gamma_x(t)\sin\theta_1 + k_1\gamma_y(t)\cos\theta_1)\cos\theta_1 \quad (8)$$

$$\dot{\theta}_1 = \frac{\begin{array}{c}(u_1 \sin\theta_1 + v_1 \cos\theta_1)(\dot{u}_1 \cos\theta_1 - \dot{v}_1 \sin\theta_1 + \ddot{\gamma}_x(t) + k_1(u_1 \cos\theta_1 + v_1 \sin\theta_1 - \dot{\gamma}_x(t))) + \\ (u_1 \cos\theta_1 + v_1 \sin\theta_1)(-\dot{u}_1 \sin\theta_1 + \dot{v}_1 \cos\theta_1 - \ddot{\gamma}_y(t) + k_1(-u_1 \sin\theta_1 + v_1 \cos\theta_1 - \dot{\gamma}_y(t)))\end{array}}{(u_1^2 - v_1^2)}$$

This system is a system of dynamical airplane that is a leader and been given a control so that it can pass the desired trajectory. The initial and final condition of the state variables (x_1, y_1, θ_1) are known. The solution of this system of differential equations use numerical approximation by substituting the desired path in system (8).

IOP Conf. Series: Journal of Physics: Conf. Series **1108** (2018) 012041 doi:10.1088/1742-6596/1108/1/012041

4. The Control Design of the Following Agent

In this section, Design control of the follower using geometry approach. Figure (1) shows three dubins cars. Where d_1 and d_2 are the distance of agents to the leader. ϕ_1 and ϕ_2 are the orientation of agents to position of the leader.

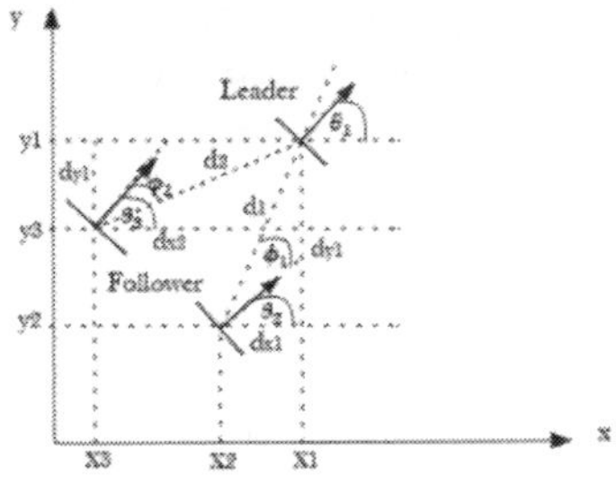

Figure 1. Three Dubins Airplane.

4.1 The Control Design of The First Follower

From the figure (1), one obtain

$x_1 - x_2 = d_1 \cos(\theta_2 + \phi_1)$ and,

$y_1 - y_2 = d_1 \sin(\theta_2 + \phi_1)$

Differentiating the equation above with respect to time, one obtains,

$$\dot{x_1} - \dot{x_2} = -d_1 \sin(\theta_2 + \phi_1)\left(\dot{\theta_2} + \dot{\phi_1}\right)$$
$$\dot{y_1} - \dot{y_2} = d_1 \cos(\theta_2 + \phi_1)(\dot{\theta_2} + \dot{\phi_1}) \tag{9}$$

Then from the above equation is obtained:

$$\dot{x_2} = \dot{x_1} + d_1 \sin(\theta_2 + \phi_1)\left(\dot{\theta_2} + \dot{\phi_1}\right)$$
$$\dot{y_2} = \dot{y_1} - d_1 \cos(\theta_2 + \phi_1)(\dot{\theta_2} + \dot{\phi_1}) \tag{10}$$

After that, by substitution equation $\dot{x_2} = u_2 \cos\theta_2 + v_2 \sin\theta_2$ dan $\dot{y}_2 = -u_2 \sin\theta_2 + v_2 \cos\theta_2$

One obtain;

$$u_2 \cos\theta_2 + v_2 \sin\theta_2 = \dot{x_1} + d_1 \sin(\theta_2 + \phi_1)\left(\dot{\theta_2} + \dot{\phi_1}\right)$$
$$-u_2 \sin\theta_2 + v_2 \cos\theta_2 = \dot{y_1} - d_1 \cos(\theta_2 + \phi_1)(\dot{\theta_2} + \dot{\phi_1})$$

To get control of the u_2 and v_2 elimination done so obtained,

$$u_2 = \dot{x_1} \cos\theta_2 - \dot{y_1} \sin\theta_2 + d_1\left(\dot{\theta_2} + \dot{\phi_1}\right)\sin\left((\theta_2 + \phi_1) + \theta_2\right)$$
$$v_2 = \dot{x_1} \sin\theta_2 + \dot{y_1} \cos\theta_2 - d_1\left(\dot{\theta_2} + \dot{\phi_1}\right)\cos\left((\theta_2 + \phi_1) + \theta_2\right)$$

And from u_2 and v_2 obtained,

$$w_2 = \frac{(u_2 - \dot{x_1}\cos\theta_2 + \dot{y_1}\sin\theta_2)d_1\left(\cos\left((\theta_2 + \phi_1) + \theta_2\right)\right) + (v_2 - \dot{x_1}\sin\theta_2 - \dot{y_1}\cos\theta_2)d_1 \sin\left((\theta_2 + \phi_1) + \theta_2\right)}{d_1\left(\sin\left((\theta_2 + \phi_1) - \theta_2\right)\right)d_1\left(\cos\left((\theta_2 + \phi_1) + \theta_2\right)\right)}$$

Then, this control ($u_2, v_2,$ and w_2) is substituted to the system (1) with $i = 2$. Thus obtained a system of the differential equations of the first follower:

$$\dot{x}_2 = \left(\dot{x_1}\cos\theta_2 - \dot{y_1}\sin\theta_2 + d_1\left(\dot{\theta_2} + \dot{\phi_1}\right)\sin\left((\theta_2 + \phi_1) + \theta_2\right)\right)\cos\theta_2 + \left(\dot{x_1}\sin\theta_2 + \dot{y_1}\cos\theta_2 - d_1\left(\dot{\theta_2} + \dot{\phi_1}\right)\cos\left((\theta_2 + \phi_1) + \theta_2\right)\right)\sin\theta_2$$

$\dot{y}_2 = -\big(\dot{x}_1\cos\theta_2 - \dot{y}_1\sin\theta_2 + d_1(\dot{\theta}_2+\dot{\phi}_1)\sin((\theta_2+\phi_1)+\theta_2)\big)\sin\theta_2 + \big(\dot{x}_1\sin\theta_2 + \dot{y}_1\cos\theta_2 - d_1(\dot{\theta}_2+\dot{\phi}_1)\cos((\theta_2+\phi_1)+\theta_2)\big)\cos\theta_2$

$$\dot{\theta}_2 = \frac{(u_2-\dot{x}_1\cos\theta_2+\dot{y}_1\sin\theta_2)d_1(\cos((\theta_2+\phi_1)+\theta_2))+(v_2-\dot{x}_1\sin\theta_2-\dot{y}_1\cos\theta_2)d_1\sin((\theta_2+\phi_1)+\theta_2)}{d_1(\sin((\theta_2+\phi_1)-\theta_2))d_1(\cos((\theta_2+\phi_1)+\theta_2))} \quad (11)$$

The system in equation (11) is a system of dynamical airplane which is first follower and been given a control so that it can be controlled dependent on the leader.

4.2 The Control Design of the Second Follower

From the figure (1) is obtained,
$x_1 - x_3 = d_2\cos(\theta_3 - \phi_2)$ and,
$y_1 - y_3 = d_2\sin(\theta_3 - \phi_2)$
Using similar steps such as in 4.1 one may design the control of the second follower. Thus obtain a system of differential equations of the first follower.

$$u_3 = \dot{x}_1\cos\theta_3 - \dot{y}_1\sin\theta_3 + d_2(\dot{\theta}_3+\dot{\phi}_2)\sin((\theta_3+\phi_2)+\theta_3)$$
$$v_3 = \dot{x}_1\sin\theta_3 + \dot{y}_1\cos\theta_3 - d_2(\dot{\theta}_3+\dot{\phi}_2)\cos((\theta_3+\phi_2)+\theta_3)$$

And from u_3 and v_3 is obtained,

$$w_3 = \frac{(u_3-\dot{x}_1\cos\theta_3+\dot{y}_1\sin\theta_3)d_2\cos((\theta_3+\phi_2)+\theta_3)+(v_3-\dot{x}_1\sin\theta_3-\dot{y}_1\cos\theta_3)d_2\sin((\theta_3+\phi_2)+\theta_3)}{d_2\sin((\theta_3+\phi_2)+\theta_3)\,d_2\cos((\theta_3+\phi_2)+\theta_3)}$$

Then, this control ($u_3, v_3,$ and w_3) is substituted to the system (1) with $i = 3$. System of the differential equations of the second follower:

$\dot{x}_3 = \big(\dot{x}_1\cos\theta_3 - \dot{y}_1\sin\theta_3 + d_2(\dot{\theta}_3+\dot{\phi}_2)\sin((\theta_3+\phi_2)+\theta_3)\big)\cos\theta_3 + \big(\dot{x}_1\sin\theta_3 + \dot{y}_1\cos\theta_3 - d_2(\dot{\theta}_3+\dot{\phi}_2)\cos((\theta_3+\phi_2)+\theta_3)\big)\sin\theta_3$

$\dot{y}_3 = -\big(\dot{x}_1\cos\theta_3 - \dot{y}_1\sin\theta_3 + d_2(\dot{\theta}_3+\dot{\phi}_2)\sin((\theta_3+\phi_2)+\theta_3)\big)\sin\theta_3 + \dot{x}_1\sin\theta_3 + \dot{y}_1\cos\theta_3 - d_2(\dot{\theta}_3+\dot{\phi}_2)\cos((\theta_3+\phi_2)+\theta_3)\cos\theta_3$

$$\dot{\theta}_3 = \frac{(u_3-\dot{x}_1\cos\theta_3+\dot{y}_1\sin\theta_3)d_2\cos((\theta_3+\phi_2)+\theta_3)+(v_3-\dot{x}_1\sin\theta_3-\dot{y}_1\cos\theta_3)d_2\sin((\theta_3+\phi_2)+\theta_3)}{d_2\sin((\theta_3+\phi_2)+\theta_3)d_2\cos((\theta_3+\phi_2)+\theta_3)} \quad (12)$$

Such as is the case in the previous system(11), the system in equation (12) is a system of dynamical airplane that role as second follower and been given a control so that it can be controlled dependent on the leader.

5. Numerical Simulation

In this section, some numerical simulations which is to illustrate the system(8),(11) and (12) are reported. For illustration, a desired path is the following parametric curve:

$$\gamma_x(t) = \frac{3}{49}t^3 - \frac{25}{29}t^2 + 5t + 3$$
$$\gamma_y(t) = \frac{2}{99}t^3 - \frac{15}{500}t^2 + 5t + 4 \quad (13)$$

The problem in this section is to design the control for three airplanes, so the three airplane move to follow the desired path from the initial position ($t = 0$ seconds) to the end($t = 25$ seconds). First of all is designing motion control of the leader airplane for track a desired path. Design control of the leader airplane used the tracking error dynamics. Figure 2 below shows the trajectory of the leader airplane tracing the desired path by using the method. From figure 3 above shows that the trajectory of the leader airplane can track the desired path from the position 0 seconds to the end position 25 seconds with a small enough distance. This means that the leader airplane can move from one area (the starting position) to another area (final position), hope of results numerical simulations are the first follower airplane and the second follower airplane also move to follow the trajectory of the leader airplane from the start position ($t = 0$ seconds) to the end ($t = 25$ seconds).Trajectory error between the trajectories of the leader airplane with desired path can be seen in Figure 3.

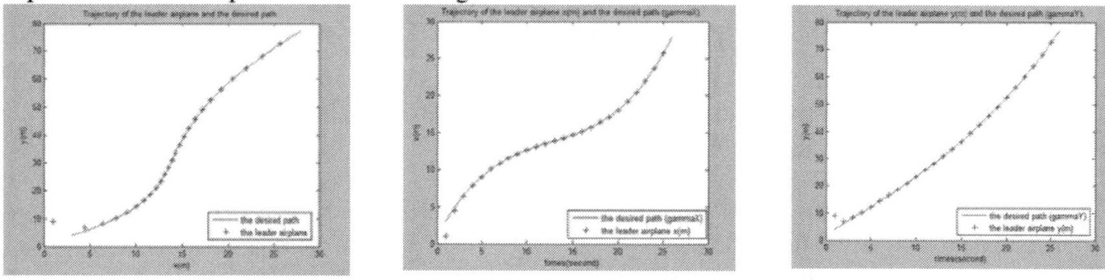

Figure 2. Trajectory of the leader airplane and the desired path.

From the results of simulations can be seen that the tracking aircraft that serves as a leader originally indeed outside the path, but in the 5th second tracking planes can already be on the desired trajectory.

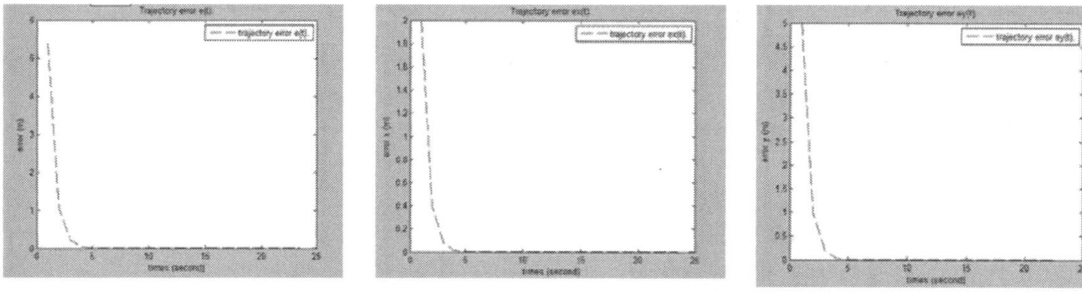

Figure 3. Trajectory error between the trajectory of the leader airplane and desired path.

Based on the results of the simulations the above error is getting smaller over time.This indicates that the method used is quite effective because the airplane was able to do tracking on the path until the end.

6. Conclusion

From the numerical simulation result above, it can be seen that the tracking error of the path of the leader airplane tracing a desired path is sufficiently small and the distance between the path of leader airplane and the desired path is preserved. A geometry approach for formation control of a group of airplane is investigated in this paper. A small error that indicates that the method used is a fairly effective method. In the future works, we will discuss the movement control of model swarm consisting of more three airplanes with a specific geometry formation.

References

[1] Miswanto, Pranoto, H.Muhammad, D Mahayana 2015 The Control Design of Ship Formation with the presence of a Leader *International Journal of Robotics and Automation (IJRA)* vol. 4 No.1 March pp 53-62

[2] V Gazi, B Fidan, Y S Hanay, and M Ilter Koksal 2007 Aggregation, Foraging and formation control

of swarm with non-Holonomic Agents UAing potential Functions anda sliding Mode Technique *Turk, J. Elec. Engin.* Vol.15 No.2

[3] Miswanto, I Pranoto, and H Muhammad 2006 A model of swarm movement with the presence of a leader *Proceeding of the International Conference on matehematics and natural Sciences* pp 740-2 (Bandung: ITB)

[4] Gazi and R Ordonez 2007 Target Tracking Using Artificial Potentials and Sliding Mode Control *International Journal of Control* Vol.80 No.10 pp 1626-35

[5] D Wang and G Xu 2000 Full state tracking and internal Dynamics of Nonholonomic Wheeled Mobile Robots *Proceedings of the American control Conference Chicago* Illinois pp 3274-8

[6] C Y Tzeng and J F Chen 1999 Fundamental Properties of Linear Ship Steering dynamic models, *Journal of Marine Science and Technology* Vol.7 No.2 pp 79-88

[7] A P Aguir and J P Hespanha 2003 Position Tracking of underactuated Vehicles *Proceedings of the American control Conference Denver* Colorado pp 1988-93

[8] A Behal, D M. Dawson, B Xian and P Setlur 2001 Adaptive tracking control of underactuated Surface Vessels *Proceedings of the IEEE International Conference on Control Application Mexico City*

[9] Miswanto, I Pranoto, J Naiborhu, S Achmadi 2012 Formation Control of Multiple Dubin's Car System with Geometric Approach *IOSR Journal of Mathematics (IOSRJM)* **1** pp 16-20

Creativity of Secondary Mathematics Teacher within Solve the 2-Dimentional Figure Problem Based On Teaching Experiences

P Wijayanti
Faculty of Mathematics and Natural Science, Universitas Negeri Surabaya, Indonesia

Abstract: Creativity is an ability produced from the cognitive activities to get ideas in solving problems by combining the concepts that have been mastered to mark by fluency, flexibility, novelty, and elaboration. This study aims to get description about the creativity of the junior mathematics teachers and senior mathematics teachers on the problems on plane figures. This study is an explorative study with qualitative approach. The instruments used in this study are the researcher itself as the main instrument supported with the written assignments on the problems on plane figures, and the valid and reliable interview guidelines. Data collection was conducted by doing task-based interviews. There are two subjects in this study which are junior mathematics teachers and senior mathematics teachers. The result of this study shows that junior secondary mathematics teachers in solving the problems on plane figures more creative than senior secondary mathematics teachers in the aspect of fluency, flexibility, novelty, and elaboration.

1. Introduction

Creativity is a term that can't be released from human life every day, for example a great farmer think about how to solve his problem, that is process the waste material so it can be used by the other lifes around him. Then, creativity is also a word that always follows an artistic masterpieces or a new science invention or unreal innovative product, for example Basuki Abdulah's painting, creating electric car and or the others. Sometimes creativity is defined as an exceptional ability that brought from born or obtained when adult, especially for artists and scientist ([1]:16).

Creativity is defined different by many experts according to each viewpoint. Alexander [2] defined creativity as a mentality process that involve creation of a concept and new ideas, or seeing the new relation between every concept and ideas that has been exist. In line with the above opinion, Munandar [3] explained that creativity is an ability to seeing many solution to a problem possibility. Furthermore, Semiawan ([4]:7) also explained that creativity is an ability to give new ideas and apply it in problem solving. Based on that some opinion, basically creativity is a person ability to produce something new, both in new creation or combination with some things that has been exist.

If we examine the meaning of creativity above, then everyone could be a creative human. The result showed that style of parents in raised their child and house environment play an important role to stimulate creativity ([1]:55). Effort to cultivate the creativity in school as described in Constitution Number 20 Year 2003 ([5]:4), about National Education System, explained that national education aims to develop potential learners to be a human who is faithful and devoted to God Almighty, noble, healthy, knowledgeable, capable, creative, independent and become a democratic state citizen, and responsible. To make one of those national educational purpose reached, i.e.to develop learners ability or potential became a creative human, educational process need to pay attention how to improve learners creative thinking by not only emphasizing upgrading to get ideas, but also encourage the desire and develop bravery to do it.

Besides, in Ministerial Regulation Number 41 ([6]:6) about Standard Process also mentioned that learning process in every units of primary and secondary education must be interactive, inspiring, delightful, challenging, and motivating the learners to actively participate and provides enough space for initiatives, creativity, and independence in accordance with talent, interest, physical, and psychological development of learners. But according to observation in the field, found that the most learning process, especially math, still not yet develop student creativity. It was because generally the teacher only emphasizing the quantity of the math not the quality of math.

Based on the description above can be concluded that the emphasis in education develops student creativity through educational process. To reach it, education system also must be conditioned to build character of the student that could raise student creativity. Teacher has a very important role in order to realize the aims of national education, one of those is creative student.

According to some theories, everyone has a creative ability, but most people don't know how to become creative. According to Evans [7], latest study showed that people can be taught to be more creative. As well as Sebastian's opinion, creativity doesn't appear because of genetic, but because of habit and skill which can be trained and developed. It is shown trough Sebastian's experience in dozen years leading some company, apparently the result of workers personality for IQ rarely change but in work attitudes and creativity can be raised after honed. Beside that, Weisberg ([1]:17) explained that "creativity rooted in past experience and sourced in the same process of thinking which we all use everyday". Both of that opinion explain that someone creativity might be affected by his experience in order to faced problem in their life. Therefore can be presumed that teaching experience of teachers affected their creativity. The fact in field, there is a teacher have long teaching and there is teacher just teach. According to teaching experience of that teacher, teacher can be divided into two, that are senior teacher and junior teacher. Senior teacher is a teacher that have long teaching experience, while junior teacher is a teacher that still have less experience in teaching.

Open ended maths problem is a one means which can be used to train and developed creativity or creative thinking. According to Plucker and Makel ([8]:52),"..., *Divergen Thinking tests ask for multiple responses to either figural or verbal prompts, and responses are scored for fluency, flexibility, originality, and elaboration of ideas.*" So divergen thinking can be seen from fluency, flexibility, originality, and elaboration of ideas. Geometry is a one part in mathematics that gave much problems which to solve it using divergen thinking. Subjects of geometry in Middle School based with two-dimentional figure material. Because of that, two-dimentional figure material is one of the subjects of geometry that gave much spaces to train and develop creativity. According to above description, then to reached one of national education purposes, it is important to do a study about how maths teacher's creativity in Middle School within solve the two-dimensional figure problem is reviewed from the different teaching experience. Therefore, the purposes of this study are (1) to describe junior maths teacher in Middle School creativity within solve two-dimentional figure problem and (2) to describe senior maths teacher in Middle School creativity within solve two-dimentional figure problem.

2. Literature Review

2.1. Creativity

IOP Conf. Series: Journal of Physics: Conf. Series **1108** (2018) 012042 doi:10.1088/1742-6596/1108/1/012042

There are many literatures show many options which can be used to expressed and defined creativity. Some people said that creativity is a kind of special thinking, while the other people said that creativity is moreover to produce a product. Stein said, "Creativity is a process that results in a novel work that is accepted as useful, tenable, or satisfying by a group of people at some point in time" ([9]:148). Creativity was defined more emphasizing to process, so creativity is a process produce a product by transfigure from the existing products and that product unique, having some purposes, and worth for the creator that marked with newly, useful, and satisfy for a group which having a same vision is some times.

That thing in line with Healy and Schifter opinion ([10]:104), "creativity as the ability to take existing objects and combine them in different ways for new purposes". Therefore creativity is an ability to create a new combination according to existing concepts with do an action of combining some element which have not combined yet before. Besides, creativity is also an ability to find new relations and seeing something according to perspective or viewpont that hasn't exist before. Pames [4] also explained that creative ability could be called trough problem that spurring five creative attitude, that are fluency, flexibility, originality, elaboration, and sensitivity. Fluency is an ability to express the same ideas to solve the problem. Flexibility is an ablility to produce any ideas to solve a problem outside the common category. Originality is an ability to give an unique or exceptional response. Elaboration is an ability to develop ideas in details to realize that idea become true. Sensitivity is an ability to catch and produce problem as a response to a situation.

Furthermore, the characteristics of creative person are having some new ideas, brave to be different, create a thought that still doesn't popular, doesn't be afraid to try, and doesn't be afraid of failure ([11]:10).

According to decriptions above, so in this study, creativity was defined as an ability which produce from cognitive activity to get an idea within solve the problem by combining concepts that have been mastered which marked by fluency, flexibility, novelty, and elaboration. According to Guilford (in [7]) fluency related to ability to easily produce as much as possible ideas. Flexibility refers to the ability to generate many kinds of ideas. Then Evans also stated that flexibility is also one of basic from novelty or originality. In solving problems, this flexibility related to an ability to try various approaches. Then the meaning of flexibility in this study is an ability resulting from cognitive activity to get an idea or thought in order to solving problems by doing combinations from concepts that have been mastered which is marked by diversity of ideas or strategies that are used to find the answer from the problems that faced. Evans [7] stated Creativity earns in new ideas and discoveries. Any definition of creativity must be included the element of novelty. And then elaboration --- the ability to supply details to complete a given outline or skeleton figure. That state including the meaning that elaboration is an ability to develop or detailing something in detail. Therefore in this study the meaning of elaboration is an ability resulting from cognitive activity to get an idea or thought in order to solving the problems by doing combination from concepts that have been mastered which is marked by ability to develop or expand an idea in order to get response from the problems faced.

2.2. Math Teacher

Republic Indonesia Constitutions Number 14 Year 2005 explained that teacher is a professional workers with the main task which are educating, teaching, guiding, directing, training, assessing, and evaluating learners in early childhood education in figureal education way, basic education, and middle education. Then teacher have a role as professional workers in primary education, middle education, and early childhood education in figureal way which appointed in accordance with the legislation. Based on understanding and position teacher in above constitutions, teacher is the human figure that take position and has a role in eduation. Then function of the teacher is to improve human dignity and has a role as learning agent that has a fuction to improve the quality of national education.

Therefore teachers has very important role in order to determining educational process. A professional teacher expected to produce qualified graduates. Professionalism of teacher as a spearhead of curriculum implementations in the class needs to get an attention (Depdiknas, 2005). Teacher has a responsibility to seeing everything that happen in the class in order to help the student development process. Delivery of learning materials is only one of the various activity in study as a dynamic process in every phase and student development process.

2.3. The Different Teaching Experience
According to years of service or the length of the predicate teachers teach, then according to Ministry Regulation PAN Number 16 Year 2009 about Teacher Functional Position and Teacher Credit Number can be divided in 4 group, namely first teacher, young teacher, medium teacher, and main teacher. Furthermore, in this study the meaning of the differences of experience teaching, maths teachers are classified into 2, namely junior math teacher and senior math teacher. Junior math teacher is a math teacher who has been teaching for less than or equal to 8 years since first teaching certificate was taken out from authorized party. Senior math teacher is a math teacher who has been teaching for more than 8 years since first teaching certificate was taken out from authorize party.

2.4. Creativity Math Teacher of Middle School within solve Two-Dimentional Figure Problems
Something important that need to remember that creativity is also awakened or produced ideas or concepts and also combine ideas that has been exist to produce something new. Creativity is one of the parts that integral from human and with that creativity which separate the human with creatures in the world. Futhermore Rogers (in Plucker in [12]) said "The very essence of the creative is its novelty, and hence we have no standard by which to judge it." That stated explain that something basic to see creativity is novelty, but there is no standard yet that could assess them. Because of that, in this study novelty is defined as new for teacher who became a subject reasearch.

Furthermore, creativity in this reasearch related with how to solve the faced problems. Guilford ([9]:156) stated that "...problem solving and creative thinking are closely related. The very definitions of those two activities show logical connections. Creative thinking produces novel outcomes, and problem solving involves producing a new response to a new situation, which is a novel outcome". The state explained that there are relationship between creativity and problem solving because problem solving become an aspect in creative. In line with Middle School Maths Teacher in order to solving the two-dimentional figure problem, then according to theory that has explained, can be concluded some definitions as follows. Middle School math teacher creativity within solve the two-dimentional figure problem is an ability resulting from cognitive activity maths teacher to get an idea or thought in order to solving the two-dimentional figure problem by doing combination from concepts that have been mastered which marked by fluency, flexibity, novelty, and elaboration. Maths teacher's fluency within solve the two-dimentional figure problem are ideas diversity or strategy that math teacher used to finding answer from two-dimentional figure problem given. Maths teacher's novelty within solve the two-dimentional figure problem is the given answer or idea/strategy that maths teacher use to get answer that hasn't exist before. Maths teacher's elaboration within solve the two-dimentional problem in this study, there are two things, namely maths teacher elaboration to find the another answer from the two-dimentional figure problem given and maths teacher elaboration from the face problem to classroom learning.

3. Result And Discussion

3.1. Creativity Of Middle School Junior Math Teacher to Solve the Two-Dimentional Figure Problem
Creativity of Middle School junior math teacher in two-dimentional figure problem solving, that is divided exactly-6 cm isosceles triangle into two parts that have a same area, shown in the following description. Middle School Junior maths teacher's fluency showed by diversity of the answer within solve the two-dimentional figure problem according to the result of the division the name of the figures which obtained, that is two right triangles isosceles which congruent, two random triangles which its pedestal and its height is equal, a trapezodial isosceles, and random quadrangle, a right isosceles triangle, a circle and a figure with n-facets and two figures with n-facets.

IOP Conf. Series: Journal of Physics: Conf. Series **1108** (2018) 012042 doi:10.1088/1742-6596/1108/1/012042

Middle School Junior maths teacher's flexibity to solve the two-dimentional figure problem showed by diversity idea that is used to finding an answer in divide right-6 cm isosceles triangle, that created triangle heavy line, created random figure that have 9 cm^2 area, and created random figure which consist of 36 swaths.

Middle School Junior maths teacher's novelty to solve two-dimentional figure problem showed by the answer of how to divide triangle into two parts which has the same area that had never been though before, namely one part of the divide results are random figure which have 9 cm^2 area and the divide results of the second figure are two figure with n-facet, and idea or method that never thought before, namely create 9 cm^2 area random figure and create random figure that consists of 36 swaths on swaths paper with 0,5 cm of size. The novelty within this is new according to Middle School's mathematics junior teacher, because that answer is newly found when working and previously thinking or did by that teacher.

Elaboration of Middle School's mathematics junior teacher shown when solve the problem of two-dimensional figure and development that teacher did based on tyhe problem of two-dimensional figure that already finished and can applied on learning in class. First, there is elaboration of main idea to subs idea for obtaining some other alternative answer based on the previously answer, that is development of idea from trapezoidal figure to circle and triangle figure. Second, there is elaboration of idea after finishing two-dimensional figure that teacher did by making 4 open questions which can applied on learning in class.

3.2. Creativity of Middle School Senior Maths Teacher to Solve the Two-Dimentional Figure Problem

Creativity of Middle School senior math teacher in two-dimentional figure problem solving, that is divide right-6 cm isosceles triangle into two parts that have a same area, shown in the following description. Middle School Senior maths teacher's fluency is shown by diversity of answers to solve the problem of two-dimensional figure based on obtained names of the figures divided result, there are two congruent right isosceles triangle, two figures of n-facet, and two incongruently triangle but have the same pedestal, length, and high.

The flexibility of Middle School's mathematics senior teacher shown by diversity of ideas that use to find answer on dividing right 6-cm isosceles triangle, that angle is divided by line of right triangle, making heavy line of triangle, making random figure that consist of 36 swaths with 0,5 cm of size and curve line. The novelty of Middle School's mathematics senior teacher within solve the problem of two-dimensional figure is shown by answer to dividing triangle into two parts that have same area that not yet thinkable previously, there are two figures of n-facet, and idea/strategy that not yet thinkable previously, there is making random figure that consist of 36 swaths with 0.5 cm of size.

Elaboration of Middle School's mathematics senior teacher is shown when solve the problem of two-dimensional figure and development that teacher did based on problem of two-dimensional figure that already finished and can applied on learning in class. First, there is elaboration of main idea to the other ideas for obtaining some other alternative answer based on the previously answer, there are development divided line from straight line to curve line. Second, there is elaboration of idea after finishing the problem of two-dimensional figure that teacher do by making 2 open questions that can be applied on learning in class, that are 1 geometry question and 1 arithmetic.

4. Conclusion

The result of this study showed that (1) middle school's junior maths teacher creativity within solve the two-dimentional figure problem, cover (a) fluency, is shown by there are diversity in answer to solve the two-dimentional figure problem based on the result of the division figures obtained; (b) flexibity, is shown by there are diversity in idea that used to find answer within divide a right-6 cm isosceles triangle into two parts with the equal area; (c) novelty, is shown by new answers and ideas/strategies that have never been thought before in solving the two-dimentional figures problem; and (d) elaboration, is shown by develop idea to create random figure with 9 cm^2 area to trapezodial figure, right isosceles triangle, and circle; and (2) middle school's senior maths teacher creativity within solve the two-dimentional figure problem, cover (a) fluency, is shown by there are diversity in answer to solve the two-dimentional figure problem based on the name of the result of the division figures obtained; (b) flexibity, is shown by there are diversity in idea that is used to find answer within divide a right-6 cm isosceles triangle into two parts with the equal area; (c) novelty, is shown by new answers and ideas/strategies that have never been thought before in solving the two-dimentional figures problem; and (d) elaboration, is shown by develop idea to create line divider from straight line into curved line.

5. References

[1] Chia, Eng-Hock 2000 *Anda Juga Bisa Menjadi Kreatif: Menerapkan Pemikiran Kreatif dalam Kehidupan Sehari-hari* (Jakarta: Prestasi Pustakaraya)
[2] Alexander 2009 Tahap Kreativitas: Bermain-main Dengan Aljabar http://sigmetris.com didownload 6 Mei 2010 jam 07.47
[3] Munandar, Utami 2009 *Pengembangan Kreativitas Anak Berbakat* (Jakarta: Pusat Perbukuan & Rineka Cipta)
[4] Semiawan, Conny, Munandar A S, & Munandar, S C U 1984 *Memupuk Bakat Dan Kreativitas Siswa Sekolah Menengah* (Jakarta: Gramedia)
[5] Peraturan Menteri Pendidikan Nasional Undang-undang Nomor 20 tahun 2003 tentang *Sistem Pendidikan Nasional* (Lampiran) (Jakarta: BSNP)
[6] Peraturan Menteri Pendidikan Nasional Undang-undang Nomor 41 Tahun 2007 tentang *Standar Proses Untuk Pendidikan Dasar dan Menengah* (Lampiran) (Jakarta: BSNP)
[7] Elson, Robert http://www.articledashboard.com/ Article/Definition-of-Creativity/683674 didownload pada 15 Oktober 2009 jam 19.28
[8] Evans, James R 1991 *Creative Thinking In the Decision and Management Sciences* (Ohio: South-Western Publishing Co)
[9] Idris, Noraini 2006 Creativity In The Teaching And Learning Of Mathematics: Issues And Prospect *Malaysia: Masalah Pendidikan 2006* pp 103-13
[10] Kaufmann, G, Helstrup, T, & Teigen, K H 1995 Problem Solving And Cognitive Processes (CPS: Linking Creativity And Problem Solving by Scott G. Isaksen) http://www. cpsb.com/study/articles/creative-problem-solving/CPS-LinkingCandPS.html download pada 22 Pebruari 2010 jam 21.00
[11] Nasution, S 2003 *Metodologi Penelitian Naturalistik Kualitatif* (Bandung: Tarsito)
[12] Kaufman, J C dan Sternberg R J 201 *The Cambridge Handbook of Creativity* (New York: Cambridge University Press)
[13] Soenarno, Adi 2006 *Creativity Games* (Yogyakarta: Andi Offset)

IOP Conf. Series: Journal of Physics: Conf. Series **1108** (2018) 012043 doi:10.1088/1742-6596/1108/1/012043

The effect of virtual lab and gender toward students' creativity of physics in senior high school

G Gunawan[1,3], N M Y Suranti[1], N Nisrina[1], L Herayanti[2], and R Rahmatiah[1]

[1]Physics Education Study Program, University of Mataram, Jln. Majapahit No 62 Mataram 83125, Indonesia
[2]Physics Education Study Program, IKIP Mataram, Jln. Pemuda No 59A Mataram 83125, Indonesia
[3]Corresponding author: gunawan@unram.ac.id

Abstract. Creativity is one of the success goals in the learning process. One of the efforts to develop and improve creativity is through computer-based learning. This study aims to investigate the effect of virtual laboratory on student creativity which consists of verbal, numerical and figural creativity in physics. The increase of students' average score was compared based on their gender. This quasi-experimental study used a pre-test and post-test control group design conducted at four different schools with 51 male and 51 female students. The data were obtained based on creativity tests (essay form) that have been validated by experts. The tests of creativity can improvement differences that were done by calculating the average difference of N-gain score. The results showed that the female student's scores have higher than male students. In the aspect of verbal creativity, male and female students have the equal score relatively. Whereas, in the aspect of numerical and figural creativity that the female students have higher scores than the male students.

1. Introduction

The advanced development of information and communication technology in this 21st century bring the significant effect on everyday life, especially education. One of the benefits of information and communication technology in education is computer-based learning media.Learning media was used to support the learning which covers to collect, to present, and to process experimental data [1]. Shyr[2] also revealed that the computer-based learning has grown in schools to replace traditional methods. In year, the virtual learning environment is widespread in high-level education, not only for presenting the subject matter but also for facilitating the communication in learning.

Physics as part of science education has a strong connection with the existence of learning media. Some concepts in physics, especially for the abstract concept, actually it cannot be separated from the touching of learning media. This is caused by the various abstract concepts that often difficult for teachers to visualize and deliver to students verbally. Moreover, most students also argue that while studying the abstract concepts of physics, they find the difficulty in terms of mastering the material, proofing the concept in real terms, and further regarding analyzing it in real life. This activity indirectly trains students' creativity to perform every step of scientific work. Unfortunately, many school laboratories have the limit experimental tools. This condition causes the experimental activity cannot be implemented optimally. As a result, the development of students creativity becomes poorly trained. Students need effective learning media that can support their activities for physics experiment virtually. Students can still train and develop their creativity as when doing real experiments. Therefore, the presence of learning media is necessary for physics teaching.

One of the effective learning media is a virtual laboratory. The virtual laboratory can guide students to experiment as well as doing real experiments in general. The existence of virtual laboratories is also intended to gain students' experience for solving difficult problems and situations that they faced. The virtual laboratories in physics learning can increase the verbal and figural creativity higher than conventional learning [3], and encourage the increasing of problem-solving skills of students [4]. Gorghiu [5] stated that the virtual laboratory is very helpful to increase the interest, motivation and learning skills.

A virtual laboratory is also useful in providing opportunities for students to learn by doing, developing thinking skills, and problem-solving skills [6]. In another study, the use of project-based learning model that is supported by virtual media can improve students' creativity [7]. The use of virtual laboratory in college learning is also proven to improve generic science skills of physics teacher candidates [8], as well as improving the critical thinking disposition of pre-service physics teachers, especially on truth-seeking and open mindedness indicators [9].

Creativity is the student's ability to follow learning activity in order to discover and use the new ideas that are unusual but still logical and rational. According to Gunawan et al. [10], creativity is an essential component of global competition in the 21st century. Therefore, the innovation in the learning is needed to help students more creative, including with the use of computer technology. Horng et al. [11] revealed that one of the goals of creative instruction is to create a student-centered learning environment with multimedia-assisted that can encourage students to think creatively in imagining freely and understanding the relationship of concept with real life.

In physics learning, the gender of students has little effect on their creativity. Gender is a sociocultural and psychological dimension of male and female [12]. Philbin et al. [13] stated that male and female have different learning styles. Honigsfeld and Dunn [14] also revealed that male and female students have different learning styles in many ways. Furthermore, Stadler et al. [15] in their research on gender stated that male and female students have differences in the meaning of physics that can affect the learning outcomes. This is due to the learning of physics associated with the real context in accordance to students daily life. This conformity involves three aspects: (1) the linkage between the learning content and daily life, (2) the pattern of body language understanding, and (3) the touching events of feelings and emotions. Gunawan et al. [16] on their research in three different high schools also found that the use of virtual laboratory in physics learning can enhance the verbal and figural creativity of students, both male and female. Female students have higher verbal creativity than male students, while male students have higher figural creativity than female students.

This study aims to investigate the effect of the virtual laboratory on student creativity which includes verbal, numerical and figural creativity. The result of student creativity improvement then compared by gender. Several previous studies on gender have been widely discussed but few compare gender based on student creativity. Most research results on gender conclude that there are differences between male and female students. This difference becomes the primary task of teachers to understand better the character of each student's gender in accepting and understanding the learning.

2. Method

This research used quasi-experimental study with pre-test and post-test control group design conducted in four different senior high schools. The sampling technique was purposive sampling. There were 102 students involved in the study, consist of 51 female students and 51 male students. Students were selected based on initial capability data as well as the proportion of female and male student ratios in each school. The treatment provided was the use of a virtual laboratory to compare verbal, numerical, and figural creativity based on students' gender. The creativity test used in the form of a description test consists of verbal, numerical, and figural creativity tests that have been validated by experts.

The obtained data then analyzed with simple statistics to determine the average N-gain score and t-test. N-gain was used to avoid misinterpretation of increasing creativity of students, both male, and female. Meanwhile, the t-test was intended to calculate the significance of differences in the increase in creativity between male and female students. Before the t-test, the homogeneity and normality tests of the data were carried out.

3. Results and Discussion

This study was conducted to examine the effectiveness of a virtual laboratory in four different high schools. The differences in the physics creativity enhancementare further compared to gender; male and female. The virtual lab used in learning comes with features program that helps students to experiment according to the goals. The virtual laboratory was equipped with a control tool that allows students to think and be creative to determine one of several solutions in solving the given problem.

Students were given tests before and after learning physics with a virtual laboratory. The results of data analysis showed that there was the difference in creativity improvement based on the student's gender which reviewed from the pre-test and the post-test of student physics creativity. Pre-test of creativity was given before the use of virtual laboratories in learning, to know the initial creation of the students. The results of the homogeneity test showed that the data of male and female students were homogeneous. This shows that male and female students have the same prior knowledge. The normality test also indicated that the data were normally distributed.

Furthermore, students follow the learning with the use of virtual laboratories. The last of the lesson, students were given a final test of creativity to find out the result and the increase of student creativity. Observations were made during the lesson. Enhancement of the student creativity was known by using the N-gain test. Figure 1 below presents the resulting study about the mean scores difference on pretest, posttest, and N-gain score of the students by their gender.

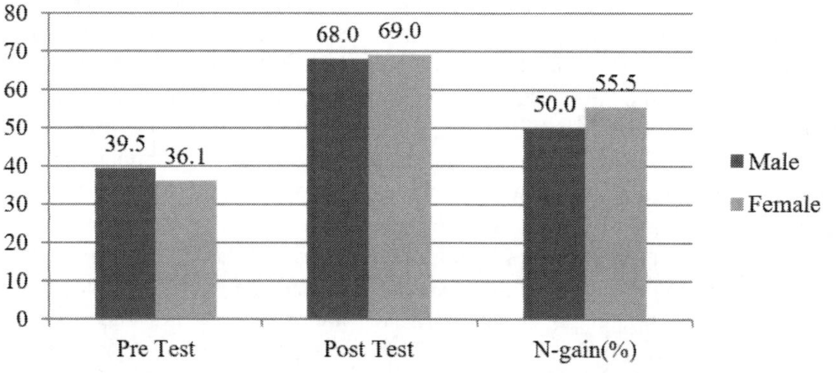

Figure 1. The difference in average score of creativity of students.

Figure 1 showed that the average score of the initial test of creativity for male students was 39.5 and for female students was 36.1. These pre-test scores were not significantly different. This may be due to students who did not understand the lesson well and only used their initial knowledge to solve the problem. After having treatment using a virtual laboratory in learning, students'creativity test results was increased. Based on Figure 1 mainly the post-test chart, it could be seen that the average score of the final test of creativity for male students was 68.0 and 69.0 for female students. The final test score showed that female students have higher creativity than male students, although not significantly different. It is assumed that male and female students have different abilities to collect information that they studied. Studying with a virtual laboratory could help students; for both male and female students to develop and accumulate their creativity. This result relates to Çelik et al. [17] that stated learning with virtual programs can enhance learners' understanding. According to Syyr [2], virtual laboratories have been able to help teachers adjust learning and enable students to develop their ideas and identify problems.

IOP Conf. Series: Journal of Physics: Conf. Series **1108** (2018) 012043 doi:10.1088/1742-6596/1108/1/012043

The increased creativity of male and female students was known based on the results of N-gain test. Based on Figure 1, it could be seen that the male student's score was 50.0% and the female student's score was 55.5%. The acquisition of N-gain scores of both groups students showed that the increased creativity included in the medium category. Male students had a lower creativity increase than female students but did not differ significantly. This statement was reinforced by the results of hypothesis testing with t-test that $t_{count} < t_{table}$ (1.25<1.99), so it could be concluded that there was no significant difference between the creativity of male students and the creativity of female students. The results of this study were in line with the research results of Suprapto et al.[18] that stated gender does not affect students' creative thinking ability. Afriana, et al. [19] in their research also found that the increase in science literacy in the male and female classes was not significant. In contrast to the results of this study, Bacharach et al. [20] reported that the gap in science achievement depends on the gender of the students. Asis& Nurdin [21] in their study also concluded that based on the terms of reference and mental rotation, the dominant male subject used spatial skills, while the dominant female subjects used logical reasoning.

The results of the students' creativity tests were divided into tests of verbal, numerical, and figural creativity. The results of the test compared to the increase based on students' gender. The differences enhancement in the verbal, numeric, and figural creativity of students based on their gender are shown in Figure 2.

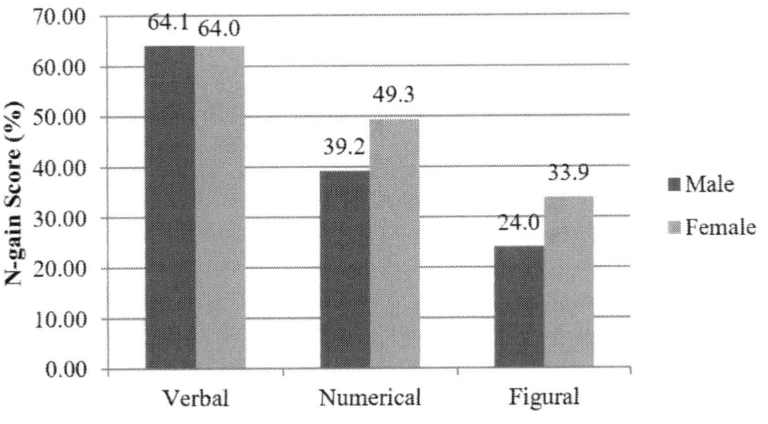

Figure 2. The differences in each aspect of students' creativity.

Figure 2 shows that there was the differences between the creativity of male and female students in each aspect. In verbal creativity, male students get an increase at 64.1 while female students score slightly lower at 64.0. This increase is not significantly different which is reinforced by the results of the hypothesis testing with a t-test that $t_{count} < t_{table}$ (0.24<1.99).In numerical creativity, male students get an increase in the value of creativity 39.2 while female students score 49.3 higher which is not significantly different. This was indicated by the t-test result which is $t_{count} < t_{table}$ (1.73<1.99).Moreover, also in figural creativity, male students score 24.0 while female students score 33.9. Female students have a higher improvement score than male students, but not significantly different. Itis showed by the results of the t-test that is $t_{count} < t_{table}$ (1.67<1.99).

Based on Figure 2, it could be seen that in general the increase of female students' creativity was higher than male students, although the score of both increases was almost the same on the test of verbal creativity. Next, will be discussed one by one for the results of each type of creativity. First, the type of creativity that ishas the highest increase was verbal creativity, both for male and female students. This suggested that the use of virtual laboratories could help students, both male,and female, to think divergingly by combining ideas about a problem verbally. This enhancement was reflected in the fluency, flexibility, originality, and elaboration capabilities that demonstrated by the students. This result proved that male and female students have a good enough ability in forming ideas through words, as well as directing the focus problem on mastering communication in writing. According to Dalgarno et al. [22], the virtual experiments have been able to improve students' skills in deductive reasoning, hypothesis formation, andeffective testing through experiments.From the gender aspect, Skaalvik & Skaalvik [23] mentioned that male students have a higher self-concept, performance expectancy, intrinsic motivation in mathematics than female students, whereas female students have the higher intrinsic motivation to learn the language than male students.

Numerical creativity got the second highest increasing after verbal creativity. Female students have a higher increase than male students, although not significantly different. It was assumed that female students have better abilities associated with numbers, structured thinking, mathematical logic, as well as more precision on basic mathematical calculations. Also, Voyer & Voyer [24] stated that female makes more effort than male in learning to use mathematical calculations. In his research, Fennema [25] concluded that male students tend to have higher mathematical abilities than female. Loori [26] also revealed that women in learning tend to use intrapersonal intelligence while men prefer to use logic and mathematical intelligence. Both male and female students alike experienced an increase in numerical creativity in the moderate category. This statement showed that the use of virtual labs could help students to associate new ideas in mathematical logic quite well. The results of this study were in line with the statement of Oidov et al. [6] who argued that with virtual activities, students could perform numerical measurements and evaluations of the process being explored, interpret data (information / facts), and write formulas and formulate the fundamental laws of physics.

The type of creativity that has the lowest increase was the figural creativity, for both male and female students. These results in line with Wu et al. [27] that concluded that student has significantly higher scores on real-world problems, and significantly lower in figural tasks, and on verbal assignment there was no different for the groups. Female students have a higher increase than male students, although not significantly different. The increase of figural creativity of female students was in the medium category, while the increase of figural creativity of male students was in a low category. This result suggested that the use of virtual laboratories, female students were able to use better abilities in divergent thinking to form ideas by combining patterns of shapes or images to solve a problem. Thisis supported by the research of Cheung & Lau [28] that suggested female students have better figural creativity, which includes figural fluency, figural flexibility, figural uniqueness, and figural unusualness.

4. Conclusion

The use of a virtual laboratory in physics learning could improve students' creativity, namely on verbal, numerical, and figural creativity. This increase could be seen from the score of N-gain of each student's creativity, both for male students and female students. The highest student creativity improvement was in verbal creativity, then numerical creativity and the lowest increasing was in figural creativity. In general, female students have a higher level of creativity than male students, although statistically, the ability difference did not differ significantly. In verbal creativity, the increase for male and female students was similar. In numerical and figural creativity, female students have a higher increase than male students.

As for suggestions that can be given for further research, mainly the use of virtual laboratories in learning needs to be preceded by explanation of the function of the program features as well as the opportunity for the students to try several times so that the students will more familiar before using it in the primary learning session. The use of virtual laboratories should be supported by an inquiry-based worksheet where the work steps can be independently adjusted by the students to achieve the learning objectives.This will further encourage the creativity and new ideas of students than if the worksheets have been prepared before by the teacher. The results of the study recommend for further research to find out in detail at which stage the students' creativity develops, both in male and female as well as comparisons at each cognitive level of students after the learning is done.

5. References

[1] Finkelstein N D, Adams W K, Keller C J, Kohl P B, Perkins K K, PodolefskyN S, ... and LeMaster R 2005When learning about the real world is better done virtually: a study of substituting computer simulations for laboratory equipment*Physical Review Special Topics-Physics Education Research***1** 010103

[2] Shyr W J 2010Enhancement of PLC programming learning based on a virtual laboratory*World Transactions on Engineering and Technology Education*. **8** 196

[3] Gunawan G, Harjono A, Sahidu H and Herayanti L 2017 Virtual laboratory of electricity concept to improve prospective physics teachers' creativity*JurnalPendidikanFisika Indonesia.***13** 102

[4] Gunawan G, Harjono A, Sahidu H and Herayanti L 2017 Virtual laboratory to improve students' problem-solving skills on electricity concept *JurnalPendidikan IPA Indonesia.***6**257

[5] Gorghiu L M, Gorghiu G, Alexandrescu T and Borcea L 2009 Exploring chemistry using virtual instrumentation-challenges and successes*Research and Innovations in Intergrating ICT in Education***1**371

[6] Oidov L, Tortogtokh U andPurevdagva E 2012 Virtual laboratory for physics teaching*InInternational Conference on Management and Education Innovation IPEDR***37** 319

[7] Gunawan G, Sahidu H, Harjono A and Suranti N M Y 2017 The effect of project based learning with virtual media assistance on student's creativity in physics *CakrawalaPendidikan***36** 167

[8] Gunawan G, Setiawan A and Widyantoro D H 2013Model virtual laboratory fisika modernuntukmeningkatkanketerampilangeneriksainscalonguru*JurnalPendidikandanPembelajaran***20** 25

[9] Gunawan G and Liliasari L 2012 Model Virtual laboratory fisika modern untukmeningkatkandisposisiberpikirkritiscalon guru*JurnalCakrawalaPendidikan***31**185

[10] Gunawan G, Harjono A,Sahidu H and Nisrina N 2018 Improving students' creativity using cooperative learning with virtual media on static fluida Concept *Journal of Physics: Conference Series***1006** 012016

[11] HorngJ S, Hong J C, ChanLin L J, Chang S H and Chu H C 2005 Creative teachers and creative teaching strategies*International Journal of Consumer Studies***29** 352

[12] Santrock J W 2008 *PsikologiPendidikan*(Jakarta: KencanaPrenada Media Group)

[13] Philbin M, Meier E, Huffman S andBoverie P 1995 A survey of gender and learning styles*Sex Roles***32**485

[14] Honigsfeld A and Dunn R 2003 High school male and female learning-style similarities and differences in diverse nations*The Journal of Educational Research.***96** 195

[15] Stadlert H, Duit R and Benke G 2000 Do boys and girls understand physics differently? *Phys. Educ***35** 417

[16] Gunawan G, Suranti N M Y, Nisrina N, Ekasari R R and Herayanti L 2017Investigating students creativity based on gender by applying virtual laboratory to physics instruction. *Advances in Social ScienceEducation and Humanities Research***158** 303

[17] Çelik H, Sari U andHarwanto U N 2015 Evaluating and developing physics teaching material with algodoo in virtual environment: archimedes' principle *International Journal of Innovation in Science and Mathematics Education***23** 40

[18] Suprapto, Zubaidah Sand Corebima A D 2018Pengaruhgender terhadapketerampilanberpikirkreatifsiswapadapembelajaranbiologi*JurnalPendidikan: Teori, Penelitian, &Pengembangan***3** 325

[19] Afriana J, Permanasari Aand Fitriani A 2016Penerapanproject based learningterintegrasiSTEM untukmeningkatkanliterasisainssiswaditinjaudari gender *JurnalInovasiPendidikan IPA***2** 202

[20] Bacharach V R, BaumeisterA A and FurrR M 2003 Racial and gender science achievement gaps in secondary education*The Journal of Genetic Psychology***164** 115
[21] Asis M and NurdinArsyad 2015Profilkemampuanspasialdalammenyelesaikanmasalahgeometrisiswa yang memilikikecerdasanlogismatematistinggiditinjaudariperbedaan gender*JurnalDayaMatematis***3** 78
[22] Dalgarno B, Bishop A G and BedgoodJr D R 2012 The potential of virtual laboratories for distance education science teaching: reflections from the development and evaluation of a virtual chemistry laboratory *Proceedings of The Australian Conference on Science and Mathematics Education(formerly UniServe Science Conference)***9**
[23] Skaalvik S and Skaalvik E M 2004 Gender differences in math and verbal self-concept, performance expectations, and motivation*Sex Roles***50**241
[24] Voyer D and Voyer S D 2014 Gender differences in scholastic achievement: a meta-analysis*psychological bulletin***140** 1174
[25] Fennema E 1976 Influences of selected cognitive, affective and educational variables on sex-related differences in mathematics learning and studying*ERIC* 68
[26] Loori A A 2005 Multiple intelligences: a comparative study between the preferences of males and females*Social Behavior and Personality: An International Journal***33** 77
[27] Wu C H, Cheng Y, IpH M and McBride-Chang C 2005 Age differences in creativity: task structure and knowledge base *Creativity Research Journal***17** 321
[28] Cheung P C and Lau S 2010 Gender differences in the creativity of Hongkongschool children: comparison by using the new electronic Wallach–Kogan creativity tests *Creativity Research Journal.***22** 194

Acknowledgments
Authors deliver sincere gratitude to the Ministry of Research, Technology, and Higher Education of Republic of Indonesia for awarding the Research Grant in 2015-2017. The team also thank to all parties involved during the process and publication of results of this study.

IOP Conf. Series: Journal of Physics: Conf. Series **1108** (2018) 012044 doi:10.1088/1742-6596/1108/1/012044

Classification of Schizophrenia Data Using Support Vector Machine (SVM)

T V Rampisela[1] **and Z Rustam**[2]

[1,2]Mathematics Department, University of Indonesia, Indonesia

Abstract. Schizophrenia is a severe and chronic mental disorder. This disorder is marked with disturbances in thoughts, perceptions, and behaviours. Due to these disturbances that can trigger Schizophrenics to commit suicide or attempt to do so, Schizophrenics have a lower life expectancy than the general population. Schizophrenia is also difficult to diagnose as there is no physical test to diagnose it yet and its symptoms are very similar to several other mental disorders. Using Northwestern University Schizophrenia Data, this research aims to distinguish people who are Schizophrenics and people who are not. The data consists of 392 observations and 65 variables that are demographic data and clinician-filled Scale for the Assessment of Positive and Negative Symptoms questionnaires. Classification method used is machine learning with Support Vector Machines (SVM). Simulations are done with different data and percentage of training data. In each simulation, accuracy is measured. Model performance validation and evaluation are done by averaging ten times Hold-Out Validations that were done. In conclusion, SVM successfully classified Schizophrenia data with final accuracy of 90.1%. Furthermore, SVM with linear kernel and Gaussian kernel reached an accuracy of 95.0% in at least one simulation in classifying Schizophrenia data.

1. Introduction

Schizophrenia is a severe and chronic mental illness characterized by disturbance in thinking, perception, and behaviour. In general, Schizophrenia symptoms are divided into positive, negative, and cognitive symptoms. Positive symptoms are additional brain activities that are not supposed to exist, for example hallucination and delusion while negative symptoms are the opposite: things that are supposed to exist but are not present, namely apathy and lack of emotion. Cognitive symptoms are symptoms related to disturbances in memory and difficulty in concentrating [1]. All these symptoms of Schizophrenia can affect the quality of life and productivity significantly [2].

Schizophrenics' life expectancy are 2-3 times lower than the general population [2], as 10% committed suicide. Furthermore, 20-40% Schizophrenics attempted suicide at least once [3]. Anyone could suffer from Schizophrenia even though the cause of it has not been determined for sure [2]. This is exacerbated by the difficulty of detecting Schizophrenia as there is no radiology, laboratory, or psychometric test that can be used to diagnose this illness yet and the similarities of Schizophrenia symptoms with several mental illnesses, e.g. Major Depression and Bipolar [1]. Hence, an easy, objective, and accurate method to diagnose Schizophrenics is needed so that they can get the suitable treatment. One of the methods is through machine learning.

Detecting Schizophrenia by making use of psychological evaluation results is considered as a classification problem in machine learning. Machine learning learns the patterns produced from data training and the results from the learning process would then be used to predict data testing [4]. A few past researches have attempted to apply machine learning in diagnosing Schizophrenia or

Schizophrenia-related problems. There are several machine learning methods used in those studies, namely Fisher Linear Discriminant Analysis [5], Linear Discriminant Analysis and k-Nearest Neighbour [6], Random Forest, Support Vector Classification, Elastic Net, as well as Least Absolute Shrinkage and Selection Operator [7]. However, there are still not many studies on the application machine learning in Schizophrenia-related problems.

The Support Vector Machines (SVM) method is a binary classification method that aims to create a model with good generalization ability [8] with an optimum global solution [9]. SVM creates a hyperplane that separates two target values such that margin, the nearest distance between data and hyperplane, is maximized. In this study, SVM is chosen as it can generalize well and has good accuracy [8]. SVM itself has been used to solve classification problems in numerous studies, for example in intrusion detection system [10], policyholders satisfactory [11], insolvency prediction in insurance companies [12], and brain cancer classification [13].

2. Data

Northwestern University Schizophrenia Data [14] is used in this research. This research uses clinical data from the database, i.e. the group data, demographic data (gender, dominant hand, ethnic, race, and age), as well as Scale for the Assessment of Negative Symptoms (SANS) dan Scale for the Assessment of Negative Symptoms (SAPS) questionnaires data. The data consists of 392 observations and 65 variables. There are 171 data with the label "Schizophrenics" and 221 with "non-Schizophrenics".

3. The use of kernel function

The kernel function is defined as follows:

$$K(x_i, x_j') = \varphi(x_i) \cdot \varphi(x_j)$$

where $\varphi(x)$ is a function that maps $x \in R^n$ to feature space. Each time $\varphi(x_i) \cdot \varphi(x_j)$ appears in classification algorithm, it can be replaced by $K\left(x_i, x_j'\right)$ [15]. By using kernel function, it is hoped that the data can be separated linearly in a higher dimension.

In this study, three different kernels [16] are used:

- Linear kernel, $K(x_i, x_j') = x_i \cdot x_j$
- Polynomial kernel, $K\left(x_i, x_j'\right) = \left(x_i \cdot x_j + 1\right)^p$
- Gaussian kernel, $K\left(x_i, x_j'\right) = \exp\left(-\gamma\left\|x_i - x_j\right\|^2\right)$, $\gamma = 1$ [17].

4. Support Vector Machines (SVM)

Let the pattern classified be a set of m row vectors, $A_i\ (i = 1,2,\dots,m) \in R^n$, where $A_i = (A_{i1}, A_{i2}, \dots, A_{in})^T$. Let $y_i \in \{1, -1\}$ be the class of the i-th pattern. On the case with data that can be separated linearly, $w \in R^n$ and $b \in R$ are determined such that: $A_i w \geq 1 - b$ for $y_i = 1$ and $A_i w \leq -1 - b$ for $y_i = -1$ [18].

The plane defined by:

$$w^T x + b = 0 \tag{1}$$

is located between two planes defined by $w^T x + b = 1$ and $w^T x + b = -1$. The plane in equation (1) separates each class from another by a margin of $\frac{1}{\|w\|}$ on each side. The SVM model is obtained by maximizing the margin and is equivalent to the following problem [18]:

$(SVM\ 1)\ \min\limits_{w,b} \frac{1}{2} w^T w$

subject to:

$$A_i w \geq 1 - b \text{ for } y_i = 1 \text{ and } A_i w \leq -1 - b \text{ for } y_i = -1 \tag{2}$$

However, when the two classes cannot be fully separated linearly, there will be error in satisfying the inequality in equation (2) on several patterns. Hence, equation (2) can be modified to:

$$A_i w + q_i \geq 1 - b \text{ for } y_i = 1 \text{ and } A_i w - q_i \leq -1 - b \text{ for } y_i = -1$$
$$q_i \geq 0, i = 1,2,\dots,m \qquad (3)$$

where q_i denotes the error variable of the i-th data sample. This classification method is considered as one with soft margin as it classifies data with some errors. Soft margin depends on the value of q_i. Classification of the sample for testing is obtained by determining the sign of $w^T x + b$ [18].

Formulation of SVM with *soft margin* and linear kernel is given by:

$$(SVM\ 2)\ \min_{w,b,q} Ce^T q + \frac{1}{2} w^T w$$

subject to:

$$A_i w + q_i \geq 1 - b \text{ for } y_i = 1$$
$$A_i w - q_i \leq -1 - b \text{ for } y_i = -1$$
$$q_i \geq 0, i - 1,2,\dots,m$$

where e is a vector which all its entries are 1 and the scalar C is a trade-off parameter. A larger value of C means that the model would prioritize the minimization of q, while a smaller one would prioritize the minimization of w [18].

5. Parameter Optimization

In this study, several parameters are optimized by using the grid search method. The grid search method finds one by one the combination of parameters that produces the optimum model [19]. The parameters that are optimized are:

- $p = 2,3,4$ on the polynomial kernel, and
- C on the interval $[10^{-3}, 10^3]$ by logarithmic scale.

6. Model Performance Validation

To validate the model performance, Hold-Out Validation is done. In Hold-Out Validation, the data is randomly separated into two parts: training and testing data. The model evaluation will be used on the testing data. Computationally, this validation method is easy and fast [20].

Hold-Out Validation is done on different percentages of data in this study. To overcome the weakness of Hold-Out Validation, which is very dependent on the data used for training and testing [20], simulations are done 10 times with a different set of data for each percentage of data used.

7. Model Performance Evaluation

Model performance evaluation in this study is done by measuring accuracy, which is how often the classification model makes a correct prediction. Accuracy is the ratio of the number of correct predictions and the total number of predictions [20]. A classification model performance is said to be good if the accuracy of the model is high [21].

8. Simulation Results and Analysis

In this section, results and analysis of classification of Schizophrenia data with SVM will be covered. The software MATLAB R2017a was used to form the SVM with soft margin model.

8.1. Classification of Schizophrenia data using SVM with linear kernel

The result of Schizophrenia data classification using SVM with linear kernel can be seen below.

Table 1. Accuracy of Schizophrenia data classification using SVM with linear kernel.

	Accuracy of the i-th simulation (%)

Percentage of Training Data (%)	1	2	3	4	5	6	7	8	9	10
10	90.1	89.2	89.0	85.8	88.7	86.1	88.7	88.4	89.0	86.4
20	92.0	89.5	87.9	89.5	88.5	89.5	88.5	88.2	88.9	87.9
30	90.2	90.9	88.0	88.7	88.7	88.7	87.6	88.0	89.5	87.6
40	90.7	91.1	89.0	87.3	89.0	88.6	88.1	88.1	89.0	87.3
50	90.3	91.8	89.3	89.3	88.8	86.7	88.3	88.8	89.3	87.8
60	90.4	93.6	91.1	88.5	88.5	90.4	88.5	88.5	90.4	88.5
70	89.0	94.9	91.5	89.0	87.3	90.7	87.3	89.0	91.5	90.7
80	91.1	94.9	92.4	89.9	86.1	88.6	89.9	86.1	92.4	89.9
90	92.5	**95.0**	92.5	85.0	**82.5**	87.5	90.0	85.0	**95.0**	92.5

Table 2. Final accuracy of Schizophrenia data classification using SVM with linear kernel.

Percentage of Training Data (%)	Accuracy (%)
10	88.1 ± 1.47
20	89.0 ± 1.22
30	88.8 ± 1.09
40	88.8 ± 1.27
50	89.0 ± 1.39
60	89.9 ± 1.68
70	**90.1 ± 2.30**
80	90.1 ± 2.79
90	89.8 ± 4.48

It can be seen from Table 1 that the SVM model with linear kernel that produces the best accuracy for classification of Schizophrenia data is the one with 90% training data with an accuracy of 95.0%. On the other hand, the worst-performing model is also obtained from the one with 90% training data with an accuracy of 82.5%. This can happen as the Hold-Out Validation method is very dependent on the data used for training and testing.

Based on the final accuracy of Schizophrenia data classification using SVM with linear kernel (Table 2), in general, the more percentage of training data used in building the model, the more accurate is the final accuracy. The SVM model with linear kernel that gives the best final accuracy for the classification of Schizophrenia data is the one that uses 70% training data with a mean of 90.1%, whereas the worst one is the model that uses 10% training data with a mean of 88.1%.

8.2. Classification of Schizophrenia data using SVM with polynomial kernel

It can be seen from Table 3 that the SVM model with polynomial kernel that produces the best accuracy for classification of Schizophrenia data is the one with 90% training data with an accuracy of 92.5%. However, the model that has the worst performance is also obtained from the one with 90% training data with an accuracy of 68.0%.

Based on the final accuracy of Schizophrenia data classification using SVM with polynomial kernel (Table 4), in general, the more percentage of training data used in building the model, the more accurate is the final accuracy. The SVM model with polynomial kernel that gives the best final accuracy for the classification of Schizophrenia data is the one that uses 90% training data with a mean of 87.0%, whereas the model that gives the worst performance uses 30% training data with a mean of 79.1%.

Table 3. Accuracy of Schizophrenia data classification using SVM polynomial kernel

	Accuracy of the i-th simulation (%)									
Percentage of Training Data (%)	1	2	3	4	5	6	7	8	9	10
10	90.1	**68.0**	76.2	83.0	82.2	85.3	84.7	76.8	69.1	79.3
20	77.7	75.8	77.4	75.2	84.7	78.7	82.2	82.5	76.1	81.2
30	70.9	79.3	77.8	73.5	82.9	80.0	81.5	80.7	81.8	82.2
40	75.0	78.4	79.2	74.6	86.4	80.1	84.3	81.8	79.7	79.2
50	75.0	79.6	78.6	81.6	83.2	83.7	83.7	75.0	86.2	80.1
60	85.4	89.2	78.3	79.0	82.8	84.1	82.2	77.1	82.8	84.1
70	87.3	80.5	91.5	84.7	82.2	87.3	83.1	78.0	91.5	90.7
80	91.1	82.3	87.3	83.5	82.3	87.3	82.3	86.1	87.3	78.5
90	**92.5**	87.5	**92.5**	77.5	80.0	87.5	90.0	82.5	90.0	90.0

Table 4. Final accuracy of Schizophrenia data classification using SVM with polynomial kernel

Percentage of Training Data (%)	Accuracy (%)
10	79.5 ± 7.07
20	79.1 ± 3.29
30	79.1 ± 3.96
40	79.9 ± 3.67
50	80.7 ± 3.75
60	82.5 ± 3.61
70	85.7 ± 4.77
80	84.8 ± 3.68
90	**87.0 ± 5.24**

8.3. Classification of Schizophrenia data using SVM with Gaussian kernel

It can be seen from Table 5 that the SVM model with gaussian kernel that produces the best accuracy for classification of Schizophrenia data is the one with 90% training data with an accuracy of 95.0%. On the other hand, the worst-performing model is also obtained from the one with 90% training data with an accuracy of 82.5%. Similar to the case with linear kernel, this can happen because the Hold-Out Validation method relies heavily on the data used for training and testing.

Table 5. Accuracy of Schizophrenia data classification using SVM with Gaussian kernel

	Accuracy of the i-th simulation (%)									
Percentage of Training Data (%)	1	2	3	4	5	6	7	8	9	10
10	88.1	89.2	87.5	88.7	88.7	88.7	88.7	88.4	88.4	88.4
20	89.2	89.5	86.9	89.5	88.5	89.5	88.5	88.2	87.9	87.9
30	90.5	88.7	88.0	88.7	88.7	88.7	87.6	88.0	89.5	87.6
40	90.7	91.1	87.7	88.6	89.0	88.6	88.1	88.1	89.0	87.3
50	90.3	91.8	89.3	89.3	88.8	86.7	88.3	88.8	89.3	87.8
60	90.4	93.6	91.1	88.5	88.5	90.4	88.5	88.5	90.4	88.5
70	89.0	94.9	91.5	89.0	87.3	90.7	87.3	89.0	91.5	90.7
80	91.1	94.9	92.4	89.9	86.1	88.6	89.9	86.1	92.4	89.9
90	92.5	**95.0**	92.5	85.0	**82.5**	87.5	90.0	85.0	**95.0**	92.5

Based on the final accuracy of Schizophrenia data classification using SVM with Gaussian kernel (Table 6), in general, the more percentage of training data used in building the model, the better is the final accuracy. The SVM model with Gaussian kernel that performs for the classification of Schizophrenia data is the one that uses 70% training data with a mean of 90.1%, whereas the worst one is the model that uses 10% training data with a mean of 88.5%.

Table 6. Final accuracy of Schizophrenia data classification using SVM with Gaussian kernel

Percentage of Training Data (%)	Accuracy (%)
10	88.5 ± 0.444
20	88.6 ± 0.855
30	88.6 ± 0.892
40	88.8 ± 1.22
50	89.0 ± 1.39
60	89.9 ± 1.68
70	**90.1 ± 2.30**
80	90.1 ± 2.79
90	89.8 ± 4.48

9. Conclusion

Machine learning can solve the classification problem of Schizophrenia data using the Northwestern University Schizophrenia Data. There are three types of simulations done in this study: SVM with linear, polynomial, and Gaussian kernel. In each simulation, the parameters of the models are optimized with grid search method, while the validation and evaluation of model performance are done by Hold-Out Validation and calculating the accuracy.

The SVM method with linear and Gaussian kernel has a similar performance, but better than the one that uses polynomial kernel. With linear and Gaussian kernel, SVM managed to classify Schizophrenia data with a final accuracy of 90.1%. Additionally, in classifying Schizophrenia data, SVM with linear kernel and Gaussian kernel successfully reached an accuracy of 95.0% in at least one simulation.

For future researches, it is recommended to use other dataset, kernels, or evaluation and validation methods. By doing this research, it is hoped that the results could help the medical field in predicting whether someone suffers from Schizophrenia or not, so that they can receive the right treatment for their illness.

10. References

[1] American Psychiatric Association 2013 *Diagnostic and statistical manual of mental disorders, fifth edition* (Arlington, VA: American Psychiatric Association)

[2] World Health Organization 2018(online) *Schizophrenia Key Facts* http://www.who.int/en/news-room/fact-sheets/detail/schizophrenia

[3] Schizophrenia.com 2010(online) *Schizophrenia symptoms and diagnosis of Schizophrenia* http://schizophrenia.com/szfacts.htm

[4] Witten I H and Frank E 2005 *Data mining: practical machine learning tools and techniques, second* edition (Burlington: Elsevier Inc)

[5] Neuhaus A H, Popescu F C, Bates J A, Goldberg, T E, and Malhotra, A K 2013 Single-subject classification of schizophrenia using event-related potentials obtained during auditory and visual oddball paradigms *European Archives of Psychiatry and Clinical Neuroscience* **263**(241)

[6] Ahn M, Hong J H, and Jun S C 2012 Feasibility of approaches combining sensor and source features in brain-computer interface *Journal of neuroscience methods* **204**(1) 168–78

[7] Hettige N C, Nguyen T B, Yuan C, Rajakulendran T, Baddour J, Bhagwat N, Bani-Fatemi A, Voineskos A N, Mallar Chakravarty M, De Luca V 2017 Classification of suicide attempters in schizophrenia using sociocultural and clinical features: A machine learning approach *Gen Hosp Psychiatry* **47** 20-8

[8] Cortes C and Vapnik V N 1995 Support-vector networks *Machine Learning* **20**(3) 273–97

[9] Burges C J C and Crisp D J 1999 Uniqueness of the SVM solution *Neural Information Processing Systems* **12** 223-9
[10] Rustam Z and Zahras D 2018 Comparison between Support Vector Machine and Fuzzy C-Means as classifier for intrusion detection system *Journal of Physics: Conference Series* **1028**(1)
[11] Rustam Z and Ariantari N P A A 2018 Support Vector Machines for classifying policyholders satistfactorily in automobile insurance *Journal of Physics: Conference Series* **1028**(1) 012005
[12] Rustam Z and Yaurita F 2018 Insolvency prediction in insurance companies using Support Vector Machines and Fuzzy Kernel C-Means *Journal of Physics: Conference Series* **1028**(1), 012118
[13] Panca V and Rustam Z 2017 Application of machine learning on brain cancer multiclass classification *AIP Conference Proceedings* **1862**(1) 030133
[14] Wang L, Kogan A, Cobia D, Alpert K, Kolasny A, Miller M I, and Marcus, D 2013 Northwestern University Schizophrenia Data and Software Tool (NUSDAST) *Frontiers in Neuroinformatics* **7** 25
[15] Witten I H, Frank E, and Hall M A 2011 *Data mining: practical machine learning tools and techniques, third edition* (Burlington: Elsevier Inc)
[16] Han J and Kamber M 2006 *Data mining concepts and techniques, second edition* (San Francisco: Elsevier Inc)
[17] Xue H, Yang Q, and Chen S 2009 *The top ten algorithms in data mining* (New York: Chapman and Hall/CRC)
[18] Jayadeva, Khemchandani R, and Chandra S 2007 Twin Support Vector Machines for pattern classification *IEEE* Transactions *on Pattern Analysis and Machine Intelligence* **29**(5) 905–10
[19] Lin C-J, Hsu C-W, and Chang C-C 2003 *A Practical Guide to Support Vector Classification*
[20] Zheng A 2015 *Evaluating machine learning models: a beginner's guide to key concepts and pitfalls* (Sebastopol, CA: O'Reilly Media, Inc)
[21] Tan P N, Steinbach M, and Kumar V 2006 *Introduction to Data Mining* (Boston: Pearson Education)

Acknowledgments
The authors would like to express appreciation for the support of data collection and sharing for this project, which were funded by NIMH grant 1R01 MH084803.

The Effect of rGO Mass Composition on The Performance of Activated Carbon/rGO Supercapacitor Electrode Based on Coconut Shell (*Cocos nucifera*)

S H Intifadhah[1], L Rohmawati[1,3], W Setyarsih[1] and Tukiran[2]

[1]Departement of Physics, Universitas Negeri Surabaya, Surabaya
[2]Departement of Chemistry, Universitas Negeri Surabaya, Surabaya
[3]Corresponding author:lydiarohmawati@unesa.ac.id

Abstract. Activated carbon (AC) is composed of reduced graphene oxide (rGO) and has been synthesized from natural material that is coconut shell (*Cocos nucifera*) by dry mixing method. The purpose of this research was to know the effect of adding rGO mass on the electrode to the specific capacitance value. The rGO mass variation used was 10-25 wt%. AC/rGO electrodes were tested by cyclic voltammetry to determine their specific capacitance values at a scan rate of 100 mV/s with Sodium Sulphate (Na_2SO_4) 1 M in the electrolyte solution and a potential range at 0-1 Volt. The electrode by 90 wt% AC composition and 10 wt% rGO had a higher specific capacitance value of 6.26 Fg^{-1} than other electrode mass compositions, as well as the activated carbon individual electrodes having a specific capacitance value of 3.38 Fg^{-1}. Therefore, it could be concluded that the specific capacitance of the electrodes could be increased by the addition of rGO material.

1. Introduction

Supercapacitor is an energy storage system which is capable to provide high power storage in a short time [5,9,19] and has a cell capacitance range of 0.043 F - 2,700 F [13]. Generally, a supercapacitor consists of a pair of electrodes, a separator and an electrolyte solution [1]. Determining the electrical properties of the supercapacitor requires selection of the right material as a supercapacitor electrode. Electrodes are electrical conductors that have the ability to absorb and release electrons from their surfaces. The storage process occurring on the electrode surface will affect the specific capacitance value [15]. Materials that have already begun to be used as supercapacitor electrode materials include activated carbon, *graphene* and reduced Graphene Oxide [3,7,19]. Reduced Graphene Oxide (rGO) is a *graphene* oxide material in which the carbon atoms of *graphene* undergo oxidation and reduction processes. In the process of oxidation, there are several oxygen and hydrogen atoms bonded to carbon atoms called graphene oxide. Meanwhile, in the process of reduction, the release of several bonds of graphene oxide occurs so that it obtaines structure that almost resembles *graphene*. The result of this reduction process is called reduced graphene oxide (rGO). The difference between *graphene*, *graphene oxide* and rGO is the content of the O. *Graphene* atomic group is only composed of carbon atoms while *graphene oxide* and rGO have an atomic O group. The O group on *graphene oxide* is more (41-50%) than rGO (13-22 %) [14]. The synthesis of *graphene* and activated carbon composites as supercapacitor electrodes is performed by ultrasonification, hydrothermal carbonation and chemical activation resulted in a specific capacitance of 210 Fg^{-1}. The capacitance of the *graphene* composite and the activated carbon is greater than the capacitance of the pure activated carbon electrode of 76 Fg^{-1}. It suggests that with the addition of graphene can improve the process of storing the charge of the electrode [19]. In subsequent studies, the activated carbon/reduced Graphene Oxide (rGO) composites were used as electrodes through hydrazine method [7]. The synthesis of the composite electrode was performed by using different variations of rGO mass composition i.e. 10 - 50 wt% rGO which, in the mass of 20 wt% rGO, obtained the best capacitance value of 181 Fg^{-1} compared to the mass of the other electrodes. The activated carbon/ rGO composite electrode also has a higher capacitance value than the pure activated carbon electrode [7]. A research [6] concerning the manufacture of rGO from coconut shell was then compiled with ZnO as a supercapacitor electrode, in which the composite mass variation was rGO-ZnO (1: 1, 1: 2 and 2: 1). The result of the research obtained the best capacitance value on rGO-ZnO composite with the ratio of 1: 2 composition of 0.613906 Fg^{-1}, where the process of voltametric characterization at scan rate 100 mVs^{-1} by using electrolyte solution KOH 1 M [6]. Based on the aforementioned researches, the material used was relatively expensive [7,19], so this research proposed an innovation by using simpler and easier method in making activated carbon /rGO (reduced Graphene Oxide) composite without the need of expensive solvent. The method of making electrode in this research used dry mixing method. Dry mixing method is a method of mixing the material in the form of a powder with a certain length of time without using solvent and it is done in the outside air and used when the material does not easily experience oxidation in the outside air [8].

2. Materials and Methods

2.1. Materials

The material used was from the natural coconut shell (*Cocos nucifera*). All chemicals used were pure analys and used without purification, including Sodium Oxide (NaOH 0.5M) as chemical activator, Chloride Acid (HCl 1M), Sodium Sulphate (Na_2SO_4 1M) as electrolyte solution, Poly Ethylene Glycol (PEG) 4000 as a binder and deionize water.

2.2. Synthesis of Activated Carbon

The activated carbon is synthesized from coconut shell using dehydration method, carbonation and chemical activation. The coconut shell was dehydrated in ± 7 days then carbonized at 400 ^{0}C and smoothed. Afterwards sieved on 200 mesh sieve then activated using 0,5 M NaOH activator and soaked for 24 hours with carbon mass ratio: NaOH of 1: 3. Subsequently, the carbon bath was calcined at 800 ^{0}C for 5 hours, after which it was washed with 0,5 M HCl and deionized water to a pH of 7 [12]. The activated carbon powder is then dried at 110 ^{0}C for 1 hour to remove the water content in the powder.

2.3. Synthesis of reduced Graphene Oxide (rGO)

Reduced Graphene Oxide (rGO) is synthesized from coconut shell with dehydration, carbonation and calcination. The coconut shell was dehydrated for ± 7 days then carbonized at 400 ^{0}C. After it was calcined with a temperature of 400 ^{0}C for 5 hours then it crushed into powder. Then, the rGO powder was washed with deionize water until the powdered sludge was obtained [10]. To remove the water content, it was then dried with a temperature of 110^{0}C for 1 hour.

2.4. Synthesis of Supercapacitor Electrode

The supercapacitor electrode was made by dry mixing method on two main ingredients, activated carbon and reduced Graphene Oxide (rGO). The activated carbon/rGO composite electrode was composed of several mass variations i.e. 10 wt% to 25 wt% rGO. Each of the activated carbon powder and rGO was mixed simultaneously with Poly Ethylene Glicol (PEG) 4000 as a powder binder at a ratio of 1: 1 [18]. The mixture was crushed by using alu mortar for 1 hour to obtain a homogeneous powder mixture. Furthermore, the electrode powders were compressed using a pressure of 1.5×10^6 Pa to form a pellet-like electrode. Compaction method is a process of powder compaction into a sample with a certain shape according to the mold [8].

3. Instruments

The activated carbon powder and reduced Graphene Oxide (rGO) were characterized using X-Ray Diffraction with an angle of $2\theta = 10^0$-50^0 of 0,5-1,0 grams to understand the phase and material structure. XRD test result was presented in graphic characteristic of XRD. The activated carbon /rGO electrode was characterized by Cyclic Voltemetry to know the electrochemical characteristics of the specific capacitance from the electrode. Cyclic Voltemetry (CV) was tested on all variations of composition of activated carbon/rGO mass with electrolyte Sodium Sulphate (Na_2SO_4 1M) solution, scan rate 100 mVs^{-1} and potential range at 0 - 1 Volt resulting voltamogram curve. The analysis of the electrode conductivity properties used Electrochemical Impedance Spectroscopy (EIS) method in the frequency range 100 kHz - 100 mHz with the result of Nyquist plot.

4. Result and Discussions

The results of the characterization of X-Ray Diffraction on the activated carbon and reduced Graphene Oxide (rGO) synthesized from the coconut shell (*Cocos nucifera*) is shown in Figure 1. The test used powder in which each sample was as much as 0,5 - 1,0 gram by using angle (2θ) in the range 10^0 - 50^0.

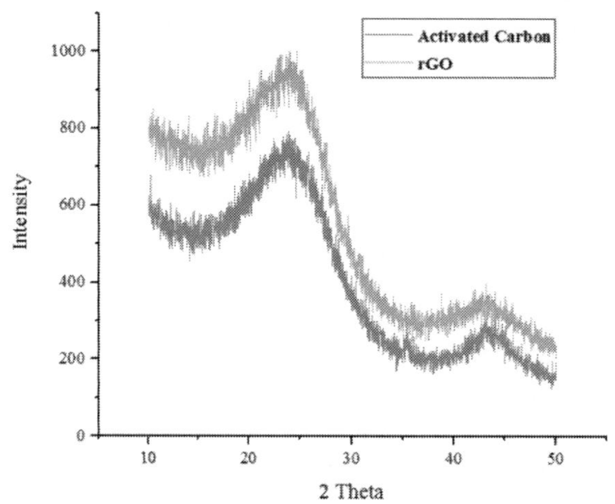

Figure 1. XRD patterns of activated carbon and reduced Graphene Oxide (rGO).

The XRD patterns in Figure 1 is the characteristic of activated carbon and rGO. The two wide diffraction peaks were at $2\theta = 20$ - 30^0 and $2\theta = 40$ - 50^0 indicating that the phase was amorphous, composed by carbon bonds irregularly so it could be used as adsorbent material [17]. The activated carbon from the coconut shell in this study showed that the diffraction peaks were at $2\theta = 35^0$, 43^0 and 45^0. While the pattern on the reduced Graphene Oxide (rGO) shows the diffraction peak at $2\theta = 24^0$ and 43^0 where there was the same peak characterization at $2\theta = 24^0$ and 44^0 [4]. The rGO phase formed was semi crystalline which was proved to have a higher intensity compared to activated carbon. The XRD intensity value was influenced by the crystallinity rate of a material, if the material was increasingly crystalline, the intensity got higher [17]. The results of the characterization of both samples showed the existence of a stacked peak and wide, whereas the difference was at the diffraction peak ie the 2θ angle formed and the intensity level.

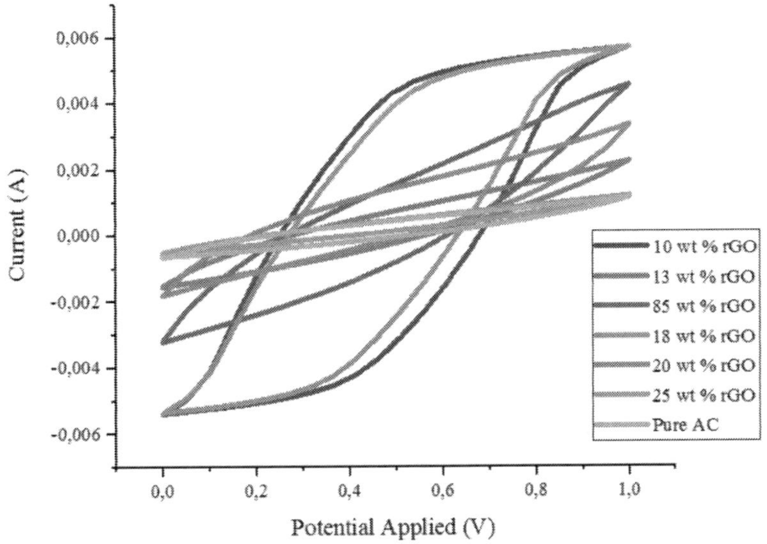

Figure 2. CV curve of pure AC electrode and composite electrode of activated carbon/rGO at different mass composition using scan rate 100 mVs^{-1} with potential range 0-1 V in 1M Na_2SO_4 electrolyte solution.

Cyclic voltametry was the most commonly used characterization test for determining the capacitance performance of the electrodes and the quantitative value of the electrode specific capacitances [5,19]. The CV curve of the activated carbon/rGO electrode was shown in Figure 2 at all variations of the mass composition using a scan rate of 100 mVs^{-1} and a potential range of 0 - 1 V in a 1M Na_2SO_4 electrolyte solution. The CV curve obtained had a different shape and area due to the capacitance value held on each of the different electrodes. The specific capacitance values of electrodes 10 wt%, 13 wt%, 15 wt%, 18 wt%, 20 wt% and 25 wt% rGO were obtained in the order of 6.26 Fg^{-1}, 5.27 Fg^{-1}, 3.87 Fg^{-1}, 4.24 Fg^{-1}, 3.71 Fg^{-1} and 3.88 Fg^{-1}. Meanwhile, the pure electrode capacitance value of pure activated carbon was smaller than all composite electrodes with 3.37 Fg^{-1}. The electrode capacitance value could be increased by the addition of rGO material [7], then the rGO material from the coconut shell could be used as composite material on activated carbon. The network structure was made by the incorporation of rGO material into the activated carbon electrode. It bridged the pores which were connected with the activated carbon particles conductively. Thus, the rapid electrolyte ion transport process was facilitated in the electrode [7]. Furthermore, it was found that the specific capacitance value decrease was caused by the presence of increasingly rGO composition. The decreasing capacitance value was caused by the composition of excessive rGO material. Most of the surface of the activated carbon was covered by rGO, which the ion diffusion process would be made in the pores of the electrode obstructed and the obtained low capacitance value would be affected [7]. If the pores of the electrodes were closed to the addition of too much composite material, then the electron absorbtion process would be inhibited. The magnitude of the capacitance value of the electrode was affected by that [3]. Therefore, the best electrode mass composition with the highest capacitance value was 90 wt% activated carbon composition and 10 wt% rGO.

The activated carbon/rGOelectrode 90-10 with the best capacitance value was tested by EIS or Electrochemical Impedance Spectroscophy in order to determine the conductive behavior of the electrode meterial [19].

IOP Conf. Series: Journal of Physics: Conf. Series **1108** (2018) 012045 doi:10.1088/1742-6596/1108/1/012045

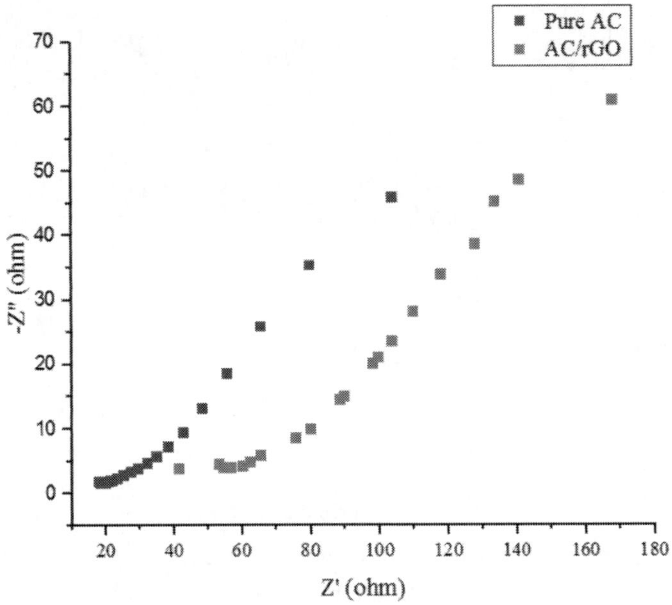

Figure 3. Nyquist plot of pure activated carbon and activated carbon / rGO electrode with a frequency range of 100 kHz - 100 mHz.

The Nyquist plot of the activated carbon electrode and the activated carbon/rGO electrode was shown in Figure 3. Figure 3 is an Electrochemical Impedance Spestroshopy (EIS) analysis with a frequency range of 100 kHz - 100 mHz. Nyquist plot consisted of component Z ' is a real impedance and Z " is an imaginary impedance. The Nyquist EIS plot in the drawing also consisted of 3 parts, the semicircle, the almost vertical line and the vertical line. Therefore, through the plot Nyquist, the obtained electronic resistance value (R_e), which was the lowest point and the charge transfer resistance (R_{ct}), were the difference of the highest value with the lowest value [2]. In Fig. 3 (a) the Nyquist plot was obtained from the activated carbon/ rGO composite electrode with R_{ct} of 13.0193 Ω and R_{tot} of 54.3093 Ω. Meanwhile, the pure activated carbon electrode obtained R_{ct} value of 2 Ω and R_{tot} of 18.9614 Ω. R_{tot} which were the total of R_e and R_{ct} [2]. In addition, Figure 3 shows that the diameter of the semicircle plot on the composite electrode was greater than that of the pure activated carbon electrode. When the diameter of the semicircle increased, the load transfer resistance (R_{ct}) [11] became higher. Therefore, the value of the quantitative charge transfer resistance in the activated carbon/rGO electrode was higher than the one of the pure activated carbon electrodes.

5. Conclusion

The best specific capacitance value was obtained at composite electrode with 90 wt% activated carbon composition and 10 wt% rGO ie 6.26 Fg^{-1}, while the specific value of pure active carbon electrolactivity was smaller at 3.38 Fg^{-1}. Therefore, the specific capacitance value on the activated carbon electrode could be increased with the addition of rGO mass.

6. References

[1] Aslani M 2012 Electrochemical Double Layer Capacitors (Supercapacitors) (Stanford: Stanford University)

[2] Dhika M A 2012 Pembuatan Material Anoda Lithium Titanate ($Li_4Ti_5O_{12}$) dan Studi Pengaruh Ketebalan Elektroda terhadap Performa Elektrokimia Baterai Ion Lithium. *Report of Universitas Sumatera Utara.*

[3] Drobny D, Tychyna S, Maletin Y A, Stryzhakova N and Zelinsky S 2013 Methods for Manufacturing Carbon Electrodes for Supercapacitors. *Pros and Cons. Proceedings Of The International Conference Nanomaterials: Applications And Properties* 1-2. Sumy State University.

[4] Fu C, Zhao G, Zhang H and Li S 2013 Evaluation and Characterization of Reduced Graphene Oxide Nanosheets as Anode Materials for Lithium-Ion Batteries. *Int. J. of Electrochemical Science* **8** pp 6269 - 6280.

[5] Huang K J, Zhang J Z, Liu Y and Liu Y M 2015 Synthesis of Reduced Graphene Oxide Wrapped-Copper Sulfide Hollow Spheres as Electrode Material for Supercapacitor *Int. J. of Hydrogen Energy* 40(32) pp 1-10.

[6] Kurniawan A F 2016 *Sintesis Komposit Grafena Oksida Tereduksi (rGO) Hasil Pembakaran Tempurung Kelapa Tua dengan Seng Oksida (ZnO) sebagai Superkapasitor* (Institut Teknologi Sepuluh Nopember : Surabaya)

[7] Li H, Pan L, Nie C, Liu Y and Sun Z 2012 Reduced Graphene Oxide and Activated Carbon Composites for Capacitive Deionization. *J. of Material Chemistry* 31 15556–15561.

[8] Nayiroh N 2018 (Online) Metalurgi Serbuk http://nurun.lecturer.uin-malang.ac.id/wpcontent/uploads/sites/7/2013/03/METALURGI-SERBUK.pdf.

[9] Peng X Y, Liu X X, Diamond D and Lau K T 2011 Synthesis of Electrochemically-Reduced Graphene Oxide Film With Controllable Size and Thickness andits Use in Supercapacitor *J. Carbon.* 49(11) pp 3488-3496

[10] Putra G B 2017 *Sintesis Elektroda Superkapasitor dengan Proses Eksfoliasi Kimia dan Pengaitan* Fe^{3+} *pada Grafena Oksida Tereduksi dari Tempurung Kelapa Tua* (Institut Teknologi Sepuluh Nopember: Surabaya)

[11] Ribeiro D, Souza C and Abrantes J 2015 Use of Electrochemical Impedance Spectroscopy (EIS) to Monitoring The Corrosion of Reinforced Concrete *IbraconSturctures and Material Journal.* **8** pp 529-546

[12] Rohmawati L and Setyarsih W 2017 Potensi Bahan Alam sebagai Elektroda Superkapasitor *Jurnal Fisika Prosiding Semnas PPM.* 325-329

[13] Sahay K and Dwivedi B 2009 Supercapacitors Energy Storage System for Power Quality Improvement: An Overview *J. of Electrical Systems* **5**(4) pp 1-8

[14] Sebastian S 2015 (online) Graphenea https://www.azonano.com/*Graphenea online acses.*

[15] Sharma P and Bhatti T 2010 A Review on Electrochemical Double-Layer Capacitors. J. *Energy Conversion and Management* **51**(12) pp 2901-2912

[16] Suwandana R F and Susanti D 2015 Analisis Pengaruh Massa reduktor Zinc terhadap Sifat Kapasitif Superkapasitor Material Graphene. *Jurnal Teknik ITS.* **4** pp 95-100.

[17] Tang Y-b, Liu Q and Chen F-y 2012 Preparation and Characterization of Activated Carbon from Waste Ramulus Mori, *Chemical Engineering Journal* **203** pp 19-24.

[18] Wati G A 2015 *Kapasitansi Elektroda Superkapasitor dari Tempurung Kelapa* (Universitas Negeri Surabaya : Surabaya)

[19] Zheng C, Zhou X, Cao H, Wang G dan Liu Z 2014 Synthesis of Porous Graphene/Activated Carbon Composite with High Packing Density and Large Specific Surface Area for Supercapacitor Electrode Material. *Journal of Power Sources* **258** 290-296.

Acknowledgement
Thanks to the Direktorat Riset Pengabdian Masyarakat Direktorat Jendral Penguatan Riset dan Pengembangan Kementrian Riset, Teknologi, dan Pendidikan Tinggi in the fiscal year 2017 trough UNESA Rector's decision number 522/UN38/HK/LT/2017 and fiscal year 2018 number 252/UN38/HK/LT/2018.This research has been supported by Department of Physics Universitas Negeri Surabaya. The research process was conducted at the Materials Physics Laboratory of Universitas Negeri Surabaya, Laboratory of Materials and Metallurgy and Chemical Engineering Laboratory of

Institut Teknologi Sepuluh Nopember. This research is also supported by lecturers and students of Universitas Negeri Surabaya with motivation, suggestion and assistance.

Isolation and characterization of silicate-solubilizing bacteria from paddy rhizosphere (*Oryza sativa* L.)

A Sulizah*[1,1], Y S Rahayu[1,2] and S K Dewi[1,3]

[1] Department of Biology, Universitas Negeri Surabaya, Indonesia.
[1]anissulizah@mhs.unesa.ac.id, [2]yunirahayu@unesa.ac.id, [3]saridewi@unesa.ac.id

Abstract. Silicate is known as a beneficial element that has important role for growth and development of paddy. Paddy is the greatest accumulator of Si up to 10% of its dry weight, but this plant can't absorb that element by itself. The uses of paddy rhizosphere bacteria as a silicate solubilizing agent should be studied furthermore. This study aims to isolate and characterize silicate bacteria from paddy rhizosphere that solubilize silicate optimally in *Bunt and Rovira* medium. The isolation, qualitative, quantitative test and characterization of silicate bacteria up to 39 characters, which is include of colony morphological, cell morphological and physiological-biochemical test. This study was conducted on February-June 2018 in Microbiology Laboratory and Plant Physiology Laboratory, Faculty of Mathematics and Natural Sciences, State University of Surabaya. Five isolates silicate solubilizing bacteria were found OS4, OS5, OS7, OS12 and OS13. The highest Solubilizing Index was gained by OS7 on 1,10, while the highest silicate concentration was solubilized by OS12 on 1,053 ppm in *Bunt and Rovira* broth.

1. Introduction

Optimalization for the growth of paddy plant as a stable Indonesian food could be through by nutritional selection [1]. Three groups of element divided by its function were essential, beneficial and toxic element. Essential element has its function to enhance growth from all type of stress in plant so that has main role in growth and development phase. Toxic element capable to inhibit growth and development of plant, while beneficial element influence growth and development stage in some plants, mainly found in monocots [2]. Silica as beneficial element has function for its cell strength in growth and development of paddy, which is accumulated 10% of its dry weight [3]. Epidermis layer strength by silica gel, make paddy is more resistant from any phytopathogenic, such as bacteria and fungi [4]. Desilication of silica-dioxide was mayor problem in soil, influences the availability of silica in paddy plant is lower than the occurrence of silicate in nature 220-240 billion/tons [1]. Silica-dioxide chelation found higher around roots, which included of main inhibition factor the uptake of nutrient by paddy through lateral root [22]. Rhizosphere bacteria play a vital role in the functioning of plants by influencing their physiology and development in rhizosphere [5]. Symbiosis of rhizosphere between rhizosphere bacteria influences avaibility nutrient uptake by organic acid productions [20]. Although the one of role rhizosphere bacteria as a silica-weathering agent had been recognized, the isolation and characterization of silicate-solubilizing bacteria in paddy rhizosphere should be studied furthermore.

2. Method

2.1. Isolation of Paddy Rhizosphere Bacteria
Sample of rhizosphere was uprooted from six rice fields, Bangkingan, Jeruk, Kapasan, Sawo, Sumber Makmur and Siwalan Makmur. Ten grams of rhizosphere were diluted by 0,85% NaCl, shaken two minutes by vortex. The serial dilutions up to 10^{-7} which the composition was 1 ml of sample into 9 ml of 0,85% NaCl, each dilution continuously shaken two minutes by vortex. The diluted inoculum is then added to a molten *Bunt and Rovira* Agar in a Petri dish, mixed and allowed to solidify. The isolates paddy rhizosphere bacteria incubated at 30^0C for 4 days in *Bunt and Rovira* Agar [6,7].

2.2. Characterization of Silicate-Solubilizing Bacteria
Colony characteristics such as shape, elevation, margin, texture, pigmentation and optic were examined. Cell morphology characteristics such as cell shape, cell arrangement, Gram staining and spore formation. Physiological and biochemical test of silicate solubilizing bacteria include of catalase test, motility, indole production, H_2S production, ornithine test, acid fast staining, *Methyl Red* (MR), *Voges Proskauer* (VP), urease test, nitrate reduction, citrate utilization, fermentation of glucose, fructose, sucrose, lactose, starch, mannitol, growth at pH 3, 7, 10 and temperature 25^0C, 30^0C, 45^0C [8,9,10,11,12,13].

2.3. Silica Solubilizing Activity in Bunt and Rovira Agar
The selected bacterial isolates from paddy rhizosphere were subjected to a silica solubilizing test in *Bunt and Rovira* Agar, incubated at 30^0C for 4 days. Silica source (quartz) were air dried and passed through 325 mesh-sieve. The medium contains of (g/750 mL): 20 g glucose; 20 g agar; 1 g peptone; 1 g yeast extract; 0,5 g $(NH_4)_2SO_4$; 0,4 g K_2HPO_4; 0,1 g $MgCl_2$; 0,01 g $FeCl_3$; 250 ml soil extract; 750 ml aquades; pH 6,6-7,0 [11]. Isolates that produces a clear zone was recognized has its capability to solubilize 0,25% quartz in *Bunt and Rovira* Agar. Clear zone from each isolates was measured by Solubilizing Index [14].

2.4. *Silica Solubilizing Activity in Bunt and Rovira Broth*
The solubility of silica was investigated in 100 ml *Bunt and Rovira* broth, which each of isolates supplemented by 0,25% quartz. Source of silica (0,25% quartz) were add separately into *Bunt and Rovira* broth. One ml of the bacteria cell 10^8 cfu/ml was inoculated into *Bunt and Rovira* broth for 7 days. After incubation periods, the culture was centrifuged at 10.000 rpm for 15 minutes to remove supernatant from debris. One ml of supernatant was added reagents into urine container and analyzed by silico-molybdate's method [11,15].

2.5. *pH Measurement in Bunt and Rovira Broth*
pH was measured before and after incubation in *Bunt and Rovira* broth. Before its sterilization, the pH *Bunt and Rovira* broth was measured by pH roll (6,6-7,0). After seven days incubation, each pH culture measured, then compared against pH before incubation periods [11].

3. Results and discussion
Sixteen isolates were isolated from paddy rhizosphere, then five of sixteen isolates have a capability to solubilize 0,25% quartz in *Bunt and Rovira* Agar. Characterizations of colony morphology silica solubilizing bacteria included of shape, margin, elevation, texture, pigmentation and optic were explained on Table 1.

Table 1. Colony morphology of silicate solubilizing bacteria.

Isolat	Shape	Margin	Elevation	Texture	Pigmentation	Optic
OS4	Round	Entire	Raised	Smooth and glossy	Creamy	Opaque
OS5	Irregular and spreading	Irregular	Raised	Smooth	Milky white	Opaque
OS7	Irregular and	Undulate	Flat	Smooth and glossy	Creamy	Opaque

	spreading					
OS12	Irregular and spreading	Irregular	Flat	Smooth	Milky white	Opaque
OS13	Irregular and spreading	Undulate	Flat	Smooth and glossy	Milky white	Opaque

Five isolates were characterized in terms of cell morphology (cell shape and arrangement), physiological-biochemical test (catalase test, motility, indole production, H_2S production, ornithine test, acid fast staining, MR, VP, urease test, nitrate reduction, citrate utilization, fermentation of glucose, fructose, sucrose, lactose, starch, mannitol, growth at pH 3, 7, 10 and temperature 25^{0}C, 30^{0}C, 45^{0}C).

Table 2. Characterization of silicate solubilizing bacteria.

Characteristic	OS4	OS5	OS7	OS12	OS13
Cell shape	Bacil	Bacil	Bacil	Bacil	Bacil
Cell arrangement	Streptobacil	Streptobacil	Streptobacil	Streptobacil	Streptobacil
Gram	-ve[a]	-ve	-ve	-ve	-ve
Spore	-	-	+	+	+
Catalase	+[b]	+	+	+	+
Motility	+	+	+	+	+
Indole production	+	+	+	-	+
H_2S	-	-	-	+	-
Ornithine	+	+	-	+	+
Acid Fast	+	+	+	+	+
MR	-	-	-	+	+
VP	-	-	-	+	-
Urease	-[c]	-	-	-	-
Nitrate	-	+	+	+	+
Citrate utilization	+	+	-	+	+
Glucose	++[d]	++	++	++	++
Fructose	++	++	++	++	++
Sucrose	++	++	++	++	++
Lactose	-+[e]	--	--	--	-+
Starch	++	++	++	++	++
Mannitol	--[f]	++	--	++	--
pH 3	+	-	+	+	+
pH 7	+	+	+	+	+
pH 10	-	+	+	+	+

25^0C	+	+	+	+	+
30^0C	+	+	+	+	+
45^0C	+	+	+	+	+

[a] Gram negative.
[b] positive test.
[c] negative test.
[d] represents acid and gas formulation.
[e] represents no acid, gas formulation.
[f] represents no acid and gas formulation.

The characterization was included of colony, cell morphological, physiological and biochemical test belong to silicate bacteria characteristic. Colony morphological showed that irregular and spreading, round shape; undulate, entire, regular margin; flat, raised elevation; smooth, smooth and glossy texture; creamy, milky white pigmentation; opaque optic. Cell morphology of five isolates was found bacil, streptobacil, Gram negative, spore or no spore formation. Physiological and biochemical test in silicate solubilizing bacteria include of catalase positive test, motile, did not produce urease enzyme, highly ability to ferment glucose, fructose, sucrose, starch, capable to growth at temperature 25^0C, 30^0C, 45^0C, optimal growth at neutral pH 7, positive or negative test in H_2S production, ornithine, MR, VP, nitrate reduction, citrate utilization, fermentation of lactose, mannitol, growth at pH 3 and 10 [8,9,10,11,12,13].

Silica solubilizing activity were examined on *Bunt and Rovira* Agar and Broth. Each of five isolates were determined its capability in *Bunt and Rovira* Agar by Solubilizing Index (SI), while the other activity of silica solubilized determined by *Bunt and Rovira* broth by concentration of silica.

Table 3. Silica solubilizing activity of five isolates in *Bunt and Rovira* medium.

Isolates	Solubilizing Index	Concentration of Silica (ppm)
OS4	1,08	0,745
OS5	1,07	0,796
OS7	1,10	0,662
OS12	1,05	1,053
OS13	1,05	0,663

Each of five isolates were found its capability in solubilize silica by *Bunt and Rovira* medium. The highest solubilizing index was gained by OS7, vary size of clear zones showed that each of five isolates capable to solubilize silica by their own activities. Vasanthi [7] explained that evaluation media of silica solubilizing bacteria by using *Bunt and Rovira* Agar contains silica source 0,25% showed a high capability silica solubilization. Inoculant in *Bunt and Rovira* Agar optimally growth at range pH 6,7-7,0; incubated at 30^0C for 4 days. Many factors can influence halozone formation in media agar, such as substrate, inoculant size, volume (thickness) of agar layer, medium composition, pH and incubation temperature [16,17,18,19].

Five isolates paddy rhizosphere bacteria solubilize 0,25% quartz determined by silico-molybdate's method, while the concentration of silica was higher than control (0,599 ppm). The highest silica solubilization was 1,053 ppm by OS12 in *Bunt and Rovira* broth. Decrease of pH was compared between before incubation at 6,8 and after incubation period in *Bunt and Rovira* broth at range pH 3,3-4,1 related to mechanism of bacteria solubilizing silica by organic acid production. Organic acid productions by silica solubilizing bacteria break the silicon-dioxide bond (O-Si-O) in quartz, so the soluble silica will release. Role of organic acid by producing of cation (H^+) help to dissolves silicate minerals. Vary acid organics such as acetic acid and citrate acid that found in quartz solubilization take a vital role as protonation molecule, cation barrier. Such organic acid include of carboxylate acid group, categorized to weak acid which is easily ionized by cation. Cation's excess can influence pH in silica solubilization. Excess of cation pull the anion hydroxide (OH^-) as a quartz substituent through organic acid production by bacteria. Chelation of anion in order to solubilize quartz in soluble form, monosilicic acid ($Si(OH_4)$) absorb by paddy plant through lateral roots [20,21,22].

4. Conclusions

Five paddy rhizosphere isolates were found solubilize silica optimally in *Bunt and Rovira* Agar and broth. The highest capability of silica solubilization in *Bunt and Rovira* Agar was gained by solubilizing index of OS7 by 1,10, while the highest silica solubilizing activity in *Bunt and Rovira* broth were found in OS12 with silica concentrate 1,053 ppm. Characteristic of paddy rhizosphere were identified as silicate bacteria.

5. References

[1] Husnain, Sri R and Ibrahim A 2011 *Pengelolaan hara silika pada tanah pertanian di Indonesia* (Bogor: Balai Penelitian dan Pengembangan Tanaman Pertanian) p 238

[2] Sahebi M, Mohamed M H, Abdullah S N A, Mohamed Y R, Pariza A, Tengoua F F, Jamaludin N W A and Mohamed S 2015 Importance of silicon and mechanisms of biosilica formation in plants *Hindawi Publishing Corporation* **2015** pp 1-16 http://dx.doi.org/10.1155/2015/396010

[3] Makarim A K, Suhartatik E and Kartohardjono A 2007 Silikon: hara penting dalam produksi padi *Jurnal Iptek Tanaman Pangan* **2** pp 195-204

[4] Snyder G H, Matichenkov V V and Datnoff L E 2006 Handbook of plant nutrition Chapter 19 Silicon (Internet) http://www.siliforce.com/pdf/7c/Snyder%20Mati%20Les%20effets%20du%20Silicium.pdf

[5] Sumarsih 2003 *Mikrobiologi dasar ilmu tanah* (Yogyakarta: UPN Veteran Press) p 238

[6] Balai Besar Penelitian dan Pengembangan Sumberdaya Lahan Pertanian 2007 *Metode Analisis Biologi Tanah* (Bogor: Balai Besar Penelitian dan Pengembangan Sumberdaya Lahan Pertanian) ed R Saraswati, E Husein et al pp 15-17

[7] Vasanthi N, Saleena L M and Raj S A 2013 'Evaluation of media for isolation and screening of silicate solubilizing bacteria *Int. Curr. Res.* **5** pp 406-408

[8] Abhilash, Ghosh A, Pandey B D and Sarkar S 2015 Microbial variants from iron ore slimes: mineral specificity and pH tolerance *Indian J Microbiol* **55** pp 30-39 http://dx.doi.org/10.1007/s12088-015-0544-6

[9] Osman A G 2009 Study of some characteristics of silicate bacteria *J.Sc. Tech* **10** pp 27-33

[10] Saha S 2013 *Biobenefication of bauxite ore through bacterial desilication* Thesis (India: National Institute of Technology Rourkela) pp 39-62

[11] Santi L P and Didiek H G 2017 Solubilization of silicate from quartz mineral by potential silicate solubilizing bacteria *Menara Perkebunan* **85** pp 95-102 http://dx..doi.org/10.22302/iribb.jur.mp.8v5i2.247

[12] Zhang X, Sun D and Lin H 2011 Screening and desilication of a silicate bacteria *Advanced Materials Research* **236-8** pp 253-257 10.4028/www.scientific.net/AMR.236-238.253

[13] Zhou H, Zeng X, Liu F, Qiu G and Hu Y 2006 Screening, identification and desilication of a silicate bacterium *J. Cent. South Univ. Technol.* **13** pp 337-341

[14] Premono E, Moawad M and Vlek P L G 1996 Effect of phosphate solubilizing *Pseudomonas putida* on the growth of maize and its survival in the rhizhosphere *Journal of Indonesian Crop Science* **11** 13-23

[15] Samin and Susanna T S 2006 *Prosiding PPI-PDIPTN* (*Yogyakarta*) pp 56-64

[16] Brady M S and Katz S E 1990 Factors influencing optimalization of diffusion assays for antibiotics *J Assoc Off Anal Chem.* **73** pp 202-205

[17] Mujahid T Y, Syed A S, Abdul W, Javeria M, Nuzhat A and Tanyeer A 2015 Effects of different physical and chemical parameters on phosphate solubilization activity of plant growth promoting bacteria isolated from inidgenous soil *Journal of Pharmacy and Nutrition Sciences* **5** pp 64-70

[18] Smith R A, Calviello C M, DerMarderosian A and Palmer M E 2000 Evaluation of antibacterial activity of belizean plants: an improved method *Pharmaceutical Biology* **38** pp 25-29 https://doi.org/10.1076/1388-0209(200001)3811-BFT025

[19] Suriani S, Soemarno and Suharjono 2013 Pengaruh suhu dan pH laju pertumbuhan lima isolate bakteri anggota genus *pseudomonas* yang diisolasi dari ekosistem sungai tercemar deterjen di sekitar kampus Universitas Brawijaya *J-PAL.* **3** pp 58-62

[20] Pittman E D and Lewan M D 1994 *Organic acids in geological processes* (Berlin: Springer Verlag) p 148

[21] Bennett P C, Melcer M E, Siegel D and Hassett P. 1988. The dissolution of quartz in dilute aqueous solutions of organic acids at 25^{0}C *Geochimica et Cosmochimica Acta* **52** pp 1521-1530 https://doi.org/10.1016/0016-7037(88)90222-0

[22] Ma J F and Naoki Y 2015 A cooperated system of silicon transport in plants *Trends in Plant Science* **8** pp 1-8 http://dx.doi.org/10.1016/j.tplants.2015.04.007

IOP Conf. Series: Journal of Physics: Conf. Series **1108** (2018) 012047 doi:10.1088/1742-6596/1108/1/012047

Graceful Labelling of Edge Amalgamation of Cycle Graph

D E Nurvazly[1] and K A Sugeng[1]
[1]Department of Mathematics, Faculty of Mathematics and Natural Sciences, Universitas Indonesia, 16424, Indonesia.

Abstract. There are various topics in graph theory. The theory of labelling is one of the topics in graph theory. The theory of labelling was introduced firstly by Alex Rosa in 1967. Alex Rosa introduced the labelling of β which then by Golomb referred to as the graceful labelling. In addition, there is also a $\hat{\rho}$ labelling which is a modification of the graceful labelling. Graceful labelling itself has many variations such as even graceful labelling, odd graceful labelling, and odd-even graceful labelling. In 2009 Lu used $\Theta(C_m)^n$ to denote the graph made from n copies of C_m (cycle that has m vertices) that share an edge. $\Theta(C_m)^n$ graph also known as book graph that has n pages and m vertices. In this paper, the researchers discussed about graceful labelling, $\hat{\rho}$ labelling, and odd-even graceful labelling of $\Theta(C_3)^n$ graph.

1. Introduction

There are various topics in graph theory, the theory of labelling is one of the topics in graph theory. The theory of labelling was first introduced by Alex Rosa in 1967. Alex Rosa introduced the labelling of β which then by Golomb referred to as the graceful labelling. Let G be a graph with $|E| = n$, and the mapping $f: V(G) \to \{0,1,\dots,n\}$ is an injective function then the vertex labelling of graph G is called graceful labelling if for each$uv \in E$ is labeled with $|f(u) - f(v)|$ and each edge label has a different value [1]. A graph is called graceful if it has graceful labelling.

In addition, there is also a labelling $\hat{\rho}$ which is a modification of graceful labelling. A graph with an n edges is said to have labelling $\hat{\rho}$ if it has a mapping of $f: V(G) \to \{0,1,\dots,n+1\}$ which is an injective function and induces a bijective function of edge labelling $f^*: V(E) \to \{1,2,\dots,n\}$ or $f^*: V(E) \to \{1,2,\dots,n-1,n+1\}$ where each edge$uv \in E$ with vertex$u, v \in V$has $f^* = |f(u) - f(v)|$[1].

Another modification of graceful labelling is the odd-even graceful labelling. Sridevi, Navaneethakrishnan, Nagarajan and Nagarajan define the odd-even graceful labelling on the graph with n edges as a mapping $f: V(G) \to \{1,3,5,\dots,2n+1\}$ which is an injective function where when each edge$uv \in E$ is given the label $|f(u) - f(v)|$, the edge label is $\{2,4,6,\dots,2n\}$. A graph that has an odd-even graceful labelling is referred to as an odd-evengraceful graph [4].

HuiChuan Lu defines the graph $\Theta(C_m)^n$as a graph made from n copies of C_m that share an edge[2]. Graf $\Theta(C_m)^n$ is also known as a book graph (a n page book with m −polygonal pages).
Definition 1. A$\Theta(C_3)^n$graph is a graph made from n copies of C_3 that share an edge.

2. Main Result

IOP Conf. Series: Journal of Physics: Conf. Series **1108** (2018) 012047 doi:10.1088/1742-6596/1108/1/012047

In the section, it would be shown that A $\mathbf{\Theta}(\boldsymbol{C_3})^{\boldsymbol{n}}$ had several variations of graceful labellings, such as graceful labelling, $\widehat{\boldsymbol{\rho}}$ labelling, and odd-even graceful labelling.
Theorem 1.A$\Theta(C_3)^n$ graph has graceful labelling.
Proof. Let the notation of vertices of A $\Theta(C_3)^n$ graph was shown in Figure 1.

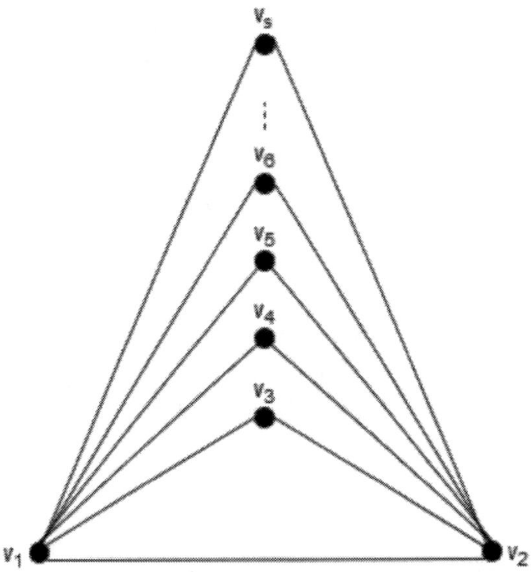

Figure 1. Vertices notation of a$\Theta(C_3)^n$ graph

Following notation in Figure 1, it could be seen that the set of vertices $V(\Theta(C_3)^n) = \{v_1, v_2, v_3, \dots, v_s\}$, and the set of edges $E(\Theta(C_3)^n) = \{v_1v_2, v_1v_3, \dots, v_1v_s, v_2v_3, v_2v_4, \dots, v_2v_s\}$. Thus, the number elements of V and E (respectively) were n and $2s - 3$, or $|V| = s$ and $|E| = 2s - 3$, where s was the number of C_3 in the graph.

Let defined function f for vertices $\Theta(C_3)^n$ graph as follows:

$$f(v_1) = 0 \quad (1)$$

$$f(v_i) = 2i - 3, \text{ for } i = 2, 3, 4, \dots, s \quad (2)$$

Based on the function f which gives labels to all vertices, it could be seen that each vertex had distinct label and forming a set $\{0,1,2,\dots,|E|\}$. Thus, the function$f$ that is defined at equation (1) and (2), was injective function from V to set $\{0,1,2,\dots,|E|\}$. Each edge $uv \in E$ had label which was derived from function$f$as$f'(uv) = |f(u) - f(v)|$.

Thus, function f' at edges$\Theta(C_3)^n$ could be formulated as follows:

$$f'(v_1v_2) = |(f(v_1) - f(v_2)| = |0 - 1| = 1 \quad (3)$$

$$f'(v_2v_j) = |f(v_2) - f(v_j)| = |1 - (2j - 3)| = 2j - 4, \; j = 3, 4, 5, \dots, s \quad (4)$$

$$f'(v_1v_j) = |f(v_1) - f(v_j)| = |0 - (2j - 3)| = 2j - 3, \; j = 3, 4, 5, \dots, s \quad (5)$$

The function f'that was defined by equation 3 – 5 gave each edge had distinct label. The labels forming a set $\{1,2,\dots,|E|\}$. Hence, the function f and the induced function f'provided distinct values for edges and also the labellings for vertices set had distinct values in $\{0,1,2,\dots,|E|\}$. It could be concluded that$\Theta(C_3)^n$graph was graceful. ■

Note that Theorem 1 was also proved by Solairaju, A., & N. Abdul A [3] by a different name of a graph that was The Merging Graph $N_{**} C_4$.

The following graph in Figure 2 was graceful labelling of $\Theta(C_3)^3$ graph.

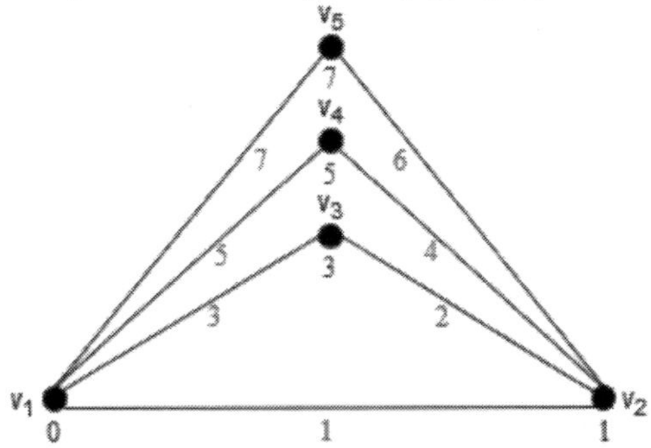

Figure 2. Graceful labelling of $\Theta(C_3)^3$ graph

The following Figure3shows a graceful labelling of $\Theta(C_3)^4$ graph.

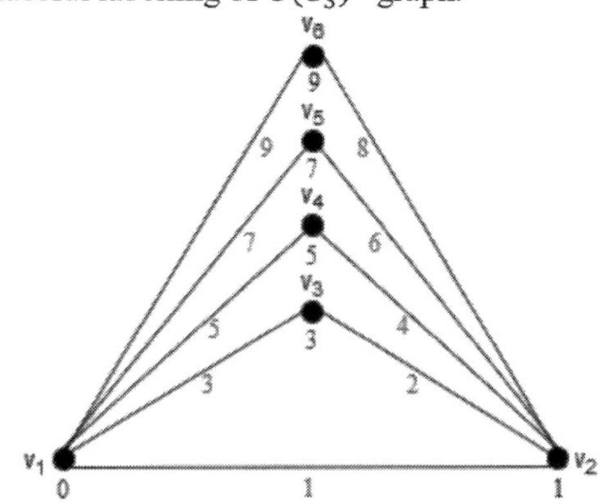

Figure 3. Graceful labelling of $\Theta(C_3)^4$ graph

Theorem 2. A $\Theta(C_3)^n$ graph has $\hat{\rho}$ labelling.
Proof. Let the notation of vertices of A$\Theta(C_3)^n$ graph as was shown in Figure 1.
Thus$V(\Theta(C_3)^n) = \{v_1, v_2, v_3, \dots, v_s\}$, and $E(\Theta(C_3)^n) = \{v_1v_2, v_1v_3, \dots, v_1v_s, v_2v_3, v_2v_4, \dots, v_2v_s\}$ and $|V| = s$ and $|E| = 2s - 3$.
Defined a function$f$of vertices $\Theta(C_3)^n$ graph as follows:

$$f(v_1) = 1 \quad (6)$$

$$f(v_i) = 2i - 2, \text{ for} i = 2, 3, 4, \dots, s \quad (7)$$

Based on the function $f$which was defined at Equation (6) and (7), it could be seen that each vertex had distinct label and forming a set $\{1,2,\dots,|E|+1\}$. Thus, the function f was an injective function from V to set $\{1,2,\dots,|E|+1\}$. Each edge $uv \in E$ had a label from function$f'$by$f'(uv) = |f(u) - f(v)|$.

Function f' of$\Theta(C_3)^n$could be formulated as follows:

$$f'(v_1v_2) = |(f(v_1) - f(v_2)| = |1 - 2| = 1 \quad (8)$$

$$f'(v_2v_j) = |f(v_2) - f(v_j)| = |2 - (2j - 2)| = 2j - 4, \; j = 3,4,5,\dots,n \quad (9)$$

$$f'(v_1v_j) = |f(v_1) - f(v_j)| = |1 - (2j - 2)| = 2j - 3, \; j = 3,4,5,\dots,n \quad (10)$$

It could be seen that the labelling edges that defined by Equation (8) – (10) were different and forming a set $\{1,2,\dots,|E|+1\}$. Hence, the function f and the induced function f'provided distinct values for edges and also the labellings for vertices set had distinct values in $\{1,2,...,|E|+1\}$. Hence, $\Theta(C_3)^n$ graph has a $\hat{\rho}$ labelling. ■

The following graph in Figure 4 was $\hat{\rho}$ labelling of $\Theta(C_3)^3$ graph.

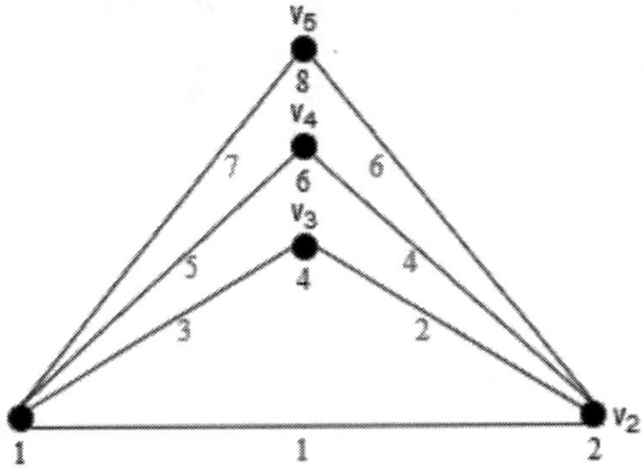

Figure 4. $\hat{\rho}$ labelling of $\Theta(C_3)^3$ graph

The following graph in Figure5 was $\hat{\rho}$labellingof$\Theta(C_3)^4$ graph.

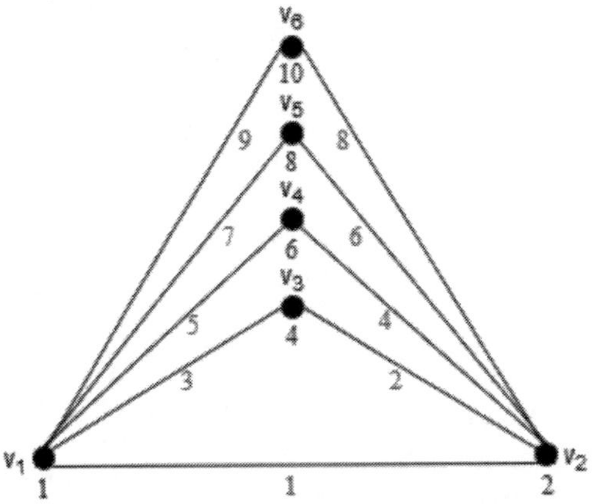

Figure 5.$\hat{\rho}$labellingof$\Theta(C_3)^4$ graph

Theorem 3.A$\Theta(C_3)^n$ graph is odd-even graceful graph.

Proof. Let the notation of vertices of A$\Theta(C_3)^n$ graph was shown in Figure 1.

Using notation in Figure 1, it was known that $V(\Theta(C_3)^n) = \{v_1, v_2, v_3, \dots, v_s\}$, and a set of edges $E(\Theta(C_3)^n) = \{v_1v_2, v_1v_3, \dots, v_1v_s, v_2v_3, v_2v_4, \dots, v_2v_s\}$. Thus, $|V| = s$ and $|E| = 2s - 3$.

Defined function f which gave labels to the vertices of $\Theta(C_3)^n$ graph as follows:

$$f(v_1) = 1 \quad (11)$$

$$f(v_i) = 4i - 5, \text{ for} i = 2, 3, 4, \dots, s \quad (12)$$

Based on the function f which was defined at equation (11) – (12), it could be seen that each vertex had different label that forming a set $\{1,3,7,11,...,2|E|+1\}$. Thus, the function f was an injective function from V to set $\{1,3,7,11,...,2|E|+1\}$. Each edge $uv \in E$ had label by function$'$. The function f' was induced from function f by $f'(uv) = |f(u) - f(v)|$.

Thus, we had function f' labelling at edges $\Theta(C_3)^n$ as follows:

$$f'(v_1v_2) = |(f(v_1) - f(v_2)| = |1 - 3| = 2 \quad (13)$$

$$f'(v_2v_j) = |f(v_2) - f(v_j)| = |3 - (4j - 5)| = 4j - 8, \; j = 3, 4, 5, \dots, s \quad (14)$$

$$f'(v_1v_j) = |f(v_1) - f(v_j)| = |1 - (4j - 5)| = 4j - 6, \; j = 3, 4, 5, \ldots, s \tag{15}$$

Based on function f' that defined inequation (13) – (15), it could be seen that each edge had distinct label that forming a set $\{2,4,6...,2|E|\}$. It could be seen that function f'provide distinct values at each edges that form a set $\{2,4,6...,2|E|\}$. Hence the function f and the induced function f' provide distinct values for edges and also the labellings for vertices set had distinct values in $\{2,4,6...,2|E|\}$. Hence, $\Theta(C_3)^n$ graph was an odd-even graceful graph. ■

Figure 6 was an odd-even graceful labelling of $\Theta(C_3)^4$ graph.

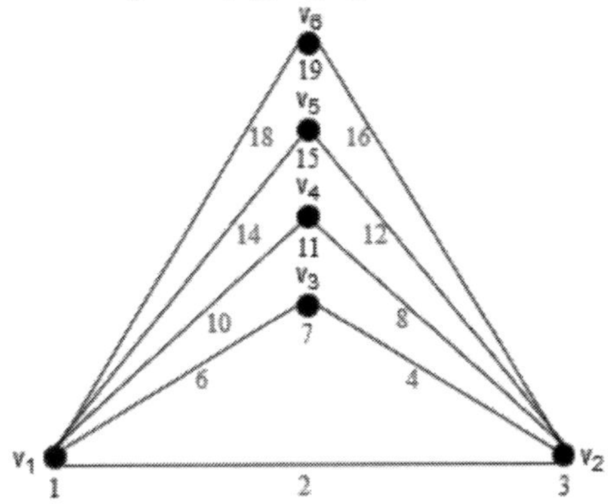

Figure 6. Odd-even graceful labellingof$\Theta(C_3)^4$ graph

3. Conclusion

$\boldsymbol{\Theta(C_3)^n}$graph had a graceful labelling, $\hat{\rho}$ labelling, and odd-even graceful labelling.

4. References

[1] Gallian J A 2017 A Dynamic Survey of Graph Labelling. *The Electronic Journal of Combinatorics*, #DS6

[2] Lu HC 2009. On Large harmonious graphs. *Ars Combin***91**

[3] Solairaju A and Abdul N 2013 The Gracefulness of The Merging Graph $N_{**} C_4$. *International Journal of Engineering and Science***2**pp 22-24

[4] Sridevi R,NavaneethakrishnanS, Nagarajan A, NagarajanK 2012 Odd-Even Graceful Graphs. *J. Apl. Math. & Informatics***30**pp5-6.

Acknowledgments

The Research was Supported By PITTA UI Research Grant 2018 Number 2276/UN2.R3.1/HKF.05.00/2018.

IOP Conf. Series: Journal of Physics: Conf. Series **1108** (2018) 012048 doi:10.1088/1742-6596/1108/1/012048

A Responsive Web-Based QR Code for Laboratory Clearance Form

N Rochmawati[1, *], Y Anistyasari[1], D F Suyatno[1] and I F Kurniawan[1]

[1]Department of Informatics Engineering, Universitas Negeri Surabaya, Surabaya 60231, Indonesia
Corresponding author: naimrochmawati@unesa.ac.id

Abstract. Engineering faculty of UNESA has a policy for students who will graduate to collect a form called Laboratory Clearance Form. Laboratory Clearance Form is a letter stating that student has no indemnity in labs. To have Laboratory Clearance Form, student should get the signature approval many sub heads of the laboratories in the department. It is a problem if the department has a lot of laboratories. Department of PKK, one of department in UNESA has 21 laboratories so their students should ask each 21 sub head laboratories to sign their forms. The process takes a long time. The second problem is sometimes there are students who cheat at the signatures. To avoid cheating, Laboratory Clearance Form should be completed with QR code. In this paper, we propose the design of the application relating to Laboratory Clearance Form using QR code. This application use framework Bootstrap which supports responsive web technology. By using the framework, the application can be accessed by mobile phone. The outcome is expected a design of application to ease student who will graduate having their Laboratory Clearance Form in the Departments of Engineering faculty of UNESA

1. Introduction

Over the years, technology has been growing [4]. Jobs that are usually done manually, can now be done by the app. Recent years, human needs are increasingly complex. everything is online. As well as education, the information related to education needs to be online as it is faster and more efficient. Engineering faculty of UNESA has a policy for students who will graduate to collect a form called Laboratory Clearance Form. Laboratory Clearance Form is a letter stating that student has no indemnity in labs. Student should get the signature approval from many sub heads of the laboratories in the department. Because there are a lot of laboratories in technical faculty of UNESA, it takes time to proceed Laboratory Clearance Form. QR code is one of technology which can be used to prevent duplicating something we protect. It also can be used to prevent student from doing cheating in applying Laboratory Clearance Form. It is effective to use because QR code can be recognized with mobile phone. Nowadays, almost everyone has mobile phone. Therefore student can proceed Laboratory Clearance Form easier without having to chase sub heads of the laboratories to find the signatures. In this paper, we propose the design of the application relating to Laboratory Clearance Form using QR code. This application use framework Bootstrap which supports responsive web technology.

IOP Conf. Series: Journal of Physics: Conf. Series **1108** (2018) 012048 doi:10.1088/1742-6596/1108/1/012048

Figure 1. Example of QR Code

2. Literature

2.1. QR Code

QR code is an image of a matrix barcode that stores data in two dimensions. Data is presented as square dots with specific pattern in both horizontal and vertical dimensions. QR codes can be used in items' identifications, objects' tracking, general marketing and advertisements [1]. QR code is now being widely used in a variety of businesses [2]. QR code seems to be applied not only in business but almost in everywhere, QR Code is developed by automotive company from Japan in 1994 by Denso Wave [3], [4]. The advantages of QR Code can keep some information. Contact information such as phone numbers, contact names and emails can also store in QR code labels. Information such as text, images, even videos can also store in QR code. International Standards Organization and International Electro technical Commission (ISO/IEC) have published two standard since year 2000 about QR Code. QR code example is shown in Figure 1.

2.2. Responsive Web

Responsive Web Design is a terminology introduced by web designer Ethan Marcotte on an article published by A List Apart website. In the last four years, there is an increase for more than 45% of the use of mobile phones for accessing the web [5]. Certainly many people access website through mobile phones. The problem is when website accessed through different devices can lead a broken design [6] that is why responsive web is needed. The broken design happened such as hidden critical links, hidden components as well as hidden important information [7]. In general, Responsive web is method to create web which is the UI could be easily seen and use by any size of screen. Responsive Web design can respond environment based on platform size. It can be resized depend on the size of device [8].

2.3. PHP

PHP was invented for the first time by Rasmus Lerdoff in 1994. He wrote the code using C programming language to build websites for many clients. Next, PHP was quickly following by other web developerts and continue to improve PHP [9].

2.4. PostgreSQL

PostgreSQL is a database management system that is quite popular in the world. For more than 15 years developed, postgresql and has proven to have a good reputation. PostgreSQL has complete features and has the same quality as Oracle and SQL Server although it is free, those features are streaming replication, aggregate function, etc. The latest version of PostgreSQL is version 10.

2.5. Web Application

Web application is a computer software that is encoded in a programming language that supports web-based software such as HTML, JavaScript, CSS, Ruby, Python, Php, Java and other programming languages.

2.6. Laravel

Laravel is a PHP MVC framework developed by Taylor Otwell in 2011 [10] and now has reached version 7 and with the slogan laravel "Php That Does Not Hurt. Code Happy & Enjoy The Fresh Water". People will feel that php programming is fun and easy because there are so many features that will help us in coding using this laravel framework. Beta version was released in June 2011 and built from scratch. Some features on that version is Eloquent ORM, routing based on closure function, helper for form, validation, authentication, etc.

3. Methods

Waterfall model, the most used system development life cycle, is used to design and develop the software. Waterfall model has five steps that shown in the Figure 2.

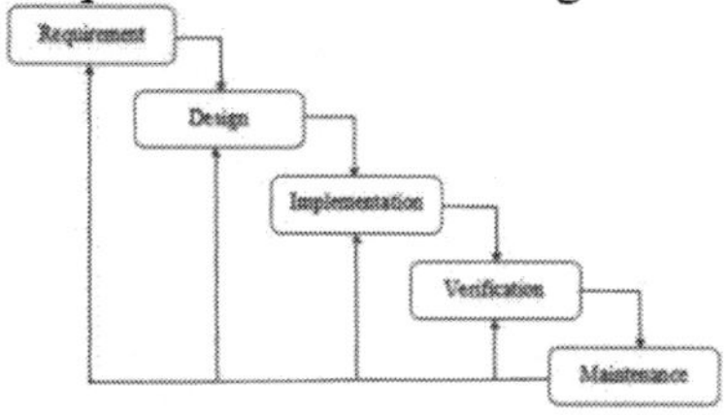

Figure 2. Waterfall Model

3.1. Requirement

Before using this system, collage students have to ask all signatures of sub head of laboratories and head of laboratory by face to face. But by using this system, collage students do not require to ask all signatures face to face. System will help students to get Laboratory Clearance Form without difficult steps, those are:

- Student has to submit the letter in submit form.
- Student must upload needed documents (depend on the lecturers).
- All of kasublab and kalab approve the submitted letter.
- Student downloads the letter.
- Lecturer can validate the letter using QR code in the printed letter from the student.

Students cannot manipulate the letter because every student has different unique QR code.

3.2. Flow Map

Flow map is the extended diagram from flowchart that represents the flow of the system. The flow map of the proposed system can be seen in Figure 3.

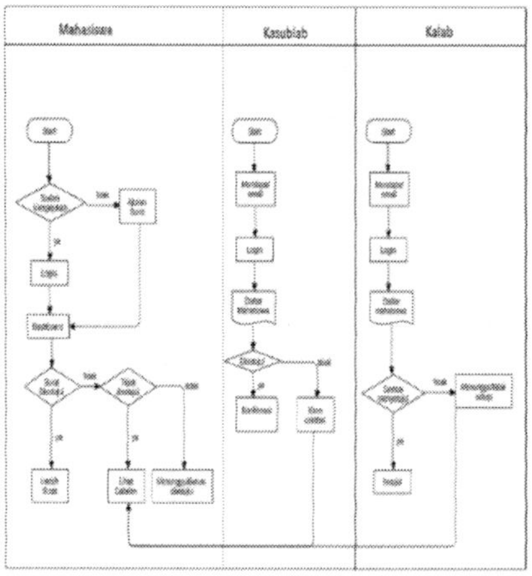

Figure 3. Flowmap of the proposed system

3.3. PDM (Physical Data Model)

PDM is used to represents the database of the system. PDM of this application is shown in Figure 4.

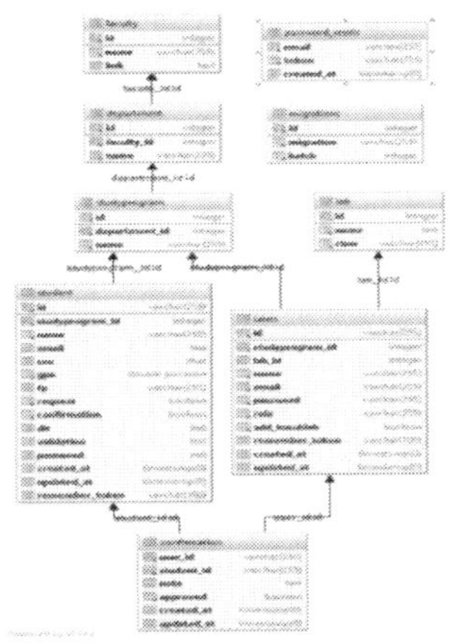

Figure 4. Physical Data Model

3.4. PDM (Physical Data Model)

DFD is used to make people understand about the flow of data or information through system. DFD of this application is shown in Figure 5.

3.5. Implementation

Applications we use in this paper are totally free and responsive web-based application. The software we used are:

- Linux Operating System
- PostgreSQL Database Server
- Apache Web Server
- PHP Web Programming Language
- HTML, CSS, JavaScript (jQuery, DataTables and Vue.js)
- Bootstrap Framework
- Laravel Framework
- Notepad++, HeidiSQL, GitKraken and Visual Studio Code

All software can be downloaded for free. This research used Bootstrap framework for responsive web and Laravel framework to ease developing system

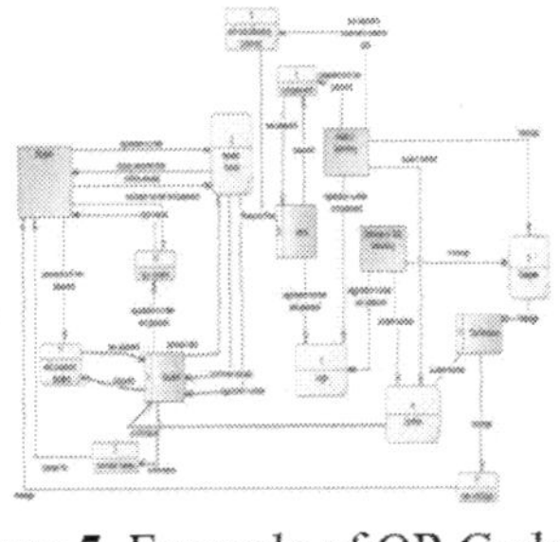

Figure 5. Example of QR Code

3.6. Verification and Validation

In this step, we used white-box and black-box testing before implementing it. White-box testing is to find the wrong groove source code algorithm and black-box testing for searching the wrong detail, user interface etc. We used those testing methods for searching some bugs and then we fixed it up.

3.7. Maintenance

IOP Conf. Series: Journal of Physics: Conf. Series **1108** (2018) 012048 doi:10.1088/1742-6596/1108/1/012048

In this step, we used white-box and black-box testing before implementing it. White-box testing is to find the wrong groove source code algorithm and black-box testing for searching the wrong detail, user interface etc. We used those testing methods for searching some bugs and then we fixed it up.

4. Results and Discussion

The result of this research is application to create Laboratory Clearance Form. To create license from this application, the student needs to request license by accessing this web application and filling request form with their registration number. If it success, each sub head laboratory and head laboratory will receive notification from the system about student who request the license. If all of sub head laboratories confirm, the head of laboratory will receive notification. After the head of the laboratory confirms, the student then can download the Laboratory Clearance Form.

4.1. Laboratory Clearance Form

Laboratory Clearance Form is generated by the system with pdf format. it contains QR code for validation which is generated when all sub head laboratory and head laboratory confirm. Laboratory Clearance Form. Can be seen in Figure 6.

Figure 6. Laboratory Clearance Form

4.2. QR Code Reader

The image of the generated QR code can be read to validate the user. The view of QR code Reader is shown in Figure 7.

Figure 7. A Laboratory Clearance Form

4.3. Responsive Web

This application can be accessed from both computer and smartphone. The GUI of Web Responsive can be seen in Figure 8 and 9.

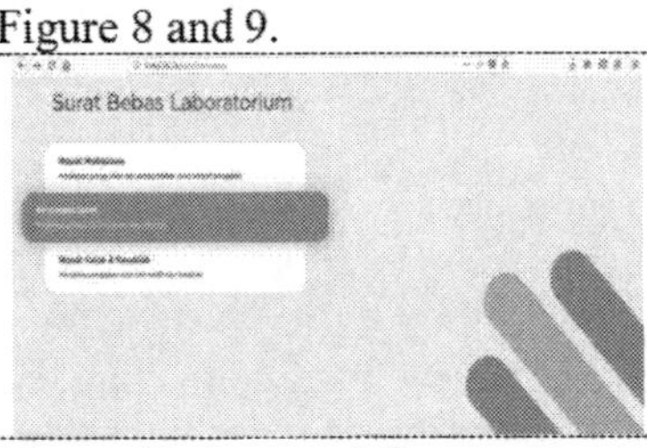

Figure 8. Accessed from computer

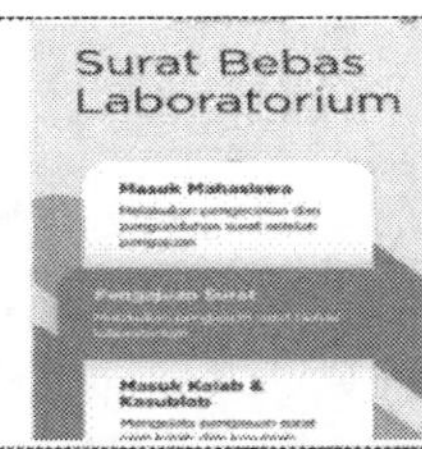

Figure 9. Accessed from mobile phone

IOP Conf. Series: Journal of Physics: Conf. Series **1108** (2018) 012048 doi:10.1088/1742-6596/1108/1/012048

4.4. Email Notification

Once the student requests Laboratory Clearance Form, each of the sub laboratory will receive message from the system. The view of email notification can be seen in Figure 10.

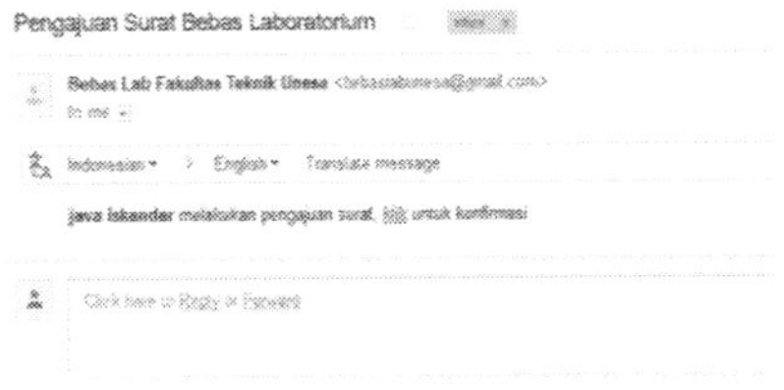

Figure 10. Email Notification

5. Conclusion

The result of this research is application to create Laboratory Clearance Form. To create license from this application, the student needs to request license by accessing this web application and filling request form with their registration number. If it success, each sub head laboratory and head laboratory will receive notification from the system about student who request the license. If all of sub head laboratories confirm, the head of laboratory will receive notification. After the head of the laboratory confirms, the student then can download the Laboratory Clearance Form.

References

[1] Bani-Hani R M, Wahsheh Y A, and Al-Sarhan M B 2014 Secure QR code system *10th Int. Conf. Innov. Inf. Technol.* pp 1–6

[2] Sutheebanjard P and Premchaiswadi W 2010 QR-Code Generator *Knowl. Eng. 2010 8th Int. Conf. ICT* pp 89–92

[3] Kieseberg P, et al 2010 QR code security *Proc. 8th Int. Conf. Adv. Mob. Comput. Multimed. - MoMM '10* p 430

[4] Narayanan A S 2012 QR Codes and Security Solutions *Int. J. Comput. Sci. Telecommun,* **3** (7), pp 1–4.

[5] Subic N, Krunic T, and Gemovic B 2014 Responsive web design – Are we ready for the new age ? *J. Appl. Knowl. Manag,* **2** (1) pp 93–103

[6] Mazzei J 2012 [Online] Responsive Web Design https://www.linkedin.com/in/jennifermazzei/

[7] Lestari D M, Hardianto D, and Hidayanto A N 2014 Analysis of user experience quality on responsive web design from its informative perspective *Int. J. Softw. Eng. its Appl,* **8** (5) pp. 53–62

[8] Rochmawati N, Buditjahjanto I G P A, Putra R E, and Wicaksono A Y 2018 A Responsive Web-Based QR Code for Inventory in the Laboratory of Informatics, UNESA *IOP Conf. Ser. Mater. Sci. Eng.,* **288** (1).

[9] Severance C 2012 Inventing PHP: Rasmus lerdorf *Computer,* **45** (11) pp 6–7

[10] Shihan M N I 2014 *Laravel 4 Docs (Updated for 4 . 1 . x)* (New York: Lean Publishing)

Virtual Laboratory to Improve Students' Conceptual Understanding in Physics Learning

G Gunawan[1,*], N Nisrina[1], N M Y Suranti[1], L Herayanti[2], and R Rahmatiah[1]

[1]Physics Education Study Program, Universitas Mataram, Jln. Majapahit No 62 Mataram 83125, Indonesia
[2]Physics Education Study Program, IKIP Mataram, Jln. Pemuda No 59A Mataram 83125, Indonesia

*Corresponding Author: gunawan@unram.ac.id

Abstract. Students' conceptual understanding is the main outcome of all education in the world. Students must be able to understand the concept well in order to apply their knowledge in daily life. One of the efforts to improve the conceptual understanding is through computer-based learning. This study aims to improve the students' conceptual understanding of physics by using the virtual laboratory. The participants of this study were students from three different senior high schools. The students were divided into three experimental groups and three control groups. The research instrument used was multiple choice forms. The results indicated that the use of virtual laboratory gave a positive effect on students' conceptual understanding. The final test result showed that the experimental group got a higher score than the control group. Data analysis was based on the number of students' correct answers in every cognitive aspect. The result of the analysis in every cognitive aspect showed that in each school, students were able to answer the questions correctly on aspects of C1 (remembering), C2 (knowing), C3 (applying), and C4 (analyzing). While on the aspects of C5 (evaluating) and C6 (creating) still needed to be improved.

1. Introduction

Physics is one of the subjects in school taking an important role in shaping the mindset of students to become a qualified human. That goal will be achieved if students can understand the concept of physics well and correctly, then applying the concept in everyday life. Hence in physics, students are trained to think in order to study a phenomenon logically and mathematically. Therefore, students are expected to have a good conceptual understanding so the goals of physics learning can be achieved as well.

Student's conceptual understanding is reflected in the results of physics learning evaluation. Silaban [1] argued that conceptual understanding is the students' ability in recording and back transferring of some information from learning, which can be used in solving, analyzing, and interpreting the problem. Many studies reveal that conceptual understanding is one of success learning indicator. Teachers as educators who have a direct role in student learning evaluation often complain about the achievement of students' physics learning results. Many students have not gained optimal learning outcomes and have not improved significantly. In this case, the quality and quantity of physics learning results are certainly influenced by many factors, one of them is the innovation in

IOP Conf. Series: Journal of Physics: Conf. Series **1108** (2018) 012049 doi:10.1088/1742-6596/1108/1/012049

learning, especially the use of the excellent learning media. Teachers as facilitators must be able to present the appropriate learning media to facilitate them to visualize abstract physics concepts [2].

Innovations in learning to support students' understanding are essential for educators [3]. One of the innovations that can solve the problem of students conceptual understanding is the utilization of computer-based media. According to Tüysüz [4], the computer-based media use in laboratory activities can increase student's interest in learning in the classroom and also help them to improve learning outcomes. Laboratory activities are essential in physics learning. Through experimental activities, students' openness attitude and curiosity can be improved [5], and also practice their scientific attitudes. Jufri et al. [6] stated that the activities of physics experiments that are not supported by good facilities and infrastructure could have a negative impact on student learning outcomes. Therefore, computer-based experimental activities through virtual laboratories can be selected to visualize the physics concepts in order to make students understand the subject easily.

Many activities can be shown in a virtual laboratory, which covers presenting the practices and exercises, tutorials, games, simulations, inventions, and troubleshooting. This virtual laboratory provides an opportunity to use computer animation in learning that is oriented to microscopic representations. Students can experiment over and over again and also visualize the abstract processes that are impossible to see or imagine in a real laboratory. Some studies generally suggest that the use of virtual laboratories has a good impact on learning outcomes. Gunawan & Setiawan [7] found that the use of virtual laboratories can enhance conceptual understanding in modern physics courses, especially on the quantum theory of electromagnetic radiation and matter, which are abstract concepts that are difficult to visualize in real experiments. In addition, Nisrina et al. [8] also reveal that the cooperative learning models combined with virtual media can improve students' conceptual understanding in each sub-material of static fluid.

Based on the description that has been stated, this study uses the virtual laboratory media as an alternative to improve students' conceptual understanding. Virtual laboratories are used to help students in carrying out the experimental activities as well as in real laboratories. This study aims to improve students' physics concept understanding through the virtual laboratories. Students' conceptual understanding of physics in this study covers all aspects of cognitive according to Bloom's taxonomic revision by Krathwohl [9], which include remembering, understanding, applying, analyzing, evaluating, and creating.

2. Method

This quasi-experimental research with pre-test and post-test control group design was conducted at three different schools. The sample was chosen by purposive sampling technique, with total 195 samples that spread in three experimental groups and three control groups. The experimental group was given a learning treatment with the use of virtual laboratory media, while the control group was given the conventional learning treatment. The virtual laboratory consists of several computer simulations used for physics experiments. Learning in each session has been carried out for 90 minutes. Each student learns to use a virtual laboratory for 5 to 6 times.

Furthermore, the level of conceptual understanding of students was compared to each sample school. Student conceptual data are gained from multiple choice test instruments that have been tested for validity, reliability, differentiation, and difficulty. The test instrument included six cognitive aspects, namely C1 (remembering), C2 (understanding), C3 (applying), C4 (analyzing), C5 (evaluating), and C6 (creating). The conceptual understanding comparison of experimental groups and control groups in each school was performed with a simple analysis of the students' conceptual understanding test results and the number of students' correct answers on every cognitive aspect.

3. Results and Discussion

Conceptual understanding can be interpreted as an understanding of the concepts and the application in daily life. Conceptual understanding is not only about understanding simple concepts but also can be described as the ability to know, understand, apply, classify, generalize, synthesize, and conclude the information obtained [10]. In this study, the virtual laboratory is used to improve conceptual understanding of physics concepts conducted at three different schools. The result of this study showed that there were differences in conceptual understanding in three schools. Figure 1 shows the comparison of the students' conceptual understanding.

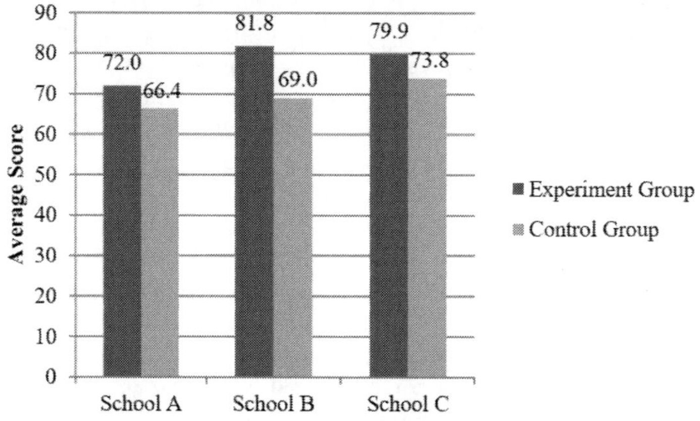

Figure 1. Comparison of students' conceptual understanding in each school

Figure 1 shows that in each school, the experimental group students have a higher score of a mean conceptual test than the control group students. At school A, the experimental group gets a higher score of 5.6% than the control group. While in school B, the experimental group gets a higher score of about 12.8% than the control group. Moreover, at school C, the experimental group get a higher score of about 6.1% than the control group. These results indicate that students' conceptual understanding of using virtual laboratories was better than conventional learning models. The findings of this study are supported by Gunawan et al. [11] who stated the virtual laboratories improve the students' conceptual understanding, both for males and females.

Virtual laboratories help students to develop physics and logic inference [12], enhance verbal and figural creativity [13], [14] and improve the ability to solve physics problems [15]. Wiyono et al. [16] state that the use of interactive multimedia can improve the physics conceptual understanding, especially abstract concepts. Interactive multimedia can help students to understand abstract physics concepts that are difficult to explain by the teacher. Furthermore, Zacharia & Anderson [17] revealed that the use of virtual media could help students visualized problems and solutions, and also can foster positive attitudes toward physics. Surely it is indirectly a positive impact on students' conceptual understanding of physics concepts.

The results of students 'conceptual understanding tests were also analyzed based on the number of students' correct answers in each cognitive aspect. It is intended to know the level of conceptual understanding of the students obtained. Figure 2 shows the percentage comparison of the number of students' correct answers in every cognitive aspect. In each aspect, the experimental group also get a higher percentage than the control group, that are C1 (remembering) by 11.1%, C2 (understanding) by 1.1%, C3 (applying) difference by 7.7%, C4 (analyzing) by difference of 14.4%, C5 (evaluate) by 7.4% difference, and C6 (creating) by 2.7% difference. This result implied that the use of virtual laboratories in learning was able to provide better conceptual understanding in every cognitive level of students compared to the use of conventional learning models.

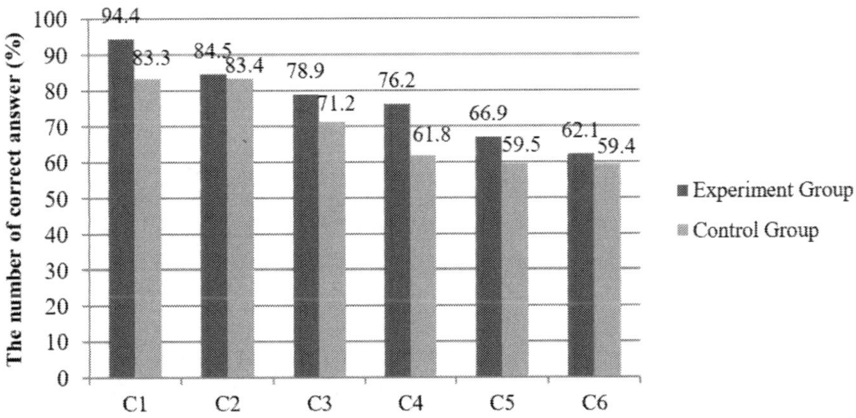

Figure 2. Comparison of the number of students' correct answers in every cognitive aspect

The percentage of the correct answers that the students successfully worked was on the cognitive aspect C1, that is by acquiring 94.4% for the experimental group and 83.3% for the control group. This result suggested that most experimental group and control group students can answer the questions in the remembering category correctly. The questions for category remembering were still relatively easy cause this was the lowest category. Problems with the remembering category did not involve a high-level thinking process. It just recognized or recalled a term, title, and formula without any claimed to understand or use it. In the C1 aspect, students just needed to take a simple knowledge of their memory. Therefore, it should be that the questions in this category can be answered well and correctly by the students.

Furthermore, for the questions in C2 (understanding) and C3 (applying) categories, most students in the experimental group as well as the students in the control group were still able to respond appropriately. Cognitive aspects of C2 and C3 were still relatively easy. Cognitive aspects C2 is the ability of a person to understand something, once something was known and remembered, and can see it in various ways. Understanding is a higher level of thinking ability than remembering. In the C2 aspect, students are required to determine the meaning of teaching messages, including verbal, written, and graphical communications. In the aspect of the C3 category, students are required to use procedures that are understood before in certain situations. The cognitive aspect of C3 is a higher level aspect of understanding (C2) because students must be able to collect and select appropriately a particular concept, law, proposition, rule, idea, or method known and understood in order to be applied to a new situation.

The percentage of students' correct answers on the cognitive aspects of C4 (analyzing) and forward begins to fall from previous cognitive aspects. In the cognitive aspects of C4, the experimental group get the correct answers percentage of 76.2% and 61.8% for control group. Nevertheless, this percentage is still quite good because more than half students can answer the C4 cognitive aspects correctly. The cognitive aspects of C4 and next belong to a high cognitive aspect. Students need to start using their high order thinking skills. In the analyzing aspect, students are required to break down the information that has been understood before into their constituent parts and to detect how the parts are interconnected with each other.

In the C5 (evaluating) and C6 (creating) questions, the percentage of students' correct answers decreases to below 70% and 60%. This is clearly due to the high-level thinking ability of C5 and C6 aspects. In C5 aspect, students are required to make a consideration of particular situations, values, ideas, or methods based on existing criteria and standards. The cognitive processes in this category involve interpreting, considering, examining, deciding, arguing, and criticizing. While in C6 (creating), students are required to create and combine several elements into a unified form. This C6 cognitive aspect creates some elements together to build a wholly logical and functional and organizes them into new patterns or structures. Therefore, it is necessary to have excellent high-level thinking skills in order to understand the cognitive aspects of C5 (evaluating) and C6 (creating).

Analysis of students' answers on each aspect of cognitive showed that students' ability in answering conceptual understanding test was higher in the lower cognitive aspects, and otherwise was getting weaker on higher cognitive aspects. In this study, students' conceptual understanding of the cognitive aspects of C1 (remembering) to C6 (creating) overall was good. This was indicated by the percentage of students' correct answers on every cognitive aspect got more than 50%, with the percentage of the experimental group that was higher than the control group. This mean, more than half the students in the experimental group as well as the control groups had a good conceptual understanding. The use of virtual laboratories in experimental group learning increased students' motivation to learn so they had better conceptual understanding than the control group. The findings of this study are in line with the findings of Gunawan et al. [11] study which reveals that the more interesting the learning media used, the more students' learning motivation. Thus, students' physics conceptual understanding is improved as well. By having a conceptual understanding, students can solve their problems during the learning process. Zacharia [18] stated that the use of virtual experiments could improve students' conceptual understanding compared to real experiments in the laboratory. The use of virtual laboratory can remove the tedious part of the real experimental activities [19]. It can help students understand the higher levels of cognitive analysis, synthesis, and evaluation.

Based on the descriptions that have been put forward can be concluded that the use of virtual laboratory is very suitable for learning cause it can help students improve their conceptual understanding. Some studies have found that not all models or learning media can improve the conceptual understanding in all cognitive aspects. Rais & Ardhana [20] found that the application of project-based learning model is appropriate for teaching material in the higher cognitive domain such as application, analysis, synthesis, and evaluation. Conversely, the model is less appropriate to use in teaching lower cognitive domain. In addition to conceptual learning, it is also important to consider the good interaction between teacher and student and between student and student. Students' conceptual understanding will develop well if the good interactions occur during learning. The better the interaction that occurs, the better the conceptual understanding of the students.

4. Conclusion

Based on the result analysis and discussion; it can be concluded that the virtual laboratory gave the positive effect on student conceptual understanding. In each school, the experimental group students had the higher average score of conceptual understanding tests than the control group students. The result of conceptual understanding analysis on each cognitive aspect also showed that the experimental group students were better in answering the questions correctly than the control group students. The percentage of students' correct answers was highest on the C1 (remembering) and the lowest on C6 (creating). Students were able to answer questions with cognitive aspects of C1 (remembering), C2 (understanding), and C3 (applying), and C4 (analyzing); while the cognitive aspects of C5 (evaluating) and C6 (creating) still needed to be improved again. As for suggestions for further research that is to combine virtual laboratory media with the approach or other learning strategies in order to obtain maximum results.

5. References

[1] Silaban B 2014 Hubungan antara penguasaan konsep fisika dan kreativitas dengan kemampuan memecahkan masalah pada materi pokok listrik statis *J. Penelit. Bid. Pendidik.* **20** 65

[2] Suranti N M Y, Gunawan G and Sahidu H 2016 Pengaruh model project based learning berbantuan media virtual terhadap penguasaan konsep peserta didik pada materi alat-alat optik *J. Pendidik. Fis. Teknol.* **2** 73

[3] Ye L, Wong, N S, and Ho J W Y 2016 Design development and evaluation of biochemistry virtual laboratory for blended learning *GSTF Inter. J. Educ.* **3** 40

[4] Tüysüz C 2010 The effect of the virtual laboratory on students' achievement and attitude in chemistry *Inter. Online J. Educ. Sci.* **2** 37

[5] Gunawan G and Liliasari L 2012 Model virtual laboratory fisika modern untuk meningkatkan disposisi berpikir kritis calon guru *J. Cakrawala Pendidik.* **31** 185

[6] Jufri A W, Gunawan G and Purwoko A A 2012 Pemetaan kompetensi siswa sma pada bidang ipa di kabupaten sumbawa *J. Kependidik.* **11** 87

[7] Gunawan G and Setiawan A 2010 Using virtual laboratory to increase students' understanding on modern physics *In Proceeding The 4th International Seminar on Science Education, Bandung: Program Pendidikan IPA SPs UPI*

[8] Nisrina N, Gunawan G and Harjono A 2016 Pembelajaran kooperatif dengan media virtual untuk peningkatan penguasaan konsep fluida statis siswa *J. Pendidik. Fis. Teknol.* **2** 66

[9] Krathwohl D R 2002 A revision of bloom's taxonomy: an overview *Am. J. Psychol.* **122**

[10] Hermawanto, Kusairi S and Wartono 2013 Pengaruh blended learning terhadap penguasaan konsep dan penalaran fisika peserta didik kelas x *J. Pendidik. Fis. Indones.* **9** 67

[11] Gunawan G, Nisrina N, Suranti N M Y, Ekasari R R, and Herayanti L 2018 The effect of virtual labs toward students' conceptual understanding of physics based on gender *Adv. Soc. Sci., Educ. Hum. Res.* **173** 128

[12] Gunawan G, Setiawan A and Widyantoro D H 2013 Model virtual laboratory fisika modern untuk meningkatkan keterampilan generik sains calon guru *J. Pendidik. Pembelajaran (JPP)* **20** 25

[13] Gunawan G, Harjono A, Sahidu H and Nisrina N 2018 Improving students' creativity using cooperative learning with virtual media on static fluida concept *J. Phys.: Conf. Series* **1006** 012016

[14] Gunawan G, Sahidu H, Harjono A and Suranti N M Y 2017 The effect of project based learning with virtual media assistance on student's creativity in physics *Cakrawala Pendidik.* **36** 167

[15] Gunawan G, Harjono A, Sahidu H and Herayanti L 2017 Virtual laboratory to improve students' problem-solving skills on electricity concept *J. Pendidik. IPA Indones.* **6** 257

[16] Wiyono K, Setiawan A and Paulus C T 2012 Model multimedia interaktif berbasis gaya belajar *J. Pendidik. Fis. Indones.* **8** 74

[17] Zacharia Z and Anderson O R 2003 The effects of an interactive computer-based simulation prior to performing a laboratory inquiry-based experiment on students' conceptual understanding of physics *Am. J. Phys.* **71** 618

[18] Zacharia Z C 2007 Comparing and combining real and virtual experimentation: an effort to enhance students' conceptual understanding of electric circuits *J. Comput. Assisted Learn.* **23** 120

[19] Herga N R and Dinevski D 2012 Using a virtual laboratory to better understand chemistry-an experimental study on acquiring knowledge in information technology interfaces (ITI) *Proceedings of the ITI 2012 34th International Conference* 237

[20] Rais M and Ardhana W 2014 Project-based learning vs pembelajaran dengan metode ekspositori dalam menghasilkan kemampuan belajar teori perancangan mesin *J. Pendidik. Pembelajaran* **20** 33

Acknowledgments
Authors deliver sincere gratitude to the Ministry of Research, Technology, and Higher Education of Republic of Indonesia for awarding the National Strategic Research Grant in 2015-2017. The team also thank to all parties involved during the process and publication of results of this study.

IOP Conf. Series: Journal of Physics: Conf. Series **1108** (2018) 012050 doi:10.1088/1742-6596/1108/1/012050

A mathematical model of TB control with vaccination in an age-structured susceptible population

Dipo Aldila*, **Zahra Alya Sari Ryanto, and Alhadi Bustamam**

Department of Mathematics, Universitas Indonesia, Kampus UI Depok, 16424 Depok, Indonesia

E-mail: aldiladipo@sci.ui.ac.id

Abstract. A new mathematical model for the transmission dynamics of tuberculosis (TB) with the intervention of vaccination in an age-structured susceptible population is designed and analyzed in this article. The model is constructed as an SEIR-based system of ten-dimensional ordinary differential equation. Each population is then further classified according to its age-class; children (<15 years old) and adult (15-65 years old). In the susceptible population, the classification goes even further by taking vaccination criteria into account. Mathematical analysis of the equilibrium points and its local stability is performed, both analytically and numerically, to help understand the possibility of the situation in the field for long-term behavior. We also show the form of the basic reproduction number as the spectral radius of the next-generation matrix. The basic reproduction number will become the threshold parameter to handle the existence of equilibrium points. Numerical simulations of the model are done for various scenarios to provide a better understanding of the analytical results. We can conclude that the vaccination strategy is successful in suppressing the spread of TB among the human population.

1. Introduction and model construction

Tuberculosis (TB) is a contagious disease caused by a bacillus –commonly *Mycobacterium tuberculosis*. The disease is transmitted by active TB patients through tiny droplets of liquid that comes out of the patient's body when the patient speaks, coughs, or sneezes. The tiny droplets of liquid are then scattered in the air and inhaled by other individuals [1].

In some cases, TB infection can progress directly into active TB. However, in a large number of cases, TB infection initially enters a latent phase, in which the bacillus is made 'asleep' or inactive by the immune system of the infected individual. The development of TB infection into active TB is highly dependent on the immunity of the infected individual. Latent TB infection may progress to active TB if the infected individual experiences a decreased immunity, for example, due to HIV [1]. Such changes can occur at any time –from several months to several decades after the initial infection [2].

To end TB epidemiology, WHO recommends the use of Bacille Calmette-Guérin (BCG) vaccination as a prevention act against the disease. The BCG vaccine contains an attenuated *Mycobacterium bovis* [3]. BCG vaccine protection lasts up to 15 years since vaccination [4]. Although the efficacy of this vaccine continues to be controversial, until now, BCG vaccine is still the only vaccine used to prevent TB disease. Its use is considered sufficient to counteract the chronic form of TB and suppress the actual death toll [5].

According to some researches, BCG vaccination administered at a younger age had a more significant impact than the vaccine given to individuals older than 15 years old [6]. The distinctive characteristics of TB disease in different age groups are not only seen in individual response to BCG vaccine. Children (1-15 years) were noted to have lower levels of morbidity and mortality rate than adults (15-65 years) [7]. Also, adults have a higher probability of transmitting the disease and entering the active TB phase immediately after the initial infection. The study of the interaction dynamics between age, immunity, and its effects on the evolution of pathogens is essential for the development of future vaccination strategies [8].

Many mathematical models have been constructed to investigate the dynamics of TB disease –be it models focusing on the process of relapse and reinfection [9, 10], immigration impacts [11], vaccinations [12, 13], as well as models that discuss the influence of the age class [14]. However, the model constructed by Yu, Mingtao, & Sanling (2017) [14] hadn't distinguished the latent, infected, and recovered population according to its age groups –hence, the model was unable to explain the reinfection process.

Therefore, in this article, we construct a mathematical model that accommodates the vaccination process and age-group classification in TB transmission. To accommodate the latent phase contained in TB disease, this model adopts the SEIR epidemic model, in which each compartment is further divided by age-group. The susceptible subpopulations are then also divided according to the vaccination criteria. In this model, we assume that the vaccine efficacy in children is 100%; hence children who are currently under the protection of BCG vaccine cannot be infected with TB. We also assume that the vaccine efficacy in adults will decrease and that only adults can transmit TB. The transmission diagram of this model can be seen in Figure 1.

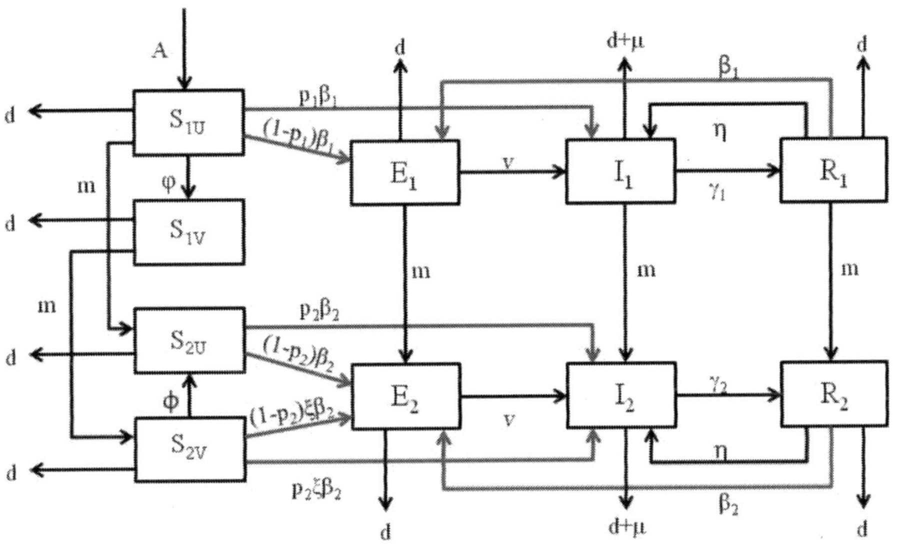

Figure 1: Transmission diagram of TB vaccination model with age-class.

The system of ordinary differential equation governing the model is given by:

$$\begin{aligned}
\frac{dS_{1_U}}{dt} &= A - (I_2\beta_1 + d + m + \varphi)\, S_{1_U},\\
\frac{dS_{1_V}}{dt} &= \varphi\, S_{1_U} - (d+m)\, S_{1_V},\\
\frac{dS_{2_U}}{dt} &= mS_{1_U} + \phi\, S_{2_V} - (I_2\beta_2 + d)\, S_{2_U},\\
\frac{dS_{2_V}}{dt} &= mS_{1_V} - (\xi\, I_2\beta_2 + d + \phi)\, S_{2_V},\\
\frac{dE_1}{dt} &= \beta_1 I_2\left((1-p_1)\, S_{1_U} + R_1\right) - (d+m+v)\, E_1,\\
\frac{dE_2}{dt} &= \beta_2\left((1-p_2)\,(\xi\, I_2 S_{2_V} + I_2 S_{2_U}) + I_2 R_2\right) + mE_1 - (d+v)\, E_1,\\
\frac{dI_1}{dt} &= I_2\beta_1 p_1 S_{1_U} + \eta\, R_1 + vE_1 - (d+m+\mu+\gamma_1)\, I_1,\\
\frac{dI_2}{dt} &= \beta_2\,(\xi\, I_2 p_2 S_{2_V} + I_2 p_2 S_{2_U}) + mI_1 + \eta\, R_2 + vE_2 - (d+\mu+\gamma_2)\, I_2,\\
\frac{dR_1}{dt} &= I_1\gamma_1 - (I_2\beta_1 + d + \eta + m)\, R_1,\\
\frac{dR_2}{dt} &= mR_1 + I_2\gamma_2 - (I_2\beta_2 + d + \eta)\, R_2.
\end{aligned} \tag{1}$$

2. Model analysis

In this section, we analyze the existence of equilibrium points and basic reproduction number from system 1. System 1 has two equilibrium points, i.e., disease-free equilibrium (DFE) and endemic equilibrium (EE). The DFE point is given by:

$$DFE = (S_{1_U}, S_{1_V}, S_{2_U}, S_{2_V}, E_1, E_2, I_1, I_2, R_1, R_2) = (S^*_{1_U}, S^*_{1_V}, S^*_{2_U}, S^*_{2_V}, E^*_1, E^*_2, I^*_1, I^*_2, R^*_1, R^*_2),$$

where $E^*_1 = E^*_2 = I^*_1 = I^*_2 = R^*_1 = R^*_2 = 0$, $S^*_{1_U} = \frac{A}{d+m+\varphi}$, $S^*_{1_V} = \frac{\varphi A}{(m+d)(d+m+\varphi)}$, $S^*_{2_U} = \frac{(d^2+dm+d\phi+m\phi+\phi\varphi)Am}{(m+d)(d+m+\varphi)(d+\phi)d}$, $S^*_{2_V} = \frac{Am\varphi}{(m+d)(d+m+\varphi)(d+\phi)}$. In the disease-free state, the total population can be obtained by summing up the total healthy subpopulations, i.e., $N = S_{1_U} + S_{1_V} + S_{2_U} + S_{2_V} = \frac{A}{d}$, which means that the total population in the disease-free state is neither affected by the age-group classification nor by the vaccination process. However, the proportion of each healthy subpopulations are indeed affected by the said factors.

The ratio of the un-vaccinated individuals to the vaccinated individuals is given by:

$$\frac{S_{1_U} + S_{2_U}}{S_{1_V} + S_{2_V}} = \frac{d^3 + 2\,d^2m + d^2\phi + dm^2 + 2\,dm\phi + m^2\phi + m\phi\,\varphi}{(d+\phi+m)\,d\varphi}.$$

By dismissing the age-group classification ($m = 0$), we have

$$\frac{S_{1_U} + S_{2_U}}{S_{1_V} + S_{2_V}} = \frac{d}{\varphi}. \tag{2}$$

Vaccination is a recommended intervention to prevent the spread of TB. Therefore, the ratio of unvaccinated individuals to the vaccinated individuals is expected to be zero. Based on equation 2, it is seen that the ratio of the susceptible human who do not use vaccination and who use vaccination can tend to zero if the life expectancy of human increases or vaccination process increases.

The second equilibrium point is the one that describes a condition where all of the subpopulations in the system 1 is positive. The said point cannot be shown explicitly in this article. However, in the simplified case of system 1, i.e., when $(m = 0, p_1 = p_2 = p, \gamma_1 = \gamma_2 = \gamma, \beta_1 = \beta_2 = \beta)$, the EE point can be shown explicitly as a function of I in the following form:

$$EE = (S_1^* = f_1(I), S_2^* = f_2(I), S_3^* = f_3(I), E^* = f_4(I), R^* = f_5(I), I^*)$$

where

$$\begin{aligned}
S_1^* =& \frac{A}{I\,\beta + d + \varphi},\\
S_2^* =& \frac{\varphi\, A}{(\phi + d)\,(I\,\beta + d + \varphi)},\\
S_3^* =& \frac{\phi\,\varphi\, A}{(I\,\beta + d)\,(I\,\beta + d + \varphi)\,(\phi + d)},\\
E^* =& \frac{I\,\beta\,\epsilon}{(I\,\beta + d)\,(I\,\beta + d + \varphi)\,(d + v)\,(\phi + d)\,(I\,\beta + d + \eta)},\\
\epsilon =& \gamma\,\beta^2\,(d + \phi)\,I^3 - \beta\,(d + \phi)\,(A\beta\,p - A\beta - 2\,d\gamma - \gamma\,\varphi)\,I^2 + (A\beta\,\left(2\,d^2 + d\eta + 2\,d\phi + \eta\,\phi + \phi\,\varphi\right) +\\
&\gamma\,d\,(d + \varphi)\,(d + \phi) - A\beta\,p\,(2\,d^2 + d\eta + 2\,d\phi + \eta\,\phi + \phi\,\varphi))I - A\,\left(d^2 + d\phi + \phi\,\varphi\right)\,(p - 1)\,(d + \eta),\\
R^* =& \frac{\gamma\, I}{I\,\beta + d + \eta},
\end{aligned}$$

and I^* as a solution of the fourth degree characteristic polynomial that has a general form as follows:

$$I(A_0 + A_1\, I + A_2\, I^2 + A_3\, I^3) = 0, \tag{3}$$

where

$$\begin{aligned}
A_0 =& A\,\left(d^2 + d\phi + \phi\,\varphi\right)\,(dp + v)\,(d + \eta)\,\beta - d\,(d + \varphi)\,(d + v)\,(d + \phi)\,(d^2 + d\eta + d\gamma + d\mu +\\
&\eta\,\mu),\\
A_1 =& A\,(dp + v)\,\left(2\,d^2 + d\eta + 2\,d\phi + \eta\,\phi + \phi\,\varphi\right)\,\beta^2 - (d + \phi)\,(3\,d^4 + 2\,d^3\eta + 3\,d^3\gamma + 3\,d^3\mu +\\
&3\,d^3 v + 2\,d^3\varphi + 2\,d^2\eta\,\mu + 2\,d^2\eta\,v + d^2\eta\,\varphi + 2\,d^2\gamma\,v + 2\,d^2\gamma\,\varphi + 3\,d^2\mu\,v + 2\,d^2\mu\,\varphi + 2\,d^2 v\varphi +\\
&2\,d\eta\,\mu\,v + d\eta\,\mu\,\varphi + d\eta\,v\varphi + d\gamma\,v\varphi + 2\,d\mu\,v\varphi + \eta\,\mu\,v\varphi)\beta,\\
A_2 =& A\,(d + \phi)\,(dp + v)\,\beta^3 - (d + \phi)\,(3\,d^3 + d^2\eta + 3\,d^2\gamma + 3\,d^2\mu + 3\,d^2 v + d^2\varphi + d\eta\,\mu + d\eta\,v +\\
&d\gamma\,v + d\gamma\,\varphi + 3\,d\mu\,v + d\mu\,\varphi + dv\varphi + \eta\,\mu\,v + \mu\,v\varphi)\beta^2,\\
A_3 =& -\beta^3\,(d + \phi)\,\left(d^2 + d\gamma + d\mu + dv + \mu\,v\right).
\end{aligned}$$

Please note that this endemic equilibrium point will exist if there is at least one positive root from polynomial in equation 3, which lead us to condition

$$I_1^* + I_2^* + I_3^* + I_4^* = -\frac{A_2}{A_3} \geq 0.$$

By choosing a parameter set, i.e., $A = \frac{10000}{65}$, $d = \frac{1}{65}$, $\mu = 0.0025$, $\beta = 3.15 \times 10^{-5}$, $m = \frac{1}{15}$, $v = 6$, $\eta = 0.00341$, $\gamma = 0.25$, $\phi = \frac{1}{15}$, $p = 0.1$, $\varphi = 0$, such that we have $-\frac{A_2}{A_3} = 110.06 \geq 0$, the endemic equilibrium is given by $(S_1, S_2, S_3, E, I, R) = (2374.91, 28.95, 30.26, 63.41, 1536.36, 5716.47)$.

Next, we analyze the basic reproduction number ($\mathcal{R}_0$) of system 1. Basic reproduction number is defined as the expected number of new cases caused by one initial infection in a closed population during one infection period [15]. The $\mathcal{R}_0$ of system 1 is given by the spectral radius of the spectral radius of the next-generation matrix of system 1. Please see [15] for further detail about how to construct $\mathcal{R}_0$ using this approach. The basic reproduction number of system 1 is given by

$$\mathcal{R}_0 = \frac{mv\beta_1 A(1-p_1)\omega_2}{(d+v)(d+m+v)(d^2+d\eta+2dm+d\mu+d\gamma_1+\eta m+\eta\mu+m^2+m\mu+m\gamma_1)\omega_1} + \frac{mv\beta_2 A(1-p_2)\left(d\varphi\xi+d^2+dm+d\phi+m\phi+\phi\varphi\right)(d+\eta)}{(d+m)(d+\phi)d(d+v)\omega_1} + \frac{\left(d^2+2d\eta+dm+\eta^2+\eta m+\eta\gamma_1\right)m\beta_1 p_1 A}{(d^2+d\eta+2dm+d\mu+d\gamma_1+\eta m+\eta\mu+m^2+m\mu+m\gamma_1)\omega_1} + \frac{(d+\eta)\beta_2 p_2 Am\left(d\varphi\xi+d^2+dm+d\phi+m\phi+\phi\varphi\right)}{(d+m)(d+\phi)d\omega_1} \quad (4)$$

where

$$\omega_1 = (d+m+\varphi)\left(d^2+d\eta+d\mu+d\gamma_2+\eta\mu\right),$$

$$\omega_2 = 2d^3+4d^2\eta+3d^2m+d^2\mu+d^2v+d^2\gamma_1+2d\eta^2+4d\eta m+2d\eta\mu+2d\eta v+2d\eta\gamma_1+dm^2+dm\mu+ dmv+dm\gamma_1+\eta^2 m+\eta^2\mu+\eta^2 v+\eta m^2+\eta m\mu+\eta mv+\eta m\gamma_1+\eta v\gamma_1.$$

In a simpler situation, where no age-class is implemented ($m = 0, p_1 = p_2 = p, \gamma_1 = \gamma_2 = \gamma, \beta_1 = \beta_2 = \beta$), the $\mathcal{R}_0$ is simply given by:

$$\mathcal{R}_0^* = \frac{(1-p)(K_1+K_2)v}{(v+d)(\gamma+d+\mu)} + \frac{p(K_1+K_2)}{\gamma+d+\mu},$$

where $K_1 = \frac{\beta A}{d+\varphi}$, $K_2 = \frac{\beta\phi\varphi A}{d(d+\varphi)(\phi+d)}$.

To show how the $\mathcal{R}_0$ determines the existence of endemic equilibrium point, we choose a set of parameter, i.e., $A = \frac{10000}{65}$, $d = \frac{1}{65}$, $\mu = 0.0025$, $\beta_1 = 2.75 \times 10^{-5}$, $\beta_2 = 3.15 \times 10^{-5}$, $m = \frac{1}{15}$, $v = 6$, $\eta = 0.00341$, $\gamma_1 = 0.496$, $\gamma_2 = 0.25$, $\phi = \frac{1}{15}$, $p_1 = 0.025$, $p_2 = 0.05$, $\varphi = 10^{-3}$, $\xi = 0.3$, such that $\mathcal{R}_0 = 1.20312 > 1$. From that set of parameter, the endemic equilibrium point is given by:

$$EE = (1575.858, 19.206, 3305.377, 14.707, 4.230, 20.327, 46.039, 530.024, 228.267, 4162.359).$$

3. Simulation of the autonomous system

The system 1 is simulated using the parameter values in Table 1 to illustrate the theoretical results contained in this article.

Table 1: Model parameters.

Par	Definition	Value	Par	Definition	Value
A	Birth rate	10000/65	α	Vaccine efficacy	0.7 [17]
φ	Vaccination rate	1	ξ	Vaccine failure ratio	0.3
m	Maturity rate	1/15 [14]	v	Transition rate	6 [14]
β_1	Children infection rate	2.75×10^{-5} [7]	γ_1	Children recovery rate	0.496 [16]
β_2	Adults infection rate	3.15×10^{-5}	γ_2	Adults recovery rate	0.25 [18]
p_1	Children rapid infection ratio	0.025 [8]	η	Relapse rate	0.00341 [19]
p_2	Adults rapid infection ratio	0.05 [16]	d	Natural death rate	1/65
ϕ	Vaccine effect reduction rate	1/15 [4]	μ	Disease-induced death rate	0.0025 [20]

The first simulation is given in Figure 2 to see how $\mathcal{R}_0$ change respect to the change of other parameters in model 1. Sensitivity analysis of the basic reproduction number ($\mathcal{R}_0$) in equation 4 is performed on the parameters that affect vaccine performance, i.e., vaccination rate (φ), vaccine effect reduction rate (ϕ), and vaccine failure ratio (ξ).

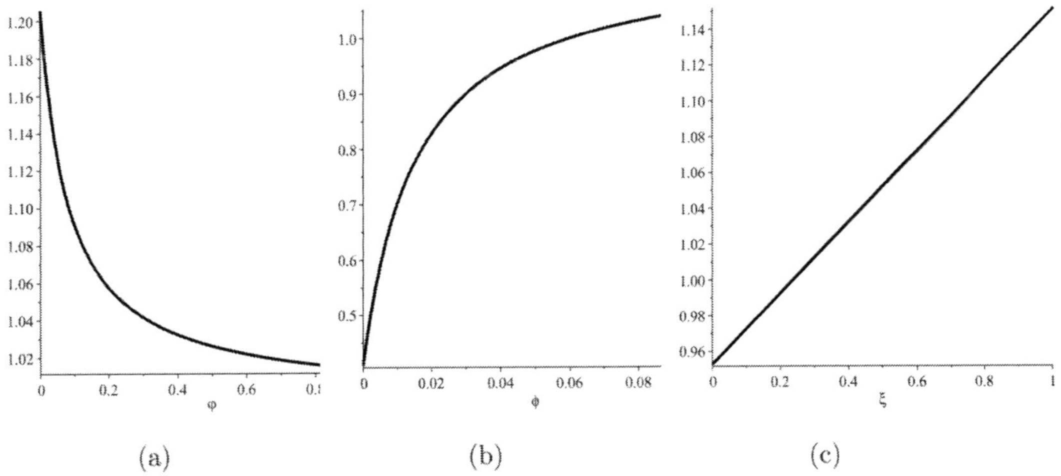

Figure 2: Basic reproduction number ($\mathcal{R}_0$) sensitivity analysis curve refers to (a) vaccination rate, (b) vaccine effect reduction rate, and (c) vaccine failure ratio. The y-axis is the $\mathcal{R}_0$.

From Figure 2, we can see that $\mathcal{R}_0$ has a negative correlation with φ and a positive correlation both with ϕ and ξ. It can be seen that enlarging φ will reduce $\mathcal{R}_0$. In the other hand, enlarging ϕ or ξ will increase $\mathcal{R}_0$. This results next will be illustrate in how the change of φ and ϕ impact the dynamic of model 1.

From the autonomous simulation with the variation of φ value in Figure 3(a), we can see that:

- Although there is indeed a vaccination process in the population, the total population in S_{2_V} eventually decreases. That is because the vaccinated adults will lose its vaccine effect, 15 years after the vaccination. For the same reason, the total population in S_{2_U} eventually increases.
- The dynamics in the infected subpopulations (E_1, E_2, I_1, I_2, R_1dan R_2) apparently more affected by other factors, e.g., transition rate (v) and recovery rate (γ). The variation of φ value does not seem to affect the said subpopulations significantly. However, the total individuals in each subpopulation seem slightly greater when the vaccination rate is lower.

Aside from the distinctive vaccination rate used in the simulation, in Figure 3(b), we can see that the total healthy population tends to increase while the total infected population tends to decrease. When $\varphi = 10$, the total population keeps changing until it reaches $DFE = (15, 1860, 6614, 1511, 0, 0, 0, 0, 0, 0)$ and when $\varphi = 0.05$, the total population keeps changing until it reaches $EE = (1093, 666, 4237, 522, 2, 12, 19, 39, 97, 2983)$.

In Figure 4(a), we can see that the significant changes only happen on the susceptible adults subpopulations. This is because the process of vaccine effect reduction only happens on the said subpopulations. Observe that the increase of total individual in S_{2_U} and the decrease of the total individual in S_{2_V} is slightly slower when the vaccine offers protection for as long as 20 years than when the vaccine protection only lasts for ten years. Also, the shorter vaccine protection period

IOP Conf. Series: Journal of Physics: Conf. Series **1108** (2018) 012050 doi:10.1088/1742-6596/1108/1/012050

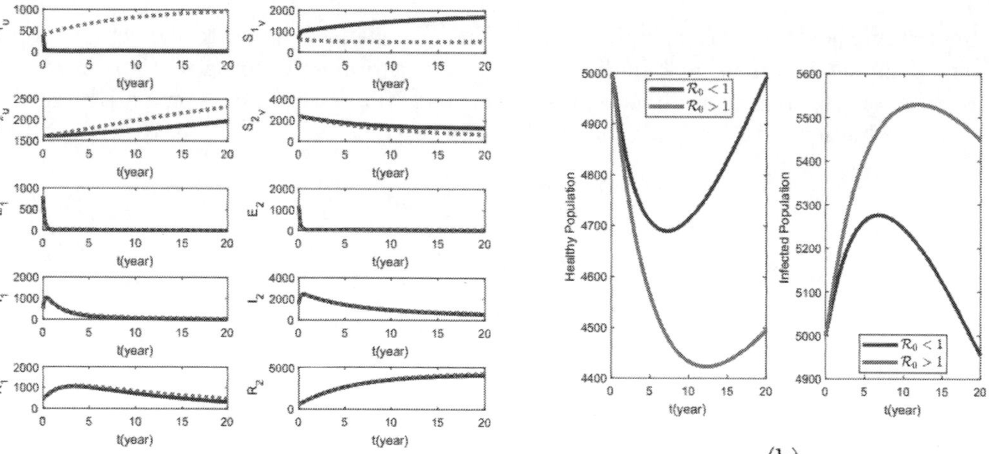

(b)

(a)

Figure 3: (a) The dynamics of each compartment in model 1 and (b) the dynamics of total healthy $(S_{1_U} + S_{1_V} + S_{2_U} + S_{2_V})$ and infected $(E_1 + E_2 + I_1 + I_2 + R_1 + R_2)$ population, with $\varphi = 10$ for $\mathcal{R}_0 = 0.99846$ (blue), $\varphi = 0.05$ for $\mathcal{R}_0 = 1.12655$ (red), and $t \in [0, 20]$ years.

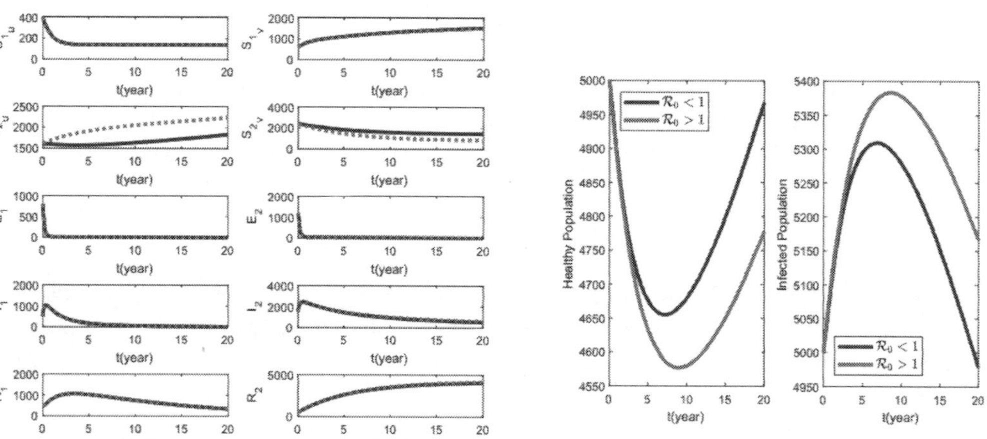

(b)

(a)

Figure 4: (a) The dynamics of each compartment in model 1 and (b) the dynamics of total healthy $(S_{1_U} + S_{1_V} + S_{2_U} + S_{2_V})$ and infected $(E_1 + E_2 + I_1 + I_2 + R_1 + R_2)$ population, with $\phi = 1/20$ for $\mathcal{R}_0 = 0.97713$ (blue), $\phi = 1/10$ for $\mathcal{R}_0 = 1.05281$ (red), and $t \in [0, 20]$ year.

results in more infected and recovered adults. It can also be calculated that when $\phi = 1/20$, the system 1 tends to approach $DFE = (142,\ 1733,\ 6358,\ 1767,\ 0,\ 0,\ 0,\ 0,\ 0,\ 0)$ as $t \to \infty$, while when $\phi = 1/10$, it tends to approach $EE = (142,\ 1728,\ 5669,\ 989,\ 1,\ 5,\ 1,\ 119,\ 5,\ 1326)$ as $t \to \infty$.

4. Conclusion

In this article, a mathematical model of the spread of TB disease involving the age-classes in susceptible compartments has been developed. The model has two equilibrium points, i.e., the disease-free equilibrium and the endemic equilibrium point. The basic reproduction number ($\mathcal{R}_0$) has been constructed using the next-generation matrix approaches. From the analytical study on the system, we know that other than administering vaccination, TB transmission can be reduced by increasing life expectancy. However, the autonomous simulation shows that the best way to reduce TB transmission is by working on its performance, involving its vaccination rate, vaccine protection period, and vaccine efficacy.

Acknowledgements

We thank all reviewers for their constructive comments and suggestions. This research was financially supported by Universitas Indonesia, with PITTA research grant scheme 2018 (Number ID: 2252/UN2.R3.1/HKP.05.00/2018).

References

[1] Crofton J, Horne N and Miller F 2009 *Clinical Tuberculosis* 3rd edn (Oxford: Macmillan Education)
[2] Song B, Castillo-Chavez C and Aparicio J P 2002 Tuberculosis models with fast and slow dynamics: The role of close and casual contacts *Mathematical Biosciences* **180** pp 187-205
[3] McShane H 2011 Tuberculosis vaccines: Beyond bacille Calmette-Guérin *Philosophical Transactions of the Royal Society B: Biological Sciences* **366** pp 2782-2789
[4] Barreto M L, Pereira S M and Ferreira A A 2006 BCG vaccine: efficacy and indications for vaccination and revaccination *Journal de Pediatria* **82** pp 4554
[5] Luca S and Mihaescu T 2013 History of BCG vaccine *Mædica* **8** pp 53-58
[6] World Health Organization 2017 *Report on BCG Vaccine Use for Protection against Mycobacterial Infections including Tuberculosis, Leprosy, and Other Nontuberculosis Mycobacteria (NTM) Infections*
[7] World Health Organization 2015 *Global Tuberculosis Report 2015*
[8] Donald P R, Marais B J and Barry C E 2010 Age and the epidemiology and pathogenesis of tuberculosis *The Lancet* **375** pp 1852-1854
[9] Bhunu C P, Garira W, Mukandavire Z and Zimba M 2008 Tuberculosis transmission model with Chemoprophylaxis and treatment *Bulletin of Mathematical Biology* **70** pp 1163-1191
[10] Feng Z, Castillo-Chavez C and Capurro A F 2000 A model for tuberculosis with exogenous reinfection *Theoretical Population Biology* **57** pp 235-247
[11] Jia Z W, Tang G Y, Jin Z, Dye C, Vlas S J, Li X W, Feng D, Fang L Q, Zhao W J and Cao W C 2008 Modeling the impact of immigration on the epidemiology of tuberculosis *Theoretical Population Biology* **73** 437-448
[12] Liu S, Li Y, Bi Y and Huang Q 2017 Mixed vaccination strategy for the control of tuberculosis: A case study in China *Mathematical Biosciences and Engineering* **14** pp 695-708
[13] Tasmi, Aldila D, Soewono E and Nuraini N 2016 Mathematical model for transmission of tuberculosis in badger population with vaccination *AIP Conference Proceedings* **1723**
[14] Yu Z, Mingtao L and Sanling Y 2017 Analysis of transmission and control of tuberculosis in Mainland China, 2005-2016, based on the age-structure mathematical model *International Journal of Environmental Research and Public Health* **14** p 1192
[15] Diekmann, Hesterbeek and Roberts 2010 The construction of next-generation matrices for compartmental epidemic models *Journal of The Royal Society*
[16] Blower S M, McLean A R, Porco T C, Small P M, Hopewell P C, Sanchez M A and Moss A R 1995 The intrinsic transmission dynamics of tuberculosis epidemics *Nat Med* **1** pp 815-821
[17] Roy A et al 2014 Effect of BCG vaccination against Mycobacterium tuberculosis infection in children: systematic review and meta-analysis *British Medical Journal* **349** pp 1-11
[18] Ncube R T, Takarinda K C, Zishiri C, Boogaard W, Mlilo N, Chiteve C, Siziba N, Trinchán F and Sandy C 2017 Age-stratified tuberculosis treatment outcomes in Zimbabwe: are we paying attention to the most vulnerable? *Public Health Action* **7** pp 212-217
[19] Millet J P, Shaw E, Orcau A, Casals M, Miró J M and Caylá J A 2013 Tuberculosis recurrence after completion treatment in a European city: reinfection or relapse? *PLOS ONE* **8**
[20] World Health Organization 2013 *Global Tuberculosis Report 2013*

IOP Conf. Series: Journal of Physics: Conf. Series **1108** (2018) 012051 doi:10.1088/1742-6596/1108/1/012051

Skill Analysis of Students' Creative Thinking In Implementation Of Problem Based Learning With Plastic Waste Handling Context

F Eny [1], **R Momo** [2], **and S Wahyu** [2]

[1] Student of Chemistry Science Education School of Postgraduate Studies, Indonesia Education University
[2] Lecturers of Chemistry Education In Indonesia Education University

Abstract : The objective of this research is analysing the implementation of problem based learning (PBL) towards students' creative thinking ability in plastic waste handling context. According to the rules of High School Competency Standard, students must have critical thinking skills, creative, innovative and be able to collaborate and communicate. One of the competencies that will be developed is the ability to think creatively which includes: the ability to think smoothly, flexibility, originality, elaborative and evaluative in plastic waste handling context. The research method was descriptive qualitative research. The subjects of this research were 65 students consisting of 2 classes of Grade X which are X IPA 2 and X IPA 3 in SMAN 1 North Tambun Bekasi Regency in year lesson 2017/2018. The instruments used in this study included written test, observation sheet, student work sheet, questionnaire and interview guideline. The data results were obtained from Pretest and Posttest and being processed and analysed using N-Gain. The creative thinking skill has improved with N-Gain 0.64 which is in the medium category, it shows that ability of mastery of concepts and skills of creative thinking can be developed through PBL.

1. Introduction

Learning is a set of actions designed to support the learning process of students. According to the National Education System Act (UUSPN) No. 20 of 2003[1], learning is the process of students' interaction with educators and learning resources in a learning environment. A learning will be success when a positive behavior changes occur in students in accordance with objective learning plan. In the revised 2013 curriculum, the competencies required to improve the skills and abilities that can be applied to face the challenges of life in the 21st century.

Competencies to be achieved by students after graduating secondary education must acquire skills to live with a positive attitude with creative and critical thinking, communicative , and collaborative with honesty and openness based on the potential processes and products of science. Chemistry as a part of natural science is gained and developed on the basis of experiments in seeking answers to the question of what, why, and how natural phenomena especially relating to the composition, structure, properties, transformation, dynamics and energy of matter. Chemistry as a process or method of investigation involves thinking, reasoning, formulating problems, conducting experiments and observations, analyzing data and concluding to obtain products [2], [3].

In the framework of mastery of 21st century skills of chemistry in high school regarded not only for the transfer of knowledge and skills *(transfer of knowledge and skills)* students, but also establish a high-level thinking skills (analysis, synthesis, critical, creative and innovative) through scientific work. Knowledge, skills, ability to think and act on chemistry learning will equip students for living in the community [4].

In relation to the above, it is necessary to strive for a learning that is able to activate the ability of students, especially those which related to life skills of the 21st century that one of them is the creative thinking skills in chemistry learning, especially in terms of solving a real problem in everyday life. In the 2013 curriculum, the highlight of instructional models are the activity and creativity, inspiring, fun and initiate, student-centered, authentic, contextual and meaningful learning, among others: Problem Based Learning Model. According to Tan [5], Problem-based learning is a learning innovation since the thinking skills of students truly optimized through group work or team systematically, so that students can empower hone, test, and develop the ability to think continually.

The learning strategy with PBL is expected to hone the creative thinking ability of students in chemical learning that is related to daily life, especially in the case of handling of plastic waste. Nowadays plastic waste becomes the main problem because of the increasing number of people, activation of life pattern, social and economic level and technological progress. Therefore, one of the environmental damage has become a major problem of plastic waste [6]. This research will be conducted to analyze the implementation of PBL in students creative thinking ability in plastic waste handling context. Creative thinking skills include: the ability to think smoothly, flexibility, originality, elaborative and evaluative [7]. Table 1 shows the indicators of achievement for the creative thinking skills.

Table 1. Aspects of creative thinking.

No	Aspects of creative thinking	Indicators
1	Well-thinking	Ask a lot of questions
2	Elaborative thinking	Able to develop and explain in detail about the idea
3	Flexible thinking	Reveal many ideas about a problem
4	Original thinking	Give consideration based on his own point of view
5	Evaluative thinking	Have a rationale in a decision taken

(Munandar [8])

The *PBL* phase consists of five stages: **Stage 1:** Organizing students to the problem, **Stage 2:** Organizing students to learn, **Phase 3:** Guiding independent inquiry and group, **Phase 4:** Developing and presenting the works and exhibitions, **stage 5:** Analyzing and evaluating the problem-solving process [9].

2. Method

The method used in this research was qualitative descriptive research. According Sukardi ([10]: 163) descriptive research was a research method that describe the object or subject in carefully in accordance with what is with the aim to describe the systematic facts and characteristics of the object studied appropriately. This research was conducted at SMAN 1 North Tambun of Bekasi Regency in class X, those are X IPA 2 and X IPA 3. The number of students will be classified to high, medium and low group based on the chemistry daily score. Data collection techniques were test and non-test techniques. The instruments used in this study include written test, observation sheet, student work sheet, questionnaire and interview guideline. The results of this study were obtained from pretest and posttest and being processed and analyzed using N-Gain and SPSS formula.

3. Results and Discussion

At the beginning of PBL, students made direct observations to garbage dumps in groups so the students can actually found problems about plastic waste. Students propose questions about plastic waste written on the work sheet prepared. Those questions will be the basis to formulate a problem that will be solved through group discussion. The next step was organize students to learn and guide independent and group investigations. At this stage, students were conduct a group discussions and trace the information to complete the formulation of predetermined problems. The last stage of PBL was present the work in the form of ideas about plastic waste containers. Then, the problem-solving process will be analyzed.

Although the learning activities were conducted in groups, the assessment was done individually through pre-test and post-test of improvement of students creative thinking ability. Differences in pre-test and post-test values of students measured by N-Gain with the criteria of increasing the score of N-Gain as follows:

Table 2. N-Gain Score Increase Criteria.

N-Gain Score	Interpretation
$(\langle g \rangle) > 0.7$	High
$0.3 < (\langle g \rangle) \leq 0.7$	Medium
$(\langle g \rangle) \leq 0.3$	Low

Table 3 shows the average score of students pre-test and post-test for creative thinking skills.

Table 3. The average value of Pre-Test and Post-test of Creative Thinking.

No	Creativity	Average		N - Gain
		Pre-test	Post-test	
1	Creative Thinking	28,0	69,1	0,64

Based on Table 3 and the criteria for increasing the score of N-Gain shows that the students' thinking skills have improved after the implementation of problem based learning. The increase was categorized as a moderate increase, indicated in the N - Gain of 0.64 . Regarding the creative thinking ability of our students will discuss in more detail as follows:

3.1 Generally Creative Thinking

The overall creative thinking skill has improved with N-Gain 0.64 which was in the medium category. This can be seen from the following figure:

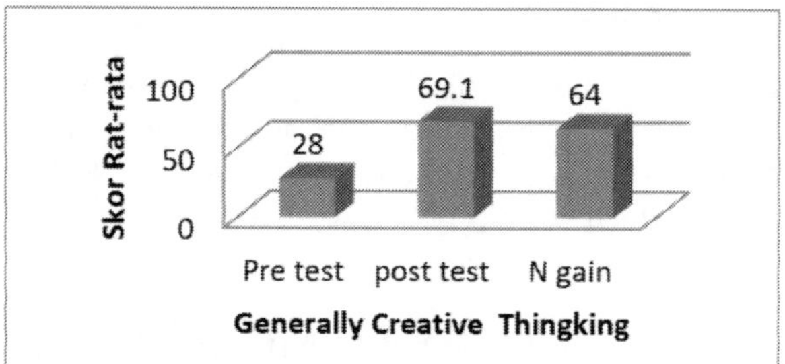

Figure 1. N- Gain the ability of Creative Thinking in its entirety.

In the figure, the overall creative thinking skill has improved significantly.

3.2 Creative thinking by Group Category

Complex thinking is called high-level thinking consisting of critical thinking, creative thinking, problem solving and decision making (Tanwil [11]). So, creative thinking is similar with high-level thinking and problem-solving process. Based on the category of creative thinking, the improvement in high group is bigger than medium and low groups. This is due to the mastery of the concept in the high group is stronger than the medium and low groups. This is shown in the following figure:

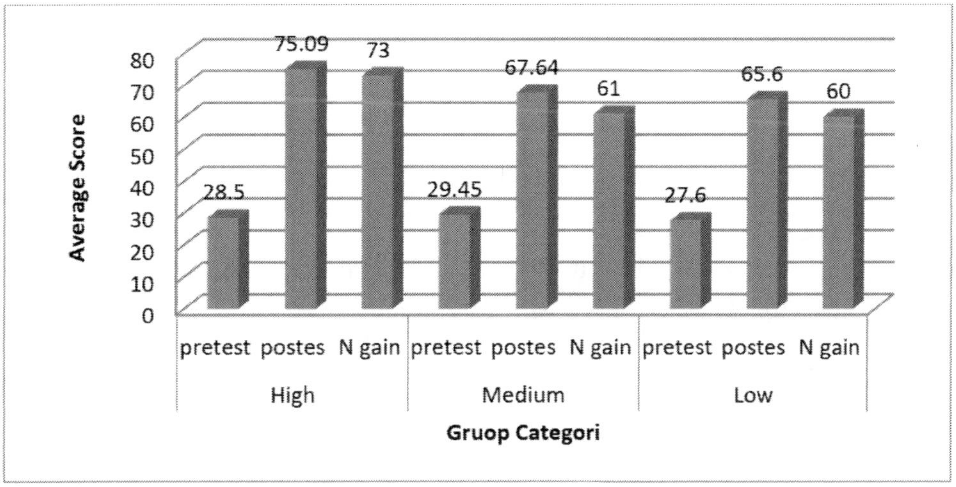

Figure 2. Creative thinking by Group Category.

3.3 Creative thinking by Indicators

The creative thinking skills of students based on the indicators were measured by the average value of creative thinking ability on the average of pre-test and post-test and N-Gain values on handling of plastic waste, it can be seen in figure 3.

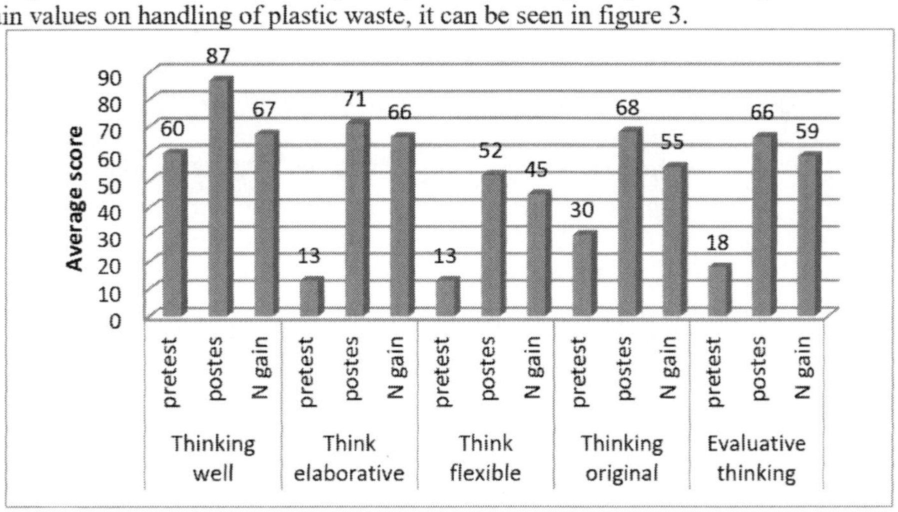

Figure 3. The average of pretest, posttest and N-Gain based on indicators of creative thinking.

Based on the value of N-gain in each indicator of creative thinking, it has increased especially in the ability to think smoothly that is the ability of learners to ask questions with N-gain value is 0.67. Whereas in the ability to think flexibility is to reveal the idea of a problem has the lowest N-gain. This was because there are still many ideas and those were not original, it can be seen in the ability of original thinking that has a value of N-gain above slightly from the ability to think flexibility of 0.55. Nevertheless, overall

IOP Conf. Series: Journal of Physics: Conf. Series **1108** (2018) 012051 doi:10.1088/1742-6596/1108/1/012051

creative thinking skills have improved on each indicator based on N-gain value in the medium category.

4. Conclusion

Based on the results, it can be concluded that problem-based learning can improve the ability of creative thinking. PBL can be recommended and applied to the learning process of chemistry that has suitability characteristics between learning materials with PBL stages.

5. References

[1] Permendiknas RI UU No. 20 2003 *Sistem Pendidikan Nasional* (Jakarta: Menteri Pendidikan Nasional)
[2] Permendikbud No. 20 2016 tentang *Standar Kompetensi lulusan* (Jakarta: Kementrian pendidikan dan kebudayaan)
[3] Permendikbud No. 21 2016 tentang *Standar Isi* (Jakarta: Kementrian Pendidikan dan kebudayaan)
[4] BSNP 2010 *Paradigma Pendidikan Adad XXI* (Jakarta: Departemen Pendidikan Nasional)
[5] Tan, O S 2003 *Problem-based Learning Innovation :* Using Problems to Power Learningin 21 Century (Singapore: Thomson Learning)
[6] Setyowati 2013 Pengetahuan dan prilaku ibu rumah tangga dalam pengelolaan sampah plastik *Jurnal Kesehatan masyarakat Nasional* Vol.7 no 12 pp 562-6
[7] Wahyu W, dkk 2015 Using Problem Based Learning to Improve Students'Creative Thingking Skills on Water Purification *Proceeding of International seminar on Mathhematics, Science and Computer Sciense Education* (MSCEIS 2015)
[8] Munandar, U 2012 *Pengembangan Kreativitas Anak Berbakat* (Jakarta: Rineka Cipta)
[9] Rusmono 2012 *Startegi Pembelajaran dengan Problem Based Learning* (Jakarta: Ghalia Indonesia)
[10] Sukardi 2003 *Metodologi Penelitian Pendidikan* (Yogyakarta: Bumi Aksara)

IOP Conf. Series: Journal of Physics: Conf. Series **1108** (2018) 012052 doi:10.1088/1742-6596/1108/1/012052

Fuzzy Support Vector Machines based on Adaptive Particle Swarm Optimization for Credit Risk Analysis

M D Murjadi*[1] and Z Rustam[1]

[1] Department of Mathematics, Universitas Indonesia, Indonesia.
*Corresponding author: melina.dewi@ui.ac.id

Abstract. Every bank has loaning activities. Banks have several criteria for determining whether credit will give or not because every credit loan has a risk that the credit cannot return. In other words, banks need to analyze the credit applicant before granting the loan. Credit loan is a case of binary classification. The classification from applicant's data might be helpful for the bank in consideration whether the applicant will return the loan or not. Support Vector Machines (SVM) is a classification technique based on structural risk minimization, which is effective for binary classification. This method developed into Fuzzy Support Vector Machines (FSVM), which is able to minimize the influence of outlier in finding the best hyper plane. Adaptive Particle Swarm Optimization (APSO) is an extension of Particle Swarm Optimization (PSO). In APSO-based FSVM, APSO used to determine the fuzzy score by finding the class canter of each attribute that may give the highest accuracy. The result of this study is APSO-based FSVM can give the highest accuracy for each process. The highest rate of accuracy is 75,67%, which used APSO-based FSVM with 70% of training data and linear kernel.

1. Introduction

In this era, credit risk analysis is essential. In credit loaning, creditor and debitor make an agreement for some things, like how much loan will be granted and how long the installment period is. After the loan is given, the debitor has a responsibility to pay the loan as in the agreement. However, the fact is that many debitors fail to pay the loan. The creditor is at risk that the debitor may not be able to fulfill the agreement. This is why banks have some criteria to determine which applicants will be granted the loan [1]. These criteria can be translated into data that can be used for credit risk analysis. This problem is about classification of which applicants will fulfill the agreement and which applicants will not fulfill the agreement.

Some research has been done to identify which model is efficient for classification and prediction in credit risk analysis with statistical method like multiple discriminant analysis [2], classification trees [3], linear programming[4], KNN classifiers [5], and the others. However, these models fail to fulfillmultivariate normality assumption for independent variabels. Recent research for this problem is developed on artificial intelligence methods like genetic algorithm[6], Support Vector Machines (SVM) [7], Artificial Neural Network (ANN) [8], and other methods[9].

SVM is a machine learning method for classification that first introduced by Vapnik in 1992. SVM has been widely applied in classification cases such as for digit handwriting recognition [10], voice identification [11], face detection [12], text classification [13], cancer classification [14], protein classification [15], credit risk data classification [7], and so forth. The development of the SVM technique continues as its application in measuring credit score inference [16], cancer classification by selection of Artificial Bee Colony features [17], prediction of stock movements [18], interference detection systems [19], and predictors of bankruptcy [20].

In 2002, Lin and Wang developed the SVM formula into Fuzzy Support Vector Machines (FSVM) by adding fuzzy values to each data point [21]. In this way, the FSVM method can classify unrecognized data on the SVM method. In 2014, Punniyamoorthy and Sridevi used the FSVM method to analyze credit data using Fuzzy C-Means (FCM) to assign fuzzy values [9]. In 2014, Almasi et al. modified the FSVM method by using Adaptive Particle Swarm Optimization (APSO) method to assign fuzzy values for each data point. This modification can handle data with noise attributes that it have not been handled by the previous FSVM [22].

2. Support Vector Machines

This section reviews the mathematical formulation of SVM. Assume a training data set $\boldsymbol{S}$ with the labeled training points $(\boldsymbol{x_i}, \boldsymbol{y_i})$ where every training data $\boldsymbol{x_i} \in \boldsymbol{R_n}$belongs to a class labeled by $\boldsymbol{Y_i} \in \{\mathbf{1}, -\mathbf{1}\}$for $\boldsymbol{i} = \mathbf{1}, \dots, \boldsymbol{n}$. The optimal hyperplane can be obtained by solving the quadratic optimization problem as in equation (1).

$$\min \frac{1}{2}\|\boldsymbol{w}\|^2 + C\sum \xi_i$$

$$\text{subject to} y_i(\boldsymbol{w}^T\boldsymbol{x_i} + b) \geq 1 - \xi_i \text{and} \xi_i \geq 0 \; \forall \; i = 1, \dots, n \quad (1)$$

The parameter $\boldsymbol{\xi_i}$ is a slack variable. The parameter $\boldsymbol{C}$ is defined by user as a regularization parameter to control the model complexity. For non linear data, data can be mapped to a higher dimensional feature space where data can be separated by a hyperplane. This can be done by using the kernel trick. The primal problem in equation (1) can be transformed to its dual form as in equation (2).

$$\max\left(-\frac{1}{2}\sum_{i=1}^{n}\sum_{j=1}^{n}\alpha_i\alpha_j y_i y_j K(x_i, x_j) + \sum_{i=1}^{n}\alpha_i\right)$$

$$\text{subject to} \sum_{i=1}^{n}\alpha_i y_i = 0 \text{ and} 0 \leq \alpha_i \leq C, \forall i = 1, \dots, n \quad (2)$$

The kernel function is written as $\boldsymbol{K}(.,.)$. Some of the kernel functions are shown in Table 1. In solving equation (2), $\boldsymbol{\alpha_i} > \mathbf{0}$ is the support vector.

Table 1. The kernel functions

Name	**Kernel function**
Linear Kernel	$K(x_i, x_j) = {x_i}^T x_j$
Polynomial Kernel	$K(x_i, x_j) = (t + {x_i}^T x_j)^d$
RBF Kernel	$K(x_i, x_j) = \exp\left(-\frac{\|x_i - x_j\|^2}{\sigma^2}\right)$
Sigmoid Kernel	$K(x_i, x_j) = \tanh(\beta_0 x_i^T x_i + \beta_1)$

3. Fuzzy Support Vector Machines

While SVM is effective for binary classification, SVM is sensitive to outlier. In 2002, Lin and Wang developed the extension method of SVM called Fuzzy Support Vector Machines. In this method, every point on training data will be given fuzzy membership values$\boldsymbol{s_i}$. Assume a training data set $\boldsymbol{S}$ with the labeled training point $(\boldsymbol{x_i}, \boldsymbol{y_i}, \boldsymbol{s_i})$ where every training data $\boldsymbol{x_i} \in \boldsymbol{R_n}$belongs to a class labeled by $\boldsymbol{Y_i} \in \{\mathbf{1}, -\mathbf{1}\}$for $\boldsymbol{i} = \mathbf{1}, \ldots, \boldsymbol{n}$. The optimal hyperplane can be obtained by solving the quadratic optimization problem as in equation (3).

$$\min \frac{1}{2} \|\boldsymbol{w}\|^2 + C \sum s_i \xi_i$$

$$\text{subject to} y_i(\boldsymbol{w}^T \boldsymbol{x}_i + b) \geq 1 - \xi_i \text{and} \xi_i \geq 0 \; \forall \, i = 1, \ldots, n \tag{3}$$

The difference between SVM and FSVM is on the terms$\boldsymbol{s_i \xi_i}$. In FSVM, the measure of error has different weights. The FSVM can also be solved by its dual form.

3.1. Generation of membership.

The fuzzy membership can be obtained by setting it as a function of the distance between the data point and its fixed class center. Data is divided into two classes, that is positive class and negative class. The class center is defined by the mean vector of the attributes as in equation (4) and (5),

$$\boldsymbol{x}_+ = \frac{1}{n_+} \sum_{x_i \in C^+} \boldsymbol{x}_i \tag{4}$$

$$\boldsymbol{x}_- = \frac{1}{n_-} \sum_{x_i \in C^-} \boldsymbol{x}_i \tag{5}$$

where $\boldsymbol{x}_+$ is the mean from positive class and $\boldsymbol{x}_-$is the mean from negative class. Meanwhile, n_+ and $n_-$are the total observations from class C^+andC^-respectively. Then, the maximum radius from the positive class and the negative class are defined as in equation (6) and (7).

$$\boldsymbol{r}_+ = max|\boldsymbol{x}_+ - \boldsymbol{x}_i| \; for \; \boldsymbol{x}_i \in C^+ \tag{6}$$

$$\boldsymbol{r}_- = max|\boldsymbol{x}_- - \boldsymbol{x}_i| \; for \; \boldsymbol{x}_i \in C^- \tag{7}$$

The fuzzy membership s_i is defined as a function of mean and radius from each class as in equation (8) (Almasi, Gooqeri, Asl, dan Tang, 2015).

$$s_i = \begin{cases} 1 - \frac{\|x_+ - x_i\|}{r_+ + \delta} & if \; y = 1 \\ 1 - \frac{\|x_- - x_i\|}{r_- + \delta} & if \; y = -1 \end{cases} \tag{8}$$

where δ is a positive constant with a small value to avoid in the case where $s_i = 0$.

4. Adaptive Particle Swarm Optimization

In the first step of the PSO algorithm, a number of particles are generated. Each particle has speed and position. At each iteration, the speed and position of each particle are calculated and updated according to equation (9) and (10):

$$\boldsymbol{v}_{k+1}^i = w\boldsymbol{v}_k^i + c_1 r_1 (\boldsymbol{p}_k^i - \boldsymbol{x}_k^i) + c_2 r_2 (\boldsymbol{p}_k^g - \boldsymbol{x}_k^i) \tag{9}$$

$$\boldsymbol{x}_{k+1}^i = \boldsymbol{x}_k^i + \boldsymbol{v}_{k+1}^i \tag{10}$$

where$\boldsymbol{x}_k^i$ is the i-th particle position on the k-th iteration and $\boldsymbol{v}_k^i$ is the velocity of the i-th particle on the k-thiteration. The best position of a particle at each iteration is denoted by $\boldsymbol{p}_k^i$ and the global best particle position is denoted by$\boldsymbol{p}_k^g$. The cognitive parameters and social parameters of particles are denoted by c_1 and c_2 of nonnegative value, r_1 and r_2 are random numbers between 0 and 1. The optimal solution will be obtained through the iteration process. The coefficient of inertia is denoted by w at interval [0,1]. To improve the searching ability of PSO, w is defined as in equation (11) [23]:

$$w(i) = w_{max} - \frac{w_{max} - w_{min}}{iter_{max}} \times i \tag{11}$$

In the early stages, the nature of the exploration is more necessary than the nature of exploitation so that it takes the large value of cognitive parameters and small values of social parameters. When approaching the final stage, the nature of exploitation is more necessary than the nature of the exploration so that the large value of social parameters and small values of cognitive parameters are required. Therefore, the values of cognitive parameters and values of social parameters need to be adapted so that the PSO algorithm is modified into Adaptive Particle Swarm Optimization (APSO). Equations (12) and (13) are the formulas used to convert the values of cognitive parameters and values of social parameters adaptively.

$$If\ c_1^{final} < c_1, then\ c_1 = \left(c_1^{final} - c_1^{initial}\right)\left(\frac{iter}{iter_{max}}\right) + c_1^{initial} \quad (12)$$

$$If\ c_2^{final} > c_2, then\ c_2 = \left(c_2^{final} - c_2^{initial}\right)\left(\frac{iter}{iter_{max}}\right) + c_2^{initial} \quad (13)$$

where c_1 is the cognitive parameter and c_2 is the social parameter.

5. FSVM based on APSO

In general, the value of fuzzy membership is obtained through the distance function between the data points and the class center. The development of fuzzy membership valuation method is based on this concept. Although with this method, FSVM can handle noise and outlier, this concept has not been able to solve noise problem in attribute. In the APSO-based FSVM method, APSO is used to change the attributes of a positive class center and a negative class center such that the error generalization can be minimized. After APSO convergence, the function of fuzzy membership assignment is made based on the center position of the refined class [22].

APSO is used to change the position of the class center attribute. By modifying the position of the class center of each class, each data point can achieve different weights that are relevant to its class point. Therefore, the upper and lower limits for each class attribute are set, and APSO is used to change the class center of each attribute within these limits. The structure of each particle in APSO can be seen in Table 2. Each particle consists of two parts, the first part is used to locate the optimal positive class center attribute and the second part is used to locate the optimal negative class center attribute.

Table 2. The structure of each particle APSO

	Parameter 1		Parameter m	Parameter m+1		Parameter 2m
Minimum value	$x_{+i1} - \alpha$	...	$x_{+im} - \alpha$	$x_{-i1} - \alpha$	...	$x_{-im} - \alpha$
	x_{+i1}	...	x_{+im}	x_{-i1}	...	x_{-im}
Maximum value	$x_{+i1} + \alpha$	...	$x_{+im} + \alpha$	$x_{-i1} + \alpha$	...	$x_{-im} + \alpha$

Here is the algorithm for FSVM based on APSO [22]:

1. Initialize FSVM model parameters
2. Initialize APSO population
3. Generate fuzzy memberships based on initial class centers
4. Train FSVM with initial model parameters
5. Evaluate the cost function
6. Update the initial population based on the procedure of APSO until the optimization approach terminate
7. Use the global best of the APSO as the class centers

6. Computational experiments

The data used in this paper is the German Credit version of the numerical data obtained through the UCI repository. This data is obtained from Prof. Dr. Hans Hofmann in 2000. The data has 24 attributes in numerical version. The attributes are the status of existing checking account, duration (in months), credit history, purpose, credit amount, savings account, length of work, installment rate of disposable income, gender, marital status, other debtors, length of residency, property owned, age (in years), other installment plans, housing, number of existing credits at this banks, job, number of people being liable to provide, telephoneand foreign worker [24]. German Credit data was processed several times by modifying the amount of training data. The amount of training data used is 10%, 20%, 30%, 40%, 50%, 60%, 70%, 80%, and 90%. Each group of training data is processed with SVM, FSVM, and FSVM programs based on APSO. MATLAB R2015a is used for these experiments.

The model obtained from each method is used to process the testing data. The programs use linear kernel and polynomial kernels with $\boldsymbol{d} = \mathbf{2}$ and $\boldsymbol{d} = \mathbf{3}$. The number of particles used for APSO is 10 and the number of maximum iterations is 100. The cognitive and social parameters are given as follows: $c_1^{initial} = 1.6$, $c_1^{final} = 0.7$,$c_2^{initial} = 0.9$, $c_2^{final} = 1.7$.The value of α is 0.1 [22]. Predicted results are evaluated by confusion matrix method.

The results of the SVM, FSVM, and FSVM based on APSO can be seen in Table 3. The highest accuracy rate is 75.67% which is obtained from FSVM based on APSO with training data 70% and linear kernel. The results show that FSVM based on APSO has better generalization performance than SVM and FSVM.

Table 3. The experiment results

Kernel	Method	Training data								
		10%	**20%**	**30%**	**40%**	**50%**	**60%**	**70%**	**80%**	**90%**
Linear	SVM	33.33%	41.63%	35.00%	44.33%	42.80%	35.75%	38.67%	36.50%	34.00%
	FSVM	46.89%	70.13%	69.71%	69.50%	72.00%	69.75%	72.67%	72.50%	69.00%
	FSVM - APSO	75.26%	71.25%	71.29%	72.00%	73.40%	74.00%	75.67%	74.50%	70.00%
Poly ($d = 2$)	SVM	66.56%	64.62%	62.71%	62.50%	62.80%	60.75%	63.67%	60.50%	61.00%
	FSVM	66.67%	64.88%	62.57%	63.17%	61.80%	61.00%	63.67%	63.50%	67.00%
	FSVM - APSO	69.00%	66.37%	63.00%	63.83%	62.20%	61.50%	66.00%	67.50%	67.00%
Poly ($d = 3$)	SVM	70.22%	69.63%	68.29%	68.17%	66.00%	68.25%	69.00%	66.00%	61.00%
	FSVM	70.33%	69.50%	68.57%	68.17%	67.00%	68.00%	68.67%	65.50%	60.00%
	FSVM - APSO	70.56%	69.75%	68.57%	68.83%	68.00%	69.25%	69.67%	67.00%	61.00%

7. Conclusions

The main difference between SVM and FSVM method is the fuzzy membership. Therefore, choosing a proper function to generate fuzzy membership is the main problem of FSVM. APSO method is used to find the purified positive and negative class centers. The effects of outliers from each attributes are reduced. From this experiment, the FSVM based on APSO can achieve the highest accuracy among SVM and FSVM.

8. References

[1] DexuHand Miao We 2016Financial Exclusion and Inclusive Finance *China Economist.***11**(3)pp 64–76
[2] Pinches GE and Mingo K A 1975The role of subordination and industrial bond ratings *The Journal of Finance.***30**(1)pp 201–06
[3] Makowski P 1985 Credit scoring branches out *Credit World.* **5**(1)pp 30–37
[4] Glover F1990 Improved linear programming models for discriminant analysis *Decision Sciences.***21**(4)pp 771–85
[5] Henley W Eand Hand DJ1996 A k-nearest-neighbor classifier for assessing consumer credit risk *The Statistician.***45**pp 77–95

[6] Chen MC and Huang SH2003Credit scoring and rejected instances reassigning through evolutionary computation techniques *Expert Systems with Applications.***24**(4)pp433–41

[7] Huang Z, et al. 2004 Credit rating analysis with support vector machines and neural networks: a market comparative study *Decision Support Systems.***37**(4)pp 543–558

[8] Lai K K, Yu L, Wang S and Zhou L 2006 Neural network meta learning for credit scoring *Intelligent Computing*, Springer Berlin Heidelberg pp 403–08

[9] Punniyamoorthy MandSridevi P 2016 Identification of a standard AI based technique for credit risk analysis *Benchmarking: An International Journal* **23**(5)pp 1381–90

[10] Cortes C and Vapnik VN 1995Support vector networks *Machine Learning***20**(3) pp 273–97

[11] Schmidt M 1996 Identifying speaker with support vector networks *Interface '96 Proceedings*, Sydney

[12] Osuna E, Freund R andGirosi F 1997Training support vector machines: an application to face detection*IEEE Conference on Computer Vision and Pattern Recognition*pp 130–136

[13] Joachims T 1997 Text categorization with support vector machines*Technical report* LS VIII Number 23, University of Dortmund

[14] Guyon I, et al. 2002Gene Selection for Cancer Classification using Support Vector Machines *Machine Learning* **46**(1–3)pp 389–422

[15] Leslie, et al. 2004 Mismatch String Kernels for Discriminative Protein Classification *Bioinformatics* **20**(4)pp 467–76

[16] Rustam Z and Yaurita F 2017Application of Support Vector Machines for Reject Inference in Credit Scoring*The 3rd International Syposium on Current Progress in Mathematics and Sciences AIP Proceeding Conference*,Bali,Indonesia

[17] Nadira T and Rustam Z 2017Classification of Cancer Data Using Support Vector Machines with Features Selection Method Based On Global Artificial Bee Colony*The 3rd International Syposium on Current Progress in Mathematics and SciencesAIP Conference,* Bali, Indonesia

[18] Rustam Z, Vibranti D F, and Widya D 2017 Predicting the Direction of Stock Price Movement using Support Vector Machines and Fuzzy Kernel C-Means*The 3rd International Symposium on Current Progress in Mathematics and SciencesAIP Conference,* Bali, Indonesia

[19] Olivera N and Rustam Z 2017 Comparison of Fuzzy Robust Kernel C-Means and Support Vector Machines for Intrusion Detection Systems using Modified Kernel Nearest Neighbor Feature Selection*The 3rd International Syposium on Current Progress in Mathematics and SciencesAIP Conference,* Bali,Indonesia

[20] Rustam Z andYaurita F 2018 Insolvency Prediction in Insurance Companies Using Support Vector Machines and Fuzzy Kernel C-Means*Journal of Physics: Conference Series 1028(1), 012118*

[21] Lin C F and Wang S D 2002Fuzzy support vector machines *IEEE Transactions on Neural Network***13**(2)pp 464–71

[22] Almasi O N, et al. 2015 A New Fuzzy Membership Assignment Approach for Fuzzy SVM Based on Adaptive PSO in Classification Problems *Journal of Mathematics and Computer Science***14**pp 171–82

[23] Shi Y and Eberhart R 1998A modified particle optimizer *Proceedings of the 1998 IEEE World Congress on Computational Intelligence*pp 69–73

[24] Lichman M 2013 UCI Machine Learning Repository [http://archive.ics.uci.edu/ml] Irvine, CA: University of California, School of Information and Computer Science

A Comparison of the Bayesian Method under Symmetric and Asymmetric Loss Functions to Estimate the Shape Parameter *K* of Burr Distribution

I Fithriani[1,*], A R Hakim[2] and M Novita[1]

[1] Department of Mathematics, Universitas Indonesia, Depok 16424, Indonesia
[2] Undergraduate Program in Mathematics, Universitas Indonesia, Depok 16424, Indonesia

*Corresponding Author: ida.fithriani@gmail.com

Abstract. Burr distribution is one of the most important types of distribution in Burr system and has gained special attention. It has an important role in various disciplines, such as reliability analysis, life testing, survival analysis, actuarial science, economics, forestry, hydrology and meteorology. Thus, the parameter estimation for Burr distribution becomes an important thing to do. The frequentist approach using the maximum likelihood method is the most commonly used way to estimate the parameters of a distribution. In this paper we considered using the Bayesian method to estimate the shape parameter k of Burr distribution using gamma prior which is a conjugate prior. The Bayes estimate for the shape parameter k is obtained under the squared-error loss function (SELF) which is one of the symmetric loss function and the precautionary loss function (PLF) which is one of the asymmetric loss function. Through a simulation study, the comparison was made on the performance of the Bayes estimate for the shape parameter k under these two loss functions with respect to the mean-squared error (MSE) and the posterior risk.

1. Introduction

Burr distribution was first introduced in 1942 by Irving Wingate Burr. Burr distribution was originally known as Burr Type XII distribution, one of the twelve types of the continuous distributions in Burr system. It was discussed in more detail by Burr [1]. Let X be a random variable having Burr distribution with two positive parameters k and c, then it has a distribution function defined by

$$F(x) = 1-(1+x^c)^{-k}, \quad x \geq 0. \tag{1}$$

Therefore, its probability density function is defined by

$$f(x) = \mathrm{d}F(x)/\mathrm{d}x = kcx^{c-1}(1+x^c)^{-k-1}, \quad x > 0. \tag{2}$$

The parameters k and c of Burr distribution are shape parameters [2].

In the past Burr distribution used in various fields of science including reliability analysis [2,3], life testing [4,5], economics [6], forestry [7], survival analysis [8], actuarial science [9], hydrology [10], and meteorology [11]. To find out more about the characteristics of the data having Burr distribution, the parameters of Burr distribution need to be estimated. The parameter estimation of Burr distribution through the frequentist approach is common. Wingo [5], Wang, Keats and Zimmer [3] and Gupta, Gupta and Lvin [4] considered the problem of estimation using the maximum likelihood method. However, the Bayesian approach received more attention. Papadopoulos [12] considered using the Bayesian method to estimate the parameter k under the squared-error loss function (SELF) when the parameter c is known. Evans and Ragab [13] assumed that both k and c are unknown and applied the Bayesian method to estimate them under the squared-error loss function. Moore and Papadopoulos [14] obtained the Bayes estimate for the parameter k under various symmetric loss functions. Al-Noor and Al-Ameer [15] compared the Bayesian method under various symmetric and asymmetric loss functions and the other estimation methods to estimate the parameter k by employing the mean-squared error (MSE) of the estimate obtained.

In this paper we considered comparing the Bayesian method under the squared-error loss function (SELF) which is a symmetric loss function and the precautionary loss function (PLF) which is an asymmetric loss function to estimate the unknown parameters k of Burr distribution when the parameter c is known. We considered using the gamma prior distribution as the prior distribution of the parameter k because it is a conjugate prior, so the posterior distribution obtained is also a member of gamma distribution family. We obtained the Bayes estimate for the parameter k in more detail mathematically. To compare these two loss functions in estimating the parameter k, we obtained the posterior risk of the estimate under each loss function. A numerical comparison of the estimates was made through a simulation study in terms of the mean-squared error (MSE) and the posterior risk.

2. The gamma prior distribution of the parameter *k*

We consider using prior distribution which is a conjugate prior distribution. A prior distribution is called a conjugate prior distribution for a parameter if and only if the posterior distribution obtained is from the same distribution family as that prior distribution [16]. One of the prior distributions which is a conjugate prior distribution is the gamma prior distribution. Let K be a random variable having the gamma prior distribution with the prior density function defined as

$$\pi(k)=\frac{1}{\Gamma(\alpha)\beta^{\alpha}}k^{\alpha-1}\mathrm{e}^{-k/\beta},\quad k>0. \tag{3}$$

Bayesian statisticians frequently write that $\pi(k)$ is proportional to $k^{\alpha-1}\mathrm{e}^{-k/\beta}$; that is, $\pi(k)\propto k^{\alpha-1}\mathrm{e}^{-k/\beta}$. We can see that K has the gamma distribution with parameters $\alpha,\beta>0$. According to Hogg and Craig [17], the parameter estimates for α and β, by using the method of moment, are

$$\hat{\alpha}=\bar{X}^2/S^2 \quad\text{and}\quad \hat{\beta}=S^2/\bar{X}, \tag{4}$$

respectively, where $\bar{X}=\sum_{i=1}^{n}X_i/n$ and $S^2=\sum_{i=1}^{n}(X_i-\bar{X})^2/n$.

3. The posterior distribution of the parameter *k*

Let $X_1,X_2,\ldots,X_n$ denote a random sample from Burr distribution with unknown parameter k, where the parameter c is known, then the joint probability density function, or likelihood function, of $X_1,X_2,\ldots,X_n$ given $K=k$, denoted by $L(x_1,x_2,\ldots,x_n|k)$, is

$$L(x_1,x_2,\ldots,x_n|k)=\prod_{i=1}^{n}f(x_i|k)=k^nc^n\left(\prod_{i=1}^{n}\frac{x_i^{c-1}}{1+x_i^c}\right)\left[\prod_{i=1}^{n}(1+x_i^c)\right]^{-k}. \tag{5}$$

From equation (3) and (5), the joint probability density function of $X_1,X_2,\ldots,X_n$ and K is

$$L(x_1,x_2,\ldots,x_n|k)\pi(k)=\frac{1}{\Gamma(\alpha)\beta^{\alpha}}k^{\alpha+n-1}c^n\left(\prod_{i=1}^{n}\frac{x_i^{c-1}}{1+x_i^c}\right)\exp\left\{-\left[\frac{1}{\beta}+\sum_{i=1}^{n}\ln(1+x_i^c)\right]k\right\}. \tag{6}$$

Therefore, the posterior density function of the parameter k, denoted by $\pi(k|x_1,x_2,\ldots,x_n)$, is

$$\pi(k|x_1,x_2,\ldots,x_n)=\frac{L(x_1,x_2,\ldots,x_n|k)\pi(k)}{\int_0^{\infty}L(x_1,x_2,\ldots,x_n|k)\pi(k)\mathrm{d}k}. \tag{7}$$

By substituting equation (6) to equation (7), we obtain

$$\pi(k|x_1,x_2,\ldots,x_n)=\frac{1}{\Gamma(\alpha+n)}\left[\frac{1}{\beta}+\sum_{i=1}^{n}\ln(1+x_i^c)\right]^{\alpha+n}k^{\alpha+n-1}\exp\left\{-\left[\frac{1}{\beta}+\sum_{i=1}^{n}\ln(1+x_i^c)\right]k\right\};\qquad(8)$$

that is, it is the probability density function of the gamma distribution with the parameters

$$\alpha^*=\alpha+n\quad\text{and}\quad\beta^*=\left[\frac{1}{\beta}+\sum_{i=1}^{n}\ln(1+x_i^c)\right]^{-1}.\qquad(9)$$

This shows that the gamma prior distribution is a conjugate prior distribution for the parameter k.

4. The Bayes estimate for the parameter *k*

To find the Bayes estimate for the parameter k, we need a loss function denoted by $\mathscr{L}(k,\hat{k})$. The Bayes estimate is obtained by finding $\hat{k}$ minimizing the posterior risk defined by

$$R(\hat{k})=E[\mathscr{L}(K,\hat{k})\,|\,x_1,x_2,\ldots,x_n].\qquad(10)$$

In this paper we consider using two different loss functions, i.e. the squared-error loss function (SELF) which is a symmetric loss function and the precautionary loss function (PLF) which is an asymmetric loss function. We use the notation $\hat{k}_{\text{SELF}}$ to denote the Bayes estimate for the parameter k under the SELF and the notation $\hat{k}_{\text{PLF}}$ to denote the Bayes estimate for the parameter k under the PLF.

4.1. The Bayes estimate for the parameter k under the SELF

In this case the squared-error loss function (SELF) is defined by

$$\mathscr{L}(k,\hat{k}_{\text{SELF}})=(k-\hat{k}_{\text{SELF}})^2.\qquad(11)$$

Then, from equation (10), we obtain the posterior risk

$$R(\hat{k}_{\text{SELF}})=E[(K-\hat{k}_{\text{SELF}})^2\,|\,x_1,x_2,\ldots,x_n]\qquad(12)$$

and $\hat{k}_{\text{SELF}}$ minimizing the posterior risk $R(\hat{k}_{\text{SELF}})$ is

$$\hat{k}_{\text{SELF}}=E(K\,|\,x_1,x_2,\ldots,x_n);\qquad(13)$$

that is, $\hat{k}_{\text{SELF}}$ is the mean, or the first moment, of the posterior distribution of the parameter k [14]. Therefore, we obtain $\hat{k}_{\text{SELF}}=\alpha^*\beta^*$; that is,

$$\hat{k}_{\text{SELF}}=(\alpha+n)\left[\frac{1}{\beta}+\sum_{i=1}^{n}\ln(1+x_i^c)\right]^{-1}.\qquad(14)$$

4.2. The Bayes estimate for the parameter k under the PLF

In this case the precautionary loss function (PLF) is defined by

$$\mathscr{L}(k,\hat{k}_{\text{PLF}})=(k-\hat{k}_{\text{PLF}})^2/\hat{k}_{\text{PLF}}.\qquad(15)$$

Then, from equation (10), we obtain the posterior risk

$$R(\hat{k}_{\text{PLF}})=E[(K-\hat{k}_{\text{PLF}})^2/\hat{k}_{\text{PLF}}\,|\,x_1,x_2,\ldots,x_n]\qquad(16)$$

and $\hat{k}_{\text{PLF}}$ minimizing the posterior risk $R(\hat{k}_{\text{PLF}})$ is

$$\hat{k}_{\text{PLF}}=[E(K^2\,|\,x_1,x_2,\ldots,x_n)]^{1/2};\qquad(17)$$

that is, $\hat{k}_{\text{PLF}}$ is the square root of the second moment of the posterior distribution of the parameter k [18]. Therefore, we obtain $\hat{k}_{\text{PLF}}=\beta^*\sqrt{(\alpha^*+1)\alpha^*}$; that is,

$$\hat{k}_{\text{PLF}} = \sqrt{(\alpha+n+1)(\alpha+n)}\left[\frac{1}{\beta}+\sum_{i=1}^{n}\ln(1+x_i^c)\right]^{-1}. \tag{18}$$

5. The posterior risk of the Bayes estimate for the parameter *k*

5.1. The posterior risk of the Bayes estimate for the parameter k under the SELF

The posterior risk in equation (12) can be written as

$$R(\hat{k}_{\text{SELF}}) = (\hat{k}_{\text{PLF}})^2 - (\hat{k}_{\text{SELF}})^2. \tag{19}$$

By substituting equation (14) and (18) to equation (19), we obtain

$$R(\hat{k}_{\text{SELF}}) = (\alpha+n)\left[\frac{1}{\beta}+\sum_{i=1}^{n}\ln(1+x_i^c)\right]^{-2}. \tag{20}$$

5.2. The posterior risk of the Bayes estimate for the parameter k under the PLF

The posterior risk $R(\hat{k}_{\text{PLF}})$ in equation (16) can be written as

$$R(\hat{k}_{\text{PLF}}) = 2(\hat{k}_{\text{PLF}} - \hat{k}_{\text{SELF}}). \tag{21}$$

By substituting equation (14) and (18) to equation (21), we obtain

$$R(\hat{k}_{\text{PLF}}) = 2\left[\sqrt{(\alpha+n+1)(\alpha+n)}-(\alpha+n)\right]\left[\frac{1}{\beta}+\sum_{i=1}^{n}\ln(1+x_i^c)\right]^{-1}. \tag{22}$$

5.3. The comparison of the posterior risk of the Bayes estimate for the parameter k under the SELF and the PLF

Based on the posterior risk in equation (19) and (21), we obtain the ratio

$$\frac{R(\hat{k}_{\text{SELF}})}{R(\hat{k}_{\text{PLF}})} = \frac{(\hat{k}_{\text{PLF}})^2 - (\hat{k}_{\text{SELF}})^2}{2(\hat{k}_{\text{PLF}} - \hat{k}_{\text{SELF}})} = \frac{\hat{k}_{\text{SELF}} + \hat{k}_{\text{PLF}}}{2}. \tag{23}$$

From equation (23), we obtain that

- if $\hat{k}_{\text{SELF}} + \hat{k}_{\text{PLF}} < 2$, we get $R(\hat{k}_{\text{SELF}}) < R(\hat{k}_{\text{PLF}})$; that is, the Bayesian estimation for the parameter k under the SELF gives better result than under the PLF;
- if $\hat{k}_{\text{SELF}} + \hat{k}_{\text{PLF}} = 2$, we get $R(\hat{k}_{\text{SELF}}) = R(\hat{k}_{\text{PLF}})$; that is, the Bayesian estimation for the parameter k under the SELF gives result as good as under the PLF; and
- if $\hat{k}_{\text{SELF}} + \hat{k}_{\text{PLF}} > 2$, we get $R(\hat{k}_{\text{SELF}}) > R(\hat{k}_{\text{PLF}})$; that is, the Bayesian estimation for the parameter k under the PLF gives better result than under the SELF.

6. Simulation Study and Results

In this section we perform a simulation study to compare the performance of each estimate of the parameter k numerically in terms of the mean-squared error (MSE) and the posterior risk. The simulation study is made by the following steps.

- We determine the values of the shape parameters k and c of Burr distribution. We consider c = 0.5, 2.0 and k = 0.1, 0.5, 1.0, 1.5, 2.0.
- We determine the sample size n. We consider using the sample size n = 10, 50 and 100.
- We determine the number of replication, i.e. L = 1000.
- We generate the random samples of size n from a uniform distribution (U) over the interval (0,1), say $u_1, u_2, \ldots, u_n$. Then, we convert them to the samples having Burr distribution with the parameters k and c through the inverse of the distribution function in equation (1), i.e.

$$x_i = [(1-u_i)^{-1/k} - 1]^{1/c}, \quad i = 1, 2, \ldots, n. \tag{24}$$

- By assuming that the parameter c is known, we find the Bayes estimate for the parameter k under the SELF and the PLF according to equation (14) and (18), where the parameters α and β of the gamma prior distribution are estimated by $\hat{\alpha}$ and $\hat{\beta}$ in equation (4).
- We make a comparison by calculating the MSE of the estimate according to equation

$$\mathrm{MSE}(\hat{k}) = \sum_{s=1}^{L} (\hat{k}_s - k)^2 / L, \tag{25}$$

where $\hat{k}_s$ is the estimate of the parameter k at the s-th replication.

- In addition to using the MSE, we also make a comparison by calculating the posterior risk according to equation (20) and (22), or equation (19) and (21).

The results of the simulation study are summarized and tabulated in Table 1 below.

Table 1. The MSE and the posterior risk of the estimates for the parameter k of Burr distribution

c	k	n	MSE		Posterior Risk	
			$\mathrm{MSE}(\hat{k}_{\mathrm{SELF}})$	$\mathrm{MSE}(\hat{k}_{\mathrm{PLF}})$	$R(\hat{k}_{\mathrm{SELF}})$	$R(\hat{k}_{\mathrm{PLF}})$
0.5	0.1	10	0.0015572	0.0018519	0.0013717	0.0107700
		50	0.0002100	0.0002185	0.0002107	0.0020218
		100	0.0001104	0.0001130	0.0001035	0.0010093
	0.5	10	0.0398777	0.0471927	0.0340126	0.0533953
		50	0.0052962	0.0055429	0.0053338	0.0101740
		100	0.0025734	0.0026272	0.0025690	0.0050303
	1.0	10	0.1565890	0.1877870	0.1393640	0.1081660
		50	0.0216592	0.0225742	0.0211730	0.0202580
		100	0.0105273	0.0107741	0.0103350	0.0100883
	1.5	10	0.1835370	0.2156420	0.2538360	0.1478210
		50	0.0516081	0.0539030	0.0480314	0.0304910
		100	0.0243331	0.0248663	0.0231992	0.0151107
	2.0	10	0.2421290	0.2722450	0.4053530	0.1876190
		50	0.0882654	0.0922875	0.0853002	0.0406391
		100	0.0372423	0.0378557	0.0406266	0.0200066
2.0	0.1	10	0.0018893	0.0022305	0.0014299	0.0108862
		50	0.0002288	0.0002387	0.0002131	0.0020321
		100	0.0001031	0.0001053	0.0001030	0.0010071
	0.5	10	0.0451579	0.0533056	0.0341517	0.0520079
		50	0.0063109	0.0066031	0.0053816	0.0101787
		100	0.0025913	0.0026525	0.0025797	0.0050371
	1.0	10	0.0649506	0.0788826	0.1028850	0.0889534
		50	0.0213653	0.0223466	0.0209623	0.0199810
		100	0.0104318	0.0107050	0.0103143	0.0100411
	1.5	10	0.0667565	0.0567915	0.1485480	0.1056750
		50	0.0377020	0.0386422	0.0443179	0.0289185
		100	0.0211745	0.0214849	0.0224884	0.0147853
	2.0	10	0.3449830	0.2932900	0.1730210	0.1123570
		50	0.0566033	0.0544790	0.0708417	0.0364830
		100	0.0314448	0.0310473	0.0379675	0.0191825

Based on the tabulated MSE values, the larger sample size we use, the smaller MSE value we obtain, under the SELF as well as under the PLF. We also obtain that under these loss functions, when the true values of the parameter k are smaller, the MSE values are also smaller. In general, the estimates for the parameter k under the SELF have smaller MSE values than under the PLF.

Based on the tabulated posterior risks, the larger sample size we use, the smaller posterior risk we obtain, under the SELF as well as under the PLF. We also obtain that under these loss functions, when the true values of the parameter k are smaller, in general, the posterior risks are also smaller. In general, the estimates for the parameter k under the SELF have smaller posterior risks than under the PLF when the true values of the parameter k are less than 1. Otherwise, when the true values of the parameter k are more than or equal to 1, the estimates under the PLF have the smaller posterior risks.

7. Conclusion

The Bayesian estimation for the shape parameter k of Burr distribution under the squared-error loss function (SELF), which is a symmetric loss function, is found to be superior compared to the precautionary loss function (PLF), which is an asymmetric loss function, in terms of the mean-squared error (MSE) for all sample sizes and all true values of the parameter k. However, in terms of the posterior risk, the estimation under the SELF is found to be superior in turn with the PLF; that is, the SELF gives better results than the PLF when the true values of the parameter k are small (less than 1).

8. References

[1] Burr I W 1942 Cumulative frequency functions *Annals of Mathematical Statistics* **13** 215–32
[2] Lewis A W 1981 The Burr distribution as a general parametric family in survivorship and reliability theory applications *Ph.D. Thesis* Department of Biostatistics University of North Carolina at Chapel Hill
[3] Wang F K, Keats J B and Zimmer W J 1996 Maximum likelihood estimation of the Burr XII parameters with censored and uncensored data *Microelectron. Reliab.* **36** pp. 359–62
[4] Gupta P L, Gupta R C and Lvin S J 1996 Analysis of failure time data by Burr distribution *Communications in Statistics–Theory and Methods* **25** pp. 2013–24
[5] Wingo D R 1983 Maximum likelihood methods for fitting the Burr XII distribution to life test data *Biometrical Journal* **25** 77–84
[6] McDonald J B 1984 Some generalized functions for the size distribution of income *Econometrica* **52** 647–63
[7] Lindsay S R, Wood G R and Woollons R C 1996 Modelling the diameter distribution of forest stands using the Burr distribution *Journal of Applied Statistics* **23** 609–20
[8] Githany M E and Al-Awadhi S 2002 Maximum likelihood estimation of Burr XII distribution parameters under random censoring *Journal of Applied Statistics* **29** 955–65
[9] Burnecki K, Härdle W and Weron R 2004 An introduction to simulation of risk processes *Encyclopedia of Actuarial Science* 1–7
[10] Shao Q, Wong H, Xia J and Ip W 2004 Models for extremes using the extended three parameter Burr XII system with application to flood frequency analysis *Hydrological Sciences Journal* **49** 685–702
[11] Usta I and Kantar Y M 2013 Analysis of the Burr Type XII distribution for estimation of wind speed distributions *Proc. on Science and Engineering* (Thailand: I-SEEC) p 302–8
[12] Papadopoulos A S 1978 The Burr distribution as a failure model from a Bayesian approach *IEEE Transactions on Reliability* **27** 369–71
[13] Evans I G and Ragab A S 1983 Bayesian inferences given a type-2 censored sample from Burr distribution *Communications in Statistics–Theory and Methods* **12** 1569–80
[14] Moore D and Papadopoulos A S 2000 The Burr Type XII distribution as a failure model under various loss functions *Microelectronics Reliability* **40** 2117–22
[15] Al-Noor N H and Al-Ameer H A A 2014 Some estimation methods for the shape parameter and reliability function of Burr Type XII distribution / comparison study *Mathematical Theory and Modeling* **4** 63–77
[16] Klugman S A, Panjer H H and Willmot G E 2012 Loss models from data to decisions 4th ed (Canada: John Wiley & Sons, Inc.)
[17] Hogg R V and Craig A T 1995 Introduction to mathematical statistics 5th ed (United States of

IOP Conf. Series: Journal of Physics: Conf. Series **1108** (2018) 012053 doi:10.1088/1742-6596/1108/1/012053

America: Prentice Hall, Inc.)
[18] Norstrom J G 1996 The use of precautionary loss function in risk analysis *IEEE Transactions on Reliability* **45** 400–3

Acknowledgments
This work is supported by Hibah PITTA 2018 funded by DRPM Universitas Indonesia No.2308/UN2.R3.1/HKP.05.00/2018.

IOP Conf. Series: Journal of Physics: Conf. Series **1108** (2018) 012054 doi:10.1088/1742-6596/1108/1/012054

Analysis of Hypertension Disease using Logistic and Probit Regression

D I Ruspriyanty[1] and A Sofro[1,*]

[1]Mathematics Department, Universitas Negeri Surabaya, Ketintang, Surabaya 60231, East Java, Indonesia
*Corresponding Author. Email: ayuninsofro@unesa.ac.id

Abstract. Hypertension (high blood pressure) is when blood pressure has reached or exceeded 140 mmHg (systole) and 90 mmHg (diastole). Suspected factors of the disease are age, obesity, education, partner status, and occupation. In this paper, we analyze models for determining high-risk factors of hypertensive disease. The results show that education and partner status are a significant factor affecting the disease. The Akaike Information Criterion value (AIC) of the logistic model is 118.64 and the probit model is 118.79. Therefore, the performance of a logistic model is better than probit due to providing the smaller value of AIC.

1. Introduction

Hypertension is a dark killer (the silent killer) because it includes a deadly disease, because it is accompanied by symptoms that are often regarded as an ordinary disorder so that the sufferer is late to realize the coming of hypertension. The sign of hypertension are headaches, palpitations, difficulty breathing after working hard or lifting a heavy burden, fatigue, blurred vision, ringing ears (tinnitus), and nosebleeds [12]. Death due to hypertension ranks higher than other causes [3].

In 2013 showed that 25.8% of Indonesia's population suffers from hypertension. According to the Central Java Provincial Health Office, in 2013, Semarang City ranks first for hypertension cases with hypertension prevalence of 55.6% [4]. Hypertension is an increase in systolic blood pressure of more than 140 mmHg and diastolic blood pressure of more than 90 mmHg in two measurements with an interval of five minutes in a rest/quiet state [9]. Factors that influence the occurrence of hypertension is divided into 2 groups, namely uncontrollable factors and factors that can be controlled. Uncontrollable factors such as age, sex, genetics, race and controllable factors such as exercise habits, diet, alcohol consumption, salt, coffee, and stress.

Based on the case above, we describe the hypertension disease as a dependent variable and there are five independent variables used. Thus, we use logistic and probit regression method. Logistic regression is one of the non-linear regression methods that can be used to find the response variable relationship of dichotomous (nominal or ordinal in two categories)or polychotomous (with nominal or ordinal scale with more than two categories) with one or more predictor variables [1]. The probit regression is also a nonlinear regression used to analyze the relationship between one variable response (dependent variable) with some predictor variable (independent variable), where the response variable is a qualitative data dichotomy that is 1 to state the existence of a characteristic and 0 to express the non-existence a characteristic [6].

In this research, we will analyze the factors that influence hypertension disease with logistic and probit regression. The data is collected from Budi Artiyaningrum in 2015 [4]. The status of disease from the observations become the dependent variable, i.e if the subject is hypertension, the case is categorized one, otherwise zero. In this paper, we will discuss the method, i.e logistics and probit regression. How to estimate parameter and analyze the disease data using both of those approaches.

2. Method of Research

2.1 Logistic Regression

Logistic regression is one of the non-linear regression methods that can be used to find the relationship of dichotomous (nominal or ordinal in two categories) or polychotomous (with nominal or ordinal scale with more than two categories) with one or more predictor variables [1]. If the response variable consists of two categories (dichotomy/binary), i.e $y = 1\,(success)$ and $y = 0\,(fail)$, logistic regression method can be applied is binary logistic regression [2], then the response variable y follows the Binomial distribution with probability function:

$$f(y_i) = \pi(\boldsymbol{x}_i)^{y_i}(1-\pi(\boldsymbol{x}_i))^{1-y_i}, (1)$$

where $y_i = 0; 1$, $\pi(\boldsymbol{x}_i)$ is the probability of success, $1-\pi(\boldsymbol{x}_i)$ is the probability of failure, and $i = 1, 2, \dots, n$ is the observation index. Logistic regression models used are [5]:

$$\pi(\boldsymbol{x}) = \frac{e^{\beta_0+\beta_1x_1+\beta_2x_2+\cdots+\beta_px_p}}{1+e^{\beta_0+\beta_1x_1+\beta_2x_2+\cdots+\beta_px_p}}, \quad (2)$$

where p is the number of independent variables. The logistic regression model in equation (2) is transformed using the logit transformation obtained:

$$g(\boldsymbol{x}) = \log\left(\frac{\pi(x)}{1-\pi(x)}\right) = \beta_0 + \beta_1x_1 + \beta_2x_2 + \cdots + \beta_px_p. (3)$$

Then, it will explain the estimation of the $\boldsymbol{\beta}$ parameter in logistic regression by using Maximum Likelihood Estimation (MLE).

2.2 Parameter Estimation

Maximum Likelihood Estimator (MLE) is one of the methods used to estimate/predict parameters on logistic regression. The method estimates the $\boldsymbol{\beta}$ parameter by maximizing the likelihood function provided that the data must follow a certain distribution. If the dependent variable (y) follows the Binomial distribution with the probability of success $\pi(\boldsymbol{x}_i)$ and the probability of failure $1-\pi(\boldsymbol{x}_i)$ thus the likelihood function for the logistic regression model is:

$$\begin{aligned} L(\boldsymbol{\beta}) &= \textstyle\prod_{i=1}^{n} f(y_i; \boldsymbol{\beta}) \\ &= \textstyle\prod_{i=1}^{n} \pi(\boldsymbol{x}_i)^{y_i}\,[1-\pi(\boldsymbol{x}_i)]^{1-y_i}. (4) \end{aligned}$$

Mathematically it would be easier if equation (4) is brought to the log-likelihood function and maximizes the log-likelihood. The following is the log-likelihood function:

$$\ell(\boldsymbol{\beta}) = \textstyle\sum_{j=0}^{p}\left[\sum_{i=1}^{n} y_i x_{ij}\right]\beta_j - \sum_{i=1}^{n} \ln\left[1 + exp\left(\sum_{j=0}^{p}\beta_j x_{ij}\right)\right]. (5)$$

Equation (5) is differentiated to $\boldsymbol{\beta}$ to obtain the following equation:

$$\frac{\partial\ell(\boldsymbol{\beta})}{\partial\beta_j} = \textstyle\sum_{i=1}^{n} y_i x_{ij} - \sum_{i=1}^{n} x_{ij}\pi(\boldsymbol{x}_i) = 0 \qquad, j = 0,1,2,\dots,p, (6)$$

where p is the number of predictor variables with

$$\pi(\boldsymbol{x}_i) = \frac{e^{\sum_{j=0}^{p}\beta_j x_{ij}}}{1 + e^{\sum_{j=0}^{p}\beta_j x_{ij}}} \qquad, j = 0,1,2,\dots,p\,.$$

To find the derivatives of equation (6) which are equated with zero, it often does not get explicit results so that the using the numerical method is the Newton Raphson method to optimize equation (6).

2.3 Probit Regression

The probit model is a non-linear model used to analyze the relationship of response/dependent variable with several predictor/independent variables where the response variable is qualitative data dichotomy that is 0 and 1 [6]. The probit regression method uses the cumulative distribution function Normal (Normal cumulative distribution function) to explain the function of the equation. Since the response/dependent variable is dichotomous/binary, the response/dependent variable (y) follows the Binomial distribution with the probability function as follows:

$$f(y_i; \pi_i) = {\pi_i}^{y_i}(1-\pi_i)^{1-y_i},$$

with $y_i = 0; 1$, π_i is the probability occurrence of the i for $y_i = 1$, and $1 - \pi_i$ is the probability occurrence of the i for $y_i = 0$. The transformation function in the probit regression model is the cumulative function of the Normal distribution as a link function in GLM.

$$P(y_i = 1|x_i) = \Phi(\boldsymbol{x}_i'\boldsymbol{\beta}) = \int_{-\infty}^{\boldsymbol{x}_i'\boldsymbol{\beta}} \phi\,(z)dz, \quad (7)$$

where $\Phi(.)$ is the cumulative function of the Normal distribution and $\phi(.)$ is the Normal distribution probability function.

$$\Phi(g(\boldsymbol{x})) = \frac{1}{\sqrt{2\pi}} \int_{-\infty}^{g(x)} e^{-\frac{z^2}{2}} dz\,.$$

In general, the probit regression model can be expressed as follows:

$$\pi_i = \Phi(Z_i) = \Phi(\beta_0 + \beta_1 x_{i1} + \beta_2 x_{i2} + \cdots + \beta_p x_{ip} + \varepsilon_i).$$

Since probit model is related to the cumulative function of Normal distribution, it can be written probit model as follows:

$$Z_i = \beta_0 + \beta_1 x_{i1} + \beta_2 x_{i2} + \cdots + \beta_p x_{ip} + \varepsilon_i\,.$$

To obtain an expectation of the probit value (Z_i), then the inverse of the normal cumulative distribution function can be obtained:

$$Z_i = \Phi^{-1}(\pi_i) = \beta_0 + \beta_1 x_{i1} + \beta_2 x_{i2} + \cdots + \beta_p x_{ip} + \varepsilon_i\,.$$

Then, it will explain the estimation of the $\boldsymbol{\beta}$ parameter in probit regression by using Maximum Likelihood Estimation (MLE).

2.4 Parameter Estimation

Maximum Likelihood Estimator (MLE) is one of the methods used to estimate/predict parameters of a known distribution model. The likelihood function for the probit regression model can be written as follows:

$$L(\boldsymbol{\beta}) = \prod_{i=1}^{n} f(y_i; \pi_i)\,. \quad (8)$$

The transformation function in the probit regression model is the cumulative function of the Normal distribution so that the equation (8) becomes:

$$L(\boldsymbol{\beta}) = \prod_{i=1}^{n} \left(\Phi(\boldsymbol{x}_i'\boldsymbol{\beta})\right)^{y_i} \left(1 - \Phi(\boldsymbol{x}_i'\boldsymbol{\beta})\right)^{1-y_i},$$

where $\Phi(.)$ is the cumulative function of the Normal distribution. The log likelihood obtained as follows:

$$\ell(\boldsymbol{\beta}) = \sum_{i=1}^{n} y_i \log\left(\Phi(\boldsymbol{x}_i'\boldsymbol{\beta})\right) + \sum_{i=1}^{n} (1 - y_i) \log\left(1 - \Phi(\boldsymbol{x}_i'\boldsymbol{\beta})\right). \quad (9)$$

The parameter value of $\boldsymbol{\beta}$ can be obtained by maximized the log-likelihood function of finding the derivative of $\boldsymbol{\beta}$. To find the derivatives of equation (9) which are equated with zero, it often does not get explicit results so that the using the numerical method is the Newton Raphson method to optimize equation (9).

2.5 Testing of Parameter Significance

The model obtained needs to be tested whether the independent variables contained in the model have a significant relationship with the dependent variable. The test is a simultaneous test and a partial test [5]. The simultaneous test is performed to find out the significance of $\boldsymbol{\beta}$ coefficient to the overall response/dependent variable, with the following test hypothesis:

$H_0 : \beta_1 = \beta_2 = \cdots = \beta_p = 0$

H_1 : there is at least one $\beta_j \neq 0$ with $j = 1,2, \ldots, p$

G test statistic or likelihood ratio test, namely:

$$G = -2\ln\left(\frac{\left(\frac{n_1}{n}\right)^{n_1}\left(\frac{n_0}{n}\right)^{n_0}}{\sum_{i=1}^{n}\widehat{\pi_i}^{y_i}(1-\widehat{\pi_i})^{(1-y_i)}}\right), \quad (10)$$

where $n_1 = \sum_{i=1}^{n} y_i$, $n_0 = \sum_{i=1}^{n}(1 - y_i)$, and $n = n_1 + n_0$. The G test statistic follows the Chi-square distribution so that reject H_0 if $G > \chi^2_{(\alpha,p)}$ with p is the number of independent variables or the number of parameters in the model at a significant level α or *p-value*$< \alpha$ [5].The partial test is to determine the effect of $\boldsymbol{\beta}$ coefficient individually by comparing the standard error. The hypothesis is as follows:
$H_0 : \beta_j = 0$
$H_1 : \beta_j \neq 0$ with $j = 1,2, \dots, p$
Wald test statistics [10]:

$$W = \frac{\widehat{\beta}_j}{\widehat{SE}(\widehat{\beta}_j)} \quad (11)$$

W test statistic follows the standard normal distribution with the rejection region: reject H_0 if value $|W| > Z_{(1-\alpha)/2}$or *p-value*$< \alpha$. $\widehat{\beta}_j$ is the value of the regression parameter estimation and $\widehat{SE}(\widehat{\beta}_j)$ is the standard error [2].

2.6 Best Model Selection Criteria

AIC is the model selection criteria considering the number of parameters and the smaller the AIC value the better the model. According to [4] Akaike Information Criterion (AIC) is defined as follows:

$$AIC = -2\ln L(\hat{\beta}) + 2k \quad (12)$$

where $L(\hat{\beta})$ is the likelihood value, k is the number of parameters.

3. Experiment Set Up

3.1 Dataset

The data used is obtained from the results of previous research with the title of factors related to the incidence of uncontrolled hypertension in patients who perform routine checks at Kedungmundu Public Health Center [4].

3.2 Operational Definition of Research Variable

The dependent variable is hypertension disease. The dependent variable is categorical, there are only two categories of response 1 states a person suffering from hypertension and response 0 states someone who does not suffer from hypertension. Independent variables are: age is the length of respondent's life calculated from birth to the last birthday and it is a continuous variable expressed in units of years, obesity is overweight as a result of excessive body fat accumulation. How to know the level of obesity is by using Body Mass Index (BMI) [8].
$BMI = \frac{Weight\ (kg)}{Height\ (m)^2}$, education is the level of school that has been followed by the respondents which are divided into two categories are 1: low (no school, elementary, junior high), 0: high (high school, college), partner status is the condition of respondents based on the presence and absence of a life partner (husband/wife) in everyday life which is divided into two categories are 1: no couple (single, widower, widow, separate), 0: there is a couple (married), and occupation is the condition of respondents who have jobs or do not have a job in everyday life which is divided into two categories are 1: not work (housewife), 0: work (self-employed, laborers, farmers, employees).

4. Result and Discussion

4.1 Factors Affecting of Hypertension Disease Using Logistic Regression

The results simultaneously with the value of test statistic $G$is 15.35 which is more than critical value from Chi-square with alpha 0.05 and degree of freedom 5. We also get p-value is 0.008968 which is less than alpha five percent. Thus there is a strong evidence to reject H_0. So it can be concluded that there are at least one of the independent variables significant on hypertension disease. Meanwhile, the result of parameters on the partial test performed using software R is shown in table 1.

Table 1. Partial Test of Logistic Regression Parameter

Independent Variable	Coef.	SE	z_{value}	$p - value$	Exp(Coef.)
Age	-0.01226	0.03335	-0.368	0.7132	0.989
Obesity	0.06350	0.06210	1.023	0.3065	1.066
Education	1.74148	0.74705	2.331	0.0197 *	5.706
Partner Status	1.29371	0.58716	2.203	0.0276 *	3.646
Occupation	-0.42108	0.49787	-0.846	0.3977	0.656
(intercept)	-2.62926	2.30484	-1.141	0.2540	0.072

*) significant at $\alpha = 5\%$

Based on the result of partial test logistic regression parameter in table 1 above shows that the value of test statistic W for education is 2.331 and the value of test statistic W for partner status is 2.203 which are more than critical value from $Z_{(1-\alpha)/2}$ with alpha 0.05. We also get the p-value of education is 0.0197 and p-value of partner status is 0.0276 which are less than alpha five percent. Thus the independent variables are education (x_3) and partner status (x_4) significant to the model. So it can be concluded that education and partner status are the factors that influence hypertensive disease because it has the $p - value < \alpha$ with significance level $\alpha = 5\%$. Odds ratio performed to determine the magnitude of the influence of each independent variable is significant as follows: respondents with no school and primary/junior high school have a 5.706 times greater risk of developing hypertension compared to senior high school/college education and respondents with status single/widower/widow/separate has a risk 3.646 times greater suffering from hypertension compared with a married status.

Logistic regression model is following:

$$\pi(x) = \frac{e^{-2,62926-0,01226x_1+0,06350x_2+1,74148x_3+1,29371x_4-0,42108x_5}}{1+e^{-2,62926-0,01226x_1+0,06350x_2+1,74148x_3+1,29371x_4-0,42108x_5}}.$$

4.2 Factors Affecting of Hypertension Disease Using Probit Regression

The results simultaneously with the value of test statistic $G$is 15.20 which is more than critical value from Chi-square with alpha 0.05 and degree of freedom 5. We also get p-value is 0.009513 which is less than alpha five percent. Thus there is a strong evidence to reject H_0. So it can be concluded that there are at least one of the independent variables significant on hypertension disease. Meanwhile, the result of parameters on the partial test performed using software R is shown in table 2.

Table 2. Partial Test of Probit Regression Parameter

Independent Variable	Coef.	SE	z_{value}	$p - value$
Age	-0.007738	0.020460	-0.378	0.7053
Obesity	0.037465	0.037046	1.011	0.3119
Education	1.034035	0.434989	2.377	0.0174 *
Partner Status	0.782626	0.352979	2.217	0.0266 *
Occupation	-0.236106	0.299571	-0.788	0.4306
(intercept)	-1.540951	1.383035	-1.114	0.2652

*) significant at $\alpha = 5\%$

Probit regression model obtained is as follows:

$$\pi_i = \Phi(Z_i) = -1.540951 - 0.007738x_1 + 0.037465x_2 + 1.034035x_3 + 0.782626x_4 - 0.236106x_5.$$

Based on the result of partial test probit regression parameter in table 2 above shows that the value of test statistic W for education is 2.377 and the value of test statistic W for partner status is 2.217 which are more than critical value from $Z_{(1-\alpha)/2}$ with alpha 0.05. We also get the p-value of education is 0.0174 and p-value of partner status is 0.0266 which are less than alpha five percent. Thus the independent variables are education (x_3) and partner status (x_4) significant to the model. Based on the model of probit regression above shows that the significant factors of the disease are: respondents with low education (no school and primary/junior high) have a higher risk of hypertension with a Z score of 1.034035 compared with higher education (high school/college graduation) and respondents who have no partner status (single/widower/widowed/separate) have a higher risk of hypertension with Z score of 0.782626 compared with the status of a couple (married).

4.3 Best Model Selection Criteria

The results of AIC value is shown in Table 3.

Table 3. AIC Value of Logistic and Probit Regression

Model	AIC
Logistic Regression	118.64
Probit Regression	118.79

The smaller AIC value, the better model. The AIC value of the logistic model is smaller than the one. Thus, from Table 3, it shows that logistic regression model is better than probit regression model.

5. Conclusion

Based on the results of analysis and discussion in this study can be concluded that hypertension disease is influenced by factors are education and partner status at a significance level of 5%. The logistic regression model is better than the probit regression model because the AIC value of logistic regression is the smaller than another model.

References

[1] Alan A 1990 *Categorical Data Analysis* (Canada: John Wiley & Sons Inc)
[2] Amr I A 2010 *Applying Logistic Regression Model to the Second Primary Cancer Data* (Egypt: Ain Shams University Press) pp 1-20
[3] Bambang H 2011 *Hipertensi: The Silent Killer* (Jakarta: Perhimpunan Hipertensi Indonesia)
[4] Budi A 2015 *Faktor-Faktor yang Berhubungan dengan Kejadian Hipertensi Tidak Terkendali pada Penderita yang Melakukan Pemeriksaan Rutin di Puskesmas Kedungmundu Kota Semarang Tahun 2014* (Semarang: Universitas Negeri Semarang)
[5] David W H and Stanley L 2000 *Applied Logistic Regression* (New Jersey: John Wiley & Sons, Inc)
[6] Evy W 2013 *Model Regresi Probit untuk Mengetahui Faktor-Faktor yang Mempengaruhi Jumlah Penderita Diare di Jawa Timur* (Surabaya: Universitas Negeri Surabaya)
[7] Hamparsum B 2000 Akaike's Information Criterion and Recent developments in informational complexity *Journal of Mathematical Psychology* **44**(1) pp. 62–91
[8] Kemenkes RI 2012 *Cara Mencegah dan Mengatasi Obesitas* (Jakarta: Direktorat Bina Gizi Masyarakat)
[9] Kemenkes RI 2014 *Pusdatin Hipertensi. Infodatin* pp. 1-7
[10] Saroje K S and Habshah M 2010 Importance of Assessing the Model Adequacy of Binary Logistic Regression *Journal of Applied Sciences* **10**(6) pp. 479-486
[11] Sumarni H A 2016 *Emisi Transportasi* (Makassar: Penebar Plus+)
[12] Vitahealth 2006 *Hipertensi* (Jakarta: PT. Gramedia Pustaka Utama)

IOP Conf. Series: Journal of Physics: Conf. Series **1108** (2018) 012055 doi:10.1088/1742-6596/1108/1/012055

The Existence of 1-Error-Correcting Pair for Maximum Distance Separable (MDS) Non-Generalized Reed Solomon (GRS) code and Almost Maximum Distance Separable (AMDS) code

M Ridha and K A Sugeng*

Department of Mathematics, Faculty of Mathematics and Natural Science
Universitas Indonesia
Email: muhammad.ridha@sci.ui.ac.id, kiki@sci.ui.ac.id

Abstract. One of the topics for decoding linear codes is an error-correcting pair. In this paper, it would be proved that the non-GRS MDS code and the AMDS $[n, 1, n-1]$ code have 1-error-correcting pair. It will also show that the AMDS code $[n, n-1, 1]$ has no error-correcting pair.

1. Introduction

In coding theory, singleton bound is an upper bound on the size of an arbitrary linear code with length n, size M, and minimum distance d. Let C be an $[n, k, d]$ code. The Singleton Bound says that $d \leq n - k + 1$. If this equality holds,$d = n - k + 1$, then the code is called MDS code or Maximum Distance Separable code. If $d = n - k$, then the code is called almost Maximum Distance Separable code or AMDS code [1].

One of the most commonly used MDS code is the Generalized Reed Solomon (GRS) code. Let n be an arbitrary integer such that $1 \leq n \leq q$. Let $\boldsymbol{a} = (a_1, a_2, \ldots, a_n)$ be an n-tuple of mutually distinct elements of $\mathbb{F}_q \cup \{\infty\}$. Let $\boldsymbol{b}$ be an n-tuple of nonzero elements of $\mathbb{F}_q$. Let k be an arbitrary integer such that $0 \leq k \leq n$. C is a generalized Reed-Solomon code, denoted by $GRS_k(\boldsymbol{a}, \boldsymbol{b})$, is generated by:

$$G = \begin{pmatrix} b_1 & b_2 & \cdots & b_n \\ b_1 a_1 & b_2 a_2 & \cdots & b_n a_n \\ \vdots & \vdots & \ddots & \vdots \\ b_1 a_1^{k-1} & b_2 a_2^{k-1} & \cdots & b_n a_n^{k-1} \end{pmatrix} [1].$$

One of the decoding methods for linear codes is error-correcting pairs. Let C be an $\mathbb{F}_q$-linear code of length n. Let A, B be subspaces of $\mathbb{F}_{q^m}^n$. Then (A, B) is called a t-error -correcting pair (t-ECP) for C if the following conditions are satisfied:

(1) $(A * B) \perp C$,
(2) $k(A) > t$,
(3) $d(B^{\perp}) > t$,
(4) $d(A) + d(C) > n$ [2].

*) Corresponding author

In 2015, Márquez-Corbella and Pellikan[2] characterized MDS code that had an error-correcting pair. The result shown that an MDS code of minimum distance $2t+1$ had a t-ECP if and only if it was a generalise Reed-Solomon code. This paper did not show the existence of error-correcting pair for MDS code which is not a GRS code and almost MDS code. Therefore, in this research, it studied about the existence of error-correcting pair for non-GRS MDS code and AMDS code.

2. Error-Correcting Pair for MDS Non-GRS code

In 1989, Roth and Lempel constructed a non-Reed-Solomon Type of MDS codes. He uses a subset of $\mathbb{F}_q$ to construct this code; this subset is called (m,t,δ)-set. A subset S of $\mathbb{F}_q$ with size m is called an (m,t,δ)-set in $\mathbb{F}_q$ if there is an element $\delta \in \mathbb{F}_q$ such that there is no sum of t elements of S equal to δ [3].

Theorem 2.1 Let n and k be integers with $k \geq 3$ and $k+3 \leq n \leq 1+2$. Let G be generator matrix of $[n,k]$ code,

$$G = \begin{bmatrix} 1 & 1 & \cdots & 1 & 0 & 0 \\ \alpha_0 & \alpha_1 & \cdots & \alpha_{n-3} & 0 & 0 \\ \vdots & \vdots & \ddots & \vdots & \vdots & \vdots \\ \alpha_0^{k-3} & \alpha_1^{k-3} & \cdots & \alpha_{n-3}^{k-3} & 0 & 0 \\ \alpha_0^{k-2} & \alpha_1^{k-2} & \cdots & \alpha_{n-3}^{k-2} & 0 & 1 \\ \alpha_0^{k-1} & \alpha_1^{k-1} & \cdots & \alpha_{n-3}^{k-1} & 1 & \delta \end{bmatrix}$$

where α_i, $i=0,1,\dots,n-3$, are distinct elements of $\mathbb{F}_q$ and $\delta \in \mathbb{F}_q$. Thus$G$ does not generate a GRS code, and$G$ generates an MDS code if and only if α_i form an $(n-2,k-1,\delta)$-set in $\mathbb{F}_q$ [3].

In the following examples, it will be given a [6,3,4] MDS code which is a non-GRS code. This MDS code has 1-ECP with the following explanation.

Let $\boldsymbol{\alpha} = \{1,2,3,4\} \subseteq Z_7$. So, $\boldsymbol{\alpha}$ is a(4,2,1)-set. By Theorem 2.1,

$$G = \begin{bmatrix} 1 & 1 & 1 & 1 & 0 & 0 \\ 1 & 2 & 3 & 4 & 0 & 1 \\ 1 & 4 & 2 & 2 & 1 & 1 \end{bmatrix} \approx \begin{bmatrix} 1 & 0 & 0 & 1 & 4 & 5 \\ 0 & 1 & 0 & 4 & 6 & 3 \\ 0 & 0 & 1 & 3 & 4 & 6 \end{bmatrix}$$

was a generator matrix of non-GRS MDS code C, with parameters[6,3,4]. Then G was a parity check matrix of $C^{\perp}$. Therefore $C^{\perp}$ is a set of all $\boldsymbol{c} \in Z_7$ such that $G\boldsymbol{c}^T = 0$[1].

Thus,

$$\boldsymbol{c} = (c_1,\ c_2,\ c_3, 6c_1 + 6c_2 + 6c_3, 5c_1 + 6c_2 + 6c_3, 3c_1 + 2c_2 + 1c_3).$$

Let B be an [6,1] code, where for all $\in B$, $\boldsymbol{b} = [b,b,b,b,b,b]$, $b \in Z_7$. Therefore, it could be obtained that $C^{\perp} * B = C^{\perp}$ or $(C^{\perp} * B) \perp C$. Since $C^{\perp}$ was an MDS code with $n(C^{\perp}) = 6$ and $k(C^{\perp}) = 3 > 1$, then $d(C^{\perp}) + d(C) = 8 > 6 = n$. Since the parity check matrix of $B^{\perp}$ was

$$H = [1 \quad 1 \quad 1 \quad 1 \quad 1 \quad 1]$$

and 2 was the smallest integer such that every two columns of H are linearly dependent, then $d(B^{\perp}) = 2 > 1$. Thus, (A,B) were 1-error-correcting pair for C or C has 1-ECP.

The example above could be generalised to be the following theorem, where it could show that the MDS non-GRS code in Theorem 2.1 had 1-ECP.

IOP Conf. Series: Journal of Physics: Conf. Series **1108** (2018) 012055 doi:10.1088/1742-6596/1108/1/012055

Theorem 2.2 Let C be an MDS non-GRS code as in Theorem 2.1 and B is $[n,1,n]$ code with generator matrix

$$G_B = [1 \quad 1 \quad \cdots \quad 1].$$

Then $(C^{\perp}, B)$ are 1-error-correcting pair of C.

Proof. To show that $(C^{\perp}, B)$ were 1-ECP of C, it might be shown that the following conditions were satisfied.

1. $(C^{\perp} * B) \perp C$
2. $k(C^{\perp}) > 1$
3. $d(B^{\perp}) > 1$
4. $d(C^{\perp}) + d(C) > n$

Now, it would be proved that the four conditions above were satisfied.

1. Let

$$g_i(x) = \sum_{j=0}^{k-1} g_{ij} x^j \triangleq \prod_{0 \le j \le k-1 : j \ne i} (x - \alpha_j), \qquad 0 \le i \le k-1.$$

Let $\lambda_{ij} = g_i(\alpha_j)$, $0 \le i, j \le k-1$.
Let

$$a \triangleq \prod_{i=0}^{k-1} (\alpha_{n-3} - \alpha_i) \ne 0$$

and

$$b \triangleq \delta - \sum_{i=0}^{k-1} \alpha_i$$

Then, $H = [-A^T | I]$ was a generator matrix of $C^{\perp}$ [3],

$$-A^T = \begin{bmatrix} -\frac{\lambda_{0k}}{\lambda_{00}} & -\frac{\lambda_{1k}}{\lambda_{11}} & \cdots & -\frac{\lambda_{(k-1)k}}{\lambda_{(k-1)(k-1)}} \\ -\frac{\lambda_{0(k+1)}}{\lambda_{00}} & -\frac{\lambda_{1(k+1)}}{\lambda_{11}} & \cdots & -\frac{\lambda_{(k-1)(k+1)}}{\lambda_{(k-1)(k-1)}} \\ \vdots & \vdots & \ddots & \vdots \\ -\frac{\lambda_{0(n-3)}}{\lambda_{00}} & -\frac{\lambda_{(k-1)k}}{\lambda_{11}} & \cdots & -\frac{\lambda_{(k-1)(n-3)}}{\lambda_{(k-1)(k-1)}} \\ -\frac{1}{\lambda_{00}} & -\frac{1}{\lambda_{11}} & \cdots & -\frac{1}{\lambda_{(k-1)(k-1)}} \\ -\frac{b+\alpha_0}{\lambda_{00}} & -\frac{b+\alpha_1}{\lambda_{11}} & \cdots & -\frac{b+\alpha_{k-1}}{\lambda_{(k-1)(k-1)}} \end{bmatrix},$$

Thus, for every $\boldsymbol{c} \in C^{\perp}$ and $\boldsymbol{b} = [k \quad k \quad \cdots \quad k]^T \in B$,

IOP Conf. Series: Journal of Physics: Conf. Series **1108** (2018) 012055 doi:10.1088/1742-6596/1108/1/012055

$$c * b = \begin{bmatrix} -\frac{a_k k \lambda_{0k}}{\lambda_{00}} - \cdots - \frac{a_{n-3} k \lambda_{0(n-3)}}{\lambda_{00}} - \frac{a_{n-2} k}{\lambda_{00}} - \frac{a_{n-1} k (b + \alpha_0)}{\lambda_{00}} \\ -\frac{a_k k \lambda_{1k}}{\lambda_{11}} - \cdots - \frac{a_{n-3} k \lambda_{(k-1)k}}{\lambda_{11}} - \frac{a_{n-2} k}{\lambda_{11}} - \frac{a_{n-1} k (b + \alpha_1)}{\lambda_{11}} \\ \vdots \\ -\frac{a_k k \lambda_{(k-1)k}}{\lambda_{(k-1)(k-1)}} - \cdots - \frac{a_{n-3} k \lambda_{(k-1)(n-3)}}{\lambda_{(k-1)(k-1)}} - \frac{a_{n-2} k}{\lambda_{(k-1)(k-1)}} - \frac{a_{n-1} k (b + \alpha_{k-1})}{\lambda_{(k-1)(k-1)}} \end{bmatrix}.$$

Therefore, $(C^{\perp} * B) = C^{\perp}$ or $(C^{\perp} * B) \perp C$.

2. Since $n - k \geq 3$, then $k(C^{\perp}) = n - k \geq 3 > 1$.
3. Since $G_B = [1 \quad 1 \quad \cdots \quad 1]$ was the parity check matrix of $B^{\perp}$ and 2 was the smallest integer such that every two columns of H were linearly dependent, then $d(B^{\perp}) = 2 > 1$ [1].
4. Since C and $C^{\perp}$ are MDS codes, then $(C) = n - k + 1$ and $d(C^{\perp}) = n - (n - k) + 1 = k + 1$. Therefore, $d(C^{\perp}) + d(C) = k + 1 + n - k + 1 = n + 2 > n$.

Since the four conditions above were satisfied then $(C^{\perp}, B)$ are 1-ECP of C.

3. Error-Correcting Pair for MDS Non-GRS code

An $[\boldsymbol{n}, \boldsymbol{k}, \boldsymbol{d}]$ code is called almost MDS code if $\boldsymbol{d} = \boldsymbol{n} - \boldsymbol{k}$. In this section, it will be shown the existence of error-correcting pair for Almost MDS code.

3.1. The existence of 1-ECP for $[n, 1, n-1]$ *code*

Let C be a linear code over Z_2 with a generator matrix $G = [1 \quad \cdots \quad 1 \quad 0]$. Then C is an $[n, 1, n-1]$ almost MDS code, $n \geq 4$. Thus,

$$H = \begin{bmatrix} 1 & 1 & 0 & \cdots & 0 \\ 1 & 0 & 1 & \cdots & 0 \\ \vdots & \vdots & \vdots & \ddots & \vdots \\ 0 & 0 & 0 & \cdots & 1 \end{bmatrix} \approx \begin{bmatrix} 1 & 1 & 0 & \cdots & 0 \\ 0 & 1 & 1 & \cdots & 0 \\ \vdots & \vdots & \ddots & \ddots & \vdots \\ 0 & 0 & \cdots & 1 & 1 \end{bmatrix}$$

is a parity check matrix of C and generator matrix of $C^{\perp}$[1].

Let A was an $[n, n-2]$ linear code over Z_2 with generator matrix $G_A = [C|D]$, $D^T = [0 \quad \cdots \quad 0 \quad 1]$, and $C = [c_{ii}]$, with$c_{ii}, c_{i(i+1)} = 1$ and zero was elsewhere. Let B was an $[n, 2]$ linear code over Z_2 with generator matrix

$$G_B = \begin{bmatrix} 1 & 1 & \cdots & 1 & 0 \\ 0 & 0 & \cdots & 0 & 1 \end{bmatrix}.$$

Thus,

$$A * B = \begin{bmatrix} a_1 \\ a_1 + a_2 \\ \vdots \\ a_{n-3} + a_{n-2} \\ a_{n-2} \\ a_{n-2} \end{bmatrix} * \begin{bmatrix} b_1 \\ b_1 \\ \vdots \\ b_1 \\ b_1 \\ b_2 \end{bmatrix} = \begin{bmatrix} a_1 b_1 \\ a_1 b_1 + a_2 b_1 \\ \vdots \\ a_{n-3} b_1 + a_{n-2} b_1 \\ a_{n-2} b_1 \\ a_{n-2} b_2 \end{bmatrix}$$

Therefore, $(A * B) \perp C$.

Since G_B was the parity check matrix of $B^{\perp}$ and 2 was the smallest integer such that every two columns of H were linearly dependent, then $d(B^{\perp}) = 2 > 1$. Let $G_A' = [I|A']$ be a reduced row echelon form of G_A and a generator matrix of A, then $(A')^T = \begin{bmatrix} 1 & 1 & \cdots & 1 \\ 1 & 1 & \cdots & 1 \end{bmatrix}$. Since the parity check matrix of A is $H_A = (-(A')^T|I)$, then $d(A) = 2$. So,

$$d(A) + d(C) = 2 + n - 1 = n + 1 > n.$$

Since $(A * B) \perp C, k(A) = n - 2 \geq 4 - 2 > 1$, $d(B^{\perp}) > 1$, and $d(A) + d(C) > n$, then C had1-ECP.

3.2. The existence of ECP for $[n, n-1, 1]$ *code*
In this section, it would be seen that the $[n, n-1, 1]$ AMDS code did not have an error-correcting pair. Let C was a linear code over $\mathbb{F}_q$, then

$$St(C) = \{x \in \mathbb{F}_q^n | xC \subseteq C\}$$

the stabilizer of C [4].

Theorem 3.1 Let (A, B) be a pair of $\mathbb{F}_q$-linear codes in $\mathbb{F}_q^n$. Then,

$$k(A * B) \geq k(A) + k(B) - k\big(St(A * B)\big)$$

[4].

Theorem 3.2 Let C be $[k, k-1, 1]$ linear code over $\mathbb{F}_q$. Let $G = (I_k|A)$ be the generator matrix of C, where I_k is a$k \times k$ identity matrix and A is $(k-1) \times 1$ matrix with exactly one zero entry.

$$A = \begin{pmatrix} -a_{11} \\ \vdots \\ -a_{m1} \\ 0 \\ -a_{(m+2)1} \\ \vdots \\ -a_{(k-1)1} \end{pmatrix},$$

Then C does not have an error-correcting pair.

Proof. Let $C^{\perp}$ be the dual code of C, then

$$G_{C^{\perp}} = (a_{11} \quad \cdots \quad a_{m1} \quad 0 \quad a_{(m+2)1} \quad \cdots \quad a_{(k-1)1} \quad 1).$$

The generator matrix was $C^{\perp}$. So, Stabilizer of $C^{\perp}$ was a linear code with generator matrix

$$G_{St} = \begin{pmatrix} 1 & \cdots & 1 & 0 & 1 & \cdots & 1 & 1 \\ 0 & \cdots & 0 & 1 & 0 & \cdots & 0 & 0 \end{pmatrix}.$$

It used Theorem 3.1, $3 \geq k(A) + k(B)$. Since $k(A) \geq t$, then $k(A)$ and $k(B)$ respectively were 2 and 1.

By singleton bound, it was obtained

$$d(A) \leq n - 2 + 1 = n - 1.$$

Thus,

$$d(A) + d(C) \leq n - 1 + 1 = n$$

Thus, C did not have an error-correcting pair.

4. Conclusions

Let $\boldsymbol{C_1}$ be an MDS non-GRS code that satisfied Theorem 2.1. Then $(\boldsymbol{C_1^\perp}, \boldsymbol{B})$, $\boldsymbol{B}$ was a linear code with generator matrix $\boldsymbol{G_B} = [\mathbf{1} \quad \mathbf{1} \quad \cdots \quad \mathbf{1}]$, were 1-error-correcting pair of $\boldsymbol{C_1}$.

It was known that there is an AMDS code that has 1-error-correcting pair. Let $\boldsymbol{A}$ be an $[\boldsymbol{n}, \boldsymbol{n}-\mathbf{2}]$ code, for $\boldsymbol{n} \geq \mathbf{4}$, over $\boldsymbol{Z_2}$. Let $\boldsymbol{G_A} = [\boldsymbol{M}|\boldsymbol{N}]$ was the generator matrix of $\boldsymbol{A}$, $\boldsymbol{N} = [\mathbf{0} \quad \cdots \quad \mathbf{0} \quad \mathbf{1}]^T$, and $\boldsymbol{M} = [\boldsymbol{m_{ij}}]$, with $\boldsymbol{m_{ii}}, \boldsymbol{m_{i(i+1)}} = \mathbf{1}$ and zero was elsewhere. Let B was an $[\boldsymbol{n}, \mathbf{2}]$ linear code over $\boldsymbol{Z_2}$ with generator matrix

$$\boldsymbol{G_B} = \begin{bmatrix} \mathbf{1} & \mathbf{1} & \cdots & \mathbf{1} & \mathbf{0} \\ \mathbf{0} & \mathbf{0} & \cdots & \mathbf{0} & \mathbf{1} \end{bmatrix}.$$

Let $\boldsymbol{C_2}$ was an $[\boldsymbol{n}, \mathbf{1}, \boldsymbol{n}-\mathbf{1}]$ linear code over $\boldsymbol{Z_2}$ with generator matrix

$$\boldsymbol{G} = [\mathbf{1} \quad \cdots \quad \mathbf{1} \quad \mathbf{0}].$$

Then, $(\boldsymbol{A}, \boldsymbol{B})$ were 1-error-correcting pair of $\boldsymbol{C_2}$. But not all AMDS codes had error-correcting pair. It was shown that $[\boldsymbol{k}, \boldsymbol{k}-\mathbf{1}, \mathbf{1}]$ AMDS code over $\mathbb{F}_q$ with generator matrix $\boldsymbol{G} = (\boldsymbol{I_k}|\boldsymbol{A})$,

$$\boldsymbol{A} = \begin{pmatrix} a_{11} \\ \vdots \\ a_{m1} \\ \mathbf{0} \\ a_{(m+2)1} \\ \vdots \\ a_{(k-1)1} \end{pmatrix},$$

did not have an error-correcting codes.

5. References

[1] Pellikaan R, Wu X, Bulygin S, and Jurrius, R 2017*Codes, Cryptology and Curves with Computer Algebra*(Inggris: Cambridge University Press)

[2] Márquez-Corbella I and Pellikaan R2015 A Characterization of MDS Codes that Have an Error Correcting Pair. *Applications of Computer Algebra, ACA 2015*, 40 issue C, 224-245

[3] Roth M.R and Lempel A1989 A Construction of Non-Reed-Solomon Type MDS Codes*IEEE Transactions on Information Theory***35** issue 3 pp 655-657

[4] Mirandola D andZémor G2015 Critical Pairs for the Product Singleton Bound*IEEE Transactions on Information Theory***61** issue 9 pp 4928-4937

Acknowledgments

Part of this research is funded by PITTA-UI Research Grant 2018 No 2277/UN2.R3.1/HKP.05.00/2018.

Investigating the Pre-service and In-service Science Teachers' Conceptions of A 'teacher' at the Beginning of Their Initial Teacher Training Program

J Marlizayati[1] and S Hardimah[2]

[1,2]Science Education Academic Group, Sultan Hassanal Bolkiah Institute of Education, Universiti Brunei Darussalam, Jalan Tungku Link, BE 1410, Brunei Darussalam

Abstract. This study is an on-going research that aims to explore pre-service and in-service teachers' conception of 'a teacher' at the beginning of their initial teacher training program. This particular study administered an open ended questionnaire that consists of 4 questions to 13 pre-service science teachers and 7 in-service science teachers during the first session of the initial teacher training programme. The open-ended questionnaire explores the teachers' conception of what makes a teacher, a good teacher, an excellent teacher and a 21st century teacher. A thematic analysis was conducted by two researchers in order to ensure trustworthiness of the emerging themes that emerged found from the data. One of the potential contributions from the findings of this research is the discourse of how pre-service and in-service science teachers' construct their conceptions on the different dimensions of a teacher i.e. from good to a 21st century teacher.

1. Introduction

This study is driven with aims to bring benefit and improvement in preparing both pre-service and in-service science teachers during their teacher training. The findings of this on-going study are hoped to unpack the discourse among these prospective science teachers on their conceptions of what makes a teacher, a good teacher, an excellent teacher and a 21st century teacher.

Loughran [1] indicated that prospective teachers enter teacher preparation with experiences they have experienced as students. This can be considered as students' perspective whereas once they enter teacher preparation, the prospective teachers are challenged to view their thinking from teachers' perspective. These prospective teachers' perspective could be shaped by their personal identity and belief [2, 3, 4, 5]. In science education, teachers' practices are expansively explained as influenced by their beliefs [2] Hence, recognition of the prospective teachers' belief would help to provide teacher educators to openly discuss and intentionally confront and challenge as well as define and redefine the prospective teachers' views on what constitute good and effective teaching and practices.

It is essential that not only the beliefs of the prospective teachers to be acknowledged but also the role of identity and their knowledge about teaching as well as subject and pedagogical content knowledge [6, 7, 8, 9, 10]. Teachers are inclined to focus on their subject knowledge which are then reflected on how they carry out their roles and hence their identities [7]. Thus it is central to see the relationship between teachers' practices and identities [11].

Hence, this study is hoped to better understand teachers' conceptions of what makes a 'teacher' through exploring their views at the onset of joining the programme. This will be a continuous study by accessing their conceptions during the middle and towards the end of the programme.

There are two main research questions formulated for this study which are:

Research question 1: What are the pre-service and in-service teachers' conceptions of different categories of 'a teacher'?
Research question 2: What are the dominant themes for the different categories of 'a teacher'?

2. Methodology

The study administered an open-ended questionnaire to 20 science teachers during the first session of their initial teacher training programme. Their responses to the questionnaire were used as the main source of data. This study employed a qualitative and quantitative approach to analyse the data. In terms of qualitative approach, a thematic analysis was used. As to quantitative approach, a simple tally of the recurrence of a theme was utilised.

2.1. Participants

20 science teachers were involved in this study i.e. 13 pre-service science teachers and 7 in-service science teachers. The in-service science teachers have a range of one to four years of teaching experience in schools.

2.2. Questionnaire

The questionnaire consisted of 2 main questions. The first main question consisted of 3 sub-questions. These questions were developed to elicit the science teachers' view on what makes (i) a teacher, (ii) a good teacher, (iii) an excellent teacher and (iv) 21st century teacher.

3. Findings

8 themes emerged from the overall data collected. There were (i) delivery of necessary content, (ii) awareness of students' capabilities and academic achievement, (iii) ability to explain concepts, (iv) pedagogical versatility, (v) memorisation, (vi) effort and commitment, (vii) sense of responsibility for future generation and (viii) becoming an inspiration and role model.

Some of the themes can only be found under one or two categories of what makes a teacher. Hence the findings are divided into two sections. The first section is focusing on each theme across the different categories of what makes a teacher. As to the second section discussed the dominant themes found under what makes (i) a teacher, (ii) a good teacher, (iii) an excellent teacher and (iv) 21st century teacher. Table 1 below is a summary of the findings

Table 1. Summary of the findings.

	A Teacher	A Good Teacher	An Excellent Teacher	A 21st Century Teacher
Delivery of Necessary Content	14	6	5	2
Awareness of Students' Capabilities and Academic Achievement	2	6	4	0
Ability to Explain Concepts	2	5	1	0
Pedagogical Versatility	0	5	11	13
Memorisation	3	0	0	0
Effort and Commitment	0	3	7	0
Sense of Responsibility for Future Generation	1	1	9	12
Becoming an Inspiration and Role Model	1	0	6	0

3.1. First section

As can be seen from Table 1, the different themes based on the responses from the questionnaire will be elaborated further in this section. Furthermore, the frequency of the themes occurrence will also be described.

3.1.1. Delivery of necessary content

Responses that indicated the role of teachers was to deliver or transmit or transfer necessary knowledge in the context of syllabus were categorised as delivery of necessary content. 14 out of 20 science teachers mentioned the importance of having the ability to deliver subject content knowledge as part of being a teacher. 6 out of 20 science teachers reiterated this ability that makes a good teacher. As for what makes an excellent teacher, only 5 responses highlighted this theme again. While only 2 science teachers mentioned this ability for what makes a 21st century teacher.

3.1.2. Awareness of students' capabilities and academic achievement

Responses that indicated teachers monitored their students' development especially in terms of academic achievement were categorised as awareness of students' capabilities and academic achievement. Only 2 respondents mentioned this trait for being a teacher. As for a good teacher, 6 respondents highlighted this trait. While only 4 respondents mentioned this trait for an excellent teacher and none for 21st century teacher.

3.1.3. Ability to explain concepts

Responses that indicated teachers that are able to explain concepts in a simplified way to ensure students understand were categorised under ability to explain concepts. There were only 2 science teachers mentioned the ability to explain concepts that makes a teacher and there were 5 science teachers highlighted this trait for a good teacher. As for an excellent teacher, only one science teachers reiterated this ability and no one mentioned this trait for what makes a 21st century teacher.

3.1.4. Pedagogical versatility

Responses that indicated the different ways of teaching such as encouraging collaboration among students or project based task or differentiated teaching were categorised as pedagogical versatility. 5 out of 20 science teachers mentioned about this ability for a good teacher. There were 11 and 13 science teachers who indicated this trait for an excellent and 21st century teacher respectively.

3.1.5. Memorisation

Responses that stated that the teacher only delivers content from textbook and encouraged students to memorise information were categorised as memorisation. There were only 3 respondents mentioned that a teacher required his or her students to memorise but none were available for the other categories.

3.1.6. Effort and commitment

Responses that stated teachers that extend their work hours outside of school hours in terms of preparation of teaching resources, lesson planning and taking care of students' welfare and willingness to sacrifice time and money for the students' benefit were categorised as effort and commitment. The respondents' data demonstrated that only for a good teacher (3 respondents) and an excellent teacher (7 respondents) would put more effort into resource making and extend their work hours.

3.1.7. Responsibility for future generation

Responses that indicated that teachers who help in shaping students' behaviour and mind set by equipping them with skills that applicable in the real world and instilling morale values were categorised as responsibility for future generation. Only one respondent mentioned about a teacher and a good teacher must feel responsible for the development of their students, not only in terms of academic but also life in general as they are the future generation of the country. While 9 respondents mentioned about this trait for an excellent teacher and 12 respondents for a 21st century teacher.

3.1.8. Becoming an inspiration and role model

Responses that indicated that teachers should inspire students and become a role model for the students were categorised as becoming an inspiration and role model. Only one respondent mentioned under what makes a teacher should be a role model and an inspiration to their students. While 6 respondents mentioned about this for an excellent teacher. While no respondent for both: a good teacher and a 21st century teacher.

3.2. Second section
As can be seen from Table 1, the dominant themes of the questionnaire under the different categories of teacher is discussed below.

3.2.1. What makes a teacher?
The dominant theme under this category was delivery of necessary content to the students. This demonstrated that the pre-service and in-service teachers put emphasis on the content knowledge for 'a teacher' category.

3.2.2. What makes a good teacher?
The dominant theme under this category was the ability to explain concepts to students. With comparison with the finding of 5.2.1, the pre-service and in-service teachers viewed 'a good teacher' as someone that is not only has a good content knowledge but also someone who has the ability to explain concepts to students.

3.2.3. What makes an excellent teacher?
The dominant themes under this category were teachers' invested effort and commitment in teaching their students and be an inspiration and role model to their own students. In contrast to the first two categories, the pre-service and in-service teachers viewed 'an excellent teacher' as someone who is committed in their teaching and hold the responsibility of being a role model to the students.

3.2.4. What makes a 21st century teacher?
The dominant themes under this category were the teachers should have pedagogical versatility and have a sense of responsibility for future generation that went beyond the confinement of the classroom. As compared to the finding of the 'excellent teacher' category, the pre-service and in-service teachers had broaden the scope for 'a 21st century teacher' in that this type of teacher is not only as an inspiration to students in the classroom but to view these students as the country future generation. Furthermore, the finding also emphasised the importance of pedagogical skills in teaching.

4. Discussion

The finding leads to a general image of what constitutes and influences these teachers' conception. Among them is their understanding of the importance of subject content knowledge and also the need to have pedagogical content knowledge. Theme on becoming inspiration and role model leads to these teachers' awareness on the role of their identity as science teachers.

As stated in the second section of the finding, the responses from the pre-service and in-service science teachers indicated that a 'teacher' and a 'good teacher' are teachers who are able to deliver content to their students and be able to explain concepts to their students. This finding contradicts with the finding of another study done by Fajet, Bello, Leftwich, Mesler and Shaver [12]. In their study, they investigated traits of a good teacher and they found that the least mentioned theme was the subject content knowledge of the teacher. We stipulated that these particular traits of subject content knowledge and ability to explain concepts in our study might be closely linked to the nature of science itself. The subject matter is usually presented as a factual and theory laden body of knowledge. This can specifically be seen with the pre-service teachers where their exposure to science teaching was based on their experience in school and university level. Nonetheless, the in-service science teachers were also inclined to these themes for what makes a teacher and a good teacher. Thus the nature of science being presented might have some influence in their conceptions of a teacher and a good teacher.

In contrast to the science teachers' responses to what makes 'an excellent' teacher and 'a 21st century' teacher, the focus shift more towards (i) social responsibility such as being a role model or an inspiration and shaping country's future generation and (ii) pedagogical content knowledge. Taking role model as related to identity, this concurs with the literature review discussed above on the relationship between teachers' identities and content knowledge in shaping their practices. The idea of being 'an excellent' teacher and 'a 21st century' teacher went beyond the limits of the classroom and school. In terms of pedagogical versatility, as a 21st century teacher, technology played an important role in the classroom. With the advancement of technology, it affords 'a 21st century teacher' to be more creative and innovative in their teaching so that students can be engaged during the lesson.

5. Conclusion

Several trends can be seen from our data analysis that are highly dependent on the different categories of 'teacher'. Nature of science might played a major role in the response pattern observed among the pre-service and in-service science teacher in their conception of 'a teacher' and 'a good teacher.' As to 'an excellent teacher' and 'a 21st century teacher', social responsibility and pedagogical content knowledge played a major role in their responses.

As this study is an on-going research, it is hoped that the future data collection will be able to tap into more in-depth insights from these pre-service and in-service teachers. In turn, this in-depth data will provide better understanding on their developmental progress in becoming a teacher to 21st century teacher.

6. References

[1] Loughran J 2006 *Developing a Pedagogy of Teacher Education: Understanding Teaching and Learning about Teaching*. (New York: Routledge)
[2] Bryan, L 2012 *Research on science teacher beliefs Second International Handbook of Science Education* (Dordrecht: Springer)
[3] Bullough, R 1991 *J. Teacher Ed* **42** 43
[4] Deemer S 2004 *Educational Research* **46** 73
[5] Pajeres M 1992 *Review of Educational Research* **62** 307
[6] Beauchamp C and Thomas L 2011 *International J of Educational Research* **50** 6
[7] Ben-Peretz M 2011 *Teaching and Teacher Education* **27** 3
[8] Clandinin D and Connelly F 1996 *Educational Researcher* **25** 24
[9] Gess-Newsome J 1999 *Secondary teachers' knowledge and beliefs about subject matter and their impact on instruction Examining pedagogical content knowledge* (The Netherlands: Kluwer Academics)
[10] Richardson V 1996 *The role of attitudes and beliefs in learning to teach Handbook of research on teacher education* (New York: Simon & Schuster Macmillan)
[11] Flores M and Day C 2006 *Teaching and teacher* **22** 219
[12] Fajet W, Bello M, Leftwich A, Mesler J L and Shaver A N 2005 *Teaching and Teacher Education* **21** 717

A mathematical model for chemotherapy paradoxical reaction in Tuberculosis transmission

Febriana Tri Rahmawati[1], **Alhadi Bustamam**[1], **and Dipo Aldila**[*1]
[1]Department of Mathematics, Universitas Indonesia, 16424 Depok, Indonesia
E-mail: *aldiladipo@sci.ui.ac.id

Abstract. An SEIR with the age-class model is analyzed in this article to understand how paradoxical reaction in Tuberculosis (TB) treatment impact the success of TB control program. Mathematical model analysis of equilibrium points and the basic reproduction number of the model are analyzed. A numerical experiment about the dependency of basic reproduction number respect to the change of other parameters is given to give an illustration of how the paradoxical reaction might impact the success of chemotherapy intervention on TB. From the numerical investigation, we find that the paradoxical reaction plays an important role to determine the TB will coexist or disappear.

1. Recent facts about Tuberculosis (TB)

Tuberculosis (TB) is a contagious disease caused by the bacterium *Mycobacterium tuberculosis*. The disease spreads with an airborne intermediate and is transmitted from an infected person to a healthy person through touch or sneezing [1]. The transmission process occurs through small water particles released when infected people are coughing or sneezing and attached to the healthy person. When bacteria enter the body of a healthy person, there are two infection conditions, namely the situation when the bacteria are still passive, and when bacteria directly become active in the body. Passive bacterial conditions are also referred to as latent TB infection conditions. This latent condition does not show the visible characteristics of TB [2]. TB symptoms that appear are fever, cough, shortness of breath, chest pain, and malaise [3].

TB prevention can be done by administering the Bacille Calmette-Guerin (BCG) vaccine, which until now is believed to reduce the potential for this disease. The primary purpose of this vaccine is to ward off bacterial infections. BCG vaccine is most useful for protecting infants and children, with an active period of approximately 15 years. TB treatment should include early and advanced stages. Treatment in the early stages is intended to kill bacteria effectively and given for two months. Advanced treatment is a crucial stage to kill the remaining bacteria in the body so that patients can recover and prevent recurrence. If the procedure is done in an orderly manner, then the standard of 6 months of the patient can be healed. The main anti-TB drugs commonly used are rifampicin and isoniazid [4].

Sometimes chemotherapy doesn't always cure. Chemotherapy in patients with tuberculosis also has several problems, including paradoxical reactions. Paradoxical reaction (PR) during chemotherapy for tuberculosis (TB) has been defined as clinical or radiological worsening of preexisting TB lesions or the development of new lesions in a patient who has already received anti-TB therapy for several days and whose condition has been reported to be improving. This paradoxical reaction usually occurs after initial treatment for 1-2 months after the beginning of the treatment period [5]. Symptoms that typically appear are high fever, deterioration of chest radiographs, and swollen lymph nodes [6]. Paradoxical reactions can be caused by an abnormal immune response or when the immune system decreases, but it is not known precisely what causes it [7]. Paradoxical reactions are more straightforward to attack children who do not have immunity against tuberculosis. Children with paradoxical reactions may show symptoms of TB more clearly than children without paradoxical reactions [5].

The spread of TB disease can be modeled mathematically. One commonly used mathematical model is the SIR epidemic model discovered by Kermack and McKendrick in 1927. In the range of TB disease, an epidemic model suitable to use is the SEIR epidemic model which is the development of the SIR epidemic model. Several articles have previously discussed the model of TB disease spread, i.e., with age-class divisions [8] and with the effect of vaccines [2].

Unlike the articles mentioned earlier, in this article, we construct the mathematical model of TB involving two age class, i.e., children and adult classes. We also include the paradoxical reaction of TB intervention which impacts in a prolongation of the duration of human to be recovered from TB.

This paper is organized as follows. In section 2, we present the construction of the mathematical model. In section 3, we analyzed the model and followed with numerical experiments in section 4. Finally, in section 5, we give some conclusions and discussions.

2. Mathematical model derivation

To analyze the impact of the paradoxical reaction of chemotherapy in TB control program, we will use a deterministic SEIR model with two age class in each compartment ($i = c$ denote children, while $i = a$ denote adult class). We divide the human population in susceptible compartments (S_c, S_a), infected by TB but not yet infectious, let call it as exposed TB (E_c, E_a), infected compartments (I_c, I_a), and recovered compartments (R_c, R_a). The mathematical model is given below which constructed based on followed assumptions.

$$
\begin{aligned}
\frac{dS_c}{dt} &= \theta - ((\rho_c + (1-\rho_c))\,\beta_c I_a + m + \mu\ )\,S_c, \\
\frac{dS_a}{dt} &= mS_c - ((\rho_a + (1-\rho_a))\,\beta_a I_a + \mu\ )\,S_a, \\
\frac{dE_c}{dt} &= ((1-\rho_c)\,S_c + R_c)\,\beta_c I_a - (m + \mu + \varepsilon\ )\,E_c, \\
\frac{dE_a}{dt} &= ((1-\rho_a)\,S_a + R_a)\,\beta_a I_a + mE_c - (\mu + \varepsilon\ )\,E_a, \\
\frac{dI_c}{dt} &= \rho_c\beta_c S_c I_a + \varepsilon\,E_c + \eta R_c - (q_c\gamma_1 + (1-q_c)\,\gamma_2 + m + \mu)\,I_c, \\
\frac{dI_a}{dt} &= \rho_a\beta_a S_a I_a + \varepsilon\,E_a + \eta R_a + mI_c - (q_a\gamma_1 + (1-q_a)\,\gamma_2 + \mu)\,I_a, \\
\frac{dR_c}{dt} &= (q_c\gamma_1 + (1-q_c)\,\gamma_2)\,I_c - (\beta_c I_a + m + \mu + \eta)\,R_c, \\
\frac{dR_a}{dt} &= (q_a\gamma_1 + (1-q_a)\,\gamma_2)\,I_a + mR_c - (\beta_a I_a + \mu + \eta)\,R_a,
\end{aligned}
\tag{1}
$$

Some explanation about the infection and treatment term in the model (1) is given as follows. Since infection of TB mainly only caused by adult people [9], the infection process only occurs when susceptible or recovered compartments made successful contact with infected adult compartment (I_a). When susceptible human successfully gets infected, ρ_i proportion of them will go directly to infectious compartment caused by the fast infection progression, while the rest of them, $1-\rho_i$, goes to the exposed compartment. The chemotherapy treatment is given in the rate of γ_j, when $j=1$ denote the recovery rate without paradoxical effect, while $j=2$ is the recovery rate with paradoxical effect. We assume that the proportion of successful chemotherapy treatment as a constant parameter, let denote it as q_i. Therefore, the recovery rate without paradoxical effect is given by $q_i\gamma_1 I_i$, while $(1-q_i)\gamma_2 I_i$ gives the recovery rate with paradoxical reaction. Please note that since the paradoxical reaction to chemotherapy intervention in TB will prolong the infection period, we have that $\gamma_2 < \gamma_1$.

3. Theoretical analysis of the model

There are two equilibrium points of the model (1), i.e., the disease-free equilibrium point which given by

$$\Omega_1 = \left(S_c = \frac{\theta}{m+\mu}, S_a = \frac{m\theta}{\mu\,(m+\mu)}, E_c = 0, E_a = 0, I_c = 0, I_a = 0, R_c = 0, R_a = 0\right), \quad (2)$$

while the endemic equilibrium point is not in simple form to be shown. However, in a simpler model of system (1), i.e., when there are no age class involved in to the model $(m=0, \rho_a=\rho_c=\rho, q_a=q_c=q, \beta_a=\beta_c=\beta)$, the endemic equilibrium point is given by

$$\Omega_2 = \left(S^* = \frac{\theta}{\beta\, I^*+\mu}, E^* = \frac{K\,(I^*)^3 + L\,(I^*)^2 + MI^*}{\varepsilon\,(\beta\, I^*+\mu)\,(\beta\, I^*+\eta+\mu)}, R^* = \frac{(q\gamma_1 + (1-q)\gamma_2)\, I^*}{\beta I^* + \eta + \mu}\right), \quad (3)$$

with

$$\begin{aligned} K &= \beta^2 q\gamma_1 + (1-q)\beta^2\gamma_2 + \beta^2\mu, \\ L &= (1-q)2\,\beta\,\mu\,\gamma_2 - \beta^2\rho\,\theta + 2\,\beta\,\mu\, q\gamma_1 + \beta\,\eta\,\mu + 2\,\beta\,\mu^2, \\ M &= \mu^2 q\gamma_1 + (1-q)\mu^2\gamma_2 - \beta\,\eta\,\rho\,\theta - \beta\,\mu\,\rho\,\theta + \eta\,\mu^2 + \mu^3, \end{aligned}$$

and I^* is taken from the positive root of the two-degree polynomial in the form of :

$$A_2(I^*)^2 + A_1(I^*) + A_0 = 0, \quad (4)$$

with

$$\begin{aligned} A_2 &= \beta^2\mu\,(q(\gamma_1-\gamma_2)+\varepsilon+\mu+\gamma_2), \\ A_1 &= (\mathcal{R}_0^* - \mathcal{K})(\mu\,(\mu+\varepsilon)\,(q(\gamma_1-\gamma_2)+\eta+\mu+\gamma_2)) \\ A_0 &= (1-\mathcal{R}_0^*)(\mu^2\,(\mu+\varepsilon)\,(q(\gamma_1-\gamma_2+\eta+\mu+\gamma_2)) \end{aligned}$$

where $\mathcal{R}_0^* = \frac{\beta\,\theta\,(\mu\,\rho+\varepsilon)(\eta+\mu)}{\mu^2(\mu+\varepsilon)(q\gamma_1-q\gamma_2+\eta+\mu+\gamma_2)}$. and $\mathcal{K} = \mu\left(1 + \frac{\mu^3(q(\gamma_1-\gamma_2)+\epsilon+\mu+\gamma_2)}{\mu^2(\mu+\varepsilon)(q\gamma_1-q\gamma_2+\eta+\mu+\gamma_2)}\right)$. The existence of the endemic equilibrium Ω_2 given in the following theorem.

Theorem 1. *Endemic equilibrium Ω_2 will*

(i) have a unique positive endemic equilibrium ($\Omega_2 \in \mathbb{R}_4^+$) whenever $\mathcal{R}_0^* > 1$, and

(ii) have two positive endemic equilibrium ($\Omega_2 \in \mathbb{R}_4^+$) whenever $\mathcal{K} < \mathcal{R}_0^* < 1$ & $A_1^2 - 4A_2A_0 \geq 0$.

The proof of Theorem 1 is simple. Theorem 1(i) appears from the consequence that polynomial 4 will have exactly one positive root of I^* when the multiplication of the root is negative, i.e., $\frac{A_0}{A_2} < 0$ which led us into a condition of $\mathcal{R}_0^* > 1$. In the other hand, theorem 1(ii) will have two positive root if and only if $\frac{A_0}{A_2} > 0$, $\frac{-A_1}{A_2} > 0$, and the discriminant of 4 should be non negative to guarantee the non-imaginary roots which gave us $\mathcal{K} < \mathcal{R}_0^* < 1$ and $A_1^2 - 4A_2A_0 \geq 0$.

Next, we analyze the basic reproduction number ($\mathcal{R}_0$) of model (1) using the next-generation matrix approach. Please see [10] for more example about the use of next-generation matrix method to determine $\mathcal{R}_0$. The basic reproduction matrix of model (1) is the spectral radius of the next-generation matrix K_s which taken by

$$K_s = -B\Sigma^{-1}C, \tag{5}$$

where B, Σ and C are in the following form.

$$\boldsymbol{B} = \begin{bmatrix} 0 & 1 & 0 & 0 & 0 & 0 \\ 0 & 0 & 0 & 0 & 1 & 0 \\ 0 & 0 & 0 & 0 & 0 & 1 \end{bmatrix}, \quad \boldsymbol{C} = \begin{bmatrix} \rho_c\beta_c S_c & 0 & 0 \\ \rho_a\beta_a S_a & 0 & 0 \\ (1-\rho_c)\,\beta_c S_c + \beta_c R_c & \beta_c I_a & 0 \\ (1-\rho_a)\,\beta_a S_a + \beta_a R_a & 0 & \beta_a I_a \\ -\beta_c R_c & -\beta_c I_a & 0 \\ -\beta_a R_a & 0 & -\beta_a I_a \end{bmatrix}, \tag{6}$$

$$\boldsymbol{\Sigma} = \begin{bmatrix} -q_c\gamma_1 - (1-q_c)\,\gamma_2 - m - \mu & 0 & \varepsilon & 0 & \eta & 0 \\ m & -q_a\gamma_1 - (1-q_a)\,\gamma_2 - \mu & 0 & \varepsilon & 0 & \eta \\ 0 & 0 & -m-\mu-\varepsilon & 0 & 0 & 0 \\ 0 & 0 & m & -\mu-\varepsilon & 0 & 0 \\ q_c\gamma_1 + (1-q_c)\,\gamma_2 & 0 & 0 & 0 & -\eta-m-\mu & 0 \\ 0 & q_a\gamma_1 + (1-q_a)\,\gamma_2 & 0 & 0 & m & -\eta-\mu \end{bmatrix}.$$

Although the $\mathcal{R}_0$ of the complete model (1) can not shown analytically in this paper, the simpler form of $\mathcal{R}_0$ for a simpler model, i.e., when $m = 0, \rho_a = \rho_c = \rho, q_a = q_c = q, \beta_a = \beta_c = \beta$ is given by

$$\mathcal{R}_0 = \frac{\beta\,\theta\,(\mu\,\rho + \varepsilon)\,(\eta + \mu)}{\mu^2\,(\mu + \varepsilon)\,(q\gamma_1 - q\gamma_2 + \eta + \mu + \gamma_2)}. \tag{7}$$

Please note that this $\mathcal{R}_0$ is the same with $\mathcal{R}_0^*$ which become the requirement to guarantee the existence of endemic equilibrium Ω_2.

4. Numerical experiments

From the results in the previous section, it is indicated that $\mathcal{R}_0$ hold a vital role to determine whether the TB will exist or vanish from the field, which also stated from many authors in other epidemiological models [11, 12, 13, 14]. They found that the disease will exist whenever $\mathcal{R}_0 > 1$ and will vanish when $\mathcal{R}_0 < 1$. Therefore, the numerical experiment in this section will be shown to see how the $\mathcal{R}_0$ impact the dynamic of the model (1) and how it fluctuated depend on many parameters.

For the following simulations, we use set of parameters : $\theta = \frac{10000}{65}$, $\beta_c = 1 \times 10^{-5}$, $\beta_a = 2.73 \times 10^{-5}$, $\gamma_1 = 0.3$, $\gamma_2 = 0.1$, $\rho_c = 0.1$, $\rho_c = 0.7$, $\eta = 0.0341$, $m = \frac{1}{15}$, $\epsilon = 6$, $\mu = \frac{1}{65}$. The first simulation is performed to see how q_c and q_a determine the magnitude of $\mathcal{R}_0$. From Figure 1, we can see that enlarging the proportion who do not get a paradoxical effect of TB chemotherapy (q_a and q_c) will reduce $\mathcal{R}_0$. With this relation in hand between q_a and q_c determining $\mathcal{R}_0$, we can determine how large is q_a or q_c to achieved the disease-free equilibrium condition. For an example, let $q_c = 0.5$, therefore it is needed $q_a = 1$ so that $\mathcal{R}_0 = 0.8746568876 < 1$ while if

$q_a = 0.8$ it is needed $q_c = 0.7$ so that $\mathcal{R}_0 = 0.9981619276 < 1$. To confirm how the dynamic of susceptible compartments and infected compartments will change respect to the change of q_c and q_a, the autonomous simulation of model (1) is given in Figure 2.

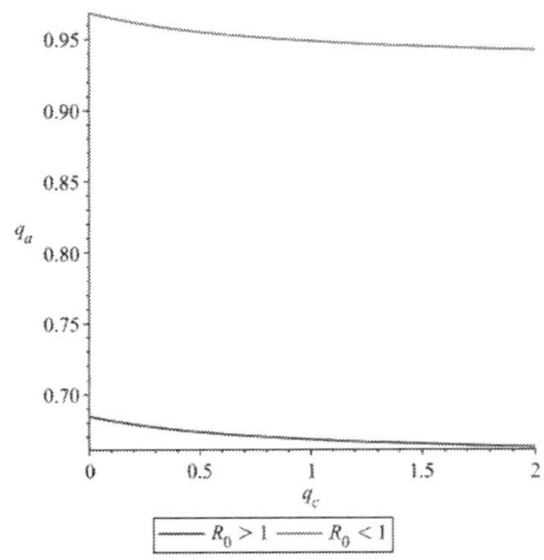

Figure 1: Sensitivity of $\mathcal{R}_0$ respect to paradoxical parameters of chemotherapy (q_a and q_c).

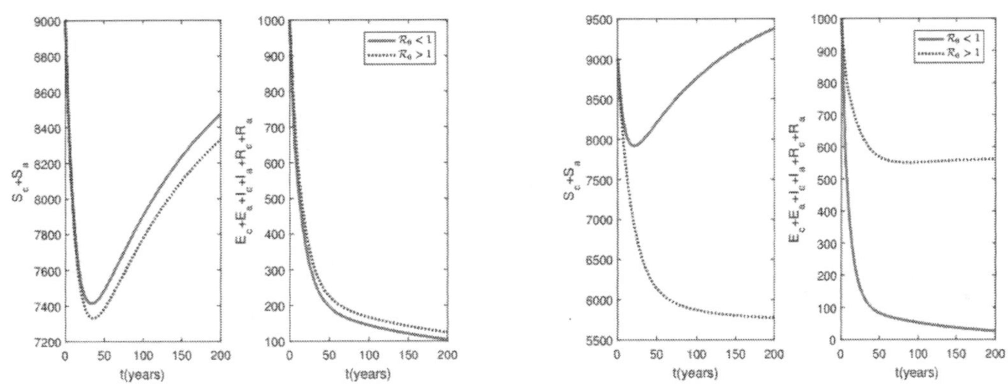

(a) (b)

Figure 2: (a) The dynamic of susceptible compartments and infected compartments in system (1) with the fixed value of $q_a = 0.8$ and the changed value of $q_c = 0$, unsuccess chemotherapy treatment in children (blue) and $q_c = 0.7$ (red). (b) The dynamic of susceptible compartments and infected compartments in system (1) with the fixed value of $q_c = 0.5$ and the changed value of $q_a = 0.6$ (blue) and $q_a = 1$, success chemotherapy treatment in adult (red).

5. Conclusions and discussion

In this paper, a mathematical model is made to look at the effect of the paradoxical reactions of chemotherapy on TB, which play an essential role in determining TB to coexist or disappear. From the model (1), we can get two equilibrium point, i.e., the disease-free equilibrium point and the endemic equilibrium point, and the basic reproduction number $\mathcal{R}_0$. We find that two conditions have to be fulfilled for the endemic equilibrium point to exist. Based on the numerical simulation of the autonomous model, we see that Paradoxical reactions can make people with tuberculosis a longer time in their recovery. It appears that the smaller the proportion of individuals affected by the paradoxical reaction will increase $\mathcal{R}_0$ and make the total susceptible individuals decrease as the total number of infected individuals increases due to longer treatment duration. Then, the higher the proportion of individuals who are not exposed to a paradoxical reaction will decrease $\mathcal{R}_0$ and make total susceptible individuals accrue together with reduced total infected individuals due to no treatment duration as long as the individual is affected by a paradoxical reaction. Further development of the model could be involved an optimal control problem to see how the limitation of the budget for TB chemotherapy will impact the optimal intervention that can be implemented in the field.

6. Acknowledgments

This research is funded by Universitas Indonesia with PITTA research grant 2018 scheme with contract Number ID: 2252/UN2.R3.1/HKP.05.00/2018.

References

[1] World Health Organization 2018 Tuberculosis Fact sheet Reviewed January 2018
[2] Liu S, Li Y, Bi Y and Huang Q 2017 Mixed vaccination strategy for the control of tuberculosis: A case study in China *Mathematical Biosciences and Engineering* **14(3)** 695-708
[3] Sudoyo A W, Setiyohadi B, Alwi I, Simadibrata M and Setiati S 2009 *Buku Ajar Ilmu Penyakit Dalam (vol 3)* (Jakarta: Interna Publishing)
[4] Kementrian Kesehatan Republik Indonesia *Pedoman Nasional Pengendalian Tuberkulosis* 2014 Kementrian Kesehatan RI Jakarta
[5] Olive Carole, Mouchet Franoise, Toppet Vronique, Haelterman Edwige and Levy Jack 2013 ParadoxicalReactionDuring TuberculosisTreatment inImmunocompetentChildren:ClinicalSpectrumand RiskFactors *The Pediatric Infectious Disease Journal* **32** 446-449
[6] Centre for Disease Control and Prevention 2003 *Treatment for Tuberculosis* CDC United States
[7] Bloch S, Wickremasinghe M, Wright A, Rice A, Thompson M, & Kon O M 2009 Paradoxical reactions in non-HIV tuberculosis presenting as endobronchial obstruction *European Respiratory Review* **18(114)** 295-299
[8] Zhao Y, Li M, & Yuan S 2017 Analysis of Transmission and Control of Tuberculosis in Mainland China 20052016 Based on the Age-Structure Mathematical Model *International Journal of Environmental Research and Public Health* **14(10)** 1192
[9] Donald P R, Marais B J and Barry C E 2010 Age and the epidemiology and pathogenesis of tuberculosis The Lancet 375 pp 1852-1854
[10] Diekmann O, Heesterbeek J A P, and Roberts M G 2010 The Construction of Next-Generation Matrices For Compartmental Epidemic Models *J. R. Soc. Interface* **7** 873-885
[11] Aldila D., Agustin M. R., 2018 A Mathematical Model Of Dengue-Chikungunya Co-Infection In A Closed Population. *J. Phys.: Conf* Ser 974 012001(1)-13
[12] Aldila D, Nuraini N, Soewono E 2014 Optimal control problem in preventing of swine flu disease transmission *Applied Mathematical Sciences* (69-72) pp 3501-3512
[13] Bustamam A, Aldila D, Yuwanda A 2018 Understanding Dengue Control for Short-Term and Long-Term Intervention with a Mathematical Model Approach *Journal of Applied Mathematics* **2018** 9674138(1)-14
[14] Putri Y E, Rozi S, Tasman H, Aldila D 2017 Assessing the effect of extrinsic incubation period (EIP) prolongation in controlling dengue transmission with wolbachia-infected mosquito intervention *AIP Conf* Proc Vol **1825** 020019(1)-9

IOP Conf. Series: Journal of Physics: Conf. Series **1108** (2018) 012058 doi:10.1088/1742-6596/1108/1/012058

Fraction Magnitude Knowledge: Students Strategy in Estimating Fractions

I K Amalina[1,*], Y Fuad[1], and I D Agustina[2]
[1]Program Pascasarjana Department of Mathematics, Universitas Negeri Surabaya, Surabaya, East Java, Indonesia
[2]Sekolah Menengah Pertama LabSchool Universitas Negeri Surabaya, Surabaya, East Java, Indonesia
*Corresponding author: ijtihadiamalina16070785040@mhs.unesa.ac.id

Abstract. Understanding concepts of fraction magnitude is necessary and inevitable steps of mastering fractions. One way to assess understanding of fraction magnitude is through magnitude estimation. Every student has different strategies for solving mathematics problem. These strategies show how deep students understanding, especially in estimating fractions. This study examined students' strategies in estimating fractions based on fraction magnitude knowledge. The sample was the students of VIIB class of a school in Surabaya, 20 boys and 8 girls. All students had to complete a diagnostic's test which was the fraction estimation test. There were 9 students with the high ability in estimating fractions, 8 students with medium ability in estimating fraction, and 11 students with low ability in estimating fraction. Three volunteer students with high ability in estimating fraction who had different and interesting strategies were chosen as our respondents. The semi-structured interview was utilized to investigate the process, strategy and students' conceptual of fraction magnitudes. the strategy used by the first and the third students are the concept of equivalence fraction and convert fraction into decimal. Moreover, the strategy used by the second student is applying the concept of part-whole, but he was only fluent in estimating unit fraction. For further study, there is need to categorize the logical stage of students' strategy in estimating fractions.

1. Introduction

Fractions have been used not only in mathematics but also in everyday life. In addition, understanding fractions are important to success in various professions such as sciences and technology. Fraction is one of the topics that must be mastered by students of 7th grades [1]. Understanding the concept of fractions and how to use them are fundamental to the development of mathematics and higher concepts in mathematics [2-6,].

The concept of fractions is difficult for students in Indonesia, America and Australia. The main source of the difficulty is because students do not know about the fraction magnitudes [2, 6, 7, 8]. There are some errors of 10 and 11 years old students in Belgium in understanding the concept of fractions [2]. The error indicates that the students do not understand the relation between the numerator and the denominator, or the students do not have a good ability about the fraction magnitudes. An understanding of fraction magnitude is a necessary and unavoidable stage of understanding about fractions [6]. Difficulties to understand the fractions can affect students' opportunity to learn advanced topics in mathematics [2, 4]. Therefore, it is very important for students to have the understanding of fractions correctly [2].

Understanding of fraction magnitudes can be assessed through the ability to compare fractions, the ability to estimate fractions on a number line, and the ability to represent fractions [9-11]. The magnitude comparison task and the number line estimation task both assess central component of mathematical competence [12]. Number line estimation assess children representation of fraction magnitude. Making placements on a number line is one form of basic numerical magnitude [13].

Students are said to understand fractions when achieving conceptual and procedural understanding [2, 4]. Conceptual understanding of fractions is the knowledge of a concept, nature, and how it is derived [14]. This understanding includes fractions consisting of numerators and denominators, its magnitude is determined by numerator-denominator relation, and that these magnitudes can be ordered on a number line [4]. While procedural understanding is the ability to perform a series of actions to solve the problem [2, 15].

Students use strategy in solving problems, including estimating fractions on a number line. The strategy chosen depends on the characteristics of the questions and the learners. Therefore not only different students use different strategies on the same problem, but also one student can use different strategies on similar problems or same problems at different times. The students' strategies use will change over time. Students with greater experience and knowledge used more adaptive and efficient strategies [16].

This research will be studied strategy used by students in fraction magnitudes that is estimating fractions on the number line which analyzed based on conceptual and procedural capability. Students with the high score will be chosen based on the interesting and different strategy used.

2. The Number Line Fraction Estimation

A fraction is a number that can be represent by an ordered pair of whole numbers $\frac{a}{b}$ (or a/b), where $b \neq 0$ [17, 18]. Fraction also can be defined as part of whole [19]. Fraction can be represented on a number line. Fraction $\frac{1}{b}$ can be located on a number line by defining a number line with the interval 0 to 1 as a whole and partitioning into b equal parts, each part is of length $\frac{1}{b}$ and the number $\frac{1}{b}$ is located at the right endpoint of the part with its left endpoint at 0.

The number line also can be used in estimating fraction. Estimating fraction on a number line is used to involve students in demonstrating their understanding of the concept of fraction as a part of whole. Besides, it also demonstrating their understanding about fraction magnitude [9, 11]. In estimating fractions, students are engaged in multiple steps in estimating a location for a given fraction by using strategies, such as identifying a benchmark fraction (example $\frac{1}{2}$) and comparing it to the given fraction to figure out which side of the number line the given fraction belongs to (example: less than $\frac{1}{2}$ or greater than $\frac{1}{2}$) [13].

3. Method

This research is conducted on 7th grade students of a school in Surabaya, selected the students of VIIB class. This class consists of 20 boys and 8 girls, chosen purposively from three available classes. All students have completed a test namely fraction estimation task for 30 minutes. There were five fractions including unit and non-unit fractions that have been checked and estimated in the line number by students which are $\frac{1}{6}, \frac{5}{7}, \frac{7}{121}, \frac{84}{216}$, and $\frac{110}{312}$. The test results were categorized into high, medium, and low ability in estimating fraction based on procedural and conceptual understanding. Three students with high ability in estimating fraction, different, and interesting strategies were chosen as subjects to analyze deeply about their abilities. The semi-structured interview was conducted on three subjects for 20 minutes to examine students' conceptual understanding deeply, process and strategy used. Table 1 shows the indicators of conceptual and procedural understanding in estimating fractions.

Table 1. Indicators of Conceptual and Procedural Understanding in Estimating Fractions.

	Procedural Understanding	**Conceptual Understanding**
Understanding the problem	Mention the information in the problem (including known and asked in the problem)	Explain the information in the problem (including known and asked in the problem)
Devising a plan	Determine the procedure and strategy used to answer the problem	Explain the argument of using procedure and strategy
Carrying out the plan	Carry out the procedure and strategy that has been determined to solve the problem	
Looking back	Recheck the consistency of the use of the procedure and strategy at each stage of problem solving	

4. Results

The fraction estimation task was analyzed based on procedural and conceptual understanding then categorized into high, medium, and low ability. Based on the analysis results there were 9 students with the high ability in estimating fractions, 8 students with medium ability in estimating fraction, and 11 students with low ability in estimating fraction. Therefore most students had low ability in estimating fractions. The difficulties were they did not understand the problem, they were not got the similar problem before and they did not understand the concept of fraction. From 9 students with the high ability in estimating fraction, there were three students who had the different and interesting strategy in estimating fraction and will be chosen as subjects. The first student is students with the higher score, the second and third students are students with the second and third rank of the scores.

4.1. The Strategy Used by Student 1

Student 1 estimated and checked correctly four from five fractions in the number line. From his answered, he used the concept of the decimal. Figure 1 shows the resulting test of student 1.

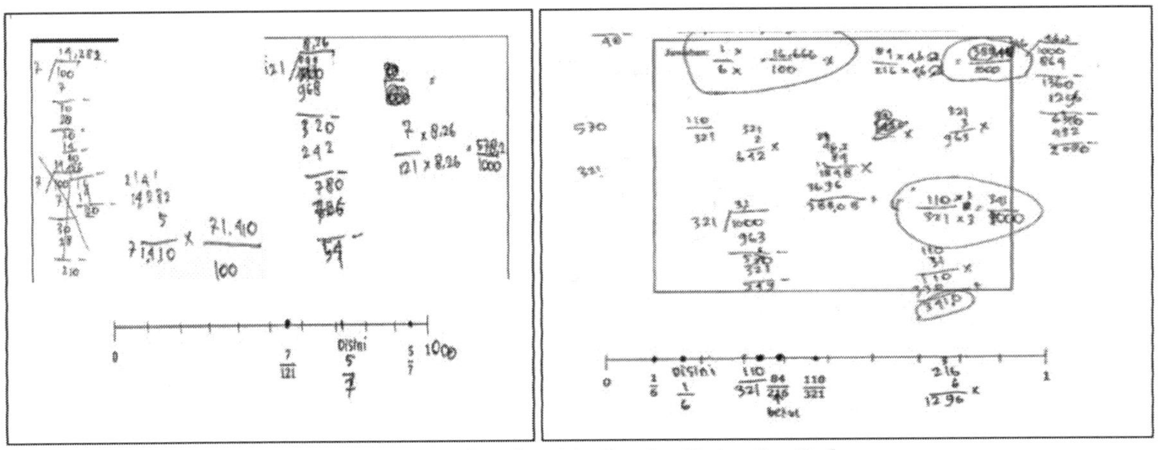

Figure 1. The Answer of Student 1 in Fraction Estimation Task

IOP Conf. Series: Journal of Physics: Conf. Series **1108** (2018) 012058 doi:10.1088/1742-6596/1108/1/012058

From the resulting test, he converted the fraction into decimal. He used that strategy fluently. It can be seen from the process he was written in the answer sheet. He used division concept to convert the fraction into decimal. It means he had a better procedural understanding. But for making it sure, the above answer was justified with the result of the interview as follows.

Q : Could you tell me what is the information on the problem as you know?
S1 : There are a number line and fractions
Q : What did you do with these fractions?
S1 : I'll check them and estimate the location in the number line
Q : Could you tell me what did you know from this number line?
S1 : The number line is between 0 to 1, it was divided into 10 parts. And these parts are decimals

From the result of the interview, student 1 was able to mention and explain the information in the problem. He understood the instruction, he stated that he will check and estimate all known fraction in the number line. He also understood the number line is divided into 10 parts, therefore the value of these parts must be between 0 to 1 or must be decimal.

The strategy used by student 1 is converted the fraction into the decimal. It was justified with the result of the interview as follow

Q : How did you check and estimate them?
S1 : I used the concept of the decimal. For example $\frac{7}{121}$, because 121 is more than 100 so I used 1000 for the denominator. So 121 is multiplied by a number and I get 1000, suppose this number is a and I multiplied 7 with a. Just like finding the equivalent fraction.

From the interview's result, he used the concept of the decimal and equivalent fraction fluently. He also understood the argument of the strategy used. Therefore, student 1 satisfied all the indicator of procedural and conceptual understanding.

4.2. The Strategy Used by Student 2

Student 2 has been answered the question in the answer sheet but it was not clear enough. Figure 2 shows the answer of student 2 in the fraction estimation task.

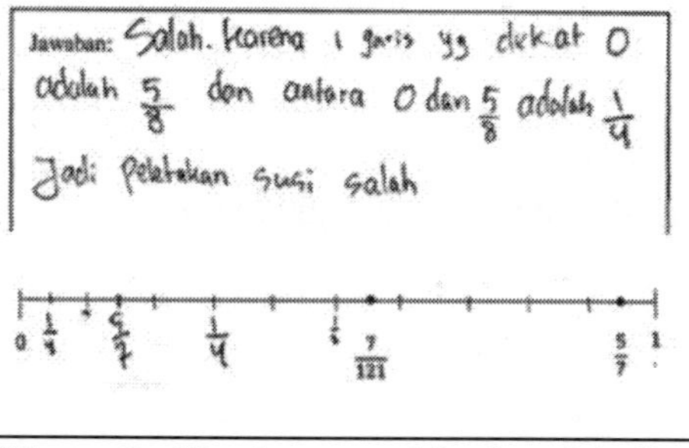

The location is wrong because $\frac{5}{8}$ in near 0, and between 0 and $\frac{5}{8}$ is $\frac{1}{4}$

Figure 2. The Answer of Student 2 in Fraction Estimation Task

From the resulting test, it can not be analyzed because student 2 just wrote the few information. Therefore it needed interview for deep analysis. A part of the interview with student 2.

Q : Could you mention the information in the problem?
S2 : There are fractions in the number line, and I will check and estimate them in the number line
Q : How did you estimate and check them?
S2 : I just imagined the number line as a cake, and it will be divided into some parts. The number line is between 0 to 1. I divided 1 by 2 becomes $\frac{1}{2}$, and divided $\frac{1}{2}$ by 2 becomes $\frac{1}{4}$ until I got the fraction that I will estimate.
Q : So, what did your answer in the answer sheet means?

S2 : I mean between 0 and $\frac{1}{4}$ is $\frac{5}{7}$, but I can not estimate fraction that has big numerator and denominator.

From the result of the interview, student 2 understand what are known and asked in the problem. He stated that the instruction is to check the known fraction and estimate the location in the number line. The strategy used by student 2 is the concept of part-whole. He imagines a number line as a whole and divided it into 2 parts, it becomes $\frac{1}{2}$, and $\frac{1}{2}$ is divided into 2 part becomes $\frac{1}{4}$, and so on. He divided a whole becomes 2 smaller parts until he gets the fraction that will be estimated. It means he knows the concept and strategy used to estimate fraction. But he had difficulty in procedural understanding. He could not apply the procedure of that concept. It can be seen in the fraction $\frac{5}{7}$, he stated that the location of $\frac{5}{7}$ is between 0 and $\frac{1}{4}$. He also could not apply that concept in the fraction with large numerator and denominator.

4.3. The Strategy Used by Student 3

Student 3 used converting the fraction into the decimal to estimate location of the fraction in the number line. In the answer sheet, he estimated four fractions from five fractions fluently. Figure 3 shows the resulting test of student 3.

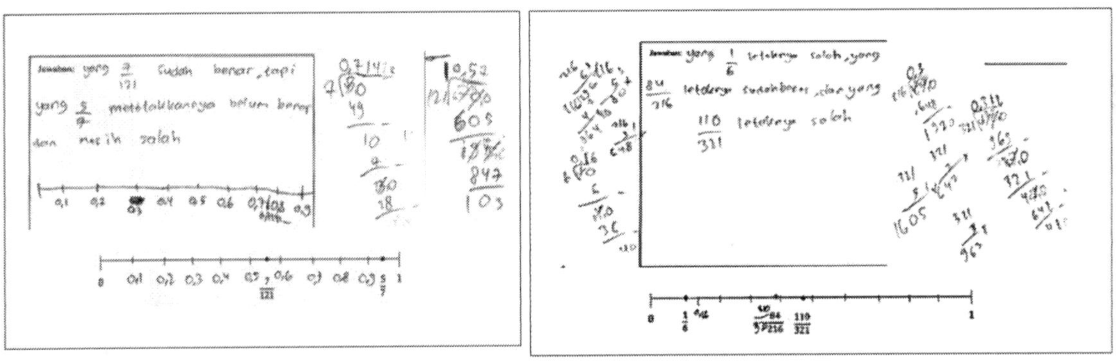

Figure 3. The Answer of Student 3 in Fraction Estimation Task

Based on the resulting test, the strategy used by student 3 is convert fraction into decimal. He used division to convert them. He divided the numerator by the denominator and used rounding concept. The above answer was justified with the result of the interview as follows.

Q : Did you know what are known and asked about the problem?
S3 : Fraction that has been located in the number line. And the instruction is to check and to estimate the location of the fraction in the number line
Q : What strategy did you use?
S3 : I represent fractions into decimals.
Q : How did you do?
S3 : Divided the numerator by the denominator
Q : Why you used that strategy?
S3 : I don't know, I just applied what my teacher did

Student 3 understood the information in the problem. He explained that he has to check the location of fractions that have been put in the number line and estimated them. He used the same strategy as student 1 but he did not find the equivalent fraction. He divided the numerator by the denominator to represent the fraction into the decimal. He did that strategy fluently but he did not know the concept of that strategy. He just applied the procedure without knowing the concept.

5. Discussion

Students use different strategies to estimate the fractions on the number line. Student with greater understanding including knowledge and experience of the fraction magnitudes, the students will use an efficient and adaptive strategy [16]. This is contrary to the results of research. The first student has a good procedural and conceptual understanding but the student uses an inefficient strategy. The strategy used in estimating fractions on a number line takes a lot of time. But by using the strategy, it can be seen that the student understands the concepts related to fractions such as equivalent fractions and decimal. The first student use inefficient strategy because he has no knowledge about other strategies.

Students are said to understand the fraction magnitudes when they have conceptual and procedural understanding in estimating fractions [2, 4]. The second student has only a conceptual understanding. So based on the theory, the second student has not understood the fraction magnitudes [2, 4]. However, the second student uses a more efficient strategy than the first student. The strategy used is the basic concept of fractions (part-whole). This is because the student experience is lacking in using the strategy, therefore he did not use the procedure fluently.

6. Conclusion

Based on the result, we conclude that the strategy used by the first and the third students are the concept of equivalence fraction and convert the fraction into the decimal. The third student faced difficulty in conceptual understanding. He just applied the procedure without knowing the concept. Moreover, the strategy used by the second student is applying the concept of part-whole, but he was only fluent in estimating unit fraction. For using the strategy students' experience and knowledge are needed.

7. Acknowledgements

This research reported herein was supported by Junior High School of LabSchool Universitas Negeri Surabaya. The opinions expressed herein do not necessarily reflect the position, policy, or endorsement of the supporting agency.

8. References

[1] The Indonesian Ministry of Education and Culture 2016 *Regulations of the Ministries of Education and Culture Number 24* (Jakarta: State Secretariat)

[2] Gabriel F 2016 *Understanding Magnitudes to Understand Fractions*. APMC, **21** (2), 36—40.

[3] Hansen N, Rinne L, Jordan N C, Ye A, Resnick I and Rodrigues J 2017 *Co-Development of Fraction Magnitude Knowledge and Mathematics Achievement from Fourth through Sixth Grade*. Learning and Individual Differences, **60**, 18—32.

[4] Mou Y, Li Y, Hoard M K, Nugent L D, Chu F W, Rounder J N and Geary D C 2016 *Developmental Foundations of Children's Fraction Magnitude Knowledge*. Cognitive Development, doi: 10.1016/j.cogdev.2016.05.002

[5] Siegler R S, Ducan G J, Davis-Kean P E, Duckworth K, Claessens A, Engel M, Susperreguy M I and Chen M 2012 *Early Predictors of High School Mathematics Achievement*. Psychological Science. **23**, 691—697.

[6] Siegler R S, Fazio L K, Bailey D H and Zhou X 2013 *Fractions: The New Frontier for Theories of Numerical Development*. Trends in Cognitive Science, **17** (1), 13—19.

[7] Ridwan M 2015 *Peningkatan Hasil Belajar Membandingkan Pecahan Menggunakan Media Garis Bilangan Limbah Triplek pada Siswa Kelas 3 SDN Batok 01 Kecamatan Gemarang Kabupaten Madiun*. Indonesian Digital Journal of Mathematics and Education. 2(3), 134—143.

[8] Siegler R S and Pyke A A 2013 *Developmental and Individual Differences in Understanding of Fractions*. Developmental Psychology, **49**(10), 1994—2004.

[9] Booth J L, Newton K J and Twiss-Garrity L K 2014 *The Impact of Fraction Magnitude Knowledge on Algebra Performance and Learning*. Journal of Experimental Child Psychology, **118**, 110—118.
[10] DeWolf M, Bassok M and Holyoak K J 2016 *A Set for Relational Reasoning: Facilitation of Algebraic Modelling by a Fraction Task*. Journal of Experimental Child Psychology, http://dx.doi.org/10.1016/j.jecp.2016.06.016.
[11] Lacey W H 2017 *The Role of Magnitude Process and Working Memory for Learning Algebra* (Houston: University of Houston)
[12] Schneider M, Thompson C A and Rittle-Johnson B 2017 *Cognitive Development from a Strategy Perspective: A Festschrift for Robbert S. Siegler* (London: Psychology Press)
[13] Namkung M 2014 *Cognitive Predictors of Calculations and Number Line Estimation with Whole Numbers and Fractions* (Nashville, TN: Vanderbilt University)
[14] Rittle-Johnson B and Schneider M 2014 Developing Conceptual and Procedural Knowledge of Mathematics. In Kadosh R C, and Dowker A (Ed). *Oxford Handbook of Numerical Cognition* (Oxford: Oxford University Press).
[15] Rittle-Johnson B, Siegler R S and Alibali M W 2001. Developing Conceptual Understanding and Procedural Skill in Mathematics: An Iterative Process. Journal of Educational Psychology, **93**(2), 346—362.
[16] Fazio L K, DeWolf M and Siegler R S 2015 Strategy Use and Strategy Choice in Fraction Magnitude Comparison. Journal of Experimental Psychology, **42**(1), 1—16.
[17] Musser G L, Peterson B E and Burger B F 2013 *Mathematics for Elementary Teachers: A Contemporary Approach* (New Jersey: Willey)
[18] Wu H 2012 *Fraction* (Barkeley: University of California)
[19] Walle J A V D, Karp K S and Bay-Williams J M 2013 *Elementary and Middle School Mathematics* (New Jersey: Pearson)

IOP Conf. Series: Journal of Physics: Conf. Series **1108** (2018) 012059 doi:10.1088/1742-6596/1108/1/012059

Application of Learning Model (LC) 7E with Technology Based Constructivist Teaching (TBCT) and Constructivist Teaching (CT) Approach as Efforts to Improve Student Cognitive Ability in Static Fluid Concepts

O Miadi[1,*], I Kaniawati[2] and T R Ramalis[3]

[1] Physics Education Department, Universitas Pendidikan Indonesia, INDONESIA.
[2,3] Physics Education Department, Universitas Pendidikan Indonesia, INDONESIA.

*Corresponding Author: oktamiadi@student.upi.edu

Abstract. This study aims to get a representation of the students' cognitive abilities improvement on the concept of static fluid as a result of the application of learning cycle 7E with Technology Based Constructivist Teaching (TBCT). This research uses quasi experiment method with Pre-test Post-test Two Equivalent Group Design. The subjects were class XI students of academic year 2017-2018 as many as 56 students divided into 2 classes, 29 students of class XI IPA 1 as experimental class and 27 students of class XI IPA 2 as control class. Measurement of students' cognitive abilities used a multiple choice instrument test on the concept of static fluid. The data were collected using pre-test and post-test. The results showed that the learning of physics with learning model LC 7E with TBCT approach can significantly improve students' cognitive abilities on the concept of static fluid with n-gain value of 0.56 categorized medium compared to physics learning with learning model LC 7E with CT approach with n-gain value of 0.51 categorised medium. Hypothesis test results showed that there were no significant differences in students' cognitive abilities on the static fluid concept between students who received learning model LC 7E with TBCT approach with students who received learning model LC 7E with CT approach with a significance value of 0,217. Although there were no significant differences, the improvement in cognitive abilities of students in the experimental class was high compared to the control class.

1. Introduction

The educational objectives and development of the current quality of education should shift from simply meeting the value and graduation requirements to a broad understanding balanced with high cognitive and high-level skills. Physic learning should include the cognitive, psychomotor, and affective aspects supported by technology as a provision to continue education at a higher level and to develop science and technology in the future. Some efforts to realize the goal and the development of quality education that includes these aspects are the proper instructional learning one of them learning model LC. The LC model is one of the learning models that also use the constructivist approach as education in the world [1].

Learning cycle is a constructivism learning model developed by Robert Karplus in the Science Curriculum Improvement Study (SCIS) from the University of California, Berkeley in the 1970s [2]. In the beginning, the learning cycle model consists of 3 phases and is called learning cycle 3E consisting of exploration phase, concept introduction, and concept application. In 1980, Rodger W. Bybee developed a learning cycle learning model into 5 phases: engagement, exploration, explanation, elaboration and evaluation [3]. Each "E" phase sequentially provides a learning experience to students in connecting prior knowledge with a new concept [4]. These five phases are then known as the 5E learning cycle. In its development, 5E learning cycle model has changed into 7E learning cycle. The addition of cycles in the learning cycle 7E is elicited and extended. In the 7E learning cycle, engagement is developed into eliciting and engaging. Elaboration and evaluation are developed to be elaborating, evaluating and extending so that the 7E learning cycle has elicit, engage, explore, explain, elaborate, evaluate and extend [5].

Previous studies conducted by Parasuma and Srikanta [6] in his research entitled Effect of Constructivist Teaching (CT) and Technology Based Constructivist Teaching (TBCT) on the Academic Achievement of Secondary School Student stated that learning with TBCT and CT approach can help learners in improve learning outcomes of learners, but TBCT is more effective than CT approach. According to Powell [7], technology is used to facilitate or mediate the learning process. Demirici [8] reveals that the use of simulation and animation for abstract topics makes learners more active in learning, providing opportunities for learners to build difficult concepts easier. The success of learning cycle 7E learning has been proven to improve cognitive learning outcomes, and students' generic science skills [9], can improve learning outcomes, critical thinking skills and students' science process skills [10], student learning result [11], can improve students 'critical thinking skills [12], can improve students' understanding and achievement [13]. And can improve the ability to understand on the students who get the learning model of learning cycle 7E with approach of Technology Based Constructivist Teaching (TBCT) significantly higher than the students who get the learning model of learning cycle 7E with Constructivist Teaching (CT) approach [14]. Pre-test and post-test Chemistry Achievement Test (CAT) values indicate that 5E learning cycle and lecturer methods are increasing [15].

The role and influence of the media are integrated with the 7e learning cycle learning model where each phase has simulation and video activities

2. Experimental Method

2.1 Research Method

The method used is Quasi Experiment. In quasi experiments researchers should make use of existing intact groups [16]. The research design chosen was Pre-test Post-test Two Equivalent Group Design [17]. This design is done by giving the pre-test to know the cognitive ability of the student, then carry out the learning activity in the experimental class using learning cycle 7E with TBCT approach and the learning activity in the control class by using learning cycle 7E with CT approach. Then both classes are given the post-test with the same problem. The design in question is shown in the table 1.

Table 1. Pre-test Post-test Two Equivalent Group Design Research.

Classes	Pretest	Treatment	Posttest
Experiment	O_1	X_1	O_1
Control	O_1	X_2	O_1

Where: O_1 = Cognitive Ability Test, X_1 = Treatment was given to groups using Learning Cycle 7E with TBCT approach, X_2 = Treatment was given to groups using Learning Cycle 7E with CT approach.

2.2 Sample of Research

The population in this study includes all students of class XI SMA at Cerenti, Kuantan Singingi Regency in the even semester of 2017-2018 school year. The sample in this study were students from two classes of XI who were selected by purposive sampling, namely a method or sampling technique carried out with a specific purpose that is in accordance with the characteristics of the research subjects needed in this study, this is done because the whole class has the same characteristics (not there is a superior class) also students are taught by the same teacher [18]. So, it is obtained class XI IPA 1 as the experimental class and class XI IPA 2 as the control class.

2.3 Variable of Research
Variables in this study consist of independent variables and dependent variables. The independent variable is the instructional model of learning cycle 7E with TBCT approach, while the dependent variable is students' cognitive ability.
2.4 Instrument of Research
To obtain data and information needed in this research, researcher makes a set of research instrument. The test used is a cognitive ability test used to measure students' cognitive enhancement of the concept of static fluid. A multiple choice test of cognitive ability is 20 items with five choices (A, B, C, D, and E). The test forms used in the pre-test and the post-test are the same.

2.5 Data Collection Techniques
Data collection is done to obtain the information needed in order to achieve the research objectives. In this study, the data collection techniques are the means used to obtain empirical data that can be used to achieve research objectives. The data collection techniques conducted in this study is a test of cognitive ability. Giving pre-test aims to see the ability of students before getting treatment (treatment) learning activity learning cycle 7E with TBCT approach, while post test to see the results achieved by students after getting treatment.

2.6 Data Analysis
2.6.1 Enhancing Cognitive Ability
To see the improvement of students' cognitive abilities, an analysis of the n-gain score $< \bar{g} >$ was carried out. Normalized gain scores can be expressed by the formula (1). [19].

$$< \bar{g} >= \frac{post\ test\ score - pre\ test\ score}{\max score - pre\ test\ score} \quad (1)$$

Categories of students' cognitive abilities can be seen in table 2.

Table 2 Category of Enhancing Students' Cognitive Ability.

Percentage	Category
$0.00 < <\bar{g}> \leq 0.30$	Low
$0.30 < <\bar{g}> \leq 0.70$	Medium
$0.70 < <\bar{g}> \leq 1.00$	High

2.6.2 Hypothesis Testing
Testing the hypothesis in this study using a two-tailed t-test (2-tailed) with a significant level of α = 0.05, with the formula (2).

$$t_{count} = \frac{\bar{X}1 - \bar{X}2}{\left(\frac{(n_1-1)S_1^2 + (n_2-1)S_2^2}{n_1+n_2-2}\right)\sqrt{\left(\frac{1}{n_1}+\frac{1}{n_2}\right)}} \quad (2)$$

Where: X_1 = N-gain average experiment class, X_2 = N-gain average control class, n_1 = number of experimental class students, n_2 = number of control class students, S_1 = experimental class variant, S_2 = control class variant.

With the testing criteria: if $t_{count} > t_{table}$ then Ho is rejected at the significant level $\alpha = 0.05$ and the degree of freedom dk = $n_1 + n_2 - 2$, so Ha is accepted.

H_o = $\mu_1 = \mu_2$ there is no average difference between the two classes
H_a = $\mu_1 > \mu_2$ the average experimental class is greater than the control class
Where: μ_1 = the average experimental class, μ_2 = the average control class.

3. Result and Discussion
3.1 Result
3.1.1 Enhancing Cognitive Ability
This study was conducted in the even semester of the academic year 2017-2018 for three meetings. In the experimental class, the learning model of learning cycle (LC) 7E with Technology Based Constructivist Teaching (TBCT) approach was applied, while the control class was given the treatment

of learning cycle learning model (LC) 7E with Constructivist Teaching (CT) approach. Prior to the treatment, first test of the two classes to determine students' early cognitive abilities on the concept of static fluid. Then apply the LC 7E learning model with TBCT approach to experiment class and LC 7E learning model with CT approach to control class. After the treatment process is completed, the activity ends with a final test that is intended to determine students' cognitive improvement on the concept of static fluid. Research activities were purposed to obtain data about the initial test score and the final test of students' cognitive abilities. Based on preliminary test score data and the final test can be determined improvement of students' cognitive abilities through the calculation of n-gain $< \bar{g} >$.

The average pre-test score of cognitive ability of experimental class students before learning is 28 and the average post-test score of 68 with n-gain value $<\bar{g}>$ of 0.56 with medium criterion. While the average initial test score cognitive ability of control class students before the learning of 27 and the average final test score of 65 with n-gain value $< \bar{g} >$ of 0.51 with medium criteria. The n-gain value $< \bar{g} >$ in Hake is included in the medium criterion. This suggests that students' cognitive abilities of the Static Fluid concept are increased by moderate category after treatment. The value of n-gain $< \bar{g} >$ experimental class given by LC 7E learning model with TBCT approach is higher than control class given by LC 7E learning model with CT approach. Data cognitive ability of students can be seen in the Table 3.

Table 3. Result of pre test, post test, and n-gain values of cognitive ability of experimental class and control class.

Classes	**Pre test averages**	**Post test averages**	**n-gain $< \bar{g} >$**
Experiment	28	68	0.56
Control	27	65	0.51

Increased cognitive abilities of students on each aspect of cognitive can be known from the test scores obtained by students on each item given in the initial test and final test. The number of questions used consists of 20 items of multiple choices with 5 choices (A, B, C, D, and E). There are four cognitive aspects that are made in the research instrument made on the concept of Static Fluid ie cognitive aspects of C1 (remember), C2 (understand), C3 (applying), and C4 (analysing). The pre-test score, post-test, and the n-gain value $< \bar{g} >$ of the experimental class on each cognitive aspect can be seen in Figure 1.

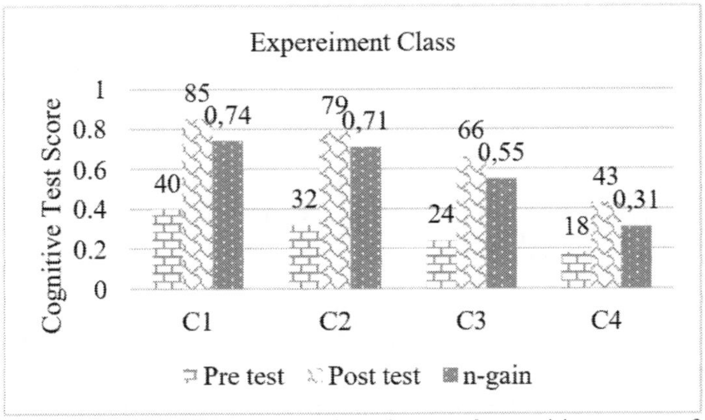

Figure 1. Pre-test and post-test score, and n-gain values each cognitive aspect of the experimental class. Remember (C1), understand (C2), applying (C3), analysing (C4)

The average cognitive abilities of students in the experimental class in each aspect of the diagram increase. The ability to remember C1 has the highest increase with n-gain of 0.74 which is included in the high category. This shows that the experimental class students are able to remember the concept of hydrostatic pressure, Pascal law, and Archimedes law well. The ability of students to analyse C4 has the lowest increase with n-gain of 0.31 which belongs to the medium category. This means that the ability of experimental class students to analyse the concept of static fluid is still low. Pre-test and post-test score, and n-gain values each cognitive aspect of the control class are shown in figure 2.

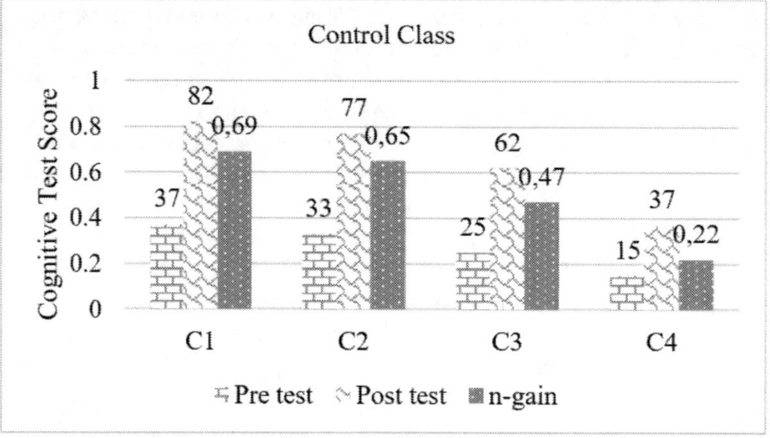

Figure 2. Pre-test and post-test score, and n-gain values each cognitive aspect of the control class. Remember (C1), understand (C2), applying (C3), analysing (C4).

Similar to the experimental class, the average pre-test, post-test, and n-gain scores of control students in each aspect of the diagram also increased. The ability to remember C1 has the highest increase with n-gain of 0.69 which is included in the medium category. This shows that the control class students are quite capable of remembering the concepts of hydrostatic pressure, Pascal law, and Archimedes' law well. The ability of students to analyse C4 has the lowest increase with n-gain of 0.22 which belongs to the low category. This means that the ability of control class students to analyse the concept of static fluid is still low.

To see the differences in cognitive enhancement of each aspect between the experimental class and the control class can be seen in the Table 4.

Table 4. Comparison of n-gain values of cognitive ability of experimental class and control class on each aspect of cognitive ability.

Classes	n-gain			
	C1	C2	C3	C4
Experiment	0.74	0.71	0.55	0.31
Control	0.69	0.65	0.47	0.22

Table 3 shows that cognitive abilities in both classes are equally improved but the experimental class has a higher increase than the control class in every aspect of cognitive ability.

3.1.2 Hypothesis testing

Before testing the hypothesis, it is necessary to test the normality and homogeneity of the distribution of data on the average score of n-gain cognitive abilities obtained by students to determine whether the sample used in the study came from a population that was normally distributed and homogeneous. Based on the results of the calculation, data is obtained as shown in the Table 5.

Table 5. Normality and Homogeneity Test Results Related to Students' Cognitive Ability in Experiment and Control Classes.

Classes	**N**	**Normality Test**		**Homogeny Test**	
		Sig	**Category**	**Sig**	**Category**
Experiment	29	0.10	Normal	0.92	Homogeneous
Control	27	0.42	Normal		

Table 5 shows that from the normality test related to students' cognitive abilities by using data on average n-gain scores in the experimental class total 29 students and control class 27 students, the significance values of normality were 0.10 and 0.42 respectively. This value indicates that the sample comes from a normally distributed population. Homogeneity test is done after it is known that the sample comes from a population that is normally distributed. The result of homogeneity test using the data of the average score of n-gain cognitive abilities obtained a significance value of 0.92. This significance value is greater than the significance level ($\alpha = 0.05$) so that this value indicates that both samples have the same (homogeneous) variant.

Based on the normality and homogeny test which states that the data is normally distributed and homogeneous, the hypothesis testing can be done using parametric tests using t-test (independent samples test) which is processed with the help of SPSS Statistic 24 data processing software. The results of hypothesis testing can be seen in Table 6.

Table 6. Hypothesis Test Results for Increasing Students' Cognitive Ability.

	t test (independent samples test) cognitive abilities of experimental and control class students	
	Sig. (*2-tailed*)	**Significance level (α)**
Equal variances assumed	.217	0.05
Equal variances not assumed	.218	

Based on the results of hypothesis testing shown in table 6, Equal variances assumed with the value Sig. (2-tailed) of 0.217, this value is greater than the significance level (α) of 0.05. Then according to the basis of decision making in the Independent Sample T-Test test it can be concluded that Ho1 is accepted and H11 is rejected. This means that the hypothesis of this study is that there is no significant difference in students' cognitive abilities on the static fluid concept between students who get 7E learning cycle learning with a TBCT approach with students who get 7E learning cycle learning with a CT approach.

3.2 Discussion

3.2.1 Enhancing Cognitive Ability

Based on the results of the tests performed in Table 2, there was an increase from the initial test to the final good test in the experimental class and control class. The n-gain value in the experimental class is 0.56 while in the control class is 0.51, this value is included in the medium criterion. This shows that the improvement of cognitive ability in experimental class given by LC 7E learning model with TBCT approach is higher than control class given by LC 7E learning model with CT approach.

The cognitive abilities of students in every aspect of cognitive are quite good, as evidenced by the improvements that occur in every aspect that is tested against the students. In Figure 1, the difference between the pre-test and post-test average of the experimental class students is shown. Students' cognitive abilities experienced the highest increase in C1 of 0.74 and the lowest increase in C4 that is 0.31. This shows that students find it difficult to solve the problem along with the level of cognitive abilities that are tested.

Figure 2 also occurs in the same case as the experimental class, where there is a difference of pre-test and post-test grade values of the control class students. Students' cognitive abilities experienced the highest increase in C1 of 0.69 and the lowest increase in C4 is 0.22. This shows that students' cognitive abilities on the cognitive aspects of C4 (analysis) are still low. The low cognitive abilities in C4 were due to lack of continuous training on aspects of analysis. This is seen when students have finished experimenting based on LKS then it is considered able to analyse the concepts of static fluid or has finished doing tasks based on existing applications in learning. However, the graph shows that the overall cognitive abilities of the experimental class students are higher than the control class in each dimension of cognitive abilities tested. This is caused by learning in the experimental class assisted with PhET programs and learning videos that explain and exemplify concepts related to static fluid phenomena while the control class must test experiments and analyse them by searching in textbooks.

In table 3, it can be seen that the increase of n-gain value of each aspect of the experimental cognitive ability of the experimental class is higher than the control class. This shows that the experimental class given by LC 7E learning model with TBCT approach is significantly more influential than control class given by LC 7E learning model with CT approach. This is consistent with research conducted by Parasuma and Srikanta [6] that CT and TBCT Interventions help students to perform better in post-test. The TBCT approach is more effective than the CT approach to improve student achievement. Resti [14] showed that there was an increase in comprehension ability with a normalized gain score $< \bar{g} >$ for the experimental class of 0.51, moderately higher than the control class of 0.33.

3.2.2 Hypothesis testing

In general, the improvement of students' cognitive abilities between the experimental and control classes did not differ significantly, both the experimental class and the control class both experienced an increase in cognitive abilities in the medium category. This does not mean that there is no difference at all but there are differences but not significant. This shows that students' cognitive abilities can be improved by applying the 7E learning cycle with both TBCT and CT approaches.

Although there were no significant differences, the improvement in cognitive abilities of students in the experimental class (TBCT) was higher compared to the control class (CT). This is due to technological assistance in the form of PhET programs in the application of 7E learning cycle learning models students become facilitated in training cognitive abilities because students become easier to find and to determine answers to questions in learning. As stated by Munir [20] that simulation on a computer is useful to explain a complicated relationship about a concept. Kocakaya et al. [21] also carried out research using computer assisted instruction whose use was applied to the 7E learning cycle learning model proven to improve the ability to understand prospective teacher students. Parallel findings were also obtained from Suci et al. [22] which states that the application of learning based on PhET simulation and optical KIT using direct learning models can complete student learning outcomes. According to Alessi and Trollip [23] "ICT-based learning has many advantages. One of them is that the advantage of using the time used becomes more effective, the subject matter becomes more accessible, attractive, and inexpensive.

4. Conclusion

Based on the analysis and discussion of students' cognitive abilities on the concept of static fluid, it can be concluded as follows:

Overall students' cognitive abilities improved with moderate criteria. Improvement of cogitative ability of experimental class students given by LC 7E learning model with TBCT approach is equal to 0.56 higher than control class given LC 7E learning model with approach of CT equal to 0.51.

All aspects of cognitive ability have improved. The cognitive aspects of C1 in the experimental class were 0.74 and control classes 0.69, C2 in the experimental class of 0.71 and control classes 0.65, C3 in the experimental class 0.55 and control classes 0.47, and C4 in the class experiments of 0.31 and control classes 0.22.

Hypothesis testing results show that there is no significant difference in students' cognitive abilities on the static fluid concept between students who get 7E learning cycle learning with the TBCT approach with students who get 7E learning cycle learning with a CT approach with Sig. (2-tailed) of 0.217, this value is greater than the significance level (α) of 0.05. Although there were no significant differences, the improvement in cognitive abilities of students in the experimental class was high compared to the control class.

5. References

[1] Ergin I 2012 *Lat. Am. J. Phys. Educ.* **6** 1
[2] Wena M 2009 *Strategi Pembelajaran Inovatif Kontemporer* (Jakarta: Bumi Aksara)
[3] Bybee R 2006(online) The BSCS 5E Instructional Model: Origins, Effectiveness and Applications. Article http://www.bscs.org/pdf/bscs5eexecsummary.pdf

[4] Kurnaz M A and Calik M 2008 *J. Phys. Teach. Educ.* **5** 1
[5] Eisenkraft A 2003 A proposed 7E Model Emphasizes "Transfer of Learning" and the Importance of Eliciting Prior Understanding, Reprinted with Permission from "the Sciences Teacher". *Vol 70 No.6 Copyright © by the National Science Teacher Association (NSTA)*
[6] Parasuma D and Srikanta 2014 *Res.Paper* **3** 10
[7] Powell R G and Powell D L 2010 *Classroom Communication and Diversity Enhancing Instructional Practices* (New York: Rutledge. Second Edition)
[8] Tuysuz C 2010 *Int. Onl. J. Educ. Sci. ISSN: 1309-2707*
[9] Apriani D et al 2012 Pembelajaran Learning Cycle 7E dan Keterampilan Generik Sains Stiews Terhadap Hasil Belajar *Universitas Negeri Semarang. ISSN No 2252-6609*
[10] Balta N and Sarac H 2016 *Eur. J. Educ. Res.* **5** 61
[11] Siribunnam R and Tayraukham S 2009 *J. Soc. Sci.* **5** 279
[12] Hartono 2013 *J. Pendidik. Fisika Indones.* **9** 58
[13] Damar S Y 2013 *The Effect of The Instruction Based on The Epistemologically and Metacognitively Improved 7E Learning Cycle on Tenth Grade Students' Achievement And Epistemological Understandings in Physics* (Turkey: Thesis of Middle East Technical University)
[14] Warliani R 2016 *Math. Sci. Comp. Sci. Educ.* doi: 10.1063/1.4983961
[15] Jack G U 2017 *Educ. Res. Review* **12** 7
[16] Fraengkel J R, Wallen N E and Hyun H H 2012 *How to Design and Evaluate Research in Education*. (New York: McGraw-Hill Companies, Inc)
[17] McMillan J H and Schumacher S 2006 *Research in Education: Evidence-Based Inquiry* (Boston: Pearson Education)
[18] Sugiyono 2013 *Metode Penelitian Kuantitatif Kualitatif dan R&D* (Bandung: Alfabeta)
[19] Hake R 1999(onlinea) *Analyzing Change / Gain Score* (http://lists.asu.edu/cgi-bin/wa?A2=ind9903&L=area-d&P=R6855)
[20] Munir 2012 *Multimedia Konsep dan Aplikasi Dalam Pendidikan* (Bandung: Alfabeta)
[21] Kocokaya S and Gonen 2010 *Turkish Onl. J. Dist. Educ.* **11** 206
[22] Prihatiningtyas S, Prastowo T and Jatmiko B 2017 *Eduscope, Januari* **3** 2
[23] Sutrisno 2011 *Pengantar Pembelajaran Inovatif* (Jakarta: Gaung Persada Press)

Investigating the Development of Reasoning Abilities Among Bruneian Physics Students After 1 Year Exposure to Cambridge International Advanced Level Program

J Marlizayati[1], A H Nor'Arifahwati[2] and A P Rahiman[3]

[1]Science Education Academic Group, Sultan Hassanal Bolkiah Institute of Education, University Brunei Darussalam, Jalan Tungku Link, BE 1410, Brunei Darussalam
[2]Learning Sciences and Technology, Sultan Hassanal Bolkiah Institute of Education, University Brunei Darussalam, Jalan Tungku Link, BE 1410, Brunei Darussalam
[3]Design and Creative Industries Department, Faculty of Arts and Social Sciences, University Brunei Darussalam, Jalan Tungku Link, BE 1410, Brunei Darussalam

Abstract. The aim of this study is to investigate the development of different reasoning abilities among Bruneian physics students after undergoing one year of Cambridge International Advanced level ('A' level) programme. This particular study administered pre and post questionnaires, namely, verbal critical reasoning, spatial ability and abstract reasoning. These questionnaires were distributed among 'A' level physics students from four out of six, sixth form centres. There were 173 students (73 male and 100 female students) who participated in the study. Non-parametric tests (i.e. Wilcoxon Signed-Rank Test and Mann Whitney U) were conducted on the pre-test and post-test scores of verbal critical reasoning, spatial ability and abstract reasoning and their corresponding gain scores due to violation of normality and presence of outliers in the data. Wilcoxon Signed-Rank Test revealed that there were significant differences in the spatial reasoning score and abstract reasoning score from pre-test to post-test. Meanwhile Mann Whitney U test indicated that male students outperformed female students in abstract reasoning during pre-test and post-test. Thus for this particular sample, there is an increase in students' abstract reasoning abilities and spatial reasoning abilities. A further investigation might be needed in exploring the reasoning abilities scores of the students with their 'A' level results to make a stronger claim for this study.

1. Introduction

There is a growing interest in using psychometric test as one of the means for job recruitment amongst most large companies and small-medium enterprises (SMEs). Hence it is no surprise the occurrences of this phenomenon are also observed in higher educational institutions such as colleges and universities. For instance, The Graduate Record Examination (GRE) is an admission requirement for many graduate schools in the United States and in other English-speaking countries. This test measure verbal reasoning, quantitative reasoning, critical thinking, and analytical writing skills that are not related to any specific field of study. Another similar type of test also been meted out among United Kingdom's universities such as University of Leeds and Cardiff University. They adapted UKCAT content and preparation as part of admission requirement. UKCAT is mainly a psychometric test which include basic arithmetic, reading and writing ability, along with character, and personal and social attitudes. In addition, Israel higher education also has used standardized test called psychometric entrance exam which heavily weighed for university admission as a large body of research indicating that the Psychometric Entrance Test has a high predictive ability. Hence our study

IOP Conf. Series: Journal of Physics: Conf. Series **1108** (2018) 012060 doi:10.1088/1742-6596/1108/1/012060

is utilising psychometric tests to determine Bruneian physics students' reasoning abilities that focus on verbal reasoning ability, spatial ability and abstract reasoning.

In this section, the definitions used for verbal critical reasoning, spatial reasoning ability and abstract reasoning will be elaborated. As the instruments used in this study were adopted from Newton and Bristoll [1], the definitions stated by these authors were used.

1.1. Verbal critical reasoning
Verbal critical reasoning is one's ability to understand, manipulate, interpret analytically and logically make sound conclusions based on information provided [1]. It is important to note that verbal reasoning is not similar to reading, however, mastering critical reading is a requirement for verbal reasoning as mentioned by Burton, Welsh, Kostin and VanEssen [2].

1.2. Spatial ability
Spatial reasoning ability is defined as one's ability to visualize and manipulate two-dimensional or three-dimensional shapes or patterns [1]. It encompasses matching shape, rotating shape, combining shapes, viewing cube in three-dimensions, and manipulating of other solid shapes in two and three dimensions.

1.3. Abstract reasoning
Abstract reasoning is very much related to general intelligence where one's ability to see relationships and co-relationships (i.e. symbols and patterns) with no pre-requisite knowledge on language or mathematics [1].

1.4. Previous studies
Colom, Contreras, Arend, Leal and Santacreu [3] discussed several meta-analyses on gender differences in verbal ability and spatial ability performances. It was commonly found that females tend to out-perform their males counterpart in verbal ability test while the males tend to outperform the females in spatial ability test. However, their study found that males outperformed females in both the verbal reasoning test and spatial ability test which contradicted the meta-analyses trend on gender differences on verbal reasoning ability. They contributed the reason of this finding was due to the verbal reasoning tasks that they used in their study required spatial ability to solve it.

Most of the literature mentioned that spatial reasoning is one of the cognitive domains that has a consistent trend where male students tend to outperformed female students [4-6]. In Ganley, Vasilyeva and Dulaney's study [4], they found that spatial ability especially mental rotation ability was more dominant in male students as compared to female students. In the case of abstract reasoning, very limited research papers that looked into the area of physics or science subjects for secondary students can be found. Nonetheless the instrument used to measure the abstract reasoning ability for this study was based on Spearman's two-factor theory of intelligence [7].
Based on the discussion of the literature above, this study was guided by two research questions which are as follow:

RQ 1: Are there any differences among the students' performance on verbal critical reasoning, spatial ability and abstract reasoning tests from pre-test to post-test?
RQ 2: Are there any differences between male and female students' performances of the three tests?

2. Method
This particular study administered pre and post questionnaires, namely, verbal critical reasoning, spatial ability and abstract reasoning. These questionnaires were distributed among 'A' level physics

students from four out of six, form six centres. There were 173 students (73 male and 100 female students) who participated in the study. The pre-test were administered at the beginning of the year (February) before the students officially started their 'A' level course. The post-test was conducted towards the end of the year, which was in November, after they sat for their final year examination. Each school agreed to a specific date to administering the pre and post-test. The test papers were collected at the end of the each session.

During the pre-test, 391 students involved in the study. However, only 173 students sat for both pre and post-test. Students who had not sat for the post-test were not included in the data analysis of this study. The consent form and the three tests were bundled together and were given identification numbers. The identification numbers were coded accordingly to the students' name which can be found in the consent form. This is the coding procedure to ensure anonymity of the students. The same procedure was also applied during the post-test session.

3. Data analysis

A non-parametric tests were conducted on the pre-test and post-test score of verbal critical reasoning, spatial ability and abstract reasoning and their corresponding gain scores. By inspecting the boxplot of the pre-test score of verbal critical reasoning and spatial ability, there were several outliers observed and none for abstract reasoning pre-test score. In terms of normality, the pre-test score of verbal critical reasoning and spatial ability showed violation to normality by graphical (histogram and Q-Q normal plot) and numerical (Shapiro Wilk test) inspection using SPSS. While pre-test score of abstract reasoning only violated normality via numerical inspection (Shapiro Wilk). Similar patterns were observed in the post-test score of the three tests. While for the gain score of the three tests, the boxplot showed several outliers for the gain score of spatial ability and abstract reasoning and none for the gain score of verbal critical reasoning. In terms of normality, the gain score of spatial ability and abstract reasoning showed violation to normality via graphical (histogram) and numerical (skewness and kurtosis and Shapiro Wilk test) inspection using SPSS. While the gain score of verbal critical reasoning only showed violation of normality via numerical inspection. Thus Wilcoxon Signed-Rank Test is used to analyse the data to answer research question 1 (RQ 1) and Mann Whitney U Test to answer research question 2 (RQ 2).

4. Findings

This section will describe the findings of the students' performances on verbal reasoning test, spatial reasoning test and abstract reasoning test. Table 1 and Table 2 below indicated the significant p value (below 0.05) of Wilcoxon Signed Rank test and Mann Whitney U test respectively.

Table 1. Wilcoxon Signed Rank test for spatial ability and abstract reasoning ability.

	Spatial_total_post-test Spatial_total_pre-test	**Abstract_total_post-test Abstract_total_pre-test**
Z	-5.982^{b}	-6.812^{b}
Asymp. Sig. (2-tailed)	.000	.000

a. Wilcoxon Signed Ranks Test

b. Based on negative ranks.

Table 2. Mann Whitney U test for spatial ability and abstract reasoning ability for pre-test and post-test result.

	Spatial_total_pretest	Spatial_total_post	Abstract_total_pretest	Abstract_total_post
Mann-Whitney U	2899.000	2938.500	2969.000	2940.500
Wilcoxon W	7949.000	7988.500	8019.000	7990.500
Z	-2.314	-2.194	-2.100	-2.189
Asymp. Sig. (2-tailed)	.021	.028	.036	.029

a. Grouping Variable: Sex

4.1. Students' verbal reasoning performance
There were 38 students who managed to score less in their post-test than in their pre-test and 85 students managed to score higher in post-test compared to their pre-test. While 50 students had the same score in pre-test and post-test. However, Wilcoxon signed rank test showed no significant difference in the verbal reasoning score from the pre-test to the post-test.

4.2. Students' spatial reasoning performance
There were 43 students who scored less in their post-test compared to their pre-test scores and 113 students scored higher in their post-test compared to their pre-test. While 17 students had the same score in pre-test and post-test. A Wilcoxon signed-rank test showed that over the span of a year upon the exposure of doing 'A' level programme, there is a significant difference in the spatial reasoning score from the pre-test (Mdn = 35) to the post-test (Mdn = 38), $Z = 5.85$, $p < .001$, $r = .45$

4.3. Students' abstract reasoning performance
There were 44 students who scored less in their post-test compared to their pre-test scores and 112 students scored higher in their post-test compared to their pre-test. While 17 students had the same score in pre-test and post-test. A Wilcoxon signed-rank test showed that over the span of a year upon the exposure of doing 'A' level programme, there is a significant difference in the abstract reasoning score from the pre-test (Mdn = 14) to the post-test (Mdn = 15), $Z = 6.81$, $p < .001$, $r = .52$

4.4. Students' performances based on gender
A Mann Whitney U test was conducted on the students' gain score in comparing between male and female students achievement. It was found that there was no statistical significant difference in students gain score in terms of gender. However, Mann Whitney U test showed that there were a significant difference in spatial ability test score and abstract reasoning test score for the pre-test and post-test between male and female students.

A Mann Whitney U test indicated that for spatial ability pre-test score, male students (Mdn = 36) performed better compared to the female students (Mdn = 34), $U = 2899$, $p = .021$, $r = .18$. In the case of the post-test, Mann-Whitney test indicated that for spatial ability score, male students (Mdn= 38) also performed better compared to the female students (Mdn = 37), $U = 2939$, $p = .028$, $r = .17$.

On the other hand, for abstract reasoning pre-test score, Mann Whitney U test indicated that the male students (Mdn = 15) performed better than female students (Mdn = 13), $U = 2969$, $p = .036$, $r = .16$. While for the post-test score, male students (Mdn = 16) also performed better compared to the female students (Mdn = 15), $U = 2941$, $p = .029$, $r = .17$

5. Discussion

As discussed in the finding section, the verbal critical reasoning result did not indicate a significant improvement from pre-test to post-test. Furthermore there was no gender differences in the sample's performance for verbal reasoning. This finding seemed to contradict with most of the literature in terms of gender differences in the verbal reasoning performance score. Unlike our finding for the spatial ability score, it was in line with most of the literature [3,4] where male students performed better than the female counterparts in the pre-test and post-test. However there were no gender differences observed in terms of gain score from pre-test to post test for verbal critical reasoning, spatial ability and abstract reasoning.

As for the reasoning abilities development, the students demonstrated growth in terms of spatial ability and abstract reasoning ability over the course of one year of undergoing Cambridge International Advance Level Programme. Caution need to be taken in interpreting these results as there could be other factors that might contribute to the differences of the pre-test and post-test result. Hanifah et al [8] demonstrated this possibility of other contributing factors in their studies. A further investigation might be needed in exploring the reasoning abilities scores of the students with their 'A' level results to make a stronger claim for this study.

6. Conclusion

Our study indicated that there were significant differences in the physics students' score for spatial ability and abstract reasoning but not for verbal critical reasoning ability after undergoing one year of Cambridge International Advance Programme. As to gender differences in performance, it was observed that male tend to performed better than female in spatial ability test and abstract reasoning test. This finding concurred with the reported literature [4-6].

References

[1] Newton P and Bristoll H 2009(online) *Spatial ability practice test 1 Psychometric Success* (www.psycometric-success.com)
[2] Burton N W, Welsh C Kostin I & Essen T. 2009. *Toward a definition of verbal reasoning in higher education.* ETS Research Report Series **2**.
[3] Colom R, Contreras J, Arend I, Garcia Leal O and Santacreu J 2004 *The Psychological Record* **54** 365
[4] Ganley C M, Vasilyeva M and Dulaney A 2014 *Child Development* **85** 1419
[5] Miller D I and Halpern D F 2011 *Proc. of the Ann. Meeting of the Cognitive Science Society* **33**(33)
[6] Linn C M and Petersen C A 1985 *Child Development* **56** 1479
[7] Psychometric success 2007(online) *Abstract Reasoning Tests* (http://www.psychometric-success.com/aptitude-tests/abstract-reasoning-tests.htm)
[8] Hanifah 2018 *J. Phys. Conf. Ser.* **947** 012014

Mathematical modeling and numerical simulation of tuberculosis spread with diabetes effect

Maulana Malik, Monica Larasati, Dipo Aldila*
Department of Mathematics, Universitas Indonesia, 16424 Depok, Indonesia
E-mail: *aldiladipo@sci.ui.ac.id

Abstract. Tuberculosis (TB) is an infectious disease which the cases continue to increase because of it co-infection with another disease, for instance, is with diabetes. To understand the effect of diabetes on the spread of TB, a mathematical model to describe the correlation between TB and diabetes will be analyzed in this article. The idea is by dividing the population between those who have diabetes and who does not have diabetes. From this model, we obtained the Basic Reproduction Number that becomes an indicator where TB can be endemic or not in a population. Through an analysis of the sensitivity of the basic reproduction number and numerical simulations, we conclude that diabetes has a significant effect on the spread of TB.

1. Introduction

Tuberculosis (TB) is an infectious disease caused by the bacteria *Mycobacterium Tuberculosis* [1]. The main organ that is attacked by TB is the lungs. TB is transmitted through fluid from the respiratory system of TB-infected patients. Transmission can be through coughing or sneezing from an infected individual. Until now, TB is one of the diseases that can not be cured. TB alone is one of the top 10 causes of death in the world, wherein 2016, 10.4 million people are exposed to TB and 1.7 million die due to TB [2].To treat tuberculosis, prevention can be done by using vaccines, such as the BCG vaccine (*Bacille Calmette-Guerin*) and can also be done by treatment using Drugs of Tuberculosis (OAT).

Although prevention and treatment of Tuberculosis (TB) continue throughout the world, it does not reduce the incidence of TB cases every year. Many factors can make TB cases continue to increase, including increased air pollution, malnutrition, coinfection with HIV, and coinfection with Diabetes [3].

Diabetes is a chronic disease that occurs when the pancreas does not produce enough insulin or when the body can not effectively use the insulin it produces [4]. Diabetes can be caused by hereditary factors or appear because of the individual's lifestyle. The appearance of diabetes is higher in heredity. The study of medical science reveals that individuals who have relatives or families with a history of Diabetes have a three times greater possibility of having Diabetes compared to individuals without a positive family history [4].

Diabetes is one of the dominant factors in declining human health levels in different parts of the world. By 2014, as many as 442 million people and by 8.5% of adults aged 18 and older have diabetes. By 2015, diabetes is the cause of 1.6 million individual deaths worldwide. WHO predicts diabetes will be the cause of death at the level -7 on the order of death rates in 2030 [5].

Currently, several epidemiological studies have shown that Diabetes is positively associated with TB [6]. Where Diabetes makes a person's risk of getting TB 3 times bigger. About 15% of TB cases worldwide may be associated with Diabetes. The history of co-infection between DM and TB has been started since 1800. This co-infection was recognized in 1883, when Windle [7] said TB appeared in more than 50% of diabetes and when it came to the conclusion that TB, 2 to 3 times more common in people who have diabetes. Studies show that diabetes suppresses the immune response, which ultimately facilitates Mycobacterium tuberculosis infection. Chances are people with TB disease will die, or relapse is higher if they also have Diabetes [7].

Previously, there have been some researches that describe mathematical models for the spread of TB, such as the mathematical model by Kim Soyoung and others (2018) [10] that discusses the spread of TB and the intervention strategies for mitigating tuberculosis, using the S-E-I/L model, which the latent compartment in this paper is divided into two, that is high risk latent (E) and low risk latent (I), but in Kim model, there isn't recovered compartment. But on research by Jinhui Zhang and others (2014) [11] about the mathematical model for Tuberculosis, which the model is similar to Kim. In the model of Zhang, there is recover or treatment (T) compartment, so the model is S-L-I/J-T. But in the paper by Zhang, there is only one latent compartment, whereas the infectious compartment is divided into two, which is infectious and hospitalized (J) and infectious but not hospitalized (I). To optimize capturing the latency dynamics in models of Tuberculosis transmission can be seen through research by Romain Ragonnet and others (2017) [12], in Romain research there are six kinds of models of transmission between susceptible, latent, and infected compartment. One of them explains the influence of fast propagation and slow propagation of the *Mycobacterium Tuberculosis*, from which the susceptible compartment will transfer into infected compartment if it includes fast propagation and will transfer into latent compartment if it is slow propagation. Based on some of these mathematical models, to describe tuberculosis transmission in a mathematical model, this article will construct S-E-I-R model, where between the susceptible, latent, and infected compartments the effect of fast propagation and slow propagation is used.

Meanwhile, a mathematical model that describes the transmission of TB in people who have diabetes is shown in research by Moualeu and others (2012) [4] about "Analysis of The Impact of Diabetes on The Dynamical Transmission of Tuberculosis", which will be the basic model of this article. Since there is a possibility that diabetes can be transmitted vertically from new-born, we accommodate this fact to include a new infection from new-born into the model. This is the main difference of the model that will be discussed in this article with the model in [4].

In the next section, the mathematical model analysis for a simple TB SEIR model without DM will be analyzed and followed with the analysis of the complete model where DM and DM infection in newborn included into the model in the next section. Some conclusions are given in the fourth section.

2. Mathematical model and analysis of tuberculosis without diabetes effect

In this section, we analyze an SEIR model for Tuberculosis (TB) with a deterministic approach. Let S indicate the susceptible compartment, E and I indicate the exposed and infectious compartment, respectively, while R indicate the recovered compartment. The transmission and transition process of Tuberculosis in this article can be seen in 1

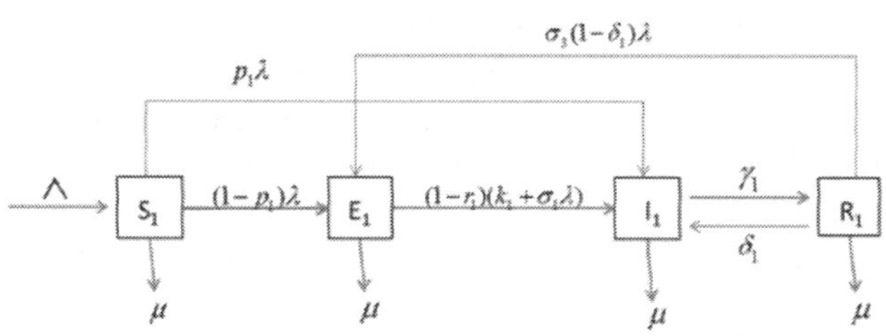

Figure 1. Interaction diagram on SEIR model in 1. Please note that the red curve is the infection term, while blue curve is the transition, natural death and new-born term.

Some explanation needs to be explained regarding the Fig. 1. Only infected compartment

(I) that could transmit the disease, while E is an infected compartment, but not yet capable of spreading TB. After getting infected by an infected individual with rate of infection (λ), individual from S compartment will transferred into I with rate of p_1 (fast propagation), while the rest, $1 - p_1$, transferred into E (slow propagation). The transition from E to I depend on internal factor (k_1) and external factor (infection). Since there is a prevention intervention to suppress the development of TB bacteria with *Chemoprophylaxis*, denoted the proportion is r_1, only people who do not follow the treatment that can be transferred into I compartment from E. If someone has been infected by *Mycobacterium Tuberculosis*, then the body will form an immune system against the bacteria, which in this paper called protective factors and denoted by σ. There is a recovery rate denoted by γ_1 and relapse rate denoted by δ_1. With this assumption, the model of Tuberculosis spread with *Chemoprophylaxis* effect is given by :

$$\frac{dS_1}{dt} = \Lambda - (\mu + \lambda)\, S_1 \tag{1a}$$

$$\frac{dE_1}{dt} = (1 - p_1)\,\lambda\, S_1 + \sigma_3\,(1 - \delta_1)\,\lambda\, R_1 - (1 - r_1)\,(\sigma_1\lambda + k_1)\, E_1 - \mu\, E_1 \tag{1b}$$

$$\frac{dI_1}{dt} = p_1\lambda\, S_1 + (1 - r_1)\,(\sigma_1\lambda + k_1)\, E_1 + \delta_1 R_1 - (\mu + \gamma_1)\, I_1 \tag{1c}$$

$$\frac{dR_1}{dt} = \gamma_1 I_1 - \sigma_3\,(1 - \delta_1)\,\lambda\, R_1 - (\mu + \delta_1)\, R_1, \tag{1d}$$

with $\lambda = \frac{\beta I_1}{N}$.

Taking the right-hand side of model 1 equal to zero and solve it respect to all variables, obtained that the model had two equilibrium point, i.e., the trivial equilibrium, let call it as disease-free equilibrium which given by

$$DFE_1 = \left(\frac{\Lambda}{\mu}, 0, 0, 0\right), \tag{2}$$

and the second un-trivial equilibrium is the endemic equilibrium, which given by

$$EE = (S_1^*, E_1^*, R_1^*, I_1^*) \tag{3}$$

with

$$S_1^* = \frac{\Lambda N}{N\mu + \beta I_1} \tag{4a}$$

$$E_1^* = -\frac{\left(-g_1\left(-\Lambda p_1 + g_2 I_1\right) I_1\beta^2 - \left(\mu\left(\left(-1 + g_1\right)\mu + \gamma_1 g_1 - g_5\right) I_1 + \Lambda p_1 g_3\right) N\beta + N^2\mu^2 g_4\right) I_1 N}{\left(-I_1 g_1\beta + g_3 N\right)\left(N\mu + \beta I_1\right)\left(r_1 - 1\right)\left(\beta I_1\sigma_1 + Nk_1\right)} \tag{4b}$$

$$R_1^* = \frac{N I_1\gamma_1}{-\beta I_1\delta_1\sigma_3 + \beta I_1\sigma_3 + N\mu + N\delta_1} \tag{4c}$$

and $g_1 = (\delta_1 - 1)\sigma_3, g_2 = \gamma_1 + \mu, g_3 = mu + \delta_1, g_4 = \gamma_1 + \mu + \delta_1, g_5 = \gamma_1 + \delta_1$.

The I_1^* is a positive root from three–degree polynomial in the form of

$$A_0 + A_1(I_1) + A_2(I_1^2) + A_3(I_1^3) = 0 \tag{5}$$

where

$A_0 = N^3\mu^3 k_1 r_1 + N^3\mu^2\delta_1 k_1 r_1 + N^3\mu^2\gamma_1 k_1 r_1 + \Lambda N^2\beta\mu^2 p_1 + \Lambda N^2\beta\mu\delta_1 p_1 - \Lambda N^2\beta\mu k_1 r_1 - \Lambda N^2\beta\delta_1 k_1 r_1 - N^3\mu^4 - N^3\mu^3\delta_1 - N^3\mu^3\gamma_1 -$
$N^3\mu^3 k_1 - N^3\mu^2\delta_1 k_1 - N^3\mu^2\gamma_1 k_1 + \Lambda N^2\beta\mu k_1 + \Lambda N^2\beta\delta_1 k_1$

$A_1 = -N^2\beta\mu^2\delta_1 k_1 r_1\sigma_3 - \Lambda N\beta^2\mu\delta_1 p_1\sigma_3 + \Lambda N\beta^2\delta_1 k_1 r_1\sigma_3 + N^2\beta\mu^3\delta_1\sigma_3 + N^2\beta\mu^3 r_1\sigma_1 + N^2\beta\mu^2\delta_1\gamma_1\sigma_3 + N^2\beta\mu^2\delta_1 k_1\sigma_3 +$
$N^2\beta\mu^2\delta_1 r_1\sigma_1 + N^2\beta\mu^2\gamma_1 r_1\sigma_1 + N^2\beta\mu^2 k_1 r_1\sigma_3 + \Lambda N\beta^2\mu p_1\sigma_3 - \Lambda N\beta^2\mu r_1\sigma_1 - \Lambda N\beta^2\delta_1 k_1\sigma_3 - \Lambda N\beta^2\delta_1 r_1\sigma_1 - \Lambda N\beta^2 k_1 r_1\sigma_3 -$
$N^2\beta\mu^3\sigma_1 - N^2\beta\mu^3\sigma_3 - N^2\beta\mu^2\delta_1\sigma_1 - N^2\beta\mu^2\gamma_1\sigma_1 - N^2\beta\mu^2\gamma_1\sigma_3 + N^2\beta\mu^2 k_1 r_1 - N^2\beta\mu^2 k_1\sigma_3 + N^2\beta\mu\delta_1 k_1 r_1 + N^2\beta\mu\gamma_1 k_1 r_1 +$
$\Lambda N\beta^2\mu\sigma_1 + \Lambda N\beta^2\delta_1\sigma_1 + \Lambda N\beta^2 k_1\sigma_3 - N^2\beta\mu^3 - N^2\beta\mu^2\delta_1 - N^2\beta\mu^2\gamma_1 - N^2\beta\mu^2 k_1 - N^2\beta\mu\delta_1 k_1 - N^2\beta\mu\gamma_1 k_1$

$A_2 = -N\beta^2\mu^2\delta_1 r_1\sigma_1\sigma_3 + \Lambda\beta^3\delta_1 r_1\sigma_1\sigma_3 + N\beta^2\mu^2\delta_1\sigma_1\sigma_3 + N\beta^2\mu^2 r_1\sigma_1\sigma_3 - N\beta^2\mu\delta_1 k_1 r_1\sigma_3 - \Lambda\beta^3\delta_1\sigma_1\sigma_3 - \Lambda\beta^3 r_1\sigma_1\sigma_3 + N\beta^2\mu^2\delta_1\sigma_3 +$
$N\beta^2\mu^2 r_1\sigma_1 - N\beta^2\mu^2\sigma_1\sigma_3 + N\beta^2\mu\delta_1\gamma_1\sigma_3 + N\beta^2\mu\delta_1 k_1\sigma_3 + N\beta^2\mu\delta_1 r_1\sigma_1 + N\beta^2\mu\gamma_1 r_1\sigma_1 + N\beta^2\mu k_1 r_1\sigma_3 + \Lambda\beta^3\sigma_1\sigma_3 -$
$N\beta^2\mu^2\sigma_1 - N\beta^2\mu^2\sigma_3 - N\beta^2\mu\delta_1\sigma_1 - N\beta^2\mu\gamma_1\sigma_1 - N\beta^2\mu\gamma_1\sigma_3 - N\beta^2\mu k_1\sigma_3$

$A_3 = -\beta^3\mu\delta_1 r_1\sigma_1\sigma_3 + \beta^3\mu\delta_1\sigma_1\sigma_3 + \beta^3\mu r_1\sigma_1\sigma_3 - \beta^3\mu\sigma_1\sigma_3$

This three–degree polynomial will give at least one positive root if and only if the sum of all the roots, i.e $-\frac{A_2}{A_3}$ is greater or equal to 0. Using the set of parameters given by : $\beta = 5, k_1 = 0.0033, \gamma_1 = 2, \sigma_1 = 1.5, \sigma_3 = 1, p_1 = 0.5, \delta_1 = 0.0485, r_1 = 0, \mu = 1/65, \Lambda = 1000/65, N = 1000$, we find that $-\frac{A_2}{A_3} \approx 19.3723308$ is greater than 0 and lead us into existence of endemic equilibrium given by $EE = (S_1 = 31.56562298, E_1 = 326.1612647, I_1 = 131.1112551, R_1 = 511.1618574)$.

Next, using the next-generation matrix approach [9], we define the basic reproduction number of TB model 1 as :

$$\mathcal{R}_0 = \frac{(\mu p_1 + k_1(1 - r_1))\,(\mu + \delta_1)\,\beta\,\Lambda}{\mu^2\,(\mu + k_1(1 - r_1))\,(\mu + \delta_1 + \gamma_1)\,N}. \tag{6}$$

This basic reproduction number determine the endemic level of disease in many mathematical models of disease spread. In this article, we find that the DFE is locally asymptotically stable if $\mathcal{R}_0 < 1$. If $\mathcal{R}_0 > 1$, then the TB will exist and spread among the population, and will persist if $\mathcal{R}_0 < 1$. Since $\frac{\partial \mathcal{R}_0}{\partial p_1} = \frac{\beta\mu}{(\mu+\gamma_1)(k_1(1-r_1)+\mu)} > 0$, we have that the $\mathcal{R}_0$ will increase when p_1 increase. So as when $\frac{\partial \mathcal{R}_0}{\partial k_1}$, we get that $\frac{\partial \mathcal{R}_0}{\partial k_1} = \frac{(1-r_1)(1-p_1)\beta\mu}{(\mu+\gamma_1)(k_1(1-r_1)+\mu)^2} > 0$, therefore, $\mathcal{R}_0$ will increase when k_1 increase. This means that the TB will get more infectious when more people experience the fast propagation of TB and when more people transition to infected because of an internal factor. However, when $\frac{\partial \mathcal{R}_0}{\partial \gamma_1} = \frac{k_1\beta(r_1-1)-\beta\mu p_1}{(\mu+\gamma_1)^2(k_1(1-r_1)+\mu)} < 0$, we have that the $\mathcal{R}_0$ will decrease when γ_1 increase. That means that the more treatment obtained, The spread of TB will decreased.

3. Mathematical model and analysis of tuberculosis without diabetes effect

Since many medical reports stated that diabetes plays an important role in determining the spread of TB [7] as we explain in the first section, in this section we expand the TB model 1 with involving the diabetes population. The idea is to separate each compartment in model 1 into two compartments, i.e. with and without diabetes. With this assumption, the transmission diagram in Fig. 1 become

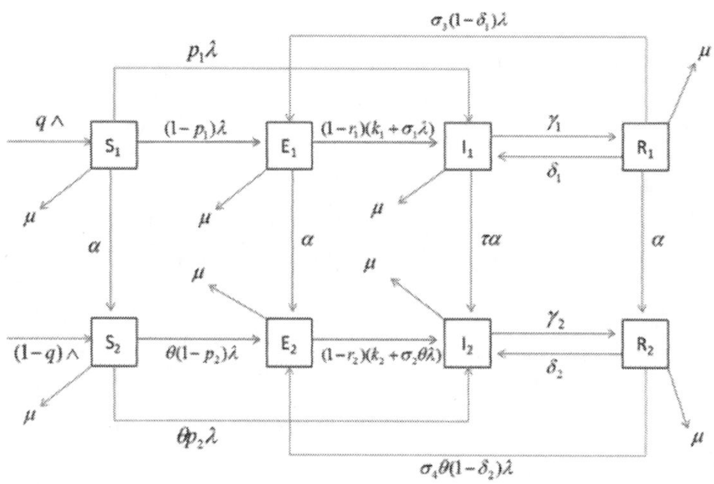

Figure 2. Transmission diagram of TB with diabetes effect. Please note that the index $i = 1$ determine the population without diabetes , while $i = 2$ is the population with diabetes.

There are several additional assumptions in Fig. 2, except what already explain in the previous section for model 1. Each individual in all of the compartment can acquire diabetes with rate α. Since diabetes makes the probability of individually contracted by TB 3 times bigger, every rate of infection (λ) multiplied by an enhancement factor denoted by θ. There is individual which born with diabetes genes denoted the proportion is q. With this transmission diagram in hand, the TB model with the effect of diabetes is given by :

$$\frac{dS_1}{dt} = q\Lambda - (\mu + \alpha + \lambda)\, S_1 \tag{7a}$$
$$\frac{dE_1}{dt} = (1-p_1)\,\lambda\, S_1 + \sigma_3\,(1-\delta_1)\,\lambda\, R_1 - (1-r_1)\,(\lambda\,\sigma_1 + k_1)\, E_1 - (\mu+\alpha)\, E_1 \tag{7b}$$
$$\frac{dI_1}{dt} = p_1\lambda\, S_1 + (1-r_1)\,(\lambda\,\sigma_1 + k_1)\, E_1 + \delta_1 R_1 - (\alpha\,\tau + \mu + \gamma_1)\, I_1 \tag{7c}$$
$$\frac{dR_1}{dt} = \gamma_1 I_1 - \sigma_3\,(1-\delta_1)\,\lambda\, R_1 - (\mu + \delta_1 + \alpha)\, R_1 \tag{7d}$$
$$\frac{dS_2}{dt} = (1-q)\,\Lambda + \alpha\, S_1 - (\lambda\,\theta + \mu)\, S_2 \tag{7e}$$
$$\frac{dE_2}{dt} = \theta\,(1-p_2)\,\lambda\, S_2 + \sigma_4\theta\,(1-\delta_2)\,\lambda\, R_2 + \alpha\, E_1 - (1-r_2)\,(\lambda\,\theta\,\sigma_2 + k_2)\, E_2 - \mu\, E_2 \tag{7f}$$
$$\frac{dI_2}{dt} = \theta\, p_2\lambda\, S_2 + (1-r_2)\,(\lambda\,\theta\,\sigma_2 + k_2)\, E_2 + \tau\,\alpha\, I_1 + \delta_2 R_2 - (\mu + \gamma_2)\, I_2 \tag{7g}$$
$$\frac{dR_2}{dt} = \gamma_2 I_2 - \sigma_4\theta\,(1-\delta_2)\,\lambda\, R_2 + \alpha\, R_1 - (\mu + \delta_2)\, R_2. \tag{7h}$$

The disease-free equilibrium of model (9) is given by

$$DFE_2 = \left(\frac{q\Lambda}{\mu+\alpha}, 0, 0, 0, \frac{\Lambda\,(\mu(1-q)+\alpha)}{\mu\,(\mu+\alpha)}, 0, 0, 0\right). \tag{8}$$

The endemic equilibrium is not easy to be determine because of the complexity of the model. Although the existence of endemic equilibrium is difficult to be shown analytically, the existence of endemic equilibrium still can be shown numerically. Using the set of the parameter in Table 1, the endemic equilibrium of model 7 is given by $EE = (S_1 = 110.9967987, E_1 = 79.30910195, I_1 = 6.826041731, R_1 = 118.0735319, S_2 = 57.88899758, E_2 = 151.9062284, I_2 = 75.46260318, R_2 = 399.5366963)$

The basic reproduction number of TB model 7 cannot be shown explicitly in this thesis. Using the formula of the next-generation matrix, as $\mathbf{K_L} = -\mathbf{N}\mathbf{\Sigma}^{-1}\mathbf{M}$ we get the basic reproduction number of TB model 7 is $\rho(\mathbf{K_L})$. With

$$\mathbf{N} = \begin{bmatrix} 1 & 0 & 0 & 0 & 0 & 0 \\ 0 & 1 & 0 & 0 & 0 & 0 \end{bmatrix}, \mathbf{M} = \begin{bmatrix} \frac{p_1\beta\, q\Lambda_1}{N(\mu+\alpha)} & \frac{p_1\beta\,\epsilon\, q\Lambda_1}{N(\mu+\alpha)} \\ \frac{\theta\, p_2\beta\,\Lambda_1(-\mu\, q+\alpha+\mu)}{N\mu\,(\mu+\alpha)} & \frac{\theta\, p_2\beta\,\epsilon\,\Lambda_1(-\mu\, q+\alpha+\mu)}{N\mu\,(\mu+\alpha)} \\ \frac{(1-p_1)\beta\, q\Lambda_1}{N(\mu+\alpha)} & \frac{(1-p_1)\beta\,\epsilon\, q\Lambda_1}{N(\mu+\alpha)} \\ \frac{\theta\,(1-p_2)\beta\,\Lambda_1(-\mu\, q+\alpha+\mu)}{N\mu\,(\mu+\alpha)} & \frac{\theta\,(1-p_2)\beta\,\epsilon\,\Lambda_1(-\mu\, q+\alpha+\mu)}{N\mu\,(\mu+\alpha)} \\ 0 & 0 \\ 0 & 0 \end{bmatrix}$$

$$\mathbf{\Sigma} = \begin{bmatrix} -\tau\,\alpha-\mu-\gamma_1 & 0 & (1-r_1)\,k_1 & 0 & \delta_1 & 0 \\ \tau\,\alpha & -\mu-\gamma_2 & 0 & (1-r_2)\,k_2 & 0 & \delta_2 \\ 0 & 0 & -(1-r_1)\,k_1-\mu-\alpha & 0 & 0 & 0 \\ 0 & 0 & \alpha & -(1-r_2)\,k_2-\mu & 0 & 0 \\ \gamma_1 & 0 & 0 & 0 & -\mu-\delta_1-\alpha & 0 \\ 0 & \gamma_2 & 0 & 0 & \alpha & -\mu-\delta_2 \end{bmatrix},$$

IOP Conf. Series: Journal of Physics: Conf. Series **1108** (2018) 012061 doi:10.1088/1742-6596/1108/1/012061

Table 1. Estimation of parameter values

Parameter	Estimatied	Parameter	Estimated
Λ	1000/65	N	1000
β	0.5	μ	1/65
α	0.009	τ	1.1
ϵ	1.1	θ	3
p_1	0.5	k_2	0.8
k_1	0.0033	k_2	0.0066
σ_1	1.5	σ_2	1
σ_3	1	σ_4	0.8
δ_1	0.0485	δ_2	0.5
γ_1	2	γ_2	3
r_1	0	r_2	0
q	0.5		

Next, we would like to see how the basic reproduction number changed respect to the various value of parameters that related to diabetes (α and q) and its effect on the dynamic of the total population.

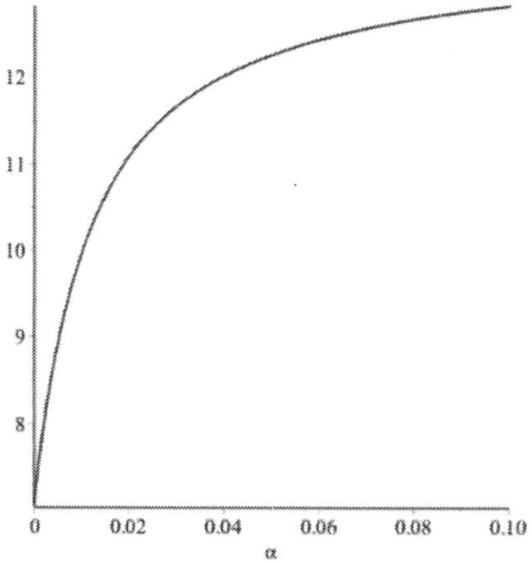

Figure 3. Basic Reproduction Number ($\mathcal{R}_0$) sensitivity analysis curve refers to acquiring diabetes rate (α). The y-axis is the $\mathcal{R}_0$

Fig. 3 shows a positive correlation with α. It describes when more and more individual with TB acquiring diabetes, the spread of TB will increases. The influence of α is affected by the value of the enhancement factor θ. Hence, the value of θ must be greater than one. Using the set of parameters in Table 1 and replace the value of θ to 1.5, the impact of the various value of α towards susceptible and infected populations could be seen in Fig. 4.

It can be seen in Fig 4 that greater the value of α will increase the magnitude of $\mathcal{R}_0$. Fig. 4 shows that the greater the value of α will make the total population of susceptible decreases, this is caused by a decrease in the value of $\mathcal{R}_0$. It is known that when the value of $\mathcal{R}_0$ increase, the number of infected populations will also increase. So the greater the value of α can cause the total infected population to increase.

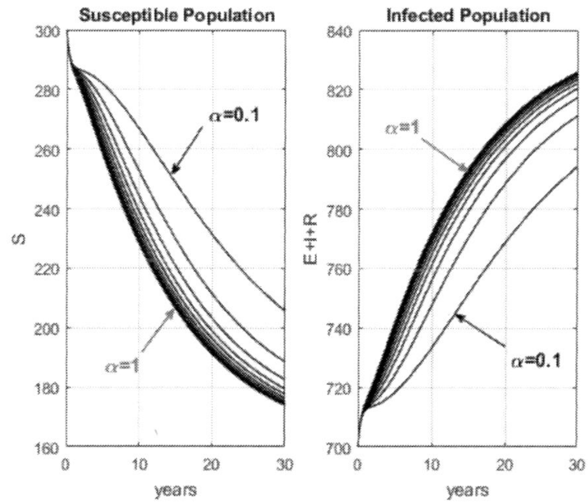

Figure 4. The changed of the value of $\alpha = 0.1, 0.2, .., 1$ to the populations of susceptible (left) and infected (right).

Using the set of parameters in Table 1 (without q), we choose multiple values of q to generate different of $\mathcal{R}_0$. Be obtained $\mathcal{R}_0(q = 0.05) = 1.3150, \mathcal{R}_0(q = 0.47) = 1.0020, \mathcal{R}_0(q = 0.7) = 0.8307$, as we can see when the value of q increase, the the value of $\mathcal{R}_0$ decreases. Fig. 5 describe the impact of various value of q towards susceptible and infected populations.

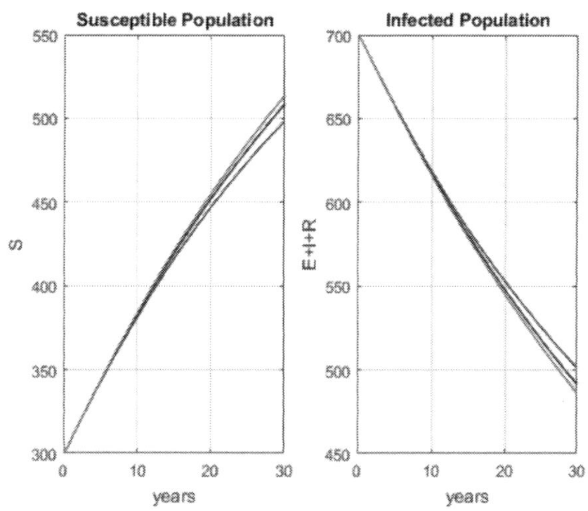

Figure 5. The changed of the value of q to the populations of susceptible (left) and infected (right).

The curve in Fig. 5 represents the value of $\mathcal{R}_0$, the red line signifies the value of $\mathcal{R}_0 = 0.8307$, the black line signifies the value of $\mathcal{R}_0 = 1.0020$, and the blue line signifies the value of $\mathcal{R}_0 = 1.3150$. Fig. 5 reflect, more and more individuals are born in a state of having diabetes it can lead to increased spread of TB. The greater the value of q will make the value of $\mathcal{R}_0$ decreases, the total of susceptible population increases and the total of the infected population decreases.

4. Conclusions

In this paper, a mathematical model of the spread of TB disease with the influence of diabetes has been constructed. From the model of TB with diabetes, we can get two equilibrium point, i.e., the disease-free equilibrium point and the endemic equilibrium point, and the basic reproduction number $\mathcal{R}_0$. In this paper, it is also constructed a simpler model, the mathematical model of the spread of TB disease without diabetes to understand the behavior of TB for some parameters which difficult to analyze in the complete model. Based on analytic studies and numerical simulations, it is known that the more individuals are infected with fast propagation, and as more individuals become infected due to the endogenous reactivation, the rate of TB spread becomes larger. Whereas, as more treatment is done or obtained, it can lead to a decrease in the spread of TB. We also find that the more individuals get diabetes can accelerate the spread of TB. However, when many individuals are born healthy without diabetes, it can reduce the spread of TB. So it can be deduced that diabetes affects the spread of TB.

Acknowledgements

We thank to all reviewers for their constructive comments and suggestions. This research was financially supported by Universitas Indonesia, with PITTA research grant scheme 2018.

References

[1] Nathella P K and Babu S 2017 Influence of diabetes mellitus on immunity to human tuberculosis *Immunology* **152** 13-24
[2] World Health Organization 2018 *Tuberculosis*
[3] Dooley K E and Chaisson R E 2009 Tuberculosis and diabetes mellitus: convergence of two epidemics *The Lancet Infectious Diseases* **9** 737-746
[4] Moualeu D, Bowong S, Tewa J, and Emvudu Y 2012 Analysis of The Impact of Diabetes on The Dynamical Transmission of Tuberculosis *Mathematical Modelling of Natural Phenomena* **7** 117-146
[5] World Health Organization *Genetics and Diabetes*
[6] World Health Organization 2016 *Tuberculosis & Diabetes*
[7] Oscarsson P N and Silwer H 1958 Incidence of Pulmonary Tuberculosis Among Diabetics *Acta Med Scand* **335** 23-48
[8] Verhulst F 1990 *Nonlinear differential equations and dynamical systems* Berln: Springer-Verlag
[9] Diekmann, Hesterbeek and Roberts 2010 The construction of next-generation matrices for compartmental epidemic models *Journal of The Royal Society*
[10] Kim S, Reyes A A, and Jung E 2018 Mathematical model and intervention strategies for mitigating tuberculosis in the Philippines *Journal of Theoretical Biology* **443** 100-112
[11] Zhang J, Li Y, and Zhang X 2015 Mathematical modeling of tuberculosis data of China *Journal of Theoretical Biology* **365** 159-163
[12] Ragonnet R, Trauer J M, Scott N, Meehan M T, Denholm J T, and Mcbryde E S 2017 Optimally capturing latency dynamics in models of tuberculosis transmission. *Epidemics* **21** 39-47

IOP Conf. Series: Journal of Physics: Conf. Series **1108** (2018) 012062 doi:10.1088/1742-6596/1108/1/012062

Classification of Osteoarthritis Disease Severity Using Adaboost Support Vector Machines

T R Adyalam[1], Z Rustam[1] and J Pandelaki[2]

[1]Department of Mathematics, Universitas Indonesia, Kampus UI Depok, Depok 16424, Indonesia
[2]Department of Radiology, Universitas Indonesia, Kampus UI Salemba, Jakarta 10430, Indonesia

Abstract. Osteoarthritis (OA) is a condition when the joint is painful due to mild inflammation that arises due to friction of the ends of the joint bone. OA is the most chronic disease and joint disability in elderly people. One way to prevent this disease is to do early detection using machine learning for classification. In this study, it was used Adaptive Boosting (AdaBoost) and Support Vector Machines (SVM) together as classifiers. The purpose of this study was to see whether AdaBoost SVM could produce good accuracy with SVM as comparison. Tests were conducted using 10% until 90% data training. Polynomial and RBF kernel were used with number of AdaBoost cycle. The highest accuracy value of SVM was 75% in 90% training data, while the highest accuracy value of AdaBoost SVM was 85,714% in 80% training data. Therefore, it could be that AdaBoost can improve the performance of SVM in classification of OA disease severity.

1. Introduction

Osteoarthritis (OA) is a long-term chronic disease characterized by the deterioration of cartilage in joints which results in bones rubbing together and creating stiffness, pain, and impaired movement [1]. A joint is an area of the body where two different bones are met. OA is the most common chronic disease and joint disability in all ages but usually in elderly people. This disease can affect any joint, but it occurs most often in knees, hips, lower back and neck, small joints of the fingers and the bases of the thumb and big toe. Common risk factors include increasing age, obesity, previous joint injury, overuse of the joint, weak thigh muscles, and genes [2].

In normal joints, a firm and rubbery material which called cartilage covers the end of each bone. Cartilage provides a smooth, gliding surface for the joint motion and acts as a cushion between the bones. In OA, the cartilage breaks down so that causing pain, swelling and problems moving the joint [3]. The severity of OA divided into three categories which are (1) not severe – suspicious narrowing of the joint and potential osteophytes, (2) severe – definitive osteophytesand potential narrowing of the joint gaps on the weight bearing side, and (3) very severe – multiple osteophytes, definitive narrowing of the joint gaps, and potential bone deformities [4].

The progressive development of osteoarthritis is a slow and multi factorial process. The onset of clinical symptoms (pain and progressive loss of function) is usually preceded by molecular and cellular changes in the joint that result in the loss of homeostasis. The diagnosis of osteoarthritis by laboratory tests, X-rays, and a magnetic resonance imaging (MRI) scan [2]. Effective antiosteoarthritic drugs and biological interventions should maintain or restore joint homeostasis and structural integrity. Since the key molecular regulators of these processes remain poorly understood, thus rendering the development of effectiveness and safe treatments still a major challenge [5]. There is no cure of it, but treatments are available to manage symptoms. Long-term management of the disease will include several factors, managing symptoms included stiffness and swelling, improving joint mobility and flexibility, maintaining a healthy weight, and getting enough of exercise [2].

One way to prevent this disease can be done by doing early detection using machine learning. The detection of OA is important. The earlier treatment can prevent the destruction of cartilage and bone [3]. Machine learning is a branch of science that allows computers to learn based on existing data. Machine learning uses mathematical algorithms implemented as computer programs to identify patterns in large datasets, and to iteratively improve in performing this identification with additional data [6]. One type of this learning is supervised learning for classification. In this study, it was used Adaptive Boosting (AdaBoost) and Support Vector Machines (SVM) together as classifiers.

2. Methods

2.1. Supervised Learning

Supervised learning used targeted training data so that each training data was a pair of data $\{(\mathbf{x}_1, y_1), \ldots, (\mathbf{x}_m, y_m)\}$. This learning aimed to form a model that could provide prediction results in accordance with the target for a testing data. Training data was used to construct predictive models to applied into testing data.

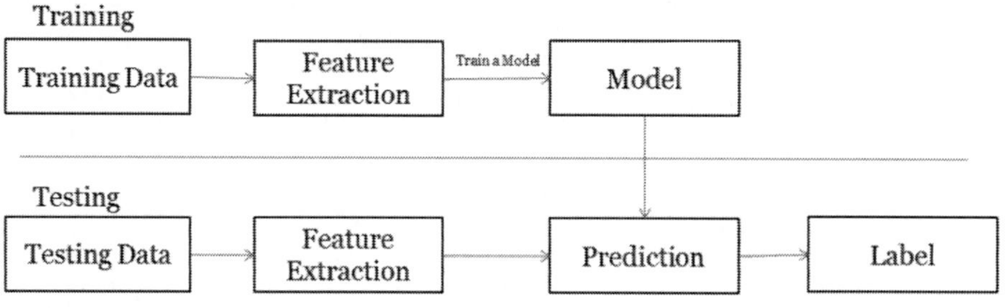

Figure 1. How supervised learning works.

Supervised learning could provide discrete predictions called classification. Classification was the arrangement of the system in groups or classes according to established rules or standarts.

2.2. Support Vector Machines (SVM)

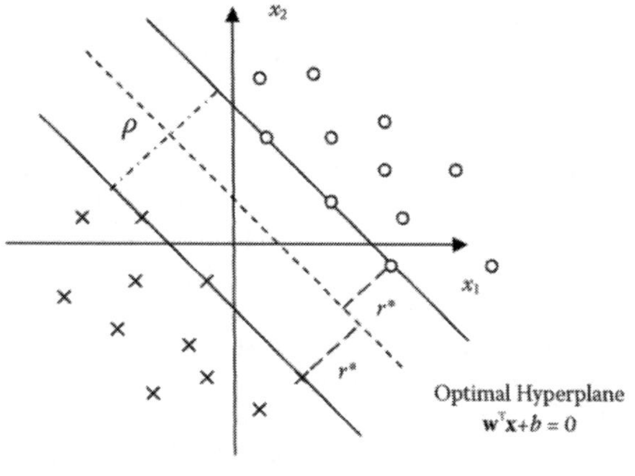

Figure 2. Optimal hyperplane illustration on SVM (Source: [7]).

SVM was one of supervised learning methods developed by Vapnik in 1992 for classification. SVM aimed to resolve the classification problem by forming a hyperplane that maximize the margin by dividing the two classes of data [8]. Margin was the closest distance between hyperplane to the nearest point of each class (support vectors). In SVM, the optimal separating hyperplane is determined by giving the largest margin separation ρ between different class [9].

Given a dataset $D = \{(\mathbf{x}_1, y_1), \dots, (\mathbf{x}_m, y_m)\}, \mathbf{x}_i \in \mathbf{X},\ y_i \in \mathrm{Y} = \{-1, +1\}$, SVM resolved the following mathematics model:

$$\min_{\mathbf{w},b} \frac{1}{2} ||\mathbf{w}||^2 \quad \text{s.t. } y_i(\mathbf{w}^T\mathbf{x}_i + b) \geq 1, \quad i = 1, \dots, n \tag{1}$$

For misclassification error cases, it was added C parameter and slack variable so that SVM mathematics model becomes:

$$\min_{\mathbf{w},b} \frac{1}{2} ||\mathbf{w}||^2 + C \sum_{i=1}^{n} \xi_i \quad \text{s.t. } y_i(\mathbf{w}^T\mathbf{x}_i + b) \geq 1 - \xi_i, \quad \xi_i \geq 0, \quad i = 1, \dots, n \tag{2}$$

with $C > 0$.

For linearly non-separable cases, it was added kernel function. Kernel function could be defined as

$$K(\mathbf{x}_i, \mathbf{x}_j) = \phi(\mathbf{x}_i)^T . \phi(\mathbf{x}_j) \tag{3}$$

There were various kernel functions i.e. [10]:

1) Linear kernel : $K(\mathbf{x}_i, \mathbf{x}_j) = \mathbf{x}_i^T \mathbf{x}_j$
2) Polynomial kernel : $K(\mathbf{x}_i, \mathbf{x}_j) = (\mathbf{x}_i^T \mathbf{x}_j + 1)^d$
3) RBF kernel : $K(\mathbf{x}_i, \mathbf{x}_j) = \exp\left(-\gamma \left|\left|\mathbf{x}_i - \mathbf{x}_j\right|\right|^2\right), \gamma > 0$

2.3. Adaptive Boosting (AdaBoost)

AdaBoost introduced by Yoav Freund and Robert Schapire in 1995. This method aimed to maintain a weight distribution $\mathbf{w}$ of base classifier (SVM is a base classifier in this paper) iteratively [11]. AdaBoost was an ensemble method that improved the classification results by constructing a set of classifier and combining it.Given a dataset$D = \{(\mathbf{x}_1, y_1), \dots, (\mathbf{x}_m, y_m)\}, \mathbf{x}_i \in \mathbf{X},\ y_i \in \mathrm{Y} = \{-1, +1\}$, this method performs base classifier training iteratively as many cycles$t = 1,2, \dots, T$. Initial weight vector $\mathbf{w}^1$ in this training was arranged the same as follows:

$$\mathbf{w}_i^1 = \frac{1}{m}, \quad i = 1, 2, \dots, m \tag{4}$$

At each round, the weight vector would be updated until it obtained the right result.

The base classifier worked to find hypothesis $h_t = \{-1, +1\}$, for $\mathbf{w}_i$. The quality of hypothesis $h_t(\mathbf{x}_i)$measured by training error ε_t as follows:

$$\varepsilon_t = \sum_{i=1}^{m} \mathbf{w}_i^t, \quad y_i \neq h_t(\mathbf{x}_i) \tag{5}$$

Training error was calculated from the trained weights vector. If $\varepsilon_t > 0{,}5$, then the weighting process was stopped and iteration was not continued.

After the hypothesis h_t accepted, AdaBoost would determine the weight of hypothesis h_t, α_t. It was obtained $\alpha_t \geq 0$ if $\varepsilon_t \leq 0{,}5$, and the value of α_t would be increase since ε_t was decreased. Thus, it was formulated as follows:

$$\alpha_t = \frac{1}{2}\ln\left(\frac{1-\varepsilon_t}{\varepsilon_t}\right) \tag{6}$$

Then, the weight vector $\mathbf{w}_i^t$ were updated to

$$\mathbf{w}_i^{t+1} = \frac{\mathbf{w}_i^t\, exp\{-\alpha_t y_i h_t(\mathbf{x}_i)\}}{Z_t} = \frac{\mathbf{w}_i^t}{Z_t} \times \begin{cases} exp\{-\alpha_t\},\ y_i = h_t(\mathbf{x}_i) \\ exp\{\alpha_t\},\ y_i \neq h_t(\mathbf{x}_i) \end{cases}$$
$$\sum_{i=1}^{m} \mathbf{w}_i^{t+1} = 1 \tag{7}$$

With Z_t was a normalisation constants that made $\sum_{i=1}^{m} \mathbf{w}_i^{t+1} = 1$ so $\mathbf{w}_i^{t+1}$ could be distributed.

The hypothesis resulted $H(\mathbf{x})$ based on the number of weights of T hypothesis of the base classifier as follows:

$$H(\mathbf{x}) = \text{sign}(\sum_{t=1}^{T} \alpha_t h_t(\mathbf{x})) \tag{8}$$

This algorithm was based on Schapire & Singer in 1999 [12].

Table 1. AdaBoost algorithm

AdaBoost
1. **Input**: Dataset $D = \{(x_1, y_1), \ldots, (x_m, y_m)\}$, a Base Classifier algorithm, the number of cycles T
2. **Initialize**: the weights of training samples: $\mathbf{w}_i^1 = 1/m$, for all $i = 1, 2, \ldots, m$.
3. **Do for** $t = 1, \ldots, T$
(1) Use the BaseClassifier algorithmto get hypotesis h_t, on the weighted training samples.
(2) Calculate the training error of h_t : $\varepsilon_t = \sum_{i=1}^{m} \mathbf{w}_i^t,\ y_i \neq h_t(\mathbf{x}_i)$.
(3) If $\varepsilon_t > 0{,}5$; then stop.
(4) Set weight for the hypotesis h_t : $\alpha_t = \frac{1}{2}\ln(\frac{1-\varepsilon_t}{\varepsilon_t})$.
(5) Update the weights of training samples: $\mathbf{w}_i^{t+1} = \frac{\mathbf{w}_i^t exp\{-\alpha_t y_i h_t(x_i)\}}{Z_t} = \frac{\mathbf{w}_i^t}{Z_t} \times \begin{cases} exp\{-\alpha_t\},\ y_i = h_t(x_i) \\ exp\{\alpha_t\},\ y_i \neq h_t(x_i) \end{cases}$, where Z_t is a normalization constant and $\sum_{i=1}^{m} \mathbf{w}_i^{t+1} = 1$.
4. **Output**: $H(x) = sign(\sum_{t=1}^{T} \alpha_t h_t(x))$.

2.4. AdaBoost SVM

AdaBoost SVM was AdaBoost method with SVM as the base classifier. Algorithm of this method was similar with algorithm in Table 1. AdaBoost performed the hypothesis weighting of SVM method to obtain a better accuracy. At each cycle, the weight in misclassification error were increased, while the weight on the already well-classified were decreased to reduce the potential weighted back in the next cycle. This process was to do predict the class (label) of hypothesis h_t [11].

Table 2. AdaBoost SVM algorithm.

AdaBoost SVM
1. **Input**: Dataset $D = \{(\boldsymbol{x}_1, y_1), \ldots, (\boldsymbol{x}_m, y_m)\}$, SVM algorithm, the number of cycles T
2. **Initialize**: the weights of training samples: $\boldsymbol{w}_i^1 = 1/m$, for all $i = 1, 2, \ldots, m$.
3. **Do for**$t = 1, \ldots, T$
(1) Use SVM algorithmto get hypotesish_t, on the weighted training samples.
(2) Calculate the training error ofh_t : $\varepsilon_t = \sum_{i=1}^{m} \mathbf{w}_i^t,\ y_i \neq h_t(\mathbf{x}_i)$.
(3) If$\varepsilon_t > 0{,}5$; then stop.
(4) Set weight for the hypotesish_t : $\alpha_t = \frac{1}{2}\ln(\frac{1-\varepsilon_t}{\varepsilon_t})$.
(5) Update the weights of training samples:$\boldsymbol{w}_i^{t+1} = \frac{w_i^t exp\{-\alpha_t y_i h_t(x_i)\}}{Z_t} = \frac{w_i^t}{Z_t} \times \begin{cases} exp\{-\alpha_t\},\ y_i = h_t(\boldsymbol{x}_i) \\ exp\{\alpha_t\},\ y_i \neq h_t(\boldsymbol{x}_i) \end{cases}$,where Z_t is a normalization constant and $\sum_{i=1}^{m} \boldsymbol{w}_i^{t+1} = 1$.
4. **Output**: $H(\boldsymbol{x}) = sign(\sum_{t=1}^{T} \alpha_t h_t(\boldsymbol{x}))$.

3. Results

3.1. Used Parameters

Table 3. Used parameters of SVM and AdaBoost.

	Parameter	Value	Explanation
SVM	Kernel	Polynomial ($d = 2$) RBF ($\gamma = 0.01$)	[13], [14]
AdaBoost	T	10	Based on numerical simulation

3.2. Numerical Results

The data used for training was OA patient data which was checked by MRI T2Map [15]. Based on those parameters used, the accuracy of SVM result would be compared between SVM with polynomial kernel (polynomial SVM) and SVM with RBF kernel (RBF SVM). The OA data was divided into training data and testing data. We constructed a model using training data and obtained accuracy by testing the model using testing data. Tests were conducted using 10% data training (90% data testing) until 90% data training (10% data testing).

Table 4. Results of OA severity classification using polynomial SVM and RBF SVM.

Polynomial SVM		RBF SVM	
Training Data	Accuracy (%)	Training Data	Accuracy (%)
10%	34	10%	41,333
20%	31,111	20%	44
30%	39,250	30%	41,667
40%	40	40%	46
50%	49,412	50%	51,765
60%	57,143	60%	64,286

70%	60	70%	68
80%	62,858	80%	68,572
90%	**65**	90%	**75**

Based on the result above, it was shown that the highest accuracy value of polynomial SVM was 65% in 90% training data, while the highest accuracy value of RBF SVM was 75% in 90% training data. Thus, RBF SVM accuracy was better than polynomial SVM accuracy.

The accuracy result would be compared between RBF SVM and AdaBoost SVM (with kernel RBF). The purpose of this study was to examine whether AdaBoost SVM could produce better accuracy with SVM as comparison.

Table 5. Results of OA severity classification using SVM and AdaBoost SVM

RBF SVM		AdaBoost SVM	
Training Data	Accuracy (%)	Training Data	Accuracy (%)
10%	41,333	10%	53,333
20%	44	20%	55,556
30%	41,667	30%	58,333
40%	46	40%	65
50%	51,765	50%	70,588
60%	64,286	60%	71,429
70%	68	70%	80
80%	68,572	80%	**85,714**
90%	**75**	90%	80

Based on the result above, it was shown that the highest accuracy value of RBF SVM was 75% in 90% training data, while the highest accuracy value of AdaBoost SVM was 85.714% in 80% training data. It was also shown that AdaBoost SVM increased the accuracy of SVM. Thus, it could be concluded that AdaBoost could improve the performance of SVM by maintains the weight of distribution.

4. Conclusion

In this study, AdaBoost SVM were using in the classification of Osteoarthritis (OA) disease severity with SVM as comparison. Based on numerical simulation, the highest accuracy value of SVM was 75% in 90% training data using RBF kernel, while the highest accuracy value of AdaBoost SVM was 85,714% in 80% training data using RBF kernel and number of cycles T = 10. Thus, AdaBoost could maintain the distribution weight of SVM iteratively and increasing its accuracy. AdaBoost SVM had better accuracy than SVM in the classification of Osteoarthritis disease severity. For further research, it can be used other base classifier in classifying the disease.

5. References

[1] World Health Organization (WHO)2013 (online)*Osteoarthritis*(www.who.int/medicines/areas/priority_medicines/BP6_12Osteo.pdf)

[2] Arthritis Foundation 2007 (online) What is Osteoarthritis? (https://www.arthritis.org/about-arthritis/types/osteoarthritis/what-is-osteoarthritis.php)

[3] Bijlsma JW, Berenbaum F andLafeber FP 2011 Osteoarthritis: an update with relevance for clinical practice*The Lancet***377**pp. 2115–2126

[4] Kellgren J and Lawrence J 2000 Radiological assessment of osteoarthrosis*Ann Rheum Dis***16** pp. 494-452

[5] Stampella A, Monteagudo S and Lories R 2018 Wnt signaling as target for the treatment of osteoarthritis *Best Practice & Research Clinical Rheumatology***xxx**pp. 1-9
[6] Lynch CM, Abdollahi B, Fuqua J D, de Carlo A R, Bartholomai J A, Balgemann R N, van Berkel V H, and Frieboes H B 2017 Prediction of Lung Cancer Patient Survival via Supervised Machine Learning Classification Techniques *International Journal of Medical Informatics***108**pp. 1-8
[7] Xue H, Yang Q and Chen S 2009 *SVM: Support Vector Machines*(USA: Taylor & Francis Group)
[8] Panca V and Rustam Z 2017 Application of Machine Learning on Brain Cancer Multiclass Classification *The American Institute of Physics (AIP) Conference***1862**(1)
[9] Kim H C, Pang S, Je H M, Kim D, and Bang S Y 2003 Constructing support vector machine ensemble*Pattern Recognition***36**pp. 2757-2767
[10] Bishop C M 2006 *Pattern Recognition and Machine Learning*(New York: Springer)
[11] Li X, Wang L and Sung E 2008 AdaBoost with SVM-based Component Classifiers Engineering *Applications of Artificial Intelligence***21**pp. 785-795
[12] Schapire R E and Singer Y 1999 Improved Boosting Algorithms Using Confidence-rated Predictions *Machine Learning***37**pp. 297-336
[13] Rustam Z and Zahras D 2018 Comparison between Support Vector Machine and Fuzzy C-Means as Classifier for Intrusion Detection System *Journal of Physics: Conference Series***1028**
[14] Rustam Z and Ariantari N P A A 2018 Support Vector Machines for Classifying Policyholders Satisfactorily in Automobile Insurance *Journal of Physics: Conference Series***1028**
[15] Pamela C 2017 *Korelasi Nilai T2Map dengan Ketebalan Kartilago Lutut Pasien Osteoartritis pada Sekuen Proton Density-Weighted dengan Menggunakan Magnetic Resonance Imaging 1,5 TeslaTesis* Jakarta: Universitas Indonesia

Acknowledgment
This research is funded by Universitas Indonesia via PITTA Research grant 2018 scheme.

IOP Conf. Series: Journal of Physics: Conf. Series **1108** (2018) 012063 doi:10.1088/1742-6596/1108/1/012063

Visualization Profile of Junior High School Students in Solving Geometry Problems Viewed from Gender Differences

M M L Puloo[1,*], D Juniati[1], and P Wijayanti[1]

[1]Postgraduate Program Department of Mathematics, Universitas Negeri Surabaya, Surabaya, East Java, Indonesia
*Corresponding author: melisapuloo16070785022@mhs.unesa.ac.id

Abstract. In solving mathematical problems, visualization has an important role in developing thinking, mathematical understanding, helping the transition of concrete thinking to abstract. This study aimed to describe the visualization process of junior high school students in solving geometry problems viewed from gender differences. The description of the student visualization process was based on the aspects of *generation, inspection*, *transformation*, and *use*. The methods of data collection in this study were conducted through test and interview. The result of the research showed that in *generation* aspect, male and female students made the picture according to the information of the problem which was given by relating it to the knowledge they had. In *inspection* aspect, male and female students applied the rules of Pythagoras theorem to determine one of its elements. In *transformation* aspect, female student changed the pattern of visual ideas into square shapes by translating the elements, while male student changed the pattern of visual ideas into triangular shapes. In *use* aspect, female student used a picture to check the result of the operation, while the male student did not use a picture to check the result of the operation. Thus, this visualization process was expected to provide information for teachers about the visualization process of junior high school students either male or female so as to assist teachers in designing learning.

1. Introduction

In learning mathematics, especially in solving mathematical problems, students are expected to be able to manage the cognitive to become more complex. Dealing with solving mathematical problems, visualization has an important role in the process of solving mathematical problems. According to [1], the important roles of visualization in the process of solving mathematical problems are developing thinking, mathematical understanding, and helping the transition of concrete thinking to abstract. Visualization can help teachers to see students' understanding of the given problem [2]. And it is needed by the students to manage the interpretation and manipulate information in the mastery of *mathematical word problem solving* [3].

According to [4] visualization refers to mental processes that describe visual or spatial information. Visualization is also defined as the process of building and transforming visual images and all representations of spatial characters that can be used in drawing, building or manipulating (images and symbolic) with pencil and paper [5]. Furthermore, according to [6], visualization is a process that is the result of the creation, interpretation, reflection of images, diagrams, in our minds embodied in paper or technology tools with the aim of describing and communicating information, mind about developing idea that is not known in advance to gain a higher understanding. In addition, [7] defining visualization in mathematics is as the process of forming images (mentally) using pencil and paper and effective implementation of the use of images that have been formed for mathematical discovery and understanding.

Because of the importance of visualization in solving problems, as according to [8] using diagrams or images can help students to understand the problem before they find solutions to the problem, mainly solve word-problem in geometry material. Therefore, students' visualization skill in solving problems is important to teachers' concerns, especially the visualization skill of junior high school students. Based on the research [9], the researchers advised the teachers of Junior High School to develop mathematical visualization skills, especially about geometry using various contexts. This is because the junior high school students based on Piaget cognitive development stage are in the formal operation stage (12 years and over), and at this stage the children have started to think abstractly and the children's understanding is needed as capital to learn mathematics in advanced level. In addition, many studies have also revealed students' difficulties in visual problems, difficulties encountered such as understanding problems, producing or drawing accurate diagrams that fit the problem situation, reading the graph correctly [6, 10, 11, 12].

In solving problems, each individual has different abilities, especially when viewed through sex differences. It is in line with the opinion of [13] that gender differences gave differences in spatial performance in geometry. Meanwhile [14] reveals that women's abilities are better than men's in terms of verbal and visual solution methods whereas in terms of quantitative reasoning men's abilities are better than women's. Further [15] explains that the development of cognitive sources is influenced by biological factors and experience. The intended cognitive source refers to the ability and speed of processing information, biological factor refers to gender, and experience refers to the frequency of a person in processing information. Thus, it can be said that the number of experiences of a person in processing visual problems will speed up a person in processing problems, and gender can affect the process of visualization of a person in solving the problem. [14] and [15] show that gender influences visual problems and methods of verbal and visual solutions of a person so that the researcher is interested to know more about the process of visualization of male and female junior high school students.

2. Method

This research was descriptive research with qualitative approach. Description of the visualization process in this study was viewed from the aspect of the visualization process presented by [16] including generation, inspection, transformation, and use. Furthermore, by qualitative approach, the data that had been collected from research subjects were arranged, simplified, organized, described in the form of words and concluded. The subject in this study was a group of students of grade VIII. A group of students was given a mathematics ability test to determine the mathematical ability of each subject. From the results of the mathematics ability test obtained, then 1 male subject and 1 female subject who had equivalent mathematical ability were selected. Furthermore, both selected subjects were given a visualization test. The results of the visualization tests obtained were then searched in depth through semi-structured interviews linked to every aspect of visualization.

3. Results

In this research, visualization process is a mental process for building and transforming visual ideas (images) in order to translate, communicate, and develop ideas in mind through the aspects of *generation, inspection, transformation*, and *use*. Therefore, the visualization process of male and female students described is seen from the aspect of *generation, inspection, transformation*, and *use*. The male and female students had equivalent mathematical abilities based on the results of the mathematical ability tests obtained.

3.1. Generation Aspect

3.1.1. Male Student

The first step of the male student in answering the problem was given by carefully reading the information in the given problem and marking the important thing in the problem that would determine the image to be formed by it. A picture of male student's answer result in describing the information of the given problem shown in Figure 1 below.

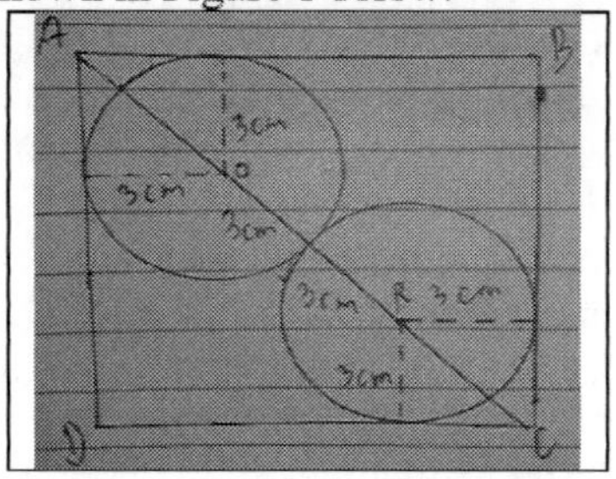

Figure 1. The Answer of Male Student in The Visualization Test

Having been searched further through the interview, the reason for describing the position of the circle as seen in the worksheet, the male student revealed that because in the problem, it was said that all the sides of the square intersected with the circle. If it was made in a row of circles in a square then there was a square side that did not intersect with the circle. He also revealed if it had to intersect with the circle in a lined position, then it formed a rectangle. The male student in his reasoning also related it to the characters of the square.

3.1.2. Female Student

The first step taken by female student in answering the given problem was identifying the information of the given problem by imagining the position of the two circles, in which the fingers of both hands formed a circle as an example of the two intended circles. Then the female student began to describe the position of the circle that might be based on the information from the given problem. A picture of female student's answer result in describing the information of the given problem shown in Figure 2 below.

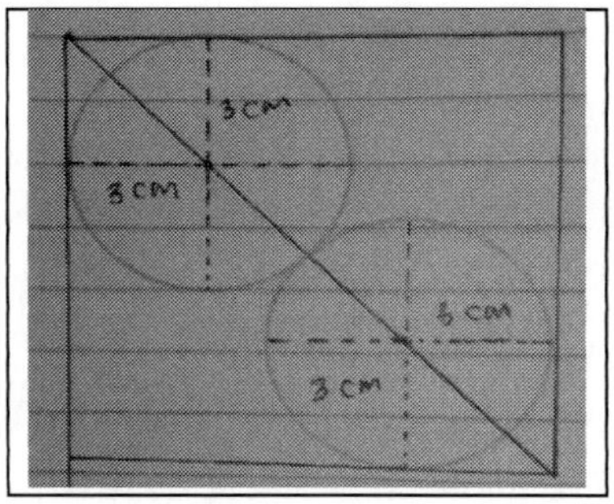

Figure 2. The answer of Female Student in Visualization Test

After being searched through the interview, the reason for drawing the position of the circle as depicted in the worksheet, the student revealed if the position of the circle lined then it formed a rectangle, whereas in the problem, the requested one was a square. The student revealed the character of the square namely all sides were equal in length.

3.2. Inspection Aspect

3.2.1. Male Student

In Figure 1, the male student determined his solving strategy by observing the center point of each circle and connecting it to one of the square vertices to form diagonal. The same was done for both circles. To find the diagonal length, the male student applied the rule of Pythagoras theorem. After being searched through an interview, the reason for applying Pythagoras's theorem rule, he revealed that by forming the diagonal the image formed was a right-angled isosceles triangle. He then tried to check the shape of a triangle he made by marking the triangles: 3 cm for the base and 3 cm for the height and he obtained √18 for the aslant side. Then he mentioned that the quadrate of aslant side of the triangle was equal to the sum of quadrate of the right angle so that his strategy was correct.

3.2.2. Female Student

The female student determined the solving strategy using Pythagoras theorem by visualizing the visual objects that were made into small rectangular shapes and then creating a diagonal line like Figure 3 below. She then checked the visual form created by marking the size of the small square created. She then applied Pythagoras theorem rule to determine the square diagonal and obtained the result of $3\sqrt{2}$.

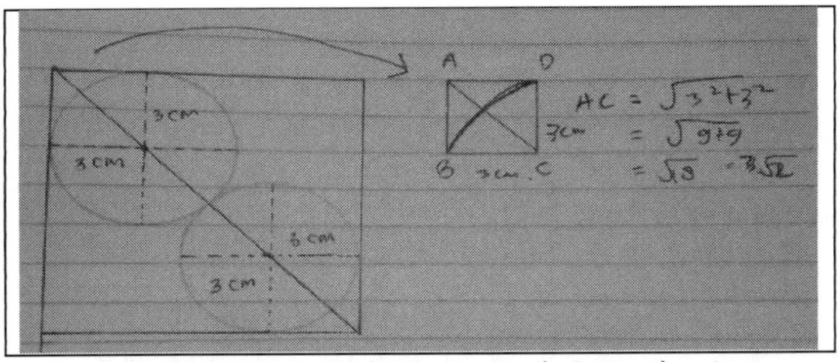

Figure 3. Female Student's Answer in Inspection Aspect

After being searched through an interview, the reason for using the Pythagoras theorem, the female student revealed because if a diagonal was created on a square forming a right triangle, Pythagoras theorem could be applied.

3.3. Transformation Aspect

3.3.1. Male Student

To get a solution, the male student made a pattern change on the image created by forming a new right triangle. He revealed that the aslant side was formed from the diagonal made in the inspection process which obtained √18 cm with a diagonal formed by connecting both points of the circle so that the result obtained was 6 cm. Then he summed up those elements namely $\sqrt{18} + 6 + \sqrt{18} = 6 + 6\sqrt{2}$. Next, to get the right angle that was not yet known he symbolized it with variable x. Then he used Pythagorean Theorem to obtain the value of $x^2 = 54 + 36\sqrt{2}\ cm^2$. He revealed that the result of x^2 obtained was the result of the square area in question. However, in this process he did not make a change or give a different name to the formed structure. The picture made by the male student to get a solution shown in Figure 4 below.

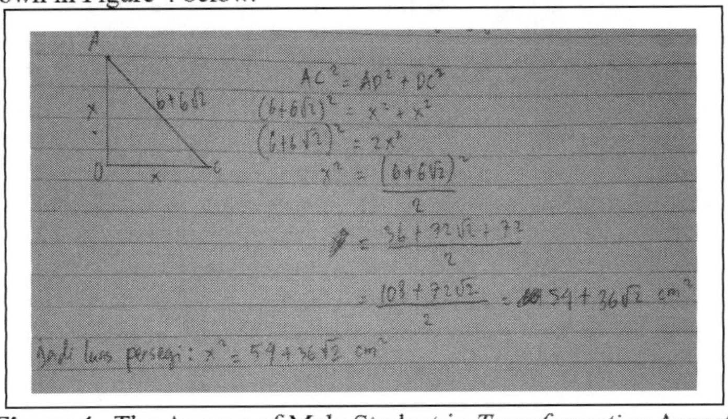

Figure 4. The Answer of Male Student in *Transformation* Aspect

3.3.2. Female Student

To obtain a solution, the female student made a change on the picture she made into a new square shape obtained by translating on the length size on the new square side. On the length size of the new square side she revealed that the length side of the new square could be shifted from the original square diagonal part obtained on the inspection aspect namely $3\sqrt{2}$ cm. Furthermore, she did translation on the size of the radius of the circle. The female student revealed that the distance between the centre point of the circle to the side of the square was equal to the distance between the auxiliary points that had been made to the square side so that the distance of the auxiliary point to the square side was 3 cm. To obtain the length of the square side, she added up the elements that had been obtained, that was $3 + 3\sqrt{2} + 3 = 6 + 3\sqrt{2}$. Next, to get the square area, she used the square area formula, which was the square of the side length, that was $6 + 3\sqrt{2}$, so that it was obtained $54 + 36\sqrt{2}\ cm^2$. In addition, she also assigned a different name to the new square formed with square PQRS. A picture of the answer from the female student in *transformation* aspect shown in Figure 5 below.

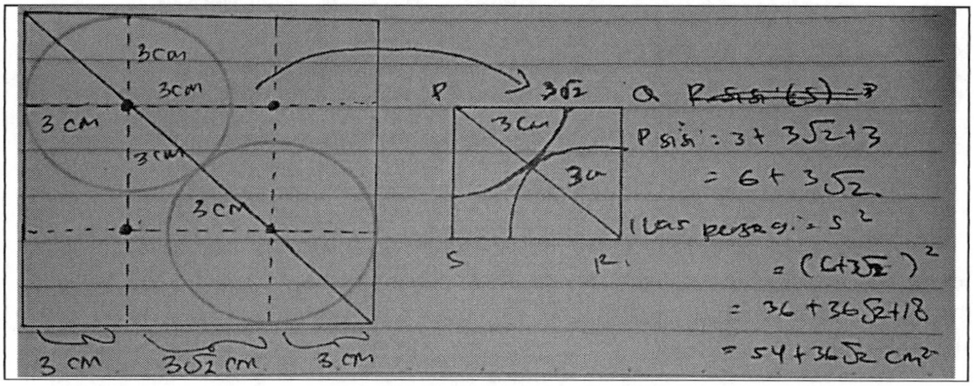

Figure 5. The Answer of Female Student in *Transformation* Aspect

3.4. Use Aspect

3.4.1. Male Student

Further, at the end of the solution, the male student did not write how to check the result of the operation obtained and when being searched through an interview, the male student did not check the picture obtained to check the result of operation. He just revealed that his calculation was correct while checking the calculation.

3.4.2. Female Student

At the end of the solution, the female student did not write how to check the correctness of the result obtained but when being searched in the interview the student revealed that to know if the result of the square area obtained was correct, then the correctness of the result of the length side of square obtained also needed to be checked. Then she connected the picture with the operation result of length side of square. She connected it by labeling the result of length side of square $6 + 3\sqrt{2}$ in the picture and checking the correctness of the result of length side of square by relating it to the Pythagoras theorem and the right triangle. She revealed that if the sum of quadrate of the square side was equal to the diagonal quadrate formed then the result was definitely correct.

4. Conclusion and Discussion

Based on the results obtained, it can be concluded on visualization process on female student: aspects of *generation, inspection, transformation,* and *use* have been carried out by the subject in solving geometry problems, while the male student only does three aspects: aspects of *generation, inspection and transformation* in solving geometry problems. [14] Reveals that women's abilities are better than men's in terms of verbal and visual solution methods. This is in accordance with the result obtained about the visualization process in female student, the four aspects of visualization had been done in solving the problem of geometry by the female student while the male student only did three aspects of visualization in solving geometry problems.

[15] Explains that the development of cognitive sources is influenced by biological factors and experiences. The intended cognitive source refers to the ability and speed of processing information, biological factor refers to gender, and experience refers to the frequency of a person in processing information. Thus it can be said that the number of a person's experiences in processing visual problems will speed up a person in processing problems, and gender can affect the process of visualization of a person in solving the problem. It is also in accordance with the results of the research that both male and female students had high mathematical skills as proven by the TKM scores obtained by both of them so as to show the effect on their good visualization process because of their large learning experience especially in the visual problem. Finally, the results of the visualization process are expected to provide information for teachers about the visualization process of junior high school students, either male or female, so as to help teachers in designing learning.

5. References

[1] Lavy, I 2006 Dynamic Visualization and The Case of 'Stars in Cages' *Proceeding 30th Conference of The International Group for The Psychology of Mathematics Education* **4** pp 25-32 (Prague : PME)

[2] Nunokawa, K 2005 Mathematical Problem Solving and Learning Mathematics: What We Expect Students To Obtain *Journal of Mathematical Behavior* pp 325-340

[3] Geary, D C 2004 Mathematics and Learning Disabilities *Journal of Learning Disabilities* pp 4-15

[4] Carden, J & Cline, T 2015 Problem solving in mathematics: the significance of visualisation and related working memory *Journal Educational Psychology in Practice* pp 1-12

[5] Huang, C H 2015 Calculus Students Visual Thinking of Definite Integral *American Journal of Educational Research* vol 3 No 4 pp 476-482

[6] Arcavi, A 2003 The Role Visual Representation in The Learning of Mathematics *Educational Studies In Mathematics* vol 52 issue 3 pp 215-241

[7] Pachemska, T A 2016 *Visualization of The Geometry Problems in Primary Math Education* **8** pp 33-37

[8] Kashefi, H 2015 Visualisation in Mathematics Problem Solving Meta-Analysis Research *E-Proceeding of The International Conference on Social Science Research* pp 576-585

[9] Utomo, E S, Juniati, D, and Siswono, T Y E 2017 Mathematical Visualization Process of Junior High School Students in Solving a Contextual Problem Based on Cognitive Style *AIP Conference Proceedings 1868,* 050011

[10] Andriyani, et al 2018 The Blind Student's Interpretation of Two-Dimensional Shape in Geometry *J. Phys : Conf.ser.947* 012055

[11] Poch, A L 2015 Students' Understanding of Diagrams for Solving Word Problems: A Framework for Assessing Diagram Proficiency *Article Teaching Exceptional Children.* pp153-162

[12] Diezmann, C M 2000 The Difficulties Students Experience in Generating Diagram for Novel Problems *Proceedings 25th Annual Conference of the International Group for the Psychology of Mathematics Education* pp 241-248

[13] U, Hanifah et al 2018 Student's Spatial Performance : Cognitive Style and Sex Differences *J. Phys : Conf. Ser. 947* 012014

[14] Mainali, B R 2014 *Investigating The Relationships Between Preferences, Gender, and High School Students Geometry Performance* (Florida: University of Central Florida)

[15] Santrock, J W 2009 *Educational Psychology 5th Edition* (New York : The McGraw Hill)

IOP Conf. Series: Journal of Physics: Conf. Series **1108** (2018) 012063 doi:10.1088/1742-6596/1108/1/012063

[16] Makina, A & Wessels, D 2010 The role of visualisation in developing critical thinking in mathematics *Journal perspectives in Education University of South Africa* Vol 28 no 1 pp 24-33

IOP Conf. Series: Journal of Physics: Conf. Series **1108** (2018) 012064 doi:10.1088/1742-6596/1108/1/012064

Identifying of factor associated with parkinson's disease subtypes using random forest

E Latifah[1,*], **S Abdullah**[1] **and S M Soemartojo**[1]

[1] Departement of Mathematics, Universitas Indonesia, Depok, 16424, Jawa Barat, Indonesia
*Corresponding author: esti.latifah@sci.ui.ac.id

Abstract. Parkinson's disease (PD) is a neurodegenerative disorder that caused by the result of lack of dopamine in a specific area of the brain called bangsal ganglia. It is a long term degenerative disorder of the central nervous system that mainly effect the motor system that has some impacts such as difficulty in speech, problem in swallowing, and dressing, trouble with handwriting or even doing some activities, and tremor. Based on this problem, researchers use the Parkinson's Progression Markers Initiative (PPMI) database to classify subtypes: Tremor Dominant (TD) and Postural Instability Gait Difficulty (PIGD). Identifying the factors of Parkinson's disease subtypes is crucial in understanding the appropriate therapy for Parkinson's disease patient. Furthermore, it gives some characteristics of patient that is classified into TD or PIGD. Classification method is used to identify the factors of parkinson's disease subtypes on 207 patient with PD and 47 variables obtained from Movement Disorder Society-Unified Parkinson Disease Rating Scale (MDS-UPDRS) part II and part III in event V12. The result is PD patient who is classified to PIGD class have the lower value in constancy of rest tremor, rest tremore amplitude (RUE), tremor, rest tremor amplitude (LUE), and postural tremor of right hand than PD patient with TD and the higher value in postural stability, walking and balance, and freezing than PD patient with TD.

1. Introduction

Parkinson's disease (PD) is a disorder of brain function that results in the degeneration of nerve cells in a specific area of the brain called basal ganglia. Degeneration of these nerve cells is due to dopamine degradation in substansia nigra pars compacta (SNC) and striatumcorpus. Dopamine decrease causes the brain activity cannot function normally that has impact to some symptoms on motor and nonmotor aspects of the body. Notice that Parkinson's disease is the most common neurodegenerative disease after Alzheimer's [11].

Parkinson's disease affects millions of the world's population or about 1% of the world's population [8]. Based on WHO data, the incidence of Parkinson's disease in Asia shows that 1.5 to 8.7 cases per year in China and Taiwan, whereas in Singapore, Wakayama and Japan there are 6.7 to 8.3 cases per year. It happens in the range of age 60 to 69. The prevalence of Parkinson's disease in Indonesia is 876,665 inhabitants [8]. The deaths of Parkinson's patients are usually not caused by Parkinson's disease itself but rather because of secondary infection [5].

There are two subtypes of Parkinson's disease: a subtype that has tremor dominant (TD) and postural instability gait disorder (PIGD) [7]. Both of these Parkinson's subtypes differ in clinical, imaging, genetic, and pathological features [6]. Parkinson with tremor dominant subtype (TD) exhibits slower progression and has a better prognosis than the postural subtype of instability gait disorder (PIGD) [15]. Parkinson's disease with TD and PIGD have different forms of motoric and non-motoric symptoms. Therefore, researchers want to examine the motor factors of Parkinson's disease that influence the subtype of Parkinson's disease using classification method that is random forest so that proper and efficient therapy can be done.

2. Data and Variables

Parkinson's Progression Markers Initiative (PPMI) data base is a study to verify the development of Parkinson's disease. The study was conducted on 145 patients with Parkinson's disease obtained from the PPMI database. The data were taken under the Movement Disorder Society-Unified Parkinson's Disease Rating Scales (MDSUPDRS) Part 2 and Part 3. Respondents selected for this study were Parkinson's patients with the following criteria:

- Patients with Parkinson's disease who have been undergoing treatment for 36 months (V12).
- Patients observed are only sufferers of motor disorders.

The following variables are used in this study:

1. Response Variable (Y)
   Y is the ratio of certain variables in the MDS-UPDRS part 2 and part 3. The mean of MDS-UPDRS items 2.10, 3.15a, 3.15b, 3.16a, 3.16b, 3.17a, 3.17b, 3.17c, 3.17d, 3.17e, and 3.18 is divided by the mean of MDS-UPDRS items 2.12, 2.13, 3.10, 3.11, and 3.12. If the ratio is greater than or equal to 1.15, then the patient is classified with TD. If the ratio is less than or equal to 0.90, then the patient is classified with PIGD [14].
2. Predictor Variable (X)
   Variable predictors consist of :
   - Data MDS-UPDRS Part 2 (13 variables). The data is a test of the motor aspects of daily life of people with Parkinson's such as speaking, eating, writing, and so on.
   - Data MDS-UPDRS Part 3 (34 variables). The data is a test performed physically by medical personnel such as response when spoken to, hand and foot movements, and so on.

3. Classification Methode

In this section the statistical methods developed and used in this study will be described.

3.1 Decision Tree

Decision tree is one of the methods used to classify a sample of unknown class data into existing classes. The use of the decision tree method depends on the response variable. If the response variable is a continuous variable, then the method used is the regression tree. If the response variable is a categorical variable, then the method used classification tree. However, in this study will only be focused on the classification problem. There are many algorithms for building a decision tree, such as ID3 (Iterative Dichotomiser 3) [16] and its further development C4.5 [17], and CART (classification and regression tree) [1].

In growing decision tree, it begins at the root node. The node that contains all the data to build a tree. From the root node will be formed branches and other nodes called parent node. Parent node itself will divide and generate new node is called child node that consist of right node and left node. Child node that can not splitting anymore is called terminal node. To understand the concept, see the ilustratin in Figure 1.

Moreover, each parent node will split into exactly two child nodes (right node and left node) based on a particular predictor variable. This process is called binary partitioning. Next, the node separation process is repeated on every child node (The child node do splitting successfully will be a parent node and so on) until the splitting process cannot be done anymore. The process of formation is known as binary recursive partitioning.

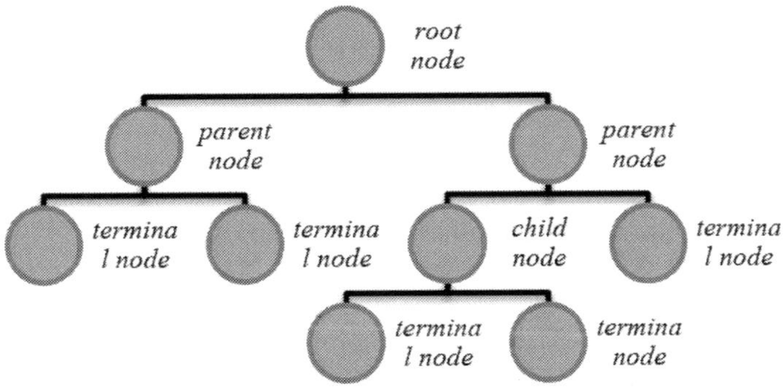

Figure 1. Structure of decision tree.

Splitting is the process of dividing the parent node into two child nodes through certain splitting rules. Parent node is divided into two child nodes based on the goodness of split criterion. The splitting criterion must fulfill the characteristic that the grouping results in the child node are more homogeneous than the preceding nodes.

In the classification and regression tree (CART), separation criteria are based on Gini index [1]. Suppose that there are n observations. These observations will be grouped into two classes (i.e. $j = A, B$). The classification rule is based on a number of M predictor variables (notation x_m, $m = 1, \ldots, M$). At each node (notation node t), the Gini index is expressed as follows,

$$Gini(t) = 1 - \sum_j p_j^2$$

with,

- $p_j = \frac{N_j(t)}{N_t}$ is the proportion of the number of objects that enter into class j on a node t.
- $N_j(t)$ is the number of objects that enter into class j on a node$t$.
- N_t is the total number of objects on a node t.

The Gini index describes the non-homogenous child nodes of the node t due to the splitting based on x_m variabel. The Gini index has a range of values between 0 and 1. The lower value of Gini index, the greater the homogeneity of a node so that the better process of separating objects into existing classes.

It takes a comparison of the non-homogeneous levels between the parent nodes (before splitting) with the non-homogeneous levels of the child node (after splitting) to determine how well an predictor variable is in splittig a node. A predictor that produces the maximum difference of Gini index between the parent node and child node to be selected as the best splitter. The difference of Gini index can be calculated as follows,

$$\Delta G(t, m) = G(t_P) - p_R G(t_R) - p_L G(t_L)$$

with,

- $\Delta G(t, m)$ is the difference of Gini index value.
- $G(t_P)$ is the Gini index valuein parent node.
- $G(t_R)$ is the Gini index value on the right node.
- $G(t_L)$ is the Gini index value on the left node.
- p_R dan p_L is the proportion of the number of objects on right and left node.

The splitting process will continue until there is no increase in homogeneity in the child node is generated.

Unfortunately, decision trees are prone to overfitting [12]. Overfitting occurs when the algorithm continues to grow the tree to reduce the error for the training data, yet this results in an increased error for the test data. The training data is a set of data that is used to build the tree. The test data is a set of data that is used to evaluate the predictive accuracy of the tree.

IOP Conf. Series: Journal of Physics: Conf. Series **1108** (2018) 012064 doi:10.1088/1742-6596/1108/1/012064

3.2 Random Forest

The random forest is an ensemble approach that offers a robust alternative to the decision trees [2]. In this method, the forming structure is similar to the decision tree method but the way of working of the two methods are quite different. There are two important steps in this method. First, the sample is obtained by taking a random sample with the replacement of data. It is called bootstrapping [2]. Second, the certain size of candidate variable is randomly taken of all predictors that used for splitting a node. It is called random feature selection [2]. As a result, the two steps above will produce many trees to form a forest with the differences of size and shape of the trees.

Suppose that there are n observations and consist of M predictor variables so that random forest is built using the following step [3]:

1. Do n random sampling of a random sample with replacement in training data. The sample generated from this step is called the bootstrap sample.
2. By using a bootstrap sample, the tree is built until it reaches its maximum size. At each node, the split selection is performed by taking $p \ll M$ of the predictors randomly. This step is called random feature selection. The best splitter is chosen from that candidat variable p.
3. Repeat steps 1 and 2 until$K$ trees are built so that the forest is formed.

The method based on bootstrapping and random feature selection is to reduce uncertainty in model predictions. In the random forest method, multiple decision trees are fitted, where each tree is fitted to a random subset of the data, and bootstrap sampled with replacement. However, to obtain a good bootstrapping result is not only through with replacement, but also be done sampling without replacement if the sample size is large enough [4].

Within a single tree, splits are determined from a random subset of predictors sampled at each node, as a means of reducing correlation among trees. The number of predictors randomly selected for each node is constant at $\sqrt{M}$ [2], where M is the total number of predictor variables. The decision on the best split is based on the CART algorithm [1] as previously described in the decision tree section. There is no pruning is performed in the growing process of each tree.

Once a tree is grown, a prediction is made for each of the cases in the out-of-bag (OOB) data. The OOB data are a subset of the bootstrap sample that is not included in the model building. For each tree K where case i is in the OOB data, class j is assigned for case i if i is predicted to be in that class. The class with the majority of assignments will be the assigned class for case i. The random forest accuracy is then calculated as the proportion of OOB cases that are correctly classified. Hence, the random forest method is appealing for its internal cross-validation embedded in the procedure.

To find the most significant variable, random forest is used Mean Decrease Gini (MDG). MDG is a measurement to find variable importance in the model. The larger MDG value of a variable predictor, the variable is more importance. MDG can be calculated as follows [3]

$$MDG_m = \frac{1}{k}\sum_t [\Delta G(t,m) I(t,m)]$$

with,

- $\Delta G(t,m)$ is the decreasing Gini index value for x_m.
- $I(t,m)$ is indicator function. $I(t,m) = \begin{cases} 1, & x_m \text{ splitting at node } t \\ 0, & \text{others} \end{cases}$
- $k = 1,2,\ldots,K$ is the number of trees formed in random forest.

4. The results for modelling data by Random Forest

The random forest method is run by using software R-version 3.4.3 (free download). Then, the following output is obtained from running the data on software R.

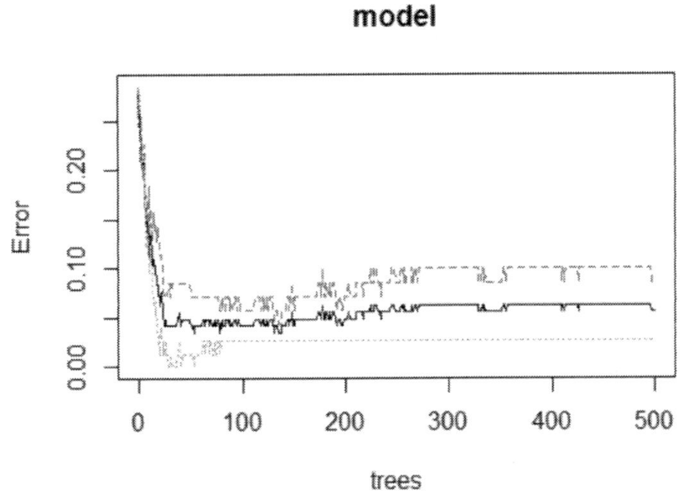

Figure 2. Error model plot with 500 trees. The red is misclassification of PIGD class plotting, the green is misclassification of TD class plotting, and the black is OOB error plotting.

Figure 2 shows the misclassifcation (error) for different classes. The red plot describes the error of the sample classified into the PIGD class, but it is wrongly predicted. In other words, the predicted PIGD sample error goes to the TD class. The error value in this sample is more constant when the tree is formed by 300 trees. In the picture, if we built a number of 300 to 500 trees, then the error is about 8%. The green plot describes the error of the predicted TD sample which goes wrongly to the PIGD class. This plot is different from PIGD sample error plot. When the tree is formed by 100 trees, the error value will be constant. If more trees are built (i.e. 100 to 500 trees), then the TD sample error is about 2%. Furthermore, the black plot is OOB error. From the picture shows if the more trees are formed, the OOB error will be more constant. When a tree is formed around 300 to 500 trees, the OOB error is about 5%. It means that random forest able to predict the patient with PD in the amount of 95%.

Afterward, Table 1 is given to find out the accuracy of the model.

Table 1. The results of confusion matrix on Parkinson's disease from software R.

		Predicted class	
Actual		PIGD	TD
Class	PIGD	64	6
	TD	2	73

Table 1 shows the confusion matrix on Parkinson's disease. The accuration of the model can easily be obtained from the confusion matrix. The accuration model is about 94,5%. It means that the model able to classify Parkinson's disease correctly.

```
Call:
 randomForest(formula = VAR.RESPON ~ ., data = trainParkins
on,      ntree = 500, mtry = 6, importance = TRUE)
               Type of random forest: classification
                     Number of trees: 500
No. of variables tried at each split: 6

        OOB estimate of  error rate: 5.52%
```

Figure 3. Output R.

The random forest model depends on *ntree* and *mtry*. *ntree* is how many trees will be built, while *mtry* is the number of candidate variables used to split each node. Figure 3 shows that in this model, the tree to be built is 500 trees and the number of variables used for splitting is 6 variables for each node.

The random forest will produce an OOB error. From the Figure 2, the value of OOB error approach its OOB error plotting. The OOB error is about 5.52%. It can be interpreted that the resulting accuracy for the prediction sample is 94.48%.

Thereafer, Figure 4 shows that the rank of all predictors. The constancy of rest variable has the higest MDG value overall predictors, whereasin the second place is rest tremor amplitude (RUE) variable. Tremor variable is in the third place after two variables above. In the next place is rest tremor mplitude (LUE) and postural stability variable. Further informations can be seen in the picture below.

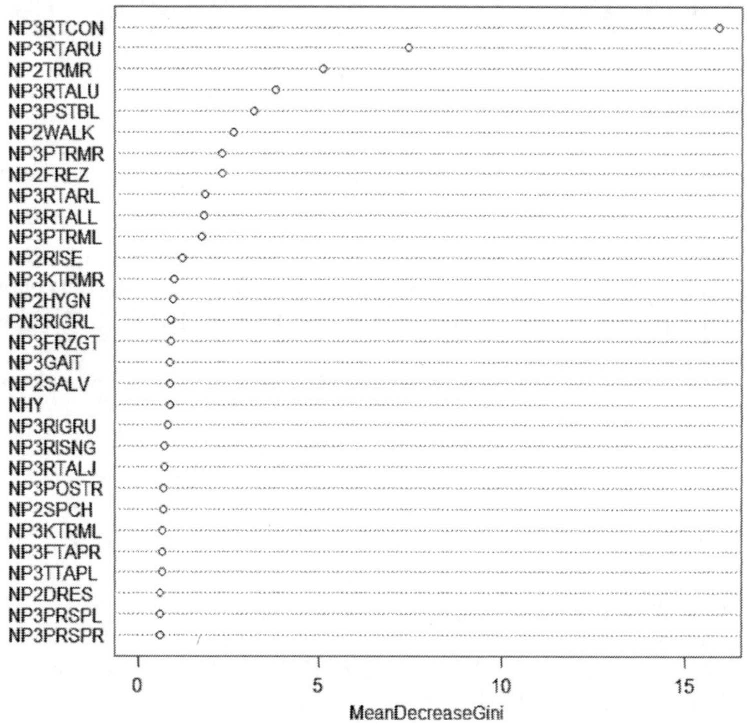

Figure 4. Variable importance plotting using MDG.

5. Conclusion

Data of 207 patients with PD were analysed to determine factors that associated with Parkinson's disease subtypes. The factors of PD subtypes have been identified. There are some importance variables based on MDG with the value > 5 such as constancy of rest tremor, rest tremore amplitude (RUE), tremor, rest tremor amplitude (LUE), postural stability, walking and balance, postural tremor of right hand, freezing. Those variabeles are the factors that influence Parkinson's disease subtypes. PD patient who is classified to PIGD class have the lower value in constancy of rest tremor, rest tremore amplitude (RUE), tremor, rest tremor amplitude (LUE), and postural tremor of right hand than PD patient with TD and the higher value in postural stability, walking and balance, and freezing than PD patient with TD.

6. Acknowledgments

This research is funded by Universitas Indonesia via PITTA 2018.

7. Reference

[1] Breiman L., et al 1984 *Classification and Regression trees* (Wadsworth: Wadsworth International Group)

[2] Breiman L 2001 Random Forests *J. of Machine Learning*, **45** pp 5–32.

[3] Breiman L and Cutler A https://info.salford-systems.com/an-introduction-to-random-forests-for-beginners (accessed: Jan 20th 2018).

[4] Buja A and Stuetzle W 2006 Observations on Bagging *J. of Statistica Sinica* **16** pp 323-351

[5] Joesoef A A 2007 *Parkinson's Disease : Basic Science* in *Parkinson's Disease & Other Movement Dissorder* (Jakarta: Pustaka Cendikia Press)

[6] Marras C and Lang A 2013 Parkinson's Disease Subtypes: Lost in Translation? *J. Neurol Psychiatry*, **84**(4)

[7] McDermott M P, et al 1995 Factors Predictive of The Need for Levodopa Therapy in Early, Untreated Parkinson's Disease *Archives of neurology* **52**(6) pp 565-570.

[8] Noviani, et al 2010 Hubungan antara Merokok Dengan Penyakit Parkinson Di RSUD Prof. Dr. Margono Soekarjo Purwokerto *Journal of health*, **4**.

[9] Parkinson's Foundation http:/www.parkinson.org (accessed: June 3rd 2018).

[10] Parkinson's Progressive Markers Initiative https://www.ppmi.org (accessed: April 5th 2018).

[11] PERDOSSI 2008 *Modul Gangguan Gerak Penyakit Parkinson*.

[12] Rokach L and Maimon O 2014 *Proactive data mining with decision trees* (New York: Springer Science & Business Media)

[13] Abdullah S 2017 *Statistical Methods for Modelling Fall and Symptoms Progression in Patients with Early Stage of Parkinson's disease* (Brisbane: Queensland University of Technology)

[14] Stebbins G T, et al 2013 How to Identify Tremor Dominant and Postural Instability/Gait Difficulty Groups with The Movement Disorder Society Unified Parkinson's Disease Rating Scale: Comparison with The Unified Parkinson's Disease Rating Scale *Movement Disorders* **28**(5) pp 668-670.

[15] Thenganatt M A and Jankovic J 2014 Parkinson Disease Subtypes *JAMA neurology*, **71**(4) pp 499-504.

[16] Quinlan J R 1986 Induction of Decision Trees *J. of Machine Learning*, **1** pp 81-106.

[17] Quinlan, J, R 1993 *C4.5: Program for Machone Learning* (New York: Spriger)

Analysis Dengue Fever and Malaria Cases using Generalized Multivariate Conditional Autoregressive Model

A Sofro

Mathematics Department, Universitas Negeri Surabaya, East Java, Indonesia

E-mail: ayuninsofro@unesa.ac.id

Abstract. An important topic in epidemiology is dengue fever. This disease is a mosquito-borne infection found in tropical and sub-tropical regions around the world. Another disease infected by mosquito is malaria. The developed methods still offer a lack information how to accommodate a cross correlation between two diseases for different cities. Hence we will provide a natural framework involving a cross correlation for multivariate spatial effect,i.e generalized multivariate conditional autoregressive model (GMCAR). We also focus on analysing the number of infected cases for two diseases in East Java Indonesia as the response variables. Therefore, in this paper we will propose a multivariate count data modelling with GMCAR to analyse the data. The result shows that the proposed model provides a better performance compare a model without cross correlation.

1. Introduction

Dengue fever and malaria are diseases which infected by mosquitoes and typically occurred in tropical and subtropical regions. The dengue virus causes dengue fever can develop into dengue hemorrhagic fever. Africa, the Americas, the Eastern Mediterranean, and South-east Asia and the Western Pacific regions and more 100 countries are endemic of the disease.

Meanwhile, another a mosquito-borne infectious disease of humans caused by parasitic protozoans is malaria. Similar with dengue fever, coma can develop in malaria cases. Areas of the Americas, many parts of Asia and Africa is also becoming high-risk regions for the disease.

Because both of two diseases come from mosquitoes, hence it is more reasonable if we analyze two diseases simultaneously. Hence, there are multiple diseases in a different area which need to analyze involving a correlation in multiple factors. The correlation is expected to be able to accommodate several types of correlations, i.e the spatial correlation for each disease across regions, dependence among multiple diseases within the same region and cross spatial correlation among multiple diseases in different regions.

The diseases have been analyzed massively recent year. Many of the researcher have tried to explore either dengue fever or malaria cases using visualization data, involving climate factor, see [1] [2], [3], [4], [5] and [6]. The investigation has improved with combining malaria disease in the analysis, see [7] and [8]. Unfortunately, most of their approaches provide a lack of information in terms of cross-correlation within and between two diseases.

The spatial correlation in the same region has been developed since 1995, see [9], [10], [11], [12], [13], [14], [15]. Meanwhile, the correlation within and between two disease in the different regions (cross correlation) has been provided by [16].

We will focus on analysing the number of infected cases due to two diseases in each city as dependent variables. Thus, in this paper, we will discuss how to model the diseases using multivariate Poisson regression with accommodating the cross correlation. We begin the investigation from the method with the simple model to the complex for multivariate cases.

2. Method

In this section, we will explore the some approaches start from the simple one. i.e Poisson regression.

2.1. Poisson Regression Model

In modelling disease incidence rates, there are a definition which needs to understand. The most common indicator used for comparing regional death rates is the standardized mortality ratio (SMR) in place i which is defined by

$$SMR_i = \frac{z_i}{E_i},$$

where z_i is observed number of deaths in place i and E_i is the expected number of deaths based on reference mortality rates applied to the regional demographic structure. Also E_i is defined as

$$E_i = \sum_g r_g \times p_{g,i},$$

where r_g is a standard mortality rate (i.e national mortality rate) of demographic group g and $p_{g,i}$ is regional population size specific to demographic group g in place i. The demographic group g is usually determined by age or sex-age attributes.

In a generalized linear model (GLM), there is a link function h connecting mean from the dependent variable with independent variables ($\mathbf{U}$), i.e.

$$\begin{aligned} E\left(\frac{z_i}{E_i}\right) &= h\left(\frac{\mu_i}{E_i}\right) \\ &= \mathbf{U}_i^T\boldsymbol{\beta} \\ &= \beta_0 + \beta_1 U_{i1} + \beta_2 U_{i2} + \ldots + \beta_k U_{ik}. \end{aligned}$$

So we can write the above equation as follows

$$h\left(\frac{\mu_i}{E_i}\right) = \beta_0 + \beta_1 U_{i1} + \beta_2 U_{i2} + \ldots + \beta_k U_{ik}, \quad i = 1, \ldots, n, \tag{1}$$

where n is the number of observations and k is the number of predictors (covariates). Function h is called the link function. We can also rewrite (1) as

$$\frac{\mu_i}{E_i} = h^{-1}(\mathbf{U}_i^T\boldsymbol{\beta})$$

There are two link functions that can be used in Poisson regression. The first is the identity link and the second is the log link. The form of identity link function as follows :

$$\frac{\mu_i}{E_i} = \mathbf{U}_i^T\boldsymbol{\beta}.$$

IOP Conf. Series: Journal of Physics: Conf. Series **1108** (2018) 012065 doi:10.1088/1742-6596/1108/1/012065

The log link is

$$\log\left(\frac{\mu_i}{E_i}\right) = \mathbf{U}_i^T\boldsymbol{\beta}.$$

If we use the log link, the relationship between mean and independent is described below.

$$\mu_i = E_i e^{\mathbf{U}_i^T\boldsymbol{\beta}}.$$

The log link function is a more popular one because the link guarantees that the value of independent variable is non negative. In term of Poisson regression, the common link function is the log link function and the model is :

$$\mu_i = E_i \exp(\beta_0 + \beta_1 U_{i1} + \beta_2 U_{i2} + ... + \beta_k U_{ik}) = E_i e^{\mathbf{U}_i^T\boldsymbol{\beta}}. \quad (2)$$

Hence, the probability density function of Poisson regression can be written as follows

$$P_r(z_i;\boldsymbol{\beta}) = \frac{e^{-\mu_i}\mu_i^{z_i}}{z_i!}$$

where $\mu_i = E_i \exp(\mathbf{U}_i^T\boldsymbol{\beta})$ is the mean and $\boldsymbol{\beta}$ are unknown parameters. Meanwhile, the mean and the variance for Poisson regression model is defined as

$$\mu_i = E_i \exp(\mathbf{U}_i^T\boldsymbol{\beta}) = Var(z_i).$$

The likelihood function of Poisson regression is

$$L(\boldsymbol{\beta} \mid \mathbf{z}) = \prod_{i=1}^{n} \frac{e^{-\mu_i}\mu_i^{z_i}}{z_i!}. \quad (3)$$

2.2. Model of Neighbourhood for Poisson Regression

The Poisson regression with neighbourhood-based spatial effect can be specified as follows

$$\begin{aligned} z_i \mid s_i &\sim Poisson(\mu_i) \\ \mu_i &= E_i \exp(U_i^T\boldsymbol{\beta} + si) \\ \log(\mu_i) &= \log(E_i) + U_i^T\boldsymbol{\beta} + s_i, \quad i = 1..., n \end{aligned}$$

where the random effect s_i is assumed to have general conditional autoregressive structure which can be defined as

$$s_i|s_j \sim \mathcal{N}(\sum_{i\sim j} b_{ij}s_j, \sigma_i^2).$$

The component b_{ij} is the weight of each other observation on the mean of s_i and σ_i^2 is a variance for s_i. As an example, if state i has m neighbours and $b_{ij} = \frac{1}{m}$ for every state that is a neighbour and 0 otherwise, then the conditional mean of a state's observation is the mean of all neighbours' observations. Based on Brook's lemma, we can write the joint distribution as follows

$$\mathbf{s} \sim \mathcal{N}(\mathbf{0}, [\tau(\mathbf{D} - \alpha W)]^{-1}), \quad (4)$$

where $\mathbf{D} = Diag(m_i)$, m_i is the number of neighbors of region i and $B = \mathbf{D}^{-1}W$. Meanwhile W denotes the adjacency matrix ($w_{ii} = 0$ and $w'_{ii} = 1$ if $i \sim j$ or shared a common boundaries, 0 otherwise). To fulfill requirement of non singular covariance matrix, we define $|\alpha| < 1$.

2.3. Multivariate Conditional Autoregressive (MCAR) Model for multivariate Poisson Regression

In this section, we will extend the model in equation (4). Therefore, the model can be expressed as follows

$$\begin{aligned} z_{ip} \mid s_{ip} &\sim Poisson(\mu_{ip}) \\ \mu_{ip} &= E_i \exp(U_{ip}^T \boldsymbol{\beta}_p + s_{ip}) \\ \log(\mu_{ip}) &= \log(E_{ip}) + U_{ip}^T \boldsymbol{\beta}_p + s_{ip}, \quad i = 1..., n, \quad p = 1, ..., P \end{aligned} \tag{5}$$

where n is the number of area and P is the number of disease. Meanwhile $\mathbf{s}$ is assumed to have multivariate conditional model (MCAR) which is the extension model from CAR model in the previous subsection.(Carlin and Banerjee, 2003; Gelfand and Vounatsou, 2003.) The specified model of MCAR(α, Λ) can be written as following

$$\mathbf{s} \sim N(\mathbf{0}, [(\mathbf{D} - \alpha W) \otimes \Lambda]^{-1}). \tag{6}$$

We define $\mathbf{s} = (s_1', ..., s_p')$ and $s_k = (s_{1k}, ..., s_{nk})'$ with $k = 1, .., p$ and Λ is positive definite covariance which accomodate nonspatial correlation among variables at any region. In this case, let $\alpha \in (-1, 1)$, $\mathbf{D} = Diag(m_i)$ and $B = \mathbf{D}^{-1}W$, W denotes the adjacency matrix ($w_{ii} = 0$ and $w_{ii}' = 1$ if $i \sim j$, 0 otherwise).

To undestand easily, we give an illustration how the model work. Suppose two diseases in each county, $p = 2$ and define $\mathbf{s}_1' = (s_{11}, ..., s_{n1})$ and $\mathbf{s}_2' = (s_{12}, ..., s_{n2})$. MCAR$(\alpha, \Lambda)$ model has covariance matrix that can be written as

$$\Sigma = \begin{pmatrix} (\mathbf{D} - \alpha W)\Lambda_{11} & (\mathbf{D} - \alpha W)\Lambda_{12} \\ (\mathbf{D} - \alpha W)\Lambda_{21} & (\mathbf{D} - \alpha W)\Lambda_{22} \end{pmatrix}^{-1}. \tag{7}$$

Carlin and Banerjee (2003). Meanwhile, the covariance matrix of MCAR$(\alpha_1, \alpha_2, \Lambda)$ model has been specified as

$$\Sigma^{-1} = \begin{pmatrix} R_1'R_1\Lambda_{11} & R_1'R_2\Lambda_{12} \\ R_2'R_1\Lambda_{21} & R_2'R_2\Lambda_{22} \end{pmatrix} = \begin{pmatrix} R_1' & 0 \\ 0 & R_2' \end{pmatrix} (\Lambda \otimes I_{n \times n}) \begin{pmatrix} R_1 & 0 \\ 0 & R_2 \end{pmatrix} \tag{8}$$

where $R_j'R_j = \mathbf{D} - \alpha_j W$ and $|\alpha_j| < 1$ for $j = 1, 2$. We define R_j as the Cholesky decomposition of $\mathbf{D} - \alpha_k W$ and also R_j is an upper triangular matrix. In this case, we need to take $|\alpha_1| < 1$ and $|\alpha_2| < 1$ to produce positive definite covariance matrix. For that reasons, this approach provide some challenges. One of them is difficulty to define the covariance matrix. Hence, in the next subsection we will explore the generalized model using directly specify the covariance matrix.

2.4. Generalized Multivariate Conditional Autoregressive (GMCAR) Model For Multivariate Poisson regression

The multivariate model in this subsection has the same typical with equation (5). The difference is how to define the covariance matrix. To have a good understanding of model, we offer an illustration in two different variables dependent, say $p = 2$. Suppose the conditional distribution $s_1|s_2 \sim N(\mathbf{A}s_1, [(\mathbf{D} - \alpha_1 W)\tau_1]^{-1})$ and $s_2 \sim N(\mathbf{0}, [(\mathbf{D} - \alpha_2 W)\tau_2]^{-1})$. We define α_1 and α_2 as a the smoothing parameters for the two types of dependent variables. Since we have $E(s_1|s_2) = \mathbf{A}s_1$, where $\mathbf{A} = \eta_0 I + \eta_1 W$ and $Var(s_1|s_2) = [(\mathbf{D} - \alpha_1 W)\tau_1]^{-1}$. Then, we assume that its element of $\mathbf{A}$ matrix are

$$a_{ij} = \begin{cases} \eta_0 & \text{if } \ j = i \\ \eta_1 & \text{if } \ j \in n_i, \ \text{ region } i \text{ and } j \text{ shared a common boundary} \\ 0 & \text{otherwise} \end{cases} \tag{9}$$

where η_0 and η_1 are the bridging parameters associating s_{i1} with s_{i2} and $s_{j2}, j \neq i$ respectively. We denote the model as GMCAR$(\alpha_1, \alpha_2, \eta_0, \eta_1, \tau_1, \tau_2)$. Hence, if model without involving the cross correlation ($\eta_1 = 0$) and $\alpha_1 = \alpha_2 = \alpha$ then GMCAR$(\alpha, \eta_0, \eta_1 = 0, \tau_1, \tau_2)$ is exactly the same as MCAR(α, Λ).

3. Result and Discussion

The data consist the numbers of infected cases due to dengue fever and malaria each city at East Java Indonesia in 2010. The secondary data is taken from statistics center East Java. East Java province has 29 districts and 9 cities, thus the total of data is 38 observations. The summary of the data is the minimum cases for dengue fever and malaria are 18 and 0 respectively. Meanwhile the maximum cases of dengue fever is 3379 and for malaria is 428 cases.

In terms of selection a proper order two diseases in GMCAR, we need to obtain several steps. The first is calculate estimated of $\hat{s_{i1}} = log(z_{i1}|E_{i1})$ and $\hat{s_{i2}} = log(z_{i2}|E_{i2})$. The second is decide the linearity of the conditional GMCAR (say, *dengue*|*malaria*),

$$E(s_1|s_2) = As_2 = A(\eta_0, \eta_1)s_2 = (\eta_0 I + \eta_1 W)s_2. \tag{10}$$

The third is obtaining $\hat{\eta_0}$ and $\hat{\eta_1}$ by minimizing $(\hat{s_1} - A(\eta_0, \eta_1)\hat{s_2})'(\hat{s_1} - A(\eta_0, \eta_1)\hat{s_2})$.Finally, we plot $A(\hat{\eta_0}, \hat{\eta_1})\hat{s_2}$ versus $\hat{s_1}$. Thus, we repeat this process for the reverse order (here, *malaria*|*dengue*). Meanwhile, we can explore plot log of SMR and its predicted to assist in selecting a proper order

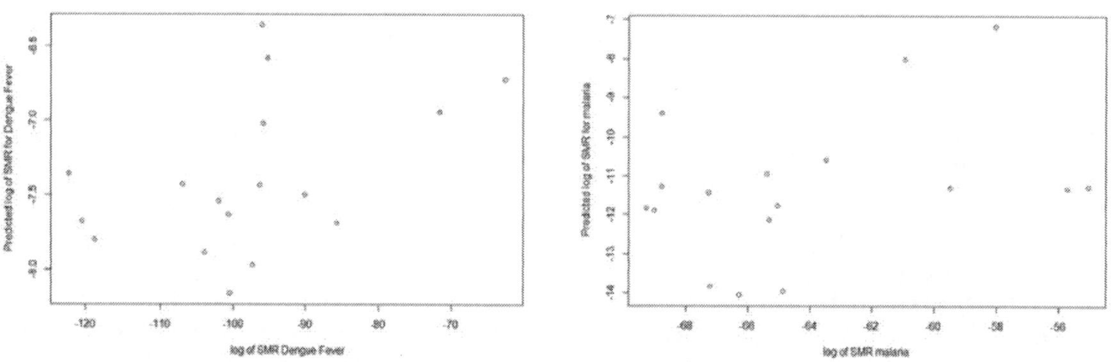

Figure 1. From leftside to righside plot *dengue*|*malaria* with sample correlation 0.4454 and *malaria*|*dengue* with sample correlation 0.3643 respectively.

by involving sample correlation of *dengue*|*malaria* and *malaria*|*dengue* in Figure 1. Hence, in this case, the data provides more evidence to choose linearity *dengue*|*malaria* due to providing the highest sample correlation.

In order to understand the performance of several models, we use hierarchical modeling extension of AIC called Deviance Information Criterion (DIC). DIC is based on the posterior distribution of deviance statistic which has a formula as follows

$$DIC = \hat{D} + PD$$

where $\hat{D}$ is denoted as the posterior expected deviance. Meanwhile, PD is defined as the effective number of parameters.

In Table 1, it shows the value of DIC form some different models. The first model is GMCAR$(\alpha_1, \alpha_2, \eta_0, \eta_1, \tau_1, \tau_2)$. The model provides a full model with conditioning order

($dengue|malaria$). The second model is GMCAR ($\eta_1 = 0$) or we can say as MCAR(α, Λ). The third model is the reverse order from full model where the conditional order is $malaria|dengue$. From the table, it shows that full model (GMCAR full) provides the smallest of DIC values

Model	$\bar{D}$	PD	DIC
GMCAR (full)	247.70	5.392	253.10
GMCAR ($\eta_1 = 0$)	247.30	11.010	258.30
GMCAR (full, reverse order)	242.50	11.730	254.20

Table 1. Model Comparison using DIC Statistics, Dengue Fever and Malaria analysis in East Java Indonesia

among all of the comparison models. It means that model with conditional $dengue|malaria$ offers the best model. The reverse order full model is also providing a relatively good performance as well comparing the reduced model (GMCAR ($\eta_1 = 0$)) as we expect.

4. Conclusion

In this paper, involving a spatial effect via GMCAR for modelling multivariate diseases is more appropriate. It is because the method is involving the bridging factor between multivariate responses.

References

[1] Fareed N, Ghaffar A and Malik TS 2016 Spatio-Temporal Extension and Spatial Analyses of Dengue from Rawalpindi, Islamabad and Swat during 2010 - 2014 *Climate* vol. **4**(2) pp.23
[2] Mustapha E O, Emeribe A U and Nasir I A 2017 Survey of malaria and anti-dengue virus IgG among febrile HIV-infected patients attending a tertiary hospital in Abuja, Nigeria *HIV/AIDS - Research and Palliative* vol. **9** pp. 145-151
[3] Lin H, Tian L, Yang W, Sun J and Liu Q 2009 Time series analysis of dengue fever and weather in Guangzhou, China *BMC Public Health* vol. **9** pp. 395
[4] Abbasi A, Abbas K, Arooj S, Habib N, Aziz W and Ashaq A 2016 Dengue Fever: A Statistical Analysis Regarding Awareness about Dengue among University Students in Azad Kashmir *Journal of Healthcare Communications* ISSN 2472-1654 vol. **2** no. 1: 1
[5] Morin C W, Comrie A C, and Ernst K 2013 Climate and Dengue Transmission: Evidence and Implications *Environmental Health Perspect*
[6] Mutheneni S R, Morse AP, Caminade C, Upadhyayula S M 2017 Dengue burden in India: recent trends and importance of climatic parameters *Emerging Microbes and Infections* vol. **6**, pp. e70
[7] Das S and Abha T 2016 Predictive analysis of dengue and malaria *International Conference on Computing, Communication and Automation (ICCCA) IEEE*
[8] Rao MR, Padhy RN and Das MK 2016 Prevalence of dengue viral and malaria parasitic co-infections in an epidemic district, Angul of Odisha, India: An ecoepidemiological and cross-sectional study for the prospective aspects of public health, *Journal of Infection and Public Health* vol. **9** issue 4 pp. 421-428
[9] Besag J and Kooperberg C 1995 On Conditional and Intrinsic Autoregressions *Biometrika* vol. **82** pp 733-746
[10] Knorr L and Besag J 1998 Modelling risk from a disease in time and space *Statistics in Medicine* vol. **17** pp 2045-2060
[11] Sun T, Kim H and He Z Q 2000 Spatio-temporal interaction with disease mapping, *Statistics in Medicine* vol. **19** pp 2015-2035
[12] McNab Y C and Dean C B 2001 Autoregressive spatial smoothing and temporal spline smoothing for mapping rate *Biometrics* vol. **57** pp 73-85
[13] Bernadinelli L, Clayton D and Montomoli C 1995 Bayesian estimates of disease maps: how important are priors?*Statistics in Medicine* vol. **14** pp. 2411-2431
[14] Wall M M 2004 A close look at the spatial structure implied by the CAR and SAR models *Journal of Statistical Planning and Inference* vol. **121** pp. 311 324
[15] Renato and Krainski E 2009 Neighborhood Dependence in Bayesian Spatial Models *Biometrical Journal* vol. **51** pp. 851869
[16] Jin X, Carlin B P and Banerjee S 2005 Generalized hierarchical multivariate CAR models for areal data *Biometrics* vol. **61**(4) pp. 950-61

Predicting Final Result of Football Match Using Poisson Regression Model

H R Azhari[1], Y Widyaningsih[2] and D Lestari[3]

[1,2,3] Departement of Mathematics, Universitas Indonesia, Kampus Baru UI, Depok, 16424, West Java, Indonesia.
(Email: lubishafis666@gmail.com, widyaningsihyekti@gmail.com, dian.lestari@sci.ui.ac.id)

Abstract. In any sport competition, there is a strong interest in knowing which team shall be the champion at the end of the championship and one of the sports is football. Football match predictions are of great interest to fans and sports press. In the last few years, it has been the focus of several studies. In this paper, the researchers propose Poisson regression model to predict the final result of football matches. The researchers predicted the average goals scored by each team by assuming that the number of goals scored by a team in a match followed a univariate Poisson distribution. Poisson regression model was formulated from four covariates: the goal average in a match, the home-team advantage, the team's offensive power, and the opponent team's defensive power. The methodology was applied to the 2017-2018 English Premier League. The results obtained using this model had a fairly good accuracy.

1. Introduction

Football is one of the most popular sports in the world which has its own charm compared to other sports. Football fans enchant in the tense and exciting moments of goals, especially for the last-minute goals, the final result, the match intensity, and the final rank. For example, the dark horse team Leicester City has won the 2014-2015 English Premier League. They surprisingly beat all the other strong teams. Real Madrid have also won three champions league title in a row, where no team has been able to do it before. This phenomenon reflects one of the most charming part of football that is complexity which makes the game result hard to be predicted.

Several papers are found in literature considering football score prediction applied to championship leagues such as the English Premier League ([1], [2], [3]), the Norwegian Elite Division [4] , the Brazilian Championship [5]. Lee [1] considered a Poisson regression to predict the number of goals from football team, where the average reects the strength of the team, the quality of the opposition and the home advantage (if it is the home team). The independence between the goals scored by the two teams was assumed and the methodology was applied to the 1995-1996 English Premier League. Brillinger [5] modeled the probabilities of win, tie and loss through an ordinal-value model and applied the model to the Brazilian Series A championship. Karlis and Ntzoufras [3] applied the Skellam's distribution to model the difference of goals between home and away teams. The authors illustrated the model using the 2006–2007 English Premier League. Koopman and Lit [6] developed a statistical model to predict the games of the 2010–2011 and 2011–2012 English Premier Leagues, assuming a bivariate Poisson distribution with coefficients that stochastically changed the intensity over time. Koopman and Lit [6] developed a statistical model to predict the games of the 2010–2011 and 2011–2012 English

Premier Leagues, assuming a bivariate Poisson distribution with coefficients that stochastically changed the intensity over time. An issue dealing with the papers cited above is that none of them considers the home team factor to calculate the probabilities of interest. Dixon and Coles [7] presented the result that 46% of the matches was won by the home team, 27% were draws and in 27% the home team lost.

In this paper, the researchers modeled the number of goal scored by each team in a match by a Poisson distribution, whose average reflected the strength of the attack and defense of the team and effect of being playing at home. The model was applied to the 2017-2018 English Premier League. The Definetti measure (DeFinetti [8]) was used to quantify the model predictive quality.

2. Model Construction

In this paper, the researchers assumed that the number of goals in each match followed the Poisson distribution so they can start constructing the regression models based on the assumption. The Poisson regression formula for this paper can be represented as:

$$Y \sim Poisson(\lambda) \qquad (1)$$

$$Y = X\beta \qquad (2)$$

From the formula, it can be seen that Y is a vector of dependent variable that consists of the home goals and away goals in games, X is a matrix of explanatory variables that records the home and away teams corresponding to the games, β is a vector containing the parameters, Offence and Deffence of the model. There were 20 teams participating in the 2017-2018 English Premier League and each of 20 teams had its offence parameter and defense parameter. Meanwhile, each of the times appeared as either a home team or an away team. Thus, it can be said that $Y = (y_{a,b}^1, y_{b,a}^1, y_{a,b}^2, y_{b,a}^2, \dots, y_{a,b}^n, y_{b,a}^n)^T$ and $\beta = (O_{afcbou}, O_{ars}, \dots, O_{westham}, D_{afcbou}, D_{ars}, \dots, D_{west}, \delta)^T$, where $y_{a,b}^i$ is the number of goals scored by team a versus team b in game-i, O_j and D_j stands for the offence and defense parameter of team j and δ explains home advantage. The status of a team in a variable incidence takes value 1 if it participates in the i-match and takes value 0 if it does not participate. For example, Arsenal versus Westham in Emirates Stadium with score 2-1. The vector $Y = [2\ 1]^T$ and first row of matrix X is $[0\ 1\ \dots\ 0\ 0\ 0\ \dots\ 1\ 1]$ because arsenal's score with home advantage and the second row is $[0\ 0\ \dots\ 1\ 0\ 1\ \dots\ 0\ 0]$. The reserachers considered $\beta = (O_{afcbou}, O_{ars}, \dots, O_{westham}, D_{afcbou}, D_{ars}, \dots, D_{west}, \delta)^T = (\beta_1, \beta_2, \dots, \beta_p)$ and this time p equally to 41. Then the log-likelihood function for β is given as follows:

$$l(\beta) = \sum_{i=1}^{n} \left\{ y_i \sum_{j=1}^{p} \beta_j x_{ji} - exp\left(\sum_{j=1}^{p} \beta_j x_{ji} \right) - log(y_i!) \right\} \qquad (3)$$

From the function, it can be seen that y_i is the number of goals scored in the-i observation, while x_{ji} is the component in matrix X in the j-column and the i-observation.

Predictions

Consider that a game between teams a and b will occurs at home stadium of team a. Denote the probability of win, draw, and defeat (loss) of team a by P_w, P_d and P_l respectively. These probabilities are given as follows:

$$P_w = P(Y_a > Y_b) = \sum_{i=1}^{\propto} \sum_{j=0}^{i-1} P(X = \lambda_a) P(Y = \lambda_b) \qquad (4)$$

$$P_d = P(Y_a = Y_b) = \sum_{i=0}^{\propto} P(X = \lambda_a) P(Y = \lambda_b) \qquad (5)$$

$$P_l = P(Y_a < Y_b) = \sum_{j=1}^{\propto} \sum_{i=0}^{j-1} P(X = \lambda_a) P(Y = \lambda_b) \qquad (6)$$

IOP Conf. Series: Journal of Physics: Conf. Series **1108** (2018) 012066 doi:10.1088/1742-6596/1108/1/012066

Similiarly to Bastos and da Rosa [9], and Suzuki et al. [10], the researchers in this study calculated the de Finetti distance in order to measure the beneficence of a prediction. This distance was given by the Euclidean distance between the point corresponding to the real outcome and the one corresponding to the prediction. For this case, just assumed that the set of all possible forecasts is given by the simplex set $S = \{(P_w, P_d, P_l)\epsilon\mathbb{R}^3: P_w + P_d + P_l = 1.\ P_w \geq 0, P_d \geq 0, P_l \geq 0\}$. *The possible real outcome, including the win, draw, and loss are represented by the points (1,0,0),(0,1,0) and (0,0,1), respectively.*

The de Finetti measure (df) which is defined as:

$$df = (P_w - b_1)^2 + (P_d - b_2)^2 + (P_l - b_3)^2 \quad . \tag{7}$$

where $(b_1, b_2, b_3) \in \{(1,0,0), (0,1,0), (0,0,1)\}$. *For example, if the prediction for the game between teams a and b is (0.2,0.65,0.15) and the real outcome is (0,1,0), the de Finetti distances is* $df = (0.2-0)^2 + (0.65-1)^2 + (0.15-0)^2 = 0.185$.

For the equiprobable case, $P_w = P_d = P_l = 1/3$, *with the win of the home team (1,0,0), the de Finetti measure is given by* $df = (1/3-1)^2 + (1/3-0)^2 + (1/3-0)^2 = 2/3$. *This value is accepted as a threshold value in order to classify the prediction as acceptable or not (see the example in Suzuki et al. (2010). If* $df < 2/3$, *the predictions are considered acceptable; otherwise, If* $df > 2/3$, *the predictions are considered poor.*

3. Application

3.1 Single Match Prediction

In this section, the researchers presented the predictions of the 32th round of the English Premier League (EPL). Table 1 showed that the probabilities of win, draw, and loss for the games, the score, and de Finetti measure if the method correctly indicated the winner team as the team with higher probability of a win.

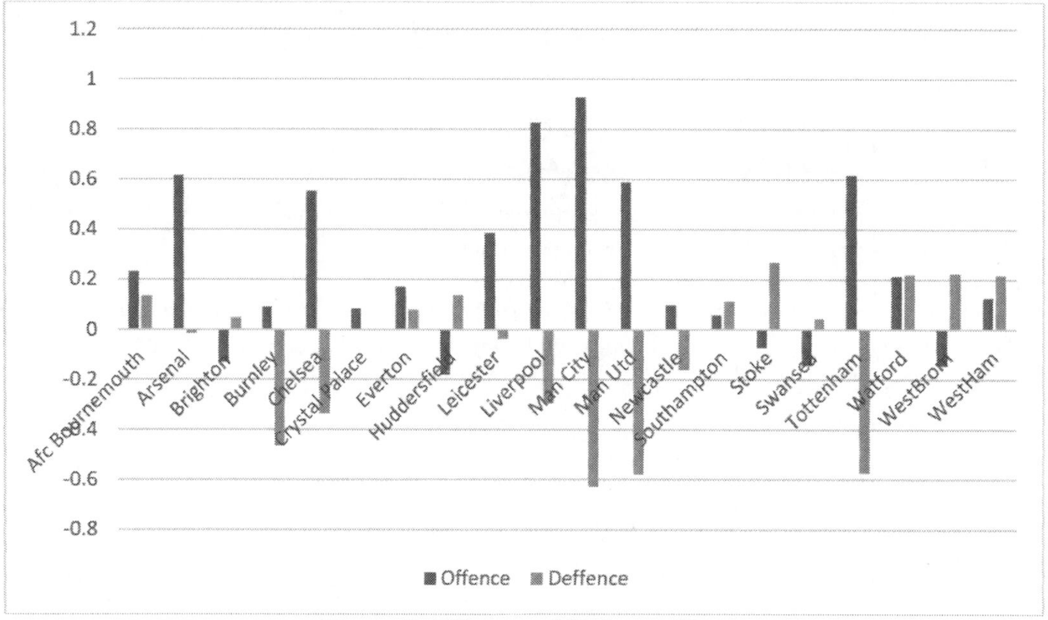

Figure 1. Offence and Defense effect

The proportion of correct prediction was 80%. The teams with an estimated probability of win higher than 0.5, were the actual winning team. Figure 1 displays the graphic of the attack and defense effect. In the graphic of Figure 1, it can be seen each attack and defense effect of each team. In offence effect, the more positive the value, the stronger the effect of the attack is. Meanwhile, in the defensive effect, the value more negative then it indicates the persistence.

Table 1. Probabilities of win, draw, and loss for each match in 32th round.

Home	Away	Probability			Score	de Finetti	Correct
		Win	Draw	Loss			
Crystal Palace	Liverpool	0,166	0,189	0,644	1-2	0,191	Yes
Brighton	Leicester City	0,401	0,302	0,297	0-2	0,746	No
Manchester United	Swansea City	0,814	0,133	0,053	2-0	0,055	Yes
Newcastle United	Huddersfield	0,607	0,238	0,154	1-0	0,235	Yes
Watford	Bournemouth	0,453	0,222	0,325	2-2	0,916	No
West Brom	Burnley	0,225	0,278	0,497	1-2	0,381	Yes
West Ham	Southampton	0,463	0,238	0,299	3-0	0,434	Yes
Everton	Manchester City	0,083	0,046	0,87	1-3	0,026	Yes
Arsenal	Stoke City	0,817	0,112	0,071	3-0	0,051	Yes
Chelsea	Tottenham	0,27	0,227	0,502	1-3	0,372	Yes

3.2 Predictions for the whole Tournament

Manchester city has the greatest attack and defense effect, so it's only natural that this team won the EPL. The smallest defense effect decreased the expected number of goals of the opposing team. In the opposite, Stoke has the worst defense effect. The researchers estimated the number of points, the number of wins, draw and loss, the number of goals for and against each team. Table 2 presents these values and it is organized by the real number of points for each team. Note that the six teams with the highest-estimated number of points are really the six best teams of the championship.

Table 2. Predictions and real values.

Team	Points		Won		Draw		Lost	
	Est.	real	Est.	real	Est.	real	Est.	real
Manchester City	105	100	34	32	3	4	1	2
Manchester United	86	81	27	25	5	6	6	7
Tottenham Hotspurs	82	77	25	23	7	8	6	7
Liverpool	81	75	24	21	9	12	5	5
Chelsea	77	70	24	21	5	7	9	10
Arsenal	69	63	21	19	6	6	11	13
Burnley	61	54	17	14	10	12	11	12
Everton	49	49	14	13	7	10	17	15
Leicester City	60	47	17	12	9	11	12	15
Newcastle United	38	44	10	12	8	8	20	18
Crystal Palace	39	44	10	11	9	11	19	16
AFC Bournemouth	41	44	11	11	8	11	19	16
West Ham United	41	42	11	10	9	12	18	16
Watford	48	41	14	11	6	8	18	19
Brighton and Hove Albion	37	40	9	9	10	13	19	16
Huddersfield Town	33	37	9	9	6	10	23	19
Southampton	33	36	7	7	12	15	19	16
Swansea City	36	33	10	8	6	9	22	21

Stoke City	26	33	6	7	8	12	24	19
West Bromwich Albion	23	31	4	6	11	13	23	19

4. Final Remarks

Through this paper, the researchers proposed a simple method with good predictive quality. It has easy implementation and low computational effort for predicting match outcomes. The researchers developed a model to estimate the probabilities of win, tie and defeat in football games. In order to calculate these probabilities, they proposed Poisson regression model in which the average of goals scored reflected the strength of attack of the team, the strength of defense of the opposing team and the home team effect. The methodology was implemented in the software R. In this paper, the researchers estimated 100 games (29th-38th round), and the accuracy for our modelling was 61%. Although the model was applied to the 2017-2018 English Premier League. Principally, it was exible and it could be easily adapted to other different tournaments. However, it would be interesting if it can count every team's chances of winning the league or the chances of being degraded. It can be seen as a direct generalization of the researchers' proposed model and it may be investigated further by adding a significant parameter which can update the model so that it gets better in terms of its accuracy.

5. References

[1] Lee A J 1997 Modeling scores in the Premier League: is Manchester United really the best? *Chance* **10** pp 15–19
[2] Everson P, Goldsmith-Pinkham P 2008 Composite Poisson Models for Goal Scoring. *Journal of Quantitative Analysis in Sports* **4** (2)
[3] Karlis D and Ntzoufras I 2009 Bayesian modelling of football outcomes: using the Skellam's distribution for the goal difference *IMA Journal of Management Mathematics* **20** pp 133–145
[4] Brillinger D R 2006 Modelling Some Norwegian Soccer Data *Advance in Statistical Modelling and Inference (Ed. V. J. Nair.) World Scientic*pp 3-20
[5] Brillinger D R 2008 Modelling Game Outcomes of the Brazilian 2006 Series a Championship as Ordinal-valued *Brazilian Journal of Probability Statistics* **22** pp 89–104
[6] Koopman SJ and Lit R 2015 A dynamic bivariate Poisson model for analysing and forecasting match results in the English Premier League *Journal of the Royal Statistical Society: Series A (Statistics in Society)* **178** pp 167–186
[7] Dixon MJ and Coles SG 1997 Modelling association football scores and inefficiencies in the football betting market *Journal of the Royal Statistical Society: Series C (Applied Statistics)* **46** pp 265-280
[8] DeFinetti B 1972 *Probability, Induction and Statictics* John Wiley, London.
[9] Bastos L S and da Rosa J M C 2013 Predicting Probabilities for the 2010 FIFA World Cup Games Using a Poisson-Gamma Model *Journal of Applied Statistics* **40** pp 1533–44
[10] Suzuki AK, Salasar LEB, Leite JG, and Louzada-Neto F 2010 A Bayesian approach for predicting match outcomes: the 2006 (Association) Football World Cup *Journal of the Operational Research Society* **61** pp 1530–39

Acknowledgements

This research was funded by Directorate of Research and Community Service of Universitas Indonesia (DRPM UI) as a grant of PITTA (Publikasi Terindeks untuk Tugas Akhir) 2018, Number: 2336/UN2.R3.1/HKP.05.00/2018.

IOP Conf. Series: Journal of Physics: Conf. Series **1108** (2018) 012067 doi:10.1088/1742-6596/1108/1/012067

Application Support Vector Machine on Face Recognition for Gender Classification

Z Rustam[1] and A A Ruvita[1]

[1]Department of Mathematics, Universitas Indonesia, Indonesia

Abstract. Face recognition system is capable of generating a variety of information about a person's identity quickly and accurately. One of them, face recognition is able to provide information about the gender (male or female) of each person. Gender classification has become an area of extensive research due to its increasing application in existing human-computer interaction (HCI) systems, advertising, biometrics, surveillance systems, content-based indexing and searching. This paper presents face recognition for gender classification using Support Vector Machine (SVM). Support Vector Machine are a system for efficiently training the linear learning machines which can be used for as powerful classification methodology. In this research, we have obtained face recognition accuracy rates for gender classification using Support Vector Machine (SVM) with different kernels. When training data were used 40 to 90 percent, SVM method with RBF kernel and also Polynomial kernel has achieved the same maximum accuracy that is 100 percent.

1. Introduction

The human face becomes an important and valuable source of information in the process of face recognition [1]. The proportion and expression of human faces are one of the most influential things to identify origins, expression, age, gender and some social information [2]. Face recognition for gender classification has a major role in a variety of potential applications. In the field of advertising, the gender classification is one of the strategies used to display ads that are relevant to the person viewing the billboard. Then in the biometric field, gender classification is used for the process of identification and access control. Furthermore, the gender classification is useful for limiting the area to only one subject, for example, when in a dormitory or train. In addition, gender classifications are also useful in content-based indexing and search, with the widespread use of electronic devices such as cameras, a large number of digital media are produced. Indexing or records of information about the number of people in the image or video, age and gender will become more noticeable with automated systems that use computer vision. In addition, for content-based searches such as finding a person's photo, classifying gender as a pre-processing step will simplify the search process and can reduce the number of searches required in the database [3]. It is therefore important to do research on gender classification in order to support its role in various fields and to produce an improved level of accuracy.

In the early stages will be collected data in the form of facial images consisting of male and female. Furthermore, the face image will be converted into a matrix form to facilitate the classification process. The classification method used is the Support Vector Machine (SVM). In the final stage will result in a comparison of the level of accuracy or success rate in the process of gender classification using Support Vector Machine (SVM) with different kernels. In this paper consists of introduction, data, related works, methodology, experiment and results, and conclusion.

1.1 Data

In this research the facial image data used comes from one benchmark face recognition is computer science research projects https://cswww.essex.ac.uk/mv/allfaces/index.html [4]. In the database, the available face images are images with the .jpg format and the face images have RGB color. Facial image data used in this study consisted of 10 male and 10 female, with each person having 10 facial images with slight expression differences. So the total image of facial images used as much as 200 images faces with a balanced composition of 100 male face images and 100 images of female faces with the number of features is 15,625.

In this research, training data and testing data were divided randomly. Training data starts from 10 percent to 90 percent. When the training data used 10 percent means the test data used is 90 percent, when the training data used 20 percent means the test data used is 80 percent, when the training data used 30 percent means the test data used is 70 percent, when the training data used 40 percent means the test data used is 60 percent, when the training data used 50 percent means the test data used is 50 percent, when the training data used 60 percent means the test data used is 40 percent, when the training data used 70 percent means the test data used is 30 percent, when the training data used 80 percent means the test data used is 20 percent, so until the training data used is 90 percent means the test data used is 10 percent.

1.2 Related Works

1.2.1 Supervised Learning

The learning method that maps input to output based on the example is called supervised learning. The input and output pairs are referred to as training data. In the process, data is labelled to facilitate customized goals or outputs. In general, supervised learning aims to predict the value of response variables or labels based on the variables or features inputted [5]. Supervised Learning is usually used in classification, approximation, control, modelling, signal processing and identification [6]. Grouping of men and women on the basis of facial recognition is a classification problem, so supervised learning is an appropriate type of learning for use in this research.

1.2.2 Support Vector Machine (SVM)

Support Vector Machines (SVM) is a learning approach originally developed by Vapnik and colleagues. Support Vector Machines (SVM) is a data-learning method based on statistical learning theory. This method belongs to the category of supervised learning used to predict both classification and regression. The SVM mechanism is to find an optimal classifier function that can separate two sets of data from two different classes. Basically, there are many separating fields that can separate data from two different classes but there is only one separating field that can optimally separate the data and minimize misclassification. The optimal classifier field is called hyperplane. This hyperplane can be found by maximizing the margin. Margin is the perpendicular distance between the hyperplane and the closest data of each class. Data from each class close to the hyperplane is called a support vector [7]. It was once discussed on [10-14]

Suppose a given set of data pairs $\{\boldsymbol{x}_i, y_i\}_{i=1}^{N}$ with x is the input vector in the form of a row vector and $x \in R^D$ so that $\boldsymbol{x}_i$ is the input vector of the sample, and dimension D (number of features), then $y \in \{-1, +1\}$ is a class label for each data sample. The point x on the hyperplane satisfies the following equation [8]:

$$f(x) = \langle \mathbf{w}, \boldsymbol{x} \rangle + b = 0 \qquad (1)$$

with w representing a vector perpendicular to the hyperplane or referred to as a normal vector, and b is a constant which is a bias parameter, $\frac{|b|}{\|w\|}$ is the perpendicular distance between the points on the hyperplane (H) to the origin, and $\|\boldsymbol{w}\|$ is the Euclidean norm of w [9]. The optimal hyperplane (H) will divide the data into positive and negative classes, assuming that all training data satisfies the following equations [8]:

$$\langle \boldsymbol{w}, \boldsymbol{x}_i \rangle + b \geq 1 \text{ for a positive sample} \tag{2}$$

$$\langle \boldsymbol{w}, \boldsymbol{x}_i \rangle + b \leq 1 \text{ for a negative sample} \tag{3}$$

The form of inequality can be rewritten as follows:

$$y_i(\langle \boldsymbol{w}, \boldsymbol{x}_i \rangle + b) - 1 \geq 0 \tag{4}$$

To get the optimum hyperplane can be done by completing the following functions:

$$min\left[\frac{1}{2}\|\boldsymbol{w}\|^2\right] \tag{5}$$

with constraints

$$y_i(\langle \boldsymbol{w}, \boldsymbol{x}_i \rangle + b) \geq 1, y_i \in \{-1,1\}, i = 1,2, \dots., N \tag{6}$$

The following are defined lagrange functions:

$$L(\boldsymbol{w}, b, \alpha) = \frac{1}{2}\|\boldsymbol{w}\|^2 - \sum_{i=1}^{N} \alpha_i [y_i(\langle \boldsymbol{w}, \boldsymbol{x}_i \rangle + b) - 1] \tag{7}$$

Here are the conditions of Karush-Kuhn-Tucker (KKT) that must be fulfilled:

$$(y_i(\langle \boldsymbol{w}, \boldsymbol{x}_i \rangle + b) - 1) \geq 0 \tag{8}$$

$$\alpha_i(y_i(\langle \boldsymbol{w}, \boldsymbol{x}_i \rangle + b) - 1) = 0 \tag{9}$$

$$\alpha_i \geq 0, i = 1,2, \dots, N \tag{10}$$

Thus, obtained the dual Lagrange form of the primal optimization problem as follows:

$$max\left(\sum_{i=1}^{N} \alpha_i - \frac{1}{2}\sum_{i=1}^{N}\sum_{j=1}^{N} \alpha_i \alpha_j y_i y_j \langle \boldsymbol{x}_i, \boldsymbol{x}_j \rangle\right) \tag{11}$$

with constraints

$$\alpha_i \geq 0 \tag{12}$$

$$\sum_{i=1}^{N} \alpha_i y_i = 0 \tag{13}$$

2. Method

There are several face recognition steps for male and female classification:

- Face Image Collection: collection of facial images consisting of male and female
- Digital Image Pre-processing: the data in the form of accumulated facial images is converted into matrix form in order to process
- Training and Classification: male or female

3. Results and Discussion

In this study used an image of a facial recognition dataset. In the dataset there are some face image folder that is faces94, faces95, faces96, and grimace. In this research using facial recognition dataset from folder faces94. The data used amounted to 200 face images. The collection of face images consists of 100 male faces and 100 female face images with JPG image format and 180 x 200 dimensions. This data collection technique is known as image acquisition. Then, do image pre-processing, the image of the face dimension 180 x 200 or portrait shaped converted into square shape with the dimensions of the image 125 x 125. After the image of the face is square, the image of the face with JPG format is converted into a form of matrix by using the algorithm built in *Matlab* program with the aim to facilitate the next process. The next stage is training, which is part of the classification process. From this training process we can get a predictive parameter and model. So, we can classify the data testing appropriately and get the smallest possible errors.

Classification is a grouping problem based on training. In this research, the classification method used is Support Vector Machine (SVM). To classify the data testing, we must get the optimum hyperplane that can separate the two classes. Suppose a sample will be classified on a particular class, the sample is in pairs (x_i, y_i), where x_i is the input and y_i is the output. Thus, the classification of two classes is a learning problem for finding a function that can map every $\boldsymbol{x_i}$ to y_i, with y_i consisting of only two values or two classes. Hyperplane is defined as follows $\boldsymbol{w} \cdot \boldsymbol{x} + b = 0$ with w is the weight vector, b is the bias parameter, and x is the point located on the hyperplane. After the classification is done, then the next stage is to evaluate. This evaluation is useful to know how well the performance of a classification program we have built. Evaluation can be seen from how much data testing is correctly classified and how much data testing is misclassified. The evaluation can be done with a confusion matrix. Then we can calculate the level of accuracy with the following formula:

$$accuracy = \frac{number\ of\ correct\ predictions}{number\ of\ predictions} x\ 100\% \tag{14}$$

The test results in this research is a level of accuracy obtained from the classification process of Support Vector Machine (SVM) by using RBF kernel and Polynomial kernel. Test results will be presented in tabular form.

The following is a trial result using all features that is 15625 features, classification process using Support Vector Machine (SVM) method with RBF kernel ($\sigma = 0.05$) and classification process using Support Vector Machine (SVM) method with Polynomial kernel ($d = 5$). Training data and testing data were divided randomly, training data diverse from 10 to 90 percent.

Table 1. Accuracy Result of Gender Classification using SVM with kernel: Polynomial, parameter: constantly 5 and using SVM with kernel: RBF, parameter: constantly 0.05, training data: diverse from 10 to 90

Training Data (%)	**Polynomial (d = 5)**	**RBF (σ = 0.05)**
	Accuracy (%)	**Accuracy (%)**
10	88.88889	85.55556
20	95.625	87.5
30	95.714	85.71429
40	100	100
50	100	100
60	100	100
70	100	100
80	100	100
90	100	100

Based on the results presented in Table 1, it can be seen when using SVM with kernel Polynomial that the best accuracy result is 100 percent obtained when the training data used is 40 to 90 percent. When training data is used 10 to 30 percent, the accuracy is 88 to 95 percent. Next, based on the results presented in Table 1, it can be seen when using SVM with kernel RBF that the best accuracy result is 100 percent obtained when the training data used is 40 to 90 percent. When training data is used 10 to 30 percent, the accuracy is 85 to 87 percent.

4. Conclusion

In this research, we have classified gender using Support Vector Machine (SVM) method with RBF kernel ($\sigma = 0.05$) and Polynomial kernel ($d = 5$). When training data were used of 10 to 30 percent, the SVM method with the polynomial kernel ($d = 5$) has better accuracy than the SVM method with the RBF kernel ($\sigma = 0.05$). Furthermore, for training data 40 to 90 percent, SVM method with RBF kernel ($\sigma = 0.05$) and also polynomial kernel ($d = 5$) has achieved the same accuracy that is 100 percent.

References

[1] O'Toole A J, Deffenbacher K A, Valentin D, Mckee K, Huff D and Abdi H 1998 *Memory & Cogn.* **26** 146
[2] Gupta R and Khunteta A 2012 *Int. J. Comp. Commun. Networking.* **1** 75
[3] Khan A, Ahmad M, Nazir M and Riaz N 2014 *Midle-East J. of Sci. Res.* **20** 1
[4] Face Recognition Database http://cswww.essex.ac.uk/mv/allfaces/index.html
[5] Cristianini N and Shawe-Taylor J 2014 *An introduction to support vector machines: and other kernel-based learning methods* (Cambridge: Cambridge University Press)
[6] Du K and Swamy M N 2016 *Neural Networks and Statistical Learning* (London: Springer)
[7] Bishop C M 2006 *Pattern Recognition and Machine Learning* (New York: Springer)
[8] Schölkopf B and Smola A J 2009 *Learning with Kernels: Support Vector Machines, Regularization, Optimization and Beyond* (Cambridge MA: MIT Press)
[9] Burges C J C 1998 *Data Min. and Knowl. Discov.* **2** 121
[10] Rustam Z and Zahras D 2018 *IOP Conf. Proceed.* **1028** 012227
[11] Rustam Z and Ariantari N PA A 2018 *IOP Conf. Proceed.* **1028** 012005
[12] Rustam Z and Yaurita F 2018 2018 *IOP Conf. Proceed.* **1028** 012005.
[13] Rustam Z, Vibranti D F, and Widya D 2017 Predicting the Direction of Indonesian Stock Price Movement using Support Vector Machines and Fuzzy Kernel C-Means The 3rd ICPMS AIP Proceeding Conference.
[14] Rustam Z and Yaurita F 2017 Application of Support Vector Machines for Reject Inference in Credit Scoring The 3rd ICPMS AIP Proceeding Conference.

Face recognition using fuzzy kernel learning vector quantization

Z Rustam[1] and Rika[1]

[1] Department of Mathematics, University of Indonesia, Indonesia

Abstract. In recent years, face recognition is widely used in various aspects as a form of technology advancement. Various studies were conducted to improve the accuracy of face recognition. In this research, Learning Vector Quantization and Fuzzy Kernel Learning Vector Quantization were used as a method of classification. The data used in this research was Labeled Face in The Wild-a (LFW-a). This database has no restrictions such as background, expression, position, and so on. Based on test results using LFW-a database, face recognition using LVQ method has highest accuracy at 89,33% and FKLVQ method has highest accuracy at 89,33% as well.

1. Introduction

Biometric is a term for measurement of physical and behavioural characteristic of humans. Physical characteristics include fingerprint, finger knuckle, iris, face, voice, etc., meanwhile behavioural characteristics include walking gait, handwriting, etc. Biometric information has been widely used for identification system in recent years due to its uniqueness in each individual. One of the most popular biometric identification system is face recognition. Compared to other biometric information, face image is easier and more flexible to acquire even when the person is unaware of being scanned.

Inspired by biological neurons, Artificial Neural Network (ANN) is a network of elements called neurons. A neuron can either inhibit or excite a signal. Input signals are inhibited or excited based on numerical weights associated with each connection to the artificial neuron [1]. In this paper, neural networks architecture namely Learning Vector Quantization (LVQ) was used to classify Labeled Faces in the Wild-a database (LFW-a). LFW-a is a database of face images designed for face recognition study without limitation such as background, expression, etc. This database already has a name label for each subject.

An LVQ network has an input layer, a competitive layer, and a linear layer. Competitive layer aims to classify input vectors to subclasses, while linear layer aims to combine subclasses into target classes. Both layers have one neuron for each class. If the competitive layer has m neuron, then the layer can learn up to m subclass which is then combined by a linear layer to form the n target class ($m \geq n$) [2].

LVQ has attracted much attention because of its simple and efficient learning process. However, LVQ has some issues such it only updates the winner neuron for each input. Moreover, it uses

Euclidean distance so that LVQ is sensitive to noise and outlier. In this paper, fuzzy membership was used to solve the first problem. To solve the second problem, kernelized distance measure was used as a replacement for Euclidean distance [3]. Thus, fuzzy kernel learning vector quantization was used to classify LFW-a database as well.

2. Classification

Classification is the process of finding a model that describes the class or category of output data, therefore it can be used to predict the class of new inputs. Classification algorithm is divided into two stages namely training and testing. In the training stage, the classification algorithm builds a model or classifier based on the learning process of training data containing input data and classes or categories of each data. The testing stage aims to find out the accuracy of the classification algorithm. At this stage, classifier classifies the testing data into classes. The accuracy of the classifier is said to be the classifier capability to correctly predict the class of testing data [4].

2.1. Learning vector quantization

LVQ is one of Artificial Neural Network (ANN) architecture. ANN is a classification method based on the principle of biological neuron network. LVQ network has 3 layers: input layer, competitive layer and linear layer [2]. Competitive layer has a winner-take-all rule, in the sense that neurons in competitive layer compete each other to become winner neurons [3].

The training process at LVQ aims to determine the weight vectors of competitive neurons, thus obtaining the training data mapping to each class on the competitive layer correctly. For each input vector, $\mathbf{x}_k$, the weight vector, $\mathbf{w}_i$, is updated with the formula as follows:

$$\mathbf{w}_i(t+1) = \mathbf{w}_i(t) + \alpha(t)h_{ik}(\mathbf{x}_k - \mathbf{w}_i(t)), \tag{2.1}$$

where $\alpha(t)$ is learning rate ($0 \le \alpha(t) \le 1$) and h_{ik} is indicative function whose value is 1 when i is the winner neuron and 0 when i is not the winner neuron [3].

The winner neuron is selected based on Euclidean distance:

$$d(\mathbf{x}_k, \mathbf{w}_i) = \|\mathbf{x}_k - \mathbf{w}_i\| = \left(\sum_{j=1}^{R} (x_{kj} - w_{ij})^2 \right)^{1/2}, \tag{2.2}$$

where R is the number of elements in the vector [2].

3. Fuzzy membership

Inspired by the Fuzzy C-Means (FCM) algorithm, the primary key of FCM is the use of the membership function found in fuzzy sets. Fuzzy C-Means Algorithm provides the identity data for each category using fuzzy membership [5]. The objective function of FCM can be formulated as follows [6]:

$$J_m = \sum_{i=1}^{c} \sum_{k=1}^{N} (u_{ik})^m \|\mathbf{x}_k - \mathbf{v}_i\|^2, \tag{3.1}$$

where:

c = number of clusters ($c \ge 2$)
N = number of data
m = degree of fuzziness ($m \ge 1$)
$\|\mathbf{x}_k - \mathbf{v}_i\|$ = Euclidean distance between $\mathbf{x}_k$ and $\mathbf{v}_i$
$X = \{\mathbf{x}_1, \mathbf{x}_2, \ldots, \mathbf{x}_n\}$ is the input data set
$V = \{\mathbf{v}_1, \mathbf{v}_2, \ldots, \mathbf{v}_c\}$ is the cluster centre set
$U = [u_{ik}]$ is the fuzzy membership.

The objective function has constraints [6]:

$$0 \le u_{ik} \le 1$$

$$\sum_{i=1}^{c} u_{ik} = 1 \qquad \forall k = 1,\dots,N$$

$$0 \le \sum_{k=1}^{N} u_{ik} \qquad \forall i = 1,\dots,c$$

The membership function, u_{ik}, was obtained as follows [6]:

$$u_{ik} = \left(\sum_{j=1}^{c} \left(\frac{\|\mathbf{x}_k - \mathbf{v}_i\|}{\|\mathbf{x}_k - \mathbf{v}_j\|} \right)^{\frac{2}{m-1}} \right)^{-1} \tag{3.2}$$

In LVQ algorithm, c denotes number of competitive neurons and $\mathbf{v}_i$ becomes $\mathbf{w}_i$ which is weight vector. The equation of u_{ik} membership becomes as follows:

$$u_{ik} = \left(\sum_{j=1}^{c} \left(\frac{\|\mathbf{x}_k - \mathbf{w}_i\|}{\|\mathbf{x}_k - \mathbf{w}_j\|} \right)^{\frac{2}{m-1}} \right)^{-1} \tag{3.3}$$

where:

c = number of competitive neurons ($c \ge 2$)

X= { $\mathbf{x}_1, \mathbf{x}_2, \dots, \mathbf{x}_n$ } is the input data set

W= { $\mathbf{w}_1, \mathbf{w}_2, \dots, \mathbf{w}_c$ } is the weight vector set

m = degree of fuzziness ($m \ge 1$)

4. Kernel

Let the set of input, X, and nonlinear mapping functions, $\phi : x \to \phi(x) \in F$, which maps x in the input space X to the feature space F which has a higher dimension. The kernel function is defined as the inner product of space F with

$$K(x, y) = \phi(x)^{\mathrm{T}} \phi(y),$$

for x, y in the X input space. With kernel K, kernelized distance function is found as follows [7]:

$$d(x, y) = \|\phi(x) - \phi(y)\| = \left(K(x, x) - 2K(x, y) + K(y, y)\right)^{1/2}. \tag{4.1}$$

The kernel used in this paper was radial basis function kernel

$$K(x, y) = \exp\left(-\frac{\|x - y\|^2}{2\sigma^2} \right) \tag{4.2}$$

where σ is the parameter in the kernel function.

5. Fuzzy kernel learning vector quantization

The fuzzy kernel algorithm was constructed with objective functions as follows [6]:

$$J_m = \sum_{i=1}^{c} \sum_{k=1}^{N} (u_{ik})^m \|\boldsymbol{\phi}(\mathbf{x}_k) - \boldsymbol{\phi}(\mathbf{w}_i)\|^2 \tag{5.1}$$

where:

c = number of competitive neurons ($c \ge 2$)

N = number of data
m = degree of fuzziness ($m \geq 1$)
$\|\boldsymbol{\phi}(\mathbf{x}_k) - \boldsymbol{\phi}(\mathbf{w}_i)\|$ = kernelized distance between $\mathbf{x}_k$ data and weight vector $\mathbf{w}_i$
$X = \{ \mathbf{x}_1, \mathbf{x}_2, \ldots, \mathbf{x}_n \}$ is the input data set
$W = \{ \mathbf{w}_1, \mathbf{w}_2, \ldots, \mathbf{w}_c \}$ is the weight vector set
$U = [u_{ik}]$ is the fuzzy membership.

The objective function has constraints:

$$0 \leq u_{ik} \leq 1$$

$$\sum_{i=1}^{c} u_{ik} = 1 \qquad \forall k = 1, \ldots, N$$

$$0 \leq \sum_{k=1}^{N} u_{ik} \qquad \forall i = 1, \ldots, c$$

By simplifying $\|\boldsymbol{\phi}(\mathbf{x}_k) - \boldsymbol{\phi}(\mathbf{w}_i)\|^2$, we get the following equation [8]:

$$\begin{aligned}\|\boldsymbol{\phi}(\mathbf{x}_k) - \boldsymbol{\phi}(\mathbf{w}_i)\|^2 &= \left(\boldsymbol{\phi}(\mathbf{x}_k) - \boldsymbol{\phi}(\mathbf{w}_i)\right)^{\mathrm{T}} \left(\boldsymbol{\phi}(\mathbf{x}_k) - \boldsymbol{\phi}(\mathbf{w}_i)\right) \\ &= \boldsymbol{\phi}^{\mathrm{T}}(\mathbf{x}_k)\boldsymbol{\phi}(\mathbf{x}_k) - \boldsymbol{\phi}^{\mathrm{T}}(\mathbf{x}_k)\boldsymbol{\phi}(\mathbf{w}_i) - \boldsymbol{\phi}(\mathbf{x}_k)\boldsymbol{\phi}^{\mathrm{T}}(\mathbf{w}_i) + \boldsymbol{\phi}^{\mathrm{T}}(\mathbf{w}_i)\boldsymbol{\phi}(\mathbf{w}_i) \\ &= K(\mathbf{x}_k, \mathbf{x}_k) - K(\mathbf{x}_k, \mathbf{w}_i) - K(\mathbf{w}_i, \mathbf{x}_k) + K(\mathbf{w}_i, \mathbf{w}_i) \\ &= 2 - 2K(\mathbf{x}_k, \mathbf{w}_i) \end{aligned} \tag{5.2}$$

By substituting equation (5.2) into equation (5.1), the objective function was obtained as follows [9]:

$$J_m = 2\sum_{i=1}^{c}\sum_{k=1}^{N} (u_{ik})^m \left(1 - K(\mathbf{x}_k, \mathbf{w}_i)\right) \tag{5.3}$$

The kernelized membership function, u_{ik}, was obtained as follows:

$$u_{ik} = \left(\sum_{j=1}^{c} \left(\frac{1 - K(\mathbf{x}_k, \mathbf{w}_i)}{1 - K(\mathbf{x}_k, \mathbf{w}_j)} \right)^{\frac{1}{m-1}} \right)^{-1} \tag{5.4}$$

Next, find the minimum distance between input vector, $\mathbf{x}_k$, and weight vector, $\mathbf{w}_i$, as follows

$$d^2(\mathbf{x}_k, \mathbf{w}_i) = K(\mathbf{x}_k, \mathbf{x}_k) - 2K(\mathbf{x}_k, \mathbf{w}_i) + K(\mathbf{w}_i, \mathbf{w}_i) \tag{5.5}$$

By simplifying the formula (5.5) using gradient descent, the equation $\mathbf{w}_i(t+1)$ can be formed as

$$\mathbf{w}_i(t+1) = \mathbf{w}_i(t) - \alpha(t) h_{ik}(t) \left(\frac{\partial d^2(\mathbf{x}_k, \mathbf{w}_i)}{\partial \mathbf{w}_i} \right)$$

$$\mathbf{w}_i(t+1) = \mathbf{w}_i(t) - \alpha(t) h_{ik}(t) \frac{2}{\sigma^2} K\left(\mathbf{x}_k, \mathbf{w}_i(t)\right)\left(\mathbf{x}_k - \mathbf{w}_i(t)\right) \tag{5.6}$$

and $h_{ik}(t)$ was calculated as follows:

$$h_{ik}(t) = \left(\frac{u_{ik}}{\max_{1<i<c} \{u_{ik}\}} \right)^{1 + \frac{t^{1/2}}{c}}$$

6. Experimental result

The database used in this paper was LFW-a database. It consists of 30 subjects with each subject having 10 face images. This database was designed for studying the problem of unconstrained face recognition. All face images in this database have different backgrounds, expressions, and face positions. Some subjects in the database also wear glasses and hats. Each face image in this database has been labelled according to the person's name in the image. The face image in this database was a grayscale image with .jpg format.

A digital image can be defined as a two-dimensional function, $f(x,y)$, where x and y are the plane coordinates and the value of $f(x,y)$ is the intensity or gray level of the image at that point. A digital image, $f(x,y)$, sized $m \times n$ pixels can be represented as a matrix containing m rows and n columns. A row vector can also be a representation of an image. The first row of $f(x,y)$ becomes n first element of the vector, the second row becomes the next n element, and so on to form a $1 \times mn$ row vector [10]. Each face image was converted into a row vector to be used as an input vector on the algorithm. The LFW-a database was divided into two parts: training data and testing data.

6.1. LVQ experiment

The experiment using Learning Vector Quantization method was performed to identify the name of subject in face images in LFW-a database. The experiment was done using 10% training data from the total number of databases, where 1 out of 10 face images of each subject is taken for training and 9 face images from each subject were used for testing. In addition, there are also experiments with 20% training data, 30% training data, 40% training data, 50% training data, 60% training data, 70% training data, 80% training data and 90% training data. The results of experiment with various number of training data as follows:

Table 1. Accuracy percentage of LVQ experiment results.

Training Data	Accuracy	Running Time (s)
10%	51.85%	12.72
20%	75.83%	12.05
30%	79.52%	11.95
40%	76.11%	16.95
50%	89.33%	16.86
60%	89.17%	24.02
70%	87.78%	26.40
80%	86.67%	48.84
90%	80.00%	50.77

As seen in Table 1, the highest accuracy obtained was 89.33% with 50% training data. The running time for Learning Vector Quantization is fast. For 50% training data, the running time is only 16.86 seconds.

6.2. FKLVQ experiment

The experiment was also conducted with Fuzzy Kernel Learning Vector Quantization classification method. Fuzzy Kernel Learning Vector Quantization method involves kernel parameter which in this case is the sigma (σ) of radial basis function kernel. The experiment was done using 50% training data, so 5 images of each subject face were used for training and 5 remaining images were used for testing. Experiment was performed using different sigma values (0,0001; 0.001; 0.05; 0.1; 1; 5; 10; 50; 100; 1000). Here are the results obtained:

Table 2. Accuracy percentage of FKLVQ experiment results

Training Data	Sigma (σ)	Accuracy	Running Time (s)
50%	0.0001	13.33%	1593.70
50%	0.001	78.00%	1628.14
50%	0.05	86.67%	1646.44
50%	0.1	89.33%	1612.30
50%	1	82.00%	1589.86
50%	5	78.00%	1592.72
50%	10	76.67%	1551.97
50%	50	76.67%	1559.38
50%	100	76.67%	1556.69
50%	1000	76.67%	1538.64

As seen in Table 2, the highest accuracy obtained was 89.33% with 50% training data and sigma 0.1. In this case, Fuzzy Kernel Learning Vector Quantization has longer running time than Learning Vector Quantization.

7. Conclusion
This paper studies face recognition to identify the identity of subject in LFW-a face image database. Learning Vector Quantization was used due to its simple and efficient learning process. To cover the weakness of this method such as its sensitivity to noise and outlier, fuzzy and kernel are implemented in the classification method. Based on the experiment using Learning Vector Quantization classification method to classify the LFW-a database, the accuracy reaches the maximum of 89.33% using 50% training data. In the experiment of Fuzzy Kernel Learning Vector Quantization classification method on LFW-a database, only 50% training were used but different sigma (σ) parameter values (0,0001; 0.001; 0.05; 0,1; 1; 5; 10; 50; 100; 1000) were implemented in the experiment. Accuracy reaches the maximum of 89.33% at the value of $\sigma = 0.1$.

8. References
[1] Engelbrecht A P 2008 *Computational Intelligence: An Introduction* (Chichester: John Wiley & Sons)
[2] Jun Y and Dian-Hong W 2005 Sequential face recognition based on LVQ networks *Proc. Of 2005 IEEE Int. Workshop on VLSI Design and Video Technology* pp 284-7
[3] Zhang D, Chen S and Zhou Z 2004 Fuzzy-kernel learning vector quantization *Advances in Neural Networks-ISNN 2004 Lecture Notes in Computer Science* pp 180-5
[4] Han J and Kamber M 2006 *Data Mining: Concepts and Techniques* (San Francisco: Morgan Kaufmann)
[5] Wulan A, Jannati M V, Rustam Z and Fauzan A A 2016 Application kernel modified fuzzy c means for gliomatosis cerebri *2016 12th Int. Conf. on Mathematics, Statistics, and Their Applications (ICMSA)*
[6] Zanaty E 2012 Determining the number of clusters for kernelized fuzzy C-means algorithms for automatic medical image segmentation *Egyptian Informatics J.* **13** pp 39-58

IOP Conf. Series: Journal of Physics: Conf. Series **1108** (2018) 012068 doi:10.1088/1742-6596/1108/1/012068

[7] Lv H and Wang W 2005 Handwritten digit recognition with kernel-based LVQ classifier in input space *Advances in Neural Networks-ISNN 2005 Lecture Notes in Computer Science* pp 203-8
[8] Rustam Z and Frederica Y 2018 Insolvency prediction in insurance companies using support vector machines and fuzzy kernel c-means *J. of Physics: Conf. Series* **1028** 012118
[9] Chiang H, Lee T and Lee J 2011 On-road speed sign recognition using fuzzy kernel-based learning vector quantization *Proc. 2011 Int. Conf. on System Science and Engineering*
[10] Gonzalez R C and Woods R E 2008 *Digital Image Processing* (New Jersey: Pearson Education)

IOP Conf. Series: Journal of Physics: Conf. Series **1108** (2018) 012069 doi:10.1088/1742-6596/1108/1/012069

Developing a Local Instruction Theory for Learning the Concept of Solving Quadratic Equation Using Babylonian Approach

A D Fachrudin[1], R I I Putri[2], A W Kohar[3], S Widadah[4]

[1]Mathematics Education Department, STKIP PGRI Sidoarjo, INDONESIA.
[2]Postgraduateprogramme on Mathemaitics education, UniversitasSriwijaya, INDONESIA.
[3]Faculty of Mathematics and Natural science, Universitas Negeri Surabaya, INDONESIA.
[4]Mathematics Education Department, STKIP PGRI Sidoarjo, INDONESIA.

[1]Corresponding Author's e-mail: dh4nyy@gmail.com

Abstract. The purpose of this research is to develop a set of learning instructions and investigate how the Old Babylonian Geometric Method: Naïve Geometry can support students' understanding about the concept of solving quadratic equations. This study aims to investigate on how students relate the Babylonian Geometric approach with the solving of the quadratic equation especially on how student bring their geometric solution into algebraic form. This research was conducted at a state junior high school in Indonesia. Design research was chosen as the method to reach the purpose. The instructional activities designed to achieve the learning objectives in this research consists of several activities, namely manipulating geometric form to solve the problem, using the Babylonian geometric method to solve the problem, linking geometric problems to algebra, and finding common formulas to solve quadratic equations. Through these activities, students acquire the idea of solving quadratic equations using algebraic completing perfect square method by manipulating and reshaping a rectangle into a square. However, our important finding is that only students who have high mathematics ability who reach the learning objective until the last stage of activity, which is to reinvent the common algebraic formula in solving the quadratic equation.

1. Introduction

In this research, we developed a local instruction theory for learning the concept of solving quadratic equation using a historical phenomenology. This was inspired by Radford's study namely Second Degree Equations in the Classroom: A Babylonian Approach. The Old Babylonian geometric approach, preserved in a tablet known as BM 13901 Problem I, used to build the concept of quadratic equation through a progressive instruction which starts with the use of manipulative and evolves through an investigative problem-solving process that combines the numerical and geometrical experiences [1]. But, we try to modify the approach that has used before. Here we also use geometric approach preserved in BM 13901 Problem II to build the learning instruction for learning the concept of the quadratic equation. Furthermore, we present some reasons that led us to conduct this study.

Some researchers have conducted studies on learning quadratic equation and noted that it is an important and interesting research object [1-4]. Some of those studies indicate that there are many students make errors in term of solving the quadratic equation. The errors, for instances, around inability to uncover the relationship between variables and roots, determining the appropriate quadratic equation for a given set of roots and vice versa [5]. In addition, O'Connor and Norton [6] found and categorized learners' errors into four patterns: lack of fractional reasoning, lack of conceptual understandings related

IOP Conf. Series: Journal of Physics: Conf. Series **1108** (2018) 012069 doi:10.1088/1742-6596/1108/1/012069

to quadratic form and algebraic process, algebraic misconception, and inefficient procedural techniques of finding solutions. Zakaria & Maat [4] argue that such kinds of errors and difficulties may be caused by lack of emphasis by teachers on understanding the language of mathematics and the skills needed by the students. These show the importance of improving the quality of learning quadratic equation.

Effort on supporting students' understanding on solving quadratic equations have been conducted through a variety of approaches, such as by employing technology-assisted learning using spreadsheet [7] and applying constructivist-oriented learning with inquiry approach [8]. Another approach which may be used is employing historical perspectives of quadratic equation. This is due to the reasons that historical development of a science is very important not only from a cultural and humanistic perspective but also because it facilitates understanding the concept of a topic [9]. Hence integration of aspects of history in learning will be very helpful, either for students or teachers. Thus, problems from history mathematics are arguably potential as sources of context to make a meaningful mathematics learning.

In general, researchers suggest to integrate mathematics learning with the history of mathematics [1,10]. Radford states that the historical construction of mathematical concepts can supply us with a better understanding of the ways in which our students construct their knowledge of mathematics [11]. Moreover, Grugnetti explains that by knowing the history of how a concept was invented or developed, will support and improve the skills about that concept [12]. In particular, based on the historical perspective, the concept of solving quadratic equations was built by the geometric foundation, such as by employing the so-called 'naïve geometry'[13]. Therefore, using naïve geometry to develop the learning instruction for learning the concept of the quadratic equation is regarded helpful for students in tackling the quadratic equation problems.

2. Theoretical Background

2.1. Naïve Geometry

The Old Babylonian geometric method is a geometric method that can be used to solving a quadratic equation. This method was identified by J. Høyrup and he called it Naïve Geometry [14]. In order to show the method, let we discuss one of the Babylonian problems, problem II of a tablet preserved at the British museum and known as BM 13901. The statement of the problem, that is to find the length of the square's side, is the following:

> My confrontation inside of the surface I have torn out: 14`30°

Meanwhile, the statement of solution appears in the tablet is the following:

> My confrontation inside of the surface I have torn out: 14`30°. 1 the wasitum; You pose. The moiety (half) of 1 you break, 30' and 30' you make span; 15' to 14`30° you append: $14`30°15'$ makes $29°30'$ equilateral. $30'$ Which you have made span to $29°30'$ you append; 30 the confrontation. (Høyrup, 1990b)
>
> note: $14`30° = 870, 30' = \frac{1}{2}, 15' = \frac{1}{4}, 14`30°15' = 870\frac{1}{4}$.

The statement of the problem is, "My confrontation inside of the surface I have torn out: 14`30 °. 1 the wasitum? Høyrup [10] explained that the "Confrontation" is a side of the square and the "surface" is a square. The side is not a simple side, but as the side along with us as a side provided with a canonical projection that forms, along with the side (rectangle form) [1]. The "wasitum" is means something going out, including something projecting from a building [14]. The problem simply is to find the value of the square's side if it is known that the area of the square minus its side is 870. To get a better understanding of the statement in the tablet, Høyrup [14] interpret the geometric shapes as shown below.

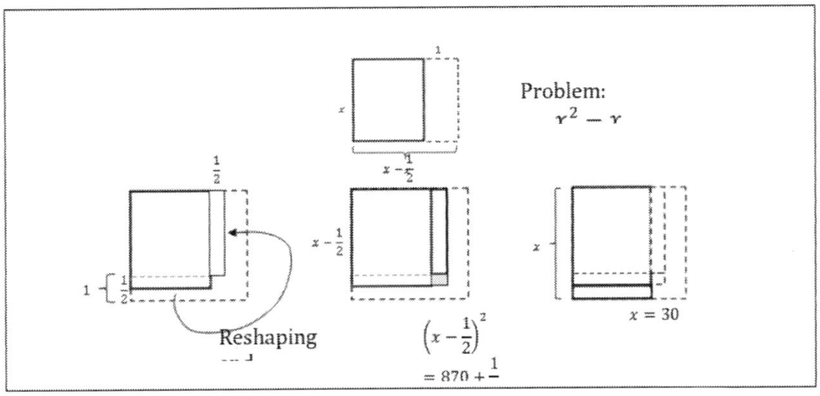

Figure 1. Geometric Interpretation of BM 13901 Problem II

Figure 1 shows that the problems and resolution steps written by the Babylonian mathematicians look simpler if interpreted into the algebraic symbol. The basic idea of the method used by Babylonian mathematicians on the problem found in BM 13901 No.2, i.e. reshaping and completing rectangle into a perfect square. Geometric interpretation also makes the problem is easier to understand. The scribes did not know any algebraic symbol, but based on the geometric and algebraic interpretations presented in figure 1, it can be concluded that they knew the formula of quadratic equations and how to find solutions.

2.2. Local Instruction theory

Gravemaijer [11] states that Local Instruction Theory (LIT) is a description of the conjecture of learning activity on a particular topic. Gravermeijer & Cobb [16] stated that LIT consists of conjecture a learning process also switch alleged guesswork about the possibility of facilities that can support the learning process. Gravemeijer [15] explains that LIT is a theory that offers a plan which can be used as guidance by teachers to choose learning activities and compose HLT.

3. Method

The core of this type of research is formed by classroom teaching experiments that centre on the development of instructional sequences and the local instructional theories that underpin them [15]. There are three main phases undertaken in this research, preliminary design, teaching experiment, and retrospective analysis [15]. In the first phase, we formulated a Hypothetical Learning Trajectory (HLT) for learning the concept of quadratic equation that is made up of learning goals for students, planned instructional activities and a conjectured learning process in which one anticipates how students' thinking and understanding could evolve when the instructional activities are used in the classroom [17].

3.1. Participants

On teaching experiment, the participants were 32 students consisting of 15 male students and 17 female students. Meanwhile, on the pilot experiment involving 5 students from the same school.

3.2. Data collection and analysis

Data were collected through video documentation, student work photos, field notes, and interviews. Then, we conducted triangulation to test the HLT that had been developed.HLT were tested through two phases, pilot experiment, and teaching experiment by analysing the students' work and interviewing.

4. Result

In preliminary design, we developed four instruction activities to be implemented in the classroom or we call it HLT. The following is a brief explanation of our HLT. In the first meeting, learning started by introducing the context and steps must be followed so that students can solve the problem using geometric method (ignoring algebraic procedure). The second problem, students have to solve the second problem (determining the length and width of a rectangle) using the same method they used in the first meeting. However, the step needed is the opposite way. In first meeting (first activity), to solve the problem they must reshape square into a rectangle but in second meeting (second activity) is the opposite. Then, in third and fourth meeting the problem used is similar (determining the side of a square), they also need to reshaping rectangle into a square but the difference is that concrete number numbers are given neither for the area nor base of the rectangle. In the last activity, we focus on how student brings the geometric procedure to the algebraic form and understand the concept of solving quadratic equation through Naïve Geometry.

The teaching sequences

Here, we try to show the brief explanation and analysis of teaching sequences developed and students work (in teaching experiment).

1) Part 1. Manipulating geometric form to solve the problem

In part 1, students are presented with the following problem:

The area of a rectangular garden is 32 ha. Around the garden, there are 120 Palm Trees which the distance of each tree is 20 meters. If the owner of the garden wants to put up an iron fence on the front side of the garden, help the owner to determine the length of the fence has to be purchased!

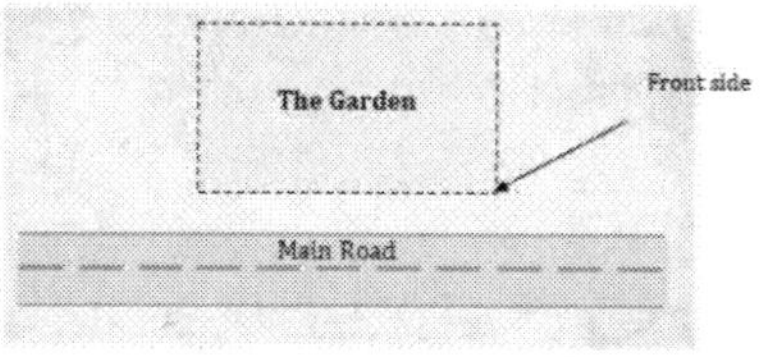

Figure 2. Problem 1

Students are asked to solve the problem using any method. Most all groups use the trial and error method to solve the problem. By using a cutting paper as props, the teacher shows them the technique of Babylonian approach or known as Naive Geometry.

Then, the following is the second problem:

What should the length and width of a rectangle if the area is 84 units and the circumference is 40 units?

By solving the problem, students understand that by performing geometric manipulation of naive geometry methods, the length of the rectangle is obtained by adding the square side formed at the beginning with the cut sides of the square (disposed to match the area of the desired rectangle). Because the square's side formed at the beginning is 10 units and the square side that removed is 4 units, the rectangle dimensions are 10 + 4 = 14 (length) and 10-4 = 6 (width) (see figure 3).

Here is the last problem of part 1:

On the last problem, to help students achieve a better understanding of the deeper level of naive geometry, we give a problem that leads students confronted with a conflict if the area to be disposed of is not square (if expressed by an integer).In this case, some groups can solve it well while others failed.

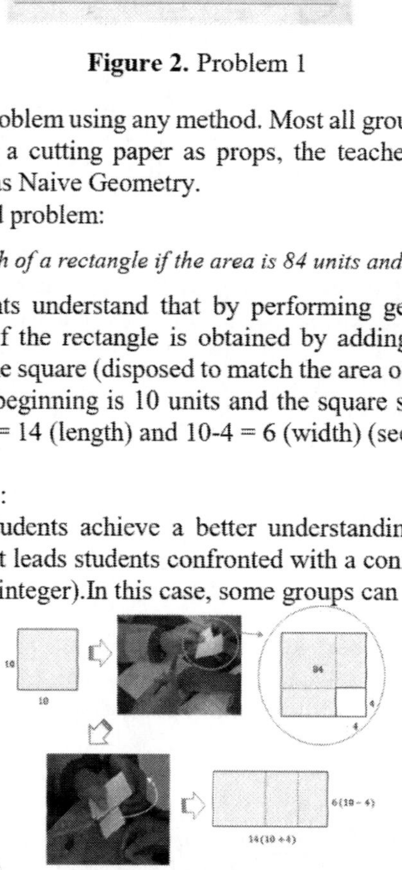

Figure 3. Students' Geometric Manipulation

2) Part 2. Using the Babylonia geometric method to solve the problem
In part 2, the students are presented with a problem that needs different use of naive geometry method. The problem is the following:

Determine the dimensions of a rectangle if known that the area is 117 unit and the difference between its length and width is 4 units?

In this case, the geometric manipulation performed by the students to solve the problem is in accordance with the conjecture on the HLT. The steps they use are almost the same as Geometric Interpretation of BM 13901 Problem II (see figure 1). But, they met some problems when we asked to work on a description of the steps to follow in order to solve this type of problem.

3) Linking geometric problems to algebra, and finding common formulas to solve quadratic equations
In part 3, the first problem must be solved by the students is to determine the side of the square if some of the area, which is rectangular in shape (width is 2 unit, length is the same with the square side), is removed so that the area becomes 24 unit area. The idea of giving this problem is to build student understanding that the geometry problems they face are a quadratic equation problem (that is to solve $x^2 - 2x = 24$). By the naive geometry manipulation method, the geometric form becomes a simple algebraic equation, $(x - 1)^2 = 25$, that can be solved easier without involving complex algebraic manipulation.

The second problem given in this part is the same type as the previous. The difference is we did not provide a concrete number. In this part, we gave the following problem to the students.

The length of a square is 'x' units. If a part of its regions, thatis rectangular shape with the width b unit and the same length as the square side, isremoved, its area becomes c units. Determine a) The algebraic form of the problem. b) A formula or steps (in the form of algebraic symbols) to determine the length of the sides of the square! (Use naive geometry method).

This section guides students to link the geometry problems they solved with algebra where students focus on how to interpret geometric manipulation into algebraic symbols. The following is the result of student work in determining the side of a square (value of x) if the shaded area is c (see figure 4).

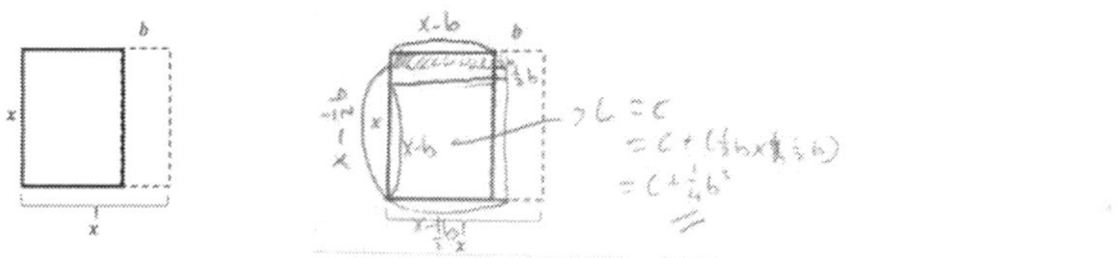

Figure 4. Problem in fourth activity and the student's work

5. Discussion and Conclusion

Based on the implementation of the HLT in teaching experiment, the use of Naïve Geometry method can be an alternative method for learning the concept of the quadratic equation because the conjecture in HLT that has been made of is quite in accordance with the result of teaching experiment. In common, the learning instruction we developed based on Old Babylonian Geometric Method: Naïve Geometry can support students' understanding of the concept of solving quadratic equations. Based on that learning conducted in the second cycle of teaching experiment, we also conclude that through these activity students can realize the idea of Naïve Geometry can be used to solve quadratic equations and finding a general form to solve quadratic equations. Through the geometric idea (reshaping into a square), also supports students to perform symbolic operations are meaningful because they are familiar with the context involved. However, from the implementation of learning activities, we found that only students with high mathematics ability who reach the learning objective until the last stage of activity, which is to reinvent the common algebraic formula in solving the quadratic equation.

In common, through activities that have been designed in HLT, students learned to understand algebraic problems and how to solve them through geometric approaches that they are familiar (contextual). Thus the students get a chance to learn how to find the solution of quadratic equations meaningfully. Implementation of HLT this time can be concluded in line with Freudenthal statement [18] that learning will happen when it is meaningful for students.

In general, we recommend teachers to integrate the history of mathematics in learning activities, which is in line with other researchers' recommendations [5,19]. On the other hand, we recommend that similar research will be conducted for other learning topics. Hopefully, it can help and enrich teachers' references in using the history of mathematics in learning.

References

[1] L. Radford and G. Gue 2000 Second Degree Equations in the Classroom : A Babylonian Approach. In V. Katz (ed.) Using history to teach mathematics. *An international perspective* pp. 69-75 (Washington: The Mathematical Association of America)

[2] J. A. Ezquerro, M. A. Hernández, and N. Romero 2010 *J. Comput. Appl. Math.*, vol. **234(4)** 960

[3] L. H. Lian and W. T. Yew 2012 Assessing Algebraic Solving Ability : A Theoretical Framework *International Education Studies* **5(6)** 177

[4] E. Zakaria and C. Author 2010 Analysis of Students ' Error in Learning of Quadratic Equations *International Education Studies* **3(3)** 105

[5] Setiawan, D 2014 Diagnosis of student difficulties in solving quadratic equations through mapping and scaffolding provided. Retrieved from http://karya-ilmiah.um.ac.id/index.php/disertasi/article/view/37707

[6] O'Connor, B. R., & Norton, S 2016 Investigating Students' Mathematical Difficulties with Quadratic Equations. *Mathematics Education Research Group of Australasia.* pp 552-559

[7] Nobre, S., Amado, N., & Ponte, J. P. D 2015 *Proc of the 12th International Conf on Technology in Mathematics Teaching* 474

[8] Nasution, A. S 2017 *Asian Journal of Natural & Applied Sciences* ***6***

[9] P. Yuste 2010 Learning mathematics through its history *Procedia - Social and Behavioral Sciences* **2(2)** 1137

[10] Fauvel, J., and Van Maanen, J. 2002 *The History in mathematics education* (Dordrecht Netherlands: Kluwer Academic Publishers)

[11] L. Radford 1996 Netherlands, The roles of geometry and arithmetic in the development of algebra: Historical remarks from a didactic perspective. In Approaches to algebra 39 Springer

[12] L Grugnetti 2000 The history of mathematics and its influence on pedagogical problems, *Paleontol. Soc*

[13] D. French 2002 Teaching and learning algebra Teaching and Learning Algebra 205

[14] M. Thought, and T. Sat 1990 Algebra and Naive Geometry. An Investigation of Some Basic Aspects of Old Babylonian Mathematical Thought *Altorientalische Forschungen* **17 (1)** 27-69

[15] K. Gravemeijer 2004 *Math. Think. Learn.* **6(2)** 105

[16] J. V. D. Akker, K. Gravemeijer, S. McKenney, and N. Nieveen 2006 *Educational design research* (London: Routledge, 107-118)

[17] M. a Simon 1995 Reconstructing mathematics pedagogy from a constructivist perspective, *J. Res. Math. Educ.* **26(2)** 114

[18] H. Freudenthal 1991 *Revisiting mathematics education: China lectures* (Nethersland: Springer)

[19] Y. Õ. Cemalettin, B. Özüm, and Z. Bahar 2011The views of the teacher and students in regards to the use of the history of mathematics in the teaching of fractal subject *Procedia - Social and Behavioral Sciences* **15** 868

Acknowledgment

We would like to thank the Ministry of Higher Education and Research for its support in funding our research, head master and mathematics teacher of junior high school state 1 (SMPN 1) of Palembang.

IOP Conf. Series: Journal of Physics: Conf. Series **1108** (2018) 012070 doi:10.1088/1742-6596/1108/1/012070

Optimal Portfolio Selection with Regime-Switching Hamilton-Jacobi-Bellman (HJB) Equation and Maximum Value-at-Risk (MVaR) Constraint

F Setyani[1], M Novita[2], M Malik[3]

[1,2,3]Department of Mathematics, Universitas Indonesia, Depok, 16424, Indonesia
Email: [1]fathia.setyani@sci.ui.ac.id, [2]mila.novita@sci.ui.ac.id, [3]maulana.malik@sci.ui.ac.id

Abstract. The forming of portfolio is necessary to determine the decision of the best investment so as to investors can identify the securities and determine the allocation of asset to obtain an optimal portfolio. The problem in forming optimal portfolio is the determination of the proportion which is allocated at investment assets in order to maximize expected return with certain risks. The model contains regime-switching market models, which states are interpreted as the states of economy. The risk measuring instrument used is VaR and MVaR is defined as the maximum value of the VaRs in all economy states. The optimal proportion formula is sought by using the stochastic optimal control theory with the aim of maximizing the discounted utility of consumption over a finite time horizon. We use regime-switching Hamilton-Jacobi-Bellman (HJB) equation and then derive a system of coupled HJB equation corresponding to the economy states. Lagrange multiplier method is used to solve the optimization problem with the constraint. We apply Kuhn-Tucker conditions due to MVaR is an inequality function, so that we derive the optimal investment and the optimal consumption. Finally, numerical examples are investigated, and the effect of parameter on the optimal investment and on the optimal consumption are studied.

1. Introduction

The optimal portfolio allocation problem is quite importance in finance. The problem is to search the best allocation of wealth among some kinds of securities. The pioneer Markowitz [1] provided a mathematical model to formulate the optimal portfolio allocation problem and developed the mean-variance approach for these problem. Furthermore, the pioneering development of the optimal portfolio allocation problem in a continuous-time framework is provided by Merton [2,3]. Afterwards, many researchers extended Merton's model.

There could be substantial changes in the economic condition over a period of time. The introduction of regime-switching effect is important in economics and finance. The market parameters, such as interest rate of bank bond, the appreciation rate, and volatility rate of risky asset, are modulated by the financial market state which is often modeled by a Markov chain. Neglecting the regime-switching effect may lead to underestimation of the risk portfolio and suboptimal results. Hamilton [4] pioneered the econometric applications of regime-switching models. Some papers with regime-switching models in finance include [7] and [8] for consumption–investment problems. Usually, investment–consumption problems focus on maximizing utility from consumption. The investor wants to maximize his/her total discounted utility over period of time in which he/she invests. In this paper, the risk is measured by using Value-at-Risk (VaR). In each state regime-switching model, we constrain a VaR value of the portfolio in a short time duration. Maximum Value-at-Risk (denoted by MVaR) is defined as the maximum value of the VaRs in all economy states. We consider the optimal investment-consumption problem subject to a maximum Value-at-Risk constraint.

This paper is structured as follows. Section 2 describes the price dynamics of the model. Section 3 determines the optimal portfolio problem and formulates the MVaR constraint. Section 4 derives the regime-switching HJB equation and present the method of the Lagrange. Section 5 determines the value function and derives the optimal solutions. In Section 6, we make numerical experiments to investigate the effect of many parameters on the optimal investment and on the optimal consumption. Finally, we conclude the paper in the last section.

2. Model formulation

In this section, we shall consider a continuous-time financial model in which two assets are traded continuously over a finite time horizon $\mathcal{T} = [0, T]$, where $T \in (0, \infty)$, one is risk-free asset B and the other one is a risky asset P. We shall consider the situation when there are two states in the Markov chain. Denote $X = \{X(t)\}_{t\in\mathcal{T}}$ as a continuous-time Markov chain with a finite state space $\boldsymbol{x} = (x_1, x_2)$. The state of X are interpreted as different states of an economy. Following [5], we shall represent the state $\boldsymbol{x}$ as a finite set of unit vectors $\boldsymbol{\varepsilon} = \{\boldsymbol{e_1}, \boldsymbol{e_2}\}$, where $\boldsymbol{e_1} = (1,0)'$ dan $\boldsymbol{e_2} = (0,1)'$, where $'$ represents the transpose of a matrix or a vector. Suppose that the Markov chain has a generator $Q = (q_{ij})_{2\times2}, i, j = 1,2$. The price process of risk-free asset B is governed by

$$dB(t) = B(t)r_i dt, \quad B(0) = b_0, i = 1,2 \tag{1}$$

where r_i denotes the instantaneous market interest rate with $r_i > 0, i = 1,2$. Suppose μ_i and σ_i denote the appreciation rate and the volatility rate of the risky asset with $\mu_i > r_i$ and $\sigma_i > 0$ for each $i = 1,2$. Let $W(t)$ denote a standard Brownian motion on $(\Omega, \mathcal{F}, \mathcal{P})$. The price process of risky asset P is assumed to be

$$\mathrm{d}P(t) = P(t)(\mu_i \mathrm{d}t + \sigma_i \mathrm{d}W(t)), \quad P(0) = p_0, \quad t \in \mathcal{T}, i = 1,2 \tag{2}$$

We assume that the market agent can invest into the risk-free asset B and the risky asset P. Let $\pi_i(t)$ denotes the amount of wealth allocated to the risky asset P at time t and state i. Let $c_i(t)$ denotes the consumption rate of an economic agent at time t and state i. Define $\mathrm{u}(t) = (c_i(t), \pi_i(t))$ as our control process. Suppose $\{S(t)\}_{t\in\mathcal{T}}$ denotes the net wealth process of the economic agent with initial wealth $S(0) = s_0$ and initial state $X(0) = x_0$, then

$$\mathrm{d}S(t) = [\pi_i(t)(\mu_i - r_i) + S(t)r_i - c_i(t)]\mathrm{d}t + \pi_i(t)\sigma_i \mathrm{d}W(t) \tag{3}$$

3. Determination of optimal portfolio problem and VaR constraint

In this section, we will discuss about the optimal portfolio problem and VaR constraint.

3.1. Forming the Objective Function Using Stochastic Optimal Control Theory

Let $U(t, c_i) : \mathcal{T} \times [0, \infty) \to \mathcal{R}$ denote a utility function such that for each $t \in \mathcal{T}$. $U(t, c_i)$ is strictly increasing and strictly concave. Let δ denotes a discount factor and a positive constant. Suppose the economic agent needs continuous consumption over the time horizon $\mathcal{T}$ and he/she uses the portfolio process $\pi_i(t)$ available at initial wealth $S(0) = s_0$ and initial state $X(0) = x_0$. So, the expected discounted utility of the agent is defined as

$$J(s_0, x_0, c_i(t), \pi_i(t)) = E\left[\int_t^T U(\tau, c_i(\tau))\mathrm{d}\tau | S(0) = s_0, X(0) = x_0\right] \tag{4}$$

3.2. Value-at-Risk (VaR) Constraint

In this sequel, we shall present the MVaR constraint of the problem. We assume that the portfolio is adjusted frequently so that the interval from $\boldsymbol{t}$ to $\boldsymbol{t + h}$ is small, where $\boldsymbol{h > 0}$. It is reasonable to assume that there is no trading in between constraint revaluation and that the consumption is approximately constant in the little time interval $[\boldsymbol{t}, \boldsymbol{t + h}]$ since in reality, one can only adjust the portfolio discretely over time, and the portfolio/consumption decision is made at the beginning of the time horizon. That is, $\boldsymbol{\pi(\tau)=\pi(t)}$ and $\boldsymbol{c(\tau)=c(t)}$ for all $\boldsymbol{\tau\in[t,t+h]}$. We also assume that there is no regime switching in the little time interval. In other words, $\boldsymbol{X(\tau)=X(t)}$, for all $\boldsymbol{\tau\in[t,t+h]}$.

First, we define the following two quantities:

$$\theta_i(\tau)=\theta_i(t)=\frac{\pi_i(t)(\mu_i-r_i)-c_i(t)}{-r_i}\text{ and }Y(\tau)=\exp[-r_i\tau]S(\tau)\ ,i=1,2,\dots,N,\tau\in[t,t+h].$$

To define VaR constraints in the sequel, we regulate (3) and get $\boldsymbol{\mathrm{d}S(t)-r_iS(t)\mathrm{d}t=-r_i\theta_i(t)\mathrm{d}t+\pi_i(t)\sigma_i\mathrm{d}W(t)}$. Put $\boldsymbol{Y(t)=exp[-r_i\tau]S(t)}$, then we get

$$\mathrm{d}Y(t)=-\exp[-r_i\tau]r_i\theta_i(t)\mathrm{d}t+\exp[-r_i\tau]\pi_i(t)\sigma_i\mathrm{d}W(t)$$

$$\leftrightarrow\int_t^{t+h}\mathrm{d}Y(\tau)=\int_t^{t+h}-\exp[-r_i\tau]r_i\theta_i(t)\mathrm{d}\tau+\int_t^{t+h}\exp[-r_i\tau]\pi_i(t)\sigma_i\mathrm{d}W(\tau)$$

$$\leftrightarrow\mathrm{d}S(t+h)=\exp[-r_ih]\big(S(t)-\theta_i(t)\big)+\theta_i(t)+\pi_i(t)\sigma_i\exp[-r_i(t+h)]\int_t^{t+h}\exp[-r_i\tau]\mathrm{d}W(\tau)$$

Mean and variance of $\mathrm{d}W(t)$ are $E\big(\mathrm{d}W(t)\big)=0$ and $Var\big(\mathrm{d}W(t)\big)=\mathrm{d}t$. We define the discounted net loss of the portfolio over $[t,t+h]$ as $\Delta_iS(t,h)=S(t)-exp[-r_ih]S(t+h),\forall i=1,2$. It can be viewed as a function of the random variable $S(t+h)$ with mean and variance:

$$\mathrm{E}[\Delta_iS(t,h)]=\theta_i(t)(1-\exp[-r_ih])\text{ and }\mathrm{Var}[\Delta_iS(t,h)]=\sigma_i^{\,2}\pi_i^{\,2}(t)(2r_i)^{-1}(1-\exp[-2r_ih])$$

Now, we define the VaR value of $\boldsymbol{\Delta_iS(t,h)}$ with the probability level $\boldsymbol{\beta}$ as $\boldsymbol{VaR(t,h,i,\beta)}$. Since $\boldsymbol{\Delta_iS(t,h)}$ is normally distributed under the condition $\boldsymbol{X(t)=\varepsilon_i}$ and we assume that there is no short selling, then we derive

$$\mathcal{P}[\Delta_iS(t,h)\geq VaR(t,h,i,\beta)|X(t)=\varepsilon_i]=\beta$$

$$\leftrightarrow VaR(t,h,i,\beta)=\theta_i(t)(1-\exp[-r_ih])-\Phi^{-1}(\beta)\sqrt{\frac{\sigma_i^{\,2}\pi_i^{\,2}(t)}{2r_i}(1-\exp[-2r_ih])},i=1,2\qquad(5)$$

We define the MVaR of the portfolio with the probability level $\boldsymbol{\beta}$ over the time horizon $\boldsymbol{[t,t+h]}$ as

$$MVaR(t,h,\beta)=\max_{i=1,2}VaR(t,h,i,\beta)$$

In other words, MVaR is the maximum value of the VaRs of the portfolio in a short time duration over different states of economy. Then, the constraint of restricting the MVaR at level $\boldsymbol{R}$ is

$$MVaR(t,h,\beta)\leq R$$

This is equivalent to the following constraint:

$$\boldsymbol{\theta_i(t)(1-\exp[-r_ih])-\Phi^{-1}(\beta)\sqrt{\frac{\sigma_i^{\,2}\pi_i^{\,2}(t)}{2r_i}(1-\exp[-2r_ih])}\leq R}\qquad(6)$$

The portfolio optimization problem with MVaR constraint is then formulated as

$$\max_{\pi_i(t),c_i(t)}J=\max_{\pi_i(t),c_i(t)}\mathrm{E}\left[\int_t^T U\big(\tau,c_i(\tau)\big)\mathrm{d}\tau\right]\qquad(7)$$

subject to

$$\mathrm{d}S(t)=[\pi_i(t)(\mu_i-r_i)+S(t)r_i-c_i(t)]\mathrm{d}t+\pi_i(t)\sigma_i\mathrm{d}W(t)$$

$$\theta_i(t)(1-\exp[-r_ih])-\Phi^{-1}(\beta)\sqrt{\frac{\sigma_i^{\,2}\pi_i^{\,2}(t)}{2r_i}(1-\exp[-2r_ih])}\leq R$$

4. Regime-switching HJB equation

In this section, we shall derive a regime-switching HJB equation for the value function described in (7). We shall also derive a system of coupled HJB equations corresponding to the regime-switching HJB equation. Then. we shall derive the optimally conditions for solving the optimal problem.

First, we consider the following optimal value function as

$$V(t,s,x)=\sup_{\pi_i,c_i} E\left[\int_t^T U\big(\tau,c_i(\tau)\big)d\tau|S(t)=s,X(t)=x\right] \tag{8}$$

Let $V_i=V(t,s,\varepsilon_i)$ for each $i=1{,}2$ and let $\mathbf{V}=(V_1,V_2)'$. Hence, the vector $\mathbf{V}$ of the value function as different regimes satisfies the following systems of coupled HJB equations:

$$\sup_{\pi_1,c_1}\left[U\big(t,c_1(t)\big)+\frac{\partial V_1}{\partial t}+[\pi_1(t)(\mu_1-r_1)+S(t)r_1-c_1(t)]\frac{\partial V_1}{\partial s}+\frac{1}{2}[(\pi_1(t)\sigma_1)^2]\frac{\partial^2 V_1}{\partial s^2}+\langle \boldsymbol{V},Q\varepsilon_1\rangle\right]=0 \tag{9}$$

and

$$\sup_{\pi_2,c_2}\left[U\big(t,c_2(t)\big)+\frac{\partial V_2}{\partial t}+[\pi_2(t)(\mu_2-r_2)+S(t)r_2-c_2(t)]\frac{\partial V_2}{\partial s}+\frac{1}{2}[(\pi_2(t)\sigma_2)^2]\frac{\partial^2 V_2}{\partial s^2}+\langle \boldsymbol{V},Q\varepsilon_2\rangle\right]=0 \tag{10}$$

with terminal condition $V(T,s,\varepsilon_i)=0$ for $i=1{,}2$ and subject to constraint:

$$\theta_i(t)(1-\exp[-r_ih])-\Phi^{-1}(\beta)\sqrt{\frac{\sigma_i{}^2\pi_i{}^2(t)}{2r_i}(1-\exp[-2r_ih])}\le R$$

In order to derive the optimal investment strategy and consumption rate, we introduce the Lagrange functions and tackle the optimization problem (9) and (10) as follow for $i=1{,}2$:

$$\begin{aligned}&\mathfrak{L}_i\big(\pi_i(t),c_i(t),\lambda(t)\big)\\&=\left(U\big(t,c_i(t)\big)+\frac{\partial V_i}{\partial t}+[\pi_i(t)(\mu_i-r_i)+S(t)r_i-c_i(t)]\frac{\partial V_i}{\partial s}+\frac{1}{2}[(\pi_i(t)\sigma_i)^2]\frac{\partial^2 V_i}{\partial s^2}+\langle \boldsymbol{V},Q\varepsilon_i\rangle\right)\\&+\lambda(t)\left(\left(\left(\frac{\pi_i(t)(\mu_i-r_i)-c_i(t)}{-r_i}\right)(1-\exp[-r_ih])-\Phi^{-1}(\beta)\sqrt{\frac{\sigma_i{}^2\pi_i{}^2(t)}{2r_i}(1-\exp[-2r_ih])}\right)-R\right)\end{aligned} \tag{11}$$

These are the three conditions that must be fulfill in order to achieve the optimal process for $i=1{,}2$:

$$\nabla_{\pi_i(t),c_i(t)}\mathfrak{L}_i\big(\pi_i(t),c_i(t),\lambda(t)\big)=0 \tag{12}$$

$$\left(\frac{\pi_i(t)(\mu_i-r_i)-c_i(t)}{-r_i}\right)(1-\exp[-r_ih])-\Phi^{-1}(\beta)\sqrt{\frac{\sigma_i{}^2\pi_i{}^2(t)}{2r_i}(1-\exp[-2r_ih])}\le R \tag{13}$$

$$\lambda(t)\left(\left(\left(\frac{\pi_i(t)(\mu_i-r_i)-c_i(t)}{-r_i}\right)(1-\exp[-r_ih])-\Phi^{-1}(\beta)\sqrt{\frac{\sigma_i{}^2\pi_i{}^2(t)}{2r_i}(1-\exp[-2r_ih])}\right)-R\right)=0 \tag{14}$$

Rearranging (12), (13), and (14), we derive $\pi_{i_{opt}}(t)$ and $c_{i_{opt}}(t)$ for $i=1{,}2$.

$$\pi_{i_{opt}}(t)=\begin{cases}-\dfrac{(\mu_i-r_i)\dfrac{\partial V_i}{\partial s}}{\sigma_i^2\dfrac{\partial^2 V_i}{\partial s^2}}, & \text{unconstrained MVaR}\\[2ex] \dfrac{R-\dfrac{c_i(t)}{r_i}(1-\exp[-r_ih])}{\left(-\dfrac{(\mu_i-r_i)(1-\exp[-r_ih])}{r_i}-\Phi^{-1}(\beta)\sqrt{\dfrac{\sigma_i^2}{2r_i}(1-\exp[-2r_ih])}\right)}, & \text{constrained MVaR}\end{cases} \tag{15}$$

and

$$\frac{\partial U}{\partial c} = \frac{\partial V_1}{\partial s}, \frac{\partial U}{\partial c} = \frac{\partial V_1}{\partial s} \tag{16}$$

5. Determination of value function and optimal solutions

In this section, we shall consider the situation when there are two states in the Markov chain. The purpose of this portfolio is to maximize return, the return function that will be used is in the form of a utility function ($U(t, c_i(t))$) and the economic agent has a power utility function for consumption:

$$U(t, c_i(t)) = \exp[-\delta t] c_i(t)^\gamma, \ \ \delta > 0, \ \ 0 < \gamma < 1, i = 1,2 \tag{17}$$

Here δ represents a discount factor for consumption, γ represents the investor's preference, and it is assumed to be a positive constant. Moreover, we assume that the rate matrix Q of the Markov chain is $Q = \begin{pmatrix} -p & p \\ p & -p \end{pmatrix}$, where p is a positive real constant. In this case, the value functions for two economics states and by substituting $\pi_{i_{opt}}(t)$ and $c_{i_{opt}}(t)$ into (9) and (10) satisfy the following coupled HJB equations:

$$\frac{\partial V_1}{\partial t} + \exp[-\delta t] c_{1_{opt}}^\gamma(t) + \left[\pi_{1_{opt}}(t)(\mu_1 - r_1) + s r_1 - c_{1_{opt}}(t)\right] \frac{\partial V_1}{\partial s} + \frac{1}{2} \pi_{1_{opt}}^2(t) {\sigma_1}^2 \frac{\partial^2 V_1}{\partial s^2} + p(V_2 - V_1) = 0 \tag{18}$$

and

$$\frac{\partial V_2}{\partial t} + \exp[-\delta t] c_{2_{opt}}^\gamma(t) + \left[\pi_{2_{opt}}(t)(\mu_2 - r_2) + s r_2 - c_{2_{opt}}(t)\right] \frac{\partial V_2}{\partial s} + \frac{1}{2} \pi_{2_{opt}}^2(t) {\sigma_2}^2 \frac{\partial^2 V_2}{\partial s^2} + p(V_1 - V_2) = 0 \tag{19}$$

Following the approach in [3], we assume that the value function is of the following form:

$$V_i = V(t, s, e_i) = \exp[-\delta t] h_i(t, s) s^\gamma, \qquad i = 1,2$$

This form is in line with the form of the power utility function. As in [6], we neglect the derivatives of $h_i, i = 1,2$ with respect to v and obtain

$$\frac{\partial V_i}{\partial s} = \gamma \exp[-\delta t] h_i(t, s) s^{\gamma - 1}, \qquad \frac{\partial^2 V_i}{\partial s^2} = \gamma(\gamma - 1) \exp[-\delta t] h_i(t, s) s^{\gamma - 2},$$
$$\frac{\partial V_i}{\partial t} = \exp[-\delta t] h_i'(t, s) s^\gamma - \delta \exp[-\delta t] h_i(t, s) s^\gamma \tag{20}$$

where h_i' represent the derivative of h_i with respect to t. Substitute (20) to (18) and (19), also substitute (20) to (15), we derive the optimal proportion of risky asset ($\pi_{i_{opt}}(t)$) and the optimal consumption ($c_{i_{opt}}(t)$) for $i = 1,2$

$$\pi_{i_{opt}}(t) = \begin{cases} \dfrac{(\mu_i - r_i)s}{\sigma_i^2(1-\gamma)}, & \text{unconstrained MVaR} \\[2ex] \dfrac{R - \dfrac{c_i(t)}{r_i}(1 - \exp[-r_i h])}{\left(-\dfrac{(\mu_i - r_i)(1 - \exp[-r_i h])}{r_i} - \Phi^{-1}(\beta)\sqrt{\dfrac{{\sigma_i}^2}{2 r_i}(1 - \exp[-2 r_i h])}\right)}, & \text{constrained MVaR} \end{cases} \tag{21}$$

and

$$c_{i_{opt}}(t) = \frac{s}{\exp\left[-\int_0^t \left[\frac{1}{2}\frac{(\mu_i - r_i)^2 \gamma}{{\sigma_i}^2 (1-\gamma)^2} + \frac{r_i\gamma - \delta}{(1-\gamma)}\right] \mathrm{d}s\right] \int_t^T \exp\left[\int_0^t \left[\frac{1}{2}\frac{(\mu_i - r_i)^2 \gamma}{{\sigma_i}^2 (1-\gamma)^2} + \frac{r_i\gamma - \delta}{(1-\gamma)}\right] \mathrm{d}s\right] \mathrm{d}s},$$

6. Simulation of proportion and consumption calculation in optimal portfolio

In this section, we shall conduct numerical experiments to provide many special effects between parameters as follows. Suppose the investor has some specific values for the model parameters: $T = 25$, $R = 5$, $\beta = 0.99$, $\delta = 0.2$, $\gamma = 0.5$, $r_1 = 0.15$, $r_2 = 0.1$, $\mu_1 = 0.2$, $\mu_2 = 0.15$, $\sigma_1 = 0.4$, $\sigma_2 = 0.6$. We substitute all of those value to (21). In this example, we assume that there are two regimes in the economy, where regime 1 represents a bull market, a market in which share prices are rising or encouraging buying and regime 2 represents a bear market, a market in which prices are falling or encouraging selling. According to [9], the stock returns are higher in a bull market, so $\mu_1 > \mu_2$. [10] found that stock volatility is higher in a bear market; thus $\sigma_1 < \sigma_2$.

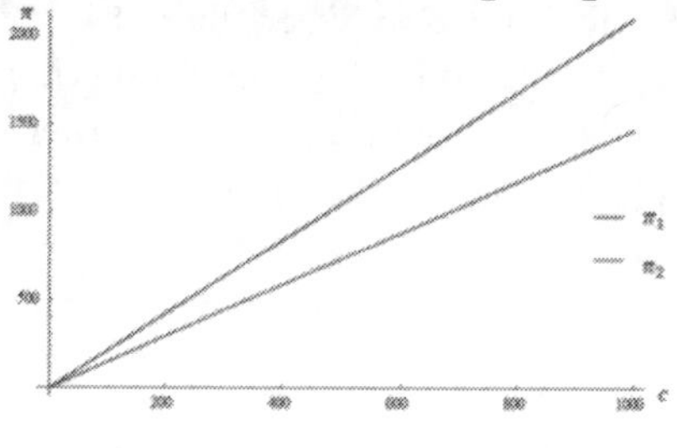

Figure 1. $\pi_{1_{opt}}$ and $\pi_{2_{opt}}$ for $R = 5, t = 5$

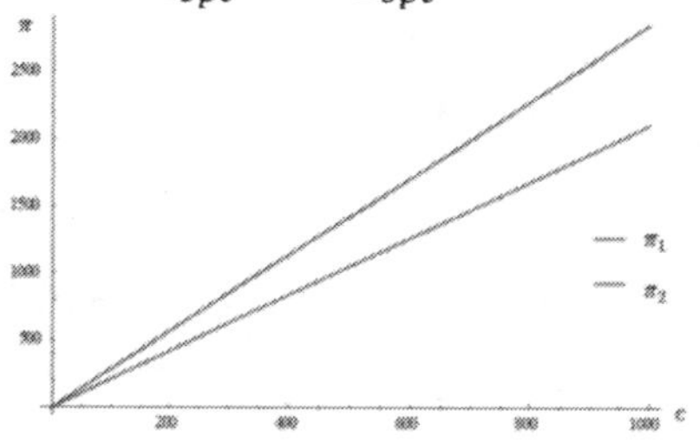

Figure 2. $\pi_{1_{opt}}$ and $\pi_{2_{opt}}$ for $R = 5,\ t = 12$

From the graph, we conclude that for the same consumption rate, the allocation of wealth for risky asset in a bull market is bigger than the allocation of wealth for risky asset in a bear market. But, in a bear market, the amount of consumption will be bigger than the amount of consumption in bull market.

7. Conclusions

We consider the optimal portfolio selection problem subject to a maximum Value-at-Risk constraint when the price dynamics of risky asset are governed by a Markov-modulated GBM. The model also contains regime-switching market modes. The optimal selection problem is formulated as a constrained utility maximization problem over a finite time horizon. We derive a system of coupled HJB equations for the problem. Moreover, with the help of Lagrange multiplier method, we derive the optimal investment and the optimal consumption. By numerical simulation, we conclude that the allocation of wealth for risky asset in bull market is bigger rather than the bear market.

Acknowledgments

This work is supported by Hibah PITTA 2018 funded by DRPM Universitas Indonesia No.2283/UN2.R3.1/HKP.05.00/2018

8. References

[1] Markowitz H 1952 Portfolio Selection *J. Financ.* **7** pp 77-91
[2] Merton R C 1969 *The Review of Economics and Statistics* pp 247-57
[3] Merton R C 1971 *J. Econ. Theory* **3** pp 373-413
[4] Hamilton J D 1989 *Econometrica* **57(2)** pp 357-84
[5] Elliot R J, Aggoun L and Moore J B 1995 (Berlin, Heidelberg, New York: Springer)
[6] Yiu K F C 2004 *J. Econ. Dyn. Control* **28** 1317-34
[7] Yiu K F C, Liu J, Siu T K and Ching W K 2010 *Automatica* **46** 979-89
[8] Hu F and Wang R 2017 *App. Math. Comp.* **313** pp 103-18
[9] Hamilton J D and Lin G 1996 *J. Appl. Econometrics* **11** pp 573–93
[10] French K R, Schwert G W and Stambaugh R F 1987 *J. Financ. Econ.* **19** pp 3–29.

Face Recognition to Identify Look-Alike Faces using Support Vector Machine

Z Rustam[1,2] and R Faradina[1,3*]

[1]Department of Mathematics, Universitas Indonesia, Depok, 16424, Indonesia
*Corresponding Author. [2]rustam@ui.ac.id; [3]ridhani.faradina@sci.ui.ac.id

Abstract. Face recognition has been one of the most interesting technology to study for many researchers. It allows a huge number of face images to be recognized in just a short amount of times, rather than recognizing each image individually through a normal human's eyes. Its idea is generally based on the assumption that each individual has a unique identity that can be distinguished between one another. However, in the real world, there are individuals who have similar faces to each other. They are referred as "look-alike" faces. This research was conducted to recognize look alike faces. Each image is represented by the features it contains. The classification method used in this research was Support Vector Machine, with the implementation of kernel. Two types of kernel used in this research were the Radial Basis Function kernel and the polynomial kernel. Results showed that SVM with Radial Basis Function kernel obtained higher accuracies.

1. Introduction

Humans have the ability to recognize faces. However, relying on only human's ability to recognize faces when dealing with large data such as hundreds to thousands and maybe even millions of facial images might cause errors. Thus, researchers have been trying continuously to construct face recognition algorithms that can produce high accuracy for various issues related to face recognition [1,3].

Face recognition is basically done on the assumption that each individual has a unique identity that can be distinguished from one individual to another. However, in the real world, there are individuals who have similar faces to each other. They are referred to as "look-alike" individuals [4].

In general, human faces look similar because of three reasons namely biological reasons such as twins and siblings, certain situations such as lighting, pose, expression or makeup, and because of plastic surgery that can turn a human face similar to other human faces in particular parts such as mouth, nose, chin or lips [4-5].

This research was conducted to recognize look alike faces. By doing so, real-world applications such as searching for missing person, criminals or fugitives can become easier. By recognizing look alike faces, the running process will save much more time because unqualified faces can be removed as they are not needed to be used for further processes. Because the data on the actual problem are usually very large, the reduction of such data can minimize the cost. A database for look-alike faces has already been made by Lamba et al. in 2011 [4], of which it became the database used in this research.

Classification method used in this research is Support Vector Machine (SVM). SVM is one of the machine learning methods introduced by Cortes and Vapnik in 1995 [6]. SVM is often used for the classification phase in facial recognition because of its high accuracy in general [1]. Initially, SVM is used for problems with only two classes. However, SVM has been developed for multiclass classification.

SVM is a great method to classify data that can be linearly separated. However, linearly separable data are rarely found in reality. So, the data need to be processed first in order for it to be linearly separable. This is where the kernel trick comes in. This research used the implementation of kernel in SVM with two types of kernel, which are the Radial Basis Function kernel and the polynomial kernel.

This paper is organized as follows: In Section 2, the methods used in this research are explained including the basics of Support Vector Machine and the implementation of kernel in SVM. Section 3 explains the data used in this research. Experiment results are portrayed in Section 4 while Section5 conveyed the conclusions.

2. Method

In this section, the methods used for this research will be explained. First, the concepts and basic formula for Support Vector Machine are first explained. After that, this section explains about the implementation of kernel in SVM.

2.1. Support Vector Machine

Support Vector Machine (SVM) is one of the most widely used methods of supervised learning machine to analyse and identify patterns [7]. For classification cases, SVM is also believed to produce high accuracy. SVM was first introduced to the world by classifying two classes. However, along with the development of researches over years, SVM has now been developed to be able to perform multiclass classification [2, 8-9]. Few examples of the application of SVM are insolvency prediction in insurance companies [10], policy holders satisfactorily classify automobile insurance [11], and intrusion detection system classification [12].

Basically, the Support Vector Machine algorithm constructs a hyper plane to separate two different classes. The hyper plane chosen by the algorithm is the one that maximizes the minimum distance between the hyper plane and the training data. The minimum distance is commonly referred to as margin [6].

The hyper plane formed by SVM is defined as follows [13]:

$$f(x) = \boldsymbol{w}^T \cdot \boldsymbol{x} + b \tag{1}$$

where $\mathbf{w}$ is a set of weights and b is the bias.

The optimization problem of SVM can be summarized as:

$$\min \frac{1}{2} \|\boldsymbol{w}\|^2 \tag{2}$$

subject to:

$$y_i(\boldsymbol{w}^T \boldsymbol{x}_i + b) \geq 1,\ \forall i = 1,2,\dots,N \tag{3}$$

The above optimization problem in equation (2) with the constraint in equation (3) was considered as a quadratic programming problem. Therefore, one way to solve this was to find the dual form of the equation. We then introduced Lagrange multiplier$\lambda_i, i = 1,2,\dots,N$. Then, Lagrange function was formed as follows:

$$L(\boldsymbol{w}, b, \lambda) = \frac{1}{2} \|\boldsymbol{w}\|^2 - \sum_{i=1}^{N} \lambda_i (y_i(\boldsymbol{w}^T \boldsymbol{x}_i + b) - 1) \tag{4}$$

By finding the derivatives of equation (4) with respect of $\boldsymbol{w}$ and b equal to zero and then substituting the results back to the equation produced the dual form of equation (2), stated as:

$$\max\left(\sum_{i=1}^{N} \lambda_i - \frac{1}{2} \sum_{i=1}^{N} \sum_{j=1}^{N} \lambda_i \lambda_j y_i y_j (\boldsymbol{x}_i \cdot \boldsymbol{x}_j)\right) \tag{5}$$

subject to:

$$\sum_{i=1}^{N} \lambda_i y_i = 0,\ \lambda_i \geq 0 \tag{6}$$

By solving equation (5) with constraints expressed in equation (6), we could find the optimal parameters and therefore the optimal hyper plane could be formed.

The above SVM discussion was accurate for linearly separable data. However, it was rare or even impossible to find such perfectly linearly separable data in this world. To overcome this, we could map the data to a higher space such that they could be separated linearly in the new space. This was called the kernel trick.

2.2. Implementation of kernel in SVM.

In general, SVM could produce high accuracy for classification with data that could be separated linearly. However, in reality, such data were rarely found. So we needed an additional method that could overcome the problem of non-linear data when using SVM. One way that could be used was by implementing kernel in SVM.

The kernel function could act as a link between non-linear data and the SVM algorithms. The idea of kernel function was to map data in the input space to a new feature space that had a higher dimension such that the data could be separated linearly in the new feature space.

Suppose $\boldsymbol{x_1}, \boldsymbol{x_2}, \ldots, \boldsymbol{x_n} \in \mathbb{R}^n$ was a set of the original data in $\mathbb{R}^n$. There was a function ϕ that mapped the data in $\mathbb{R}^n$ to a new feature space with a higher dimension$F$.

$$\phi: \mathbb{R}^n \to F$$

The kernel function was defined as follows [13]:

$$K(\boldsymbol{x}, \boldsymbol{y}) = \phi(\boldsymbol{x}) \cdot \phi(\boldsymbol{y}) \tag{7}$$

and the distance between two kernelled data were defined as:

$$d(\boldsymbol{x}, \boldsymbol{y}) = \|\phi(\boldsymbol{x}) - \phi(\boldsymbol{y})\| = \sqrt{K(\boldsymbol{x}, \boldsymbol{x}) - 2K(\boldsymbol{x}, \boldsymbol{y}) + K(\boldsymbol{y}, \boldsymbol{y})} \tag{8}$$

There were several types of kernel, but this research used two types that had been commonly used for SVM, which were the Radial Basis Function (RBF) kernel and the polynomial kernel. The RBF kernel was defined as:

$$K(\boldsymbol{x}, \boldsymbol{y}) = exp\left(-\frac{\|x-y\|^2}{2\sigma^2}\right) \tag{9}$$

and the polynomial kernel was defined as [14]:

$$K(\boldsymbol{x}, \boldsymbol{y}) = (\boldsymbol{x} \cdot \boldsymbol{y} + 1)^d \tag{10}$$

The idea of kernelled SVM remained the same as the original SVM, but the work was done when all the data had been mapped into a new feature space such that the data became linearly separable. So, the hyper plane was now defined as [6]:

$$f(\boldsymbol{x}) = \boldsymbol{w}^T \phi(\boldsymbol{x}_i) + b \tag{11}$$

The optimization problem of kernelled SVM was:

$$min \frac{1}{2} \|\boldsymbol{w}\|^2 \tag{12}$$

subject to:

$$y_i(\boldsymbol{w}^T \phi(\boldsymbol{x}_i) + b) \geq 1 - \xi_i,\ \forall i = 1,2,\ldots,N \tag{13}$$

By using Lagrange multiplier and forming the Lagrange function, we could find the dual form for the optimization of the kernelled SVM as:

$$\max\left(\sum_{i=1}^{N} \lambda_i - \frac{1}{2} \sum_{i=1}^{N} \sum_{j=1}^{N} \lambda_i \lambda_j y_i y_j K(\boldsymbol{x}_i, \boldsymbol{x}_j)\right) \tag{14}$$

subject to:

$$\sum_{i=1}^{N} \lambda_i y_i = 0, \ \lambda_i \geq 0 \tag{15}$$

By solving equation (14) with constraints expressed in equation (15), we could find the optimal parameters for the optimal hyper plane. Suppose λ^* was the solution, then we could find that

$$\boldsymbol{w} = \sum_{i=1}^{N} \lambda_i^* y_i \phi(\boldsymbol{x}_i) \tag{16}$$

Substituting $\boldsymbol{w}$ in equation (16) to $f(\boldsymbol{x})$ in equation (11) could result finding b as follows:

$$b = \frac{1}{n_S} \sum_{i\in S}^{n} \left(y_i - \sum_{i\in S}^{N} \lambda_i^* y_i K(\boldsymbol{x}_i, \boldsymbol{x}_j)\right) \tag{17}$$

3. Data

The data used in this research were collected from the Look-Alike Faces Database (LAF) which was created by Lamba et al. in 2011 [4]. From the database, there were 50 subjects or people who had 5 images of their genuine face and 5 look-alike images. So in total, 500 face images were used in this research.

The images from the database had already been separated into two different categories by Lamba et al. [4], which were the genuine face images and the look-alike face images. The genuine face images consisted of the faces of well-known or famous people such as the celebrities, ranging from Western, Eastern, and Asian origins. While the look-alike face images consisted of ordinary people from all over the world. The genuine face images were used for the training data set, while the look-alike images were used for the testing dataset. So, the ratio of the training data set and the testing dataset was 5:5 for each subject.

Images in the Look-Alike Faces Database were varied in colour and size. Some images were in colour form (RGB) and some were in grayscale form. All the images were converted into grayscale form for this research. The reason of doing that was particularly because computational time needed would be much reduced.

4. Experiment Results

Experiment for face recognition to identify look-alike faces as proposed in this research was done and the results were summarised in table 1 and table 2 below. Each table contained the results of face recognition to identify look-alike faces using Support Vector Machine. Table 1 shows the result when Radial Basis Function kernels with various σ were used, whereas table 2 shows the result when polynomial kernels with various d were used.

As seen in table 1, the experiments done using Support Vector Machine with implementation of RBF kernel resulted in similar accuracies which were 94% except for $\sigma = 0.00001$ which obtained 74.4%. Therefore, the next step was to analyse the results in the amount of time needed which was the running time. The lowest running time for this case was 91.75 seconds with $\sigma = 1$. So, the best result for face recognition to identify look-alike faces using Support Vector Machine with RBF kernel in this research was obtained when $\sigma = 1$ was used.

For the results of Support Vector Machine with polynomial kernel for the face recognition to identify look-alike faces, the highest accuracy was also 94%. The highest accuracy while using Support Vector Machine with polynomial kernel was obtained four times in this experiment, which were for $d = 7$ to $d = 10$. Further analysis is to see the running time and choose the lowest one. The lowest running time for this case is 74.796875 seconds with $d = 7$. So, the best result for face recognition to identify look-alike faces using Support Vector Machine with polynomial kernel in this research was obtained when $d = 7$ was used. It could be seen that for the ones that obtained the highest accuracy, which were from $d = 7$ to $d = 10$, the time needed became higher as the parameter increased.

IOP Conf. Series: Journal of Physics: Conf. Series **1108** (2018) 012071 doi:10.1088/1742-6596/1108/1/012071

Table 1. Results Obtained Using Support Vector Machine with Σ as the Parameter of the Radial Basis Function Kernel

σ	Accuracy (%)	Running time (seconds)
0.00001	74.4	126.390625
0.0001	94	126.671875
0.001	94	126.65625
0.01	94	125
0.1	94	126.859375
0.5	94	126.328125
1	94	91.75
10	94	314.5625
100	94	126.265625
1000	94	126.0625

Table 2. Results Obtained Using Support Vector Machine With *d* As The Parameter Of The Polynomial Kernel

d	Accuracy (%)	Running time (seconds)
1	74.4	92.109375
2	2	160.96875
3	2	127.09375
4	2	141.421875
5	2	134.140625
6	0.4	1113.828125
7	94	74.796875
8	94	82.296875
9	94	92.921875
10	94	312.671875

5. Conclusion

This research proposes face recognition to identify look-alike faces using Support Vector Machine. Two types of kernel function have been implemented, which are the Radial Basis Function (RBF) kernels and the polynomial kernels. The data used are 500 face images taken from the Look-Alike Face database (LAF).

Based on the results obtained, the highest accuracy using Support Vector Machine with RBF kernel is 94%, with 91.75 seconds as the lowest running time. For the polynomial kernel, the highest accuracy is also 94% with 74.796875 seconds needed. For comparison, the highest accuracy obtained in this research by using two different kernels are the same, but the lowest running time needed for the face recognition to finish by using polynomial kernel is lower than the running time needed when RBF kernel was used. In conclusion, for face recognition to identity look-alike faces, SVM has successfully produced high accuracies.

IOP Conf. Series: Journal of Physics: Conf. Series **1108** (2018) 012071 doi:10.1088/1742-6596/1108/1/012071

6. References

[1] Guo G, Li S and Chan K 2000 Face Recognition by Support Vector Machines *Proc. 4th. IEEE International Conference on Automatic Face and Gesture Recognition* pp 196−201

[2] Li J, Zhao B, Zhang H and Jiao J 2009 Face Recognition System Using SVM Classifier and Feature Extraction by PCA and LDA Combination *Int. Conf. on Computational Intelligence and Software Engineering* pp 1−4

[3] Wei J, Jian-qi Z and Xiang Z 2011 Face Recognition Method Based On Support Vector Machine And Particle Swarm Optimization *Expert Systems with Applications* **38** pp 4390−4393

[4] Lamba H, Sarkar A, Vatsa M, Singh R and Noore A 2011 Face Recognition For Look-Alikes: A Preliminary Study *Int. Joint Conf. on Biometrics (IJCB)* pp 1−6

[5] Moeini A, Faez K, Moeini H and Safai A M 2017 Open-Set Face Recognition Across Look-Alike Faces in Real-World Scenarios *Image and Vision Computing* **57** pp 1−14

[6] Cortes C and Vapnik V 1995 Support-Vector Networks *Machine Learning* **20** pp 273−297

[7] Vapnik V N 1999 An Overview of Statistical Learning Theory *IEEE Transactions on Neural Networks* **10**(5) pp 988−999

[8] Ko J and Byun H 2003 Multi-Class Support Vector Machines with Case-Based Combination for Face Recognition *Int. Conf. on Computer Analysis of Images and Patterns* pp 623−629

[9] Prakash J S, Vignesh K A, Ashok C and Adithyan R 2012 Multi Class Support Vector Machines Classifier for Machine Vision Application *Int. Conf. on Machine Vision and Image Processing (MVIP)* pp 197−199

[10] Rustam Z and Yaurita F 2018 Insolvency Prediction in Insurance Companies Using Support Vector Machines And Fuzzy Kernel C-Means *Journal of Physics: Conference Series* **1028**(012118)

[11] Rustam Z and Ariantari N P A A 2018 Support Vector Machines for Classifying Policyholders Satisfactorily in Automobile Insurance *Journal of Physics: Conference Series* **1028**(012005)

[12] Rustam Z and Zahras D 2018 Comparison Between Support Vector Machine and Fuzzy C-Means as Classifier for Intrusion Detection System *Journal of Physics: Conf. Series* **1028**(012227)

[13] Bishop C M 2006 Pattern Recognition and Machine Learning (New York: Springer)

[14] Liu L, Shen B and Wang X 2014 Research On Kernel Function of Support Vector Machine *Advanced Technologies, Embedded and Multimedia for Human-centric Computing* pp 827−834

A Comparison of O'Sullivan Penalized Spline and Penalized Spline Based Truncated Power Basis Methods to Predict Ozone Concentrations

R Fitri[1], Y Widyaningsih[2] and D Lestari[3]

[1, 2, 3] Department of Mathematics, Universitas Indonesia, Depok, 16424, Indonesia

Abstract. This paper aims to predict ozone concentrations based on the effects of ultraviolet light. Spline regression is piecewise polynomials that connect join points called knots. In spline regression, parameter estimation was fit by OLS (Ordinary Least Square) method. However, the OLS method leads to over parameterized and in the plot of estimated regression curve is fluctuated when using too much knots. Therefore, it needs an additional constraint which contains smoothing parameter, so that results an ideal fit. This parameter estimation method is known as PLS (Penalized Least Square) method. Spline regression using PLS method is called as penalized spline regression. O'Sullivan (1986) introduced a class of penalised splines based on B-spline basis functions. OPS (O'sullivan Penalized Spline) is a direct generalisation of smoothing splines that latter arises when the maximal number of B-spline basis functions included. One of the top performance measures the predicted regression curve that can be used is MSE (Mean Square Error). Results showed that the OPS method had a smaller MSE and GCV than the PSTPB (Penalized Spline Based Truncated Power Basis) method, so the use of the OPS method for predicting ozone against ultraviolet light was suitable to use.

1. Introduction

Among others, spline is one included in nonparametric regression. There are two bases in spline regression, namely B spline and truncated power bases. The advantage of B spline base is that we can cope with high order spline, many knots, and the knot is too closed [8]. In spline regression, parameter estimation is fit by OLS (Ordinary Least Square) method. However, the OLS method leads to over parameterized and in the plot of estimated regression curve will be fluctuated when using too much knots. Therefore, it needs an additional constraint which contains smoothing parameter, so that results an ideal fit. This parameter estimation method is known as PLS (Penalized Least Square) method. Spline regression using PLS method is called as penalized spline regression. O'Sullivan uses a relatively large number of knots. To prevent over fitting, a penalty on the second derivative restricts the edibility of the A major problem of any smoothing technique is the choice of the optimal amount of smoothing, in our case the optimal weight of the penalty. Smoothing splines have a special place in nonparametric regression. O'Sullivan penalised splines become a direct generalisation and closer approximation of smoothing splines.

Ozone is one of many gases in stratosphere. Although the concentration of ozone is relatively small, it plays an important role for life on Earth due to its ability to absorb ultraviolet radiation (UV) from the sun. In the course of the last 20 years we have often heard about "the depletion of the ozone layer" and the so-called "ozone hole", which have been considered as a threat to our health and environment [5]. Ozone concentrations, however, are not constant, and fluctuate quite a bit from day to day, depending on many factors. Therefore, this research aims to predict ozone concentrations based on the effects of ultraviolet radiation. We use OPS and PSTPB methods to create a relationship curve between ozone and ultraviolet radiation. Then from both methods we compare which smaller MSE. Minimum MSE and GCV of both methods are used to predict ozone.

2. Smoothing Splines

The idea behind smoothing splines is to combine measures of the smoothness of a function and how well it fits the data. For the goodness of fit for the function, we use the residual sum of squares as a criterion [1]. Non-smoothness, like noise and rapid changes, can be suppressed by minimizing the penalty $\int (f''(x))^2 dx$ the integrated second derivative. Together these two criteria formulate the smoothing spline technique as minimization problem

$$PLS(\lambda) = \sum_{i=1}^{n} (y_i - f(x_i))^2 + \lambda \int [f''(x)]^2 dx \tag{1}$$

where the estimated solution is $\hat{f_\lambda}(x) = \arg\min PLS(\lambda)$. As $\lambda \to 0$ we impose no penalty and end up with a very close fit, but the resulting curve could be very noisy as it follows every detail in the data. As $\lambda \to \infty$ the penalty dominates and the solution converges to the OLS line, which is as smooth as you can get (the second derivative is always 0), but may be a very poor fit. Amazingly, it can be shown that minimizing $PLS(\lambda)$ for a fixed λ over the space of all continuous different functions leads to a unique solution, and this solution is a natural cubic spline with knots at the data points. A key point is the smoothing parameter λ, which controls this trade-off between goodness-of-fit and roughness.

Theorem 2.1 Given interpolation data $(t_i, y_i)_{i=0}^{n}$. Among all functions $f \in C^2[a, b]$ which interpolate (t_i, y_i), the natural cubic spline is the smoothest, where smoothness is measure through $\int [f''(x)]^2 dx$.

Proof. We need to prove that $\mu(f) \geq \mu(S)\ \forall f \in C^2[a, b]$ introduces $g(x) = S(x) - f(x)$, $g(x) \in C^2[a, b]$, $g(t_i) = 0$, $i = 0, \dots, n$ Inserting this yields

$$\mu(f) = \int_a^b (S''(x) - g''(x))^2 dx = \mu(S) + \mu(g) - 2\int_a^b S''(x) g''(x) dx \tag{2}$$

Since $\mu(g) > 0$, we have proved our result if we can show that

$$\int_a^b S''(x) g''(x) dx = 0 \tag{3}$$

We have that,

$$\int_a^b S''(x) g''(x) dx = g'(x) S''(x)|_b^a - \int_a^b g'(x) S'''(x) dx \tag{4}$$

First part on the right hand size is zero since $z_0 = z_n = 0$. Second part we split an integral over each subdomain

$$-\int_a^b g'(x) S'''(x) dx = -\sum_{i=0}^{n-1} \int_{t_i}^{t_{i+1}} g'(x) S'''(x) dx = -\sum_{i=0}^{n-1} 6a_i \int_{t_i}^{t_{i+1}} g'(x) dx$$

$$= -\sum_{i=0}^{n-1} 6a_i g(x) \Big|_{t_i}^{t_{i+1}} = 0$$

The role of the smoothing parameter in penalized smoothing is to control the smoothness of the fitted curve. In order to compute the optimal value of the smoothing parameter λ selection criteria are considered in this paper generalized cross validation (GCV) [6]. The GCV method is computationally simpler and very well used in the literature about smoothing splines. We consider two versions of GCV error, one for the smoothing splines and other for Penalized splines. The GCV method consists of selecting λ so that minimized

$$GCV(\lambda) = \frac{1}{n} \frac{\sum_{i=1}^{n} (y_i - \hat{f}(x_i))^2}{[1 - n^{-1} tr(\boldsymbol{I} - \boldsymbol{H})]^2} = \frac{MSE(\lambda)}{[n^{-1} tr(\boldsymbol{I} - \boldsymbol{H})]^2} \tag{5}$$

where $\boldsymbol{H}$ is hat matrix in the case of the smoothing splines.

3. B-spline

The estimate $\hat{f}_\lambda(x)$ is a cubic spline, meaning the function is a piecewise cubic polynomial. It is divided into intervals associated with a knot sequence

$$a = \tau_1 \leq \tau_2 \leq \cdots \leq \tau_{k-1} \leq \tau_k = b \tag{6}$$

The spline can be represented in terms of a basis, which give the term B-spline. The function $S(x)$ is specified by a coefficient or weight vector $\boldsymbol{w} = (w_1, w_2, \ldots, w_n)^T$ and the basis matrix $\mathrm{B(x)} = [B_{1,\mathrm{d}}(x), \ldots, B_{n,d}(\mathrm{x})]$, where $B_{i,d}(\mathrm{x})$ are i basis functions. The polynomial spline is given by

$$S(x) = \sum_{j=1}^{n} B_j(\mathrm{x}) w_j \tag{7}$$

A B-spline on an interval $[a, b]$ has order $m = d + 1$, where d is the degree of the polynomials, and number of internal knots k. An augmented knot sequence τ_1 is defined by placing m equal boundary knots on the end points and such a knot sequence can support a basis of order l up to $l \leq m$

$$a = \tau_1 = \tau_2 = \cdots = \tau_m < \tau_{m+1} < \cdots < \tau_{k+m} < \tau_{K+m+1} = \cdots = \tau_{K+2m} = b \tag{8}$$

Each duplication of the boundary knots results in the loss of one continuous derivative. The number of basis function B_i supported by the augmented sequence is $k + m$. This can be shown by counting the parameters needed to be specified. We will have $k + 1$ regions multiplied with m function parameters per region, but we must subtract the parameters specified by the continuity constraints given by k internal knots multiplied by $m - 1$ constraints on derivatives per knot. So the number of parameters and basis functions are

$$(k + 1).m - k.(m - 1) = k + m \tag{9}$$

The basis is found by the Cox-de Boor recursion formula for $j = 1, \ldots, k + 2m - 1$:

$$B_{j,1}(x) = \begin{cases} 1, & \text{if } \tau_j < x \leq \tau_{j+1} \\ 0, & \text{otherwise} \end{cases} \tag{10}$$

$$B_{j,m}(x) = \frac{x - \tau_j}{\tau_{j+m-1} - \tau_j} B_{j,m-1}(x) + \frac{\tau_{j+m} - x}{\tau_{j+m} - \tau_{j+1}} B_{j+1,m-1}(x) \tag{11}$$

If the knot sequence is uniform, the basic functions become shifted copies of another [3].

4. O'sullivan Penalized Spline (OPS)

The paper *On semi parametric regression with O'Sullivan penalized splines* (Wand and Ormerod, 2008) introduces penalized splines using quintile-based, non-uniform knots [7]. The penalty is based on the integral of B-spline basis functions, as introduced by O'Sullivan, and the spline estimate satisfies natural boundary conditions. The cubic B-spline basis functions $B_1, \ldots, B_{K+4}$ are defined by a knot sequence given as (10) and (11). The knots k_i are chosen to be the $\frac{i}{K+1}$th sample quintile of the unique data points x_i. The design matrix B has entries $B_{ij} = B_j(x_i)$ and the $(K+4) \times (K+4)$ penalty matrix $\boldsymbol{\Omega}$ is defined by

$$\boldsymbol{\Omega}_{ij} = \int_a^b B_i''(x) B_j''(x) dx \tag{12}$$

The estimate $\hat{f}(x; \lambda)$ is the minimizer of

$$S(\boldsymbol{w}) = (\boldsymbol{y} - \boldsymbol{Bw})^T (\boldsymbol{y} - \boldsymbol{Bw}) + \boldsymbol{\lambda w^T \Omega w}, \tag{13}$$

Giving

$$\hat{f}(x; \lambda) = \boldsymbol{B}\widehat{\boldsymbol{w}}, \text{ where } \widehat{\boldsymbol{w}} = \left(\boldsymbol{B^T B} + \boldsymbol{\lambda \Omega}\right)^{-1} \boldsymbol{B^T y}, \tag{14}$$

Where $\boldsymbol{B}_x = [\boldsymbol{B}_1, \ldots, \boldsymbol{B}_{K+4}]$ dan $\boldsymbol{\lambda} > \boldsymbol{0}$ is the smoothing parameter.
Wand and Ormerod (2008) comment that the cubic smoothing spline arises in the special case $k = n$ and $\tau_{i+4} = x_i$ for $1 \leq k \leq n$ provided that x_i is distinct.

The estimate $\hat{f}(x; \lambda)$ satisfies the natural boundary condition, a constraint on the derivatives, meaning that,

$$\hat{f}''(a; \lambda) = \hat{f}'''(b; \lambda) = \hat{f}''(a; \lambda) = \hat{f}'''(b; \lambda) = 0 \tag{15}$$

IOP Conf. Series: Journal of Physics: Conf. Series **1108** (2018) 012072 doi:10.1088/1742-6596/1108/1/012072

which implies that $\hat{f}(x;\lambda)$ is approximately linear over the intervals $[\boldsymbol{a}, k_5]$ and $[k_{K+4}, b]$. The linearity is exact if $k_5 = \min(x_i)$ and $k_{K+4} = \max(x_i)$.
The penalty matrix $\boldsymbol{\Omega}$ can be computed in R by the second derivative design matrix $\boldsymbol{B}''$ and is given for a cubic basis

$$\boldsymbol{\Omega} = (\widetilde{\boldsymbol{B}}'')^T \text{diag}\,\mathbf{c}\,(\widetilde{\boldsymbol{B}}'') \tag{16}$$

Where $\widetilde{\boldsymbol{B}}''$ is the $\times\ (K+4)$ matrix with entries $B_j''(\tilde{x}_i)$, $\tilde{x}_i$ is from the vector

$$\tilde{x} = \left(k_1, \frac{k_1 + k_2}{2}, k_2, k_2 \frac{k_2 + k_3}{2}, k_3, \ldots, k_{K+7}, \frac{k_{K+7} + k_{K+8}}{2}, k_{K+8}\right) \tag{17}$$

and $\mathbf{c}$ is the $3(K+7) \times 1$ vector

$$\mathbf{c} = \left(\frac{1}{6}(\Delta k)_1, \frac{4}{6}(\Delta k)_1, \frac{1}{6}(\Delta k)_1, \ldots, \frac{1}{6}(\Delta k)_{K+7}, \frac{1}{6}(\Delta k)_{K+7}, \frac{4}{6}(\Delta k)_{K+7}\right) \tag{18}$$

where $(\boldsymbol{\Delta k})_k = k_{k+1} - k_k, 1 \le k \le K+7$.

This result is given by applying Simpson's rule over each of the inter-knot difference using the second derivative design matrix to calculate the integrals defining $\boldsymbol{\Omega}_{ij}$ since each function $B_j''B_j''$ is piecewise-quadratic, Simpson's rule will calculate the integral exactly. The greatest advantages with O'Sullivan splines is the direct use of the B-spline basis functions to calculate the penalty matrix, giving a better approximation to the smoothing spline penalty. The function $\boldsymbol{f}(\boldsymbol{x})$ is approximated by a weighted set of basic functions, as a parallel to finite element methods.

5. Penalized Spline Based Truncated Power Basis (PSTPB)

There are several options for the penalization criteria, but the easiest to implement is to choose a C such that $\sum_{i=1}^{K} b_i^2 < C$, where b_i refers to the weight of each linear function. This is an excellent minimization criteria, because it reduces the overall effect of individual piecewise functions and avoids over-fitting the data. This minimization criteria is more formally stated as minimizing the equation $\sum_{i=1}^{n}(y_i - \hat{y}_i)^2$ subject to $\boldsymbol{\beta}^T \boldsymbol{D} \boldsymbol{\beta} \le C$, Where

$$\boldsymbol{D} = \begin{pmatrix} 0_{(p+1)\times(p+1)} & 0_{(p+1)\times K} \\ 0_{K\times(p+1)} & I_{K\times K} \end{pmatrix} \tag{19}$$

Using Lagrange multipliers, that this is equivalent to choosing ß to minimize [4]. Ruppert and Carroll (2003) used truncated polynomials as basic functions. In particular, with truncated polynomials of degree p based on K inner knots $a < u_1 < \cdots < u_K < b$, the penalized spline estimator is defined as the solution to the penalized least squares criterion

$$\sum_{i=1}^{n}(y_i - \hat{f}(x_i))^2 + \boldsymbol{\lambda}^{2p}\boldsymbol{\beta}^T\boldsymbol{D}\boldsymbol{\beta} \tag{20}$$

With $\hat{f}(x) = \{\alpha_0, \alpha_1 x, \ldots, \alpha_p x^p, \beta_1(x-u_1)_+^p, \ldots, \beta_k(x-u_k)_+^p\}$. This has the solution

$$\widehat{\boldsymbol{\beta}} = (\boldsymbol{X}^T\boldsymbol{X} + \lambda^{2p}\boldsymbol{D})^{-1}\boldsymbol{X}^T\boldsymbol{y} \tag{21}$$

The term is called a roughness penalty because it penalizes fits that are too rough, thus yielding a smoother result. The amount of smoothing is controlled by λ , which is therefore usually referred to as a smoothing parameter.

6. Application Data

The data used in this research are UV index data and total data of layer columns downloaded from AURA satellite for Bandung area at position 6.9^0 LS $107{,}5^0$ BT. OMI is an existing sensor on the AURA satellite launched in 2004. In the OMI sensor there is a total number of ozone layer and UV index. Then performed data processing by extracting data for each UV index data and total ozone columns obtained, then drawing a graph in the form to see patterns on the data. This pattern is made to look at the UV and ozone indices. The linkage of the UV index and the number of ozone columns in each season is expected to index the UV and ozone colour bias of Bandung in each season in 2013. The total ozone column data has Dobson Unit (DU), while the UV index has no units. Ozone concentration is very instrumental in the inhibition of solar UV radiation that reaches the surface of the earth. The change of the concentration of ozone in every minute reduces the intensity of solar UV radiation. Using the OPS method and PSTPB method to create a curve of the relationship between ozone and ultraviolet radiation. Then from both methods we compare which minimum MSE and the results showed that,

Table 1. Result of Comparison between OPS and PSTPB

	GCV	λ	MSE
OPS	6.548	8.823	0.00358
PSTPB	7.098	12.755	0.00987

Based on Table 1, it can be seen that the result of comparison between OPS and PSTPB shows that OPS has a smaller MSE of 0.00358 than PSTPB with MSE of 0.00987, where the λ obtained from the OPS method is less than PSTPB, and so is the GCV.

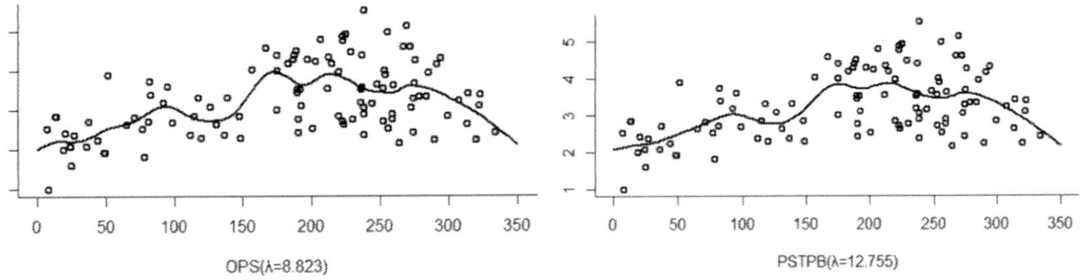

Figure 1. The plot of estimated regression above is the curve resulting from the OPS method with λ=8.823 and the curve which is obtained from the PSTPB method with λ=12.755.

7. Software

All calculations in this research were performed on the 3.3.1 version of statistical software R project, making use of the spline package for the B-splines basis construction. The Penalized splines approach was implemented through our own routine considering Eilers and Marx's definition and M.P. Wand and J.T. Ormerod. Smoothing parameters were chosen by our own routine based on the Generalized Cross Validation method. Smoothing Splines approximation was made by means of a function available in the spline library (smooth spline).

8. Conclusion

We have compared two methods namely OPS (O'sullivan Penalized Spline) and PSTPB (penalized spline truncated power base). Both methods have been compared in this paper to estimate a series of unobserved smooth curves from discrete noise observations. The comparison result between OPS and PSTPB shows that OPS has a smaller MSE that is 0.00358 compared to PSTPB with MSE 0.00987 where λ obtained from the OPS method is less than PSTPB, of which is similar to GCV. As the basis of this result, we can conclude that OPS and PSTPB spacing the splines lose their smooth control as the number of knots increases. Both inherited approaches can improve matches that give a mean squared error with respect to fine samples. On the other hand, PSTPB has numerical complexity that makes calculations easier and very insensitive to the choice of nodes, so it is enough to select a large number of vertices with equal distance. OPS methods are appropriate to avoid instability in numerical methods and at the same time have variations in the optimal smoothing rate to be found. Calculations using OPS are slightly more complicated than PSTPB methods. The OPS method has added flexibility due to the use of quartile based nodes, which provide more fundamental functions to areas with higher data density. This ensures that the computer power is not wasted in areas with small data and high uncertainty. Strategically placed knots, can reduce the number of basic functions. Another important feature of O'Sullivan splines is nature's boundary properties divided by spline alignment. So the use of OPS methods to predict ozone for ultraviolet light is suitable for use.

IOP Conf. Series: Journal of Physics: Conf. Series **1108** (2018) 012072 doi:10.1088/1742-6596/1108/1/012072

9. References

[1] Eubank R L 1999 *Nonparametric Regression and Spline Smoothing* (New York: Marcel Dekker)
[2] Eilers P H C and Marx B D 2010 Splines, knots, and penalties *Wiley Interdisciplinary Reviews Computational Statistics* **2**(6) pp 637–653
[3] De Boor C 2001 A Practical Guide to Splines Revised Edition *Applied Mathematical Sciences* chapter 27 (New York: Springer-Verlag)
[4] Ruppert D, Wand M P and Carroll R J 2009 Semiparametric Regression During 2003-2007 *Electronic Journal of Statistics* **3**(2009) pp 1193-1256
[5] Heriksen 2009 (Online) Ozone and UV-Radiation http://www.uio.no/
[6] Hastie T J and Tibshirani R J 1990 *Generalized Additive Models* London: Chapman and Hall
[7] Wand M P and Ormerod J T 2008 On Semiparametric Regression with O'Sullivan Penalized Splines *Australian & New Zealand Journal of Statistics* **50**(2) pp 179–198
[8] Khurotul L 2014 Election the Best B Spline Regression Model for Solar UV Radiation Estimation Based On Optimal Order and Point Knot *Thesis* University of Maulana Malik Ibrahim Malang

Acknowledgment

This research is funded by Universitas Indonesia via PITTA Research grant 2018 scheme.

Team Assisted Individualization (TAI) in Mathematics Learning Viewed from Multiple Intelligences

I I Siregar[1*], Budiyono[2] and I Slamet[3]

[1,2] Department of Mathematics Education, Faculty of Education and Teacher Training University of Sebelas Maret, Indonesia
[3] Department of Mathematics Education, Faculty of Education and Teacher Training University of Sebelas Maret, Indonesia

*Corresponding author: indraivantisiregar@gmail.com

Abstract. In this era, cooperative learning is very important in learning mathematics. The aim of this study is to find the effect of cooperative learning type TAI on the students' mathematics achievement viewed from multiple intelligences. This research is quasi experiment. The population of this study is the students of 8th grade Wonogiri district in the academic year 2016/2017. The sample of this research was taken using stratified cluster random sampling technique. The data was collected through the achievement and multiple intelligence tests. Data were analyzed in two ways, ANOVA with unequal cell and scheffe test. The result shows that this learning method has significant value 0.003; it means that students' mathematics achievement using TAI had successfully given different effects on mathematics learning achievement. While, in multiple intelligences category it has significant value 0.043, it means that there is a different mathematics learning achievement in multiple intelligences. There is no interaction between multiple intelligent and learning models. TAI learning model can be used by the teachers in mathematics learning, which is expected to improve the mathematics achievement.

1. Introduction

Educating student of the 21st century demands a different approach, a deviation from traditional way of teaching [1]. In order to meet the demands of students, education must change with ever-changing student [2]. Applying the effective teaching approaches can help the teachers to support students achieving learning objectives [3]. Mathematics is a science that underlies the development of modern technology and has an important role in developing the human mind power, so that mathematics is given in every level of education from primary school to high school.

According to Slavin [4] the difficulties in learning sciences are not sufficiently solved by conventional method, but can be assisted by cooperative learning method. One of the causes of the poor quality of students' in mathematics is in classroom learning, teachers are too oriented to procedural and mechanistic things such as teacher-centered learning, mathematical concepts often delivered informatively, and students are trained to complete without a deep understanding [5]

Several approaches that can be done in the classroom, namely: (1) Competition, which involve students in a fight that can determine who win and who lose; (2) Individualism, which ask students to work independently of a particular assigned task and strive alone to achieve the learning goal; and (3) Cooperation, which is to work together in a group and each member of the group is ensure to contribute and understand the task being done [6]. Cooperative learning gives more space and opportunities for students to discuss, solve problems, create solutions, provide ideas and help each other [7]. Through the use of these cooperative learning strategies, the teacher-researcher found that students' motivation to learn increases tremendously and this may have contributed to them showing higher achievement [8].

There have been many studies proving that cooperative learning could improve the quality of the learning process. Cooperative learning indicated that there is a significant difference in achievement than individual learning [7]. The cooperative learning with STAD strategy combine with TGT showed better student learning outcomes than other learning strategies that include conventional learning strategies [3]. Fui was found that the use of cooperative learning does increase students' academic achievement, help students to achieve a better understanding of physics concept and increases students; motivation to learn[8]. The result of Effandi study indicate that the cooperative learning approach resulted a higher achievement than the traditional teaching approaches and also indicate that the cooperative learning approach increase attitude towards mathematics [9].

Cooperative learning has many variants, one of them are Team Assisted Individualism (TAI). TAI is one of the simple cooperative learning model involving heterogeneous groups. There are several steps of cooperative learning model type TAI (1)Curriculum Materials, students are given the task to study the individual learning materials that have been prepared by the teacher, (2) Placement Test, students are given individual quizzes to get a basic score or an initial score. The results of these tests are used to determine the making of groups, (3) Teams, students consist of several groups where each group consists of 4-5 students with different abilities of either skill level (high, medium, and low). If possible group members come from different races, cultures, tribes and gender equality, (4) Team Study, student learning outcomes are individually discussed in groups. In group discussions, each group member checks each other's answers, (5) Teaching Group, students are facilitated in making summaries, directing, and affirming learning materials that have been studied, (6) Fact Test, students are given individual tests, (7) Team Scores and Team Recognition, teachers reward groups based on the value of improving individual learning outcomes from baseline scores to subsequent quiz scores, (8) Whole Class Unit, the teacher reminded the material to be learned in the next meeting. Teacher closes learning with messages of motivation and prayer.

Each component of cooperative learning model type TAI brings benefit to the teachers, students, high groups and low groups who work together completing the academic tasks, namely: the clever students take the responsibility to help the weak group. Thus, they can develop their abilities and skills together.

The application of this method has been proven to have a significant effect on the learning. The ability of the students of the Department of Forestry and Environmental Sciences taught by TAI type cooperative learning model is better than the students who are taught using conventional learning, with TAI strategies get an average 60.09 and conventional strategies get 53.67 [10]. The mathematics communication ability has a significant correlation with the cooperative learning type TAI, because the class using the cooperative learning, the students are expected to help each other, to mutually discuss and argue, to sharpen the knowledge they have at the current time, and to overcome the gap in the understanding among students [11].

As outlined by Gardner [12] there are nine different types of intelligences that are examined in this study, namely verbal-linguistic, logical-mathematical, visual-spatial, musical, interpersonal, intrapersonal, kinesthetic, naturalistic and existential. Every child has strong and weak sides no matter what his intelligence is, on condition that only one or two of these intelligence areas are used, the students whose intelligences types don't take place among the ones used at school cannot be able to improve their brilliant sides, complete learning in a long period without enjoying or managing to learn [13]. According to Thobroni [14] multiple intelligences is a dual ability to solve problems encountered in life. But in this study type of multiple intelligences we will limit to three types namely logical mathematical, Linguistic intelligence, Interpersonal intelligence.

Logical mathematical intelligence is the understanding and use of logical structures, including patterns and relationships, statements a proposition through experimentation, quantification, conceptualization and classification [15]. Linguistic intelligence is a kind of language capacity to achieve an aim and capability to use language learning ability in terms of writing and speaking effectively [16]. Interpersonal intelligence is the capacity of understanding, distinguishing and welcoming the emotions, aspirations and need of surrounding people [17]. Research conducted by Samsudi [18] showed a relationship between multiple intelligence and the preferred science teaching. There is a significant difference between the experimental group that taught according to the multiple intelligences theory and the control group which taught according to the traditional method; and this difference has been found to be in favor of the experimental group [13].

Based on the background, the purpose of the researcher is to find the effect of cooperative type of TAI on student mathematics achievement viewed from multiple intelligences in grade VIII in Wonogiri.

2. Method

This research used a quasi-experimental research with factorial design 2x2. The aim of this research was to know the influence of two independent variables to dependent variable. The independent variables in this research were models learning which TAI as an experimental class and direct learning as control class, and then three multiple intelligence logic mathematics, linguistics and interpersonal intelligence categorize by using questionnaires. While the dependent variable was the student's mathematics learning achievement. The population of this research was all the students of eight grades in Junior High School in

Wonogiri in 2016/2017, and the samples were three schools in Wonogiri that categorized into high, moderate and low school. The sample was taken by using cluster random sampling. This research used three methods in collecting the data. They were documentation method used to get student's profile data, the test method used to obtain student's mathematics learning achievement data that contains the indicators of geometry materials in the form of multiple choice questions and using questionnaires to categorize the multiple intelligences of students. This research hypothesis used two ways analysis of variance (two – way ANOVA) with unequal cells which the research data should be qualified for the normality and the homogeneity.

3. Result and Discussion

3.1 Decryption of Student Score in TAI model and Direct Learning

Table 1. The following is the result of students' value in cooperative model type TAI and direct learning

Statistics	Value Statistics	
	TAI	Direct Learning
Total Sample	80	81
Lowest Value	70.710	64.844
Highest Value	76.273	70.345
Average Value	73.84	67.63
Standard Deviation	13.314	12.476

3.2 Categorization of Students' Multiple Intelligences in Each Learning Model

The instrument to categorize the multiple intelligences of students is using questionnaires. The questionnaire used 15 items for each intelligence category. At the end, the students were categorized into a particular type of intelligence which was seen from the most dominant score among the three scores of questionnaires.

Table 2. Categorization of Student Multiple Intelligence

Learning Method	Intelligent	N
TAI	Logic Mathematics	24
	Linguistics	25
	Interpersonal	31
Direct Learning	Logic Mathematics	28
	Linguistics	28
	Interpersonal	25

3.3 Normality and Homogeneity Test

Prior to the two-way ANOVA test, normality and homogeneity were performed the data with significance level 5% by using SPSS. The result of normality test of student's mathematics learning achievement showed that from experimental class and control class the sig. Value > 0.05 that were 0.904>0.05, and 0.623> 0.05. This indicated that the sample of students' mathematics achievement test came from a normality distribution population.

The result of homogeneity test of student's mathematics learning achievement between experiment class and control class showed the sig. Value>0.05 that was 0.155>0.05. This indicated that the population has the same or homogeneous variance.

3.4 Two Ways ANOVA

Table 3. The Result of Two-Ways ANOVA with Unequal Cells

Tests of Between-Subjects Effects

Dependent Variable: score

Source	Type III Sum of Squares	df	Mean Square	F	Sig.
Corrected Model	3733.147[a]	5	746.629	4.768	.000
Intercept	794956.716	1	794956.716	5076.203	.000
METHOD	1388.780	1	1388.780	8.868	.003
INTELLIGENT	1005.861	2	502.930	3.211	.043

METHOD * INTELLIGENT	1048.508	2	524.254	3.348	.083
Error	24273.710	155	156.605		
Total	833089.000	161			
Corrected Total	28006.857	160			

3.5 Estimated Marginal Mean

Table 4. Method of Learning

Method	Mean	Std. Error
TAI	73.491	1.408
Direct Learning	67.594	2.392

Table 5. Multiple Intelligence

Interpersonal Level	Mean	Std. Error
Logic mathematics	72.402	1.741
Linguistics	66.987	1.722
Interpersonal	72.239	1.682

3.6 Post Hoc Test with Scheffe on Multiple Intelligence

Table 6. Post Hoc Test with Scheffe on Multiple Intelligence

Intelligence	Intelligence	Sig.
Logic Mathematics	Linguistics	.010
Linguistics	Interpersonal	.028
Interpersonal	Logic Mathematics	.038

Based on the calculation in Table 3, the effect of learning model on mathematics learning has obtained value of F = 8.868 with sig. Value 0.003<0.05 so, H_0 rejected and H_a accepted. It can be concluded that there was difference effect using learning model in mathematics learning achievement. In experimental class in Table 4 had mean score 73.491and in control class had mean score 67.594 so it concluded that TAI is greater than direct learning in student's mathematics learning achievement. The effect of interpersonal intelligence on mathematics learning achievement has a value of F obtained of F=3.211 with sig. Value 0.043 < 0.05 so, H_0 rejected and H_a accepted. It can be concluded that there was a difference between logic mathematics, linguistics, interpersonal in multiple intelligence, so it will be continued with scheffe test. Based Table 6 result using Post Hoc Test with Scheffe, the result of logic mathematics intelligence differed significantly with and linguistics intelligence, in Table 5 logic mathematics had mean score 72.402 and linguistics intelligence score 66.987 so it can be concluded that learning method effectively improve students' logic mathematics intelligence than linguistics intelligence. Then, the result of linguistics intelligence differed significantly with and interpersonal intelligence, in Table 5 linguistics had mean score 66.987 and interpersonal intelligence score 72.239 so it can be concluded that learning method effectively improve students' interpersonal intelligence than linguistics intelligence. The result of interpersonal intelligence differed significantly with and logic mathematics intelligence, in Table 5 interpersonal had mean score 72,239 and logic mathematics intelligent score 72.402 so it can be concluded that learning method effectively improve students' logic mathematics intelligence than interpersonal intelligence. In Table 3 showed the sig. Value of method intelligent 0.083, so H_0 accepted and H_a rejected. It can be concluded that there was no interaction between method of learning and multiple intelligence to mathematics learning achievement.

In direct learning, teachers give explanations and narratives to the students verbally, while students listen carefully, and record the principal expressed by the teacher. The TAI learning model is a learning model that forms a heterogeneous small group with different backgrounds and ways of thinking to help each other in need. Suppose a responsible student guides a weak or clever student, otherwise it can increase student participation in small groups. The learning model of TAI provides better learning achievement compared to direct learning model. Based on two way variance analysis with unequal cell for row effect obtained H_0 rejected. This shows that there is influence of learning model on mathematics learning achievement. The learning model of TAI provides better mathematics learning achievement than direct learning. This is in accordance with research conducted by Slavin [19], he compared this kind of learning with direct learning, TAI can improve learning achievement and solve math problems.

The second result of this study is that students with mathematical-logical intelligence have better mathematics learning achievement than students with linguistic intelligence and interpersonal intelligence; students with interpersonal intelligence have better mathematics learning achievement than students with linguistic intelligence. Based on a two-way variance analysis with unequal cells for column effects obtained H_0 rejected. This shows that there is the influence of multiple intelligences (mathematical-logical, linguistic and interpersonal intelligence) on mathematics learning achievement. According to Gardner [12] because different terms of intelligence, then one individual with the other is not the same ability to solve encountered problems. To find out which multiple intelligences provide better mathematics learning achievement, it is necessary to compute average between columns. From the comparative test average between the columns obtained the conclusion (1) logical mathematical intelligence provides better learning achievement of mathematics linguistic intelligence. The results of this study are consistent with studies conducted by Endang [20] that conclude learners with mathematical categories provide better mathematical learning achievement than linguistic categories (2) interpersonal intelligence gives higher achievement than linguistic intelligence, because students with interpersonal intelligence like work in the group and easy to interact so that students can gain understanding from other students. Students with linguistic intelligence only have the ability to explain in good language, (3) logic mathematics intelligence provides higher achievement than interpersonal intelligence, because logical mathematical intelligence students tend to work with numbers, patterns, solve problems and analyze any situation either in group or alone learning activities.

4. Conclusion

Based on the result and data analysis, it can be concluded that there were differences in learning achievement between classes that were studied with TAI a direct learning strategy. This suggests that learning strategies affect the learning achievement. Further test result (Scheffe), showed the multiple intelligences category, students' mathematics achievement with logical language intelligence have better results than those with language intelligence category and students' mathematical achievement with language intelligence has better results than those with interpersonal intelligence category. TAI learning model can be used by the teachers in mathematics learning, which is expected to improve the mathematics achievement.

5. References

[1] Vinita A and Hema G 2016 *The Int. J. of Indian Psychology* **3** 121
[2] Denise J 2013 *J. of Education and Work* **27** 220
[3] Hasmyati and Suwardi 2018 *Conf. Series. J. of Physics* 1028(2018)012090
[4] Slavin R E 2008 *Cooperative Learning Teori, Riset dan Praktik* (Bandung: Nusa Media)
[5] I Putu Hendra W 2018 *Conf. Series. J. of Physics* 1028(2018)012134
[6] Johnson D W, Johnson R T and Stanne M B 2000 *Cooperative Learning Method:Meta Analysis*
[7] Adeneye O A A, Alfred O F and Samuel A O O 2012 *Int. J. of Mathematics Trends and Technology* **3** 7
[8] Fui fong HO and Hong Kwen BOO 2007 *Asia Pasific Forum on Science Learning and Teaching* **8** 1
[9] Effandi Z, Lu Chung C and Md. Yusuf D 2010 *J. of Sosial Sciences* **6** 272-5
[10] La Rudi 2017 *Int. J. of Education and Research* **5** 69
[11] Georgina M T 2015 *J. of Education and Practice* **6** 27
[12] Gardner H 1993 *Multiple Intelligences: The teory in practice* (New York: Basic Book)
[13] Sibel G Y 2013 *Int. J. on New Trends in Education and Their Implication* **4** 27
[14] Thobroni M 2011 *Belajar dan Pembelajaran: Pengembangan* (Yogyakarta: Ar-Ruzz Media
[15] Armstrong T 2003 *The multiple intelligence of reading and writing: making the words come alive.* (New York: ASCD)
[16] Gundesli F 2006 *Multiple intelligence theory and primary education institutions' potential effects on management structure.Unpublished Master's Project* (Kahramanmaraş Sutcu Imam University: Public Department of Management)
[17] Tan 2006 *Teaching Planning and Evaluation* (Ankara: PegemA Syf)
[18] Samsudin M A 2015 *J. of education and Learning* **9** 53-59
[19] Slavin R E, Madden N A and Leavey M 1984 *J. of Educational Psychology* **76** 813

[20] Endang H 2013 *Efektivitas Model Pembelajaran Kooperatif Tipe Team Assisted Individualization (TAI) dan Problem Based Learning (PBL) Pada Prestasi Belajar Matematika Ditinjau Dari Multiple Intelegences Siswa SMP Kabupaten Lampung Timur Tahun Pelajaran 2012/2013* (UNS: Unpublised)

Analysis of difficulties in mathematics learning on students with deictic gesture type in problem-solving HOTS algebra test

H Zulfa[1], D R S Saputro[2], and Riyadi[3]

[1] Student of Sebelas Maret University, Faculty of Teacher Training and Education, Sebelas Maret University,
36 A Ir. Sutami Street, Surakarta, Central Java, 57126, INDONESIA
[2,3]Lecturer in Sebelas Maret University Jl Ir. Sutami no 36 A, Surakarta, Central Java, INDONESIA

E-mail: hariratuz@gmail.com

Abstract. HOTS is a real-life based assessment of everyday life, but in practice, the students are having trouble in completing the HOTS issues. Learning difficulty is also influenced by gesture type. Edward (2008) explains that gesture with speaking skill is data resources about thinking in mathematics. Besides, gesture is part of integral idea in mathematics. McNeill classifies the gesture into 4 types, namely iconic, metaphoric, deictic, and beat. The researcher focuses on the type of deictic gesture that is the type of gesture to show object, event, or point out people. This study aims to describe the difficulty of learning mathematics of students with a type of deictic gesture in the completion of Algebra materials especially in solving HOTS. This research type is descriptive qualitative research. The results showed a total of 3.22 % difficulties of number fact skill, 7.52 % difficulties of arithmetic skill, 35.50 % difficulties of information skill, 30.10% difficulties of language skill, 23.70 % difficulties of visual-spatial skill.
Keywords : deictic gesture, learning mathematics, algebra

1. Introduction

Students are able to solve a problem only if they understand the problem and apply their knowledge into that new situation. This is known as high order thinking skills (HOTS) [1]. Assessment of the 2013 Curriculum is adapted from international standard assessment models which hopefully encourage students with high-level thinking to think broadly and deeply about the matter [2]. Bloom's Revision of Taxonomy is used to measure high-order thinking skills (HOTS). Anderson & Krathwohl classified the cognitive process to measure high-level thinking skills by category of analyze, evaluate and create [3]. According to Trends in International Mathematics and Science Study (TIMSS) survey in 2015, Indonesian mathematics achievement ranked 45th out of 50 countries [4]. By the Organization for Economic Co-operation and Development (OECD) in 2015 using the Program for International Student Assessment (PISA), Indonesia stayed in the 64th from 72 nations [5]. According to Totok Suprayitno, the head of Badan Penelitian dan Pengembangan (Balitbang) Kemendikbud, in 2015 Indonesia's mathematics quality scored 386 which is under 490 as the mean score [6]. Applying HOTS question according to PISA will increase it [7]. The high-level thinking ability should have been begun by schools in Indonesia in order to meet the demands of the 21st century which need more developed skills [8].

The results of National Examination of Senior High School in 2015/2016 in Tulungagung obtained from the BSNP PAMER program show that the absorptive capacity of the students in mathematics was low. SMAN I Rejotangan Senior High School was one of one of them with Dual Variables Linear Equation System (DVLES) at 33.56% which is under the mean score of the city which scored 67.13%, province 58.08% or even national with 49.85%[9]. DVLES material had the lowest absorption score, so the students must have difficulties on it. Many high school students are not able to find the alternative solution through the DVLES problem [10]. The difficulty is coming from the lack understanding of Linear Equation System (LES) concepts that ultimately inhibit the further learning process. Kazemi (2002) said that students tend to be reckless in the calculation process. As the example + sign is understood as – sign [11].They are also confuse in substituting the variable with the x or y which will be a blunder in multiple choice question [12]. Specifically, mistakes are found because students did not change the sign while adding or subtracting with another equation, merging the variable with Constanta, reversed in writing the division result and adding or subtracting number before divide it [13].

In dual coding theory, Pavio has been stated that information someone got is processed by verbal (text and voice) and non verbal (diagram, picture and animation) channel [14].By studying DVLES, students will increase their visualization, critical thinking, intuition, perspective, problem-solving, conjecturing, deductive reasoning, logical argument and proofing skill.

The difficulty of learning mathematics is not only because ineffective learning but may also by differences in gesture type. Roth & Welzel (2001) stated that gesture become a vital resource in learning mathematics [15]. Gesture is important to support a heavy, general and abstract material explanation [16].McNeill classified the gesture into 4 types namely Iconic, Metaphoric, Deictic and Beat classified by the body language differences as the evidence that body are involved in thinking and speaking about ideas [17].

Gesture shows how someone thinks mathematics. Besides, gesture is part of integral idea in mathematics[18].Students with deictic gesture are tend to point or show a body gesture rather than explain a complex thing [19]. By studying it, we will understand the student mind. Roth (2002) stated that deictic gesture is an integral part of mathematics idea communication, but there is only a few educational study about the role of gesture in learning especially for abstract fields like science and mathematics.[20]

Many students have not acquired the basic skills needed.[21]Ineffectiveness in learning process will adversely affect the mathematics acquiring skills. Understanding the difficulties students' experienced in improving mathematics skills will help them in the learning process. Researcher will analyze students' difficulties with deictic gesture type on the algebra HOTS with LES material. Like Nathan, this study focused on the difficulties in mathematics learning in number fact, arithmetic, information, language, visual-spatial skill [22]. Inspired by both Nathan and Roth, this study wants to analyze the difficulty of learning mathematics at students with deictic gesture type in LESHOTS problem-solving.

2. Research Methods

The research used qualitative method with descriptive approach. In this research, it is formed a description of difficulty learning student mathematics with deictic gesture at solving LES HOTS issue. This method was carried out by analyzing oral tests and interviews with students. Subjects in this study were students with deictic gesture based on low (S1), medium (S2) and high (S3) achievement in SMAN I Rejotangan Tulungagung regency. Selection of subject in this research used purposive sampling technique. The classification was based on previous semester achievement and consultation with teacher while the deictic gesture data was taken from the McNeill gesture test. The data in this study was based on the results of the written exam on LES HOTS. Interviews were processed in such a way so it could be concluded that the learning difficulties of students with the deictic gesture of each student's ability. Research step according to Moelong (2007) are: (1) pre activity, (2) activity, (3) data analysis, (4) evaluation and report [23]. To ensure that the data is valid, data credibility test was completed using triangulation technique.

3. Research Results and Discussion

The data collection was conducted by giving Mc Neil gesture type test to 38 students of Senior High School at SMAN I Rejotanganto categorize students' gesture type. The result is on the table 1.

Table 1. Gesture Type Test Result

Gesture	Frequency	Percentage
Iconic	8	21,05 %
Methaporic	7	18,42 %
Deictic	14	36,84 %
Beat	9	23,69 %
Total	38	100 %

Table 2. Total Number of Difficulties Mathematics Skill

Hots Question	Types of difficulties mathematics skill					Total
	Number fact skill	Arithmetic skill	Information skill	Language skill	Visual-spatial skill	
1	0	0	14 (15.05%)	10 (10.75%)	7 (7.50%)	31 (33.30%)
2	3 (3.22%)	4 (4.30%)	9 (9.70%)	7(7.50%)	6 (6.45%)	29 (31.20%)
3	0	3 (3.22%)	10 (10.75)	11 (11.83)	9 (9.70%)	33 (35.50%)
Total	3 (3.22%)	7 (7.52%)	33 (35.50%)	28 (30.10%)	22 (23.70%)	93 (100%)

Table 2 shows the students difficulties on each mathematics skill from the sample in solving LES HOTS. The highest number of difficulties was in question 3. Meanwhile, the lowest number of difficulties was in question 2. In addition, information skill is the most difficult skill.

3.1. Analysis of problem-solving HOTS Question 1

Cognitive Levels: Analysis and Evaluation

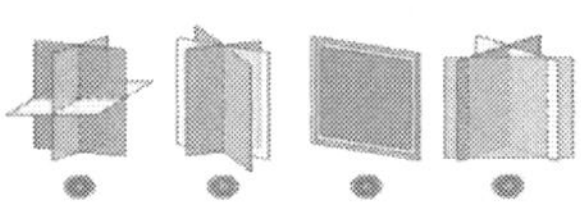

There are 4 three solid figures which each of the equation is a plane figure in the solid figure. Each plan can intersect each other in many ways which produce different set completion. Which figure with only one solution is? Explain it!

Figure 1. Illustration of solid figure

S1on the solution of the HOTS1is in cognitive level of analysis and evaluation. He answers A which is correct, but the arguments given are not correct. Based on the interviews and answers analysis on the solution of the HOTS question 1, he analyzes the problem correctly, but he is failed at the evaluation stage. In *information skill*, there are difficulties in problem-solving planning. Actually he is too lazy to write down the answer. He is also unable to remember the formula and understanding the concept of linear equation system. As the result, he cannot provide a good reasoning. In *language skill,* there are difficulties in understanding question. In *visual spatial skill,* S1 could not manipulate LES in form of picture. According to Polya, understanding problem is the first phase of problem-solving process [24]. During that phase, first of all, students need to know the story from the information given and the objective of the problem.

Design A	:	Yes, because it is collinear.
Design B	:	No, because not all of the shape is collinear.
Design C	:	No, because it is not collinear.
Design D	:	Yes, because there are shapes which collinear each other.

IOP Conf. Series: Journal of Physics: Conf. Series **1108** (2018) 012074 doi:10.1088/1742-6596/1108/1/012074

Figure 2. The answer ofS1with low ability on HOTS question 1

Figure 3 shows S2 answers all question correctly but the arguments given in each answer are less precise. This means that in the category of analysis, he has difficulties in information skills on problem-solving planning. S2 is less precise in writing information about questions and arguments written on each answer option. In figure 4, S3 answers it all correctly but he does not provide an argument in every answer. This means that on the category of cognitive and analysis he has difficulties in the information skills on the problem solving planning. It shows that there is oblivious information. S3 are also unable to provide a reason related to the answer.

Design A (Yes) because the shape is perpendicular.
Design B (No)
Design C (No) because the shape is not perpendicular.
Design D (No) because the shape is not perpendicular

Figure 3. The Answer of S2withModerate Ability on HOTS Question 1

Design A(Yes)
Design B (No)
Design C (No)
Design D (No)

Figure 4. The Answer of S3with High Ability on HOTS Question 1

3.2. Analysis of problem-solving HOTS Question 2

Cognitive Level: Evaluation

Look at Figure 5 below!

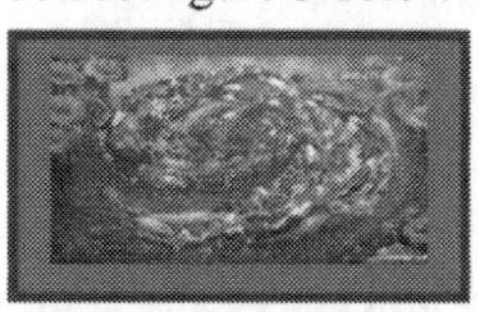

Mr. Doni has a picture frame illustrating koi fish. Its circumference is 44 cm. If the width is 6 cm lower than the length, find the measurement of the frame!

Figure 5. Illustration of picture frame

The question of HOTS 2is on the cognitive levels evaluation toS2. He encountered difficulties in solving. Below is the answer of S2who has moderate ability with deictic gesture types [25].

$(2x + 2y) = (44)$
$l + w = 22$
$l = 22 - w$
$w + 6 = 22$
$w = 14$

Figure 6. The Answer of S1 with Moderate Ability on HOTS Question 2

$2x + 2y = 44 : 2$
$x + y = 22$
$x = y + 6$
$x = 14$
$y + 6 + y = 22.$
$2y = 16$
$y = 8$
So, the length is 8 cm and the width is 14 cm.

Figure 7. The Answer of S2WithLow Ability on HOTS Question 2

Seeing Figure6, S1 answer incorrectly. He did not finish the work until the end. He has difficulties in understanding the category of HOTS question 2. In computational and mathematical working procedure, S1 responded incorrectly. He did not write the information received on the question. He was failed in substituting the length and width of the picture into symbols. It looks like he had done it carelessly. This means that S1 has difficulties on arithmetic, information, language and visual-spatial skill. As a result student was failed to carry out the problem-solving correctly. Students are faced with difficulties in problem-solving due to their lack in visualizing mathematics problem and concept [26]. The affect of mathematics difficulties is cumulative because the development of mathematics skills is based on the previous skills [27].

Figure 7shows that the solution given by S2of the HOTS question 2 is on evaluation aspect. He has difficulties in problem-solving planning. S2did not write what was known and asked. The answer became unstructured and made many errors in solving. Based on interviews and answers analysis on S2, evaluation started by analyzing the questions appropriately but at that stage, he could not complete the answer correctly. S2 was able to analyze the matter, but S2was too lazy to write a long answer. This is the cause of the answer to be unstructured so there were many mistakes. It means that the student was in trouble with *information skills*. The subject S2 also has difficulties in *arithmetic skill*. The planning procedure of the calculation experienced by subject S2 is not accurate. Furthermore, he is weak in number fact skill. He wrote: $2x + 2y = 44: 2 \ = x + y = 22$, means thathe made incorrect calculations, although the end result was in line with the correct answers. Incompetency in conceptual knowledge of division caused error in procedural skill. According to Narayanan, students were lack in arithmetic and procedure knowledge as a result from weak conceptual understanding [25].Besides, S3has answered it right.

3.3. Analysis of problem-solving HOTS Question 3

Cognitive Level: Analysis

Karno made 3 towers which constructed by rectangles, trapezoids and hexagons. The first tower has 35 cm height. Second tower has 34 cm height. Third tower has 49 cm height. If the rectangle has 2 cm in height, which one of the following tower has more rectangle? Give a logic reasoning!

Figure 8. Tower

The answer ofS1 is correct, but the argument given is not correct, it is inconsistent between the answer and the question. This means that S1has difficulties in mathematics skills such as information skill and language skill. The S1 has difficulties in manipulate tower picture with his visual-spatial skill. According to Garderen, deficiency in visual-spatial skill might cause difficulty in differentiating and relating information meaningfully [28].In another situation, the answer of S2 is wrong. It seems that S2 has difficulties in understanding the problem so that the answer is incorrect. S2 has difficulties in manipulate geometrical shape. EspeciallyS2 has difficulties in visual-spatial skill.

The first tower has more rectangles than the other tower. It is because the first tower has 4 rectangles, second tower has 3 rectangles and the third tower has only one rectangle.

Figure 9. The Answer of S3WithHigh Ability on HOTS Question 3

From the figure 9, it seems that the solution of HOTS 3 given by S3 is on analysis level. S3 has difficulties in visual-spatial skill as he is being fooled by the picture. Based on interviews and answers of S3, he started by analyzing questions incorrectly.S3 has difficulties in the visual-spatial skill which is about understanding the problem and the picture. A visual abnormality child is often unable to distinguish Geometry form [29]. Furthermore, S3 also has difficulties in language skill.S3 concluded that the first tower has the most rectangle because he was unable to realize that he needs to find how much rectangle in the third tower using Triple Variables Linear Equation System (TVLES). Factors that cause he experienced this difficulties is lack understanding of the problem.

Based on the results of the analysis, this study found some difficulties for students. These show evidence of earlier studies that show that students have much difficulty in understanding of LES especially if the problem is served in picture. This problem is caused by students' inability to transform the information on the picture into mathematics words [30].This is strengthen by previous research conducted by PPG Matematika in 2002 about the difficulties faced by students in 5 province, showing that there are lack of skill and conceptual understanding related to the algebra operation. There are also difficulties in understanding the problem, error in operation and also translating the story problem and picture into mathematics word [31].Furthermore, this research also shows difficulties of students who have the deictic gesture type in answering HOTS with dominant material of LES. The difficult parts of this are language, information and visual spatial skill with various causes. One of the causes is students' limited understanding of information given. So that, students solved the HOTS questions hardly. This research is also strengthened by Tambychik & Meerah's study which showed that the weaknesses of many students are at number fact skill, visual-spatial skills, and information skill [26].

4. Conclusion

This study concludes that students with deictic gesture type have difficulties in mathematics problem-solving HOTS Geometry due to incompetency in acquiring many mathematics skills. They are mostly weak in communicating their mind. They like to show a gesture rather than explain by verbal.

It was found that a total of 93 type of difficulties on mathematics skill which has been identified in the problem solving process involving HOTS. It is also found a total of 3.22% difficulties of number fact skill, 7.5% of arithmetic skill, 35.5% of information skill, 30.10% of language skill, 23.70% of visual-spatial skill. Information skill was found to be the most critical mathematics skill. In HOTS problem with cognitive level analysis and evaluation category, student with high and moderate ability are experiencing difficulties in information skill, while student with low ability in information and visual-spatial skill. In HOTS problem with cognitive level evaluation category, students with high ability only have few difficulties which is found in the last step, students with moderate abilities have in information, number fact and arithmetic skill, while student with low ability have difficulties in information, language, visual-spatial and arithmetic skill. In HOTS problem-solving with cognitive analysis category, students with high ability have difficulties in language and visual-spatial skill, students with moderate ability in visual-spatial skill, while students with low ability have difficulties in information, language and visual-spatial skill. Therefore, teachers should be able to recognize the difficulties faced by students in solving problems related to HOTS. Besides, teacher often give students many exercise to stimulate their creativity in solving the problem. Furthermore, the researcher suggests to study the other type of gesture to compare the students ability between each type of gesture type classified by McNeill, especially on the SPL material.

5. References

[1] Krulick S & Rudnick J A 1999 Innovative Tasks to Improve Critical and Creative Thinking Skills *Journal Developing Mathematical Reasoning in Grades* K-12: 138-145

[2] Direktorat Pembinaan SMA 2015 *Penyusunan Soal Higher Order Thinking Skill's Sekolah* Menengah *Atas (*Jakarta: Direktorat Jenderal Pendidikan Menengah Kementerian Pendidikan danKebudayaan)

[3] Anderson L W and Krathwohl D R 2001 *A Taxonomy for Learning, Teaching, and Assessing: A Revision of Bloom's Taxonomy of Educational Objectives* (Allyn & Bacon Boston: MA PearsonEducation Group)

[4] https://m.bernas.id/50899-peringkat-berapakah-indonesia-di-timss.html

[5] De Lange 2016 Mathematical Literacy for Living from OECD-PISA Perspective *Tsukuba Journal of Educational Study in Mathematics* **25**: 13-35

[6] Kemdikbud 2016 *Peringkat dan capaian PISA Indonesia mengalami Peningkatan*(Jakarta: Dirjen Pendidikan Menengah Kementerian Pendidikan dan Kebudayaan)

[7] Nizam 2016 *Hasil Penilaian Seminar Puspendik* (Jakarta: Pusat Penilaian Pendidikan Badan Penelitian dan Pengembangan Kementerian Pendidikan dan Kebudayaan)

[8] Conklin W 2012 *Higher-order thinking skills to develop 21st century learners* (Huntingon Beach:Shell Education Publishing, Inc)

[9] Kemendiknas Aplikasi Pamer UN 2016
[10] Huntley M, Marcus R, Kahan J and Miller J L 2004 Investigating high scholl student's reasoning strategies when they solve linear equation *Journal of Mathematical Behaviour***26** (2) 115-139
[11] Kazemi E 2002 Exploring test performance in mathematics the questions children's answers raise *Journal of Mathematical Behaviour***21** (2) 203-224 Retrieved from the ScienceDirect database
[12] Clement J 1982 Algebra word problem solutions: thought processes underlying a common misconception *Journal for Research in Mathematics Education***13** (1) 16-30 Retrieved from theJSTOR database
[13] Noguire D L and Tall D 2008 Procedural embodiment and magic in linear equation *Educational Studies in Mathematics***67** (1) 3-18 Retrieved from the JSTOR database
[14]Paivo A 1991 Dual coding theory: retrospect and current status *Canadian Journal Psychology* 45255-287
[15] Roth W M & Wezel M 2001 From activity to gestures and scientific language *Journal of Researchin Science Teaching* 38 103-136
[16] Corts D P & Pollio H R 1999 Spontenous production of figurative language and gesture in college lectures *Metaphor and Symbol***14** (2) 81-100
[17]McNeill D 1992 *Hand and mind: What gestures reveal about thought* (Chichago: Chichago University Press)
[18] Hostetter A B, Edward& Alibali M W 2008 Visible embodiment: Gestures as simulated action *Psychonomic Bulletin & Review* 15 495-514
[19] Kok I D, Hough J, Schlangen & Kopp S 2016 Deictic gesture in coaching interactions *Procedings of the Journal on Multimodal Analyses Enabling Artifical Agents in Human Machine Interaction***16** : 12-14
[20] Roth W M & Lawless D 2002 Scientific investigations, metaphorical gestures, and the emergence of abstract scientific concepts *Elseiver Learning and Instruction***12** (2002) 285-304
[21]Ali M N & Desa R 2004 MeningkatkanKemahiranMengingatSifir 3,6 dan 9 di Kalangan MuridPemulihan Tahap 2 *Jurnal Kajian Tindakan Pelajar PSPK. KementerianPendidikanMalaysia*102 – 112
[22] Nathan M JAlibali M W 2007 Teacher's Gestures as a means of Scaffolding Student's Understanding: Evidence from an Early Algebra Lesson Research in the Learning Sciences (NJ:Erlbaum)
[23] Moleong L J 2007 *Metodologi Penelitian Kualitatif* (Bandung: PT Remaja Rosdakarya)
[24] Polya G 1981 *Mathematical Discovery on Understanding Learning and Theacing Problem Solving*(New York: John Wiley and Sons)
[25]Narayanan L M 2007 *Analysis of error in addition and subtraction of fraction among form*(Universiti Malaya: Kertas Projek Sarjana Pendidikan)
[26]Tambychik T Meerah T S M & Aziz Z 2010 Mathematics Skills Difficulties: A Mixture of Intricacies. Procedia *Social and Behavioral Science***7**(C):142-151
[27] Carnine M 1997 Instructional design in Mathematics for student with learning disabilities *Journal of Learning Disabilities* 30:130-141
[28] Garderen D V 2006 Spatial Visualization, visual imagenary and mathematical problem solving ofstudents with varying abilities. *Journal of Learning Disabilities* **39** (6): 496-506
[29] Lerner J W 1988 *Learning Disabilities: Theories, Diagnosis and Teaching Strategies* (London: Houghton Miffin Company)
[30]Newman M A 1977An analysis of sixth-grade pupils' errors on written mathematical task *Victorian Institute for Educational Research Bulletin***39**: 31-43
[31]Wardani I G A K 2004 Penelitian Tindakan Kelas (Jakarta: Universitas Terbuka)

6. Acknowledgments
The authors would like to thank to the adviser Dr. Riyadi M.Si dan Dr Dewi Retno Sari Saputro, S.Si, M.Kom, my parents, my little brother and all friends of PPS UNS 2016 generation.

The Role of Gesture to Mathematics' Fraction Base Concept Instruction

E Surahmi[1,2,*], I K Budayasa[2], and R Ekawati[2]

[1] Fakultas Keguruan dan Ilmu Pendidikan, Universitas Madura, Jalan Raya Panglegur, Pamekasan, Jawa Timur, Indonesia, 69371
[2] Program Pascasarjana Pendidikan Matematika, Universitas Negeri Surabaya, Kampus Ketintang, Surabaya, Indonesia, 60231

*Corresponding Author: emasurahmi25@gmail.com

Abstract. Gesture is a body movement as a form of repesentasi in thinking and talking about the ideas that are owned by someone and expressed through body movement. Gesture has an important role for a teacher of mathematics learning, especially on the basic concept of fractions. Abstract mathematics makes teachers think how to communicate mathematical concepts to students in order to be more easily understood. The purpose of this study is to describe the role of gesture for elementary school teachers in Pamekasan. This research is descriptive research with qualitative approach. The results of this study show the type of gesture; (1) gesture point, as a gesture used to designate objects by using fingers, hands or stationery. (2) representational gestures, as gestures with hand or arm movements representing multiple objects, actions, concepts and relationships, (3) writing gestures, integrating teacher speech with hand movements in writing. The role of each gesture; (a) the gesture points; directing students' attention, keeping their eyes fixed for long periods of time, (b) representational gestures; helps to simulate actions and perceptions expressed in body movement and (c) writing gestures; can help students focus on understanding mathematical concepts, symbols, and mathematical notations.

1. Introduction

Gestures are defined as arm and hand movements that correspond to speech [1]. Gestures are all body movements that are integrated with speech or action to communicate something [2]. Teachers' gesture in the learning process have a very important role to attract students' attention on learning materials. The body movements of teachers when teaching mathematical concepts, partly because of their consciousness but not infrequently without their noticing [3]. The teacher uses limb movements to interact with students in learning. The use of body movements by these teachers proves that the body is involved in the interaction of learning. As it is argued that to convey ideas in learning, the body of a speaker is involved in it [4]. Morsella & Krauss's research [5] shows that speakers do more gestures when describing objects that weigh on spatial working memory. Gestures often occur when speakers express information involving spatial information [6]. Gestures are performed because they have an important role in cognition [5]. Gestures arise when information is difficult to conceptualize [7]. Gestures can reduce cognitive load in all systems from verbal information systems to spatial information systems [6]. Gestures are part of the communication used to learn better and reduce errors.

Gesture has an important role in mathematics learning so it is interesting to note, this is in accordance with expert opinion, gesture is often used in the context of learning [8-10]. Kelly et al. [11] stated that teachers often use gestures to help students who have difficulty learning math concepts. Alibali et al [12] argues that teachers use gestures when explaining mathematical concepts and the use of gestures by teachers will increase when conveying new concepts in mathematics learning. Thompson [9] says that the gestures teachers use in learning have an impact on students' understanding of the material being taught. So it can be said that gestures have a very important role to help students understand.

According to McNeill [1] there are two types of gestures: propositional gestures and non-propositional gestures. The propositional gesture is a gesture that has a component of the main picture whereas a non-propositional gesture is a speech gesture. Propositional gestures are divided into three: (1) an iconic gesture, is a gesture that contains a formal relationship closely related to what content is discussed semantically. Iconic gesture presents a picture of a concrete entity or action (2) a metaphoric gesture, a gesture that presents abstract ideas without physical form, and (3) a deity gesture that points to objects, events, or people. Non-propositional gestures are divided into two: (a) the beat gesture is a gesture in the form of rhythmic motion and is often repeated from both vertical and horizontal hands and fingers and (b) a cohesive gesture. Furthermore McNeill [5] categorizes gestures into four main categories, namely iconic gestures, metaphoric, deictic, and beat.

Alibali & Nathan [3] modified the McNeill classification of gestures [1] in mathematics learning in the classroom. Alibali & Nathan [6] classifies gestures into three types, namely (1) pointing gestures, (2) representational gestures, and (3) writing gestures. Pointing gesture to none other than deistic gestures on the McNeill classification [1]. The representational gestures include iconic gestures and metaphoric gestures. Gestures write similar or go in the iconic symbolic gesture subcategory. In line with Shein's opinion [13], the results of research on gestures consist of three; (b) representational gestures, as gestures with hand or arm movements representing multiple objects, actions, concepts and relationships, (c) gestures (gestures), gestures that are used to show objects by using fingers, hands and /or stationery; writing, as what the teacher is saying integrates with hand movements in writing.

Research on the role of teachers' gestures in mathematics and science learning has not been the focus of educational research, although teachers' gestures have an important role in learning teaching process [12]. Research on the role of teachers' gesture in mathematics learning is needed in the future [4,6]. Based on these facts, researchers conducted this research on the role of teachers' gesture in learning mathematics. Teaching materials delivered by the teacher is the basic concept of fractions. Researchers suspect the role of teachers' gesture can help students in understanding the basic concepts of fractions. In this study, the classification of gestures used refers to the classification developed by [3,13]. The gestures studied were: (1) pointing gesture, (2) representational gesture, and (3) writing gestures. This is because the teacher gesture has an important role in learning that can help students in understanding the concepts of mathematics, symbols, mathematical notations and ideas in mathematics.

2. Method

In this study, a descriptive qualitative is used to describe teacher's gesture role in instructional mathematics. The subject in this study is an elementary school teacher of mathematics, grade 3, who teaches material about base concept of fraction. The Collected data is obtained during process of teaching-learning in the classroom. The researcher is a main instrument as the researcher herself who design the research, collect and analyze the data, draw a conclusion and report the result of the study. To strengthen data of the study, a supporting instrument is needed, namely: camera, it is used to record gestures audio visually during the process of teaching-learning occur in the classroom. Techniques used in analyzing the data: (1) data reduction, it is a process of selecting, focusing, simplifying all obtained raw data, (2) serving data, it is an activity to serve the result of reduction in narrative form which enables to draw conclusion (3) drawing conclusion and verification is a process of providing conclusion concerning result of estimation, and evaluation involves meaning searching and explanation of given.

3. Result and Discussion

In this study the material presented about the basic concept of fractions, which means as follows; the fractional number can be written as $\frac{a}{b}$, with $a\ dan\ b$ integers, where $b,\ b \neq 0$ with a number whose symbol consists of an integer pair a and b, a is called numerator and a is called denominator. There are three concepts in the fraction; namely part to-whole concept, division concept and ratio concept [14]. The focus of the fractional basic concept material described by the teacher is part-to-whole concept. Material of fraction number is important to be taught by virtue of its characteristic, i.e. abstract. In early age, students tend to meet the growing in level of pre-concrete operational. Due to this case, the teachers needed to consider how deliver the material easily to be understood the students by utilizing teacher's gesture in explaining, as what Alibali, M.W & Nathan [4] said that gesture played a leading role in communication of teaching. Hence the present study focuses the role of gesture employed by the teacher as the communication form in delivering the material which refers to the classification developed by Alibali & Nathan [3] and Shain [13], that is to classify the gesture into three: pointing gesture, representational gesture, and writing gesture. This following is the explanation:

3.1 The role of pointing gesture

The opinion of Alibali, M.W & Nathan [4] said pointing gestures are often used along with speech, and they manifest speakers' indexing of speech content to objects, locations, or inscriptions In the physical environment. By "physical environment" we mean the setting for the interaction (e.g., a classroom, tutoring session, or experimental session) including the interlocutors (e.g., students, teacher, experimenter); the focal tasks; and the representations, notational systems, tools, and technological resources that are used.

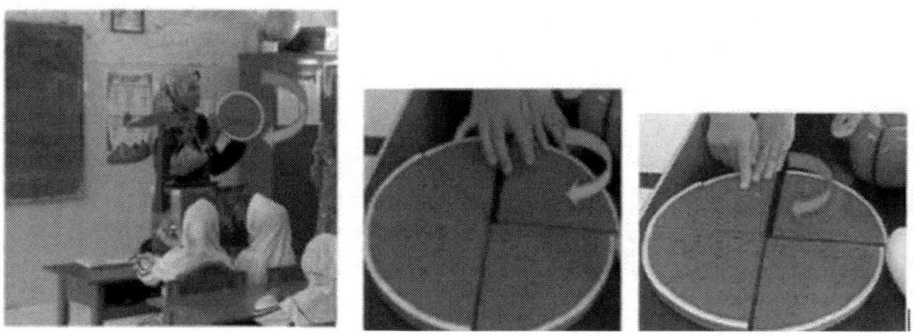

Category 1 Category 2

Figure 1. Teacher utilizes pointing gesture in category 1 and 2

The teacher uses gestures to point out in explaining the basic concept of fractions as part of the whole, the example used a on the whole piece of watermelon. Category (1) indicates that the teacher using pointing gestures to chunks $\frac{1}{4}$ which is accompanied verbally with meaning that the number of watermelon pieces there are 4 equal parts of 1 part overall. The number of 4 pieces of the piece as 1 whole, using the teacher's fingers pointing to the object that aims to provide abstract ideas $\frac{1}{4}$ into a concrete form.

Category (2), indicating that the teacher explains fraction $\frac{1}{8}$ by using pointing gesture involving the fingers on the objects of watermelon snippet and followed the saying that the number watermelon pieces there are 8 equal parts of 1 whole. 8 parts of the piece are 1 whole, 1 is the numerator and 8 is the denominator. Categories 1 and 2 show that what teachers do in explaining material to students using pointing gestures can lead to fixations on speakers and listeners. Fixation is a point in the field of gaze where the eye remains staring at it in a relatively long time [15] and aims to show location, check and direct attention.

This is in accordance with the opinion of Alibali & Nathan [3] which states that motion pointing is used to indicate something (object, person, event, location). Most pointed movements are generated from gestures by the fingers / hands or objects carried (for example, pencils, markers or chalk) that serve as a pointer. In accordance with the opinion of Shein [13] states that in particular, seeing the pointed movement refers to artifacts and visual representations related to tasks (eg, worksheets, whiteboards, and manipulatives).

3.2 The role of representational gestures

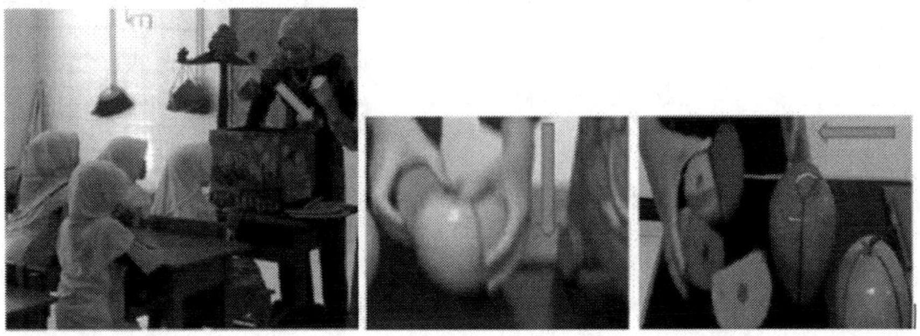

Category 1 Category 2 Category 3

Figure 2. Teacher utilizes representational gesture in category 1, 2 and 3

The teacher uses representational gestures in explaining the basic concept of fractions is part of the whole. Category 1 is a representational gesture used by teachers by involving open-handed gestures into the air to simulate the object of cognition in the visual image of a mathematical idea that is 1 part of the whole apple, then divided into 2 equal parts ie in category 2, the teacher moves the right and left hands to air with a second opening hands, right hand as $\frac{1}{2}$ part apple and left hand as another $\frac{1}{2}$ part apple, namely 2 part from 1 whole of apple. In category 3, the teacher moves his gesture as his action previously by opining his hands going up in the air, his right hand moves forward and his left hand goes up in the air but its position is not parallel to his right hand, this gesture can simulate apple from $\frac{1}{2}$ part which is previously divided 2 with the same size, that is as visual reflection of mathematics notion 2 part from $\frac{1}{2}$ which every part has size $\frac{1}{4}$

Representational gesture is utilized by the teacher in category 1, 2 and 3 of this study reveals that teacher's gesture is accompanied with or without his utterances. This regard agrees with Thompson's opinion [11] he asserts that gesture can be accompanied with or without utterances. Representational gesture is accompanied with utterances occur when the subject is thinking or trying to explain something either for himself or for another person. Besides, Representational gesture can contribute to listener's comprehension by assisting listener to simulate action and perception expressed in speaker's gesture [3]. This fact indicates that gesture is not always accompanied with utterances, this case agrees with Alibali & Nathan's outlook [4] they claim that speaker employs Representational gesture when thinking. Accordingly, gesture is used not only to explain notions but also to express notion [12] which can come about without using utterances at all.

This statement is strengthened by Alibali, M.W & Nathan [4] which state evidence from representational gestures suggests that explaining mathematical thinking involves simulations of actions on mathematical objects, simulations of visual images of mathematical ideas (often mental images of inscriptions), and simulations of the real-world situations that mathematical problems address. Speakers produce such gestures when they think and speak about mathematical ideas, and indeed they may intentionally produce such gestures in order to facilitate thinking about such ideas or to promote effective communication about such ideas.

3.3 The role of writing gesture

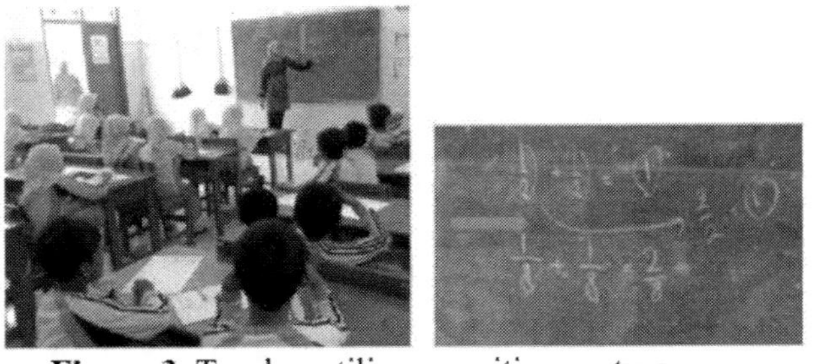

Figure 3. Teacher utilizes a writing gesture

Teachers using writing gestures to explain the basic concept of fractions are part of the whole by involving motion, arms, shoulders and fingers using chalk to write on the board followed by speech as an emphasis in explaining. Teacher writes $\frac{1}{2}$ which constitutes as a form 2 part from 1 a whole of apple having similar size, and it looks the teacher puts the used writings on the blackboard $\frac{1}{2}+\frac{1}{2}=\frac{2}{2}$ or 1 while giving an explanation if 2 part $\frac{1}{2}$ the apple is combined to be 1 part of whole apple, where 1 indicates numerator and 2 is called as denominator. Then the teacher rewrite on the blackboard for fraction $\frac{1}{8}$ which is accompanied with brief utterances that $\frac{1}{8}$ is 8 part from 1 a whole watermelon exemplified previously and then the teacher writes $\frac{1}{8}+\frac{1}{8}=\frac{2}{8}or\frac{1}{4}$ on the blackboard by giving explanation if 2 part $\frac{1}{8}$ cuts of watermelon when it is combined will be $\frac{1}{4}$ part of cuts of watermelon.

Figure 3 is a form of writing gesture which occurs when one's action leaving a permanent sign in a new media (such as: a working paper, a blackboard representing a visual thing). Shein [13] states that a writing gesture is focused on writing or drawing which is followed by utterances.

4. Conclusion

The role of gestures used by teachers in mathematics learning on the basic concepts of fractions: (1) Pointing gesture, teachers use fingers to point objects in communicating abstract ideas $\frac{1}{4}$ and $\frac{1}{8}$ in concrete form followed by utterance. (2) The representational gesture involves the motion of the second open hand to the air in an open position to simulate the object present in the cognition of the visual image, 1 being the whole apple, $\frac{1}{2}$ and $\frac{1}{4}$, with or without being followed by utterance (3) writing gesture, the teacher involves hand gestures, shoulders, arms and fingers in writing on the blackboard and greeting to write fractions $\frac{1}{2}$ and $\frac{1}{4}$. Each gesture has its own role. Pointing gestures designate to attract the learners' attention which creates gaze where one's eyes remain watch the teacher's gesture very closely for a long time gaze, representational gestures help simulate actions and perceptions expressed in gestures, and writing gestures have a role that can help students focus in understanding mathematical concepts, symbols, mathematical notations and ideas in mathematics. In this case, research on the role of teacher' gesture in mathematics learning is needed in the future [3,12]

References

[1] McNeill D 1992 *Hand and mind: What gestures reveal about thought* Chicago: University of Chicago Press

[2] Becvar A, Hollan J and Hutchins E 2008 *Representational Gestures as Cognitive Artifacts for Developing Theories in a Scientific Laboratory* Ackerman, M.S., (eds) Resources, Co-Evolution and Artifacts: Theory in CSCW pp. 117-143

[3] Alibali M W and Nathan M J 2007 *Teachers' gestures as a means of scaffolding students understanding*: Evidence from an early algebra lesson. In R. Goldman, R. Pea, B. Barron, & S. J. Derry (Eds.), Video research in the learning sciences (pp. 349–365) Mahwah, NJ: Erlbaum

[4] Alibali M W and Nathan M J 2012 Embodiment in Mathematics Teaching and Learning: Evidence From Learners and Teachers Gestures *Journal of the Learning Sciences* **21**(2) pp. 247-286

[5] Morsella E and Krauss R M 2004 The Role of Gestures in Spatial Working Memory and Speech *American Journal of Psychology* **117**(3) pp. 411-424

[6] Hostetter A B and Alibali M W 2008 Visible Embodiment: Gestures as Simulated Action *Psychonomic Bulletin & Review* **15**(3) pp. 495-514

[7] Hostetter A B, Alibali M W, and Kita S 2007 I See It in May Hand's Eye: Representasional gestures are Sensitive to Conceptual Demands *Language and Cognitive Processes* **22**(3)

[8] Goldin-Meadow S, Kim S, and Singer M 1999 What the Teacher's Hands Tell the Student's Mind About Math *Journal of Educational Psychology* **91**(4) pp. 720-730

[9] Thompson J M 2014 Teachers Perceptions of Other Teachers' Spontaneous Hand Gesturing in The EFL Classroom *Novitas-ROYAL (Research on Youth and Language)* **8**(2) pp. 119-135

[10] Neill S 1991 Classroom Nonverbal Communication (London: Routledge)

[11] Kelly S D, Manning S M, and Rodak S 2008 Gesture Gives a Hand to Language and Learning: Perspectives from Cognitive Neuroscience *Developmental Psychology and Education Language and Linguistics Compass* **2**(4) pp. 569-588

[12] Alibali M W, Young A G, Crooks N M, Yeo A, Wolfgram M S, Ledesma I M., Nathan M J, Church R B and Knuth E J 2013 Student Learn More When Their Teacher Has Learned to Gesture Effectively *Gesture* **13**(2) pp. 210-233

[13] Shein P P 2012 Seeing With Two Eyes: A Teacher's Use of Gestures in Questioning and Revoicing to Engage English Language Learners in the Repair of Mathematical Errors *International Journal of Research in Mathematics Educations* **43**(2) pp. 82-222

[14] Albert B, Bennet N, and Tet N 2004 *Mathematics for Elementary Teachers,* (New York: Mc. Graw Hill Compay)

[15] Flevares L M and Perry M 2001 How Many Do You See? The Use of Nonspoken Representations in First-Grade Mathematics Lessons *Journal of Educational Psychology*, **93**(2) pp. 330-345

Acknowledgments

With best reverence to Lembaga Pengelola Dana Pendidikan Educational Fund Management Institution (LPDP), Indonesian Endowment fund for Education, Ministry of Finance of Republik Indonesia. I am highly indebted to LPDP, that which is funding this publication and giving me precious support for studies. I deeply appreciate the allocation of fund to participate in my experimental studies where gave me insightful spirit and strong motivation to produce another and more useful publication. Hopefully the strong mission of LPDP to improve human capital competency in Indonesia will be soon achived.

IOP Conf. Series: Journal of Physics: Conf. Series **1108** (2018) 012076 doi:10.1088/1742-6596/1108/1/012076

Risk Prediction Model Occurrence of Obstructive Sleep Apnea (OSA) Risk in Chronic Obstructive Pulmonary Disease patient. Case Study: Cipto Mangunkusumo Hospital

I R Anisa[1], T Siswantining[1,*], T Kamelia[2] and D Sarwinda[1]

[1] Departement of Mathematics, FMIPA Universitas Indonesia, Kampus UI Depok, Depok 16242, Indonesia. [2] Pulmonology Division, Internal Medicine Department, Universitas Indonesia/Dr Cipto Mangunkusumo Hospital, Jakarta, Indonesia

*Corresponding author: titin@sci.ui.ac.id.

Abstract. The complication occur in Chronic Obstructive Pulmonary Disease (COPD) patient is nocturnal hypoxemia. Obstructive Sleep Apnea (OSA) and COPD can occur at the same time which cause two times uncomfortable while breathing. The purpose of this research is to determine prediction model occurrence OSA risk in COPD patient based on factor affecting the risk of OSA occurring in COPD patients. Data used in this research is primary data from COPD patients who is diagnosed by doctor at RSCM by interviewing them using Berlin questionnaire and physical examination such as measuring the circumference of neck and waist. This study uses non-probability sampling i.e. purposive sampling method. Sample of this research is 111 patients with COPD. This research uses binary logistic regression to predict model occurrence of OSA risk in COPD patients. This study shows that waist circumference and COPD Assessment Test (CAT) 2 questionnaire (COPD patients with severe degree) are significant factor of OSA on COPD patient. In addition, COPD patients with severe degree are 4.39 times greater risk suffer from OSA than mild to moderate COPD patients and each centimetre increase of waist circumference has higher risk of OSA. Accuracy of our model is estimated using classification table with cut point at 0.5 and its accuracy is 73.9%.

1. Introduction

COPD and OSA are two common pulmonary diseases. Chronic Obstructive Pulmonary Disease (COPD) is a preventable and treatable pulmonary disease indicated by persistent respiratory system and airflow barriers caused by airway or alveolar abnormalities that are usually caused by hazardous particles and gases [1]. To determine the severity of COPD symptoms experienced by patients, it can be calculated based on the questionnaire that is COPD Assessment Test (CAT) and modified Medical Research Council (mMRC)[1].

COPD is the fifth leading cause of death in the world (in 2002) and cause 5% of death in the world (in 2005). It is also known that almost 90% of COPD patients' death occur in low and middle-income countries [2]. Nocturnal hypoxemia is a complication suffered by COPD patients where oxygen intake decreases at night. This condition is exacerbated if the patient has Obstructive Sleep Apnea (OSA) [3].

On the other hand, OSA is a sleeping disorder caused by occluded airway. Normally, the soft tissue in the upper respiratory tract, including the tongue and the throat wall, will relax during sleep. In people with OSA, when the tissue relaxes, it blocks the airway, which causes temporary pauses in breathing. These pauses are brief and may only last a few seconds (at least 10 seconds) but it can happen several times during an hour of sleeping [4]. In China, man workers aged 30-60 years old suffer from OSA is 4.1% [5]. However, the prevalence for OSA populations overall ranged from 9% to 38% and is higher in men. In some elderly groups, it reaches 90% in men and 78% in women [6].

Unfortunately, COPD and OSA can happen at the same time, known as Overlap Syndrome, causing two times uncomfortable condition while breathing [3]. Two recent researches observe the prevalence of COPD patients with positive OSA is 65,9% in lung rehabilitation [7] and 10,8% in Korea [8].

Polysomnography (PSG) is widely used to diagnose OSA. While the severity degree of OSA is calculated by Apnea Hypopnea Index (AHI). AHI is divided into 3 severity classification, which are 5-14 for low degree of OSA, 15-30 for middle degree, and more than 30 for high degree of OSA [9]. While, the Berlin questionnaire as a tool for capturing COPD patients at risk for OSA and Berlin questionnaire used is a validated questionnaire for the population of Indonesia [10].

Model to estimate the risk of OSA in COPD patients is not invented yet, both in Indonesia and other countries. Therefore, this study attempts to determine estimation model of OSA on COPD patients by observing COPD patients in Rumah Sakit Cipto Mangunkusumo (RSCM).

2. Method

Regression analysis has become an integral component of the analysis of data pertaining to describe the relationship between the response variable and one or more predictor variables [11].

2.1. Binary Logistic Regression

Binary logistic regression is a statistical technique used to analyse the relationship of response variables that have two categories i.e. success and failure. Therefore, in prediction variable data type can be a nominal, ordinal, interval nor ratio.

Independent variable shows by $\mathbf{x}_{ji} = (x_{1i}, x_{2i}, \dots, x_{pi})$ and dependent variable Y_i consists of two categories: success and failure, denoted $y_i = 1$ (success) and $y_i = 0$ (fail), let Y_i as Bernoulli distributed with these functions.

$$f(y_i) = \pi(\mathbf{x}_{ji})^{y_i}[1 - \pi(\mathbf{x}_{ji})]^{1-y_i}, y_i = 0,1 \quad (1)$$

where probability of success is $\Pr(Y_i = 1|\mathbf{x}_{ji}) = \pi(\mathbf{x}_{ji})$ and probability of failure is $\Pr(Y_i = 0|\mathbf{x}_{ji}) = 1 - \pi(\mathbf{x}_{ji})$. Regression model that formed $\mathrm{E}(Y_i = 1|\mathbf{x}_{ji}) = \pi(\mathbf{x}_{ji})$ is as follows.

$$\pi(\mathbf{x}_{ji}) = \frac{e^{(\beta_0+\beta_1 x_{1i}+\beta_2 x_{2i}+\cdots+\beta_p x_{pi})}}{1 + e^{(\beta_0+\beta_1 x_{1i}+\beta_2 x_{2i}+\cdots+\beta_p x_{pi})}} \quad (2)$$

In a logistic regression model required a connecting function in accordance with the logistic regression model is the logit function. Logit transformation as a function of $\pi(\mathbf{x}_{ji})$ stated as follows:

$$\text{logit}\,[\pi(\mathbf{x}_{ji})] = g(\mathbf{x}_{ji}) = ln\left(\frac{\pi(\mathbf{x}_{ji})}{1-\pi(\mathbf{x}_{ji})}\right) \quad (3)$$

2.2. Parameter Estimator

Commonly used method for parameter logistic regression model is Maximum Likelihood Estimation (MLE). MLE estimated β parameter by maximizing likelihood function and requiring data must follow a distribution. Each observation is assumed mutually independent so that likelihood function is the multiplication of the probability function of each observation as follows.

$$l(\boldsymbol{\beta}) = \prod_{i=1}^{n} f(y_i) = \prod_{i=1}^{n} \pi(\mathbf{x}_{ji})^{y_i}[1 - \pi(\mathbf{x}_{ji})]^{1-y_i} \quad (4)$$

For $i = 1,2, \dots, n$, $\mathbf{x}_{ji} = (x_{1i}, x_{2i}, \dots, x_{pi})$ is an independent variable for the i-th observation and y_i is a dependent variable on binary i-th with values 0 and 1.

Mathematically, $\boldsymbol{\beta}$ parameter will be evaluated by maximizing the logarithm function probability as known as log likelihood defined as follows.

$$L(\boldsymbol{\beta}) = \sum_{i=1}^{n} \{y_i \ln[\pi(\mathbf{x}_{ji})] + (1-y_i)\ln[1-\pi(\mathbf{x}_{ji})]\} \tag{5}$$

The maximum value of $\boldsymbol{\beta}$ is obtained through the $l(\boldsymbol{\beta})$ to $\boldsymbol{\beta}$ and the result is zero.

$$\frac{\partial L(\boldsymbol{\beta})}{\partial \beta_j} = \sum_{i=1}^{n} x_{ji}[y_i - \pi(\mathbf{x}_{ji})] = 0 \quad ,j = 0,1,2,\dots,p \tag{6}$$

The value of $\boldsymbol{\beta}$ obtained from equation (6) called maximum estimated likelihood and denoted as $\hat{\boldsymbol{\beta}}$. Thus, prediction value for logistic regression model is $\hat{\boldsymbol{\pi}}(\mathbf{x}_{ji})$, where value from equation (2) is calculated using $\hat{\boldsymbol{\beta}}$ and $\mathbf{x}_{ji}$.

2.3. Parameter Significance Testing

Significance test consists of two stages: the significance test of model parameters simultaneously and test the significance of the model parameters partially.

1. **Simultaneously test**

   Simultaneous test is used to determine the significance of $\boldsymbol{\beta}$ parameter to dependent variable together. The hypothesis is used as follows.

   $H_0: \beta_1 = \beta_2 = \cdots = \beta_i = 0$

   H_1: at least $\beta_1 \neq 0$; $i = 1,2,\dots,p$

   Tests on these parameters is done through G test statistics are stated as follows.

$$G = -2\ln\left[\frac{L_0}{L_p}\right] \tag{7}$$

   L_0 = likelihood value without independent variable

   L_p = likelihood value with p independent variable

   The G test statistic is a Likelihood Ratio Test where the G value follows the Chi-Square distribution with the degrees of freedom as much as the independent variable p. The decision is obtained from the comparison between the test statistic value of G and χ^2tables. H_0 is rejected if the test statistic value $G > \chi^2_{(\alpha,p)}$.

2. Partially test

   Partial test is a variable test performed one by one using Wald (W) test statistic. The partial test hypothesis is as follows.

   $H_0: \beta_i = 0$

   $H_1: \beta_i \neq 0$; $i = 1,2,\dots.,p$

   the test statistic used is as follows.

$$W = \frac{\hat{\beta}_i}{SE(\hat{\beta}_i)} \tag{8}$$

   with,

   $\hat{\beta}_i$ = estimator β_i

   $SE(\hat{\beta}_i)$ = standard error β_i

2.4. Coefficient Interpretation

Interpretation of the parameter coefficients is done to determine the trend / functional relationship between independent variables with dependent variables and show the effect of changes in value on the dependent variable and shows the effect of changes in the value of the variable concerned.

Logistic Regression with Binary Independent Variable quantifying with 0 and 1. In general the value of y is continuous, but in this case the variable y becomes discrete, where the value is 0 and 1 so that the value of y is transformed like a probability value called the odds ratio. As for the logistic regression of continuous variables is to assume a linear logit function of the independent variable. The interpretation is any increase in c unit independent variables, the risk of y = 1 is equal to $e^{c\beta_1}$ times bigger.

Odds ratio is defined.

$$\mathrm{OR} = e^{\beta_1} \tag{9}$$

2.5. Classification Table

Table classification was conducted to determine the accuracy of the prediction modelClassification table is the result of cross-classification between the dependent variable (Y) with a new dichotomous variable, where value is derived from statistical probability estimates $\hat{\pi}(\mathbf{x}_{ji})$.

Table 1. Classification Table using Logistic Regression Model with Cut Point (c)

Prediction	Observation		Total
	$y = 1$	$y = 0$	
$\hat{y} = 1$	n_{11}	n_{10}	$n_{1\cdot}$
$\hat{y} = 0$	n_{01}	n_{00}	$n_{0\cdot}$
Total	$n_{.1}$	$n_{.1}$	n

Classification accuracy was measured with high sensitivity and specificity. Sensivity is the ability to predict success events correctly. While specificity is the ability to predict the occurrence of failure and true. Sensitivity and specificity are calculated as follows.

$$\text{Sensitivity} = \frac{n_{11}}{n_{11} + n_{01}} x\ 100\%$$

$$\text{Specificity} = \frac{n_{00}}{n_{00} + n_{10}} x\ 100\%$$

3. Results and Discussion

3.1 Data

The population in this study were all patients of COPD at Cipto Mangunkusumo hospital. The sample selection for this research uses non-probability sampling with purposive sampling type, where the sample is not taken randomly but rather determined by the researcher based on certain criteria. The number of samples in this study were 111 patients who had been diagnosed with COPD by doctors at RSCM by meeting inclusion and exclusion criteria.

Table 2. Inclusion and Exclusion Criteria in Patients with COPD

Inclusion Criteria	Exclusion Criteria
COPD patients with age ≥ 25 years	Patients refuse to be a research sample
Willing to take the study by interviewed by COPD Assessment Test (CAT) questionnaire and modified Medical Research Council	Pregnant mother

(mMRC) questionnaire Willing to interviewed using Berlin questionnaire for early screening Obstructive Sleep Apnea (OSA) Willing to do a physical examination by measuring the circumference of the neck and abdominal circumference	Patients with Struma (presence of swelling in the neck)

3.2 Numerical Result

Characteristics of patients with COPD in this study are based on the risk of OSA, gender, age, body mass index (BMI), CAT questionnaire, mMRC questionnaire, smoking history, neck circumference and waist circumference. By gender showed that most respondents were male as many as 82 people (74%) while the female respondents 29 people (26%) with an average age of respondents was 59.77 years, the majority of respondents in this study experienced shortness of breath are 75 respondents (68%), a total of 39 respondents (35%) in this study has moderate COPD, 36 respondents (33%) with mild COPD and 36 respondents (32%) of COPD severity. Body Mass Index (BMI) COPD patients in RSCM average overweight is 23.74 and the majority of respondents had a history of smoking, as many as 73 respondents (66%).

Based on questionnaires Berlin showed that the respondents were at risk of OSA totaling 47 respondents (47%) and respondents who are not at risk of OSA amounted to 64 respondents (58%) with a mean neck circumference and waist circumference was 37.5 cm and 91.47 cm.

Parameter significance testing consists of two stages: simultaneous test and partial test. In the simultaneous test p-value obtained by $0{,}0002449016 < \alpha = 0{,}05$, then H_0 rejected so that it can be concluded that the independent variables are used together influence on the risk of OSA. In the partial test in Table 3 that at $\alpha = 0{,}05$ seen independent variables have a significant effect, namely age, CAT questionnaire 2 and waist circumference.

After performing a significance test parameters, further suitability test will be conducted using a model of Hosmer & Lemeshow test to see whether model fit to the data. In Hosmer & Lemeshow test obtained chi-square value is 3.9829 with p-value of $0.8587 > 0.05$ then H_0 is not rejected so it can be concluded that the model used can explain the data.

Based on data analysis using the R software, found that a significantly influential factor is 2 or CAT questionnaire can be categorize as COPD severity and waist circumference bigger than it should. The results of parameter estimates are presented in Table 4.

Based on the results of parameter estimation in Table 2, the logistic regression model equation can be written as follows.

$$\hat{\pi}(\mathbf{x}) = \frac{\exp(-6{,}59615 + 1{,}47908x_1 + 0{,}06409x_2)}{1 + \exp(-6{,}59615 + 1{,}47908x_1 + 0{,}06409x_2)}$$

Based of the equation above, COPD patients with severe degree are 4.39 times greater risk suffer from OSA than mild to moderate COPD patients and each centimetre increase of waist circumference has higher risk of OSA. Accuracy of our model is estimated using classification table with cut point at 0.5 and its accuracy is 73.9%.

Table 3. Results of Parameter Estimation of Logistic Regression Model

Variable	Coefficient	*p-value*
(Constant)	-7,77206	0,02444
Gender	-1,11772	0,09797

Body Mass Index (BMI)	0,02482	0,73195
Age	0,03828	0,04052
Smoking History	0,23752	0,70713
mMRC questionnaire	-0,21011	0,71167
CAT 1 questionnaire	0,49683	0,40077
CAT 2 questionnaire	2,02619	0,00478
Neck circumference	-0,04305	0,70683
Waist circumference	0,05791	0,04443

Table 4. Results of Logistic Regression Model Parameter Estimation

Variable	Estimate	p-value
(Constant)	-6,59615	0,000474
CAT 2 questionnaire	1,47908	0.002308
Waist circumference	0,06409	0,001523

4. Conclusion

Based on data analysis, there are 111 patients who have been diagnosed with COPD by a doctor at the RSCM. From the risk factors that has been studied by previous researchers such as age, gender, body mass index (BMI), smoking history, mMRC questionnaires, CAT questionnaires, neck circumference and waist circumference the factors significantly affecting the risk of OSA in COPD patients are waist circumference and CAT 2 questionnaires or patients with severe COPD. From the analysis of the data obtained and the logistic regression model coefficients interpretation that the variables that affect significantly the risk of OSA in patients COPD with weight and waist circumference degree. In this study it was found that a more severe degree of COPD patients at risk of OSA 4.39 times greater than COPD patients with mild to moderate degree, and every increase of 1 cm in waist circumference at risk of OSA.

5. References

[1] Global Initiative for Chronic Obstructive Lung Disease (GOLD) 2017 *Pocket Guide to COPD Diagnosis, Management, and Prevention* (Global Initiative for Chronic Obstructive Lung Disease, Inc.)

[2] World Health Organization 2016 *Chronic Obstructive Pulmonary Diseases* (World Health Organization)

[3] Owens R L and Malhotra A 2010 Sleep-Disordered Breathing and COPD: The Overlap Syndrome *Respiratory Care* **55**(10) pp. 1333–1346

[4] American Sleep Association 2017 *Obstructive Sleep Apnea-Symptoms, Causes & Treatments* (America: American Sleep Association)

[5] Marry S M, Lamm B, Lauder I J, Tsang K W T, Chung K, Mok Y, and Lam W 2001 A community study of sleep-disordered breathing in middle-aged Chinese men in Hongkong *Chest Journal* **119**(1) pp. 62-69

[6] CV Senaratna 2017 Prevalence of Obstructive Sleep Apnea in the general population : A systematic review *Sleep Medicine Review* **34 pp.** 70-81

[7] Soler X, Gaio E, Powell F L, Ramsdell J W, Loredo J S, Malhotra A, and Ries A L 2015 High prevalence of Obstructive Sleep Apnea in Patients with Moderate to Severe Chronic

Obstructive Pulmonary Disease *Annals of the American Thoracic Society* **12**(8) pp. 1219-1225

[8] Choi K, Thomas R J, Kim J, Lee S K, Yoon D W, and Shin C 2017 Overlap Syndrome of COPD and OSA in Koreans *Medicine* **96**(27) p. 7241

[9] American Academy of Sleep Medicine 2008 *Obstructive Sleep Apnea* (New York: American Academy of Sleep Medicine)

[10] Kamelia T 2017 Validation and Realiability Study to Screening Obstructive Sleep Apnea for Indonesian Population. *Sleep Medicene* **40**(1) p. 152

[11] Hosmer D W and Lemeshow S 2013 *Applied Logistic Regression (Third Edition)* (United States of Amerika: John Wiley & Sons, Inc)

Acknowledgments
This research is fully supported by PITTA Universitas Indonesia Grant 2018

IOP Conf. Series: Journal of Physics: Conf. Series **1108** (2018) 012077 doi:10.1088/1742-6596/1108/1/012077

Content analysis of 13 dimensions to support student teachers' PCK in the environmental chemistry textbooks

D G Ramadhani[1,*], S B Utomo[2], and N Y Indriyanti[2]

Department of Chemistry Education, Postgraduate Program, Universitas Sebelas Maret

Author's Email :[1]gielramadhani@student.uns.ac.id, [2]sbukim98@staff.uns.ac.id, [3]nurma.indriyanti@staff.uns.ac.id

Abstract. Environmental chemistry courses in the undergraduate program is an important subject to link many concepts of chemistry and environmental sciences. Content analysis was used by the researchers to determine textbooks compatibility using 13 dimensions of content knowledge. We examined three textbooks often used as references to discuss environmental chemistry issues in the courses. We specify the analysis in the chapter on wastewater. This chapter is a very important chapter to develop trainee teachers' competencies either as teacher or scientist. We labeled each book with codes of R, S, or T with Kappa coefficient of 82.7, which means no significant differences in perception about observed aspect among researchers. The textbooks only present material-related subjects. Not all dimensions in the content knowledge contained in the textbooks, only dimensions related to basic knowledge.

1. Introduction

Environmental chemistry is the scientific study of chemical and biochemical phenomena occurring in nature. It can be defined as the study of the sources, reactions, transports, effects, and the fate of chemicals in the air, soil, and water environment and effect of human influences on them. It is an interdisciplinary science includes the atmospheric, aquatic and earth chemistry, and relies heavily on analytical chemistry, and other fields of science [1]. Environmental chemistry is important to be studied by pre-service chemistry teachers to study issues related to global environmental issues, and hopefully, they can explain the issue of environmental issues with the concept of chemistry [2].

Wastewater discharged from the households, industries, and various places generally contain harmful substances to human health and environment. Characteristics of wastewater vary depending on the locations, populations, industries, land usages, groundwater level, and level of separation between storm water and sanitary water. Liquid waste is divided into three categories: domestic wastewater include: liquid waste from the kitchen, bathroom, laundry and others; sanitary wastewater includes: domestic wastewater, commercial, office, and similar facilities; and industrial wastewater comes from industry. The properties of industrial wastewater vary depending on the raw materials used, the use of water in the process, and the additives used during the production process[3].
Environmental pollution, especially wastewater, currently is not just a national problem but also a global issue. The problems caused by environmental pollution cannot be separated with toxicology (the understanding of the adverse chemical effects of living organisms). In the last decades, heavy metal toxicology, become the great concern in terms of the health effects [4]. From a toxicological

point of view, heavy metals can be divided into two types. The first is the essential heavy metal, where their presence in a certain amount is needed by living organisms, but excessive amounts can cause intoxication. Examples of these heavy metals are Zn, Cu, Fe, Co, Mn, and so forth. The second is the unimportant or toxic heavy metal, where their presence in the body is unimportant or may even be toxic, such as Hg, Cd, Pb, Cr, As, and others [3].

The chemistry teachers should try to explain the environmental issues using chemical approach. The 21st-century learning requires the teachers to connect their knowledge with the actual phenomena or issues, so students can understand and apply it. The environmental issue became the world's discussion on the next 10 or even 50 years as humanity progresses [5]. The teachers should have the good content knowledge to explain to their students according to the demands of 21st-century learning [6]. The chemistry teachers should help students to understand the content of scientific knowledge. In line with this, the chemistry teachers must have sufficient academic competence, such as the mastery and understanding of the content as well as the pedagogical skills. Content and pedagogical knowledge should be integrated into learning to create new knowledge [7]. It is defined as Pedagogical Content Knowledge (PCK). Shulman (1986, 1987) stated the PCK is a special knowledge possessed by teachers on how to teach certain content to specific learners with the strategies to achieve understanding [8]. PCK can also be described how the teachers teach the certain subject by accessing what they know about the subject matter, what they know about the learner, what they know about the curriculum related to the subject, and what they believe as the good way of teaching for certain context. Thus, PCK is very important in the learning process and should be owned by the chemistry teachers.

Textbooks used in higher education chemistry lectures should be evaluated if there any content that can improve the ability of pre-service teachers' PCK. Textbooks used in lectures, especially environmental chemistry, discuss the concepts of chemistry related to global environmental issues should be analysed how much the contents can improve the pre-service chemistry teachers' PCK [9]. The content analysis was used to assess the content of the textbooks. Content analysis is a technique to analyse and understand textbooks. It is the investigative technique attempts to describe the textbooks objectively, systematically and quantitatively [10]. According to H.D. Laswells, this analysis is as called the quantitative semantics. Researchers can learn a lot about the community by analysing newspapers, magazines, television programs, or music. The researchers who use content analysis does not study people, but their communication. Content analysis of the textbooks is important because the textbook is one of the main components in learning, and very influential in the development of pre-service teachers' knowledge. In this research, the content in the textbooks refers to the 13 dimensions of PCK [11].

2. Method

The content analysis was used in this research, data were analyzed using qualitative and quantitative analysis. Three books were used as the samples, those were most widely used references to study the environmental chemistry[12]. The chapter on wastewater was the focus of the analysis. This chapter was very important to develop pre-service teachers' competencies either as teacher or scientist. Those books were issued by publishers from Yogyakarta, Jakarta, and New York. Those books then labeled with R, S, and T to simplify the analysis process. Content analysis was conducted based on the 13 PCK dimensions proposed by Bucat: (1) Knowing more "facts"; (2) Understanding the role of models and theories in chemistry; (3) Alternate between the macro world and related submicroscopic models; (4) Quality of images at the sub-micro level; (5) Understanding the language of chemistry; (6) Operate at multiple levels of explanation, rationalization and prediction; (7) Memory banks of episodes; (8) Ability to distinguish between demonstrable knowledge and arbitrary knowledge; (9) Appreciation of the sources of our knowledge; (10) Recognition of the place and role of chemistry in society; (11) Understanding what chemists do; (12) Knowing what we do not know; and (13) Interlinking learning[11]. The contents from those books were analyzed to find out what extent the 13 PCK dimensions were supported. Content analysis was performed by researchers and researchers' assistants, where reliability and validity were concluded from the Kappa coefficient. The researchers divide the 13 dimensions of PCK and then looked out what the extent the contents of the books can improve the pre-service chemistry teachers' PCK. The results of the analysis from the researchers then compared to find out the interpreter reliability [13]. The reliability was determined by using following formula:

$$k = \frac{\Pr(a) - \Pr(e)}{1 - \Pr(e)}$$

$\Pr(a)$ =The percentage of consistent measurements between readers
$\Pr(e)$ =The percentage of consistent measurement changes between readers
k = The Kappa Coefficient [14]

Table 1. Interpretation of the Kappa Coefficient.

Kappa.	Interpretation
<0	Poor agreement
0.00-0.20	Slight agreement
0.21-0.40	Fair agreement
0.41-0.60	Moderate agreement
0.61-0.80	Substantial agreement
0.81-1.00	Almost perfect agreement

3. Result and Discussion

Table 2. 13 dimensions of content knowledge in environmental chemistry text books.

No	Content Knowledge Dimension	R	S	T
1	Knowing more "facts"	(32)	(20)	(29)
2	Understanding the role of models and theories in chemistry;	(60)	(43)	(48)
3	Alternate between the macro world and related submicroscopic models	(1)	(1)	(2)
4	Quality of images at the sub-micro level	(1)	(1)	(2)
5	Understanding the language of chemistry	(37)	(19)	(28)
6	Operate at multiple levels of explanation, rationalization, and prediction	-	-	-
7	Memory banks of episodes	(9)	(3)	(3)
8	Distinguish between demonstrable knowledge and arbitrary knowledge	-	-	-
9	Appreciation of the sources of our knowledge	-	-	-
10	Recognition of the place and role of chemistry in society	(8)		(9)
11	Understanding what chemists do	-	-	(8)
12	Knowing what we do not know;	(34)	(23)	(29)
13	Interlinking learning	(10)	(5)	(7)

We found differences in content knowledge in the three books. In table 2, in the R, S, and T books have differences in content knowledge. The R and T books have the same qualities of content knowledge which was almost good, while the S book has the lowest content knowledge. There were differences in the dimensions contained in the R, S, and T books. There were dimensions existed only on the T, or only on the R alone. For the alternate between macroscopic world and related submicroscopic models were found in all books. The heart of chemistry is our use of models/theories at the level of atom molecules to make sense of observable phenomena.

Alternate Between the Macroscopic World and Related Submicroscopic Models

One of the PCK dimensions became the point needed to be described was the relationship between macroscopic and submicroscopic levels. In the R and S there was the explanation as follows:

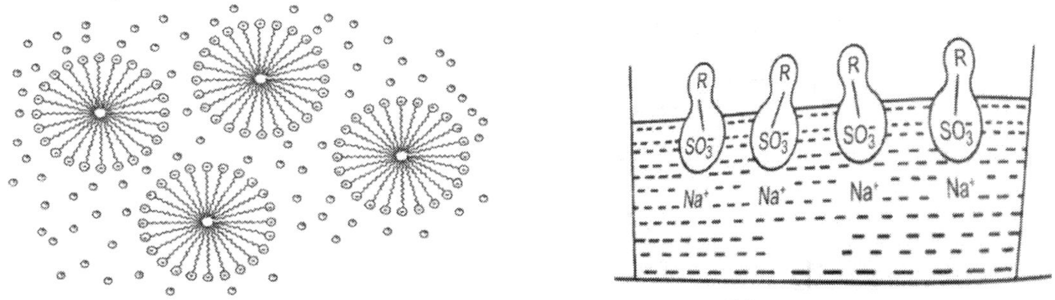

(a) (b)

Figure 1. Interaction between soap with oils, fats, and other water-insoluble organic materials. (a) Illustration in the R and S books. (b). Illustration in the T book.

At the macroscopic level, the soaps and detergents ability to remove fats and oils were well known. At the sub-microscopic level, soaps were the salt of high fatty acids, such as sodium stearate, $C_{17}H_{35}COO\text{-}Na^+$. Soap washing actions were generated from its emulsifying power and the ability to reduce water surface tension. This concept can be explained using the dual nature of soap anions. The stearate consisted of carboxyl ions as the heads, with long hydrocarbons as the tail. In the presence of oil, fat, and other insoluble organic matter, the "tails" dissolve in organic matter, while the "heads" remain in the aqueous solution. Soap emulsifies or suspends organic matter in water. Soap anions form particles of micelle colloid particles, which the hydrocarbon "tails" anions are clustered within small colloidal particles and the "heads" of the carboxylic anions are located on the surface of colloidal particles. In the T book—which uses alkyl benzene sulfonate—, and in the R and S—which use sodium stearate ($C_{17}H_{35}COO\text{-}Na+$)—, the process of interactions were described in Fig. 1a and 1b. Both useful to explain macroscopic phenomena and illustrate them in sub-microscopic levels.

What chemists do?

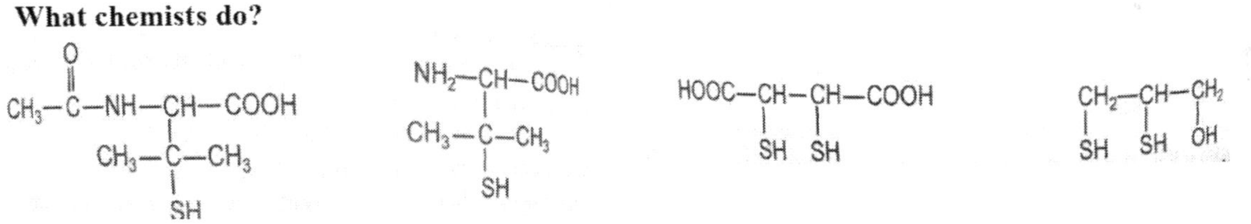

N-acetyl-DL-penicillamine D-penicillamine Dimercaptosuccinic acid British Anti-Lewisite

Figure 2. Chemotherapy Agents to Treat Mercury Poisoning

Dimensions of "What chemist do" only existed in the T book, which explains how a chemist should do if faced with certain phenomena. The T book provided illustration about mercury poisoning. Chemists were expected to know the right treatment for each case. Some of the chemotherapy agents used to treat mercury poisoning, such as N-acetyl-DL-penicillamine, or British Anti-Lewisite (BAL) were shown in Figure 2. These agents enter the cells, convert the intracellular methyl mercury into a mobile complex that ready to diffuse out of the cells, and excreted through urine and feces. The British Anti-Lewisite (BAL) was very effective to deal with inorganic mercury poisoning but cannot be used to treat methyl mercury poisoning. The T was the only book gave the dimension "what chemists do" by providing information on how the chemists deal with pollutants. The other books presented only the dangers without describing how the chemist's act. The R book described the treatment of mercury poisoning, as the chemists' way to act when the same events occur.

The result of the reliability analysis obtained the Kappa coefficient of .827 and p-value of 0.000. This result means p-value <alpha. The alpha of .05 was used in this research. This result means significant Kappa test result, so there were no perception differences of observed aspects among researchers [13] [14].

Some concern regarding the teachers' PCK was obtained from our analysis. The 13 dimensions of PCK were the basic things to be mastered by teachers to cope with the challenges of the 21st-century learning and as the basis to become teachers who teach their students well. The analyzed textbooks have supported the majority of the PCK's dimensions, but there were several PCK not yet covered in the textbooks. The operating at multiple levels of explanation, rationalisation, and prediction; distinguish between demonstrable knowledge and arbitrary knowledge; recognition of the place and role of chemistry in society; and understanding what chemists do by lecturers during the learning process needs to be added to the S book [11] [16]. Those dimensions are rarely found in chemistry textbooks, especially on the topic of environmental chemistry [17].

Addition of the operate at multiple levels of explanation, rationalisation, and prediction help the students to understand multiple representations and can connect as macroscopic, symbolic, and submicroscopic levels. Distinguish between demonstrable knowledge and arbitrary knowledge have relationships with the bank of memories to determine the provable knowledge or unnecessary results of research implied with phenomena in the chemistry [11]. Recognition of the place and role of chemistry in society requires teachers' knowledge to relate events in social context and their relation to the chemistry[18]. Appreciation of the sources of our knowledge helps the teachers to recall what is known and try to discuss more deeply, this dimension give students more cognitive conflict [15]. And "what chemist do" provide education about what we should do in facing cases or events as the chemist [12], [19].

Overall, the R and T books have almost the same quality of contained PCK while the S book contained very little PCK. The R book has more characteristics on the understanding of symbolic and chemical reactions while the T book has the power to gain a chemical context in society.

Conclusion

The textbooks only present material-related subjects. Not all dimensions of PCK contained in the textbooks. There were only dimensions related to basic knowledge such as: knowing more "facts"; understanding the role of models and theories in chemistry; understanding the language of chemistry but lacking in additional knowledge; understanding and chemistry in context. While the useful dimensions to enhance thinking and understanding skills, such as: operating at multiple levels of explanation, rationalisation, and prediction; distinguish between demonstrable knowledge and arbitrary knowledge; appreciation of the sources of our knowledge; and understanding what chemists do was not contained in the textbooks. Thus, while they can be used as the reference to studying environmental chemistry, the students are expected to use additional sources for dimensions not yet contained in the textbooks. There were no differences in perception about observed aspects among the researchers.

Acknowledgments
Thanks to Institute of Research and Community Service (LPPM), Universitas Sebelas Maret, Surakarta, Indonesia for funding this research through The Research Group Maintenance; International Publication Grant 2018.

References
[1] Greene J W 1996 Incorporation of Pollution Prevention Principles into Chemical Science Education (Thesis of the Master of Science degree, School of Public Health, Department of Environmental and Industrial Health, University of Michigan)
[2] Manahan S E. 2001 *Fundamentals of Environmental Chemistry* (CRC Press LLC)
[3] Stigliani W M 2013 *Critical Reviews in Analytical Chemistry Environmental Chemistry Education : An Issues Course In Environmental Chemistry* pp. 37–41
[4] Development Core Team 2011 *A Language and Environment for Statistical Computing*, vol. 1, no. 2.11.1.
[5] Al-Jaro M, Asmawi A and Hasim Z 2017 *Arab World English J.* **88** 264
[6] Wood E 2003 *Math. Educ.* **7** 49
[7] Shulman L S 1986 *Educational Researcher* **15** 4
[8] Jing-Jing H 2014 *Int. J. Educ. Res.* **2** 411
[9] Fernandez C 2014 *Probl. Educ. 21st century* **60** 79
[10] Gilbert J K and Treagust D 2009 *Multiple Representations in Chemical Education* (New York, N Y: Springer)
[11] Bucat R 2004 *Chem. Educ. Res. Pract.* **5** 215
[12] Yost M 1973 *J. Res. Sci. Teach.* **10** 317
[13] Gisev N, Bell J S and Chen T F 2013 *Res. Soc. Adm. Pharm.* **9** 330
[14] Thompson W D and Walter S D 1988 *J. Clin. Epidemiol.* **41** 949
[15] Baddock M and Bucat R 2008 *Int. J. Sci. Educ.* **30** 1115
[16] Kusumaningrum I A, Ashadi A and Indriyanti N Y 2017 *J. Phys. Con. Ser.* **895** 012042
[17] Yavuz S 2016 *J. Educ. Train. Stud.* **4** 118
[18] Hesse J J and Anderson C W 1992 *J. Res. Sci. Teach.* **29** 277
[19] Kahveci A 2010 *Int. J. Sci. Educ.* **32** 1495

IOP Conf. Series: Journal of Physics: Conf. Series **1108** (2018) 012078 doi:10.1088/1742-6596/1108/1/012078

Red Small Dot Segmentation for Early Warning of Diabetic Retinopathy

O S Riza[*1] **and H Tjandrasa**[2]

[1,2] Department of Informatics, Sepuluh Nopember Institute of Technology (ITS) Surabaya, Indonesia
*Corresponding author: ozzysecio@gmail.com, handatj@its.ac.id

Abstract. Diabetic retinopathy is a medical condition occurs to the retina due to diabetes mellitus and leading cause of blindness. Early detection of diabetic retinopathy is very necessary. Therefore, an automated system to detect abnormalities in retinal fundus image is expected to avoid further damage to the retina. In this study, we proposed an automated system to detect red small dots in retinal fundus images. The method is the combination of Tyler Coye and morphological supremum of opening algorithm to enhance blood vessel segmentation for red small dot detection. The system consists of three parts, the first part is the process to segment the dark area of retinal fundus image after eliminating bright regions including optic disc and exudates. The second part is the process to segment the blood vessel, the third part is to obtain the red small dot segmentation by subtracting the result of the first with the second part. The test performance evaluated using accuracy, sensitivity, and specificity. The result is 99.69%, 70.88% and 99.73% respectively.

1. Introduction

Diabetic retinopathy is a disease caused by complication of diabetes, which causes damage to the retina and can result in blindness. It could be divided into two clinical groups: proliferative diabetic retinopathy and non-proliferative diabetic retinopathy, the earliest clinical symptom found in diabetic retinopathy. Diabetic retinopathy could be seen with the development of microaneurysm, hemorrhages, hard exudates, soft exudates, and neovascular in the retina [1]. Microaneurysm and small haemorrhage categorized as red small dot.

Microaneurysm is the earliest clinical sign of diabetic retinopathy, characterized by the emergence of pockets on the blood vessel and appear as small red and dark round dots. Hemorrhages occur because of blood leakage in the blood vessel in the retina. Exudates appear because there are lipid or leaky fat in the abnormal blood vessel. Increasingly the area of microaneurysms, haemorrhage, and exudates indicates that the disease is getting worse and can lead to blindness.

The manual detection of red small dots could be time consuming and weariness in observation. The process is not easy because of the complexity of the retina structure. Therefore, it is needed to develop an automatic system for red small dot detection. Previous researches on segmenting red small dots in retinal fundus images are performed using various approaches. [2] detecting microaneurysms using shade correction in pre-processing step to eliminate illumination variation. Candidate extraction done using bilinear top-hat transformation. A similar approaches was adopted by [3] and increase the use of region growing to the algorithm. [4] initial set of candidates using a Gaussian Matched Filter then classifies this set using Tree Ensemble classifier to reduce the number of false positive.

Before detecting red small dots, blood vessel need to be removed because it has similar intensity value with red small dots. [5] perform smoothing operation on retinal images using mathematical morphology. The enhanced image then segmented using K-means clustering. Many researches using Coye algorithm to detect blood vessel in retinal fundus images [6,7]. Other studies use the supremum of opening with linear structuring element at different orientations [8,9]. Both Coye algorithm and supremum of opening provide good result for detecting blood vessel in retinal fundus images, but there are still parts of blood vessel that are not detected by one algorithm but detected on the other. This study proposed the combination of Tyler Coye and morphological supremum of opening algorithm to enhance blood vessel segmentation for red small dot detection.

This paper is organized as four sections, the first section describes the background of diabetic retinopathy and the previous researches on segmentation of red small dots in retinal fundus images. The second section consist of research method used in this study. The third section explains the experimental results and discussion. The last section is conclusion of the study.

2. Red small dot segmentation approach

2.1. Contrast-limited Adaptive Histogram Equalization (CLAHE)

Histogram equalization is a method that improve the image contrast by equalizing the histogram to enhance particular characteristics [8,9]. The histogram equalization process in adaptive histogram equalization is carried out on small region or tiles of the image instead of the entire image. The neighbouring region are smoothed utilizing bilinear interpolation. In contrast-limited adaptive histogram equalization (CLAHE), the histogram of each region is calculated first, then a clip limit for clipping histogram is achieved. Furthermore, the cumulative distribution function uses to transform the grayscale mapping.

2.2. Supremum of Opening

Supremum of Opening identifies the elongated structure of blood vessels with linear structuring element at different orientations as seen in Figure 1. Then applied the opening process to the image using different structuring elements. Furthermore, taken the supremum of the opening process result.

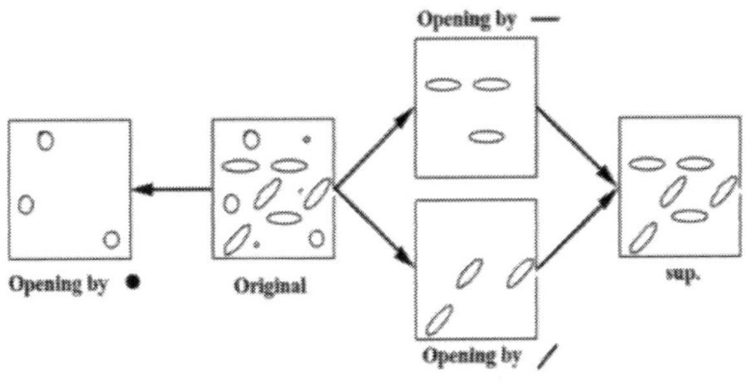

Figure 1. Supremum of Opening [9]

2.3. Tyler Coye Algorithm

Tyler Coye algorithm [12] is used to detect blood vessel in retinal fundus image. It uses principle component analysis (PCA) of Lab colour model for converting the image into grayscale. The contrast enhances using Contrast-Limited Adaptive Histogram Equalization (CLAHE). Then, the background is excluded by subtracting the average filtered image. The binary image computes by isodata thresholding and finally, the smaller components are removed by the pixel size.

3. Research Method

Segmentation of red small dots consist of three parts; the first part is the process to segment the dark area of retinal fundus image after eliminating bright regions including optic disc and exudates. The second part is the process to segment the blood vessels, the third part is to obtain the red small dot segmentation by subtract the result of the first and the second part.

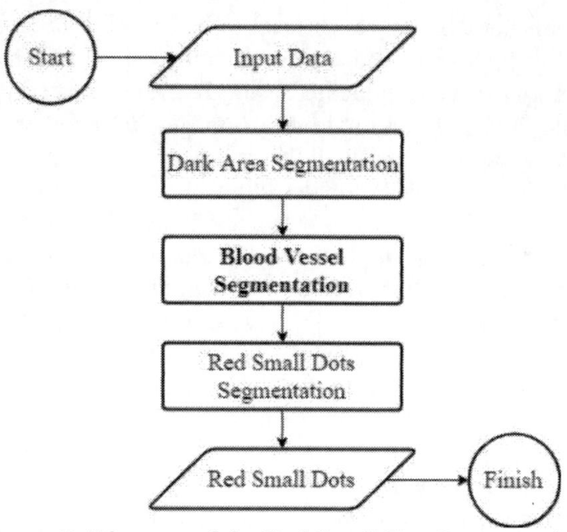

Figure 2. Diagram of the Red Small Dot Segmentation

3.1. *Dark Area Segmentation*

Dark area segmentation process eliminating bright regions including optic disc and exudates. This part consists of the binary image of retinal vessel and red small dots. The segmentation method is given by the following steps:

a. Resize retinal fundus image into 480 x 640 pixels.
b. Convert the RGB color retinal image into green channel image.
c. Find the image regional minima of the result from the second step.
d. Multiply regional minima with the green channel image.
e. Apply morphological reconstruction with the marker from the result of multiply image regional minima with green channel and the green channel image as the mask then enhance the image.
f. Threshold the result of the previous step to get the dark area of the image.

3.2. *Blood vessel segmentation*

The process to get the blood vessel segmentation is using the combination of Tyler Coye algorithm and morphological supremum of opening. The blood vessel segmentation method is described by the following steps:

a. Convert RGB color retinal image into grayscale using Coye algorithm
b. Enhance the contrast of an image by applying CLAHE.
c. Apply the morphological opening and closing of the enhanced image.
d. Eliminate the background by subtracting the result from the third step with the enhancement image.
e. Find local minima of the gray channel image.
f. Detect blood vessels using supremum of opening with linear structuring element at different orientation.
g. Apply morphological reconstruction by dilation to obtain the blood vessels with the output of elimination background as the marker and the result of supremum of opening as the mask.
h. Threshold the image from the seventh step to get the blood vessel segmentation.
i. Remove small pixels smaller than 90.

3.3. *Red small dot segmentation*

The procedure to get the red small dot segmentation is solved by subtract the binary image of the blood vessel segmentation from the dark area segmentation to get the red small dots segmentation. Only select the red small dot objects with major axis length less than 5 pixel.

IOP Conf. Series: Journal of Physics: Conf. Series **1108** (2018) 012078 doi:10.1088/1742-6596/1108/1/012078

3.4. *Performance Measure*

The results of segmentation are evaluated using Sensitivity, Specificity, and Accuracy values using criteria of true positive (TP), true negative (TN), false positive (FP), and false negative (FN) with the following equations:

$$Sensitivity = \frac{TP}{TP+FN} \quad (1)$$

$$Specificity = \frac{TN}{TN+FP} \quad (2)$$

$$Accuracy = \frac{TN+TP}{TN+FP+TP+FN} \quad (3)$$

4. Experimental Result

This work uses the retinal fundus images taken from DIARETDB1 database as the input images and randomly selected. The data has dimensions of 1500 × 1152 pixels. To reduce the size and speed up the process of segmentation, the image is reduced to 480 × 640 pixels. Figure 3-5 show the result of each steps, start from the resized colour images, the enhanced images, the result of dark area segmentation, the result of blood vessel segmentation, then the result of red small dot segmentation.

The performance result compares with segmentation of red small dots using Tyler Coye algorithm only is given by Table 1. Compare to the result of segmentation using Tyler Coye algorithm only, proposed method give better result.

Table 1. Red Small Dot Segmentation result using Tyler Coye algorithm and proposed method

Method	Accuracy (%)	Sensitivity (%)	Specificity (%)
Tyler Coye algorithm	99.65	52.52	99.69
Proposed Method	99.69	70.88	99.73

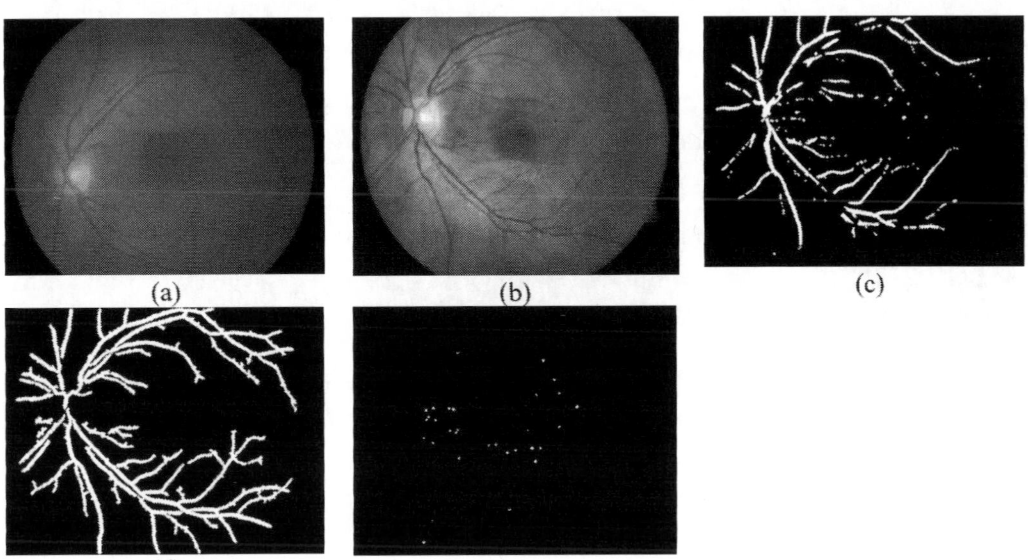

(a) (b) (c)

(d) (e)

Figure 3. The results of each steps, (a) resized colour image, (b) enhanced image, (c) dark area segmentation result, (d) blood vessel segmentation result, (e) red small dot segmentation result.

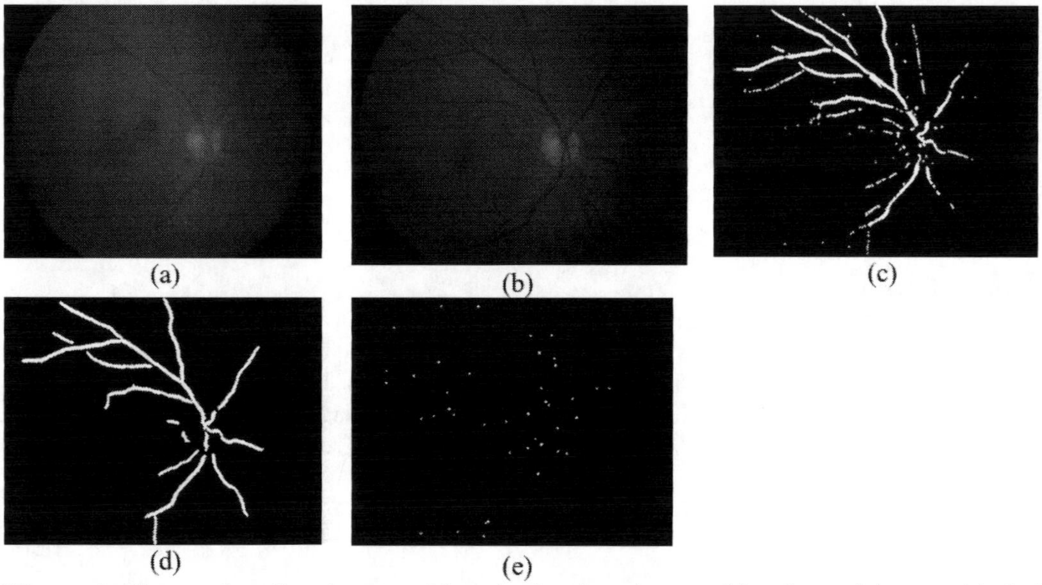

Figure 4. The results of each steps, (a) resized colour image, (b) enhanced image, (c) dark area segmentation result, (d) blood vessel segmentation result, (e) red small dot segmentation result.

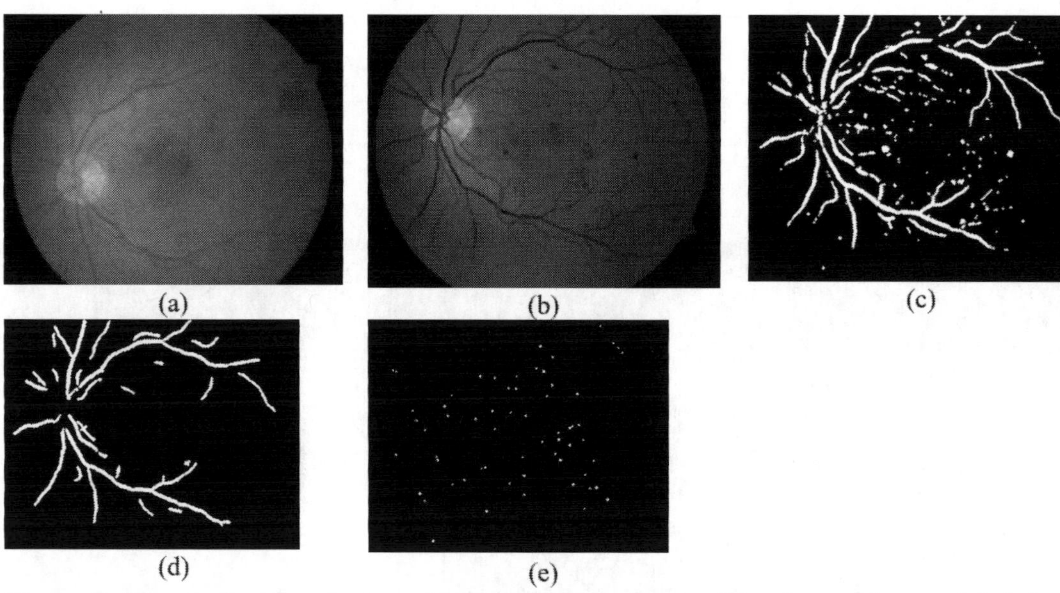

Figure 5. The results of each steps, (a) resized colour image, (b) enhanced image, (c) dark area segmentation result, (d) blood vessel segmentation result, (e) red small dot segmentation result.

5. Conclusions

Red small dot segmentation of the retinal fundus images is proposed in this work. The red small dot detection is implement by several steps, including gain dark area by removing bright areas of exudates and optic disk, detecting the blood vessels, then separating the red small dots with the blood vessels. The methods developed by combining Tyler Coye and morphological supremum of opening algorithm to enhance blood vessel segmentation for red small dot detection. The experiment results show that compared with the result using Tyler Coye only, the proposed method give better result. The average accuracy is 99.69%, the average sensitivity is 70.88%, and the average specificity is 99.73%. With the improvement of the red small dot segmentation model which can fully eliminate the blood vessel, it is expected that the performance can be increased.

Reference

[1] Xu Y, Li Z, Yang Z and Zhang D 2017 *IEEE Access* **5** 8502
[2] Spencer T, Olson J A, McHardy K C, Sharp P F and Forrester J V 1996 *Comput. Biomed. Res.* **29**(4) 284
[3] Forrester J V1997 *R. Coll. Ophthalmol* **11** 622
[4] Habib M M, Welikala R A, Hoppe A, Owen C G, Rudnicka A R and Barman S A2017 *Informatics Med. Unlocked* **9**(May) 44
[5] Hassan G, El-Bendary N, Hassanien A E, Fahmy A, Abullahm S, and Snasel V 2015 *Procedia Comput. Sci.* **65** 612
[6] Bandara A M R R and Giragama P W G R M P B 2018 2017 IEEE International Conference on Industrial and Information Systems (ICIIS) 1–5,.
[7] Furtado P, Travassos C, Monteiro R, Oliveira S, Baptista C and Carrilho F 2017 *IEEE EMBS Int. Conf. Biomed. Heal. Informatics, **BHI 2017*** 25
[8] Hajer J, Kamel H and Noureddine E 2006 *Inf. Commun. Technol. 2006. ICTTA '06. 2nd* **1** 1130
[9] Saraswati S, Tjandrasa H and Amaliah B 2016 *Deteksi Mikroaneurisma Pada Citra Fundus Retina Mata Retinopati Diabetes* (Surabaya: Institut Teknologi Sepuluh Nopember)
[10] Pisano E D *et al* 1998 *J. Digit. Imaging* **11**(4) 193
[11] Reza A M 2004 *J. VLSI Signal Process. Syst. Signal Image. Video Technol* **38**(1) 35

Students' Translation Ability of Mathematical Representations (Symbolic and Visual) Based on Their Learning Styles

I S Putra*[1], Masriyah[2] and R Sulaiman[3]
[1, 2, 3] Department of Mathematics,Universitas Negeri Surabaya, Indonesia
*Corresponding author: imamputra16070785033@mhs.unesa.ac.id; masriyah@unesa.ac.id; radensulaiman@unesa.ac.id.

Abstract. This qualitative study aims to describe students' ability of translation between mathematical representations (visual and symbolic) of visual, auditory or kinesthetic learners. The students' translation ability between mathematical representations was seen from the process undertaken during the translation, namely: unpacking the source, preliminary coordination, constructing the target, and determining equivalence. The participants of this study were three junior high school students with different learning styles, having the similar mathematical ability. The participants were given a translation ability test between mathematical representations and followed by semi-structured interviews. Then, the data were analysed by step of data condensation, data display and conclusion. The results of this study showed that at the unpacking the source stage, they are able to unpack the representations of a given resource by identifying the information contained in the source representations. At the preliminary coordination stage, they can prepare a formula or strategy as a first step to determine the target representations. In the constructing the target, they can transfer information obtained from the source representations to form a target representations. At the determining equivalence, they consider the target representations made by the representations of the given source.

1. Introduction

Representation is one of the mathematical skills that should be mastered by students in learning mathematics [1] due to it is very important for students and related to communication skills and problem solving closely [1-5]. Representation is defined as any configuration of characters, images, concrete objects etc., that can symbolise or "represent" something else [6-8]. In psychology representation can be viewed as the process of modeling concrete things in the real word into abstract concepts or symbols [9]. Many experts divide the form of mathematical representation into several types such as graphs, images, diagrams, words, manipulative and symbol forms [9-12].

In the learning process, students not only used one representation but also required to use multi-representations [13,14], because the ability of students in using various forms of mathematical representation can affect mathematical abilities [15], students can understand broader mathematical concepts and solve mathematical problems in various forms of representation [14,16].

Besides being able to use various representations, students are also required to change the form of representation [17, 18]. The ability of students to change the form of mathematical representation can be known by doing the translation. Translational ability between representations is very important in mathematics learning [19], because with translation skills students can express mathematical ideas [19] and also students have access to a broader set of strategies in problem solving [20, 21].

Translation between mathematical representations is the process of changing a given mathematical representation (source representation) into another requested mathematical representation (target representation) [17, 20-22]. Translation and the translation process are psychological forms (mental processes), intellectual and cognitive processes that are transformed into source mathematical representations into target representational forms [22]. The translation of the mathematical representation is done through four stages: unpacking the source, preliminary coordination, constructing target, and determining equivalence [21].

Many researchers have examined translation between representations, but are limited to only a few representations. There was a study of the translation between verbal representations and symbols [20], translational representations between verbal representations and graphs [18], translational representations between visual and verbal representations [17]. The findings of some previous studies showed that each number of students made mistakes in transcribing between representations [17, 18,20]. This study will make the students' translation of symbolic and visual representations of the material of straight-line equations.

Previous research has been proven that students who have different learning styles (visual, auditory and kinesthetic) will choose mathematical representations that suit their needs in solving mathematical problems [23] and students have different learning styles have different translation skills and representations skills in learning [8]. This study focuses on students who have visual, auditory, and kinesthetic learning styles

2. Methods

This research was a descriptive research with a qualitative approach that aimed to describe the ability of students in visual, auditory, and kinesthetic learning styles in translating between mathematical representations. Subjects in this study consisted of 3 subjects taken from 42 students in one school of Surabaya. The three selected subjects had different learning styles: visual, auditory and kinesthetic. The data used in this study were collected on March 2018. The instruments was used in the study consisted of learning style test, math ability test, and translation ability test between mathematical representations, as much as two matter taken from equation of straight-line equation. The learning style questionnaire and mathematics ability test was used to determine the subjects in the research, while the translation ability test was used to identify ability of students in doing translation between representations. The translation ability test between mathematical representation is shown in Table 1.

Table 1. Translation ability test between mathematical representations.

Translations from Symbol to Visual Test	Translations from Visual to Symbol Test
Sketch the graph of the line equation follow $2x + y = 4$.	Determine the equation of the graph follow. 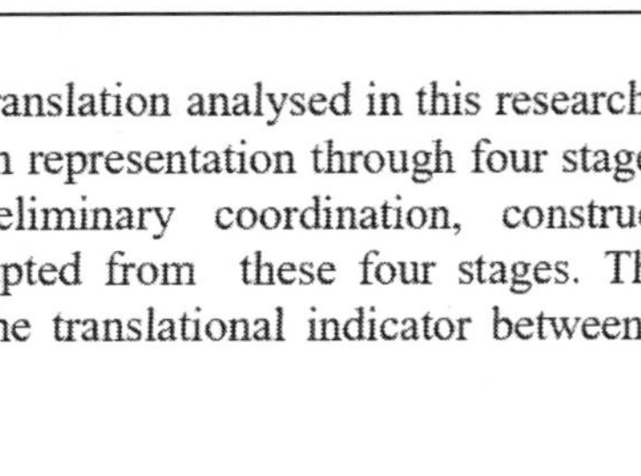

The ability of the students in conducting the translation analysed in this research was the ability of the students in performing the translation between representation through four stages of translation of mathematics, i.e., unpacking the source, preliminary coordination, constructing target, and determining equivalence [14]. The research adapted from these four stages. The indicators were created based on Indonesia's student culture. The translational indicator between the mathematical representations was described in Table 2.

Table 2. Indicators of Translation between Mathematics representations.

Stages of Translation	**Indicator**
Unpacking The Source	Identifying the form of visual representation/symbol representation, and mentioning the existing information in the form of visual representation or symbol representation and visual representation/representation symbol to be searched.
Preliminary Coordination	Determining the strategy/plan for the initial formation of target representations based on existing information on visual representation or symbol representation.
Constructing the Target	Transferring information that exists in the visual representation/symbol representation in to the initial strategy/plan of formation (Formulation).
	Form a visual representation or symbol requested in the question.
Determining Equivalency	Looking again at the conformity between the visual representation/symbol representation formed by the visual representation or symbol representation in the question.

Adapted from [17]

3. Results

3.1. Translations from Symbol Representations to Visual Representations

3.1.1. Visual Learner

The translation ability of visuals learner from symbol representations to visual representations shows in Figure 1.

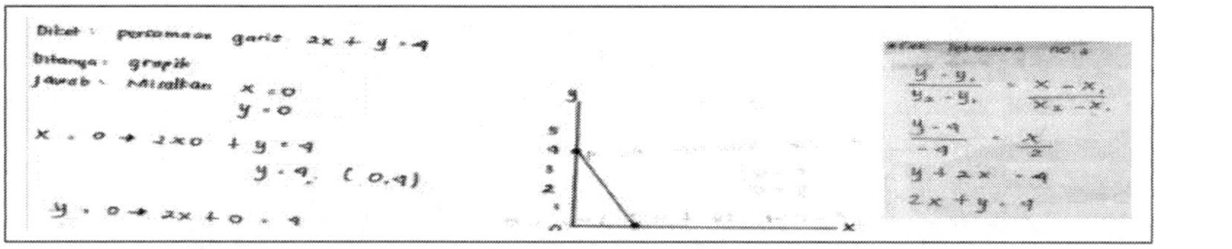

Figure 1. Translation from Symbol Representations to Visual Representations by Visual Learner

In the unpacking the source stage the visual learner student could unpack the symbol representation by identifying and mentioning the existing information on the symbol representation of the equation $2x + y = 4$. In the preliminary coordination stage the students replaced the variables x and y variables with 0. In the stage student was able to transfer information into the plan of visual representation formation by substituting $x = 0$ and $y = 0$ into the line equation so that students obtain the coordinate points from the graph $(0, 4)$ and $(2, 0)$, then the students draw a line equation graph. At determining equivalency stage the student adjusted the visual representation with the symbol doing the truth check by recreating the line equation of the graph formed.

3.1.2. Kinesthetic Learner

The translation ability of kinesthetic learner from symbol representations to kinesthetic representations shows in Figure 2.

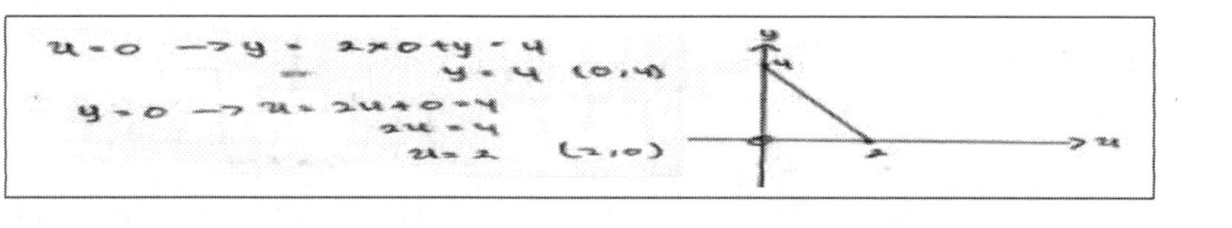

Figure 2. Translation from Symbol Representations to Visual Representations by Kinesthetic Learner

The translations performed by kinesthetic learner was the same as visual learner students. In the unpacking stage the kinesthetic learner unpacked the symbol representation by identifying and mentioning the existing information symbol representation of the line equation $2x + y = 4$. In the preliminary coordination stage the student replaced the variables x and y variables with 0. In the stage student was able to transfer information into the plan of visual representation formation by substituting $x = 0$ and $y = 0$ into the line equation so that student obtained the coordinate points from the graph $(0 , 4)$ and (2.0), then the student drawn a line equation graph. At this stage the student adjusted the visual representation with the symbol doing the truth check by recreating the line equation of the graph formed.

3.1.3. Auditory Learner
The translation ability of auditory learner from symbol representations to auditory representations shows in Figure 3.

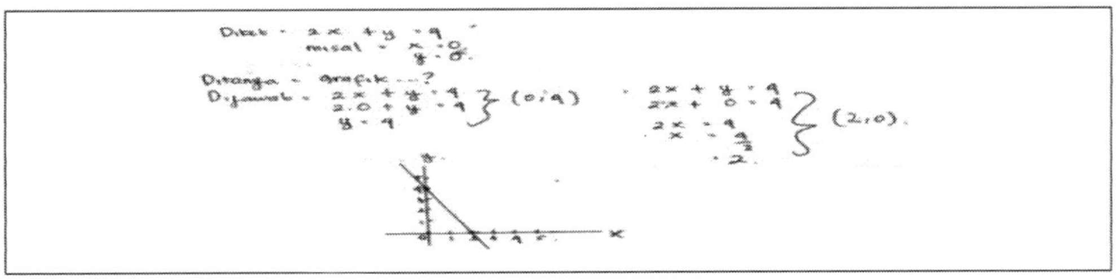

Figure 3. Translations from Symbol Representations to Visual Representations by Auditory Learner

Based on Figure 1-3 in the unpacking the source stage, preliminary coordination, and constructing the target of auditory learner perform the same process with visual and kinesthetic learning styles, but which differentiate among the three students at the determining equivalence stage. In the determining equivalence stage the auditory-style students re-check the coordinate points obtained as revealed by the students in the following interview:

ISP *Why do you believe in your answer?*
SA *Because this equation through 2 coordinates, so I believe that two coordinate are correct.*
ISP *How do you re-examine your answer?*
SA *I checked these two coordinate points correctly or not and I also checked the graphs that I made were appropriate or not.*

3.2. Translations from Visual Representations to Symbol Representations
3.2.1. Visuals Learner
The translation ability of Visuals Learner from visual representations to symbol representations shows in Figure 4.

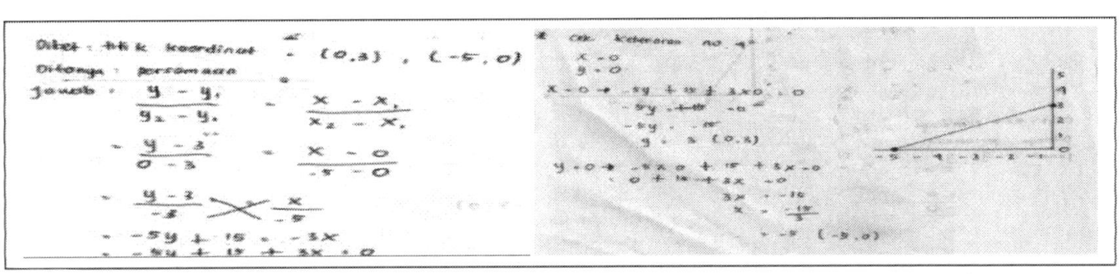

Figure 4. Translations from Visual Representations to Symbol Representations by Visuals Learner

In the unpacking the source stage, student identified the line graph that became the source representation to know and understand the information contained in the graph. So student could name information that is known from the graph. At the preliminary coordination stage, the student used the formula $\frac{y-y_1}{y_2-y_1} = \frac{x-x_1}{x_2-x_1}$ to determine the equation of the given line. In the stage of constructing the target student transfered the information they have that is two point coordinates through which the line into the formula used to determine the equation of the given line. The third subject entered the known coordinate e.g., point $(0, 3)$ and point $(-5,0)$ into the formula $\frac{y-y_1}{y_2-y_1} = \frac{x-x_1}{x_2-x_1}$ by substituting $x_1 = 0, x_2 = -5, y_1 = 3$ and $y_2 = 0$ to obtain the $-5y + 3x = -15$ equation which is the symbolic representation of the given line graph. At the determining equivalence stage, the student learned visually to recreate the graph from the graph equation -5$y + 15 + 3x = 0$ formed.

3.2.2. *Kinesthetic Learner*

The translation ability of kinesthetic learner from visual representations to symbol representations shows in Figure 5.

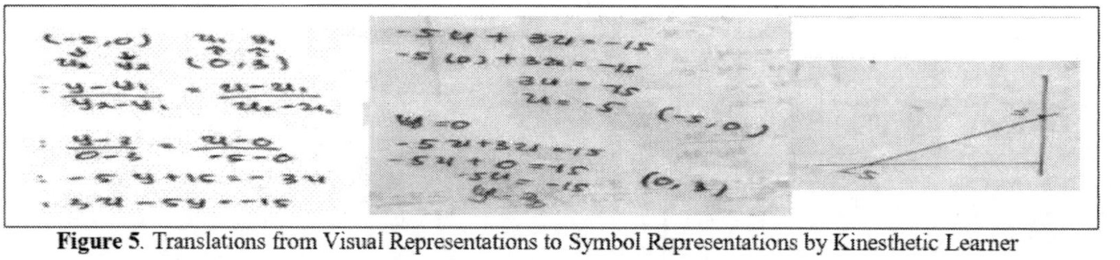

Figure 5. Translations from Visual Representations to Symbol Representations by Kinesthetic Learner

The translations of kinesthetic learner was the same as the translations of visual learner. In the unpacking the source stage, student identified the line graph that became the source representation to know and understand the information contained in the graph. So student could name information that is known from the graph. At the preliminary coordination stage, the student used the formula $\frac{y-y_1}{y_2-y_1} = \frac{x-x_1}{x_2-x_1}$ to determine the equation of the given line. In the stage of constructing the target student transferred the information that is two point coordinates through which the line into the formula used to determine the equation of the given line. The third subject entered the known coordinate points e.g. point (0, 3) and point (-5,0) into the formula $\frac{y-y_1}{y_2-y_1} = \frac{x-x_1}{x_2-x_1}$ by substituting $x_1 = 0, x_2 = -5, y_1 = 3$ *and* $y_2 = 0$ to obtain the $3y - 5x = -15$ equation which is the symbolic representation of the given line graph. At the determining equivalence stage, the visual learner to recreate the graph from the grais equation $3y - 5x = -15$ formed.

3.2.3. *Student in Auditory Learning Styles*

The translation ability of auditory Learning Styles student from visual representations to symbol representations shows in Figure 6.

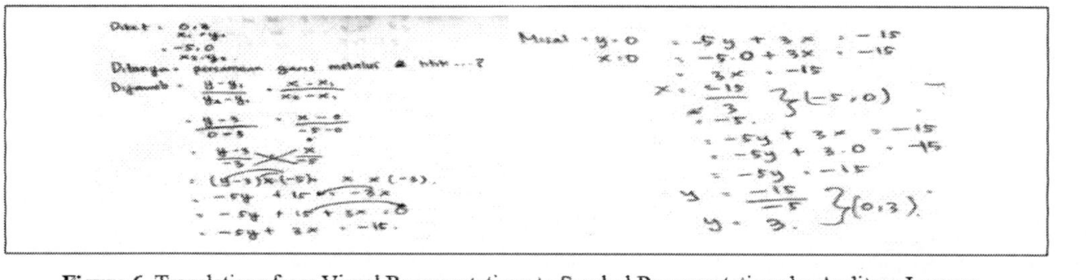

Figure 6. Translations from Visual Representations to Symbol Representations by Auditory Learner

Based on Figure 4-6 in the unpacking the scourge stage, preliminary coordination, and constructing the target of Auditory Learner performed the same process with visual and kinesthetic learner, but which differentiate among the three students at the determining equivalence stage. At the determining equivalence stage the auditory learner re-entered the known coordinate points $(-5{,}0)$ and (0.3) into the $-5y + 3x = -15$ equation.

4. Discussion

Based on the result, in general, the three students were able to translate between visual representations into symbols and symbols to visuals. To form a target representation the three students dismantled the source representation by identifying and mentioning the initial information contained in the representation of the source provided. It was consistent with research that reveals that the initial step to form a target representation of students had to dismantle the source representation by identifying to know the initial information needed to form a charger representation [21]. Furthermore, the three subjects made the initial formulation to form a target representation at the preliminary coordination stage. The three subjects were able to form the target representation requested based on the initial formulation made previously. In translating between visual representations to symbols the three subjects formed different representations such as students learning visual learning style making -5y + 15+ 3x = 0, students with kinesthetic learning style made 3y-5x = -15, while students with auditory learning style made -5y + 3x = -15, while in translating symbols into visuals the three students formed the same graph. Similarities and different forms of representation made by students based on their understanding of mathematical concepts that had been known before. It was based on the theory which states that students who had different learning styles will choose the form of representation according to their understanding and needs [9, 23].

Next, the students did the final step in translating between representations, namely determining equivalence. At this stage we could see the difference in the translation process carried out through the translation stages. In translating visual representations to symbols, the visual and kinesthetic learners re-created the representation of the visual representation of the form, while the auditory-style students only checked the two coordinate points generated, because the line could only be make two coordinate points , the student was able to perform the last stage in pass translation. In the translation of symbol representation into visuals, the auditory learners re-entered the known coordinate points (-5.0) and (0.3) into the line equation. Whereas students studied visual and kinetic recreating the graph from the line equation formed. This stage showed that the third students were able to translate representations well. It was based on previous research which showed that students with good math skills will translate well [9, 20, 21].

The results of this research indicated that the translation ability between student representations influenced by an understanding of the mathematical concepts that have been studied. The better the students' mathematical abilities, the better the translation between representations [9, 20,21].

5. Conclusion

Based on the result of the study, we concluded that the students' of visual, kinesthetic and auditory styles are able to make translation between mathematical representation (visual and symbolic) through the translation process (1) Unpacking the sources; (2) Preliminary coordination; (3) Constructing the target; (4) Determining equivalence. At the unpacking the source stage, they are able to unpack the representations of a given resource by identifying the information contained in the source representations. At the preliminary coordination stage, they can prepare a formula or strategy as a first step to determine the target representations. In the constructing the target, they can transfer information obtained from the source representations to form a target representations. At the determining equivalence, they consider the target representations made by the representations of the given source.

Acknowledgment

This research was supported by SMP Lab school UNESA Surabaya, especially for 8th grades as participants and we are very appreciate to Mr. Andi as a teacher in this class who helped the research.

6. References

[1] National Council of Teacher of Mathematics 2009 *Principle and standards for school*

mathematics (Reston: NCTM)
[2] Sabirin M 2014 *JPM IAIN Antasari* **1** 33
[3] Sierraa M G and Tirado M M 2015 *Int. J. Math. Educ. Sci. Tech.* **46** 700
[4] Stylianou D A 2010 *J. Math. Teach. Educ.* **13** 325
[5] Selling S K 2016 *J. Math. Behav.* **41** 191
[6] Gagatsis A and Elia I D 2004 *The effects of different modes of representation on mathematical problem solving* Proc. of the 28th Conference of the Int. Group for the Psychology of Math. Educ, 2 447–454
[7] Goldin G A and Shteinghold 2001 System of representation and the development of mathematics Conceps. In couco, Albert A. (Ed) *The role of representation in school Mathematics yearbook* (Reston: VA NCTM)
[8] Salkind and Hjalmarson 2007 *Mathematical Representation* (George Mason University)
[9] Hwang W Y, Chen N S, Dung J J and Yang Y L 2007 *Educ. Tech. Soc.* **10** 191
[10] Bustanul R 2017 *Int. Educ. Studies* **10** 20
[11] Chahine I 2011 *J. Math. Educ.***4** 47
[12] Chigeza P 2013 *Translating Between and Within Representations Mathematics as Lived and Interactions* Proc. of the 36th annual Conf. of the Math. Educ. Res. Group of Australia (Melbourne: VIC MERGA)
[13] Cai J and Lester F K 2005 *J. Math. Behavior* **24** 221
[14] Tripathi P N 2008 *Math. Teach. in Middle School* **13** 438
[15] Mandur K, Sadra I W and Suparta I N 2013 *Contribution of Connection Ability Representation Ability, and Mathematical Disposition on Student Learning Achievement in Private High School Manggarai Regency* E-Journal Program Pascasarjana Univ. Pend. Ganesha Program Studi Matematika
[16] Pape S J and Tchoshanov M A 2001 *Theory into Practice* **40** 118
[17] Bosse J, Adu-Gyamfi and Cheetam M 2011 *Int. J. Math.Teach. Learn.* **15** 1
[18] Rahmawati S, Subanji P, Hidayanto S, Rahmad and Anwar 2017 *Int. Elec. J. Math.* **12** 367
[19] Bal A P 2015 *J. Math. Behav.* **197** 582
[20] Molina M, Domingo SR and Cañadas MC 2017 Int. J. Scie. Math. **15** 1137
[21] Bosse M J, Adul-Gyamfi K and Chandler K 2014 *Int. J. Math. Teach. Learn.* **28** 3
[22] Bosse M J, Adu-Gyamfi K and Stiff LV 2012 *School Science dan Mathematics* **112** 159
[23] Dunn R and Dunn K 1993 *Teaching Secondary Students Through Their Individual Learning Styles* (Boston: Allyn and Bacon)

IOP Conf. Series: Journal of Physics: Conf. Series **1108** (2018) 012080 doi:10.1088/1742-6596/1108/1/012080

The Interpersonal Intelligence Profile of Seventh-Grade Students in Mathematics Learning

C A Dien[1*] and D U Wustqa[2]

[1]Graduate Program of Mathematics Education, Yogyakarta State University, Indonesia
[2]Department of Mathematics Education, Yogyakarta State University, Indonesia
[*]Corresponding Author. E-mail: cendekiaaddien@gmail.com

Abstract. Interpersonal intelligence is one of the intelligences in Gardner's multiple intelligences. This intelligence is related to respond information, understand information, and build social connections/interactions to others. The purpose of this quantitative descriptive study was to describe the interpersonal intelligence of seventh-grade students in mathematics learning based on communication, empathy, cooperation and conflict resolution aspects. In the present study, data were obtained through the questionnaires. The questionnaires were administered to 210 seventh grade students aged 14 to 15 years old, who were randomly chosen from within private junior high schools of a sub-district in Yogyakarta. The result of this study revealed that 9.52% had interpersonal intelligence with "very high" category, 46.67% had interpersonal intelligence with "high" category, 40% had interpersonal intelligence with "moderate" category, 1.90% had interpersonal intelligence with "low" category and 1.90% had interpersonal intelligence with "very low" category. Therefore, it leads to sum up that the average of interpersonal intelligence of seventh grade students in mathematics learning was in the high category.

1. Introduction

Individual success is not only assigned by knowledge and technical ability but also promoted by soft skills consisting of intrapersonal intelligence and interpersonal intelligence. This is in line with the result of [1], which revealed that 75% of long-term success in workplaces depends on people skills, while only 25% is dependent on technical knowledge. To put it another way, people skills or people smart as the element of interpersonal intelligence [2,3] occupy the largest part of a person's fruitfulness.

Interpersonal intelligence in Gardner's multiple intelligence theory among eight intelligences such as visual-spatial, linguistic-verbal, logical-mathematical, bodily-kinesthetic, musical, interpersonal, intrapersonal and naturalistic [4], is defined as the ability to regard and make distinctions among other people, including their moods, temperaments, motivations, and intentions [5]. [6] added that interpersonal intelligence relies on social intelligence–the ability and skills of a person in creating relationships, build relationships and maintain social relations so that both parties are in a win-win situation or mutually beneficial. Similarly, [7] stated that interpersonal intelligence involves the intelligence to understand others and process through interaction with them. In this study, interpersonal intelligence refers to respond information, understand information, and build social connections/interactions with others. In addition, the focus area of interpersonal intelligence lies on four aspects, including communication, empathy, cooperation and conflict resolution.

Interpersonal intelligence plays a vital role in today's education, particularly in mathematics learning. Many related studies have indicated that interpersonal intelligence could influence on the success of language learning [8,9], mathematics learning [10,11], nursing [12]. That is to say, interpersonal intelligence obviously gets involved in learning activities. The interaction of students illustrates this point. Students need to interact more with many friends regarded as smarter students, or teachers, to ask for help in solving mathematics problem. Through these interactions, students will be helped to communicate their difficulties in understanding the lessons, in order to be able to accomplish immediately their difficulties. Moreover, it will also allow for the transfer of knowledge between teacher and student or among students and other students. Thus, the well-developed interpersonal intelligence will facilitate students in each the learning activity.

The need to be able to have good interpersonal intelligence is critical for students. Interpersonal intelligence links well with the present day learning objectives of communication and collaboration as well as critical thinking and problem solving [13]. Furthermore, good interpersonal intelligence is necessary for the development of creativity, as outlined by [14] that socialization can make changes and developments toward a person's creativity. Improving interpersonal skills helps individuals enrich relationships, cope better at social environment and especially when dealing with difficult or challenging individuals as well [15]. Based on the above descriptions, the aim of this study is to describe the interpersonal intelligence of seventh-grade students in mathematics learning, based on communication, empathy, cooperation and conflict resolution aspects.

2. Method

The present study was a quantitative descriptive. It was aimed to describe the interpersonal intelligence of seventh-grade students in mathematics learning, based on communication, empathy, cooperation and conflict resolution aspects. This research was conducted during the second term of 2017/2018 in private junior high schools of a sub-district in Yogyakarta. In this study, the participants were 210 seventh grade students, ranging in age from 14 to 15 years who were randomly chosen. Data were gathered from participants with the questionnaires, then they were asked to indicate responses on a five-point likert scale as follows: 1 = never, 2 = rarely, 3 = sometimes, 4 = often, and 5 = always. Questionnaire assessment criteria refer to Table 1.

Table 1. Questionnaire Assessment Criteria of Interpersonal Intelligence

Type of Statement	Response Score				
	Never	Rarely	Sometimes	Often	Always
Positive statements	1	2	3	4	5
Negative statements	5	4	3	2	1

Data analysis of questionnaire was carried out by calculating questionnaire scores, determining the category of measurement result, and classifying the students into several categories. Table 2 was used to determine the category of student's interpersonal intelligence [16].

Table 2. Measurement Scale of Students' Interpersonal Intelligence Category

No.	Score Interval (X)	Category
1	$M_i + 1.5\ S_i < X$	Very High
2	$M_i + 0.5\ S_i < X \leq M_i + 1.5\ S_i$	High
3	$M_i - 0.5\ S_i < X \leq M_i + 0.5\ S_i$	Moderate

4	$M_i - 1.5\ S_i < X \le M_i - 0.5\ S_i$	Low
5	$X \le M_i - 1,5\ S_i$	Very Low

Labels:
X : students' interpersonal intelligence score in each respondent
M_i : mean score
S_i : standard deviation

3. Result and Discussion

According to the results of calculation, average score of seventh-grade students' interpersonal intelligence indicator based on each aspect, including communication, empathy, cooperation and conflict resolution are presented from Table 3 to Table 5.

Table 3. Average of Students' Interpersonal Intelligence Indicator Based on Communication Aspect

Indicators of Communication Aspect	Average of Indicators
Easy to interact	3.41
Active listening	3.83
Persuading/Influencing others	2.90
Open-mindedness	3.70
Assertiveness	2.88

Table 2 shows the indicator of active listening in communication aspect had highest average score with average score of communication aspect was 3.34. It means that seventh-grade students paid full attention to the teacher's explanation about the mathematics lessons.

Table 4. Average of Students' Interpersonal Intelligence Indicator Based on Empathy Aspect

Indicators of Empathy Aspect	Average of Indicators
Sensitive towards others feeling	3.34
Giving attention to others	3.34
Understanding the wishes of others	3.35

In Table 3, the indicator of understanding the wishes of others in empathy aspect had highest average score with average score of empathy aspect was 3.34. It means that seventh-grade students with good comprehension of mathematics lessons would teach their friends who had difficulty in learning mathematics.

Table 5. Average of Students' Interpersonal Intelligence Indicator Based on Cooperation Aspect

Indicators of Cooperation Aspect	Average of Indicators
Participating in team	3.67
Appreciating to team member	3.70
Responsible in team	3.40

In Table 4, the indicator of appreciating team member in cooperation aspect had highest average score with average score of cooperation aspect was 3.59. It means that seventh-grade students engaged actively and worked together in discussion, problem solving, and task management.

Table 6. Average of Students' Interpersonal Intelligence Indicator Based on Conflict Resolution Aspect

Indicators of Conflict Resolution Aspect	Average of Indicators
Discreet in resolving conflict	3.47
Discussing the problems	3.56

Table 6 illustrates the indicator of discussing the problems in conflict resolution aspect had highest average score with average score of conflict resolution aspect is 3.56. It means that seventh-grade students can receive the diversity of opinions and tend to avoid conflict.

Based on Table 2-5, the highest average aspect of all aspects is on cooperation aspect and the lowest average aspects of all aspects are on communication and empathy aspects. It showed that students with high interpersonal prefer to learn and work in group than individual. When working interactively with others, students learn to inquire, hand out ideas, clarify differences, problem-solve, create and build new understandings [17].

Table 7. Students' Interpersonal Intelligence Category

No.	Score Interval (X)	Category	Frequency	Percentage (%)
1	$160 < X$	Very High	20	9.52
2	$133.33 < X \leq 160$	High	98	46.67
3	$106.67 < X \leq 133.33$	Moderate	84	40
4	$80 < X \leq 106.67$	Low	4	1.90
5	$X \leq 80$	Very Low	4	1.90

Table 7 demonstrates that there were 20 students (9.52%) had interpersonal intelligence with "Very High" category, 98 students (46.67%) had interpersonal intelligence with "High" category, 84 students (40%) had interpersonal intelligence with "Moderate" category, 4 students (1.90%) had interpersonal intelligence with "Low" category, and 4 students (1.90%) had interpersonal intelligence with "Very Low" category. Therefore, it revealed that the average of interpersonal intelligence of seventh grade student in mathematics learning was in the high category.

The characteristics of students who have high interpersonal intelligence include (1) able to create and maintain social relationship, (2) able to empathize with other people and understand others, (3) able to solve problems in social relationship, (4) able to realize verbal and non-verbal communication, (5) having communication skill [6]. Besides, [10] found that seventh grade students with high interpersonal intelligence are at third level of creative thinking which means that they are creative.

4. Conclusion

Based on the result and discussion, it comes to the conclusion that the average of interpersonal intelligence of the seventh grade student in mathematics learning was on the high category. Of the four aspects of interpersonal intelligence, the highest average aspect was on cooperation aspect and the lowest average aspects were on communication and empathy aspects. The results and findings of this research may be advantageous in giving some insight to the teachers, the educational practitioners, and researchers of the mathematics education field. For instance, use of the appropriate and innovative mathematics teaching-learning models and approaches in terms of students' interpersonal intelligence, relationship among interpersonal intelligence and other intelligences, relationship among interpersonal intelligence and significant variable in mathematic learning.

5. References

[1] Klaus P 2010 Communication Breakdown *California Job Journal* **28** pp 1-9

[2] Contreras M 2013 *Interpersonal Skills For Entrepreneurs* (Denmark: Bookboon)

[3] Amstrong T 2014 *You Are Smarter Than You Think: A Guide to Your Multiple Intelligences* (Mumbai: Jaico Publishing House)
[4] Gardner H 2006 *Multiple Intelligences* (New York: Basic Books)
[5] Gardner H 2011 *Frames of Mind* (New York: Basic Books)
[6] Safaria T 2005 *Interpersonal Intelligence: Metode Pengembangan Kecerdasan Interpersonal Anak* (Yogyakarta: Amara Books) p 253
[7] Advani V and Hema G 2016 Effect of Interpersonal Intelligence Based Teaching Strategies on Students Academic Achievement *International Journal of Indian Psychology* **3** pp 119-128
[8] Behjat F 2012 Interpersonal and Intrapersonal Intelligences: Do They Really Work in Foreign-Language Learning? *Procedia-Social and Behavioral Sciences* **32** pp 351-355
[9] Mostafavi B B M, Akbari O and Masoominezhad F 2012 The Relationship Between Interpersonal, Visual-Spatial Intelligences and Technical Translation Quality *Journal of Research in Peace, Gender and Development* **2** pp 176-184
[10] Kuncorowati R H, Mardiyana and Saputro D R S 2017 Mathematics Creative Thinking Level Based on Interpersonal Intelligence *J. Phsy.: Conf. Ser.* **943** 012005
[11] Gerhana M T C, Mardiayana and Pramudya I 2017 The Experimentation of Learning Models Viewed from Interpersonal Intelligence *J. Phys.: Conf. Ser.* **909** 012064
[12] Bhana V M 2014 Interpersonal Skills Development in Generation Y Student Nurses: A Literature Review *Nurse Education Today* **34** pp 1430-1434
[13] Russel L 2011 *The Accelerated Learning Fieldbook: Translated by M. Irfan Zakkie* (Bandung: Nusa Media)
[14] Magno C 2008 Explaining the Creative Mind *The International Journal of Research and Review* **3** pp 10-19
[15] Lindsey N S and Rice M L 2015 Interpersonal skills and education in the traditional and online classroom environments *Journal of Interactive Online Learning* **13** pp 126-136
[16] Azwar S 2010 *Penyusunan Skala Psikologi* (Yogyakarta: Pustaka Pelajar) p 163
[17] Chiriac H E 2014 Group work As an Incentive for Learning – Students' Experiences of Group Work *Frontiers in Psychology* **5** pp 1-10

Parameter estimation for the Lomax distribution using the E-Bayesian method

A Fitrilia[1,2]**, I Fithriani**[1,3] **and S Nurrohmah**[1,4]

[1] Department of Mathematics, Faculty of Mathematics and Natural Sciences, Universitas Indonesia, Depok 16424, Indonesia.

Email : [2]aldilafitrilia@gmail.com, [3]ida.fithriani@gmail.com, [4]snurrohmah@sci.ui.ac.id

Abstract. In this paper, we will find the parameter estimator from Lomax distribution on one of censored data, that is right censored data type II. Lomax distribution is also called Pareto type II distribution. The parameter will be estimate is the shape parameter with the assumption of scale parameters (β) has been known. The parameter estimation method used in this study is E-Bayesian estimation method. E-Bayesian estimation is an expectation of Bayes estimation, in order to obtain Bayes estimation expectations is by calculating the mean of Bayes estimators. E-Bayesian estimation method is used to estimate failure rate. The estimation will use prior Gamma as conjugate prior distribution from Lomax distribution and Loss function will be used is balanced squared error loss function (BSELF). Thus, the main purpose of this study is to find the parameter estimator of the Lomax distribution on the right censored data type II using the E-Bayesian method. The final result of this study, we get the likelihood function from Lomax distribution on the right censored data type II and the parameter estimator from Lomax distribution on the right censored data type II using the E-Bayesian method.

1. Introduction

The Lomax distribution was first introduced by K. S. Lomax in 1954 [1]. Lomax distribution is a probability distribution used in business, economics, and actuarial science. The Lomax distribution has two parameters and denoted by $\text{Lomax}(\alpha,\beta)$ where α is a form parameter and β is the scale parameter. Lomax distribution has probability density function (pdf) that defined as

$$f(x;\alpha,\beta)=\frac{\alpha}{\beta}\left(1+\frac{t}{\beta}\right)^{-(\alpha+1)}, \quad x>0, \quad (\alpha,\beta>0) \tag{1}$$

and cumulative distribution function (cdf) that defined as

$$F(x;\alpha,\beta)=1-\left(1+\frac{x}{\beta}\right)^{-\alpha}, \quad x>0, \quad (\alpha,\beta>0) \tag{2}$$

The survival function or reliability function of the Lomax distribution defined as the probability of an object can survive until time t is

$$S(t) = 1 - F(t) = 1 - \left(1 - \left(1 + \frac{t}{\beta}\right)^{-\alpha}\right) = \left(1 + \frac{t}{\beta}\right)^{-\alpha}, \qquad t > 0 \tag{3}$$

The hazard function of the Lomax distribution represents the probability of a component failing at time t is

$$h(t) = \frac{f(t)}{S(t)} = \frac{\frac{\alpha}{\beta}\left(1 + \frac{t}{\beta}\right)^{-(\alpha+1)}}{\left(1 + \frac{t}{\beta}\right)^{-\alpha}} = \frac{\alpha}{\beta}\left(1 + \frac{t}{\beta}\right)^{-1}$$

$$h(t) = \frac{\alpha}{\beta + t}, \qquad t > 0 \tag{4}$$

Mean and variance of Lomax distribution is

$$\mu = \frac{\beta}{\alpha - 1}, \qquad \alpha > 1 \tag{5}$$

and

$$\sigma^2 = \frac{\alpha\beta^2}{(\alpha - 1)^2(\alpha - 2)}, \qquad \alpha > 2 \tag{6}$$

Since the parameters of α and β are unknown, parameter estimation is necessary, one of the paramereter estimation that can be used is the E-Bayesian estimation method. E-Bayesian estimation method are used to estimate failure rates, furtheremore they are suitable for censored data with small sample sizes and high reliability [2]. Estimated parameters using the Bayesian approach, which to obtain E-Bayesian estimate of α (i.e. expectation of the Bayes estimates of α) is obtained by calculating the mean of the Bayes estimator.

2. Balanced square error loss function (BSELF)

The estimator of a function, using BSEL is actually a combination of the maximum likelihood estimator of a function and the Bayes estimator using the SEL. Ahmadi et al. [3] introduced a balanced loss function (BLF) derived from Zellner [4] in the form

$$L^{q}_{\rho,\omega,\delta_0}(\Lambda(\alpha), \delta) = \omega q(\alpha)\rho(\delta_0, \delta) + (1 - \omega)q(\alpha)\rho(\Lambda(\alpha), \delta) \tag{7}$$

where $q(\cdot)$ is a suitable positive weight function and $\rho(\Lambda(\alpha), \delta)$ is any Loss function when estimating $\Lambda(\alpha)$ with δ. The parameter δ_0 is the selected prior estimator of $\Lambda(\alpha)$, obtained for example from the maximum likelihood criterion, and $0 \leq \omega < 1$. If $\rho(\Lambda(\alpha), \delta) = (\delta - \Lambda(\alpha))^2$ and $q(\alpha) = 1$ are substituted in equation (5), obtained BSELF, given as

$$L_{\omega,\delta_0}(\Lambda(\alpha), \delta) = \omega(\delta - \delta_0)^2 + (1 - \omega)\big(\delta - \Lambda(\alpha)\big)^2 \tag{8}$$

The corresponding Bayes estimator of the function $\Lambda(\alpha)$ is given as

$$\delta_{\omega,\Lambda,\delta_0}(\mathrm{x}) = \omega\delta_0 + (1 - \omega)E(\Lambda(\alpha)|\mathrm{x}) \tag{9}$$

IOP Conf. Series: Journal of Physics: Conf. Series **1108** (2018) 012081 doi:10.1088/1742-6596/1108/1/012081

3. Likelihood function from Lomax distribution on the right censored data type II

In the right censored data type II, there are r observations of n samples observed, and the study is stopped after the rth failure that occurred. In this censored, r is determined before the data is collected. The joint probability density function of right censored data type II of $X_1, X_2, \dots, X_r$ for $r < n$ is

$$f(x_1, x_2, \dots, x_r) = \frac{n!}{(n-r)!}\left[\prod_{i=1}^{r} f(x_i)\right][S(x_r)]^{n-r} \tag{10}$$

then the joint probability density function of Lomax distribution on the right censored data type II is

$$f(x_1, x_2, \dots, x_r) = \frac{n!}{(n-r)!}\alpha^r \frac{1}{\beta^r \prod_{i=1}^{r}\left(1+\frac{x_i}{\beta}\right)} \frac{1}{\prod_{i=1}^{r}\left(1+\frac{x_i}{\beta}\right)^{\alpha}} \frac{1}{\left(1+\frac{x_r}{\beta}\right)^{\alpha(n-r)}}$$

$$f(x_1, x_2, \dots, x_r) = \frac{n!}{(n-r)!}\alpha^r v(\underline{x};\beta)e^{-M\alpha} \tag{11}$$

Thus the likelihood function of the Lomax distribution on the right censored data type II can be written as

$$L(\underline{x}|\alpha,\beta) = \frac{n!}{(n-r)!}\alpha^r v(\underline{x};\beta)e^{-M\alpha} \tag{12}$$

where $\underline{\mathbf{x}} = (\boldsymbol{x_1}, \boldsymbol{x_2}, \dots, \boldsymbol{x_r})$, $\boldsymbol{v(\underline{x};\beta)} = \dfrac{\mathbf{1}}{\boldsymbol{\beta^r \prod_{i=1}^{r}\left(1+\frac{x_i}{\beta}\right)}}$

and $\boldsymbol{M \equiv M(\underline{x};\beta) = \sum_{i=1}^{r} \ln\left(1+\frac{x_i}{\beta}\right) + (n-r)\ln\left(1+\frac{x_r}{\beta}\right)}$

4. Bayesian estimation of Lomax distribution on the right censored data type II

Bayes estimation is estimation method that combines prior information with information obtained from sample data. Prior information is obtained from the distribution of parameters, while the information from the data is summarized in the likelihood function. The combination of prior information and information from the data will produce posterior information (posterior distribution). The results are expressed in terms of this posterior distribution which will be the basis in the Bayes method for estimating the parameters.

From (12), assuming β is known, the maximum likelihood estimator of the parameter α is given as

$$\ln L(\underline{x}|\alpha) = \ln\frac{n!}{(n-r)!} + r\ln\alpha + \ln v(\underline{x};\beta) + (-M\alpha)$$

$$\frac{d}{d\alpha}\ln L(\underline{x}|\alpha) = 0$$

so that the maximum likelihood estimator of the parameter α is obtained

$$\hat{\alpha}_{ML} = \frac{r}{M} \tag{13}$$

To search for Bayes estimator, we need to find the prior distribution first. According to Box and Tiao [5], the conjugate prior distribution, refers to the reference model analysis, especially in the creation of likelihood function. So in the defining of conjugate prior always thought about the defining of the prior distribution pattern that has the conjugate form with the probability density function of the likelihood construction. According to Castillo [6], If $\mathcal{F}$ is the class of the sampling distribution $p(y|\theta)$ and $\mathcal{P}$ is the class of the prior distribution for θ, $p(\theta)$, then class $\mathcal{P}$ is said to conjugate for $\mathcal{F}$ if the posterior probability $p(\theta|y)$ has the same distribution form with the prior probability function $p(\theta)$ for all $p(y|\theta) \in \mathcal{F}$. In this study, Gamma conjugate prior is used to find Bayes estimators. Likelihood function of the Lomax distribution on the right censored data type II is

$$f(\underline{x}|\alpha,\beta) = \frac{n!}{(n-r)!}\alpha^r v(\underline{x};\beta)e^{-M\alpha} \tag{14}$$

Assuming β is known, we can use the Gamma distribution as an conjugate prior distribution of α with shape and scale parameter a and b respectively and its probability density function given by

$$g(\alpha|a,b) = \frac{b^a}{\Gamma(a)}\alpha^{a-1}e^{-b\alpha}, \quad \alpha > 0 \tag{15}$$

The likelihood function and the probability density function of Gamma distribution has the same functional form where $r = a - 1$ and $b = M$. Therefore, the Gamma Distribution is used as the conjugate prior distribution of the Lomax distribution on the right censored data type II. Furtheremore, we will find the posterior distribution density function

$$q(\alpha|\underline{x}) = \frac{f(\underline{x}|\alpha,\beta)g(\alpha|a,b)}{\int_0^\infty f(\underline{x}|\alpha,\beta)g(\alpha|a,b)d\alpha} = \frac{\alpha^{r+a-1}e^{-(b+M)\alpha}(b+M)^{r+a}}{\Gamma(r+a)}$$

$$q(\alpha|\underline{x}) = \kappa\alpha^{r+a-1}e^{-(b+M)\alpha}, \quad \alpha > 0 \tag{16}$$

with $\kappa = \frac{(b+T)^{r+a}}{\Gamma(r+a)}$. Posterior probability density function in (16) is the probability density function of Gamma distribution with parameters $r + a$ and $b + T$. Since the resulting posterior distribution is Gamma distributed, just like the prior distribution which Gamma distributed, it is proven that the Gamma distribution is the conjugate prior of the Lomax distribution on the right censored data type II. The Bayes estimator of α with balanced square error loss function (BSELF) can be given as

$$\hat{\alpha}_B(a,b) = \omega\delta_0 + (1-\omega)E(\Lambda(\alpha)|\mathbf{x})$$

with δ_0 is maximum likelihood estimator and $E(\Lambda(\alpha)|\mathbf{x})$ is Bayes estimator with square error loss function.

$$E(\Lambda(\alpha)|\mathbf{x}) = \int_0^\infty \Lambda(\alpha)\, q(\Lambda(\alpha)|\mathbf{x})d\alpha = \frac{(r+a)}{(b+M)} \tag{17}$$

Then the Bayes estimator for α is obtained as

$$\hat{\alpha}_B(a,b) = \omega\left(\frac{r}{M}\right) + (1-w)\frac{(r+a)}{(b+M)} \tag{18}$$

IOP Conf. Series: Journal of Physics: Conf. Series **1108** (2018) 012081 doi:10.1088/1742-6596/1108/1/012081

5. E-Bayesian estimation of Lomax distribution on the right censored data type II

E-Bayesian estimation methods are used to estimate failure rates. E-Bayesian estimation is an expectation of Bayes estimation, in order to obtain Bayes estimation expectations is by calculating the mean of Bayes estimators.

Suppose that life an object is Lomax distributed on the right censored data type II with a joint probability density function is (11). Let the prior distribution of α be the conjugate $-\Gamma(a,b)$ distribution with the probability density function is (15) where $\Gamma(a)=\int_0^\infty x^{a-1}e^{-x}dx$ is a function of Gamma, $a>0, b>0$, and a and b are hyperparameters. According to Han [7], the hyperparameters a and b must be choose to guarantee that $g(\alpha|a,b)$ is a decreasing function of α. The derivative of $g\,(\alpha\,|\,a,b)$ with respect to α is

$$\frac{dg(\alpha|a,b)}{d\alpha}=\frac{b^a}{\Gamma(a)}[(a-1)\alpha^{a-2}e^{-b\alpha}+\alpha^{a-1}(-b)e^{-b\alpha}]$$

$$\frac{dg(\alpha|a,b)}{d\alpha}=\frac{b^a}{\Gamma(a)}\alpha^{a-2}e^{-b\alpha}[(a-1)-b\alpha] \tag{19}$$

Thus, for $0<a<1, b>0$, the prior $g\,(\alpha\,|\,a,b)$ is a decreasing function of α. Assuming that the hyperparameters a and b in $g\,(\alpha\,|\,a,b)$ are independent and $\pi(a,b)=\pi_1(a)\pi_2(b)$. The E-Bayesian estimator of α (the expectation of the Bayes estimator of α) can be written as

$$\hat{\alpha}_{EB}=E(\alpha|\underline{X})=\int_{\varrho}\int\hat{\alpha}_B(a,b)\pi(a,b)dadb \tag{20}$$

where ϱ is the set of all possible values of a and b where the prior density function is a derivative of α. $\hat{\alpha}_B(a,b)$ is the Bayes estimator of α given by (18). For more detail see Han [2].

The E-Bayesian estimates of α are derived depending on three different distributions of the hyperparameters a and b. These distribution are used to investigate the impact of the different prior distributions on E-Bayesian estimation of α. The prior distribution used must be conjugate to make it easier to find the posterior distribution. If the prior distribution is not conjugate, we will have difficulty to find the posterior distribution.The following distributions of a and b may be used

$$\pi_1(a,b)=\frac{2(c-b)}{c}, 0<a<1, 0<b<c \tag{21}$$

$$\pi_2(a,b)=\frac{1}{c}, \quad 0<a<1, 0<b<c \tag{22}$$

$$\pi_3(a,b)=\frac{2b}{c^2}, \quad 0<a<1, 0<b<c \tag{23}$$

We can obtained the E-Bayesian estimate of α based on $\pi_1(a,b), \pi_2(a,b)$, and $\pi_3(a,b)$ by using (18) and (21) in (20), (18) and (22) in (20), and (18) and (23) in (20) respectively with $\hat{\alpha}_{EB}$ is the E-Bayesian estimator of α, $\hat{\alpha}_B(a,b)$ is the Bayes estimator of α given by (18), and a, b is hyperparameters. The corresponding E-Bayesian estimate of α are respectively.

For $\pi_1(a,b)$,

$$\hat{\alpha}_{EB1}=\int_{\varrho}\int\hat{\alpha}_B(a,b)\pi_1(a,b)dbda=\int_0^1\int_0^c\hat{\alpha}_B(a,b)\pi_1(a,b)dbda$$

$$=\int_0^1\int_0^c\left[\omega\left(\frac{r}{M}\right)+(1-w)\frac{(r+a)}{(b+M)}\right]\times\frac{2(c-b)}{c^2}dbda$$

$$\hat{\alpha}_{EB1} = \omega\left(\frac{r}{M}\right) + \frac{2}{c}(1-w)\left(r+\frac{1}{2}\right)\left(\frac{c+M}{s}\ln\left(\frac{c+M}{s}\right)-1\right) \tag{24}$$

For $\pi_2(a,b)$,

$$\hat{\alpha}_{EB2} = \int_{Q}\int \hat{\alpha}_B(a,b)\pi_1(a,b)dbda = \int_0^1\int_0^c \hat{\alpha}_B(a,b)\pi_2(a,b)dbda$$
$$= \int_0^1\int_0^c \left[\omega\left(\frac{r}{M}\right) + (1-w)\frac{(r+a)}{(b+M)}\right]\times\frac{1}{c}dbda$$
$$\hat{\alpha}_{EB2} = \omega\left(\frac{r}{M}\right) + \frac{(1-w)}{c}\left(r+\frac{1}{2}\right)\ln\left(\frac{c+M}{M}\right) \tag{25}$$

For $\pi_3(a,b)$,

$$\hat{\alpha}_{EB3} = \int_{Q}\int \hat{\alpha}_B(a,b)\pi_1(a,b)dbda = \int_0^1\int_0^c \hat{\alpha}_B(a,b)\pi_3(a,b)dbda$$
$$= \int_0^1\int_0^c \left[\omega\left(\frac{r}{M}\right) + (1-w)\frac{(r+a)}{(b+M)}\right]\times\frac{2b}{c^2}dbda$$
$$\hat{\alpha}_{EB3} = \omega\left(\frac{r}{M}\right) + \frac{2}{c}(1-w)\left(r+\frac{1}{2}\right)\left(1-\frac{M}{c}\ln\left(\frac{c+M}{c}\right)\right) \tag{26}$$

6. Conclusion

In this paper, the E-Bayesian method is used to estimate the α parameter of the Lomax distribution in the right censored data type II. Since E-Bayesian estimation is an expectation of Bayes estimation, in order to obtain Bayes estimation expectations is by calculating the mean of Bayes estimators, it is necessary to find the Bayes estimation first. Loss function used in this study is balanced square error loss function (BSELF) which is a combination of the maximum likelihood estimator of a function and the Bayes estimator using the square error loss function (SELF).

7. References

[1] K. S. Lomax 1954 *J. Am. Stat. Association* **49** 847
[2] Ming Han 2008 *App. Math. Model.* **33** 1915
[3] Jafar Ahmadi, M. J. Jozani, Eric Marchand & Ahmad Parsian 2008 *J. Stat. Plan. Inference* 1180
[4] Arnold Zellner 1994 *Statistical Desicion Theory and Methods V* 337-390
[5] G. E. P. Box & G. C. Tiao 1973 Bayesian inference in statistical analysis (Philippines: Addison-Wesley Publishing Company, Inc)
[6] Enrique D. Castillo 2007 Process optimization a statistical approach (New York: Springer Science + Business Media, LCC)
[7] Ming Han 1997 The structure of hierarchical prior distribution and its applications 31-40
[8] Essam K. Al-Hussaini & Mohamed Hussein 2011 *Open J. of Stat* 28-38
[9] Hassan M. Okasha 2013 *J. Egyptian Math. Soc.* 489
[10] Hesham M. Reyad & Soha O. Ahmed 2014 *Int. J. Adv. Math. Sci.* **3** 108-120

Acknowledgments

We would like to thank the DRPM Universitas Indonesia for supporting in our research with Hibah PITTA UI 2018.

IOP Conf. Series: Journal of Physics: Conf. Series **1108** (2018) 012082 doi:10.1088/1742-6596/1108/1/012082

Verifiable Image Secret Sharing Using Matrix Projection

N F Azzahra[1*] and K A Sugeng[1*]

[1]Department of Mathematics, Faculty of Mathematics and Natural Sciences, Universitas Indonesia, Indonesia
*Corresponding Author. nurfathiyafrdn@gmail.com; kiki@sci.ui.ac.id

Abstract. Secret sharing scheme is a technique to share secret data into n pieces based on a simple (k, n) threshold scheme. However, the problem with Shamir's secret sharing scheme is that they do not provide any way to verify that the dealer was honest and the shares were indeed valid. This problem also occurr in the Thien and Lin's image secret sharing or the Li Bai's construction using matrix projection. On the other hand, a developed protocol for secret sharing called verifiable secret sharing allows every participant to validate their received piece to confirm the authenticity of the secret. Therefore, this paper discussed a proposed scheme based on verifiable secret sharing, in which the matrix projection is used to create image shares and a public matrix from watermark image. The secrets were represented in elements of a square matrix. The watermark image was used for verifiability where the reconstructed watermark image verifies the accuracy of the reconstructed secret image.

1. Introduction

Secret sharing scheme is a technique to share secret data into n pieces based on a simple (k, n) threshold scheme, in which participants will easily reconstruct the secret if there are minimum k pieces while knowledge of any $k - 1$ pieces will not be able to decipher the secret [5]. An extension to Shamir's secret sharing scheme on images proposed by Thien and Lin image secret sharing [6]. In which, the generated image shares or shadow images are smaller than the secret image. Another extension of Shamir's scheme also applied in Li Bai's method of generating an image shares using matrix projection [1, 2]. However, there is no guarantee in both schemes that they prevent dishonest dealer or shareholders.

On the other hand, a developed protocol for secret sharing called verifiable secret sharing allows every participant to validate their received piece to confirm the authenticity of the secret.Wang *et al.* developed a method for sharing a secret image with verification, in which an m x m watermark image is used to verify the accuracy of the reconstructed image [7].

2. Secret Sharing Scheme

Secret Sharing Scheme is a technique to secure secret data in which the data will be divided into shares, with each participant received one share, and the secret data will only be revealed if put together according to the predefined combinations [5].

2.1. Shamir's Secret Sharing Scheme [5]

Shamir developed a scheme to divide data D into n shares $D_1, D_2, \ldots, D_n$ such that it takes k pieces or more of D_i to determine D, and the knowledge of $k - 1$ pieces of D_i will not be able to decipher the secret. This is called (k, n) threshold scheme.Shamir's secret sharing scheme is using the idea of polynomial interpolation in which given k different points with i = 1, 2, …,k, there is only one polynomial $q(x)$ of degree k – 1 such that $q(x_i) = y_i$ for each value of i. When D is assumed to be a number, then to divide D intoD_i, choose $q(x) = a_0 + a_1x + \cdots + a_{k-1}x^{k-1}$ where $D = a_0$, andD_1 = $q(1)$, …, $D_i = q(i)$, …, $D_n = q(n)$. The knowledge of k pieces of D_i will determine the coefficient of polynomial $q(x)$, and the value of $D = q(0)$ will be able to be calculated. On the other hand, knowledge of k – 1 pieces of D_i will not be enough to calculate the value of D.

2.2. Thien and Lin's Image Secret Sharing Scheme [6]

An extension to Shamir's secret sharing scheme is published by Thien and Lin in 2002, in which when the secret data using secret sharing scheme is an image, then the secret image will be divided into n pieces of shadow images which in a smaller size of the secret image.

Image secret sharing is using a polynomial of degree k – 1 to generate n pieces of shadow images from secret image with $m \times m$ pixels.The pixel value is denoted by I. Since the grey level from a pixel value is between 0 and 255, then a prime number p = 251 will be used for later, where 251 is the biggest prime number on the range of the grey level. The polynomial is $q(x) = a_0 + a_1x + \cdots + a_{k-1}x^{k-1}$, where $D = a_0$, and $D_1 = q(1)$, …, $D_i = q(i)$, …, $D_n = q(n)$, with each piece of D_i is a shadow image.In this method, the size of the shadow images is $1/k$ of secret image.

2.3. Li Bai's Secret Sharing Scheme using Matrix Projection [1]

Li Bai developed (k, n) threshold Shamir by representing the secret data into square matrix S, and the secret matrix will be shared to all participants using matrix projection. The advantage of this scheme is the secret data is more protected.For an $m\times k$ matrix A with rank k ($m \geq k > 0$), an $m \times m$matrix $\mathbb{S}$ is matrix projection of matrix A, with $\mathbb{S} = A(A^TA)^{-1}A^T$ or denoted by $\mathbb{S}$ = proj (A).

2.4. Li Bai'sImage Secret Sharing Scheme using Matrix Projection [2]

In the same year, Li Bai developed an image secret sharing, where this method divides a secret colour image into n pieces of image shares such that any k image shares will be able to reconstruct the secret image. This image secret sharing scheme is using Thien and Lin's image secret sharing, in which the main essence is using a polynomial of degree k – 1 to construct n image shares from an $m \times m$ secret image, with the pixel value denoted by I.

The polynomial can be written as follows:

$$S_{i,t}^{(j)}(c) = (I(i, k(t-1)+1) + I(i, k(t-1)+2) + \ldots + I(i, k(t-1)+k)\, c_j^{k-1} \bmod 251$$

where $1 \leq t \leq \left\lceil \frac{m}{k} \right\rceil$ dan $1 \leq i \leq m$.

In this method, the size of image shares is $1/k + 1/m$ of secret image.

2.5. Verifiable Secret Sharing Scheme

Verifiable Secret Sharing Scheme is a cryptographic protocol in which its existence ensures each participant could verify their received piece of share without knowing the secret [3].

Wang proposed verification method in secret sharing scheme in which an $n \times n$ watermark image will be used with the secret image to construct shadow images. The accuracy of watermark image will determine the reliability from shadow images before reconstructing the secret image. If the reconstructed image using watermark image accurate, then the reconstructed shadow images is also proven to be accurate. Therefore, the secret image could be verified [7].

3. Verifiable Image Secret Sharing Scheme using Matrix Projection

The problem with reliable secret sharing scheme using matrix projection published by Li Bai is they do not provide a way for participants to validate their received image shares, where as there is a possibility of a disturbance in transmission or problem that come off by dealing with dishonest dealers or shareholders. An extension to secret sharing scheme developed by Wang *et al.* in which an $m \times m$

IOP Conf. Series: Journal of Physics: Conf. Series **1108** (2018) 012082 doi:10.1088/1742-6596/1108/1/012082

watermark image will be shared with image shares, and by determining the accuracy of watermark image, then the reconstructed secret image could be verified.

This paper will modify the image secret sharing scheme using matrix projection by Li Bai with the use of the watermark image in Wang et al.'s scheme to resolve the problem stated above.

The scheme is presented as below.

3.1. Construct Image Shares

- Determine the (k, n) threshold andthe matrix of secret image S
- Construct a random $m \times k$ matrix A of rank k where $m > 2(k-1)-1$
- Calculate projection matrix $\mathbb{S} = proj(A) \pmod p$ and remainder matrix $R = (S - \mathbb{S}) \pmod p$
- Choose n of$k$ x 1 random vector $\boldsymbol{x}_j$, and$n$of $c_j$for $1 \le j \le n$
- Computeshare vector $\boldsymbol{v}_j = (A \times \boldsymbol{x}_j)$mod $p$for $1 \le j \le n.p$ is a prime number
- Use Thien and Lin image secret sharing, the remainder matrix R is used to create shares as follows: $G_j = [g_1^{(j)} \quad g_2^{(j)} \quad \cdots \quad g_{\lceil \frac{m}{k} \rceil}^{(j)}]$
  in which$g_t^{(j)}(i) = (I(i, k(t-1)+1) + \ldots + I(i, k(t-1)+k)\, c_j^{k-1} \bmod 251$
  where $1 \le t \le \lceil \frac{m}{k} \rceil$ dan $1 \le i \le m$
- Select watermark image W and compute remainder matrix $R_w = (W - \mathbb{S}) \pmod p$
- Divide image shares $Sh_j = [\ \boldsymbol{v}_j G_j]$ to participants and make matrix R_w public

3.2. Reconstruct Secret Image

- Select k image sharesfrom any$k$participants,namely $Sh_1, Sh_2, \ldots, Sh_k$
- Construct remainder matrix R using Thien and Lin and compute $B = [v_1 \quad v_2 \quad \cdots \quad v_k]$
- Compute projection matrix$\mathbb{S} = (proj(B)) \pmod p$
- Construct remainder matrix R using Thien and Lin image secret sharing, by using G_j as follows:

$$R = \begin{bmatrix} r_1^{(1)} & r_2^{(1)} & \cdots & r_y^{(1)} \\ r_1^{(2)} & r_2^{(2)} & \cdots & r_y^{(2)} \\ \vdots & \vdots & \ddots & \vdots \\ r_1^{(x)} & r_2^{(x)} & \cdots & r_y^{(x)} \end{bmatrix}$$

in which $r_y^{(x)} = [I(x,\ k(y-1)+1) \quad I(x,\ k(y-1)+2) \quad \ldots \quad I(x,\ k(y-1)+k)]$

where$1 \le x \le m$, dan $1 \le y \le \lceil \frac{m}{k} \rceil$

- Calculate secret image $S = (\mathbb{S} + R) \pmod p$
- Determinevalue of watermark image $W = ((\mathbb{S} + R_w) \pmod p))$
- Divide image shares $Sh_j = [\ \boldsymbol{v}_j G_j]$ to participants and make matrix R_w public

3.3. Example

For further understanding, given the examples of construction and reconstruction process below.

- Choose secret image S, and (2, 4) threshold scheme, with $p = 251$ and construct random matrix A.

$$\text{Let}S = \begin{bmatrix} 2 & 5 & 2 & 3 \\ 3 & 6 & 4 & 5 \\ 4 & 7 & 4 & 6 \\ 1 & 4 & 1 & 7 \end{bmatrix} \text{ and } A = \begin{bmatrix} 3 & 7 \\ 6 & 1 \\ 2 & 5 \\ 6 & 6 \end{bmatrix}.$$

- Calculate the matrix projection $\mathbb{S}$ using the formulas stated above.

$$\mathbb{S} = (proj(A))(\text{mod } p) = (A\,(A^TA)^{-1}A^T)(\text{mod } p) = \begin{bmatrix} 1 & 221 & 201 & 214 \\ 221 & 189 & 186 & 16 \\ 201 & 186 & 43 & 76 \\ 214 & 16 & 76 & 20 \end{bmatrix}$$

- Then, calculate remainder matrix $R = (S - \mathbb{S})(mod\ p) = \begin{bmatrix} 1 & 35 & 52 & 40 \\ 33 & 68 & 69 & 240 \\ 54 & 72 & 212 & 181 \\ 38 & 239 & 176 & 238 \end{bmatrix}$

- After that, choose 4 vectors $\boldsymbol{x}_j$ with size 2 ×1, and 4 values of c_j with $1 \le j \le 4$

$\boldsymbol{x}_1 = \begin{bmatrix} 9 \\ 5 \end{bmatrix}$, $\boldsymbol{x}_2 = \begin{bmatrix} 4 \\ 4 \end{bmatrix}$, $\boldsymbol{x}_3 = \begin{bmatrix} 9 \\ 8 \end{bmatrix}$, $\boldsymbol{x}_4 = \begin{bmatrix} 3 \\ 2 \end{bmatrix}$ and $c_1 = 1, c_2 = 2, c_3 = 3, c_4 = 4$

- Compute share vector $\boldsymbol{v}_j = (A \times \boldsymbol{x}_j)\text{mod } p$ for $1 \le j \le 4$

$$\boldsymbol{v}_1 = \begin{bmatrix} 62 \\ 59 \\ 43 \\ 84 \end{bmatrix}, \boldsymbol{v}_2 = \begin{bmatrix} 40 \\ 28 \\ 28 \\ 48 \end{bmatrix}, \boldsymbol{v}_3 = \begin{bmatrix} 83 \\ 62 \\ 58 \\ 102 \end{bmatrix}, \boldsymbol{v}_4 = \begin{bmatrix} 23 \\ 20 \\ 16 \\ 30 \end{bmatrix}$$

- Use Thien and Lin image secret sharing with the remainder matrix R is used to create shares as follows: $G_j = [g_1^{(j)} \quad g_2^{(j)} \quad \cdots \quad g_{\lceil \frac{m}{k} \rceil}^{(j)}]$

For $t = 1 \rightarrow g_1^{(j)}(i) = (I(i, 1) + I(i, 2)\,c_j) \text{ mod } 251$

For $t = 2 \rightarrow g_2^{(j)}(i) = (I(i, 3) + I(i, 4)\,c_j) \text{ mod } 251$

For $j = 1 \rightarrow G_1 = [g_1^{(1)} \quad g_2^{(1)}]$ $\qquad j = 3 \rightarrow G_3 = [g_1^{(3)} \quad g_2^{(3)}]$

$j = 2 \rightarrow G_2 = [g_1^{(2)} \quad g_2^{(2)}]$ $\qquad j = 4 \rightarrow G_4 = [g_1^{(4)} \quad g_2^{(4)}]$

Calculate G_1 :

$g_1^{(1)}(i) = (I(i, 1) + I(i, 2)c_1) \text{ mod } 251$, and $g_2^{(1)}(i) = (I(i, 3) + I(i, 4)c_1) \text{ mod } 251$

For $i = 1$:

$g_1^{(1)}(1) = (I(1, 1) + I(1, 2)\,c_1) \text{ mod } 251 = (1 + 35(1)) \text{ mod } 251 = 36 \text{ mod } 251 = 36$

$g_2^{(1)}(1) = (I(1, 3) + I(1, 4)\,c_1) \text{ mod } 251 = (52 + 40(1)) \text{ mod } 251 = 92 \text{ mod } 251 = 92$

Following the same pattern for $i = 2, 3, 4$, the value will be obtained as follows:

$g_1^{(1)}(2) = 101$, $g_2^{(1)}(2) = 58$, $g_1^{(1)}(3) = 126$, $g_2^{(1)}(3) = 142$,

$g_1^{(1)}(4) = 26$, $g_2^{(1)}(4) = 163$

Thus, the matrix $G_1 = \begin{bmatrix} 36 & 92 \\ 101 & 58 \\ 126 & 142 \\ 26 & 163 \end{bmatrix}$

Using the same steps, the matrices will be obtained as follows:

$$G_2 = \begin{bmatrix} 71 & 132 \\ 169 & 47 \\ 198 & 72 \\ 14 & 150 \end{bmatrix} \qquad G_3 = \begin{bmatrix} 106 & 172 \\ 237 & 36 \\ 19 & 2 \\ 2 & 137 \end{bmatrix} \qquad G_4 = \begin{bmatrix} 141 & 212 \\ 54 & 25 \\ 91 & 183 \\ 241 & 124 \end{bmatrix}$$

- Select watermark image W and compute remainder matrix $R_w = (W - \mathbb{S})\ (\text{mod } p)$

Say $W = \begin{bmatrix} 50 & 100 & 21 & 14 \\ 22 & 76 & 200 & 54 \\ 1 & 91 & 45 & 7 \\ 24 & 66 & 96 & 120 \end{bmatrix}$, then $R_w = \begin{bmatrix} 49 & 130 & 71 & 51 \\ 52 & 138 & 14 & 38 \\ 51 & 156 & 2 & 182 \\ 61 & 50 & 20 & 100 \end{bmatrix}$

- Using calculations above, matrix share $Sh_j = [\ \boldsymbol{v}_j \ G_j]$ for $j = 1, 2, 3, 4$ is obtained as follows:

$$\text{Sh}_1 = \begin{bmatrix} 62 & 36 & 92 \\ 59 & 101 & 58 \\ 43 & 126 & 142 \\ 84 & 26 & 163 \end{bmatrix}, \text{Sh}_2 = \begin{bmatrix} 40 & 71 & 132 \\ 28 & 169 & 47 \\ 28 & 198 & 72 \\ 48 & 14 & 150 \end{bmatrix}, \text{Sh}_3 = \begin{bmatrix} 83 & 106 & 172 \\ 62 & 237 & 36 \\ 58 & 19 & 2 \\ 102 & 2 & 137 \end{bmatrix}, \text{Sh}_4 = \begin{bmatrix} 23 & 141 & 212 \\ 20 & 54 & 25 \\ 16 & 91 & 183 \\ 30 & 241 & 124 \end{bmatrix}$$

IOP Conf. Series: Journal of Physics: Conf. Series **1108** (2018) 012082 doi:10.1088/1742-6596/1108/1/012082

Image share Sh_1 will be shared to participant 1, image share Sh_2 will be shared to participant 2, and so on. Remainder matrix R_w will be a public share.

- Select 2image sharesfrom Sh_1, Sh_2

$$Sh_1 = \begin{bmatrix} 62 & 36 & 92 \\ 59 & 101 & 58 \\ 43 & 126 & 142 \\ 84 & 26 & 163 \end{bmatrix} \qquad Sh_2 = \begin{bmatrix} 40 & 71 & 132 \\ 28 & 169 & 47 \\ 28 & 198 & 72 \\ 48 & 14 & 150 \end{bmatrix}$$

- Construct matrix B using first column of share 1 and share 2, such that $B = [v_1 \quad v_2]$.

$$v_1 = \begin{bmatrix} 62 \\ 59 \\ 43 \\ 84 \end{bmatrix} \qquad v_2 = \begin{bmatrix} 40 \\ 28 \\ 28 \\ 48 \end{bmatrix} B = \begin{bmatrix} 62 & 40 \\ 59 & 28 \\ 43 & 28 \\ 84 & 48 \end{bmatrix}$$

- Compute matrix projection $\mathbb{S} = (proj(B))(\text{mod } p)$

$$\mathbb{S} = \begin{bmatrix} 1 & 221 & 201 & 214 \\ 221 & 189 & 186 & 16 \\ 201 & 186 & 43 & 76 \\ 214 & 16 & 76 & 20 \end{bmatrix}.$$

- Construct matrix G with elements from share 1 and 2 without the first column

$$G_1 = \begin{bmatrix} 36 & 92 \\ 101 & 58 \\ 126 & 142 \\ 26 & 163 \end{bmatrix} \qquad G_2 = \begin{bmatrix} 71 & 132 \\ 169 & 47 \\ 198 & 72 \\ 14 & 150 \end{bmatrix}$$

Then, use G to construct remainder matrix R with formulas stated before.

$r_y^{(x)} = [I(x,\ k(y-1)+1) \quad I(x,\ k(y-1)+2)]$

In which value of I obtained using Gauss-Jordan from:

$$\begin{pmatrix} 1 & c_1 \\ 1 & c_2 \\ \vdots & \vdots \\ 1 & c_j \end{pmatrix} \begin{pmatrix} I(x, 2(y-1)+1) \\ I(x, 2(y-1)+2) \end{pmatrix} = \begin{pmatrix} G_1\,(x,y) \\ G_2\,(x,y) \\ \vdots \\ G_j\,(x,y) \end{pmatrix} (\text{mod } p)$$

where $1 \le c_1 = j \le 2$, $1 \le x \le 4$, and $1 \le y \le 2$

$x = 1$: For $y = 1$, $G_1(1,1) = 36$, $G_2(1,1) = 71$

$$\begin{pmatrix} 1 & 1 \\ 1 & 2 \end{pmatrix} \begin{pmatrix} I(1,1) \\ I(1,2) \end{pmatrix} = \begin{pmatrix} 36 \\ 71 \end{pmatrix} (\text{mod } 251) \rightarrow \begin{pmatrix} 1 & 1 & 36 \\ 1 & 2 & 71 \end{pmatrix} (\text{mod } 251)$$

$$\rightarrow \begin{pmatrix} 1 & 1 & 36 \\ 0 & 1 & 35 \end{pmatrix} (\text{mod } 251) = \begin{pmatrix} 1 & 0 & 1 \\ 0 & 1 & 35 \end{pmatrix} \rightarrow \text{So, } r_1^{(1)} = [1 \quad 35]$$

For $y = 2$, $G_1(1,2) = 92$, $G_2(1,2) = 132$

$$\begin{pmatrix} 1 & 1 \\ 1 & 2 \end{pmatrix} \begin{pmatrix} I(1,3) \\ I(1,4) \end{pmatrix} = \begin{pmatrix} 92 \\ 132 \end{pmatrix} (\text{mod } 251) \rightarrow \begin{pmatrix} 1 & 1 & 92 \\ 1 & 2 & 132 \end{pmatrix} (\text{mod } 251)$$

$$\rightarrow \begin{pmatrix} 1 & 1 & 92 \\ 0 & 1 & 40 \end{pmatrix} (\text{mod } 251) = \begin{pmatrix} 1 & 0 & 52 \\ 0 & 1 & 40 \end{pmatrix} \rightarrow \text{So, } r_2^{(1)} = [52 \quad 40]$$

Following the same pattern, the value will be determined as follows:

$r_1^{(2)} = [33 \quad 68]$ $\qquad r_1^{(3)} = [54 \quad 72]$ $\qquad r_1^{(4)} = [38 \quad 239]$

$r_2^{(2)} = [69 \quad 240]$ $\qquad r_2^{(3)} = [212 \quad 181]$ $\qquad r_2^{(4)} = [176 \quad 238]$

Thus, the remainder matrix $R = \begin{bmatrix} 1 & 35 & 52 & 40 \\ 33 & 68 & 69 & 240 \\ 54 & 72 & 212 & 181 \\ 38 & 239 & 176 & 238 \end{bmatrix}$

- Construct secret matrix $S = (\mathbb{S} + R)$ (mod p) and Verify using watermark image W, calculate $W = ((\mathbb{S} + R_w)\ (\text{mod } p))$.

$$S = \begin{bmatrix} 2 & 5 & 2 & 3 \\ 3 & 6 & 4 & 5 \\ 4 & 7 & 4 & 6 \\ 1 & 4 & 1 & 7 \end{bmatrix} \text{ and } W = \begin{bmatrix} 50 & 100 & 21 & 14 \\ 22 & 76 & 200 & 54 \\ 1 & 91 & 45 & 7 \\ 24 & 66 & 96 & 120 \end{bmatrix}$$

As stated above, the proposed scheme constructs image shares by the secret image itself, and the watermark image will use the same projection matrix to build a public matrix. Afterward, the image shares are used to reconstruct the secret image, and the public matrix is used to determine the watermark image. Then, verify the accuracy of the watermark image using PSNR. An expected result would show high PSNR values, which show the accuracy of the watermark image, and so does the reconstructed secret image. If the watermark image is accurate, then the reconstructed secret image is accurate and valid.

4. Result and Implementation

Using the algorithm stated above, with (2, 4) threshold scheme, Figure 1 shown the secret image S, and Figure 2 shows the watermark image W. Each figure is in size 256 ×256 pixels.

Figure 3-6 below was image shares obtained by using the construction process, with each size of the image shares are 256 ×129 pixels. Figure 3 was representing image share 1, figure 4 was representing image share 2, and so on.

Figure 1. The Secret Image S

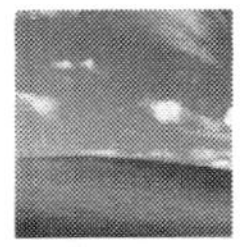

Figure 2. TheWatermarkImage$W$

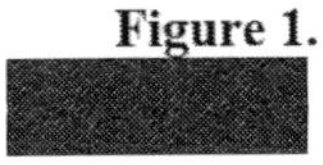

Figure 3. Representing image share 1

Figure 4. Representing image share 2

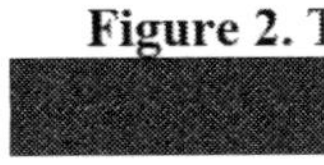

Figure 5. Representing image share 3

Figure 6. Representing image share 4

5. Conclusion

In order to prevent dishonest dealer and provide a way for participants to validate their receivedimage share, this paper presented a verifiable image secret sharing scheme using matrix projection, in which the matrix projection was used to create image shares and a public matrix from watermark image. The watermark image was used for verifiability, where the reconstructed watermark image verifies the accuracy of the reconstructed secret image.

6. References

[1] Bai L 2006 A Strong Ramp Secret Sharing Scheme Using Matrix Projection*Proc. of The 2006 Int. Symp.On a World of Wireless, Mobile and Multimedia Networks* (Washington DC: IEEE Computer Society) pp 652-656

[2] Bai L 2006 A Reliable $(\boldsymbol{k},\boldsymbol{n})$ Image Secret Sharing Scheme*2nd IEEE International Symposium on Dependable, Autonomic and Secure Computing* (Philadelphia: IEEE Computer Society) pp 31-36

[3] Chor B,Goldwasser S,Micali SandAwerbuch B 1985 Verifiable Secret Sharing and Achieving Simultaneity in the Presence of Faults*26th Annual Symp. on Found. of Comp. Sci.*pp383-395

[4]Ridayanti F 2013SkemaPembagian Data Rahasia yang MenggunakanMatriksProyeksipada Citra Digital (Skripsi)*FakultasMatematikadanIlmuPengetahuanAlamUniversitas Indonesia*

[5] Shamir A 1979 How to share a secret *Communications of the ACM***22**(11)pp 612–613

[6] Thien C C and Lin J C 2002 Secret image sharing *Computers & Graphics***26**(5)pp765-770

[7] Wang Z H, Chang C C, Tu H N, and Li M C 2010*Journals of Informations Hiding and Multimedia Signal Processing***2**(1)pp78-90

Acknowledgments

Part of this research was funded by PITTA-UI 2018 Research Grant [Project Number = 2277/UN2.R3.1/HKP.05.00/2018].

IOP Conf. Series: Journal of Physics: Conf. Series **1108** (2018) 012083 doi:10.1088/1742-6596/1108/1/012083

Students' Mathematisation in Solving Mathematical Literacy Problems with Space and Shape Contents

Lestariningsih[1], S M Amin[1], A Lukito[1], M Lutfianto[2]

[1]Department of Mathematics, Universitas Negeri Surabaya, Indonesia
[2]STKIP Al-Hikmah, Surabaya, Indonesia

Abstract. This paper aimed to describe four undergraduate students' mathematisation from Surabaya and Sidoarjo in solving mathematical literacy problems with space and shape contents. This study used descriptive research with a qualitative approach. Data were collected by giving the mathematical literacy problem with space and shape contents and interviews. Next, data were analysed by studying the four mathematisation processes: formulating, using, interpreting, and evaluating. The results of this study pointed out that students'' mathematisation in solving mathematical literacy problems with space and shape contents were as follows 1) formulating contextual problems into mathematical problems by verbally describing, using mathematical symbols that have been understood, and using appropriate modelling, 2) building or using direct mathematical objects (facts, concepts, principles, and procedures) and indirect (mathematical reasoning) to obtain mathematical solutions, 3) summarising the results obtained from the counting operations that have been performed, and the last 4)stating that the mathematical solution obtained is correct by raising logical reasons. Students' mathematisation was a principal process in solving mathematical literacy problem with space and shape contents, so it was important to emerge students' mathematisation in the mathematical classroom or daily activities.

1. Introduction

Problem solving skills in mathematics especially contextual mathematics problems require mathematisation. Mathematisation is not only used when someone makes a model or a mathematical representation of a problem with a real-life context but also when the process of solving the problem and interpreting it into a real-life context [1, 2]. Freudenthal [3] declares that mathematisation means making things more mathematical. Also, mathematisation is not only an act of mathematicians but also becomes the activities of learners in understanding the daily situations using a mathematical approach. Mathematisation is divided into two kinds, namely horizontal mathematisation and vertical mathematisation [4]. Mathematisation can be different for each learner in solving a problem.

Contextual problem is also used in Programme for International Student Assessment (PISA). Characteristics of problems which used in PISA is arranged in different domains require not only vertical mathematisation but also horizontal mathematisation. Mathematisation in PISA refers to Freudenthal's idea in Realistic Mathematics Education (RME) [5]. Some researchers conclude that mathematical cycles (mathematical modelling) in PISA are identical with horizontal mathematical and vertical mathematisation cycles in RME [6,7].

Mathematisation not only as part of the process of solving mathematical problems or contextual problems but also as part of a learning process and teaching practice which makes mathematisation capability to be owned by students [8,9]. But, many students have difficulty to do mathematisation when solving contextual problems. The students' mathematisation processes were not even and/or consistent even for the problem with same content topic [10].

Mathematisation proposed by Roux [11] is the application of concepts, procedures, and methods developed in mathematics to an object or knowledge in either mathematics or other disciplines. Meanwhile, Fosnot [12] says that mathematisation is a constructive process including observing patterns in special cases, analysing the reasons for something happening, stating in some form of generalisation, and looking for flexibility in strategy making or proof. Whereas OECD [13] stated that mathematisation can involve transforming a problem defined in the real world to a strictly mathematical form (which can include structuring, conceptualising, making assumptions, and/or formulating a model), or interpreting or evaluating a mathematical outcome or a mathematical model in relation to the original problem. Then, mathematisation in this study is the activity of transforming a problem expressed in a real life context into a mathematical model or representation, then the completion of a mathematical model or representation is interpreted into a real life context. The real world is defined as everything that is outside mathematics, like everyday life and the environment [14].

The steps of mathematisation proposed by Lynch [15] were (1) considering the real-world problem, (2) organizing based on mathematical contexts, (3) paring down to the most important aspects, (4) using mathematical skills and solving problem, and (5) considering the solution toward terms of the real situation. OECD [13] declared the steps of mathematisation which were (1). formulate, (2) employ, (3) interpret, and (4) evaluate. The steps of mathematisation which used in this study were (1) formulating a problem of real world context into mathematical problem, (2) using the facts, concepts, procedures and mathematical reasoning to obtain mathematical solution of the problem mathematically, (3) interpreting the mathematical solution to the real world context to the initial problem, and (4) evaluating solution to real world context to the problem.

Mathematisation is also one of the seven fundamental mathematical capabilities required by the individual in solving mathematical literacy problem [16]. Mathematical literacy problem is expressed mainly in the form of contextual problems or mathematical problems using real world context because of their structure emphasizes the need to develop the capacity of a person in using mathematics in a variety of contexts. Mathematical literacy problem can be used to measure the level of mathematical literacy that is owned by someone. Mathematical literacy describes a person's ability to reason mathematically and use the concepts, procedures, facts and mathematical tools to explain and predict phenomena. Mathematical literacy is an essential ability for young people to prepare for life in a modern society in the 21st century [16].

Problem in PISA is based on aspects of mathematical content and context. The context indicates the location or places that exist in the problem and mathematical content is a mathematical topic used in problems. There are four categories of content used in PISA, namely, (1) change and relationships, (2) shape and space, (3) quantity, (4) uncertainty and data [17]. This study uses shape and space content contained in PISA. Shape and space content include phenomena about the visual world that involves patterns, properties of objects, positions and orientations, representations of objects, encoding visual information, navigation, and dynamic interactions associated with real forms. We use shape and space content because this content includes aspects of mathematical content in mathematics present in the curriculum.

Previous study about mathematisation showed that student's specific forms of mathematisation. It was associated with the mathematisation of motions which were (1) deductive reasoning, (2) alternative mathematical methods, and (3) geometrical diagrams [11]. Other study about mathematisation of Irish students showed that their ability to mathematics (that is the ability to transfer a realistic situation into a mathematical situation) was poor hence there was no evidence of any significant gender difference in test performance, and they find difficulty when applying mathematical knowledge in unfamiliar situations [15].

Based on the previous explanation, the purpose of this study was to describe students' mathematisation in solving mathematical literacy problems with space and shape contents.

IOP Conf. Series: Journal of Physics: Conf. Series **1108** (2018) 012083 doi:10.1088/1742-6596/1108/1/012083

2. Method

This study used descriptive research with qualitative approaches. Participants in this study were four undergraduate students of mathematics education with high mathematics ability. The selection of these participants was made by using mathematics skills test. Before the mathematics skills tests were used, validity was performed to maintain the quality of the problems, namely construct validity, content validity, face validity, and item validity.

Then to know the ability of mathematisation, students were asked to solve mathematical literacy problems with space and shape content by adapting the PISA problems using Indonesian context (see Table 1). Problems were designed as two problems given at different times for triangulation needs. Mathematical literacy problems that have been designed to be validated through expert review include the suitability of content, construct, and face.

After students finished mathematical literacy problems, we conducted interviews to explore mathematisation data. Interview data were analysed using qualitative data analysis consist of three stages, namely data condensation, data presentation, and drawing conclusion.

Table 1. Mathematical literacy problems with space and shape contents.

First Problem	Second Problem
Susan wants to make a solid from small cubes. Here is a picture of a small cube (Figure 1) and a solid that build from small cubes (Figure 2). She uses glue to fit the cubes to be solid. First, Susan glued eight cubes to make a solid as shown in figure B. Then Susan wants to make a solid with a length of 6 small cubes, a width of 5 small cubes, and a height of 4 small cubes. She wants to use as little as possible a small cube with a hole in the solid. How many cubes are Susan needed to make this? Explain! 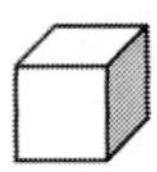 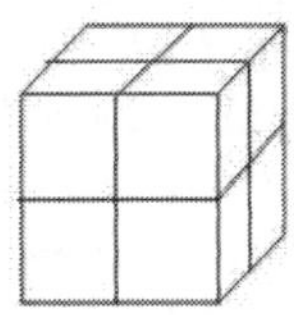 **Figure1**. Small Cube.**Figure 2**. Solid.	A cake trader "*TerangBulan*" sells two kinds of *terangbulan* with the same thickness but different in sizes (See Figure 3). The small *terangbulan* has a diameter of 30 cm, and the price is Rp 30.000, - and the big *terangbulan* has a diameter of 40 cm, and the price is Rp 40.000,-. Which *terangbulan* is cheaper? Explain!  **Figure 3**. *TerangBulan*.

3. Results and Discussion

3.1. Students' mathematisation on formulating a problem of real world context into a mathematical problem

Regarding to the first step, students identified the mathematical aspects of real life context in the problems by presenting what was known in the form of verbal descriptions. In the first problem, students wrote that Susan wanted to make the solid that looked likea solid object in the inner part (see Figure 4) while in the second problem, the students wrote the diameter and the price of the small and big *terangbulan* (see Figure 5).

Figure 4. Verbal Description of First Problem.

Furthermore, students also identified the important variables according to real situation problems by stating the image information in the problem using mathematical statements or symbols that have been understood. In the first test, students declared information in the form of Figure 1 (Small Cubes) and Figure 2 (Solid from Small Cubes) into a description that was on the problem there was one figure of small cube and 1 drawing the composition of a small cube shaped block while in the second problem. Students declared the information that was obtained by writing the small *terangbulan* as the first *terangbulan* and the big *terangbulan* as the second *terangbulan*. Figure 5 also illustrates those students using symbols in the second problem.

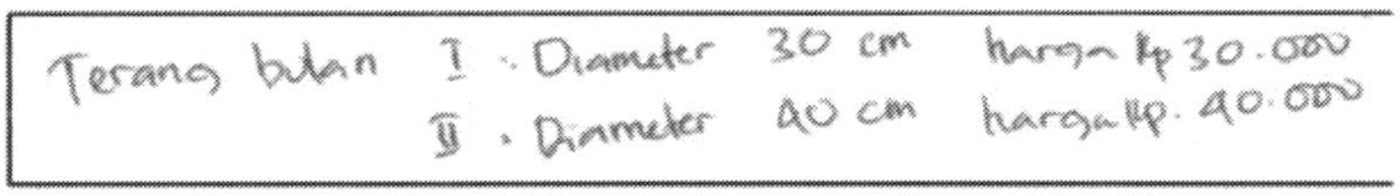

Figure 5. Verbal Description of the Second Problem.

Then, students represented the situation mathematically by using appropriate modelling to clarify the problem and facilitated themselves in solving the problem, as well as identified the variables that were important in the context of the problem to determine the mathematical aspects. In the first test, students made a visualization of the information. *"Susan wants to make a solid that looks like a solid with full of cubes includes in the inner part. It has length 6 small cubes, width 5 small cubes, and height 4 small cubes"* as a representation of a build composed of several small cubes to clarify the problem and facilitated them in solving the problem. In the second test, students made visualisation of small *terangbulan* and big *terangbulan* as the representation of the first circle and the second circle respectively.

Students identified the important variables in the context of the problem to determine the mathematical aspect of the first problem by calculating the length, width and height by marking using the "dots". Then, it was known that the length was 6 small cubes, the width was 5 small cubes, and the height was 4 small cubes (see Figure 6). They said, *"This one is 6 in length, 5 small cubes in width and 4 small cubes in height (while recalculating the dot on the drawing in the answer sheets)." While from the second problem, students knew that the radius is half of the diameter by stating, "The radius is half of the diameter, and counts the radius by* $30 \div 2 = 15$".

Moreover, from the data analysis, it was obtained that the main categories and other categories are seen as subcategories. The main category that emerges is the way students formulate mathematical literacy problem with the content of shape and space into mathematical problems with three subcategories, namely (1) students formulate mathematical literacy problem into mathematical problems by using verbal description, (2) students formulate mathematical literacy problem into mathematical problems using mathematical statements or symbols that have been understood, and (3) students formulate mathematical literacy problem into mathematical problems using the appropriate modelling to clarify the problem and facilitate themselves in solving the problem.

3.2. Students' mathematisation in using the facts, concepts, procedures and mathematical reasoning to obtain a mathematical solution of the problem mathematically

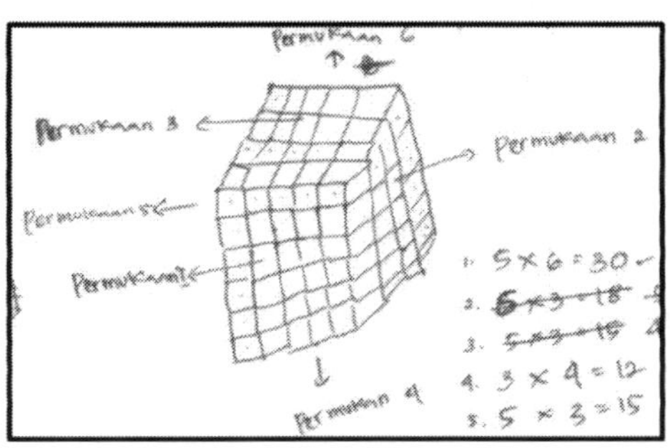

Figure 6. Solid from Cubes.

Regarding to the second step, students designed a strategy to find a mathematical solution by planning to use the concept of cube and circle arrangement to represent the context contained in the problem. Students also designed to apply principles that fit with the context of the problem, namely the surface area of solid and the area of the circle. For the second problem, they said, *"The area of the circle=* $\pi \times r^2$". Then, students planed strategies to find mathematical solutions by constructing facts by applying the concept of sequence when using labels to express the context contained in the problem. They used labels ranging from first surface to sixth surface on each constructed surface drawn respectively to represent the front surface, the right surface, the top surface, the lower surface, the left surface, and the back surface (see Figure 6). They said, *" In this problem, the front side is likened surface 1, hold the side here (right) is likened surface 2, top side is likened surface 3, bottom side is likened subsurface 4, and the left side is likened surface 5, and the back side is likened surface 6."*

Students used the concepts of cube, rectangle, and circle to represent the context contained in the problem. They also used the procedure of addition, multiplication, and division as well as doing mathematical reasoning (deductive reasoning using the associative properties of addition and multiplication) to obtain a mathematical solution. Figure 7 shows the mathematical reasoning mentioned.

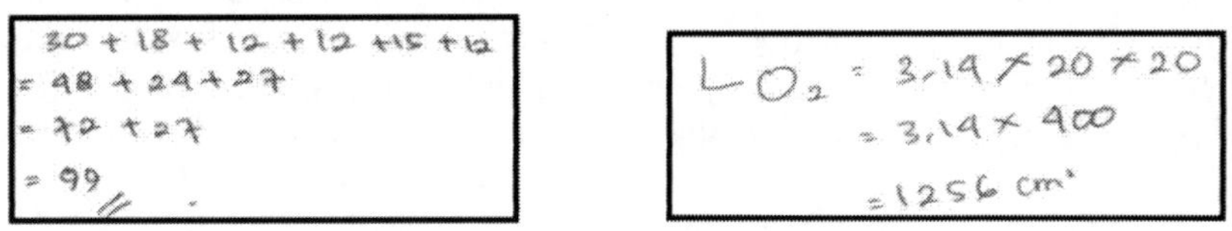

Associative properties of addition associative properties of multiplication

Figure 7. The Associative Properties of Addition and Multiplication

In the step of using facts, concepts, principles, procedures and mathematical reasoning to derive mathematical solutions from mathematical problems, it was obtained a main category and two subcategories. The main category that emerged was students constructing or using mathematical learning objects to obtain mathematical solutions. Then, two subcategories of data could be created in the convergence interpretation. First, students constructed or used mathematical learning objects directly in mathematical literacy to obtain mathematical solutions. Second, students used indirect mathematics learning objects to obtain mathematical solutions that are by using mathematical reasoning like the associative properties of addition and multiplication.

3.3. Students' mathematisation on interpreting the mathematical solution to the real world context to the initial problem

In the step of interpreting mathematical solutions into real world contexts in the original problem, students concluded the results obtained from the counting operations that have been done on the first and second problem. In the first problem, they concluded by stating the number of small cubes needed to make the solid. They said, *"Susan will use 99 small cubes to make the solid block."* Then, in the second problem, the students concluded the results obtained from the counting operations that have been done by stating that *terangbulan* that cost cheaper was *terangbulan* which had size 40 cm. They said, "In conclusion, I think the cheaper *terangbulan* is 40 cm in size".

3.4. Students' mathematisation onevaluating solutions to real world context to the problem

In the last step, students evaluated the compatibility between mathematical solutions in the context of real world problems by stating that the mathematical solutions obtained were correct because students were already using the appropriate procedures. In the first problem, students counted the number of small cubes of each surface then added. In the second problem, students calculated the area of both *terangbulan* then used the comparison to get a mathematical solution. Furthermore, students explained the reasons why mathematical results or conclusions were appropriate or not appropriate to the context of the problem by stating that the results or conclusions obtained are suitable with the context of problem because students used concepts, principles, and procedures to find out. In the first problem, they said, *"The results or conclusions obtained are suitable with the context of problem because Wealready counted the number of small cubes on each side and sum up them"*. Then, in the second problem, they said, *"We used formula to get areas of circles then we compared them"*.

Moreover, findings on mathematisation of students with high mathematics ability in solving mathematical literacy was in line with study conducted by Lynch [15] because mathematical literacy problems was using realistic context for students. Students with high mathematics skills also did mathematisation by using facts, concepts, principles, procedures and mathematical reasoning to obtain mathematical solutions [15,16]. However, it was found that some important categories when students did mathematisation namely the way students formulate mathematical literacy problems using shape and space content into mathematical problems, and students constructing or using mathematical learning objects to get mathematical solutions.

4. Conclusions

It was important to highlight that mathematisation is a fundamental process in solving mathematical literacy problems because through mathematisation it can be known (1) the process of formulating real world context problems into mathematical problems, (2) the use of mathematical learning objects to obtain mathematical solutions, (3) the summary of the results, and (4) the evaluation of the results. The result of study showed that the students did the mathematisation by (1) identifying the mathematical aspect in real situation problem, (2) identifying/raising the important variables according to real situation problem, (3) representing the situation mathematically by using the appropriate modelling, (4) planning the strategy to find mathematical solutions, (5) applying mathematical facts/rules/algorithms/structures or mathematical strategies when looking for solutions, (6) interpreting mathematical solutions in real life context, (7) evaluating the fit between mathematical solutions in the context of real-world problems, and (8) explaining the reasons why results or mathematical conclusions are appropriate or not inappropriate with the context of the problem.

Mathematical literacy and PISA problems with change and relationships, quantity, uncertainty and data content were recommended to be used. It was recommended to develop a more complete description on students' mathematisation by applying various problems. Moreover, further studies could also use problems that were developed by researchers.

References

[1] Gravemeijer K and Terwel J 2000 *J. Curr. Studies***32** 777
[2] Lestariningsih L, Amin SM, Lukito A andLutfianto M 2018 *Proc. of University of Muhammadiyah Malang's 1st INCOMED, ASSEHR* **160** p. 291
[3] Freudenthal H 1968 *Educational Studies in Mathematics* **1** 3
[4] Wake G 2014 *Educational Studies in Mathematics***86** 271
[5] OECD 2003*The PISA 2003 Assessment Framework: Mathematics, Reading, Science and Problem Solving Knowledge and Skills*(Paris: OECD Publishing)
[6] Kaiser G and Willander T 2005 *Teaching Mathematics and Its Applications***24** 48
[7] De Lange J 2006 *Tsukuba J. Educ. Study in Mathematics* **25**13
[8] Biccard P and Wessels D 2015 *Pythagoras***36** 294
[9] Biccard P and Wessels D 2017 *African Journal of Research in Mathematics, Science and Technology Education* **21**61
[10] Murata A and Kattubadi 2012 The Journal of Mathematical Behavior **31** 15
[11] Roux S 2010*Early Science and Medicine***15** 319
[12] Fosnot C T 2005 *The Constructivist***16** 1
[13] OECD 2016 *PISA 2015 Assessment and Analytical Framework: Science, Reading, Mathematic and Financial Literacy* (Paris: OECD Publishing).

[14] Siregar N F 2016 *Logaritma* **4** 17
[15] Lynch P E 2011 *Mathematisation and Irish students: The ability of Irish second-level students to transfer mathematics from the classroom to solve authentic, real-life problems* (Doctoral dissertation, National University of Ireland Maynooth).
[16] Blum W and Ferri R B 2009 *J. Math. Modelling and Application* **1** 45
[17] Stacey K 2011 *J.* Math. *Educ. (Indo-MS_JME)* **2** 1

Acknowledgment
We would like to thank Ministry of Higher Education and Research on a doctoral dissertation research grant.

Analysing PISA-like assessment test measuring Scientific Literacy using Three-Parameter Logistic (3PL) of IRT–2018

S Rohmah[1*], I Kaniawati[2] and T R Ramalis[2]

[1]Student of Postgraduate School, Universitas Pendidikan Indonesia, Indonesia
[2]Postgraduate School, Universitas Pendidikan Indonesia, Indonesia
*Corresponding Author. Email: siti.rohmah@student.upi.edu

Abstract. This study investigated the multiple-choice of scientific literacy resembling to PISA framework assessment test to assess senior high school students' literacy acquisition in physics by applying Item Response Theory (IRT) with three-parameter logistic. The multiple-choice instrument composed by 15 items test measuring three aspects of the scientific literacy assessment in PISA 2015 framework such as Context, Knowledge and Competencies. Three-parameter logistic (3PL) used to analysing test are difficulty, discriminatory, and guessing parameters of the instrument items were fitted by IRT model, using EIRT. Data was gathered on 32 students in science program of student high school in Bandung region. The IRT framework reveals item-level information and indicates appropriate ability ranges for the test. With the result that these methods can apply to diagnose and evaluate the features of items at various scientific literacy aspects of test takers.

1. Introduction

The PISA-like assessment test developed by Maulana and Suhandi in 2016, is known standard multiple-choice test and Likert test of scientific literacy for physics course at high school student. The test consists of two types of test. The first type of test be made up of 15 items of multiple choices test that covers three domain of scientific literacy framework developed PISA which is context domain, knowledge domain, and competencies domain. The second type of test be made up of 18 statement that cover the fourth domain which is attitude domain [8]. All domains of scientific literacy are important to measure student's proficiency to learn about physics as part of science. Previous research that using this test has been demonstrated classical test assessment methods: the item difficulty index and the item discriminatory index [1]. However, the framework of classical test theory (CTT) for test assessment has some important limitations. For instance, the item parameters depend on the ability distribution of examinees and the ability parameters depend on the set of test items. To overcome these shortcomings, the item response theory (IRT) was introduced [2–4].

Therefore, the purpose of this study is to explore PISA-like assessment test based on the framework of IRT. We will first present the key concepts of IRT, focusing on the three-parameter logistic (3PL) model used in the study (Sec. 2). Section 3 is the research method that used by the test from tenth grade high school students that study physics. Section 4 (results and discussion) will be discussed some limitations of CTT also advantages of IRT, and the significance of using 3PL-IRT analysed by our data. Last, in Sec. 5, we will summarise what we did and found in this study.

2. Item Response Theory

The IRT framework rests on the assumption that the performance of an examinee on a test item can be predicted from the item's typical features and the examinee's latent traits—often called abilities or person parameters. The relationship between the examinees' performance on an item and their ability is described by an item characteristic function (or item response function), which quantitates how the probability of a correct response to a specific item increases with the level of an examinee's ability. The graph of this relationship is known as the item characteristic curve (ICC). The empirical ICCs in prior published research, of relevance to IRT, tend to be S-shaped or sigmoidal. As the ability increases, the empirical ICC rises slowly at first, more sharply in the middle, and again slowly at very high levels of ability. In its early days, the normal ogive function was commonly used to model the ICC, while nowadays the logistic function is a popular alternative, as shown in Eq. (1). An example of the ICC from our data, established by fitting a logistic function, is shown in Fig. 1.

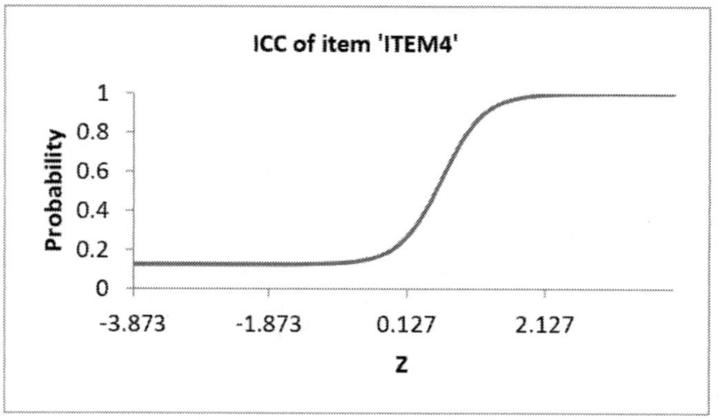

Figure 1. Item Characteristic Curve (ICC) of Item 4 in the PISA-Like Assessment Test, Modelled from Data on Tenth Grade Students (N = 32)

In item response function models, the ability parameter is usually standardised to zero mean and unit standard deviation (SD), and is commonly denoted by the Greek letter theta (θ). Theoretically, the ability can range from $-\infty$ to ∞, but its values in practice are usually in the interval -3 to 3. The PISA-like assessment test consists of multiple-choice questions with responses sometimes based on guessing, and the three-parameter logistic (3PL) model of IRT is appropriate for the investigation of the related item parameters (i.e., difficulty, discriminatory, and guessing parameters). In the item response function, the probability of answering item number i correctly for an examinee with the ability θ is given by

$$P_i(\theta) = c_i + (1 - c_i)\frac{1}{[1 + \exp\{-Da_i(\theta - b_i)\}]} \quad (1)$$

where b_i is the item difficulty parameter (θ at the inflection point of the ICC), and a_i is the slope of the ICC at that inflection point, called the item discriminatory parameter. The lower asymptote level c_i of the ICC, which corresponds to the probability of a correct answer at very low-ability levels, is referred to as the *pseudo-chance level* or sometimes as the guessing parameter. The constant D normally equals 1.7, and is used to make the logistic function as close as possible to the normal ogive function. The 3PL model can be reduced to the two-parameter logistic (2PL) model by setting $c = 0$. The 2PL model is most plausible for open ended questions, in which responses are rarely guesses. Moreover, this can be further reduced to a one-parameter logistic (1PL) model by considering only the b parameter at which the probability of a correct response is 0.5, while holding a fixed [2,7–8]. The special cases with $D = 1.0$, $c = 0$, and $a = 1.0$, are known as Rasch models [5].

In this study, we applied the EIRT: Item Response Theory applications using Excel add-in created by a collaborative effort between Université du Luxembourg and Université Laval (Québec, Canada). IERT has integrated well-known models, namely, the 1PLM, 2PLM, and 3PLM to estimate the appropriate parameters. Also included in the program is a generalization of the logistic model to accommodate multiple-choice items. One possible generalization is Bock's nominal model. This model is appropriate with nominal data, for instance, with a four-option multiple-choice item. EIRT also provides classical test theory statistics such as Cronbach's alpha [6].

3. Data Collection

The 15-item PISA-like assessment test was validated by researcher in physics education that used to measure student's scientific literacy in physics class with topic static fluid. The test was revised based on suggestions from the professors. We applied the assessment to 32 student high school in Bandung. These students had learned static fluid concepts through lectures and integrated demonstrations and group discussions by teacher. The participants took 30 min to complete the assessment. The collected data were analysed using the 3PL model in IRT and Instrument information function plots.

4. Result and Discussion

In applying IRT to the data gathered on tenth grade high school student in Bandung, we used the add-ins on Microsoft program called EIRT to fit the three-parameter logistic (3PL) models, one for each PISA-like assessment test item. We assumed a single ability, named the scientific literacy ability, which represents the latent traits in each student that affect performance in the scientific literacy. Each logistic model is determined by identifying its three parameters: discrimination a, difficulty b, and guessing c. These identified parameters are shown in Table 1 for the 15 items test, categorised by physic concepts in static fluid. Moreover, the item difficulty index (*diff*) and discrimination using the point-biserial coefficient (r_{pbs}) from the CTT framework are included as the last columns of Table 1. The criterion ranges of the item parameters are shown by interval bounds in square brackets.

Table 1. The Model Parameters Identified in IRT Analysis, Namely, Discrimination a, Difficulty b, And Guessing c, for the 15 Items in PISA-Like Assessment Test Categorized by Static Fluid Concepts, Along with the Item Difficulty Index (*Diff*) and the Discrimination (*disc*) from CTT Analysis

Static Fluid Concept	Item	Discrimination		Difficulty		Guessing
		IRT (a)	*CTT-(disc)*[a]	IRT (*b*)	CTT (*diff*)[b]	IRT (*c*)[c]
		[0,3]	[0, 0.4]	[-3,3]	≥0.18	[0.12, 0.18]
Pascal's law	1	0,595	0,000	0,622	0,500	0,166
	5	0,623	0,093[d]	3,182	0,313	0,194
	8	0,628	0,164	-0,435	0,625	0,168
	10	0,677	0,073	1,352	0,406	0,166
	12	0,719	0,062	-1,398	0,750	0,167
Archimedes principle	3	1,005	0,239	-0,542	0,656	0,164
	7	1,546	0,200	0,476	0,438	0,154
	11	0,884	0,056[d]	31,790[d]	0,188	0,180
	14	0,931	0,022[d]	-1,637	0,813	0,167
Capillarity	2	1,443	0,263	-1,257	0,813	0,164
	4	3,275	0,274	0,631	0,281	0,124
	13	0,607	0,000	0,683	0,500	0,173
Viscosity and Stoke's law	6	0,843	0,225[d]	3,966	0,188	0,167
	9	0,214	0,489[d]	-3,993	0,750	0,171
	15	0,773	0,015[d]	1,863	0,344	0,173

[a] in classical test theory, the item discrimination is the point-biserial correlation coefficient between the item scores (0 or 1) and the total test scores

[b] in classical test theory, the item difficulty is the percentage of correct response, so a low value indicates a high item difficulty

[c] in classical test theory, there is no formal definition of guessing chance. Guessing chance of a multiple-choice question is often estimated as (1/ number of choices), which is 20% for every item in five multiple choice tests. This estimation implies the probability of choosing each choice is equal.

[d] outside the criterion range.

The item difficulty b is the ability θ at the inflection point of ICC. In the logistic model, at this point, the probability of correct answer is $(1 + c)/2$, midway between the asymptote levels, as seen by

IOP Conf. Series: Journal of Physics: Conf. Series **1108** (2018) 012084 doi:10.1088/1742-6596/1108/1/012084

substituting $\theta = b$ in Eq. (1). When $c = 0$, as in the 1PL and 2PL models, the probability of a correct answer is 0.5 and this could be used to identify b. Parameter b is named "difficulty" because a harder test item requires higher ability b for probability 0.5 of a correct answer. The criterion range for b is chosen to [-3,3]. Clearly, an item with b close to −3 is very easy, while b close to 3 is very difficult for the sample population of examinees.

The results in Table 1 show that item 6 was the most difficult question for the group of students, with the maximum b = 3,9. Item 6 has only 18% of the students correctly answered item 6 (*diff* = 0,18). Since only a few students from the high-ability group could answer the test correctly, item 6 clearly presented the most difficulty to the normal-ability students. Although difficulty b from the IRT analysis and the *diff* from the CTT analysis differ in their theoretical interpretations, they tend to be in good agreement. For example, item 5 (b = 3,1) and item 15 (b = 1,8) were somewhat difficult for the students, according to IRT. Their correct response rates were only 31% and 34%, respectively, indicating them as hard items according to the *diff* values also. Further agreement was found in item 14, whose CTT-*diff* indicated an easy question, with about 81% correct answers, consistent with difficulty b = −1,6. However, the two measures of difficulty seem to disagree on item 4. Viewed through CTT, item 4 appeared to be quite difficult items with correct answer rates 28%, but their respective b values were 0,6 quite easy for the difficulty parameter of IRT. As discussed earlier, this is one downside of the CTT approach, whose results depend on the examinees' ability. The guessing parameter c of an item represents the probability that an examinee with very low-ability level answers correctly. This may relate to the attractiveness of the answer choices and/or the guess behaviour of the examinees. Its value is equal to the level of a lower asymptote for the ICC, and it ranges from 0 to 1. Typically, c should be less than 0.3 [7]. Table 1 shows that, overall, the very low-ability students had less than a 30% chance of choosing the correct option for most the PISA-like assessment test items.

To show how the probability of correct response for a specific item depends on the ability of an examiner, we build the item characteristic curve (ICC). The ICC of a well-designed question should have a sigmoidal S shape [2,7]. Then the probability of a correct response would consistently increase with ability, and a high slope at the inflection point would indicate sharp separation by ability around that point. As shown by the solid line in Fig. 2, item 4 of the PISA-like assessment test mostly agrees with these criteria in our data. It has high discrimination power (a = 3.275) for separating examinees at medium-ability level near b = 0,63, and very low-ability students have a 28% chance to correctly answer the item (c = 0,12).

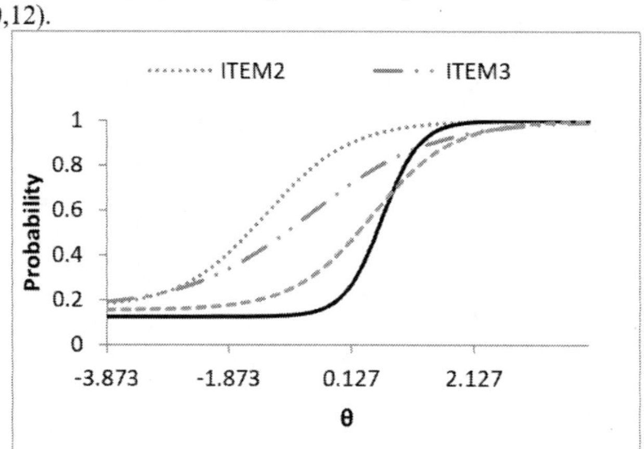

Figure 2. The ICCs for items 2, 3, 4, and 7 in the PISA-like assessment test.

Overall, the results show that each PISA-like scientific literacy assessment test item has proper discrimination power at its specific difficulty (or ability level). Clearly, the steepness of the curve demonstrates the capability of the item to discriminate in the ability domain between examinees who understand the concept targeted by the item and those who do not. In general, a set of test items or questions should cover a range of ability domains in which the test takers are expected to differ. To determine how well the assessment does in testing adoption of the physics concept by the examinees at their various ability levels, the test information function was investigated in IRT framework.

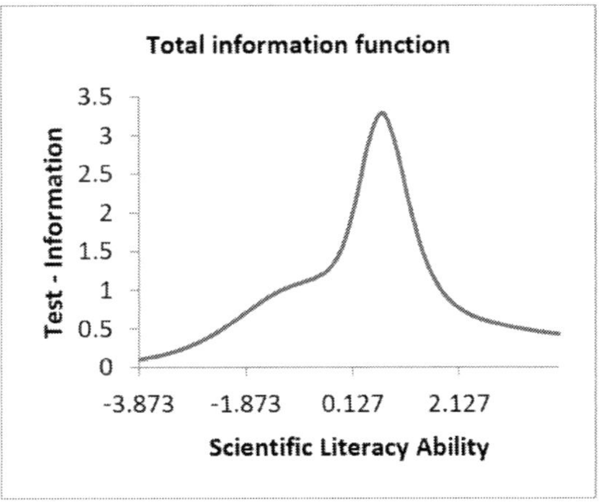

Figure 3. Test information curve of the 15-item PISA-like across scientific literacy ability levels of the tenth-grade high school students tested.

The information function for a test at θ, denoted I(θ), is defined as the sum of the item information functions at θ:

$$I(\theta) = \sum_{i=1}^{n} I\,(\theta) = \sum_{i=1}^{n} \frac{[P_i(\theta)']^2}{[P_i(\theta)][Q_i(\theta)]} \quad (2)$$

where I(θ) is the item information function of item i, P(θ)' 0 is the derivative of P(θ) with respect to θ for item i, and $Q_i(\theta)$ is $1 - P_i(\theta)$ [2,7]. This feature is not available in CTT. For example, the point biserial coefficient for an item is influenced by all items in the test. A plot of I(θ) for the 15-item PISA-like assessment test across ability levels of scientific literacy is shown in Fig. 3. The information curve peaks sharply to its maximum value 3,75 at θ = 1.12. This indicates that the test provides information about the scientific literacy ability of student most effectively when the examinees have abilities roughly in the range from 0.1 to 2.1 (medium to high). For cases with scientific literacy abilities less than 0, the PISA-like test provides very little information that would distinguish their differences, while the results of the 15-item test are highly sensitive to ability differences around ability 1.1. In general, the purpose of the test decides what type of information curve would be desired. For example, a flat curve over the whole range of abilities indicates that the component items are sensitive to variations at different ability levels, so the test information obtained by summing the item information becomes evenly distributed. This is desired if the test serves to assess a wide variety of abilities. In contrast, a sharply peaked curve sensitively reports differences of the test takers only around that peak, elsewhere it acts like a pass or fail threshold. This may be desirable when the purpose of the test is to provide eventual pass or fail labels. Test developers can benefit from these item and overall test information curves, to revise tests with such considerations of purpose in mind.

5. Conclusion

Results in this study indicate that the 15-item scientific literacy test, with 5 choices per item, is useful for testing students' literacy acquisition in physics when the ability of the examinees is from medium to high. It can be applied as a pass or fail threshold instrument at a somewhat high-ability value (θ ≈ 1.1). This insight is clearly provided by the test information function. Items 5 and 6 are useful for separating examinees at high-ability levels. However, the item and ability parameters of scientific literacy reported in this study only pertain to the test responses by high school student that study physics in static fluid concept. Overall, the approach and findings of the current study may be used to develop and improve testing and enhance its sensitivity and effectiveness within a given range of abilities. Test developers can analyse item and ability parameters using IRT. Moreover, using IRT with the item parameters held constant, the same group of students can be tested before and after instruction to determine the learning gains in ability. The current results and approach can directly benefit anyone who uses the assessment, to gain improved accuracy of diagnosis. Other than that number of student participant should be added more to increase better precision of assessment test.

6. References

[1] Maulana A and Suhandi A 2017 *AIP Conf. Proc.* **1848**(050004)
[2] Hambleton K, Swaminathan H and Rogers H 1991 *Fundamentals of Item Response Theory* (Newbury Park, CA: Sage Publications Inc.)
[3] Hambleton R K and Jones R W 1993 *Educ. Meas.* **12**(38)
[4] MacDonald P and Paunonen S 2002 *Educ. Psychol. Meas.* **62**(921)
[5] Kolen M J and Brennan R L 2004 *Test Equating, Scaling, and Linking Methods and Practices, 2nd ed.* (New York, NY: Springer Science and Business Media)
[6] Valois P, Houssemand C, Germain S and Abdous B, 2011 *Proc. Soc. Behav. Sci.* **15**(552)
[7] Harris D 1989 *Educ. Meas.* **8**(35)
[8] Organisation for Economic Co-Operation and Development (OECD) 2013 *PISA 2015 Draft Science Framework* (Paris: OECD)

IOP Conf. Series: Journal of Physics: Conf. Series **1108** (2018) 012085 doi:10.1088/1742-6596/1108/1/012085

A Sequence Convergence of 1 −Dimensional Subspace in a Normed Space

Manuharawati[1,*], D N Yunianti[2], and M Jakfar[3]

[1, 2, 3] Department of Mathematics, Faculty of Mathematics and Natural Science, Universitas Negeri Surabaya, Indonesia
*Corresponding author: manuhara1@yahoo.co.id

Abstract. In this paper, the researchers will be introduced the concept of a sequence convergence of 1 −dimensional subspaces (lines) in a normed space and shall discuss some properties of it. Furthermore, it will be proved a continuity property of angles among subspaces in inner product spaces. Finally, the notion of limit of a sequence of 2 −dimensional subspaces (planes) in a normed space is studied. The researchers also obtain a result which describe how the convergent of a sequence of lines is associated to the convergent of a sequence of planes in a normed space.

1. Introduction

A sequence in the metric space and its properties has been widely used in investigating functional continuity [1], discussing fixed point theorems [2], discussing the central limit theorem in statistics [3], and discussing the numerical regression convergence [4].

The sequence in the metric space can be vector sequence, function sequence, matrix sequence, real numbers sequence, and so on. This has been discussed in some research conducted by many researchers:[5-13] discussing the convergence of vector sequences and their properties.

The application of vector sequences' convergence are widely studied by [14] [15]. The two research have discussed the angle between vectors in the inner product space and the concept of convergence of vector sequences in the inner product space are used to prove the continuous properties of the angular function between two vectors in the inner product space. In addition to both research, there are still many mathematicians who discuss the angle between subspace and its properties in the inner product space such as [3] [16] and [17]. However, neither of which shows the classification of the continuous properties of the angular function between the two subspaces in the inner product space.

The question arises: "Can we show the continuity of the angular function between two vector subspaces in the inner product space?" This will be answered if we have the convergence concept of subspace sequences from vector space. Readers may also agree if the concept is researched. It will also be obtained many applied to be produced.

In this paper, the researchers will define the convergence of sequence inone dimensional subspaces. This is a new concept introduced in the discussion of the convergence of sequence. The researchers also will discuss the basic properties of it. As a consequence of this definition, the property of continuity from the angle between the two subspaces will be shown. At the end, we will generalize the concept of convergence these subspaces in two-dimensional subspaces.

Before discussing the convergence concept of vector subspaces sequences, we first recall the definition of convergence of vector sequences as follows:

IOP Conf. Series: Journal of Physics: Conf. Series **1108** (2018) 012085 doi:10.1088/1742-6596/1108/1/012085

"a vector sequence (u_n) in normed space is said convergent to a vector u if $(\|u_n - u\|) \to 0$".

The definition of vector sequences convergence cannot be derived directly to define the convergence of subspace sequences. For example, a one-dimensional subspace can not only be replaced byu_n and u respectively as the basis of its subspace sequence, because when the definition is used, the convergence concept of subspaces sequences dependent on basis selection. Thus, by using the definition, there are subspaces sequences in normed space are convergent and divergent depending on the basis selection. For example $X = \mathbb{R}^2$, $U_n = span\{(1,0)\} = U \subset X$. If we take the basis of U_n where $\{u_n = (1,0)\}$for every n and the basis of U is $\{(1,0)\}$, then the sequence (U_n) converges to U because $(\|u_n - u\| = 0) \to 0$. But if we take the basis ofU_n is $\{u_n = (2,0)\}$ and $\{(1,0)\}$is the basis of U, then the sequence $(U_n)$is divergent to U because $\|u_n - u\| = \|(2,0) - (1,0)\| = \|(1,0)\| = 1 \nrightarrow 0$.

To overcome these weaknesses, it is necessary to construct the definition of convergent in finite dimensional subspace in normed space that independenton basis selection. According to the survey conducted by the author (journals, magazines, proceedings, national and international seminars), the concept of convergence in finite dimensional subspace in normed space has beenresearched by none.

If we choose $u$as the basis of one dimensional subspace $U = span\,\{u\}$, then for each $\alpha \in R$, αu is also the basis of U. Therefore, we can make the basis become the unit basis which the norm is 1 by multiplying the basis vector by $\frac{1}{\|u\|}$, or$\frac{u}{\|u\|}$. However, the one dimension subspace has two unit basis, that are$\frac{u}{\|u\|}$ and $\frac{-u}{\|u\|}$. Geometrically, it can be presented in Figure 1.

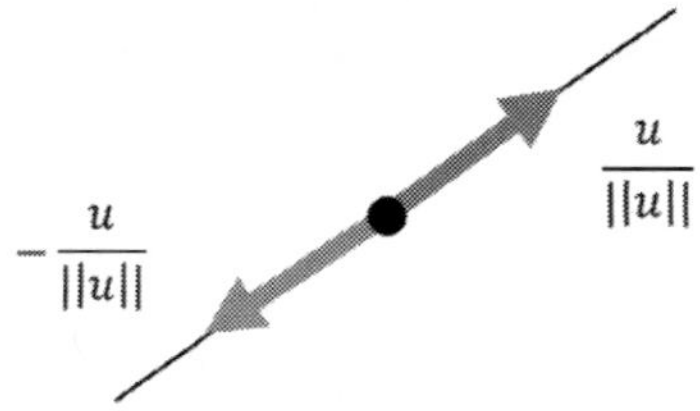

FIGURE 1. Representation of unit basis of a line.

By considering that the definition of $\|u_n - u\| \to 0$ for $n \to \infty$, it can be modified by

$$\left\|\frac{u_n}{\|u_n\|} - \frac{u}{\|u\|}\right\| \to 0$$

or

$$\left\|\frac{u_n}{\|u_n\|} + \frac{u}{\|u\|}\right\| \to 0$$

for $n \to \infty$.

However, if the result of modification is used as a definition, then the concept of convergence of line sequence is still dependent on thebasis selection.

For example, let $x = \mathbb{R}^2$, $U_n = span\{(1,0)\} = U \subset X$. If we take the base U_n for every n where $u_n = (1,0)$ and the basis of U is $(1,0)$, then sequences (HU_n) converges to U because $\left\|\frac{u_n}{\|u_n\|} - \frac{u}{\|u\|}\right\| = 0 \to 0$ for $n \to 0$. But if we take the basis of U_n is

$$u_n = \begin{cases} (1,0), & \text{if } n \text{ even} \\ (-1,0), & \text{if } n \text{ odd} \end{cases}$$

and the basis of U is $(1,0)$, then sequence (U_n) divergent to U, since both $\left\|\frac{u_n}{\|u_n\|} - \frac{u}{\|u\|}\right\|$ and $\left\|\frac{u_n}{\|u_n\|} + \frac{u}{\|u\|}\right\|$ divergent to 0 for $n \to 0$. So to resolve it will be modified again to be

$$\min\left\{\left\|\frac{u_n}{\|u_n\|} - \frac{u}{\|u\|}\right\|, \left\|\frac{u_n}{\|u_n\|} + \frac{u}{\|u\|}\right\|\right\} \to 0$$

for $n \to \infty$.

Next, Let X is a normed space. A 1-dimensional subspace of a normed space X is said a line in X, and 2-dimensional subspace of X is said a plane in X.

2. A Sequence of Lines

Let $\boldsymbol{X}$ is a normed space. Base on the introduction, we get the notion of limit of a sequence of lines in $\boldsymbol{X}$ (1-dimentional subspace of a normed space $\boldsymbol{X}$) and it will be the focus on this section.

Definition 2.1 Let (U_n) is a sequence of lines in $X$with$U_n = span\{u_n\}$ and $U = span\{u\}$ is a line in X. A sequence$(U_n)$is said to be **converge** to U, or U is is said to be a **limit** of (U_n), if and only if

$$\min\left\{\left\|\frac{u_n}{\|u_n\|} - \frac{u}{\|u\|}\right\|, \left\|\frac{u_n}{\|u_n\|} + \frac{u}{\|u\|}\right\|\right\} \to 0$$

as $n \to \infty$.

If a sequence has a limit, we say that the sequence is convergent; if it has no limit, we say that the sequence is divergent.We will sometimes use the symbolism $U_n \to U$, which indicates the intuitive idea that the lines U_n "approach" the line U as $n \to \infty$

The following theorem convinces us that our definition make sense.

Theorem 2.2. *The definition 2.1 satisfies homogeneity property. It means, The definition is dependent of the choice of bases for U and U_n for every $n \in \mathbb{N}$.*

Proof.

For every $n \in \mathbb{N}$, if u_n change with another basis of U_n, that is $\alpha_n u_n$ with $\alpha_n \in \mathbb{R}$ and u were replaced by$\beta u$with$\beta \in \mathbb{R}$, then there were two cases. If $\alpha_n \beta \geq 0$ then we find that

$$\left\|\frac{\alpha_n u_n}{\|\alpha_n u_n\|} - \frac{\beta u}{\|\beta u\|}\right\| = \left\|\frac{u_n}{\|u_n\|} - \frac{u}{\|u\|}\right\| \text{ and } \left\|\frac{\alpha_n u_n}{\|\alpha_n u_n\|} + \frac{\beta u}{\|\beta u\|}\right\| = \left\|\frac{u_n}{\|u_n\|} + \frac{u}{\|u\|}\right\|$$

On the other hand, if$\alpha_n \beta < 0$ then we get

$$\left\|\frac{\alpha_n u_n}{\|\alpha_n u_n\|} - \frac{\beta u}{\|\beta u\|}\right\| = \left\|\frac{u_n}{\|u_n\|} + \frac{u}{\|u\|}\right\| \text{ and } \left\|\frac{\alpha_n u_n}{\|\alpha_n u_n\|} + \frac{\beta u}{\|\beta u\|}\right\| = \left\|\frac{u_n}{\|u_n\|} - \frac{u}{\|u\|}\right\|.$$

Consequently,

$$\min\left\{\left\|\frac{u_n}{\|u_n\|} - \frac{u}{\|u\|}\right\|, \left\|\frac{u_n}{\|u_n\|} + \frac{u}{\|u\|}\right\|\right\} = \min\left\{\left\|\frac{\alpha_n u_n}{\|\alpha_n u_n\|} - \frac{\beta u}{\|\beta u\|}\right\|, \left\|\frac{\alpha_n u_n}{\|\alpha_n u_n\|} + \frac{\beta u}{\|\beta u\|}\right\|\right\}$$

■

Now, it will begin by establishing some basic properties of convergent sequences of lines since it will be needed in this section.

Fact 2.3. *If $U_n = span\{u\} = U$ then $U_n \to U$.*

Proof.

It is known that

$\left\|\frac{u_n}{\|u_n\|} - \frac{u}{\|u\|}\right\| = \left\|\frac{u}{\|u\|} - \frac{u}{\|u\|}\right\| = \|0\| = 0 \to 0$ as $n \to \infty$. Hence, $U_n \to U$.■

Theorem 2.4. *Let $U_n = span\{u_n\}$ and $U = span\{u\}$. If $U_n \to U$then*

$$\left\|\frac{u_n}{\|u_n\|} - \frac{u}{\|u\|}\right\| \left\|\frac{u_n}{\|u_n\|} + \frac{u}{\|u\|}\right\| \to 0$$

for $n \to \infty$.

Proof.

If $U_n \to U$ then, by Definition 2.1, it was obtained

$$\min\left\{\left\|\frac{u_n}{\|u_n\|}-\frac{u}{\|u\|}\right\|,\left\|\frac{u_n}{\|u_n\|}+\frac{u}{\|u\|}\right\|\right\}\to 0$$

as$n\to\infty$. Since$\left\|\frac{u_n}{\|u_n\|}+\frac{u}{\|u\|}\right\|\le\left\|\frac{u_n}{\|u_n\|}\right\|+\left\|\frac{u}{\|u\|}\right\|=2$ (Triangle Inequality) andalso$\left\|\frac{u_n}{\|u_n\|}-\frac{u}{\|u\|}\right\|\le$ 2, then

$$\left\|\frac{u_n}{\|u_n\|}-\frac{u}{\|u\|}\right\|\left\|\frac{u_n}{\|u_n\|}+\frac{u}{\|u\|}\right\|\le 2\min\left\{\left\|\frac{u_n}{\|u_n\|}-\frac{u}{\|u\|}\right\|,\left\|\frac{u_n}{\|u_n\|}+\frac{u}{\|u\|}\right\|\right\}^2\to 0$$

as$n\to\infty$.■

Theorem 2.5. *Let$U_n=span\{u_n\}$ and $U=span\{u\}$. $U_n\to U$if and only if there exists $a_n\in\mathbb{R},a_n\neq 0$ for every $n\in\mathbb{N}$such that*

$$\left\|\frac{u_n}{\|u_n\|}-\frac{a_nu}{\|a_nu\|}\right\|\to 0$$

as $n\to\infty$.

Proof.

($\Rightarrow$) If $U_n\to U$ then it was known that

$$\min\left\{\left\|\frac{u_n}{\|u_n\|}-\frac{u}{\|u\|}\right\|,\left\|\frac{u_n}{\|u_n\|}+\frac{u}{\|u\|}\right\|\right\}\to 0$$

as $n\to\infty$. So, There exists$a_n\in\mathbb{R},a\neq 0$with$a_n\in\{1,-1\}$such that

$$\left\|\frac{u_n}{\|u_n\|}-\frac{au}{\|au\|}\right\|\to 0$$

as $n\to\infty$.

($\Leftarrow$) Let there exists$a_n\in\mathbb{R},a_n\neq 0$ such that

$$\left\|\frac{u_n}{\|u_n\|}-\frac{a_nu}{\|a_nu\|}\right\|\to 0$$

as $n\to\infty$, then we have two cases, If$a_n>0$ then it was obtained

$$\left\|\frac{u_n}{\|u_n\|}-\frac{u}{\|u\|}\right\|=\left\|\frac{u_n}{\|u_n\|}-\frac{a_nu}{\|a_nu\|}\right\|\text{ and }\left\|\frac{u_n}{\|u_n\|}+\frac{u}{\|u\|}\right\|=\left\|\frac{u_n}{\|u_n\|}+\frac{a_nu}{\|a_nu\|}\right\|,$$

if$a_n<0$ then it was known that

$$\left\|\frac{u_n}{\|u_n\|}-\frac{u}{\|u\|}\right\|=\left\|\frac{u_n}{\|u_n\|}+\frac{a_nu}{\|a_nu\|}\right\|\text{ dan }\left\|\frac{u_n}{\|u_n\|}+\frac{u}{\|u\|}\right\|=\left\|\frac{u_n}{\|u_n\|}-\frac{a_nu}{\|a_nu\|}\right\|.$$

Finally,

$$\min\left\{\left\|\frac{u_n}{\|u_n\|}-\frac{u}{\|u\|}\right\|,\left\|\frac{u_n}{\|u_n\|}+\frac{u}{\|u\|}\right\|\right\}=\min\left\{\left\|\frac{u_n}{\|u_n\|}-\frac{a_nu}{\|a_nu\|}\right\|,\left\|\frac{u_n}{\|u_n\|}+\frac{a_nu}{\|a_nu\|}\right\|\right\}\to 0$$

as $n\to\infty$. So, $U_n\to\infty$. ■

Theorem 2.6. (Uniqueness of Limit) *Let$U_n=span\{u_n\}$ and $U=span\{u\}$. If $U_n\to U$and $U_n\to V$ then $U=V$.*

Proof.

Let $U_n\to U$. By applying Theorem 2.5, then we can find $a_n\in\mathbb{R},a_n\neq 0$ for every $n\in\mathbb{N}$ such that

$$\left\|\frac{u_n}{\|u_n\|}-\frac{a_nu}{\|a_nu\|}\right\|\to 0$$

as $n\to\infty$. Since $U_n\to V$then according to Theorem 2.5, there exists $b_n\in\mathbb{R},a_n\neq 0$ such that

$$\left\|\frac{u_n}{\|u_n\|}-\frac{b_nv}{\|b_nv\|}\right\|\to 0$$

as $n\to\infty$.

Next, by using Triangle Inequality, it was obtained that

$$\left\|\frac{a_nu}{\|a_nu\|}-\frac{b_nv}{\|b_nv\|}\right\|\le\left\|\frac{u_n}{\|u_n\|}-\frac{a_nu}{\|a_nu\|}\right\|+\left\|\frac{u_n}{\|u_n\|}-\frac{b_nv}{\|b_nv\|}\right\|\to 0$$

as $n\to\infty$.

On the other hand, it was known that

$$\left\|\frac{a_n u_n}{\|a_n u_n\|}-\frac{b_n v}{\|b_n v\|}\right\|=\left\|\frac{a_n u_n}{|a_n|\|u_n\|}-\frac{b_n v}{|b_n|\|v\|}\right\|=\left\|\frac{u_n}{\|u_n\|}-\frac{|a_n| b_n v}{|b_n| a_n\|v\|}\right\|=\left\|\frac{u}{\|u\|}-\frac{(a_n b_n) v}{\|(a_n b_n) v\|}\right\|$$
$$=\left\|\frac{u}{\|u\|}\pm\frac{v}{\|v\|}\right\|.$$

Therefore, $\left\|\frac{u}{\|u\|}-\frac{v}{\|v\|}\right\|\rightarrow 0$or $\left\|\frac{u}{\|u\|}+\frac{v}{\|v\|}\right\|\rightarrow 0$. Hence, $\left\|\frac{u}{\|u\|}-\frac{v}{\|v\|}\right\|=0or\left\|\frac{u}{\|u\|}+\frac{v}{\|v\|}\right\|=0$. It means that $u=cv$ for some $c\in\mathbb{R}, c\neq 0$. So, $U=span\{u\}=span\{cv\}=span\{v\}=V$. ■

Theorem 2.7. *Let $U_n=span\{u_n\}$ and $U=span\{u\}$. $U_n\rightarrow U$if and only if for every $a\in\mathbb{R}, a_n\neq 0$, there exists $a\in\mathbb{R}, a_n\neq 0$ such that*

$$\left\|\frac{a u_n}{\|a u_n\|}-\frac{a_n u}{\|a_n u\|}\right\|\rightarrow 0$$

as $n\rightarrow\infty$.

Proof.

($\Rightarrow$) If $U_n\rightarrow U$ then by using Theorem 2.5, there were existed $a_n\in\mathbb{R}, a_n\neq 0$, such that

$$\left\|\frac{u_n}{\|u_n\|}-\frac{a_n u}{\|a_n u\|}\right\|\rightarrow 0$$

as $n\rightarrow\infty$. It meant that for all $a\in\mathbb{R}, a\neq 0$, if$a>0$ then

$$\left\|\frac{a u_n}{\|a u_n\|}-\frac{a_n u}{\|a_n u\|}\right\|=\left\|\frac{u_n}{\|u_n\|}-\frac{a_n u}{\|a_n u\|}\right\|\rightarrow 0,$$

If $a<0$, then there were exists $b_n\in\mathbb{R}, b_n\neq 0$, that is$b_n=-a_n$, such that

$$\left\|\frac{a u_n}{\|a u_n\|}-\frac{b_n u}{\|b_n u\|}\right\|=\left\|-\frac{u_n}{\|u_n\|}-\frac{-a_n u}{\|-a_n u\|}\right\|=\left\|\frac{u_n}{\|u_n\|}-\frac{a_n u}{\|a_n u\|}\right\|\rightarrow 0$$

as $n\rightarrow\infty$.

($\Leftarrow$) Let for every $a\in\mathbb{R}, a\neq 0$ there were exists $a_n\in\mathbb{R}, a_n\neq 0$ such that

$$\left\|\frac{a u_n}{\|a u_n\|}-\frac{a_n u}{\|a_n u\|}\right\|\rightarrow 0$$

as $n\rightarrow\infty$. There were two cases, if$a>0$ then it was obtained

$$\left\|\frac{u_n}{\|u_n\|}-\frac{a_n u}{\|a_n u\|}\right\|=\left\|\frac{a u_n}{\|a u_n\|}-\frac{a_n u}{\|a_n u\|}\right\|\rightarrow 0,$$

or if$a<0$ then it was obtained that

$$\left\|\frac{u_n}{\|u_n\|}-\frac{-a_n u}{\|-a_n u\|}\right\|=\left\|\frac{u_n}{\|u_n\|}+\frac{a_n u}{\|a_n u\|}\right\|=\left\|\frac{a u_n}{\|a u_n\|}-\frac{a_n u}{\|a_n u\|}\right\|\rightarrow 0,$$

as $n\rightarrow\infty$. So, by applying Theorem 2.5, $U_n\rightarrow U$. ■

3. Relationship to Angle in an Inner Product Space

Let $(X,\langle\cdot,\cdot\rangle)$ is an inner product spaces. Inspired by the notion of limit of a sequence of lines in X, it could be described the continuity property of angle between two lines.

Theorem 3.1. (Continuity Property) *Let $U_n, V_n, U$and Vare1-dimentional subspaces of a normed space $X$and $\theta(U,V)$ is an angle between $U$and V. If $U_n\rightarrow U$ and $V_n\rightarrow V$then $\theta(U_n,V_n)\rightarrow\theta(U,V)$.*

Proof.

Since $U_n\rightarrow U$ and $V_n\rightarrow V$ then by applying Theorem 2.5, it was obtained and found $a_n, b_n\in\mathbb{R}$, $a_n\neq 0$, and$b_n\neq 0$ for every $n\in\mathbb{N}$ such that

$$\left\|\frac{u_n}{\|u_n\|}-\frac{a_n u}{\|a_n u\|}\right\|\rightarrow 0\text{and}\left\|\frac{b_n v}{\|b_n v\|}-\frac{v_n}{\|v_n\|}\right\|\rightarrow 0 \text{ as } n\rightarrow\infty.$$

It was known that

$$\left|\left\|\frac{u_n}{\|u_n\|}-\frac{v_n}{\|v_n\|}\right\|-\left\|\frac{a_n u}{\|a_n u\|}-\frac{b_n v}{\|b_n v\|}\right\|\right|\leq\left\|\frac{u_n}{\|u_n\|}-\frac{a_n u}{\|a_n u\|}+\frac{b_n v}{\|b_n v\|}-\frac{v_n}{\|v_n\|}\right\|$$
$$\leq\left\|\frac{u_n}{\|u_n\|}-\frac{a_n u}{\|a_n u\|}\right\|+\left\|\frac{b_n v}{\|b_n v\|}-\frac{v_n}{\|v_n\|}\right\|\rightarrow 0$$

as $n \to \infty$. Therefore, $\left\|\frac{u_n}{\|u_n\|} - \frac{v_n}{\|v_n\|}\right\| \to \left\|\frac{a_n u}{\|a_n u\|} - \frac{b_n v}{\|b_n v\|}\right\|$ as $n \to \infty$. Hence, by the definition of angle between two subspaces of a normed space ([17]), it was obtained that

$$\theta(U_n, V_n) = \arccos\left|\frac{\langle u_n, v_n\rangle}{\|u_n\|\|v_n\|}\right| = \arccos\left|\frac{1}{2}\left(2 - \left\|\frac{u_n}{\|u_n\|} - \frac{v_n}{\|v_n\|}\right\|^2\right)\right|$$
$$\to \arccos\left|\frac{1}{2}\left(2 - \left\|\frac{a_n u}{\|a_n u\|} - \frac{b_n v}{\|b_n v\|}\right\|^2\right)\right| = \arccos\left|\frac{\langle a_n u, b_n v\rangle}{\|a_n u\|\|b_n v\|}\right|$$
$$= \arccos\left|\frac{\langle u, v\rangle}{\|u\|\|v\|}\right| = \theta(U, V)$$

as $n \to \infty$. ■

Relationship between the notion of limit of a sequence of lines in X and angle between vectors in X ([18] and [19]) is described in the following theorem.

Theorem 3.2. *Let $U_n = span\ \{u_n\}$ and $U = span\ \{u\}$ are1-dimentional subspaces of a normed space Xand $\theta_{Thy}(u, v)$ is an angle between two vector u and v. $U_n \to U$ if and only if $|\cos\theta_{Thy}(u_n, u)| \to 1$ as $n \to \infty$.*

Proof.

In [18] and [19], angle between u_n *and* u, $\theta_{Thy}(u_n, u)$, is defined as

$$\cos\theta_{Thy}(u_n, u) = \frac{1}{4}\left[\left\|\frac{u_n}{\|u_n\|} + \frac{u}{\|u\|}\right\|^2 - \left\|\frac{u_n}{\|u_n\|} - \frac{u}{\|u\|}\right\|^2\right].$$

In inner product spaces X, for $n \to \infty$, if $\left\|\frac{u_n}{\|u_n\|} + \frac{u}{\|u\|}\right\|^2 \to 0$ then $\left\|\frac{u_n}{\|u_n\|} - \frac{u}{\|u\|}\right\|^2 \to 4$, and if $\left\|\frac{u_n}{\|u_n\|} - \frac{u}{\|u\|}\right\|^2 \to 0$ then $\left\|\frac{u_n}{\|u_n\|} + \frac{u}{\|u\|}\right\|^2 \to 4$. Thus,

$$\min\left\{\left\|\frac{u_n}{\|u_n\|} - \frac{u}{\|u\|}\right\|, \left\|\frac{u_n}{\|u_n\|} + \frac{u}{\|u\|}\right\|\right\} \to 0 \leftrightarrow \frac{1}{4}\left[\left\|\frac{u_n}{\|u_n\|} + \frac{u}{\|u\|}\right\|^2 - \left\|\frac{u_n}{\|u_n\|} - \frac{u}{\|u\|}\right\|^2\right] \to 1. \blacksquare$$

4. A Sequence of Planes

Theorem 2.7 was used to get the notion of limit of a sequence of planes in X (2-dimentional subspace of a normed space X).

Definition 4.1. Let (U_n) is a sequence of planes in X with $U_n = span\{u_{1_n}, u_{2_n}\}$ and $V = span\{v_1, v_2\}$is a plane in X. A sequence(U_n) is said to be converge to V if and only if for every, $b \in \mathbb{R}$, there exists $a_n, b_n \in \mathbb{R}$such that

$$\left\|\frac{(au_n + bv_n)}{\|au_n + bv_n\|} - \frac{(a_n u + b_n v)}{\|a_n u + b_n v\|}\right\| \to 0$$

As $n \to \infty$.

Note that $a, b \in \mathbb{R}$ in Definition 4.1 are arbitrary. It means that our definition is independent of the choice of basis for U_n and V. So, our definition make sense. Also, if $U_n$and V in Definition 4.1 are a 1-dimensional subspace of X, then the definition is equivalent with Theorem 2.7.

The following theorem describe a relation between the convergent of a sequence of lines in X and the convergent of a sequence of planes in X.

Theorem 4.2. Let $U_n = span\{u_n\}$, $V_n = span\{v_n\}$, $U = span\{u\}$, $V = span\{v\}$with u and v are linearly independent, and $B_n = span\{u_n, v_n\}$ and $B = span\{u, v\}$. If $U_n \to U$and $V_n \to V$ then $B_n \to B$.

Proof:

It was easy to prove this theorem by using Theorem 2.7, Definition 4.1, and Triangle Inequality.■

Remark 4.3. As the reader, it might have realized that the Definition 2.9 may also be generalized to define the limit of a sequence of m-dimentional subspace of a normed space. Let X is a normed space with $\dim X > m$. Let (V_n) is a sequence of m-dimentional subspaces of X with $V_n = span\{v_{1_n}, v_{2_n}, \ldots, v_{m_n}\}$ and $V = span\{v_1, v_2, \ldots, v_m\}$ is an m-dimentional subspace of X with$1_n, 2_n, \ldots, m_n \in \mathbb{N}$. A sequence$(V_n)$ is said to be converge to V if and only if for every $a_1, a_2, \ldots, a_m \in \mathbb{R}$there exists$b_{1_n}, b_{2_n}, \ldots, b_{3_n} \in \mathbb{R}$ such that

$$\left\| \frac{\left(\sum_{i=1}^{m} a_i u_{i_n}\right)}{\left\|\sum_{i=1}^{m} a_i u_{i_n}\right\|} - \frac{\left(\sum_{j=1}^{m} b_{j_n} v_j\right)}{\left\|\sum_{j=1}^{m} b_{j_n} v_j\right\|} \right\| \to 0$$

as $n \to \infty$.

Note that $a_1, a_2, \ldots, a_m$ in Definition 4.3 were arbitrary. It meant that the definition was also invariant under any change of basis for V_n and V. So, the definition made sense.

5. References

[1] Kreyszig E 1978 Introductory Functoinal Analysis with Applications (New York: John Wiley and Son)Komlos J 1972 *Every Sequence Converging to O Weakly in L_2 Contain An Unconditional Convergence Sequence* (Budapes: Publishing House of The Hunggarian Academi of Sci.)

[2] Davey B A and Priestly H A 2002 *Introduction to Lattices and Order* (Cambridge: Cambridge University Press)

[3] Voit J 2003 The Statistical Mechanics of Financial Markets *Springer-Verlag* **124**

[4] Isakov V 2006 SciencesInverse Problems for Partial Differential Equations (Second Edition *American Institute of Mathematical***127**

[5] Altman M 1957 A Fixed Point Theorem in Bannach Space*Bul. Acad. Polon. Sci.n Cl. III.***5** 19-22

[6] Alexits 1961 *Convergence Problems of Orthogonal Series* (Budapes: Publishing House of The Hunggarian Academi of Sci.)

[7] Powell 1976 Some Global convergence Properties of a Variable Metric Algorithm for Minimization without Exact Line Searches*SIAM-AMS Proceeding* **9**

[8] Parashar S D and Choudhary B 1994 Sequence Spaces Defined By Orlicz Functions*Indian J. Pure Appl. Math***25** (4) 419-428

[9] Quarteroni A, Sacco R, and Saleri F 2007 *Numerical Mathematics* (New York:Springer Berlin Heidelberg)

[10] Manuharawati and Yunianti D 2013a *Integral Henstock di dalam Ruang Metrik Kompak Lokal dari Fungsi Bernilai Vektor (*Surabaya: Laporan Penelitian, LPPM Unesa Surabaya).

[11] Manuharawati and Yunianti D 2013b Kriteria Chaucy dari Suatu Fungsi bernilai Vektor Pada Sel di Dalam Ruang Metrik Kompak Lokal*Prosiding Seminar Nasional Matematika dan Aplikasinya, Unair*

[12] Jakfar M 2018 Ruang Fungsi Terintegralkan p Berbobot *kubik* **3**(1) 60-67

[13] Jakfar M, Gunawan H, and Idris M 2015 *Hasil Kali Dalam Berbobot pada Ruang $L^p(X)$, Prosiding Semnastika 2015 Seminar Nasional Matematika dan Pendidikan Matematika* 298-304 (Surabaya: Unipress Unesa)

[14] Diminnie C R, Andalafte E Z and Freese R W 1988 Generalized Angles and a Characterization of Inner Product Spaces *Houston J. Math.***14** 475-480

[15] Gunawan H, Lindiarni J, Neswan O 2008 P-, I-, g-, and D-Angles in Normed Spaces*ITB J. Sci.***40 A** 24-32

[16] Risteski I B and Trenčevski K G 2001 Principal values and principal subspaces of two subspaces of vector spaces with inner product *Beitr. Algebra Geom* **42** 289–300

[17] Gunawan H, Neswan O and Setya-Budhi W 2005 A Formula for Angles between Subspaces of Inner Product Spaces*Beitr¨age zur Algebra und Geometrie Contributions to Algebra and Geometry***46** 311-320

[18] Thürey V 2009 Angles and polar coordinates in real normed spaces *arXiv:*0902.2731v2.

[19] Manuharawati and Jakfar M 2017 Thy-angle in an n-Normed *Space Far East Journal of Mathematical Sciences (FJMS)* **102**(5) 979-994

IOP Conf. Series: Journal of Physics: Conf. Series **1108** (2018) 012086 doi:10.1088/1742-6596/1108/1/012086

Delay's propagation of *citrus tristeza virus* on citrus plant

D A Zulfiyana[1], Y Fuad[1*] and B P Prawoto[1]

[1]Department of Mathematics, Universitas Negeri Surabaya, Indonesia
[*]Corresponding Author, Email: yusuffuad@unesa.ac.id

Abstract. Citrus plant is one of the biggest production fruit plants in the world as well as in Indonesia. Citrus plant can be infected with a disease by virus, such as *Citrus Tristeza Virus* (CTV) propagated by *Toxoptera citricida*. Based on the well-known SIR-SI model, it was derived the CTV propagation model of the citrus plant. The obtained model involves citrus plant which consists of subpopulation susceptible, infected, and recovered. While the vector's population were subpopulation susceptible and infected insects. Then two equilibrium points were determined for the model with and without delay, i.e. the disease–free equilibrium and the endemic equilibrium points. Hence the obtained model, which were analysed around two equilibrium points with and without delay, were asymptotic stable. Moreover, the basic reproduction number (ρ) was determined for obtaining the parameter value m. The numerical simulation with and without delay were demonstrated and the responses of the model were asymptotic stable. By applying time delays, it is concluded that the spread of CTV on citrus plant can be inhibited by accelerating the harvest before it is infected to the virus, so that citrus fruits can be harvested more better than without giving any delay.

1. Introduction

Plants have a very important role for life as a producer of oxygen, producer of food and the fulfillment of all kinds of needs of living things. One of the largest fruit production plants in the world is citrus. In 2014 citrus plantation production occupies the position of the soil with a total of 70.86 million tons [1]. Citrus plants may develop viral illnesses, such as Citrus Tristeza (CTV) [2].

Healthy citrus plants can be infected by the virus because it is infested by insects which previously infected citrus plants that have been infected. One of the insects that spread the virus is aphids (Toxoptera citricida) [3]. Toxoptera citricida proliferates depending on the temperature between $20° C - 30° C$ and the capacity increases to 0.4 when the temperature is $25° C$. Female Toxoptera citricida can produce a population of over 4,400 eggs within 3 weeks without any natural predators [4].

It has been interested to study the model of the spread of plant virus that describes the interaction between citrus plants and virus-carrying insects, and one was inspired by an article [5]. The purpose of this study is to discuss CTV dispersion model with and without delay, by assuming the virus would be transmitted by contact between citrus and insect plants using Holling Type II. Then the study demonstrates also the stability and simulations of CTV deployment model in citrus plants with and without delay. It is necessary to obtain an alternative to control the spread of the virus in citrus plants. Such alternatives can be taken into consideration for appropriate action in handling any problem posed by the CTV deployment.

2. Mathematical Modelling for The Propagation of CTV on Citrus Plant

Epidemic model is utilized to determine the spread of a disease in certain areas. The SIR epidemic model was first introduced by [6]. The mathematical model was applied in the spread of the virus in citrus plants involves the population of citrus plant K and insect population N. Based on the well-known SIR-SI model, it was derived the CTV propagation model of the citrus plant. The obtained model involved citrus plant population which consists of subpopulation susceptible $S(t)$, infected $I(t)$, and recovered $R(t)$, respectively. While the vector's population were subpopulation susceptible insect $S_X(t)$ and infected insect $I_Y(t)$, respectively. The interactions between citrus plants and insects influence the spread of viruses in citrus and insects. These interactions implement Holling Type II through two ways, namely virus transmission from insects to citrus plants and transmission of virus from citrus plants to insects. Thus, a compartment diagram of the propagation of CTV on citrus plants is depicted as follows:

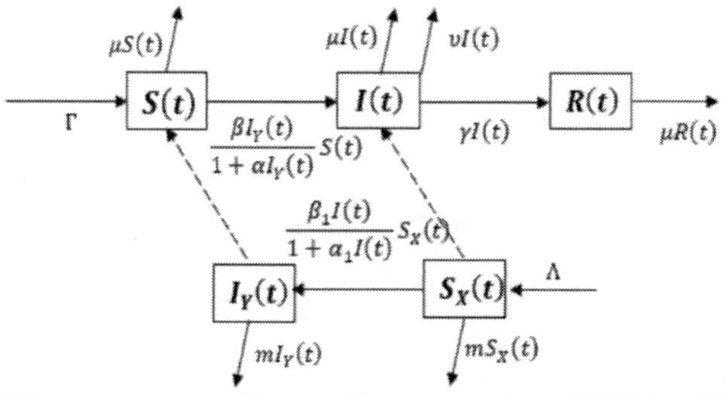

Figure 1. The compartment diagram for propagation of CTV on Citrus Plant

The mathematical model of the diagram, for propagation of CTV on citrus plant without any delay, is stated as follows:

$$
\begin{aligned}
\frac{dS(t)}{dt} &= \Gamma - \mu S(t) - \frac{\beta I_Y(t)}{1+\alpha I_Y(t)} S(t) \\
\frac{dI(t)}{dt} &= \frac{\beta I_Y(t)}{1+\alpha I_Y(t)} S(t) - \upsilon I(t) - \mu I(t) - \gamma I(t) \\
\frac{dR(t)}{dt} &= \gamma I(t) - \mu R(t) \\
\frac{dS_X(t)}{dt} &= \Lambda - \frac{\beta_1 I(t)}{1+\alpha_1 I(t)} S_X(t) - mS_X(t) \\
\frac{dI_Y(t)}{dt} &= \frac{\beta_1 I(t)}{1+\alpha_1 I(t)} S_X(t) - mI_Y(t)
\end{aligned} \tag{1}
$$

Meanwhile, the propagation of CTV on citrus plant is relating to some delays. Assuming that delay time is only occured on susceptible citrus plants, infected and infected insects because with these delays the occured CTV may inhibit its spreading. The model is formulated as follows:

$$
\begin{aligned}
\frac{dS(t)}{dt} &= \Gamma - \mu S(t) - \frac{\beta I_Y(t-\tau_1)}{1+\alpha I_Y(t-\tau_1)} S(t-\tau_1) \\
\frac{dI(t)}{dt} &= \frac{\beta I_Y(t-\tau_1)}{1+\alpha I_Y(t-\tau_1)} S(t-\tau_1) - \omega I(t) \\
\frac{dR(t)}{dt} &= \gamma I(t) - \mu R(t) \\
\frac{dS_X(t)}{dt} &= \Lambda - \frac{\beta_1 I(t)}{1+\alpha_1 I(t)} S_X(t) - mS_X(t) \\
\frac{dI_Y(t)}{dt} &= \frac{\beta_1 I(t-\tau_2)}{1+\alpha_1 I(t-\tau_2)} \left(\frac{\Lambda}{m} - I_Y(t-\tau_2)\right) - mI_Y(t)
\end{aligned} \tag{2}
$$

where, $\Gamma = \mu K + \upsilon I(t)$; $\Lambda = mX + mY$; K = total population of citrus plants; N = total population of insects; $S(t)$ = population of susceptible citrus plant; $I(t)$ = population of citrus plants infected by CTV; $R(t)$ = population of recovered citrus plant; $S_X(t)$= population of susceptible insect; $I_Y(t)$= population of insects infected by CTV; β = infection rate of citrus plants caused by insects; β_1 rate of insect infections caused by citrus plants; α = constant saturation of citrus plants caused by insects; α_1 = constant saturation of insects caused by citrus plants; μ = natural mortality rate in citrus plant; m = natural rate of death on insects; γ = recovered rate of citrus plant; Λ = rate of population increase of insects (birth or immigration; Γ= rate of population growth of citrus plants; and υ = death rate of infected citrus plants due to the virus.

2.1 Stability of Equilibrium

Definition 1.

Given first-order differential equations $\dot{x} = f(x,t)$, and $x(t,x_0)$ are system solutions with initial values $x(0) = x_0$. A vector $\bar{\boldsymbol{x}}$ that satisfies $\boldsymbol{f}(\bar{\boldsymbol{x}}) = 0$ is called the equilibrium point. The equilibrium point $\bar{\boldsymbol{x}}$ is stable if for $\varepsilon > 0$ there is $\delta > 0$, so for $\|x_0 - \bar{\boldsymbol{x}}\| < \delta$ apply $\|x(t,x_0) - \bar{\boldsymbol{x}}\| < \varepsilon$, for $t > 0$. An equilibrium point $\bar{\boldsymbol{x}}$ is said to be asymptotically stable if $\bar{\boldsymbol{x}}$ is stable and there is $\delta_1 > 0$, so for $\|x_0 - \bar{\boldsymbol{x}}\| < \delta_1$ imply $\lim_{t\to\infty}\|x_0 - \bar{\boldsymbol{x}}\| = 0$. In addition the stability of the equilibrium point of the system can be determined by considering the real part of eigen values of the characteristic equation [7].

3. Result and Discussion

3.1 The Analysis of Equilibrium Point

The parameter values used in the propagation of CTV on citrus plants are $(K; N; \beta; \beta_1; \alpha; a_1; \mu; \gamma; \Lambda; \upsilon) = (1000; 40; 0{,}001; 0{,}001; 0{,}2; 0{,}1; 0{,}01; 0{,}01{,}20, 0{,}2)$. The analysis of equilibrium point of the propagation of CTV on citrus plant is by making the right-hand side of each equation (2) equal to zero. Thus, the equilibrium point is $\tilde{E} = (\tilde{S}, \tilde{I}, \tilde{R}, \tilde{S_X}, \tilde{I_Y})$, where

$$\tilde{S} = \frac{\Gamma(1+\alpha\tilde{I_Y})}{\mu(1+\alpha\tilde{I_Y})-\beta\tilde{I_Y}}\ ; \tilde{I} = 0 \text{ atau } \tilde{I} = \frac{\beta\beta_1\tilde{S_X}-\omega}{m\omega\alpha_1+\alpha\beta_1\omega\tilde{S_X}}; \tilde{R} = \frac{\gamma\tilde{I}}{\mu}$$

$$\tilde{S_X} = \frac{\Lambda}{\left(\frac{\beta_1\tilde{I}}{1+\alpha_1\tilde{I}}+m\right)}; \tilde{I_Y} = \frac{\beta_1 I(t)}{m(1+\alpha_1 I(t))}\tilde{S_X}$$

By assuming two conditions, namely $\tilde{I} = 0$ or $\tilde{I} = \frac{\beta\beta_1\tilde{S_X}-\omega}{m\omega\alpha_1+\alpha\beta_1\omega\tilde{S_X}}$, two equilibrium points are obtained as follows.

1. The disease–free equilibrium point is $E^0 = (S^0, I^0, R^0, S_X^0, I_Y^0)$.
   Where $S^0 = \frac{\Gamma(1+\alpha\tilde{I_Y})}{\mu(1+\alpha\tilde{I_Y})-\beta\tilde{I_Y}}\ ; I^0 = 0; R^0 = \frac{\gamma I^0}{\mu};\ S_X^0 = \frac{\Lambda}{\left(\frac{\beta_1 I^0}{1+\alpha_1 I^0}+m\right)}\ ; I_Y^0 = \frac{\beta_1 I^0 S_X^0}{m(1+\alpha_1 I^0)}$.
   Note that $\Gamma = \mu K + \upsilon I(t)$. Substituting all parameter values, it is obtained that
   $$E^0 = (S^0;\ I^0;\ R^0;\ S_X^0;\ I_Y^0) = (K; 0; 0; N; 0)$$
2. Assuming $S \neq 0, I \neq 0, R \neq 0, S_X \neq 0, I_Y \neq 0$. The endemic equilibrium point is
   $$E^* = (S^*;\ I^*;\ R^*;\ S_X^*;\ I_Y^*)$$
   Where $S^* = \frac{\Gamma(1+\alpha I_Y^*)}{\mu(1+\alpha I_Y^*)-\beta I_Y^*}; I^* = \frac{\beta\beta_1 S_X^*-\omega}{m\omega\alpha_1+\alpha\beta_1\omega S_X^*}; R^* = \frac{\gamma I^*}{\mu}; S_X^* = \frac{\Lambda}{\left(\frac{\beta_1 I^*}{1+\alpha_1 I^*}+m\right)}\ ; I_Y^* = \frac{\beta_1 I^*}{m(1+\alpha_1 I^*)}S_X^*$.
   Obviously, the endemic equilibrium point depends on I^*, and it is assumed that $I^* = 2{,}66$. Substituting the parameter values, then the endemic equilibrium point is $E^* = (S^*, I^*, R^*, S_X^*, I_Y^*) = (994{,}68; 2{,}66; 2{,}66; 39{,}23; 0{,}33)$.

3.2 Basic Reproduction Number

A basic reproduction numbers (ρ) is used to determine the extent of disease spread. The value ρ is determined by using the next generation matrix F. The matrix F consists of a matrix consisting of $\mathcal{F}_i$ indicating a new infection rate in different compartments and V consisting of $\mathcal{V}_i$ showing the rate of displacement from one compartment to another.

$$F = \begin{pmatrix} 0 & \frac{\beta S^0}{1+\alpha Y^0} \\ \frac{\beta_1\Lambda}{m(1+\alpha_1 I^0)} & 0 \end{pmatrix}, V = \begin{pmatrix} \omega & 0 \\ 0 & m \end{pmatrix}$$

The basic reproduction numbers is as follows: $\rho = \sqrt{\frac{\beta\beta_1\Lambda K}{m^2\omega}}$

Substituting the parameter values to the basic reproduction number, then it is obtained that if $\rho < 1$, then the virus does not spread for $m > 0.30$. Otherwise, an endemic state will occur whenever $m < 0.30$.

3.3 Linearization Model Without Delays

1. The Jacobian matrix of the linier model around disease free equilibrium point is

$$\mathcal{J} = \begin{pmatrix} -\mu - \frac{\beta I_Y^0}{1+\alpha I_Y^0} & \upsilon & 0 & 0 & -\frac{\beta S^0}{(1+\alpha I_Y^0)^2} \\ \frac{\beta I_Y^0}{1+\alpha I_Y^0} & -\omega & 0 & 0 & \frac{\beta S^0}{(1+\alpha I_Y^0)^2} \\ 0 & \gamma & -\mu & 0 & 0 \\ 0 & -\frac{\beta_1 S_X^0}{(1+\alpha_1 I^0)^2} & 0 & -\frac{\beta_1 I^0}{1+\alpha_1 I^0} - m & 0 \\ 0 & \frac{\beta_1 S_X^0}{(1+\alpha_1 I^0)^2} & 0 & \frac{\beta_1 I^0}{1+\alpha_1 I^0} & -m \end{pmatrix} \quad (3)$$

The linear model around free-disease equilibrium point $E^0 = (S^0;\ I^0;\ R^0;\ S_X^0; I_Y^0)$ is therefore given as follows:

$$\begin{aligned} \frac{dS}{dt} &= \left(-\mu - \frac{\beta I_Y^0}{1+\alpha I_Y^0}\right) S + \upsilon I - \frac{\beta S^0}{(1+\alpha I_Y^0)^2} I_Y \\ \frac{dI}{dt} &= \frac{\beta I_Y^0}{1+\alpha I_Y^0} S - \omega I + \frac{\beta S^0}{(1+\alpha I_Y^0)^2} I_Y \\ \frac{dR}{dt} &= \gamma I - \mu R \\ \frac{dS_X}{dt} &= \left(-\frac{\beta_1 I_Y^0}{(1+\alpha_1 I^0)^2}\right) I - \left(\frac{\beta_1 I^0}{1+\alpha_1 I^0} + m\right) S_X \\ \frac{dI_Y}{dt} &= \frac{\beta_1 I_Y^0}{(1+\alpha_1 I^0)^2} I + \frac{\beta_1 I^0}{1+\alpha_1 I^0} S_X - m I_Y \end{aligned} \quad (4)$$

2. The Jacobian matrix of the linier model around endemic equilibrium point is

$$\mathcal{J} = \begin{pmatrix} -\mu - \frac{\beta I_Y^*}{1+\alpha I_Y^*} & \upsilon & 0 & 0 & -\frac{\beta S^*}{(1+\alpha I_Y^*)^2} \\ \frac{\beta I_Y^*}{1+\alpha I_Y^*} & -\omega & 0 & 0 & \frac{\beta S^*}{(1+\alpha I_Y^*)^2} \\ 0 & \gamma & -\mu & 0 & 0 \\ 0 & -\frac{\beta_1 S_X^*}{(1+\alpha_1 I^*)^2} & 0 & -\frac{\beta_1 I^*}{1+\alpha_1 I^*} - m & 0 \\ 0 & \frac{\beta_1 S_X^*}{(1+\alpha_1 I^*)^2} & 0 & \frac{\beta_1 I^*}{1+\alpha_1 I^*} & -m \end{pmatrix} \quad (5)$$

The linear model around the equilibrium endemic point $E^* = (S^*;\ I^*;\ R^*;\ S_X^*;\ I_Y^*)$ is given as follows:

$$\begin{aligned} \frac{dS}{dt} &= \left(-\mu - \frac{\beta I_Y^*}{1+\alpha I_Y^*}\right) S + \upsilon I - \frac{\beta S^*}{(1+\alpha I_Y^*)^2} I_Y \\ \frac{dI}{dt} &= \frac{\beta I_Y^*}{1+\alpha I_Y^*} S - \omega I + \frac{\beta S^*}{(1+\alpha I_Y^*)^2} I_Y \\ \frac{dR}{dt} &= \gamma I - \mu R \\ \frac{dS_X}{dt} &= \left(-\frac{\beta_1 S_X^*}{(1+\alpha_1 I^*)^2}\right) I - \left(\frac{\beta_1 I^*}{1+\alpha_1 I^*} + m\right) S_X \\ \frac{dI_Y}{dt} &= \frac{\beta_1 S_X^*}{(1+\alpha_1 I^*)^2} I + \frac{\beta_1 I^*}{1+\alpha_1 I^*} S_X - m I_Y \end{aligned} \quad (6)$$

3.4 Simulation the model with and without delays

Consider the equation (1), our numerically simulations are demonstrated with two different values $m = 0{,}5$ and $m = 0.25$. Furthermore, the simulations are shown as follows.

1. Simulation of the model, with and without delays, around the disease–free equilibrium point.

Consider the disease-free equilibrium point $E^0 = (1000; 0; 0; 40; 0)$ with $K = 1000$, $N = 40$, and the initial value $(945; 55; 0; 35; 5)$, $\tau_1 = 40$, $\tau_2 = 18$ and $m = 0{,}5$. The responses are given as follows in figure 2.

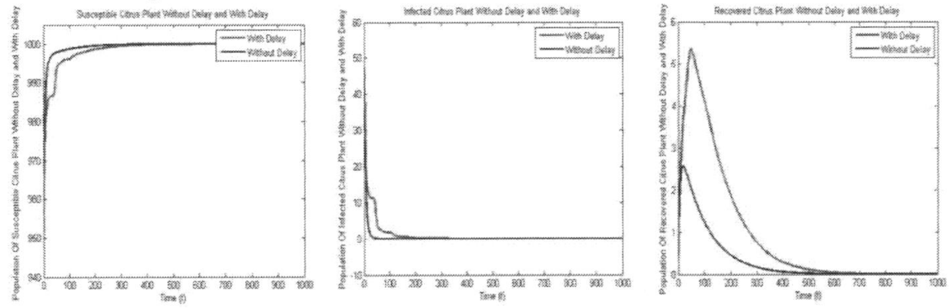

(a) (b) (c)

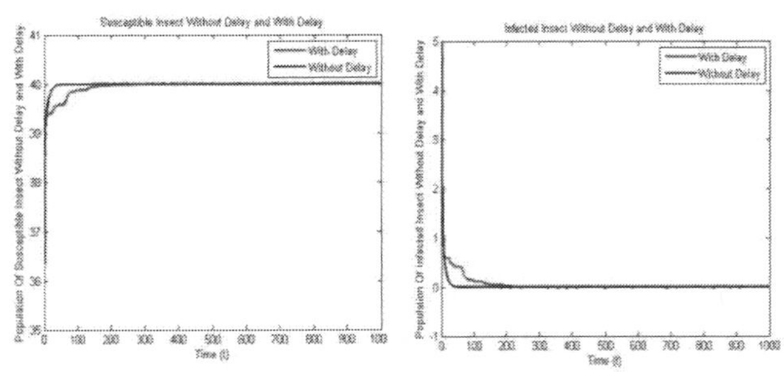

(d) (e)

Figure 2. (a) Susceptible Citrus Plants, (b) Infected Citrus Plants, (c) Infected Citrus Plants, (d) Susceptible Insects, and (e) Insects Infected with $m = 0.5$

2. Simulation of the model, with and without delays, around the endemic equilibrium point.
 Consider the endemic equilibrium point $E^* = (994{,}68; 2{,}66; 2{,}66; 39{,}23; 0{,}33)$ with $K = 1000$, $N = 40$, the initial value $(945; 55; 0; 35; 5)$, $\tau_1 = 40$, $\tau_2 = 18$ and $m = 0{,}25$. The responses of the model are given as follows.

(a) (b) (c)

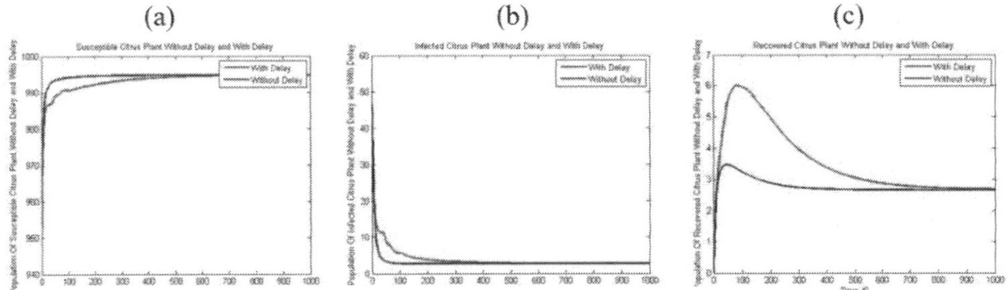

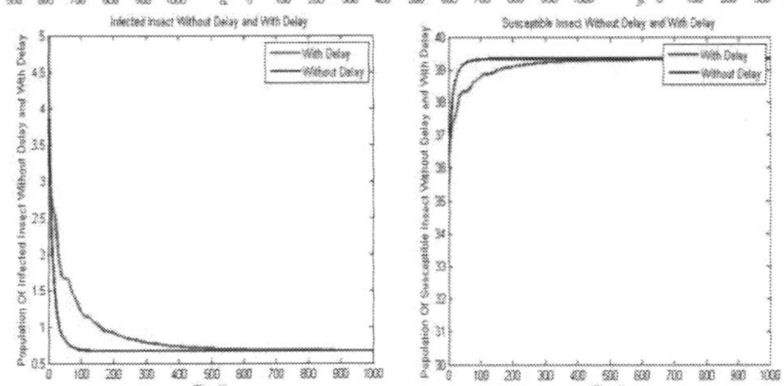

(d) (e)

Figure 3. (a) Susceptible Citrus Plants, (b) Infected Citrus Plants, (c) Infected Citrus Plants, (d) Susceptible Insects, and (e) Insects Infected with $m = 0.25$

Based on Figure 2 and Figure 3, it demonstrates the model responses with and without delays blue and red responses, respectively. The red graph of the citrus plant population with some delays could provide more advantageous than without delays (blue graph). Therefore, it can be concluded that the CTV deployment of citrus crops can be inhibited by accelerating the harvest before citrus plants are infected with the virus, so that the harvesting of citrus fruit can be more better than without delays.

3.5 Stability Analysis

1. Stability of the model around the disease–free equilibrium point $E^0 = (S^0;\ I^0;\ R^0;\ S_X^0;\ I_Y^0)$. Based on the matrix $\mathcal{J}$ in the equation (4) and by substituting the parameter value, it is obtained that $\lambda_1 = -0{,}01$, $\lambda_2 = -0{,}01, \lambda_3 = -0{,}50$, $\lambda_4 = -0{,}12$ and $\lambda_5 = -0{,}60$. Hence, the model is asymptotically stable
2. Stability of the model around endemic equilibrium point $E^* = (S^*;\ I^*;\ R^*;\ S_X^*;\ I_Y^*)$. Based on the matrix $\mathcal{J}$ in the equation (5) and by substituting the parameter value, it is obtained that $\lambda_1 = -0{,}01$, $\lambda_2 = -0{,}38$, $\lambda_3 = -0{,}09$, $\lambda_4 = -0{,}02$, and $\lambda_5 = -0{,}25$. Therefore the model is also asymptotically stable.

4. Conclusion

The CTV propagation models with and without delays had been well-derived based on the model SIR – $S_X I_Y$. Two obtained equilibrium points, i.e. the disease-free equilibrium point $E^0 = (1000, 0, 0, 40{,}0)$ and the endemic equilibrium point $E^* = (994{,}68; 2{,}66; 2{,}66; 39{,}23; 0{,}77)$, are utilized to liniearize and analyze the model. The stability of the model without any delay addressed that the model is asymptotically stable, as well as the model with some delays. In addition, the basic reproduction number (ρ) is significantly involved to determine the parameter value m, and it is obtained that $m \geq 0.30$ and $m < 0.30$ for the disease-free and endemic conditions, respectively. The simulation of models with and without delays, with $m = 0{,}5$ for disease free and $m = 0{,}25$ for endemic state, emphasized that both populations are asymptotically stable. By applying some delays, it can be concluded that the CTV deployment of citrus plants can be inhibited by accelerating the harvest before the citrus plants are infected with the virus, so that the citrus fruits harvesting can be better than without delays.

5. References

[1] The Statistics Portal 2016 (Online) Global Fruit Production in 2016, by Variety (in Milion on Metric Tons) https://www.statista.com/statistics/264001/worldwide-production-of-fruit by-variety/

[2] Silva G, Marques N and Nolasco G 2012 The Evolutionary Rate of Citrus Tristeza VirusRanks Among the Rates of The Lowest RNA Viruses *Journal of General Virology* **93** 419-429

[3] Mawassi M, Mietkiewska E, Gofman R, Yang G and Bar-Joseph M 1996 Unusual Sequence Relationships Between Two Isolates of Citrus Tristeza Virus *Journal of General Virologi* **77** 2359-2364

[4] Halbert S E and Brown L G 2011 *Brown Citrus Aphid, Toxoptera citricida (Kirkaldy) (Insecta: Hemiptera: Aphididae)* (Florida: University of Florida)

[5] Jackson M and Chen-Charpentie Benito M 2016 Modelling Plant Virus Propagation with Delays *Journal of Computational and Applied Mathematics* **309** 611-621.

[6] Zaman G, Yong H K and Jung I H 2008 Stability Analysis and Optimal Vaccanation of an SIR Epidemic Model *J. Biosystems* 400-249.

[7] Olsder G J and Woude J W V D 2004 *Mathematical System Theory* (Netherland: VVSD)

The Application of Different Curing Models to the Strength Development of High Calcium Fly Ash Geopolymer Incorporating Limestone

A Wardhono

Department of Civil Engineering, Universitas Negeri Surabaya, Indonesia

Corresponding atuthor: ariewardhono@unesa.ac.id

Abstract. One of the issues in geopolymer specimen is the requirement for heat curing treatment temperature. The use of a material with high calcium content is one alternative to overcome this problem. However, the addition of calcium tends to alter the basic geopolymer matrix being similar to the basic matrix in normal concrete, which requires a water addition during curing process. This research aims to investigate the effect of different curing models to the strength development of fly ash geopolymer specimens with limestone inclusion. Class C fly ash was used as primary material. The blended sodium silicate and 10 Molar NaOH were used as alkaline activator. Three different curing models were applied, i.e. 24-hours water immersion, watering once a day, and watering twice a day, respectively. The results show that the addition of 30% limestone under watering once a day curing models gives the best result with the strength of 32.54 MPa. The limestone inclusion significantly affects the strength development of geopolymer mortar strength. It also slightly increases the water requirement during curing process to achieve its structural integrity. This is might be attributes to the changes in geopolymer matrix due to the addition of limestone.

1. Introduction

The use of geopolymer as an alternative material of concrete without cement has been widely studied [1-3]. Geopolymer, which was coined by Davidovits [4], is a material based on fly ash or other material containing silica and aluminate compounds reacted with an alkaline activator. This material is more environmentally friendly material compared to that normal Portland Cement (PC)-based concrete. According to Davidovits [5], the production of 1 ton cement also releases approximately 1 ton of CO_2, which leads to the concern of global warming problem. In addition, geopolymer does not require a water immersion during curing process due to the matrix characteristic of geopolymer, which is different from the PC concrete. The geopolymer matrix is a polymer compound consisting of silicate and aluminate forming the Si-O-Al geopolymer matrix bonds [5]. However, the main issue of geopolymer is the need for high temperature during the curing process. Geopolymer are not be able to achieve its structural integrity at normal temperatures. This high temperature is required to speed up the process of silicate (Si) and aluminate (Al) reaction in forming the geopolymer matrix. This might attribute to the low calcium content in fly ash material [6-8]. The use of a high calcium compound material is one alternative to overcome this problem [9-10].

However, the limestone addition in geopolymer will promote a new reaction between silicate and calcium. The silicate contents in geopolymer will divide between the two reactions: the reaction between silicate and aluminate forms a geopolymer matrix, and the reaction between silicate and calcium forms C-S-H gel, similar to that matrix in PC concrete [11-12]. The appearance of the second reaction raises a question as to whether the geopolymer incorporating limestone needs water during the curing process due to the matrix similarity to that of PC concrete.

This paper reports the investigation results on the effect of different curing models to the strength development of fly ash geopolymer specimens with limestone inclusion. Class C fly ash was used as primary material. The blended sodium silicate and 10 Molar sodium hydroxide (NaOH) were used as alkaline activator. The ratios between fly ash and limestone addition were 1.0:0.0, 0.9:0.1, 0.8:0.2, 0.7:0.3, 0.6:0.4, and 0.5:0.5, respectively. Three different water curing models were applied, i.e. 24-hours under water immersion, watered once a day, and watered twice a day. The strength developments of geopolymer specimens were measured in accordance with ASTM Standard [13].

2. Research Methodology

2.1. Materials

High calcium fly ash from Paiton coal power plant Indonesia was used as fully cement replacement material. It also has a low silicate (12.63%) and high ferrite (51.74%) contents. The chemical compositions of high fly ash are listed in Table 1. While, limestone or calcium carbonate ($CaCO_3$) was obtained from limestone material distributor. Both material compositions were identified by X-Ray Fluorescence (XRF) test using PANalytical test equipment type Minipal 4.

Table 1. Chemical composition of high calcium fly ash and limestone (mass %).

Materials	SiO_2	Al_2O_3	Fe_2O_3	CaO	MgO	K_2O	TiO_2	Mn_2O_3	SO_3
Fly ash class C	15.02	5.31	52.68	17.97	0.05	1.73	0.86	0.39	0.18
Limestone	-	-	0.29	98.42	-	-	-	-	-

The CaO content of 17.97% (> 10%), the total content of SiO_2+Al_2O_3+ Fe_2O_3 of 73.01% (> 50%), and the SO_3 content of 0.18% categorized the fly ash as high calcium class C fly ash in accordance with ASTM C618 [14]. A blended sodium hydroxide (NaOH) solution and sodium silicate was used as activators.

2.2. Mix design

The mix design details of high calcium class fly ash geopolymer incorporating limestone (LFAG) mortars are shown in Table 2.

Table 2. Mix proportions of fly ash geopolymer incorporating limestone (LFAG) mortars

Mixture	Portland Cement	Lime stone	Fly ash	Fine sand	Sodium silicate	NaOH 10 M	Water	Curing models
PC	1	-	-	3.0	-	-	0.45	CM1-24h
LFAG 1	-	0.1	0.9	3.0	0.21	0.14	-	CM1-24h, CM2-1X, CM3-2X
LFAG 2	-	0.2	0.8	3.0	0.21	0.14	-	CM1-24h, CM2-1X, CM3-2X
LFAG 3	-	0.3	0.7	3.0	0.21	0.14	-	CM1-24h, CM2-1X, CM3-2X
LFAG 4	-	0.4	0.6	3.0	0.21	0.14	-	CM1-24h, CM2-1X, CM3-2X

Mixture	Portland Cement	Lime stone	Fly ash	Fine sand	Sodium silicate	NaOH 10 M	Water	Curing models
LFAG 5	-	0.5	0.5	3.0	0.21	0.14	-	CM1-24h, CM2-1X, CM3-2X

2.3. Curing models
Three different curing models were applied for all LFAG mortar specimens. Curing Model 1 (CM1-24h) was applied by immersing the LFAG mortar for 24-hours until prior to be testing. Curing model 2 (CM2-1X) was performed by watering the LFAG specimens once a day, while Curing model 3 was carried out by watering the LFAG specimens twice a day. For control specimen (PC), Curing Model 1 was applied.

2.4. Testing mortar specimens
The LFAG mortar specimens were prepared by 5 x 5 x 5 cm^3 cube steel molds. The strength of LFAG mortars were tested by compressive strength test. Universal Testing Machine (UTM) equipment was used in accordance with ASTM C109 [13]. Three mortar geopolymer cubes were tested at the age of 28 days after casting.

3. Results and Discussions

3.1. Strength development of LFAG mortars
Table 3 shows the strength development of LFAG mortar specimens for all mixes at 28 days. All geopolymer incorporating limestone specimens exhibit a higher compressive strength compared to that PC-cement based concrete. The highest compressive strength of 32.54 MPa was achieved by LFAG3 specimen with 30% limestone addition by watering the specimen once a day.

Table 3. Compressive strength of LFAG mortars at 28 days with different curing models

Mixture	Compressive strength at 28 days (MPa)		
	Curing CM1-24h	Curing CM2-2X	Curing CM3-1X
PC	13.96	13.80	12.97
LFAG1	15.82	16.11	16.65
LFAG2	18.14	19.26	20.50
LFAG3	28.44	30.50	32.54
LFAG4	16.51	18.92	19.81
LFAG5	15.96	16.20	17.70

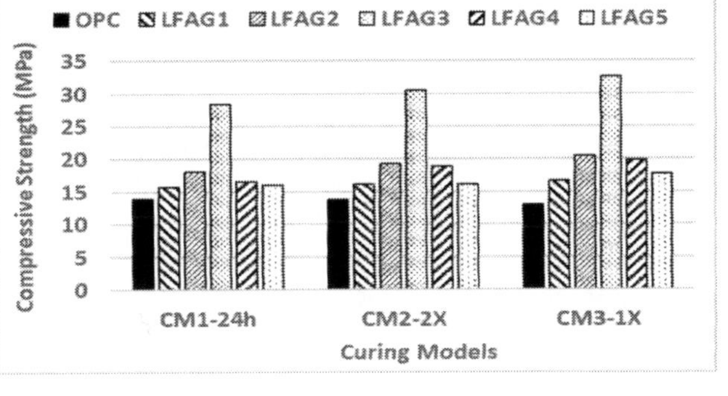

Figure 1. The effect of limestone on strength development of LFAG specimens at different curing models at 28 days

Figure 1 demonstrates the effect of limestone inclusion on strength properties of LFAG mortar specimens under with three different curing models, i.e. CM1-24h, CM2-2X, and CM3-1X. The results show that the addition of limestone significantly affects the strength of LFAG specimens. All curing models exhibits a similar strength development patterns along with the limestone inclusion. The highest compressive strength was achieved by LFAG3 with the strength of 28.44 MPa, 30.50 MPa, and 32.54 MPa in all curing models, respectively. However, the further addition of limestone more than 30% tends to reduce strength of LFAG. This might be attributed to the excess un-reacted calcium, which form a calcium hydroxide compounds. According to Sinsiri et al. [15], the calcium hydroxide compound tends to have a concrete expansion and creates a crack, which might lead to the strength reduction at further calcium inclusion. In addition, despite LFAG1 (strength of 15.82 MPa) and LFAG2 (strength of 15.96 MPa) specimens demonstrate the lowest compressive strength test result, LFAG 1 and LFAG 2 still exhibit a slightly better strength performance compared to that PC mortar control.

3.2. Effect of curing models on strength development

The effect of different curing models on strength development of LFAG mortar specimens are shown in Figure 2.

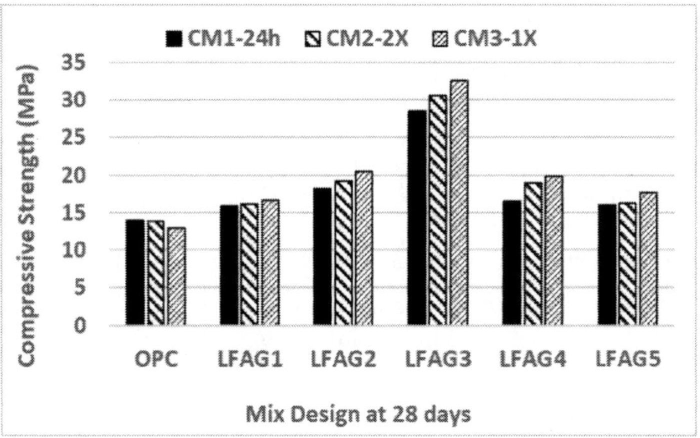

Figure 2. The effect of curing models on strength development of LFAG mortars at 28 days

Figure 2 demonstrates that the duration of curing process does not significantly affect the strength development of geopolymer incorporating limestone. The longer duration of curing process tends to inhibit the strength development of LFAG specimens. Although the limestone inclusion significantly affects the strength development of LFAG, all mixes under curing model CM1-24h (water immersion for 24-hours) display the lowest compressive strength with the lowest strength of 15.82 MPa. The reduction of water curing duration from 24-hours (CM1-24h) to twice a day watering (CM2-2X) improve the strength of LFAG specimens. A further shorter water curing duration shows a significant increase of strength from 28.44 MPa (LFAG3, CM1-24h) to 32.54 MPa (LFAG3, CM3-1X) with increase of 12.60%. Longer water curing duration, particularly under water immersion curing models, tends to inhibit the strength development of high calcium fly ash geopolymer. According to Rangan [16], water plays no significant role in the chemical reaction in geopolymer mixture. It only contributes to the workability of the geopolymer mixture. This is in contrary to the chemical reaction of water in PC control.

4. Conclusions

The effect of different curing models on strength development of high calcium fly ash geopolymer mortar specimens incorporating limestone were experimentally investigated at 28 days age. The main conclusions that may be drawn based on this research are, (1) The inclusion of limestone significantly affects the strength development of high calcium fly ash geopolymer specimens cured at normal temperature, (2) The optimum strength was achieved by 30% limestone inclusion at 28 days. Further limestone addition tends to reduce the strength of geopolymer incorporating limestone, (3) The use of water curing models does not significantly govern the strength properties of geopolymer incorporating limestone, and (4) The highest compressive strength was obtained by curing model 2 that was performed by watering the LFAG specimens once a day. The longer curing process tends to inhibit the strength development of geopolymer incorporating limestone.

5. References

[1] Ahmaruzzaman M 2010 *J. Progr. Energy Comb. Sci.* **36(3)** 327-63
[2] Law D W, Adam A, Molyneaux T, Patnaikuni I, and Wardhono A 2015 *Materials Struct.* **48(3)** 721-31
[3] Gunasekara C, Setunge S, and Law D 2017 *ACI Struct. J.* **144(3)** 743-52
[4] Davidovits J 1991 *J. Thermal Anal.* **37** 1633-56
[5] Davidovits J 1994 *World Resource Rev.* **6(2)** 263-78
[6] Vijay K, Kumutha R., and Vishnuram B G 2010 *Inter. J. Physical Sci.* **5(9)** 1419-23
[7] Wardhono A, Law D W, and Strano A 2015 *Procedia Eng.* **125** 650-56
[8] Kong D L G and Sanjayan J G 2010 *Cement Conc. Res.* **40(2)** 334-39
[9] Temuujin J, van Riessen A, and Williams R 2009 *J. Haz. Materials* **167** 82-88
[10] Sakulich A R, Anderson E, Schauer C, and Barsoum M W, *Const. Build. Materials* **23** 2951-57
[11] Wardhono A, Law D W, Sutikno, and Dani H 2017 *AIP Conf. Proc.* **1887** 020030
[12] Yip C K, Lukey G C, Provis J L, and van Deventer J S J 2008 *Cement Conc. Res.* **38(4)** 554-64
[13] ASTM Std. 2004 **ASTM C109**
[14] ASTM Std. 2004 **ASTM C618**
[15] Sinsiri T, Chindaprasirt P, and Jaturapitakkul C 2010 *Int. J. Min., Metal. Materials* **17** 683-90
[16] Rangan BV 2010 Proc. Inter. Workshop on Geopolymer Cement and Concrete

Acknowledgment
The authors would like to express their sincere thanks to Universitas Negeri Surabaya for providing the support for this research project.

Revealing Student's Multiple-Misconception on Electric Circuits

W Widodo[1], L Rosdiana[1], A M Fauziah[1], Suryanti[2]

[1]Department of Science Education, Universitas Negeri Surabaya, Indonesia
[2]Department of Primary Teacher Education, Universitas Negeri Surabaya, Indonesia

E-mail:wahonowidodo@unesa.ac.id

Abstract. Quick identification of misconception on classroom setting is an important step to gain deep understanding on physics. The study aims at developing simple tools to determine students' misconceptions and to reveal students' misconception about electric circuits. This research was conducted in the form of a survey study. Initially, Multiple Misconception Revealing Test (MMRT) with some characteristics, i.e. arousing curiosity on students to solve using of several concepts, potent to reveal one or more student's misconception(s), allow the students to explain their answers, as well as allow to be followed up, was developed. The test was administered to 60 of higher education students (prospective science teachers) and data were analyzed both quantitatively and qualitatively. The results shows that only 17% students have right conception about electrical circuit, the others have one of three type of multiple misconception. About 7% students hold Type 1 multiple misconception (cannot ditinguish between serial and parallel principles); 43% students have the Type 2 (misunderstand circuit principles as well as cannot ditinguish between serial and parallel principles); and 33% students have the Type 3 (misunderstand circuit principles, misunderstand circuit components, and cannot ditinguish between serial and parallel principles). The conclusion is that there is no single student misconception on classroom, particularly on electrical circuits; and the MMR test can reveal three kinds of multiple misconception on it.

1. Introduction

It is known that students often have ideas and knowledge about nature that are substantially different from the views of scientists [1], called misconceptions. Misconceptions are encountered at various levels of educational units and cultural backgrounds [2][3]. Magnusson, Bole and Templin [1] state that experts may still have a misconception encoded in the neural networks of their brains that must be inhibited in order to answer scientifically. Misconceptions may deeply penetrate into students' minds and be resisted to change [4][5]. Misconception is one of the major obstacles in studying physics [6], and thus leads students to the failure when studying physics [1].

Misconceptions occur on a variety of physics topics, including electricity. The misconceptions include consumption model, current collision model, equalize or exchange current with voltage, closed circuit model without including voltage source [7]. In addition, there are students who misunderstand the series and parallel circuits, including the current and voltage properties of the circuit [4][8]. Misconceptions, including in electrical circuits, are not single in one concept, but tend to be interconnectedness between concepts [2]. Also, students tend to have misconceptions including more than a concept or multiple misconceptions.

IOP Conf. Series: Journal of Physics: Conf. Series **1108** (2018) 012088 doi:10.1088/1742-6596/1108/1/012088

Remedial of misconceptions can be done, but not easy. It is not only the students' misconception tend to be multiple, but also they tend to hold on to the explanations they have formulated in their minds. This remediation requires conceptual change in students. These are cognitive conflicts and analogies [9], including using educational-technology [10].

In the context of prospective science teachers, misconceptions about the electricity they may have must be remediated. To teach physics effectively, misconceptions must be spotted and overcome [11]. This remedy is conducted in a relevant lecture. However, this remedy requires a quick way to uncover any misconceptions that may occur to them.

Various methods of uncovering misconceptions have been developed, among others: concept inventory, test with Certaintly Response Index (CRI) or two tier test [12], which later developed into 3 tiers [13][6], and 4 tiers [9]. They require a sufficiently long time to administer test, analyze, and determine misconceptions for students. In addition, the test is in the form of multiple choice, generally each item is only related to a single misconception. Therefore, these methods are not suitable for classroom diagnostic test purposes, such as a simple way to reveal students' misconceptions fastly as well to be handled easily by teacher.

In order to classroom diagnostic tests for misconceptions, the idea is: do not need many test items, test items that engaging curiosity on student (to solve), on "inclusive concept" so that requiring students to use of several concepts, as well as allow to be followed up. However, whether this idea can reveal one or more student's misconception(s), this needs to be examined.

2. Method

This research was conducted in the form of a survey using Multiple Misconception Revealing Test (MMRT) completed with selected interview. MMRT was administered to 30 of prospective science teachers at one of the state universities in Indonesia who take Electricity and Magnetism course, and interviews are conducted on 9 selected students in response to MMRI. The data were analyzed descriptively by categorizing data, crosscheck with students' notes and results of interview.

The MMRT as a data collection tool is developed according to the ideas above, with the following details:

1. the test is about open ended question;
2. problems developed in parallel circuits, allowing students to use a variety of relevant concepts;
3. problem is on the level of creating, ie in the form of constructing [14] with certain rules. Problem solving by drawing, so easily recognizable right or wrong quickly (in the context of diagnostic test) and expected arousing curiosity on student to solve; and
4. problems are equipped with a level of confidence and an explanation of why they answer like that (the idea of three tiers test). See Figure 1.

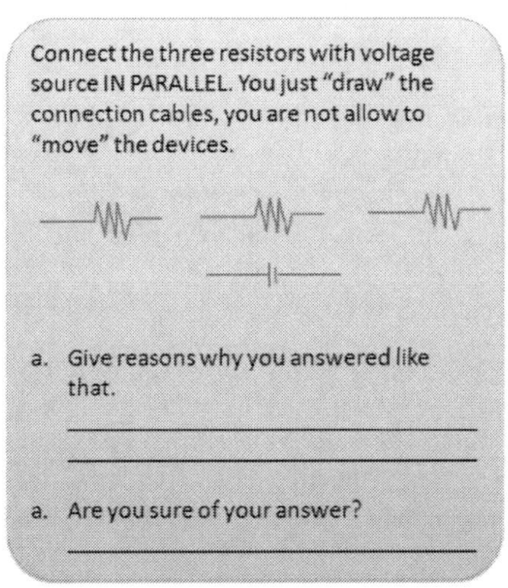

Figure 1 The MMRT of Electric Circuit

Data obtained from MMRT were analyzed using the approaches which require the definition of scientifically complete response (nomothetic) and the classification of explanations in certain categories (ideographic) [4]. See Figure 2.

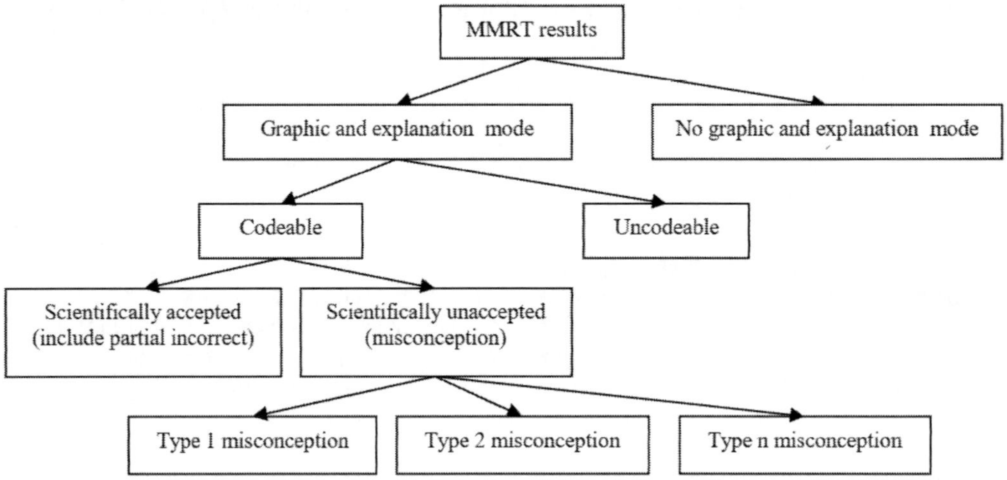

Figure 2 Analysis of MMRT Result

3. Results and Discussion

As the analysis of the data was completed, the findings obtained show that only 7% answers were in the scientifically accepted category. Because all participants answered with confidence (pictures and explanations), the rest (93%) experienced misconceptions. The results of further analysis show, misconceptions that occur can be categorized into 3 types of misconception. See Figure 3

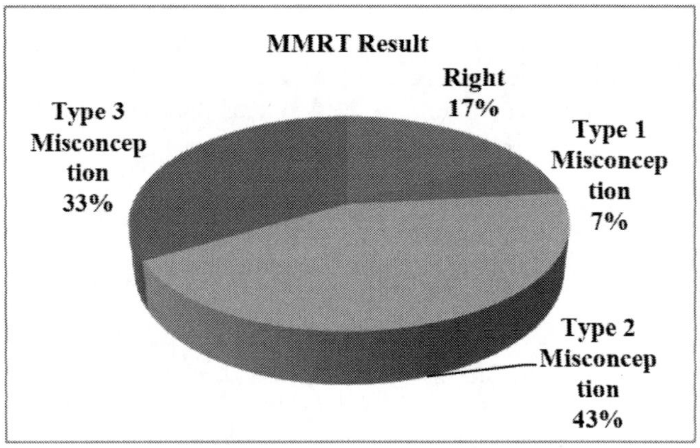

Figure 3 MMRT Results Analysis

Almost all textbooks and explanations in the class use a picture of a parallel circuit like Figure 4 (a). It seems that students hold this strong in their cognitive structure. This is evident from the student's writing, "... this matter is not the same as in the book, it makes me ...".

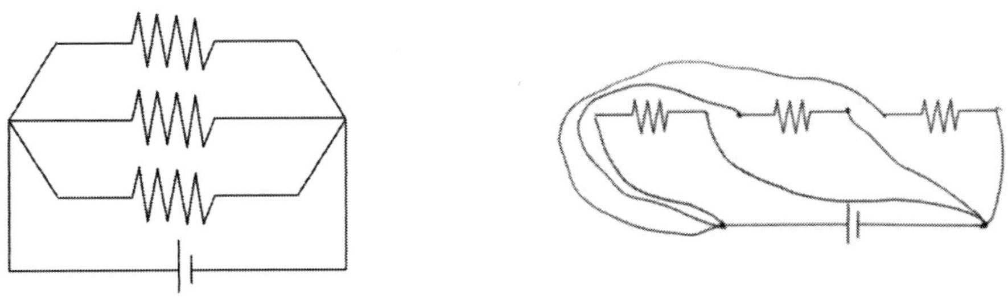

(a) (b)

Figure 4 (a) Parallel circuits commonly described in books, and (b) correct answers

The mental picture in his/her mind of parallel circuits makes it memorized, so that students are unable to recognize the essence of parallel circuits. Generally, misconception tests on electrical circuits do not extend up to this point, so these results cannot be confirmed with other studies. However, it is known that one will have a mental picture of something, in this case a parallel circuit. The mental picture is useful because it is the simplest and most imaginable one, but it apparently makes the person misunderstood of the essence, in this case is parallel circuit.

The misconceptions that occur to them are not only in one concept but in some other aspects, including: the inappropriateness of distinguishing the essence of series and parallel circuits, the inaccuracy in understanding closed circuits, and the inaccuracy in understanding of the components in a closed circuit. So, it turns out this misconception is layered on each student, so it tends to be multiple misconception. Multiple misconceptions that occur can be classified into 3 categories, as follows.

1. Type 1 Misconception

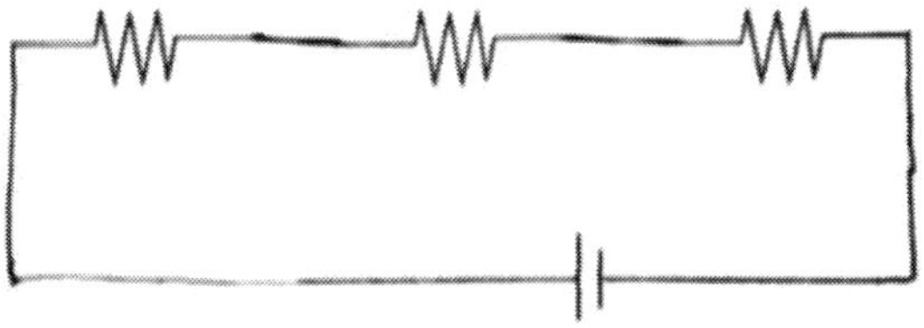

Figure 5 Example of Type 1 Misconception

On Type 1 Misconception, they can't distinguish between series and parallel circuits, but they have a correct understanding of closed circuits. According to them, "... according to the book, if there are three resistor lined horizontally, the circuit is like that ...". Apparently, the process of thinking: **"The type of circuit depends on how the resistors are placed with each other"** becomes misconception on electric circuits. Such misconceptions have never appeared in previous research.

2. Type 2 Misconception

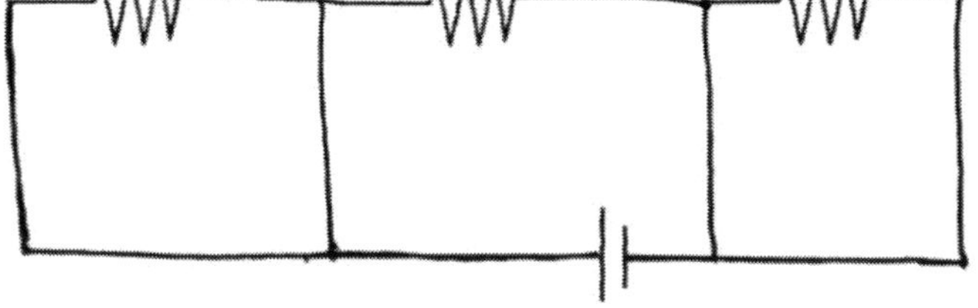

Figure 6 Example of Type 2 Misconception

On Type 2 Misconception, they cannot ditinguish between serial and parallel principles. They also misunderstand about circuit principles. Acording to student's notes, "...I try to imitate the drawing in the book..". The misconception closes to "unipolar model" [15]

3.Type 3 Misconception

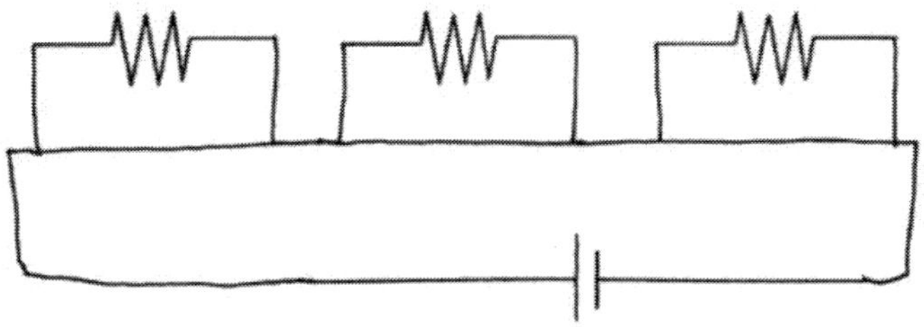

Figure 7 Example of Type 3 Misconception

On Type 3 Misconception, they cannot ditinguish between serial and parallel principles. They also misunderstand about circuit principles as well as circuit components. Again, mental images "in accordance with the book" seem to have been held in their minds and defeated the essence of electrical circuits. The misconception closes to "Scientifically Unacceptable" chategory [4].

4. Conclusion

To conclude, there are two main findings of the study, that are:

1. MMRT can reveal students misconception relatively quickly, so it can be a diagnostic tool in the learning process.
2. On the subject of research, only 17% have right conception about parallel circuit, the rest tend to have multiple misconceptions. There is the misconception not encountered in the previous studies: the type of circuit depends on how the resistors are placed with each other.
3. There are 3 kinds of multiple misconception on electric circuits. The standard drawing of parallel circuits in the book seems to make students not recognize the essence of parallel circuits.It is needed to be further develop of MMRT on another subject.

References

[1]. Magnusson, S.J., Bole, R.A., Templin, M. 1994. Conceptual Development: Re-examining Knowledge Construction in Science. *Annual Meeting of American Education Research Association*, LA, April 5-8, 1994. 14-15

[2]. Berg E 1991 *Miskonsepsi Fisika dan Remediasi* (Salatiga: UKSW)

[3]. Pablico J R 2010 *Misconceptions on force and gravity among high school students.* LSU Master's Theses. 2462. http://digitalcommons.lsu.edu/gradschool_theses/2462.

[4]. Küçüközer H And Kocakülah S 2007 *J. of Turk. Sci. Educ.* **4(1)**

[5]. Osborne R. and Freyberg P 1985 *Learning Science: The Implication of Children's Science* (NH: Heinemann)

[6]. Gurcay, Deniz; Gulbas and Etna 2015 *Res. in Sci. and Tech. Educ.,* **33(2)** 197-217

[7]. Berg E and Grosheide W 1997 *Sch. Sci. Rev.,* **78(284)** 89-94

[8]. Lee Y and Law N 2001 *Int. J. of Sci. Educ.,* **23(2)** 111-149

[9]. Dreyfus A, Jungwirth E, Eliovitch R 1990 *Sci. Educ.,* **74(5)** 555-569

[10]. Ross, John A, Bruce, Catherine D 2009 *Int. J. of Math. Educ. in Sci. and Tech.* **40(6)** 713-727

[11]. Ozkan, Gulbin, Sezgin Selcuk, Gamze 2013 *Asia-Pacific Forum on Sci. Learning and Teaching*, **14 (1)**.

[12]. Hasan S, Bagayokod D, and Ella L K 1990 *Phys. Educ.* **34(5)**

[13]. Cohen R., Eylon B and Ganiel U 1983 *Am. J. of Phys.* **51(5)**.

[14]. Anderson L W and Krathwohl D R 2001 *A Taxonomy for Learning, Teaching, and Assessing: a Revision of Bloom's Taxonomy of Educational Objectives* (New York: Longman).

[15]. Driver R, Squires A, Rushworth P and Wood-Robinson V 2015 *Making Sense of Secondary Science* (London: Routledge)

IOP Conf. Series: Journal of Physics: Conf. Series **1108** (2018) 012089 doi:10.1088/1742-6596/1108/1/012089

Numerical analysis of fluidal flow in heat exchangers using finite element method to reduce exhaust emission level in air

A Fatahillah [1], Dafik[2], B P Prawoto[3], A R Lazuardi[4]

[1,2,4]Mathematics Educational Study Program, University of Jember, Indonesia
[3]Department of Mathematics, Universitas Negeri Surabaya, Indonesia

Abstract. Air pollution is one of the problems faced by all countries around the world. In Indonesia, air pollution is mostly caused by exhaust gas of vehicles and industrials emission gas. POPKA factory which becomes one of the biggest industrial factories in Indonesia uses Heat Exchanger engine to cool down the emission of gas. Although the simulation oh Heat Exchanger has been widely used, it is considered as steady in under several conditions. Accordingly, these conditions produce less resemble with actual condition. Furthermore, the calculation method which is used does not give a high accuracy, therefore another method in calculating the formed mathematical modeling is necessarily employed. This research created a new model of calculation method to overcome the problems which could display the effectiveness of Heat Exchanger type intercooler in cooling down the residual materials of POPKA factory and could be used to manage the residual emission into the air. It is aided MATLAB software in calculation process, FLUENT and GAMBIT software in the stimulation, and Gauss-Seidel method to know the effectiveness of the calculation.

1. Introduction

Air pollution is not a new problem which has been faced all countries in the world. As the data published by the University of Washington, it can be seen that almost all countries are experiencing air pollution [1]. Both rich and poor countries face the problem of air pollution. Indonesia as one of the developing countries also cannot escape from air pollution problems. In Indonesia, the largest air pollution contributor is vehicle exhaust emissions which is about 60-70%, followed by industrial exhaust emissions which is 10-15%, and the rest is contributed by other burnings [2]. Air pollution mostly occurs in some cities which have dense population and large industrial factories such as Jakarta, most cities in Java and Borneo. Figure 1 shown he number of deaths due to air pollution in some countries by University of Washington below.

IOP Conf. Series: Journal of Physics: Conf. Series **1108** (2018) 012089 doi:10.1088/1742-6596/1108/1/012089

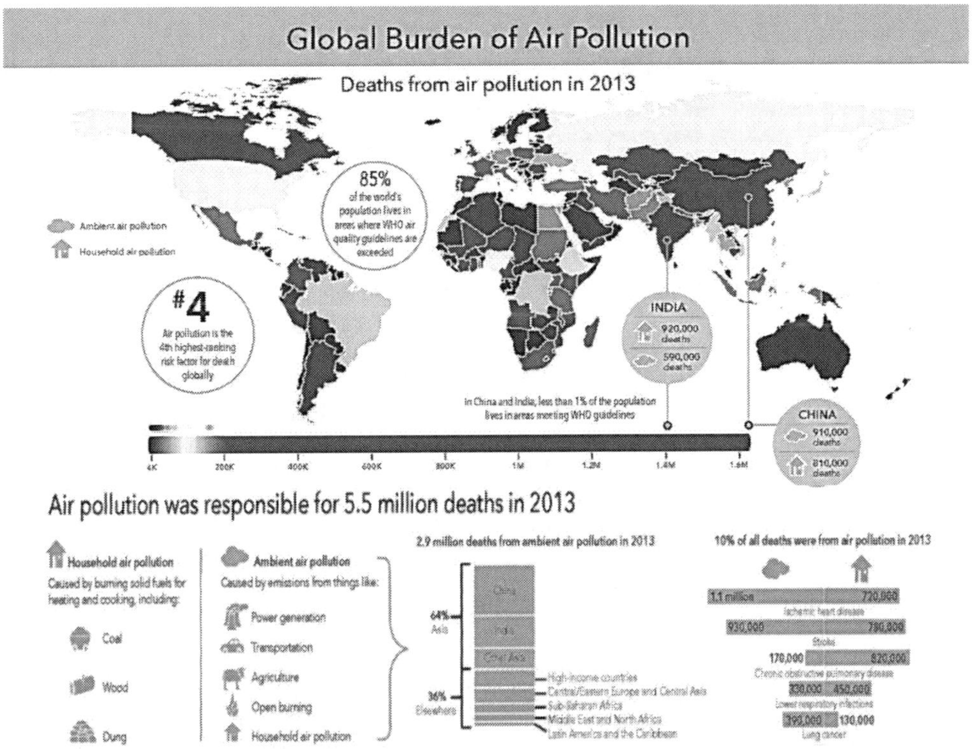

Figure1. The number of deaths due to air pollution in some countries by University of Washington
(Source: http://www.healthdata.org/infographic/global-burden-air-pollution)

POPKA factory is one of urea fertilizer manufacturing factories in Indonesia which uses Heat Exchanger type intercooler with shell and tube. Exhaust gases such as CO_2 (carbon dioxide), CO (carbon monoxide), SO_x (sulfur oxide), and NO_x (nitrogen oxide) are cooled by Heat Exchanger, and then discharged into the air. The purpose of exhaust gas emission is to reduce the effects of global warming and air pollution. CFD simulation has been widely used in Heat Exchanger; however the calculation is still limited. Some calculation factors on the Heat Exchanger are always considered as steady condition, thus so many factors such as ambient temperature and flow velocity are ignored. This will affect the simulation result. In addition, the calculation method used does not have a high accuracy, so other methods in calculating the formed mathematical modeling is highly needed.

Jung and Assanis [4] built a numerical model based on the thermal resistance concept. This model could effectively predict the heat exchange performance, and the calculation results were in good agreement with the experimental data. Some issues that will be examined in this research is analyzing and completing the analysis model of fluidal flow in the heat exchanger numerically using finite element to reduce exhaust emission level in air, with the formulation of the problem as follows is: How does the mathematical model of heat transfer in the fluidal of heat exchanger type shell and tube during the cooling down emission gas, How does the ambient temperature material affect the cooling down emission gas in the heat exchanger type shell and tube using the finite element method, How does the velocity of fluidal flow affect the cooling down emission gas in the heat exchanger type shell and tube using the finite element method, and How the effectiveness of the finite element method in analyzing the heat transfer of fluidal in the heat exchanger type shell and tube when the cooling down emission gas.

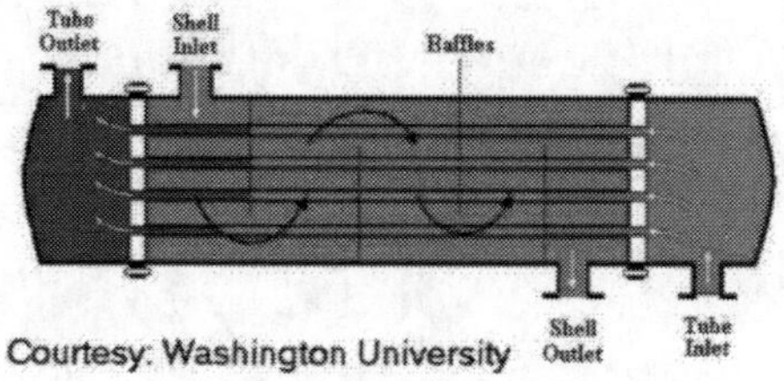

Figure 2. Heat Exchanger Shell and Tube type

Figure 2 show that Heat Exchanger Shell and Tube type. Based on the formulation of problem above, the main goal of this research is knowing the mathematical model of heat transfer in fluidal-type heat exchanger shell and tube current the cooling process of emission gas, determining the effect of speed fluidal flow to the cooling process of emission gas in the heat exchanger shell type tube using the finite element method, determining the effect of ambient temperature material to the cooling process of emission gas in the heat exchanger type shell and tube using the finite element method, knowing the effectivity of the finite element method in analyzing the heat transfer fluidal of heat exchanger type of shell and tube current the cooling process of emission gas.

The basic equation of fluidal flow model in this heat exchanger using the energy equation by differentiating each item will obtained:

$$\nabla^2 \mu C_k + \left(\frac{\nabla^2 \mu_t C_k}{\tau_k}\right) - \nabla \rho q u \phi - \frac{\partial \rho C_k}{\partial t} = -2\nabla^2 u \mu_t \quad (1)$$

with ρ = mass density (1.1), u = velocity of x-axis (0.5), Φ = velocity of y-axis (0), and μ = viscosity of substances (1000). In these equations, Gk represents the generation of turbulence kinetic energy due to the mean velocity gradients [5], Ck are the advection parameters and q is the amount of heat on Heat Exchanger, calculated by :

$$G_k = \frac{\mu_t}{\rho} \frac{\partial u_i}{\partial x_j}\left(\frac{\partial u_i}{\partial x_j} + \frac{\partial u_j}{\partial x_i}\right) \quad (2)$$

$$q = m\, c\, \Delta t \quad (3)$$

$$Ck = \rho\, q\, u \quad (4)$$

Furthermore, the equation will be derived using the quadratic approach and integrate of each tribe.

2. Research Method

Finite element method is one of approach method of numeric that bases on the problems at every part of element that is named by finite element [3]. The first step that must be done in this research is collecting data and supporting the theory in order to get to design the heat transfer model of fluidal flow in heat exchanger based on the speed of fluidal flow and ambient temperature. The next is completion the heat transfer models of fluidal flow in heat exchanger based on the speed of fluidal flow and ambient temperature using finite element method. Then created computer program with MATLAB and FLUENT and testing program. The next step in this research is analyzing the relationship between the level of environmental temperature and the speed of fluidal flow to the long process of emission gas cooling. The final step in this research is evaluating the results of testing and validation to produce the heat transfer of fluidal flow analysis in heat exchanger.

3. Result and Discussion

In this simulation process used MATLAB software that will help researchers in describing a large temperature of emission gas formed at each point of the domain. From a global matrix equation that has been formed will be simulated with MATLAB program and the result will be analyzed for some simulations. The first simulation is the process of emission gas cooling by the exact method. The result of simulation can be seen in Figure 3.

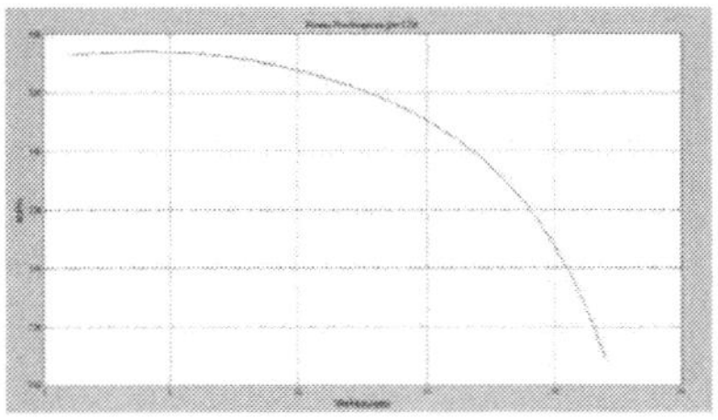

Figure 3. The Cooling Process of emission gas for 20 hours

In figure 3 shown the movement of temperature decrease on each global node throughout the domain. It can be seen that the decreasing of temperature is inversely to the longer it takes. The longer time of the cooling process will decrease the gas temperature. It is based on the Law of Cooling Newton where the longer it takes the object to release or receive the heat is inversely to the sum of temperature required of objects to receive or release the heat. Newton's law of cooling suggests that the intensity of energy transfer in the form of heat depends on a difference of temperatures of the interacting physical systems [6].

$$\frac{Q}{t} = \frac{h\,A\,\Delta T}{l} \quad (5)$$

where ΔT is logarithmic mean temperature difference in countercurrent arrangement of shell and Tube heat exchangers [7]. And Q is amount of heat, t is time, h is thermal conductivity, A is surface area, and *l* refers to length of the object.

In this second simulation will be simulated a process of cooling emission gas to the influence of the difference temperature in the environment around the system using the usual method and the Gauss-Seidel method. In this second simulation temperature that will be used is 290 Kelvin, 293 Kelvin and 296 Kelvin as a comparison, while for a time, domain length and the initial temperature gas of the first simulation. The simulation results can be seen in Figure 4a-4b

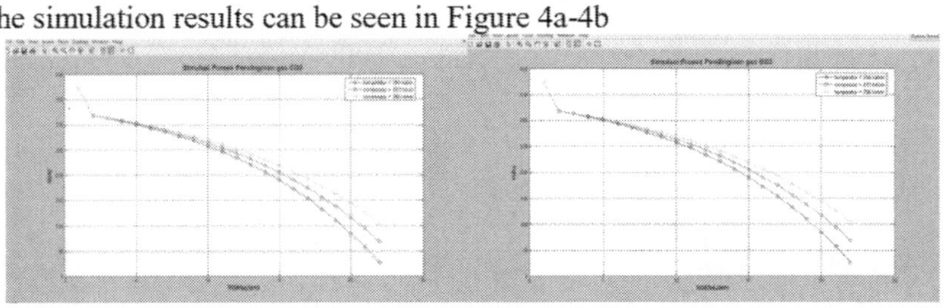

a b

Figure 4. The cooling process of emission gas based on temperature of around the system
a) exact method, b) Gauss – Seidel method

From the graph in Figure 4a and 4b, there is no difference between simulation using exact method and Gauss Seidel method. The lower ambient temperature will make the cooling process of emission gas is faster and conversely. The simulations which then simulated a cooling process of emission gas to the difference of speed fluidal flow using the usual method and the Gauss-Seidel method. In the second simulation, the flow rate that will be used is 1 m/s, 1.5 m/s and 2 m/s as a comparison, while for a time, domain length and the initial temperature gas of the first simulation. The simulation results can be seen in Figure 5a-5b.

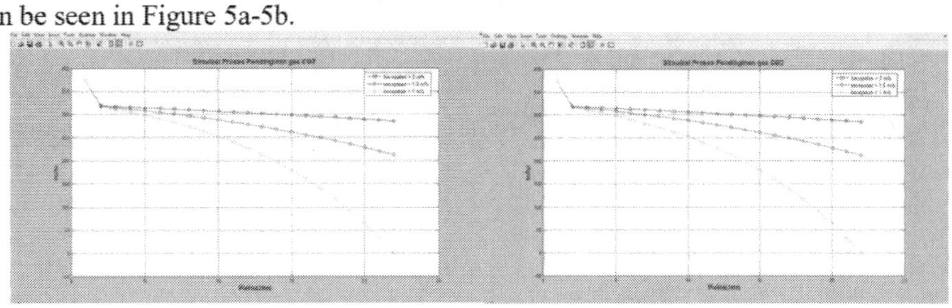

a b

Figure 5. The cooling process of emission gas based on initial velocity
a) exact method, b) Gauss – Seidel method

Based on the graph in figure 4a and 4b, there is no difference between simulation using exact method and Gauss Seidel method. The lower flow rate will make the cooling process of emission gas is faster and conversely.

In this simulation FLUENT software used to describe the examined object in GAMBIT software first and then simulation with FLUENT software. This simulation is done to show the pattern of fluidal flow in the heat exchanger based on the initial of different temperatures and speeds in order to know the effectiveness of heat exchangers in the cooling process of emission gas. In this first simulation will be described a pipe of heat exchanger in the two-dimensional form. The variable that will be compared is the influence of emission gas in the different ambient temperature to the pipe of Heat Exchanger. The ambient temperature of emission gas that will be compared is 296K, 293K, and 290K. The simulation results can be seen in Figure 6a-6c.

Based on the2-dimensional visualization of the simulation in Figure 6a-6c can be seen that initial temperature of emission gas is very high, which at the end of the inlet pipe is red colored. But the color starts to turn yellow, green and blue at the end of the exit pipe. It indicates that the temperature of emission gas is decrease of the initial inlet pipe to the end of the exit pipe. Based on the three pictures, the pipe of heat exchanger with ambient temperature of 290K is the most rapid cooling where the red part is meager and dominated by yellow and slightly green and blue colors. While the pipe of heat exchanger with the ambient temperature of 296K ambient temperature is the most slow cooling where the red color indicates the amount of high temperatures are the largest among of the three pipe. While the pipe with the ambient temperature of 293K is more slowly cooling than the pipe with the ambient temperature of 290K but is faster than the pipe with ambient temperature of 296K. We can see the number of high temperature on the inlet pipe shown the red color is larger than the number of pipe with an ambient temperature of 290K but is fewer than a pipe with ambient temperature of 296K. From the three pipe, we can see that at the end of the inlet pipe, high-temperature gas represented red is slowly changing into yellow, green, and finally to blue color towards the end of the pipe out. It is accordance with the MATLAB simulation that the gas temperature has decreased even with variation of ambient temperature.

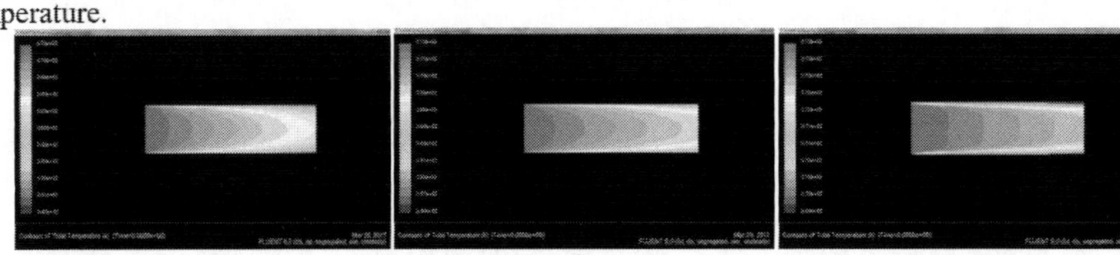

a b c

Figure 6. The cooling process with the ambient temperature: a) 290K, b) 293K, and c) 296K

In the second simulation will be illustrated the pipe of Heat exchangers in the 3 dimensional shape. The variable that will be compared is the varying fluidal velocity of the heat exchanger pipe. Fluidal velocity that will be compared are 1 m/s, 1.5 m/s and 2 m/s. The result of simulation can be seen in Figure 7a-7c.

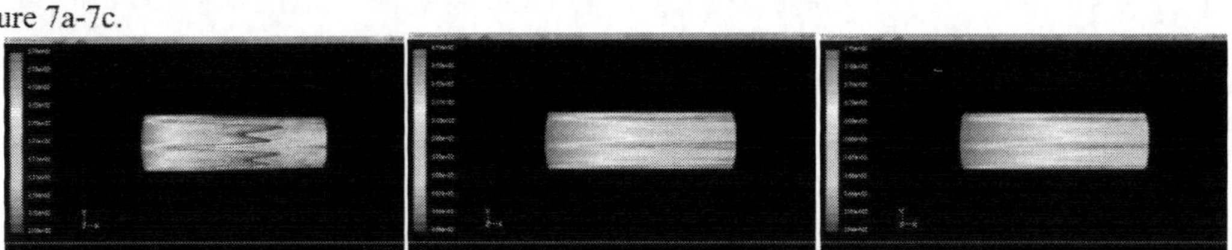

a b c

Figure 7. The cooling process with the speed: a) 1 m/s, b) 1.5 m/s and c) 2 m/s

Based on the 3-dimensional visualization of the simulation in Figure 7a-7c can be seen that the initial temperature of emission gas is very high, which at the end of the inlet pipe is colored red. But the color starts to turn yellow, green and blue at the end of the exit pipe It indicates that the temperature of emission gas is decrease of the initial inlet pipe to the end of the exit pipe. Based on the three pictures, the pipe of heat exchanger with the fluidal speed of 1 m/s is the most rapid cooling because it dominated by blue color. In the pipe of heat exchanger with the fluidal speed of 2 m/s has the slowest process of cooling down due to there is domination of red, yellow, and green in most. Meanwhile on the pipe with a fluidal speed of 1.5 m/s has the slower cooling process than the pipe at a speed of 1 m/s which the number of red color indicates the high temperature at the inlet is bigger than the pipe with a fluidal speed of 1 m/s and faster than pipe with a fluidal speed of 2 m/s with the biggest of high temperature shown in much red color on the pipe. On the three pipe, we can see that at the end of the inlet pipe, high-temperature gas is represented in red is slowly changing into yellow, green, and finally to blue color towards the end of the pipe out. It is suitable with the simulation of MATLAB that the gas temperature is decreased although the speed of fluidal flow is different.

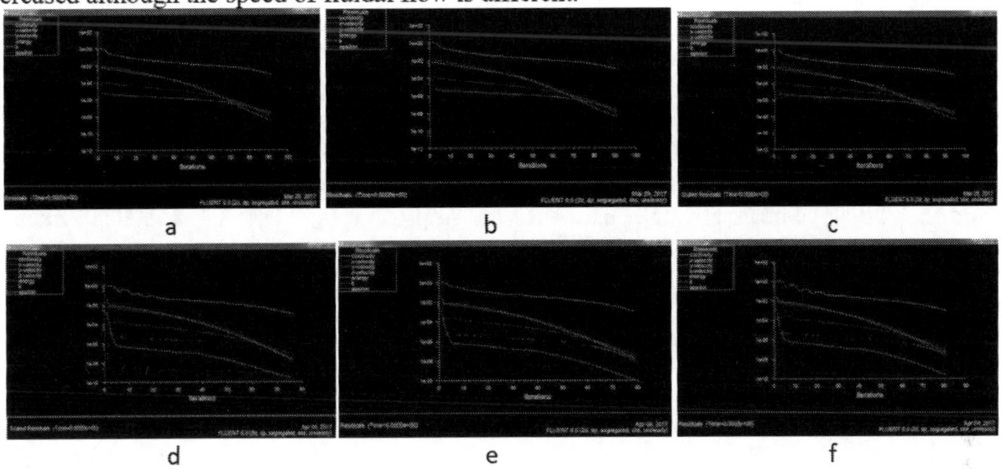

Figure 8. The iteration chart of cooling process with the ambient temperature: a) 290K, b) 293K, c) 296K and with the initial rate: d) 1 m/s, e) 1.5 m/s, f) 2 m/s

The final step in the mathematical modeling is analyzing and making a conclusion. In this section will be deduced from the error relative calculation using Gauss-Seidel method and iteration graphs using FLUENT. In Figure 2 and 3, we can see that there is no big difference between the graph using the Gauss-Seidel method and the usual methods. It indicates that the value of the calculation error is very small. Based on the calculation using MATLAB, there is an error relative calculation using the finite element method is 0.0026848 to the influence of ambient temperature, and 0.0028581 for the effect of fluidal speed with the limit values of tolerance 0.01. It is also strengthened with the iterations chart using FLUENT with the convergent completion in figure 8a-8f.

In Figure 8a shows that the iteration convergent when the iteration is 91st. In the figure 8b shows that the iteration convergent when the iteration is 92nd. While the figure 8c shows that the iteration convergent when the iteration is 95th. It indicates that differences in the ambient temperature also affects the iteration in cooling process of CO_2 gas. The bigger ambient temperature in Heat Exchanger cause the number of iterations required is also growing and conversely.

In Figure 8d shows that the iteration convergent during the 76th iteration. In the figure 8e shows that the iteration convergent during the 79th iteration. While the figure 8f shows that the iteration convergent when the iteration is 81st .It indicates that the difference in the fluidal speed also affect the iterations in cooling process of emission gas. The bigger fluidal speed in Heat Exchanger causes the number of iterations required is also growing and conversely.

4. Conclusion and Suggestion

Based on research conducted, it can be concluded that the bigger ambient temperature around the system cause the longer cooling process of emission gas and the higher flow rate of emission gas cause the longer cooling process of emission gas. Therefore, to minimize the exhaust gas emission level the plant is expected to condition the ambient temperature of the engine and adjust the fluid flow rate while inside the engine so that the cooling process does not last long and the exhaust emissions are not too high. The calculation using the finite element method is effective because it has a relative error under 0.01.

Based on research conducted, we expected that the calculation can use the other method and the simulation can also use the other software than FLUENT and MATLAB. For the next research still required the other factor than the flow rate and the ambient temperature, and for the precise gas may be more focused on each type of exhaust emissions.

5. References

[1] Kreith F 1998 *The CRC Handbook of Thermal Engineering* (USA: CRC Press)
[2] Samuel, S and Parlindungan M 2011 Analysis of Froude Number On Tour Boat at Jatiluhur Reservoir With Computational Fluid Dynamic Approach (Case Study KM Jasatirta) *Jurnal Kapal* **32**(3) pp. 112-118
[3] Widodo B 2011 Mathematical Modeling and Numerical Solution of Iron Corrosion Problem Based on Condensation Chemical Properties *Australian Journal of Basic and Applied Sciences* **5**(1) pp. 79-86
[4] Jung D H and Assanis D N 2006 Numerical Modeling of Cross Flow Compact Heat Exchanger with Louvered Fins using Thermal Resistance Concept *SAE Paper* 2006-01-0726
[5] Huang Y 2014 Multiscale Thermal Analysis Approach For The Typical Heat Exchanger Automotive Cooling System *International Communications in Heat and Mass Transfer* **59** pp. 78-87
[6] Davidzon M I 2012 Newton's law of cooling and its interpretation *International Journal of Heat and* Mass *Transfer* **55** pp. 5397–5402
[7] Jiaqiang E 2018 Performance Enhancement of a Baffle-Cut Heat Exchanger of Exhaust Gas Recirculation *Applied Thermal Engineering* **134** pp. 86-94

Acknowledgement
We would like to thank the University of Jember, Indonesia, which give us a grant and chance to disseminate an intellectual output of our research.

The Effectiveness of ARICESA-based Learning Material on Students' Motivations

R Setiani[1*]

[1]STKIP PGRI Tulungagung
[*]corresponding author: rahyusetiani@gmail.com

Abstract. This research aims to describe the effectiveness of ARICESA (Attention, Relevance, Inquiry, Confidence, Enjoyment, Satisfaction, and self-Assessment)-based learning material to improve students' motivation on the basic concept of natural science. This research was conducted in 6 weeks by employing a pre-experimental research, that is one group pre-test and post-test design. The data to seek for the effectiveness was obtained through students' motivation questionnaire that consists of 5 aspects (choice of tasks, effort, persistence, self-confidence, and achievement). Then, the data that were obtained were analysed descriptive and inferential statistics. The mean scores of students' motivation before joining learning process with ARICESA-based learning material (initial motivation) is still in "very less" criteria in each aspect of motivation. The mean scores of students' motivation after following the learning (final motivation) increase in "good" criteria. The increase was shown quantitatively with a normalized gain (N-gain) mean of 0.65 in the medium criteria. The dependent t-test results showed p-value (significance) of 0.000 (<0.05). This means that there is a significant difference from the beginning and the end of students' motivation in an undergraduate program of primary school teacher education for basic science concept learning.

1. Introduction

The prospective teachers should have good competence with a various of 21st-century skills. "Assessment and Teaching of 21st-Century Skills" (ATC21S) categorizes 21st-century skills into 4 categories, one of them are ways of thinking [1]. The ways of thinking category consist of creativity, innovation, critical thinking, and metacognition [1], [2]. In the ways of thinking category, motivation and succeed confidence become capital to achieve these skills [1]. This shows that the learning motivation to support the achievement of 21st-century skills.

Learning motivation has a positive correlation with the level of student self-sufficiency, which became one of 21st-century skills [1], [3], and the competence of graduates referring to Indonesian National Qualification Framework. Learning motivation becomes the main factor of learning independence [4]. One of the intrapersonal resources as a principal to achieve learning independence is the motivation to learn [5]. When students have high motivation, then the student will spend a lot of time to study independently [6], [7]. Intrinsic and extrinsic motivations are adaptive that encourage and sustain learning independence [8]. The higher level of the learning motivation, the higher level of learning self-sufficiency.

The creativity that is one of the 21st-century skills is also influenced by learning motivation. Learning motivation has a significant causality relationship between creative thinking ability [9]–[11]. Creative thinking skills emerge from learning motivation [12], [13]. The motivation to learn motivation can improve creativity [9], [12]. Learning motivation is a major factor in creative thinking [14].

Learning motivation is the main principal of students to achieve 21st-century skills such as self-sufficiency and creativity [1], [5], [14]. The challenges of achieving 21st-century skills are not supported by the students' motivation. The preliminary study that has conducted in an undergraduate program of primary school teacher education in STKIP PGRI Tulungagung, showed that the students' motivation is still very low. This fact is supported by PISA findings [15], TIMMS 2015 [16], and UNESCO [17] that indicating low student' motivation. The low student' motivation also indicated from those who are not motivated in the educational environment in school [18]–[20]. Based on the description above, the researcher develops a natural sains learning materials that are able to fulfill the criteria of validity, practicality, and effectiveness. Generally, the purpose of this research is to describe the effectiveness of ARICESA-based learning material to improve students' motivation on the basic concept of natural science.

ARICESA-based learning is carried out based on the ARICESA model syntax. The ARICESA model (Attention, Relevance, Inquiry, Confidence, Enjoyment, Satisfaction, and self-Assessment) is the result of the expansion of the ARCS model by adding an inquiry process (I), creating an enjoyment atmosphere (E), and self-assessment (A) that has fulfilled the content and construct validity [21]. The ARCS model (Attention, Relevance, Confidence, and Satisfaction) is an excellent synthesis of motivational research, but the weakness of ARCS is to only focus the source of motivation on the individual as a unit of analysis. Whereas there are external factors that can also affect learning motivation, such as the presence of collaborative factors on social presence, the structure of experience that causes learning to flow, and environmental conditions [22]. I as an Inquiry phase can encourage learning that provides an active role for students while learning. This inquiry model is also able to bridge the needs of students to the space of social interaction and active involvement to generate motivation and self-determination [23], [24]. E as an enjoyment phase that makes it clear that learning must be in the form of joyful learning. The learning environment (in terms of the physical structure of the classroom environment and equipment and materials to be used in the classroom) must be designed appropriately to increase learning motivation [25]. A as an Assessment phase that encourages self-assessment of students themselves as a continuation of learning that supports inquiry. If it is associated with the forming of learning motivation, then self-assessment is more integrated with the achievement of mastery goals aspects [26].

2. Methods

The research subjects of the implementation of ARICESA-based learning material on basic science concepts (Force and Motion, Temperature, Heat, Energy and Changes, Light, and Sound) are 34 students who are 2nd graders of the undergraduate program of primary school teacher education, STKIP PGRI Tulungagung. The development of learning material are adapted from Plomp design [27]. The developed learning material in this research consists of the lesson plan, student worksheet, and learning book. The stages in research and development method include preliminary research, prototyping stage, and assessment phase. However, this research is limited to the stage of prototyping stage. The learning material that is implemented in this research has passed the validity test stage with "very valid" category.

This research was conducted in 6 weeks by employing a pre-experimental research, that is one group pre-test and post-test design [28]. The experimental design was used to investigate the effectiveness of ARICESA-based learning material. The one-group pretest-posttest has been performed by means of a group of students subject to treatment and the dependent variables observed (measured) to assess the effect of the treatments [28]. The treatment in question is related to learning by applying ARICESA-based learning material which measured the learning motivation on the basic concept of science. The learning process at the time of limited implementation is done by the lecturers of the Basic Concepts of Science.

The data to seek for the effectiveness was obtained through students' motivation questionnaire that consists of 5 aspects (choice of tasks, effort, persistence, self-confidence, and achievement). Student motivation data was obtained by completing questionnaire before (initial motivation) and after (final motivation) the implementation of ARICESA-based learning material. Analysis of students' motivation to describe students' self-condition about learning motivation level with implementation of ARICESA-based learning material, analyzed descriptively qualitative with formula percentage of agreement. Students said to be complete if the score of learning motivation of at least 3.25 and classical completeness is achieved when 85% of students complete. The completeness of the indicators using questionnaires motivation to learn. Students are said to master the indicator when the minimum value of 3.25 and the completeness of the indicator is achieved if 75% of students have been completed. The data that were obtained were analyzed descriptive and inferential statistics. The scores of students' motivation are categorized based on the following criteria: very less, less, good, and very good. The differences in students' motivation scores before and after the treatment were analyzed using N-gain. The significant of motivation improvement were analyzed using dependent t-test.

3. Result and discussion

The descriptions of student learning motivation were done in two ways, using N-gain and paired t-test. The use of N-Gain is intended to find out the criteria of increasing motivation to learn basic concepts of science students. While the use of paired t-test is intended to know the significance of the increase. Table 1 shows descriptive analysis of students' learning motivation questionnaire of each indicator based on excellent category (EX), good (G), enough (E), and less (L). It also shows the average score of learning motivation in each indicator. In summary, the description of Table 1 is presented in Figure 1 and Figure 2.

Table 1. The analysis results of learning material questionnaire.

Indicator of Motivation	Categories	Initial Motivation			Final Motivation		
		∑	%	x	∑	%	x
Choice of Tasks	EX	0	0.00	1.92	14	41.18	3.95
	G	0	0.00		20	58.82	
	E	13	38.24		0	0.00	
	L	21	61.76		0	0.00	
Effort	EX	0	0.00	1.86	17	50.00	3.94
	G	0	0.00		17	50.00	
	E	11	32.35		0	0.00	
	L	23	67.65		0	0.00	
Persistence	EX	0	0.00	1.89	15	44.12	3.88
	G	0	0.00		19	55.88	
	E	16	47.06		0	0.00	
	L	18	52.94		0	0.00	
Self-Confidence	EX	0	0.00	1.97	18	52.94	3.96
	G	0	0.00		16	47.06	
	E	16	47.06		0	0.00	
	L	18	52.94		0	0.00	
Achievement	EX	0	0.00	1.89	13	38.24	3.83
	G	0	0.00		21	61.76	
	E	12	35.29		0	0.00	
	L	22	64.71		0	0.00	

∑: number of student; x: average score

Figure 1 shows that the implementation of the ARICESA-based learning material can increase the average score of learning motivation in all learning motivation indicators. Based on the results of the initial motivation and final motivation, the implementation of ARICESA-based learning material can increase the number of students to a higher level of motivation on each indicator, which is summarized in Figure 4.2.

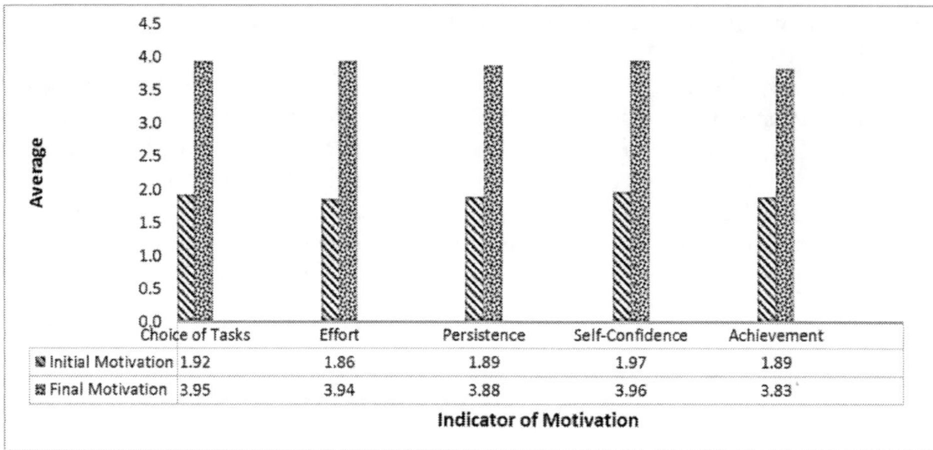

Figure 1. The average of learning motivation score.

Figure 2 shows that all students before attending the learning process with ARICESA-based learning material is still in enough category and less good in each indicator of motivation. No student has a good and excellent motivation category. Implementation of the ARICESA-based learning material can increase the number of students in the higher category. All students after learning have a motivation to learn science in good and excellent category. No student still has a category of motivation is good enough or less good.

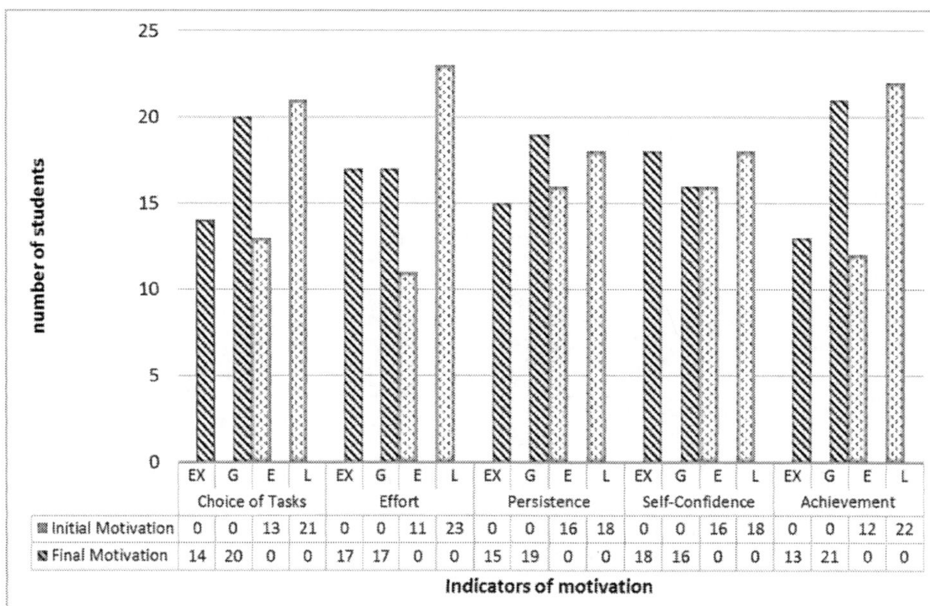

Figure 2. The number of students on each indicator of learning motivation.

Table 2 shows that the completeness of the final learning motivation indicator the basic science of students has reached 100% in each indicator. The level of achievement is already above the minimum threshold set that is 75%. It is much improved from the completeness of the student's initial motivation on each indicator. The results of initial and final motivational measurements showed an increase in motivation with a normalized gain average of 0.65 in the medium category.

Table 2. Completed indicator and N-gain indicators of learning motivation.

Indicator of Motivation	Initial Motivation			Final Motivation			N-gain	
	Score	Completed Indicator		Score	Completed Indicator		<g>	Criteria
Choice of Tasks	1.92	0.00%	Not completed	3.95	100.00%	Completed	0.66	Medium
Effort	1.86	0.00%	Not completed	3.94	100.00%	Completed	0.66	Medium
Persistence	1.89	0.00%	Not completed	3.88	100.00%	Completed	0.64	Medium
Self-Confidence	1.97	0.00%	Not completed	3.96	100.00%	Completed	0.66	Medium
Achievement	1.89	0.00%	Not completed	3.83	100.00%	Completed	0.62	Medium
Average	1.91	0.00%	Not completed	3.91	100.00%	Completed	0.65	Medium

The testing of difference of initial and final learning motivation was done to see the significance of improvement of learning motivation after applied ARICESA-based learning material. The test of difference of learning motivation was used paired t-test. Assumptions in paired t-test analysis are the normality and homogeneity of variance between groups [29]. Data analysis used to determine data normality was Kolmogorov-Smirnov test, whereas to test the homogeneity of variance between groups was used Levene's test [29]. Both testing techniques are assisted by SPSS 18 software. Test results of normality of learning motivation by using Kolmogorov-Smirnov Test obtained p-value or statistical significance equal to 0.194 for initial motivation measurement and 0.200 for final motivation measurement. Both have values greater than 0.05; that means learning motivation follow normal distribution function. While the results of homogeneity test of data variance of learning motivation by using Levene's test obtained p-value or statistical significance of 0.186. The value is greater than 0.05; that means the variance of initial and final learning motivation are homogeneous.

Based on the normality and homogeneity test results obtained that the data has followed the normal distribution function and the variation of the data is also homogeneous, so testing the difference between learning motivation were done by paired t-test. This test was conducted to determine whether there are differences in initial motivation and final motivation. The result of paired t-test analysis on learning motivation of basic science concepts is presented in Table 3.

Table 3. The result of paired t-test analysis on learning motivation of basic science concepts.

	t	df	Sig (2-tailed)	Description
Initial and final motivation	-26.722	33	0.000	H_0 Rejected

Table 3 shows that the paired t-test statistic value for initial motivation and final motivation is -26.722 and has p-value (significance) of 0.000 (<0.05). Because of the significance value of the test result is less than 0.05; then the conclusion is H0 rejected or H1 accepted. This means that there is a significant difference between initial and final learning motivation of students.

Based on the data findings, it can be stated that the developed ARICESA-based learning materials has qualified effectiveness in terms of student's learning motivation. In other words, the ARICESA-based learning materials is an effective tool for improving the learning motivation. This is in accordance with the basic component characteristics of the ARICESA model that is the ARCS motivation model and the Inquiry model. Application of ARCS model in learning can increase learning motivation [30]–[32]. Similarly, the results of the application of inquiry models that have been done by previous researchers have proven to improve learning motivation [32]–[34]. The combination of the two characteristics of the model into the ARICESA model eventually proved to be able to increase the motivation to learn. The joyful learning (enjoyment) component included in the ARICESA model also makes students focus, persistent, and make optimal efforts in understanding learning materials [35]. That is, students are increasingly motivated to learn with a fun learning atmosphere [36].

4. Conclusion

Based on the findings during the research it can be concluded that the ARICESA-based learning material are effective to improve students' motivation on basic science concepts in undergraduate program of primary school teacher education. That have shown from N-gain for learning motivation the medium category with an average score of 0.65 and paired t-test results for learning motivation indicate a significant difference between before and after the implementation of ARICESA-based learning materials. Based on the conclusion of the research result, it is given some suggestions for further research in order to pay attention to the design of classroom management, especially in group work, related to the number of members, supporting equipment, and the allocation of learning time, and also need to be tested consistency of improvement of learning motivation.

5. References

[1] Griffin P, Care E, and McGaw B 2012 The Changing Role of Education and Schools *Assessment and Teaching of 21st Century Skills Springer Science+Business Media* 1–15.

[2] Wagner T 2010 *The Global Achievement Gap* (New York: Basic Books)

[3] Ananiadou K and Claro M 2009 21st century skills and competences for new millennium learners in OECD countries *OECD Publishing* **41**

[4] Mahmoodi M H, Kalantari B and Ghaslani R 2014 Self-Regulated Learning (SRL), Motivation and Language Achievement of Iranian EFL Learners *Procedia - Soc. Behav. Sci.* **98** 1062–1068

[5] Zhang W 2017 Using classroom assessment to promote self-regulated learning and the factors influencing its (in)effectiveness *Front. Educ. China* **12** (2) 261–295

[6] Tagawa M 2008 Physician Self-Directed Learning and Education *Kaohsiung J. Med. Sci.* **24** (7) 380–385

[7] Kim K -J 2009 Motivational Challenges of Adult Learners in Self-Directed E-Learning *J. Interact. Learn. Res.* **20** (3)317–335

[8] Pintrich P R 1999 The role of motivation in promoting and sustaining self-regulated learning," *Int. J. Educ. Res.* **31** 459–470

[9] Chen J K and Chen I S 2012 Critical creativity criteria for students in higher education: Taking the interrelationship effect among dimensions into account *Qual. Quant.* **46** (4) 1057–1075

[10] Trianggono M M 2017 Analisis Kausalitas Pemahaman Konsep Dengan Kemampuan Berpikir Kreatif Siswa Pada Pemecahan Masalah Fisika *J. Pendidik. Fis. dan Keilmuan(JPFK)* **3** (1) 1–12

[11] Runco M A and Chand I 1995 Cognition and creativity *Educ. Psychol. Rev.* **7**(3) 243–267

[12] Cropley D and Cropley A 2010 Recognizing and fostering creativity in technological design education *Int. J. Technol. Des. Educ.* **20**(3) 345–358

[13] d'Inverno M and Luck M 2012 Creativity Through Autonomy and Interaction," *Cognit. Comput.* **4** (3) 332–346

[14] Hamza T S and Hassan D K 2016 Consequential creativity: Student competency and lateral thinking incorporation in architectural education *Int. J. Technol. Des. Educ.* **26** (4) 587–612

[15] PISA 2016 *PISA 2015 Results in Focus* (OECD)

[16] Mullis I, Martin M, Foy P and Hooper M 2016 *TIMSS 2015: International Results in Mathematics and Sience* (Boston: TIMSS & PIRLS International Study Center)

[17] UNESCO 2010 *Current challenges in basic science education* (Paris: UNESCO - Education Sector)

[18] Fortus D and Vedder-Weiss D 2014 Measuring students' continuing motivation for science learning *J. Res. Sci. Teach.***51** (4) 497–522

[19] Cetin-dindar A 2015 Student Motivation in Constructivist Learning Environment *EURASIA J. Math. Sci. Technol. Educ.* **12** (2) 233–247

[20] Ng B L L, Liu W C and Wang J C K 2016 Student Motivation and Learning in Mathematics and Science: A Cluster Analysis *Int. J. Sci. Math. Educ.* **14**(7) 1359–1376

[21] Setiani R, Resbiantoro G, and Wanda A 2018 The Validity of ARICESA-Based Learning Materials in Basic Science Concept for Student of Primary School Teacher Education *Proceedings of the 1st Annual International Conference on Mathematics, Science, and*

Education (ICoMSE 2017) **218** 62–70.

[22] Robinson D G 2013 Learner Motivation Systems: A Framework for the Analysis and Design of Learner Motivation *Framework for design learner motivation*, Eastern Virginia Medical School

[23] Williams D R, Brule H, Kelley S S and Skinner E A 2018 Science in the Learning Gardens (SciLG): a study of students' motivation, achievement, and science identity in low-income middle schools *Int. J. STEM Educ.* **5** (8) 1–14

[24] Ng B L L, Liu W C and Wang J C K 2016 Student Motivation and Learning in Mathematics and Science: A Cluster Analysis *Int. J. Sci. Math. Educ.* **14** (7) 1359–1376

[25] Surucu A and Ozdemir H 2013 Comparison of the chemistry learning motivations of the science and primary school teacher candidates *Anthropologist* **16** (3) 671–676

[26] McMillan J H and Hearn J 2008 Student Self-Assessment: The Key to Stronger Student Motivation and Higher Achievement *Educ. Horizons* **87** (1) 40–49

[27] Plomp T 2013 Educational Design Research: An Introduction *Educational Design Research Part A: An introduction*, T. Plomp and N. Nieveen, Eds. (Enschede: Netherlands Institute for Curriculum Development (SLO)) 10–51

[28] Creswell J W 2012 *Educational research: Planning, conducting, and evaluating quantitative and qualitative research, 4th ed.* **4**. (Washington, D.C.: SAGE Publications)

[29] Peers I S 1996 *Statistical Analysis for Education and Psychology Researchers*

[30] Andrade H and Du Y 2007 Student responses to criteriareferenced self-assessment *Assess. Eval. High. Educ.* **32** (2) 159–181

[31] Malik S 2014 Effectiveness of arcs model of motivational design to overcome non completion rate of students in distance education *Turkish Online J. Distance Educ.* **15** (2) 194–200

[32] Lee I C 2010 The Effect of Learning Motivation, Total Quality Teaching and Peer-Assisted Learning on Study Achievement: Empirical Analysis from Vocational Universities or Colleges' students in Taiwan *J. Hum. Resour. Adult Learn.* **6** 56–73

[33] Wang P H, Wu P L, Yu K W and Lin Y X 2015 Influence of Implementing Inquiry-based Instruction on Science Learning Motivation and Interest: A Perspective of Comparison," *Procedia - Soc. Behav. Sci.* **174** 1292–1299

[34] Megawati D and Manurung S R 2016 Pengaruh Model Pembelajaran Inquiry Training dan Motivasi Terhadap Hasil Belajar Fisika Siswa Effect of Inquiry Learning Model and Motivation on Physics Outcomes Learning *J. Pendidik. Fis.* **5** (1) 1–6

[35] Singh U 2013 Technology Integration in Teacher Education : an Infusion or a Delusion? *J. Arts, Sci. Commer.* **IV** (3) 147–155

[36] Meyer D Z, Antink M A, Nabb K A, Connell M G and Avery L M 2013 A Theoretical and Empirical Exploration of Intrinsic Problems in Designing Inquiry Activities *Res. Sci. Educ.* **43** (1) 57–76

IOP Conf. Series: Journal of Physics: Conf. Series **1108** (2018) 012091 doi:10.1088/1742-6596/1108/1/012091

Critical Thinking Disposition of Prospective Science Teachers at IKIP Mataram, Indonesia

H Fitriani[1], M Asy'ari[1], S Zubaidah[2], and S Mahanal[2]

[1]Institut Keguruan dan Ilmu Pendidikan (IKIP) Mataram, Jl. Pemuda No. 59A, Mataram 83125, Indonesia
[2]Universitas Negeri Malang, Jl. Semarang No.5, Kota Malang, Jawa Timur 65145, Indonesia
Email: muhammadasyari@ikipmataram.ac.id

Abstract. This study aims to identify prospective teachers' critical thinking disposition by gender that consist of seven components: 1) inquisitiveness, 2) self-confidence, 3) truth-seeking, 4) open-mindedness, 5) analyticity, 6) systematic, and 7) maturity. The survey methodology was used in this study. Research sample consisted of 100 students in Biology Education Program, Faculty of Science and Mathematics Education IKIP Mataram, Indonesia. Data on critical thinking dispositions were collected using The California Critical Thinking Disposition Inventory (CCTDI). The results showed that prospective science teachers' critical thinking disposition generally neither strength nor weakness category (score $3.1 \leq X \leq 3.9$). Based on the result, it can be stated that prospective science teachers' critical thinking disposition is neither strength nor weakness and there are differences in critical thinking disposition based on prospective science teachers' gender on the components: 1) inquisitiveness (p= 0.001); 2) self-confidence ($t_{(100)}$= 2.331, p= 0.022); 3) open-mindedness ($t_{(100)}$= 4.531, p= 0.000); and 4) maturity (p= 0.004).

1. Introduction

Critical thinking ability teaches humans to know better about thinking process, because it is a reasonable and reflective thinking focused on deciding what to believe or do [1]. Critical thinking is a controlled assessment with a logical purpose related to evidence, field, concept, method, and criterion [2]. One of the traits of a critical thinker is an ability to regulate his own knowledge [3]. An ideal critical thinkers have high curiosity, broad knowledge, trust in reason, flexible, open-minded to critical, honest in facing personal bias, wise in making judgments, willing to reconsider, formulate problems clearly, orderly in complex matters, diligently search for relevant information, right in the selection of criteria, focus on investigation, and persistent in the investigation [4]. Critical thinking requires the student to actively and skillfully conceptualize, apply, analyze, synthesize, and/or evaluate information gathered or resulting from observation, experience, reflection, communication reasoned, or as a guide to beliefs and actions [5] [6]. The process of inquiry can build students' critical thinking skills [7] and educational demands also require students to build their own knowledge [8].

Critical thinking is often regarded as a general goal of learning in higher education but it is not specifically learned through the provision of relevant teaching materials to achieve these goals so that learning of students' critical thinking skills in a comprehensive manner cannot be achieved. Good critical thinkers combine empowerment of critical thinking skills by maintaining a consistent critical thinking disposition [9] in finding problem-solving solutions [10]. The critical thinking disposition is important for learning because the cognitive component of critical thinking is not enough to be a critical thinker, critical thinking disposition also plays an important role and should be learned [1]. Disposition is a tendency of someone to do something, in certain circumstances, for example, someone tends to be open to alternative solutions to problems encountered from outside himself or consider other solutions of solutions owned. Critical thinking disposition is an attribute or habit of mind that is integrated into one's beliefs or actions [11] to effectively solve problems and make decisions as a product of thinking [9]. The three basic components of critical thinking as thought dispositions are willingness, sensitivity, and ability [12]. Learners should build their own knowledge [13] through inquiry processes, both independently and mentally [14] which in order to perform these processes requires the willingness, sensitivity, and ability to solve the problems encountered, thus thinking critically as a skill component lifelong learning [15] can be done well. Furthermore, [16] describes the critical thinking disposition of seven characteristics, namely; curiosity, open-minded, systematic, analytic, truth-seeking, critical thinking, confidence, and maturity.

The results of several studies show the positive impact of critical thinking on learning achievement [17] [18] and improving the performance of Special Education Integration Program (SEIP) teachers [19]. Furthermore, the results of the study [20] showed a positive and very significant correlation between students' critical thinking dispositions and their clinical performance with hemodialysis patients. The general procedure undertaken to teach students critical thinking is the integration of process skills in learning. Process skills or learning through discovery are basic skills for an inquiry [21] because in it students are taught to connect their prior knowledge or concrete knowledge of students with new learning concepts. The ability of discovery and inquiry independently is more important than the achievement of the concept that students learn [22]. Yuksel and Alci explained that the disposition or attitude of students to think critically is an important initial condition to be able to think critically and help students predict the tendency of better skills in making mature decisions [23]. The critical thinking disposition can also encourage students to train their critical thinking skills [24]. In line with this opinion, there is a positive and significant relationship between critical thinking dispositions and critical thinking skills [25] [26].

Critical thinking skills are one of the basic capital or intellectual capital that is very important for everyone and is a fundamental part of human maturity so it is important to be taught at every level of education. Critical thinking ability is used as a basis for analysing an argument and the emergence of an interpretation in developing a consistent and logical reasoning pattern, the ability to understand assumptions, formulate problems, make deductions and make decisions appropriately and can be developed through the process of learning activities or behaviours during the process. Gender is one of the factors that can influence one's behaviour [27].

Gender is a generic term that refers to male and female [28] that make up the psychology and social role of a person [29] so that it impacts on how individuals think, behave, and perceive a phenomenon within the self [30]. Based on the description, male and female may differ in the context of thinking skills and thinking dispositions, [31] suggesting that male in thinking are more analytical and flexible than female, while female are less skilled in abstract and logical thinking [32]. Further [29] states that male and female have no distinction in conceptual understanding, but male are superior in solving problems rather than female. On the other hand, [33] mentions that there is no significant correlation between critical thinking between male and female. On the other hand, [34] state that critical thinking dispositions differ according to the sex that supports female students.

This study aims to identify differences in critical thinking disposition of prospective teachers in Biology Education Studies Program, Faculty of Science and Mathematics Education IKIP Mataram, Indonesia that specifically based on gender, consisting of seven components: 1) inquisitiveness, 2) self-confidence. 3) truth-seeking, 4) open-mindedness, 5) analyticity, 6) systematics, and 7) maturity [35] [36]. The results of this study also present an intersection between each component that becomes a special characteristic of prospective male and female teachers' critical thinking disposition.

2. Method

The survey methodology was used in this study to identify students' critical thinking disposition by giving The California Critical Thinking Disposition Inventory (CCTDI) questionnaire compiled by [35] [36]. CCTDI is the principal tool for surveying the dispositional aspects of critical thinking containing 75 items of statements for the 7 critical dispositional components of thinkers: 1) inquisitiveness (10 items), measuring one's intellectual curiosity level; 2) self-confidence (9 items), refers to a person's level of trust in reasoning; 3) truth-seeking (12 items), refers to the someone's spirit to seek the truth, dare to ask questions, and be honest and objective in the quest for problem solving faced, even if the findings do not support the interests or opinions of someone previously formed; 4) open-mindedness (12 items), refers to the disposition of open-mindedness and one's tolerance to different views and sensitivity to the possibility of one's bias; 5) analyticity (11 items), refers to one's awareness of potentially problematic situations, anticipates possible outcomes or consequences, and prioritizes the application of the reason and use of evidence even if the problem is difficult to resolve; 6) systematics (11 items), referring to the disposition of an organized, orderly, focused, and diligent investigation; and 7) maturity (10 items), referring to a reflective assessment of an individual. CCTDI is designed for use in adults in general. Scores for each of the seven dispositions can range from 1 to 6. "Score 3 or below shows a weak disposition. Score between 3.1 until 3.9 does not indicate the strength or weakness of the respondent's critical thinking disposition. A score of 4.0 to 4.9 shows a minimal disposition of critical thinking disposition's mean score and scores above 5.0 show strong thinking dispositions [36]. Cronbach alpha for the seven components of CCTDI disposition ranges from 0.71, while the total reliability of 75 statement items in CCTDI ranges from 0.90 to 0.91 [35] [36] further, Kokdemir in [23] reported the value of the total reliability of all components in a CCTDI is 0.8 (reliable category).

The sample in this research is 100 prospective teacher students (50 male and 50 female) of biology education wheres selected purposively. The prospective teachers' critical thinking disposition then categorized using Table 1 below:

Table 1. Category of Student Critical Thinking Disposition

Skor	Category
$X \geq 5$	Strong
$4.0 \leq X \leq 4.9$	Enough
$3.1 \leq X \leq 3.9$	Neither Strength nor Weakness
$X \leq 3.0$	Week

Adopted from [36]

Prospective teachers' critical thinking disposition was analyzed using SPSS for Windows software. The analysis results include t-test and Mann-Whitney to identify differences in each component of the prospective teachers' critical thinking disposition and the correlation between each component of the prospective teachers' critical thinking disposition.

3. Result and Discussion

The result of prospective teachers' critical thinking disposition questionnaire is classified by gender subsequently, independent sample t-test and Mann-Whitney test are conducted to determine the difference between prospective teachers' critical thinking disposition. The results of the students' differences analysis are briefly presented in Table 2.

Table 2. The t-test and Mann-Whitney test results of prospective teachers' critical thinking disposition based on gender.

					Independent Sample t-Test		
Dispositions	**Gender**	**N**	**Mean**	**SD**	**df**	**T**	**p**
Self-Confidence	Male	50	3.0330	.37810	100	2.331	.022
	Female	50	2.8552	.38455			
Open-Mindedness	Male	50	2.9230	.32115	100	4.531	.000
	Female	50	2.6360	.31216			
					Mann-Whitney Test		
Dispositions	**Gender**	**N**	**Mean**	**SD**	**ΣN**		**p**
Inquisitiveness	Male	50	40.47	.47406	100		.001
	Female	50	60.53				
Maturity	Male	50	43.50	.36480	100		.004
	Female	50	57.50				

Based on Table 2 it can be seen that the critical thinking disposition of female prospective teacher is better than male in composite inquisitiveness (mean = 60.53), and maturity (mean = 57.50), while male are better on the self-confidence component (mean = 3.0330) and open-mindedness (mean = 2.9230).

IOP Conf. Series: Journal of Physics: Conf. Series **1108** (2018) 012091 doi:10.1088/1742-6596/1108/1/012091

Correlations between each component of male and female critical thinking disposition and the correlation of the disposition of all samples were also analyzed to determine which disposition components interacted with each other. The disposition of analyticity becomes the most positively correlated component with other disposition components both from gender review (Table 3) and generally (Table 4).

Table 3. Correlation of students' critical thinking disposition component based on Gender

Dispositions	Gender	N	p	Annotation
Inquisitiveness (I)	Male	50	$.013^{A}; .025^{M}$	50^{AM}
	Female	50	$.040^{C}; .000^{T}; .012^{A}; .017^{M}$	50^{CTAM}
Self-Confidence (C)	Male	50	$.021^{M}$	50^{M}
	Female	50	$.040^{I}; .024^{A}; .002^{M}$	50^{IAM}
Truth-Seeking (T)	Male	50	$.006^{A}$	50^{A}
	Female	50	$.000^{I}; .033^{A}$	50^{IA}
Open-Mindedness (O)	Male	50	$.005^{S}$	50^{S}
	Female	50	$.001^{A}$	50^{A}
Analyticity (A)	Male	50	$.013^{I}; .006^{T}$	50^{IT}
	Female	50	$.012^{I}; .024^{C}; .033^{T}; .001^{O}; .001^{M}$	50^{ICTOM}
Systematicity (S)	Male	50	$.005^{O}$	50°
	Female	50	-	-
Maturity (M)	Male	50	$.025^{I}; .021^{C}$	50^{IC}
	Female	50	$.017^{I}; .002^{C}; .001^{A}$	50^{ICA}

Table 4. The correlation of prospective teachers' critical thinking disposition components

Dispositions	N	p	Annotation
Inquisitiveness (I)	100	$.000^{T}; .001^{A}; .000^{M}$	100^{TAM}
Self-Confidence (C)	100	$.038^{A}; .045^{S}; .002^{M}$	100^{ASM}
Truth-Seeking (T)	100	$.000^{I}; .001^{A}; .013^{M}$	100^{IAM}
Open-Mindedness (O)	100	$.006^{A}; .022^{S}$	100AS
Analyticity (A)	100	$.001^{I}; .038^{C}; .001^{T}; .006^{O}; .006^{M}$	100^{ICTOM}
Systematicity (S)	100	$.045^{C}; .022^{O}$	100^{CO}
Maturity (M)	100	$.000^{I}; .002^{C}; .013^{T}; .006^{A}$	100^{ICTA}

Critical thinking is the process of seeking, producing, evaluating, analysing, synthesizing and conceptualizing information as a guide to the development of one's thinking skills with the awareness and the ability to use that information creatively [23]. The critical thinking disposition is the realm of one's consciousness to think critically [36]. The results showed that female prospective teacher was better than male in composite inquisitiveness (mean = 60.53), and maturity (mean = 57.50), while male was better on self-confidence components (mean = 3.0330) and open-mindedness (mean = 2.9230). Different results are presented [23] which suggest that males are superior in inquisitiveness and systematics components than female. In line with the results of the study, [31] stated that male in thinking are more analytical and flexible than female, while female is less skilled in abstract and logical thinking [32]. Male are better at logical thinking than female but are better at controlling emotions [37] so that in accordance with the results of the female is superior in the maturity component. Critical thinking that requires precise, objective, and reflective attributes [38] [39] [40] requires good emotional control in order to make good decisions.

Critical thinking disposition components such as truth-seeking ($p = .118$), analyticity ($p = .800$), and systematics ($p = .663$) were not significantly different between male and female. In general, prospective teachers' critical thinking disposition of both male and female are categorized as neither strength nor weakness ($3.1 \leq X \leq 3.9$) so it is necessary to be developed. People can have abilities but are sometimes not used [38] [3]. In this stage of the abilities, disposition plays an important role, because to be a successful critical thinker, the critical thinking skills that constitute cognitive skills are insufficient and require the development of a critical thinking disposition. The development of critical thinking and other high-level thinking require good plan through the learning process by applying various models that are appropriate to students' conditions and the material being studied [41] [42].

4. Conclusion

The prospective science teachers' critical thinking disposition generally should be improves. The critical thinking disposition of female and male prospective teacher varies in several components, namely inquisitiveness, maturity, self-confidence, and open-mindedness. The female prospective teacher was better than male in composite inquisitiveness and maturity, while male was better on self-confidence and open-mindedness components. The disposition component of Analyticity has a dominant correlation in all components of critical thinking disposition of male and female so it is important to be learned or integrated into the learning process in the classroom.

5. References

[1] Ennis R H 1996 *Critical Thinking* (New York: Prentice-Hall)
[2] Harkreader H. 2000 *Fundamental of nurse caring and clinical Judgment* (Philadelphia: WB Saunders Company)
[3] Facione PA 1991 *Critical Thinking: A statement of Expert Consensus for Purposes of Educational Assessment and Instruction* (Millbrae CA: California Academic Press)
[4] Gul R B, Cassum S, Ahmad A, Khan S, Saeed T & Parpio Y 2010 Enhancement of critical thinking in curriculum design and delivery: A randomized controlled trial for education. *Procedia Social and Behavioral Sciences* **2** 3219-3225.
[5] Scriven M and Paul R 1996 Defining Critical Thinking: A Draft Statement for the National Council for Excellence in Critical Thinking Available on http://www.criticalthinking.org/University/univlibrary/library.nclk.
[6] Phan H 2010 Critical Thinking As a Self-Regulatory Process Component in Teaching And Learning *Psycothema* **(22)** 289-292
[7] Arends R 2012 *Learning to Teach* Ninth Edition (New York: McGraw-Hill)
[8] Kemendikbud 2013 *Permendikbud No 73 Tahun 2013 Penerapan Kerangka Kualifikasi Nasional Indonesia (KKNI) Bidang Pendidikan Tinggi* (Jakarta: Kemendikbud)
[9] Facione N C and Facione P A 2007 *The critical thinking disposition inventory CCTDI A test of critical thinking disposition* (Cambridge: California Academic Press)
[10] Kabeel AR and Eisa S A M 2016 The Correlation of Critical Thinking Disposition and Approaches to Learning among Baccalaureate Nursing *Students Journal of Education and Practice* **7** (32) 91-103.
[11] Profetto-McGrath J, Hesketh K L, Lang S and Estabrooks CA 2003 A study of critical thinking and research utilization among nurses *Western Journal of Nursing Research* **25** (3) 322-337
[12] Perkins D N, Jay E and Tishman S 1993 Beyond abilities: A dispositional theory of thinking *Merrill-Palmer Quarterly* **39** (1) 1-21
[13] Ibrahim M and Wahyusukrtiningsih 2014 *Model Pembelajaran Inovatif Melalui Pemaknaan, (Belajar Perilaku Positif dari Alam)* (Surabaya: Unesa University Press)
[14] Nur M 2011 *Modul Keterampilan-keterampilan Proses dan Hakikat Sains* (Surabaya: PSMS Unesa)
[15] Hudgins B B and Edelman S 1988 Children's Self Directed Critical Thinking: A Model for Its Analysis and Two Examples *Journal of Educational Research* **81**(5) 262-273
[16] Mahmoud H G 2012 Critical Thinking Dispositions and Learning Styles of Baccalaureate Nursing Students and its Relation to Their Achievement *International Journal of Learning & Development* **2**(1) 398-415.
[17] Collins K M and Onwuegbuzie A J 2000 Relationship between critical thinking and performance research methodology courses. Paper presented at the Annual Conference of the Mid-South Educational Research Association (United States: ERIC)
[18] Prayogi S and Asy'ari M 2013 Implementasi Model PBL (Problem Based Learning) untuk Meningkatkan Hasil Belajar dan Kemampuan Berpikir Kritis Siswa *Jurnal Prisma Sains* **(1)1** 79-87
[19] Sulaiman T, Kuppusamy S K, Ayub A F M and Rahim S S A 2016 Relationship between Critical Thinking Disposition and Teaching Efficacy Among Special Education Integration Program Teachers in Malaysia. *2nd International Conference and Workshop on Mathematical*

Analysis
[20] Mohamed HA and Mohammed S S 2016 Relationship between critical thinking disposition of nursing students and their performance for patients on hemodialysis *IOSR Journal of Nursing and Health Science* **5** (6) 45-53
[21] Nur M 2011 *Modul Keterampilan-keterampilan Proses dan Hakikat Sains* (Surabaya: PSMS Unesa)
[22] Ausubel D P 1962 A Subsumption Theory of Meaningful Verbal Learning and Retention. *The Journal of General Psychology* – (66) 213-244
[23] Demirhan E and Köklükaya AN 2014 The critical thinking dispositions of prospective science teachers *Procedia-Social and Behavioral Sciences* **116** 1551-1555
[24] Nieto A M and Saiz C 2011 Skills and Dispositions of Critical Thinking: Are They Sufficient? *Anales the Psicologia* **27** (1) 202-209
[25] Giancarlo C A & Facione N C 1994 *A study of the critical thinking disposition and skill of Spanish and English-speaking students at Camelback high school* (Millbrae, CA: The California Academic Press)
[26] Facione N, Facione P 1997 *Critical Thinking Assessment in Nursing Education Programs: An Aggregate Data Analysis* (Millbrae CA: The California Academic Press)
[27] Funder DC 1987 Errors and mistakes: Evaluating the accuracy of social judgment *Psychological Bulletin* **101** (1) 75-90
[28] Fin L S and Ishak Z 2012 A Priori Model of Students Academic Achievement: The Effect of Gender as Moderator *Procedia-Social and Behavioral Sciences* **65** 1092-1100
[29] Fuad N M, Zubaidah S, Mahanal S, and Suarsini E 2017 Improving Junior High Schools' Critical Thinking Skills Based on Test Three Different Models of Learning *International Journal of Instruction* ***10*(1)** 101-116
[30] Santrock J W 2011 *Educational Psychology Fifth Edition* (Dallas: University of Texas)
[31] Dagun S M D 1992 *Maskuline dan Feminisme: Perbedaan Pria dan Wanita dalam Fisiologi, Psikologi, Seksual, Karier dan Masa Depan* (Jakarta: PT. Rineka Cipta)
[32] Krutetskii V A 1976 *The Psychology of Mathematical Abilities in School children* (Chicago: University of Chicago Press)
[33] Taghva F, Rezaei N, Ghaderi J and Taghva R 2014 Studying the relationship between critical thinking skills and students' educational achievement (Eghlid Universities as Case Study) *International Letters of Social and Humanistic Sciences* **25** 18-25
[34] Yenice N 2011 Investigating pre-service science teachers' critical thinking dispositions and problem solving skills in terms of different variables *Educational Research and Reviews* **6** 497-508
[35] Facione P A and Facione N C 1994 *Holistic Critical Thinking Rubric* (Millbrae CA: California Academic Press)
[36] Facione P A, Facione N C and Giancarlo C A 2000 *The California Critical Thinking Disposition Inventory; Test Manual: 2000 Update* (Millbrae CA: California Academic Press)
[37] Lippa R A 2005 *Gender, nature, and nurture* (London: Routledge)
[38] Ennis R 2011 Critical thinking: Reflection and perspective Part II *Inquiry: Critical thinking across the Disciplines* **26** (2) 5-19
[39] Paul R, and Elder L 2006 *Critical thinking: Learn the tools the best thinkers use, concise edition* (Pearson Prentice Hall, Upper Saddle River, NJ)
[40] Facione P 2011 *Critical Thinking. What It Is and Why Its Counts* (Measured Reason and The California Academic Press)
[41] Yusnaeni, Corebima A D, Susilo H, & Zubaidah S 2017 Creative thinking of low academic student undergoing search solve create and share learning integrated with metacognitive strategy *International Journal of Instruction* **10(2)** 245-262
[42] Mahanal S, Zubaidah S, Bahri A, & Dinnurriyah S M 2017 Improving students' critical thinking skills through Remap NHT in Biology classroom *Asia-pasific forum on science learning and teaching* **17(2)** article 11 19

IOP Conf. Series: Journal of Physics: Conf. Series **1108** (2018) 012092 doi:10.1088/1742-6596/1108/1/012092

The Effect of Yacon Tubers as Prebiotics and Sinbiotics against *Salmonella typhimurium* IFO 12529 on *Rattus norvegicus*

L Yuanita[1], P R Wikandari[1], Isnawati[2], W B Sabtiawan[3] and D A P Sari[3]

[1]Department of Chemistry,Universitas Negeri Surabaya, Indonesia
[2]Department of Biology, Universitas Negeri Surabaya, Indonesia
[3]Department of Science, Universitas Negeri Surabaya, Indonesia
Corresponding Author:lennyyuanita@unesa.ac.id

Abstract. This study is to obtain the role of yacon tubers through storage and boiling variations as prebiotics and sinbiotics against *Salmonella typhimurium* pathogenic bacteria (*in-vivo*). There are 80 of *Rattus norvegicus* that were divided into 10 groups: five of them by yacon tubers prebiotic supplements and five other feed by yacon tubers sinbiotics with *Bifidobacterium longum* Reuter ATCC 15707, infected on days 22 and 23. *Salmonella typhimurium* IFO 12529 infections is on days 29 and 30. Total length of yacon tubers supplements is 40 days. There are four kinds of treatment for yacon tubers, namely, T1 (without storage and boiling), T2 (without storage, 60 minute boiling), T3 (14 day storage, no boiling), T4 (14 day storage, 60 minute boiling). The results of the study: 1) The use of prebiotic yacon tubers inhibits growth or decreases the amount of *Salmonella typhimurium*, and it is more effective to use yacon tubers sinbiotics along with *Bifidobacterium longum*, 2) Among the yacon tubers treatment supplements, the T1 is the most effective treatment for decreasing the amount of *Salmonella typhimurium*, with or without *Bifidobacterium longum*.

1. Introduction

The yacon tubers [*Smallanthus sonchifolia* (Poepp.et Endl.) H.Robinson] are known as a source of FOS prebiotic bioactive compounds (fructooligosaccharide); containing 70-80% dry weight saccharide, consisting of 350.1 fructose, glucose 158.3, sucrose 74.5, FOS (GF2-GF9) 206.4, 13.5 mg inulin, also chlorogenic acid antioxidant 48.5 mg / kg dry weight [1]. FOS is the oligosaccharides of the fructosyl unit binding β-2,1 with glucose, 2-9 degree of polymerization (glucose- (fructose) n). The main components of FOS are 1-kestose (GF2), nystose (GF3), and 1-b-D-fructofuranosylnystose (GF4). Chlorogenic acid is a polyfenol compound that can be used as a specific growth substrate, intestinal bacteria include Lactobacillus and *Bifidobacterium* strains, because they can metabolize chlorogenic acid into caffeine and quinic acid [2].

One of the factors that affect the levels of bioactive compounds FOS and chlorogenic acid, is the activity of enzymes. The fructan 1-exohydrolase enzyme catalyzes the breakdown of FOS to fructose and glucose [3]. Miyaguchi et al [4] revealed that the FOS's yacon will drop when it is heated in 100°C after it is peeled and cut. After 6 days storage in the shade, the concentration of yacon tubers' FOS decreased from 50-62 to 36-48% dry weight, while the concentration of fructose, glucose, sucrose increased from 29-34 to 48-52% [5]. Six days sun heating decreased FOS into 29-44% and increasing the saccharide level into 45-51% dry weight. Chlorogenic acid is easily oxidized and polymerizes to form quinone compounds that cause brown color. In the case of boil tubers, there is a significant decrease of chlorogenic acid content, it isomerized with 4-O-caffeoylquinicacid (4-O-CA) and 5-O-caffeoylquinicacid (5-O-CA) [6][7].

Gastric acid and gastrointestinal enzymes that can reach the large intestine cannot digest FOS and inulin, but they are fully fermented to the cecum and colon by some bacterial strains such as *Bifidobacterium* and *Lactobacillus*, selectively to stimulate its growth; this is due to the production of β-fructofuranosidase / fructanase [8]. Sghir et al. [9] study showed that *Lactobacillus* and *Bifidobacterium* use FOS for its growth, but *Bifidobacterium* grows faster in FOS than *Lactobacillus*.

Fermentation of intestinal microflora to FOS-inulin will simultaneously produce various short chain fatty acids (SCFA), CO_2, CH_4, H_2, lactic acid, thus decreasing the pH of the large intestine, which will inhibit the growth of pathogenic bacterial groups such as *Salmonella thypii*, reducing the formation or accumulation of carcinogenic compounds and harmful metabolites. Tanner et al [10] found that the addition of FOS greatly inhibited the growth of *Salmonella enterica* Typhimurium N-15 in the PolyFermS porcine model, and correlated with increased production of SCFA, especially acetate and propionate.

One way to stabilize and enhance the effect of probiotics is by combining probiotics and prebiotics, called synbiotics. According to Suskovic et al. [11], synbiotics increase the survival of probiotic bacteria, decrease the amount of *Clostridia*, *Enterobacteriaceae* and *E. coli*. Gmeiner et al. [12] obtain synbiotic *L. acidophilus* 74-2 with FOS, increase SCFA and β-galactosidase activity that has a role in the hydrolysis of galactooligosaccharide.

The result showed that through in vitro method yacon tubers as a source of FOS prebiotics and synbiotics along with *Lactobacillus acidophilus* IFO 13951, *Bifidobacterium longum* Reuter ATCC 1570, can inhibit the growth of *Salmonella typhimurium* IFO 12529 [13]. The purpose of this research is to find yacon tubers' role through variations of storage and boiling time as prebiotics and synbiotics with *Bifidobacterium* against *Salmonella typhimurium* on *Rattus norvegicus*.

2. Method

There are four groups of yacon tubers involving T_1 (no storage and no boiling), T_2 (no storage, with boiling 60 minutes), T_3 (14 days of storage, without boiling), and T_4 (14 days of storage, with boiling 60 minutes). There are 5 feed treatments involving standard (S), the T1, the T2, the T3, and the T4 (containing 10% FOS = 250 g yacon / 1000 g). 80 tested animals, wistar *Rattus norvegicus* strain, are divided into 10 groups, the first five groups without the addition of *Bifidobacterium longum* Reuter ATCC 15707 and the second five groups with the addition of *Bifidobacterium longum* Reuter ATCC 15707. Each animal is placed in individual cages and drinks in ad libitum. 4-day environmental adaptation with standard feed, and then with the feed containing the treatment yakon for 17 days. On the 22nd and 23rd days, an infection is performed by oral isolates *Bifidobacterium longum* Reuter ATCC 15707 @ 1/2 x2x 1010 cfu (suspension in 1 ml of saline 1:3 w/v and containing 3% $NaHCO_3$ newly made) to 5 groups [14][15]. At days 29 and 30, all of an animal model to infected *Salmonella typhimurium* IFO 12529 isolates orally @ 1 / 2x2x108 cfu (suspension in 1 ml of saline 1:3 w/v and containing 3% of newly-created $NaHCO_3$) [14][15]. After an incubation period of ten days, the tested animal is killed. Feces collected on the first day and the tenth incubation period. Then analyzed the animals model: 1) *Bifidobacterium* and *Lactobacilllus* in colon (cfu), and 2) *Salmonella* feces at the beginning and end of incubation (cfu). Total plate count is used to count the bacteria in this research. MRS De Man, Rogosa and Sharpe broth / media (MRS) are used for *Lactobacillus acidophilus*, MRS-cysteine HCl 0.05% for *Bifidobacterium longum*, while bismuth sulfite media for *Salmonella typhimurium*. The result of measurement has been analyzed quantitatively.

3. Results and Discussion

Yuanita et al. [16] results showed that increasing storage (0,7,14 days) and boiling time (0, 30, 60 minutes) would decrease of FOS and increase of saccharides concentrations. Yacon tubers contain FOS as prebiotic, which trigger a probiotic population growth of lactic acid bacteria, especially *Bifidobacterium* and *Lactobacillus* that naturally found in *Rattus norvegicus* colonic. This is due to FOS can be ingested through hydrolysis as a result of β-fructofuranosidase of probiotic microorganisms, produced glucose and fructose. The result of fermentation reaction of glucose fructose lowers pH

bowel/colon because the resulting acetic, propionic, butyric and lactic acids. According to Gibson and Roberfroid [17], the fermentation reaction stoichiometry described as follows:
$1\ C_6H_{12}O_6 \rightarrow 1.5\ CH_3COOH + 0.33\ CH_3CH_2COOH + 0.15\ CH_3(CH_2)_2COOH + 0.33\ CH_3CHOHCOOH + 0.33\ CO_2 + 0.33\ H_2O$

At the end of the treatment, it is obtained the number of *Bifidobacterium* and *Lactobacillus*on colon bacteria (Table 1 and 2), and *Salmonella* in the feces of *Rattus norvegicus* (Table 3).

Table 1 The average number of *Bifidobacterium* and *Lactobacillus* of *Rattus norvegicus* colon (10^{-3}) cfu.

Group	Feed Treatment	*Bifidobacterium*	*Lactobacillus*	Total
1	S	0.60	1.80	2.40
2	T1	33.40	38.60	72.00
3	T2	19.80	20.80	40.60
4	T3	7.20	11.80	19.00
5	T4	7.80	12.60	20.40

S: standard feed, T_1= fresh and without boiling; T_2=fresh and 60 minutes boiling time; T_3= 14 days storage time and without boiling; T_4= 14 days storage time and 60 minutes boiling time.

Table 2 The average number of *Bifidobacterium* and *Lactobacillus* of *Rattus norvegicus* colon with infection of *Bifidobacterium longum Reuter ATCC 15707* (10^{-3}) cfu.

Group	Feed Treatment	*Bifidobacterium*	*Lactobacillus*	Total
6	S + *B. longum*	12.70	11.90	24.60
7	T1 + *B. longum*	39.50	50.45	89.95
8	T2 + *B. longum*	24.80	50.50	75.30
9	T3 + *B. longum*	11.70	17.05	28.75
10	T4 + *B. longum*	14.80	21.70	36.50

S: standard feed, T_1= fresh and without boiling; T_2=fresh and 60 minutes boiling time; T_3= 14 days storage time and without boiling; T_4= 14 days storage time and 60 minutes boiling time.

The average number of *Bifidobacterium* on the data Tables 1 and 2 can be described as follows in Figure 1, while the average number of *Lactobacillus* in Figure 2.

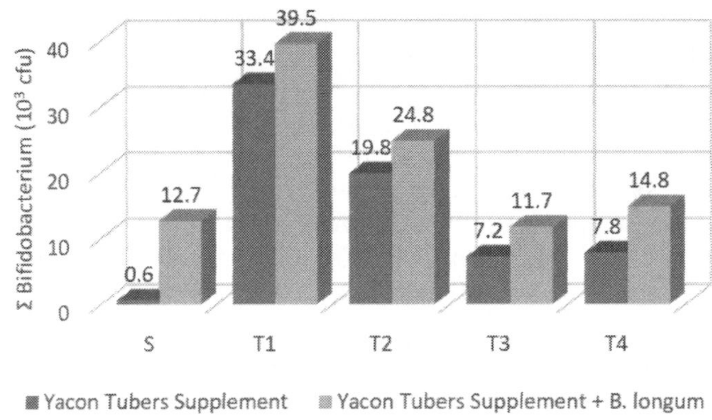

Figure 1. The influence of yacon tubers supplement against average number of *Bifidobacterium* colonic of *Rattus norvegicus*.

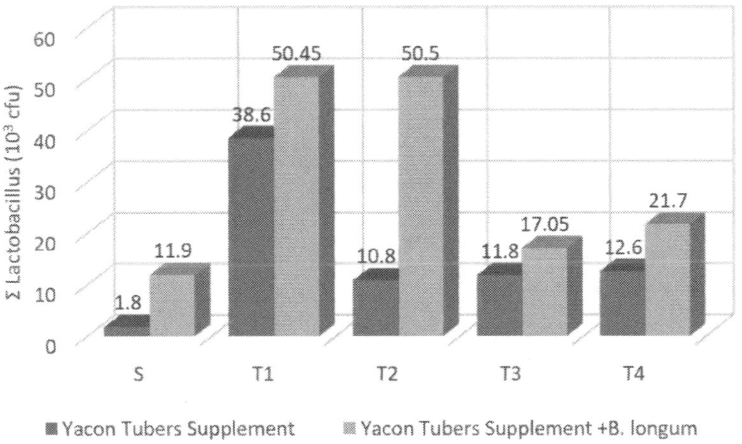

Figure 2. The influence of yacon tubers supplements against average number of *Lactobacillus* colonic of *Rattus norvegicus*.

From the results in Table 1 and 2, it can be argued that:

1. The use of yacon tubers synbiotic supplement with *Bifidobacterium longum* is more effective against a number of colon *Bifidobacterium* and *Lactobacillus* compared to treatments with the yacon tubers prebiotic supplement alone. The combination of probiotics and non-digestible carbohydrates (FOS) may be a way of stabilization and /or improvement of the probiotic effect. The use of probiotic that orally administered can improve immune status by increasing circulating and local antibody levels, endurance live probiotic colon bacteria, increasing the activity of useful enzymes that degrade of prebiotic and supression of pathogenic microorganism in intestinal tract [11][12].
2. The use of yacon tubers as prebiotic or sinbiotic supplement, increasing the number of beneficial colon bacteria *Bifidobacterium* and *Lactobacillus* compared to treatments with the feed standards. This is because yacon tubers is a source of FOS serve as nutrients for the colon bacteria which have fructan 1-exohydrolase. Likewise, saccharides and chlorogenic acid of yacon tubers are useful for intestinal bacteria growth.
3. On the usage of yacon tubers as prebiotic or synbiotic, fresh and no boiling (T1) treatment most effective. May be due to the highest of FOS and chlorogenic acid on the T1 treatment yacon tubers. Chlorogenic acid can be used as a specific growth substrate, not degraded or absorbed in the upper digestive tract, but can be hydrolyzed by the intestinal bacteria produce quinic acids and hydroxycinnamate acids i.e. caffeic acids [2].
4. The yacon tubers supplement of T4 treatment (14 days storage and boiling 60 minutes) produced colon bacteria of *Rattus norvegicus* more higher than T3 yacon tubers supplement (14 days of storage, without boiling). May be due to glucose and fructose content in yacon tubers resulted in the T4 treatment more higher than T3, cause hydrolyzed during storage and boiling. Glucose and fructose also a source of carbon for growth of *Bifidobacterium* and *Lactobacillus*.
5. The amount of *Lactobacillus* colonic more higher than the *Bifidobacterium*, also against the standard feed groups. *Lactobacillus* have optimum growth at pH 6.5, while *Bifidobacterium* at pH 5-5.5. pH range of colon is 5.5 to 7, its mean that pH colonic is more suitable for the growth of *Lactobacillus* than *Bifidobacterium*. This sensitivity includes temperature, pH, the presence of nutrients according to their needs, and the presence of other compounds that interfere with its growth. Research of Sghir et al[9] showed that *Lactobacillus* and *Bifidobacterium* using FOS for its growth, but the *Bifidobacterium* grown faster on FOS than *Lactobacillus*. Then on infection of *Bifidobacterium longum,* competition for FOS limiting nutriens occurs by *Bifidobacterium*; while *Lactobacillus* using saccharides as a source of carbon for growth. *Bifidobacterium* is more affected than *Lactobacillus*,

population growth is limited, even rapidly entering the stationary phase and subsequent phase of death.

The average number of *Salmonella of Rattus norvegicus* feces through infections of *Salmonella typhimurium IFO 12529* on the usage of yacon tubers as prebiotic and synbiotic are provided in Table 3 and Figure 3.

Table 3. The average number of *Salmonella* of *Rattus norvegicus* feces through infections *of Salmonella typhimurium IFO 12529* (10^{-3} cfu).

Group	Feed Treatment	Feces	
		Early Treatment *	End Of Treatment *
1	S	infinite	16.40
2	T1	infinite	1.20
3	T2	infinite	1.80
4	T3	infinite	5.40
5	T4	infinite	7.60
6	S + *b. longum*	infinite	19.20
7	T1 + *b. longum*	infinite	0.60
8	T2 + *b. longum*	infinite	0.80
9	T3 + *b. longum*	infinite	3.60
10	T4 + *b. longum*	infinite	4.20

S: standard feed, T_1= no storage and boiling; T_2= no storage, boiling 60minutes; T_3= 14 days of storage, without boiling; T_4= 14 daysof storage, boiling 60 minutes. * Early treatment: after 1 day, end of treatment: after 10-day treatment.

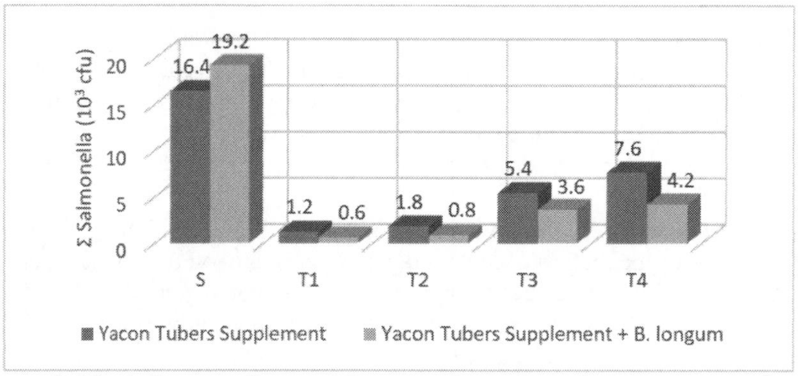

Figure 3. The influence of yacon tubers supplements against average number of Salmonella of *Rattus norvegicus*feces.

Table 3 showed that:

1. The use of yacon tubers synbiotic with *Bifidobacterium longum* is more effective inhibiting the growth of *Salmonella* than as prebiotics.
2. Yacon tubers as beneficial supplements inhibit the growth of *Salmonella*. This indicates the role of the yacon tubers FOS as prebiotic. The high levels of FOS is followed by growth of bacteria to intestinal generated high acidity; low pH will inhibit the growth of pathogenic bacteria.
3. T1 yacon tubers supplement has ability to suppress the growth of *Salmonella* in the feces of Rattus norvegicus. Moreover, similarly *Salmonella* bacteria not found in the cecum and colon [18]. This is due to the high concentration of FOS in fresh yacon tubers as a source of nutrients for colon bacteria.Fermentation of intestinal microflora against FOS simultaneously will produce short chain fatty acids, CO_2, CH_4, H_2, lactic acid, that is lowering the pH of the large intestine; by consequence,

it will inhibit the growth of pathogenic bacteria group. Undegraded or unhydrolyzed chlorogenic acid also suppress the growth of *Salmonella*. According to Karunanidhi et al. [19], several Gram-positive (*Staphylococcus aureus, Streptococcus pneumoniae, Streptococcus mutans* and *Bacillus subtilis*) and Gram-negative (*Salmonella typhimurium, Shigella dysenteriae*, and *Escherichia coli*) bacterial pathogens have been reported to be highly susceptible to chlorogenic acid.

4. Increasing of average number of colonic bacteria, decreasing of *Salmonella*, except the T4 treatment. It is due to the highest amount of glucose on the T4 yacon tubers. The glucose can be used for the growth of *Salmonella* through fermentation process.

4. Conclusion

To conclude, there are two main findings of the study, that usage of yacon tubers prebiotic inhibits the growth or decrease of the number of *Salmonella typhimurium*, but more effective in the use of yacon tubers synbiotic with *Bifidobacterium longum* and Among the supplements of treatment yacon tubers, the T1, fresh and no boiling yacon tubers supplement, is the most effective for lowering the number of *Salmonella typhimurium*, with or without a *Bifidobacterium longum*.

Acknowledgement
International Development Bank supported this research for the Development and Upgrading of Seven Universities in Improving the Quality and Relevance of Higher Education in Indonesia.

5. References

[1] Lachman J, Fernandez E C and OrsakM 2003*Plant Soil Environ.***49(6)** 283.
[2] Couteau D, Mc Cartney AL, Gibson G R, Williamson G and Faulds CB 2001*J. Appl. Microbiol.* **90(6)** 873.
[3] Shiomi N, Benkeblia, N S, Onodera O T, Takahashi N, Fujishima M, Yoshihira T and Kosaka S 2007 *J. Appl. Glycosci.* **54** 187.
[4] Miyaguchi Y and Inoue E 2012 *Nippon Shokuhin Kagaku Chinese Story Kaishi.* **59(4)** January.
[5] GraefeS, HermannM, Manrique I, Golombek S and Buerkert A 2006 *Field Crops Research***86** 157.
[6] Kan S, Cheung MW M, Zhou Y and Ho W S 2014 *J. Food Sci.* **79(2)** C147.
[7] Xu B and Chang S K C 2008*J. Agric. Food. Chem.***56(16)** 7165.
[8] Swennen K, Caurtin C M and Delcour A 2006 *Food Sci. Nutr.***46**459.
[9] Sghir A, Chow J M and Mackie R I 1998 *J Appl Microbiol.***85**769.
[10] Tanner SA., Chassard C, Berner A Z and Lacroix C 2014*Gut. Pathogens***6**44.
[11] Suskovic J, KosB and Matosic S J G 2001 *Food Biotechnol. Technol.* **39(3)** 227.
[12] GmeinerM., Kneifel W, Kulbe K D, Wouters R, De Boever P, Nollet L and VerstraeteW 2000 *Appl. Microbiol. Biotechnol.***53**219.
[13] Yuanita L, Wikandari P R and Isnawati 2015 *The 5th Inter. Conf. Indones. Chem. Soc. Proc.* Samarinda, East Kalimantan. August, 30-31, 2015.
[14] Ten Bruggencate SJ M, Bovee-Oudenhoven IM J, Lettink MLG W and Van d M R 2005*J. Nutr.***135**(4) 837.
[15] Boove OIMJ, Termont DSML, Heidt P J and Van d M 1997*Gut.***40**497.
[16] Yuanita L, Wikandari P R and Sabtiawan W B 2017 *Adv. Sci. Lett.***23(12)** 11982.
[17] Gibson G R and Roberfroid M B 1995*J. Nutr.***125**1401.

[18] Yuanita L, Wikandari P R and Isnawati 2016 Yacon Tuber FOS [*Smallanthus sonchifolia* (Poepp.et Endl.) H. Robinson] as Prebiotik and Sinbiotik to inhibit the growth of pathogenic bacteria. *Research Report* PUPT-IDB. LPPM UNESA.
[19] Karunanidhi, Thomas R, Van B A and Neela V 2013 *BioMed. Res. Inter.* **2013** 1.

Survey Study for The Usage of Solar Energy at Household By The Employees of Al-Shifa Medical Complex in Gaza Strip

H J El-Khozenadar[1] and F F El-batta[2]

[1] Electrical Engineering Department,Engineering College,Islamic University of Gaza,P. O. Box 108, Palestine
[2] Electrical Appliances Department, Ministry of Health, Gaza, Palestine

*Corresponding Author: hkhozondar@iugaza.edu

Abstract. Providing residences of Gaza with electric power for 24 hours/day is one of the key problems that facing decision-making bodies in energy sector. Most recently, residence in Gaza started to adopt solar energy system as a replacement or as a complementary to the current energy sources. In this paper, we present a survey results that measures the success of using solar energy at homes in Gaza Strip. The study population is the employees of Al-Shifa Medical Complex (about 2000 employees). The questionnaires are distributed to a sample of 10% of the total population. Collected data is analyzed using SPSS. Results show that gender and job titles have no effect of the decision of adopting solar energy systems. However, income has a direct effect on the decision of adopting solar energy systems.

1. Introduction:

Photovoltaic, PV, is a device that converts light from into electrical energy. The main components of solar energy system are Solar Panels, Inverter, Battery Pack, and Charge Controller [1]. For residential use, there are several types of solar systems available in the market to consider. While these are based on different factors like grid connection, material type, battery, and placement, there are three main types of solar power systems: stand- alone systems, grid-connected systems and solar PV hybrid system [2]. Gaza Strip has been suffering from a chronic deficit in electricity since 2006. Around two million people are living on less than 4 hours of electricity a day because of the 61% electricity deficit. Several alternatives, including the use of electric generators, solar panels and uninterruptible power supply units were initiated to overcome electricity shortages, particularly in houses, hospitals and schools. That problem started in 2006, when Israeli forces occupation bombed electricity-generating station located in the central Gaza Strip [3]. The problem is increased due to the tensions between Hamas and Fatah, over custom tax revenue, funding of Gaza, and political authority. Gaza is supplied by three primary sources: Israeli power line (120 MW), Egypt power line (37MW if all lines worked) and the Palestine Electric Company (PEC) power line, and the said quantity depends on the amount of fuel available for the production of electricity [3]. On average, the PEC provides approximately 80 MW or more than 50% of its full capacity of 120 MW. However, PEC currently can provide Gaza only by 54MW.In addition; Israeli line is reduced to 70MW out of its full capacity due to the above-mentioned reasons.

Thus, Gaza Strip receives only 161 MW from all the aforementioned sources. Moreover, Gaza's electricity demands vary seasonally. During the summer and winter months it requires 440 MW. During the rest of the year, it requires 380 MW. Then, irrespective of the seasonal needs, there is a significant deficit between 220 to 280 MW through the course of the year, and this deficit causes recurring interruption of electrical supply to homes and economic and service facilities [3].

In this way, the electricity crisis in Gaza has become one of the outstanding problems that affect various aspects of the lives of the Palestinian citizens. This crisis is also considered as particularly acute due to its impact on the economic and social welfare in the Gaza Strip. In 2012, Ministry of Health started to use solar energy system in several clinics and hospitals, e.g., Children's incubator in ALNASER Hospital, and as a backup electrical source in intensive care unit and operations rooms [4]. Ministry of Education and Higher Education also started to equipped several schools with solar systems, e.g., Bashir Al Rais secondary school for girls and then rolled similar projects to this day. Recently, the Ministry of Education and Higher Education and the Palestine Investment Fund (PIF) signed an agreement to develop and install solar power systems on 500 public schools in west bank and Gaza strip with a total capacity of 35 MW and an investment volume of 35 million dollar [5]. Ministry of Agriculture and the Ministry of the Interior, the local municipalities, universities, factories, stores, and private companies are following the same steps. Most recently, people started to use the system to provide electricity to their private homes or apartments. Gaza Electricity Distribution Corporation (GEDCO) started to encourage people to install their own solar system at their houses by providing solar systems to their customers with installment system to be collected with the monthly bill [6]. In the last few years, little research have been established to study the implementation of solar energy in Palestine and in Gaza Strip in particular. Some studies [7 -13] proposed alternative sources of renewable energy such as solar, wind and sea wave energy. Other researchers (e.g. [14]) developed plans for the construction of solar power plants in the Gaza Strip. In 2016, Adel Juaidi [9] and others analyzed the current energy sector situation in Palestine and highlighted the status of the potential renewable energy (RE) as an essential future energy source in Palestine.

In a previous work by the authors, the motivation for the local institutions to install solar energy is studied [15]. The results of analyzing the collected data show that the local institutions chose to use solar energy regardless to the price due to its environmental value. The results did not depend on the attributes of the institutions and the authors recommend using solar energy as another source of energy [15]. However, this study does not include other sectors in the local society, e.g., household, small shops, and hotels. In this work, authors expand their work to include households. Authors could not find a good reference to the list of people of install solar system at their houses, they decided to choose a close society consists of the workers in Al-Shifa Medical Complex to study the usage of solar system at houses in Gaza strip.

2. Research Problem

The people in Gaza are suffering from power outage. Students at schools and universities have to do their homeworks in 3 hours per day period when electricity is on. Similarly, professors and teachers should use these 3 hours to do all their work duties. As it obvious, this is impossible. Thus, people in Gaza started to adopt solar systems at their houses. The focus of this work is to study the implementation of solar systems by households in Gaza Strip. This research attempts to answer the following main question: "Can solar energy be an alternative to conventional energy for domestic use in Gaza Strip?" In specific, the research problem will answer the following questions: (1) Is the environmental value affects the domestic user's decision to adopt solar energy system?; (2) Is the cost of energy affects the domestic user's decision to adopt solar energy system?; (3) Is solar energy market affects the domestic user's decision to adopt solar energy system?; (4) Are the attributes of domestic users (gender, job title, and income) affect the decision to adopt solar energy system?

Importance of Research: The importance of research stems by providing an accurate information about the use of solar energy at houses by surveying people working at Al-Shifa Medical Complex in Gaza Strip.

Research Objectives: The main objectives of this research are to find out the useful of implementing solar energy system for domestic use in Gaza Strip. The specific objectives of the research can be summarized in the following points; (1) Shine the light on the domestic use of solar energy systems in Gaza Strip, (2) Examine the role of the available solar energy environmental value in the decision of domestic users to adopt this technology, (3) Examine the impact of solar energy marketing in Gaza Strip and its effect on domestic users' decision to adopt solar technology, (4) Find out the effect of the price of the solar energy system on the decision of domestic users to adopt the solar energy system, (5) Find out the effect of different attributes of domestic users (age, sex, income, geographic region, type of house, and house ownership) on adopting solar technology, and (6) Propose solutions and recommendations to avoid most of the gaps and obstacles that affect the willingness of local inhabitants to adopt solar technology.

Variables of Research: Figure 1 defines the independent variables and dependent variable.

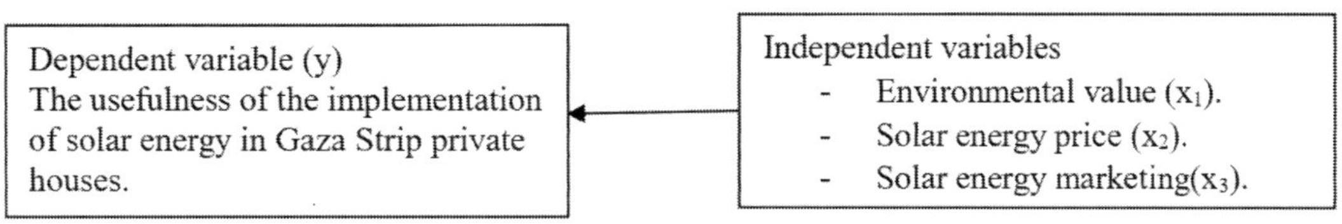

Figure 1. The relation between dependent and independent variables.

Hypothesis: The significance level is measured at $\alpha \leq 0.05$ [16]. (1) There is statistically significant differences in the attitudes of the employees of Al-Shifa Medical Complex in Gaza regarding the success of applying the solar energy technology (environmental benefit, solar energy, market value, total) due to gender variable; (2) There is statistically significant differences in the attitudes of the employees of Al-Shifa Medical Complex in Gaza regarding the success of applying the solar energy technology (environmental benefit, solar energy, market value, total) due to the variable of the job title., (3) There is statistically significant differences in the attitudes of the employees of Al-Shifa Medical Complex in Gaza regarding the success of applying the solar energy technology (environmental benefit, solar energy, market value) due to the variable income level.

3. Methodology

In this section, we present the actions and steps taken in the field study, which addresses the methodology, the community of the study, and the sample of the study. Moreover, we clarify the tools used in the study, their steps, and statistical methods used to analyze the data to reach results and achieve the objectives of the study.

3.1 Study Approach

The researchers followed the analytical descriptive approach that tries to answer the basic question in the science and nature of the phenomenon discussed, including analysis of the phenomenon, its environment, and the relationship between its components.

3.2 The study population

All employees at Al-Shifa Medical Complex, which is 1819 employees (549 doctors, 650 nurses, 370 administrators, 250 technicians)

3.3 The study sample

The sample of the study was about 10% of the total population. 200 employees of Al-Shifa Medical Complex is randomly chosen to fill the questionnaire. 58% of the study sample are male and 42% are female. Regarding the job title, the sample has the following distribution: 29.5% of the sample works as administrators, 28.5% of the sample are nurses, 15.5% of the sample are programmers or accountants, 13% of the sample are doctors, 8.5% of the sample are technicians and 5% of the sample are engineers.

3.4 The study Tools

The researchers used the questionnaire titled "usefulness of implementing solar energy at Gaza Strip houses". The questionnaire has two main sections. The first section includes personal data for each participant. The second part has 22 items of the questionnaire on a five-point scale: strongly agree, agree, not certain, disagree, and strongly disagree. The second part is divided to three main parts; (1) The First part: Environmental benefit and included 8 items, (2) The second part: The adoption of solar energy and included 9 items, and (3) The third part: The market value of solar energy and included 5 items.

3.4.1 Statistical Analysis. The researchers used the statistical methods of statistical analysis for the Social Sciences (SPSS).The following statistical methods have been used; (1) Descriptive statistics, including the percentage, arithmetic mean, standard deviation and relative arithmetic weight. This is mainly used to determine the frequency of categories of a variable and to describe the variables of the study, (2) Pearson correlation coefficient is to verify the reliability of internal consistency between the items of the questionnaire and the total score of the questionnaire, (3) Cronbach's Alpha coefficient is to determine the stability of the scales, (4) Spearman Brown Split Half is to determine the reliability of the test after changing test length, and (5) One Samples T test is to determine whether the average response level has reached a specific mean of 2 or not.

3.4.2 Research limits. The study identified the following limits; (1) Objective limit: This study is limited to studying the implementation of solar power in the Gaza Strip houses, (2) Spatial limit: study confined in Gaza Strip, (3) Sample limit: The study included only employers at Al-Shifa Medical Complex, and (4) Time limits: This study has conducted in the last quarter of 2017.

4. Research Results and analysis

Results on the main question: Can solar energy be an alternative to conventional energy for domestic use in Gaza Strip? In order to answer this question, the percentage, mean and relative weight of the items of the questionnaire are calculated. The results are exhibited in tables 1, 2 and 3 for the three main parts.

Environmental benefit: After calculating the arithmetic mean, standard deviation and relative weight for environmental benefit, we found that the highest ranked item is item 8: "If the cost of energy is equal to solar and conventional energy, you will choose solar energy" ranked first with a relative weight of 73.10%. and The lowest ranked item is item 6: "You are aware of the environmental risks resulting from the use of traditional energy sources" ranked sixth with a relative weight of 67.60%. The relative weight of the environmental benefit of applying solar technology by the employees of Al-Shifa Medical Complex in Gaza is 69.43%.

The adoption of solar energy: After calculating the arithmetic mean, standard deviation and relative weight of the adoption of solar energy, we realized that the top ranked item is item 11: "The government subsidy, if any, for the prices of solar energy that drives you to adopt it" ranked first with a relative weight of 74.20%. while the lowest ranked item is: item12: "The cost of solar energy is less than the cost of conventional energy in the long term" ranked seventh with a relative weight of 67%. The relative weight of the adoption of solar energy technology in Al-Shifa Medical Complex in Gaza is 67.93%.

Market Value: After calculating arithmetic mean, standard deviation and relative weight, we found that the top ranked item is item 20: "You see the need for specialized governmental centers to research renewable green energy" ranked first with a relative weight of 77%. While the lowest ranked item is item 18: "You see the future of solar energy to replace the place of traditional energy" ranked fourth with a relative weight of 68.70%. The relative weight of the success of applying the solar energy technology (market value) of the employees of Al-Shifa Medical Complex in Gaza is 72.04%.

Results related to the first hypothesis: To validate the hypothesis, the independent sample t test was used to determine the differences between the two variables. Results of the analysis show that there are no statistically significant differences in the application of solar energy technology (environmental benefit, solar energy, market value, total) by Al-Shifa Medical Complex employees in Gaza due to gender variable.

Results related to the second hypothesis: To validate the hypothesis, the One Way ANOVA test was used to determine the differences between the two variables. The results of the analysis indicates that there are no statistically significant differences at the level of significance of 0.05 in the success of applying solar energy technology (environmental benefit, solar energy, market value, total) in the employees of Al-Shifa Medical Complex in Gaza due to the variable of the job title.

Results related to the third hypothesis: To validate the hypothesis, the One Way ANOVA test was used to determine the differences between the two variables. Results exhibits that there are statistically significant differences in the degree of significance of the application of solar energy technology (environmental benefit, solar energy, market value, total) byAl-Shifa Medical Complex employees in Gaza due to variable income level. To illustrate the differences, the Scheffe' Test and the following results are obtained:

For Environmental benefit, Scheffe' Test presents that there are statistically significant differences at the level of significance of 0.05 among the categories of income level (2000 - 4000, less than 2000) for the category of income level from 2000 NIS to 4000 NIS.

Regarding Adoption Solar Energy, Scheffe' Test shows that there are statistically significant differences at the level of significance of 0.05 among the categories of income level (2000 - 4000, less than 2000) for the category of income level from 2000 NIS to 4000 NIS.

And for Market value, Scheffe' Test presents that there are statistically significant differences at the level of significance of 0.05 among the categories of income level (2000 - 4000, less than 2000) for the category of income level from 2000 NIS to 4000 NIS.

6. Conclusion

In this work, authors studied the reasons behind adopting solar energy systems as an alternative source of energy at households in Gaza Strip. The study conducted on a sample of 200 employees at Al-Shifa medical complex. Different factors are considered in the study including gender, income and job titles. In addition, the study addresses different reasons, which might be the motivation for the households to use solar energy. These reasons are the environmental factor, marketing and prices. The sample members do not use solar energy as a full replacement for traditional energy. The results of analyzing the collected data from the questionnaire using SPSS show that the gender and job titles have no effects on the decision of adopting solar energy. However, income level has impact on the decision of adopting solar energy system. This make sense because the adoption of solar energy needs enough budget no matter what job title or sex a person have.

References

[1] Roos C 2009 *Solar Electric System Design, Operation and Installation* (Olympia: Washington State University Extension Energy Program)

[2] Parimita M K R 2016 *Solar Photovoltaic System Applications* (London: Springer)

[3] PalThink for Strategic Studies 2014 *Case Study: Renewable Energy in the Gaza Strip: Short, Mid, and Long-Term* (Gaza: PalThink for Strategic Studies)

[4] Ministry of health 2017(online) Ministry of health (*http://www.moh.gov.ps*)

[5] Ministry of high education 2017 (online) *ministry of high education* (http://www.mohe.ps/: http://www.mohe.ps/)

[6] GEDCO 2017 (online) Gaza Electricity Distribution Corporation (http://www.gedco.ps/)
[7] Ouda M 2001 Prospects of Renewable Energy in Gaza Strip *China: Energy Research and Development Center* pp. 1-6
[8] Naim A N 2010 Potential of Solar Pumping in Palestine *International Water Technology Conference (IWTC)* pp. 1-11
[9] AdelJuaidi F A 2016 An overview of renewable energy potential in Palestine *Renewable and Sustainable Energy Reviews* pp. 943–960
[10] Abu-Hafeetha M F 2009 *Planning for Solar Energy as an Energy Option for Palestine* (Nablus: An-Najah National University)
[11] Aydi J Y 2011 The solar Energy potential of gaza strip *Global Journal of researches* pp. 47-51
[12] Elnaggar M, Edwan E, and Ritter M 2017 Wind Energy Potential of Gaza Using Small Wind Turbines: A Feasibility Study *Energies* **8**(18) pp. 1-13
[13] Yaseen B T 2009 Renewable Energy Applications in Palestine *DISTRES* pp. 52-65
[14] Abualtayef M 2016 Solar Voltaic Energy *GAZA* **11**
[15] El-Khozenadar H J and El-batta F F 2018 *Solar Energy as an alternative to conventional energy in Gaza Strip: Questionnaire based study* (Nablus: An-Najah National University)

IOP Conf. Series: Journal of Physics: Conf. Series **1108** (2018) 012094 doi:10.1088/1742-6596/1108/1/012094

Students' perceptions toward online learning and face-to-face learning courses

S Bali[1,*] and M C Liu[2]

[1, 2] Department of Education and Human Potentials Development, National Dong Hwa University, Taiwan

*Corresponding author: sandrotuab@gmail.com

Abstract: The implications of online learning versus face-to-face learning have been discussed for several years in higher education. This study examined the issues of student perception toward online learning and face-to-face learning in the context of social presence, social interaction, and satisfaction in Indonesia Open University, Taiwan branch (n=107). The comparison of the online group and the face-to-face group conducted to explore student-learning perceptions regardless of the course delivery method and the online environment. The result of this study indicate that face-to-face learning perception was higher than online learning in term of social presence, social interaction, and satisfaction. However, there is no statistically significant difference in learning preference found among level of student. Meanwhile, some students were very comfortable in online learning since it led them to the chance to being innovative by using computer technology.

1. Introduction

Online education has definitely moved into higher education with new programs being added continuously. The blended synchronous learning (a "blend" of online and traditional approaches) mode is also gaining and developing currency in higher education, while its effects on students' and instructors' experiences are yet to be fully explored. With the rapid expansion of the transnational education market, more and more universities join the ranks of transnational education providers or expand their transnational education offering, such as distance learning or blended learning, including in Indonesia, such as in Indonesia Open University.

Traditionally, online learning perceived as lack interactivity compared to face-to-face learning. It is mainly due to the lack of social presence, lack of social interaction, and lacks of students' satisfaction. However, online learning has been promoted as being more cost effective and convenient than traditional educational environments as well as providing opportunities for more learners to continue their educations.

Previous research have investigated some research regarding to the students perception and satisfaction toward online learning and face-to-face learning. Fortune, Spielman, and Pangelinan [1] investigated 156 students who took and enrolled in either an online learning section or face-to-face learning of the Recreation and Tourism course at multicultural university in Northern California, United States. They found that no statistically significant difference in learning preference was found between those enrolled in the two different learning modes.

Kemp and Grieve [2] found that a result showed that undergraduate Psychology students (n = 67) at Australian university preferred to complete activities face-to-face rather than online. However, their study argued that online and face-to-face activities can lead to similar levels of academic performance, but that students would rather do written activities online but engage in discussion in person. Another study found by Tratnik [3] indicated that there are significant differences in student satisfaction levels between online and face-to-face learning of English as a foreign language. Students taking the face-to-face course were generally more satisfied with the course on several dimensions than their online counterparts.

Although online learning continues to grow rapidly, it still remains at an early stage of development. Consequently, developers and deliverers of online learning need more understanding of how students perceive and react to elements of blended learning, since student perception and attitude is critical to motivation and learning, along with how to apply these approaches most effectively to enhance learning [4]. Therefore, the significance of this research lies on exploring the Indonesia Open University students' perception toward online and face-to-face courses. More specifically, the purpose of this research were to explore the students' perception toward online and face-to-face courses in term of social presence, social interaction, and satisfaction. Besides, this study also examine the difference of online learning and face-to-face learning perception among different level of students.

2. Literature Review

2.1. Social Presence

One of the most important factors in online learning is social presence. Social presence can be defined as the degree of salience between two communicators using a communication medium due to the differing perceptions of medium of its online learning [5, 6]. Social presence also can be interpreted as the degree to which a person is perceived as "real" in mediated communication, which contribute to the social climate in the classroom experience [7]. In regard to that, Garrison, Anderson, and Archer [8] also argued that social presence refers to Community of Inquiry (CoI) framework, namely the students' ability to project themselves socially and emotionally as "real people" in an online learning community which is support cognitive and affective objectives of learning in a community of learners.

As Salmon [5] argued, the importance of social presence is to cultivate and build the successful of communication and group which has something to do with cognitive presence. As Gunawardena [7] argued, social presence is necessary to increase the communication both in traditional and technology-based classrooms. Instructor must be able to create and maintain the educational arrangements, while students enjoy that arrangement. If the social presence is low, it means the interaction in learning is also low.

2.2. Social Interaction

Moore [9] demonstrated that interaction is one of the most important components of teaching and learning experiences. Interaction is the most important thing when it comes to face-to-face learning [10] or even in online learning [11]. Inevitably, interaction that uses the social aspect must be apply in learning in order to improve student learning by enhancing student knowledge.

According to Tu [10], social interaction is fundamental to the explanation of the relationship between social presence and the social learning theory. When social interaction becomes part of the classroom dynamics, therefore classrooms become active places [12]. Also, social interaction of learning which is very essential for the successful learning provides and support productive and meaningful learning. Besides, social interaction also promotes learning engagement which has been identified as positively affecting achievement of learning outcomes [13].

2.3. Satisfaction

Similar to social presence and social interaction, satisfaction in courses is an important thing in learning. When comparing satisfaction with online and face-to-face courses, many researcher argued that learners tend to be more satisfied with face-to-face interactions [14]. In regard to that, Gunawardena and Zittle [15] also argued interactions between learner and instructor also contribute to satisfaction. According to

Lo [16], satisfaction deals with three parts, namely satisfaction with instructor's directions and support, satisfaction with own commitment to learning, and satisfaction with course policies as well.

There are several studies have been conducted to measure the level of student satisfaction in traditional and online environments. However, this study solely focused on social presence, social interaction, and satisfaction dimension, because these three dimension were suitable and amenable to conduct with participants' background.

Table. 1 The dimensions of online and face-to-face learning by different studies

Spears (2012)	Fortune , M., Spielman, M., and Pangelinan, D. (2011)	Gray, J., and DiLoreto, M. (2016)	In this Study
a. Social Presence	a. Learning Environment	a. Course Structure/ Organization	a. Social Presence
b. Social Interaction	b. Face to Face	b. Learner Interaction	b. Social Interaction
c. Collaborative Learning	c. Technology	c. Student Engagement	c. Satisfaction
d. Satisfaction	d. Learning	d. Instructor Presence	
	e. Preferences	e. Student Satisfaction	
	f. Collaboration		

3. Method

3.1. Context and Participants

Participants in this study (n=107) were undergraduate students pursuing a bachelor degree in Indonesian Open University, Taiwan branch, which consisted of three major namely Management, English Literature and Communication. They were enrolled in blended education learning for some courses on the field of humanities and social sciences. The data were collected from four group of students level as shown in Table 2.

Table 2. Summary of Sample demographics (n=107)

Background	Subtotal	
	n	%
Gender		
Male	7	6.54
Female	100	93.46
Level		
Freshman (1)*	28	26.17
Sophomore (2)*	35	32.71
Junior (3) *	27	25.23
Senior (4) *	17	15.89
Total	107	100

*Years of studying

3.2. Instrument

The survey in this study included into three instruments, which has twenty-five items and three dimensions. The dimension were applied into the scale. Social Presence Scale developed by Spears [17], Social Interaction Scale was developed by Picciano [18] and Sudents Satisfaction Scale [17, 19]. The reliability of the full survey (composed of three subscales) was .973 for online learning, and .977 for face-to-face learning.

The Social Presence Scale was developed by Spears [17] was used to measure social presence, social interaction, collaborative learning, and satisfaction in online and face-to-face courses. The Social

Interaction Scale was developed by Picciano [18] to examine performance in an online course in relationship to student interaction and sense of presence in the course. Similar to Social Presence scale, the students Satisfaction Scale also developed by Spears [17] and by Gray and DiLoreto [19], which was designed to measure student engagement, student Satisfaction, and perceived learning in online learning environments. In regard to those scale, this study only adopted and modified the scale and items which are amenable.

3.3. Data Analysis
The validity and reliability in this study using the SPSS software. All factors and items considered to be correlated. Meanwhile, earlier was mentioned that the questionnaire given to the participants were adopted and modified which also had been translated into Indonesian language. In order to translate the questionnaire from English to Indonesian language, one expert was involved in editing or rewording each statement. The construct validities and reliabilities of the of the Social Presence Scale, Social Interaction Scale and Students Satisfaction Scale both in online learning and in face-to-face learning can be described in table 3 by the Pearson product-moment correlation coefficient. Next, the comparison of mean and standard deviation also shown in table 4, and finally, the examination of online learning and face-to-face learning perception among different level of students also examined through ANOVA test, in order to portray the different perception of online learning and face-to-face learning among the different level of student.

4. Results and Discussion
4.1. Students Perception toward Online Learning in term of Social Presence, Social Interaction and Satisfaction
This study employed three instruments to measure students' perception toward online learning and face-to-face learning. All of the scale used a five-point Likert scale. Although these scales have been used in previous studies of online learning, some minor modifications to the wording of specific scale items as discussed previously were made. Thus reliability for each variable used in this study was analyzed by calculating Cronbach's alpha. The Cronbach's alpha for the online learning was .973 and for face-to-face learning was .977. Carmines and Zeller [20] indicated that a Cronbach's alpha of at least .800 should be achieved for widely used.

In this study, the level of analysis also involved obtaining descriptive statistics. As shown in table 3, the face-to-face learning show significantly higher than the online learning. Students rated face-to-face learning higher in all of dimension. For social presence in face-to-face learning (M=33.29, SD=.5.48), than online learning (M=31.51, SD=.69). With regard to social interaction, there was a statistically different between face-to-face learning (M=24.20, SD=4.34) and online learning (M=22.77, SD=.4.28). The result also indicated that there were a significance different in the students' perception of face-to-face learning and online learning in term of satisfaction. The satisfaction rated (M=44.07, SD=7.48). Those results supported hypothesis in this study: face-to-face learning led to more positive perceptions, higher levels, and stronger sense compared to online learning. The main reason why face-to-face learning is higher than online learning is due to the lacks of social presence and social interaction itself toward online learning.

Table 3. The comparisons of Online Learning among Social toward Social Presence, Social Interaction, and Satisfaction

Dimension	Variable	Mean	Standard Deviation
Social Presence	Online learning	31.51	5.55
	Face-to-face learning	33.29	5.48
Social Interaction	Online learning	22.77	4.28
	Face -to-face learning	24.20	4.34
Satisfaction	Online learning	41.63	7.62
	Face-to-face learning	44.07	7.48

To examine the difference of online learning and face-to-face learning perception, ANOVA test also used in in this study. Table 4 describes the students' perception of online learning and face-to-face learning among different level in the university. Based on ANOVA test, there is no significantly different among university students at different levels of online learning and face-to-face learning, the university students level are freshman, sophomore, junior, and senior.

Table 4. The comparisons of online and face-to-face learning among the different level of study

		Sum of Squares	df	Mean Square	F	Sig.
Online Learning	Between Groups	2.76	3	.92	2.13	.10
	Within Groups	44.32	103	.43	1.31	
	Total	47.09	106			
Face-to-face learning	Between Groups	1.71	3	.57	1.31	.27
	Within Groups	44.84	103	.43		
	Total	46.55	106			

The result showed that p value > .05, meanwhile .10> .05, and .27> .05.

4.2. Discussion

The purpose of this study was to explore the students' perception toward online and face-to-face courses in term of social presence, social interaction, and satisfaction. In addition, this study also aimed to examine the difference of online learning and face-to-face learning perception among different level of students. Many previous research identified the satisfaction of face-to-face learning is higher than online learning, this research also argued that face-to-face learning are more amenable compared to online learning in term of social presence, social interaction, and satisfaction. However, the findings of this study argued that there is no significantly perception of online learning and face-to-face learning among different level in the university (freshman, sophomore, junior, and senior).

Our findings demonstrated that level in the university does not have influence to the online learning and face-to-face learning. With respect to learning perspectives, there was no statistically significant difference between the participants enrolled in the online and face-to-face learning. Even though face-to-face learning perceived more satisfy, many chose online learning over face-to-face classes for the convenience and ease of time and for the opportunity to work when they wanted instead of when they had to [21]. In addition, online education has efficient cost, and it allows learners to work at their own pace to complete their full degree. Therefore, instructor must have a concrete understanding of the theory and the practice in order to be effective, whether in online learning or in face-to-face learning in order to experience class participation [22].

5. Conclusions

Universities now have the ability to provide blended learning or distance learning opportunities through online classes. Therefore, a survey instrument was distributed to students to explore students perception toward online and face-to-face courses in term of social presence, social interaction, and satisfaction. Moreover, this study also aimed to examine the difference of online learning and face-to-face learning perception among different level of students. Even though online learning perceived as less social interaction, lacking social presence, and synchronicity in communication, online learning actually has some advantages to the students.

5.1. Limitation of the Study

The participants in this study were just from one university on the field of humanities. Adding more sample from different university should be conducted on the field of natural science as well. Most of the participants were female (n=107, male=7, female=100). Therefore this study doesn't compare based on gender, which is very common in education field. Moreover, this study also was based upon students' perception rather than field of observation in face-to-face learning or analyzing the online learning video.

6. References

[1] Fortune M, Spielman M and Pangelinan D 2011 Students' perceptions of online or face-to-face learning *and* social media in hospitality, recreation and tourism *Journal of Online Learning and Teaching* **7**(1) pp 1-16
[2] Kemp N and Grieve R 2014 Face-to-face or face-to-screen? undergraduates' opinions and test performance in classroom vs. online learning *Educational Psychology* **5** pp. 1-14
[3] Tratnik A 2017 Student satisfaction with an online and a face-to-face Business English course in a higher education context *Journal Innovations in Education and Teaching International* **15**(1) pp. 1-10
[4] Koohang A and Durante A 2003 Learners' perceptions toward the web-based distance learning activities/assignments portion of an undergraduate hybrid instructional model *Journal of Informational Technology Education* pp 105-113
[5] Salmon G 2004 *E-moderating: The Key to Teaching and Learning Online* (New York: Taylor and Francis Group)
[6] Short J, Williams E, and Christie B 1976 *The social psychology of telecommunications* (London: John Wiley and Sons)
[7] Gunawardena C 1995 Social presence theory and implications for interaction and collaborative learning in computer conferences *International Journal of Educational Telecommunications* **1**(2) pp. 147-166
[8] Garrison D, Anderson T, and Acher, W 2000 Critical inquiry in a text-based environment: Computer conferencing in higher education *The Internet and Higher Education* **1** pp. 87–105
[9] Moore M G 1993 *Three types of interaction. In Harry K, John M and Keegan D (Eds.) Distance Education: New Perspectives* (New York: Routledge)
[10] Tu C H 2000 On-line learning migration: From social learning theory to social presence theory in a CMC environment *Journal of Network and Computer Applications* **23**(1) pp. 27-37
[11] Jung I, Choi S, Lim C, and Leem J 2002 Effects of different types of interaction on learning achievement, satisfaction and participation in Web-based instruction *Innovations in Education and Teaching International* **2** pp 153–162
[12] Hurst B, Wallace R and Nixon S 2013 The impact of social interaction on student learning *Reading Horizons* **52**(4) pp. 375-398
[13] Carini R, Kuh G, and Klein S 2006 Student engagement and student learning: Testing the linkages *Research in Higher Education* **1** pp. 1–32
[14] Hiltz S R 1993 *The virtual classroom: learning without limits via computer networks* (Norwood, New Jersey: Ablex Publishing Corporation)
[15] Gunawardena C N and Zittle F J 1997 Social presence as a predictor of satisfaction within a computer-mediated conferencing environment *The American Journal of Distance Education* **11**(3) pp. 8-26
[16] Lo C 2010 How student satisfaction factors affect perceived learning *Journal of the Scholarship of Teaching and Learning* **10**(1) pp. 47-54
[17] Spears L R A 2012 (online) *Social presence, social interaction, collaborative learning, and satisfaction in online* and *face-to-face courses* (Doctoral dissertation, Iowa State University) Retrieved from http://lib.dr.iastate.edu/cgi/viewcontent.cgi?article=3983andcontext=etd.
[18] Picciano G A 2002 Beyond student perceptions: issues of interaction, presence, and Performance in an online course *Journal of Asynchronous Learning Networks* **1** pp. 21-40
[19] Gray J and DiLoreto M 2016 The effects of student engagement, student satisfaction, and perceived learning in online learning environments *International Journal of Educational Leadership Preparation* **11**(1) pp. 1-20

[20] Carmines E G and Zeller R A 1979 *Reliability and validity assessment* (California: Sage Publications)
[21] Cuthrell K and Lyon A 2007 Instructional strategies: What do online students prefer? *MERLOT Journal of Online Learning and Teaching* **3**(4) pp. 357-362
[22] Artino A R 2010 Online or face-to-face learning? Exploring the personal factors that predict students' choice of instructional format *The Internet and Higher Education* **13**(4) pp. 272-276

Is f(x) unique? Prospective teachers' conceptual and procedural knowledge on a definite integral problem

A H Rosyidi*, A W Kohar

Departement of Mathematics, FMIPA, Universitas Negeri Surabaya, Kampus Unesa Ketintang, Surabaya, Indonesia, 60231
*Corresponding email: abdul.haris@unesa.ac.id

Abstract. This is a descriptive-explorative research which investigates the extent to which prospective teachers who completed calculus courses involve their procedural or conceptual knowledge in relation to the concept of definite integral through a problem-solving task. Data were collected from 30 prospective teachers' work on a problem-solving task related to Integral. Data were analysed by categorizing the prospective teachers' work into whether they considered $f(x)$, as an integrand, meet the property of uniqueness or not. Results on initial test point out that more than a half of prospective teachers ($n = 17$) answered that the integrand of $f(x)$ is unique, which means there is only one $f(x)$ which satisfy the equation, while the remaining 13 prospective teachers answered $f(x)$ is not unique. The reasons of those who answered $f(x)$ is unique were found around procedural errors related to technique of integration and misunderstanding of using fundamental theorem of calculus. More specifically, there was no prospective teachers who related their performance with the concept of definite integral as area-relation and limit of Riemann sums. Results on the test after reflection activity points out that the prospective teachers seemed aware of their mistakes in finding $f(x)$, proven by the increasing number of responses arguing that $f(x)$ is unique as well as increasing number of variety of functions that can be possibly generated.

1. Introduction

According to Hiebert and Lefevre [1], the main concepts of integral theory are that the definite integral of a function is the limit of Riemann sums, the integral-area relation and the fundamental theorem of calculus. However, research has found learners' limited understanding of underlying concepts related to those three concepts. For example, while learners could perform routine procedures for finding the area under a curve, the learners rarely could explain their procedures [2]. Also, many learners were found to be able to perform procedures and basic techniques of integration, but they failed to understand what behind the procedures [3]. Furthermore, research also found that learners could compute the area under a given curve and could apply the Fundamental Theorem of Calculus to evaluate a definite integral, but they do not know why area under a curve is represented by the definite integral or why the theorem yields the expected result [4]. Along the same lines, Sealey [3] found that some learners in his study understood that the given word problem should be solved using integral concept which utilize the representation of area under curve, but they experienced difficulties on graphing appropriate curve to get the correct representation. In relation to the relationship betweeen symblic notation and its contextual meaning, Mahir reported that while learners show manipulation of concepts on variable which is limited to algebraic symbol, they do not concern on the possible contextual meaning of such algebraic manipulation [2]. Thus, regarding knowledge of integration, all the preceding integral-related findings indicates learners' conceptual knowledge, i.e. knowledge of concepts, which are abstract and general principles [5], is not as good as learners' procedural knowledge, i.e. knowledge of procedures, i.e. a series of steps, or actions done to accomplish a goal [5].

There is a consensus empirically proven by evidence regarding mathematics learning arguing that conceptual knowledge supports procedural knowledge, as well as vice versa [6]. In this regard, Mahir found that the learners who have conceptual understanding on integral theory, indicated by succesfully solving problem related to the concepts of Fundamental theorem of calculus, integral-area relation, and integral of function as the algebraic sum of areas simulataneously, could also have good procedural performance, indicated by succesfully solving problems related to integral techniques. However, it does not mean that learners always use conceptual approach when doing conceptual task in character, as found by Engelbrecht et al [7] . Rather, they often use procedural approach which lead to the failure of obtaining solution or the longer steps written to get solution.

Findings on the lack of understandings become warrants for not only students studying at school levels and undergradute level, but also for those who are prepared to be future teachers. This is because many findings also indicate prospective teachers' limited knowledge regarding undergraduate mathematics topics, such as mathematical proof [8], derivatives [9], concept of limit [10] and of course integral calculus [4-5]. Accordingly, capturing the prospective teachers' understanding on the integral-related concepts within their studies is required as one of tools of evaluation for the improvement of courses programs related to calculus offered in the curriculum at university level.

The aim of this study is to investigate the extent to which prospective teachers who completed calculus course involve their procedural or conceptual knowledge in relation to the concept of definite integral through a problem-solving task.

2. Procedural and conceptual knowledge on definite integral

Literature has much indicated that learners often evaluate definite integrals by applying Fundamental Theorem of Calculus (FTC). FTC could not only connect the definite and indefinite integral, but also providing an efficient method for evaluating definite integrals [11].

Conceptually, Thompson and Silverman [12] asserts that students should not only conceive definite integral as the area under a curve, but also must interpret it as the quantities being accumulated as being created by accruing incremental bits formed multiplicatively. In the context of area under curve, this describe Riemann sum. In this case, Sealey [4] provides a conceptual knowledge need to be understood by a learner. Through his research related to uncovering students' understanding of definite integral from Riemann sum, a learner should figure out that such knowledge has some layers, namely the Product layer, summation layer, limit layer, and function layer to thoroughly understand the meaning of $f(b) = \lim_{n\to\infty} \sum_{i=1}^{n} f(x_i) \triangle x$. Radmehr and Drake [13] have conceptualised four types of knowledge regarding integral, two of which are conceptual knowledge and procedural knowledge. Conceptual knoweldge consist of knowlegde of classifications and categories, knowledge of principles and generalizations, and knowledge of theories, models, and structures. Regarding definite integral, this is related the knowledge about the relationship between definite integral and Riemann sums, integral-area relationship, and FTC. Meanwhile, procedural knowledge consist of knowledge of subject-specific skills and algorithms, knowledge of subject-specific techniques and methods, and knowledge of criteria for determining when to use appropriate procedures. In relation to definite integral, this is related to the knowledge of classifying an integrand according to its form, knowledge of different methods for finding areas (e.g. with respect to x and y), knowing that it is easier to find the definite integral using FTC rather than sums when antiderivate can be found in terms of elementary functions.

In this study, the concern is around prospective teachers' understanding on definite integral related to the first part of FTC: "If f is continues at every point in $[a, b]$ and F is any antiderivative of f on $[a, b]$, then $\int_a^b f(x)dx = F(b) - F(a)$" (Thomas, p.277) [14], the uniqueness of definite integral resulted from the left part of the equation within such theorem, and the extent to which they relate their responses to the concept of integral-area relationship or even the limit of riemann sums.

3. Methods

3.1. Participants and Instruments

Participants of this study is 30 prospective teachers who were studying at bachelor of mathematics education program at Universitas Negeri Surabaya, Surabaya, Indonesia, having completed some basic mathematics courses such as calculus and advanced calculus. Aside the researchers themselves, the main instrument is a problem-solving task involving the concept of calculus integral. The indicator of the problem-solving task is to find out the possible integrand of $f(x)$ if both the value of definite integral associated with such $f(x)$ and the interval of x are given. See Figure 1.

Given that $f(x)$ is a continuous function satisfying the equation

$$\int_{2}^{5} f(x)dx = -6$$

1. Find $f(x)$
2. Is $f(x)$ unique? Give your reasons

Figure 1. The problem-solving task of definite integral

The prospective teachers were then provided a week after solving such task to reflect on whether they need to revise their work and the extent to which they carried out any changes. This activity is to clarify the problem-solving processes they undertook when solving the task in the initial test.

3.2. Analysis on prospective teachers' procedural and conceptual knowledge

Data were analysed by coding each of the responses into an appropriate procedural error and conceptual errors, as well as procedural and conceptual knowledge correctly performed by prospective teachers. Also, the responses were also categorized into whether $f(x)$ is unique or not. Thus, they were summarised into a table. To describe, we exemplify the responses of student teachers' proof schemes on those two tasks by presenting examples of each of the levels. To maintain th validity and reliability of the analysis, the coding was first carried out by the second author, then it was checked and clarified by the first author. Furthermore, the authors discussed the possible explanation of each of the codings to be presented by comparing the findings with theoretical review discussing conceptual and procedural knowledge on definite integral, as discussed in the preceding section.

4. Results and discussion

4.1. Variety of functions of f(x): Conceptual and Procedural Performance

Table 1. Variety of functions of $f(x)$

Performances	**Constant function**		**Linear function**		**Quadratic function**		**Unknown**	
	Initial test	After reflection*	Initial test	After reflection*	Initial test	After reflection*	Initial test	After reflection*
Conceptual error	5	2	0	0	0	0	4	1
Procedural error	1	2	0	0	0	0	5	3
Correct performance	12	21	2	16	1	6	0	0

*The number of response in total for each of the results of initial test and after reflection is more than 30 because at after reflection, particularly, some participants give more than one form of functions.

Table 1 presents the changes of responses regarding the forms of function at initial test and the test after reflection ditributed into whether they are correct or conceptually/procedurally incorrect. In general, most of the responses at initial test in relation to the type of functions are recorded as contant function and unknown function, which means the prospective teachers only concerned on a simple polynomial function that can represent the $f(x)$. After reflection, the prospective teachers seemed aware of their mistakes in finding $f(x)$, proven by the increasing number of correct answer. With regard to prospective teachers' performances, it was found that a half of them found correct $f(x)$ and the remaining found incorrect $f(x)$ at initial test. Conceptual errors were found around two types: (1) the assumption that $f(x)$ is only a constant function, and (2) the assumption that $f(x)$ is the result of definite integral. In detail, the first conceptual error was found from 5 prospective teachers and the second error was found from 4 prospective teachers. Figure 1 shows these two errors.

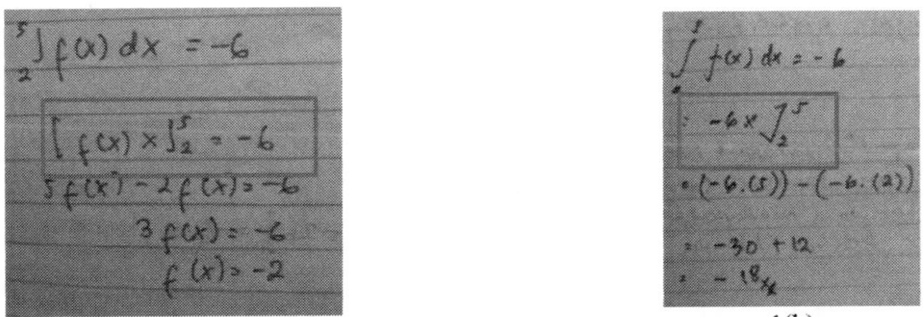

1(a) 1(b)

Figure 2. Conceptual error in finding $f(x)$

The prospective teacher whose response is shown at figure 2(a) seemed did not consider the various possible of $f(x)$. Instead, he assumes that $f(x)$ is a constant function such that it can follow one of theorems in integral, which is if k constant, then $\int k\, dx = kx + c$. Meanwhile, figure 2(b) shows a prospective teacher who unable to differentiate between the result of an integral operation and the integrand so that -6 becomes a constant function that can changes become $-6x$ after an integration process. In conclusion, figure 2(a) and 2(b) shows that the prospective teachers was unable to use appropriate application of FTC conceptually.

Procedural errors, on the other hand, were found around three types: (1) result of integral is assumed as the part of integrand, (2) inappropriate use of distributive law in operating integrand, and (3) mistakes in writing up the integrand: no brackets for more than one terms in the integrand. Figure 2(a), 2(b), and 2(c) respectively point out these three errors.

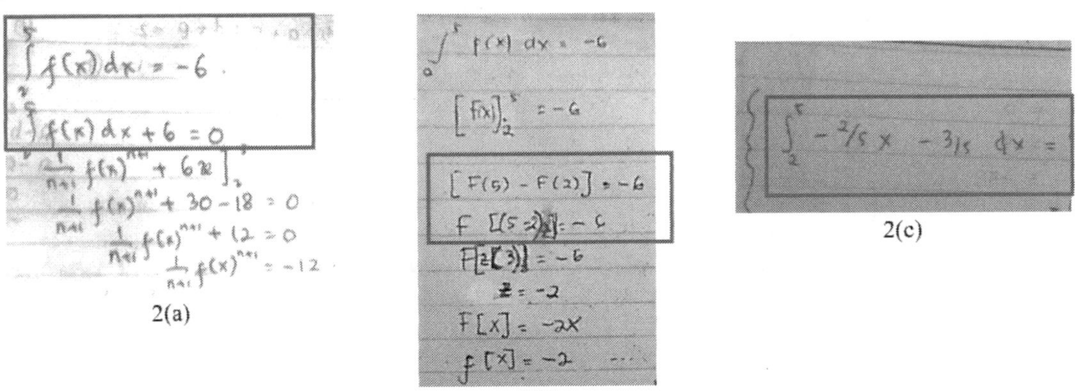

2(a) 2(b) 2(c)

Figure 3. Procedural error in finding $f(x)$

Procedurally, response on figure 3(a) indicates an error since the prospective teacher at this figure moved the result on integral in the right side to the left side as many learners do when carying out an algebraic procedure, then assume it as part of the integrand that can be then integrated with a formula of elementary integral. Figure 3(b) shows that $F(x)$ is not viewed as a unity of symbol representing an integrated function of $f(x)$. Instead, it is viewed as F times x so that the prospective teacher having this response factorize the form of $F(5) - F(2)$ becomes $F(5 - 2)$. Meanhile, figure 3(c) indicates an incorrect writing of terms in an integrand, which leads the integral does not make sense symbolically. The response should have been written as $\int_2^5 \left(-\frac{2}{5}x - \frac{3}{5}\right) dx$, instead of $\int_2^5 -\frac{2}{5}x - \frac{3}{5}\, dx$.

Correct responses are also performed by the prospective teachers in this study. The responses ranges from $f(x)$ in the form of constant function to quadratic function. Figure 3(a) and 3(b) respectively shows these two types of responses.

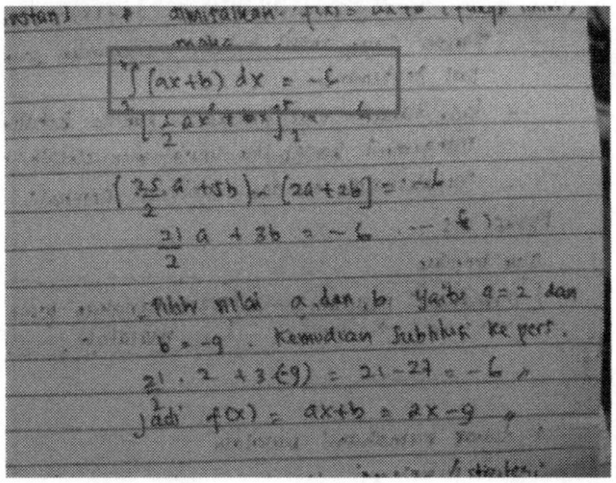

4(a)

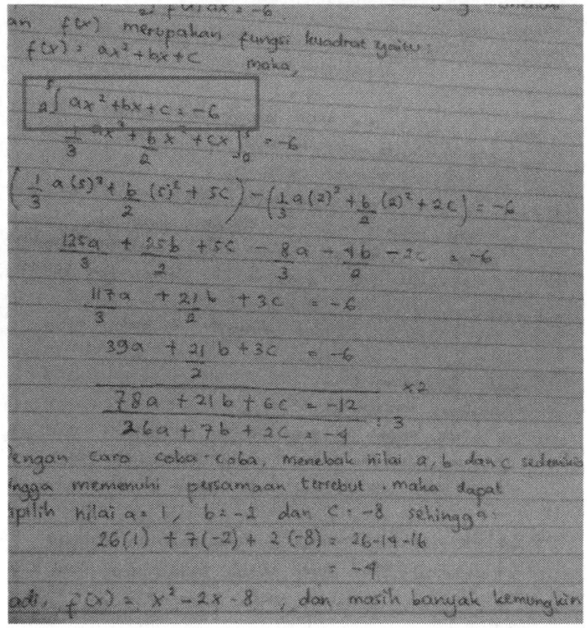

4(b)

Figure 4. Correct performance of finding $f(x)$

Th responses on figure 4(a) and 4(b) show that the linear function and the quadratic function is respectively supposed as $ax + b$ and $ax^2 + bx + c$. Then, these are integrated by applying a formula of elementary integral within FTC. In the end, each variable of a, b, and c is found by selecting integers which can substitute a, b, and c to satisfy the integral result equation.

4.2. The uniqueness of integrand

In the initial test, regardless of whether the $f(x)$ found by the participants is correct or not, there were 17 prospective teachers assuming that the integrand of $f(x)$ is unique, which means there is only one $f(x)$ which satisfies the equation, while the remaining 13 prospective teachers answered $f(x)$ is not unique (more than one possible forms). Table 1 shows distribution of prospective teachers' responses on the problem-solving task on initial test and after reflection.

Table 2. Prospective teachers' response on the uniqueness of integrand

Uniqueness of	Correct		Incorrect	
integrand	Initial test	After reflection	Initial test	After reflection
Unique	7	3	10	0
Not unique	8	25	5	2

Table 2 shows that in initial test the number of responses of correct and incorrect responses are equally distributed, in which the responses of integrand as a unique function increases and those as not unique function decreases from initial test to the test after reflection. Thus, in general, the reflection activity made the prospective teachers become more aware of their errors they undetook during the initial test. The following section describe some examples of findings examplifying each of entries in table 1.

4.2.1. Integrand as a unique function

Findings indicate some reasons for those assuming the integrand as a unique function. The reasons of those who answered $f(x)$ is unique were found around procedural errors related to technique of integration and missundertanding of using fundamental theorem of calculus. The latter finding, particularly, is related to the prospective teachers's perceived assumption that since the result of a definite integral is unique, the integrand is unique as well. In other words, the converse of a statement that "for $f(x)$ continuous function, if the integral result for the operation of $\int_a^b f(x)dx$ is k, then k is unique", is assumed correct. The other finding is related to the failure of finding f(x) procedurally which leads prospective teacher conclude that $f(x)$ is unique, of course without any supporting argument.

4.2.2. Integrand as not unique functions

For those who answered that $f(x)$ is not unique, none of their answer was indicated to show possible various algebraic form of $f(x)$ which represent the integrand. Instead, students were found to let $f(x)$ as contant function, linear function, and quadratic function, instead polynomial function which has degree more than 2 or other types of form of functions. More specifically, there was no prospective teachers who related their performance with the concept of definite integral as area-relation. Thus, it indicate prospective teachers' lack of conceptual understanding of definite integral. However, once the prospective teachers revised their work, most of them (28), primarily for those who answered $f(x)$ is unique, changed their answer to the assumption that $f(x)$ is not unique, and only 2 prospective teachers who kept their answers, i.e. $f(x)$ is unique. Regarding problem-solving stages, prospective teachers experienced the stages of devising a plan as the most difficult steps to determine $f(x)$. This is because the prospective teachers found difficulties in determining the possible form of $f(x)$ which can be included in the integral calculation. Also, at 'understanding the problem' stages, prospective teachers were also found to have misunderstanding of the meaning of 'unique' for $f(x)$.

The findings in this study reveals that no prospective teachers who considered the concepts of integral-area relationship or even Riemann sums as the approach in finding $f(x)$. This is aligned with previous report arguing that learners tend to focus on the application of the first part of FTC, instead of such two approaches in solving a definite integral problem [3-4]. Therefore, the results of this study encourage future empirical research to further study about the impact of conceptual understanding on procedural fluency, particularly about integral concept, and vice the versa. Also, the study also may consider further studying the performance of prospective teachers regarding the factual knowledge, procedural knowledge, conceptual knowledge, and metacognitive knowledge as suggested by Radmehr and Drake [8] and how its impact on their students. The latter suggestion is related to the mathematical knowledge for teaching a particular topic that a teacher should hold [15].

5. Conclusion

To highlight, this present study found that the number of prospective teachers who answered that the integrand of $\boldsymbol{f(x)}$ is unique, which means there is only one $\boldsymbol{f(x)}$ which satisfy the equation, is more that those who answered that $\boldsymbol{f(x)}$ is not unique for a given definite integral problem with the result of integral and the interval of the integral are both given. The classification of the preservice teachers' work into conceptual error or procedural error for some of their responses, in this case, maybe somewhat superficial. This is due to the finding of both procedural and conceptual errors which appears, which lead to the questions of which one of those errors dominantly occurs. Thus, we need to further explore by interviewing the person having the responses to obtain some clarification.

6. References

[1] Hiebert J and Lefevre P 1986 Conceptual and procedural knowledge in mathematics: An introductory analysis, In Hiebert J *Conceptual and procedural knowledge: The case of mathematics* 1-27 (Hillsdale, NJ: Lawrence Erlbaum Associates).

IOP Conf. Series: Journal of Physics: Conf. Series **1108** (2018) 012095 doi:10.1088/1742-6596/1108/1/012095

[2] Mahir N 2009 Conceptual and procedural performance of undergraduate students in integration *International Journal of Mathematical Education in Science and Technology* **40**(2) 201-211.

[3] Sealey V 2006 Definite integrals, Riemann sums, and area under a curve: What is necessary and sufficient *Proceedings of the 28th annual meeting of the North American Chapter of the International Group for the Psychology of Mathematics Education* Vol. **2** (1991) 46-53.

[4] Sealey V 2014 A framework for characterizing student understanding of Riemann sums and definite integrals *The Journal of Mathematical Behavior* **33** 230-245.

[5] Canobi K H 2009 Concept–procedure interactions in children's addition and subtraction *Journal of experimental child psychology* **102**(2) 131-149.

[6] Rittle-Johnson B, Schneider M, & Star J R 2015 Not a one-way street: Bidirectional relations between procedural and conceptual knowledge of mathematics *Educational Psychology Review* **27**(4) 587-597.

[7] Engelbrecht J, Harding A, & Potgieter M 2005 Undergraduate students' performance and confidence in procedural and conceptual mathematics *International Journal of Mathematical Education in Science and Technology* **36**(7) 701-712.

[8] Rosyidi A H & Kohar A W 2018 Student Teachers' Proof Schemes on Proof Tasks Involving Inequality: Deductive or Inductive? *Journal of Physics: Conference Series* **947**(1) 012028.

[9] Siyepu S W 2013 An exploration of students' errors in derivatives in a university of technology *The Journal of Mathematical Behavior* **32**(3) 577-592.

[10] Denbel D G 2014 Students' Misconceptions of the Limit Concept in a First Calculus Course *Journal of Education and Practice* **5**(34) 24-40.

[11] Radmehr F & Drake M 2017 Exploring students' mathematical performance, metacognitive experiences and skills in relation to fundamental theorem of calculus *International Journal of Mathematical Education in Science and Technology* **48**(7) 1043-1071.

[12] Thompson P & Silverman J 2008 The concept of accumulation in calculus. In Carlson M & Rasmussen C (Eds.), *Making the connection: Research and teaching in undergraduate mathematics education* pp. 43–52 (Washington, DC: Mathematical Association of America).

[13] Radmehr F & Drake M 2017 Revised Bloom's taxonomy and integral calculus: unpacking the knowledge dimension *International Journal of Mathematical Education in Science and Technology* **48**(8) 1206-1224.

[14] Thomas G B, Finney R L, Weir M D, & Giordano F R 2010 *Thomas' calculus 12th edition* (Boston: Addison-Wesley).

[15] Siswono T Y E, Kohar A W, & Hartono S 2018 Designing Tasks to Examine Mathematical Knowledge for Teaching Statistics for Primary Teachers *Journal of Physics: Conference Series* **947**(1) 012008.

Epitope Mapping of Capsid Protein L1 from Human Papillomavirus to Development Cervical Cancer Vaccine through Computational Study

E R Purnama[1], V D Kharisma[2]

[1]Biology Departement, Faculty Mathematic and Natural Sciences, Universitas Negeri Surabaya
[2]Biology Departement, Faculty Mathematic and Natural Sciences, Universitas Brawijaya

[1]Corresponding author: erlixpurnama@unesa.ac.id.

Abstract. Human papillomavirus (HPV) is a virus that plays an important role in the occurrence of cervical cancer, HPV is a member of the family papovaviridae, genus papillomavirus. The HPV gene is composed of two parts: early and late gene. Early proteins in HPV include E1, E2, E4, E5, E6, and E7 Late gene expressed from the late promoter acting as coders for L1 and L2 protein capsid. The L1 protein has a conserved region composed of cysteine and lysine residues, both of which have involved in the binding process between virions and host receptors. Previous research has shown that vaccines can be developed based on epitopes that have conserved areas. This study is important to identify conserved protein sequences in L1 of HPV capsid, predict epitope mapping of B cells and antigenicity in the conserved region of L1 HPV capsid, as well as the similarity of amino acid residues of epitope composers with surface receptors of human body cells. The conserved areas were identified in L1 HPV as a potential epitope of B cells based on epitope mapping analysis of positions 23-46 and 97-119 with EGRGQPLGGSGHPNDDE DRDKQ and RHNGGPGPSGSSQFNKPYWAQGN peptides and each had a peptide length of 22-mer and 23-mer. The 97-119 epitope has a high antigenicity score and the similarity of the low amino acid residue sequence to the cell surface receptor of the human body. So the 23-mer RHNGGPGPSGSSQFNKPYWAQGN peptide can be used as a reference for the development of cervical cancer prevention vaccine.

1. Introduction

HPV infection has increased since 1960 due to the increased cases of cervical cancer and the development of genital warts into carcinomas. HPV types 6 and 11 were found to be 35% in genital warts whereas HPV types 16 and 18 found 63% in carcinoma [1]. HPV is composed of the double strand DNA (dsDNA) genome and two types of capsid proteins L1 and L2. HPV is a non-enveloped virus and replicates in the nucleus of an infected host cell [2]. It is generally classified into two types of high risk (HR) and low risk (LR). The type of HR often causes cervical cancer, such as HPV types 16, 18, 31, 33, 35, 39, 45, 51, 52, 56, 58, 59, and 68 while the LR type only triggers benign, like HPV type 6, 8, 7, 11, and 15 [3]. The HPV gene is composed of two parts, early and late gene. The early proteins in HPV include E1, E2, E4, E5, E6, and E7. Then, a late gene expressed by the late promoter acts as a protein-coding of L1 and L2 capsid [4]. In addition, L1 protein HPV has a conserved region composed by cysteine and lysine residues, this both residues have involvement in the process of binding between virions with host receptors [5]. B cell epitope mapping is a method to predict the region of proteins, that can be recognized and related to cell B response [6]. The epitope is a specific area of the part of the antigen binding to the paratope of the antibody and can also recognized by T and B cells, the epitope can be either self-epitope or outside [7]. Based on this statement, the development of epitope conserve based vaccine for maximum cervical cancer prevention is necessary because to provide more protection against HPV infection by adding other HPV types that have been identified and not yet used in previous vaccine making, the importance of protein sequence identification conserved and predicted properties of antigenicity and epitope mapping on the L1 protein HPV underlying this study.

2. Method

2.1. Identification of Conserved Region on L1 HPV

Sequence data obtained from NCBI (*www.ncbi.nlm.nih.gov*). The search in a biological database is done by using keyword "L1 Human Papillomavirus", these sample collected from all high-risk strains of virus available in the database, then after the samples obtained successfully stored in FASTA format on a notepad. This study used MEGA 5.05 software to perform alignment of L1 HPV protein sequences and identification of conserved protein sequences as has been done in previous studies [8]. After the alignment sequence process is complete then the next step is protein modeling and 3D alignment, where this method aims to perform 3D construction and molecular simulation of conserved domain on L1 capsid HPV.

2.2. Prediction Potential B Cell Epitop

This method has functions to determine or predict the epitope of B cells and their antigenicity. Epitope and antigenicity predictions are analyzed through facilities on the IEDB online webserver (*www.iedb.org*) [8, 9; 10]. After the prediction of the B cell epitope, the peptide obtained was then further analyzed using basic local alignment tools protein (BLASTp) (*http://blast.ncbi.nlm.nih.gov/blast.cgi*). This aims to compare the similarities with the protein sequences contained in the human body.

3. Results and Discussion

3.1. The Position of Conserved Region on L1

L1 HPV protein has a conserved region composed of cysteine and lysine residues, both of which have involved in the binding process between virions and host receptors [5]. The samples of L1 sequence from HPV HR type is obtained from the NCBI database were 13 types and 869 strains. The alignment process used the MEGA 5 program and produced a conserved protein sequence with a 173-mer sequence length, had residual C (cysteine) of 8 and 6 K (lysine) residues, then analyzed protein modeling on the web server Swiss Model for obtaining 3D protein structures.

The 3D model protein produced by the Swiss-Model server shows the identified query protein sequence value of 77.03% according to the HPV L1 capsid template consisting of 5 chains with 3ofl PDI ID. Peptide generated from the modeling process that has a protein sequence length of 148-mer. The peptide 3D protein model consists of alpha-helix, beta-sheet, and coil displayed on the surface, using PyMol software. 3D sequence structure obtained by homology modeling on web server *www.swissmodel.expasy.org*, then visualized protein structure and 3D alignment using PyMol software [8]. The result of 3D alignment (Figure 1) shows that the peptide (red color) shown as a sphere match with chain A (green) in HPV L1 capsid template is shown in cartoon form, it indicates the present position of conserve sequence on L1 HPV.

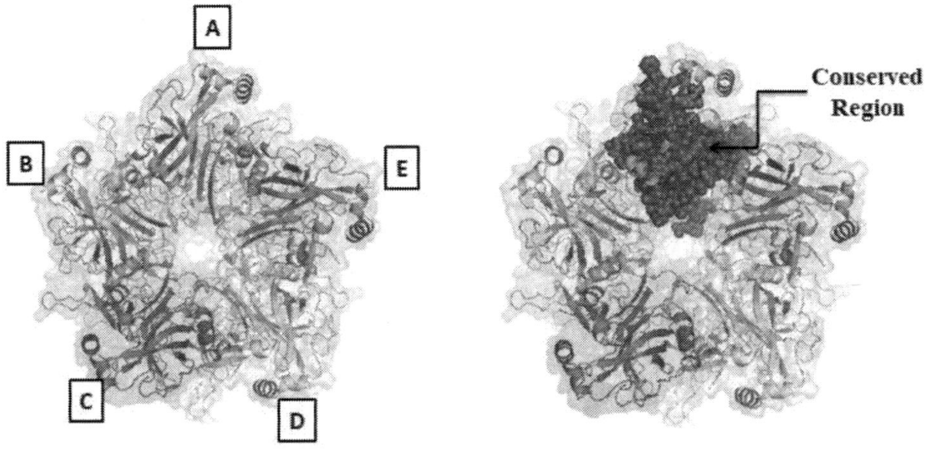

Figure 1. Molecular visualization 3D structure of L1 protein. Pentamer structure L1 template 3ofl (left). The position of conserved amino acid residues after 3D alignment (right).

3.2. B Cell Epitope Prediction

Scores of the results of linear epitope prediction analysis of BepiPred method from IEDB show that there is a predicted protein sequence position of the B-cell epitope that is 25-46 sequence with 22-mer sequence length and 97-119 sequence with 23-mer length. Epitope mapping graph using BepiPred method of IEDB webserver output analysis shows that peptides with protein sequence positions 23-46 and 97-119 with the threshold of 0.510 are included in the yellow region. Peptide position 23-46 with protein sequence EGRGQPLGGSGHPNDDEDRDKQ and position 97-119 with protein sequence RHNGGPGPSGSSQFNKPYWAQGN predicted as an epitope of B cells in chain A, then displayed in its 3D structure with the selection of staining structures in the form of spheres and cartoon using PyMol software to describe the position of epitope. The peptide region is shown in the PyMol software, in spheres structures dyed red, yellow, and blue sequentially showing the conserved region, position peptides 23-46, and 97-119. The coloring of cartoon structures in green, cyans, and magenta sequentially depicts staining chains A, B, and E (Figure 2).

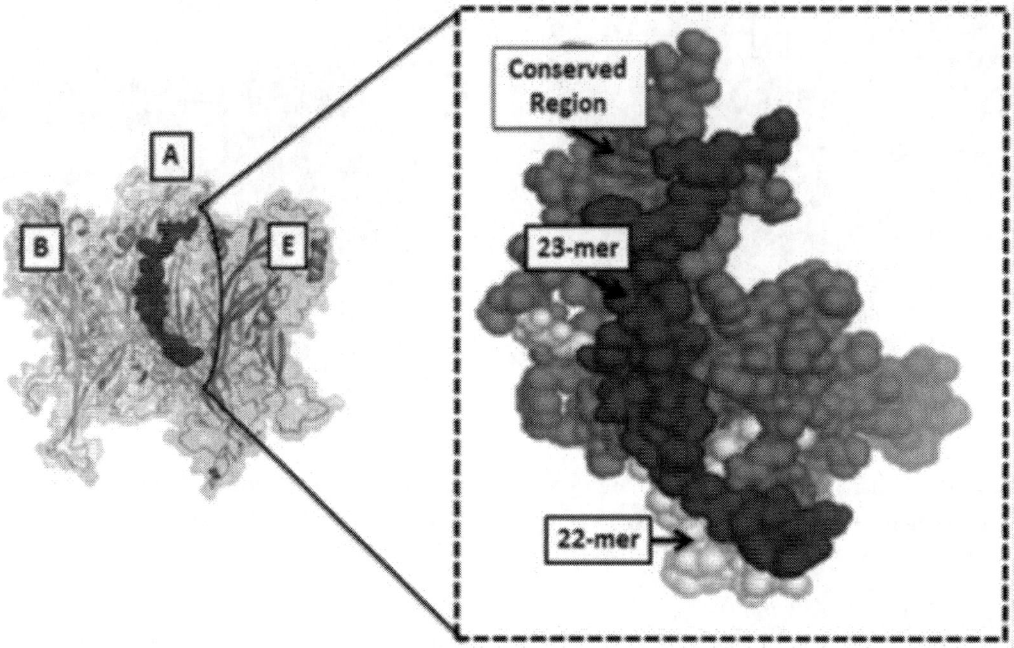

Figure 2. Epitope mapping on L1 HPV protein. Epitope visualization in chain A, B, E (left) with cartoon structure in each chain. Visualization of the epitope (yellow and blue) in conserved sequence area (red) with spheres structure (right).

Antigenicity is a property possessed by antigens that allow antigens to trigger B cell responses to produce specific antibodies, and the nature of this antigen also refers to immunogenicity [11]. Peptides of 22-mer and 23-mer lengths are potential as epitopes, then compared their antigenicity values using the Kolaskar & Tongaonkar method on the IEDB web server and displayed in the PyMol software. The Kolaskar & Tongaonkar method is a computational protein antigenicity prediction method based on physicochemical properties and experimental data, with the accuracy of up to 75% [12]. The predicted antigenicity score using the Kolaskar & Tongaonkar method showed that the peptide of 23-mer sequence number 97-119 has a high antigenicity score of 22-mer peptide position 25-46 (Figure 4) and shown in PyMol in spheres structure (peptide) as well as cartoon (L1 chain A protein) and clarified its position using the surface structure (Figure 6). It shows 23-mer peptide with protein sequence RHNGGPGPSGSSQFNKPYWAQGN predicted to trigger the formation of adaptive immune response by B cell.

Analysis of basic local alignment sequence tool (BLAST) in the vaccine design stage serves to perform data comparisons with proteins in the human body, it is done to avoid the autoimmune response from the body of patients who will receive the vaccine, on reading the results of BLAST maximum score of 70% query sequence similarities with proteins contained in the human body [10]. The results of the RHNGGPGPSGSRHNGGPGPSGSSQFNKPYWAQGN protein analysis with 23-mer length showed very low similarity score with the surface receptor of a human body cell that is >40. So the epitope can be used as a reference for the development of HPV vaccine. This research shows that bioinformatics is a very useful tool in the vaccine design process and useful for analyzing the interactions of genes and analysis of herbal mediated cell apoptosis [8].

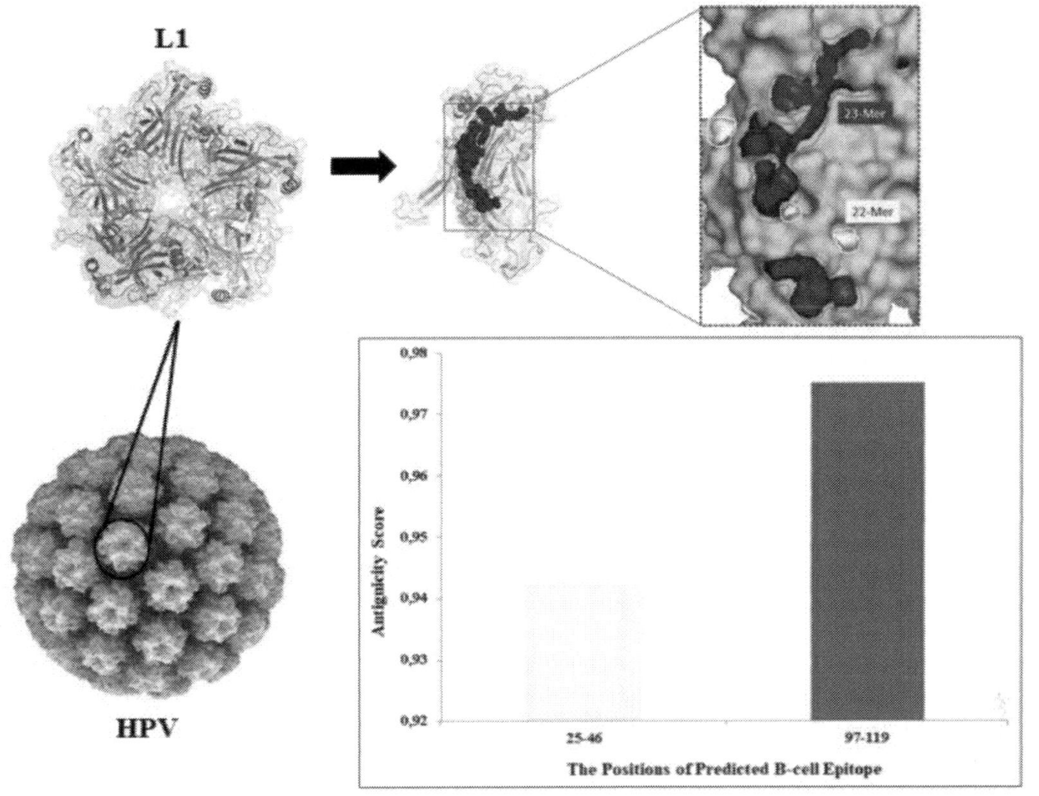

Figure 3. Prediction of Kolaskar & Tongaonkar antigenicity scores and peptide position visualization in L1 chain protein A. An antigenicity score at sequence no. 25-46 is lower than 97-119. Peptides with yellow and blue spheres are peptides of 22-mer and 23-mer, with positions 25-46 and 97-119. The cartoon structure is a L1 capsid protein that is colored green by the coloring of the chain and clarified its position in surface structure of the PyMol software.

4. Conclusion

The positions of conserved protein sequence potentially B cell epitope 23-46 and 97-119 position with EGRGQPLGGSGHPNDDEDRDKQ and RHNGGPGPSGSSQFNKPYWAQGN peptide and each have a peptide length of 22-mer and 23-mer. The 97-119 epitope has a high antigenicity score and the similarity of the low amino acid residue sequence to the cell surface receptor of the human body. So the 23-mer RHNGGPGPSGSSQFNKPYWAQGN peptide can be used as a reference for the development of cervical cancer prevention vaccine.

5. References

[1] Chang Y, Brewer N T, Rinas A C, K Schmitt, and J S Smith J S 2009 Evaluating the impact of human papillomavirus vaccines *Vaccine* **27**(32) 4355-62.

[2] Clifford G M, Gallus S, Herrero R, Munoz N, Snijders P J, Vaccarella S, Anh P T, Ferreccio C, Hieu N T, Matos E, Molano M, Rajkumar M, Ronco G, Sanjose S, Shin H R, Sukvirach S, Thomas J O, Tunsakul S, Meijer C J, and Franceschi S 2005 Worldwide distribution of human papillomavirus types in cytologically normal women in the International Agency for Research on Cancer HPV prevalence surveys: a pooled analysis *Lancet* **8** 366-991.

[3] Baseman JG and Koutsky L A 2005 The epidemiology of human papillomavirus infections *J. Clin. Virol.* **32** S16-S24.

[4] King J A, Sonsma A J, Vriend J H, Sande M, Feltkamp C M, Boot H J, and Koopmans G P M 2015 Genetic Diversity in the Major Capsid L1 Protein of HPV-16 and HPV-18 in the Netherlands *PLoS ONE* **11**(4) e0152782

[5] Dasgupta J, Bienkowska-Haba M, Ortega M E, Patel H D, Bodevin S, Spillmann D, Bishop B, Sapp M, and Chen X S 2011 Structural basis of oligosaccharide receptor recognition by human papillomavirus *J. Biol. Chem*. **286** 2617–2624.
[6] Nakagawa M, Kim H K, and Moscicki B A 2004 Different Methods of Identifying New Antigenic Epitopes of Human Papillomavirus Type 16 E6 and E7 Proteins *Clin. Immunol.* **11**(5):889-896.
[7] Parham P 2005 The immune system (New York: Garland Science Publishing).
[8] Loly S S, Widodo N, Djati S M, and Utomo D K 2012 Epitope mapping of gp350/220 conserved domain of epstein barr virus to develop nasopharyngeal carcinoma (npc) vaccine *J. Bioinfor.* **8**(10) 479-482.
[9] Karimatul H, Fitriyah, Ardyati T, Deocaris C, and Widodo 2016 Polytope Prediction for Dengue Vaccine Candidate Based on Conserved Envelope Glycoprotein of Four Serotypes of Dengue Virus and Its Antigenicity *J. Pure App. Chem.* **5**(2) 101-107.
[10] Widodo. Metode Desain Vaksin (Pendekatan Bioinformatika). http://science.lecture.ub.ac.id/files/2012/04/METODE DESAIN-VAKSIN.pdf.
[11] Wickelgren I 2004 Policing the immune system *Science* **306** 596-599.
[12] Kolaskar S A and Tongaonkar C P 1990 A semi-empirical method for prediction of antigenic determinants on protein antigens.

Pre-service physics teachers' understanding on total lunar eclipse: A response of supermoon on January 31st 2018

N Suprapto*, U A Deta, N A Lestari, A Kholiq, A S Adam, H Mubarok, and M N R Jauhariyah

Department of Physics, Universitas Negeri Surabaya, Indonesia.

*Corresponding Author's E-mail: nadisuprapto@unesa.ac.id

Abstract. At the end of January 2018, the spectacular phenomena of astronomy that some people call super-blue-blood moon happened on the night of January 31st or the early morning of February 1st. These phenomena are consequences of a total lunar eclipse. In Indonesia, these three terms of moon become a controversy. Therefore it needs to be more investigated. Furthermore, this study has two research questions: (1) to what extent do the Indonesian PSTs' existing knowledge and reasoning of the lunar eclipse phenomena? and (2) to what extent do the Indonesian PSTs' existing knowledge and reasoning of the supermoon, blue moon, and blood moon?. A qualitative research method was considered in helping the researcher to construct some meanings. There were eight teen participants involved in an interview and a written questionnaire process. The PSTs' reasoning was categorised into four levels: dissatisfied (naïve), intelligible, plausible, and fruitful. The findings indicated that PSTs have some reasoning about the lunar eclipse, however, they unfamiliar with the super-blue-blood moon. It means their reasoning reached the level of plausible and fruitful according to the first research question and at the level of 'dissatisfied' and 'intelligible' for the second research question.

1. Introduction

At the end of January 2018, the spectacular phenomena of astronomy that some people call supermoon, blue moon, and blood moon happened on the night of January 31st or the early morning of February 1st. These phenomena are consequences of a total lunar eclipse. It occurs when the moon is in its closest position to the earth (perigee). In Indonesia, these three terms of moon become a controversy [1]. A supermoon happens where the moon is in 'perigee' position to the earth, blue moon indicating it is the second full moon in a month, and the blood moon is signifying the moon radiates blood red. The first full moon of January happened on the night of January 1st or the early morning of January 2nd, according to our location [2]. It was noted that the study about the moon had been considered by some authors, i.e. [3-9].

To response this phenomenon, as physics researchers we captured the perception of public or citizen about the phenomena. Pre-service physics teachers (PSTs) as a group are appropriate to explain these phenomena. They have already learnt about concepts of astronomy, including eclipse since primary school, indeed, through earth science course in the university. Specifically for PSTs, they will teach students scientific phenomena like a solar eclipse and lunar eclipse to their students.

IOP Conf. Series: Journal of Physics: Conf. Series **1108** (2018) 012097 doi:10.1088/1742-6596/1108/1/012097

Furthermore, this study has two research questions: (1) To what extent do the Indonesian PSTs' existing knowledge and reasoning of the lunar eclipse phenomena? (2) To what extent do the Indonesian PSTs' existing knowledge and reasoning of the supermoon, blue moon, and blood moon?

2. Method

A qualitative research method was considered in helping the researcher to comprehend better the process of constructing meaning. The study used an interview and a written questionnaire. There eight teen participants in this study which forms the focus of this report. The participants were referred to by pseudonyms with popular Indonesian nickname, i.e. Rani (Rn), Hesy (Hs), Harun (Hr), Handal (Hn), Nabil (Nb), etc. The qualitative data from the results of semi-structured interviews were transcribed and analysed by reading the responses to each interview question [10-11]. The PSTs' reasoning was categorised into four levels: dissatisfied (naïve), intelligible, plausible, and fruitful.

3. Result and Discussion

RQ 1: To what extent do the Indonesian PSTs' existing knowledge and reasoning of the lunar eclipse phenomena?

Based on the RQ 1 data findings, There is the different opinion about lunar eclipse among PSTs'. Many of them have the lowest level or in the dissative (naïve) level. These opinions are shown as the statements and figure below:

> *"Sun, moon and earth are taken place in one straight line".* [Hs]
>
> *"Blue moon, super moon and blood moon happens where sun, moon and earth are taken place in one straight line".* [Kn]

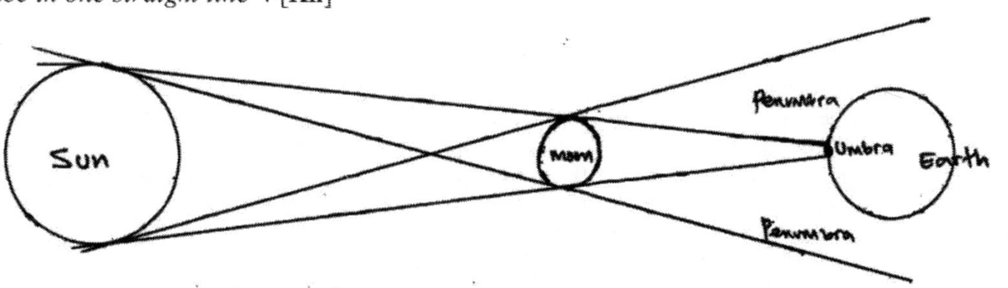

Figure 1. Kn's opinion about lunar eclipse illustration.

> *"The Blue moon happens where the earth, moon and sun are taken place in one straight line".* [Zk]

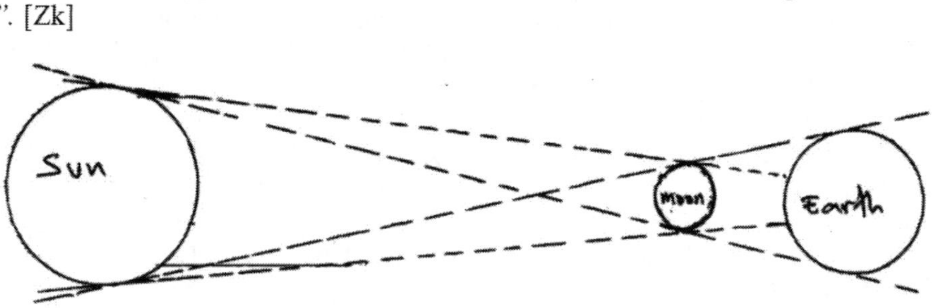

Figure 2. Zk's opinion about lunar eclipse illustration.

In the next level (intelligible level), some reasoning shows inappropriate statement and illustration but better than the previous level. These opinions show in the statements and illustrations below:

> *"The distance of earth and moon is closer".* [Ll]

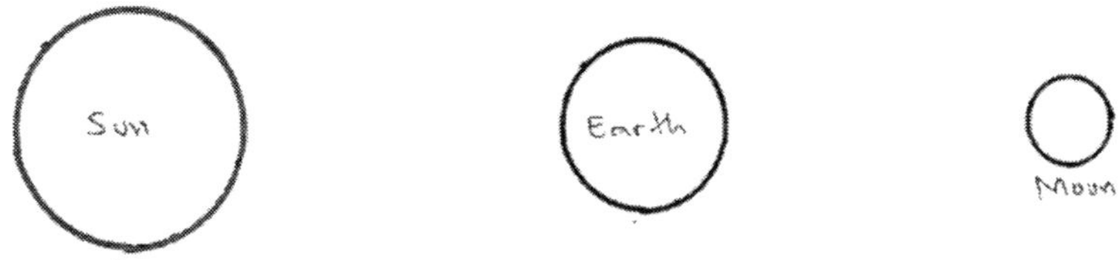

Figure 3. Ll's opinion about lunar eclipse illustration.

"As far as I know, lunar eclipse happened when the earth is blowing light from the sun to the moon. It means the earth position between the sun and the moon". [Al]
"The position of earth and moon is closer". [Fd]

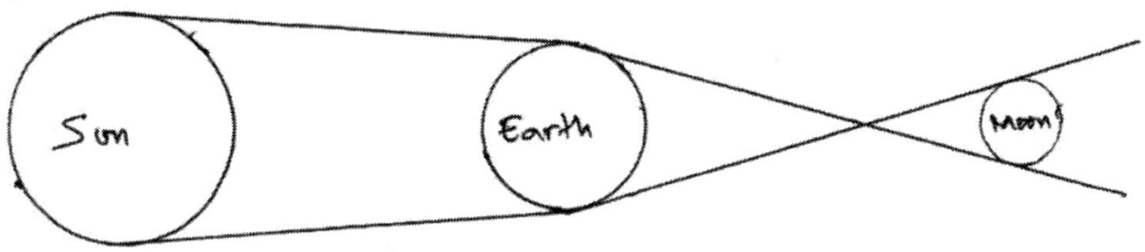

Figure 4. Fd's opinion about lunar eclipse illustration.

"The lunar eclipse happened when moon, earth and sun in a straight line". [Vr]

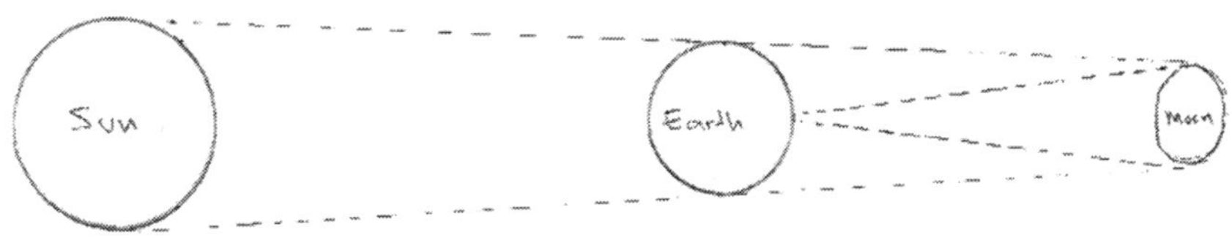

Figure 5. Vr's opinion about lunar eclipse illustration.

In the plausible level, the PSTs' opinion about lunar eclipse getting better, but not in a perfect description. These opinions' show at the statements and illustrations below:

"All type of lunar eclipse (Full moon, blue moon and blood moon) happened in rare time and the position of earth between sun and earth". [El]

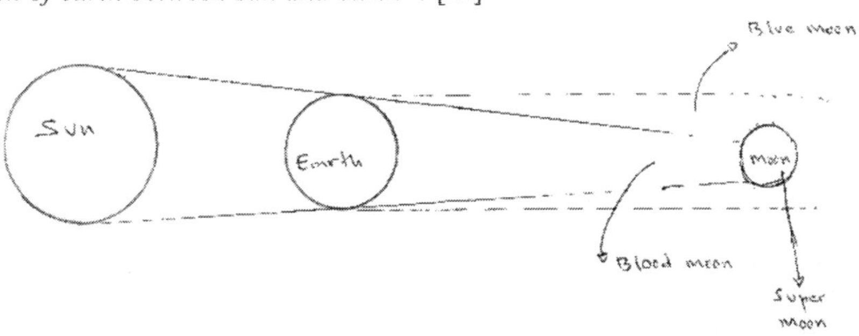

Figure 6. El's opinion about lunar eclipse illustration.

"lunar eclipse is a phenomenon when the sunlight that should be reflected by the moon can't reach the moon surface or can't be reflected". [Hr]

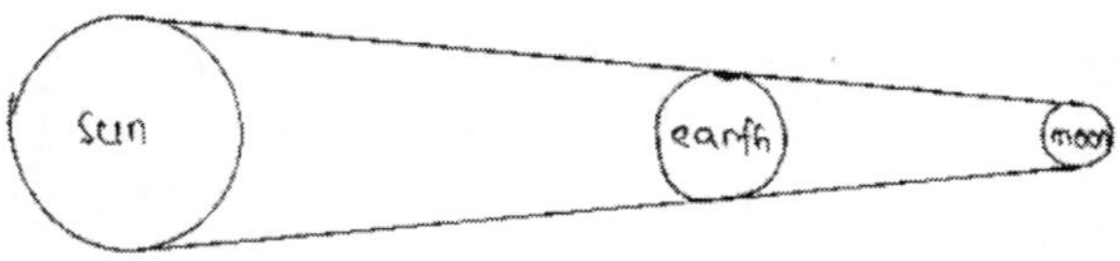

Figure 7. Hr's opinion about lunar eclipse illustration.

> *"The lunar eclipse occurs when the sun of the earth and the moon are in one straight line".* [Cd]

Regarding RQ 1, the findings also indicated that PSTs have some reasoning about the lunar eclipse. It means their reasoning reached the level of plausible and fruitful according to the first research question. The following are the example of scientific quotations (fruitful level) which performed by them:

> *"The lunar eclipse occurs when part or all of the moon's cross-section is covered by the shadow of the earth. It happens when the earth is between the sun and the moon in the same straight".* [Jl]

> *"I know about lunar eclipse phenomenon that states if the solar, the earth, and the lunar are taken place at one straight line. Because it happens, it can cause the lunar doesn't have light sometimes".* [Rn]

> *"Supermoon, blue moon and blood moon happen where sun, earth and moon that taken place at one straight line. The difference between of super moon and the common eclipse of the moon is the moon closest to earth, so the moon looks bigger than usual. And why scientific given name 'blood moon' because the colour of the moon is red".* [Hn]

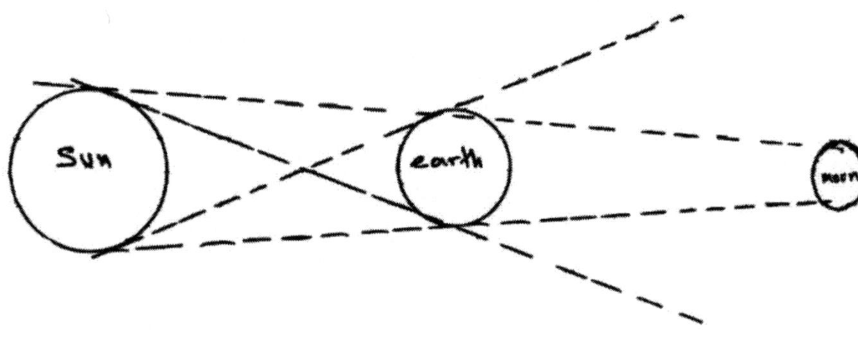

Figure 8. Hn's opinion about lunar eclipse illustration.

RQ 2: To what extent do the Indonesian PSTs' existing knowledge and reasoning of the supermoon, blue moon, and blood moon?

Based on the RQ 2, there are several reasoning of the Indonesian PSTs, which include four level categories. The first level is dissatisfied (naïve),

> *"A supermoon becomes the size of moon clearly, when the lunar eclipse happened, the general people said that "the moon was eaten Buto (Buto is a giant in the java legend)".* [Zk]
> *"Actually, I don't know the reason, because the super moon happens it just because of the position of the moon and sun. And we can see the supermoon several positions in the earth".* [Dn]
> *"I think the black moon does not exist because cloud covers the moon. But the blue moon and supermoon exist, this phenomena caused the rotation of the earth."* [Sk]

"I really believe that super moon, blue moon and blood moon is one of the marvellous phenomena". [Rn]
"Supermoon, blue moon and blood moon exists. There is a myth of those phenomena; if there is a pregnant woman in a solar eclipse, then the family must give food to their neighbor and people's pure, to protect the fetus". [Hs]

In this level, PSTs' reasoning about super-blue-blood moon is direct to myth, there is a correlation about super-blue-blood moon with the social condition, especially Javanese people.
For the second level is an intelligible category, there are several sample reasoning of PSTs' about the super-blue-blood moon.
"Supermoon happens when moon seems bigger than usual; I believe that supermoon exists because I ever see the real condition of the moon is in the biggest condition". [St]
"Supermoon happened because moon look bigger than usual, and blood moon happened because the colour of the moon is red". [Lk]
"Supermoon happens when the coordinate of the moon is near to the earth". [Gb]

Most of PSTs' reasoning on this level is understand about the supermoon and the colour of the blood moon.
For the third level is a plausible category, this is the several sample reasoning of PSTs' about the super-blue-blood moon.
"Supermoon happens when moon seems bigger than usual, the colour of blood moon is red, and the colour of the blue moon is bluish".[Nl]
"supermoon is the condition of the moon that has bigger size than usual, and blood moon is the colour of the moon is reddish. Same like blue moon, the colour of the moon is bluish". [Mt]
"supermoon is an anomaly, which the size of the moon is biggest. A blood moon is a condition, where the colour of the moon is reddish. And the blue moon, the colour of the moon is bluish".[Nb]

Most of PSTs' reasoning assumes that the blue moon is the colour of the moon that can see from the earth.
And for the fourth level is a fruitful category, this is the several sample reasoning of PSTs' about the super-blue-blood moon.
"supermoon is real because the moon's orbit isn't perfectly round, it's eclipse, so sometimes it's in the closest point to the earth. Blue moon isn't bluish; it is just terminology of 2 full moons in a month. A blood moon is real because it's light is filtered by molecules in the atmosphere".[Al]
"Supermoon happened when the position of the moon is very near with the earth. Blue moon is when two full moons happen in the same calendar month. The blood moon is the colour of the moon is red".[Ls]
"super moon happened when the position of the moon is closest with the earth. A blood moon happens when the lunar eclipse happens totally, and the colour is red. Blue moon is not the colour of the moon, but the period of lunar eclipse happen twice in one month".[St]

Almost PSTs' reasoning on this level is perfect, about the supermoon and a blue moon, but for blood moon, there are several PSTs' have reasoned that the colour of the moon is red

To sum up, the following section demonstrated they unfamiliar with the supermoon, blue moon, and blood moon. It means their reasoning reached the level of plausible and fruitful according to the first research question and at the level of 'dissatisfied' and 'intelligible' for the second research question. Meanwhile, regarding the second research question, the following are the example of a quotation by PSTs,

"Super blue blood moon is one of the phenomena that happen to the lunar phase. I believe it and really believe it will happen..." [Rn]
"Supermoon happens when moon seems bigger than usual; I believe that supermoon exists because I ever see the real condition of the moon is in the biggest condition". [St]

On the other hands, the following are the scientific conception created by scientists: "(1) Super moon happened when the moon's perigee - its closest approach to Earth in a single orbit - coincides with a full moon. In this case, the supermoon also happens to be the day of the lunar eclipse", "(2) A blue moon is when two full moons happen in the same calendar month; lunar eclipses occur when the moon passes into Earth's shadow", "(3) the term blood moon, there is a light coming into the penumbra area. If a glimpse of the moon is there, it will not make the colour of the moon become reddish or the so-called blood moon. Only if the moon passes entirely to the earth's umbra, the moon's colour becomes red, and the reason is that the light is so dimly lit" [4-9].

4. Conclusion

Finally, the conclusion can be derived from this study include: pre-service physics teachers performed enough understanding of the lunar eclipse phenomena with some levels of reasoning from naïve to fruitful. In contrast, their prevailing knowledge and reasoning of the supermoon, blue moon, and blood moon in the area of dissatisfied (naïve) to intelligible.

Acknowledgements

This paper is part of the spontaneous study to encounter the special phenomena of a lunar eclipse. Therefore, the authors would like to thank all pre-service physics teachers especially "Junior Level" who took part in the study via the course "Philosophy of Science". Besides, thanks to M M Yazid for contributing to data collection.

References

[1] Putri G S 2018 *Kontroversi istilah "super moon", "blood moon", dan "blue moon"* Retrieved from https://sains.kompas.com/read/2018/01/30/210600623/kontroversi-istilah-super-moon-blood-moon-dan-blue-moon
[2] Emspak J 2018 *Super blue blood moon 2018: Here are the best photos and videos* Retrieved from https://www.space.com/39208-super-blue-blood-moon-guide.html
[3] Nissani M 1994 *Science Activities: Class. Proj. & Curr. Ideas* **31** 26
[4] Guy M and Young T 2010 *Science Activities: Class. Proj. & Curr. Ideas* **47** 75
[5] Brown T and Brown K 2017 *Science Activities: Class. Proj. & Curr. Ideas* **54** 48
[6] Abell S, Martini M and George M 2001 *Int. J. Sci. Educ.* **23** 1095
[7] Barnett M and Morran J 2002 *Int. J. Sci. Educ.* **24** 859
[8] Subramaniam K and Padalkar S 2009 *Int. J. Sci. Educ.* **31** 395
[9] Ogan-Bekiroglu F 2007 *Int. J. Sci. Educ.* **29** 555
[10] Creswell J W and Plano Clark V 2011 *Designing and conducting mixed methods research (2nd ed.)* (Thousand Oaks, CA: Sage)
[11] Suprapto N, Lin H-H, Ku C-H and Chang T-S 2018 "Unless you can explain": Voices from graduate students and their Professor regarding the importance of science communication courses. Paper submitted to *The Qualitative Report*

Application of *Acetobacter Xylinum* in The Production of *Nata De Legen* From Palmyra Palm Neera

M T Asri[1,*], N Ducha[1], E Ratnasari[1], and A Bashri[1]

[1]Biology Department, Universitas Negeri Surabaya, Indonesia

*Corresponding Author: mahananiasri@unesa.ac.id

Abstract. Legen (neera) is produced from palmyra palm (*Borassus flabellifer* Linn.) neera that favored local people, but storability of legen not durable, this palmyra neera fermentation is relatively quick so it doesn't have high selling value. A technology is needed in the form fermentation to produce *nata de legen* using *Acetobacter xylinum*. *Nata de legen* is made by fermentation using legen medium with the addition of sucrose and food grade ZA. *Nata de legen* production was carried out on 15 medium samples which were repeated twice. *Nata de legen* post-harvest processing is carried out for bleaching and softening using strong base solutions. The results showed that the 15 medium samples succeeded in producing *nata de legen* with white turbid morphology, chewy, fibrous texture, incubation period 7-9 days, thickness of nata ranging between 1.05-1.55 cm, weight between 500-700 g, the surface is smooth or wavy and smells acid. The results of post-harvest processing produce white *nata de legen* and nata fibers easily removed.

1. Introduction

Palmyra Palm (Borassus flabellifer) is native to the Indian subcontinent and Southeast Asia, including Indonesia [1]. Almost all parts of palm trees can be used. Primary product of palmyra palm is nira obtained from tapping flowers, which can be drunk directly or processed into sugar, and secondary product of palmyra palm is used for craft materials [2]. All of these products are of economic value, both for farmers and sellers. The problems experienced by farmers and sellers are legen (palmyra neera) and Palmyra fruit don't last long and spoil quickly, if they are not sold within a few days they will be spoiled and thrown away. This causes loss to farmers and sellers. Application of technology is needed as an effort to use legen into another food products that have more selling value and could increase profit for farmers and sellers. The technology chosen is applying fermentation to produce nata de legen. In general, the production of nata is done by direct inoculation into liquid medium. Immobilization of cells is a technique used to trap the cells into a matrix. The use of immobilized cells for the production of nata is one alternative to the product resulted in a cell-free nata [3]. Nata is a bacterial cellulose formed from bacterial activity [4], with Acetobacter xylinum metabolizes glucose and converts it into bacterial cellulose that has unique properties including high purity, crystallinity and mechanical strength [5]. Cellulose products produced by aerobic bacteria (Acetobacter xylinum) get much attention because of their unique physicochemical properties compared to plant cellulose, including their application in the food industry [6]. Legen that has been processed into nata de legen has high selling value and potential so that it can improve the status of the surrounding community to become a business center for nata de legen with skilled and creative producers in processing nata de legen into a variety of food and beverage products.

The process of making nata de legen from palmyra is an effort to increase the economic value of legen, which so far has only been used to make legen drinks, which not everyone likes, because of its high alcohol content. Nata is a processed food product good for health for its high fiber which is useful to help digestive process of food. It also has high economic value and profitable because the production process requires simple ingredients and low cost. Nutrient content of palmyra neera consists of water (87.66%), sugar (12.04%), protein (0.36%), fat (0.36%), and ash (0.21%). The result of processing isotonic drinks with a combination technique obtained sugar content of 10.4-16.3%, pH 4.0-4.3, as well as the smell and taste of the preferred category until very preferred [7]. Different results on mesocarp of Palmyra fruit showed a water content of 77.31%, total fat 0.11%, ash 1.43%, tannin 0.08%, total carotenoids 8324.6 μg / 100 g, carotene 6217.48 μg / 100 g [8]. Another technique for preserving nata is using the drying method. Nata samples were dried using three physical drying methods such as oven, tray dryer or freeze dryer until it achieved 3-5% moisture content [9].

The problem experienced by legen farmers and sellers is that it performs natural fermentation in a short time which causes destructive legen, sour taste, and the smell of alcohol so that it does not sell and will eventually be discarded. This causes losses for farmers and sellers. One effort to reduce losses from damaged legen is to process it into other products, such as liquid or solid palm sugar which has a long production process and uses a lot of fuel but has a low selling value. Therefore, other technologies are needed to utilize legen which is spoiled into other food products that have a higher selling value and increase profits for farmers and sellers. The technology applied to increase legen value is conventional biotechnology in the form of fermentation to process legen into nata de legen using Acetobacter xylinum. The aim of this research was to study legen fermentation technology into nata de legen using the Acetobacter xylinum.

2. Method

This research was conducted using a sample of 15 which was repeated twice each time, carried out in the resident's house of Shoberoh Village, Dalegan Village in natural conditions. This research was carried out through several stages, namely multiplication of the bacterium *Acetobacter xilynum*, making *nata de legen* and post-harvest processing from *nata de legen*. Making stater is done by breeding *A. xlinum* bacteria on siwalan or legen medium with the addition of sucrose (25 $g.L^{-1}$) and foodgrade ZA (0.5 $g.L^{-1}$) placed in a sterile bottle aerobically with an incubation of 7 days at room temperature (28-30 °C). The next stage is the artificial *nata de legen* made on legen medium with the addition of sugar and foodgrade ZA with the same composition as the medium for making starter *Acetobacter xylinum*, the medium is boiled at 100 °C for 15 minutes then poured into as many sterile plastic trays 1 L/tray. The medium in the tray is left to cool and inoculated with the *A. xilinum* starter as much as 10%. Medium covered with plain brown paper. Medium that has been inoculated with starter bacteria is incubated at room temperature (28-30 °C), for 7-9 days aerobically until *nata de legen* is formed. The third stage is the harvest of *nata de legen* and post-harvest processing to get the *nata de legen* ready for consumption. Nata harvest is done by nata separated from the rest of the medium and cleaned by the lowest layer by peeling. *Nata de legen* is washed with water until clean and cut into 1x1 cm pieces. Nata pieces are soaked with water for 3 days, every day the water is replaced. On the 4th day the nata was soaked in a strong base solution (NaOH) with a concentration of 0.5% for 24 hours. Then the nata is washed with water many times until the pH becomes neutral (pH 7). After the pH of nata neutral nata is boiled until boiling for 15 minutes. Nata is ready to be processed into food/beverage products.

3. Result and Discussion

The results of this study were obtained in 3 stages. Stage 1, making starter (multiplication) of *Acetobacter xylinum*. Stage 2, making *nata de legen* and stage 3 post-harvest processing. In production of starter (multiplication) of *Acetobacter xylinum* in 15 samples, each sample was repeated 2 times. The results showed that each sample of the medium in the bottle was inoculated with *A. xylinum* bacteria can grow well on legen medium with the yield of 0.9-1.2 cm at 7 days incubation (Figure 1), and the medium shows a cloudy color. In the manufacture of this starter, no contaminated samples were found.

In the stage of making *nata de legen* which was also carried out on 15 samples which were repeated 2 times, obtained a good l result which was obtained by *nata de legen* with a supple texture, whitish brown (cream) with a weight of 800-900 g.L^{-1} media and only 1-1.5 cm in incubation 7-9 days (Table 1).

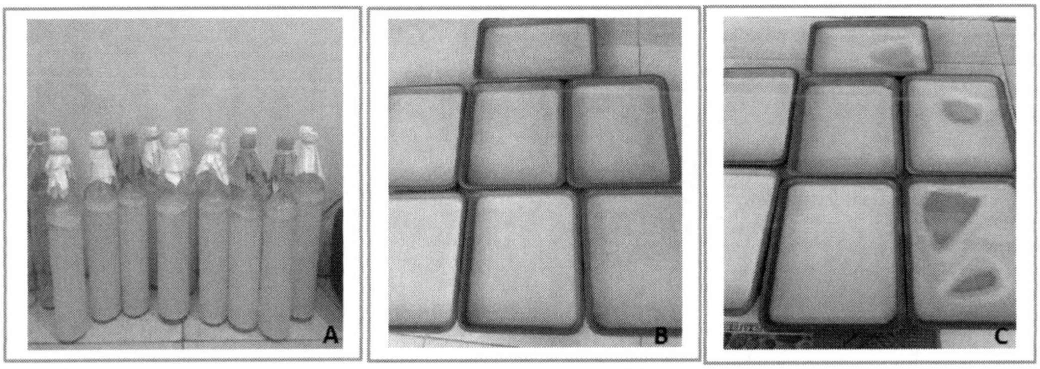

Figure 1. Stater *Nata De Legen* Containing *Acetobacter Xylinum* (A), *Nata De Legen* with A Smooth Surface (B), And Some Surfaces With Holes (C)

Result of research from *Acetobacter xylinum* application in the manufacture of starter and *nata de legen* from palmyra sap in Table 1. The application of *Acetobacter xylinum* multiplication or known as starter is carried out in the second stage, making *nata de legen*. According from the 15 samples that were repeated as many as two times, the results were shown in Table 1.

Table 1. Morphology of *Nata De Legen* Starter (*Acetobacter Xylinum*) And *Nata De Legen* Made from Legen

Sample	Starter morphology		*Nata de legen* / Liter medium						
	Solution	**Thickness of nata (cm)**	**Texture**	**Fiber**	**Appea-rance**	**Thickness (cm)**	**Weight (g)**	**Smell**	**Surface**
1	TW	1.0	Chewy	Fibrous	TW	1,2 5	643	Acid	Smooth
2	TW	0.9	Chewy	Fibrous	TW	1.0 5	500	Acid	Smooth
3	TW	1,2	Chewy	Fibrous	TW	1.45	700	Acid	Smooth
4	TW	0.9	Chewy	Fibrous	TW	1.3 5	686	Acid	Smooth
5	TW	0.9	Chewy	Fibrous	TW	1.20	629	Acid	Smooth
6	TW	0.9	Chewy	Fibrous	TW	1.3 0	657	Acid	Smooth
7	TW	1,2	Chewy	Fibrous	TW	1, 50	700	Acid	Perforated
8	TW	1.0	Chewy	Fibrous	TW	1,2 5	643	Acid	Smooth
9	TW	1.1	Chewy	Fibrous	TW	1.25	657	Acid	Smooth
10	TW	1.0	Chewy	Fibrous	TW	1.4 5	700	Acid	Smooth
11	TW	1,2	Chewy	Fibrous	TW	1.50	700	Acid	Perforated
12	TW	1.0	Chewy	Fibrous	TW	1.50	700	Acid	Perforated
13	TW	0.9	Chewy	Fibrous	TW	1.10	609	Acid	Smooth
14	TW	0.9	Chewy	Fibrous	TW	1,2 5	643	Acid	Smooth
15	TW	0.9	Chewy	Fibrous	TW	1.25	643	Acid	Perforated

Note: TW = turbid white. Measurement of nata starter was carried out at the 7th day incubation and *nata de legen* measurements performed on the 6-9th day incubation (when the medium has not left)

IOP Conf. Series: Journal of Physics: Conf. Series **1108** (2018) 012098 doi:10.1088/1742-6596/1108/1/012098

The Result from 15 samples medium nata that were selected with *Acetobacter xylinum* with morphology of nata turbid white, chewy, fibrous, 7-9 days incubation period ranges from 1.05 to 1.55 cm, weighs between 500-700 g, smells acid, and no samples were contaminated. High fiber supplement mixture of nata was reported able to reduced lipid level of consumer [10]. No medium left after 9 days incubation. So almost all mediums change to *nata de legen*. Although there are some that have perforated but uncontaminated surface (Figure 2).

Figure 2. Post-harvest processing (box-shaped nata pieces) (A) and *nata de legen* bleaching (B)

At the post-harvest stage, processing of nata is carried out so that nata can be consumed. Unprocessed nata is still in sheet form, tastes sour and the fiber is difficult to remove. The nata formed after 9 days of incubation is in sheet form. This nata is immediately harvested and washed with water to remove the acid odor. Besides that the lowest layer of nata is also removed so that the remnants of the media become nonexistent. Nata subsequently cut into cubes or rectangles with a size according to taste (generally 1x1 cm) and nata then soaked in water until the color becomes whiter and the smell of acid reduced subs 3 days and each day the water is replaced prior to the next process is softening and bleaching.

Bleaching is done by soaking the nata in 0.5% caustic soda (NaOH) solution for 24 hours. Nata washed many times until the alkaline nature of the loss (held not slippery) and his own re-neutral pH (pH 7). Soft nata with features can be torn or loose fiber when bitten. The white and soft nata are then boiled and ready for the product.

Propagation results of *Acetobacter xylinum* on the legen medium showed that the starter had been formed with characteristics already formed white chewy mass on the surface of the media in the starter bottle and the medium liquid showed turbid white. The white chewy mass is called *nata de legen*, while the rather turbid liquid below it is called starter. This stater can be used to make *nata de legen* or make it stand again with the legen siwalan media. Siwalan Legen can be used as a medium for making nata because it has a composition similar to coconut water. According to [11], palmyra legen contains total sugar 10.93 (g/100 cc); Protein 0.35 (g/100 cc); Nitrogen 0.056 (g/100 cc); Phosphorus 0.14 (g/100 cc); Iron 0.4 (g/100 cc); Vitamin C 13.25 (mg/100 cc).

The high content of legen nutrition can be converted into nata. *Nata de legen* has a smooth, white and high-fiber texture. By using the bacterium *Acetobacter xylinum*, the liquid form of legen can be converted into cellulose fiber sheets that are chewy and can be used as a mixture of drinks or packaged foods. The thick sheet of cellulose was washed, boiled and cooked in sugar syrup for food applications such as desserts, fruit cocktails and jellies [12]. The results of the multiplication of bacteria *A. xylinum* / stater were then applied to the legen siwalan medium to make *nata de legen* . In this application the results obtained with the characteristics nata nata formed its thickness ranges between 1-1, 5 cm, b closely nata ranges from 5 to 700 g, surface texture of nata is smooth, some perforated with fiber regularly, supple, turbid white and smelled acid.

The morphology of the surface of the nata caused by the cover of the tray (brown paper) that is not neat so that it sticks to the media, the area with brown paper causes *A. xylinum to* not produce nata properly. However, the areas affected by the brown paper were not contaminated because the paper was still new (so it was still sterile) besides the medium which was high in acidity (pH 3) so that it could be a selector for the microbial environment or microbes that might be on the sticky brown paper. In post-harvest processing by soaking the pieces of nata in fresh water for 3 days with water replaced every day and soaking in a strong base solution for 24 hours, it can whiten and squeeze the nata. Fiber of nata previously released when chewed. An increase in nata production has been reported in a way addition 12% sucrose concentration to the medium has influenced the formation of nata optimum results with average 46.13 g [13] and how to increase the feasibility of *nata de legen production*, it is necessary to increase the yield of nata production up to 90% [14].

4. Conclusion

Acetobactr xylinum which will be applied to the making of *nata de legen*, can be multiplied on the legen medium with the results of the starter liquid turbid white and the surface of the media formed nata with a thickness ranging between 0.9 -1.2 cm. *Nata de legen* produced from legen medium showed morphology of turbid white, supple, fibrous, incubation period of 7-9 days, with thickness of nata ranging from 1.05-1.55 cm, weight between 500-700 g, surface smooth or perforated and acidic. The results of post-harvest processing produce white *nata de legen* and fibers of nata easily removed.

5. References

[1] Kazi A A, Tandel M B, Pathak J G and Prajapati D H 2017 Potentiality of colocasia intercrop under naturally occurring palmyra palm (Borassus flabellifer L.) Journal of Tree Sciences 36(1) pp. 58-61
[2] Tambunan P 2010 The potential and policy for lontar development to increase the people income Jurnal Analisis Kebijakan Kehutanan 7(1) pp. 27–45
[3] Nugroho D A and Aji P 2015 Characterization of nata de coco produced by fermentation of immobilized Acetobacter xylinum Agriculture and Agricultural Science Procedia 3 pp. 278–282
[4] Budiyanto M A K 2002 Mikrobiologi Terapan (Malang: Universitas Muhammadiyah Malang)
[5] Halib N, Amin M C I M, and Ahmad I 2012 Physicochemical properties and characterization of nata de coco from local food industries as a source of cellulose Sains Malaysiana 41(2) pp. 205–211
[6] Esa F, Tasirin S M, and Rahman N A 2014 Overview of bacterial cellulose production and application. Agriculture and Agricultural Science Procedia 2 pp. 113–119
[7] Gafar P A and Heryani S 2012 The development of Aren sap drink processing technology by using ultrafiltration and deodorization techniques Jurnal Hasil Penelitian Industri 25(1), pp. 1-10
[8] Idayati E, Suparmo, and Darmadji P 2014 Potency of mesocarp bioactive compounds in lontar fruit (Borassus fl abeliffer L.) as a source of natural antioxidant Agritech 34(3) pp. 277-284
[9] Pa'e N, Hamid N I A, Khairuddin N, Zahan K A, Seng K F, Siddique B M, and Muhamad II 2014 Effect of different drying methods on the morphology, crystallinity, swelling ability and tensile properties of nata de coco Sains Malaysiana 43(5) pp. 767–773
[10] Mesomya W 2006 Effects of health food from cereal and nata de coco on serum lipids in human songklanakarin Journal of Science Technology 28(1) pp. 23–28
[11] Davis L S and Johnson K N 1987 Forest Management Third Edition (New York: McGrawHill Book Company)
[12] Esa F, Tasirin S M, and Rahman N A 2014 Overview of bacterial cellulose production and application Agriculture and Agricultural Science Procedia 2 pp. 113–119
[13] Widiyanti N L P M, Mulyadiharja S, Sukarta I N, and Pradnyandari N W I 2017 The effect of addition sucrose concentrations toward weight of nata de lontar (Borassus flabellifer) Linn Journal of Physics: Conf. Series 1040 (2018) pp. 1–8
[14] Kadir A W 2007 Financial analysis of processing nata lontar (from lontar sap) Jurnal Penelitian Sosial dan Ekonomi Kehutanan 4(4) pp. 341–355

IOP Conf. Series: Journal of Physics: Conf. Series **1108** (2018) 012099 doi:10.1088/1742-6596/1108/1/012099

Optimizing the Drying Temperature of Temulawak Simplicia (Curcuma xanthorrhiza Roxb.) Based on Water and Ash Content and Functional Compound

M A Anggarani[1,*], Rusijono[2], D A Maulana[3]

[1] Department of Chemistry, Universitas Negeri Surabaya, Ketintang, Surabaya, 60231, Indonesia
[2] Department of Education, Universitas Negeri Surabaya, Lidah Wetan, Surabaya, 60231, Indonesia
[3] Department of Mathematic, Universitas Negeri Surabaya, Ketintang, Surabaya, 60231, Indonesia

*Corresponding Author : mirwaanggarani@unesa.ac.id

Abstract. This study has been conducted to produce the standardization of temulawak (Curcuma xanthorrhiza Roxb). Simplicias was one of the harvest-processed forms. The parameters were optimized based on water content and ash content and functional compound (essential oil, xantorizol, oleoresin, curcumin, and curcuminoid). To get temulawak simplicias product with high quality, it has been done the optimization on the stage of drying process. For the dried by the oven, it has been conducted optimization on the drying temperatures of 70 °C, 80 °C and 90 °C. The analysis result shown that the active compound content in the simplicia product indicates that the product of temulawak simplicia with optimum quality has been obtained from the production process by oven drying method using temperature of 80 °C. The temulawak simplicia product contains essential oils, xantorizol, oleoresin, curcumin, and curcuminoid, each by 1.48%, 67.32%, 1.21%, 92.42% and 1.25%. Meanwhile, the temulawak water content and ash content are 5% and 3,16% that achieve in standard applicable of simplicias product.

1. Introduction

Temulawak (Curcuma xanthorrhiza) is a native Indonesian plant and traditionally utilized for a range of illness including liver damage, hypertension, diabetes, and cancer. Most people use the rhizome of this plant as they believe it has medicinal effect. Curcumae xanthorrhizae rhizoma consists of the dried rhizome; Curcuma xanthorrhiza Roxb. (C. xanthorrhiza D. Dietrich). The synonyms of Curcuma xanthorrhiza is also known as Javanese Turmeric or Temulawak. The rhizomes of Curcuma xanthorrhiza Roxb contain volatile oil, saponin, flavonoid and tannin. Chemistry analysis showed that the main substances of Curcuma xanthorrhiza Roxb are starch (48.18-59.64%), fiber (2.58-4.83%), volatile oil such as, phelandren, camphor, tumerol, sineol, borneol, and xanthorrhizol (1.48-1.63%), and also curcuminoid, curcumine and desmetoxicurcumine (1.6-2.2%) (Afifah, 2005). The root contains two classes of characteristic constituents; curcuminoids (1-2%), a mixture of dicinnamoylmethane which derivatives like curcumin (diferuloylmethane), monodemethoxycurcumin (feruloyl-phydroxycinnamoylmethane) and bisdesmethoxycurcumin (bis-(p-hydroxycinnamoyl) methane) [1], and other phenolic and non-phenolic diarylheptanoids [2]. Volatile oil (3-12%), composed mainly of sesquiterpenes (e.g. β-curcumene, ar-curcumene), xanthorrizol (44.5%) and a small amount of camphor (1.39%) [3].

The Material handling after harvesting time needs to be considered because it affects the quality of the processed product. The quality and safety of bio-farm product determines by the quality of raw material, post-harvest handling and processing techniques. The post-harvest handling techniques consist of the stages of sorting, washing, and drying [4], sorting/grading, packaging, labeling, and storage, either before it is sold in fresh form or after further processing. However until at a recent time there is no researcher who report in detail about the standard operating procedures (SOP) for the manufacture of processed products after harvesting.

Based on its usefulness, on this research, the content of the functional compound in temulawak include: (1) essential oils; (2) xantorizol; (3) oleoresin; (4) curcumin; (5) curcuminoid. The content of functional compound is influenced by the water content and the ash content. Functional compound, water content, and ash content become the nutritional content parameters of temulawak simplicia product.

2. Method

2.1 Material

The main material used in this research is temulawak (*Curcuma xanthorrhiza Roxb).* The optimization of post-harvest handling of temulawak was conducted using a number of apparatuses as in the previous publication [5]. It includes washer (homemade) used to optimize washing time with NESCO Lab MS-H280 Pro magnetic stirrer as a motor producing whirlpool was used in order to remove the contaminants, Bosch MUZ4DS3 slicer machine with slicing capacity of 60 g/min and the slicing thickness of 0.30 cm; J-LabTech LDO-030E (Daihan Labtech Ltd. Co.) as well as the drying machine with a range temperature of 50 °C – 250 °C and a drying time up to 60 minutes, were used to optimize rhizome drying.

2.2 Manufacturing of temulawak simplicia

Post-harvest handling of temulawak rhizome is done through three main stages covering washing, slicing, and drying. To obtain optimum rhizomes quality, it has been done the drying process by oven. Specifically, the temperature used was 70 °C, 80 °C and 90 °C. In addition, during the drying process, it was not involved the reversing, and drying process was stopped as soon as constant mass was obtained.

2.3 Water content and ash content testing of temulawak simplicia product

The testing procedure performed used the gravimetric method.

2.4 Functional compound testing of temulawak simplicia product

The testing procedure of functional compound content of temulawak simplicias products includes essential oil content using Stahl distillation methods. The testing procedure of curcuminoids, curcumin was performed using Agilent 1290 Infinity Liquid Chromatography (LC) and STM Columns. While for oleoresin and xantorizol, an analysis was performed using Pharmaspec UV-1700 UV Visible Spectrophotometer.

3. Result and Discussions

Data in Table 1 shows the Quality Requirements of temulawak Simplicia [6]. Based on data in Table 2, it is known that the drying process by using oven with temperature 80 °C has produced temulawak simplicia with water content and ash content respectively 5% and 3,16% which was obtained from drying for 1,600 minutes.

Table 1. Quality Standard of Temulawak Simplicia Issued by Materia Med Indonesia

No	Criterion	Units	Requirements
1	Appearance		
1.1	Odor	-	Normal
1.2	Taste	-	Normal
2	Water	% (b/b)	Max. 12.0
3	Ash	% (b/b)	Max. 7.0
5	Metal contamination		
5.1	Lead (Pb)	mg/kg	Max. 10.0
5.2	Copper (Cu)	mg/kg	Max. 30.0
6	Arsenic contamination	mg/kg	Max. 0.1
7	Microbe contamination		
7.1	The total of plate account	colony/g	Max. 10^6
7.2	E. coli	apna/g	Max. 10^3
7.3	Mold	colony/g	Max. 10^4
8	Aflatoxin	mg/kg	Max. 20.0

Table 2. Quality of temulawak simplicia from the drying process by oven for 1,600 minutes

No.	Parameter of quality	Drying Temperature		
		70 °C	80 °C	90 °C
1	Aroma	Normal	Normal	Normal
2	Taste	Normal	Normal	Normal
3	Water content (%)	10	5	5
4	Ash content (%)	6.02	3.16	3.16

The data has been shown that the simplicia product of this experiment has the normal aroma and taste (organoleptic testing). It can be inferred that the product has been accordance with the quality standard of temulawak simplicia. The Table 2 has been shown that the lowest water content of temulawak simplicia can be obtained when it is dried at a temperature of 80 °C. From this temperature, the water content and the ash content are categorized as lower level. It can be inferred that the product has been accordance with the quality standard of temulawak simplicia that the maximal level of water content is 12.0 % and the maximal level of ash content is 7.0 %, while the water content of this experiment is 5.0 % and the ash content of this experiment is 3.16 %.

Microbial growth in foodstuffs is closely related to the amount of water available. The amount of water in materials for microbial growth is known as water activity (aw / activity). Total microbial inhibition will occur in the food which stays less than 0.6. The water content which is contained in a food product will affect the appearance, taste, and age. The water content presented in a food product will affect the appearance and the taste of the product. According to [7] the water content in the material also determines the acceptance, freshness and durability of the material. Ash content is a parameter of product purity that is influenced by mineral elements in the foodstuff [7]. Ash content illustrates the amount of un-burnt minerals to substances that can evaporate.

On the previous data, it has been shown that the lowest water content and ash content can be obtained when it is dried at a temperature of 80 °C. Water is a major component in foods that affect appearance, texture, aroma, and taste of its ingredients. The water content in food ingredients also determines the acceptability of a food ingredient. The sensitivity of a commodity against water losses due to evaporation depends on the deficit at atmospheric pressure in the vicinity and the surface layer of the commodity.Organoleptic test is an important parameter to measure the level of consumer acceptance and its preference to the product. Organoleptic assessments are made in this hedonic test which includes color, aroma, taste, texture and general acceptance of the product.

Drying is the oldest preservation technique of agricultural products and it is essential process used all over the world for the preservation of farm product. It helps in reducing the water contains of the produced simplicia to a level below deterioration. It does not occur for a definite duration [4]. Lower water content will have a direct impact on the easily covered by mold and bacteria. It also eliminates the activity of enzymes that can decipher the active compound content. Further, it facilitates further processing, so it can be more compact, durable and easily stored [5].

Drying will never replace canning and freezing because the method does a good job of retaining the taste, appearance, and nutritive value of fresh food, but drying is an excellent way to eat delicious food. One of the biggest advantages of dried foods is that much less storage space than canned or frozen foods.

However, to obtain optimum drying temperatures in the production process of temulawak simplicia, at this stage it was also conducted an analysis of active compounds content in simplicia products which was dried using an oven temperature 70 °C; 80 °C; and 90 °C. Based on data from the product which is shown in the Table 9, it can be seen that there is the highest content of essential oil, oleoresin, xantorizol, curcuminoid and curcumin, which were obtained on the drying temperature at 80 °C.

Table 3. Active Compound of Temulawak Simplicia Resulted from Drying Temperature Optimization

No.	**Drying Temperature (°C)**	**Active Compound Content (%)**				
		Essential oil	**Oleoresin**	**Xantorizol**	**Curcuminoid**	**Curcumin**
1.	Raw material	1.89	1.23	68.67	1.28	94.37
2.	70 °C	1.50	1.21	67.40	1.25	92.57
3.	80 °C	1.48	1.21	67.32	1.25	92.42
4.	90 °C	1.29	1.00	60.05	1.00	87.98

Based on data from the active compound content of temulawak simplicia products which is shown in Table 3, it is concluded: (1) increasing the drying temperature from 70 °C to 80 °C did not cause a significant decrease in the active compound content of temulawak simplicia; (2) increasing the drying temperature from 80 °C to 90 °C caused a significant decrease in the active compound content of temulawak simplicia. Drying temperature at 80 °C claimed as the best temperature to dry temulawak to produce temulawak simplicia, with the lowest water content and ash content. Moreover, it contains an active compound with the highest content (percentage). Thereby, it can be said that the temulawak material has higher stability of increased drying temperature.

Furthermore, in order to obtain the manufacturer procedure of temulawak simplicia quality, an analysis had been performed using a standard which was established by Materia Med Indonesia (Table 2) [5]. Based on optimization result of temulawak simplicia production with optimum quality, the temperature of drying at 80 °C, contains 1.48% essential oil; 1.21% oleoresin; 67.32% xantorizol; 1.25% curcuminoid; and 92.42% curcumin.

IOP Conf. Series: Journal of Physics: Conf. Series **1108** (2018) 012099 doi:10.1088/1742-6596/1108/1/012099

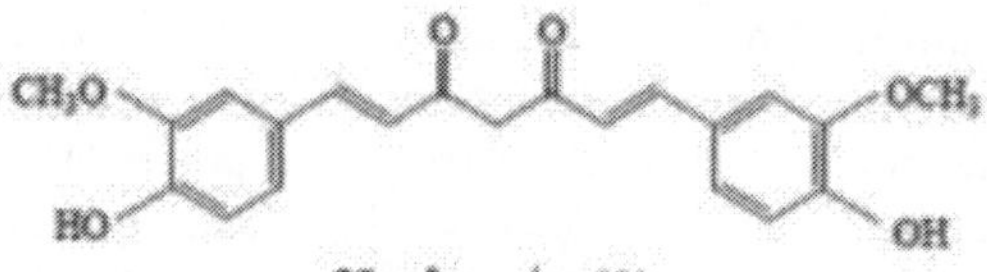
Kurkumin (1)

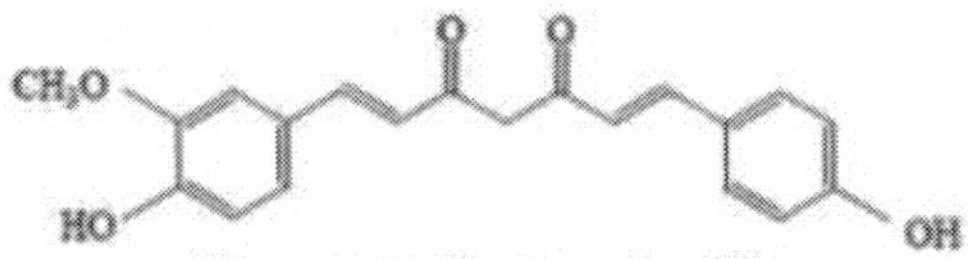
Demetoksikurkumin (2)

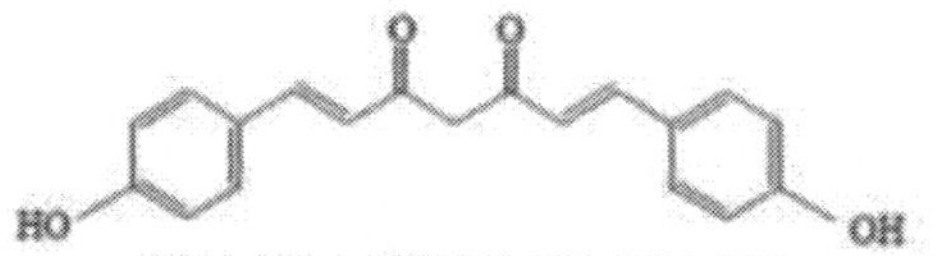
Bisdemetoksikurkumin (3)

Figure 1. Curcuminoid structure of temulawak

Curcuminoid is a fraction that responsibles to the yellow color of temulawak rhizome. Curcuminoids are insoluble in water but dissolve in acetone, alcohol, acetic acid, and alkali hydroxide. In addition, curcuminoid has a distinctive aroma and is non-toxic manner [1-2]. This curcuminoid fraction is stable against a high temperatures but it is unstable of light. Because of this nature, when curcumin is exposed to light, there will be such degradation or decomposition of the structure of cyclization of curcumin [8]. Curcuminoids and components that make up volatile oils such as xanthorrhizol are phenolic compounds that are antioxidants [1-2]. The main functions of antioxidants are used as an effort to minimize the oxidation process of fat and oil. It minimizes the occurrence of food damage, extends the life of the food industry, increases the stability of fats contained in foods and prevents the loss of the sensory and nutritional quality [6]. Natural antioxidants can protect the human body against damage caused by reactive oxygen species. It also able to inhibit the occurrence of degenerative diseases and is able to inhibit lipid peroxidation in food. Natural antioxidants generally have a hydroxy group in their molecular structure.

4. Conclusion

The result showed that the drying temperature affects the water content, the ash content, and the active compound of temulawak simplicia products. Drying temperature at 80 °C is claimed as the best temperature to dry temulawak to produce temulawak simplicia, with the lowest water content and ash content, and contain an active compound with the highest percentage. The water content of temulawak simplicia at optimal temperature of drying is 5%, and ash content of temulawak simplicia at optimal temperature of drying is 3.16%. Based on the optimization result of temulawak simplicia production with the optimum quality at 80 °C, it is resulted that the product contains 1.48% essential oil; 1.21% oleoresin; 67.32% xantorizol; 1.25% curcuminoid; and 92.42% curcumin.

5. References

[1] Ruslay S, Abas F, Shaari K, Zainal Z, Maulidiani, and Sirat H 2007 Characterization of the components present in the active fractions of health gingers (Curcuma xanthorrhizaand Zingiber zerumbet) by HPLC-DAD-ESIMS *Food Chem* **104**(3) pp. 1183-91

[2] Uehara I, Yoshida T, Mori K, Hatano T, Okumura T, Komagoe K, Fujita Y and Okuda, T 1987 Studies on inhibition mechanism of autoxidation by tanninsand flavonoids V. Radical-scavenging

effects of tannins and related polyphenolson l,l-diphenyl-2-picrylhydrazyl radical *Chem. Pharm. Bull.* **37** pp. 1919-1921

[3] Zwaving J and Hand Bos R 1992 Analysis of the essential oils of five Curcuma spp. *Flavor Fragance J.* **7** pp. 19-22

[4] Prasad J, Vijay V K, Tiwari G N, and Sorayan V P S 2006 Study on performance evaluation of hybrid drier for turmeric (Curcuma Longa L.) drying at village scale *J. Food. Eng.* **75** pp. 497-502

[5] Kusumawati N, Anggarani M A, Rusijono, Setiarso P, Muslim S 2017 Product standardization of ginger (Zingiber officinale Rosc.) and red ginger (Zingiber officinale var. Rubrum) simplicia through washing time, slice thickness and raw materials drying process optimization *International Journal on Advanced Science Engineering Information Technology* **7** pp. 15-21

[6] Raharjo M and Rostiana O 2005 Budidaya Tanaman Temulawak. Sirkuler No. 11 (Bogor: Balai Penelitian Tanaman Obat dan Aromatik. Badan Penelitian dan Pengembangan Pertanian)

[7] Winarno F G, Srikandi F dan Dedi F 1980 Pengantar Teknologi Pangan (Jakarta: Gramedia)

[8] Tonnesen H H and Karlsen J 1985 Studies on curcumin and curcuminoids alkaline degradation of curcuming *Z. Lebens, Unters, Forsc* **180** pp. 132-134

IOP Conf. Series: Journal of Physics: Conf. Series **1108** (2018) 012100 doi:10.1088/1742-6596/1108/1/012100

Comparison of Student Achievement in Agricultural Biotechnology-STEM Integrated Using Research Based Learning

H Subekti[1], A R Purnomo[2], H Susilo[3], Ibrohim[4] and H Suwono[5]

[1,2]Science Department, Faculty of Mathematics and Natural Sciences, Universitas Negeri Surabaya, Indonesia
[3,4,5]Biology Department, Faculty of Mathematics and Natural Sciences, Universitas Negeri Malang, Indonesia
[3]Corresponding author: herawati.susilo.fmipa@um.ac.id

Abstract. Biotechnology has been widely used to meet the challenges of producing products in the current industrial revolution era 4.0. One of the trends in educational field to face the challenge is to integrate Science, Technology, Engineering, and Mathematics (STEM) altogether. The purpose of this study is to compare two models of STEM Research-Based Learning, that is, Project Based Learning (PjBL) and Aesthetic Student-Teacher Elaborate Research (ASTER). Fifty five students from two classes were selected through typical case sampling; class A was about 29 students taught using PjBL, whereas Class B was 26 students taught using ASTER. This study used quantitative data which were analyzed using t-test. The results indicated that learning achievement of both classes were not quite different, with of ASTER (achievement of 48.3%) and PjBL (achievement of 45.5%). Therefore, this study suggests that further empirical studies need to be conducted on the effect of STEM integrated learning with the use of ASTER to strengthen the results of the study.

1. Introduction

Education in the 21st century increasingly emphasizes the importance of Science, Technology, Engineering and Mathematics (STEM) contributing to the development and growth of modern industrial society [1,2]. Advances in information technology and biotechnology are the two determinants of the current era [3]. In this context, the mastery of intelligence and biotechnology is one of the challenges facing a new era called the industry revolution 4.0 or the fourth world industrial revolution [4]. The existence of biotechnology has been widely used in various products throughout the 21st century [5]. Thus, preparing the quality of human resources in terms of innovation and meeting challenges of industrial revolution 4.0 is mandatory.

STEM education so called as STEM integrated learning is essential to modern education [7] and provides global competitiveness [8]. In this context, STEM integrated learning is a global movement in education, through various integration patterns to develop the quality of human resources, in line with modern-day demands [9]. Therefore, it becomes a trend in today's education [10].

To prepare competitive citizen, many countries around the world prioritize the quality improvement of science [11] and technology learning. One of the efforts to improve learning quality in Indonesia is to integrate STEM with the curriculum system in Indonesia [9]. In addition, STEM integrated learning can develop discipline knowledge, stimulate interest in STEM learning, as well as assist in future job-related skills [12]. Introducing STEM integrated learning to students is then an important component of the curriculum in the 21st century. Besides, STEM integrated learning in higher education is needed to provide more support for innovative learning, such as in biotechnology lectures.

For decades the biotechnology has grown very rapidly [13]. Plant biotechnology allows the use of new genetic elements found to plant new varieties of agricultural crops with added value [14]. This leads to increase nutrition over fruits and vegetables. Agricultural biotechnology materials are one of the basic biotechnology course materials that is trying to develop steps of integrating STEM so as to students learn to deepen knowledge of science [15]. Producing Virgin Coconut Oil (VCO), for instance, involves students to do complex thinking as STEM mandated, such as, determining variables, choosing various techniques, calculating expenditure, and learning how to use apparatus to apply for.

In addition, there is also a new challenge for the development of life-based curriculum as well as biology practices that develop capabilities by integrating the mastery of content, attitudes, characters, values, management skills, and research. Therefore, the need for a learning approach can answer global challenges. The ability to think logical, critical, and creative become necessary to analyze and solve problems given to the students. Through this *observation based learning* activity, students will be accustomed to working in networking through *collaborative learning* to enhance student creativity. The research-based learning undertaken in research activities is a Project-Based Learning Model (PjBL) and Aesthetic Student Elaborate Research (ASTER).

PjBL is a form of constructivist learning and collaboration in the learning process of using student-centered learning, which enables students to work together to solve problems and learn from each other to build their knowledge from pieces of which. PjBL is a systematic learning that actively involves students in developing their knowledge and skills [17] and can motivate students to learn independently [18]. Problem-solving can also be done through the empowerment of various skills [19]. The use of projects in lectures has several advantages such as giving students the opportunity to link statistics with everyday life [18].

Different to PjBL models which deals with learning quality improvement [19], ASTER model is a model developed by researchers referring to the mix of the local wisdom of Patrap Triloka based on the Ki Hajar Dewantara Education Idea [20]. The teaching model tries to enhance the patrap triloka values of wisdom and the form of research activities to facilitate STEM education. ASTER itself can be defined as "sensitivity (patrap) of an educator (lecturer/teacher) in giving guidance *(mentor)*or stage *in ngarso sung* tulodo, *coaching* or stage *in the middle of the game* , and the motivation or the level of self-determination of students in the learning pattern (with research) is carried out together".

This research is important as an effort to harmonize students' competence with the needs of the community through the integration of STEM and the raising idea of Ki Hajar Dewantara. That way, forming itself to be able to challenge the times will continue to be an important consideration [21]. In short, the application of STEM integrated learning emphasizes the student's thinking ability in implementing interdisciplinary integration and intellectual achievements for science, technology, engineering and mathematics [22]. Therefore, the integration of STEM in the lectures is felt with urgency in planning learning activities.

Based on the above explanation, the purpose of this study compares learning achievement between PjBL and ASTER. The question asked in this study is how the difference in learning outcomes of students with Model PjBL and Aster Model on Agricultural Biotechnology materials?

2. Method

This research used a quantitative approach with post-test-only control group design, implemented at Science Department, State University of Surabaya. Participants of this study were second-year students taking General Biotechnology course in the third semester, academic year 2017/2018 from two different co-educational classes. To be precise, class one consist of 3 males and 26 females), while class two contains 4 males and 22 females. In addition, they have relatively the same economical background and ethnic.

The instrument developed in this study is written test used to measure the learning outcomes of relevant student knowledge. The STEM items of the instrument are as follows: (1) Science, the role of tissue culture of plant (score 15); (2) Technology, protoplast combination technology (score 20) and utilization of devices water flow laminarly related to tissue culture (score 20); (3) Engineering, determination of research variables and product quality (score 20); and Mathematics, the effectiveness of manufacturing technique to produce VCO related to pricing (Score 25).

The data collected were analyzed using quantitative. In order to know the comparison results of PjBL and ASTER, the data were analysed using *t-test* with the aid of Minitab 16. Before the *t-test* was conducted, the initial data were tested for normality test using a *histogram with fit* and homogeneity test using *Levene's test*.

3. Results and Discussion

3.1. Overview of Research

Biotechnology is the fastest-growing industry in the production of commercial applications and products worldwide [13]. Thus, it has resonated with STEM's integrated learning as one of the solutions to human resource quality problems and the competitiveness of a country, both developed and developing countries [9]. Integrated learning in the STEM has a positive impact on student achievement [23] in terms of future STEM talent and competitiveness [12,7]. Descriptive statistical analysis results about both model STEM based are presented in Table 1.

Table 1. Descriptive Statistics of PjBL Model and Aster Model

Variable	N	St-Dev	Mean	Median	Minimum	Maximum
PjBL Model	29	15.95	67.1	66.0	34.0	98.0
Aster Model	26	13.80	68.2	71.0	36.0	88.0

Internationally, STEM's integrated learning can increase students' interest [24] and influence students to become technologically literate [7]. Related to the context of STEM, it is one of the key drivers of technological innovation [25]. One of the learning practices in Indonesia that can be developed is learning by integrating STEM [26]. Through this STEM integration, studies in learning will be more multidisciplinary [23]. However, the way to develop a sense of interest and curiosity with the world of disciplinary science is not an easy matter. This study then contributes to biotechnology learning that integrates this discipline so as to make students engage in critical, creative, and collaborative thinking. STEM stands for study or professional practice in science, technology, engineering, and mathematics [27]. STEM's integrated learning is an integrated education that combines scientific research, technology, engineering design, and mathematical analysis into a unified learning paradigm, including curriculum content, teaching activities, and educational policies. Integration pattern taken in this research activity is to unite engineering content (E) or engineering, and technology (T), as well as mathematical (M) learning in agricultural biotechnology. Incorporation visualization as illustrated in Figure 1.

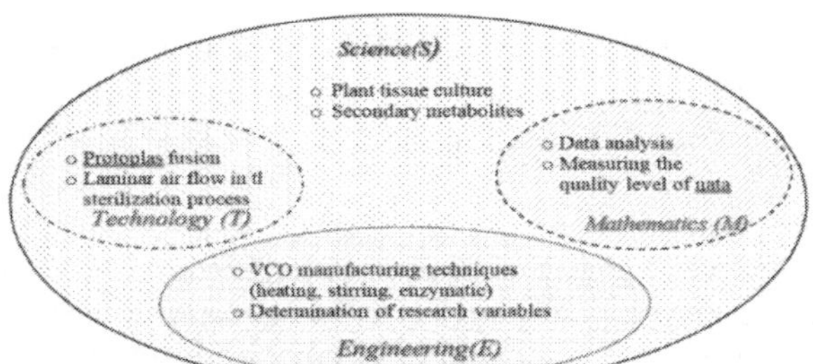

Figure 1. Incorporation Of Technology, Engineering, And Mathematical In Agricultural Biotechnology Materials

In the context of this study, we define integrated STEM as a combination of content and concepts from multiple disciplines [28]. Integration takes place in STEM disciplinary knowledge and process [28] by which agricultural biotechnology not only produces vegetables and fruits but also offers an effective way of producing new value-added products with the potential to influence nutrition and human health.

3.2. Normality Test Results

The important assumption in parametric testing is normality. Paired sample normality checks performed with (1) histogram with fit (2) Chi-square, and (3) normalized test Kolmogorov-Smirnov [29]. In this research activity, the norm test used a histogram with fit described in Figure 2 and Figure 3.

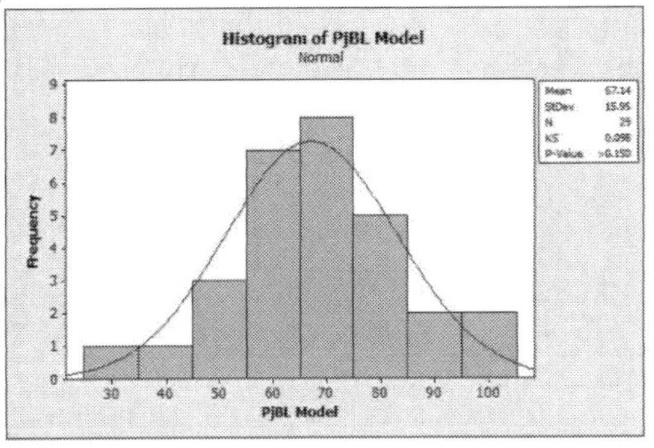

Figure 2. Histogram Test *with Fit* Normalized PjBL Model

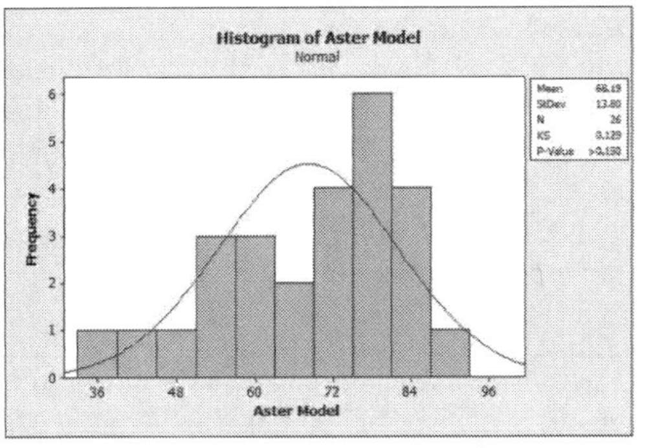

Figure 3. Histogram Test *with Fit* Aster Normalized Model

Normal data can be viewed descriptively by using the probability plot of the difference data above. Referring to figure 2 and 3, the data show a bell-shaped line with a normal distribution plot. In the upper right corner, it shows corresponding PjBL Model value $KS = 0.098$ and $p - value > 0.150 > \alpha = 0{,}05$. Likely, histogram of Aster Model the above figure shows that the result is relatively similar, with value $KS = 0{,}129$ and $p - value > 0.150 > \alpha$ =0.05. Conclusions, data support for acceptance $H_0$with a 95% confidence level.

3.3. Variance Homogeneity

Variant homogeneity test aims at checking the similarity of variances between the two populations [29]. Homogeneity can be interpreted to have the same variance, or in other words similar or nearly identical variants presented in Table 2.

Table 2. Test Uniformity Using Levene Test

	PJBL Model	Aster Model	Statistic	P-Value
F Test (normal)	28	25	1.34	0.468
Levene's Test	1	53	0.25	0.25

Analysis of testing using Levene's Test shows homogeneous data p≥0.05. Results show that F_{count} = 1,34 dengan $p - value = 0.468 \geq 0{,}05 = \alpha$. So, the data supports to accept H_0 which means both variants of homogeneous data with a 95% confidence level.

3.4. Analysis of Range

Analysis of range shows the difference between the largest and the smallest value [29], where the distant values are far greater. On the other hand, if the values are relatively close, the size of the diversity tends to be small. PjBL Model is data range $95 - 35 = 60$. Aster Model is data range $88 - 36 = 52$. This shows that the PjBL model data range is higher than the Aster Model. In other words, PjBL model values are varied more than the Aster Model.

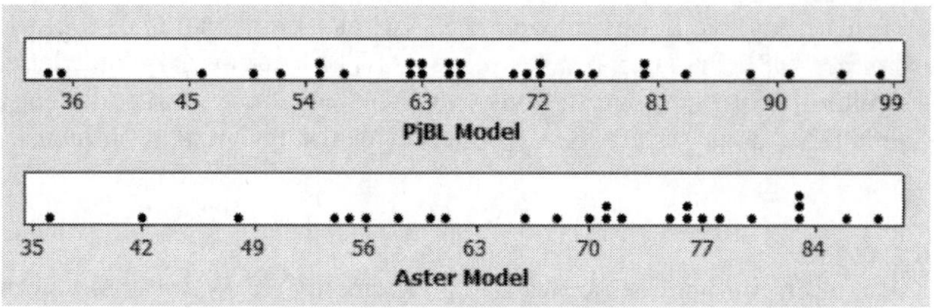

Figure 4. Dot Plot of PjBL and Aster Model of Achievement

3.5. *Achievement of Learning Indicators*

Developmental Progress in the STEM is expected to be the driving forces of the future economy and overall well-being not only for developed country, but also for developing countries [30]. Integrated learning STEMs can relate to scientific investigations, by formulating hypothesis answered through research [31]. Therefore, this skill should be considered as the benchmark for graduates.

Table 3. Average Comparison of Learning Achievement Indicators PjBL Model and Aster Model

No	Indicators	PjBL *N*= 29	ASTER N = 26
1	Describing the role of plant tissue culture from the aspect of medicinal benefits *(*Science*)*.	10.1	8.5
2	Analyzing the effectiveness of the technique in producing Virgin Coconut Oil (VCO) in accordance with pricing *(*Mathematics*)*.	16.6	19.9
3	Identifying the variables of the study and the determination of the quality of nata de corn *(*Engineering*)*.	5.9	6.8
4	Communicating the combined technique of protoplasts and tools related to the process of tissue culture cultivation (Technology).	12.8	13.1
	Percentage	**45.4**	**48.3**

The biotechnology sector is one of the key strategy that will transform into a developed nation by 2020 [13]. Biotechnology has grown rapidly, particularly in agriculture and livestock, food and pharmaceutical industry [16] through interdisciplinary approach [32]. The implications for STEM include how they can narrow the gap between education and workplace competencies [27].

4. Conclusion

This research presents the empirical evidence that the learning achievement trends in the aster model (achievement of 48.3%) are higher than PjBL (achievement of 45.5%). Although the use of these two models is considered to be ineffective in achieving the completeness of student learning (<50%), the implementation of this model is considered to be equivalent to an empirically proven PjBL model that can improve the quality of learning. The implication of this study is that the Model PjBL and Aster Model can be used as an alternative to implementing research-based learning process on agricultural biotechnology courses. It is evident that the Patrap Triloka gesture developed into the potential Aster model and can be regarded as a learning model that can enhance STEM's integrated material control.

5. Acknowledgments

I would like to thank the Ministry of Research, Technology, and Higher Education (Kemristekdikti) of Republic of Indonesia who has funded this study. Also, I thank the academic community at Universitas Negeri Surabaya for their expertise in assisting this study and the lecturers and colleagues at Doctorate Program of Biology Education, Universitas Negeri Malang for suggestions and criticisms.

6. References

[1] León J, Núñez J L and Liew J 2015 *Learn. Individ. Differ.* **43** pp 156–63

[2] Le H and Robbins S B 2016 *J. Vocat. Behav.* **95–96** 21
[3] Suwono H, Lestari U, Lukiati B and Lestari S R 2018 *Int. Conf. on Learn. Innov. (ICLI 2017)* **164** 97
[4] Kemristekdikti 2018 *Presiden Jokowi: Tantangan Kita Kedepan, Revolusi Industri 4.0* pp 1–2
[5] Bahri N M, Suryawati E, and Osman K 2014 *Eurasia J. Math. Sci. Technol. Educ.* **10** 195
[6] Widayati D T, Luknanto D, Rahayuningsih E, Sutapa G, Harsono, Sancayaningsih R P and Sajarwa 2010 Padoman Umum Pembelajaran Berbasis Riset *STAR*
[7] Kanematsu H and Barry D M 2016 *STEM and ICT Education in Intelligent Environments* (London: Springer)
[8] Christensen R 2015 *J. Sci. Educ. Technol.* **24** 898
[9] Firman H 2015 Pendidikan Sains Berbasis STEM: Konsep, Pengembangan, dan Peranan Riset Pascasarjana *Seminar Nasional Pendidikan IPA dan PKLH* (Bogor: Pendidikan IPA Program Pascasarjana Universitas Pakuan)
[10] Pellas N, Kazanidis I, Konstantinou N and Georgiou G 2017 *Educ. Inf. Technol.* **22** 2235
[11] Guzey S S, Moore T J, Harwell M and Moreno M 2016 *J. Sci. Educ. Technol.* **25** 550
[12] Lai C S 2018 *J. Educ. Sci. Environ. Heal.* **4** 110
[13] Bahri N M, Suryawati E and Osman K 2014 *J. Math. Sci. Technol. Educ.* **10** 195
[14] Ricroch A, Surinder C and Fleischer S J 2014 *Plant Biotechnology: Experience and Future Prospects* (London: Springer)
[15] Lou S, Chou Y C, Ru-Chu S and Chih-Chao C 2017 *Eurasia J. Math. Sci. Technol. Educ.* **8223** 2387
[16] Suwono H, Lestari U, Lukiati B and Lestari S R 2018 *Adv. in Soc. Sci. & Humanit. Res.* **164**
[17] Sumarni W, Wardani S and Gupitasari D N 2016 *J. Pendidik. IPA Indones..* **5** 157
[18] Mairing J P and Lorida D 2013 *J. Pendidik.* **14** 53
[19] Hernawati D, Amin M, Irawati M, Indriwati S and Aziz M 2018 *Eurasia J. Math. Sci. Technol. Educ.* **14** 2475
[20] Subekti H 2018 Model Pengajaran ASTER (Aesthetic Student-Teacher Elaborate Research): Sebuah Model Pengajaran Berbasis Kearifan Lokal Patrap Triloka Terintegrasi STEM Education Untuk Mengembangkan Keterampilan Riset, Literasi Informasi, dan Komunikasi Sains pp 1–50
[21] Ames T, Reeve E, Stewardson G and Lott K 2017 Preferred *J. Technol. Educ.* **28**
[22] Tsai H Y, Chung C C and Lou S J 2018 *Eurasia J. Math. Sci. Technol. Educ.* **14** 15
[23] Mayasari T, Kadarohman A and Rusdiana D 2014 Pengaruh Pembelajaran Terintegrasi Science, Technology, Engineering, and Mathematics (STEM) pada Hasil Belajar Peserta Didik: Studi Meta Analisis *Seminar Nasional Pendidikan IPA VI* ed W Widodo, Erman and E Sudibyo (Surabaya: Unesa University Press)
[24] Kier M W, Blanchard M R, Osborne J W and Albert J L 2014 *Res. Sci. Educ.* **44** 461
[25] Goy S C, Wong Y L, Low W Y, Nurani S, Noor M, Onyeneho N, Daniel E, Azizan S and Ginikauzoigwe A 2017 *Stud. High. Educ.* **0** 1
[26] Khaeroningtyas N, Permanasari A and Hamidah I 2016 *J. Pendidik. IPA Indones.* **5** 94
[27] Jang H 2016 *J. Sci. Educ. Technol.* **25** 284
[28] Nadelson L S and Seifert A L 2017 *J. Educ. Res.* **110** pp 221
[29] Mairing J P 2017 *Statistika Pendidikan: Konsep dan Penerapannya Menggunakan Minitab dan Microsoft Excel* ed Seno (Yogyakarta: ANDI)
[30] Al-Salami M K, Makela C J and de Miranda M A 2017 *Int. J. Technol. Des. Educ.* **27** 63
[31] Kelley T R and Knowles J G 2016 *Int. J. STEM Educ.* **3** 1
[32] Barak M and Assal M 2018 *Int. J. Technol. Des. Educ.* **28** pp 121

Improving Collaborative Critical Thinking Skills of Physics Education Students through Implementation of CinQASE Learning Model

Hunaidah*[1,2], E Susantini[1], Wasis[1], B K Prahani[3] and M A Mahdiannur[4]

[1] State University of Surabaya, Jl. Ketintang, Surabaya 60231. Indonesia
[2]Universitas Halu Oleo, Jl. H.E.A. Mokodompit, Kendari, 93231, Indonesia
[3] UIN Sunan Ampel Surabaya, Jl. A.Yani No.117, Surabaya 60237, Indonesia
[4] Universitas Kaltara, JalanTanjung Selor Hilir, Bulungan 60237, Indonesia
*E-mail:hunaidah@uho.ac.id

Abstract. The CinQASE Learning Model is a collaborative-based learning that is designed to improve students' collaborative critical thinking skills. The CinQASE Learning Model has a syntax consisting of (1) Problem Presentation, (2) Individual Work, (3) Team Work in Collaboration, (4) Class Discussion, and (5) Evaluation and feedback. The purpose of this research is to improve the collaborative critical thinking skill of physics education students through implementing CinQASE learning model. This research used one group pre-test and post-test design toward 56 physics education students of Halu Oleo University at academic year 2017/2018. The collaborative critical thinking skills of physics education students were measured by using Direct Assessment of Collaborative Critical Thinking (DACCT) with following indicators: questioning, analyzing, synthesizing, and evaluating. Data analysis technique used Paired t-test, n-gain and ANOVA test. The results showed that: (1) Average post-test score was 79.35 (2) the improvement on collaborative critical thinking skill of physics education students was at $\alpha = 5\%$, (3) Average n-gain score of collaborative critical thinking skills was 0.62 ; and (4) No difference of n-gain on collaborative critical thinking skills in all groups; and (5) Students responded in good category toward the implementation of CinQASE learning model. Therefore the CinQASE learning model has proven to be effective to improve students' collaborative critical thinking skills.

1. Introduction

College education has an important role in the industrial revolution 4.0. Especially colleges that produce educator candidates who are expected to provide prospective educators who have expertise that supports them to become successful individuals in the world of work and social life. To succeed in today's world, students need such skills that support them, such as critical thinking and problem solving, creativity and innovation, communication and collaboration [1-8]. The development of students' critical thinking skills is considered as one of the most important goals of education [3, 9-10]. Critical thinking is a process of thinking to determine what is done and what is believed [11]. Several studies have been conducted by applying various models, learning, strategies, methods, and approaches in research to improve critical thinking skills [12-23]. However, the development of critical thinking skills has been emphasized at the individual level. In fostering collaborative critical thinking skills, it is pointed to three things; they are constructivism, situasionism, and collaboration. It further develops that educator candidates must have collaborative critical thinking skills. Collaboration encourages the development of critical thinking through discussion, clarification, ideas, and evaluation of other people's ideas [24]. There is persuasive evidence that collaborative teams reach higher levels of critical thinking and store information longer, than students who secretly work as individuals [25]. The collaborative team serves as a mediating tool to develop individual critical thinking skills. Collaborative learning provides an opportunity for learners to engage in discussion, take responsibility for their own learning, thereby helps individuals become critical thinkers [26].

The preliminary study results indicated that collaborative critical thinking skills activity in UHO Faculty of Teacher Training and Education Physics education students (Kendari, Indonesia) is still low. This is because the lectures often involve the individual critical thinking skills. Lecturers need to facilitate the learning process that embodies the collaborative work that makes natural science students in Teacher Training Institution (TTI) become able to develop their thinking results. Based on recent literature studies through the searching of international databases in the field of education, critical thinking skills at the group or team level that are commonly called as collaborative critical thinking, are still not widely researched, whereas collaborative thinking skills are a vital skill that physics education students must master in TTI as well as physics teachers in supporting their increasing competence within the TPACK framework.

There are several team-based learning models that can improve critical thinking skills. However, the implementation of explicit team-based learning models has not been able to improve the collaborative critical thinking skills. The Cooperative Learning (CL) model has not yet been explicitly shown to be capable to engage in collaborative critical thinking skills [27]. Furthermore, Team Based Learning (TBL) shows a significant improvement in critical thinking skills but it is less efficient and practical because the time required for each phase is not possible to be completed in a single meeting [28]. Based on team-based learning models that have been described previously, they can implicitly train critical thinking skills but they have not yet clearly improved collaborative critical thinking skills. Therefore, in this study it is discussed the use of Collaborative in Questioning, Analysing, Synthesizing and Evaluating (CinQASE) model which has been developed and proven to be valid to improve students' collaborative critical thinking skills [29]. The results of this study are expected to be empirical evidence in the disseminate process of the CinQASE learning models' effectiveness to improve the collaborative critical thinking skills of physics education students.

2. Experimental Method

2.1 General Background of Research

The main purpose of this research is to analyse the improvement of collaborative critical thinking skills of physics education students through the implementation of CinQASE learning model. This research was conducted in physics education department at Halu Oleo University (Kendari, Indonesia), academic year 2017/2018 who takes basic physics course. The improvement physics education students' collaborative critical thinking skills through the implementation of the CinQASE learning model is determined based on: (1) Post-test scores of physics education students' collaborative critical thinking skills that must be at least in medium category, (2) A statistic increase in score between pre-test and post-test of physics education students' collaborative critical thinking skills, (3) The n-gain of physics education students' collaborative critical thinking skills must be at least on low improvement criteria, (4) The consistency of n-gain average score of physics education students' collaborative critical thinking skills in three experimental groups, and (5) the responses of physics education that must be at least in enough category.

2.2 Sample of Research

The sample in this research was 56 students of physics education at Halu Oleo University (Kendari, Indonesia), academic year 2017/2018 who takes basic physics course. All physics education students were divided into 3 groups namely class A (19 students), class B (19 students), and class C (18 students). Each group took the basic physics course at academic year 2017/2018.

2.3 Instrument and Procedures

The collaborative critical thinking skills of physics education students were measured by using a Direct Assessment of Collaborative Critical Thinking (DACCT) with indicators of valid and reliable questioning, analysing, synthesizing, and evaluating [29-31]. Meanwhile, the student response data was obtained by using a student response questionnaire that was given after the entire learning process was completed. The questionnaire response aims to determine the student's response to the instruments and learning process by applying CinQASE learning model. The physics course that was used in this research was chosen to be the one that was suitable with the characteristic of CinQASE learning model; basic physics. This study used one group pre-test and post-test design, O1 X O2 [32]. The learning began by giving pre-test (O1). Every physics education student was required to complete DACCT. After the pre-test, the lecturers applied CinQASE learning model and instruments in each group (X). The implementation of the CinQASE learning model had been conducted for seven meetings on the basic physics course. The physics learning by using the CinQASE Learning model has syntax that consists of 5 phases, they are: (1) Problem Presentation, (2) Individual Work, (3) Teamwork in Collaboration, (4) Class Discussion, and (5) Evaluation and feedback. The learning instruments consisted of: syllabus, lecture plan, student work sheet, student textbook, DACCT, and response questionnaire (valid and reliable) [29]. Each phase of the CinQASE learning model collaborative critical thinking skills in basic physics courses included: questioning, analysing, synthesizing, and evaluating. The implementation of CinQASE learning model ended by post-test (O2) by using DACCT. Each student was required to complete the DACCT on a post-test and filled out a student response questionnaire.

2.4 Data Analysis

The improvement of students' collaborative critical thinking skills was analysed by using the N-Gain equation [33]. The result of N-gain calculation was then converted by criteria: High > 0.70; Medium 0.3 - 0.70; Low <0.30. The inferential statistical tests used Parried t-test (analysis of statistical improvement) and n-gain consistency analysis of all physics education students after the implementation of CinQASE learning model that is used ANOVA test. The analysis of student response questionnaire used Guttmann scale, that was for students who answered Yes, they got one score (1) and students who answer No, they got zero (0). The percentage of student responses was converted to the following criteria: 81% - 100% = Very Good; 61% - 80% = Good; 41% - 60% = Good enough; 2% - 40% = Less Good; 0% - 20% = Not Good [29].

3. Result and Discussion

The results are presented in Table 1, Table 2, and Table 3 which is described as follows:

Table 1. The learning outcomes of physics education students' collaborative critical thinking skills.

Students' Initial	Class											
	A				B				C			
	O1	O2	<g>	C	O1	O2	<g>	C	O1	O2	<g>	C
M1	37.00	69.00	0.51	M	45.00	79.00	0.62	M	53.00	80.00	0.57	M
M2	49.00	78.00	0.57	M	57.00	89.00	0.74	H	37.00	69.00	0.51	M
M3	51.00	76.00	0.51	M	54.00	72.00	0.39	M	41.00	75.00	0.58	M
M4	45.00	70.00	0.45	M	41.00	87.00	0.78	H	37.00	69.00	0.51	M
M5	53.00	77.00	0.51	M	37.00	95.00	0.92	H	36.00	73.00	0.58	M
M6	41.00	67.00	0.44	M	30.00	70.00	0.57	M	53.00	90.00	0.79	H
M7	40.00	70.00	0.50	M	45.00	86.00	0.75	H	65.00	90.00	0.71	H
M8	45.00	80.00	0.64	M	29.00	77.00	0.68	M	47.00	81.00	0.64	M
M9	41.00	79.00	0.65	M	53.00	78.00	0.53	M	37.00	80.00	0.68	M
M10	65.00	92.00	0.77	H	45.00	83.00	0.69	M	45.00	77.00	0.58	M
M11	41.00	67.00	0.44	M	53.00	79.00	0.55	M	45.00	82.00	0.67	M
M12	45.00	90.00	0.82	H	45.00	69.00	0.44	M	57.00	90.00	0.77	H
M13	49.00	78.00	0.57	M	35.00	73.00	0.58	M	45.00	86.00	0.75	H
M14	37.00	70.00	0.52	M	35.00	74.00	0.60	M	51.00	85.00	0.69	M
M15	43.00	75.00	0.56	M	43.00	86.00	0.75	H	57.00	89.00	0.74	H
M16	39.00	81.00	0.69	M	49.00	66.00	0.33	M	45.00	79.00	0.62	M
M17	61.00	92.00	0.79	H	51.00	73.00	0.45	M	41.00	72.00	0.53	M

Students' Initial	Class											
M18	53.00	91.00	0.81	H	45.00	78.00	0.60	M	53.00	92.00	0.83	H
M19	45.00	87.00	0.76	H	37.00	80.00	0.68	M				
Mean	46.32	78.37	0.61	M	43.63	78.63	0.61	M	46.94	81.06	0.65	M

Note: O1 (Pre-test); O2 (Post-test), C (Category); <g> (n-gain); L (Low); M (Moderate); H (High)

Table 1 shows the learning outcomes of physics education students' collaborative critical thinking skills on physics learning. In all groups, the average pre-test score is still low and did not pass. This is because the students are still unfamiliar with collaborative critical thinking skills in physics. The findings are in accordance with the results of preliminary studies that the collaborative critical thinking skills of physics education students are still relatively low [29]. The opposite condition occurs in the post-test results where after the implementation of the CinQASE learning model, no students get a low score. The average value of collaborative critical thinking skills in class A, class B, and Class C are 78.37, 78.63, and 81.06 respectively. Table 1 also informs the positive results of the CinQASE learning model implementation that the n-gain of collaborative critical thinking skills in class A, class B, and class C are 0.61, 0.61, and 0.65 respectively and are in the medium category. The results of this study show that the implementation of the CinQASE learning model is proved to be effective to improve the collaborative critical thinking skills of physics education students. This is because the developed CinQASE learning model meets the validity, practicality and effectiveness to improve the collaborative critical thinking skills of physics education students [29]. This is supported by the results of the study [2-8, 34-47] that the model, method, media, approach, teaching and learning that can meet validity, practicality and effectiveness will be able to improve and achieve the learning objectives.

Table 2. Average score of physics education students' collaborative critical thinking skills indicator.

Collaborative critical thinking skills indicator	Class											
	A				B				C			
	O1	O2	<g>	C	O1	O2	<g>	C	O1	O2	<g>	C
Questioning	2.15	2.84	0.81	H	2.32	2.68	0.53	M	2.17	2.67	0.60	M
Analysing	1.73	2.63	0.71	H	1.42	2.37	0.60	M	1.67	2.61	0.71	H
Synthesizing	1.57	2.42	0.59	M	1.22	2.42	0.67	M	1.28	2.44	0.67	M
Evaluating	1.21	2.4	0.66	M	1.37	2.58	0.74	H	1.17	2.56	0.76	H

Note: O1 (Pre-test); O2 (Post-test), C (Category); <g> (n-gain); L (Low); M (Moderate); H (High)

Table 2 shows that all collaborative critical thinking skills indicator of physics education students in the pre-test are in the low category and still need to be improved. Positive results indicate an increase in collaborative critical thinking skills of physics education students, which is shown in table 2 that all indicators of collaborative critical thinking skills are in high category. This is shown through an n-gain score analysis where the n-gain value for each indicator increases. The highest increase can be seen in indicators of questioning, analysing, synthesizing, and evaluating with a range of n-gain scores from 0.53 - 0.81.

It shows that during the learning process, students have enhanced their collaborative critical thinking skills through scientific experimentation activities to build and discover facts to support their argument in collaboration. The process of collaborative critical thinking starts from the initial knowledge, verifies, and modifies can be trained by using the step technique: (1) create a story, (2) test conflict, (3) evaluate the story, and (4) develop a contingency plan. This training strategy has been validated in several samples. In collaborative critical thinking, individual decision-makers should often make choices in dynamic complex situations. This arrangement is often with incomplete, conflicting, informative, and ambiguous information. Collaboration is processes where someone looks at different aspects of a problem, explores differences and seeks solutions that transcend the limitations of his/her own knowledge of what is possible. During collaboration, a person with complementary skills interacts to create a shared understanding of a situation in a whole compared than someone who has achieved individually [48]. Effective collaboration requires special skills in terms of negotiating and listening actively. Collaborative critical thinking can be regarded as one of the factors that affect the team's ability to collaborate. The indicators of collaborative critical thinking skills in this study are adapted from indicators that have been developed with questioning, analysing, synthesizing, and evaluating components [31]. Theoretically, it is argued that a person who is accustomed to work in groups continuously will have a negative impact if he/she is required to work individually.

The results of the normality and homogeneity test of variance suggest that the pre-test, post-test, and n-gain of collaborative critical thinking skills are homogeneous and normally distributed for the whole group. Therefore, the impact of implementing the CinQASE learning model to improve collaborative critical thinking skills in the courses for the whole group used Paired t-test while the consistency test used ANOVA test. The paired t-test and ANOVA test results are presented in Table 3.

Table 3. The result of Paired t-test and ANOVA test of student's collaborative critical thinking skills.

Paired t-test, $\alpha = 5\%$						**ANOVA test, $\alpha = 5\%$**					
Class	N	Mean	t	df	p		Sum of Squares	df	Mean Square	F	Sig.
A	19	-3.20	-21.89	18	.000	Between groups	.023	2	.012	.713	.495
B	19	-3.50	-13.99	18	.000	Within groups	.870	53	.016		
C	18	-3.41	-32.46	17	.000	Total	.894	55			

Table 3 shows that the mean of collaborative critical thinking skills for classes A, B, and C are -3.20, -3.50, -3.41 and have degrees of freedom (df) 18, 18, and 17, t arithmetic -21.89, -13.99 , and -32.46 respectively. This result is significant because $p < 5\%$. The result of the calculation has a negative sign which means that there is difference in the pre-test and the post-test (improvement) of physics education students' collaborative critical thinking skill after the implementation of CinQASE learning model in all classes. Table 3 shows that F arithmetic gives F arithmetic < F table with significance level P = .495> 5%. The results show that the implementation The CinQASE learning model can significantly improve the collaborative critical thinking skills of physics education students for all groups.

This is because the CinQASE learning model has been developed by design to improve students' collaborative critical thinking skills with the following phases. (1) Activity in phase 1 (Problem presentation), (a) Lecturer motivates students to engage in team collaboratively in problem solving, (b) Lecturer gives scaffolding in the form of questions to grow up student's curiosity, (c) Lecturer provide a social environment that supports lectures by grouping students heterogeneously into groups of 6-8 collaborative people; (d) Lecturers present authentic real-life situation problems that allow various solutions to be solved; and (e) Lecturers deliver objectives of the lesson related to problems that will be solved individually and collaboratively. (2) Activity in Phase 2: Individual Work includes (a) Lecturer directs students to work individually in their group to analyse the given problems, (b) Students collect information according to the given problems from their experience individually, which will be communicated to the work team in collaboration. (3) Activities in Phase 3: Collaborative Critical Thinking Team Work includes (a) Lecturers direct students to work in pairs in collaborative groups, (b) Lecturers distribute worksheet (LKM) to each partner in the collaborative group as a guide in solving the problem (d) Students work together in a paired collaborative group and work on the provided worksheet, (e) Each pair in the collaborative group discusses the collaboration of the collaborative group, (c) the lecturer distributes the worksheet in each collaborative group to analyse the results of experiment or investigation on problems to be solved collaboratively, (f) lecturers develop social structures that encourage the emergence of appropriate behaviour to collaborate among students, (g) lecturers provoke interaction between students with the physical and social environment in analysing the problem solving . (4) Phase 4' Activities: Classroom Discussions include (a) Lecturers organize tasks to be presented by each collaborative group, (b) Each collaborative group presents its work according to the task, the lecturer facilitates other groups to discuss the results of the exposure group (c) Lecturers facilitate the growth of collaborative work between groups of students in the classroom. (5) Activities in Phase 5: Evaluation & Feedback include (a) Lecturers help students to reflect or evaluate the process and results of investigation or collaborative problem solving, (b) Lecturer guides students to draw conclusions based on work result at team work collaboration stage and class discussions, (c) Lecturers provide feedback. The results are reinforced by theoretical and empirical support that the CinQASE learning model is based on motivational theory, social constructivist theory, positive dependency theory, and cognitive psychology-theory [29,49-51].

The result of students' response analysis that was collected is concluded that the overall percentage of students' response after the implementation of CinQASE learning model was above 70%. It shows that students responded very well to the developed learning model. The responses also show that the developed and implemented collaborative critical thinking skills test instrument in the six-session learning activities can assist students in understanding the material and improve collaborative critical thinking skills during the learning activities. Therefore, the CinQASE learning model proved to be effective to improve collaborative critical thinking skills of physics education students.

4. Conclusion

The CinQASE Learning Model is a collaborative-based learning that is specifically designed to improve students' collaborative critical thinking skills. The CinQASE Learning Model has syntax that consists of 5 phases, they are: (1) Problem Presentation, (2) Individual Work, (3) Team Work in Collaboration, (4) Class Discussion, and (5) Evaluation and feedback. The result of the research proves that: (1) Average post-test value was 79.35 (high category), (2) There is improvement of physics education students' collaborative critical thinking skills at $\alpha = 5\%$, (3) N-gain average score of physics education students' collaborative critical thinking skills was 0.62 (medium category); and 4) No difference (consistent) n-gain of collaborative critical thinking skills of physics education students in all groups; and (5) Students responded well to the implementation of the CinQASE learning model. Another finding based on the results of interviews shows that the CinQASE learning model has a nurture effect to improve the collaborative skills and argumentation. Therefore the CinQASE learning model has proven to be effective to improve collaborative critical thinking skill of physics education students. The implication of this research is that CinQASE learning model can be used as alternative solution to improve collaborative critical thinking skill of physics education students in Indonesia. Further research can be done by implementing the CinQASE learning model to improve the collaborative critical thinking skills of education and pure science students.

Acknowledgments

The author's gratitude goes to the Ministry of Research, Technology and Higher Education for funding the Research (SK Number: 056/SP2H/LT/DRPM dated 30 January 2018).

5. References

[1] Griffin P and Care E 2015 *Assessment and teaching of 21st century skills* (New York: Springer)
[2] Jatmiko B, Prahani B K, Munasir, Supardi Z A I, Wicaksono I, Erlina N, Pandiangan P, Althaf R, and Zainuddin 2018 *J. Baltic Sci. Educ.* **17** 1
[3] Jatmiko B, Widodo W, Martini, Budiyanto M, Wicaksono I, and Pandiangan P 2016 *J. Baltic Sci. Educ.* **15** 441
[4] Pandiangan P, Sanjaya M, Gusti I and Jatmiko B 2017 *J. Baltic Sci. Educ.* **16** 651
[5] Prahani B K, Winata S W and Yuanita L 2015 *J. Penelit. Pend. Sains***4** 503
[6] Prahani B K, Limatahu I, Nur M and Yuanita L 2016 *Int. J. Educ. Res.* **4** 231
[7] Prahani B K, Nur M, Yuanita L and Limatahu I 2016 *Vidhya Karya***31** 72
[8] Suyidno, Nur M, Yuanita L, Prahani B K and Jatmiko B2018 *J. Baltic Sci. Educ.***17** 136
[9] Thaiposri P and Wannapiroon P 2015 *Procedia Soc. Behav. Sci.***174** 2137
[10] Forawi S A, Almekhlafi A G and Al-Mekhlafy M H 2012 *Online Submission***1** 99
[11] Geertsen H R 2003 *Teach. Sociology* **31** 1
[12] Ennis R H 1996 *Critical thinking* (The University of Illinois: Pretince Hall Inc)

[13] Burbach M E, Matkin G S and Fritz S M 2004 *College Student J.* **38**482
[14] Cheong C M and Cheung W S 2008 *Australian J. of Educ. Technol.***24**556
[15] Ennis R H 2011 *Inquiry***26**4
[16] Ernst J and Monroe M 2004 *Environmental Educ. Res.* **10** 507
[17] Facione P A 2013 *Insight Assessment* **I** 1
[18] Jenicek M 2006 *Medical Sci. Monitor* **12** 28
[19] Miri B, David B C, and Uri Z 2007 *Res. Sci. Educ.* **37**353
[20] Mundilarto and Ismoyo H 2017 *J. Baltic Sci. Educ.***16**761
[21] Popil I 2011 *Nurse Educ. Today* **31**204
[22] Siew N M and Mapeala R 2016 *J. Baltic Sci. Educ.***15**602
[23] Snyder L G and Snyder M J 2008 *J. Res. Business Educ.* **50**9
[24] Gokhale A A 1995 *J. Tech. Educ.***7** 22
[25] Johnson R T and Johnson D W 1986 *Science and Children***24** 31
[26] Totten S, Sills T, Digby A and Russ P 1991 (Garland Publishing: New York).
[27] Lavasani G M, Weisani M and Ejei J 2012 *Proc. Soc. Behav. Sci.***15** 1881
[28] Marin L M and Halpern D F 2011 *Thinking Skills and Creativity***6** 1
[29] Hunaidah 2018 *Makalah seminar hasil* (Surabaya: Pascasarjana Unesa)
[30] Grichanik M 2014 The effects of collaborative critical thinking training on trust development and effectiveness in virtual teams (*Graduate Theses and Disertations*). *http://scholarcommons,usf.edu/etd?utm_source=schokarcommons*
[31] Godrati N 2015 *Conceptualizing and measuring collaborative critical thinking on asyncronous discussion forums: challenges and possible solutions* (*Digital Culture and Education*)
[32] Fraenkel J R, Wallen N E, and Hyun H H 2012 *How to design and evaluate research in education* (8th ed.) (New York: McGraw-Hill)
[33] Hake R R 1998 *Am. J. Phys.***66**64
[34] Purnamawati, Mulbar U, and Saliruddin 2017 *WTE&TE***15** 82
[35] Tan, M., Yu, P., and Gong, F., 2016 *WTE&TE***14** 72
[36] Limatahu I, Suyatno S, Wasis, and Prahani B K *J. Phys.: Conf. Ser.***997** 0120022
[37] PrahaniB K, SupraptoN, Suliyanah, Lestari N A, JauhariyahM N R, Admoko S and Wahyuni S 2018 *J. Phys.: Conf. Ser.***997** 0120008
[38] Purwaningsih E, Suyatno S, Wasis and Prahani B K 2018 *J. Phys: Conf. Ser.***997** 0120032
[39] Sunarti T, Wasis, Madlazim, Suyidno and Prahani B K 2018 *J. Phys: Conf. Ser.***997** 0120013
[40] Madeali H, and Prahani B K 2018 2018 *J. Phys: Conf. Ser.***997** 0120029
[41] Erika F, Prahani B K, Supardi Z A I and Tukiran 2018 *WTE&TE***17** 1
[42] Hadi S A, Susantini E, and Agustini R 2018 *J. Phys: Conf. Ser.***947** 012063
[43] Sudiarman W, Soegimin W and Susantini E 2015 *J. Penelit. Pendidik. Sains***4** 636
[44] Susantini E, Isnawati and Lisdiana L 2016 *J. Sci. Educ.***2** 74
[45] Susantini E, Lisdiana L, Isnawati, Al Haq A T and Trimulyono G 2017 *Biochemistry and Molecular Bio. Educ.***45** 216
[46] Susantini E and Qomariyah N 2013 *J. Pendidik. IPA Indonesia* **2** 142
[47] Susantini E, Thamrin M, Isnawati H and Lisdiana L 2012 *J. Pendidik. IPA Indonesia* **1** 102
[48] Freeman J & Hess K P *Collaborative critical thinking* (Paper presented at 8thInternational Command and Control Research and Technology Symposium June 17 – 19, 2003, National Defense University, Washington, DC)
[49] Moreno R 2010 *Educational psychology* (New York: John Wiley and Sons Inc)
[50] Arends R I 2012 *Learning to teach* (New York: Mc. Graw-Hill Companies)
[51] Slavin E R 2011 *Educational psychology. theory and practice* (Boston: Pearson)

Effect of Earthquake's Surface Wave on Ionosphere Layer

A Santoso[1] and B Muslim[1]
[1] Researcher in Space Science Center, LAPAN, Jln. Dr. Djundjunan 133 Bandung, INDONESIA

Abstract. The lithosphere and the atmosphere/ionosphere, continue to exchange energy through various coupling mechanism. The earthquake creates energy waves, such as Direct Shock Acoustic Waves (SAWs) and Rayleigh wave induced Acoustic Waves (RAWs) and Atmospheric Gravity Waves (AGW) when the earthquakes occur at sea. If the earthquake event is large enough (Mw> 6), then SAWs, RAWs and AGWs induce the detected ionosphere plasma as an anomaly. Near the epicentrum, the Coseismic Ionospheric Disturbance (CID) diffuses at almost the same speed as the acoustic wave speed of the shock in the ionosphere. At the distant point of the epicentrum, Rayleigh waves will induce ionospheric interference. Summing up seismological information from seismo-ionospheric manifestations is a subject related to ionospheric seismology. In this study, the ionospheric response to earthquakes was analyzed using TEC data in Indonesia especially to identify surface wave using TEC data. As a case study We were used Aceh earthquake (coordinates 3.295^{0}N; 95.982^{0}E) on December 26, 2004. It is expected that the seismo-ionospheric manifestations show the behavior of CID..

1. Introduction

The source of ionospheric disorders can come from below and above. From below are lithosphere like earthquake, volcanic eruption and nuclear explosion. While the top of the space weather such as geomagnetic storms, CME and flares. Particularly related to earthquakes, ionospheric disorders are primarily caused by (1) shock acoustic waves (SAW), (2) Rayleigh Acoustic Wave (RAW) acoustic waves or also called Surface Seismic Waves and (3) gravity acoustic wave (Acoustic Gravity Wave, AGW) whose frequency falls between 0.1 mHz to 10 mHz. This is a collection of seismic waves present in the ionosphere. The phenomenon of the relationship between earthquakes and ionsofer is called seismo-ionospheric. The study of seismo-ionospheric has been widely practiced. This is due to the fact that many advanced equipment are installed (eg HF Doppler Sounding, DEMETER and GNSS) en masse to monitor ionospheric plasma disturbances caused by earthquakes both on land and in space so as to facilitate study and monitoring seismo-ionospheric disorder.

Studying these seismo-ionosphere anomalies can further be used for the detection of epicenter earthquakes and tsunami early warning. In addition, it also offers the possibility of remote sensing of seismic signals [1,2] mainly for two reasons: (1) continuity of vertical displacement on the surface, the atmosphere is then forced to move at the same vertical speed as the ground surface and (2) the conservation of kinetic energy and high exponential density air density. Further, it should be noted that it is easier to detect the diffuse waves in the ionospheric plasma, rather than the neutral atmosphere due to the radio propagation properties (plasma dispersive properties).

In this paper, we study effect of earthquake's surface wave to ionosphere using TEC data during Aceh earthquake (coordinates 3.2950N; 95.9820E) on December 26, 2004. The objective is to know

IOP Conf. Series: Journal of Physics: Conf. Series **1108** (2018) 012102 doi:10.1088/1742-6596/1108/1/012102

effect earthquake's surface wave to ionosphere. Furthermore, for the long term it can be applied for earthquake epicenter detection as well as tsunami early warning.

1.1 Types of Seismic Wave

Seismology is the study of earthquakes and seismic. Seismic waves can be defined as, "the waves of energy caused by the sudden breaking of rock within the earth or an explosion" [3]. The two main types of waves are **body waves** and **surface waves**. Body waves can travel through the earth's inner layers, but surface waves can only move along the surface like ripples on water. Earthquakes radiate seismic energy as both body and surface waves.

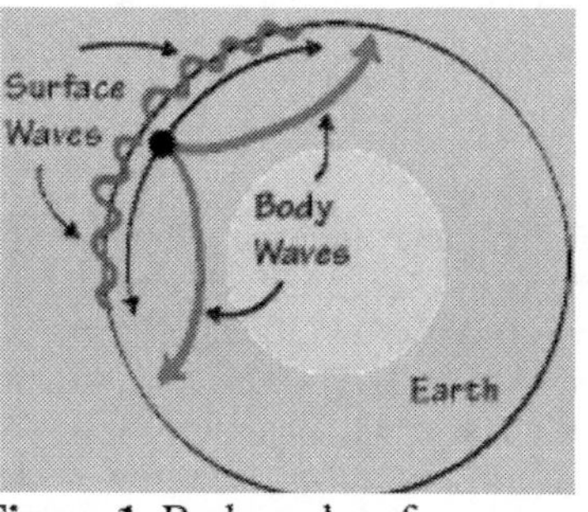

Figure 1. Body and surface waves

1.2 State Art of Seismology

Recent observations of strong ground motion accelerations is strongly dominated by Rayleigh waves [4,5] as recorded in México earthquakes from subduction earthquakes. In many previous studies, surface waves has been used to infer the shear wave velocity structure in the upper few kilometers of the crust [6]. However, in these studies some difficulties exist in terms of identifying surface waves because they require an earthquake of magnitude about 7, which produce Rayleigh waves over a broad range of periods [7]. Large earthquakes produce important surface waves, particularly in sediment-filled areas [8]. Tanaka, Yoshizawa and Osawa [9] observed that during large earthquakes, surface waves with period around 8 sec were dominant in records of sedimentary soils of Tokyo. Then, it is clear that layering is responsible for practically all the observed effects during earthquakes. The effect of local soil conditions in modifying the earthquake motions has been established for some time.

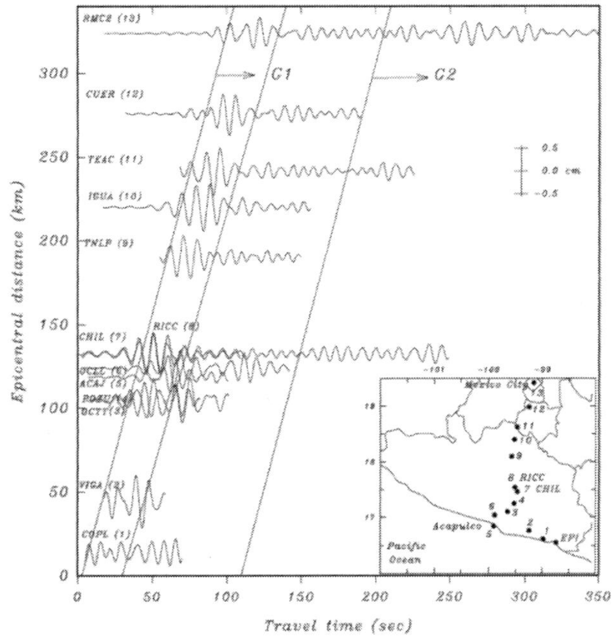

Figure 2. Vertical displacement seismograms (0.09-0.13 Hz) along the way between México City and the epicenter of the September 14, 1995 earthquake.

Based on the Figure 2, the number in parentheses shows the location of each station in the map of the bottom right angle. Significant amplitudes are detected in the prolonged records in Chilpancingo (RICC) and in Teacalco (TEAC) stations, besides the recorded in México City.

2. Method

Data used as an analysis material is TEC data from satellites crossing Indonesian territory at the time of Aceh earthquake (coordinates 3.295^0N; 95.982^0E) of December 26, 2004; making plotting of TEC data versus time and visually analyzing the TEC data anomalies around the time of the earthquake. To refine the analysis, also used index data Dst as geomagnetic storm event parameter. The Dst Index is used to verify whether anomalies detected through pure TEC are from earthquakes or from space weather sources. And then, for more detailed analysis, an FFT analysis of TEC data from each satellite is performed. Time of surface wave to ionosphere assumed as

T = distance satellite to epicentrum / 3km s^{-1} + 8 minute (3-1)

The last, it is for conclusions.

3. Results and Discussion

From TEC data, we obtained 6 satellites at MSIA observatory and 10 satellites at NGNG observatory.

IOP Conf. Series: Journal of Physics: Conf. Series **1108** (2018) 012102 doi:10.1088/1742-6596/1108/1/012102

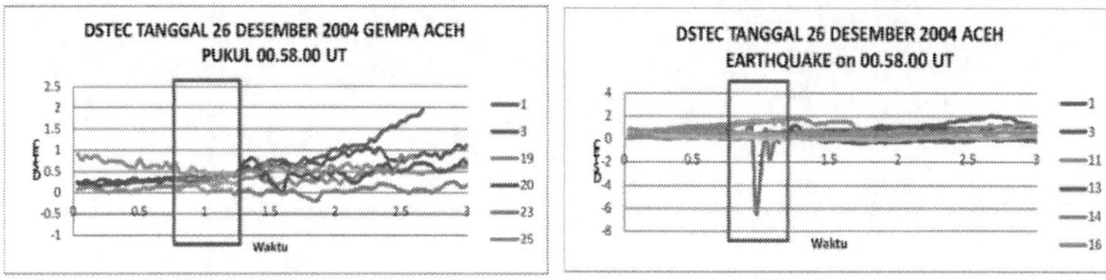

Figure 3. (a) DSTEC from MSIA observatory and (b) DSTEC from NGNG observatory on December 26, 2004

Figure 4-1, in the square visually it is not clear that there is a DSTEC pattern that is disturbed, but it is quite clear that at around 01:00 UT the pattern seems to collect all at one point and a little fluctuation DSTEC pattern. Furthermore, it is seen that the DSTEC value as if starting to experience the uptrend from that point. To ascertain whether there is any effect from the space weather, the following shows the index data Dst and the AE index.

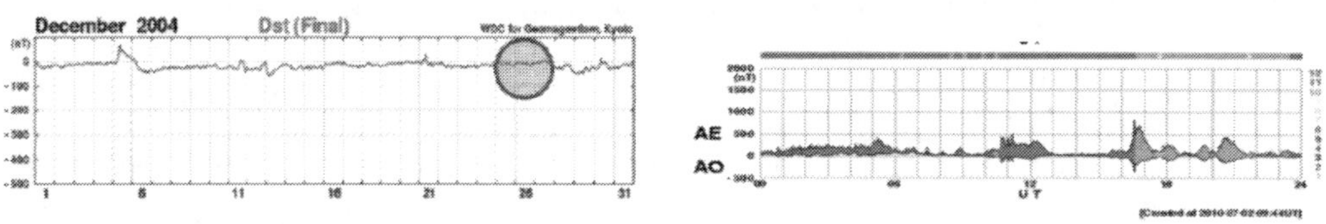

Figure 4. (Left) Dst index on December, 2004 and (Right) AE index on December 26, 2004

The Dst Index on December 26, 2004 showed no pattern caused by aerospace weather disturbance, and from the AE index also reinforced the phenomenon. Thus, the space weather of December 26, 2004 can be said to be in quiet condition and it is certain that the space weather is not a source of DSTEC pattern trigger on December 26, 2004 at around 01.00 UT.

To see more clearly the disturbance on DSTEC, FFT analysis is performed and the results are as shown in Table 1.

Table 1. The travel time pf surface wave

No	Name Observatory	Number of Satellite	Epicentrum	Coordinate of satellite	Travel time (Minutes)	
					Eq 3-1	FFT
1	MSIA	1	3.92^{0} N; 95.892^{0}E	0.57^{0} S; 101.14^{0} E	12.114	11.614
2		3		2.46^{0} S; 99.99^{0} E	12.410	13.870
3		19		4.72^{0} S; 98.97^{0} E	13.319	14.815
4		20		0.49^{0} S; 95.65^{0} E	10.339	18.940
5		23		8.92^{0} N; 94.55^{0} E	11.573	13.495
6		25		6.38^{0} N; 06.35^{0} E	14.947	19.685
7	NGNG	1		0.99^{0} S; 101.3^{0} E	12.350	15.152
8		3		2.93^{0} S; 100.14^{0} E	12.670	11.223
9		11		11.18^{0} S; 90.63^{0} E	17.538	14.881

10		13		19.94° S; 111.28° E	25.377	16.807
11		14		4.95° S; 104.61° E	15.542	8.993
12		16		2.99° N; 99.77° E	10.527	13.228
13		19		5.0° N; 99.18° E	13.530	10.504
14		20		1.18° S; 95.84° E	10.760	10.163
15		23		9.24° N; 94.33° E	11.800	12.020
16		25		6.09° N; 106.58° E	15.043	12.987

From Table 1, NGNG observatory satellite number 20, travel time value for equation 3-1 and FFT is very close.

4. Conclusion

Travel time of surface wave in about of 6.8 minute until to 27.2 minute after earthquake. The difference of the travel time of surface wave between the results of equation 3.1 with DSTEC data is supposed as the effect of moving away or approaching the satellite

5. References

[1] Lognonn E P, Garcia R, Crespon F, Occhipinti G, Kherani A, Artru-Lambin 2006 *J. Seismic waves in the ionosphere. Europhys News* **37**(4): 11e4

[2] Reddy C D 2016 Seismo-ionospheric anomalies and implications from recent GNSS observations in India and South-East Asia *Geode sy. and geodynamics* 17(1) pp. 11-18

[3] UpSEIS 2006 (online) What Is Seismology? (http://www.geo.mtu.edu/UPSeis/waves.html)

[4] Gómez-Bernal A and Saragoni R 1995 Interpretation of dynamic soil effects on México City valley using the dense accelerograph network *Proc. of the Fifth Int Conf. on Seismic Zonation,* France **1** pp. 747-754

[5] Gómez-Bernal A and Saragoni R 1996 Oscillations of the México City surface layer excited by seismic surface waves *Proc. 11th World Conf. On Earth. Eng.* **718**

[6] MacBeth C D and Burton P 1985 Upper crustal shear velocity models from higher mode Rayleigh wave dispersion in Scotland *Geophysics. J. R. Astr. Soc.* **83** pp. 519-529

[7] Ewing W M, Jardetzky W S, and Press F 1957 *Elastic waves in Layered Media* (New York: McGraw Hill)

[8] Bard P Y and Bouchon M 1980 The seismic response of sediment-filled valleys. Part 2. The case of incident P and SV waves *Bull. Seism. Soc. Am.* **70** pp. 1921-1941

[9] Tanaka T, Yoshizawa S, and Osawa Y 1980 Characteristics of strong earthquake ground motion in the period range from 1 to 15 seconds *Proc. 7th World Conf. Earthquake Eng.* **2** pp. 609-616

IOP Conf. Series: Journal of Physics: Conf. Series **1108** (2018) 012103 doi:10.1088/1742-6596/1108/1/012103

A Note On The Weighted Fourier Transforms on p –Integrable Function Spaces

Y Fuad[1,*] **and M Jakfar**[1]
[1] Mathematics Departement, Universitas Negeri Surabaya, Surabaya, Indonesia
*Corresponding Author: yusuffuad@unesa.ac.id

Abstract. We have known that the Fourier transformis well defined on L^1 space. In this paper, we shall introduce generalization of the Fourier transform on L^p spaces with $p \geq 1$ which is called by weighted Fourier Transform, and discuss its properties. The invers transform is studied and we also obtain a result which describe how the weighted Fourier transform is applied to solve generalization of the heat equation problem.

1. Introduction

Fourier Analysis studies how to analyse a range of function by expanding it to a series or integral of a certain function (which properties has been well known, as polynomial function or trigonometric function) ([1]). Fourier Analysis is a powerful tool to solve a various problems particularly in partial differential equation that has been existed in pure and applied science. Fourier Analysis also can be used to analyse digital signals and images as in [2], [3] and [4].

The theory in Fourier analysis which has been mostly used to solve problems in applied science is Fourier transform. One of its benefit is to find a solution of classical differential equation—heat equation in $\mathbb{R}$. The Fourier transform is only well defined on L^1 ([1]). One questions rise: can we find a transform which is a generalization of the Fourier transform and it is well defined not only on L^1 but also well defined on L^p for $p \geq 1$? This is a basic problem in this paper. In this paper, by using Hölder's innequality [5], we will construct a new transform which is generalisation of the Fourier transform, and also found many interesting results, which we will use to solve generalization of the heat equation problem.

Throughout the paper, the researchers declared the definition of $L^p = L^p(\mathbb{R})$ spaces as in [5]. For $1 \leq p < \infty$,

$$L^p = \left\{ f : \int_{-\infty}^{\infty} |f(x)|^p dx < \infty \right\},$$

and for $p = \infty$,

$$L^\infty = \{ f : ess\sup\{|f(x)| : x \in \mathbb{R}\} < \infty \}.$$

where $ess\sup\{|f(x)| : x \in \mathbb{R}\}$ is "essential" supremum of $|f|$. In other word, L^∞ is the space all of functions which are bounded almost everywhere.

2. Weighted Fourier Transformation For $1 \leq p < \infty$

In this sections, we shall construct the generalized Fourier transform. To define the transform on L^p, we have to use weights (look at [6] and [7]). Let us choose $w \in L^p$ where $w(x) > 0$ for all $x \in \mathbb{R}$ when $1 \leq p < \infty$. Next, it was defined the mapping $\mathcal{F}_{p,w}$ which maps a function $f \in L^p$ for $1 \leq p < \infty$ to

$$\mathcal{F}_{p,w}f(\xi) := \int_{-\infty}^{\infty} e^{-i\xi x}|w(x)|^{p-1}f(x)dx. \tag{1}$$

Throughout this paper, $\widehat{f_w}$ is called weighted Fourier transform of f.

The mapping is well defined for $f \in L^p$ because for all $x \in \mathbb{R}$, we have

$$\left|e^{-i\xi x}\big(w(x)\big)^{p-1}f(x)\right| = \left|\big(w(x)\big)^{p-1}f(x)\right|$$

which is integrable.

It follows from Hölder's innequality [5] that we get this proposition.

Proposition 2.1. *If $f \in L^p$ and $w \in L^p$ with $w(x) > 0$ for all $x \in \mathbb{R}$, then $\mathcal{F}_{p,w}f \in L^\infty$.*

Proof. The first, using Hölder's innequality, we obtained $w^{p-1}f \in L^1$. Indeed,

For$1 < p < \infty$,

$$\int_{-\infty}^{\infty}\left|\big(w(x)\big)^{p-1}f(x)\right|dx \leq \left(\int_{-\infty}^{\infty}(|w(x)|^{p-1})^{\frac{p}{p-1}}dx\right)^{\frac{p-1}{p}}\left(\int_{-\infty}^{\infty}|f(x)|^p dx\right)^{\frac{1}{p}}$$
$$= \left(\int_{-\infty}^{\infty}|w(x)|^p dx\right)^{\frac{p-1}{p}}\left(\int_{-\infty}^{\infty}|f(x)|^p dx\right)^{\frac{1}{p}} < \infty,$$

for $p = 1$,

$$\int_{-\infty}^{\infty}\left|\big(w(x)\big)^{p-1}f(x)\right|dx = \int_{-\infty}^{\infty}\left|\big(w(x)\big)^{0}f(x)\right|dx = \int_{-\infty}^{\infty}|f(x)|dx < \infty.$$

So, $w^{p-1}f \in L^1$.

Now, it could be investigated that the weighted Fourier transform was well defined in L^p. For every $f \in L^p$ with $p \geq 1$, we get

$$\begin{aligned}|\mathcal{F}_{p,w}f(\xi)| &= \left|\int_{-\infty}^{\infty} e^{-i\xi x}|w(x)|^{p-1}f(x)dx\right| \\ &\leq \int_{-\infty}^{\infty}\left|e^{-i\xi x}|w(x)|^{p-1}f(x)\right|dx \\ &= \int_{-\infty}^{\infty}\left|e^{-i\xi x}\right|\left|\big(w(x)\big)^{p-1}f(x)\right|dx \\ &= \int_{-\infty}^{\infty}\left|\big(w(x)\big)^{p-1}f(x)\right|dx < \infty \quad (\text{since } w^{p-1}f \in L^1)\end{aligned}$$

Since$|\mathcal{F}_{p,w}f(\xi)| < \infty$ for all $\xi \in \mathbb{R}$. So, $\mathcal{F}_{p,w}f \in L^\infty$. ■

IOP Conf. Series: Journal of Physics: Conf. Series **1108** (2018) 012103 doi:10.1088/1742-6596/1108/1/012103

***Remark** 2.2.*The transform on (1) was called by the weighted Fourier Transform because $\mathcal{F}_{p,w}f$ is generalization of the Fourier transform which was defined by

$$\mathcal{F}(f)(\xi) := \int_{-\infty}^{\infty} e^{-i\xi x} f(x)dx \qquad (2)$$

where $f \in L^1$. The Fourier transform is well defined only on L^1, however, the weighted Fourier transform was well defined on L^p for $1 \le p < \infty$, According to Proposition 2.1. In addition, if $p = 1$, then $\mathcal{F}_{p,w}f = \mathcal{F}f$.

The proposition below show that the weighted Fourier transform of function on $L^p$is continuous on $\mathbb{R}$ ([8])

Proposition 2.3. *If $f \in L^p$and $w \in L^p$ with $w(x) > 0$ for all $x \in \mathbb{R}$, then $\mathcal{F}_{p,w}f$ is continuous on $\mathbb{R}$.*

Proof. For all ξ and h in $\mathbb{R}$

$$\mathcal{F}_{p,w}f(\xi + h) - \mathcal{F}_{p,w}f(\xi) = \int_{-\infty}^{\infty} e^{-i\xi x}(e^{-ihx} - 1)|w(x)|^{p-1}f(x)dx,$$

Thus,

$$|\mathcal{F}_{p,w}f(\xi + h) - \mathcal{F}_{p,w}f(\xi)| \le \int_{-\infty}^{\infty} |e^{-ihx} - 1| \left|\left(w(x)\right)^{p-1} f(x)\right| dx.$$

The integral in the right sideis dominated by $2|w^{p-1}f(x)|$ and went to *0* as $h \to 0$. So, according to the Lebesgue Dominated Convergences Theorem (look at [9]), the right side must converges to 0 as $h \to 0$. Consequently, the left side also went to 0 as $h \to 0$. ∎

Some propositions below are basic properties of the weighted Fourier Transform.

Proposition 2.4. *if $f \in L^p$ and $w \in L^q$ with $w(x) > 0$ for all $x \in \mathbb{R}$, then $\lim_{|\xi| \to \infty} \mathcal{F}_{p,w}f(\xi) = 0$*

Proof. It was observed that

$$-\mathcal{F}_{p,w}f(\xi) = \int_{-\infty}^{\infty} e^{-i\xi\left(x + \frac{\pi}{\xi}\right)} |w(x)|^{p-1} f(x)dx = \int_{-\infty}^{\infty} e^{-i\xi x} \left|w\left(x - \frac{\pi}{\xi}\right)\right|^{p-1} f\left(x - \frac{\pi}{\xi}\right)dx.$$

Hence,

$$\begin{aligned}2\mathcal{F}_{p,w}f(\xi) &= \mathcal{F}_{p,w}f(\xi) - \left(-\mathcal{F}_{p,w}f(\xi)\right) \\ &= \int_{-\infty}^{\infty} e^{-i\xi x}\left(|w(x)|^{p-1}f(x) - \left|w\left(x - \frac{\pi}{\xi}\right)\right|^{p-1} f\left(x - \frac{\pi}{\xi}\right)\right)dx\end{aligned}$$

Thus

$$2|\mathcal{F}_{p,w}f(\xi)| \le \int_{-\infty}^{\infty} \left||w(x)|^{p-1}f(x) - \left|w\left(x - \frac{\pi}{\xi}\right)\right|^{p-1} f\left(x - \frac{\pi}{\xi}\right)\right| dx.$$

Since $w^{p-1}f \in L^1$ (look at the proof of proposition 2.1), then (using the continuity of norm in L^1) the right side went to 0 if $|\xi| \to \infty$. Consequently, the left side also went to 0 if $|\xi| \to \infty$. ∎

Corollary 2.5. *Let $p \ge 1$ and $w \in L^p$with $w(x) > 0$ for all $x \in \mathbb{R}$. The weighted Fourier transformmapsa functionf $\in L^p$ to a function $\widehat{f_w}$ in $C_0(\mathbb{R})$. In other words, $\mathcal{F}_{p,w}f$ is continuous and bounded.*

Note. $C_0(\mathbb{R})$ is the space of functions that are continuous and bounded in $\mathbb{R}$ with zero limit in $\pm\infty$.

Proposition 2.6. *If $f, g \in L^p$ for $1 \le p < \infty$, α and β are (real) scalars, and $w \in L^p$ with $w(x) > 0$ for all $x \in \mathbb{R}$, then $\mathcal{F}_{p,w}(\alpha f + \beta g) = \alpha \mathcal{F}_{p,w} f + \beta \mathcal{F}_{p,w} g$.*

Proof. Since integral is linear. ∎

Next, researchers would discuss a spacial case for the weighted Fourier transform properties. In this case, we take a function w as

$$w_0(x) = \sum_{i=0}^{\infty} \lambda_{(i-1,i]}(x) \min_{x\in(i-1,i]} \left(e^{-|x|}\right) + \lambda_{[i,1+i)}(x) \min_{x\in[i,1+i)} \left(e^{-|x|}\right), \quad \forall x \in \mathbb{R}.$$

It could be observed that $w_0 \in L^p$ for any $p \ge 1$ because $w_0(x) \le e^{-|x|}$ and it was known that the function k with $k(x) = e^{-|x|}$ is in L^p. It could also be investigated that $w_0' = 0$ almost everywhere.

Proposition 2.7. *If f is a differentiable function on $\mathbb{R}$, both f and its derivative f' are in L^p for $p \ge 1$, and*

$$w_0(x) = \sum_{i=0}^{\infty} \lambda_{(i-1,i]}(x) \min_{x\in(i-1,i]} \left(e^{-|x|}\right) + \lambda_{[i,1+i)}(x) \min_{x\in[i,1+i)} \left(e^{-|x|}\right), \quad \forall x \in \mathbb{R},$$

then the weighted Fourier transform of the derivative is given by

$$\mathcal{F}_{p,w_0} f'(\xi) = i\xi \mathcal{F}_{p,w_0} f(\xi), \quad \forall \xi \in \mathbb{R}.$$

Proof. By integrating of parts, it was obtained

$$\mathcal{F}_{p,w_0} f'(\xi) = \int_{-\infty}^{\infty} e^{-i\xi x} |w_0(x)|^{p-1} f'^{(x)} dx = \int_{-\infty}^{\infty} e^{-i\xi x} |w_0(x)|^{p-1} \frac{df}{dx}(x) dx =$$
$$-\int_{-\infty}^{\infty} f(x) \frac{e^{-i\xi x}|w_0(x)|^{p-1}}{dx} dx = -\int_{-\infty}^{\infty} f(x) \left(-i\xi e^{-i\xi x} |w_0(x)|^{p-1} + e^{-i\xi x} \frac{|w_0(x)|^{p-1}}{dx}\right) dx =$$
$$i\xi \int_{-\infty}^{\infty} f(x) \left(e^{-i\xi x} |w_0(x)|^{p-1}\right) dx - 0 = i\xi \int_{-\infty}^{\infty} e^{-i\xi x} |w_0(x)^{p-1}| f(x)\, dx = i\xi \mathcal{F}_{p,w_0} f(\xi). \ ∎$$

Corollary 2.8. *If $f \in L^p$ for $p \ge 1$, f has second derivatve, both its derivative f' and its second derivative f'' are in L^p, and*

$$w_0(x) = \sum_{i=0}^{\infty} \lambda_{(i-1,i]}(x) \min_{x\in(i-1,i]} \left(e^{-|x|}\right) + \lambda_{[i,1+i)}(x) \min_{x\in[i,1+i)} \left(e^{-|x|}\right), \quad \forall x \in \mathbb{R},$$

then the weighted Fourier transform of the derivative is given by

$$\mathcal{F}_{p,w_0} f''(\xi) = -\xi^2 \mathcal{F}_{p,w_0} f(\xi), \quad \forall \xi \in \mathbb{R}.$$

Proof. It followed that

$$\mathcal{F}_{p,w_0} f''^{(\xi)} = \mathcal{F}_{p,w_0} (f')'^{(\xi)(\xi)} = i\xi \mathcal{F}_{p,w_0} f'^{(\xi)} = i\xi \left(i\xi \mathcal{F}_{p,w_0} f(\xi)\right) = -\xi^2 \mathcal{F}_{p,w_0} f(\xi),$$

for every ξ in $\mathbb{R}$. ∎

3. Invers Wighted Fourier Tansform

Throughout this section, we will discuss inversion formula of the weighted Fourier transform. When we revisit the Fourier Transform on L^1, weshall find Fourier Inversion Theorem (at [1]) below:

"If $f \in L^1$ is peace wise continue, and $\mathcal{F}(f) \in L^1$ then f is continuous and

$$f(x) = \frac{1}{2\pi} \int_{-\infty}^{\infty} \mathcal{F}(f)(\xi) e^{i\xi x}\, d\xi$$

for all $x \in \mathbb{R}$."

Now, it will be generalised that the Fourier Inversion Theorem has inversion formula of the weighted Fourier Transform on L^p.

Next, one of the main result of this section: Inversion formula of the weighted Fourier Transform on L^p.

Theorem 3.1. (Weighted Fourier InversionTheorem) *If $f \in L^p$ for $1 \le p < \infty$, $w \in L^p$ where $w(x) > 0$ for all $x \in \mathbb{R}$, and f and w are piece wise continuous, and $\mathcal{F}_{p,w}f \in L^1$ then f is continuous and*

$$f(x) = \frac{1}{2\pi|w(x)|}\int_{-\infty}^{\infty} \mathcal{F}_{p,w}f(\xi)e^{i\xi x}\,d\xi$$

for all $x \in \mathbb{R}$.

Proof. Since$f \in L^p$ and $w \in L^p$, then according to Hölder's innequality, we obtained $|w|^{p-1}f \in L^1$. Base on hypothesis, f was piece wise continuous and $\mathcal{F}(|w|^{p-1}f) = \mathcal{F}_{p,w}f \in L^1$ and also the Fourier Inversion Theorem, we obtained

$$|w|^{p-1}f = \frac{1}{2\pi}\int_{-\infty}^{\infty} \mathcal{F}(|w|^{p-1}f)(\xi)e^{i\xi x}\,d\xi = \frac{1}{2\pi}\int_{-\infty}^{\infty} \mathcal{F}_{p,w}f(\xi)e^{i\xi x}\,d\xi.$$

Consequently,

$$f(x) = \frac{1}{2\pi|w(x)|^{p-1}}\int_{-\infty}^{\infty} \mathcal{F}_{p,w}f(\xi)e^{i\xi x}\,d\xi$$

for every $x \in \mathbb{R}$, and $w^{p-1}f$ was continuous in $\mathbb{R}$. Moreover, since w was piece wise continuous, then f was also wise continuous in $\mathbb{R}$. ■

Remark 3.2. The Weighted Fourier Inversion Theorem show us that weighted Fourier transform has an invers which is called by invers weighted Fourier transform. If the weighted Fourir transform is denoted by $\mathcal{F}_{p,w}$, then the invers weighted Fourier transform is denoted by $\mathcal{F}_{p,w}^{-1}$. So, if $\mathcal{F}_{p,w}^{-1}f(x) = \frac{1}{2\pi|w(x)|^{p-1}}\int_{-\infty}^{\infty} f(\xi)e^{i\xi x}\,d\xi$ then $f = \mathcal{F}_{p,w}^{-1}(\mathcal{F}_{p,w}f)$.

Corollary 3.3. *If $f \in L^p$ for $1 \le p < \infty$ and $w \in L^p$ where $w(x) > 0$ for all $x \in \mathbb{R}$, and f is piece wise smooth, then*

$$\lim_{T\to\infty} \frac{1}{2\pi|w(x)|^{p-1}}\int_{-T}^{T} \mathcal{F}_{p,w}f(\xi)e^{i\xi x}\,d\xi = \frac{1}{2}[f(x+) + f(x-)],$$

for every $x \in \mathbb{R}$.

4. Further Result

Now, we will apply the weighted Fourier transform on classical differential equation—heat equation in $\mathbb{R}$, that is

$$u_t = ku_{xx} \quad \text{, for almost every } x \in \mathbb{R}$$

with the initial condition $u(x,0) = f(x)$ where $f \in L^p$ for $p \ge 1$ and the boundary condition $u(x,t) \to 0$ and $f(x) \to 0$ as $x \to \pm\infty$.

By applying Weighted Fourier transform on both sides of the equation (both sides consider respect to x) with

$$w_0(x) = \sum_{i=0}^{\infty} \lambda_{(i-1,i]}(x) \min_{x\in(i-1,i]}\left(e^{-|x|}\right) + \lambda_{[i,1+i)}(x) \min_{x\in[i,1+i)}\left(e^{-|x|}\right), \quad \forall x \in \mathbb{R}.$$

we obtained

$$\frac{\partial \mathcal{F}_{p,w_0} u}{\partial t}(\xi, t) = -\xi^2 k \mathcal{F}_{p,w_0} u(\xi, t) \tag{3}$$

$$\mathcal{F}_{p,w_0} u(\xi, 0) = \mathcal{F}_{p,w_0} w(\xi).$$

The solution of this differential equation is

$$\mathcal{F}_{p,w_0} u(\xi, t) = \mathcal{F}_{p,w_0} f(\xi) e^{-\xi^2 kt}.$$

By using the Weighted Fourier Inversion Theorem, we get the following solution of the heat equations

$$u(x,t) = \frac{1}{2\pi |w(x)|^{p-1}} \int_{-\infty}^{\infty} \mathcal{F}_{p,w_0} f(\xi) e^{-\xi^2 kt} e^{i\xi x}\, d\xi$$

with $w \in L^p$ and

$$w_0(x) = \sum_{i=0}^{\infty} \lambda_{(i-1,i]}(x) \min_{x \in (i-1,i]} \left(e^{-|x|}\right) + \lambda_{[i,1+i)}(x) \min_{x \in [i,1+i)} \left(e^{-|x|}\right), \quad \forall x \in \mathbb{R}$$

for every $x \in \mathbb{R}$.

5. References

[1] Folland G B 1992 *Fourier Analysis and Its Applications* (Wardsworth & Brooks/Cole: Pacific Grove)
[2] Anumaka M C 2012 Analysisand Applications of Laplace/Fourier Transformationsin Electrict Circuit *IJRRAS* **12**(12)
[3] Gabor D 1946 Theory of communication *J. Inst. Electr. Eng.* **93** pp. 429-457
[4] Shannon C E 1949 Communication in the present of noise *Proc. of the Inst. Of Radio Eng.* **37** pp. 10-21
[5] Kreyszig E 1978 *Introductory Functional Analysis with Applications* (New York: John Wiley & Sons)
[6] Jakfar M 2018 Ruang Fungsi Terintegralkan p Berbobot *kubik* **3**(1) 60-67
[7] Jakfar M, Gunawan H, and Idris M 2015 *Hasil Kali Dalam Berbobot pada Ruang $L^p(X)$, Prosiding Semnastika 2015 Seminar Nasional Matematika dan Pendidikan Matematika* 298-304 (Surabaya: Unipress Unesa)
[8] Manuharawati, Yunianti D N, and Jakfar M 2018 Some basic properties of Uniformly Symmetrically Continuous (Real Valued) Functions on Metric Spaces Journal of Physics: Conference Series **947**(1) 012006
[9] Rudin 1986 *Real and Complex Analysis, 3rd ed.* (New York: McGraw-Hill)

Acknowledgement
This research was founded by FMIPA Universitas Negeri Surabaya. Researchers also would like to thank the referee for his/ her valuable comments on the early version of this paper.

IOP Conf. Series: Journal of Physics: Conf. Series **1108** (2018) 012104 doi:10.1088/1742-6596/1108/1/012104

The Analysis of Students' Mathematical Thinking based on Their Mathematics Self-Concept

N Delima[1], M A Rahmah[2], A Akbar[3]

[1]Universitas Subang, Jln. R.A. Kartini KM 3 Subang, West Java, Indonesia
[2]Universitas Subang, Jln. R.A. Kartini KM 3 Subang, West Java, Indonesia
[3]SMA N 3 Subang, Jln. Emo Kurniatmaja Subang, West Java, Indonesia

Abstract. This research aims to describe the students' mathematical thinking based on their mathematics self-concept. In addition, this study also aims to determine whether there are differences in mathematical thinking between students who have positive mathematics self-concept (MSC) with students who have negative MSC and to know how much influence MSC to the students' mathematical thinking. There are four indicators of students' mathematical thinking which is used in this study. First, Specializing, i.e. trying the problem by looking at the example, paying attention to a simple case. Second, Generalising, i.e. looking for patterns and relationships. Third, Conjecturing, i.e. predicting relationships and results. Fourth, Convincing, i.e. finding and communicating the reason why something is right. This research used a descriptive method. The population of this study is all senior high school students in Subang, West Java. Instruments in this research are the test of students' mathematical thinking and MSC questionnaire. Based on the finding of this research, generally, students who have positive MSC, have good mathematical thinking abilities. Mann-Whitney U test results showed that significantly there are differences in mathematical thinking between students who have positive MSC with students who have negative MSC. Variables of MSC give positive and big influence on the students' mathematical thinking.

1. Introduction

The objectives of mathematics education in school include formal objectives which emphasize organizing, reasoning, and shaping students' personality; and material objectives which emphasize problem-solving skill and applying mathematics ([1]). In achieving those material objectives, numerous researchers suggest that mathematics learning inside the classroom should imitate a mathematician ([2]). When a mathematician works on the mathematical problem, the activity will include exemplifying, specializing, completing, deleting, correcting, comparing, sorting, organizing, changing, varying, reversing, altering, generalizing, conjecturing, explaining, justifying, verifying, convincing, refuting. Those activities, according to [3], are several of the activities that eventually will build students' mathematical thinking skill. Therefore, the objectives of mathematics education in school will be achieved, if the learning process is able to construct a mathematical thinking skill.

There are four basic processes in the construction of mathematical thinking skill, according to [4]. Those four processes are (1) specializing, is the students' skill in finishing numerous exercises by looking at the example; (2) generalizing, is the students skill in identifying patterns and connections; (3) conjecturing, can be defined as the students' skill in predicting correlation and result; and (4) convincing or students' skill in finding and communicating reasons why something is considered as true.

IOP Conf. Series: Journal of Physics: Conf. Series **1108** (2018) 012104 doi:10.1088/1742-6596/1108/1/012104

Mathematical thinking skill is closely related with mathematical literacy proposed by PISA. Stacey, in [5], states that the framework used by PISA to measure mathematical literacy includes several elements from mathematical thinking skill such as reasoning element, modeling and making connections among ideas. For that reason, it can be said that mathematical thinking skill is one skill that supports the mastery of other disciplines outside mathematics such as science, technology, economics, and even the advancement in the field of economy.

One of the processes that shape the mathematical thinking skill is convincing. This process demands students to have sufficient self-confidence in defending and convincing others about the truth from conjecture that they have decided ([4]). According [6] report there are several aspects that relate to mathematics self-belief, such as mathematics self-efficacy (MSE), which is individual's self-belief in handling and solving mathematics problem effectively, mathematics self-concept (MSC) which is someone's self-belief in accomplishing mathematics activities by him or her own self, mathematics anxiety is the feeling of being depressed and worried when dealing with mathematics, and engagement, the presence connection between mathematical and other activity both inside and outside classroom. Marsh (in [7]) defines mathematics self-concept (MSC) as an individual's perception towards his or her own reasoning and mathematics skills.

According to the research conducted by Adegoke ([8]) written in a Journal of Studies in Education, it is found that students' accomplishment in mathematics subject has a significant and positive correlation with their mathematics self-concept (MSC). The finding is parallel with what has been revealed by Marsh & Hau, 2013; Marsh, 2015; Liu, Wang & Perkins, 2018 (in [8]) which mention that the students' academic skills have a positive and significant correlation with their academic self-concept. In the meantime, Stacey (in [5]) declares that people will need to use their mathematical thinking skill when trying to solve particular problems using mathematics. Consequently, it can be said that the students' achievement in mathematics subject is also can be identified base on their mathematical thinking skill. Therefore, it is assumed that there is a correlation between students' mathematical thinking skill and their mathematics self-concept (MSC). This research aims to elaborate the analysis of mathematical thinking skill based on the senior high school students' mathematics self-concept (MSC). In other words, the objective of this research is to uncover the students' mathematical thinking skill based on the mathematics self-concept (MSC) they already have. Moreover, this research tries to investigate whether there is a difference in their mathematical thinking skill between the students who have positive mathematics self-concept (MSC) and they who have negative MSC or no as well as to uncover how huge the influence of the mathematics self-concept towards the mathematical thinking skill.

2. Methodology

This research is a descriptive one conducted in Subang. The population for this research is all senior high school students in Subang. The selection of senior high school students as the population is because they are going to make the decision on something that is very different from what they have ever experienced before. Decision making in selecting which university or profession (for students who do not continue their study), according to [9], is influenced by the students' mathematics self-concept. Meanwhile, the sampling for this research is purposively performed based on the mean score of national examination for mathematics subject from all senior high school students in Subang.

The instruments used in this study are mathematical thinking skill test and mathematics self-concept questionnaire. The type of test used to measure students' mathematical thinking skill is essay one consisting of nine numbers with mathematical thinking skill indicators proposed by [4]. Those indicators are specializing, is the students' skill in finishing numerous exercises by looking at the example; generalizing, is the students skill in identifying patterns and connections; conjecturing, can be defined as the students' skill in predicting correlation and result; and convincing or students' skill in finding and communicating reasons why something is considered as true. In this instrument, there are two numbers that measure convincing and conjecturing indicators. The given test items are related to transformation and infinite series and line materials, with the composition one item for transformation and eight items for infinite series and line. The test instrument is considered as valid through product moment r test with 5% coefficient level and has 0.805 Cronbach alpha coefficient correlation.

In the meantime, mathematics self-concept (MSC) questionnaire is designed by modifying mathematics self-concept questionnaire used by PISA. The questionnaire used by PISA is a close-ended questionnaire with strongly agree, agree, disagree, and strongly disagree options consisting of five statements: (1) I am not good enough at mathematics, (2) I get a good score in mathematics, (3) I learn mathematics very fast, (4) I always believe that mathematics is one of the coolest subjects, and (5) I understand every mathematics material, even the most difficult one. The modification is performed through inserting numerous new statements and completing statement sentences by making them suitable for learning materials so that the number of statements in one questionnaire is thirty items; fifteen items for each transformation and infinite series and line. The mathematics self-concept (MS) questionnaire instrument is also considered as valid through Spearman r test with 5% significant level and 0.877 Cronbach alpha reliability coefficient.

3. Findings and Discussion

Data for this research consist of students' mathematical thinking skill test score and their mathematics self-concept (MSC) questionnaire score. The students' mathematical thinking skill scores are classified into three categories that low, average, and high. This classification is carried out by creating the frequency distribution table with the number of class is eleven. The following is the table of students' mathematical thinking skill score frequency distribution for each group.

Table 1. Table of Students' Mathematical Thinking (SMT) Score Frequency Distribution

No	Group	SMT Score	Number of Students	%
1	Low	$1 \leq SMT\ Score \leq 11$	11	35
2	Average	$12 \leq SMT\ Score \leq 22$	8	26
3	High	$23 \leq SMT\ Score \leq 32$	12	39

By looking at the above table 1, it can be seen that the majority of the students have high mathematical thinking skill. According to the data, the mean of the overall mathematical thinking skill score is 17.42 with 10.17 standard deviation. Meanwhile, the percentage of the students who get mathematical thinking skill score above average is 51.61%. It demonstrates that more than half of the students have relatively high mathematical thinking skill.

As it is mentioned before, mathematical thinking skill has four indicators. The following table displays the students' means score for each indicator.

Table 2. The Students' Mathematical Thinking (SMT) Mean Score for each Indicator

	Students' Mathematical Thinking (SMT) Indicators			
	Specializing	*Generalizing*	*Conjecturing*	*Convincing*
Mean	4.90	6.48	3.02	3.02
Standard Deviation	2.68	3.77	2.65	2.65

Based on Table 2, students are more capable when solving generalizing test items. These items request students to complete a line of numbers and to decide the pattern of those numbers. The generalizing aspect is an aspect of mathematical thinking skill that is naturally owned by every person. It is parallel with Gestalt's cognitive psychology theory who mentions that every person has a tendency to complete and to fill incomplete experiences to make them meaningful ([10]; [11]). The third year students of the senior high school are those who have already been familiar with the material of lines of numbers since they were in elementary school, and the more complicated version of this material then is relearned in senior high school. By doing so, when there is a recalling act of that material, the level of difficulty faced by the students will be lower compared with other types of test items.

Watson & Mason ([12]) mention that the test items containing generalizing indicator can help students to increase their ability in solving complicated problems. This statement is reinforced by [4] who demonstrate that before coming to the generalizing process, the students should step the process of specializing, conjecturing, and symbolizing as it displayed in figure 1. This phenomenon tells us that generalizing skill is a series of process that constructs mathematical thinking skill. This makes the fact that the highest students' mean score lays in the test items that measure their generalizing skill is not surprising. This data echoes the aforementioned data that uncovers more than half of the students have higher mathematical thinking skill.

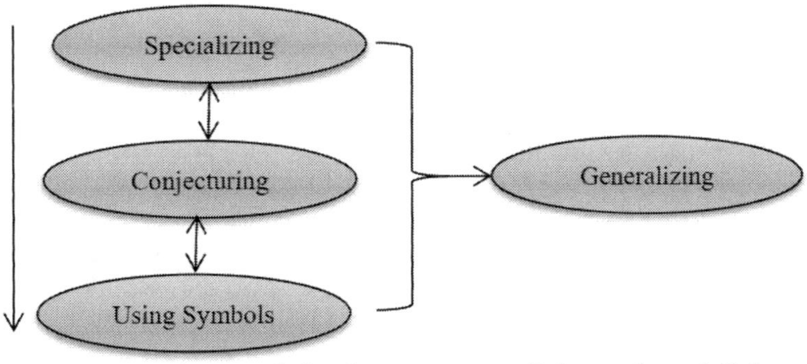

Figure 1. The correlation between generalizing and specializing

The mean score of each student's mathematics self-concept (MSC) is classified into two categories that are students with positive mathematics self-concept and students with negative mathematics self-concept. This classification is performed by creating two interval classes with the length of each class is 1.5. The following table displays the frequency distribution of students' mathematics self-concept (MSC) score in each group.

Table 3. Classification of students' Mathematics Self-Concept (MSC) and their mathematical thinking (SMT) skill

No	Group	MSC mean score	Numbers of student	SMT mean score	SMT Score standard deviation
1	Positive	$1.00 \leq \overline{MSC} < 2.50$	9	21.89	8.28
2	Negative	$2.50 \leq \overline{MSC} \leq 4.00$	22	15.59	10.70

In table 3 above, the majority of the students have negative mathematics self-concept. The students who have negative mathematics self-concept get a lower mean score of mathematical thinking compared with those who have positive mathematics self-concept. This phenomenon is parallel with the theory proposed by Marsh, 1994; Managet & Eiklend, 1998 (in [13]) that there is a significant positive correlation between students' mathematics self-concept and their achievement in mathematics subject. This mathematics self-concept is closely related with the students' self-confidence towards their mathematics mastery. There are students who still think that mathematics is difficult, and it, in turn, makes mathematics difficult for them. However, there are several students who have enough self-confidence when they are dealing with particular mathematics learning materials; in turn, this contributes to shaping their positive mathematics self-concept.

This research also uncovers the students' mathematics self-concept towards each learning material. The following table displays data about the students' mathematics self-concept for each learning material.

Table 4. Students' Mathematics Self-Concept (MSC) for each Learning Material

MSC	Materi Ajar					
	Transformation			Infinite series		
	Numbers of student	SMT mean score	SMT Score standard deviation	Numbers of student	SMT mean score	SMT Score standard deviation

Positive	7	1.00	0.58	10	22.80	8.12
Negative	24	1.08	0.78	21	13.29	9.58

SMT: Students' mathematical thinking skill

Based on the previous table 4, it can be observed that for transformation material, there is an anomaly where students with negative mathematics self-concept have higher MTS mean score compared with those who have positive mathematics self-concept. It is possibly because of the presence of a student who actually has good mathematical thinking skill, but less confident about her skill, in her daily life, she is introvert and rarely interacts with her peers. Moreover, there is also a student has good self-confidence but her mathematical thinking skill is low. In her daily life, this student is the one who is delighted to socialize with her peers and has a good artistic talent and is usually involved in the modern dance competition so that she has a high self-confidence. [14] and [15] state that the characteristic of students who have a tendency of mastering music and rhyme, tend to use their right side of their brain in thinking. [14] also mentions that those students usually have a good social life but find it is difficult to learn things that are more technical. Because of there are students whom their ways of thinking are dominated by their right side of their brain in this research, it causes although those students have a positive mathematics self-concept, their mathematical thinking skill is low. In this transformation material, students are only given one test item that measures only one indicator of mathematical thinking skill, which is specializing indicator, so that the score of this material does not represent the overall students' mathematical thinking skill. Therefore, this anomaly is not applicable to the students' mathematical thinking skill.

In the infinite series material, there is no such anomaly. It is because this material has the average level of difficulty. This material has already been known by the students since they were in their elementary school (although in the most simple form). The test items for this material measure all indicators of mathematical thinking skill. Consequently, this data reinforces the previous statement that claims, students with positive mathematics self-concept have higher mathematical thinking skill compared with those with negative mathematics self-concept.

A research indeed needs to draw the conclusion that can be generalized and be responsibly proven statistically. Therefore, in this research, it is also conducted proportion examination and examination of the difference of two means. The data used to perform both hypothesis examinations are the mathematical thinking skill score and the score of mathematics self-concept questionnaire for infinite series and line learning materials. Proportion examination was carried out to test whether the number of students with positive mathematics self-concept is significantly different from those with negative mathematics self-concept or not. The following is the data tabulation from each group.

Table 5. Number of Students in each Group Data Tabulation

		MSC Infinite Series		
		Negative	Positive	Total
SMT Infinite Series	Low	8	1	9
	Average	7	2	9
	High	6	7	13
Total		21	10	31

The above table 5 demonstrates that the majority of students with negative mathematics self-concept have low mathematical thinking skill. The second biggest number of students is the group of students with positive mathematics self-concept and have high mathematical thinking skill. The difference of the attained score between students with positive mathematics self-concept and those with negative one is not too far. The Fisher's Exact Test proportion examination was carried out to investigate the significance of the difference between both groups of students. The upcoming table shows the result of the test.

Table 6. Fisher's Exact Test

	Value	df	Exact Sig. (2-sided)
Fisher's Exact Test	4.599		.093
N of Valid Cases	31		

By giving a close attention to the previous table, it can be inferred that the proportion of students with positive mathematics self-concept and those with negative one is significant. In other words, it can be said that the number of the sample employed to test the difference of means is not too different significantly. The examination was also continued to test the data normality. The following table displays the result fo the test.

Table 7. Normality Test

	MSC Infinite Series	Kolmogorov-Smirnov[a]			Shapiro-Wilk		
		Statistic	Df	Sig.	Statistic	Df	Sig.
SMT Infinite Series	Negative	.167	21	.129	.921	21	.092
	Positive	.307	10	.008	.789	10	.011

a. Lilliefors Significance Correction

According to the table 7 above, it can be observed that the group of students who have negative mathematics self-concept has abnormal data distribution. As a result, mean difference examination was performed using U Mann-Whitney test.

Table 8. U Mann-Whitney Test

	SMT Infinite Series
Mann-Whitney U	44.500
Z	-2.562
Asymp. Sig. (2-tailed)	.010

Grouping Variable: MSC Infinite Series

The result of U Mann-Whitney uncovers that there is a significant difference in mathematical thinking skill between students with positive mathematics self-concept and those with negative mathematics self-concept. The table 8 is parallel with descriptive statistics in table 3 and table 7. The students with positive mathematics self-concept will have better mathematical thinking skill when it is being compared with those who have negative mathematics self-concept. This echoes what has been published in the PISA 2012 report that states, in every nation joint in OECD, the mathematics self-concept (MSC) is positively associated with their performance in mathematics. Furthermore, [8] also declares that there is a positive correlation between the students' mathematics achievement and their mathematics self-concept (MSC). It corresponds to the previous research reports conducted by Marsh & Hau (2103); Marsh (2105); Liu, Wang & Perkins (2008) that reveal, the students' academic achievement has a significant and positive correlation with their academic self-concept.

This research has unearthed that there is a significant difference of mathematical thinking skill between students who have positive mathematics self-concept and those who have the negative one. This difference indicates that the students' mathematics self-concept will influence the students' mathematical thinking skill. The level of influence given by the mathematics self-concept can be determined using Cohen effect size. Based on the result of the computation, it is revealed that the value of effect size from both variables is 1.039, from which can be concluded that mathematics self-concept variable positively and relatively hugely influences mathematical thinking skill. This finding is identical with research conducted by [16] that analyzed TIMSS data and found that there is a positive correlation between self-concept and mathematics subject achievement of the students in 16 different nations. Three years earlier than that research, [17] found that there is a reciprocal effect between academic achievement and academic self-concept, both variables are influencing each other, in which the change in the students' academic self-concept can cause the change in their academic achievement and vice versa. Their research then is reinforced by [18] who declared that academic self-concept holds a crucial role in increasing students' academic achievement.

The positive correlation between mathematics self-concept and mathematics achievement is also revealed in the research carried out by [19] and [20]. The students' mathematics achievement can be measured in various ways. The mathematical thinking test score that is measured in this research is one of the students' mathematics achievements so that it can be said that this research has confirmed the correlation between the students' mathematics self-concept and their mathematical thinking skill. This relationship implies that if you want to improve students' mathematical thinking skills, the teacher can improve students' mathematics self-concept first. Because it will be easier to touch the cognitive domain of students if the students' affective is understood by the teacher. Therefore, at the beginning of the lesson, the teacher should first measure the mathematics self-concept that is owned by the students, so that the delivery of material can be given using an appropriate model.

4. Conclusion

Taking into account the result of data analysis elaborated on the previous section, it can be concluded that the students with positive mathematics self-concept have higher mathematical thinking skill contrasted with those who have negative mathematics self-concept. The result of U Mann-Whitney test demonstrates that the mathematical thinking skill of the students who have positive mathematics self-concept is significantly different from those who have negative mathematics self-concept. The mathematics self-concept variable gives positive and significant influence on the mathematical thinking skill. This research provides a recent reference about the interconnectedness of students' mathematical thinking skill and their mathematics self-concept. It is highly recommended for further researchers to investigate the method to develop a learning model for the students with diverse mathematical thinking skill and mathematics self-concept.

5. References

[1] Ekawati, E. 2011 *Peran, Fungsi, Tujuan, dan Karakteristik Matematika Sekolah* [Online] Available at http://p4tkmatematika.org/2011/10/peran-fungsi-tujuan-dan-karakteristik-matematika-sekolah/

[2] Breen S. & O'Shea A. 2010 Mathematical Thinking and Task Design *Iris Mat. Soc. Buletin* **66**

[3] Mason,J. & Johnston-Wilder,S. 2004 *Designing and Using Mathematical Tasks* (UK: Tarquin)

[4] Mason,J., Burton, L. & Stacey,K 2010 *Thinking Mathematically second edition* (London: Pearson Education Limited)

[5] Stacey, K. 2006 *What is Mathematical Thinking and Why is it Important?* [Online] Available at https://www.researchgate.net/publication/254408829

[6] PISA 2012 *Mathematics Self Belief and Participation in Mathematics Related Activities* [Online] Available at http://www.oecd.org/pisa/keyfindings/PISA2012-Vol3-Chap4.pdf

[7] Githua,B.N & Mwangi,J.G. 2003 Students' Mathematics Self Concept and Motivation to Learn Mathematics: relationship and Gender Differences among Kenya's Secondary – School students in Nairobi and Rift Valley Provinces. *International Journal of Educational Development* **23** 487–499.

[8] Adegoke, B. A. 2015 The Big-Fish-Little-Pond Effect on Mathematics Self Concept of Junior School Student in Academically Selective and Non-Selective Schools. *Journal of Studies in Education* **5** (2)

IOP Conf. Series: Journal of Physics: Conf. Series **1108** (2018) 012104 doi:10.1088/1742-6596/1108/1/012104

[9] Marsh, H. W., (2010) The Structure of Academic Self Concept: The Marsh/ Shavelson Model *Journal of Psychology* **82** (4)
[10] Guberman, S. 2015 On Gestalt Theory Principles *Gestalt Theory* **27**(1) 25-44
[11] Hidayati, T.N. 2011 Implementasi Teori Belajar Gestalt pada Proses Pembelajaran *Jurnal Falasifa* **2** (1)
[12] Watson, A. & Mason, J. 1998 *Questions and Prompts for Mathematical Thinking* (Derby: Association of Teachers of Mathematics)
[13] Isiksal, M., et.al. 2009 Mathematics Anxiety and Mathematical Self-Concept: Consideration in Preparing Elementary-School Teachers *Social Behavior an Personality* **37** (5) 631-644
[14] Nurasiah 2016 Urgensi Neuroscience dalam Pendidikan (sebagai Langkah Inovasi Pembelajaran) *Al-Tadzkiyyah: Jurnal Pendidikan Islam*, **7**(1) 72.
[15] Jensen, E. 2008 *Brain-Based Learning* (Yogyakarta: Pustaka Pelajar)
[16] Wilkins, J. L. M., Zembylas, M., & Travers, K. J. 2002 Investigating correlates of mathematics and science literacy in the final year of secondary school. In D. F. Robataille & A. E. Beaton (Eds.), *Secondary Analysis of the TIMSS Results: A Synthesis of Current Research* (Boston, MA: Kluwer Academic Publishers)
[17] Marsh, H. W., Byrne, B. M., & Yeung, A. S. 1999 Causal ordering of academic self-concept and achievement: Reanalysis of a pioneering study and revised recommendations *Educational Psychologist* **34** (3) 155-167
[18] Marsh, H. W., & Martin, A. J. 2011 Academic self-concept and academic achievement: Relations and causal ordering *British Journal of Educational Psychology 81*(1) 59-77
[19] Thien, L. M., & Ong, M. Y. 2015 Malaysian and Singaporean students' affective characteristics and Mathematics performance: evidence from PISA 2012. *SpringerPlus* **4**(1) 1-14
[20] Bong, M., & Skaalvik, E. M. 2003 Academic self-concept and self-efficacy: How different are they really? *Educational psychology review* **15**(1) 1-40

Acknowledgments
The authors would like to express their gratitude to *Kemenristekdikti* (Ministry of research, technology, and higher education) of Indonesia for providing them Research Grant for this project.

Isolation and Characterization of Potassium-Solubilizing Bacteria from Paddy Rhizosphere (*Oryza sativa* L.)

R Fatharani[1,2], Y S Rahayu[1,3]

[1]Department of Biology, Universitas Negeri Surabaya, Indonesia.
[2]radhinfatharani@mhs.unesa.ac.id, [3]yunirahayu@unesa.ac.id

Abstract. Potassium is one of macronutrients required by plants. The concentrations of soluble Potassium in the soil are usually very low, more than 90% of Potassium in the soil are (K)-bearing minerals form, one of them is Feldspar which quite abundant found in Indonesia. Rhizobacteria have been found to solubilize (K)-bearing mineral into a form that plants can access. This study aims to isolate and characterize Rhizobacteria from paddy rhizosphere that potentially dissolve mineral feldspar as potassium source. This study was conducted on February-June 2018 in Microbiology Laboratory, Department of Biology, Universitas Negeri Surabaya. Seven bacteria isolates capable of solubilizing potassium were LJK 1, LJK 2, LJK 4, LBK 4, LBK 5, PSUK 4, and PSIK 6. Based on the potassium-solubilizing index, seventh isolates were capable of solubilizing feldspar in Alexandrov solid medium, the best isolate with the highest score was LJK 2 (5,25 cm in diameter), contrary with the lowest score was LBK 4 (2,37 cm). In amended Alexandrov broth, isolate suspension of LJK 2 caused maximum solubilization (6,846 mg/L), while the minimum solubilization was PSIK 6 (6,724 mg/L). All these isolates were characterized on the basis of morphological and physiological-biochemical test up to 39 characteristics.

1. Introduction

Potassium (K) is an essential nutrient that plays important roles for plant growth and development, such as activating enzyms, maintaining turgor, transporting of nutrient, and protecting plant from disease and insects. As one of the agricultural commodities, paddy needs Potassium in the generative phase for the developing and filling grain. In such, the availability of Potassium become one of the factors that affect the productivity of paddy. Unfortunately, more than 90 to 98 per cent of Potassium found in the soil are K-bearing minerals which are unavailable for plant uptake. Indonesia has some K-bearing minerals of low solubility such as feldspar and mica [1]. Soil microorganisms have been reported play a key role in ion cycling and soil fertility [2]. Many studies have known that microbes decompose silicate minerals such as feldspar and mica, transformed insoluble K into available K so that can be directly absorbed by plants [3]. In light of above facts, the objective of this research was to isolate and characterize of Potassium-solubilizing bacteria (KSB) from paddy rhizosphere and its ability to solubilize feldspar as the sole source of insoluble K in Alexandrov medium.

2. Method

2.1. Isolation of Paddy Rhizosphere Bacteria

Sample of rhizosphere was taken from four farming areas, Bangkingan, Jeruk, Sumber Makmur and Siwalan Makmur. Ten grams of rhizosphere were inoculated onto 90 ml *Alexandrov broth* for 4 days. The serial dilutions up to 10^{-7} consisted of 1 ml of sample and 9 ml of 0,85% NaCl solution. Each

IOP Conf. Series: Journal of Physics: Conf. Series **1108** (2018) 012105 doi:10.1088/1742-6596/1108/1/012105

dilution was continuously shaken for two minutes by vortex, and was inoculated on *Alexandrov* agar plate. The isolates were then incubated at 30^{0}C for 4 days.

2.2. Characterization of Potassium Solubilizing Bacterial Isolates
The isolates were cultured on Alexandrov's medium. Cell and colony morphology and physiological-biochemical characteristics were examined. Motility, catalase test, *Ziehl-Nelseen* test, *Methyl Red* test, *Voges-proskauer* test, urease test, nitrate reduction test, indole production test, ornithine utilization test, H_2S production test, citratase utilization, growth in different pH (3, 7, and 10), growth in different temperature (25^{0}C, 30^{0}C, and 45^{0}C), and starch hydrolysis were studied as physiological-biochemical characters.

2.3. Potassium Solubilizing Activity in Alexandrov Agar
The selected bacterial isolates from paddy rhizosphere were subjected to a Potassium solubilizing test in *Alexandrov agar*, incubated at 30^{0}C for 4 days. Detection of potassium solubilization by different rhizobacterial isolates based upon the ability of solubilization zone formation. The medium contains of glucose 5,0 g; agar 20 g; $MgSO_4.7H_2O$ 0,5 g; $CaCO_3$ 0,1 g; $FeCl_3$ 0,006 g; Ca_3PO_4 2,0 g; and Feldspar 3 g diluted by 1 L deionized water [4].

2.4. Potassium Solubilizing Activity in Alexandrov Broth
The solubility of Potassium was investigated in 50 ml *Alexandrov* broth, all those isolates incubated at 30^{0}C for 7 days. The growth suspension was centrifuged at 7,000 rpm for 10 minutes in the microcentrifuge to separate the supernatant from the cell growth and insoluble Potassium. One ml of the supernatant was taken and placed into 50 ml volumetric flask, added with distilled water and mixed thoroughly. The solution was fed to *Atomic Absorption Spectrometer* to determine (AAS) K content [5].

2.5. pH Measurement in Alexandrov Broth
pH was measured before and after incubation in *Alexandrov* broth. Before its sterilization, the pH of *Alexandrov broth* was measured by pH roll (6,6-7,0). After seven days incubation, each pH culture measured, then compared against pH before incubation periods.

3. Results and discussion

Twenty two KSB isolates were isolated from rhizospheric soils of paddy (*Oryza sativa* L.). Seven isolates were capable of solubilizing Feldspar in Alexandrov agar presented in Table 1. These seventh isolates colonies showed exhibiting zone of clearance indicating Potassium solubilization. The value of clear zone from each isolates examined by using Premono's selection ratio, noted as Solubilizing Index (SI) [6]. Solubilization of K from mineral Feldspar by bacterial strains also examined in Alexandrov broth using AAS (Table 1).

Table 1. Potassium solubilization values of selected bacterial isolates in Alexandrov agar

Isolates	Solubilizing Index (SI)	Solubilization AAS
LJK 1	3,03	6,803
LJK 2	5,25	6,846
LJK 4	3,00	6,784
LBK 4	2,37	6,759
LBK 5	2,63	6,770
PSUK 4	2,67	6,744
PSIK 6	3,11	6,724

The morphological colony (shape, margin, elevation, surface, pigmentation, and optic) and cell morphology of seventh bacteria are presented in Table 2.

IOP Conf. Series: Journal of Physics: Conf. Series **1108** (2018) 012105 doi:10.1088/1742-6596/1108/1/012105

Table 2. Morphological colony, cell shape, gram reaction and spore formation characteristics of Potassium Solubilizing Bacteria

Isolates	Colony characters	Cell shape	Gram reaction	Spore formation
LJK 1	Irreguler, undulate, raised, smooth, white, opaque	Coccus	-ve	+[a]
LJK 2	Irreguler, undulate, flat, smooth, white, translucent	Bacil	+ve	-[b]
LJK 4	Irreguler, undulate, raised, smooth, white, translucent	Bacil	-ve	+
LBK 4	Circular, entire, raised, smooth, creamy, opaque	Coccus	+ve	+
LBK 5	Irreguler, entire, flat, smooth, creamy, opaque	Coccus	+ve	-
PSUK 4	Irreguler, undulate, raised, smooth, white, opaque	Coccus	+ve	+
PSIK 6	Irreguler, entire, flat, smooth, white, opaque	Coccus	+ve	+

[a] positive (+)
[b] negative (-)

Physiological-biochemical characteristics test include motility, catalase test, Ziehl-Nelseen test, Methyl Red test, Voges-proskauer test, urease test, nitrate reduction test, indole production test, ornithine utilization test, H_2S production test, citratase utilization, growth in different pH (3, 7, and 10), growth in different temperature (25^{0}C, 30^{0}C, and 45^{0}C), and starch hydrolysis (Table 3).

Table 3. Physiological-biochemical characteristics of Potassium Solubilizing Bacteria

Characters	LJK 1	LJK 2	LJK 4	LBK 4	LBK 5	PSUK 4	PSIK 6
Motility	+	+	+	+	+	+	+
Catalase test	+	+	+	+	+	+	+
Ziehl-Nelseen	+	+	+	+	+	+	+
MR	+	+	+	+	+	+	+
VP	+	-	+	-	-	-	-
Urease	-	-	-	-	-	-	-
Nitrate reduction	+	+	+	+	+	+	-
Indole production	-	-	-	+	+	-	-
Ornithine utilization	+	+	+	+	+	+	+
H_2S production	+	+	+	-	-	-	-

Characters	LJK 1	LJK 2	LJK 4	LBK 4	LBK 5	PSUK 4	PSIK 6
Citratase utilization	+	-	+	+	+	+	+
Growth in pH 3	+	-	+	+	+	+	+
Growth in pH 7	+	+	+	+	+	+	+
Growth in pH 10	+	+	+	+	+	-	-
Growth in 25^{0}C	+	+	+	+	+	+	+
Growth in 30^{0}C	+	+	+	+	+	+	+
Growth in 45^{0}C	+	+	+	+	+	+	+
Glucose hydrolysis	+	+	+	+	+	+	+
fructose hydrolysis	+	+	+	+	+	+	+
Lactose hydrolysis	+	-	+	+	+	-	-
Sucrose hydrolysis	+	+	+	+	+	+	+
Mannitol hydrolysis	+	+	+	+	+	+	+
Amilum hydrolysis	+	-	+	+	-	+	+

Potassium solubilization by rhizobacterial isolates was studied on Alexandrov agar and Alexandrov broth medium. Solubilizing index was used to measure the K-solubilization zone ability of the KSB on Alexandrov agar. Seventh of KSB were found to be capable of K-solubilization, and the Solubilization Index ranged from 2,37 to 5,25 cm in diameter. Strain LJK 2 showed the highest ability to solubilize K (5,25 cm in diameter) followed by PSIK 6, LJK 1, LJK 4, PSUK 4, LBK 5, and LBK 4. Solubilization of Potassium also conducted on Alexandrov broth then fed to AAS, this aims to determine the concentration of ion K from solubilization activities by KSB after 7 days incubation period. The results exhibited that isolate LJK 2 was consistent with the previous studies on Alexandrov agar as the best isolate compared to other isolates, its dissolved K of 6,846 mg/L at 7 DAI.

There were least differences in the solubilization activity of each isolates, it is hypothesized that the release of K may be due to the various organic acids that produced by KSB, besides, the production of organic acids by each isolates have different concentration. Organic matter while solubilize minerals produced by KSB has known as oxalic acid and citric acid [4], ferulic acid and coumaric acid [7], formic acid, malic acid and acetate acid [8], tartric acid [9]. These organic acids produced by the KSB might enhanced the dissolution of K-bearing minerals by supplying protons, destabilizing surface of K-bearing and complexing Ca^{2+}, Fe^{2+}, and Al^{3+} ions [10,11,12]. In this study, feldspar ($KalSi_3O_8$) was the sole source of K-bearing, this reacts with organic acids promotes cation exchange reactions between H^+ and K^+ ions. Furthermore, K released from feldspar into available K for plant uptakes, besides, these solubilizing activity also forms a secondary mineral as kaolinite ($H_4Al_2Si_2O_9$) [1,13].

Characterization of the KSB were examined by 39 characters: morphological colony (shape, margin, elevation, surface, pigmentation, and optic) and cell morphology. Physiological-biochemical characteristics test include motility, catalase test, Ziehl-Nelseen test, Methyl Red test, Voges-proskauer test, urease test, nitrate reduction test, indole production test, ornithine utilization test, H_2S production test, citratase utilization, growth in different pH (3, 7, and 10), growth in different temperature (25^{0}C, 30^{0}C, and 45^{0}C), and starch hydrolysis. Characterization performed in this study was also conducted in previous studies to characterize Potassium solubilizing bacteria [4,14,15].

4. Conclusion

Isolation and screening processes have found seven isolates from rhizospheric soils of paddy (Oryza sativa L.). Seventh isolates were capable of solubilizing Feldspar in Alexandrov agar and Alexandrov broth. The best isolate with the highest score in Alexandrov agar was LJK 2 (5,25 cm diameter), this isolate also caused maximum solubilization in Alexandrov broth (6,846 mg/L). All these isolates were characterized on the basis of morphological and physiological-biochemical test up to 39 characteristics.

Seventh KSB in this study are able to dissolve Potassium from K-bearing mineral have both economic and environmental advantages, so that these isolates potentially to be applied as an alternative bio-fertilizer agent of Potassium, generally the bio-fertilizer can be applied to the seed, soil, or during composting. Microencapsulation engineering and fermentation engineering can be done to ensure an increase in the microbial population and reduce the potential for microbial contamination of other undesirable during storage and transportation [16].

References

[1] Pratama D and Anas S 2016 Ability of Potassium-solubilissing microbes to solubilise Feldspar and their effects on sorghum growth *Malaysian Journal of Soil Science* **20** 163-175.
[2] Bin Lian, Bin Wang, Mu Pan, Congqiang Liu and H. Henry Teng 2010 Microbial release of Potassium from K-bearing minerals by thermophilic fungus Aspergilus fumigatus *Geochimica et Cosmochimica Acta* **72** 87-89.
[3] Zhang C and Fanyu K 2014 Isolation and identification of Potassium-solubilizing bacteria from tobacco rhizospheric soil and their effect on tobacco plants *Applied Soil Ecology* **82** 18-25
[4] Prajapati K B and Modi H A 2012 Isolation and characterization of Potassium solubilizing bacteria from ceramic industry soil *CIB Tech Journal of Microbiology* **1** 8-14.
[5] Parmar P and Sindhu S S 2013 Potassium solubilization by rhizosphere bacteria: influence of nutritional and environmental conditions *Journal of Microbiology Research* **3** 25-31.
[6] Premono E M, Moawad A M, and Vlek P L G 1996 Effect of Phosphate solubilizing Pseudomonas putida on the growth of maize and its survival in the rhizosphere *Indonesian J. Crop Sci.* **11** 13-23.
[7] Setiawati T C and Laily M 2016 Solubilization of Potassium containing mineral by microorganisms from sugarcane rhizosphere *Agriculture and Agricultural Sci.Procedia* **9** 108-117.
[8] Shanware A S, Surekha A K, and Minal M T 2014 Potassium solublisers: occurrence, mechanism and their role as competent biofertilizers *Int.J.Curr.Microbiol.App.Sci* **3** 622-629.
[9] Hu X F, Chen J, and Guo J F 2006 Two Phosphate and Potassium solubilizing bacterial isolated from Tiannu Mountain, Zhijiang, China *World J. Microbiology and Biotechnology* **22** 983-990.
[10] Stillings L L, Drever S L, Brantly Y S, and Oxburgh R 1996 Rates of feldspar dissolution at pH 3-7 with 0-8 mM oxalic acid *Chemistry Geology* **132** 79-90.
[11] Meena V S, Bihari R M, Jai P V, Abhinav A, Ashok K, Kangmin K and Vivek K B 2015 Potassium solubilizing rhizobacteria (KSR): isolation, identification, and K-release dynamics from waste mica *Ecological Engineering* **81** 340-347.
[12] Styriakova I, Styriak I, Galko I, Hradil D, and Bezdicka P 2003 The release of iron-bearing minerals and dissolution of feldspar by heterotrophic bacteria of Bacillus species *Ceramics Silicaty* **47** 20-26.
[13] Ismangil and Eko H 2005 Degradasi mineral batuan oleh asam-asam organik *Jurnal Ilmu Tanah dan Lingkungan* **5** 1-17.
[14] Ahmad M S and Zargar M Y 2017 Characterization of Potassium solubilizing bacteria (KSB) in rhizospheric soils of apple (*Malus domestica Borkh.*) in temperate Kashmir *Journal of Applied Life Sci.International* **15** 1-7.
[15] Archana D S 2007 Studies on potash solubilizing bacteria *Ph.D theses* UAS Dharwad India.
[16] Gunalan 2006 Penggunaan mikroba bermanfaat pada bioteknologi tanah berwawasan lingkungan vol 32 (Universitas Sriwijaya: Majalah Sriwijaya) p 2.

IOP Conf. Series: Journal of Physics: Conf. Series **1108** (2018) 012106 doi:10.1088/1742-6596/1108/1/012106

The Influence of Sub-Microscopic Media Animation on Students' Critical Thinking Skills Based on Gender

R A Mashami[1] and G Gunawan[2]

[1]Department of Chemistry Education, IKIP Mataram, Indonesia
[2]Department of Physics Education, Universitas Mataram, Indonesia

Abstract. This study aimed to examine the influence of sub-microscopic media animation on students' critical thinking skills in terms of gender. The study used pre-test and post-test control group design. The sample consisted of experimental group and control group, each of which consisted of 27 students (12 males and 15 females). The sampling technique used was purposive sampling. Data on critical thinking skills was taken using a description test on buffer concepts. Further data were analysed using a t-test at 95% significant level. The results showed that sub-microscopic media animation influence students' critical thinking skills, both male and female. In the experimental group, students' critical thinking skills are higher than male students, which means that sub-microscopic animation media influence female students' critical thinking skills better than male students.

1. Introduction

Technology has been growing so rapidly in just a few years. For education, technological progress has a very significant influence, such as changes in teaching and learning systems, ease of access to knowledge, bridging communication between teachers and students, and maximizing learning evaluation [1,2]. Especially in science lessons, technological sophistication is very helpful in the development of learning media with the existence of various computer applications [3-6]. The use of computer technology helped students increase their curiosity and open attitude [7], as well as students' creativity in the figural and verbal aspects [8].

Learning media acts as a tool that helps teachers convey information in this subject matter to students. Information in the form of material submitted without media assistance will make it difficult for students to understand it, including in studying chemistry. Among the three levels of representation (macroscopic, sub-microscopic, and symbolic), sub-microscopic levels are more difficult to understand because sub-microscopic representations explain the structure and process at the particle (atom/molecular) level of the observed phenomenon [9-11]. The particle level cannot be observed directly or with the help of the tool. The sub-microscopic aspect can be studied with the help of instructional media featuring simulations of particle animation to make it look real and easier to understand. The learning media is called sub-microscopic media animation. The use of animation contributes to students' conceptual understanding [12] and representational competences [13]. Chemistry teachers must make much effort to create an ideal environment for teaching and learning by incorporating computer-assisted methods [14].

One of the important tasks of teachers in the classroom is not only to provide knowledge to the students but to teach them how to learn and know about how rapidly proliferating information [15]. Somehow, we are surprised at how technology, communication, and information are growing so rapidly.

Information in a place that is very far away once we can know at once immediately. It is inconceivable that such an age will be found by future students, which will surely be more sophisticated than current conditions. Such a drastic change certainly has a negative impact. Therefore, we must prepare our students from now on. Students must be able to think critically in response to information and knowledge that is increasingly unlimited. Students should be able to sort the correct or wrong information and wisely criticize the progress of the times in the future. Many critical thinking skills are one of the most important soft skills because critical thinking skills can be used in a variety of disciplines and aspects of life, including when students enter the workforce [16,17].

The world of formal education that becomes a place to grow children until adolescents even enter adulthood need to place critical thinking skills as one direction in learning achievement. The student must be trained to develop his critical thinking skills while mastering the material he or she studies. Critical thinking is the appropriate use of reflective skepticism, and that is necessarily linked with a specific area of knowledge [15]. However, and more significant for the current task, critical thinking is an element in many levels and what we expect students to achieve at the end.

Critical thinking skills do not just belong to a person, to extend that critical thinking is exercised and drills [18]. Studying the Media of Sub-microscopic Animation means training students to understand the abstract concept and to process the concept so that it becomes more real and reliable. It is expected to train critical thinking skills, because critical thinking is trying to understand a subject, thinking about it, appreciating it, understanding the strengths and limitations of it and then developing a point of view on the subject.

Sex differences in science learning are interesting to research. Based on research on mathematics learning, students' thinking ability and academic achievement were different from female [19,20]. Differences in thinking ability of male and female students were also found in physics learning [21]. The conceptual understanding of the physics reviewed in the gender aspects shows the differences between male and female [22]. There was a significant difference between creative thinking skills, particularly fluency and elaboration, between male and female students in biology learning [23]. The gender differences in chemistry learning have not been much discussed. Therefore, this study aimed to analyze the influence of sub-microscopic media animation on students' critical thinking skills in terms of gender.

2. Method

This quasi-experiment research used pre-test and post-test control group design. The study population was all students of class XI IPA in SMAN 7 Mataram which amounts to four classes. The sampling technique used was purposive sampling with the aim of obtaining samples having the same initial capability. The number of male and female samples was also adjusted so as not to differ greatly thus affecting the data. The sample consisted of experimental group and control group, each group consisted of 27 students (12 males and 15 females). The experimental group students were treated in the form of learning using sub-microscopic media animation while the control group was not given that treatment.

At the end of the lesson, the two sample groups were given a final test to obtain students' critical thinking skills data. The instrument used was a description test on buffer concepts. The description test measured seven indicators of critical thinking skills, namely (1) focus on a question, (2) answer questions, (3) judge observation reports, (4) induce or judge induced results, (5) deduce or judge deduced results, (6) make and judge value judgments, and (7) attribute unstated assumptions . Score critical thinking skills between 0 and 20. The data were analyzed statistically with the help of SPSS 23 program, including normality test, homogeneity test, and hypothesis test. In addition, the normalized gain score (N-gain) data was also calculated so that the data acquisition improves critical thinking skills on each indicator.

3. Result and Discussion

The learning was done on the concepts of buffer solution as many as six meetings. The experimental group studied the concept of buffer solution using sub-microscopic media animation, while the control group studied the concept of buffer solution without using that media. Before the lesson begins, a pre-test was performed to determine students' initial critical thinking skills. The result of the initial data normality test with Shapiro-Wilk analysis gets the significant value of each 0.200 for the experiment group and control group which means that data was normally distributed. The homogeneity test using Levene's Test obtained a value of 0.787 indicating no difference in variance between the control group and the experimental group. Furthermore, independent sample t-test yields a significant value of 0.556. Based on the data analysis it was concluded that the initial critical thinking skill of the experimental group and control group was not different before being given treatment.

Furthermore, post-test was given to the students to obtain critical thinking skill data after learning was complete. By comparing experimental and control group data it can be seen how the influence of sub-microscopic media animation on students' critical thinking skills. Data critical thinking skills were grouped into two, male student data and female student data (Table 1). Furthermore, the data were analyzed by gender at 95% confidence level.

Table 1. Data of Students' Critical Thinking Skills

Criteria	Control Group		Experimental Group	
	Male	Female	Male	Female
Amount	12	15	12	15
Lowest Score	7.83	7.42	11.25	11.67
Highest Score	14.92	17.83	16.25	19.33
Average Score	11.69	12.94	13.44	15.37

Result of post-test data analysis critical thinking skill of male student, that was: 1) normality test with Shapiro-Wilk test obtained value 0.667 for experiment group and 0.387 for control group which mean that data was normal distributed, 2) Homogeneity test using Levene's Test obtained value equal to 0.551 indicating no difference of variance between control group and experiment group, and 3) hypothesis test with an independent sample t-test yielded significant value equal to 0.036 which mean there was a significant difference between the critical thinking skill of male group of experiment group and control group. The difference between the experimental group and the control group was only in the treatment so that it can be concluded that the difference in skills was caused by the use of sub-microscopic media animation in the learning process.

The influence of sub-microscopic media animation can also be seen in increasing the score in each indicator of critical thinking skills. Figure 1 shows the critical thinking ability of male who study the higher sub-microscopic media animation in all indicators. Indicator with the highest average score in both experiment and control group was indicator 6, make and judge value judgments. The biggest difference in critical thinking skills is seen in indicator 3, judge observation reports. While the smallest difference appeared in indicator 2, answer questions.

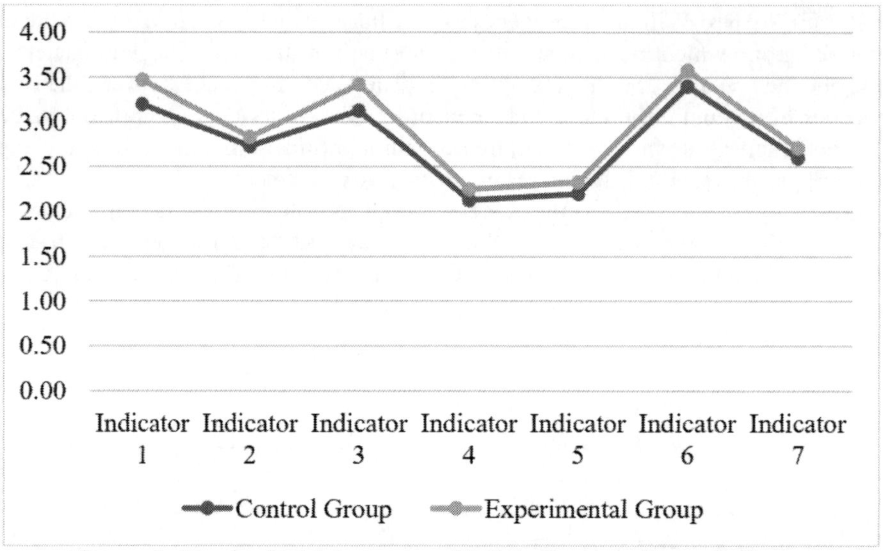

Figure 1. Graph of Male's Average N-Gain Score on Every Indicator

The same data analysis was also performed on female student data with the following results: 1) normality test with Shapiro-Wilk analysis obtained value of 0.150 (experimental group) and 0.278 (control group) which means the data of the two groups are normally distributed, 2) homogeneity test using Levene'sTest obtained a value of 0.172 indicating the control group and experimental group was homogenous, 3) hypothesis test with independent sample t-test yields significant value of 0.032 which means critical thinking skills of female students experimental group was different from the critical thinking skills of female students control group.

When analyzed based on the average increase in each indicator, the students' critical thinking skills were higher than the control group students (Figure 2). Indicators who score the highest average of both the experimental group and the control group are the indicator 6, make and judge value judgments, as well as the male students. The indicator with the largest N-gain difference was indicator 7, ie attribute unstated assumptions. While the smallest difference was in the indicator 5, deduce or judge deduced results.

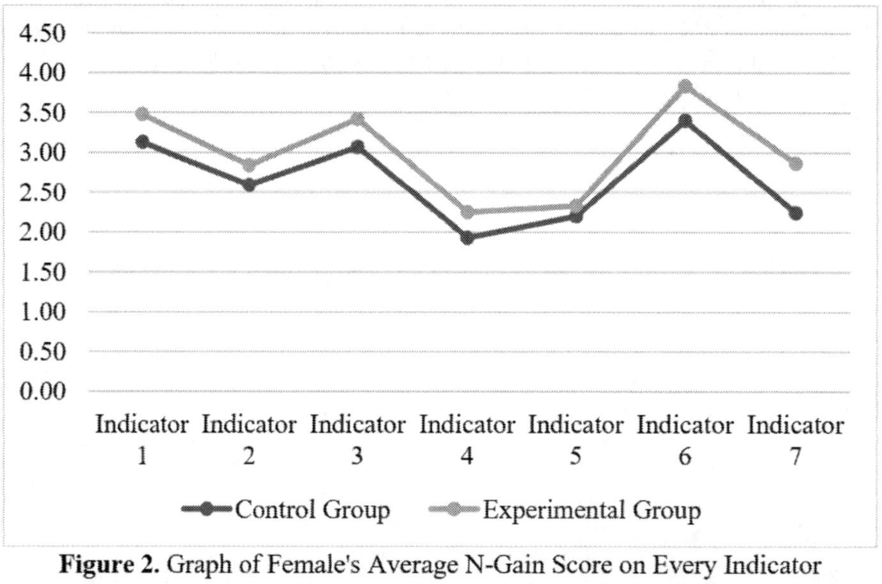

Figure 2. Graph of Female's Average N-Gain Score on Every Indicator

The result of the data analysis showed that the critical thinking skills of experiment group students are influenced by sub-microscopic media animation. Buffer solutions include abstract concepts with concrete examples. Although the phenomenon of the buffer solution can be observed visually, understanding the concept is not as simple as looking at an example. An animation is needed further to explain the molecular level of the phenomenon. Animation as one form of visualization is an important element in teaching, understanding, and making science ideas. Visualizations such as languages and other communication tools are effective in influencing one's cognitive thinking and capacity [24]. If the sub-microscopic aspect of the concept of chemistry can be well understood it will greatly assist the students in understanding the concept of chemistry as a whole [10,25]. This applies to both male and female students The knowledge gained through visualization can enhance deep understanding, hypothesize, reason, and solve problems [26]. One of the advantages of educational technology is that it can be readily used to support the development of critical thinking [27].

To strengthen the influence of sub-microscopic media animation on critical thinking skills, an analysis of male and female students data in the control group was performed. The results of the analysis of the data are: 1) the normality test with the Shapiro-Wilk analysis get the value of 0.667 (experimental group) and 0.150 (control group) which means the data of the two groups are normally distributed, 2) the homogeneity test using Levene'sTest obtained value of 0.044 which shows the variance is not homogeneous, 3) hypothesis test with independent sample t-test yields significant value of 0.019 which means there are differences in the critical thinking skills of female and male students.

Broadly speaking, critical thinking skills of female students are higher than the students' critical thinking skills. As shown in Figure 3, female students excel at all critical thinking indicators. It found on a study that female is better than male in critical thinking [28]. The effectiveness of animation use depends on material content and gender. Female students have a better ability in learning animation media than male. This is because animation can help female students manage the cognitive input they receive rather than static media [29].

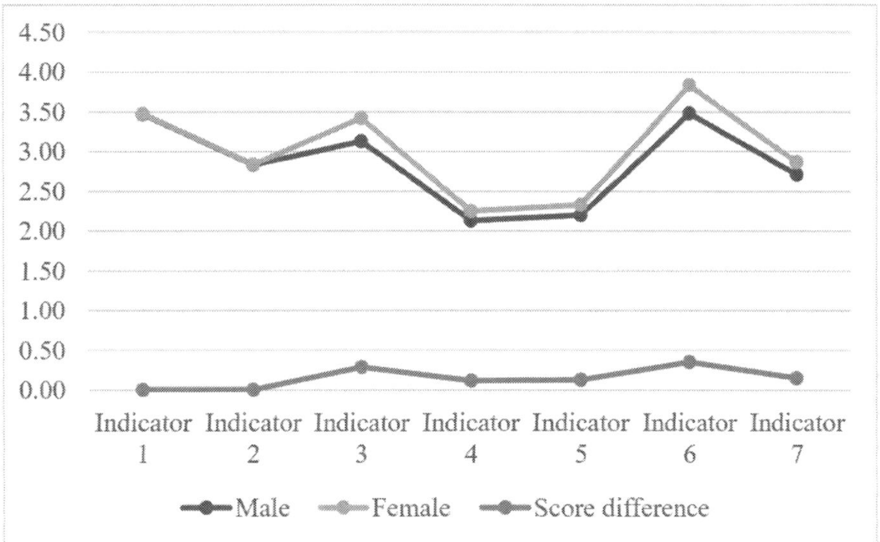

Figure 3. Comparison of Average Score By Male and Female at Experimental Group

Figure 3 also shows the difference in the scores of male and female in each indicator. The ability to focus on questions (indicator 1) and answer the question (indicator 2) of the male was almost the same as female students. The biggest difference in critical thinking lies in the ability to make and judge value judgments. This was consistent with the results of the studies, male students dare to make decisions even at risk without much consideration [30,31]. Female students were just the opposite, more careful in making decisions and considering everything more when thinking. The causes of differences in some aspects of ability between male and female include hormonal, brain structure, cognitive processes, and social processes that are considered to have a significant effect on male and male differences female [27].

Although the female has a higher average value of results than male, female tend to lack confidence in judging their own abilities. The level of self-confidence is even higher in boys, which is indicated by the perception of boys who are higher than female [31,32]. Other studies have different conclusions, the students' critical thinking skills are better than the critical thinking skills of female students [33,34]. Other research results show that the use of animation can remove the boundary between male and female abilities [35].

4. Conclusion

The conclusion obtained from this research was sub-microscopic media animation influence students' critical thinking skills, both male and female. However, female's critical thinking skills were better than male students, which means that sub-microscopic media animation had more influence on female students' critical thinking skills than male students, especially on the ability to make and judge value judgements. The ability of female students in focusing on a question and answer questions was similar to male students.

5. References

[1] Stosic L 2015 the importance of educational technology in teaching *In. J.of Cognitive Research on Sci. Engineering, and Ed.* **3**(1) pp. 159-173

[2] Widowati A 2016 'new pedagogies' of experience-based learning form in science learning *Proc. of 3rd In. Conf. on Research, Implementation, and Ed. Of Math. And Sc.* pp: 25-31

[3] Mashami R A and Raehanah R 2016 pengembangan multimedia interaktif untuk meningkatkan keterampilan berpikir kreatif mahasiswa pada materi larutan penyangga *J. Kependidikan* **15**(3) pp: 265-269

[4] Lu'mu 2017 learning media of applications design based android mobile smartphone *In. J. of Applied Engineering Research* **12**(17) pp. 6576-85

[5] Herga N R 2016 virtual laboratory in the role of dynamic visualization for better understanding of chemistry in primary school *Eurasia J. of Math., Sci. & Tech. Ed.* **12**(3) pp. 593-608

[6] Astuti F K, Cahyono E, Supartono, Van N C, and Duong N T 2018 effectiveness of elements periodic table interactive multimedia in Nguyen Tat Thanh high school *In. J. of IndonesianEd. and Teaching* **2**(1) pp. 1-10

[7] Gunawan G and Liliasari L 2012 model virtual laboratory fisika modern untuk meningkatkan disposisi berpikir kritis calon guru *J. Cakrawala Pendidikan* **31**(2) pp. 185-199

[8] Gunawan G, Harjono A, Sahidu H, and Nisrina N 2018 improving students' creativity using cooperative learning with virtual media on static fluida concept *J. of Physics: Conf. Series* **1006** (1) 012016

[9] Chittleborough G and Treagust D F 2007 the modelling ability of non-major chemistry students and their understanding of the sub-microscopic level *Chemistry Education Research and Practice* **8**(3) pp. 274-292

[10] Chandrasegaran, Treagust, and Mocerino 2007 Enhancing students' use of multiple levels of representation to describe and explain chemical reactions *School Science Review* **88**(325) p. 115

[11] Gilbert J K and Treagust D F 2009 Multiple Representation in Chemical Education *Models and Modeling in Science Education* **4**

[12] Falvo D 2008 animations and simulations for teaching and learning molecular chemistry *In. J. of Technol. In Teaching and Learning* **4**(1) pp. 68-77

[13] Mashami R A, Andayani Y, and Gunawan G 2014 pengaruh media animasi submikroskopik terhadap kemampuan representasi siswa *J. Ilmiah Pend. Kimia "Hydrogen"* **2**(1) pp. 149-152

[14] Bharathy J B 2015 importance of computer-assisted teaching & learning methods for chemistry *Sci. J. of Ed.* **3**(4-1) pp. 11-16

[15] Santos L F 2017 the role of critical thinking in science education *J. of Ed. And Practice* **8**(20) pp: 159-173

[16] Karakoc M 2016 the significance if critical thinking ability in term of education *In. J. of Humanities and Social Science* **6**(7) pp. 81-84

[17] Murawski L M 2014 critical thinking in the classroom and beyond *J. of Learning in Higher Education* **10**(1) pp. 25-30

[18] Thomas T 2011 developing first-year students' critical thinking skills *Asian Society Science* **7**(4) pp: 26-35

[19] Pargulski J R and Reynolds M R 2017 sex differences in achievement: distributions matter *Personality and Individual Differences* **104** pp. 272-278

[20] Rasyid M A, Budiarto M T, and Lukito A 2018 junior high school students' reflective thinking on fraction problem solving: in case of gender differences *J. of Phy.: Conf. Series* **947**

[21] Gunawan G, Suranti N M Y, Nisrina N, Ekasari R R, and Herayanti L 2017 investigating students creativity based on gender by applying virtual laboratory to physics instruction *Advances in Social Science, Education and Humanities Research* **158**(1) pp. 303-310

[22] Gunawan G, Nisrina N, Suranti N M Y, Ekasari R R, and Herayanti L 2018 the effect of virtual labs toward students' conceptual understanding of physics based on gender *Advances in Social Science, Education and Humanities Research* **173**(1) pp. 128-131

[23] Hilman M F, Ari W, Diana R 2018 analysis of creative thinking ability and process of creative thinking ability development of junior high school students on biology learning *Assimilation: Indo. J. of Biology Ed.* **1**(1) pp. 21-28

[24] Barbara T 2005 prolegomenon to scientific visualizations *In JK. Gilbert (Ed.), Visualization in Science Education* **1** pp. 29-42

[25] Kozma R and Russell J 2005 students becoming chemists: developing representational competence *In JK. Gilbert (Ed.) Visualization in Science Education* **7** pp. 121-145

[26] Wang and Jacobson. 2011 guest editorial - knowledge visualization for learning and knowledge management *Educational Technology & Society* **14**(3) pp. 1–3

[27] Snowman and McCown 2012 Psychology Applied to Teaching, 13th Ed. (USA: Wadsworth, Cengage Learning)

[28] Babalhayaeji F and Ghiasi M 2011 analysis of LIS research articles 1997-2007: critical thinking, gender, and education *Library Philosophy and Practice* **7**(1) pp. 1-14

[29] Wong M, Castro-Alonso J C, Ayres P, and Paas F 2015 gender effects when learning manipulative tasks from instructional animations and static presentations *Ed. Technol. & Society* **18**(4) pp. 37–52

[30] Aloka P J O and Bojuwoye O2013 gender, age and teaching experiences differences in decision-making behaviours of members of selected Kenyan Secondary School disciplinary panel *Asian Soc. Sci.* Vol. **9**(10) pp. 43-55

[31] Linda V L, Westenberg P M, and Crone E A 2008 a developmental study of risky decisions on the cake gambling task: age and gender analyses of probability estimation and reward evaluation *Developmental Neuropsychology* **33**(2) pp. 179-196

[32] Irwansyah F S, Yusuf Y M, Farida I, and Ramdhani M A 2018 augmented reality (AR) technology on the android operating system in chemistry learning *IOP Conf. Series: Materials Sci. and Engineering* **288** 012068

[33] Harish 2013 critical thinking skills among ninth standard students in relation to gender, intelligence and study habits *In. J. of Ed. and Psychological Research (IJEPR)* **2**(3) pp. 13-20

[34] Aliakbari M and Sadeghdaghighi 2011 investigation of the relationship between gender, field of study, and critical thinking skill: the case of iranian students *Proc.of the 16th Conf. of Pan-Pacific Association of Applied Linguistics* pp. 301-310 (Hongkong)

[35] Sanchez C A and Wiley J 2010 sex differences in science learning: closing the gap trough animation *Learning and Individual Differences* **20** pp. 271–275

IOP Conf. Series: Journal of Physics: Conf. Series **1108** (2018) 012107 doi:10.1088/1742-6596/1108/1/012107

Split and Conquer Method in Penalized Logistic Regression with Lasso (Application on Credit Scoring Data)

F Shofiyah[1], and A Sofro[1,*]

[1]Mathematics Department, Universitas Negeri Surabaya, Ketintang, Surabaya 60231, East Java, Indonesia

[*]Corresponding Author's Email: ayuninsofro@unesa.ac.id

Abstract. Big data is one of the biggest issues recently. We need a new approach to deal with the problem. One of a statistical strategy that can be used to solve the problem is split and conquer method. In this paper, we focus on non-Gaussian data, i.e binomial distribution. In this research will be discussed about the implementation of the method for credit scoring data. The result is there are 5 important independent variables in credit scoring data. The first variable is the percentage of total balances in credit cards and private lines of credit except real estate divided by the number of credit limits. The second variable is the age of debtor. The third variable is how many times debtor has been 30-59 days late pay in the last 2 years. The fourth variable is how many times the debtor has been late for pay 90 days or more. The last variable is how many times debtor has been 60-89 days late pay in the last 2 years.

1. Introduction

Big data is datasets that so voluminous and complex, so it is difficult to be analyzed. A statistical strategy that can be used to solve big data problems is split and conquer method [1]. The method solves the big data problem by divide big data into Q blocks, so it can be easily processed by computing. The estimator results of each block are combined in the final estimate. One example of data that requires big data analysis is credit scoring data [2].

Credit scoring is the assessment process using the historical data of the debtor [2]. The historical data is used to classify debtors who are eligible to be loaned and not. Credit scoring data can contain millions or even billions of debtor's historical records. If all data is used in the classification model, then it will be very inefficient and time-consuming [2]. For that reason, we can use the Split and Conquer method. In addition, credit scoring data requires selecting variables to identify relevant variables. The method that can be used is penalized logistic regression. Penalized logistic regression is a logistic regression that estimates coefficients by maximizing log-likelihood function plus the penalty function on coefficient size [3].

One interesting penalty function is the Least Absolute Shrinkage Selection Operator (LASSO). LASSO was proposed by Tibshirani in 1996, see [4]. Efron (2004) has been introduced a regression scheme that involved LASSO as one in it, sees [5]. LASSO penalty function will able to generate regression with a good parameter estimator and further minimize the error from the logistic regression model [6]. LASSO penalty function also plays a role in identifying which predictor variables are important for the response variable. LASSO penalty reduces the number of coefficients by making it 0. Thus LASSO can reduce the dimensions of the data credit scoring.

In this research will be analyzed big data on real secondary data credit scoring use the Split and Conquer method on penalized logistic regression with LASSO penalty. The credit scoring data consisted of 150,000 observations, 1 dependent variable dependent, and 10 independent variables.

2. Method

2.1 Logistic Regression

The logistic regression model is a model that describes the relationship between several factors (predictor variables) with dichotomous (binary) response variables. For the response variable, $y = 1$ to express the results of a successful and $y = 0$ express the result of the failed. In logistic regression, the variable response is Bernoulli distributed. The response variable is usually denoted by y and the predictor variable is symbolized by x. Suppose y_i is the i-th response variable, π_i is the i-th probability of success, and i is the observation measure, the probability function of y is:

$$f(y_i, \pi_i) = \begin{cases} \pi_i(1-\pi_i)^{1-y_i}, for\ y_i = 0,1 \\ 0 \qquad\qquad otherwise \end{cases}. \quad (1)$$

The logistic model with p predictor variable is expressed as follows

$$\pi_i = \frac{\exp(\beta_0 + \beta_1 x_1 + \beta_2 x_2 + \cdots + \beta_p x_p)}{1 + \exp(\beta_0 + \beta_1 x_1 + \beta_2 x_2 + \cdots + \beta_p x_p)}. \quad (2)$$

The logistic regression model in equation (2) is transformed using the logit transformation obtained:

$$\log\left(\frac{\pi(x)}{1-\pi(x)}\right) = \beta_0 + \beta_1 x_1 + \beta_2 x_2 + \cdots + \beta_p x_p. \quad (3)$$

To estimate logistic regression, parameter estimation method is required, that is maximum likelihood estimation

2.2 Parameter Estimation

The Maximum Likelihood Estimation / MLE method is one of the methods in parameter estimation. This method was first introduced by R.A Fisher in 1912, see [7]. Let $y_1, y_2, \dots y_n$ be a continuous random variable of size n with probability functions $f(y_i, \boldsymbol{\beta})$ and $\boldsymbol{\beta}$ are unknown parameters.
In logistic regression, the likelihood function is as follows [8]

$$L(\boldsymbol{y}; \boldsymbol{\beta}) = \prod_{i=1}^{n} [\pi_i^{y_i}(1-\pi_i)^{1-y_i}], \quad (4)$$

and the likelihood function is as follows [8]

$$\log L(\boldsymbol{y}; \boldsymbol{\beta}) = \sum_{i=1}^{n} \{y_i \log(\pi_i) + (1-y_i)\log(1-(\pi_i)\}. \quad (5)$$

Parameter estimation ($\boldsymbol{\beta}$) can be obtained by maximizing log-likelihood function.

2.3 Least Absolute Shrinkage and Selection Operator (LASSO)

LASSO method was first introduced by Tibshirani in 1996, see [4]. LASSO shrinks regression coefficients from predictor variables with high correlation with errors, being either zero or near zero [8]. The LASSO penalty function is

$$P(\boldsymbol{\beta}) = \sum_{j=1}^{p} |\beta_j|, \quad (6)$$

with $\sum_{j=1}^{p} |\beta_j| \le \lambda$, and λ is a tuning parameter that controls the depreciation of the LASSO coefficient by $\lambda \ge 0$. LASSO can shrink some coefficients toward zero even exactly zero [6], so it can select predictor variables.

2.4 Penalized Logistic Regression Model with LASSO penalty

There are several practical problems and computational problems in the regression analysis, such as multicollinearity and high dimension. To deal with this problem, we need a variable selection method. Penalized regression is a modern approach to choosing a regression model with high prediction accuracy and efficiency [7]. The penalized regression has the advantage of produce more stable results in highly correlated data and data that has a much larger number of predictors than the sample size.
The log-likelihood of Penalized logistic regression use LASSO penalty on equation (6) is as follows [9]

$$\log L_p(\boldsymbol{y};\boldsymbol{\beta}) = \sum_{i=1}^{n} -[y_i \log(\pi_i) + (1-y_i)\log(1-\pi_i)] + \lambda \sum_{j=1}^{p} |\beta_j|. \quad (7)$$

Some coefficients will be equal to zero, depend on LASSO penalty. Therefore, the LASSO performs the selection of variables.
The LASSO coefficient estimate is as follows [9]

$$\widehat{\boldsymbol{\beta}}_{LASSO} = \arg min_{\beta} \left\{ -\sum_{i=1}^{n} \{y_i \log(\pi_i) + (1-y_i)\log(1-(\pi_i)\} + \lambda \sum_{j=1}^{p} |\beta_j| \right\}. (8)$$

After estimating, then do the optimization through coordinate descent.

2.5 Optimization on Penalized Logistic Regression using Coordinate Descent
Suppose the current estimates of the parameter are $(\tilde{\beta}_0, \tilde{\beta})$. To make calculations easier, we form the approximation of the squared log-likelihood regression penalized logistics use the Taylor expansion [10], i.e

$$(\hat{\beta}_0, \hat{\beta}) = argmin \frac{1}{n} \sum_{i=1}^{n} w_i (z_i - \beta_0 - x_i^T \beta)^2 + \lambda \sum_{j=1}^{p} |\beta_j|, \quad (9)$$

where
$\tilde{\eta}_i = \tilde{\beta}_0 + x_i^T \tilde{\beta}$; $\tilde{\pi}_i = \frac{1}{1+\exp(-\tilde{\eta}_i)}$; $z_i(\tilde{\eta}_i) = \tilde{\eta}_i + \frac{y_i - \tilde{\pi}_i}{\tilde{\pi}_i - (1-\tilde{\pi}_i)}$; $w_i(\tilde{\eta}_i) = \tilde{\pi}_i(1-\tilde{\pi}_i)$.
The solution for equation (9) is

$$\hat{\beta}_j = \frac{S\left(\frac{1}{n}\sum_{i=1}^{n} w_i x_{ij} r_i + \left(\frac{1}{n}\sum_{i=1}^{n} w_i x_{ij}^2\right)\tilde{\beta}_j, \lambda\alpha\right)}{\frac{1}{n}\sum_{i=1}^{n} w_i x_{ij}^2 + \lambda(1-\alpha)}. \quad (10)$$

In the equation (10) there is a function S. The function S is soft- thresholding which has the following value

$$S(c,\gamma) = \begin{cases} c-\gamma, & if\ c > 0, and\ \gamma < |c| \\ c+\gamma, & if\ c < 0, and\ \gamma < |c| \\ 0, & if\ \gamma \geq |c| \end{cases}. \quad (11)$$

The following will explain the penalized likelihood optimization algorithm on penalized logistic regression use coordinate descent [11]

1. Initialization $(\tilde{\beta}_0, \tilde{\beta})$, set $\tilde{\eta}_i = \tilde{\beta}_0 + x_i^T \tilde{\beta}$ for $i = 1,2, \dots n$
2. Calculate $z_i = z_i(\tilde{\eta}_i)$ and $w_i = w_i(\tilde{\eta}_i)$
3. Estimate β with equation (9)
4. for $j = 1,2, \dots p$ and $r_i = y_i - \tilde{\beta}_0 - x_i^T \tilde{\beta}$,
   calculate the solution $\hat{\beta}_j$ with the equation (10)
5. Set $\tilde{\beta} = \hat{\beta}$, $\tilde{\beta}_0 = \hat{\beta}_0$
6. Repeat steps 2-4 until β convergence

The coordinate descent algorithm requires tuning parameters, then we will look for initial value tuning parameters use pathwise solution

2.6 Pathwise Solution

To select the value of λ, we first choose the grid/value range of λ, that is λ_{max} and λ_{min} value. The first step is to find λ by setting $\beta = 0$ [12]. To find the value of λ_{max} is as follows :

$$\lambda_{max} = max_j \frac{1}{n\alpha}\sum_{i=1}^{n} w_i(0)x_{ij}z(0)_i, \qquad (12)$$

where $\lambda_{min} = \epsilon\lambda_{max}$. The fixed value for ε depends, for $n < p$, then $\epsilon = 0.05$ and for $n \geq p$, then $\epsilon = 0.0001$. After it will be selected optimal tuning parameters using k-cross-validation

2.7 k-fold Cross Validation (KCV)

In k-fold cross-validation, the datasets are divided randomly into the same size set, then find for cross-validation value λ, i.e

$$CV_i(\lambda) = \ell\left(\beta_{-i}(\lambda)\right) - \ell_{-i}\left(\beta_{-i}(\lambda)\right). \qquad (13)$$

Where ℓ_{-i} is the partial log of likelihood data block $k-1$ (except the i-th data block).
Our total goodness of fit estimate, $CV(\lambda)$, is the sum of all $CV_i(\lambda)$. Then, choose λ values that maximize cross-validation $CV(\lambda)$.

2.8 Split and Conquer Methods for Penalized Regression with LASSO on Big Data

Split and Conquer method developed by Chen and Xie in 2014, see [13]. This method has advantages that are to overcome the problem of data that is too large.
n data is divided into Q data block, then we find the parameter estimation from each data block uses the penalized logistic regression with LASSO.
In this section will be discussed about the conquer method, which combines the value of $\widehat{\boldsymbol{\beta}}$ of each data block. The combined estimate of with $q = 1, \ldots .Q$ is

$$\widehat{\boldsymbol{\beta}}^{(c)} = A\left(\sum_{q=1}^{Q} A^T\,\widehat{\mathbf{S}}_q A\right)^{-1} \sum_{q=1}^{Q} A^T\widehat{\mathbf{S}}_q A\widehat{\boldsymbol{\beta}}_{q,\hat{\mathcal{A}}^{(c)}}. \qquad (14).$$

The following will explain how to calculate each of the above variables :
Defined

$$\hat{\mathcal{A}}_q = \{j: \widehat{\boldsymbol{\beta}}_{q,j} \neq 0\}. \quad (15)$$

With j is index of variable , so $j = 1,2,\ldots p$
$\widehat{\boldsymbol{\beta}}_{q,j}$ is the estimator of the q-th data block on the j-th variable. $\hat{\mathcal{A}}_q$ is the set of an index the non-zero estimator. Thus, $\widehat{\boldsymbol{\beta}}_{q,\mathcal{A}_q}$ is a sub-vector contain only the nonzero estimator [13].
Defined

$$\hat{\mathcal{A}}^{(c)} = \left\{j: \sum_{q=1}^{Q} \mathrm{I}\left(\widehat{\boldsymbol{\beta}}_{q,j} \neq 0\right) > \omega\right\}. \quad (16)$$

$\hat{\mathcal{A}}^{(c)}$ is the set of selected variables index of the combined estimator, where $\omega \in [0, Q)$ is a predefined threshold and I is an indicator function [13]. So if the index j appears more than ω times in $\hat{\mathcal{A}}_q$, then the j index is included to the set $\hat{\mathcal{A}}^{(c)}$.
So, $\widehat{\boldsymbol{\beta}}_{q,\mathcal{A}^{(c)}}$ is the estimator of the variable on q-th data block and j-th variables, where j is j which belongs to the set $\hat{\mathcal{A}}^{(c)}$.
Suppose $E = diag(v_1, \ldots, v_p)$ is a $p \times p$ matrix with $v_j = 1$ if $\sum_{q=1}^{Q} \mathrm{I}(\widehat{\boldsymbol{\beta}}_{q,j} \neq 0) > \omega$ and 0 if otherwise, and $A = E_{\hat{\mathcal{A}}^{(c)}}$ is the matrix $p \times |\hat{\mathcal{A}}^{(c)}|$, ie the submatrix of the matrix E, where the rows

IOP Conf. Series: Journal of Physics: Conf. Series **1108** (2018) 012107 doi:10.1088/1742-6596/1108/1/012107

chosen from E is the rows j contained in the set $\mathcal{A}^{(c)}$. Defined $\widehat{\boldsymbol{S}}_q = X_q^T \Sigma(\widehat{\boldsymbol{\theta}}_q) X_q$ where $\widehat{\boldsymbol{\theta}}_q = \boldsymbol{X}_q \widehat{\boldsymbol{\beta}}_q$ and $\Sigma(\widehat{\boldsymbol{\theta}}_q) = diag(\sigma(\theta_{i1}) \dots, \sigma(\theta_{iQ}))$. The value of $\sigma_i = b''(\theta_i) = \frac{e^{\theta_i}}{(1+e^{\theta_i})^2}$.

3. Experiment Set Up

3.1 Dataset

In this study, we use the Credit Scoring data (Give Me Some Credit). It is available at kaggle.com (https://www.kaggle.com/c/GiveMeSomeCredit). Credit scoring data is used to predict debtors who deserve credit and do not. In this data is available historical data about 150,000 debtors. Of the 150,000 observations, there are some missing values that must be eliminated and used in this study as many as 110290 observations.

3.2 Variable

The dependent variable is y. It has a binary value of either 1 or 0. A value of 1 means that the borrower is delinquent and has defaulted on his loans for the last 2 years, while a value of 0 means that the borrower is a good customer and repays his debts on time for the last two years. There are 10 independent variables. Variable (x_1) represents the percentage of total balances in credit cards and private lines of credit except for real estate (no installment debt such as Auto loans) divided by the number of credit limits, variable (x_2) represents the age of debtor, variable (x_3) represents how many times debtor has been 30-59 days late pay (not too bad) in the last 2 years, variable (x_4) represents the percentage of monthly debt payment, allowance, living expenses divided by monthly gross income, variable (x_5) represents the debtor income every month, variable (x_6) represents the number of credit lines and loans, variable (x_7) represents how many times the debtor has been late for pay 90 days or more, variable (x_8) represents denotes the number of mortgage and real estate loans including home equity lines of credit, variable (x_9) represents how many times debtor has been 60-89 days late pay (not too bad) in the last 2 years, and variable (x_{10}) represents the number of family members.

4. Result and Discussion

The first step, we do 'Split' procedure that is to divide data in *the* Q data block. We choose $Q = 10$ then find for λ tuning parameter value through a k-cross-validation algorithm. Before doing k-cross-validation process, we must find the value of initial value (λ) through pathways solution. Selected tuning parameter in each data block will be shown in the following table

Table 1. Optimal tuning parameters in each data block

Data block (q)	$\hat{\lambda}$
1	0.01329002
2	0.001475477
3	0.01944356
4	0.001367119
5	0.002261008
6	0.01088266
7	0.005072335
8	0.001735969
9	0.02407742
10	0.01441134

We estimate the parameters by the maximum likelihood estimation (MLE) method and the optimization use the coordinate descent. The procedure of MLE is to maximize the log-likelihood function in equation (7). After maximizing the log-likelihood, we optimize the log likelihood with the coordinate descent. We obtain the parameter estimation value for each data block, then the estimation result from Q data block is combined in final estimation use Conquer method in equation (14)

Table 2. Combined estimator

Variable	$\hat{\beta}_j$
Constant	-3.58203
x_1	1.47142
x_2	-0.00368
x_3	0.372446
x_4	0
x_5	0
x_6	0
x_7	0.612527
x_8	0
x_9	0.508872
x_{10}	0

Based on the table, we can see that some coefficients are zero. It is due to the LASSO penalty function on penalized logistic regression. LASSO shrinks coefficients on irrelevant variables. So that the irrelevant variables have a small effect on the response variable. The shrinkage of the LASSO coefficient results in a simpler regression model. The relevant variables that exist in the model are x_1, x_2, x_3, x_7 and x_9

5. Conclusion

The relevant variables present in the model are (x_1) represents the percentage of total balances in credit cards and private lines of credit except real estate divided by the number of credit limits, (x_2) represents the age of debtor, (x_3) represents how many times debtor has been 30-59 days late pay in the last 2 years, (x_7) represents how many times the debtor has been late for pay 90 days or more, and (x_9) represents how many times debtor has been 60-89 days late pay in the last 2 years.

6. References

[1] Meng C, Wang Y, Zhang X, Mandal A, and Ma P 2017 *Effective Statistical Methods for Big Data Analytics* pp. 1–22.
[2] Márquez J 2008 *An Introduction to Credit Scoring For Small and Medium Size Enterprises,*
[3] Chung Y, Shopia R, Vincent D, Andrew G and Jingchen L N 2013 *A Nondegenerate Penalized Likelihood Estimator For Variance Parameters In Multilevel Models* **78** pp. 685–709.
[4] Tibshirani R 1996 *Regression Shrinkage and Selection via the LASSO* **5**.
[5] Efron B 2004 *The Estimation of Prediction Error : Covariance Penalties and Cross-Validation* **99** pp. 619–632.
[6] Fonti V and Belitser E 2017 *Feature Selection using LASSO.*
[7] Gunes F 2015 Penalized Regression Methods for Linear Models in SAS/STAT ® *SAS Institute Inc* pp. 114.
[8] Algamal Z Y and Lee M H 2015 Applying penalized binary logistic regression with correlation based elastic net for variables selection *J. Mod. Appl. Stat. Methods.***14**(1) 168-179.
[9] Krona E 2017 A simulation study of model fitting to high dimensional data using penalized logistic regression *Bachelor Thesis Matematiska institution Mathematical Statistics Stockholm University*
[10] Yang Y and Zou H 2014 A coordinate majorization descent algorithm for ℓ1 penalized learning *J. Stat. Comput. Simul.* **84** pp. 84–95.

[11] Friedman J, Hastie T, and Tibshirani R 2010 Regularization Paths for Generalized Linear Models via Coordinate Descent *J. Stat. Softw.* **33**(1) pp. 1–24.
[12] Simon N, Friedman J, Hastie T, and Tibshirani R 2011 Regularization Paths for Cox's Proportional Hazards Model via Coordinate Descent *J. Stat. Softw.* **39**(5) pp. 1-13.
[13] Chen X and Xie M 2014 A Split-and-Conquer Approach for Analysis of Extraordinarily Large Data *Statistica Sinica* **24** pp. 1655–1684.

IOP Conf. Series: Journal of Physics: Conf. Series **1108** (2018) 012108 doi:10.1088/1742-6596/1108/1/012108

Students' Covariational Reasoning in Fraction Compare Problem

R W Ningrum[1*], **Y Fuad**[2], **and R Ekawati**[2]

[1]Program Pascasarjana, Universitas Negeri Surabaya, INDONESIA
[2]Department of Mathematics, Universitas Negeri Surabaya, INDONESIA

*Corresponding Author: rachmanianingrum16070785029@mhs.unesa.ac.id

Abstract. Covariational reasoning plays important role to mathematics education such as fraction compare word problem. This study investigates students' covariational reasoning after their studies concerning fraction word problem in first semester on fifth grade. Three students who had high mathematics ability were chosen to solve problem that involved how quantities change with focused on operation (addition, subtraction, multiplication, and division) and compare. Interviews were conducted to reveal the students' reasoning while solving covariational problems. The results emphasizes that fifth grade students were nearly unable to construct the relation of quantity that changes with the other quantity. It cause students faced difficulty in determining the operation of the problem solving, could not operation the fraction with different denumerator, and could not apply the concept of fractions. These findings suggest that learning fraction should be increased emphasis on coordinationg two quantities changing about compare magnitude and operation of fraction of change and to improve the knowledge in fraction compare.

1. Introduction

Mathematics word problem are very interesting in the research world. Several studies [1-7] showed that all level students assume that compare word problem are difficult. One of the factor causing students' difficulties in solving compare word problem is the lack of the school learning that concerns the students' reasoning. Schools place greater emphasis on procedures rather than reasoning [8]. Reasoning is one of the important components on every mathematics learning. Reasoning is the process of thinking in drawing the conclusios based on logical analysis and argument. Compare word problem needs students' reasoning to solve the problem. Covariational reasoning is one of the reasoning that students need to solve compare word problem. Covariational reasoning to be the cognitive activities involved in coordinating two varying quantities while attending to the ways in which they change in relation to each other [9,10]. For example elementary students that given questions, "if You got 6 score for the mathematics test, then you don't get additional money. If you got 7 score, then you get additional money amount Rp 1.000,00. If you got 8 score, then you get additional money amount Rp 2.000,00." It showed the students know two quantities was changes. Thats students can be said thinking by covariational reasoning.

Moreover, many researchers [9-12] had investigated about covariational reasoning, but none of them exposed covariational reasoning that related with fractions. The fraction is one of the main concepts in studying mathematics especially algebra [13-17]. In addition none of researchers exposed covariational

IOP Conf. Series: Journal of Physics: Conf. Series **1108** (2018) 012108 doi:10.1088/1742-6596/1108/1/012108

reasoning on elementary students. Therefore this study focuses to investigation on how students' reason during their studies concerning two covarying quantities in fractions on elementary students.

2. Method

This study was conducted on the five grade students of the elementary school at Mojokerto, in academic year 2017/2018 which was chosen by purposively sampling. The fifth grade class consists of 12 male and 12 female. All students were given the fraction ability test and the fraction compare problem, which was focused on arithmetic's operations (addition, subtraction, multiplication, and division). The results of fraction ability test were ranked into three levels, i.e. high, medium, and low mathematical abilities respectively.

The instrument consists of three items, and had validated by two senior lecturers of the mathematics' department of Unesa. The six students invited to complete three written problems that involve interpreting and representing how quantities vary in tandem, which were identically designed in context. Three of these students (labeling with A, B, and C) were volunteerly invited to participate in the interview sessions that concentrated on how students reason during solving covariational problems. The selection of the interview subjects was also based on diversity responses on their written test. The analysis focused on students' use of reasoning covariationally while responses to the written test, which related with five mental actions of covariational reasoning were adapted from [9]. Five mental action of covariational reasoning on compare problem were modified by researcher which appropriate with students elementary (see Table 1).

Table 1. Mental Action of Covariational Reasoning

Mental Action	Description of Mental Action	Behavior
Mental Action 1 (MA 1)	Coordinating the changing dependence of one quantity on an ongoing basis.	• Determining the value of two interrelated quantities • Verbalizing that the value of two quantities are interrelated.
Mental Action 2 (MA 2)	Coordinating the direction of change of one quantity with changes in the other quantity.	• Using the countdown to determine another quantity change • Verbalizing that the quantity of one changes depends on the other quantity.
Mental Action 3 (MA 3)	Coordinating the amount of change of one quantity with changes in the other	• Determining the results of the second change quantity • Verbalizing that the occurrence of quantity changes affects other quantity changes.
Mental Action 4 (MA 4)	Coordinating the average rate-of-change of the quantity with unfiorm increments of change in the input quantity	• Determining changes in amount of quantity and change of quantity with same increasement. • Verbalizing that the change in quantity (with same incresement) affects other quantity change.
Mental Action 5 (MA 5)	Coordinating the momentary changes of the quantity that constantly cahnages in both quantities	• Determining changes in amount of quantity and changes of quantity with changes increasement (up or down). • Verbalizing that quantity change affects other quantity changes.

Adopted from [9]

3. Results and Discussion

The fraction ability test were given by students on fifth grade. The result of fraction ability are shown in Table 2. The results of fraction ability test were ranked into three levels, i.e. high, medium, and low mathematical abilities respectively.

Table 2. Students' Fraction Test Ressults

Sex	score > 80	80 ≥ score ≥70	score < 70
Boys	3	6	3
Girls	4	5	4

Based on fraction ability test were chosen students with high mathematics ability to solve fraction compare tests. Fraction compare tests were given to know how students' covariational reasoning on compare problem. Table 3 shows the kinds of responses that seven students provided on the written test. Only two students of these high fraction ability given an acceptable solution, while four students had some missed calculation, and one other had some missed problem. During the follow up interview, the three interview subjects provided varied responses.

Table 3. Students' Fraction Compare Test Ressults

Sex	True	False
Boys 1	1	5
Boys 2	4	2
Girls 1	1	5
Girls 2	2	4
Girls 3	3	3
Girls 4	4	2
Girls 5	2	4

In this study, we only found three from five mental action that can be achieved by students. Students A provided responses that indicated three mental actions that linear with framework of covariational reasoning. He could be verbalizing that two different fraction are interrelated, one fraction of one changes depends on the other fraction, and the occurrence of fraction changes affects other fraction changes. He had accepted answer (see Figure 2). Students A stated the following statements:

I undertand how to know the difference of their money (problem at test). It needs the price of their paid. So, the first, I must known how many chicken egg they will buy for each (MA 1). Then, I could find the price of their paid for each by multiplying how many chicken egg they will buy and the price of chicken egg per kg (MA 2). After that, I could find the difference of their money (MA 3).

Student B had same reasoning with students A. He had finished with accepted answer (see Figure 3). He could states that two difference fraction are interrelated (MA1). He explained why he could find the price of chicken egg for each person (MA 2) until he got the difference of their money (MA 3), he stated as follows

I must find the price of chicken egg for each to hunt for the difference of their money (problem test). I did for multiplying the price of chicken egg per kg and the amount of chicken egg(kg), and I get that.

Student C had differ reasoning from student A and student B. Student C suppose that the difference in the problem are the difference with proportion (see Figure 4), she stated as follow.

Since the price of the chicken egg per person had been find, I can found the difference of the chicken egg per person. But after that I must subtract the price of chicken egg per person by the price of chicken egg per kg. So, I found the price of Dian is Rp 6.000,- and the price of Ani is Rp10.000,-. In this item, I had hesitant. Therefore, I had two answer. The second answer of the end question is the difference of the price of chicken egg per person.

She appeared to have difficulty in determining the results of the change quantity. She could be verbalizing that two different fraction are interrelated (MA 1) and one fraction of one changes depends on the other fraction (MA 2). But she almost missed for verbalizing the occurrence of fraction changes affects other fraction changes. She indicated that she appeared to have difficulty to determine the results of the change quantity. In order studies, students like that is defined pseudo [10].

The data analysis was focused on students' covariational reasoning, which related to three mental actions of covariational reasoning. Students were given fraction compare test. The first item test on fraction compare test as follow.

Ani buy $\frac{5}{4}$ kg chicken eggs and Dian buy $\frac{1}{2}$ kg chicken eggs. The price of the eggs is Rp 12.000,-They have $\frac{1}{4}$ kg from the shop because its celebrate shop's birthday for each.

a. How many chicken egg which bought by Ani and Dian for each before addition?
b. Which one of the chicken egg have more after addition by shop?
c. How much the difference of their money?

1. Ani membeli $\frac{5}{4}$ kg telur ayam dan Dian membeli $\frac{1}{2}$ kg telur ayam. Harga per kilogram telur adalah Rp 12.000,-. Masing-masing dari mereka mendapat tambahan $\frac{1}{4}$ kg dari toko karena toko merayakan hari ulang tahunnya

a. Berapa kg telur yang dibeli Ani dan Dian masing-masing sebelum ditambah?
b. Telur siapa yang lebih banyak setelah ditambah oleh toko? Jelaskan! (Bisa dengan menggunakan garis bilangan).
c. Berapakah selisih uang yang harus dibayar oleh Dian dibandingkan dengan uang yang dibayar oleh Ani?

Figure 1. Question 1 in Fraction Compare Test

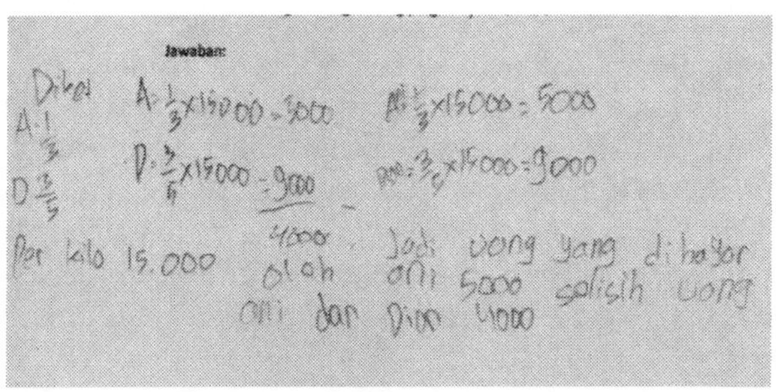

Figure 2. Subject A Answers

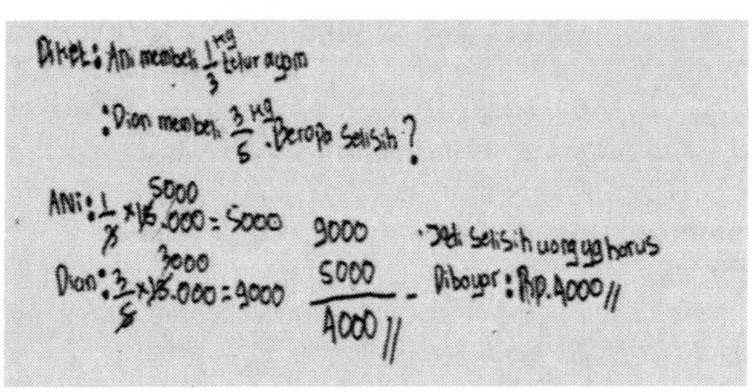

Figure 3. Subject B Answers

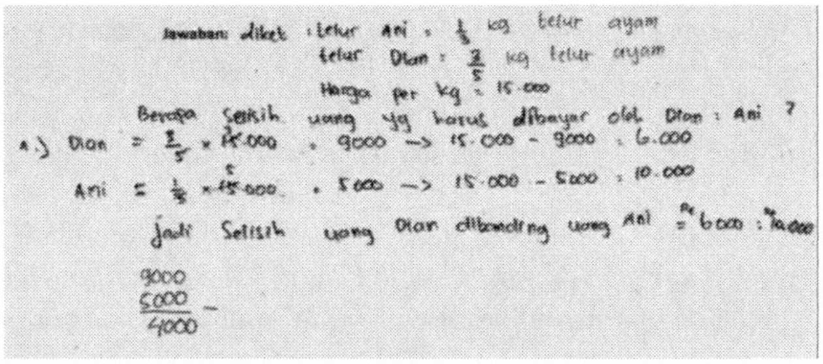

Figure 4. Subject C Answers

Based on these cases, it can be seen that students' covariational reasoning didn't occurred on all mental action. It caused the students (elementary students) did not use the fully knowlegde. In this case the students did not really understand compare problem. It seen that one students missed on her answer. She had two answer because she had hesitant. Not only one students who didn't understand compare problem. But the other students didn't fully understand compare problem. It caused the students rarely get compare problemon mathematics class on daily activities. The teacher approved it too. The teacher recognized that the students had given more conceptual knowledge for the first. She expected the students should understand compare problem if she will be given the compare problem.

4. Conclusion

Based on results, students were nearly unable to construct the relation of the quantities that change with the other quantities and worked on three mental actions. Those mental action are: coordinating the chnging dependence of one quantity on an ongoing basis, coordinating the direction of change of one quantity with changes in the other quantity, and coordinating the amount of change of one quantity with changes in the other. However, students appeared to have difficulty in determining the results of the second change quantity and have difficulty on compare problem. This findings suggest that learning in fraction and compare problem should placed increased emphasis on coordinating two quantities changing in tandem and conceptual knowledge in fraction. Further studies are needed to investigate how to develop students' reasoning in coordinating the average rate-of-change of the quantity with unfiorm increments of change in the input quantity and coordinating the momentary changes of the quantity that constantly cahnages in both quantity.

5. References

[1] Artut P D 2015 Preschool Childrens Skills in Problem Solving Mathematical Word Problem *Educational Research and Reviews* **10**(18), pp 2539—2549

[2] Boonen A J H, and Jolles J 2015 Second Grade Elementary School Students' Differing Performance on Combine, Change and Compare Word Problems *Interational Journal of School and Cognitive Psychology,* **2** 122

[3] MacDonald L R and Banes L 2017 More Than Words: Struggling Reader's Comprehension of Word Problems. *Journal of Teacher Action Research* **3**(3) pp 24-39

[4] Monroe E E and Cutler C S 2014 (online) Oral Retellig: Solution Strategy For Comparing Word Problems? Article obtained from https://www.research-gate.net/publication/242326659

[5] Sepeng P and Madzorera A 2014 Sources of Difficuly in Comprehending and Solving Mathematical Word Problems *International Journal of Education Science* **6**(2) pp 217-225

[6] Ulusoy F and Argun Z 2017 Secondary school students' representations for solving geometric word problems in different clinical interviews *International Journal of Eduacation in Mathematics, Science and Technology (IJEMST)*

[7] Mattarella-Micke A and Beilock S L 2010 Situating Math Word Problem *The Story Matters. Psychonomic Bulletin & Review* **17**(1) pp 106-111

[8] Singh P 2000 Understanding The Concepts of Proportion and Ratio Among Grade Nine Students in Malaysia *Internastional Journal oef Mathematical Education in Science and Technology,* **31**(4) pp 579-599

[9] Carlson M, Jacobs S, Laren S, and Hsu 2002 Applying Covariational Reasoning while Dynamic Events: A Framework and A Study *Journal for Research in Mathematics Education* **33**(5) pp 352-378
[10] Subanji 2011 *Teori Berpikir Pseudo Penalaran Kovariasional* (Malang: Universitas Negeri Malang)
[11] Strom A D 2006 The Role of Covariational Reasoning in Learning and Understanding Exponential Functions *Proceedings of the Twenty Eighth Annual Meeting of the North American Chapter of the International Group for the Psychology of Mathematics Education* **2** pp 624-630
[12] Zeytun A S, Çentikaya B, and Erbas A K 2010 Mathematics Teachers' Covariational Reasoning Levels and Predictions about Students' Covariational Reasoning Abilities *Education of Science: Theory and Practice* **10**(3) pp 1602-16012
[13] Kim S U, Wang P, and Michaels C A 2015 Using Explicit C-R-A Instruction to Teach Fraction Word Problem Solving to Low-Performing Asian English Learners. Reading & Writing Quarterly **31**(3) pp 253-278
[14] NMAP 2008 *Foundations for success: The final report of the National Mathematics Advisory Panel*. Washington (DC: U.S. Department of Education)
[15] Booth J L and Newton K J 2012 Fractions: Could They Really be the Gatekeeper's Doorman? *Contemporary Educational Psychology* **37** pp247—253
[16] Booth J L, Newton K J, and Twiss-Garrity L K 2014 The Impact of Fraction Magnitude Knowledge on Algebra Performance and Learning *Journal of Experimental Child Psychology* **118** pp 110—118.
[17] Siegler R S, Thompson C A, and Schneider M 2011 An Integrated Theory of Whole Number and Fraction Development *Cognitive Psychology* **62** pp 273—296

IOP Conf. Series: Journal of Physics: Conf. Series **1108** (2018) 012109 doi:10.1088/1742-6596/1108/1/012109

Nanogold's Influence on Antioxidant Activity of Green Tea Extracts in the Framework of New Essential Ingredients Discovery in Cosmetic Formulation

T Taufikurohmah[1,*], I G M Sanjaya[2] and S Tjahjani[1]

[1]Chemistry Department, Universitas Negeri Surabaya, Indonesia
*Corresponding author: titiktaufikurohmah@unesa.ac.id

Abstract. The circulation of cosmetic containing hazardous materials especially mercury has occurred in Indonesia and other countries. The impact has also been felt by the wider community. So far efforts have been made, among others, by searching for and finding new essential ingredients in cosmetic formulas. Mercury in cosmetics leads to the formation of free radicals in the body that must be resisted with antioxidants. Nanogold has been shown to have antioxidant activity. Similarly, green tea extracts are also reported to possess antioxidant activity. Their combined form in cosmetic formulas are expected to increase antioxidant activity. Thus, this study will examine antioxidant activity of nanogold combined with green tea extract. This is particularly important with regard to the application of both in cosmetic formulas. The results obtained show that antioxidant activity of green tea extract of 20 ppm are 91,16%. Antioxidant activity of green tea extract with nanogold in 5, 10, 15, 20, 25, 30 and 35 ppm are 90,63; 90,77; 90,90; 91,42; 91,56; 91,42, and 91,42 respectively. To sum up, the combination of nanogold and green tea extract gives no significant impact towards antioxidant activity.

1. Introduction

Cosmetics containing hazardous materials, especially mercury, have been found in the market and are widely distributed in both urban and rural communities. This can be known from the research related mercury contents in facial cream in 2012 where 5 of 6 facial creams were positively analyzed containing mercury [1]. The increase occurs in 2017 where 39 samples out of 45 positively contain mercury [2]. The case of mercury containing cosmetics is not only in Indonesia but covers almost all countries in Southeast Asia, India, and other countries [3].

The impact has also been felt by the public at large, especially when the termination of cosmetics [4]. Mercury is a very powerful antimicrobial material, so that at the beginning of use all acne is healed and the skin becomes smooth [5]. At the time of the discontinuation of cosmetics, acne will appear even much more. Mercury also suppresses melanocyte cells so these cells not to excrete melanin that its skin pigment [6]. In the early of use the skin show very white and satisfying users. At the time of discontinuation of cosmetics, hyperpigmentation that occurs because melanocyte cells become uncontrolled produce melanin. It does not stop here, the damage of cells and tissues continues directly due to the use of mercury [7]. Mercury is the source of free radicals in the human body. Free radicals attack all important organs and cause damage organ function [8].

So far, efforts have been made, among others, by searching for and finding new essential ingredients in cosmetic formulas. Mercury in cosmetics leads to the formation of free radicals in the body that must be resisted with antioxidants [4]. New materials as antioxidants are tested for their activity using artificial free radical 2,2-diphenyl-1-picrylhydrazyl (DPPH) [9]. DPPH includes stable artificial free radicals that are stable so that when new materials of antioxidants can reduce this free radicals was certainly also able to reduce the actual free radical. The quantitative price of the antioxidant power is indicated by the magnitude the free radical damping expressed in percent of damping [10].

Nanogold has been shown have antioxidant activity. Percent damping of nanogold to DPPH at 20 ppm was 56.85% and at 25 ppm 66.27% [11]. Using Nanogold as cosmetics is selected concentration of 20 ppm because it has a higher colloidal stability. Drops of free radicals by nanogold are still ongoing because nanogold was not damaged after give an activity. This was evidenced by the increase in the damping percentage that reaches 76.01% in measurements after one hour and becomes 94.01 after two hours. The mechanism of action of nanogold as antioxidants is different from organic antioxidants. Nanogold works as a catalyst that conducting catalytical process where its returns to original state after the catalytical process ends. Than the catalyst or nanogold in order prepare for subsequent catalytic processing without any damage [12].

The green tea is a plant containing catechins that are the dominant antioxidant compounds found in this plant [13]. Green tea extract provides a protective effect of skin damage due to oxidation, photoaging and preventing premature aging. The use of green tea extract topically can provide a smoothing or moisturizing effect on facial skin, and can reduce black stains [14]. Green tea extracts were also reported to have antioxidant activity through activity testing using DPPH [15]. Green tea extract has an antioxidant activity of 91.39% in concentrations of 193 mg / L or 193 ppm [13]. When compared to nanogold that have 56.85% damping activity in concentrations of only 20 ppm nanogold looks superior to antioxidant material [10].

To further clarify the antioxidant activity of the material it is necessary to test the activity of green tea extract and nanogold at the same concentration. In this study both will be tested its antioxidant activity separately and simultaneously. In addition, also will be conducted related research to answer the problem of how the effect of adding nanogold to antioxidant activity of green tea extract. This is particularly important with regard to the application of both in cosmetic formulas.

2. Material and Method

Materials used in this study include: Nanogold 20 ppm, made from HauCl4 base material that is reduced with sodium citrate. To obtain nanogold 20 ppm HauCl4 1000 ppm base solution of 20 ml diluted to 1000 ml. This solution is heated to boiling and added 2 g of sodium citrate. The color change from yellow, to colorless, then turns red to explain the process of reduction from Au3 + to Au + and finally to Au in colloidal form. Nanogold 20 ppm then tested its antioxidant activity.

The green extract is made by heating 1000 ml of water and incorporating 20 mg of dried green tea leaf powder. At first green tea extract pale green, after allowed to cool the color increases thick. This shows the process of extraction more perfect. The active substances of dissolved green tea in water are increasing. This green tea extract is then tested for its antioxidant activity. In this activity test required free radical artificial 2,2-diphenyl-1-picrylhydrazyl (DPPH).

The instruments used in the antioxidant activity test of both nanogold, green tea extract and the combination of both are UV-visible Spectrophotometers. The decrease in UV absorption of DPPH compounds indicates the occurrence of the free radical damping process. The difference in absorption compared with the initial absorption multiplied by 100% is the percentage of damping which also shows the quantity of antioxidant activity quantitatively. The greater the percentage of the damping the higher the antioxidant activity of antioxidant material.

The methods used to answer the problem are:

Nanogold Synthesis

The first step in this research is by heating aquades 200 mL to boiling. Adding sodium citrate as much as 0.3 gram and 1000 ml HAuCl4 solution of 7 ppm, 6 mL, 5 mL, 4 mL, 3 mL, 2 mL, 1 mL for each concentration of 35 ppm, 30 ppm, 25 ppm, 20 ppm , 15 ppm, 10 ppm and 5 ppm. The mixture is stirred using a stirrer at 500 rpm. The heating is stopped when the solution turns into a burgundy color.

Green tea extract 20 ppm

100 grams of green tea is put into a glass of chemistry. Adding aquades at 100°C until the solution is ± 1cm above the sample. Stirring green tea extract and wait for up to 15 minutes. Filter the solution

IOP Conf. Series: Journal of Physics: Conf. Series **1108** (2018) 012109 doi:10.1088/1742-6596/1108/1/012109

using a vacuum pump. Insert the filtrate into the freeze dryer tool until the dry extract is obtained. A total of 2 mg of dried extract was diluted with 100 mL warm aquades using 100 mL measuring flask. The solution is shaken until homogeneous so that the solution concentration of green tea extract 20 ppm.

Testing the Effect of Nanogold Addition on Antioxidant Activity of Green Tea Extract
DPPH powder 4 mg was put into a 100 mL measuring flask, dissolved with ethanol on a 100 mL measuring flask. Shake until homogeneous and placed in dark space. Measured using a UV-Vis Spectrophotometer instrument at a wavelength of 400-600 nm to obtain the maximum wavelength of DPPH. 5 mL solution of green tea extract 20 ppm was inserted into 7 dark bottles, then 5 mL nanogold 5, 10, 15, 20, 25, 30, 35 ppm and 100 mP DPPH solution added 10 mL. Shuffled to homogeneous and measured absorbance using the UV-Vis Spectrophotometer instrument at maximum DPPH wavelength.

3. Result and Discussion

3.1 Nanogold Synthesis and Characterization

In this study, the synthesis of gold nanoparticles was made by heating aquades to boiling, then added 0of sodium citrat. The color of the solution is changed begins with a yellow colored solution becomes colorless and then turns into dark blue and red. Addition of sodium citrate serves as a reducing agent also as a stabilizer agent so that the nanogold solution is not aggregated. The negative charge of the citrate ion will be absorbed by the nanogold surface so that the gold particles will repel each other. The reaction is as follows:

$$2Au^{3+} + 4C_6H_5O_7^{3-} + 6H_2O \rightarrow 2Au + 4C_6H_8O_7 + 3O_2$$

When gold is in ionic form, Au3 + will repel each other because of the influence of similar charge. However, after being reduced by sodium citrate to Au0, the charge of Au atoms becomes neutral so that the atoms will get closer to each other through the bonds between the metals and the longer the cluster will grow to the nanometer size. The explanation can be illustrated as follows:

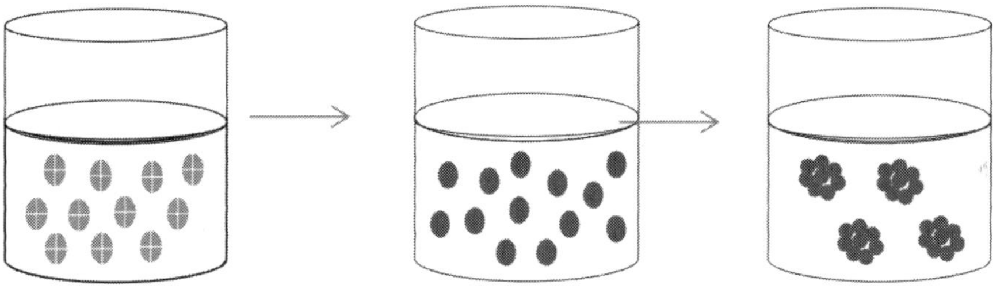

Fig 1. The Illustration of The Nanogold Cluster Growing

The color intensity resulting from nanogold synthesis of 5 ppm - 35 ppm is concentrated. This relates to the cluster diameter and the cluster distance density. The higher the concentration the narrower the distance between the clusters and the color will be stronger and the higher the nanogold concentration, the more clusters are formed so that the resulting color is stronger. The synthesis of gold nanoparticles was then characterized by UV-Vis spectrophotometer and obtained the maximum wavelength to be used to calculate the cluster diameter using the Brus equation as follows:

$$\frac{1240,6}{524} = 1,3 + \frac{14,84}{R^2}\left(\frac{1}{(0,25)^2} + \frac{1}{(0,25)^2}\right) - \frac{2,6}{6,5R}$$

Table 1. The Diameter of Nanogold cluster used Bruss Equation

No	Konsentrasi Nanogold (ppm)	λ maks (nm)	Diameter Cluster (nm)
1	5 ppm	524	20,90
2	10 ppm	521	20,95
3	15 ppm	526	20,99

4	20 ppm	524	20,90
5	25 ppm	522	20,82
6	30 ppm	527	22,05
7	35 ppm	525	20,95

3.2 Preparation and Characterization of Green Tea Extracts

The green tea extract solution begins by weighing the green tea leaf powder that has been dried and in the blender.Selection of solvent using aquades to extract green tea caused catechin compounds contained in green tea is polar and soluble in water. Catechins have more than one hydroxyl group so that the solubility in the water is greater. In addition, the aquades solvent is a safe solvent and is not toxic when applied to human skin. Analysis of green tea extract solution using UV-Vis Spectrophotometer showed that 20 ppm green tea extract solution had maximum wavelength absorption at 291 nm and absorbed value was 0.162.

3.3 Testing the Effect of Nanogold Addition on Antioxidant Activity of Green Tea Extract

The preparation of DPPH solution is by weighing a 4 mg DPPH black powder inserted into 100 mL measuring flask, dissolved with ethanol up to miniskus boundary mark. Shake until homogeneous and placed in dark space. Furthermore, it was measured using a UV-Vis Spectrophotometer instrument at a wavelength of 400-600 nm to obtain the maximum wavelength of DPPH. DPPH solution has a maximum absorption wavelength of 518 nm and absorbance of 0.758. This absorbance value will be used as an initial DPPH absorbance control.

Analysis of antioxidant activity was done by measuring the absorbance of green tea extract and nanogold tested with DPPH radicals. The absorption of DPPH will decrease and the intensity of purple color from DPPH will fade over time when reacted with antioxidants. This is because the antioxidant will contribute an atom H or Au to the DPPH radical so that the DPPH radical will become more stable. Calculation of free radical damping activity by nanogold and green tea extract at each concentration was done by entering absorbance data which have been analyzed using UV-Vis Spectrophotometer instrument into the following% damping formula:

$$\% \text{ of dumping} = \frac{\text{Abs}_{\text{DPPH}} - \text{Abs}_{\text{sampel}}}{\text{Abs}_{\text{DPPH}}} \times 100\%$$

Based on the measurements of the UV-Vis Spectrophotometer at the maximum DPPH wavelength of 518 nm, there is a% damping as shown in the following table.

Table 2. Dumping percentage of Nanogold and green tea extract.

The concentration (ppm)	**% of dumping (hours)**				
	1 mnt	**1**	**2**	**3**	**4**
Green tea extract 20 ppm	91,16	91,95	91,69	92,61	93,14
Green tea extract 20 ppm + NG 5 ppm	90,63	91,95	92,08	92,35	93,67
Green tea extract20 ppm + NG 10 ppm	90,77	91,03	91,03	91,69	93,01
Green tea extract20 ppm + NG 15 ppm	90,90	91,82	92,22	92,22	93,67
Green tea extract 20 ppm + NG 20 ppm	91,42	92,08	92,35	92,74	93,40
Green tea extract 20 ppm + NG 25 ppm	91,56	92,08	93,01	93,54	82,85

Green tea extract 20 ppm + NG 30 ppm	91,42	91,95	93,01	92,88	77,04
Green tea extract 20 ppm + NG 35 ppm	91,42	92,22	92,74	93,40	77,04

Calculation percent of DPPH free radical damping by green tea extract samples added with nanogold at concentrations of 5-35 ppm and reacted with 0.005% DPPH solution at 518 nm wavelength for 1 minute, 1 hour, 2 hours, 3 hours, 4 hours. The timing of the interaction of the sample with DPPH is very influential on the decrease of DPPH absorbance. This suggests that the longer the interaction time, the more radical muted by nanogold and green tea extract. DPPH radical damping by the addition of nanogold increased during minute 1 to 3 hours. However, it dropped dramatically when the 4th hour added nanogold concentration of 25 ppm, 30 ppm and 35 ppm. This is because nanogold particles have aggregated to inhibit antioxidant activity at high concentrations of 25 ppm to 35 ppm. The increase was only demonstrated by the addition of nanogold at a concentration of 5 ppm to 20 ppm.

Based on the graph it can be seen that the concentration of 5 ppm to 20 ppm the percent of free radicals damping or scaffolding continues to be increase. The results of optimum and best concentrations were shown at concentrations of 20 ppm having an average percentage of damping 92.4%. Decrease in antioxidant activity was shown at concentrations of 25 ppm to 35 ppm. It is caused that in the higher concentration nanogold cluster will grow larger and aggregation. This condition caused nanogold not be compete to bind the DPPH (free radicals). One way anova analysis was used in this study to test the hypothesis that the addition of nanogold had an effect on the antioxidant activity of green tea extract. The result of anova test gives significant value p <0,05 which shows that there is a real influence between nanogold variable on green tea extract.

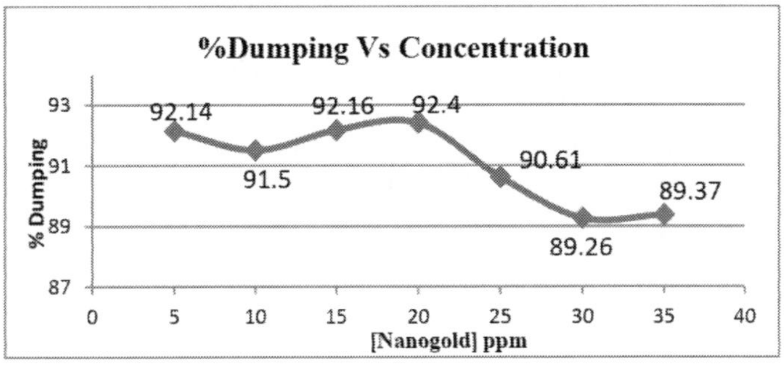

Fig 2. The percentage dumping green tea extract with nanogold

4. Conclusion

The results of antioxidant activity green tea extract 20 ppm was 91,16% that increased after 1 hour, 2 hours, 3 hours and 4 hours were 91,95%; 91.69%; 92.61% and 93.14%. The results obtained that antioxidant activity of green tea extract of 20 ppm were 91,16%. Than antioxidant activity of green tea extract with nanogold 5, 10, 15, 20, 25, 30 and 35 ppm are 90,63; 90,77; 90,90; 91,42; 91,56; 91,42, and 91,42. Nanogold various concentrations give increased the antioxidant activity of green tea extract was not significant. Nanogold various concentrations give unequal effects at the antioxidant activity of green tea extract. The effect at concentrations of 5, 10, 15 and 20 ppm nanogold decreased the antioxidant activity of green tea extract, while the concentration of 20 and 25 ppm of nanogold increased the antioxidant activity of green tea extract. In conclusion to increase the antioxidant activity in cosmetic formulas, the nanogold and green tea extract are included gradually or not simultaneously.

5. References

IOP Conf. Series: Journal of Physics: Conf. Series **1108** (2018) 012109 doi:10.1088/1742-6596/1108/1/012109

[1] Taufikurohmah A, Soegianto A, Sanjaya I G M, Baktir A, and Syahrani A 2013 Mercury exposure effects to skin tissue of mus muscullus at fibroblasts cell proliferation and collagen quantity Res. *J. Pharm. Biol. Chem. Sci.* **4**(4)

[2] Taufikurohmah T, Wardana A P, Tjahjani S, Sanjaya I G M, Baktir A, and Syahrani A 2018 The Clinical Test of Nano gold Cosmetic for Recovering Skin Damage Due to Chemicals: Special Case *J. Phys. Conf. Ser.* **947**(1)

[3] Cosmetics S L 2012 FACT SHEET: Mercury in Skin Lightening Cosmetics pp. 1–4.

[4] Tong J, Wang Y and Lu Y 2015 ScienceDirect In vitro evaluation of inorganic and methyl mercury mediated cytotoxic effect on neural cells derived from different animal species *JES* **41**(I) pp. 138–145

[5] Martinez-finley E J and Aschner M 2014 Recent Advances in Mercury Research pp. 163–171

[6] Straka E, Ellinger I, Balthasar C, Scheinast M, Schatz J, Szattler T, Bleichert S, Saleh L, Knö M, Zeisler H, Hengstschläger M, Rosner M, Salzer H, and Gundacker C 2016 Mercury toxicokinetics of the healthy human term placenta involve amino acid transporters and ABC transporters **340** pp. 34–42

[7] Skalny A V, Skalnaya M G, Bjørklund G, Nikonorov A A, and Tinkov A A 2016 Mercury as a possible link between maternal obesity and autism spectrum disorder Med. Hypotheses **91** pp. 90–94

[8] Rooney J P K 2014 The retention time of inorganic mercury in the brain — A systematic review of the evidence *Toxicol. Appl. Pharmacol.* **274**(3) pp. 425–435

[9] Of T, Activity A, Of D, and Anthocyanin T 2010 Test Of Antioxidant Activity And Determination Of Total Anthocyanin Content In Rind Of Mangosteen (Garcinia mangostana L) **15**(2) pp. 64–70

[10] Taufikurohmah T, Sanjaya I G M, Baktir A, and Syahrani A 2012 Activity Test of Nanogold for Reduction of Free Radicals, a Pre-Assessment Utilization Nanogold in Pharmaceutical as Medicines and Cosmetics **2**(12) pp. 611–617

[11] Taufikurohmah T, Sanjaya I G M, Baktir A, and Syahrani A 2014 Synthesis of Nanogold and Stability Test of This Colloidal as Essential Material in Drug, Supplement and Cosmetics **3**(5) pp. 2012–2015

[12] Alanazi F K, Radwan A A and Alsarra I A 2010 Biopharmaceutical applications of nanogold *Saudi Pharmaceutical Journal* **18**(4) pp. 179–193

[13] Anjarsari I R D 2016 Katekin teh Indonesia: prospek dan manfaatnya Indonesia tea catechin: prospect and benefits *J. Kultiv.* **15**(2) pp. 99–106

[14] Sutarna N A A H and Alatas F 2016 Bahan Aktif Pembuatan Sediaan Krim Tabir Surya **4**(2) pp. 32–35

[15] Rohdiana D and Shabri 2012 Analisis individual katekin teh hijau hasil ekstraksi dan fraksionasi kromatografi kolom *J. Penelit. Teh dan Kina* **15** pp. 81–88

Acknowledgments

Acknowledgments are directly to the Directorate General for Research and Development of the Ministry of Research and Technology, which has provided funding through the Directorate of Research and Community Service in the PUSN financing scheme for the fiscal year 2017-2018.

Exudate Segmentation in Retinal Images of Diabetic Retinopathy Using Saliency Method Based on Region

N Nur[1,*] and H Tjandrasa[2]

[1,2] Department of Informatics Sepuluh Nopember Institute of Technology (ITS) Surabaya, Indonesia

*Corresponding author: [1]nnahya16@gmail.com

Abstract. Diabetic retinopathy is a complication of diabetes that attacks the eye organs. One of the early signs of the disease is the appearance of an exudate wound that occurs because there is lipid in the abnormal blood vessels which can cause blindness if it appears around the macula. Early detection of exudates is expected to reduce the risk of blindness to diabetic retinopathy patients. One of the problems in the detection process of exudates is that the size of the object is quite small compared to the overall image. In this research, we propose the segmentation of exudates with saliency method based on region obtained using intensity thresholding. There are three main stages in this research, namely optic disc removal, location detection of exudates, and exudates segmentation. Optical disc removal is done using midpoint circle algorithm. At the detection stage of the exudates location, the image will be subdivided into smaller sub-images called patches then classified into exudate patch and exudate-free patch based on the intensity thresholding obtained from each image. For a sub-image classified as exudate patch, then it is segmented using saliency method. The evaluation method is done on the diaretDB1 dataset by calculating the value of accuracy, sensitivity, and specificity of the proposed method respectively 99.33%, 81.65% and 99.42%. The proposed method is expected to be able to detect exudates well.

1. Introduction

Diabetic retinopathy is a complication of diabetes that occurs in the eye [1]. One of the early signs of the disease is the appearance of an exudates wound that occurs because there is lipid in abnormal blood vessels and can cause blindness when its happen near of the macula [2]. Early detection of the emergence of exudates is expected to reduce the risk of blindness to diabetic retinopathy patients. Manual evaluation procedure is performed to analyze abnormalities that occur in the retinal fundus image. The diagnosis takes considerable time and the accuracy of the measurement depends entirely on the experience of the ophthalmologist [3]. Early treatment can minimize the possibility of blindness so speed and accuracy in detecting exudates becomes very important. Detection of exudates automatically by performing an analysis of the retinal fundus image is expected to be the solution to the problem.

Several previous studies have attempted to segment the exudates on the fundus image. Harangi B. et al. [4] used several preprocessing methods and mathematic morphological for candidate extraction and then applied the active contour method to the preprocessing result and combined it to obtain the segmentation result. Banerjee S. et al. [5] used the mean shift and normalized cut method to overcome oversegmentation exudates. Both methods directly segmented the whole image.

However, because the size of the exudates is quite small compared to the overall image, the segmentation process is less to give maximum results [6]. In this study, we propose to detect regions containing exudates before segmentation process. Thus, the segmentation process will focus only in specific areas and not segment the overall image. Therefore, in order to obtain a good segmentation result of exudates, the optic disc should be removed because it is not relevant to the segmentation.

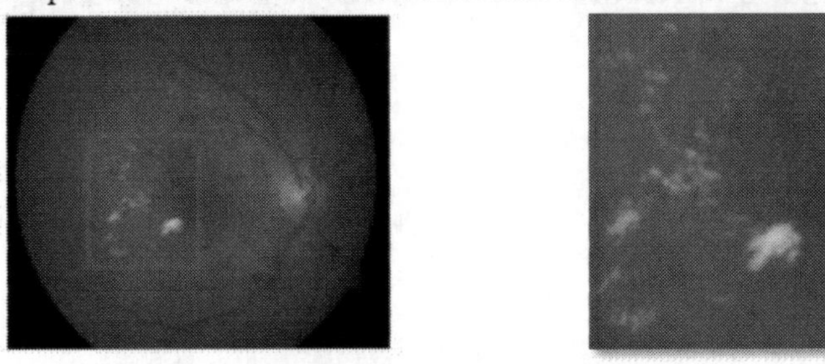

(a) Sample of fundus image (b) Zoom-in exudate

Figure 1. Retinal fundus image contain exudates

2. Methods

The proposed method involves three main stages: (1) optic disc removal; (2) exudates region detection; (3) exudates segmentation.

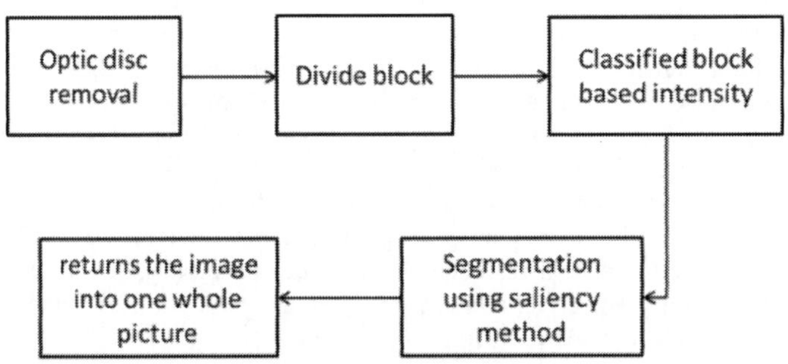

Figure 2. Procedure of proposed method

2.1 Optic disc removal

Optic disc removal is intended to avoid false detection in the segmentation stages due to the color and contrast of optic disk that almost the same as the exudates. Some optical disk segmentation methods include using active contour and Hough transform [7]. In this research, midpoint circle algorithm is used because the result of optic disc segmentation using this method is good enough with less complexity [8]. The first step to remove the optic disc is find the maximum pixel area. After that, the area with the maximum pixels is labeled and the labeled area is combined. Calculation of the area and the centroid in the picture are performed. After the area is acquired, the maximum pixel area is bordered using the midpoint circle algorithm and defined by:

$$x^2 + y^2 - r^2 = 0 \quad (1)$$

The result of the elimination process can be seen in Figure 3 (b).

2.2 Exudates region detection

In this stage, the image will be divided into several parts of smaller images using non-overlapping patch. After that, the patches classified into two category. The patches that contain exudates called the exudate patch and patch that does not contain exudates called the exudate-free patch. The area or patch is detected by using intensity thresholding because areas that containing exudates will be brighter than others [9].

It is favorable to enhance contrast of image so that the exudates can easily be differentiated from the fundus image because exudates brighter than background [10]. In this method, contrast enhancement using morphological processing is employed on the L channel of input fundus image. This contrast enhancement method uses combination of white top hat and black top hat transform of the image L.

Before intensity thresholding is applied for exudates location detection, intensity adjustment is employed to enhance intensity of exudates present in the image. Intensity adjustment is performed by mapping the intensity values in input image to new values in output image such that 1% of data is saturated at low and high intensities of input image [9]. After intensity adjustment, exudates location are detected using thresholding technique. Intensity thresholding is defined by:

$$I_{thres} = z - 0.01 \times z \tag{2}$$

where z is maximum intensity value in the intensity adjusted image I. Illustration of the classification result of each patch can be seen in the figure 3(d). Patches that classified as exudates patches are marked with a red block.

2.3 Exudates segmentation

The last step is to segment the exudates patch using the saliency method then determine a threshold value to acquire the final segmentation. Saliency method could capture the unique colors in the image [11],[12],[13]. So that, the exudates that have a high pixel value can be highlight in the salient image. In this study, the saliency method used refers to the saliency methods by [14]. The method aims to acquire the saliency map $\boldsymbol{S_{map}}$ of image I which has a length of W and width of H pixels formulated as

$$\boldsymbol{S_{map}(x,y)} = ||\boldsymbol{I_{\mu}} - \boldsymbol{I_{whc}(x,y)}|| \tag{3}$$

Where $\boldsymbol{I_{\mu}}$ is the mean pixel value of the image and $\boldsymbol{I_{whc}}$ is a Gaussian blurred vector version of the original image to remove unwanted noise and $||.||$ is the norm L2. LAB color space used in this method and L2 norm is Euclidean distance. The result of exudates patch segmentation using saliency method shown in figure 3(e). After the segmentation process, the last step is to recombine the smaller images into one whole picture as shown in figure 3(f).

2.4 Evaluation Method

The DiaretDB1 dataset [15] is used to evaluate the proposed method by calculating the value of accuracy, sensitivity, and specificity that defined by equation (4),(5), and (6).

$$\text{Accuracy} = \frac{(TP+TN)}{(TP+TN+FP+FN)} \tag{4}$$

$$\text{Sensitivity} = \frac{TP}{(TP+FN)} \tag{5}$$

$$\text{Specificity} = \frac{TN}{(TN+FP)} \tag{6}$$

where True Positive (TP) is the number of pixels that classified as exudates according to both the proposed method and groundtruth, False Positive (FP) is the number of pixels that classified as exudates using proposed method but categorized as non-exudates on the groundtruth, False Negative (FN) is the number of pixels that classified as non-exudates according to proposed method but categorized as exudates on the groundtruth, and True Negative (TN) is the number of pixels that classified as non-exudates using proposed method and also categorized as non-exudates on groundtruth.

3. Results and Discussion

Result of the proposed method can be seen in figure 3 below. Figure 3(a) is original image which is obtained from DiaretDB1 dataset and 3(b) is the result of optic disk removal. After the optic disc eliminated from the original image, then the result image divided into smaller part called patch as shown in figure 3(c). Patch size is set 32x32 pixel. After that, classification using intensity thresholding is performed to obtain exudates patch which will be used in segmentation process. Each exudates patch segmented using the saliency map and then combine all patches into one whole picture as shown in figure 3(f).

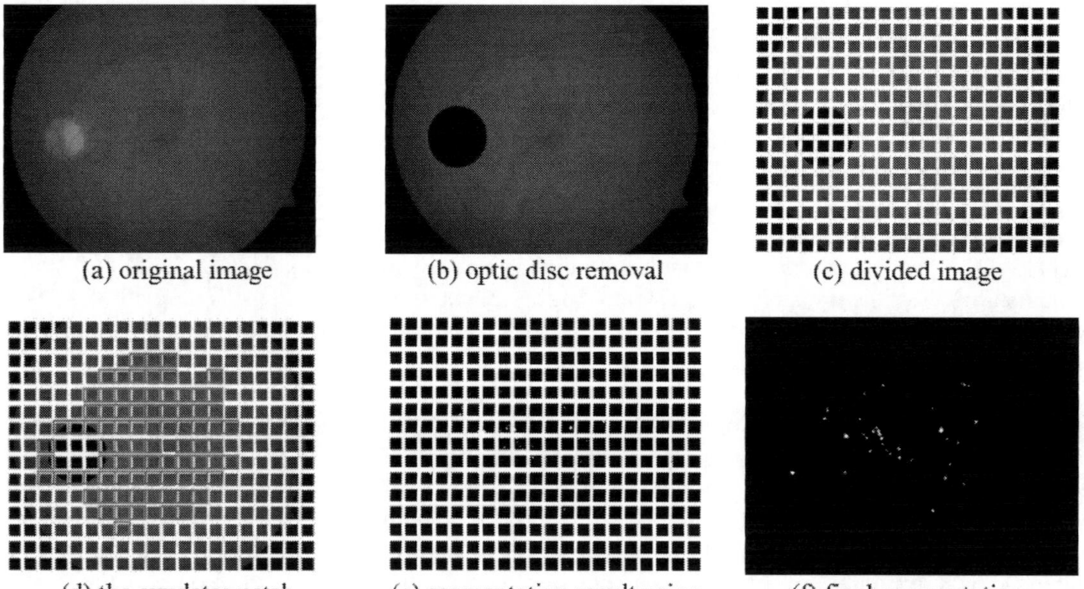

(a) original image (b) optic disc removal (c) divided image

(d) the exudates patch classification results (e) segmentation result using saliency method of each patch (f) final segmentation

Figure 3. Result of the proposed method

The result of exudates segmentation using the proposed method shown in figure 4. Figure 4 (a) is the sample of original image, 4(b) is the segmentation result using proposed method, and 4(c) is ground truth image. Detected exudates are marked with white objects that shown in figure 4(b) below.

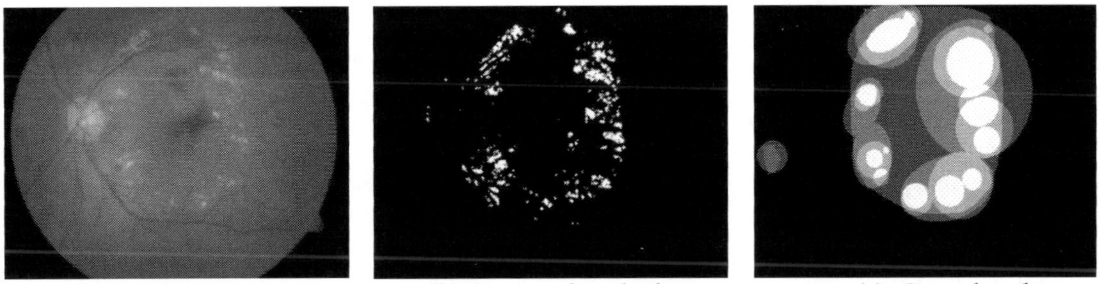

(a) Original image (b) Proposed method (c) Ground truth

Figure 4. Segmentation result using proposed method

Table 1. Exudates segmentation result

Image	Accuracy (%)	Sensitivity (%)	Specificity (%)
1	99.87	81.78	99.89
2	98.12	77.05	98.42
3	99.41	97.81	99.41
4	99.36	75.35	99.39
5	99.68	85.95	99.69
6	99.45	76.5	99.47
7	98.96	72.47	99.22
8	99.01	79.09	99.23
9	99.7	75.21	99.76
10	99.73	95.3	99.74
Average	**99.33**	**81.65**	**99.42**

Based on Table 1, the result obtained that the average accuracy value of proposed method reached 99.33%. Sensitivity and specificity value of the proposed method respectively 81.65% and 99.42%. The sensitivity value of overall images obtained is quite low because there are still remains of optic disc that is not erased perfectly. In addition, in the detection location exudates process, there are misclassified patches because of the characteristic of the image is quite dark. There are block that contain exudates but categorized as exudates-free patch and vice versa. It affects the segmentation process that causes the sensitivity level is not optimal.

4. Conclusion

In this paper, instead of directly segment the exudate on the entire image, certain areas that containing exudates are detected first. There are the main step in this paper, such as optic disk removal, exudates region detection, and exudate segmentation. Average value of accuracy, sensitivity, and specificity of the proposed method respectively 99.33%, 81.65% and 99.42% which shows that the proposed method can segment the exudate well. Segmentation results are influenced by the classification of patches so that for the future work, the segmentation result can improve with more accurate patch classification method.

5. References

[1] Tjandrasa H 2015 Classification of Non-Proliferative Diabetic Retinopathy Based on Segmented Exudates using K-Means Clustering pp. 1–8

[2] Qomariah N and Handayani T 2017 Exudate Detection in Retinal Fundus Images using Combination of Mathematical Morphology and Renyi Entropy Thresholding

[3] Kaur J and Mittal D 2017 A generalized method for the segmentation of exudates from pathological retinal fundus images *Integrative Medicine Research* **38**(1) 27–53
[4] Harangi B and Hajdu A 2014 Automatic exudate detection by fusing multiple active contours and regionwise classification *Computers in Biology and Medicine* **54** 156–171
[5] Banerjee S and Kayal D 2016 Detection of hard exudates using mean shift and normalized cut method *Integrative Medicine Research* **36**(4) 679–685
[6] Liu Q, Zou B, Chen J, Ke W, Yue K, and Chen Z 2017 A location-to-segmentation strategy for automatic exudate segmentation in colour retinal fundus images *Computerized Medical Imaging and Graphics* **55** 78–86
[7] Tjandrasa H, Wijayanti A, and Suciati N 2012 Optic Nerve Head Segmentation Using Hough Transform and Active Contours **10**(3) 531–536
[8] Tjandrasa H, Wijaya A Y, Arieshanti I, and Salyasari N D 2013 Segmentation of Hard Exudates in Retinal Fundus Images using Fuzzy C-Means Clustering with Spatial **4** 39–44
[9] Singh A, Sengar N, and Kishore M 2015 Automatic Exudates Detection in Fundus Image using Intensity Thresholding and Morphology **4** 330–334
[10] Mariko S, Ryo K, Jonathan E N, Tien Y W, Ecosse L, Jie J W 2013 Quantitative Measurement of Hard Exudates in Patients With Diabetes and Their Associations With Serum Lipid Levels *Investigative Ophthalmology & Visual Science* **54** 5544-5550
[11] Itti L, Koch C, and Niebur E 1998 A Model of Saliency-Based Visual Attention for Rapid Scene Analysis **20**(11) 1254–1259
[12] Achanta R, Hemami S, Estrada F, Sabine S, and Epfl D L 2009 Frequency-tuned Salient Region Detection (Ic) **4** 1597–1604
[13] Wei Y, Wen F, Zhu W, and Sun J 2012) Geodesic Saliency Using Background Priors
[14] Achanta R, Estrada F, Wils P, and Sabine S 2012 Salient Region Detection and Segmentation
[15] Kauppi T, Kalesnykiene V, Kamarainen J K, Lensu L, Sorri I, Raninen A, Pietila J 2007 the DIARETDB1 diabetic retinopathy database and evaluation protocol *Procedings of the British Machine Vision Conference* **12** 15.1-15.10

Acknowledgement
This publication was partially supported by Indonesia Endowment Fund for Education (LPDP) Ministry of Finance Republic Indonesia.

IOP Conf. Series: Journal of Physics: Conf. Series **1108** (2018) 012111 doi:10.1088/1742-6596/1108/1/012111

Development 3D Animated Story as Interactive Learning Media with Lectora Inspire and Plotagon on Direct and Inverse Proportion Subject

E Oktavianingtyas[1,*], F S Salama[1], A Fatahillah[1], L A Monalisa[1], T B Setiawan[1]

[1,2,3,4,5] Mathematics Education, University of Jember, Indonesia

*Corresponding Author: rvien_okta@ymail.com

Abstract. Industrial Revolution 4.0 made everything connected with technology. In this case, the use of technology on the learning process could improve TPACK skill. The aim of this research was to produce an interactive learning media in the form of 3D animated story with Lectora Inspire and Plotagon on direct and inverse proportion subject that can be accessed online on computer or android. This media was designed based on Problem Based Learning model where 3D animated story was used to illustrate the problems and explained them until the students found the direct and inverse proportion formula. This research employed Research and Development (RnD) that used ADDIE model which consist of six phases, they are analyze, design, development, implementation, and evaluation. The result showed that the interactive learning media and its user manual book were categorized very high valid category with correlation coefficient α equal to 0,86 and 0,97, percentage of the students responses was 88,44% which means that the media in practical category. Based on the test result, the comprehension of the concepts was improved. This meant that the media was effective. In general, the media was rated valid, efficient, and effective. This means the media is feasible to be used in mathematics learning.

1. Introduction

The 21st century presents itself as a challenging time demanding a particular set of skills [1]. One of skill that should be had by the teacher is TPACK. TPACK or Technological Pedagogical Content Knowledge is a framework to integrate technology in the teaching activity. TPACK has been chosen as the basis for identifying what has been described as the desired qualities of teachers of the next generation. Expert teachers of the 21st century, are those who possess the knowledge of how ICT can be used to teach [2]. Therefore the use of technology in learning activity can improve TPACK skill on teachers. Three inspects knowledge in TPACK are technological knowledge, content knowledge, and pedagogical knowledge with interaction between them [3]. Scheme of TPACK can be seen in the Figure 1.

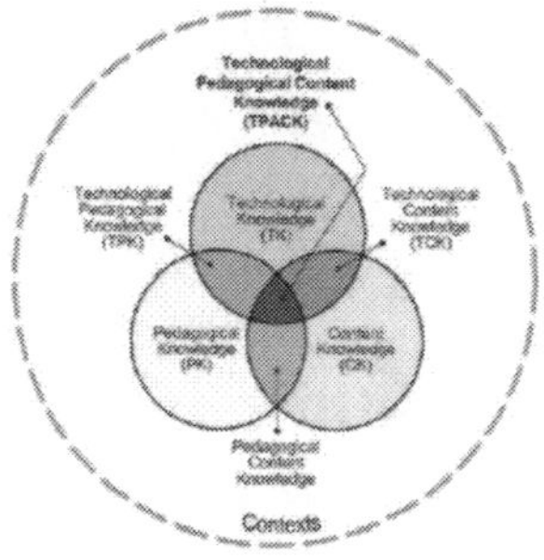

Figure 1. TPACK Scheme

In globalization era, there are some issues about the development of technology. One of them is industrial revolution 4.0. Industrial revolution 4.0 does not influence only the production itself, but also the labor market and the educational system as well [4]. Due to the fact, the use of technology in education should be developed. The aim of developing the use of technology is to improve the quality of education itself. One of the ways that can be used by the teacher to receive the industrial revolution 4.0. is using technology as a media such as the use of interactive media in the mathematics learning activity. Media information or ICT can help visualize abstract concepts and be able to engage students' active roles in learning, so it can reduce learning difficulties because mathematics is abstract materials [5].

Mathematics is one of lesson that should be understood by people. Mathematics is not only for science knowledge but it is as a tool which is used for solving the problem in daily life [6]. The use of interesting media in mathematic lesson can increase the students' desire, interest, motivation, and simulating them to learn. The application of media also makes the process of learning activity is more effective and interesting [7]. Educational technologies for the students develop a positive attitude towards mathematics, increase interest and provide benefits such as reducing a fear [8].

One of the technologies which can be used as the media is the audiovisual technology. As stated by Turan on his study, the effect of audio visual supported to learn mathematics [9]. In general, information and communication technology (ICT) has been suggested as a means to improve educational outcomes [10]. Animation story is one of the audiovisual technology which is used by the teacher as the media of learning. Akamca on his study stated that the use of cartoon in the learning process had positive impact on the students' achievement [11].

The concept of cartoon is one of the ways to expand the students' knowledge and fix misunderstanding of the students about the phenomenon of scientific. Moreover, learning math by using mathematics cartoon is more fun than conventional learning so that the process of learning activities are more effective and the students motivated to raise their achievement in mathematics learning [12]. *Software* which is used to make animation story 3D is Plotagon and Lectora Inspire is used to make 3D animated story as learning media.

One of the industrial revolution 4.0 program is IoT (Internet of Thing) or IoE (Internet of Everything) [13]. It shows that the use of internet is widely used in industrial revolution 4.0. Therefore, animation media 3D is made as online media on the website study Schoology so that the students can access it every time and everywhere.

One of the mathematics materials which is tough as difficult materials by the students is direct and diverse proportion subject. According to Raharjati, the students are difficult in differentiate direct and inverse proportion subject [14]. Therefore, it needed mathematics learning media which emphasize on how to differentiate direct and inverse proportion subject.

Based on the discussion above, the researcher wanted to do a research about developing 3D animated story as interactive learning media using *Lectora Inspire* dan *Plotagon* on direct and inverse

IOP Conf. Series: Journal of Physics: Conf. Series **1108** (2018) 012111 doi:10.1088/1742-6596/1108/1/012111

proportion subject, valid, practice, and effective in teaching and learning activities, that can be accessed by the students by online using computer or gadget

2. Methods

This kind of this research was Research and Development (RnD). This research includes qualitative research that conducted at MTS 1 Negeri Jember and the participants were 15 students from VII F. The aim of this research was to develop an interactive media with 3D animated story in Direct and Inverse proportion subject that valid, effective, and efficient. The development follow ADDIE style, they are Analysis, Design, Development, Implementation, and Evaluation [15]. In analysis phase, collected some data from analyzed curriculum, the character of the students, and the materials. These data were used as a base to design what the media looks like. In Design phase consist of determined the media, the strategy of explaining the materials, also made story board of media before create the media with Lectora Inspire software. In this phase also made an instrument of validation sheet for validation process. In development phase consist of constructed the media with Lectora Inspire and Plotagon software according to storyboard, and evaluated the media by validators. In this research developer used Lectora Inspire 16 and Plotagon 1.22. The media was validated by three people who from Mathematics Education University of Jember and MTS 1 Negeri Jember. At this phase would be determined the media in valid category or not. If the media in valid category, the step will continue to the implementation phase. In this phase also published the media by uploading on the website Schoology so the students could access it online with computer or android. In Implementation phase consist of tried the media and collected students' response data and did the test of learning achievement. In evaluation phase, analyzed the data which was gained from the questionnaire and the test result. At this phase would be determined the media in effective and efficient category or not.

According to Fanani, a good media should fulfill the aspects of quality, they are (1) validity; (2) the effectiveness; (3) efficiency [16]. Therefore in this research used three research instruments for each category, they are validation sheet that was used to measure the validity of learning media, questionnaire response that was used to know the response from the students after they applied the media to measure the efficient of media, and the test of learning achievement which was used to know the accomplishment of the aim teaching and learning activity and the effectiveness of the media. The validation sheet and the questionnaire used Likert scale and the question of test was taken from mathematics national book. The media is valid if the correlation of coefficient (α) include high or very high category is $0{,}60 \leq \alpha$. The media is effective if 80% subject fulfill the learning exhaustiveness in tryout. This research used standard value from MTS Negeri 1 Jember which is 75. Media is efficient if the average score of the questionnaire from the participants $P \geq 80\%$.

3. Results and Discussion

3.1. Learning media

This research used animation story as learning media that was made as interactive media by using software Plotagon dan Lectora Inspire. This could be accessed online by using computer or gadget. Hopefully the use of animation story could help the students in understanding direct and inverse proportion subject.

There are four menus in the beginning of media, they are materials, exercises, author and competence. The explanation of the materials and the exercises can be found on the menu of materials while the tasks are available on the menu of rehearsal. The menu of author consist of the profile of the author and the menu of competence consist of main competence and basic competence based on curriculum 2013. The figure of the media can be seen in figure 2 until figure 5.

Figure 2. The Beginning Page of Media

Figure 3. Subject Discussion Page 1

Figure 4. Subject Discussion Page 2

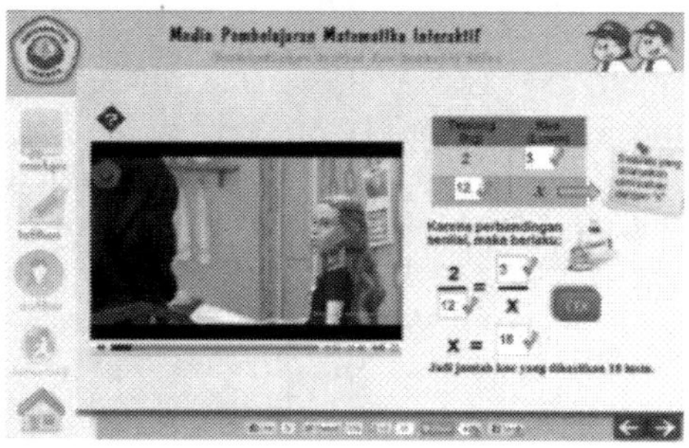

Figure 5. Exercise Page

The model used to explain the materials was Problem Based Learning where animation story had a function as visualization of the problem used. The explanation of the materials was divided into four pages. In the first page, there is an animation video that tells two friends who discuss about proportion subject. In the end of the video, there is a table that should be completed by the students. The first page can be seen on figure 3. The students could continue the discussion on the next page if they could complete all the questions on the table with right answer. In the next page, the animation video was played once again where the character in the video explained the answer on the previous page. After the video was end, there emerged next questions. It can be seen in figure 4. There are two types of exercises in each material. The exercises also were made in the form of animation video. The exercises page can be seen in figure 5.

3.2. Validity of learning media

There are six aspects of assessment on the attempt of properness learning media, they are the ease of use and navigation, cognitive load, knowledge space and information presentation, media integration, aesthetics, and overall functionality which were measured by the validator. The result of the validity by three validators in each aspect is explained in figure 6.

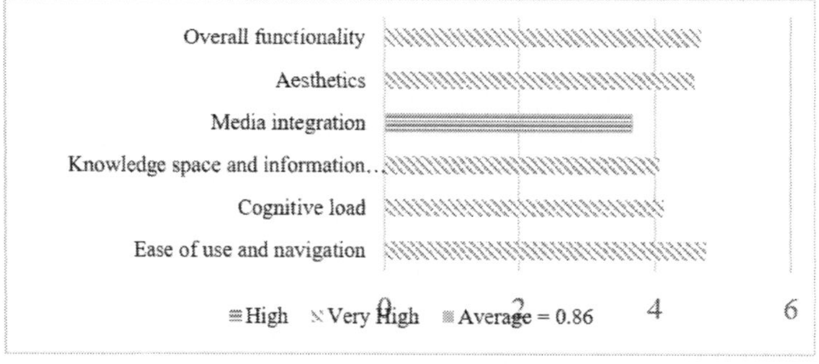

Figure 6. Validity of Learning Media

On the validity process, the researcher got the coefficient α equal to 0,86. This proved that the learning media used was valid with very high category. Therefore, this media could be used in the process of teaching and learning activities. There were two aspects measured in the manual book of learning media validation instrument, they were technique presentation and language. The result of validity in each aspect by validator showed on the figure 7.

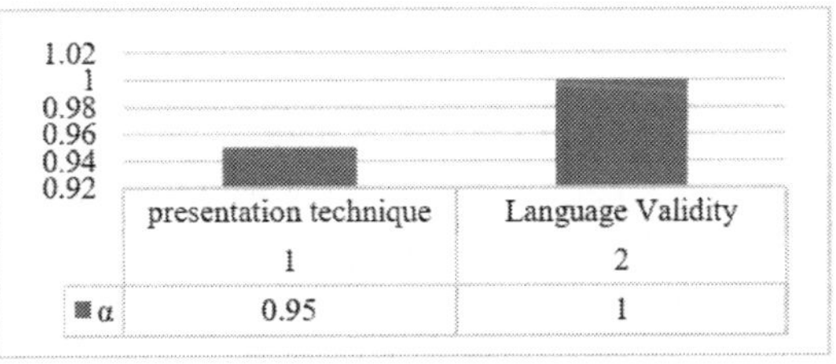

Figure 7. Validity of User Manual Book

Based on the result of the data validation book, the researcher got the total average value 0.97 from two aspects. It showed that the manual book of learning media had very high category in validity.

3.3. The efficiency of learning media

The efficiency of learning media measured based on the questionnaire which had been fulfilled by the students after applying this learning media. The result of the students' response in each aspect showed in the figure 8.

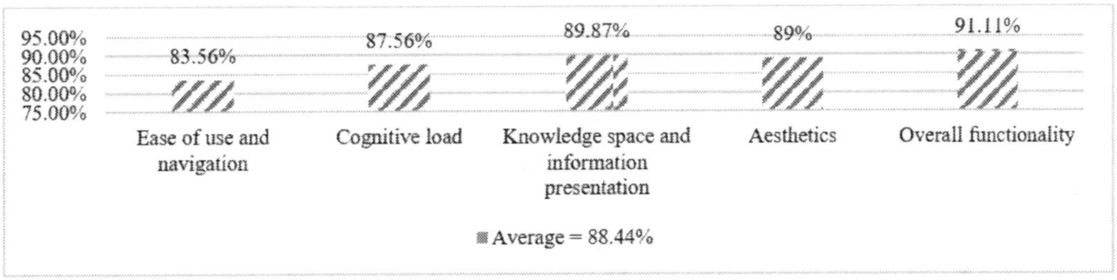

Figure 8. The Result of Questionnaire

Based on the questionnaire filled up by the students, the researcher got 88,44% the total average score from all aspects, so that the learning media developed in the category efficiency.

According to the questionnaire on the aspect of the ease of use and navigation, the subject assessed the media could be operated easily and the instruction was easy to be understood. The subjects who seldom to use computer felt that the media was easy to be operated. Although, there were some students who needed instruction in some pages. In the cognitive load, the subjects measured that the problem and the tasks offered on the video were easy to be understood. The materials were also easy to be understood and the questions were made based on the materials. in the aspect knowledge space and information presentation, the subject measured that the video using the language clearly, using right spelling, and explaining the materials clearly. The video also had good series in plot story. The materials in the video also explained systematically. in aesthetics aspects, the subjects measured that the video was interesting, the color used on the media was appropriate, the alphabet used could be read easily, and the video was clear enough. In overall functionality of all aspects, the subjects measured that this media could help them to understand the materials. Moreover, they felt happy and have an interest to study by using this media. This showed that this learning media could increase the students' interest for mathematics learning and made them feel happy so they did not feel bored in mathematics learning. The result of this study supported the research which was conducted by Turan that the effect of audio visual support to study mathematics [9]. Supriadi on his study said that learning mathematics by using math cartoon is more interesting than conventional learning so the teaching and learning activities was more effective and the students motivated to increase their achievement in mathematics learning [12].

3.4. The effectiveness of learning media

Attempting the effectiveness of learning media which was developed by the researcher was effective. It can be seen from percentage the result of the students' posttest was 80%. It can be concluded that the learning media was effective. It is supported by the research conducted by Akamca that the use of cartoon in the atmosphere of study had positive impact on the students' achievement [11]. Therefore, interactive learning media in the form of animation story measured as a feasible media that could help the students in learning direct and inverse proportion subject.

Moreover, the measurement of the effectiveness media, the result of the students' achievement was used to see preferment of the students' value after using this learning media. This research used first score as standard of comparison score which was got from the daily test by using appropriate materials on seven days before attempting the learning media had been done. The comparison between first score and posttest score from the 15 subjects could be seen in the figure 9.

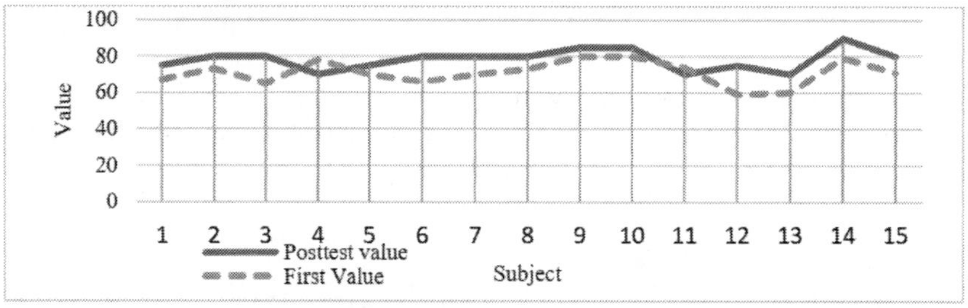

Figure 9. Collation of Scores

In the first score, from fifteen subjects there are seven students who did not get score which was appropriate with the standard score from the school. From the comparison between first score and the posttest score, the score was increase and appropriate with the standard score. It is supported by Akamca on his study who said that the use of cartoon or animation could give positive influence on the students [11].

4. Conclusion

Based on the data analysis and discussion above, it can be concluded that 3D animated story as learning media by using Problem Based Learning Model help the students in understanding the problem and also help the students to find direct and inverse proportion formula. Moreover, the learning media had high validity category with the coefficient validity 0,86 while the manual book had high validity category with coefficient validity 0.97. It proved that the media and its user manual book are proper to be used in the mathematics teaching and learning activities. In addition, the learning media had category efficient with the percentage of average score 88,84%. This proved that the students felt interesting and assisted after using this media. However, the learning media also was effective. It was proved from the percentage of posttest score, it was 80%. It can be concluded that 3D animated story as learning media helped the students in understanding the materials about direct and inverse proportion subject. The media was valid, efficient, and effective so that the media can used in the process of teaching and learning activities.

5. References

[1] Esdras P, Omar C, Daniel B, Bernd M, and Henrique R 2018 Exploring gamification tosupport manufacturing education on industy 4.0 as an enable for innovation and sustainability *Procedia Manufacturing* **21** pp. 438-445
[2] Vicente C R, Christine R, Helen D, and Sue G 2017 Integrating ICT into teacher education programs from a TPACK perspective: Exploring perceptions of university lecturers *Computers & Education* **115** pp 1-24.
[3] Matthew J K, Punya M, Mete A, and Joshua M R 2013 *The Technological Pedagogical Content Knowledge Framework for Teachers and Teacher Educators* (Canada: Commonwealth Educational Media Centre for Asia)
[4] Andrea B and Jiří T 2017 Requirements for Education and Qualification of People in Industry 4.0. *Procedia Manufacturing* **11** pp. 2195-2202
[5] Evi W, Budayasa I K, and Masriyah *Proc. Int. Conf. on Mathematics* **27** p. 274
[6] Ozdamli F 2012 Pedagogical framework of m-learning *Procedia Social and Behavioral* **31** pp 927-931
[7] Hariharan N K, Arsaythamby V, and Ng S H 2013 Perception of teacher towards media usage in teaching mathematics in secondary schools *Procedia Social and Behavioral Sciences* **112** pp. 1093-1098
[8] Burcu T 2014 Smart Board İn Mathematics Education, The Use of Cartoon Characters İmpact On Student Success *Procedia Social and Behavioral Sciences* **143** pp 809-815
[9] Burcu T 2014 The Opinions of Teacher on The Use of Cartoon Character in The Mathematics Lesson *Procedia Social and Behavioral Sciences* **141** pp. 1386-1391
[10] Keri L S and Bradley J 2017 Blogging mathematics: Using technology to support mathematical explanations for learning fractions *Computer and Education* **111** pp. 1-43
[11] Güzin Ö A, Murat E, and Hülya H 2009 Effects of computer aided concept cartoons on learning outcomes *Procedia Social and Behavioral Sciences* **1** pp. 296-301
[12] Supriadi 2008 Penggunaan Kartun Matematika dalam Pembelajaran Matematika *Jurnal Pendidikan Dasar* **6** pp. 1-6
[13] Saurabh V, Prashant A, and Santosh B 2018 Industry 4.0 - A Glimpse *Procedia Manufacturing* **20** pp. 233-238
[14] Raharjanti, Meliyana, Toto N and Sri M 2016 *Proc. Nat. Conf. on Mathematics and Its Learning* (Surakarta: University of Muhammadiyah Surakarta)
[15] Meli Y, Ihsan, and Subaer Development of Interactive Learning Media on Kinetic Gas Theory at SMAN 2 Takalar *Journal of Physics* **812** pp. 1-6
[16] Risa M, F Dafik and Arif F 2017 Pengembangan Media Pembelajaran Interaktif Online Menggunakan Edmodo Berbantuan Software Geogebra pada Materi Sistem Persamaan Linier Dua Variabel *Kadikma* **8** pp. 78-85

Acknowledgments

Praise and great gratitude to Almighty God submitted by the authors for blessings and help to complete this paper on time. On this occasion with great humility, we would like to thank you to Headmaster and mathematics teacher at MTsN 1 Jember for given us a facility to applied the media. Next for validator from the Lectures of Mathematics Education, University of Jember for evaluating the media and test instrument.

IOP Conf. Series: Journal of Physics: Conf. Series **1108** (2018) 012112 doi:10.1088/1742-6596/1108/1/012112

Machine Learning with Partially Homomorphic Encrypted Data

K Muhammad[1], K A Sugeng[1,*] and H Murfi[1]

[1]Department of Mathematics, Universitas Indonesia, Kampus UI Depok, Depok 16424, INDONESIA
E-mail: [1]khalid.muhammad@sci.ui.ac.id, [1,*]kiki@sci.ui.ac.id, [1]hendri@ui.ac.id

Abstract. Machine learning had been widely used to analyze various kinds of data, including sensitive data such as medical and financial data. A trained machine learning model can be wrapped in a web application so that people can access it easily via internet. However, if the data to be analyzed is private or confidential, this will cause a problem; the application administrator may read the input. As shown by Dowlin et al. in their remarkable paper, this kind of problem can be solved with homomorphic encryption scheme. Paillier encryption scheme is one kind of encryption scheme that has homomorphic property. In this research, we will show that one type of machine learning model can take an input encrypted by Paillier encryption scheme and produce an encrypted output that shares the same key. A machine learning model will be trained with the MNIST database of hand-written digits. This model will be tested with the test data encrypted with Paillier encryption scheme. The experiment shows that the model achieved 92.92% accuracy on the test set.

1. Introduction

Machine learning had been widely used in various fields, from computer vision, natural language processing, to medical and finance. Goodfellow et al. defined machine learning algorithm as an algorithm that is able to learn from data; machine learning enables us to tackle tasks that are too difficult to solve with fixed programs written and designed by human beings [1]. A well-trained machine learning model can be wrapped in a web application so that people can access it easily via internet, Google Translate is such an example of this. However, people won't use it to analyze a private data because the application administrator may read their input. As shown by Dowlin et al., homomorphic encryption schemes have the ability to overcome this kind of problem. It preserves the structure of the message space such that we can perform operations such as addition and multiplication over the ciphertext space [2]. With a homomorphic encryption scheme, one can encrypt his/her data in advance before he/she inputs it to the machine learning model. The machine learning model will produce an encrypted output that shares the same key. The person with the key can then decrypt it. This decrypted output will have the same value to the output that are produced by taking the unencrypted data as an input. This way, the person can still uses the machine learning model with his/her private data without worrying that the application administrator will read his/her data. Furthermore, the application administrator also can't read the output produced by the application because it is still in an encrypted form.

This research is inspired by the remarkable work of Downlin et al. [3], they trained a deep convolutional neural network model with the MNIST database [4] of handwritten digits. The trained model is then modified to what they call CryptoNets, neural networks that can take an input encrypted by a homomorphic encryption. They use YASHE' encryption scheme, a fully homomorphic encryption scheme, to encrypt the test set. The CryptoNet model achieved 98.9% accuracy when tested with the encrypted test set.

One of the encryption schemes that has a homomorphic property is Paillier encryption scheme, firstly introduced in 1999 by Pascal Paillier on his paper "Public-key cryptosystems based on composite degree residuosity classes" [5]. This encryption scheme is proven to be semantically secure against chosen-plaintext attack, moreover it is more efficient than Goldwasser-Micali cryptosystem, as well as the provably-secure RSA and Rabin schemes [6]. Paillier encryption scheme supports addition of ciphertexts and multiplication of a ciphertext with a plaintext (an unencrypted scalar). With this property in hand, we can do matrix multiplications and vector additions. Similar to what Dowlin et al. has shown, we will see that a trained shallow neural network model can be simplified to a matrix multiplication and a vector addition, hence this model can take an input encrypted by Paillier encryption scheme. We will also demonstrate this idea by training a model with the MNIST database [4].

2. Paillier Encryption Scheme

Paillier encryption scheme is an asymmetric encryption scheme. It uses (n, g) as the public-key, where n is a product of two big prime numbers p and q, and g is an element of $Z^*_{n^2}$ such that its order is a multiple of n. The two prime numbers p and q are used as a secret-key.

One version of this encryption scheme sets g equals to $n + 1$ and requires (p, q) to be a prime of equal length (in their binary representation). The requirement assures that $\phi(n) = (p - 1)(q - 1)$ has an inverse modulo n. The inverse of $\phi(n)$ is denoted by μ and used as the secret-key [5]. The scheme is summarized in Figure 1.

Key Generation:

- $p, q \in P$ with equal length
- $n = pq$
- $g = 1 + n$
- $\phi(n) = (p - 1) \cdot (q - 1)$
- $\mu = \phi(n)^{-1} \; mod \; n$ (μ is used as *private-key*)

Encryption:

- Plaintext $m < n$ ($m \in Z_n$)
- Choose $r < n$ dengan $\gcd(r, n) = 1$ randomly ($r \in Z^*_n$)
- Ciphertext $c = g^m . r^n \; mod \; n^2$

Decryption:

- Ciphertext $c < n^2$
- Plaintext $m = L(c^\lambda \; mod \; n^2) \cdot \mu \; mod \; n$

Figure 1. Paillier encryption scheme

Suppose that two plaintexts m_1 and m_2 are encrypted to $e(m_1) = g^{m_1}.r_1^n \bmod n^2$ and $e(m_2) = g^{m_2}.r_2^n \bmod n^2$ where $r_1, r_2 \in Z_n^*$. We can see that

$$e(m_1) \cdot e(m_2) \bmod n^2 = (g^{m_1}.r_1^n \bmod n^2) \cdot (g^{m_2}.r_2^n \bmod n^2) \bmod n^2$$

$$e(m_1) \cdot e(m_2) \bmod n^2 = g^{m_1}.r_1^n.g^{m_2}.r_2^n \bmod n^2$$

$$e(m_1) \cdot e(m_2) \bmod n^2 = g^{m_1+m_2}.(r_1.r_2)^n \bmod n^2.$$

Since g has order n in $Z_{n^2}^*$ and $r_1, r_2 \in Z_n^*$ then

$$e(m_1) \cdot e(m_2) \bmod n^2 = g^{m_1+m_2 \bmod n}.(r_1.r_2 \bmod n)^n \bmod n^2, \text{ where } r_1.r_2 \bmod n \in Z_n^*$$

$$e(m_1) \cdot e(m_2) \bmod n^2 = e(m_1 + m_2 \bmod n) \quad (1)$$

Furthermore, if $m_1 + m_2 < n$ then

$$e(m_1) \cdot e(m_2) \bmod n^2 = e(m_1 + m_2).$$

Because of this property, Paillier encryption scheme is called a partially homomorphic encryption scheme. This property can be generalize to addition of n plaintexts

$$e(m_1) \cdot e(m_2) \cdot \ldots \cdot e(m_n) \bmod n^2 = e(m_1 + m_2 + \cdots + m_n \bmod n).$$

Now suppose that k is a positive integer then

$$e(m_1)^k \bmod n^2 = (g^{m_1} \cdot r_1^n \bmod n^2)^k \bmod n^2$$

$$e(m_1)^k \bmod n^2 = (g^{m_1} \cdot r_1^n)^k \bmod n^2$$

$$e(m_1)^k \bmod n^2 = (g^{m_1})^k \cdot (r_1^n)^k \bmod n^2$$

$$e(m_1)^k \bmod n^2 = g^{k.m_1} \cdot \left(r_1^k\right)^n \bmod n^2$$

Since g has order n in $Z_{n^2}^*$ and $r_1 \in Z_n^*$,

$$e(m_1)^k \bmod n^2 = g^{k \cdot m_1 \bmod n} \cdot \left(r_1^k \bmod n\right)^n \bmod n^2, \text{ where } r_1^k \bmod n \in Z_n^*$$

$$e(m_1)^k \bmod n^2 = e(k \cdot m_1 \bmod n) \quad (2)$$

Furthermore, if $km_1 < n$ then

$$e(m_1)^k \bmod n^2 = e(k \cdot m_1)$$

We will make use of these two properties to build a system where a machine learning model can take an input encrypted by Paillier encryption scheme and produce an output encrypted with the same key.

3. The Machine Learning Model

First, we will train a neural network model with description given as below:

1. Input layer with 784 neurons corresponding to flattened array of size 28x28 of the MNIST image data.
2. First hidden layer: fully connected (dense) layer with 256 neurons and identity function as the activation function
3. Second hidden layer: fully connected layer with 128 neurons, also having identity function as the activation function
4. Output layer with 10 neurons corresponding to 10 classes of the MNIST database (image of handwritten digits of the integers from 0 to 9).
5. There are $(784 \times 256) + (256 \times 128) + (128 \times 10) = 234.752$ weights parameter and $256 + 128 + 10 = 394$ biases parameter that can be optimized in the training process.

This model can be written in a form of matrix multiplications as below:

$$a^{(2)} = W^{(2)}a^{(1)} + b^{(2)} \quad (3)$$

$$a^{(3)} = W^{(3)}a^{(2)} + b^{(3)}$$
$$\hat{y} = a^{(4)} = softmax\left(W^{(4)}a^{(3)} + b^{(4)}\right)$$

where:

- $a^{(1)}, a^{(2)}, a^{(3)}$, and $a^{(4)}$ respectively is a column vector of length 784, 256, 128, and 10.
- $b^{(2)}, b^{(3)}$, and $b^{(4)}$ respectively is a column vector of length 256, 128, 10.
- $W^{(2)}$, $W^{(3)}$, and $W^{(4)}$ respectively is a matrix of size 256x784, 128x256, 10x128.
- $\hat{y}$ or $a^{(4)}$ is the output of the model in the form of a column vector of length 10.

After the training phase, the weights and biases will have a fixed value, so the model can then be simplified as

$$\hat{y} = softmax\left(W^{(4)}\left(W^{(3)}\left(W^{(2)}a^{(1)} + b^{(2)}\right) + b^{(3)}\right) + b^{(4)}\right)$$
$$\hat{y} = softmax\left(W^{(4)}W^{(3)}W^{(2)}a^{(1)} + W^{(4)}W^{(3)}b^{(2)} + W^{(4)}b^{(3)} + b^{(4)}\right)$$
$$\hat{y} = softmax\left(W'^{(2)}a^{(1)} + b'^{(2)}\right),$$

where

$$W'^{(2)} = W^{(4)}W^{(3)}W^{(2)} \text{ and } b'^{(2)} = W^{(4)}W^{(3)}b^{(2)} + W^{(4)}b^{(3)} + b^{(4)},$$

is a matrix of size 10x784 and a column vector of length 10 respectively [7].

Because of the softmax activation function in the output layer, this model cannot take an input encrypted with Paillier encryption scheme yet, softmax function is not a function that only consists of addition and scalar multiplication. However, we will see that the model is able to predict the input correctly without computing the softmax function in the output layer.

Neural networks model determine the inputs' class by looking at the index of the maximum value of $\hat{y}$. On the other hand, softmax function dosen't change the order of its input, for example if the input vector is $x = (8,7,10,6,9)$, $softmax(x) = (0.086, 0.031, 0.636, 0.011, 0.234)$, we see that the elements of x and $softmax(x)$ have the same order. In general, suppose that $x = (x_1, x_2, \dots, x_n)$ and WLOG we can assume that $x_1 < x_2 < \cdots < x_n$, since e is greater than 1 we can see that $e^{x_1} < e^{x_2} < \cdots < e^{x_n}$, finally since $\sum e^{x_i} = e^{x_1} + e^{x_2} + \cdots + e^{x_n}$ is positive then $\frac{e^{x_1}}{\sum e^{x_i}} < \frac{e^{x_2}}{\sum e^{x_i}} < \cdots < \frac{e^{x_n}}{\sum e^{x_i}}$, thus we have proven that in general case the elements of x and $softmax(x)$ have the same order. We can now safely drop the softmax function in the output layer of the model. The model is now simplified to

$$\hat{y} = W'^{(2)}a^{(1)} + b'^{(2)}. \tag{4}$$

Suppose that Alice have successfully trained a model that has a structure like (3) which she then simplify to (4). She then builds a web application so that people can access it easily via internet. Bob wants to use the model with his private data $m = [m_1\ m_2 \dots m_{784}]^T$. The output that Bob looking for is $\hat{y} = Wm + b$. Because m is private, Bob will encrypt it first to $e(m) = [e(m_1)\ e(m_2) \dots e(m_3)]^T$ which he then inputs it to Alices' app. In order to get $e(\hat{y})$, the app have to compute

$$\begin{bmatrix} \left(\left(e(m_1)^{w_{1,1}} \cdot e(m_2)^{w_{2,1}} \cdot \dots \cdot e(m_{784})^{w_{784,1}}\right) \cdot b_1\right) mod\ n^2 \\ \left(\left(e(m_1)^{w_{1,2}} \cdot e(m_2)^{w_{2,2}} \cdot \dots \cdot e(m_{784})^{w_{784,2}}\right) \cdot b_2\right) mod\ n^2 \\ \vdots \\ \left(\left(e(m_1)^{w_{1,10}} \cdot e(m_2)^{w_{2,10}} \cdot \dots \cdot e(m_{784})^{w_{784,10}}\right) \cdot b_{10}\right) mod\ n^2 \end{bmatrix} \tag{5}$$

which equals to

$$\begin{bmatrix} \left(e\left(m_1 \cdot w_{1,1} + m_2 \cdot w_{2,1} + m_{784} \cdot w_{784,1} + b_1\right)\right) mod\ n^2 \\ \left(e\left(m_1 \cdot w_{1,2} + m_2 \cdot w_{2,2} + m_{784} \cdot w_{784,2} + b_2\right)\right) mod\ n^2 \\ \vdots \\ \left(e\left(m_1 \cdot w_{1,10} + m_2 \cdot w_{2,10} + m_{784} \cdot w_{784,10} + b_{10}\right)\right) mod\ n^2 \end{bmatrix}$$

$$= [e(\hat{y}_1)\ e(\hat{y}_2) \dots e(\hat{y}_{10})]^T = e(\hat{y})$$

Therefore, using equation (5) Alices' app can outputs $e(\hat{y})$ when given $e(m)$ as input. Bob can decrypt this this output to $\hat{y}$, i.e. the value that he was looking for from the trained model. This way, Bob can still obtain the value that he was looking for from Alices' model without having to give his private data to Alice. Furthermore, Alice, as the application admin, won't be able to read the output—which may be confidential as well—since it is in an encrypted form. Figure 2 gives an illustration on how this concept works.

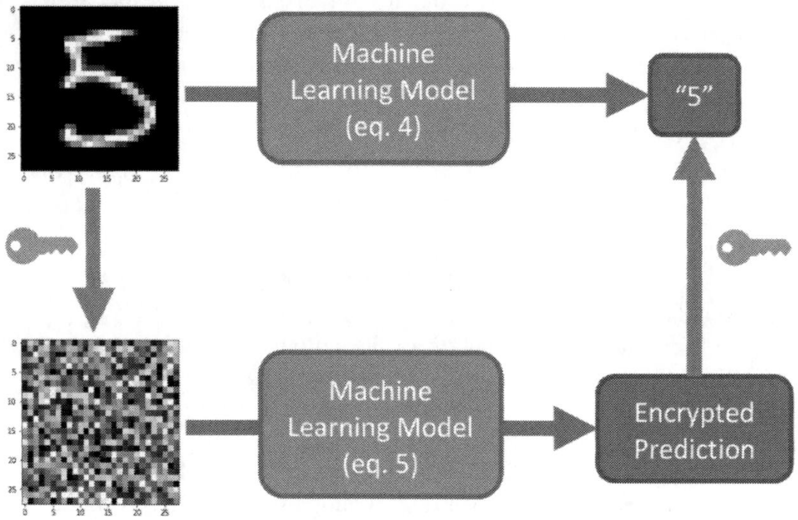

Figure 2. Summary

4. Empirical Result

We will implement the whole concept with Python programming language [8]. Three main library that we will use are Keras [9] for training the neural network model, NumPy [10] for matrix operations, and python-paillier [11] for using Paillier homomorphic encryption scheme. We use matplotlib [12] to plot the training progress.

We train the model in 150 epochs with Adam [13] as the optimizer. The best model is obtained in the 122th epoch with an accuracy of 92.92% on the test set. Figure 3 illustrates the training progress. After the training phase, we can extract the weights and biases of the trained model to NumPy arrays. We can then simplify the model to (4) using NumPy array operations. This simplified model—the model that doesn't have softmax on the output layer—gave the same output as the initial model. This confirms what we have discussed in the last section.

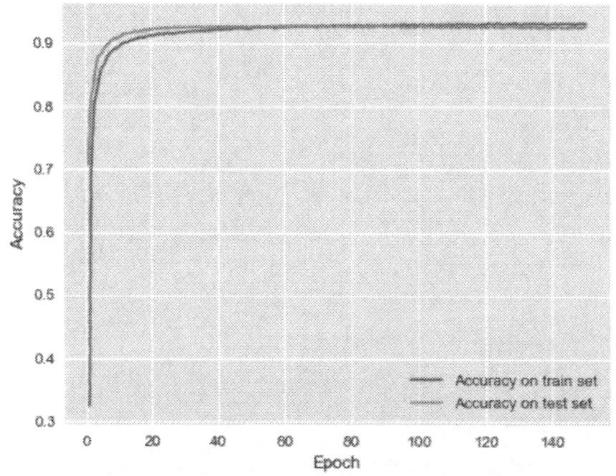

Figure 3. Training progress

The test set is encrypted pixel-by-pixel using phe, the encoding from floating point numbers to integers is handled by phe. Using equation (5), we get an encrypted prediction of the test set. After we decrypt the output of equation (5), we see that it results the same class prediction with the simplified model, it also gives a 92.92% accuracy.

5. Conclusion

Paillier encryption scheme is homomorphic over addition and scalar multiplication operation of ciphertexts. We have trained a neural network model and simplify the model to one matrix multiplication and one vector addition. By using the weights and biases of this simplified model, we can generate an encrypted prediction for private data.

Fully homomorphic encryption scheme such as FV [14] and YAHSE' [15] can be used for further research. This type of encryption allows addition and multiplication of ciphertexts, using these two operations we can compute polynomials, and by polynomials there are quite many thing we can do such as approximating functions using Taylors' theorem. Hesamifard, Takebi, & Ghasemi used approximation of ReLU function with polynomials to be used as an activation function in a neural network hidden layer [2]. Applying this concept for financial or medical data will also be interesting.

6. References

[1] Goodfellow I, Bengio Y, and Courville A 2016 Deep learning (Cambridge: MIT press)
[2] Hesamifard E, Takabi H, and Ghasemi M 2017 CryptoDL: Deep Neural Networks over Encrypted Data *Preprint arXiv:1711.05189*
[3] Dowlin N, Gilad-Bachrach R, Laine K, Lauter K, Naehrig M, and Wernsing J 2016 Cryptonets: Applying Neural Networks to Encrypted Data with High Throughput and Accuracy *International Conference on Machine Learning* pp 201-10
[4] LeCun Y 1998 The MNIST Database of Handwritten Digits http://yann.lecun.com/exdb/mnist/
[5] Paillier P 1999 Public-key Cryptosystems Based on Composite Degree Residuosity Classes *International Conference on the Theory and Applications of Cryptographic Techniques* pp 223-38 (Springer, Berlin, Heidelberg)
[6] Katz J and Lindell Y 2014 *Introduction to Modern Cryptography* (CRC press)
[7] Nielsen M A 2015 Neural Networks and Deep Learning (Determination Press) http://neuralnetworksanddeeplearning.com
[8] Van Rossum G 2007 Python Programming Language *USENIX Annual Technical Conference* **41** p 36
[9] Chollet F 2015 Keras *GitHub repository* https://github.com/keras-team/keras
[10] Oliphant T E 2006 A Guide to NumPy (USA: Trelgol Publishing)
[11] N1 Analytics 2014 python-paillier *GitHub repository* https://github.com/n1analytics/python-paillier
[12] Hunter J D 2007 Matplotlib: A 2D Graphics Environment *Computing in Science & Engineering*
[13] Kingma D P and Ba J 2014 Adam: A Method for Stochastic Optimization *Preprint arXiv*:1412.6980
[14] Fan J and Vercauteren F 2012 Somewhat Practical Fully Homomorphic Encryption *IACR Cryptology ePrint Archive*
[15] Bos J W, Lauter K, Loftus J, and Naehrig M 2013 Improved Security for a Ring-based Fully Homomorphic Encryption Scheme *IMA International Conference on Cryptography and Coding* pp 45-64 (Springer, Berlin, Heidelberg)

Acknowledgments

This research is funded by PITTA-UI 2018 Research Grant [Project Number = 2277/UN2.R3.1/HKP.05.00/2018].

IOP Conf. Series: Journal of Physics: Conf. Series **1108** (2018) 012113 doi:10.1088/1742-6596/1108/1/012113

The Potential of Sidoarjo Marine Product: *Corbula* Sp. as Therapeutic Nutrition to Improve Immune System

E T Evanda[1], Isnawati[1], and Y T Rachmad[2,*]

[1]Biology Department, Universitas Negeri Surabaya, Ketintang, Surabaya 60231, East Java, Indonesia

[2]Biology Department, Faculty of mathematics and natural science, Brawijaya University, Jl veteran, Malang, East Java, 65145, Indonesia

*Corresponding Author: yoga.rachmad@gmail.com

Abstract. The regulation of immune responses are controlled by hormone and gene regulation. Micro-mineral such as Selenium is the precursor for immune response based on glutathione hormone activity. The glutathione hormone mainly works against diseases such as HIV/AIDS, avian influenza, and cancer. The aim of this study was to explore the selenium levels and antioxidant activity marine product *Corbula* sp in Sidoarjo East Java. The *Corbula* sp. was obtained Candi district, Sidoarjo. Measurement of selenium concentration was performed by atomic absorption spectroscopy. Measurement of antioxidant activity was performed by FRAP (ferric reducing antioxidant power). The selenium measurement result showed that *Corbula* sp was contained 0.28 ppm. The antioxidant activity was identified by increasing levels at 517 nm. The study was indicated that *Corbula* sp in Sidoarjo East Java, contains selenium levels to increase antioxidant activity as improvement of immune response with a potential function as alternative nutrition.

1. Introduction

The consumption pattern is influenced by the accessibility of purchasing product, ease of access, knowledge of foodstuffs and culture [1]. Indonesians tend to choose diet with high cholesterol such as fast food, risking to a healthy condition. Immediate changes in food consumption patterns often lead to increased incidences of chronic diseases in the community [2].

One of the emerging factors of chronic disease is an increase in the number of antigens that potentially disrupt the condition of cell homeostasis. Changes in homeostasis will lead to impair immune responses in the body in the form of cell degradation. The most common antigen is free radicals. Free radicals can be formed through normal physiological activities in the body or due to the influence of exogenous species. Exogenous species may take the form of a natural compound present in the environment (ozone, NO_2, ethanol), industrial chemical compounds (carbon tetrachloride) or xenobiotics arising from metabolic activity [3]. Radicals that often arise from the results of metabolism is superoxide (O_2-1) which then subsequently dismutase into hydrogen peroxide (H_2O_2). Superoxide dismutase is a catalyst enzyme in the formation of hydrogen peroxide [4]. These compounds become indicators of the process that produces superoxide in the body. Taking into account the effect of free radical-mediated reactions, there is a strong evidence that free radicals play an important role in promoting the development of chronic events affecting health [3].

The free radicals are compounds in nature where the atom or its constituent molecule has unpaired electrons in its own independent outbound orbital and is reactive to other surrounding compounds [5]. Naturally, the presence of free radicals in cellular environment can be controlled by the equilibrium between radical and antioxidant compounds.

The 21st century is beginning of the development period research on alternative nutritional products from local resources that have not been explored optimally. The study is an attempt to find alternative solutions to overcome health problems such as free radicals. The results of Hartmann and Meisel's research (2006) show up that to 2006 there were 10 food products with the role as an ACE inhibitor, immunomodulator, cytomodulator, an opioid agonist, an opioid antagonist, anti-thrombotic, anti-microbial, mineral binder, anti-hyper-cholesterol and antioxidant [6]. With the large potential of local food products as a source of healthy nutrition, of course, Indonesia has an important role by abundant natural resources.

As an archipelagic country with 60% of its territory compound by waters, Indonesia offers a large number of marine commodities. One of them is the abundance of *Corbula* sp.; one of the many fishery products in Indonesia [7]. *Corbula* sp. consumption rate reached 10 thousand tons per year, is a source of animal protein consumed by many people. These fishery products are easy to get at a relatively cheap price, making it affordable for the whole community. *Corbula* sp. is classified as triploblastic stomata. The size of the *Corbula* sp. varies considerably between 1 -2 cm long, with a circular body and enveloped by a cupped shell [8]. The potential of *Corbula* sp. as a food product is not limited to macronutrients. Micronutrient *Corbula* sp. content is a potential to be explored and has a role in support quality of consumer health. Based on the previous research Saurabh and Sahoo [8], describe the micronutrients of fishery products that have a great role as an antioxidant precursor compound is selenium [9].

Selenium is an important mineral in protein synthesis and enzyme activity of glutathione peroxidase (GSH-Px). Selenium in the synthesis of glutathione peroxidase has a role as a catalyst in the breakdown of peroxides formed in the body into non-toxic bonds [10]. Peroxides may turn into free radicals that can oxidize unsaturated fatty acids present in cell membranes, thus damaging cell membranes. Based on these roles, it is known that selenium works with vitamin E and acts as an antioxidant. The role of vitamin E by maintaining cell membranes from free radicals by releasing hydrogen ions, selenium plays a role in breaking the peroxide into non-reactive bonds so as not to damage the unsaturated fatty acids present in the membrane, helping to maintain the integrity of the membrane and protect DNA from damage [11–13]. The integrity of cell membranes is indispensable in the immune system because the production of cytokines is largely determined by the receptors contained in the cell membrane, therefore selenium is necessary to enhance cellular immunity [12,14].

Various role of selenium that supported by the big potency of *Corbula* sp. product in Sidoarjo, this research aims to explore and analyze the content of selenium and antioxidant activity level from *Corbula* sp. as a healthy nutrition alternative for the community. The study was conducted with *Corbula* sp. exploration in the eastern coastal area of Sidoarjo, followed by a test of selenium and antioxidant activity in *Corbula* sp.

2. Methods

The study was conducted from January to June 2013 consisting of exploration and measurement of selenium concentration and anti-oxidant activity. The exploration of *Corbula* sp (*Corbula faba* Hind) was conducted in Sidoarjo east coast, at Balongbendo village where *Corbula* sp is intensively produced. Exploration was done through collection, identification, and determination over the sample.

Selenium concentration in the sample was analyzed using Shimadzu model AA 6300G atomic absorption spectrometer (Kyoto, Japan) with graphite furnace (GFA-EX 7i), a selenium hallow cathode lamp as radiation source (Hamamatsu Photon-ics, Japan) at the 196.0 nm wave length with a slit width of 0.7 nm, 23 mA current and deuterium background corrector, was used for measurement. A L'vov total pyrolitic graphite platform tube (Shimadzu) was used. The pyrolitic graphite platform was modified according to the procedure described in previous study [15]. The samples injection volume was 20 µL. Integrated absorbance (peak a rea) was used for quantitation analysis.

The pH values were measured with a Metrohm pH- meter (model: 691, Herisau, Switzerland) supplied with a glass-combined electrode. The Centurion Scientific centrifuge (model 1020D, UK) was used for centrifuging.

Antioxidant activity of Corbula sp. was analized using 1,1-Diphenyl-2-picryl-hydrazyl (DPPH) reagent, was purchased from Sigma–Aldrich (USA). Ascorbic acid and butylated hydroxytoluene (BHT) were purchased from E. Merck (India). Propyl gallate was purchased from Life Science Central Laboratories, Malang, Indonesia. Stock solutions of DPPH were prepared in methanol, and methanol buffered with acetic acid buffer (0.1 M, pH 5.5), respectively. Buffered methanol was prepared by mixing 40 ml of 0.1 M acetate buffer (pH 5.5) with 60 ml methanol. The solvents and other chemicals were of analytical grade. The reaction tubes, in triplicates, were wrapped in aluminum foil and kept at 30°C for 30 min in dark. All measurements were done under dim light. Spectrophotometric measurements were done at 517 nm using Spectronic Genesys 5 spectrophotometer [16]. The data are mean ± SD.

3. Results and Discussion

The following table and graph represent the observation data about selenium concentration and antioxidant activity.

Table 1. The results of testing selenium and protein in *Corbula* sp.

Samples	Selenium (µg/g)	Protein (mg/g)
Corbula	0.279 ± 0.09	15.05 ± 0.5

The table 1 show that *Corbula* sp. containing selenium 0.279 µg/g and protein 15.05 mg/g ($p < 0.05$)

A

B

C

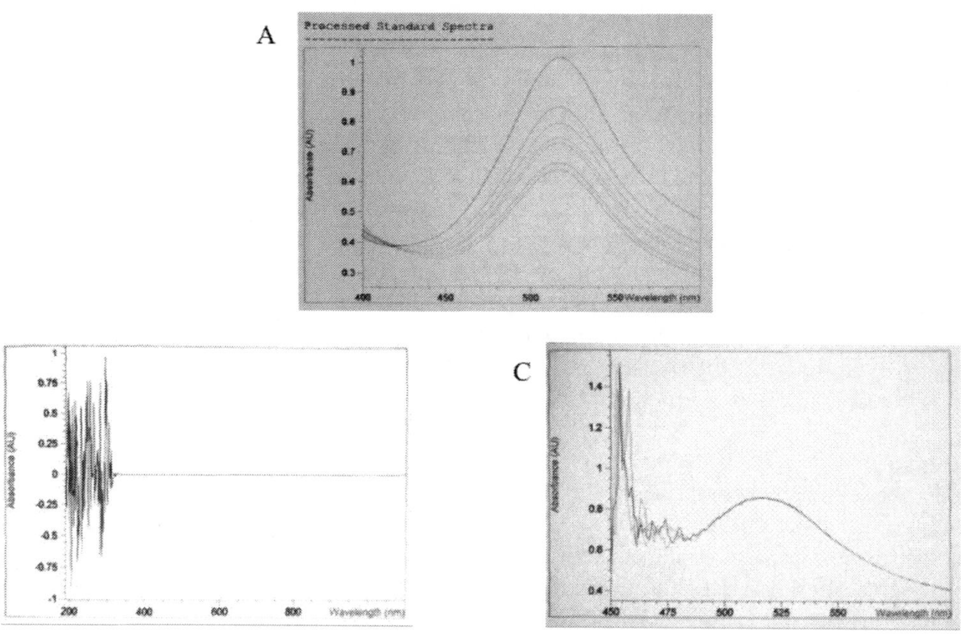

Figure 1. antioxidant activity of *Corbula* sp. A. Optimal wavelength spectrum using reagent *1,1-diphenyl-2-picrylhydrazyl. B.* antioxidant activity control (-) without *Corbula* sp. C. Antioxidant activity of *Corbula* sp. was identified by increasing levels at 517 nm

The relationship between disease and nutrition has been shown to be related one another [17]. Consumption of nutrients have an influence on the metabolic system. This fact supports that in the state of nutritional deficiency the immune system will be depressed and increase the risk of metabolic changes [1].

In conditions of immune system suppression, the glutathione level tends to decrease. Decreased glutathione levels are enhanced by nutritional supplements. Some cases of deficiency of selenium or glutamic acid, glycine, cysteine, and methionine lead to increased indications of decreased glutathione enzymes [17]. In addition, low selenium significantly decreased the activity and sensitivity of immune cells during the incubation period of infection.

Problem-solving with the efforts of the fulfillment of nutrition sources that support metabolism activity is one effort that can be carried out by society in overcoming health problem [18]. By utilizing the potential *Corbula sp* that has the content of selenium, glutamic acid, glycine, cysteine and methionine that can be used as a therapy in HIV infection. will be used by the body to produce glutathione and glutathione peroxidase [19].

The process of glutathione formation in the body by the presence of glutamic acid, glycine, cysteine, and methionine content in the *Corbula sp* through the digestive tract of proteins in general, begins in the stomach with the help of stomach acid and pepsin enzymes, then continues in the small intestine by the aid of protease enzymes and pancreatic enzymes to into free amino acids, some of which are glutamic acid, glycine, cysteine, and methionine. This free amino acid subsequently enters the blood circulation through the portal vein and is brought into the blood circulation of the organ which is then used as the synthesis of glutathione [20].

Of the four amino acids, glycine and glutamate are the ready-to-use amino acids for glutathione synthesis; but for cysteine and methionine, those will change first into cysteine [3]. The change of cysteine to cysteine is mediated by the aid of GSH-cystine transhydrogenase enzyme, while for the conversion of methionine into cysteine is mediated by several enzymes, namely: methionine adenosyl transferase, methyltransferase, SAH hydrolase, cystathionine beta-synthase and vitamin B6 aids [21]. The cysteine formed later together with glutamate will be converted to L-γ-glutamyl cysteine by the aid of the glutamate-cysteine ligase enzyme. Furthermore, L-γ-glutamyl cysteine joins the glycine to form glutathione through the intermediate enzyme glutathione synthase [13].

The glutathione that has been formed will be converted to glutathione peroxidase by the aid of selenium. Glutathione and glutathione peroxidase act as antiretroviral to antigens. An antiviral that plays glutathione occurs through 2 reactions. First, through the blockade of oxidative stress that occurs due to metabolic abnormalities. Oxidative stress plays a role in activating the transcription of cellular factor NF-B (nuclear factor-kappa B) [22]. NF-B plays a role in improving and maintaining the cellular inflammatory cell. Glutathione may interfere with the complex phosphorylation process of NF-B and active NF-B release, or active NF-B delivery process to the nucleus so that the inflammatory condition can be suppressed [23].

Glutathione plays a role in inhibiting antigen activity by activating macrophage cells. Macrophages play a role in the infection process and counteract the cytopathic effects of the antigen [1]. The role of glutathione delivered to macrophage cells by neutrophil cells can play a role in enhancing and regulating the activity of macrophages before the antigen can be overcome. The results of in vitro study note that antigen activity can decrease significantly due to the macrophages that have high glutathione expression [13]. Meanwhile, according to a recent study conducted by Hurwitz and colleagues at the University of Miami in Florida, USA for 9 months of 174 HIV-infected patients (91 patients grouped with 200 μg selenium and 84 others in the placebo group) found that selenium could maintain HIV viral load and even in some patients, selenium can lower HIV viral load averaging as much as 10,000. Selenium also increases CD4 counts in HIV-infected patients [12].

4. Conclusion

Selenium is consisted in *Corbula* sp. important for co-activator glutathione peroxidase enzyme. Consumption of selenium source has a potential role to improve immune response based on glutathione peroxidase enzyme metabolisms pathway. The abundant number of *Corbula* sp. in Sidoarjo was a potential source for health nutrition.

5. Reference

[1] Calder P C 2013 Feeding the immune system. *Proceedings of the Nutrition Society*, **72**, 299–309

[2] Dubie T, Sisay T, Zeru F, Gebru M, Muktar Y 2015 The potential application of avian egg antibodies with emphasis on immunotherapeutic and immunodiagnostic purpose **7** 145–158,

[3] Shah, Sudhir V, Radhakrishna Baliga, Mohan Rajapurkar, & Vivian a Fonseca 2007 Oxidants in Chronic Kidney Disease *Journal of the American Society of Nephrology : JASN* **18** (1) 16–28 doi:10.1681/ASN.2006050500

[4] Gough D R and Cotter T G 2011 Hydrogen peroxide: a Jekyll and Hyde signalling molecule. *Cell death & disease* **2**(e213)

[5] Kumar V and Cotran R S 2007 *Buku Ajar Patologi edisi 7 volume 2* (JAkarta: Penerbit Buku Kedokteran EGC)

[6] Whitney E and Rady R S 2011 *Understanding nutrition* ISBN 0309069491

[7] Brander K M 2007 Global fish production and climate change *Proceedings of the National Academy of Sciences of the United States of America* **104** 19709–19714

[8] Cowey C B and Sargent J R 1972 Fish Nutrition *Advances in Marine Biology* **10** 383–494

[9] Saurabh S and Sahoo P K 2008 Lysozyme: An important defence molecule of fish innate immune system *Aquaculture Research* **39** 223–239

[10] Letelier M E, Molina-Berríos A, Cortés-Troncoso J, Jara-Sandoval J, Holst M, Palma K, Montoya M, Miranda D, González-Lira V, 2008 DPPH and oxygen free radicals as pro-oxidant of biomolecules *Toxicology in Vitro* **22** 279–286

[11] Rayman M P 2012 Selenium and human health *Lancet* **379** 1256–1268

[12] Gill H and Walker G 2008 Selenium, immune function and resistance to viral infections *Nutrition and Dietetics* **65** 41–47

[13] Vunta H, Belda B J, Arner R J, Reddy C C, Vanden Heuvel J P, Prabhu K S 2008 Selenium attenuates pro-inflammatory gene expression in macrophages *Molecular Nutrition and Food Research* **52** 1316–1323

[14] Magor K E 2011 Immunoglobulin genetics and antibody responses to influenza in ducks *Developmental and Comparative Immunology* **35** 1008–1017

[15] Ng P C and Henikoff S 2006 Predicting the Effects of Amino Acid Substitutions on Protein Function *Annual Review of Genomics and Human Genetics* **7** 61–80

[16] Pyrzynska K and Pękal A 2013 Application of free radical diphenylpicrylhydrazyl (DPPH) to estimate the antioxidant capacity of food samples *Analytical Methods* **5** 4288–4295

[17] Brambilla D, Mancuso C, Scuderi M R, Bosco P, Cantarella G, Lempereur L, Di Benedetto G, Pezzino S, Bernardini R 2008 The role of antioxidant supplement in immune system, neoplastic, and neurodegenerative disorders: A point of view for an assessment of the risk/benefit profile. *Nutrition Journal* **7** 1–9

[18] Naylor R L, Goldburg R J, Primavera J H, Kautsky N, Beveridge M C, Clay J, Folke C, Lubchenco J, Mooney H, Troell M 2000 Effect of aquaculture on world fish supplies *Nature* **405** 1017–24

[19] Da-Silva I, Ribeiro A, Canal C, Trevizan L, Macagnan M, Gonçalves T, Hlavac N, de Almeida L, Pereira R 2010 The impact of organic and inorganic selenium on the immune system of growing broilers submitted to immune stimulation and heat stress *Revista Brasileira de Ciência Avícola* **12** 247–254

[20] Duggal S, Chugh T, Das, Duggal A K 2012 HIV and malnutrition: Effects on immune system. *Clinical and Developmental Immunology* **42**(3)

[21] Rayman M P 2012 Selenium and human health *The Lancet* **379** 1256–1268

[22] Vlassara, Helen, Torreggiani M, Post J B, Zheng F, Uribarri J, and Striker G E 2009 Role of

Oxidants/inflammation in Declining Renal Function in Chronic Kidney Disease & Normal Aging. *Kidney International. Supplement* **76**(114). Elsevier Masson SAS: S311. doi:10.1038/ki.2009.401.

[23] Martati T, Pancasila U 2010 Penetapan Mineral Selenium Dalam Produk Suplemen Kesehatan: Aplikasi Metode Nuklir Dalam Bidang Farmasi 28 2–3

IOP Conf. Series: Journal of Physics: Conf. Series **1108** (2018) 012114 doi:10.1088/1742-6596/1108/1/012114

Marshall-Olkin Extended Inverse Weibull Distribution and Its Application

R M Pakungwati[1], Y Widyaningsih[2]and D Lestari[3]

[1,2,3] Department of Mathematics, Faculty of Mathematics and Natural Sciences, Universtity of Indonesia, West Java 16424, Indonesia

Email: [1]rtmtiara@gmail.com, [2]widyaningsihyekti@gmail.com, [3]dian.lestari@sci.ui.ac.id

Abstract. Weibull distribution is the most popular distribution in wind speed energy literature. However, in real life, the wind speed data may not always modelled by Weibull distribution. An alternative in modeling wind speed data is the inverse Weibull distribution. Inverse Weibull distribution is modifications from the Weibull distribution with the transformation variables. Marshall and Olkin (1997) introduced an interesting method of adding a parameter to a well established distribution so we extend the Invers Weibull distribution by the Marshall-Olkin method (IWMO). The probability density function (pdf), cumulative distribution function (cdf), hazard rate, survival function, moment and quantiles of IWMO are derived. We also discuss the estimation of the model parameters by maximum likelihood. The IWMO distribution was applied to wind speed data. The results were given which illustrate the IWMO distribution and were compared to Weibull distribution and Inverse Weibull distribution. Model comparison using the log likelihood, AIC, and BIC showed that IWMO fit the data better than the other distributions.

1. Introduction

Weibull distribution has attracted the attention of statisticians working on theory and methods as well as in various fields of applied statistics [1]. Two-parameter Weibull Distribution is commonly-used and is an accepted distribution in the wind energy literature [2]; see Souloukngа et al [3] using Weibull distribution for wind speed in Faya-Largeu region and Ozay et al [4] in Alaçatı region. However, in real life, the wind speed data may not always be modeled by using the Weibull distribution. In the other words, it may not represent all wind speed characteristic encountered in nature. Therefore, the alternative distributions are used in such cases like Inverse Weibull (IW) Distribution, as introduced by Keller et al see[5]. Akgül et al.[6] used IW distribution for modelled wind speed in two different stations of Turkey, i.e. Bursa and Sakarya, and show that Invers Weibull better than Weibull distribution.

IW distribution is modification from the Weibull Distribution with the transformation variables. Let Y is a continuous random variable with Weibull distribution with parameters θ, τ. Let $F_W(y)$ is the cumulative distribution function (cdf) of Weibull distribution, and $f_W(y)$ the probability distribution function (pdf) of Weibull distribution, then [7]

$$F_W(y) = 1 - e^{-\left(\frac{y}{\theta}\right)^{\tau}}, \tag{1}$$

$$f_W(y) = \frac{\tau\left(\frac{y}{\theta}\right)^{\tau} e^{-\left(\frac{y}{\theta}\right)^{\tau}}}{y}, \tag{2}$$

With θ is scale parameters and τ is shape parameters.
The transformed variable $= \frac{1}{Y}$, has the cdf $F_{iw}(z)$ and pdf $f_{iw}(z)$

$$F_{iw}(z) = e^{-(z\theta)^{-\tau}}, \tag{3}$$

$$f_{iw}(z) = \tau\theta^{-\tau}z^{-(\tau+1)}e^{-(z\theta)^{-\tau}}, \tag{4}$$

Where $y, x, \theta, \tau > 0$
The equation (3) and (4) is cdf and pdf of IW distribution respectively. The pdf of IW distribution generally exhibits a long right [1].

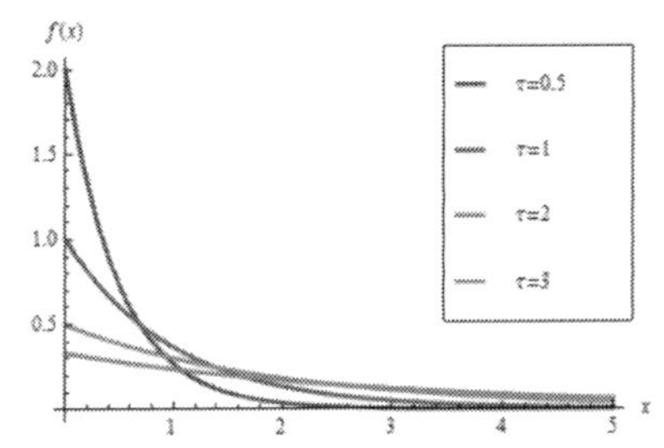

Figure 1. Plots of the pdf of the Weibull for θ = 1

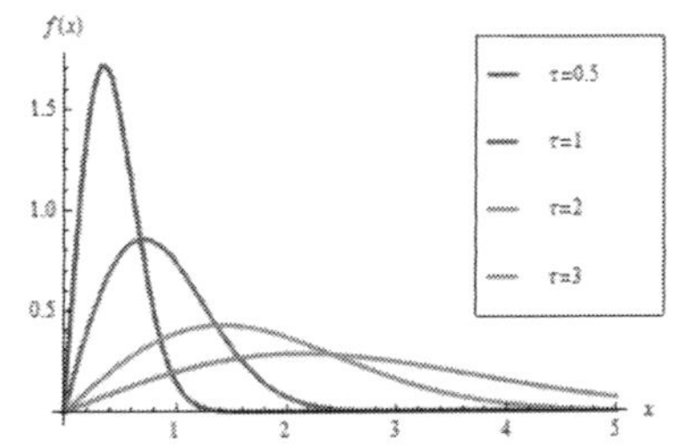

Figure 2. Plots of the pdf of the Weibull for θ = 2

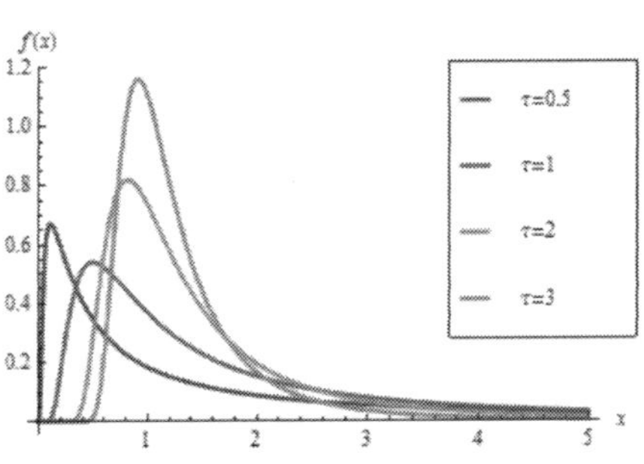

Figure 3. Plots of the pdf of the IW for $\theta = 1$

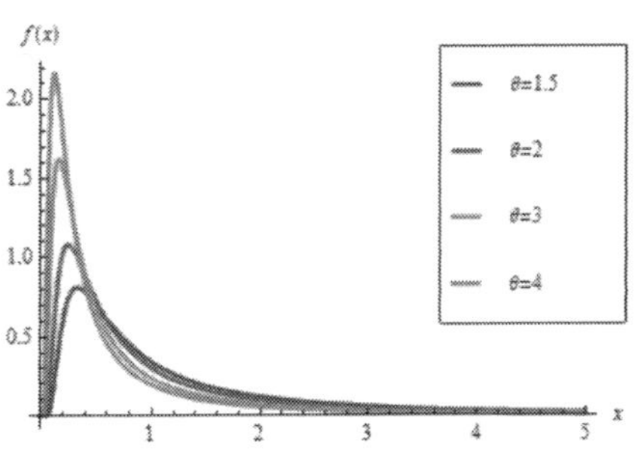

Figure 4. Plots of the pdf of the IW for $\tau = 1$

Based on figures 1 and 2, the pdf of Weibull distribution is unimodal and decreasing function and has a light tail. Based on figures 3 and 4, the pdf IW distribution is unimodal and has a heavy or long tail.

In 1995 Marshall and Olkin [8] proposed a method to expand families of distributions based on the survival function of a distribution by adding a new parameter. If $\bar{F}(x)$ denote the survival function of a continuous random variable X, then the usual device of adding a new parameter results in another survival function $\bar{G}(x)$ defined by [9]

$$\bar{G}(x,\alpha) = \frac{\alpha\,\bar{F}(x)}{1-\bar{\alpha}\,\bar{F}(x)};\ -\infty < x < \infty\,; \alpha > 0\,, \bar{\alpha} = 1-\alpha \tag{5}$$

If $G(x)$, $g(x)$and $r(x)$are the cdf,pdf and hazard rate function corresponding to $\bar{G}$, then

$$G(x,\alpha) = \frac{F(x)}{1-\bar{\alpha}(1-F(x))}\ ;\ -\infty < x < \infty\,; \alpha > 0\,, \bar{\alpha} = 1-\alpha \tag{6}$$

IOP Conf. Series: Journal of Physics: Conf. Series **1108** (2018) 012114 doi:10.1088/1742-6596/1108/1/012114

$$g(x,\alpha) = \frac{\alpha f(x)}{[1-\bar{\alpha}\bar{F}(x)]^2};\ -\infty < x < \infty\ ;\alpha > 0\ ,\bar{\alpha} = 1-\alpha \tag{7}$$

$$r(x,\alpha) = \frac{h(x)}{1-\bar{\alpha}\bar{F}(x)};\ -\infty < x < \infty\ ;\alpha > 0\ ,\bar{\alpha} = 1-\alpha \tag{8}$$

Where $h(x)$ is the hazard rate function corresponding to $f(x)$ and $f(x), F(x)$ is the pdf and cdf corresponding to $\bar{F}(x)$.

Marshall and Olkin [8] have noted that the method has a stability property, if the method is applied twice, nothing new is obtained. Marshall-Olkin extended distributions offer a wide range of behavior than the basic distributions from which they are derived. The property that the extended form of distributions can have an interesting hazard function depending on the value of the added parameter α and therefore can be used to model real situation in a better manner than the basic distribution, resulted in the detailed study of Marshall-Olkin extended family of distributions by some researchers like Okasha et al [10], Ghitany et al [11], Josse et al [12],jayakumar and Mathew [13], Pérez-Casany, M., and Casellas [14], Krishna et al [15], Gui [16], and Lemonte[17]

2. IWMO Distribution and its properties

An extended distribution with marshall-olkin is commonly used for reliability applications, such as on the Inverse Weibull extension with Marshall-Olkin and its application in reliability, see [10]. But in this paper we use slightly different constructions parameters on its Inverse Weibull distribution.

2.1. IWMO distribution

In this section we will extend the inverse weibul distribution with pdf and cdf in equations (3) and (4) using the Marshall-Olkin method. Let X be the random continuous variable that have IWMO distribution with the parameters $\theta, \tau > 0$. And α are the real constants, α > 0, by substituting equations (3) and (4) into equation (5) - (8) we get cdf, pdf, hazard rate function and survival function of IWMO respectively as follows

$$G_{iwmo}(x|\theta,\tau,\alpha) = \frac{e^{-(x\theta)^{-\tau}}}{\alpha-(\alpha-1)(1-e^{-(x\theta)^{-\tau}})} \tag{9}$$

$$g_{iwmo}(x|\theta,\tau,\alpha) = \frac{\alpha\tau\theta^{-\tau}x^{-\tau-1}e^{-(x\theta)^{-\tau}}}{[\alpha-(\alpha-1)e^{-(x\theta)^{-\tau}}]^2} \tag{10}$$

$$r_{iwmo}(x|\theta,\tau,\alpha) = \frac{\tau\theta^{-\tau}x^{-\tau-1}e^{-(x\theta)^{-\tau}}}{(1-e^{-(x\theta)^{-\tau}})(\alpha-(\alpha-1)e^{-(x\theta)^{-\tau}})} \tag{11}$$

$$\bar{G}_{iwmo}(x|\theta,\tau,\alpha) = \frac{\alpha(1-e^{-(x\theta)^{-\tau}})}{\alpha-(\alpha-1)e^{-(x\theta)^{-\tau}}} \tag{12}$$

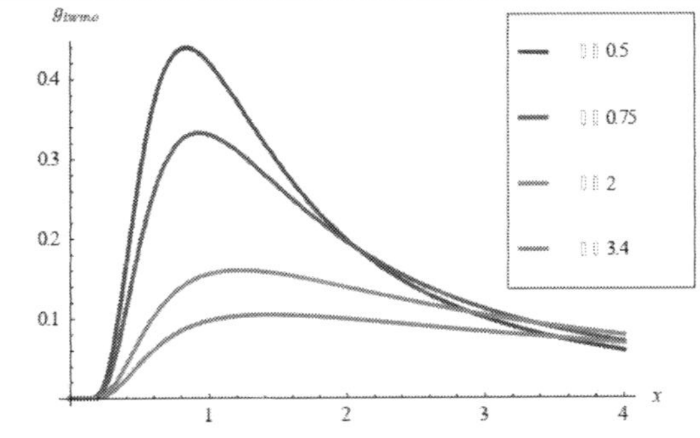

Figure 5. Plots of the pdf of the IWMO for $\theta = 0.5$ and $\tau = 1$

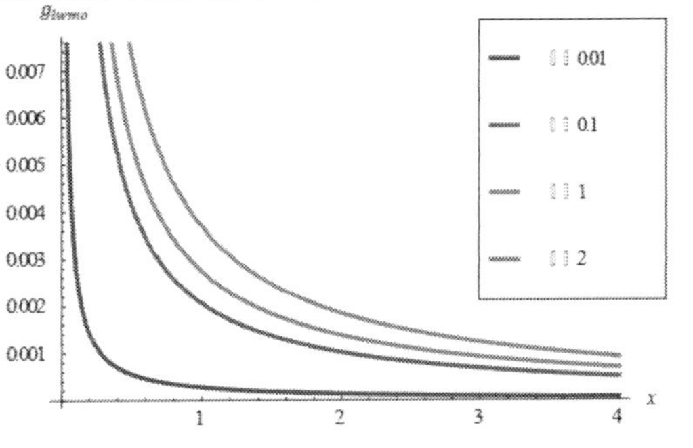

Figure 6. Plots of the pdf of the IWMO for for $\theta = 0.1$ and $\tau = 0.01$

From figures 5 and 6 we can see pdf of IWMO is decreasing and uimodal function.

2.2. Moment

In this section will discuss the kth moments of IWMO distribution. If X be the random variable thas IWMO distribution with parameters ,τ . and α is a real constant where $\theta, \tau, \alpha > 0$ then rth moments of IWMO is

$$E(X^k) = \int_0^\infty x^k \, g_{iwmo}(x)\, dx \tag{13}$$

then

$$E(X^k) = \sum_{j=0}^{\infty} \frac{1}{\alpha}\left(1-\frac{1}{\alpha}\right)^j \left(\frac{1}{\theta^\tau}(j+1)\right)^{\frac{k}{\tau}} \Gamma\left(1-\frac{k}{\tau}\right), \tau > k \tag{14}$$

We can get mean and variance by substituting $k = 1$ and $k = 2$ respectively.

2.3. Quantiles

In this section will discuss the quantiles of IWMO distribution. For a constant $p \in (0.1)$, the quantile p of a value ω_p is solution for

$$G_{iwmo}(\omega_p) = p \tag{15}$$

then the quantile p of a value ω_p

$$\omega_p = \left[\theta^\tau \ln\left(1-\frac{p-1}{p\alpha}\right)\right]^{-\frac{1}{\tau}}, \tag{16}$$

We can get median by substituting $p = 0.5$

3. Parameters Estimation

We use the Maximum Likelihood estimation method for parameters estimation. Let $X_1, X_2, \ldots, X_n$ be a random sample from IWMO distribution, then the likelihood function is given by

$$L(x|\theta,\tau,\alpha) = \frac{\tau^n \theta^{-\tau n} e^{-\sum_{i=1}^n (x_i\theta)^{-\tau}} \prod_{i=1}^n x_i^{-(\tau+1)}}{\alpha^n \prod_{i=1}^n \left(1-(1-\alpha)e^{-(x\theta)^{-\tau}}\right)^2}, \tag{17}$$

Then the logarithm of the likelihood function

$$lnL(x|\theta,\tau,\alpha) = n\ln(\theta^\tau) + n\ln\tau - n\ln\alpha - (\tau+1)\sum_{i=1}^n \ln(x_i)$$
$$-\theta^{-\tau}\sum_{i=1}^n (x_i)^{-\tau} - 2\sum_{i=1}^n ln\left(1-\left(1-\frac{1}{\alpha}\right)e^{-(x_i\theta)^{-\tau}}\right). \tag{18}$$

IOP Conf. Series: Journal of Physics: Conf. Series **1108** (2018) 012114 doi:10.1088/1742-6596/1108/1/012114

The maximum equation (18) can be obtained by taking the first partial derivatives of the log-likelihood function with respect to the three parameters, the results are

$$\frac{\partial lnL}{\partial\theta} = -\frac{\tau n}{\theta} + \theta^{-1-\tau}\tau\sum_{i=1}^{n}(x_i)^{-\tau} - 2\sum_{i=1}^{n} - \frac{\tau e^{-(x_i\theta)^{-\tau}\left(1-\frac{1}{\alpha}\right)x_i(\theta x_i)^{-1-\tau}}}{1-e^{-(x_i\theta)^{-\tau}}\left(1-\frac{1}{\alpha}\right)} = 0, \quad (19)$$

$$\frac{\partial lnL}{\partial\tau} = \frac{n}{\tau} - n\ln\theta - \sum_{i=1}^{n}x_i + \theta^{-\tau}\ln\theta\sum_{i=1}^{n}(x_i)^{-\tau} - \left(\theta^{-\tau}\sum_{i=1}^{n} - \ln x_i(x_i)^{-\tau}\right) - 2\left(\sum_{i=1}^{n} - \frac{e^{-(x_i\theta)^{-\tau}\left(1-\frac{1}{\alpha}\right)\ln(\theta x_i)(\theta x_i)^{-\tau}}}{1-e^{-(x_i\theta)^{-\tau}}\left(1-\frac{1}{\alpha}\right)}\right), \quad (20)$$

$$\frac{\partial lnL}{\partial\alpha} = -\frac{n}{\alpha} - 2\left(\sum_{i=1}^{n} - \frac{e^{-(x_i\theta)^{-\tau}}}{1-e^{-(x_i\theta)^{-\tau}}\left(1-\frac{1}{\alpha}\right)}\right) \quad (21)$$

The maximum likelihood estimates of θ, τ, and α are obtained by solving the nonlinear equations. These equations are not in closed form and the values of the parameters α, β, and θ must be found by numerical methods.

4. Wind Speed data

In this section, the wind speed data obtained from Central Java, Indonesia, namely Jragung station [18] is used to determine the best fitting distributions. The data is the monthly average wind speed miles per day from 2011 to 2015, the data consists 60 observations. The location of the jragung station is within 07 ° 09 '17.16 "south latitude and 110 ° 33' 40.26"east longitude. Statistical descriptions of the data are given in Table 1. From figure 7, we see that the data has extreme values. Table 2 is an estimate of the parameters using the maximum likelihood method.

TABLE 1. Descriptive statistics for wind speed data

Median	Mean	Variance	Skewness	Kurtosis	Min	Max
22.115	30.79	2423.225	6.130	42.893	2.74	377.09

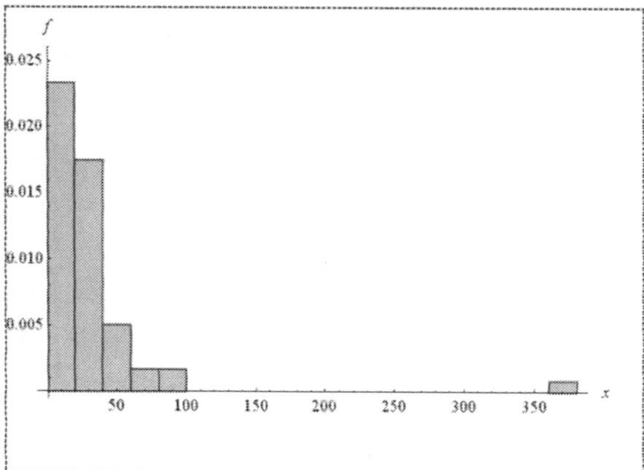

Figure 7. The histogram wind speed data

TABLE 2. Maximum Likelihood Estimation for wind speed data

Distribution	$\hat{\theta}$	$\hat{\tau}$	$\hat{\alpha}$
IWMO	0.30276	1.8707	27.36221
IW	0.08166	1.12	-

Weibull	30.6667	0.9926	-

4.1. Fitted models distributions

For the comparison for fitted models distribution IWMO with the IW and Weibull distribution,we used the log-likelihood values (log L), the Akaike information criteria (AIC) defined by -2log L + 2q, and the Bayesian information criterion (BIC) defined by -2log L +q log(n), where q is the number of estimated parameters and n is the sample size. The best model would give by the high value of log L and the lowest values from AIC and BIC. From Table 3 we have IWMO better than the others distributions.

TABLE 3. Summary of fitted distributions

Distribution	log L	AIC	BIC
IWMO	-256.766	519.533	525.816
IW	-261.565	527.129	531.318
Weibull	-265.63	535.259	539.448

4.2. Goodness of fit distribution models

For Based on Kolgomorov-Smirnov test with level of significance $\alpha = 0.05$, the value of $w_{1-\alpha} = 0.1755$. From table 4, all value of *T* is smallest than 0.1755, so that all models fit in explaining the data.

TABLE 4. Summary of fitted distributions

Distribution	T
IWMO	0.06975
IW	0.1252
Weibull	0.0978

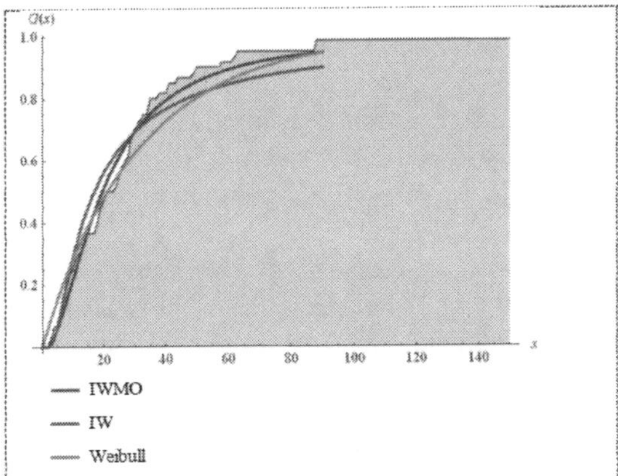

Figure 8. The empirical and fitted cumulative functions

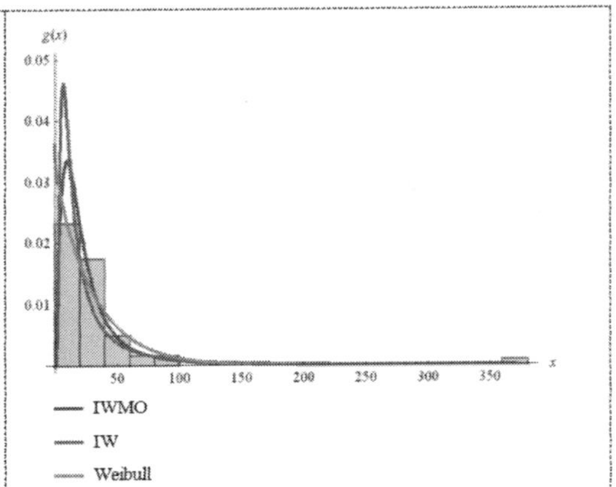

Figure 9. The pdf and histogram wind speed data

Figure 8 gives The empirical and fitted cumulative functions, and from Figure 9 the graph of pdf IWMO is not too sharp like IW but also not decreased like Weibull.

5. Conclusion

In this paper we extend the Inverse Weibull distribution by the Marshall-Olkin method (IWMO) corresponding to the Weibull distribution [7]. Some properties of our proposed model IWMO were derived. The estimation of parameters was obtained by maximum likelihood method. The results of simulation on wind speed data in Jragung Station, Indonesia show that IWMO is better than Weibull and IW distributions.

IOP Conf. Series: Journal of Physics: Conf. Series **1108** (2018) 012114 doi:10.1088/1742-6596/1108/1/012114

6. Acknowledgements
We wish to thank DRPM Univeritas Indonesia through Hibah Publikasi Internasional Terindeks untuk Tugas Akhir Mahasiswa UI (PITTA) 2018 Number: 2337/UN2.R3.1/HKP.05.00/2018

7. References
[1] Rinne H 2009 *The Weibull distribution : a handbook* (United States of America: Taylor & Francis Group, LLC)
[2] Kantar Y M, Usta I, Arik I and Yenilmez I 2018 Wind speed analysis using the Extended Generalized Lindley Distribution *Renewable Energy* **118** 1024-1030
[3] Souloukngа M H, Doka S Y, Revanna N, Djongyang N and Kofane T C 2018 Analysis of wind speed data and wind energy potential in Faya-Largeau, Chad, using Weibull distribution *Renewable Energy* **121** 1-8
[4] Ozay C and Celiktas M S 2016 Statistical analysis of wind speed using two-parameter Weibull distribution in Alaçatı region *Energy Conversion and Management* **121** 49-54
[5] Keller A Z, Kamath A R R and Perera U D 1982 Reliability analysis of CNC machine tools *Reliability Engineering* **3**(6) 449-473
[6] Akgül F G, Şenoğlu B and Arslan T 2016 An alternative distribution to Weibull for modeling the wind speed data: Inverse Weibull distribution *Energy Conversion and Management* **114** 234-240
[7] Klugman S A, Panjer H H and Willmot G E 2012 *Loss models from data to decisions (4th ed.)* (Canada: John Wiley & Sons, Inc.)
[8] Marshall A W and Olkin I 1997 A new method for adding a parameter to a family of distributions with application to the exponential and Weibull families *Biometrika* **84**(3) 641-652
[9] Jose K K 2011 Marshall-Olkin family of distributions and their applications in reliability theory, time series modeling and stress-strength analysis *Proc. ISI 58th World Statist. Congr Int Stat Inst, 21st-26th August* **201** 3918-3923
[10] Okasha H M, El-Baz A H, Tarabia A M K and Basheer A M 2017 Extended Inverse Weibull distribution with reliability application *Journal of the Egyptian Mathematical Society* **25**(3) 343-349
[11] Ghitany M E, Al-Hussaini E K and Al-Jarallah R A 2005 Marshall–Olkin extended Weibull distribution and its application to censored data *Journal of Applied Statistics* **32**(10) 1025-1034
[12] Jose K K, Naik S R and Ristić M M 2010 Marshall–Olkin q-Weibull distribution and max–min processes *Statistical papers* **51**(4) 837-851
[13] Jayakumar K and Mathew T 2008 On a generalization to Marshall–Olkin scheme and its application to Burr type XII distribution *Statistical Papers* **49**(3) 421-439
[14] Pérez-Casany M and Casellas A 2013 Marshall-Olkin Extended Zipf Distribution. *arXiv preprint arXiv:1304.4540*
[15] Krishna E, Jose K K, Alice T and Ristić M M 2013 The Marshall-Olkin Fréchet distribution *Communications in Statistics-Theory and Methods* **42**(22) 4091-4107
[16] Gui W 2013 A Marshall-Olkin power log-normal distribution and its applications to survival data *International Journal of Statistics and Probability* **2**(1) 63
[17] Lemonte A J 2013 A new extension of the Birnbaum–Saunders distribution *Brazilian Journal of Probability and Statistics* **27**(2) 133-149
[18] Data Kecepatan Angin Bulanan (km/hari) Stasiun Klimatologi Jragung. (2017,September 20). Retrieved from http://data.jatengprov.go.id/dataset/data-kecepatan-angin-rerata-bulanan-sta-klimatologi-jragung-kab-demak/resource/68a6a1d7-fc84-4879-8c6c-4f2bee4dc058

IOP Conf. Series: Journal of Physics: Conf. Series **1108** (2018) 012115 doi:10.1088/1742-6596/1108/1/012115

Comparison of SVM and FSVM for predicting bank failures using chi-square feature selection

Z Rustam[1,3], F Nadhifa[1,4] and M Acar[2,5]

[1] Department of Mathematics, Faculty of Mathematics and Natural Sciences, Universitas Indonesia, Depok 16424, Indonesia
[2] Department of Business Administration, Faculty of Economics and Administrative Sciences, Sulcuk University, Campus 42031, Konya, Turkey

E-mail: [3]rustam@ui.ac.id, [4]farah.nadhifa@sci.ui.ac.id, [5]melekacar@selcuk.edu.tr

Abstract. Bankruptcy doesn't happen suddenly, but there are early indications that can be seen by investigating the financial statement of a bank. In this research, we aim to find the best bankruptcy prediction model to give an early warning for regulators so that it can help them to prevent or lessen the negative effects on economic systems. We will be performing SVM and modification of SVM by adding fuzzy membership function called FSVM to analyze bank's health. We chose machine learning for bankruptcy prediction because it can give faster result rather than traditional statistical method. The prediction accuracy will be measured by using the dataset that consists of 65 Turkish banks of which each of them has an information of 20 financial ratios. Furthermore, to improve the accuracy prediction, we also perform chi-square feature selection (CSFS) to filter any irrelevant features of total 20 features in our dataset. CSFS can sort all 20 features based on chi-square score from the most relevant feature to the least one. After that, we will choose 5, 10, and 15 best features, so that we have four datasets to be classified into healthy and non-healthy banks. We found that using 5 features and SVM classifier gives the highest accuracy prediction, which scores 98.28%. For most cases, SVM gives better performance compared to FSVM.

1. Introduction

The bank plays a big role on economic system as they significantly contribute through the facilitation of business. Hence, the collapse of several banks can cause a huge damage to financial systems not only in a country but also globally as it had happened in the late 1990s and around 2008 [1]. There are so many caused that can triggered banks to go bankrupt such as interest rate volatility, high-risk taking, poor management, and inadequate accounting. As a consequence, the study of bankruptcy prediction of banks is becoming crucial for investors/creditors, borrowing firms, and governments as it can decrease the negative effects that can harm financial stability [2]. Constructing bankruptcy prediction model could identify banks with potential problems before they face solvency and liquidity crisis so that the danger of cascading failures of banks will be significantly reduced [3]. In other words, bankruptcy prediction model can act as an early warning system for regulators so that they can take actions before a financial crisis occurs.

To predict the bank's health, we need to investigate bank performance which can be influenced by internal factors (bank specific) and macroeconomic (external) [4]. There are various rating systems that can be used to determine bank's healthiness through analyzing their balance sheet reports. Therefore, in this research, we will be using dataset containing 20 financial ratios from 65 Turkish banks in the period of 1997—2004 issued by Banks Association of Turkey (BAT) as the variable predictor based on

CAMELS rating system [5]. In order to achieve a faster result, we chose to implement a machine learning technique rather than the traditional statistical technique. Moreover, machine learning capability for pattern recognition is very essential for this case. With two supervised machine learning approaches executed that is Support Vector Machines (SVM) and modified SVM called Fuzzy Support Vector Machines (FSVM) to construct the prediction. In previous studies, SVM has been applied by I.P.A Wardana and Z Rustam for decision-making in stock investment [6], Rustam and Frederica for insolvency prediction in insurance companies [7], Z. Rustam, D.F. Vibranti, and D. Widya for predicting the direction of Indonesian stock price movement [8]. Therefore, we know that SVM could be a great tool for building a prediction model, especially in the financial field. Furthermore, we will also be modifying SVM using fuzzy membership and in some cases, according to [9], adding fuzzy membership to SVM will minimize the error classification. Furthermore, using chi-square feature selection (CSFS) we aim to filter any irrelevant features and deduct 20 features to 15, 10, and 5 best features. Thereupon, there will be four datasets to be processed using two classifiers as mention before. As a result, we can compare each accuracy prediction and find the best model for our problem.

A brief explanation about theoretical background will be discussed in section 2. After that, section 3 will give an explanation about experimental designs for our research. Then, experimental results will be presented in section 4. And lastly, we will conclude and discuss the outcome of our research in section 5.

2. Theoretical background

2.1. Support vector machines (SVM)

Support vector machines (SVM) is a powerful method that gives high accuracy in terms of prediction because it holds structural risk minimization (SRM) principle. Moreover, according to [5] the application of SVM is relatively rare compared to neural networks and statistical method. Hence, we chose SVM as the classifier to construct a model for bankruptcy prediction. The idea of SVM is to find the best discriminant boundaries of two classes called hyperplane. In this case, the hyperplane is discriminant boundaries with the largest margin. We use [10], [11], and [12] as our main references for this section. In linearly separable case, given sample of several banks as training dataset $\boldsymbol{x_i} \in \mathbb{R}^n$ with corresponding label $y_i \in \{-1, +1\}$ $(i = 1,2, \dots, N)$, we can formulate a separator as $f(\boldsymbol{x}) = \boldsymbol{w} \cdot \boldsymbol{x} + b = 0$. Thus, maximizing hyperplane can be done by solving optimization primal problem (1) respected to constraint (2) as follows:

$$\min\left(\frac{1}{2}\|\boldsymbol{w}\|^2\right) \tag{1}$$

$$\text{s.t. } y_i(\boldsymbol{w} \cdot \boldsymbol{x_i} + b) \geq 1, \forall i = 1, \dots, N \tag{2}$$

Where $\boldsymbol{w}$ is the parameter vector and b is the bias. Therefore, the equation above can be qualified as a quadratic programming (QP) problem. We will solve the optimization problem by introducing Lagrange multipliers $\alpha_i \geq 0$ for each constraint in (2). Using Lagrange multipliers will obtain inner product form which is very useful for non-separable case later on. Hence, we have primal Lagrangian form as follows:

$$L(\boldsymbol{w}, b, \boldsymbol{\alpha}) = \frac{1}{2}\|\boldsymbol{w}\|^2 - \sum_{i=1}^{N} \alpha_i(y_i((\boldsymbol{w} \cdot \boldsymbol{x_i} + b) - 1)) \tag{3}$$

Setting the derivatives $L(\boldsymbol{w}, b, \boldsymbol{\alpha})$ respected to $\boldsymbol{w}$ and b both equal to zero, we will get

$$\frac{\partial L}{\partial \boldsymbol{w}} = \boldsymbol{w} - \sum_{i=1}^{N} \alpha_i y_i \boldsymbol{x_i} = 0 \tag{4}$$

$$\frac{\partial L}{\partial b} = \sum_{i=1}^{N} \bar{\alpha}_i y_i = 0 \tag{5}$$

Substitute both (4) and (5) to (3) will eliminate $\boldsymbol{w}$ and b so that we have dual Lagrangian form as follows:

$$\max\left(-\frac{1}{2}\sum_{i=1}^{N}\sum_{j=1}^{N}\alpha_i\alpha_j y_i y_j(\boldsymbol{x_i}\boldsymbol{x_j})+\sum_{i=1}^{N}\alpha_i\right) \tag{6}$$

$$s.t.\sum_{i=1}^{N}\alpha_i y_i=0\,,\alpha_i\geq 0\;(i=1,\ldots,N) \tag{7}$$

In real life cases, it's very rare to find linearly separable data as we assume in the previous case. Hence, we use the kernel trick method to couple with SVM to cope with non-linear separable data. Kernel trick follows the projection idea, where the dataset is mapped into an inner product space. Furthermore, we will be mapping data from the original space to another higher space using kernel function defined as $K(\boldsymbol{x}_i,\boldsymbol{x}_j)=\phi(\boldsymbol{x}_i)\cdot\phi(\boldsymbol{x}_j)$. Therefore, the data will be linearly separable without even knowing ϕ explicitly [10]. Hence substituting K to (6) we will have:

$$\max\left(-\frac{1}{2}\sum_{i=1}^{N}\sum_{j=1}^{N}\alpha_i\alpha_j y_i y_j K(\boldsymbol{x}_i,\boldsymbol{x}_j)+\sum_{i=1}^{N}\alpha_i\right) \tag{8}$$

We can find the value of α_i using SMO algorithm and assume there are $\bar{\boldsymbol{\alpha}}=(\bar{\alpha}_1,\bar{\alpha}_2,\ldots,\bar{\alpha}_N)$ as the solution of Lagrange multipliers so that we can easily obtain $\bar{\boldsymbol{w}},\bar{b}$, and $f(\boldsymbol{x})$ in Eq. (9), (10), (11) consecutively

$$\bar{\boldsymbol{w}}=\sum_{i=1}^{N}\bar{\alpha}_i y_i\boldsymbol{x_i} \tag{9}$$

$$\bar{b}=y_u-\sum_{i\in S}\bar{\alpha}_i y_i K(\boldsymbol{x}_i,\boldsymbol{x_u}) \tag{10}$$

$$f(\boldsymbol{x})=sgn\left(\sum_{i\in S}\bar{\alpha}_i y_i K(\boldsymbol{x}_i,\boldsymbol{x_u})+\bar{b}\right) \tag{11}$$

It is stated in [11] that a low polynomial or RBF kernel give satisfactory results and to outperform conventional classifiers. Hence, we'll try to use both kernels with the following formulation respectively:

$$K(\boldsymbol{x}_i,\boldsymbol{x}_j)=(\boldsymbol{x}_i\cdot\boldsymbol{x}_j+1)^d \tag{12}$$

$$K(\boldsymbol{x}_i,\boldsymbol{x}_j)=exp\left(-\frac{\|\boldsymbol{x}_i-\boldsymbol{x}_j\|^2}{2\sigma^2}\right) \tag{13}$$

2.2. *Fuzzy support vector machines (FSVM)*

We will modify SVM by adding fuzzy membership to each data in order to have weighted data attributed to each corresponding class [9]. We formulate fuzzy membership for each data in (1) and (2) [13]

$$\mu_{iy_0}=e^{-\frac{\|x_i\|^2}{\sigma^2}}\text{ for }x_i\in y_0=0 \tag{14}$$

$$\mu_{iy_1}=1-e^{-\frac{\|x_i\|^2}{\sigma^2}}\text{ for }x_i\in y_1=1 \tag{15}$$

where y_j ($j = 0,1$) is observed result or class of healthy ($y_0 = 0$) or non-healthy banks ($y_1 = 1$), $\|x_i\|^2 = \sum_{i=1}^{N} x_i^2$, $\sigma^2 = \frac{1}{N-1}\sum_{i=1}^{N}(x_i - \bar{x})^2$, and $\bar{x} = \frac{\sum_{i=1}^{N} x_i}{N}$ with N as sample size. The idea is to have weighted data by multiplying every data with corresponding fuzzy membership value.

2.3. *Chi-square feature selection*

Feature selection is one of many ways to improve data quality. This research will propose chi-square feature selection (CSFS) to filter any redundant ratios by investigating chi-square score of every ratio. CSFS can measures how dependent a ratio respected to the corresponding class. In this case, the smaller chi-square score indicates that it gives a bigger impact for determining bank's health. First, we will construct a contingency table that provides the foundation for statistical inference to see the relationship between observed variables

Table 1. Contingency table

	Class = 0	Class = 1	
C_1	$n_{10}(\mu_{10})$	$n_{11}(\mu_{11})$	n_{1*}
...	...	...	...
C_r	$n_{r0}(\mu_{r0})$	$n_{r1}(\mu_{r1})$	n_{r*}
	n_{*0}	n_{*1}	n

Chi-square score of every ratio will be measured using the following formulation [14]:

$$\chi^2 = \sum_{i=1}^{r} \frac{(n_{i1} - \mu_{i1})^2}{\mu_{i1}} + \frac{(n_{i0} - \mu_{i0})^2}{\mu_{i0}} \tag{16}$$

where n_{ij} is the number of instances that have a value of C_i (feature label) with $i = 1, \dots, r$ and class j, $n_{*j} = \sum_{i=1}^{r} n_{ij}$, $n_{i*} = n_{i0} + n_{i1}$, expected value $\mu_{ij} = \frac{n_{*j} n_{i*}}{n}$, and $n = n_{*0} + n_{*1}$ with n is the number of instances in the labeled data.

3. Experimental designs

3.1. *Dataset and variables*

Using data provided by Prof. Melek Acar [5] that she collected from the annual publication "Banks in Turkey" by the Banks Association of Turkey (BAT). This dataset has 21 solvent and 44 insolvent banks, which is 65 Turkish banks in total. Each of this bank has 20 financial ratios grouped in six components of CAMELS rating system including Capital adequacy, Asset quality, Management quality, Earnings, Liquidity, and Sensitivity to market. All 20 ratios from six components will be used as features in this study shows in Table 2 below.

Table 2. Financial ratios

Ratio	Description	Ratio	Description
CA_1	Shareholder's equity/total assets	E_2	Net profit/average shareholder's equity
CA_2	Shareholder's equity/total loans	E_3	Income before taxes/average assets
CA_3	Shareholder's equity+net profit/total assets + off-balance sheet commitments	E_4	Interest income/total operating income
AQ_1	Specific provision/loans under follow-up	E_5	Non-interest expenses/total operating income
AQ_2	Total loans/total assets	L_1	Liquid assets/total assets

AQ_3	Loans under follow-up/total loans	L_2	Total loans/total deposits
AQ_4	Specific provision/total loans	SMR_1	Trading securities/total assets
AQ_5	Permanent assets/total assets	SMR_2	FX assets/FX liabilities
M_1	Personnel expenses/average assets	SMR_3	Net on balance sheet position/total shareholder's equity
E_1	Net profit/average assets	SMR_4	Net interest income/average assets

In the preprocessing phase, we will be labeling the continuous data to qualitative form so that we can filter the features using CSFS. Based on [4] each ratio can be categorized as strong, satisfactory, fair, marginal, and unsatisfactory. Hence, we labeled all ratios with the numerical value of 1, 2, 3, 4, and 5 by dividing the original domain into five subintervals equally.

3.2. Measure for performance evaluation

The performance of the classifier will be evaluated using the following metrics:

$$Accuracy = \frac{TP + TN}{TP + FP + FN + TN} \times 100\% \tag{17}$$

where TP is the number of true positive cases, which represents instances which are correctly categorized in the 'positive' class, FN is the number of false negatives, which represents 'positive' class instances that are wrongly classified, TN is the number of true negatives, which represents instances that are correctly categorized in the 'negative' class, and FP is the number of false positive, which represents 'negative' class cases that are wrongly classified. A higher accuracy value indicates better classifier performance.

4. Experimental results

4.1. Variable selection

Based on chi-square score for every ratio that we've measured using Eq. (16), we will be choosing 5, 10, and 15 best subset features for our prediction model. Consequently, we will have a total of four datasets including all features dataset. Table 2 below shows chi-square score for every feature by order which number 1 shows smallest chi-square score (most relevant feature) to largest chi-square score (least relevant feature).

Table 3. Chi-square score for each feature

	Var	Chi		Var	Chi		Var	Chi		Var	Chi
1	AQ_4	0,299	6	AQ_1	3,0499	11	SMR_4	5,1933	16	AQ_3	8,8697
2	CA_1	0,9373	7	L_1	3,3093	12	E_5	6,2718	17	E_2	9,1636
3	E_1	1,3207	8	CA_3	3,4552	13	AQ_5	6,8119	18	M_1	9,2849
4	L_2	1,6967	9	CA_2	4,144	14	E_3	7,8713	19	E_4	11,8446
5	AQ_2	2,5892	10	SMR_1	4,8388	15	SMR_3	8,55	20	SMR_2	21,2597

4.2. Classifier accuracy

Table 3 and Table 4 below have summarized the result of our experiments. We have tried to used several parameter values and found out that highest accuracy prediction achieved while using kernel RBF with $\sigma = 0{,}05$ and polynomial kernel of degree 4 ($d = 4$). We split SVM and FSVM into four categories correspond to features subsets that we are using while performing two different kernels for each category.

Table 4. The accuracy of SVM from four different datasets using RBF kernel ($\sigma = 0{,}05$) and polynomial kernel $d = 4$

Data	Accuracy (%)			
training	20 features	15 features	10 features	5 features

	RBF	Polynom	RBF	Polynom	RBF	Polynom	RBF	Polynom
10%	79,31	65,52	84,48	68,97	84,48	74,14	**98,28**	86,21
20%	82	51,92	84,62	59,62	48,08	65,38	98,08	86,54
30%	80	68,89	84,44	71,11	77,78	75,56	95,56	95,56
40%	76,92	**87,18**	82,05	82,05	76,92	92,31	92,31	94,87
50%	75	84,38	81,25	84,38	71,88	90,63	93,75	**96,88**
60%	88,46	69,23	88,46	69,23	88,46	84,62	92,31	96,15
70%	**89,47**	73,68	**89,47**	73,68	**94,74**	**94,74**	89,47	89,47
80%	84,62	84,62	85,71	**85,71**	85,71	92,86	92,86	92,86
90%	83,33	71,43	83,33	66,67	83,33	83,33	83	83,33

Table 5. The accuracy of FSVM from four different datasets using RBF kernel ($\sigma^2 = 0{,}05$) and polynomial kernel $d = 4$

Data training	Accuracy (%)							
	20 features		**15 features**		**10 features**		**5 features**	
	RBF	Polinom	RBF	Polinom	RBF	Polinom	RBF	Polinom
10%	77,59	67,31	82,76	67,24	82,76	70,69	89,66	84,48
20%	73,08	67,31	80,77	69,23	80,77	75	90,38	76,92
30%	80	60,00	80	62,22	82,22	73,33	91,11	95,56
40%	82,05	76,92	82,05	74,36	84,62	82,05	92,31	**97,44**
50%	**84,38**	78,13	**84,38**	84,38	**87,50**	**96,88**	**96,88**	96,88
60%	80,77	73,08	80,77	**84,62**	84,62	86,62	96,15	96,15
70%	78,95	78,95	78,95	78,95	84,21	84,21	94,74	94,74
80%	76,92	**84,62**	71,43	71,43	78,57	78,57	92,86	92,86
90%	83,33	83,33	83,33	83,33	83,33	83,33	83,33	83,33

It can be seen from Table 4 that the best accuracy prediction generated by SVM is 98,28% while using 5 subset features and RBF kernel. Whereas, from Table 5 we know that FSVM gives the best accuracy prediction at 97,44% while using a polynomial kernel. We also can conclude that the accuracy prediction of both classifiers increase as we reduce the irrelevant features. In other words, selecting features using CSFS successfully increase our dataset quality so that we have a better bank failures prediction model. Furthermore, Table 6 below will be giving information for running time and memory usage of each algorithm to give a brief comparison between the two classifiers.

Table 6. Running time for each experiment with best accuracy prediction based on feature subset using SVM and FSVM

Features	SVM		FSVM	
	Running time (s)	**Memory (Kb)**	**Running time (s)**	**Memory (Kb)**
5	0,874	14666,06	0,844	14146,82
10	0,917	21127,53	0,996	15736,51
15	1,484	26548,1	0,971	22380,82
20	3,837	110136,42	1,25	30640,19
	1,778	43119,5275	1,01525	20726,085

From Table 6, we know that SVM has longer running time and two times bigger memory consumption than FSVM. However, the gap between the two classifiers in terms of running time isn't too big.

5. Conclusion

This research aimed to apply SVM and modified SVM, called FSVM to construct a bankruptcy prediction model. As we can see from our experimental results, SVM gives higher accuracy with the highest value being 98,28% but while using 10 features, FSVM achieved better accuracy prediction with the value of 96,88%. Although SVM algorithm tends to spend more time and memory usage we still conclude that SVM will be a better bank failures predictor compared to FSVM because there isn't a significant difference between the two cases. This may occur because FSVM tends to give a maximal performance in multiclass cases. For the next research, we suggest to use another modification approach, also try to find another bigger and more update dataset in order to have an even more reliable bankruptcy prediction model. We also suggest using another feature selection approach that can be used for continuous data. Then, the result of feature selection can be compared to our results by using CSFS.

6. References

[1] Hong H and Jean-Laurent V 2016 Predicting bank failure statistical technique versus intelligent technique VEAM Seminar Paper

[2] Sungjae L and Wu S C 2013 A multi-industry bankruptcy prediction model using the back-propagation neural network and multivariate discriminant analysis *Expert Systems with Applications* **40** 2941–2946

[3] Michael N 2017 Identification of failing banks using clustering with self-organizing neural networks *NTELLI 2013: The Second International Conference on Intelligent Systems and Applications* **108C** 1327–13333

[4] Desta T S 2016 Financial performance of "best african banks": Comparative analysis through CAMEL rating *Journal of Accounting and Management* **6** 1–20.

[5] Melek A, Kara Y and Baykan Ö K 2009 Predicting bank financial failures using neural networks, support vector machines and multivariate statistical methods: A comparative analysis in the sample of savings deposit insurance fund (SDIF) transferred banks in Turkey *Expert Systems with Applications* **36** 3355–3366.

[6] I P A Wardana and Z Rustam 2017 Application of support vector machines on decision-making in stock investment *Proceeding of 3rd International Symposium on Current Progress in Mathematics and Sciences*

[7] Z Rustam and Frederica Y 2017 Insolvency prediction in insurance companies using support vector machines and fuzzy kernel c-means *Journal of Physics: Conference Series* **1028** 012118

[8] Z Rustam, D F Vibranri and D Widya 2017 Predicting the direction of Indonesia stock price movement using support vector machines and fuzzy kernel c-means *Proceeding of 3rd International Symposium on Current Progress in Mathematics and Sciences*

[9] Shigeo A and Takuya I 2002 Fuzzy support vector machines for multiclass problems *European Symposium on Artificial Neural Networks* (Bruges: Belgium) p 113–118

[10] Burges C J C 1998 A tutorial on support vector machines for pattern recognition *Data Mining and Knowledge Discovery* **2** 121–167

[11] Hofmann M 2006 Support vector machines–kernels and kernel trick *Houptseminar Report*

[12] Bishop C M 2006 *Pattern Recognition and Machine Learning* (New York: Springer)

[13] Z Rustam 2008 Face recognition using fuzzy supportvectorr machines *National Mathematics Conference XIV*

[14] Shaomin W and Flach P A 2002 Feature selection withlabeledd and unlabelled data *Workshop on Integrating Aspects of Data Mining, Decision Support, and Meta-Learning* p 156–167

The creativity of Prospective Teachers in Mathematical Patterns Problem Solving Based on Emotional Intelligence

I Safitri[1,*], P Wijayanti[1], Masriyah[1]

[1]Program Pascasarjana Department of Mathematics, Universitas Negeri Surabaya, East Java

[*]Corresponding author: indasafitri16070785006@mhs.unesa.ac.id

Abstract. Creativity is the ability that must be possessed by teachers in order to stimulate the creativity of their students. Teachers must also have emotional intelligence because emotional intelligence has a positive relationship with creativity. This research aims at describing the creativity of prospective teachers in mathematical patterns problem based on emotional intelligence. The research method used was a qualitative method. Three prospective teachers, representing each type of Goleman's emotional intelligence types, were selected as the research subjects. The data were collected from the test of creativity, the mathematical patterns problem, and interviews. The prospective teacher's creativity is described based on the Torrance test of creativity. The results of this research revealed that, of the three categories of emotional intelligence, only prospective teachers who have self-motivation abilities that meet three indicators. He was able to make many solutions to the problem of mathematical patterns, was able to make variations on the solutions of mathematical pattern problems and was able to find unique solutions.

1. Introduction

Creativity has an important role in everyday life. In addition, creativity also plays a role in the progress of a nation. Creativity has the meaning of an ability to produce something [1] or find solutions to problems. There are four design of creativity task namely fluency, flexibility, originality and elaboration. For assessors who are not creativity understanding about indicators of elaboration, Torrance's recommendation is to use only three namely fluency, flexibility and originality [2]. Several studies, especially in mathematical research, many use only three indicators to know creativity or measure creativity [3] [4] [5].

Furthermore, if viewed from GCI [6], Indonesia's creativity index was still far below Thailand. Though Indonesia first developed than Thailand. Therefore, creativity enters in one of the goals of national education [7].

Schools that are an important part of the world education is expected to realize the goal of national education. The role held by this school is then submitted by the teachers. So it can be said that teachers have an important task to stimulate the creativity of learners. But the first teacher must have the creative ability as well.

The results of Vale's research revealed that the creative dimensions of teachers are still on the dimension of fluency and slightly on the flexibility dimension [3]. Based on this, it is important to know how the creativity of the prospective teachers.

There are various ways to find out how the creativity. One of them through the provision of problems that have many solutions [8] [9]. The problem in mathematics that can be used in the research was the mathematical patterns problem. In general, the pattern is identical with a result of creation or creativity. But in mathematics patterns are used as a tool to provide an understanding of mathematical concepts [4] [10].

Speaking of creativity, it must also pay attention to the matter of emotional intelligence. Emotional intelligence is the ability to reduce emotion, emotional control, the ability to use humor and play in challenges, and the ability to deal with conflicts in a positive and confident [11]. A person with good emotional intelligence can create a creation rather than a person who lacks good emotional intelligence [12]. Emotional intelligence consists of two aspects namely personal skills and social skills [13]. This study chooses to use the emotional intelligence of personal skills including self-awareness, self-control, and self-motivation. Self-awareness is knowing what is being felt in a condition that is being experienced and using it as a guide in making decisions for yourself. Self-control is overcoming emotions in such a way that positively impact the execution of tasks and able to return from emotional stress. Self-motivation is using the deepest desire to move and guide toward the goal, helping to take the initiative and act very effectively and to endure failure and frustration [14]. Because the problem used in this research is a problem that must be solved alone.

Based on this, this study discusses how the creativity of prospective teachers in making solutions to mathematical patterns problem based on emotional intelligence.

2. Method

The method used in this study was a descriptive qualitative approach to know how the creativity of prospective teachers in making solutions to mathematical patterns problem based on emotional intelligence. Especially personal skills namely self-awareness, self-control, and self-motivation.

Table 1. Personal Skill - Emotional Intelligence

Categories	Definitions
Self-awareness	knowing what is being felt in a condition that is being experienced and using it as a guide in making decisions for yourself
Self-control	overcoming emotions in such a way that positively impact on the execution of tasks and able to return from an emotional stress
Self-motivation	using the deepest desire to move and guide toward the goal, helping to take the initiative and act very effectively and to endure failure and frustration

The first step was to choose 3 research subject in a 6th class at a college in Jombang by giving a test of emotional intelligence. Furthermore, the three subjects were given problems of mathematical patterns and asked to make many solutions to the problem.

Solutions created by the subject were used to determine the creativity of the subject by using Torrance version of creativity indicators including fluency, flexibility, and originality. Finally, researchers conduct interviews with each subject. Because this research uses Torrance's creativity indicators, the highest creativity was owned by subjects that meet the indicators of originality.

Table 2. Torrance's Indicator Creativity for This Research

IOP Conf. Series: Journal of Physics: Conf. Series **1108** (2018) 012116 doi:10.1088/1742-6596/1108/1/012116

Indicators	Definitions
Fluency	the number of solutions that are true.
Flexibility	variation of strategy or in this research was a variation of pattern form which was represented in the form of a mathematical statement.
Originality	unique solution.

3. Result

The problem of mathematical pattern

1. Find a different way to calculate the sun in the image below. Then express the way that you've found the pattern and write down the mathematical formula.

2. Make a variety of sequence. Where the second term of each sequence that you create is to be of the form below.

3.1 Self Awareness

For the subject of this category when looking at the problem first, the subject decides to focus and maximize the ability to work on problems was mastering, in this case, was the first mathematical problem. As for the problem of the second mathematical pattern which is not a problem that the subject was mastered, subject to completing it in the run-up time. So the subject can only create solutions that are much less than the second problem.

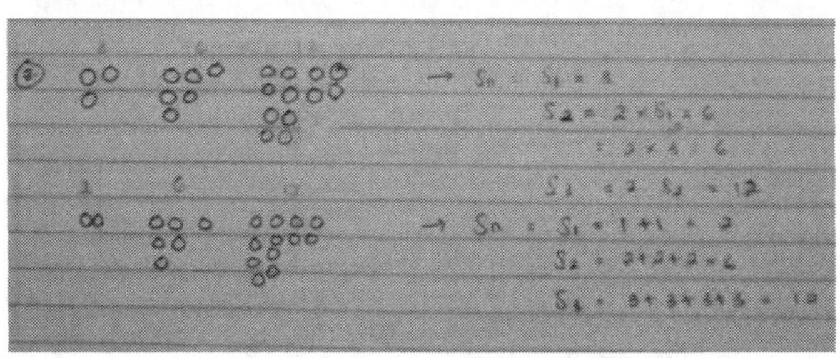

Figure 1. For the second problem, the subject did not fulfill all indicators of creativity.

For the creativity side, the subject meets two indicators of creativity for the first problem of fluency and flexibility. Fluency was seen from the number of solutions made by the subject. Flexibility is seen from the variation of representation. The subject does make many solutions to the first problem but there are only two variations of representation that can be found in the solutions made by the subject ie the representation in which it uses addition and representation operations in which it uses multiplication and addition operations. As for the problem of the second mathematical pattern, the subject did not satisfy the indicator of creativity.

3.2 Self Control

For the subject of this category, the subject also worked on the mathematical problem he first mastered, namely the first problem. After the subject felt enough to solve the first problem, the subject solved the second problem. Subjects have enough time to answer the second problem. The difference in the number of solutions to the first and second problems was not as large as the difference in the number of solutions made by subjects who entered the self-awareness category.

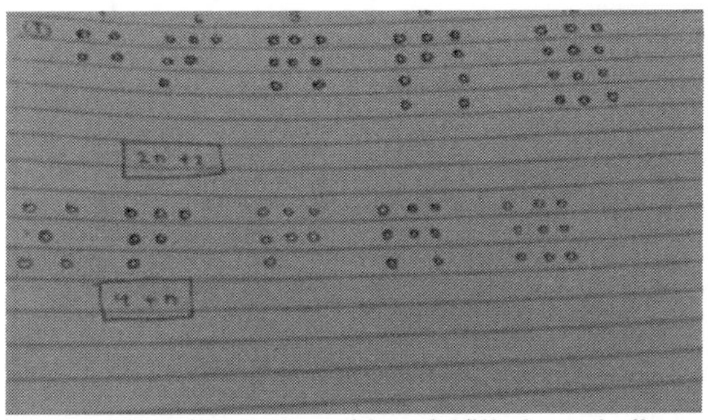

Figure 2. For the second problem, the subject fulfilled two indicators of creativity. Fluency, and flexibility.

The solutions made by the subjects in this category satisfy two creativity indicators for the first problem of fluency and flexibility and two indicators for the second problem. For the first problem, the subject did not make the solution as much as the subject in the category of self-awareness but the solution that he made also has 2 variations of representation ie the representation in which there was the addition operation, and the representation in which there were two operations namely the addition operation and the multiplication operation. As for the second problem, the subject only fulfilled two indicators of fluency and flexibility because it was only able to make two solutions to the problem of the second mathematical pattern with different representation.

3.3 Self-Motivation

For subjects of this category, take the same actions as subjects in the category of self-control. But it has differences from the side of creativity.

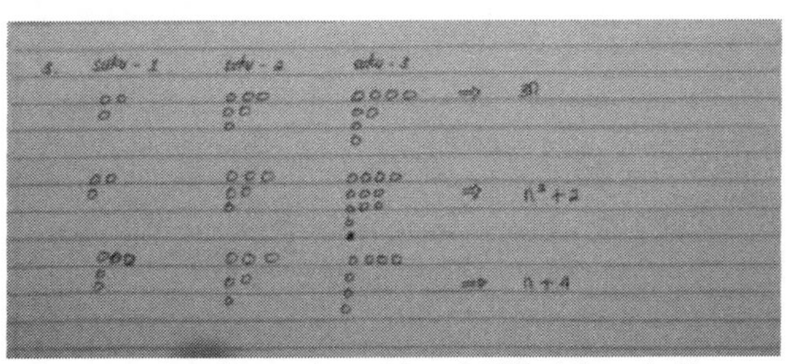

Figure 3. For the second problem, the subject fulfilled three indicators of creativity. Fluency, flexibility, and originality.

For the first problem, the subject fulfills two indicators namely fluency and flexibility. For flexibility, the subject made solutions same with another subject. For the second pattern problem, the subject fulfills three indicators namely fluency and flexibility and originality. For fluency, the subject did make many solutions. For flexibility, the subject did make three variations and for originality, the subject did make a different variation with another subject in this research.

4. Discussion

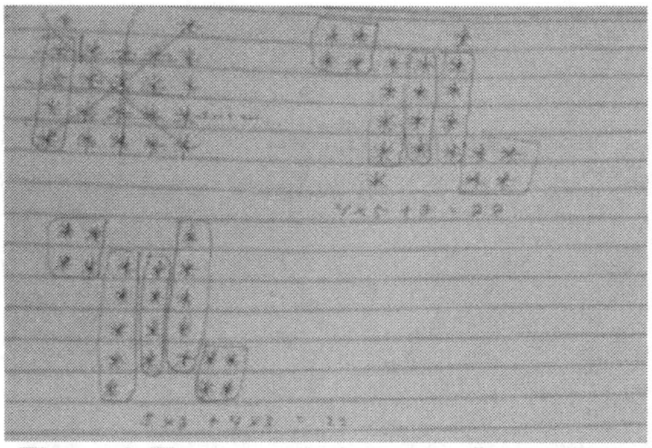

Figure 4. The most solution to the first problem

From the research results can be seen that there are differences between the subject with self-awareness category, subject with self-control category, and subject with self-motivation category. The first difference was found in the results of solutions made by the three subjects. For the subject of self-awareness category more make the solution to the first problem but not able to make a solution to the second problem. As for the subject category of self-control and self-motivation able to create solutions for two problems in this research.

The second difference in creativity. Because this research uses Torrance's creativity indicators, the highest creativity was owned by subjects that meet the indicators of originality. For the first problem, the three subjects fulfilled the two indicators of creativity namely fluency and flexibility. As for the second problem, the subject with self-awareness category only fulfilled one indicator. Then the subject with self-control category fulfilled two indicators and the subject with self-motivation category fulfilled three indicators namely fluency, flexibility and originality. Based on this it can be said that the subject with the category of self-motivation has the highest creativity.

5. Conclusion

From the above description, it can be concluded that the subjects, who meet the three indicators of creativity on the first pattern problem, were the subject in the category of self-motivation. While the rest meet the two indicators of creativity that is fluency and flexibility. This shows that motivation was important in learning especially when want to give stimulus creativity to learners by way of giving a problem which has many solutions. In addition, the above description shows that average the subject was more creative when solving problems that are more subjects mastered.

6. Reference

[1] KBBI 2006(online) Arti Kreasi (https://kbbi.web.id/kreasi accessed 20 Maret 2018)

[2] Hebert T P 2002 *E. Paul Torrance: His Life, Accomplishment, and Legacy, The National Research Center on The Gifted and Talented*

[3] Issabel V and Barbosa 2015 Mathematics Creativity in Elementary *Journal of the European Teacher Education Network, Training Teacher* **10** 101-109
[4] Issabel V 2012 Pattern Problem Solving Task as a mean to Foster Creativity in Mathematics *Proceeding of the* 36th *Conference of the International Group for the Psychology of Mathematics Education* **4** pp. 171-178
[5] Tavares D *2015* Visual Patterns and The Development of Creativity and Functional Reasoning *Journal of The European Teacher Education Network* **9** 79-40
[6] Florida R 2015(online) *The Global Creativity Index*, Martin Property Institute (https://www.google.co.id/url?sa=t&source=web&rct=j&url=http://martinprosperity.org/media/Global-Creativity-Index-2015.pdf)
[7] Permendikbud 2013(online) *UU RI No. 20 Tahun 2003 Tujuan Pendidikan Nasional* (http://kelembagaan.ristekdikti.go.id)
[8] Leikin R 2009 Evaluating Mathematical Creativity: The Interplay Between Multiplicity and Insight, *Psychological Test and Assessment Modelling* **55**
[9] Eric L M 2006 Creativity: The Essence of Mathematics *Journal for The Education of gifted* **30**(2) pp. 236-260
[10] Walle J A V D 2008 *Matematika Sekolah Dasar dan Menengah: Pengembangan Pengajaran* (Jakarta: Erlangga)
[11] Kanhai A 2014 Emotional Intelligence: A Review of Researches *European Academic Research* **2**(1)
[12] Isen A M, Kimberly A D and Gary P N 1987 Positive Affect Facilities Creative Problem Solving *Journal of Personality and Social Psychology* 52(6) 1122 – 1131
[13] Goleman D 2002 Emotional Intelligence (Jakarta: Gramedia Pustaka Utama)
[14] Goleman D 2003 Working with Emotional Intelligence (Jakarta: PT. Gramedia Pustaka Utama)

Acknowledges
This research reported herein was supported by Universitas Hasyim Asy'ari Jombang East Java. The opinions expressed herein do not necessarily reflect the position, policy, or endorsement of the supporting agency.

IOP Conf. Series: Journal of Physics: Conf. Series **1108** (2018) 012117 doi:10.1088/1742-6596/1108/1/012117

The Role of Refutational Text as a Conceptual Change Effort to Fix the Misconception on Addition and Subtraction of Integers

S Retnowati[1*], S M Amin[2], and E M Imah[2]

[1]Mathematics Education Program, Universitas Negeri Surabaya, Surabaya, Indonesia
[2]Mathematics Department of Universitas Negeri Surabaya, Surabaya, Indonesia

*Corresponding author: [1]sriretnowati16070785047@mhs.unesa.ac.id

Abstract. There are conceptual difficulties in studying operations of integers from which a poor understanding of this concept could lead to misconceptions. One way that can be done to help the students' misconceptions was through the use of conceptual change techniques. There were four conditions in the conceptual change technique and the refutational text meets of all conditions. The aim of this study was to describe the role of refutational text in conceptual changes that can fix misconceptions on addition and subtraction of integers. This research began with developing refutational text-aided learning media. Posttest pretest design was used. The pretest was given at 30 4th-grade students at the beginning of the lesson. In the teaching stage, the explanation of the concept through refutational text-aided learning media was presented. Posttest was conducted at the end of the study. Interviews were conducted twice, ie during pretest and posttest. Data analysis was done by comparing the result of answer and result of student interview at pretest and posttest. The result as a whole there is an increase in the average value ie M_{PRE} = 43 and M_{POST} = 80.1. In addition, there is an indication of a conceptual change of students in understanding the matter of addition and subtraction of integers.

1. Introduction

An integer is a part of a rational number, consisting of positive integers, zero and negative integers [1]. There is a conceptual difficulty to understand the concept of negative integers and operations using the set of integers [2,3]. Besides negative integers is a difficult subject to teach [4], one of the causes of students' difficulties in understanding negative integers is the difficulty in understanding the use of a negative sign as a type of number or as a sign of subtraction operation. This is in accordance with the principles of Vygotsky, which defines the existence of a close relationship between signs such as language or symbols with cognitive development [5].

In several studies, found that there are still errors that students do in the matter of addition and subtraction of integers. One of the causes of errors in the matter of the addition and subtraction of integers is due to a misconception of the concept of arithmetic operations on integers [4,6,7,8,9]. Based on Makonye & Fakude's research [4], it was found that 83.3% of 8th-grade students had misconceptions of the concept of addition and subtraction of integers caused by poor conceptual understanding [4]. Poor conceptual understanding can lead to errors in other parts, such as procedural errors (false in sign manipulation). This is reinforced by the finding that as many as 62.9% of students with poor conceptual understanding do procedural errors in performing addition and subtraction operations on integers [4].

Given that integers are one of the most important and challenging topics in the transition from arithmetic to algebra [10], there is an indication that student misconceptions about the matter of integers can cause difficulties in learning algebra [11]. Misconceptions about the concept of addition and subtraction on integers can have long-term effects, where adults who have been educated and taught over the years are found to have misconceptions about this concept, generally they assume that the sum of two numbers will always result in a larger number and a two-digit subtraction will always result in smaller numbers, including in the case of negative integers [12].

One method that can be used to fix misconceptions is by making conceptual changes [13,14]. The proposed teaching technique for conceptual change is teaching that can lead to cognitive conflict [15]. Cognitive conflict is a cognitive imbalance that can cause a person to discover or develop a new concept [15]. One way that can be used to achieve cognitive conflict in this research is by using refutational text [13], this is because the refutational text is the only way of teaching that aims to encourage conceptual change, where such conceptual changes can last a long time [16]. In addition, the refutational text also fulfills the four necessary steps for conceptual change can occur [17]. To achieve conceptual change, incorrect and correct concepts are presented in one text at the same time so two opposite concept can be activated in memory [18]. In the refutational text, incorrect concept (misconception) and correct concepts are very close to each other so it is quite possible that both can be activated simultaneously in working memory and make conceptual changes happen [19]. The aim of this study is to describe the role of refutational text in a conceptual change effort to fix the misconception of matter of addition and subtraction of integers. This research begins by developing ICT media that uses refutational text in it to make students easier to understand the refutational text and support the research being conducted.

2. Refutational Text and Conceptual Change

Misconceptions can be fixed by a process called conceptual change [17]. The refutational text is aim to achieving conceptual change by explicitly stating, refuting and replacing an incorrect conception [18]. In the refutational text, two opposite concepts (incorrect and correct concepts) are presented in one text at the same time. In order to be referred to as a refutational text, the text must have two essential elements [13], ie: (i) the text must clearly indicate the wrong concept which will be refuted by the refutational text; (ii) the text must clearly state that this concept is not true and why (reason) by explaining the correct concept. Part of the refutational text always contains three components, ie: misconceptions, refutation cue and explicit disapproval (clear rejection) of the misunderstanding with an emphasis on currently accepted scientific explanations [13, 16]. Conceptual change is very likely to happen in learning. Conceptual change is a change in the structure of existing concepts or knowledge to be able to assimilate or accommodate to get new concepts or knowledge. In other words, to be able to make a conceptual change, one must rearrange the existing concept structure (existing knowledge) to be able to assimilate or accommodate new knowledge or concepts [13]. There are four steps in the conceptual change process, ie: (a) Dissatisfied, (b) Intelligible, (c) Plausible and (d) Fruitful [17].

3. Methods

This research is a research development and experimental research. Developmental research is done to develop learning media using refutational text. Learning media with refutational text is used to give students ease in understanding the refutational text, thus can supporting the research conducted. Development of learning media using the Plomp development model consisting of five stages [20], ie: (a) Preliminary Investigation, (b) Design, (c) Realization, (d) Test, Evaluation and Revision, (e) Implementation. Experimental research was conducted with one group sample pretest-posttest. This experimental research is the implementation of refutational text-aided learning media that has been developed.

3.1. Refutational Text-Aided Learning Media

An ICT media with refutational text in it, in this learning media there are parts of learning from apperception stage to evaluation. Refutational text in this learning media is emphasized at the time of giving of materials. The given learning material is 8 addition and subtraction forms, consisting of 4 addition of integers and 4 of subtraction of integers. The addition and subtraction forms of integers are as follows: (i) $a + b$, (ii) $a + (-b)$, (iii) $-a + b$, (iv) $-a + (-b)$, (v) $a - b$, (vi) $a - (-b)$, (vii) $-a - b$, and (viii) $-a - (-b)$ in condition $a < b, a = b, a > b$ with $a, b \in$ integers. In each of the materials taught will be given the conclusion of the concept. Then the students are asked to try some practice questions from each section studied, the exercise questions are in the form of multiple choice and when students make mistakes in the selection of answers then the refutational text will appear to guide the subject in correcting the error. One a part of the refutational text in media is as Figure 1.

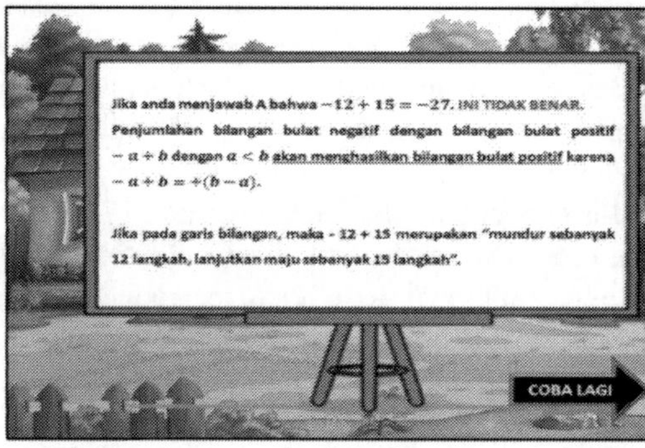

Figure 1. Figure the use of refutational text in learning media.

In Figure 1 the blackboard is a place of refutational text, when students answer correctly then there will be an explanation of the concept and they can continue to the next page. However, when students answer wrong the refutational text will appear to direct the student's mistake by telling them that the answer is wrong and explained the correct concept. The link between the conceptual change process and the use of refutational text, ie: (i) when students choose the wrong answer and a board containing refutational text appears, students will know that they are experiencing misconceptions that cause errors, (ii) in refutational text there is an explanation of the correct concept, so students are encouraged to be able to understand the new information, (iii) students are guided to be able to explain the new information in accordance with understanding that makes sense to him, (iv) by understanding refutational text, students are expected to be able to use the correct concept to solve different problems in the same material.

3.2. Pretest and Posttest

In this study pretest and posttest use the same problem, consisting of 4 questions of addition on integers and 4 problem of subtraction on integers, such: (i) $a + b$, (ii) $a + (-b)$, (iii) $-a + b$, (iv) $-a + (-b)$, (v) $a - b$, (vi) $a - (-b)$, (vii) $-a - b$, and (viii) $-a - (-b)$. Each question consists of 3 sub-items with condition $a < b, a = b, a > b$ with $a, b \in$ integers. In addition, to determine whether there is or not misconceptions experienced by students on the addition and subtraction of integers, the question of pretest and posttest is given CRI (Certainty of Response Index) scale 0 – 5 to find out the level of confidence of students in providing answers [21].

4. Results and Discussion

Participants as many as 30 students in 4th grade, consists of 15 male and 15 female students. Age between 10–11 years. All the students are physically normal. The selection of participants of 4th grade of primary school because the matter of addition and subtraction of integers involving negative integers begins to be taught at this level, so it is important to immediately correct the misconceptions that may occur in this material. This research design uses one group sample pretest-posttest. At the beginning of the study, the students were given a pretest consisting of 8 questions of the addition and subtraction of integers. After the pretest, an interview was conducted on the entire student to know the initial level of students' understanding of the addition and subtraction of integers through the answers given by the students at the time of pretest. At four subsequent meetings, students are asked to study the addition and subtraction materials of integers using refutational text-assisted learning media, students study independently. Then given the posttest by using the same test given at the time of pretest. The results of pretest and posttest are grouped into high, medium and low-value groups. The determination of high, medium and low score groups is based on the mean and standard deviation on each test.

The selection of interview subjects is based on the students who experience the change of the value group after being given the learning by using refutational text-aided instructional media. The change of value groups between the pretest and the posttest is a change in the low to moderate, low to high and moderate to high values group. If there is a decrease in the value group, one additional interview subject will be chosen for an interview. The results of the interview during the posttest will be compared with the results of the interview during the pretest to determine whether or not there is any conceptual change that occurs during the exposure of refutational text.

The aim of this study is to describe the role of refutational text in a conceptual change effort to fixing the misconception of matter of addition and subtraction of integers. Overall there is an increase in the results of pretest and posttest values, where the average pretest score is 43.0 and the average posttest score is 80.1. In the pretest still found a misconception, the existence of these misconceptions is indicated by the level of confidence of students at the time of reply, where students answer wrong with a high level of confidence [23], as in Table 1.

Table 1. The level of student confidence in answering
based on combinations of the correct or wrong answer and of low or high CRI.

	Low CRI (< 2.5)	**High CRI** (> 2.5)
Correct answer	Correct answers and low CRI indicate **"lack of knowledge"**	Correct answer and high CRI indicate **"concept comprehension is correct"**
Wrong answer	Wrong answers and low CRI indicate **"lack of knowledge"**	Wrong answer and high CRI indicate **"misconception"**

When pretest it was found that in the matter of addition and subtraction of integers there are 4% of students are at the level of "lack of knowledge", 47% of students are at the "correct comprehension" and 49% of students experience "misconceptions". This is seen from the level of student confidence over the wrong answer or the correct answer given. Misconceptions rate tends to decrease at posttest, ie stay 8% after exposure with refutational text. One of the differences in student answers when pretest and posttest are presented in Figure 2.

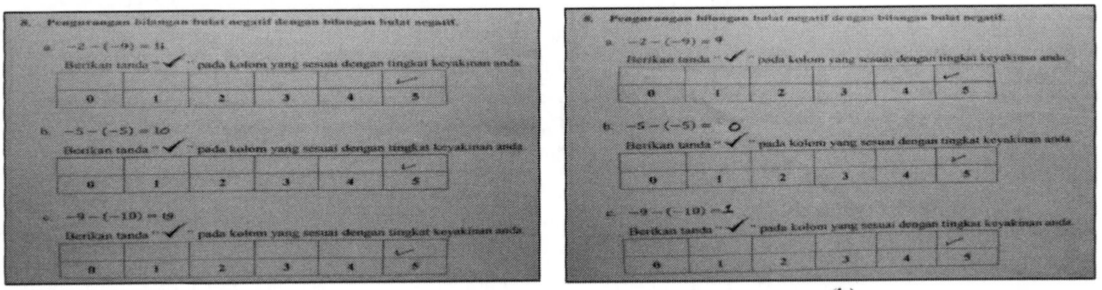

(a) (b)

Figure 2. (a) Student's answer on pretest; (b) Student's answer on posttest

In the matter of addition of positive integers with positive integers, there is no error and no misconception of all students. However, like that shown in Figure 2 in the matter of addition and subtraction materials that use negative integers, there is still a misconception. The selection of interview subjects is based on the range of scores on each test, at the time of pretest with M = 43.0 and SD = 21.18 there were 2 students with high score group, 3 medium score group students and 25 low score group students. While at posttest with M = 80.1 and SD = 20.0 there are 16 students with high score group, 10 students of medium score group and 4 students with low score group. Comparing the results of pretest and posttest, there were 6 students experiencing a change in low to moderate value group, 9 students experienced a low to high group value change and 10 students experienced moderate to the high-value group. None experienced a decline in the value group, but there were 5 students who did not experience a change in the value group of pretest and posttest.

4.1. Student Task Analysis Results

There are differences in students' answers when pretest and posttest. Analysis of student task results will be described in Table 2 and Table 3.

Table 2. Analyze the results data of the students' answers to the pretest and posttest

Addition of Integers		
Positive integers with negative integers of the form $\boldsymbol{a} + (-\boldsymbol{b})$	Negative integers with positive integers of the form $-a + b$	Negative integers with negative integers of the form $-a + (-b)$
Transcript of the answer from S_9	Transcript of the answer from S_3	Transcript of the answer from S_4
Pretest	Pretest	Pretest
(a) $5 + (-7) = 2$ ✗	(a) $-6 + 9 = -15$ ✗	(a) $-3 + (-7) = 4$ ✗
(b) $8 + (-8) = 0$ ✓	(b) $-7 + 7 = -14$ ✗	(b) $-6 + (-6) = 0$ ✗
(c) $9 + (-3) = 6$ ✓	(c) $-10 + 9 = -19$ ✗	(c) $-8 + (-2) = 6$ ✗
Posttest	Posttest	Posttest
(a) $5 + (-7) = -2$ ✓	(a) $-6 + 9 = 3$ ✓	(a) $-3 + (-7) = -10$ ✓
(b) $8 + (-8) = 0$ ✓	(b) $-7 + 7 = 0$ ✓	(b) $-6 + (-6) = -12$ ✓
(c) $9 + (-3) = 6$ ✓	(c) $-10 + 9 = -1$ ✓	(c) $-8 + (-2) = -10$ ✓

Analysis:

- Students still misconception on the addition of the forms $\boldsymbol{a} + (-\boldsymbol{b})$ with $\boldsymbol{a} < \boldsymbol{b}$. Students assume that $\mathbf{5} + (-\mathbf{7}) = \mathbf{7} - \mathbf{5} = \mathbf{2}$, but it is not. Addition of positive with negative integers equals reducing the number smaller than the larger number and the sign on the number follows the larger number.
- Students still misconception on the addition of the forms $-\boldsymbol{a} + \boldsymbol{b}$ with $\boldsymbol{a} < \boldsymbol{b}, \boldsymbol{a} = \boldsymbol{b}, \boldsymbol{a} > \boldsymbol{b}$. The student understands that the addition is equal to summing two numbers, since the sign at the beginning of the number is negative, so it will produce a larger negative number. This is not. This addition is equal to subtracting the smaller number from the larger number and the number sign following the larger number.
- Students still misconception on the addition of the forms $-\boldsymbol{a} + (-\boldsymbol{b})$ with $\boldsymbol{a} < \boldsymbol{b}, \boldsymbol{a} = \boldsymbol{b}, \boldsymbol{a} > \boldsymbol{b}$. Students understand this addition is the same as reducing the number smaller than the larger number. This is not. This addition of negative numbers with negative numbers will result in larger negative numbers, so the result will be smaller and negative, not positive.
- Overall misconceptions that occur can be corrected at posttest.

Table 3. Analyze the results data of the students' answers to the pretest and posttest

Subtraction of Integers			
Positive integers with positive integers of the form $\boldsymbol{a} - \boldsymbol{b}$	Positive integers with negative integers of the form $a - (-b)$	Negative integers with positive integers of the form $-a - b$	Negative integers with negative integers of the form $-a - (-b)$
Transcript of the answer from S_1	Transcript of the answer from S_4	Transcript of the answer from S_{21}	Transcript of the answer from S_6
Pretest	Pretest	Pretest	Pretest
(a) $0 - 4 = 4$ ✗	(a) $8 - (-10) = 2$ ✗	(a) $-5 - 8 = 13$ ✗	(a) $-2 - (-9) = -11$ ✗
(b) $8 - 8 = 0$ ✓	(b) $3 - (-3) = 0$ ✗	(b) $-6 - 6 = 12$ ✗	(b) $-5 - (-5) = -10$ ✗
(c) $9 - 8 = 1$ ✓	(c) $9 - (-7) = 2$ ✗	(c) $-10 - 7 = 17$ ✗	(c) $-9 - (-10) = -19$ ✗
Posttest	Posttest	Posttest	Posttest
(a) $0 - 4 = -4$ ✓	(a) $8 - (-10) = 18$ ✓	(d) $-5 - 8 = -13$ ✓	(a) $-2 - (-9) = 7$ ✓
(b) $8 - 8 = 0$ ✓	(b) $3 - (-3) = 6$ ✓	(e) $-6 - 6 = -12$ ✓	(b) $-5 - (-5) = 0$ ✓
(c) $9 - 8 = 1$ ✓	(c) $9 - (-7) = 16$ ✓	(f) $-10 - 7 = -17$ ✓	(c) $-9 - (-10) = 1$ ✓

Analysis:

- Students still misconception on the addition of the forms $\boldsymbol{a - b}$ with $\boldsymbol{a < b}$. Visible understanding of students stated that $\boldsymbol{0 - 4 = 4 - 0 = 4}$. This is not the case. Subtraction of positive numbers with a positive number form $\boldsymbol{a - b}$ with $\boldsymbol{a < b}$ is generating a negative number, ie $\boldsymbol{0 - 4 = -4}$.
- Students still misconception on the addition of the forms $\boldsymbol{a - (-b)}$ with $\boldsymbol{a < b, a = b, a > b}$. The student understands only that the reduction means reducing the number smaller than the larger number regardless of the sign of the number. On the number line, it produces a positive integer, a larger positive number. So it can be stated that $\boldsymbol{a - (-b) = a + b}$.
- Students still misconception on the addition of the forms $\boldsymbol{-a - b}$ with $\boldsymbol{a < b, a = b, a > b}$. Students do not yet understand that reducing negative numbers with negatives will result in larger negative numbers. Where $\boldsymbol{-a - b = -(a + b)}$.
- Students still misconception on the addition of the forms $\boldsymbol{-a - (-b)}$ with $\boldsymbol{a < b, a = b, a > b}$. Students sum up smaller numbers with larger numbers. Whereas students should reduce the numbers smaller than larger numbers.
- Overall misconceptions that occur can be corrected at posttest.

4.2. Analysis of Student Interview Results

Results of student interviews are described in Table 4.

Table 4. Data analysis of interview result with student

Interview Data	Analysis
Interview Transcript with S_2 Change of low to moderate value group. R : On the task number 3 (c), when pretest you answer $-10 + 9 = 1$, while at posttest $-10 + 9 = -1$. Do you think that is true when pretest or posttest? S_2 : When posttest. R : Why? What do you think is bigger -10 or 9? S_2 : The greater one 9. I'm just convinced of that now. R : Sure what cause? S_2 : Because earlier on the computer indicated that on the number line -10 less than 9. And there are sentences that explain also. R : So according to you, you get important information from the text that is in the learning media? S_2 : Yes. Of course. (Smiling)	Overall students who experienced an increase in the value group gave an answer indicating that there is a change in students' understanding of the addition and subtraction of integers before and after exposure to refutational text in the form of learning media. Students tend to assume that they have obtained reasonable and useful information in correcting their former misconceptions.
Interview Transcript with S_{15} Change the low to high-value group R : When pretest you state that $0 - 4 = 4$. Because 4 is bigger than 0 so the results are positive 4. S_{15} : No, its wrong. R : Why wrong? S_{15} : On the computer described if the smaller number minus the larger number the result will be negative. So the result -4. R : When you answer all the question, are you all right or wrong? S_{15} : No, there is wrong. But there are texts that help to fix.	
Interview Transcript with S_{21} Change of moderate to the high-value group. R : According to you $-5 - 8$ is how much? S_{21} : -13 R : Okay. But your answer at pretest is positive 13? S_{21} : Yes. Previously I thought if negative meeting negative would produce positive. So I answer 13. R : What do you think now about that answer? S_{21} : Yes, the answer is wrong.	

R	: How can you know that it's wrong?
S_2	: When on the computer there is an animated rabbit that jumps showing that if the negative subtraction with positive it will produce a negative number whose value is greater.

From the result of the analysis of student's task and students interview that was shown in Table 2 and Table 3, it can be seen that there is the difference between student's understanding during pretest and posttest. Students generally get information that is reasonable and interesting in fixing mistakes made before, this is in accordance with the conditions that must be met for conceptual changes [17]. Where conceptual change can occur in the presence of cognitive conflict that can be activated by presenting two opposite concepts at one time on a single text [18]. According to the four conditions expressed by Posner [17], ie: (a) Dissatisfied, (b) Intelligible, (c) Plausible dan (d) Fruitful has been fulfilled.

Dissatisfied, students know that they are experiencing misconceptions. This is given when the student answers the question wrong, then the student will know where the misconception is owned. By trying to answer again, students have a desire to correct misconceptions that occur. Intelligible, students are given new information or correct knowledge of the misconceptions that are being encountered, and from the results of interviews students simply understand the explanation of the given text. Plausible, new information provided to students makes sense. This is evident from the results of work and student interviews where students have the ability to improve the answers given. Fruitful, new information is helpful for solving future problems, this is seen when students have used the correct concept while working on posttest [17].

To further reassure the effect of refutational text in the conceptual changes of students, Effect Size calculation is done which states that $E_z = 1.8$, if $E_z < 1.00$ it can be asserted that the refutational text is strongly influencing in the conceptual changes experienced by the students [22]. Calculation of Effect Size uses a comparison of the pretest and posttest values, although there will still be other factors that cannot be measured in this study which also influence the results. However, the results given are refutational text enough give a strong role in the conceptual change effort to fix misconceptions of addition and subtraction of integers.

5. Conclusion

Refutational text plays a powerful and effective role in the effort of conceptual change to fix misconception on the matter of addition and subtraction of integers. This is seen from the Effect Size test which states that if an action is given influence more than 1 then the action gives a strong influence, wherein this study the influence of refutational text-aided learning media is $Ez = 1.8$ [22]. In addition, refutational text can also improve students' understanding matter of the addition and subtraction of integers because if there is a wrong concept will be given a signal that the concept is wrong and given the correct concept so that students can minimize the misconception in learning this material [17,18].

References

[1] Kilpatrick J Swafford and Findell B 2001 Adding it up: Helping children learn mathematics. *National research council.* [Online]. Retrieved 19 February 2018 from https://mason.gmu.edu/jsuh4/teaching/strands.htm.

[2] Altınok, A., Keşan, C., & Yılmaz, S. (2005). İlköğretim 7. Sınıf Tamsayılar Konusunun Günlük Yaşamla İlişkilendirilmesi ve Öğrenci Üzerindeki Etkisi. *XIV.* Ulusal Eğitim Bilimleri Kongresi Pamukkale Üniversitesi.

[3] Bishop J P, Lamb L L, Philipp R A, Whitacre I, Schappelle B P and Lewis M L 2014 Obstacles and Affordances for Integer Reasoning: An Analysis of Children's Thinking and the History of Mathematics. *Journal for Research in Mathematics Education*, **45**(1),19–61.

[4] Makonye J P and Fakude J 2016 A Study of Errors and Misconceptions in the Learning of Addition and Subtraction of Directed Numbers in Grade 8. *ResearchGate,* 1 – 10.

[5] Vlassis, J 2008 The Role of Mathematical Symbols in the Development of Number Conceptualization: The Case of the Minus Sign. *Philosophical Psychology*, **21**(4), 555–570
[6] Pitsi M E 2016 Identifying Common Misconceptions and Associated Errors The Grade 9 Learners Make When Adding and Subtraction. University of Johannesburg. Retrieved 20 February 2018 from https://ujcontent.uj.ac.
[7] Mohyuddin R G and Khalil, Usman 2016 Misconceptions of Students in Learning Mathematics at Primary Level. *Bulletin of Education and Research*, **38**(1), 133 – 162
[8] Fuadiah N F, Suryadi D and Turmudi 2017 Some Difficulties in Understanding Negative Numbers Faced by Students: A Qualitative Study Applied at Secondary Schools in Indonesia. *International Education Studies*, **10**(1), 157 – 168
[9] Kubar A and Cakiroglu E 2017 Prospective Teachers' Knowledge on Middle School Students' Possible Descriptions of Integers. *International Journal of Education in Mathematics, Science and Technology*, **5**(4), 279 – 294
[10] Peled I and Carraher D W 2007 Signed numbers and algebraic thinking. In J. Kaput, D. Carraher, & M. Blanton (Eds.), *Algebra in the early grades*, 303–327
[11] Vlassis, J 2001 Solving equations with negatives or crossing the formalizing gap. In M. Van den Heuvel-Panhuizen (Ed.), *Proceedings of the twenty-fifth international conference for the psychology of mathematics education*, **4**, 375 – 382
[12] Vamvakoussi X, Van Dooren W, and Verschaffel, L 2013 Educated adults are still affected by intuitions about the effect of arithmetical operations: evidence from a reaction-time study. *Educational Studies in Mathematics*, **82**(2), 323–330.
[13] Tippett C D 2010 Refutation text in science education: A review of two decades of research. *International Journal of Science and Mathematics Education*, **8**, 951 – 970
[14] Schneider M, Vamvakoussi X, and Van Dooren W 2012 Conceptual Change. In N. Seel (Ed.), *Encyclopedia of the sciences of learning*, 735–738
[15] Lem S, Onghena P, Verschaffel L, and Van Dooren W 2017 Using refutational text in mathematics education. *ZDM Mathematics Education*. doi:10.1007/s11858-017-0843-y
[16] Guzzetti B J, Snyder T E, Glass G V and Gamas W S 1992 Promoting conceptual change in science. *Reading Research Quarterly*, **28**(2), 117 – 159
[17] Posner G J., Strike K A, Hewson P W and Gertzog W A 1982 Accommodation of a Scientific Conception: Toward a Theory of Conceptual Change. *Science Education*, **8**(2), 211-227.
[18] Hynd C R 2001 Refutational texts and the change process. *International Journal of Educational Research,* **35**(7–8), 699 – 714
[19] Van den Broek, P and Kendeou P 2008 Cognitive processes in comprehension of science texts: the role of co-activation in confronting misconceptions. *Applied Cognitive Psychology*, **22**(3), 335 – 351
[20] Plomp, T 1997 *Educational Design: Introduction*. Faculty of Educational Science and Technology, University of Twente.
[21] Hasan S D Bagayoko and Kelley E L 1999 Misconceptions and Certainty Of Response Index (CRI). *Physic Education*, **34**(5), 294 – 299.
[22] Cohen, J 1988 *Statistical Power Analysis for the Behavioral Science (2nd ed)* Hillsdale N J : L. Erlbaum Associates.

Implementation of Chi Square Automatic Interaction Detection (CHAID) Method to Identify Type 2 Diabetes Mellitus in Tuberculosis Patient. A Case Study in Cipto Mangunkusumo Hospital

T Siswantining[1,*], Maghfiroh[1], T Kamelia[2], D Sarwinda[1]

[1]Department of Mathematics, FMIPA Universitas Indonesia, Kampus UI Depok, Depok 16424, Indonesia

[2]Division of Pulmonology, Department of Internal Medicine, FK Universitas Indonesia / Dr. Cipto Mangunkusumo Hospital, Jl. Diponegoro No. 71, Central Jakarta 10430, Indonesia

*Corresponding Author: titin@sci.ui.ac.id

Abstract. A Pulmonary Tuberculosis (pulmonary TB) is a chronic infectious disease caused by *Mycobacterium tuberculosis*. Chronic infection cause the body in a state of oxidative stress. In the state of stress, stress hormone production increases and can affect the increase of blood sugar levels which then trigger the occurrence of diabetes mellitus (DM). The purpose of this study is to determine the factors associated with the emergence of DM Type 2 and make a classification to characterize DM Type 2 in patients with pulmonary TB. In this research, the data used are secondary data of pulmonary TB patients obtained from Cipto Mangunkusumo Hospital (RSCM) for six years, 2012 - 2017. Chi Square Automatic Interaction Detection (CHAID) method is used to classify categorical data by dividing the data set into subgroups. The dependent variable is type 2 DM status, and the independent variables are gender, age, body mass index, level of neutrophil, level of lymphocytes, erythrocyte sedimentation rate, and anti-tuberculosis medicines such as rifampicin, isoniazid, pyrazinamide, ethambutol, and streptomicin. There are seven classes are obtained by classification using CHAID. The result from analysis of CHAID method shows that the factors related to the occurrence of type 2 DM on TB patients are age, body mass index, gender and pirazinamid (Z). Based on CHAID method, a classification of the occurrence of type 2 DM on pulmonary TB patients is obtained, which is pulmonary TB patients who are ≥ 40 years old, have body mass index ≥ 25 kg/m^2, and the most patients are male (Class-4th).

1. Introduction

Tuberculosis (TB) and diabetes mellitus (DM) are two major health problems which are epidemiologically and globally significant because they are chronically related diseases [1]. Tuberculosis (TB) is an infectious chronic disease caused by *Mycobacterium tuberculosis*. Most TB bacteria invade the lungs, but can also invade other organs, such as skin, eyes, lymph glands, bone, lining of the brain, and so on [2]. Based on WHO data in 2013, Indonesia ranks fourth after India, China and South Africa as the country with the highest incidence of TB in the world, while diabetes mellitus (DM) is a metabolic disorder disease characterized by elevated blood sugar levels and carbohydrate metabolism abnormalities, fat, and protein are caused by insulin secretion abnormalities, insulin work or both [3]. The number of DM patients worldwide is estimated at 285 million people and this number will increase up to 438 million people by 2030 [4].

Chronic infection causes the body in a state of oxidative stress. Oxidative stress is a condition where there is an imbalance between pro oxidants (free radicals) and antioxidants. When the body is in a state of stress, the production of stress hormones increases as epinephrine, cortisol, and glucagon which work synergistically. Increased production of these hormones can affect the increase in blood sugar levels (hyperglycemia). Increased blood sugar levels will affect the production of insulin from pancreatic beta cells. As a result, high blood sugar levels in prolonged time will trigger insulin resistance in muscle tissue causing diabetes [5,6].

Research conducted by [7] conducted in Nigeria that TB patients with impaired blood glucose tolerance had normal results after three months of tuberculosis treatment, while in Turkey the oral glucose tolerance test (TTGO) was examined to 58 TB patients and 23 patients with pneumonia. In TB patients, 10% of glucose intolerant and 9% had diabetes, while pneumonia patients had 17% of DM and no glucose intolerance. Both studies show that infection causes reversible glucose intolerance.

The above-mentioned study suggests that TB patients have a tendency to develop DM. This needs to be investigated in Indonesia regarding to the variables suspected to be associated with DM status in TB patients, especially in Cipto Mangukusumo Hospital (RSCM) so that the management of DM-TB patients can be improved.

Problems:

How is the classification of Type 2 DM patients based on the factors which are most significantly related to the appearance of Type 2 DM in pulmonary tuberculosis patients?

Objectives:

The objective of this research is to classify the characteristics of Type 2 DM on pulmonary tuberculosis patients based on the factors that are most significantly related.

Scope of problem:

1. This study uses medical records of RSCM taken from 2012 to 2017.
2. The patients studied were patients with pulmonary tuberculosis who are inpatient or outpatient at pulmonology clinic of RSCM.

2. Research variables

The variables that will be involved in this research are : status of type 2 diabetes mellitus on patients, which is the presence or absence of type 2 DM on Pulmonary tuberculosis patients; gender of the patients, which consists of two categories namely female and male; age of the patients with pulmonary tuberculosis, which is calculated from the time the patient is born until admitted to the hospital with the diagnosis of Type 2 DM, expressed in years; Body Mass Index (BMI) is an assessment of adult nutritional status that is measured by body weight in kilograms (kg) divided by height in squared meters (m^2);

Neutrophil level is the percentage of neutrophils in the blood of pulmonary tuberculosis patients based on the blood test results; Lymphocyte level is the percentage of lymphocyte in the blood of pulmonary tuberculosis patients based on the blood test results; Erythrocyte Sedimentation Rate (ESR) is the rate of erythrocyte cells sediment in the plasma expressed in millimeter (mm); Rifampicin (R) is an anti- tuberculosis drug which consists of two categories which are consuming or not consuming; Isoniazid (H) is an anti-tuberculosis drug which consists of two categories which are consuming or not consuming; pyrazinamide (Z) is an anti-tuberculosis drug which consists of two categories which are consuming or not consuming; Etambutol (E) is an anti-tuberculosis drug which consists of two categories which are consuming or not consuming; Streptomycin (S) is an anti-tuberculosis drug given by injection.

3. Methods

The Population in this research is patient with pulmonary tuberculosis at RSCM who were registered in medical records in 2012-2017. The number of sample is 30 patients with pulmonary tuberculosis at RSCM, and used purposive sampling. The method for data analysis is *Chi-square Automatic Interaction Detection* (CHAID).

3.1 Analysis of Chi-square Automatic Interaction Detection (CHAID) Method

CHAID is a method to classify categorical data by dividing the data set into subgroups based on dependent variable [8]. According to [9] the CHAID method is an iterative technique that tests one by one independent variables used in classification and arrange them based on the significance level of the chi-square test of the dependent variable. The results of the classification are shown in the form of a tree diagram. The best independent variable that will form the first branch in the resulting tree diagram.

Before the process of calculating the CHAID algorithm, [10] divides the independent variables into three types:

1. Monotonic Variables

The categories in these variables can be combined if they are adjacent to each other, i.e. variables whose categories indicate a sequence, such as ordinal data. For example: age, income, and level of education.

2. Free Variable

The categories in these variables can be combined even if they are not close to each other or do not pay attention to the sequence, such as nominal data. Examples: job, geographic area.

3. Floating Variable

The categories in this variable will be treated like monotonic variables, except for the missing value which can be combined with any category.

3.2 Chi-square test (χ^2)

Chi-square test is a non-parametric statistical test that can be used to test the equality of proportion, independence test, and goodness of fit test or suitability test. The chi-square test in the CHAID method according to [10] is used to determine whether the categories in the independent variables are uniform. Furthermore, the chi-square test statistic is then used to test the independence between variables, in this case it will be determined which independent variables are most significantly related to the dependent variable.

Suppose the first variable has a category b, i.e. $A_1, A_2, \ldots, A_b$ and the second variable has k categories, i.e. $B_1, B_2, \ldots, B_k$. The number of observations in the i-th category $(i = 1,2,3,\ldots,b)$ on the first variable and the jth category $(j = 1,2,3,\ldots,k)$ in the second variable is expressed by O_{ij} (Table 1).

Table 1. Structure of Chi-Square Test Data $b \times k$

Variable 2 Variable 1	B_1	B_2	B_3	...	B_k	Jumlah
A_1	O_{11}	O_{12}	O_{13}	...	O_{1k}	$O_{1.}$
A_2	O_{21}	O_{22}	O_{23}	...	O_{2k}	$O_{2.}$
A_3	O_{31}	O_{32}	O_{33}	...	O_{3k}	$O_{3.}$
-	-	-	-	...	-	-
A_b	O_{b1}	O_{b2}	O_{b3}	...	O_{bk}	$n_{b.}$
Total	n_1	n_2	n_3	...	n_k	n

Where: O_{ij} = Number of observations on the i-th row and the jth column; $n_{i.}$ = Total observations on the i-th row; $n_{.j}$ = Total observations on the j-th row; n = Total number of respondents.

The expected value for each observation is

$$E_{ij} = \frac{n_{i.}n_{.j}}{n}$$

The following are the independency test steps

1. Hypothesis

H_0: Both variables are independent

H_1: not so

2. Determine the level of significance α
3. Statistical value of the test
4. Make a decision rule
5. Make a conclusion from the outcome of the decision

The chi-square test described above is a chi-square test of independence, whereas in the CHAID method chi-square test is also used to test the equality of proportion and independence test.
To test the similarity of proportions, the hypothesis is
H_0: $p_1 = p_2$
H_1: $p_1 \neq p_2$
The next steps would be similar to the previous steps as described above.

3.3 CHAID Algorithm
The CHAID algorithm is used to separate and merge categories in the variables used in the analysis. Based on [11], the outline of this algorithm is divided into three stages, namely merging, splitting, and stopping.

a. Merging Stage
The merging stage is performed to combine categories of independent variables that have more than two categories. To perform this merger stage examined the significance of each category of independent variables to the dependent variable. The steps for the merger stage are as follows
    1. For each variable $X_1, X_2, \ldots, X_k$, create a contingency table of size 2 × J formed by a pair of categories from the category of independent variables with their dependent variables having as many as J categories.
    2. Calculate the statistical value of chi-square test on each pair of categories for each independent variable to test the equality of proportions.
    3. Pay attention to the chi-square test statistic value for each category pair, if smallest $\chi^2_{count} < \chi^2_{tabel}$, then the category is combined into a new category.
    4. Re-examine the significance of the new categories. If there are still categories of independent variables that have more than two categories then repeat steps 2 and 3, but if it is significant go to the next step.

Recalculate the statistical value of the chi-square test to determine which independent variables are most significantly related to the dependent variable. After the chi-square value is obtained, calculate the p-value of each significant independent variable.
    5. Make a Bonferroni correction for the p-value obtained in step 5.

b. Partition Stage
In this stage of separation, the variables will be partitioned or divided groups. To perform the separation, the most significant independent variables will be used as split nodes by looking at the Bonferroni corrected p-value values of the merging stages on each independent variable by choosing the smallest p-value value < α.

c. Termination Stage
Before the termination stage, repeat the merging stage to analyze subsequent subgroups. The termination will be performed if there are no more significant independent variables that indicate any difference in the dependent variables.

4. Results and Discussion
The data obtained in this study were analyzed using Chi-square Automatic Interaction Detection (CHAID) method with SPSS software version 23. The dependent variable in this study was the status of DM Type 2 in pulmonary TB patients divided into two categories, i.e. yes and no, while independent variables existed as many as 11 variables expressed in category form and determined in consultation with physician and laboratory clinical reference.

IOP Conf. Series: Journal of Physics: Conf. Series **1108** (2018) 012118 doi:10.1088/1742-6596/1108/1/012118

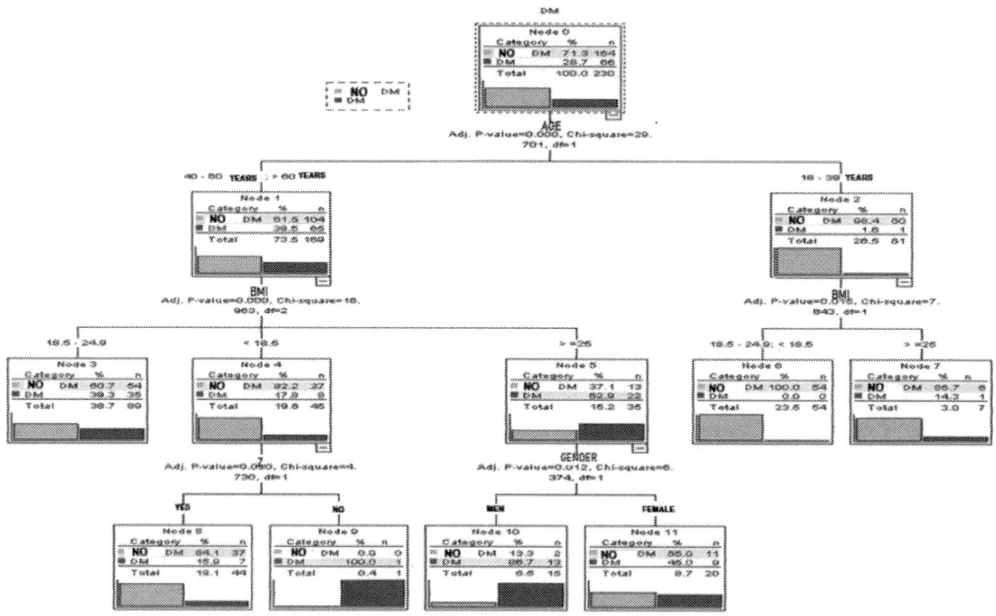

Fig 1. Result of CHAID Tree Diagram

From Figure 1 above, we found that significant factors related to the emergence of Type 2 DM in pulmonary tuberculosis patients were age, body mass index (BMI), sex, pyrazinamide (Z). The CHAID tree diagram above illustrates that pulmonary TB patients are divided into 7 classifications or groups by reading the results of the tree diagram following the top-down stopping rule, i.e. starting from the parent node, i.e. age. Then under it is a subgroup (child node) which is obtained from the division of the parent node up to the bottom end of the tree called the terminal node (Table 2).

Table 2. Classification of pulmonary TB patients

Classification Number	Characteristics
1	Pulmonary TB patients with age ≥ 40 years and body mass index (BMI) ranged from 18.5-24.9 kg / m2
2	Pulmonary TB patients with age ≥ 40 years, BMI <18.5 kg / m2, and taking pyrazinamide (Z)
3	Pulmonary TB patients with age ≥ 40 years, BMI <18.5 kg / m2, and not taking pyrazinamide (Z) drugs.
4	Pulmonary TB patients with age ≥ 40 years with BMI ≥25 kg / m2, and male sex.
5	Pulmonary TB patients with BMI ≥25 kg / m2, and female sex.
6	Pulmonary TB patients aged 18-39 years with BMI <25 kg/m^2.
7	Pulmonary TB patients aged 18-39 years with BMI ≥ 25 kg/m^2.

From the seven classifications that are formed, it can be seen in Table 3 that the number and percentage of each of the TB patients with Type 2 DM and not having Type 2 DM are as follows

Table 3. Percentage of results for each classification of pulmonary TB patients

Classification Number	Number of DM Patients	Percentage	Number of non-DM Patients	Percentage
1	35	39.33%	54	60.67%
2	7	15.91%	37	84.09%
3	1	100.00%	0	0.00%
4	13	86.67%	2	13.33%
5	9	45.00%	11	55.00%
6	0	0.00%	54	100.00%
7	1	14.29%	6	85.71%

The largest percentage of pulmonary tuberculosis patients with Type 2 DM is in the 3rd classification, which is 100%, but in that classification or group there is only one pulmonary TB patient with Type 2 DM, so the classification does not represent the population well. Therefore, to see the characteristics of pulmonary tuberculosis patients with Type 2 diabetes, the 4th classification was selected with a percentage of 86.67% with the characteristics of patients aged ≥ 40 years, having BMI ≥25 kg / m2, and male sex. This is in accordance to a study by [12] which shows that in adults aged 18-45 in China, it is known that obesity is positively correlated with the emergence of Type 2 DM in patients with pulmonary TB; the prevalence of DM incidence was higher when BMI ≥ 23.41 kg / m2. Research conducted by [13] showed that factors associated with the emergence of Type 2 DM in patients with pulmonary tuberculosis were age and body mass index (BMI). [14] in Czech and Finland shows that the number of men with Type 2 diabetes is more dominant than women.

The following is an accurate classification result obtained from data analysis using CHAID

Table 4. Accuracy of the classification results

Observed	Predicted		
	not DM	DM	Percent Correct
not DM	162	2	98.8%
DM	52	14	21.2%
Overall Percentage	93.0%	7.0%	76.5%

From Table 4 above, it was found out that the accuracy of overall pulmonary tuberculosis patient predictions using CHAID was 76.5% with prediction error of 23.5%. On the accuracy of classification results to predict Type 2 diabetes by 21.2%. This is due to limitations on research data obtained in the RSCM medical record where the number of pulmonary TB patients with Type 2 DM and the number of pulmonary TB patients with Type 2 DM were out of balance.

5. Conclusion

Based on the results of analysis by using CHAID method, the conclusion obtained is that the main factor associated with the emergence of DM Type 2 in patients with pulmonary TB is age. In addition, there are also gender factors, body mass index (BMI), and pyrazinamide (Z). The results of the classification of pulmonary tuberculosis patients with Type 2 DM based on the factors that have been obtained are patients with characteristics aged ≥ 40 years, have BMI ≥ 25 kg / m2, and male sex with percentage of accuracy is 86.67%.

Acknowledgements
This research is supported by PITTA UI 2018 Research Grant from DRPM Universitas Indonesia.

References

[1] Aregbesola A 2016 Gender difference in Type 2 Diabetes and the role of body *Annals of Clinical Biochemistry* pp. 1-24

[2] Bagozzi R P 1994 Advanced methods of marketing research. Blackwell Publishers Ltd Oxford

[3] Departemen Kesehatan Republik Indonesia 2011 *Pedoman nasional penanggulangan Tuberkulosis* (Jakarta: Direktorat Jendral Pengendalian Penyakit dan Penyehat Lingkungan)

[4] Departemen Kesehatan Republik Indonesia 2014 *Pedoman nasional penanggulangan Tuberkulosis* (Jakarta: Direktorat Jendral Pengendalian Penyakit dan Penyehat Lingkungan)

[5] Dooley K E and Chaisson R E 2009 Tubeculosis and Diabetes Mellitus: Convergen of two epidemics **9**(12) pp. 737-746

[6] Gallagher C A 2000 *An iterative approach to classification analysis*

[7] Jing Cai 2017 Association between body mass index and Diabetes Mellitus in Tuberculosis patients in China: A community based cross-sectional study *BMC Public Health* **17** p. 228

[8] Kelestimur F 2004 The endocrinology of adrenal Tuberculosis: The effects of Tuberculosis on the hypothalamo-pituitary-adrenal axis and adrenocortical function *Journal of Endocrinol Invest* **27**(4) pp. 380-386

[9] Kibirige D 2014 Endocrine dysfunction among adult patients with Tuberculosis: An African experience *Indian Journal of Endocrionology and Metabolism* **18**(3)

[10] Perkumpulan Endokrinologi Indonesia (PERKENI) 2015 *Konsensus pengelolaan dan pencegahan Diabetes Melitus Tipe 2 di Indonesia 2015* (Indonesia: PB PERKENI)

[11] Syam M S 2013 *Dukungan sosial penderita Tuberkulosis paru di wilayah kerja Puskesmas Ajangale Kebupaten Bone tahun 2013*

[12] Viswanathan V 2012 Prevalence of diabetes and pre-diabetes and associated risk factors among Tuberculosis patients in India

[13] World Health Organization and the International Union Against Tuberculosis and Lung Disease 2011 *Collaborative framework for care and control of Tuberculosis and diabetes* (Geneva: WHO Press)

[14] Sayadi M, M Zibaeenezhad, S M T Ayatollahi 2017 Simple Prediction of Type 2 Diabetes Mellitus via Decision Tree *Int Cardiovasc Res J.* **11**(2) pp. 71-76

IOP Conf. Series: Journal of Physics: Conf. Series **1108** (2018) 012119 doi:10.1088/1742-6596/1108/1/012119

Implementation of Multimodel Active Learning to Improve Basic Teaching Skills of Pre-Service Physics Teachers

N A Lestari[1,*], N Suprapto[1], U A Deta[1], and M Yantidewi[1]

[1] Department of Physics, Faculty of Mathematics and Natural Sciences, Universitas Negeri Surabaya
Jl Ketintang, Gd C3 Lt 1, Surabaya 60231, Indonesia

*Corresponding author: nuritalestari@unesa.ac.id

Abstract. Pre-service teachers must have a good basic teaching skills that are opening lesson, use of learning model, use of learning media, mastery of teaching materials, submission of teaching materials, class management, and closes lesson ability. Basic teaching skills influenced by learning material design and communicating skill. Improving pre-service's basic teaching skill can't did only by an effective learning model, but a collaborative model that can called multimodel learning. From the study known that pre-service teacher have a good score of learning design and teaching skills.

1. Introduction

Improving the quality of education can be started by increasing the basic teaching skills of prospective teachers while implementing pre-service training in universities. Increased competence of the teacher candidate is not only on the science material but also on the pedagogy. This condition relates to the teachers'role as professional educators who are not only judged from the understanding of the science concepts but also the ability to teach those concepts. Basic teaching skills that should be owned by prospective teachers (pre-service teachers) are the ability to open lessons, use of learning models, use of instructional media, mastery of teaching materials, submission of teaching material, classroom management, and closing lessons.

Improving the ability to teach students as candidates for physics teachers can be done through the application of active learning. Active learning is a learning approach that provides opportunities for students to play a more active role in the learning process (searching for information, processing information, summarizing it, and applying) by providing a learning environment that makes students not depressed and happy to carry out learning activities ([1])

Referring to Fink [2], active learning consists of two main components, namely experience, including doing and observing and dialogue, including self-dialogue and dialogue with others. Doing is an activity that refers to the learning process in which students actually do something real and directly interact with related objects. Observing activities occur where students can see and listen when other people "do something", related to what they are learning. Observing actions can be done "directly" or "indirectly." Direct observation means that students are invited to observe real activities or situations directly. Indirect observation means that students are invited to observe the situation, case study or invited to watch a movie (video). Dialogue with Self is a student activity in reflective thinking about a topic. The student can ask himself, what is or should be considered, what is felt from the topic he is studying, and "think about his own thoughts (thinking about my own thinking)". This ability to think is called metacognitive ability. Dialogue with others can be seen in the interaction between students in the class that aims to make the process of exchange of thoughts occur. This activity can be done by discussing in groups or class discussions with experts or other sources.

The improvement of teaching ability as a physics teacher candidate can be held through the application of multimodel active learning in a class that facilitates the practice of teaching (peer teaching and microteaching). The learning system was done using multimodel active which includes interactive learning models with the aim of improving the quality of student involvement during the learning process such as Problem Based Learning (PBL), Cooperative Learning type Jigsaw, and Think Pair Share (TPS), and Direct Instruction. Each model has its goal achievement that can facilitate the improvement of basic skills of teaching for pre-service teachers. PBL used to help master the basic concepts about types of learning models, Cooperative Learning (Jigsaw and TPS) used to support students in group discussions, and Direct Instruction provides examples of modeling the implementation of a learning model in class.

This research is dedicated to study the influence of the application of active learning multimodel on the ability of students as candidates for physics teacher in designing and learning implementation. Pre-service teachers are expected to have excellent and competent teaching basic skills so that later they can apply effective learning systems when teaching in schools.

2. Methods

This research uses mixed method model, that is combining both qualitative and quantitative research method using explanatory design with participant selection model (QUAL emphasized). The data obtained from the learning is analyzed qualitatively and quantitatively to describe the complete results of the research. Characteristics of this method are data retrieval, and data analysis done quantitatively as the support method and then continued with the data collection and qualitative analysis as the primary method. Qualitative methods in this study are expected to help to get a description of the variables measured such as the basic teaching skill. In general, the stages of this research as follows,

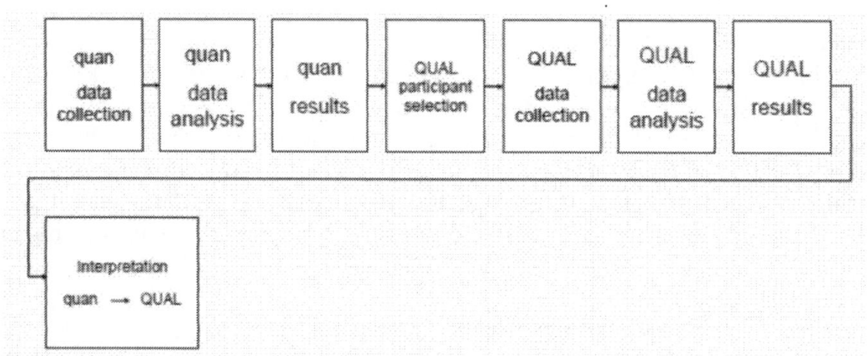

Figure 1. Design of mixed method experiment

Quantitative data retrieval research conducted after the implementation of multimodel active learning of pre-service teachers. Quantitative data in this study were obtained through tests at the end of learning which included the subject matter of the pedagogic and physics materials. Other quantitative data is in the form of the value of observations of the ability to teach physics teacher candidates which include the ability to design teaching materials and peer teaching and qualitative data is a description of learning observations using active multimodel and peer teaching observations. Besides that, the results of interviews with prospective teachers were also related to the peer teaching results that had been conducted.

In addition to getting a quantitative score, interviews and activity record to determine the success of the action also conducted. At the time of the interview session, the students were asked to state the advantages and disadvantages of the learning that had been done. Quantitative and qualitative data were then analyzed to determine their accuracy. The data analysis showed the students' teaching ability improvement after the action. These results can be used to express the effectiveness of applied learning.

3. Results and Discussion

This activity begins with the implementation of learning using active learning multimodel. Some types of learning models are implemented during the lecture process by referring to the expected goals, which can train the teaching skills of physics teacher candidates. Learning models selected include PBL (problem-based learning), Jigsaw cooperative learning and think pair share, as well as Direct Instruction.

As the final product of the learning process, students are asked to do peer teaching using learning models that have been taught in the classroom. Peer teaching is done to find out students' abilities in teaching. Before peer teaching, students were also asked to prepare lesson plans according to the model applied. Assessment of the lesson plan is carried out by looking at the completeness of the components and the suitability of the content with the objectives to be achieved. Score of lesson plan include as quantitative data. The following are quantitative data of learning design ability which obtained during the study.

Table 1. Learning Design Data

Aspect	Average score	Official Statement
Arranging learning indicators	81,25	Difficulties in determining cognitive level
Arranging learning objectives	85,00	Adjust the ABCD format (audience, behaviour, condition, degree)
Analysis of learning materials	70,00	There are still found inappropriate concepts and misconceptions
Analysis of learning models	77,50	The learning model used should not be random chosen
Arranging syntax of learning models	78,75	The stages are in sequence but the contents are incomplete
Ability to choose learning sources	81,25	Learning resources are relevant to students' needs

Based on data, known that students as candidates for physics teachers have quite good abilities in designing learning. A good lesson plan can make an effective learning because effective implementation of educational and instructional activities requires planned work ([3]). The preparation of a good lesson plan is influenced by the completeness of its components, including indicators of learning, learning objectives, teaching materials, selection of learning models and methods, syntax (step of learning model), and selection of learning resources. The lowest value obtained by students is related to the preparation of teaching materials. In the teaching material, there is still a discrepancy in the concept. An example is the static fluid material in the Archimedes's law which discusses objects that float and sink in the fluid. Figure 1 is the example of incomplete material. Students show an image of a sink object, but they do not explain the concept. Sink object has three force which included to the system, such as buoyancy force (Fa), Weight (Wb), and Normal Force (N) because the object was touched bottom surface of the glass.

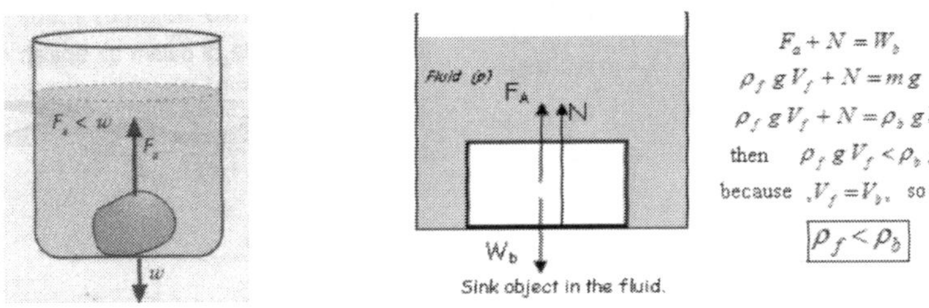

(a) (b)

Figure 2. Example of sink object from student material on the lesson plan (a)
Example of complete concept (b)

The second quantitative data are score of pre service's teaching skills. The basic teaching skills contain of opening lesson, use of learning model, use of learning media, mastery of teaching materials, submission of teaching materials, class management, and closes lesson. The data shown on Table 2.

Table 2. Description of Teaching Skill Component

Aspect	Average score	Official Statement
Opening Lesson	70,00	Less of using interesting and unusual phenomenon
Use of Learning Model	86,25	Good, because effect of modelling
Use of Learning Media	91,25	They have good creativity
Mastery of Teaching Materials	71,25	There are some misconception
Submission of Teaching Materials	88,75	The communicating skill is good
Class management	80,00	They have good interaction with students
Closes Lesson	73,75	Lack of time

Based on data of Table 2, known that students as pre-service physics teacher have highest score in using learning media (91,25). This is influenced by the high creativity of students in making and using innovative learning media. In addition, it is also supported by technological sophistication that can be utilized by the community to the fullest. In accordance with [4], that learning media can be a useful component for organizing learning activities. Students also have realized the importance of media in learning physics, so that the value obtained is optimal. Other high values are found in the ability to explain physics material. They earned an average score of 88.75 which means that the ability of students to inform teaching material is well supported by a language that can be understood by students. According to [5], students will be better able to understand and explicate how language is used as it is being acquired through interaction resourcefully, contingently, and contextually.

The ability to teach with the lowest score is to open and close the lesson. The score obtained is less than 75. Based on the results of interviews with students, the low score influenced by the timing of inappropriate activities. Students do peer-teaching with a time allocation of 1x45 minutes with certain teaching materials that done in class with a time allocation of 2x45 minutes. As a result, many students cannot complete the learning on time. Whereas for the ability to open the lesson is influenced by the lack of creativity of students in providing physical phenomena that can be a link between teaching materials and their applications. Examples delivered by students are phenomena that have often appeared during teaching activities. Students should be able to find other cases that are better and more recently the teaching material presented.

At the end of the peer-teaching stage, students are asked to conduct tests to determine their ability to master basic physics and pedagogy. Physical material tests include the fields of kinematics, dynamics, thermodynamics, optics, and magnetic electricity. While pedagogy material comprises knowledge about school curriculum, characteristics of students, learning theory, innovative learning, and assessment systems. Based on the test results, students tend to have a better pedagogy score, which is 67% and a physics score of 33%. Ideally, they should have equal value (50:50) because as candidates for physics teachers, students are required to have the ability to both teach and deliver teaching materials in physics. Figure 3 shows the results of the student test scores.

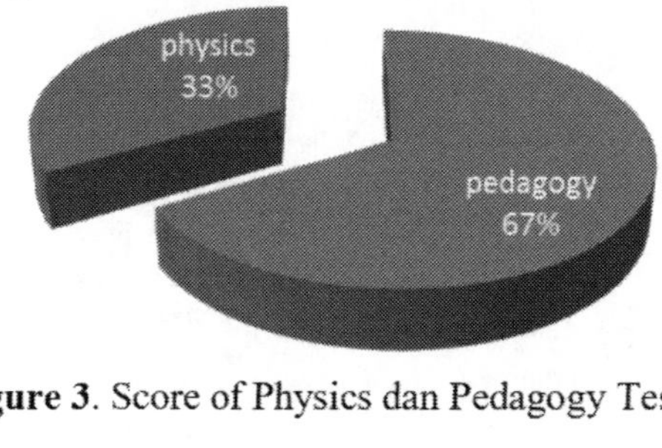

Figure 3. Score of Physics dan Pedagogy Test

Active learning that is applied is not only capable of improving students 'understanding of teaching material but also can improve students' ability to carry out learning. The ability to carry out learning is an ability that can improve through skill training. This statement supported by the research's results in [6] which can improve the ability of students who are skilled with active learning.

Furthermore, the application of multimodel through the collaboration of several types of learning models that applied in the classroom is known to improve students' teaching ability. This ability improvement is known from the results of observations when students conduct peer-teaching activities compared to the previous year's class. Students who receive learning using multimodel active learning have better abilities than last year students. This improvement is evidenced by the appearance of students when peer-teaching is neater, structured, and appropriate with the lesson plan. This is a positive impact of each model that is applied, namely (1) Problem Based Learning (PBL), helping students in analyzing the basic concepts of learning to be carried out, (2) Cooperative Learning type Jigsaw, and Think Pair Share (TPS), help students collaborate and interact socially with colleagues; and (3) Direct Instruction provides modeling learning so students know first-hand the stages in the classroom implementation.

Through active learning, students are more accessible to remember and understand teaching material. The teaching material focused on content that is the basis for students when conducting teaching practices such as learning theories and learning models. Students are not only invited to learn the material, but also develop it and form the characteristics of their learning. This condition is following by active learning goals based on Student-Centered Learning. Students get the opportunity to be actively involved during the learning process, not only emphasizing the mastery of the material but also in the development of student character (life-long learning) ([7]).

4. Conclusions

Based on the result of the analysis and discussion, it is found that the basic ability to teach physics teacher candidates is of good enough value and meets the criteria expected during the lecture process using active learning multimodel. Students can practice complete teaching methods according to the design stages of learning.

One of the interesting findings in this study is that the students as candidates for physics teacher have a dominant value in mastering the subject matter in the field of pedagogy compared to the physics materials. For suggestions, physics majors related to this problem i.e the holding of training or seminars for pre-service physics teachers need to link to how to learn effectively and efficiently. It is expected students can improve the understanding of physics and pedagogy materials. Also, further research is needed for students to obtain a comprehensive result related to the effects of innovative models in the understanding basic teaching skills.

5. References

[1] Asmani J M 2009 *Jurus-jurus Belajar Efektif untuk SMP dan SMA* (Yogyakarta: Diva Press)

[2] Fink L D 1999 Active Learning. (Online), (http://honolulu.hawaii.edu/intranet/committees/FacDevCom/guidebk/teachtip/evaluate.htm), accessed on April 17th 2018.

[3] Tok S 2010 The problems of teacher candidate's about teaching skills during teaching practice, *Procedia Social and Behavioral Sciences* **2**(1) 4142-4146

[4] Arends R I 2012 *Learning to Teach. Ninth Edition* (New York: McGraw Hill)

[5] Gilakjani A P, Ismail H N, and Ahmadi S M 2011 The Effect of Multimodal Learning Models on Language Teaching and Learning *Theory and Practice in Language Studies* **1**(10) 1321-1327

[6] Dirjen Dikti 2008 *Buku Panduan Pengembangan Kurikulum Berbasis Kompetensi Perguruan Tinggi* (Jakarta: Dirjen Dikti)

[7] Haviluddin. 2010. *Active Learning* Berbasis Teknologi Informasi. *Jurnal Informatika Mulawarman,* (Online), **5**(1) 31, (http://jurnal-media.com/tik-haviluddin.pdf, accessed on May 3th 2018).

Self-Regulated Learning Ability of Elementary School Students in Learning Mathematics

A D Astuti[1,3] and M N Wangid[2]

[1]Graduate Program of Elementary Education, Yogyakarta State University, Indonesia.
[2] Elementary Education Department, Yogyakarta State University, Indonesia.

[3]Coresponding author: ameliauny@gmail.com)

Abstract. The purpose of this study was to describe the self-regulated learning ability of elementary school students in learning mathematics. It was conducted on 284 fourth-grade elementary school students. Respondents were chosen based on the same characteristics, which are the same curriculum, grade and age range. Data were collected from students' responses using the questionnaires. This research was a quantitative descriptive with survey-design research type. The result showed that the ability of self-regulated learning students was various. Self-regulated learning ability of the students were distributed as 15% (very high category), 25% (high category), 59% (medium category), 0,4% (low category) and 0,4% (very low category). Based on the results of this study, the self-regulated learning ability of fourth grade of elementary school in learning mathematics is in the moderate category.

1. Introduction

Education and learning are two closely interconnected objects. One of the main aspects of the education world is learning activities. A good learning activity will be realized if there is synergy between teachers and students as the education stakeholders. Teachers have roles to design well the way of learning in order to provide the meaningful experiences for students in each lesson, so that can permanently contribute, both in terms of students' knowledge, behavior and thinking ability [1]. Meanwhile, the role of students is how to get good self-regulation for controlling themselves and emerge the awareness to be active-reactive in the learning process and also have the responsibility toward the success of desired goals.

Based on [2], year 1, 2, and 3 children were able to indicate that there are foundational aspects of self-regulated learning through the appropriate teacher's orientation and support. Accordingly, self-regulated learning needs to be instilled in early childhood. It is aimed to require students to take responsibility for their learning so as to have targets and learning goals and can determine the steps to achieve those goals. Good self-regulated learning can encourage success and improve students' achievement proactively with strategies for desired goals [3].

Self-regulated learning ability is very important for students to comply with every learning, especially in learning mathematics. Mathematics learning is faced with the challenges to solve the complex and non-routine problems [4]. The use of strategies, the preparation of solution steps, and strong motivation are necessary to be able to solve mathematics problems precisely. It shows that there are three phases of self-regulated learning in the learning process, including processes and beliefs that occur before beginning to learn (forethought), the process during the learning activity occurs (performance), and the process that occurs after learning (self-reflection) [5]. Components in the three phases of SRL are shown in Figure 1 below [6].

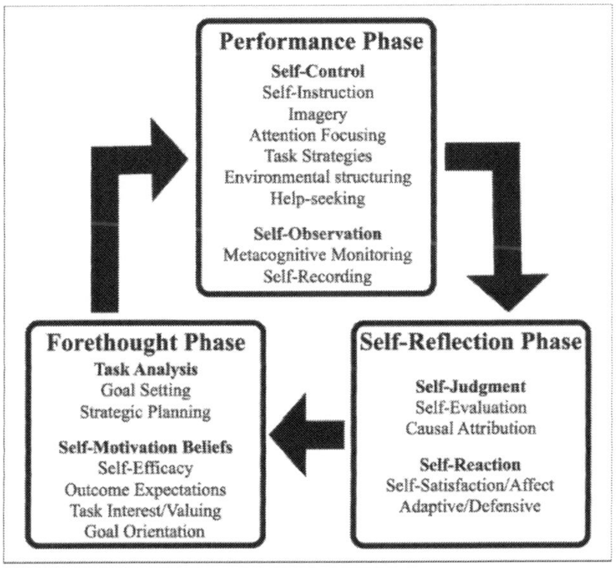

Figure 1. Self-regulated Learning Phase

In the planning phase, there are two related categories, namely task analysis, and self-motivation beliefs. Task analysis includes goal setting and strategic planning. Goals can be interpreted as the establishment or determination of the learning outcomes that an individual wants to achieve. Furthermore, strategic planning is a process and action of someone who aims and oriented for obtaining and demonstrating a skill that can be used to achieve goals. On self-motivation beliefs phase leads to how far students understand the lessons or the skills that they are learning, how far students assess the lesson or skills that they are learning, students' perceived self-efficacy and the goals or reasons students adopt for wanting to complete academic tasks have often been used to understand and explain students' motivation and concomitant effort and persistence for academic tasks [7].

On performance phase, there are two categories, namely self-control, and self-observation. Self-control in self-regulated learning is how students avoid distraction from interruption, focus on learning, and use time effectively, which were found to be important for students' learning [8]. While self-observation is students' responses that involve systematically monitoring their own performance. Observing oneself can provide information about how well one is progressing toward one's goals including self-recording to personal events and self-experimentation in order to find out the causes of the events that occur [9].

Self-reflection phase sets in efforts for comparing the obtained information through self-monitoring with predetermined standards or objectives, beliefs about the causes of strides or errors, enhancement of self-satisfaction and protection of self-image by withdrawing or avoiding opportunities to learn.

Some researchers found that the self-regulated learning ability of students on the university and high school are at the different levels. [10] showed that the understanding level of self-regulated learning of students in the department of elementary school teacher education is on a high level. Then, [11] showed that the self-regulated learning ability of 258 high school students whom 43.03% of respondents liked mathematics are 23,64% students have low self-efficacy, 57,75% students have high self-judgment and 56,60% students have a high level of anxiety toward mathematics achievement. Different self-regulated learning ability profile in each university students is demonstrated by research from [12], that there are five profiles of self-regulated learning: super self-regulators, competent self-regulators, forethought-endorsing self-regulators, performance/reflection self-regulators, and non-self-regulator or minimal self-regulators, and they significantly influence their academic achievement. Based on the explanation from some relevant studies, self-regulated learning ability had actually belonged to each individual, even since the individual is in year 1. However, the difference is the level of self-regulated learning ability of each individual. Therefore, this study was conducted to determine the profile of self-regulated learning ability of elementary school students in learning mathematics.

From the above descriptions, the purpose of this research was to describe the category of students' self-regulated learning ability in learning mathematics. This study is required to allow follow-up or suitable treatments in order to develop students' self-regulated learning ability.

2. Method

This study is a quantitative descriptive research to describe the category of students' self-regulated learning ability in learning mathematics. It was held on 284 fourth-grade of an elementary school student in Bantul District, Yogyakarta. Samples were chosen from five public elementary schools with the same characteristics, namely applying curriculum 2013, the same grade and relatively the same age range from 9-11 years. Data were collected from students' responses using the questionnaires and measured by the Likert-type scale. It is in line with the explanation of [13] that the self-regulated learning questionnaire will be better if the response's item uses scale-type. Subsequently, questionnaires were analyzed by calculating questionnaire scores, determining the category of the measurement result, and grouping the students into several categories. To determine the category of students' self-regulated learning ability, it is shown in Table 1 [14].

Table 1. Measurement scale of students' self-regulated learning ability category

Interval	Category
$M_i + 1{,}5\ S_i < X$	Very High
$M_i + 0{,}5\ S_i < X \leq M_i + 1{,}5\ S_i$	High
$M_i - 0{,}5\ S_i < X \leq M_i + 0{,}5\ S_i$	Moderate
$M_i - 1{,}5\ S_i < X \leq M_i - 0{,}5\ S_i$	Low
$X \leq M_i - 1{,}5\ S_i$	Very Low

Labels:
X : students' self-regulated learning ability score in each respondent
M_i : mean score of students' self-regulated learning ability
S_i : standard deviation of students' self-regulated learning ability

3. Result and Discussion

The result of students' ability was various on the three phases of self-regulated learning included the phase of forethought, performance, and self-reflection. The results of the average of phase forethought students' self-regulated learning ability on each phase will be demonstrated in Table 2.

Table 2. The average score of self-regulated learning ability on forethought phase

Indicator	Sub-Indicator	Average of Sub-Indicator
Task Analysis	Goal setting	2,3
	Strategic planning	2,1
Self-Motivation Beliefs	Self-efficacy	3,2
	Outcome expectations	3,2
	Intrinsic interest	2,8
	Learning goal orientation	2,5

Table 2 shows that the self-motivation beliefs indicator with the intrinsic interest sub-indicators of the forethought phase has the highest average among the other sub-indicators. Based on these two indicators, the average task analysis indicator is 2.2 and self-motivation beliefs is 2.92. In another word, in the forethought phase, students are more prominent in the self-motivation beliefs indicator than the task analysis indicator. Next, the average indicator of the performance phase will be shown in Table 3.

Table 3. The average score of self-regulated learning ability on performance phase

Indicator	Sub-Indicator	Average of Sub-Indicator
Self-Control	Imagery	2,9
	Self-instruction	3,4
	Attention focusing	3,4
	Task strategies	3,5
Self-Observation/Monitoring	Self-recording	2,9
	Self-experimentation	2,8

Table 3 shows that in the performance phase, the task strategies sub-indicator is the highest average among the other sub-indicators. From the self-control and self-observation / monitoring indicators, each is known to have an average of 3.3 and 2.85. Based on the average, it is known that the student performance phase is more prominent in the self-control indicator. Next, the average of each indicator for the third phase is shown in table 4.

Table 4. The average score of self-regulated learning ability on the self-reflection phase

Indicator	Sub-Indicator	Average of Sub-Indicator
Self-judgment	Self-evaluation	2,6
	Causal Attribution	2,5
Self-Reaction	Self-satisfaction/affect	2,9
	Adaptive/defensive	3,1

Table 4 shows that the self-reflection phase with the adaptive/defensive sub-indicator has the highest average. From self-judgment and self-reaction indicators, students are more prominent on the self-reaction indicator with an average of 3.0. The average of the self-judgment indicator is 2.55.

Based on the calculation of the average of each phase of self-regulated learning, it is known that the phase of performance is the most prominent phase between the forethought and self-reflection phase, ie with an average of 3.15, while the forethought phase shows an average of 2.68 and a self-reflection phase with an average of 2.78.

These three phases are cyclical, so they are interconnected and sustainable. Self-regulated learning will grow well if initiative emanates out of student's own sheer will to be able to awaken learning motivation in such a way that it can affect thoughts, feelings, strategies, and behaviors for the achieved goals [15]. Motivation is a major factor to initiate in action or effort that comes from inside and outside of the self, which it can be a reward or punishment [16]. Several studies have also shown that self-regulated learning and motivation are closely related to each other [17].

Then, for the levels of the category of self-regulated learning ability level of elementary school students will be demonstrated in Table 5 below.

Table 5. Self-regulated learning ability of elementary school students category

Interval Score (X)	Category	Frequency	Percentage (%)
$117 < X$	Very High	43	15
$99 < X \leq 117$	High	72	25
$81 < X \leq 99$	Moderate	167	59
$63 < X \leq 81$	Low	1	0,4
$X \leq 63$	Very Low	1	0,4

Table 5 showed that the percentage of self-regulated learning ability categories of elementary school students are 43 students (15%) with "Very High" category, 72 students (25%) with "High" category, 167 students (59%) with "Moderate" category, and 1 student (0,4%) with "Low" category as well as "Very Low" category.

The levels of self-regulated learning ability of elementary school students are affected by some factors, including personal, behavioral, and environmental factors which they are dynamic during the learning process [18]. Personal factors are related to students' self-confidence in the learning process. The perception of students' self-efficacy depends on four types that affect a person's self. They are students' knowledge, the process of metacognition, purpose, and effect. Furthermore, the environmental factor concerns the social experience of student learning through direct experience and task management or learning situations. On behavioral factor, it is related to monitor students on their own performance and systematically comparing their efforts to certain standards. In addition, students are able to respond to their own learning attitudes.

Self-regulated learning ability begins with self, it is called personal factors. Personally, it means that students must be able to observe, assess and react to their own condition properly [19]. Self-regulated learning is not a personal style or trait that can not be controlled by students, but students can control their behavior, motivation and affect, and cognition in order to improve their academic learning and performance [20].

4. Conclusion

Based on the results of this research, it is known that the level of self-regulated learning ability of fourth-grade elementary school students in learning mathematics is on the moderate category and from the three phases of self-regulated learning, students are more prominent on the performance phase. In order to develop self-regulated learning in students, teachers have an important role in providing motivation for the achievement of student learning goals [21]. In addition, knowing the level of self-regulated learning ability of students, teachers can assist them in the different treatments such as applying the appropriate model and strategies of learning-teaching, preparing and providing the proponent assignments to strengthen self-regulated learning ability of students.

References

[1] Santrock 2011 *Life-span development: perkembangan masa-hidup* (Jakarta: Erlangga)
[2] Chris J and Rosemary Hipkins 2004 Young children's emergent self-regulated learning abilities in a primary science investigation *NZARE (New Zealand Council for Educational Research Conf.* (Wellington)
[3] Jantz C 2011 Self-regulation and online developmental student success *Journal of Online Learning and* Teaching **6** 852 – 857
[4] Dhia O 2017 Self-regulated learning dalam pembelajaran matematika *MES (Journal of Mathematics Education and Science)* **2** (2) 10-16
[5] Zimmerman B J 2002 Becoming a self-regulated learner: an overview-theory into practice (London: Routledge) **41**(2) 64-70
[6] Zimmerman B J 2002 Becoming a self-regulated learner: an overview-theory into practice (London: Routledge) **41**(2) 64-70
[7] Wolters C A and Rosenthal H 2000 The Relation between students' motivational beliefs and their use of motivational regulation strategies *International Journal of Educational Research* **33** 801-820
[8] Zhu Y, Au W, and Yates G 2016 University Students' Self-control and Self-regulated Learning Blended Course *Internet and Higher Education* **30** 54-62
[9] Zimmerman B J 2000 Attaining Self-regulation: A Social Cognitive Perspective Handbook of Self-regulation Academic Press 13-39
[10] Darmiany 2016 Self-regulated learning mahasiswa Pendidikan Guru Sekolah Dasar (PGSD) tahun pertama *Jurnal Psikologi Pendidikan & Konseling* **2**(1) 72-83
[11] Marchis I, Balogh T 2010 Secondary school pupils' self-regulated learning skills *Acta Didactica* Napocensia **3** (3) 47-52
[12] Brak L B, Lan W Y, and Paton V O 2010 Profiles in self-regulated learning in the online learning environment *International Review of Research in Open and Distance Learning* **11** (1) 61-80

[13] Retnawati H 2015 Perbandingan akurasi penggunaan skala likert dan pilihan ganda untuk mengukur self-regulated learning *Jurnal Kependidikan* **45** (2) 156-167
[14] Azwar S 2010 *Penyusunan Skala Psikologi* (Yogyakarta: Pustaka Pelajar) p 163
[15] Wangid M N 2004 Peningkatan prestasi belajar siswa melalui self-regulated learning *Cakrawala Pendidikan*
[16] Zumbrunn S, Tadlock J and Danielle E 2011 Encouraging self-regulated learning in the classroom: a review of the literature Metropolitan Educational Research Consortium (MERC) (Virginia: Virginia Commonwealth University)
[17] Vrieling E, Bastiaens T and Stijnend S 2012 Effects of increased self-regulated learning opportunities on student teachers' motivation and use of metacognitive abilities Australian Journal of Teacher Education **37** 102-117
[18] Zimmerman B J 2000 Attaining self-regulation: a social cognitive perspective Handbook of Self–Regulation (CA: Academic Press) pp 13-41
[19] Wangid M N 2017 Efektivitas tutor sebaya dan pekerjaan rumah dalam meningkatkan self-regulated learning siswa *Jurnal Kependidikan* **1**(2) pp. 306-319
[20] Pintrich 1995 *New directions for teaching and learning: understanding self-regulated learning:* (San Francisco, CA: Jossey Bass Publishers)
[21] Effeney G, Carroll A, and Bahr N 2013 Self-regulated learning: key strategies and their sources in a sample of adolescent males *Australian Journal of Educational & Developmental* psychology **13** pp. 58–74

IOP Conf. Series: Journal of Physics: Conf. Series **1108** (2018) 012121 doi:10.1088/1742-6596/1108/1/012121

Improvement Performance of Clustered Shortest Geopath Routing Protocol Using Data Aggregation in Wireless Sensor Network Environments

Rozita [1], W Wibisono[2], R Anggoro[3]

[1, 2, 3] Informatics Department, Institut Teknologi Sepuluh Nopember, Teknik Kimia Street, Surabaya, Indonesia

Email: [1]rozita@aknbenk.ac.id, [2]waswib@gmail.com, [3]onggo@cs.its.ac.id

Abstract. The wireless sensor network is one of the wireless networks which consists of many sensor nodes placed in a particular region for a certain specific functionality. Sensor nodes are generally small and have limited memory, processing resources, energy and lifetime. Limitations of sensor nodes on the energy side can affect network performance such as network lifetime and delay on the event detection. There are many methods that have been done to increase network lifetime, one of which is Clustered Shortest Geopath Routing protocol (CSGP). CSGP is a method to increase network lifetime, this method the development of the shortest geopath routing protocol (SGP) and Low Energy Adaptive Clustering Hierarchy (LEACH). the cluster formation is based on the geographic information of the network while the cluster head selection is based on the position of the node closest to the clusters center point and the data sensing obtained by sensor node will be sent to the cluster head which will forward them to the sink using SGP. However, this method has the limitation on the cluster head selection which is not optimal and the density of network transmission on the SGP line so that the energy consumption of nodes in the path is greater than other sensor nodes. This research proposes a schema selection and maintenance of the cluster head to determine the adaptive path, then aggregation data to reduce the number of network transmission. The data aggregation is the process of collecting and combining useful information in a particular region. Implementation of the stages will be performed in the simulation environment by utilizing SIDNet SWANS simulator, a simulator that used for simulating the wireless sensor network.

1. Introduction

At this time where technological developments occur very rapidly, many human activities are greatly helped by the existence of technology. One of the most widely used and utilized wireless sensor networks (WSN), which can be implemented in many areas such as in the military, agriculture, environmental monitoring, health and so on [1]. WSN is one of the distributed wireless networks consisting of sinks and collections of sensor nodes scattered in certain areas, used as a medium to process the sensors, collect sensing data which then produce information according to certain characteristics so that it can be used to monitor a certain condition such as temperature, radiation, pressure, humidity, vibration, and others [2]. Each sensor node is equipped with communication system equipment so that nodes can be connected to other sensor nodes [3].

Sensor nodes have some limitations on the energy side, this limitation can affect network performance such as on the network lifetime and delay event detection. Several studies have been done to increase network lifetimes such as Low Energy Adaptive Clustering Hierarchy (LEACH) and Clustered Shortest Geopath Routing Protocol (CSGP). LEACH is one of the algorithms developed to improve network performance on the network lifetime side. The algorithm was first introduced by Heinzelman by utilizing a hierarchical or cluster-based Routing protocol [4]. Clusters can be formed by clustering. This process classifies data into multiple clusters, so data in one cluster has a high degree of similarity [5]. A shortest geopath routing protocol is one of the location-based routing protocols, forward the message to the sink with the closest distance to the sink[6].

CSGP is the development of shortest geopath routing protocol(SGP) and LEACH, the formation of clusters is based on the geographic information of the network and the selection of cluster head selection based on the position of the node closest to the center point of the cluster, the sensing data obtained by the sensor node will be sent to the cluster head, then cluster head will forward to sink data with shortest geopath routing protocol [7]. However, CSGP has not conducted further research on the selection and replacement of the cluster head so that further research is needed for the selection process and the change of cluster head, on the other hand, the transmission of data from the centralized cluster head using the nearest line to the sink can cause its solid transmission on the path so that the energy consumption of nodes in the path is greater than other sensor nodes. To reduce the number of network transmission can be done with the data aggregation process[8]. Data aggregation to reduce packet size as well as redundancy and then send packets to the required location [9]. In LEACH method the cluster head selection based on the highest residual energy, CSGP method cluster head selection based on the closest distance to the cluster center and without data aggregation that caused the high density of network traffic.

Therefore, this research proposes a schema cluster head selection to determine the adaptive path, then aggregation data to reduce the number of network transmission. Data aggregation is the process of collecting and combining useful information in a particular region [10]. Effective data collection techniques can improve energy efficiency and network life [11]. Implementation of the stages will be performed in the simulation environment by utilizing SIDNet SWANS simulator, a simulator that used for simulating the wireless sensor network [12].

2. Methodology

This method called Clustered Shortest Geopath Routing Protocol with Data Aggregation (CSGP-DA), divided by 2 phases, cluster head selection phase and data aggregation phase. Cluster head selection phase is a phase where each node performs cluster head selection based on the highest number of neighbors in a cluster. Data aggregation phase is a phase where cluster head received data from node sensor and temporarily accommodated data where priority data is medium and low priority data and then collected data will be average.

The cluster head is the selected node sensing that is responsible for forwarding messages from the cluster member to the sink. the cluster head is required in order to reduce the number of network transmission, because with the cluster head then the transmitting message will become centralized where the cluster member or sensor node does not need to forward the message directly to the sink that can cause its high transmission in the network. Sensing results from the sensor nodes will be sent to the selected cluster head which then passes the received message to the sink through the shortest path to the sink.

Firstly, the cluster head selection is performed after the heartbeat is completed, the selection parameter is based on the node that has the largest number of neighbors in a cluster. Each sensor node will find out the number of its own neighbors and the number of neighbors in a cluster, if the number of neighbor nodes is greater than the number of neighbors of each neighbor then that node will be chosen as cluster head, but if the number of neighbor's neighbor node more than the node's, then the neighbor node that will be chosen as cluster head.

To reduce the number of network transmission can use data aggregation in the cluster head so that with the reduced number of transmissions will increase network lifetime. For network scenarios that scan data such as temperature, humidity or other data that sometimes do not need to be sent at any time, data aggregation techniques can be used and then sent adaptively according to the priority of the scanned data. In this research, the experiment will be conducted on fire scenario with data temperature.

IOP Conf. Series: Journal of Physics: Conf. Series **1108** (2018) 012121 doi:10.1088/1742-6596/1108/1/012121

After the neighbor information and the sink are stored in the table, then the sensor nodes will periodically perform the sensing in accordance with the specified sampling interval, the sensing results will be grouped into several priorities according to a certain range of values that is the high priority, medium priority, and low priority. The priority data is to recognize the condition of an area where if fire condition then it is considered as a high priority and required handling faster than other priorities. Medium Priority is being considered as abnormal conditions and low priority considered as normal conditions. As for the data priority according to temperature, there is three range such as fire temperature range 40 - 110 celsius, abnormal temperature range 33-43 Celcius and normal temperature range 18-32 Celcius.

Furthermore, data aggregation phase is performed by a cluster head while the aggregation process is based on data priority level. Cluster head reduces the amount of transmission by collecting some data from the lower node and aggregate it before send to the sink node. High data priority (P1) will be sent directly to sink using SGP, whereas data with medium and low priority will be temporarily accommodated in the cluster head to a certain limit. Medium data priority (P2) are collected until the collected data reach a certain number. In this case, we use 10 data as the limit. Low data priority (P3) will be collected until collected data is equal to 20. Figure 1 shows the flowchart of the aggregation process in cluster head.

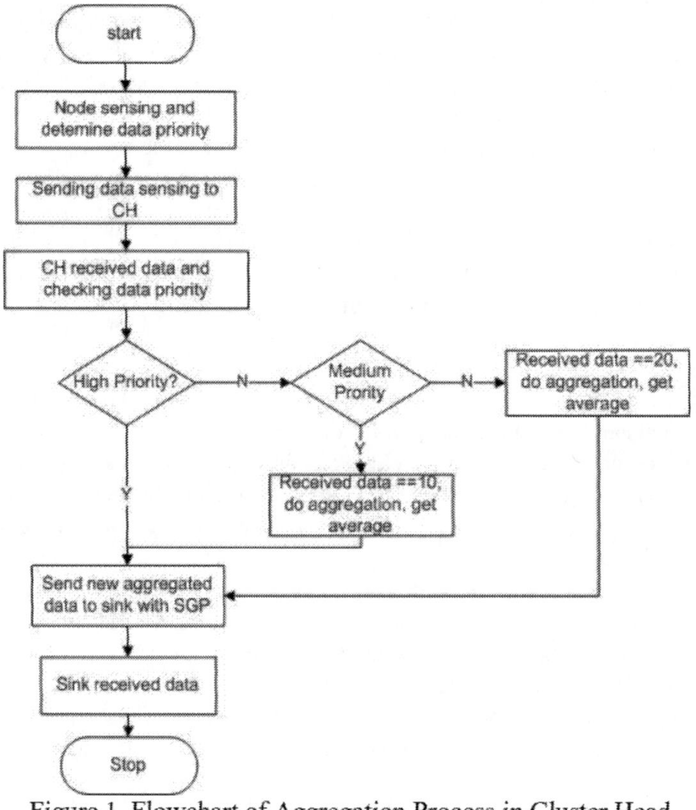

Figure 1. Flowchart of Aggregation Process in Cluster Head

3. Result and Discussion

The experiment carries out by SIDnet-SWANS simulator with 400 m2 area, 200 nodes that randomly place and 60 hours of monitoring. The power source is limited to source node battery with 3 volts and 40 mAh capacity and node sink battery with 3 volt and 80 mAh. Another experiment parameter using the default setting from SIDnet-SWANS simulator. The proposed method Clustered Shortest Geopath Routing Protocol with Data Aggregation (CSGP-DA) will be compared to the Clustered Shortest Geopath Routing Protocol without data aggregation (CSGP-WA) method and Shortest Geopath Routing Protocol without data aggregation(SGP-WA). The experiment will be run for 3 times and all result will be calculated for average value and then compare it to the other method. Proposed method CSGP-DA is tested for 3 times by using all parameter and pre-build scenario that already determined. Network lifetime is a parameter where how long lifetime of a network from nodes deployed until one of the nodes run out of power source, greater value is better. Delay event detection is a parameter to determine how long from event happens until event detected by the sink node, a lower value is better. Table 1 depicts Experiment result of CSGP-DA.

Table 1. Experiment Result of CSGP-DA

Experiment		1	2	3	Average
Network Lifetime(Minutes)		2635	2633	2650	2636
	P1	72	72	72	72
Delay Event Detection(ms)	P2	83	83	83	83
	P3	77	77	77	77

Then we need to compare proposed method CSGP-DA with CSGP without adaptive path and data aggregation (CSGP-WA) and SGP without data aggregation (SGP-WA) to know the performance of CSGP-DA. It also tested for 3 times by using same parameter and pre-build scenario. Table 2 depicts Experiment result of CSGP-WA, Table 3 depicts Experiment result of SGP-WA.

Table 2. Experiment Result of CSGP-WA

Experiment		1	2	3	Average
Network Lifetime(Minutes)		2223	2193	2218	2211.3
	P1	97	97	96	96.6
Delay Event Detection(ms)	P2	105	105	105	105
	P3	106	106	106	106

Table 3. Experiment Result of SGP-WA

Experiment		1	2	3	Average
Network Lifetime(minutes)		1159	1160	1792	1370.3
	P1	98	99	101	99.3
Delay Event Detection(ms)	P2	103	105	111	106.3
	P3	176	105	96	125.6

From the third experiment, we could compare the other performance base on average result of network lifetime and delay event detection based on data priority. Table 4 shows each average result comparison.

Table 4. Comparison Average of Network Lifetime and Delay Event Detection

Method		CSGP-DA	CSGP-WA	SGP-WA
Network Lifetime(minutes)		2636	2211.3	1370.3
	P1	72	96.6	99.3
Delay Event Detection(ms)	P2	83	105	106.3
	P3	77	106	125.6

As shown in table 4, CSGP-DA has better performance than other methods, average network lifetime of CSGP-APA is 2636 minutes that is 19,2% better than CSGP-WA that has 2211.3 minutes and 92,22% better than SGP-WA has 1370,3 minutes.

Average delay event detection of CSGP-APA is the lowest compared to other methods. This happens since with data aggregation the number of network transmission can be reduced without ignoring high data priority. Comparison Average of network lifetime shown in figure 2 and average of delay event detection shown in figure 3.

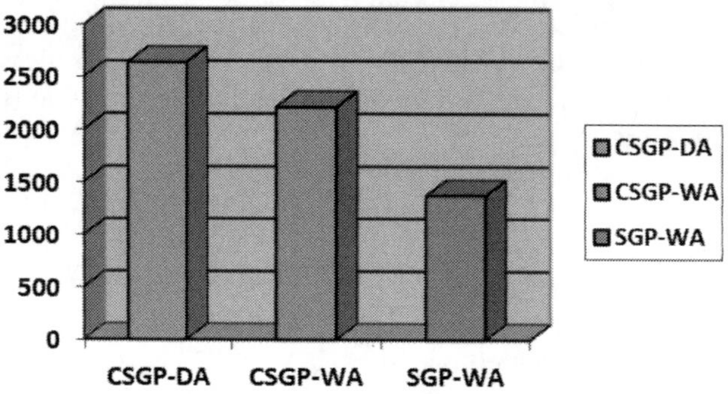

Figure 2. Comparison average of network lifetime

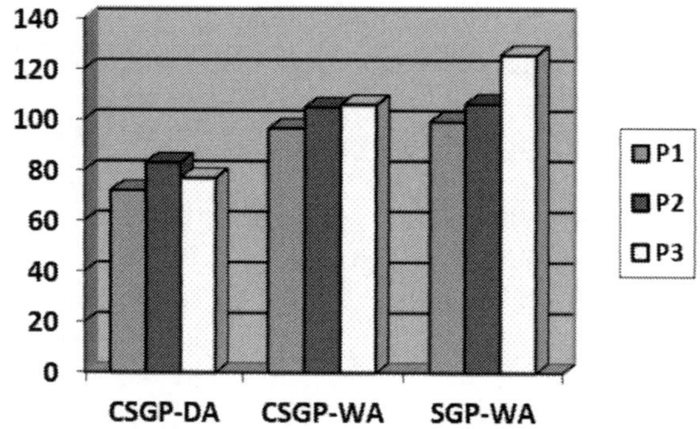

Figure 3. Comparison average of delay event detection

4. Conclusions

The results CSGP with data aggregation (CSGP-DA) method proved to increase network lifetime 19,2% compared to CSGP without data aggregation (CSGP-WA) and 92,22% compared to SGP without data aggregation (SGP-WA), then decrease delay event detection compared to CSGP without data aggregation and SGP without data aggregation. CSGP-DA is 7 ms lower than CSGP-WA and 26,3 ms lower than SGP-WA. Further research is required to handle the hole conditions in the sgp line.

5. Acknowledgment

Authors give thanks to The Indonesia Endowment Fund of Education (LPDP), Ministry of Finance, Indonesia for financial assistance and encouragement.

6. References

[1] Sohraby K, Minoli D, and Znati T 2007 *Applications of Wireless Sensor Networks*, in *Wireless Sensor Network Technology, Protocols and APlication* 38-71 (New Jersey: John Wiley & Sons, Inc.)
[2] Sivalingam K 2015 *Tutorial : Wireless Sensor Networks* (Maryland: University of Maryland, Baltimore County)
[3] Pratama I P A E and Suakanto S 2015 *Wireless Sensor Network, Pertama* (Bandung: Informatika)
[4] Heinzelman W R, Chandrakasan A, and Balakrishnan H 2000 Energy-Efficient Communication Protocol for Wireless Microsensor Networks *the Proceedings of the Hawaii International Conference on System Sciences* 1–10.
[5] Tan P N 2006 *Cluster Analysis: Basic Concepts and Algorithms*, in *Introduction to Data Mining* 125-145 (Pearson Addison-Wesley)
[6] Imieliński T and Navas J C 1999 GPS-based geographic addressing, routing, and resource

discovery *Commun. ACM* **42**(4) 86–92

[7] Astuti L D and Wibisono W 2017 Peningkatan Network Lifetime pada Wireless Sensor Network Menggunakan Clustered Shortest Geopath Routing (C-SGP) Protocol *J. Teknol. Inf. dan Ilmu Komput.* **4**(3) 148–153

[8] Rajagopalan R and Pramod K 2006 Data Aggregation Techniques in Sensor Network : a Survey *IEEE Commun. Surv. Tutorials* **8**(4) pp. 48–63

[9] Desai S, Vyas T, Nair A R, and Jain H 2017 Survey on Cluster-Based Data Aggregation in Wireless Sensor Network *Int. J. Adv. Res. Comput. Sci.* **8**(5) 311–314

[10] Randhawa S and Jain S 2017 *Data Aggregation in Wireless Sensor Networks: Previous Research, Current Status and Future Directions* (Springer US)

[11] Qayyum B, Saeed M, and Roberts J 2014 Data Aggregation in Wireless Sensor Networks with Minimum Delay and Minimum Use of Energy : A Comparative Study *2nd BCS Int. IT Conf. Mobility, Intell. Networks Smart Soc.* 1–5

[12] Ghica O C 2010 *SIDnet-SWANS Manual*, in *SIDnet-SWANS Manual* 1–50

Development of chemistry instructional materials based on Cooperative Group Investigation (CGI) to empower thinking skills

U Azizah[1,*] and H Nasrudin[1]

[1]Chemistry Department, Universitas Negeri Surabaya, Surabaya, Indonesia

*Corresponcing author: utiyaazizah@unesa.ac.id

Abstract. Thinking skill is important for students as it is the foundation for enhancing the mastery of concepts needed in the 21st century. Thinking skills that empowered in learning focuses on how students participate in solving problems. This research aims to develop chemistry instructional materials based on Cooperative Group Investigation (CGI) to empower thinking skills on colloidal system topic. The feasibility of chemistry instructional materials is evaluated based on its validity and practicality. The design of this research was four-D-model, consist of Define, Design, Develop, and Disseminate. Data of this research was analyzed by quantitative descriptive. The result of this research include: (1) The instructional materials that were developed which consist of lesson plan, students' book, worksheets, assessment, and power point media; (2) The instructional materials developed were valid with good category in construct, content, language, empirical validity, and contain principles of cooperative group investigation; (3) The teacher implemented the lesson plan well, the increased students activities were indicated by doing investigation, analyzed the result of investigation and presentation of final product that are empowering students' thinking skills. Based on the results of the research, the instructional materials of colloidal system topic can empower students' thinking skills.

1. Introduction

Education target at 21st century as the century of information technology has shifted traditionally to the knowledge which requires ability to think, and way of thinking [1, 2]. According to Piaget [3], the development of reasoning skills (empowerment process of thinking) is essential for mastery of concepts; because conceptual knowledge represents the result of a constructive process.

The instructional materials are one means of learning that can lead students to understand a concept and can improve retention of learning outcomes [4]. This matter as according to result of research to develop instructional materials for empowerment of metacognitive skills on the topic of hydrolysis and buffer solution, can improve students' activeness in learning as indicated by the improvement of students' interactions and teacher's-students interactions, student enthusiasm in learning and their eagerness to joint to the next teaching-learning activity with the same way of learning [5].

The result of the research is strengthened by result of research that implements the model of Cooperative Group Investigation (CGI) with contextual approach oriented in the learning activities of Science. It suggests that the activity has centered to students, as indicated by the improvement of students' interactions and teacher's-students interactions, teacher responses, discussion among them, teacher's-students discussion, doing investigation, presenting the report's investigators, and volunteers (student expresses their opinion and responses)[6].

Most in all concepts, there are chemistry representing the abstract concept which tends to only be comprehended better by students who have developed the ability to think the abstract. One of the learning strategies used to empower students' thinking skills is a cooperative group investigation [7, 8], where class is divided to become the groups of 5-6 students with heterogeneous members. Further students choose the topic investigated, investigate on selected topics, and present the report investigated to the whole class.

The difficulty of obtaining the quality of instructional materials, time limitation, funding, as well as expert is owned progressively against the teacher task. Therefore, this research develops and applies the instructional materials chemistry which concern on CGI oriented that can ease the task of chemistry teacher. The instructional materials consist of lesson plan, students' book, worksheets, assessment and media of power point.

2. Methods

This research represented the development research using four-D Model consisted of four stages pervading define, design, develop and disseminate [9]. In principle, the developments of instructional materials development use the framework of mind with a few appropriate adaptations.

At the first year research, it has produced the instructional materials chemistry oriented CGI, including Lesson Plan, Student Book, Worksheet, and Assessment Sheet (cognitive, psychomotor and affective) and keys, and media of power point and its research instruments. The preparations of instructional materials through the stages of define, design and develop using a series of activities: field observations, curriculum analysis, literature review, development instructional materials, and review of instructional materials for linguist and expert.

In the second year through the stages of disseminate, it was conducted trials of instructional materials result of research in first year at Senior High School Sidoarjo, Indonesia for completion of instructional materials as prototype of effective final instructional materials. The effectiveness tested with the criterion of feasibility of learning chemistry which oriented CGI, the activity of thinking skills of students, and students' responses to the learning. Before trial, it was conducted training for teachers of chemistry in terms of the application of instructional materials.The method used in training are modelling. The final result of this study is the materials oriented CGI which is expected to be implemented /applicable at school, because it was through review, expert validation, and testing.

3. Result and Discussion

In general, the data is the mean score of the feasibility of lesson plans, students' book, worksheets, assessment, and power point media that developed by validators' evaluation which is presented in Figure 1, Figure 2, Figure 3 and Figure 4. The result of validators' evaluation of lesson plan, student book, worksheets, assessment and power point media is good with the mean score of 4.1 to 4.9.

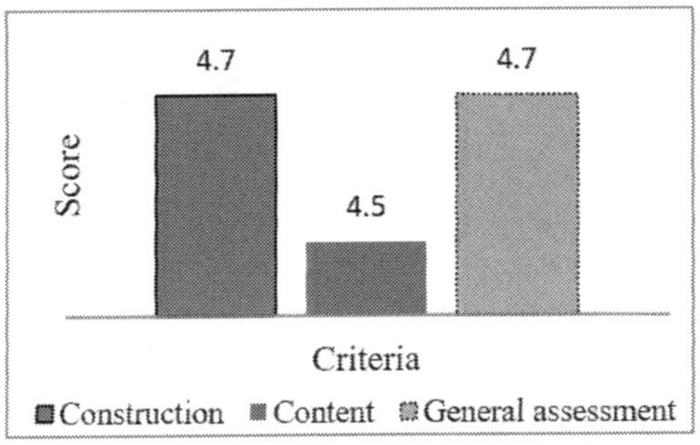

Figure 1. Validation results of Lesson Plan

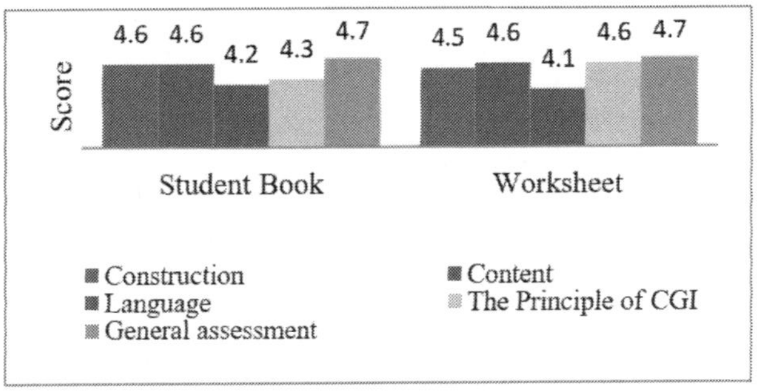

Figure 2. Validation results ofStudent Bookand Worksheet

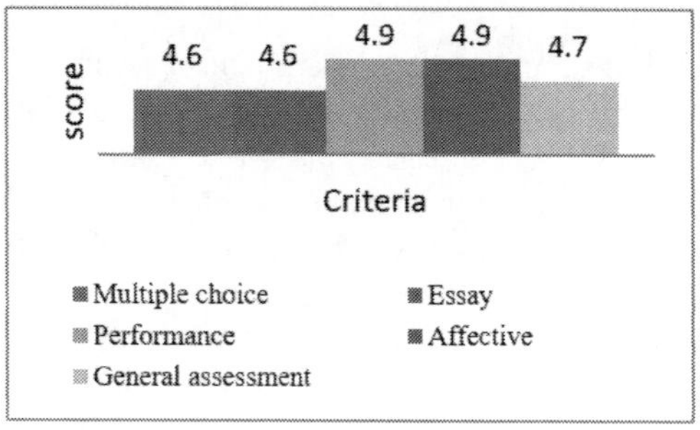

Figure 3. Validation results of Assessment

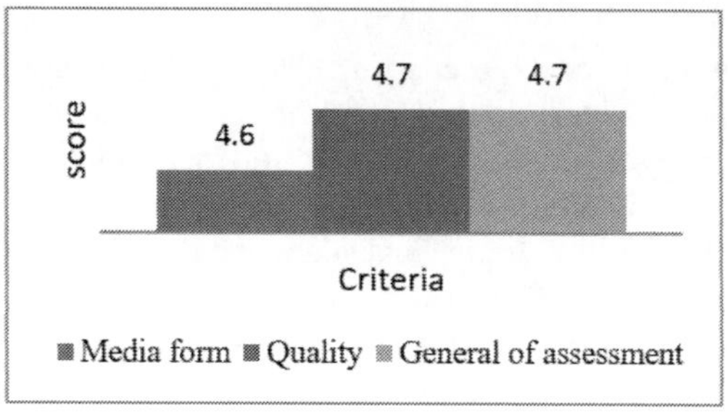

Figure 4. Validation results of Power Point

Based on the interpretation of feasibility according to the Likert Scale [10], it is said feasible if good and very good ratings. So it can be said that by Lesson Plan for colloidal system which have been developed have been feasible to be used in the learning process because they have fulfilled with the construct validity and content validity. In addition, the Student Book and worksheets for colloidal systems material which have been developed have been feasible to be used in the learning process because have fulfilled with the construct validity, content validity, language validity and according to student book group with the principle of GIC. An assessment and power point medium that has been developed in colloidal systems have also been fulfilled to be used in the learning process.

At the dissemination stage, the performance data of teachers in implementing a prototype instructional material CGI oriented includes lesson plans, student book, worksheets, assessment and media of power point on the colloidal system. They are observed by two observers using the lesson plan format which is presented in Figure 5. The teacher activity affects the activity of student thinking skills which are presented in Figure 6. In addition, the percentage of time used by students learning to do the activities supported by student responses to the Student Book and worksheets that are implemented in the teaching and learning (Figure 7).

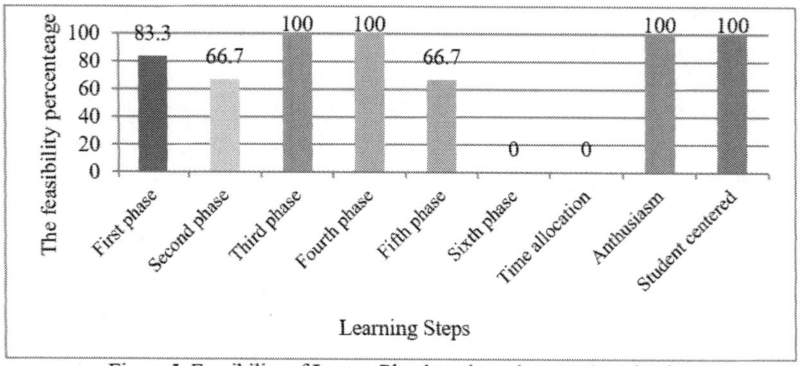

Figure 5. Feasibility of Lesson Plan based on observers' evaluation

Description of phases:

1. Communicate goals and motivate students
2. Present the information (selection of topics)
3. Organize the students in learning (Cooperative planning)
4. Guide the group to learn and work (implementation, analysis and synthesis)
5. Evaluation
6. Reward

Result of evaluation the lesson plan feasibility by observer in stage disseminate is good and very good with the feasibility percentage of 66.7% until 100% (Figure 5). Based on the score interpretation according to Likert Scale [10], the observation result is good. Implementation of teaching and learning is student-centered, and teachers act as facilitators. So that, the Lesson Plan for colloidal system which have been developed, can be implemented better, that indicate teacher have communicated the indicators, motivated students, correlated the matter with the former concept, presented the outline of information, revealed common problems and the selection of topics, explained experimental procedures in the worksheet, setting students group, assisted each group to make the transition efficiently, guided students to plan the learning procedure (cooperative planning), guided the group by using the worksheet (implementation, analysis, and synthesis), observed the activities of students in turn, evaluated students' learning outcomes through the questions and answers or each group presented the results of their work (presentation of final results), and provided feedback.

However, correlating the matter with the former concept has not been performing well. Because the matter is relatively quite a lot, so the teacher directly went to the explanation matter, without giving apperception. Similarly, at second phase, the activities of teacher reveal common problems and the selection of topics that has not been performing well. At sixth phase, teacher did not guide the student to subject matter, and gives awards to individual or group. It is because the amount of time used by student in activity minilab, chemistry laboratory, daily chemistry, and assessment cause the time allocation planned was not performing well (exceeding time planned).

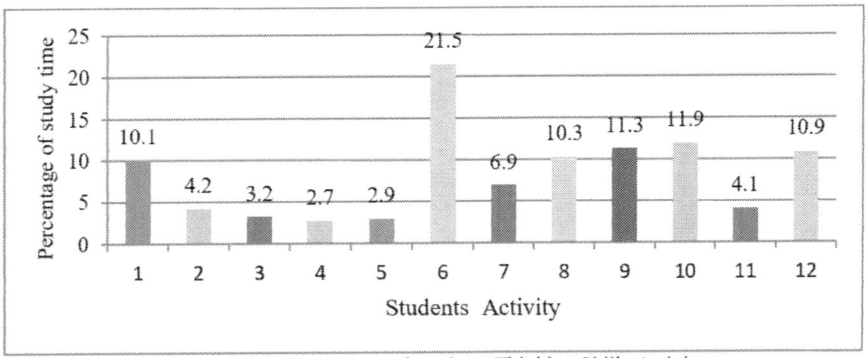

Figure 6. The Percentage of Students Thinking Skills Activity

Activities:
1. Paying attention to the explanation of teacher/students
2. Reading (student book or worksheet)
3. Writing relevant to the learning activities
4. Selecting a task in accordance with general problems
5. Planning the procedure solving task
6. Doing investigation (theoretical/observation/experiment)
7. Analyzing the result of investigation
8. Preparing investigation reports
9. Presenting the report of investigation
10. Discussing within students
11. Teacher's-students discussion
12. Volunteering (go forward, student express their Opinionand responses)

The teacher activity affects the thinking students' skills, with an active student learn in class that they only kept quiet to listen to the teacher or friends. Based on the percentage of time in learning activity (Figure 6) that the time for 80,1% used by students to be active learning in class with the details activities are reading (student book and worksheet), writing, selecting a task in according to problems, planning the procedure solving task, doing investigation (theoretical/observation/experiment), analyzing and evaluating result of investigation, compiling a report, presenting the investigation report, discussing within student, teacher's–students discussion, and volunteering (go forward, student express their opinion and responses). While 10.9 % are used to listen the teacher or friend.

In empowering students' thinking skills, teaching and learning implementation have to strive in order not to solely relate to importance transfer the information and or even the discovery information but refers to the importance of higher-order thinking skill. It has been seen on the results of the implementation of the instructional materials, in which most student used to do the investigation (theoretical/observation/experiment), analyzing and evaluating the result of investigation, compiling the report and presents the investigation report. The activity can develop the ability of abstract thinking, providing the opportunity for students to interact with friends in the investigation for the discovery of the concept, and the existence of interaction between student – teacher in expressing an idea, asking and responses question.This means that the Cooperative Group Investigation (CGI) model can train students to explore and discover concepts because it involves many students during the learning process to find their own concepts so that students can acquire knowledge and build concept independently [11, 12].

In the students' book, there are practice problems, daily chemistry, chemistry-discuss, and guessing term that guides and directs students in discovering concepts through questions. In addition, some communicative questions on the subject matter can also assist the students in constructing their own knowledge. Meanwhile, in the worksheet, the activities in mini laboratories and chemistry laboratory enable students to independently construct their own knowledge through experiments. The concepts of colloidal system material studied in the student book and worksheet were dispersion system, dispersed phase, medium dispersion, sol, aerosol, gel, emulsion, foam, lyophilic colloid, lyophobic colloid, hydrophilic, hydrophobic, Tyndall effect, Brownian motion, adsorption, coagulation, protective colloid, dialysis, condensation method, dispersion method, and peptization [13].

However, according to observers 'and researchers' notes, some students are less effective in selecting a task in accordance with problems, planning the procedure solving the task, and analyzing the result of the investigation. This happens when students conduct investigations on how to purify water and make colloid by condensation method. Some students are still wrong in selecting a task in accordance with the problem of purifying water. To purify water, students should use the following procedure: water can be purred by adding $Al_2 (SO_4)_3$. In the water, $Al_2 (SO_4)_3$ is hydrolyzed to form $Al (OH)_3$. $Al (OH)_3$ is a colloid. These can adsorb coloring matter or pollutant in water. As a result, the analysis the result of the investigation is not correct in making colloid by condensation method, because students are still confused in identifying the types of condensation reactions as follow:

(a) $FeCl_3\,(aq) + H_2O\,(l) \rightarrow Fe\,(OH)_3\,(sol) + 3HCl\,(aq)$
(b) $2H_2S + SO_2 \rightarrow 2H_2O + 3S\,(colloid)$
(c) $Na_2SiO_3(aq) + 2HCl(aq) \rightarrow 2\,NaCl(aq) + H_2SiO_3\,(colloid)$.

This means, students do not understand the difference between hydrolysis reactions, redox reactions and substitution reactions.

Nevertheless, activities in the Students' Book and worksheet can support feasibility in teaching and learning process to be more students centered so that students can independently construct their own knowledge (constructivist) while developing their skills of inquiry, through collaboration in study groups that can share their opinions. This means, students are trained to use thinking skills [14, 15]. In addition, the percentage of time used by students is supported by student responses to the Student Book and worksheets which are implemented in learning (Figure 7) that the majority of students have given the good evaluation with a percentage of approximately of 81.1% - 91.7%.

This means the student has given a positive response to the Students' Book and worksheets based on the aspects of construction including attractive physical appearance, size according to desire, and the letter used comfortably readable, interesting way of presenting the interests and concerns, and the illustrations (figures and descriptions) to support the understanding of the material. For the aspects of content, including the existence of students' book and worksheets, the subject matter is related to everyday life, the content may encourage active involvement in learning, examples of problems that presented are easy to understand, with the students' book and worksheet to learn on their own at home, interest to continue to study it, and in general students' book and worksheet have good category. For the aspects of language, including the language used is easy to understand/ communicative, and the terms and symbols used is steady and easy to understand. For the aspects which are linked to principle of group investigation cooperative, by using the Students' Book and worksheets they can construct knowledge independently, concluded the concept of matter by observation, encourage cooperation, improve knowledge, thinking processes and laboratory skills, and encourage responding to events, activities and experience.

Based on the description above, Students' Book and worksheets have been feasible to be used in meaningful learning activities that have fulfilled the empirical validity.

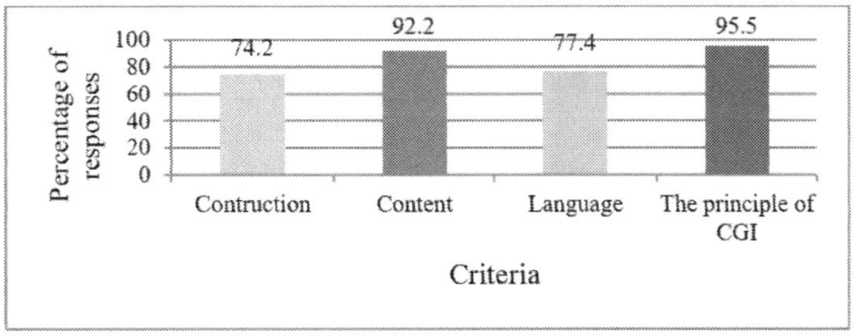

Figure 7. The percentage of student responses to student book and worksheet

Some of the comments and suggestions for improvement from students that can be used as a material consideration for revision are as follows: (a) Students' book and worksheet have goodness side, and attract students' attention, but the books need to be completed with chart/ diagram/ concept map figures, colors, chemical information of interest; a glossary at the end of the page, real concepts, content of more attention, full color and bilingual; (b) The size of the students' book and worksheet are less as according to students desire, because they are less minimize portable so that, (c) There are a few examples of questions presented in the student book which is not easy to understand, because it still needs glossary and sample questions, and (d) There are some phrases used in the students' book and worksheet which are less easy to understand, because it still needs glossary and should be in bilingual.

4. Conclusion

Based on the objectives and results of research that has been described, it can be concluded that: (1) The instructional materials were developed consist of lesson plan, students' book, worksheets, assessment and power point media; (2) The instructional materials developed were valid with good category in construct, content, language, empirical validity, and contain principles of cooperative group investigation; (3) The teacher implemented the lesson plan well, the increased students activities were indicated by doing investigation, analyzing the result investigation and presenting the final product that can empowering students' thinking skills. Based on the results of the research, the instructional materials of the colloidal system topic which developed can empower students' thinking skills.

5. References

[1] Galbreath J 1999 Preparing the 2st Century Worker: The Link Between Computer-Based Technology and Future Skill Sets*Educ. Tech.***19**(5) 14-22.

[2] Trilling B and Hood P1999Learning Technology and Education Reform in the Knowledge Age *Educ. Tech.***7**(1) 5-18

[3] Lawson A.E2000Managing the Inquiry Classroom: Problem and Solutions*The American Biology Teacher***62**(9) 641-648.

[4] Soekamto H 2002 *Prosedur Pengembangan Perangkat Pembelajaran*(Malang: Universitas Negeri Malang)

[5] Azizah Uand Nasrudin H2018 Empowerment of Metacognitive Skills through Development of Instructional Materials on the Topic of Hydrolysis and Buffer Solutions*J. Phys.: Conf. Ser.953012199*. doi: 10.1088/1742-6596/953/1/012199.

[6] Nasrudin H andAzizah U2010 Improvement Thinking Skills and Scientific Attitude Using The Implementation of "Group Investigation Cooperative Learning" Contextual Oriented at Acid, Based, and Salt Topic in Junior High School*Proceeding of The 4th International Conference on Teacher Education: Join Conference UPI & UPSI Bandung Indonesia, 8-10 November 2010.*

IOP Conf. Series: Journal of Physics: Conf. Series **1108** (2018) 012122 doi:10.1088/1742-6596/1108/1/012122

[7] Tsoi MF, Goh NK, Chia LS 2004 Using Group Investigation for Chemistry in Teacher Education*Asia-Pacific Forum on Science Learning and Teaching*5(1) 1-12
[8] ArendsRI 2007 *Classroom Instruction and Management* (New York: Mc Graw-Hill Companies,Inc.)
[9] ThiagarajanS, Semmel DSand Semmel MI1974 *Instructional Development for Training Teachers of Exceptional Children:Source Book*(Bloomington: Center for Innovtion on Teaching the Handicapped)
[10] Riduwan 2003 *Skala Pengukuran Variabel Penelitian* (Bandung:Alfabeta)
[11] Ahern-Rindell AJ 1999Applying Inquiry Based and Cooperative Group Learning Strategies to Promote Critical Thinking*College Science Teaching*28(3) 203-207
[12] Cheng, Eric CK 2011The role of self-regulated learning in enhancing learning Performance. *The International Journal of Research and Review*6(1) 1-16
[13] Timberlake KC 2012*Chemistry*(New York: Prentice Hall)
[14] Slavin RE 2009 *Educational Psychology: Theory and Practice.* Fourth Edition(Massachusetts: Allyn and Bacon Publishers)
[15] Niemi H 2002 Active learning: A cultural change needed in teacher education and schools*Teaching and Teacher Education*18(7) 763-780

IOP Conf. Series: Journal of Physics: Conf. Series **1108** (2018) 012123 doi:10.1088/1742-6596/1108/1/012123

Mathematical Writing Profile of High Social Arithmetic Ability Student in Solving Social Arithmetic Problems

M Yulianti[1,*], D Juniati[2] and S Khabibah[2]

[1]Universitas Negeri Surabaya
[2]Department of Science and Mathematics UniversitasNegeri Surabaya

[*]Corresponding author: maulidayulianti16070785031@mhs.unesa.ac.id

Abstract. In solving problems, students need to communicate their ideas. One of the media is writing in written media. Social arithmetic is one of mathematics subject that close to daily live activities. Students' understanding about social arithmetic and how students communicate their ideas in solving social arithmetic problems can be seen in their mathematical writing. This study aimed to describe how student with high social arithmetic ability mathematical writing in solving social arithmetic problems based on the use of precise mathematical language, explanation of the process, reasoning about the process to produce the right solution, connections between abstract mathematical procedures and the context of the problem. The analysis indicate that student clearly used accurate math vocabulary to communicate ideas, clearly provides the mathematical steps used to solve the problem, clearly described the reasoning used to solved the problem which includes connections between the numbers used in the strategy and the problem context, and then the final answer was clearly explained in the context of the problems. Through student mathematical writing, reader (teacher) will know that student understand the problem and the subject and also can clearly communicate their ideas.

1. Introduction

Learning mathematics is more than solving a set of problems or replicating teacher explanations. It is building a strategy for solving problems, applying various approaches, looking at whether the approach used leads to right solutions, and checking whether the solutions were acceptable [1]. In mathematics learning, students should be able to express their mathematical ideas. One of the media for students is their mathematics idea which is writing in written media.

Writing activities in learning mathematics is called writing mathematics, it is a combination of mathematics and writing [2]. There is a writing activity in solving mathematical problems, such as when students write the steps of completion and the explanation while they are solving mathematics problems.

The result of writing mathematics is a mathematical writing that consists of a visual language, symbols, or pictures to communicate mathematical ideas. Mathematical writing may also be referred to as a written description of the problem-solving process [3]. Not all types of mathematics problems can produce a written description of the process of completion or mathematical writing. To produce mathematical writing, the questions should use a question word that directs students to express their opinions, as well as explain their ideas.

Produce mathematical writing and the ability to learn in general is connected, to learn new mathematical content, to improve math problem solving, and to add knowledge-related processes [4] [5]. Student mathematical writing can also be used by teachers to understand students' understanding and misunderstanding [4] [1].

Arithmetic social is close to daily live activities, such as buying and selling activities. This term is introduced in mathematics learning since 7th grade. The problem relates to this term, is usually presented in word problem. In solving social arithmetic world problems, students have to build models or represent the problems to form appropriate calculation processes to obtain the right solution [6].

Students understanding about social arithmetic and how students communicate their ideas in solving social arithmetic problems can be seen in their mathematical writing. Based on discussion above, the researcher is eager to search and investigate how students mathematical writing in solving the problem of social arithmetic.

The writing of mathematics in solving the problem of social arithmetic is written words of visual languages, symbols, or pictures to communicate the mathematics ideas in solving stories related to the economic world. Mathematical writing can also be referred to as a written description of the problem-solving process, which is the writing that reveals how students communicate mathematical ideas, how the students representation in solving mathematical problems, and how students justify the strategies used in solving problems. There were four important aspects in mathematical writing. Those are the use of precise mathematical language, explanation of the solution process, reasoning why the solution process produced the correct answer, connections between abstract mathematical procedures and the context of the problem [7] [8][9].

2. Method

This descriptive research with a qualitative approach was designed to obtain information and explore the things that affect the incidence experienced deeply by the research subjects. This study aims to describe how students mathematical writing when communicating their ideas in solving the problem of social arithmetic.

In this research, student with high arithmetic ability was chosen from 27 students from the same environment and the level of school as research subject. The research subject was selected based on social arithmetic ability test scores. Student with score of more than 80 are categorized as student with social arithmetic ability. Furthermore, students were given the task of writing mathematics for further description of how students' mathematical writing in solving social arithmetic problems. The research subjects were interviewed to get additional information about students' mathematical writing.

All instruments that used in this research were validated by expert lecturers and piloted to several students before being given to students. The research data was collected by time triangulation technique. Therefore, subjects who have been selected were given the task of writing mathematics and an interview for several times in different time until it is obtained valid data.

3. Result and discussion

After the second triangulation, the researcher had obtained valid data. It is seen when students show consistent answers. The results of the task of writing mathematics I and II can be seen in Figures 1 and 2.

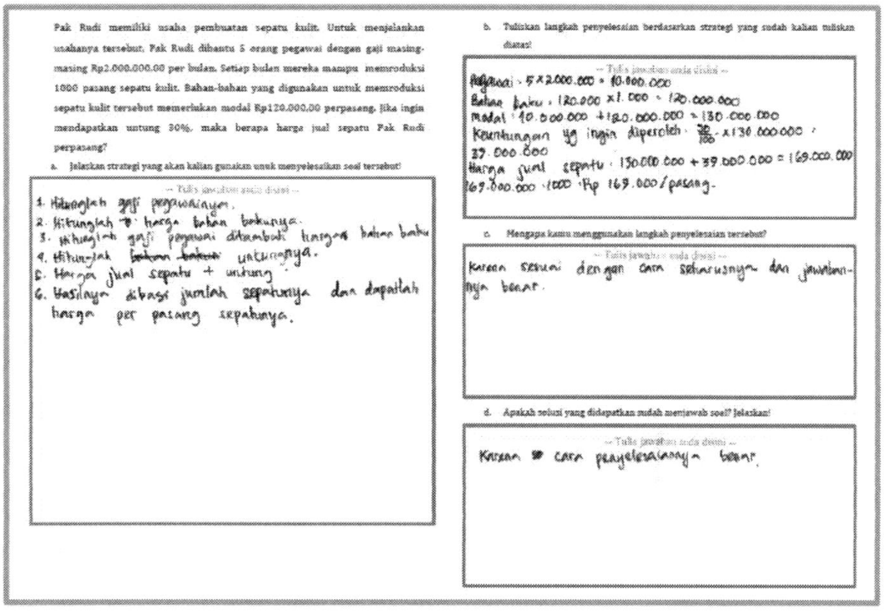

Figure 1. Student mathematical writing in first writing mathematics task.

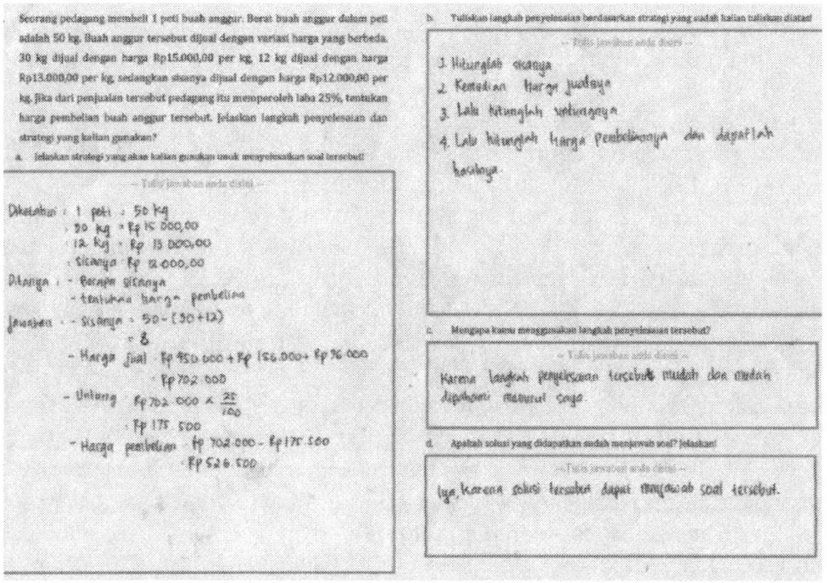

Figure 2. Students mathematical writing in second writing mathematics task.

3.1. The use of precise mathematical language

The use of precise mathematical language in students' mathematical writing can be discovered by these three indicators.

- Using mathematics language in solving problem
- Using good and correct Indonesians rules
- Use appropriate symbols or notations

While doing first writing mathematics task, students use the math language in solving the problem clearly, employ the Indonesian rules quite well and correctly and use the right symbols/ notations. Similarly, while doing second writing mathematics task, students use the language of mathematics in solving the problem clearly, use the Indonesian rule quite well and correctly and use the right symbols/notations.

Based on data triangulation on how they use mathematics language and its accuracy in student mathematical writing, it seen that student was consistent in the use of math language. So that, the data of the use of mathematical language and accuracy in the mathematical writing of students was valid. Thus, the researcher can conclude that:

- Student with high social arithmetic skills clearly use mathematical language in solving problems in their mathematical writing.
- Students with high social arithmetic skills use Indonesian rules well and correctly.
- Students with high social arithmetic skills use appropriate symbols or notations.

3.2. Explanation of the solution process

To find out how the explanation of the solution process, the indicator that must be met was "write down the explanation steps of the problem solving process". While doing first writing mathematics task, student wrote down the explanation of the steps of the completion process into the problem solving strategy. However, there is still some left information. Similarly, while doing second writing mathematics task, student wrote a description of the steps of the completion process into the settlement strategy.

Based on data triangulation on how to explain the steps of the problem solving process on the student mathematical writing, it appears that student was consistent in explaining the steps of the problem solving process. So that the data was valid. Thus, the researcher can conclude that students with high social arithmetic skills write down the explanation of the steps of the settlement process into the settlement strategy quite clearly in their mathematical writing.

3.3. Reasoning why the solution process produced the correct answer

Write down the reasons for the selection of settlement measures was the indicator to find out how the reason for selecting the completion steps used to produce the correct solution.

Students wrote down the reasons for choosing clear and confident resolution steps while working on first writing mathematics task. She declared that the settlement that she used in accordance with what she had learned and produced the right answer. Similarly, while doing second writing mathematics task, students write down the reasons for the election of the completion steps clearly and confidently. She declared that the settlement step she used was complete in accordance with what had been learned and produced the correct answer.

Based on data triangulation on why the reason for the selection of completion steps from the problem solving process on the students mathematical writing, it appears that the students was consistent in explaining the steps of writing down the reasons for the selection of the settlement measures. Thus, it can be said that the data of the reasons for the selection of completion steps of the problem solving process on the student's mathematical writing is valid. Thus, the researcher can conclude that students with high social arithmetic capability write down the reasons for choosing clear and confident resolution steps.

3.4. Connections between abstract mathematical procedures and the context of the problem

To find out how the relationship between the mathematical procedure used and the context of the question, "write an explanation of the relationship between the mathematical procedure used and the context of the question" was employed as an indicator.

While working on the first writing mathematics, students wrote an explanation of the relationship between the mathematical procedures used with the problem context quite clearly. She wrote that the settlement step used was the actual and learned way. Likewise while working on the second writing mathematics task, students wrote an explanation of the relationship between the mathematical procedures used with the context of the problem quite clearly. She wrote that the settlement step used was the actual way, done in sequence, using information from the problem, and based on the knowledge that she learned.

Based on the data triangulation on connections between abstract mathematical procedures and the context of the problem on the students' mathematical writing, it appears that students was consistent in explaining the relationship between the mathematical procedures used and the context of the questions. Thus, the researcher can conclude that students with high social arithmetic skills write an explanation of the relationship between the mathematical procedures used with the context of the question quite clearly.

4. Conclusion

The student who has high social arithmetic ability clearly used accurate math vocabulary to communicate ideas, clearly provided the mathematical steps used to solved the problem, clearly described the reasoning used to solve the problem that includes connections between the numbers used in the strategy and the problem context, and then the final answer was clearly explained in the context of the problems.

5. References

[1] Walle J AVDKerp KS and Williams MBJ 2013*Elementary and Middle School Mathematics Teaching Developmentally* (America: Pearson Education)
[2] Hebert M A and Powell S R 2016*Examining fourth-grade mathematics writing: features of organization, mathematics vocabulary, and mathematical representations. Springer Science+Business Media Dordretch*. Doi: 10.1007/s11145-016-9649-5
[3] Pugalee D K 2001 Writing, Mathematics, and Metacognition: Looking for Connections Through Students' Work in Mathematical Problem Solving *School Science and Mathematics***101**(5), May 2001.
[4] Freeman B Higgins K N and Horney M 2016 How Students Communicate Mathematical Ideas: An Examination of Multimodal Writing Using Digital Tecnologies *Contemporary Educational Technology***7**(4) 281-313.
[5] Magnant C Nasseh S and Flateby T 2016 *Mathematical Writing Assignment for Deeper Understanding and Process Writing*(Georgia Southern University)
[6] Peake C et.al 2014*Syntatic Awareness and Arithmetic Word Problem Solving in Children With and Without Learning Disabilities. Journal of Learning Disabilities*. Doi: 10.1177/0022219413520183.
[7] Denman G A 2013*Think it, show it mathematics: Strategies for explaining thinking. Huntington Beach*(CA: Shell Education)
[8] Dougherty BJ 2006 *The 'write' way: Mathematics journal prompts for algevra one*(Honolulu, HI: Curriculum Research and Development Group)
[9] Parker R and Breygfogle M L 2011learning to write about mathematics*Teaching Children Mathematics***18**(2) 90-99
[10] Craig T S 2016The role of expository writing in mathematical problem solving *African Journal of Research in Mathematics, Science and Technology Education***20** (1) 57-66
[11] De Kock W D and Harskamp EG2016 *Procedural versus content-related hints for word problem solving: an exploratory study. Journal of Computer Assisted Learning*.doi: 10.111/jcal.12148.
[12] Fuchs L S et al 2015 *Is Word-Problem Solving a Form of text Comprehension? Scientific Studies of Reading*.Doi: 10.1080/10888438.2015.1005745.
[13] Fuchs LS etal 2017*Text Comprehension and Oral Language as Predictors of Word-Problem Solving: Insights into Word-Problem Solving as a Form of Text Comprehension. Scientific Studies of Reading*. Doi:10.1080/10888438.2017.1398259.
[14] Jitendra A K et al 2016 *Is Mathematical Representation of Problems an Evidence-Based Strategy for Students With Mathematics Difficulties? Exceptional Children*. Doi: 10.1177/0014402915625062.
[15] King B Raposo D and Gimenez M 2016*Promoting Student Buy-in: using Writing to Develop Mathhematical Understanding. Georgia Educational Researcher*. Doi: 10.20429/ger.2016.130202

[16] Lubin AHoude O and De Neys W 2015Evidence for children's error sensitivity during arithmetic word problem solving *Learning and Instruction***40**(2015) p1-8
[17] Marks JL Purdy CR and Kinney LB1958 *Teaching Arithmetic For Understanding*(New York: McGRAW-HILL BOOK COMPANY, INC)
[18] MoleongLJ 2013 *MetodologiPenelitianKualitatifEdisiRevisi*(Bandung: PT RemajaRosdakarya)
[19] Monroe E andPanchyshyn R 1995 Vocabulary considerations for teaching mathematics*Childhood Education***72** 80-83, doi:10.1080/00094056.1996.10521849
[20] Peterson S SRochwerger L Brigman J and Wood K Cross-Curricular Literacy: Writing for Learning in a Science Program*ProQuest Professional Education***14**(2) 31-37.
[21] Republik Indonesia 2016*PeraturanMenteriPendidikandanKebudayaanRepublik Indonesia No.24 Tahun 2016 tentangKompetesiIntidanKompetensiDasarPelajaranPadaKurikulum 2013 PadaPendidikanDasardanMenengah* (Jakarta:Sekretariat Negara)
[22] Salkind NJ 2008*Encyclopedia of Educational Psycology*(Thousand Oaks: SAGE Publications)
[23] Siswono T Y E 2010 *PenelitianPendidikanMatematika* (Surabaya: Unesa University Press)
[24] Sugiyono2014*MetodePenelitianKombinasi Mixed Methods*(Bandung: Alfabeta)
[25] Swanson HL 2016 Word Problem Solving, Working Memory and Serious Math Difficulties: Do Cognitive Strategies Really Make a Differene? *Journal of Applied Research in Memory and Cognition*. Retrieved from http://dx.doi.org/10/1016/j.jarmac.2016.04.012
[26] Teuscher D Kulinna P H and Crooker C 2015 Writing to Learn Mathematics: An Update*The Mathematics Educator***24**(2) 56-78.
[27] Wilkie K J 2016 Student's use of variables and multiple representations in generalizing functional relationships prior to secondary school. Education Studies Mathematics. 10.1007/s10649-016-9703-x.
[28] Widoyoko E S 2016*TeknikPenyusunanInstrumenPenetilian* (Yogyakarta: PustakaPelajar)

The Influence of Contextual Learning Model and Critical Thinking to Science Literacy of High School First Year Students

K Mungawanah[1,*], Y Supriyati[2], and M A Marpaung[2]

[1] Postgraduate Program in Physics Education, Universitas Negeri Jakarta, Indonesia
[2,3] Department of Physics, State University of Jakarta, Indonesia
*Corresponding Author: khikmatulmungawanah@gmail.com

Abstract. This study aims to determine the influence of contextual learning model and the ability to think critically on science literacy of the High School first year students. This study was conducted to the first year students of SMAN 4 Kota Tangerang and SMAN 11 Kabupaten Tangerang. A Quasi-experimental method with two-way Anova (treatment by level 2x2) design was implemented in the study. 80 students were chosen as sample by using purposive sampling technique. Students on experiment class study using science, technology and society (STS) model. Meanwhile, students on control class study using contextual teaching and learning (CTL) model. The conclusion taken from the study, are as follow: student science literacy with STS model is higher than students with the CTL model; there is an interaction effect between learning model and critical thinking ability to student science literacy; student science literacy with STS model is higher compared to the students with CTL model on the group of students that possess high critical thinking ability; and student science literacy with *STS* model is lower compared to the students with CTL model on the group of students that possess low critical thinking ability.

1. Introduction

In the 21st Century, the industrial world is growing rapidly due to the advances in science and technology to cater human needs. The rapid development of industry and technology in the 21st century also causes many problems; one of them is environmental aspect. This is due to a lack of awareness on science. Humans often utilize science and technology by exploiting nature without understanding the consequences for the environment and the future of the earth.

Science is a systematic effort to create, build and organize knowledge to understand universe [1]. The presence of science shapes human behaviour and character to care and responsible for itself, society and the universe which is literally defined as science literacy. Science literacy is considered a learning outcome; the key to science literacy education is at the age of 15 because at that age, someone should have chosen the career and participate in science and technological development [2].

To assess science literacy level, Indonesia has taken PISA (Programme for International Student Assessment) test, where it was initiated by OECD (Organization Economic Co-Operation and Development) that aims to evaluate education system of 72 countries around the world [9]. Every three years, students ranged 15 years old are randomly selected to have taken the PISA test [10]. PISA-focused subjects for tests are reading, math and science. This test is diagnostic which provides useful information in the improvement of the education system [11].

The facts of PISA 2015 result showed that the average science value of OECD Country is 493, while Indonesia only reached 403[3]. The data shows that science literacy within the education in

Indonesia is still less and adequate and there is a gap in treating science education. The use of appropriate learning model is expected to be able to increase the student interest and their science literacy. One of the learning models that support students' literacy skills is contextual learning model. Contextual learning model includes technology and society model and contextual teaching and learning model.

Science learning with science, technology and society model is directed to science and technology literacy, meaning that students can understand science from the perspective of science and technology produced by them. This is in line with the fourth scientific literacy that is science, technology and community interaction intended to illustrate the impact of science on society, and to explain the implementation of science and how technology can help human [4]. The contextual teaching and learning model is a learning concept that links between the material taught by the teacher to the real-world situations of the students and encourages students to make connections between their knowledge and application in their live as family members and society [5].

In addition to positive impact of the science literacy of students, the use of science, technology and society model and contextual teaching and learning model, also requires the students to develop critical thinking ability because in both learning process the students are encouraged to dig many information analyzing the problem and answer it, and relate to the technology that exists in everyday life. John Dewey defines thinking as an active, orderly and meaningful process used to conquer the world. Meanwhile, being critical is to ask, understand and analyze. Thus, critical thinking is thinking well and organized, reflecting on the process that are part of the critical thinking [6].

The ability of critical thinking possessed by students may develop well if the implemented learning model within learning process is able to activate the students (student centre). The learning model used during learning process and critical thinking ability by each student may lead to science literacy. This is similar to the research that has been done by Ucu Cahyana et al., that there is an interaction effect between learning methods with the ability to think critically on the ability of science literacy [7]. Therefore, the writer conducted a study entitled "The Influence of Contextual Learning Model and Critical Thinking to Science Literacy of High School First Year Students".

2. Methods

This study has several objectives, among others: address the difference of science literacy among students, technology and society model and students with contextual teaching and learning model, address the interaction between learning model and critical thinking ability towards students' science literacy, address the difference in terms of students' science literacy between science, technology and society learning model with high critical thinking ability and contextual teaching and learning model with high critical thinking ability, and address the difference in terms of science literacy between science, technology and society learning model with low critical thinking ability and contextual teaching and learning model with low critical thinking ability.

The study is conducted at two schools of SMAN 4 Kota Tangerang and SMAN 11 Kabupaten Tangerang in February-May 2018. The study used quasi-experimental method with two-way Anova. There are two independent variables in this research design; they are the learning model (treatment variable) and critical thinking ability (attribute variable) and one dependent variable which are science literacy. Each independent variable is classified into two types: contextual learning model consisting science, technology and society and contextual teaching and learning model, while critical thinking ability consist of high critical thinking and low critical thinking. The following shows the design of study in Table 1:

Table 1. Research Design

Critical Thinking Ability (B)	Learning Model (A)	
	Science Technology and Society (STS) A_1	Contextual Teaching and Learning (CTL) A_2
High Critical Thinking (B_1)	$A_1 B_1$	$A_2 B_1$
Low Critical Thinking (B_2)	$A_1 B_2$	$A_2 B_2$

The target population used in this study is all students of SMAN 4 Kota Tangerang and SMAN 11 Kabupaten Tangerang on their second semester of 2017/2018 period. The population used in this study are all first year students in Science Major of SMAN 4 Kota Tangerang and SMAN 11 Kabupaten Tangerang on their second semester of 2017/2018 period. The sampling was conducted using purposive sampling technique, where two out of six class of the first year Science major of SMAN 4 Kota Tangerang were taken and two out of seven class in SMAN 11 Kabupaten Tangerang to be implemented with different learning model, which is experiment class with science, technology and society model and control class with contextual teaching and learning model. A critical thinking ability test had given in each treatment class. The values obtained from the tests are then sorted by rank, from the highest to the lowest. Based on the result, about 27% top values are taken to form the upper group and 27% of the bottom values are the lower group [7]. Sample distribution can be seen in table 2:

Tabel 2. Sample Research

Critical Thinking Ability (B)	Learning Model (A)		Total
	Science Technology and Society (STS) A_1	Contextual Teaching and Learning (CTL) A_2	
High Critical Thinking (B_1)	20	20	40
Low Critical Thinking (B_2)	20	20	40

Data collection in this study is test. Test in this study include science literacy tests on business materials, energy, momentum and impulses as well as critical thinking ability test. The data analysis technique used in this study, including analysis techniques with descriptive statistics, prerequisite analysis test and data analysis with inferential statistics.

3. Result and Discussion

At the beginning of the study, both classes were tested for critical thinking ability which aims to classify students into two parts: high critical thinking and low critical thinking. Then, the students are given pre-test to determine whether the students' initial ability in both groups of the same class or not.

Based on the result in the appendix, the mean pre-test score of the students in the experimental class before treated using science, technology and society model is 22.45 with standard deviation of 7.82, while mean pre-test score in the control class before using contextual teaching and learning model is 29.33 with standard deviation of 8.05.

Table 3. Normality Test

	Group	Total Sample	L_{score}	L_{table}	Conclusion
Pretest	A_1	40	0.072	0.140	Normal
	A_2	40	0.129	0.140	Normal
	A_1B_1	20	0.112	0.190	Normal
	A_1B_2	20	0.121	0.190	Normal
	A_2B_1	20	0.170	0.190	Normal
	A_2B_2	20	0.106	0.190	Normal
Posttest	A_1	40	0.104	0.140	Normal
	A_2	40	0.094	0.140	Normal
	A_1B_1	20	0.186	0.190	Normal
	A_1B_2	20	0.131	0.190	Normal
	A_2B_1	20	0.104	0.190	Normal
	A_2B_2	20	0.135	0.190	Normal

Based on table 3, the entire data group obtained is normal, because L_{score} smaller than L_{table}. There are homogeneity test, the result is:

Table 4 Homogeneity Test used The Fisher Test in Group A_1 and A_2

	Group	Total Sample	F_{score}	F_{table} (α=0.05)	Conclusion
Pre-test	A_1	40	1.058	3.24	Homogeneous
	A_2	40			
Post-test	A_1	40	2.09	3.24	Homogeneous
	A_2	40			

Based on table 4, the entire data group obtained is homogeneous, because F_{score} smaller than F_{table}. It is concluded that both pre-test data and post-test data of students in experiment and control class have the same population variance (homogeneous).

Table 5. Homogeneity The used The Fisher Test in Group A_1B_1, A_1B_2, A_2B_1, A_2B_2

	Group	Total Sample	F_{score}	F_{table} (α=0.05)	Conclusion
Pre-test	A_1B_1	20	0.44	7.815	Homogen
	A_1B_2	20			
	A_2B_1	20			
	A_2B_2	20			
Post-test	A_1B_1	20	3.94	7.815	Homogen
	A_1B_2	20			
	A_2B_1	20			
	A_2B_2	20			

Based on table 5, the entire data group obtained is homogeneous, because F_{score} smaller than F_{table}. It is concluded that both pre-test data and post-test data of students in experiment and control class have the same population variance (homogeneous).

The pre-test and post-test data in the experimental class and control class have been tested to be normal and homogeneously distributed, then the next stage is hypothesis test. Hypothesis test used two ways Anova, the data in hypothesis test only post-test data, with result as follows:

Table 6. Two-Way Anova Test Result

Variance Source	Db	JK	RJK	F_{score}	F_{table} (α = 0.05)	Conclusion
			Pretest			
Between A_1 dan A_2	1	171.1125	171.1125	2.743	3.12	H_0 accepted
Interaction A X B	1	2.8125	2.8125	0.045	3.12	H_0 accepted
In Group (D)	76	4740.75	62.378	-	-	
Total (T)	80	59473	-	-	-	
			Posttest			
Variance Source	Db	JK	RJK	F_{score}	F_{table} (α = 0.05)	Conclusion
Between A_1 dan A_2	1	117.6125	117.6125	6.671	3.12	H_0 rejected
Interaction A X B	1	2702.8125	2702.8125	153.299	3.12	H_0 rejected
In Group (D)	76	1339.95	17.631	-	-	
Total (T)	80	464667	-	-	-	

Based on table 6 (pretest), F_{score} is 2.743 while F_{table} is 3.12 because of F_{score} is smaller than F_{table}, it is then H_o is accepted. It can be concluded that there are no a differences in students' science literacy on the treatment of science, technology and society model and contextual teaching and learning. Based on table 6 (post-test), F_{score} is 6.671 meanwhile F_{table} is 3.12 because of F_{score} is bigger than F_{table}, it is then H_o is rejected where it can be concluded that there are differences in students' science literacy on the treatment of science, technology and society model and contextual teaching and learning.

Based on analysis pre-test data, F_{score} is 0.045 while F_{table} is 3.12 Because of F_{score} is smaller than F_{table} so H_0 is accepted and it can be concluded that there is no an interaction between the contextual learning model and the ability to think critically of the students science literacy. Based on analysis post-test data, F_{score} is 153.229 while F_{table} is 3.12 Because of F_{score} is bigger than F_{table} so H_0 is rejected and it can be concluded that there is an interaction between the contextual learning model and the ability to think critically of the students science literacy. Graph of interaction between the contextual learning model and the ability to think critically of the students science literacy, could be seen below :

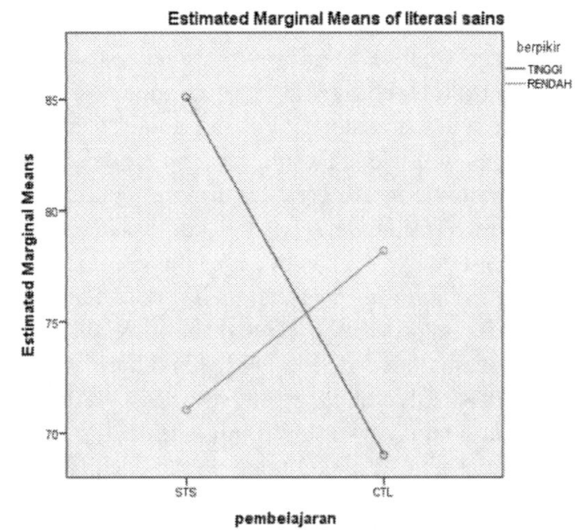

Figure 1. Graph of Interaction between The Contextual Learning Model and The Ability to Think Critically of The Students Science Literacy

Table 7. Tukey Test Result

	Group	N	Q_{score}	Q_{table}	Conclusion
Pre-test	A_1B_1	20	1.83	3.96	H_0 accepted
	A_2B_1	20			
	A_1B_2	20	1.42	3.96	H_0 accepted
	A_2B_2	20			
Post-test	A_1B_1	20	14.963	3.96	H_0 rejected
	A_2B_1	20			
	A_1B_2	20	9.798	3.96	H_0 rejected
	A_2B_2	20			

Based on table 7, hypothesis test between group A_1B_1 with A_2B_1 is explained here. Analyzed pre-test data, Q_{score} is 1.83 while Q_{table} is 3.96. Because of Q_{score} is smaller than Q_{table} so H_0 is accepted and it can be concluded that there are no a differences in students science literacy between science, technology and society learning model with high critical thinking ability and contextual teaching and learning model with high critical thinking ability. Analyzed post-test data, Q_{score} is 14.963 while Q_{table} is 3.96. Because of Q_{score} is bigger than Q_{table} so H_0 is rejected and it can be concluded there are differences in students science literacy between science, technology and society learning model with high critical thinking ability and contextual teaching and learning model with high critical thinking ability.

Based on table 7, hypothesis test between groups A_1B_2 with A_2B_2 is explained here. Analyzed pre-test data, Q_{score} is 1.42 while Q_{table} is 3.96. Because of Q_{score} is smaller than Q_{table} so H_0 is accepted and it can be concluded there are no a differences in students science literacy between science, technology and society learning model with low critical thinking ability and contextual teaching and learning model with low critical thinking ability. Analyzed post-test data, Q_{score} is 9.798 while Q_{table} is 3.96. Because of Q_{score} is bigger than Q_{table} so H_0 is rejected and it can be concluded there are differences in students' science literacy between science, technology and society learning model with low critical thinking ability and contextual teaching and learning model with low critical thinking ability.

The results showed that there are four hypotheses that have been tested such as: (1) there is a difference in terms of science literacy between students given the science, technology and society model with the students given the contextual teaching and learning model. The result data showed the average post-test in the experimental class is 78.08 while the mean post-test in the control class is 73.60. This means that the students' science literacy with science, technology and society model is higher than the students with contextual teaching and learning model.

This is because learning with science, technology and society model in the experimental class gives the students several advantages with steps included in the model that makes them become active, responsible, and able to use their knowledge to create a simple technology that makes them learning thoughtfully [8]. (2) There is an interaction between the contextual learning model and the critical thinking ability of students' science literacy. This is because the contextual learning model emphasizes the process of student involvement in finding the relationship between knowledge and its application in daily life, thus encouraging students to develop critical thinking ability. This has a positive impact on the increased literacy results of student science [12]. (3) There is difference of student science literacy between science learning model, technology and society with high critical thinking ability and learning model of contextual teaching and learning with high critical thinking ability.

The data showed that the average post-test of students who are given the science, technology and society model with low critical thinking ability is 85.10, higher than the average post-test of students with contextual teaching and learning model with high critical thinking ability is 69. This is because the science, technology and society model is a short, clear and solid learning process and may invite the students to create simple projects as the application of the subjects, while in contextual teaching and learning model, there are lots of learning model resulting to lack of focus and takes a lot of time [13]. (4) There is a difference in science literacy between science, technology and society learning model with low critical thinking and learning contextual teaching and learning model with low critical thinking ability.

The data showed that the average post-test of students who are given the model of science, technology and society with low critical thinking ability is 71.05 lower than the average post-test of students given contextual teaching and learning model with critical thinking ability low is 78. This is because in the contextual teaching and learning model, the teacher still guides the students to work on the worksheets that have been given and provides an initial explanation first [14]. This is preferable for students with low critical thinking skills.

4. Conclusion

Based on the result of the study obtained from data analysis and hypothesis testing, it can be concluded that: there is a difference in terms of science literacy between students that is given the science, technology and society model with the students that is given contextual teaching and learning model, there is an interaction between contextual learning model and critical thinking ability towards students' science literacy, there is a difference in terms of students' science literacy between science, technology and society learning model with high critical thinking ability and contextual teaching and learning model with high critical thinking ability, and there is a difference in terms of science literacy between science, technology and society learning model with low critical thinking ability and contextual teaching and learning model with low critical thinking ability.

5. References

[1] M Randy Fananta A E 2017 *Materi Pendukung Literasi Sains* (Jakarta: Kementerian Pendidikan dan Kebudayaan)

[2] Rahmawati 2013 *Analisis Literasi Sains Siswa SMP dalam Pembelajaran IPA Terpadu pada Tema Penerapan Bioteknologi Konvensional* (Bandung : Universitas Pendidikan Indonesia)

[3] Kemendikbud 2016 (online) *Gerakan Literasi Nasional*. Retrieved Februari Rabu 2017 from gln.kemdikbud: (http://gln.kemdikbud.go.id/glnsite/)

[4] Rusman 2010 *Model-Model Pembelajaran Mengembangkan Profesionalisme Guru* (Jakarta : PT Raja Grafindo Persada)

[5] Rusman L D 2012 *Kurikulum & Pembelajaran* (Jakarta : Raja Grafindo Persada)

[6] Johnson 2009 *Contextual Teaching & Learning Menjadikan Kegiatan Belajar Mengajar Mengasyikkan dan Bermakna* (Bandung : Mizan Media Utama)

[7] Ucu Cahyana A K 2017 *Relasi Kemampuan Berpikir Kritis dalam Kemampuan Literasi Sains pada Siswa Kelas IV Sekolah Dasar* (Universitas Negeri Jakarta)

[8] Gherardini M 2016 *Pengaruh Metode Pembelajaran dan Kemampuan Berpikir Kritis Terhadap Kemampuan Literasi Sains* (Jurnal Pendidikan Dasar)

[9] OECD 2003 *Assessment Framework-Mathematics, Reading, Science and Problem Solving Knowledge and Skills* (Paris : OECD)

[10] OECD 2017 *PISA 2015 Assessment and Analytical Framework Revised edition* (Paris: OECD Publishing)

[11] PISA 2009 *Assessment Framework Key Competencies In Reading, Mathematics and Science. (OECD)*

[12] Rahayuni G 2016 Hubungan Keterampilan Berpikir Kritis dan Literasi Sains Pada Pembelajaran IPA Terpadu dengan Model PBM dan STM *Jurnal Penelitian dan Pembelajaran IPA* **2**(2)

[13] Siti Muhajir E R 2015 The Differences Between STS and CTL Learning Models on Scientific Literacy and Science Achievement *Jurnal Pendidikan Matematika dan Sains* **2**(2)

[14] Suwita I K 2012 Pengaruh Model Pembelajaran STM dan CTL Terhadap Kemampuan Pemahaman Konsep Fisika dan Keterampilan Berpikir *Kritis Jurnal Penelitian Pascasarjana UNDIKSHA* **3**(5)

Acknowledgement

The authors would like to express her gratitude to Allah SWT with His blessing and grace, for the author's parents with their immense support and prayers, staffs and teachers of SMAN 4 Kota Tangerang, staffs and teachers of SMAN 11 Kabupaten Tangerang, Jakarta State University Graduate School Physics Education Program, Thesis Adviser I Prof Dr Yetti Supriyati, M.Pd, Thesis Adviser II Dr Mangasi Alion Marpaung, M.Si that has provided the author with directions, guidance, as well as opportunity to be able to finish this paper well.

IOP Conf. Series: Journal of Physics: Conf. Series **1108** (2018) 012125 doi:10.1088/1742-6596/1108/1/012125

A Comparative Study of Students and Faculty Members Views on the Importance of Educational Objectives to Join Electrical Engineering

I J El-Khozondar[1,*] and H J El-Khozondar[2]

[1]Electrical Engineering Department, Islamic University of Gaza, Gaza, Palestine
[2]Education Department, Al-Azher University, Gaza, Palestine

*Corresponding Author: hkhozondar@iugaza.edu

Abstract. Educational institutions play an important role in preparing students to serve their society. In order to prepare them properly, it is necessary to identify the desired goals to facilitate the graduates to be constructive elements in their society. The aim of this study is to compare the opinions of second-year electrical engineering students, and the views of the faculty members regarding the importance of the various educational objectives of the electrical engineering department at the Islamic University. The samples were 41 students from second year students from the electrical engineering department at the Islamic University in Gaza (42) and faculty members (17). All members of the community were contacted for a small number, but the questionnaire was retrieved from 17 male students, 10 female students, and 5 teaching staff members. The following statistical methods were used: frequency and percentage, arithmetic mean, standard deviation, relative weight, correlation coefficient, and Alpha Cronbach test. Studies indicate that there are no significant differences between the views of different groups in the educational objectives of the Department of Electrical Engineering at the Islamic University of Gaza.

1. Introduction

The goals of the academic program are the general terms that describe the objectives of the long-term academic program that the programmers seek to achieve: developing students' knowledge, skills, attitudes, attitudes, etc., after meeting the requirements for degree [1].

Therefore, the objectives of the academic program are a long-term strategy and a specific purpose that provide people something they wish to work for and are proud of when those are accomplished. This is characterized by the generality of its formulation and the general framework or reference is determined in defining the academic program's specifications: General trends of the content of the academic program, and general trends in teaching and evaluation strategies. Although the objectives of the academic program are formulated in general, they are necessary to consolidate the views of the academic program development and evaluation team and serve as a reference for differences in any component. Therefore, the objectives of the academic program provide the basis for decision making on the nature and scope of the academic program, the priority of the various activities to be implemented and the planning process of its various components, thereby helping to achieve its mission [1].

The specialization of electrical engineering (one of the branches of the Faculty of Engineering) is the specialization that gives mankind the benefit by studying communication and radar systems, control systems, measurements, electronic circuits, electromagnetic and their applications, energy transfer technologies, networks, lighting, computers etc. Because they overlap in many sectors of industry and various civil and military activities, they are almost the core of civil civilization in our time.

The research dealt with the students of the second year who have decided to choose the specialty recently before the start of teaching materials and, thus, it can be informed that the educational goals of the Department of Electrical Engineering are not being affected by the specialized materials. In order to complete the picture in many aspects, the comparison between the views of the faculty and students is performed to determine the importance of the educational objectives of the Department of Electrical Engineering. Each has a vision from its own angle.

The aim of this research is to recognize the importance of the educational objectives of the engineering department to select students' specialization in electricity from the point of view of both second year electrical engineering students and the faculty members. This study is a natural response to the university's renaissance in all aspects of education, especially the desire to achieve the conditions of the ABET[3]. It is noteworthy that of all students who enroll in the Islamic University, there are 5811 students (1759 students and 3052 students), 473 students (326 students and 148 female students) attend the Faculty of Engineering, 41 students (31 students and 10 female students) are in electrical engineering.

2. Problem of the current study

The present study attempts to answer the following main question: "Do students and faculty members agree in their views on the importance of educational goals to help students choosing the specialization of electrical engineering?" The answer to this question requires answering the following sub-questions:

a. What are the educational objectives of the highest importance for each group?
b. Do students and faculty members have the same order and priority for the educational objectives of electrical engineering specialization?

3. Objectives of the study

The objectives of this study are as follows:

a. Recognizing the importance of educational objectives for the specialization of electrical engineering from the perspective of students, faculty and parents.
b. Comparing the views of the students of the second year electrical engineering with the views of the parents and the views of the faculty regarding the importance of the educational objectives of the department.
c. Raising university education by making recommendations to develop the educational objectives of the department.

4. The Significances

The significances of this study are as follows:

a. This study concerned the study of the educational objectives of the electrical engineering department from the point of view of each of the students and teaching staff.
b. The results of this study help to enlighten the Palestinian universities, faculty members, students and those interested in the educational process with the educational goals of utmost importance from the point of view of students and teaching staff and activate them to improve the educational reality in the Palestinian universities.
c. The results of the study are expected to identify the effect of variables (student, faculty).
d. The results of this study help researchers to conduct further studies on this subject using other variables and other study sections.

5. The limitation of the study:

a. The objective limit: The study was limited to study the educational objectives of the electrical engineering department and recommendations and proposals to activate them.
b. Institutional Limit: Islamic University.
c. Human Limit: Second year students from the Department of Electrical Engineering.

6. Terminology of study :

a. The educational goal is the desired change that the educational process seeks to achieve in the behaviour of the student, a description of the educational output in the behaviour of the learner (a description of what the learner will be after passing through educational experience), and a description of the pattern of behaviour expected to occur in the personality of the learner as a result of his experience Educational or educational position.
b. The educational objective is a clear description of the behaviour to be issued by the learner at the end of the educational stage. The outcome of the learning and learning process translates into observable behaviour and measurement (Buhamama, 2008, p. 77).

7. Previous studies

The study [2] aimed to find out the extent to which the teachers of the basic stage of the skills of formulating the behavioral goals through surveying the views of educational supervisors and managers considering sex and the function of the evaluator. The study sample reached (250) supervisors and managers. The results of the study showed that the scope of the conditions for the formulation of the behavioral goals ranked first by the average of the estimates of the educational supervisors, while the field of general bases of the behavioral goals ranked first according to the averages of the managers' estimations, as such The results showed that there were no statistically significant differences between the averages of the educational supervisors' sexes and the sexes of the managers in each of the six study fields and the fields combined, except in the field of formulating the behavioral goals.

The study [3] was conducted to determine the level of satisfaction of the first batch of students who joined the Hashemite University in the first year of its foundation on the experiences gained and the services provided by the university in general fields (general experience, expertise in the field of specialization, university experience outside the classroom) Related to their future and university studies. The sample included 477 students from the first batch who spent six or more semesters. The number of respondents was 244 students from 17 departments. The arithmetical averages, standard deviations, percentages and repetitions necessary for the statement The results showed the degree of satisfaction of the students with the specific experiences in the scale. The results showed statistically significant differences at the level of significance (0.05) due to gender or college variables and interaction between them. Comparison of the level of satisfaction of the students of the Hashemite University with the degree of satisfaction of American university students with the practices that represent the experiences they have acquired outside the class work, which is the lowest level in most of its items.

The study [4] aimed to identify the views of the science teachers in the state of Bahrain on the importance of the objectives of scientific education related to the concept of scientific culture, in addition to determining the relationship between the degree of importance of the objectives and the contribution of the content of science books in their achievement from the point of view of teachers, Demographics such as: school stage - experience - gender. The study is based on a basic assumption that the teacher's awareness of the importance of the objectives of the scientific education and its priorities in his view can affect his teaching practices in the science classes. The questionnaire was applied to a sample of 306 teachers from the preparatory and secondary schools in the State of Bahrain. The objectives of the scientific education associated with the scientific culture and the importance of the contribution of the content of science books in the achievement of the average of the importance of those goals from the point of view of teachers 89.6, but on the other hand, teachers stressed the lack of current science books and the weakness of content to the objectives suggested, where The average translation of these goals and their presence in the science books from the point of view of teachers was 69.8, and the results confirm that there are no statistically significant differences in opinions related to experience. The study recommended the need to reflect the objectives of scientific education associated with the scientific culture within the content of science books.

The study et. Al Wilson [5] found that students enroll in educational institutions based on different goals, as well as colleges in various educational institutions in higher education designed their educational goals for their students in various forms so as to ensure that students are attracted to them, The first to enroll in the college is self-knowledge and personal identification, and more general goals are classified as the optimal goal at the state level and generally more general levels at the university level.

From the above we see that the previous studies confirmed: The importance of the role of the teacher in determining educational goals. However, none of the previous studies have indicated a comparison of students' views with regard to educational goals.

8. Study population and study sample

The study population consists of 41 second year electrical engineering students at the Islamic University in Gaza and 17 faculty members. The members of the study society were contacted and therefore the sample is equal to the study population. However, 17 students and 10 female students And 5 faculty members give back the questionnaires.

9. Study tool

The researchers used the questionnaire compose of 32 item in main aims related: Science and technology (6 items), social (3 items), mental abilities and communication skills (4 items), personal development (4 items), and professional development (5 items).

10. Results

Analysis of the first question: The first question states: "What are the educational objectives of the specialization of electrical engineering of the utmost importance to each group?"

To answer this question, the two researchers used the relative weight and order of this question. Table 1 illustrates this:

Table 1. Analysis of the first question

Female students		Male students		faculty		Item	number
order	Relative weight	order	Relative weight	order	Relative weight		
4	76.67	4	55.38	4	80	Understanding the nature of science and scientific methods	1
2	80.00	6	46.15	1	88	Get enough knowledge to use your computer	2
6	68.33	5	47.69	2	84	Get basic theoretical knowledge of electrical engineering	3
2	80.00	2	60.00	4	80	Ability to understand electrical networks	4
4	76.67	1	69.23	2	84	The ability to build electrical circuits	5
1	91.67	2	60.00	6	76	Obtaining accurate scientific information to keep abreast of scientific development	6
	78.89		**56.41**		**82**	**average**	

1	85.00	2	50.77	1	76	Understanding Islamic principles and values	7
2	81.67	3	44.62	3	60	Understanding the student's rights and duties towards the homeland	8
3	75.00	1	55.38	2	64	Ability to serve the community, especially the industrial sector	9
	80.56		**50.26**		**66.67**	**average**	
4	65.00	1	75.38	1	68	Learn to write reports efficiently and clearly	10
3	78.33	2	66.15	3	60	Learn logical and analytical thinking	11
1	86.67	3	60.00	3	60	To transform scientific ideas into practical projects	12
2	85.00	4	53.85	1	68	Learn the ability to innovate and create	13
	78.75		**63.85**		**64**	**average**	
3	83.33	2	58.46	3	68	Gain self-confidence	14
1	85.00	1	60.00	3	68	Gain the ability to independently and handle things	15
1	85.00	4	52.31	1	72	Gain the ability to develop and keep pace with scientific progress	16
4	73.33	2	58.46	2	64	Gain the ability to deal with all sectors of society	17
	81.67		**57.31**		**68**	**average**	
3	73.33	2	70.77	4	60	Acquire information and skills that prepare students for the job market	18
1	83.33	4	56.92	2	76	Acquiring sufficient information to enable students to enrol in higher education	19
5	68.33	5	55.38	1	80	Acquire sufficient information and skills for application in any profession	20
2	76.67	3	64.62	3	64	Acquire accurate information related to a specific occupation	21
3	73.33	1	95.38	5	56	Get a good job after graduation	22
	75.00		**68.62**		**67.2**	**average**	
1	78.97	5	59.29	4	69.57	**Total Average**	

From the previous table it is clear that each group has different views in terms of any educational goals for the specialization of electrical engineering of utmost importance. The most important and least important for each group can be summarized as follows: For the teaching staff, the relative weight of the goals of science and technology received the highest relative weight of 82%, while the goals related to mental abilities and communication skills were relatively low (64%). While the highest relative weight of the goals related to the professional development of students is 68.62%. From the students' point of view, the highest relative weight of the goals related to personal development is 81.67% while the relative weight of the goals related to the professional development of the students is 75%.

Analysis of the second question: "Do students and faculty have the same order and priority for the educational objectives of electrical engineering specialization?"To answer the question, the researchers used an independent sample test to determine the differences between the two variables.

Table 2. The number, mean, standard deviation, T test value and significance level are shown

Level of significance	T test value	Standard deviation	Arithmetic average	number		Aim
//0.055	2.001-	4.696	20.160	25	Students	First aim
		3.362	24.600	5	Faculty	
//0.773	0.293-	3.577	9.720	25	Students	Second aim
		1.414	10.000	5	Faculty	
//0.443	0.779	3.227	14.200	25	Students	Third aim
		5.630	12.800	5	Faculty	
//0.913	0.110	3.916	13.800	25	Students	Fourth aim
		2.074	13.600	5	Faculty	
//0.510	0.667	3.475	17.920	25	Students	Fifth aim
		3.114	16.800	5	Faculty	
//0.782	0.280	14.869	75.800	25	Students	Total
		12.696	77.800	5	Faculty	

* The tabular value of (T) at the degree of freedom 23 and the level of significance 0.05 = 2.069. // not significant at 0.05

The above table shows that the calculated value of T for all objectives and total is less than the T value. This means that there are no statistically significant differences at the level of significance of 0.05 in all educational objectives of engineering specialization due to the variable of students and faculty.

11. Discussion

The results indicated that the objectives of science and technology were the most important for the faculty with a weight of 82%. For students, the goals for personal development are the most important at 81.67%. And the students' opinions showed that the goals related to professional development are the most important with a weight of 68.62% since in our society the young man should do the things of the house so he seeks to develop himself professionally in order to have a suitable job He is then able to perform his role in society. When comparing the views of the groups, there were no significant differences in the importance of the educational objectives of the Department of Electrical Engineering in the Faculty of Engineering at the Islamic University. It is indicating that the objectives were prepared accurately, where experienced engineers and students were consulted in their identification.

12. Recommendations:

1. Educating parents about the importance of choosing specialization based on the educational objectives of the specialization.
2. To sensitize high school students to the objectives of the department so that they can choose the specialization based on a conscious study.
3. Expanding the study to include employers and involve them in formulating the goals of the department.

IOP Conf. Series: Journal of Physics: Conf. Series **1108** (2018) 012125 doi:10.1088/1742-6596/1108/1/012125

13. References

[1] Quiran, A WA 2014 Academic Development Center. Republic of Yemen Ministry of Higher Education and Scientific Research Council of Academic Accreditation and Quality Assurance(electronic material).
[2] Buhamamah J2008 *Journal of Education***162**pp. 76-94
[3] Abdul A and Saleh 1985*Education and teaching methods*(Dar Al Ma'arif: Cairo)
[4] Abu Q and Mahmoud 2002 *Educational Magazine***63**pp. 53-97
[5] Fadl N A W and Abukhas K A 1997The Egyptian Association for Scientific Education, The First Scientific Conference "Scientific Education for the 21st Century"*Abu Bakr-Alexandria***1** pp. 10-13
[6] Wilson RC, Gaff JG, Dienst ER, Wood L, and Bavry JL 1975*Professors and their impact on students*(John Wiley & Sons: New York)

The Validity of Teaching Material Based on Ethnoscience Batik to Increase the Ability of Scientific Literacy for Junior High School

N S Melyasari[1,*], S Suyatno[2] and W Widodo[3]

[1]Student of Science Education Program, Post-graduate of Universitas Negeri Surabaya
[2,3]Lecturer of Science Education Program, Post-graduate of Universitas Negeri Surabaya
*Corresponding author: nindysilvia@gmail.com

Abstract. Reflected from the results of pre-research about the ability of students on scientific literacy in Tulungagung, which was relatively low, the researcher has recommendation to develop a teaching material that can increase the skill of scientific literacy. In order to make it more interesting, it will be based on ethnoscience or local wisdom of Batik. The purpose of this study is to produce a valid teaching material, which will use contextual approach based on ethnoscience-Batik to increase the ability or skill of scientific literacy of learners in Junior High School. The specifications include the Learning Implementation Plan, Textbook, Student Activity Sheet (SAS), and the Scientific Literacy Test Assessment Instrument. The four major dimensions of scientific literacy that will be included in the literacy test are the aspects of scientific knowledge, scientific contexts, competencies or the process of science and attitudes. Model of the teaching material development in this research uses 4D model from Thiagarajan. Based on the assessment of three validators, the developed learning tools have considerably very valid category and can be tested for limited students in Tulungagung.

1. Introduction

The learning of the 21st century is marked by the paradigm shift from teaching to learning [1]. Learning teaches the students to learn independently to explore their own potential. At the beginning of learning, educators can arouse students' curiosity about a phenomena or issue and then formulate it in the form of a question [2]. These issues and phenomena are related to the subject matter that is learned then the students seek solution to solve the problems. The process of problem solving is tough. One guidance for the ability of solving problems from an issue or phenomenon is by using the skill of scientific literacy [3].

The expectation of Curriculum 2013 (K13) is the increasing of scientific literacy on students in Indonesia. Based on the research results of PISA 2015, Indonesia is still in the 62nd out of 70 countries. From this result, it can be said that the scientific literacy of learners is still below average. The skill of scientific literacy is very important for every human being. When a person is in school, the skill of scientific literacy correlates to learning achievement. The time when they involve in the society, literacy skill correlates with socializing and solving life problems [3].

According to DeBoer [4], the efforts to increase the skill of scientific literacy is perpetually applicable for science because it is not static. Related to the research about scientific literacy that has been done in various countries, it can be drawn a red thread that the common barrier that is faced by every school is the process of habituation, starting from learning environment that lack of literacy attitude, and systematic attitude problems faced by the learners which are different from PISA or TIMMS system. One of the methods to increase scientific literacy in Indonesia is by adding indigenous science or ethnoscience to teaching materials. The function of this ethnoscience is to make the process of learning is more interesting and attract the learners' interest since it involves the identity of their own region. Ethnoscience that rooted in the life of learners is a form of direct or contextual experience [5]. The reason for integrating ethnoscience into the learning process is to provide a new understanding that local culture can be linked to science learning, fostering a sense of love for local culture, and preserving it.

Learning that is based on ethnoscience will ease students or learners to explore the facts and phenomena that exist in society and they can be integrated with scientific knowledge [6]. Currently, research of learning based on ethnoscience to improve any skills are developed continually. The concept of teaching material that can be associated with ethnoscience is the "classification material and its changes" in VII-grade of Junior High School. The material can be associated with ethnoscience Batik because the equipment for making Batik is related to the "classification of the material". However, in general, increasing scientific literacy by using ethoscience learning is still rare. Therefore, researcher will be developing a teaching material-based on ethnoscience Batik in Tulungagung area by using material "classification materials and its changes".

2. Method

The model that is used in this study is 4D model from Thiagarajan, et al [7]. The 4D development model consists of 4 phases: (1) define, (2) design, (3) develop, and (4) desseminate. In the define phase, the things that must be done is analyzing the front-end (front-end analysis), which consists of curriculum that is used, relevant learning theories, and challenges in the future. Furthermore, doing analyzing for students that covers academic ability and level of maturity. The third phase in define are material analysis, assignment analysis, and learning objectives formulation. Moreover, in the design phase, the activities that must be done consist of making literacy test, media selection for learning process, and then followed by the preparation of teaching material based on ethnoscience Batik, which consist of the Learning Implementation Plan, Textbook, Student Activity Sheet (SAS), and the Scientific Literacy Test Assessment Instrument.

Result from the preparation of teaching material is called "the initial design" or draft I. Furthermore, draft I will be reviewed by the lecturers or mentors to be given suggestions and improvements. In the develop phase, teaching material that is being developed will go through a revision process into draft II, which later on be validated by IPA lecturer, expert lecturer, and science teacher. The purpose of validation is to obtain the feasibility for the contents of developing teaching material so that it can be used for limited trials and extensive trials. The data of validation result was analyzed quantitatively. The average result of the components that is assessed by all validators will be analyzed qualitatively on each aspect by referring to the following criteria.

Table 1. Assessment Criteria

Score interval	Category	Explanation
3,60 $\leq$ V $\leq$ 4,0	Very valid	Can be used without revision
2,60 $\leq$ V $\leq$ 3,59	Valid	Can be used with a few revision
1,60 $\leq$ V $\leq$ 2,59	Less valid	Can be used with many revisions
1,00 $\leq$ V $\leq$ 1,59	Invalid	Not yet in use and requires consultation

From the Table 1., it can be seen that the teaching material is said to be valid if it gets a score at the interval of 2,60 $\leq V \leq 3{,}59$ and particularly to be very valid if it gets a score at the interval of 3,60 $\leq V \leq 4{,}0$. However, if a developing teaching material belongs to invalid category then it should be revised, and has to be validated again.

3. Result and discussion

3.1 Result

The developing teaching material consists of (1) Learning Implementation Plan, (2) Textbook, (3) Student Activity Sheet (SAS), and (4) Scientific Literacy Test Assessment Instrument. Figure 1, Figure 2, Figure 3, and Figure 4 below are an example of an overview of developed teaching material .

Figure 1. Textbook Cover

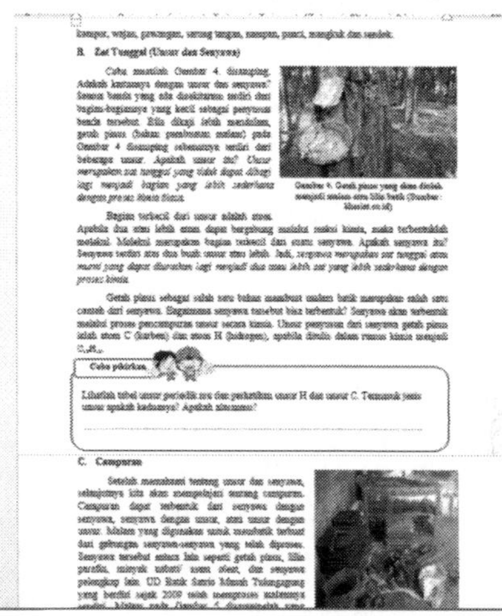

Figure 2. The Textbook content that is integrated with ethnoscience Batik

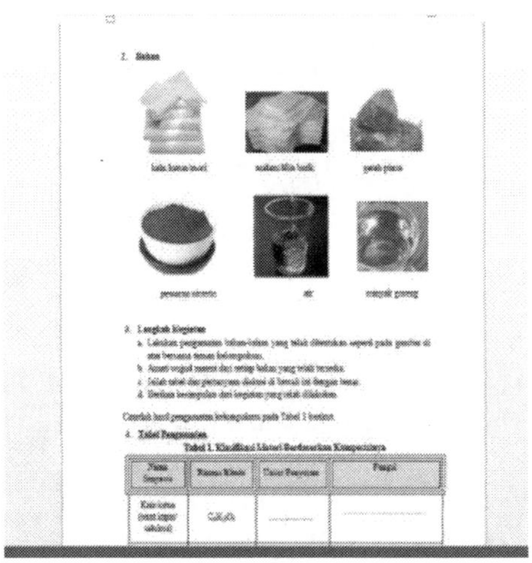

Figure 3. The content in Student Activity Sheet (SAS)

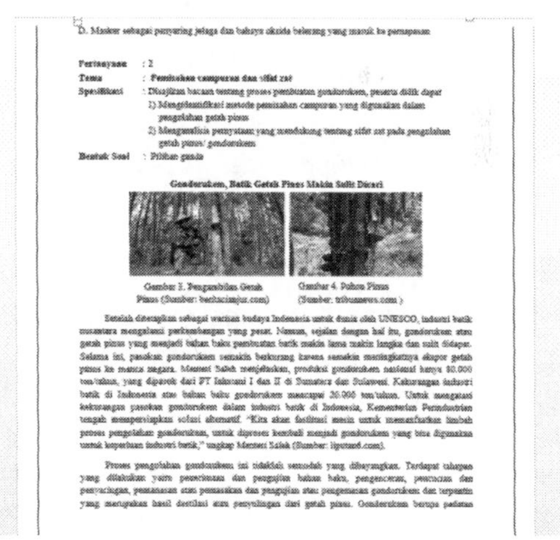

Figure 4. The example of Scientific Literacy Test

The Learning Implementation Plan refers to the Syllabus as a supporting component of teaching material. The Learning Implementation Plan consists of three meetings that are based on ethnoscience Batik. The provision of writing KD refers to Permendikbud 24 Tahun 2016. Components in Learning Implementation Plan includes school identity, subject, class/semester, subject matter, time allocation, KI, KD, Indicator of Competency Achievement, Lessons' Purpose, material summary points, learning method, media and materials, learning resources, learning steps, assessment and appraisal appendices [8].

The title of Textbook is "Material Classification and its Changes" which is supplemented by mapping concept, Basic Competence (KD), learning indicator, summary of the material that is connected with the material of making "Batik" and also changing that happened during the process of making Batik, ethnosciences information about Batik culture, examples of scientific literacy test, and bibliography [9]. Furthermore, There are three developed SAS for 3 meetings. The first meeting discusses the classification of the material based on its composition, the second meeting discusses the changes in the form of substances, and the last one discusses the activities of the mixture separation.

The SAS contains components such as title, KD, activity objectives, activity steps which is accordance with cooperative learning syntax, and bibliography [9]. Every SAS begins with a motivation that is generating a phenomena related to the material being studied. The presentation of material summaries in SAS is associated with ethnoscience Batik. Each SAS is completed with the organizing students into study groups, the guidance of the study group, guidance of the formulating problem, the materials that used for the activity, the steps of the activity, the observation table, the column data analysis, the discussion question and column for conclusion. The three SAS-based on ethnoscience Batik are complemented by key answers that make it easier for teachers to manage the learning.

The scientific literacy assessment uses 15 multiple choice questions and 7 scientific attitude questionnaires. The scientific literacy test refers to three indicators of PISA, that are explain (1) phenomena scientifically; (2) evaluate and design scientific enquiry; and (3) interpret data and evidence scientifically [10]. The three scientific literacy indicators are aligned with specific indicators on each item. Before having validation, the teaching material were reviewed by lecturers to get improvement and suggestions. It were validated by three validators (IPA lecturer, expert lecturer and science teacher). Validation result from the three validators can be seen in the following details.

Table 2. Result of validation Learning Implementation Plan

Aspect	Tendency Score	Category
The form	4	Very valid
Learning activities	3,5	Valid
Supporting learning activities	4	Very valid
Language	4	Very valid

Based on the validation result in Table 2., it can be seen that the Learning Implementation Plan has a tendency score of 4 with very valid category. On the word, the developed of Learning Implementation Plan can be used as a classroom instructional guideline with a few revisions.

Table 3. Result of Textbook

Aspect	Tendency Score	Category
Foreword	3,5	Valid
Content	3,6	Very valid
Characteristics	3,7	Very valid
Ending	3,67	Very valid
Concept explanation	4	Very valid

Based on the result of Textbook validation in Table 3., the average of tendency scores that are obtained from the validators is 3.7 with very valid category. It means that the developed of Textbook can be used as a media-based on ethnoscience Batik to increase the skill of scientific literacy at VII-grade.

Table 4. Result of Student Activity Sheet (SAS)

Aspect	Tendency Score	Category
The form	4	Very valid
Feasibility of content	3,5	Valid
Language	3,33	Valid

Table 4. describes the result of the validators assessment on the Student Working Sheet (SAS) development. The results that are obtained from these aspects have an average of 3.6 with very valid criteria. These are three SAS for 3 times meetings.

Table 5. Scientific Literacy Test Assessment Instrument

Aspect	Tendency Score	Category
Content	3,6	Very valid
Construction	3,6	Very valid
Language	3,78	Very valid

Table 5. presents validation result of scientific literacy test assessment instrument. From three aspects that are assessed, the average score obtained is 3.6 with very valid criteria. This scientific litercy test are used to assess the improvement of scientific literacy skill (after and before the learning takes place).

3.2 Discussion

The validity of teaching material uses a contextual approach (CTL) based on ethnoscience Batik in Tulugagung area. It aims to increase scientific literacy skill, has got a very valid result and can be used for limited trial. The learning uses STAD type of cooperative learning. According to Amedu [11], the cooperative model has a good effectiveness for students because the formation of learning group will be helpful for learners in understanding the materials and raising a positive communication.

The contextual approach contains 7 pillars of CTL, which are constructivisme, inquiry, questioning, learning community, modeling, reflection and authentic assessment [12]. The CTL approach teaches students to use previous experiences and build a new experiences based on existing knowledge, and connect the content that have been learned with life content in which it can be used [13]. Students are expected to find the meaning in the whole of learning process. The meaning that can be taken is the link of "classification material and its changes" with local culture or ethnoscience. The ethnoscience is a knowledge possessed by a society or tribe of nations, which is derived by a particular method and follows certain procedures, which is a part of the tradition of society and its truth can be tested empirically [14]. The scientific learning that is able to be a bridge for learners culture and scientific culture in school will increase the effectiveness of learning process.

In addition to Learning Implementation Plan, teaching material are also equipped with the Textbook. This Textbook is a set of media that is used to help the students in learning activities or autonomous activities. Textbook that can be easily used by students and teachers. The good one of Textbook also should be accordance with the times and must be easy to understand, or does not lead to multiple interpretations [15]. The developing of Textbook is expected to strengthen the scientific concept. It is strengthen by Chung and Keckler's research [16], which says that teachers can encourage the students to explore concepts, vocabulary, and scientific literacy skill from an early age with a book or the quality literature.

In addition to Learning Implementation Plan and Textbook, there are 3 items of Student Activity Sheet (SAS) that can make the students more active during the learning process. It is by working together with their groups, helping students in developing concepts, training in finding, and developing the skills of scientific process, and helping students to obtain the recording of material that have learned through the activities of observations or experiments [17]. The SAS can also be used to train the skill of scientific literacy by referring to scientific activities such as formulating problems, making research hypotheses, analyzing activity or observing data and drawing conclusions. That is in line with Genc's research [18], which says that scientific activity has effective function to improve the skills of scientific literacy and scientific attitudes.

The last one is the Instrument of Scientific Literacy Test Assessment. The instrument contains 15 items of multiple choice and 7 scientific attitude questionnaires. This test is adapted to the theme that is being studied, which is ethnoscience Batik. This assessment instrument comes with a rubric or grid of answer to facilitate the teacher in the assessment process. The literacy questions refer to the PISA 2015 form. The literacy indicators of science that is used largely consist of identifying scientific facts/ phenomena; analyzing scientific evidence/ scientific phenomena related to environmental and scientific research; and using scientific evidence/ research to formulate problems, hypothesize and conclude from scientific research. The indicators are aligned with the specific indicators on each test item.

4. Conclusion

Based on the explanation above, the validation result by the three validators indicates that the teaching material has a very valid category. The teaching material that based on ethnoscience Batik by employing a contextual approach can be tested to try on class VII grade in SMP N 1 Bandung Tulungagung. The function of this research is to increase scientific literacy skill of students on material of "classification materials and its changes".

5. References

[1] Yuliati Y 2017 *J. Cakrawala Pendas* **3** 2442-7470

[2] Daryanto 2014 *Pendekatan Pembelajaran Saintifik Kurikulum 2013* (Yogyakarta: Gava Media)

[3] eBoer G.E 2000 *J. of Research in Education* **37** 582-601

[4] Kurniawati, Ayu A, Wahyuni S, Putra, and Pramudya D A 2017 *J. of Social Science and* Humanity **7**

[5] Nurkhalisa S, Ummayah, and Fivi Fajri Dhotul 2017 *International Journal of Science and* Research **6** 2319-7064

[6] Thiagarajan S, Semmel D, and Semmel M 1974 *Instructional Development for Training Teachers of* Expectional *Children* (Minneapolis: Leader Training Institute/ Special Education, University of Minnesota)

[7] Permendikbud RI Nomor 103 2014 *Tentang Pembelajaran Pada Pendidikan Dasar dan Pendidikan Menengah* (Jakarta: Depdiknas)

[8] Departemen Pendidikan Nasional 2008 *Panduan Pengembangan Bahan Ajar* (Jakarta: Depdiknas)

[9] OECD 2013 *Assesment and Anilitycal Framework Mathematicts, Reading, Science, Problem Solving and Financial Literacy* (Paris: OECD Publising)

[10] Amedu O I 2007 *J of Education and Practice* **8**(10)

[11] Depdiknas 2007 *Panduan Pembelajaran Kontekstual Sekolah Menengah Pertama* (Jakarta: Depdiknas)
[12] Hudson C C, and Whisler V R 2007 *J of Systemics, Cybernetics and Informatics* **4** 1690-4524
[13] Sudarmin 2014 *Pendidikan Karakter Etnosains dan Kearifan Lokal* (Semarang: CV Swadaya Manunggal)
[14] Nieveen N 2010 *"Prototyping to Reach Product Quality" Design Approaches and Tools in Education and Training* (Nederlands: Kluwer Academic Publishers)
[15] Chung M H and Keckler B 2016 *Shared-Book Experience Using Science-Themed Books to Develop Scientific Literacy: An Interactive Approach with Strunggling Readers* (New York: Mercy College)
[16] Achmadi H R 1996 *Telaah Kurikulum Fisika SMU (Model Pembelajaran Konsep dengan LKS)* (Surabaya: University Press)
[17] Genc M 2015 *J of Faculty of Education* **34**(1) 141-152

Acknowledgements
Thanks to all those who helped in this research, especially to validators to validate the learning tools developed. I am also grateful to the students in SMP N 1 Bandung, Tulungagung who has helped this research.

The Argumentation Skills of Junior High School Students on Physical Changes and Chemical Changes

R N Astuti[1,*], S Suyono[2], M Nur[3]

[1]PGMI, Fakultas Tarbiyah dan Ilmu Kependidikan UIN Maliki Malang, Indonesia
[2]Department of Chemistry Education, FMIPA, Universitas Negeri Surabaya, Indonesia
[3]Postgraduate program of Science Education, Universitas Negeri Surabaya, Indonesia

*Corresponding email: riniastuti@mhs.unesa.ac.id

Abstract. The purpose of this study is to investigate the improvement of junior high school students' argumentation skills on the topic of physical changes and chemical changes. The design of this study used one group pretest-posttest design. Data were then analysed using descriptive statistics to describe the students' argumentation skills and inferential statistics to examine students' improvement. The results of the research showed that students' argumentation skills on some indicators were successfully applied: examining claims, showing evidence, making reason, and arranging contra-argument after joining the learning model which consisted of five phases, namely (1) orientation, (2) exploration, (3) pre-argument, (4) arguments and contra-argument, and (5) reflection on limited trials and extensive trials.

1. Introduction

Osborne [1] stated that the goal of science learning in the 21st century is more emphasized on the development of critical thinking skills. Developing critical thinking skills is one of the objectives of science learning in junior high schools in the 21st century [2]. This is also consistent with one of the goals to be achieved in science subjects in junior / MTs based on the Curriculum 2013 which is to develop critical thinking skills [3].

According to Paul [4], argumentation skills are part of critical thinking skills. Another similar opinion by Marttunen, Leena, Litosseliti, & Lund [5] revealed that the skill of argumentation is one of the important competencies required by students because argumentation can develop critical thinking skills. In a similar vein, Liliasari [2], argues that the most fundamental skill in developing critical thinking skills is argumentation skills.

Argumentation skill is important to be trained in the science-learning process. According to Osborne et al. [1], this is because: 1) scientists are found to get used to arguing on certain phenomena to develop and improve their scientific knowledge, 2) community also argue in the process of scientific debate, and 3) students also argue in the learning process to improve their understandings. This opinion is supported by Erduran [6] stating that argumentation skill is an important ability to be trained in science learning. Furthermore, there are three factors that lead to important argumentation skills in science learning: it can improve high-level thinking processes through social interaction activities; train students in constructing scientific arguments supported by relevant evidence through the process of learning, tasks, and modeling of teachers, and improve students' conceptual understanding. Based on these opinions, teachers need to practice argumentation skills for science learning [7]. This is because students who are trained in argumentation skills can improve their argumentation skills.

Several studies have been conducted to find a link between argumentation skill exercises and students' argumentation. Science teaching on the topic of acid and base based on argumentation significantly influenced students' argumentation skills by Sagir & Kihc [8]. Groups of students who participated in the teaching model were better in terms of argument ability compared with the group of students following the conventional learning. The results of this study are relevant to the results of Chen, Lin, Hsu, & Lee [9] research which found that students who attended physics-oriented learning had better argument ability than students who received conventional learning. The same thing was expressed by Khisfe [10] finding that students who are able to construct arguments had better argumentation skills.

The learning model that trains the argumentation skills in this research is the innovation of the QA (Question Argumentation) model developed by Chin and Osborne [11]. The innovation is done by considering learning strategies to develop students' argumentation skills in science learning developed by Simon, Erduran, & Osborne [12]. The results of research conducted by Simon et al. [12[found that a strategy of learning supporting argumentation skill in a science topic completed with six major steps in learning can significantly develop students' argumentation skills. Another result that is also obtained is the reflection step is an important step in the learning of argumentation skills. This is because the reflection step can provide an opportunity for teachers to provide feedback on the learning process and the results of the preparation of arguments by students. The results of research that have been obtained Simon et al. [12], were further developed by Dawson & Venville [13] on biology subjects. The main results of this study were derived from the development of the strategies: (1) providing a step to find evidence used in strengthening the argument, (2) provoking a counterargumentary step that requires students to retain arguments and draw up opposing arguments. Based on the advantages and disadvantages of the QA learning model and considering the argumentative learning strategy proposed by Simon et al. [12] was developed further by Dawson & Venville [13], this study resulted in a learning argumentation model that has a syntax consisting of five phases, namely (1) orientation, (2) exploration, (3) pre-arguments, (4) arguments and counterargument, and (5) reflection to improve junior high school students' argumentation skills.

2. Method

2.1. Research Design

This research is a pre-experimental quantitative studies with one-group pretest-posttest design [14]. The subjects of this research were the junior high school students of class VII in the 2016/2017 academic year at junior high school 14 Surabaya. The total number of the subjects was 30 students. This research was conducted for six meetings with the allocation time each meeting 3 x 50 minutes. The diagram of the one-group pretest-posttest design is shown in Figure 1.

O_1	X	O_2
Initial test	**Treatment**	**Final test**

Figure 1 Pretest-posttest one-group design (Fraenkel & Wallen, 2009)

Note:

O_1 :Initial argumentation skills testing for concepts physical changse and chemical changes
X :Treatment: Model of argumentation skills learning
O_2 :Final test, argumentation skills for concepts physical changes and chemical changes

Before the model of argumentation skills learning is applied, a preliminary test was firstly conducted to see the students' argumentation skills. The model of argumentation skills learning is applied for three learnings, then a final test is conducted to look at differences in student's argumentation skills between before and after learning applying the model of argumentation skills learning.

2.2. Data analysis on students' argumentation skills
Preliminary test data and student final tests were further analyzed using inferential statistical tests. The score level for claims and evidence is based on the evidence shown in support of the claim. The reason-scoring level is based on the logical reasons the students constructed to explain why the proposed evidence can support the claim and the level of the contra-argument scores based on whether the students could set up reasons that support such contra-arguments as proposed by the Zohar & Nemet [15], Acar & Patton [16], and Supeno [17]. The final rubric is used as shown in Table 1.

Table 1 Scoring guideline for students' argumentation skills

Score	Description
	Existence of claims
0	No claim
1	Claim exist, but wrong
2	Correct claim
	Existence of Evidence
0	No evidence
1	Evidence exists and supports the claims (one correct evidence)
2	Evidence exists and supports the claims (two correct evidence)
	Existence of Reasons
0	No reason
1	Reason exist, but does not align with evidence, and incorrect
2	Reason exist aligning with evidence, correct but incomplete
3	Reason exist aligning with evidence, correct and complete
	Existence of Contra-argument
0	No contra-argument
1	Contra-argument exists, but incorrect
2	Contra-argument exists and correct
3	Contra-argument exists and correct, but incorrext reasons
4	Contra-argument exists and correct, corrext reasons but incomplete
5	Contra-argument exists and correct, corrext reasons and complete

The statistical test used is the Wilcoxon Signed Ranking test (nonparametric) [18]. The hypothesis is tested as follows.

H_0: There is no difference in the student's argumentation skills before and after the learning applying model of argumentation skills learning.

H_1: There is difference in the student's argumentation skills before and after the learning applying model of argumentation skills learning.

In the testing, the hypothesis uses a significant level $\alpha = 0.05$.
H_0 rejection criteria is based on T as follows.

H_0 is accepted if $T \geq T\alpha$
H_0 is rejected if $T < T\alpha$

3. Results and discussion

Table 2 Inferential statistical test on the limited trial

Learning code	T-score	T-table	Decision
L-1	0	151	H_0 is rejected H_1 is accepted
L-2	0	151	H_0 is rejected H_1 is accepted
L-3	0	151	H_0 is rejected H_1 is accepted

Table 2 shows the result of Wilcoxon Ranking test from learning 1 to 3 (L-1, L-2, L-3), which indicates that the value of T-score < T-table with $\alpha = 0{,}05$. Therefore, H_0 is rejected, which means that the students' argumentation skills between pre- and post-learning that applying model of argumentation skills learning are not the same, in which case students who acquire argumentation skill-learning by applying model of argumentation skills learning have higher argumentation skills.

Table 3 Inferential statistical test on the extensive trial

	Class A			Class B		
Learning code	**T-score**	**T table**	**Decision**	**T-score**	**T table**	**Decision**
L-1	0	151	H_0 is rejected H_1 is accepted	0	151	H_0 is rejected H_1 is accepted
L-2	0	151	H_0 is rejected H_1 is accepted	0	151	H_0 is rejected H_1 is accepted
L-3	0	151	H_0 is rejected H_1 is accepted	0	151	H_0 is rejected H_1 is accepted

Table 3 shows the calculation of Wilcoxon Signed Ranking test statistic in class A and class B which indicates that T-score < T-table with $\alpha = 0.05$. Therefore, H_0 is rejected. This means that students' pre- and post argumentation skills applying the model of argumentation skills learning in class A and class B are not the same. These results show that students on those three learnings (L-1, L-2, L-3) have better (high) argumentation skills. Thus, science learning that brings about argumentation skills can develop students' argumentation abilities optimal, which in this case support the finding of Chen & She [19]; Sampson & Clark [21]; Zohar & Nemet [15]; Duschl & Osborne [7].

In the orientation phase, the strategy used to teach students to construct claims in this study is to use competing theories strategy. According to the strategy, students construct claims by choosing one of the statements of two controversial statements about the phenomenon to be debated [1,20]. Enhancement of student activity on the worksheet assessed by observers is to make claims using a competing theories strategy, to discuss the results of claims in limited trials and extensive trials. This is to illustrate that students have been able to construct claims as a first step in argumentation skills. This is why the ability to construct claims as an indicator of argumentation skills has improved from learning 1 to 3 after applying the model of argumentation skills learning.

In the exploration phase, the improvement of students' activity to explore concepts physical change and chemical changes, i.e. analyzing the information from the student's book, experimenting to obtain data that can be used as evidence to support the claim, discussing the data that has been obtained to be used as evidence in the limited trial and the extensive trial, investigating evidence and showing evidence used to support the claim. This is what causes the ability to demonstrate evidence as one indicator of argumentation skills improved after the applying model of argumentation skills learning. In addition, the students' abilities are found consistent with the opinions of Katchevich, Hofstein & Naaman [22] where students' argumentation skills can evolve by providing experimental opportunities to help students obtain empirical evidence that can be used in strengthening arguments. In addition to conducting experiments, proof-seeking activities can be undertaken by reviewing teaching materials [23].

Another argumentation skill assessed in this study is the ability to set up the reasons and the contra-argument. The improvement of student activity in drawing on reasons and arguments also occurs, proven by the student's ability to analyze claims, evidence, and reasons into a proper argument diagram, discuss the results of argument diagrams, and construct the counterargument (pre-argument stage). Students' ability to compose reasons and contra-arguments in limited trials and extensive trials is an illustration that students have been able to set up reasons and contra-arguments. This become an indicator of the improvement of argumentation skill after being engaged in the model of argumentation skills learning. The ability of the student in accordance with the opinion of Chin & Osborne [11] stating that students' ability to make claims, show evidence, and make reasons can be drawn through the preparation of argument diagrams to produce a complete argument. Therefore, it can be concluded that students' activities in analysing the claims, evidence, and reasons in an argument diagram and drawing up a counter-argument on applying model of argumentation skills learning can improve students' argumentation skills.

4. Conclusion

The results of this study suggest that there are significant differences in the skills of argumentation between before and after learning applying the model of argumentation skills learning on both the limited trial and the extensive trial. The syntax of the learning model consists of five phases, namely orientation, exploration, pre-arguments, arguments and counterargument, and reflection. There were also improvements of students' argumentation skills in relation to indicators: making claims, showing evidence, preparing reasons and preparing a counter-argument.

5. References

[1] Osborne J, Erduran S and Simon S 2004 *J. Sci. Edu.* **6** 915
[2] Liliasari 2009 http: //file.upi.edu.
[3] Permendikbud Nomor 68 tahun 2013 tentang Kerangka Dasar dan Struktur Kurikulum Sekolah Menengah Pertama/ Madrasah Tsanawiyah.
[4] Paul R W 1992 *J. Dev. Edu.* **2** 34
[5] Marttunen M, Leena L, Litosseliti L and Lund K 2005 *J. Edu. Res and Eval.* **4** 365
[6] Erduran S and Maria P J 2008 *Argumentation in Science Education*. (London: Springer.)
[7] Duschl R A and Osborne J 2002 *J. Stud in. Sci Edu.* **3** 39
[8] Sagir S U and Kihc Z 2012 *Eurasian. J. Phys and Chem Edu.* **2** 139
[9] Chen J, Lin H, Hsu Y and Lee H 2011 *J. Sci. Edu.* **2** 147
[10] Khisfe R 2012 *J.Res in Sci Teach.* **4** 489
[11] Chin M and Osborne J 2010 *J. The Learn. Sci.* **2** 230
[12] Simon S, Erduran S and Osborne J 2006 *J. Sci. Edu.* **2** 235
[13] Dawson M V and Venville G 2010 *J. Res. Sci. Edu.* *2* 133
[14] Fraenkel J R and Wallen N E 2009 *How to design and evaluate reseach in education*. 5th ed. (Boston: McGraw-Hill Companies)
[15] Zohar A and Nemet F 2002 *J. Res in Sci Teach.* **39** 35
[16] Acar O and Patton B R 2012 J. *Soc and Behvi Sci.* **45** 4756

[17] Supeno, Nur M dan Susantini E 2016 Model Pembelajaran Penyelesaian Masalah Argumentasi (PMA) untuk Meningkatkan Hasil Belajar Kognitif Produk, Kognitif Proses, dan Keterampilan Argumentasi Ilmiah Siswa SMK. *Unpublished dissertation.* PPS Unesa.
[18] Djarwanto 1991 Statistik Nonparametrik (Yogyakarta: BPFE Yogyakarta)
[19] Chen C H and She H C 2012 J. *Edu Tech & Soc* .**1** 197
[20] Bell P and Linn M C 2000 *Inter J. of Sci. Edu.* **8** 797
[21] Sampson V and Clark D 2011 *J. Sci. Edu.* **93** 448
[22] Katchevich D, Hofstein A and Naaman R M 2013 J. *Res in Sci. Edu.* **43** 317
[23] Wilson C D, Taylor J A, Kowalski S M and Carlson J 2010 J. *Phys. Edu.* **38** 324

The Application of Inquiry Learning to Train Critical Thinking Skills on Light Material of Primary School Students

Suryanti[1], I S Z Arifin[2], U Baginda[3]

[1]Primary Teacher Education Program of Universitas Negeri Surabaya
[2] Physics Education Program of Universitas Negeri Surabaya
[3] Graduate Program of Universitas Negeri Surabaya

*Corresponding author: suryanti@unesa.ac.id

Abstract. Critical thinking skills are seen as important skills for survival and have become one of Indonesia's educational goals. The results of initial observation of critical thinking skills in Kemangsen 2 primary school involving students of grade V, revealed that 95% students possessed score 2 (with the scale, ranging from 1 to 4). Therefore, critical thinking skills of students need to be improved by applying hands-based learning of Hands on Science; one of which is through inquiry learning. The purpose of this study is to determine the effectiveness of inquiry learning model in trained critical thinking skills of elementary students (n = 22). This research was conducted by One Group *Pretest-Posttest* Design. Data were analyzed using t-dependent tests for overall critical thinking skills and descriptive analysis for critical thinking skills for each indicator. The critical thinking skills trained in this study include evaluation, explanation, interpretation, and inference based on the results of data analysis. To sum up, inquiry learning is classified as effective in terms of training students' critical thinking skills. The critical thinking skills of students should be continuously trained on appropriate materials or topics.

1. Introduction

Critical thinking is seen as an important skill for the survival of the 21st century [1]. The skills to think critically is not only necessary in analyzing the various information but also important in the individual and social aspects [2]. Critical thinking helps to decide about a person's way of life. In everyday life, we need to have high selection skills in dealing with different situations in life [3]. All these capabilities make it possible for greater success at school, at work and at home [4]. The skills to think critically will form a critical attitude is actually meant to invite us to think clearly to help ourselves get the right knowledge and understanding [5].

Critical thinking is seen as an essential skill for 21st century survival and education experts agree that encouraging students to think critically is an essential requirement of the educational system [6]. Rahma [7] critical thinking skills are one of the most important educational goals in all levels of education. Critical thinking skills is unnatural innate in every human being since birth so it can be taught to the students [8]. Critical thinking can be taught, and not just for "talented" students but for all students [9]. The skills to think critically can be improved and developed by way of practice [10]. Students may not be able to think critically when their teachers are unable to integrate sufficient critical thinking into the practice of daily learning [11]. Teachers in this case have an important role in trained skills to think critically to students at every level of education. [12] the skills to think critically should be developed from an early age. In elementary level, critical thinking skills can be taught to fourth, fifth and sixth graders because it is in accordance with students' cognitive development at the abstract operational stage [13]. Critical thinking skills is not only a skill that needs to be taught in schools but has become one of the goals of Indonesian education should be trained.

Science is concerned with a systematic way of finding out about nature, so science is not just a collection of knowledge in the form of facts, concepts, or principles but also a process of discovery [14]. Filsaime [15] argues that critical thinking skills are needed in digging and getting to know the science so that students can master the workings taken in studying nature and solving problems related to daily life. Contextual thinking skills are contextual so that their application in science education includes a focus on tasks, problems, and issues that exist in the curriculum that require critical thinking skills [16]. From the above description, it shows that critical thinking skills can be developed on the materials that have been available in the curriculum that have characteristic fit with critical thinking skills.

Critical thinking must go through argumentation, reasoning, and inferences [22]. From the exposition, the trained aspects must contain elements of argumentation (argument evaluation), explanation and inference (inference and interpretation). The critical thinking indicators to be investigated include explanation, interpretation, inference, and evaluating arguments. [23] Inquiry learning is a dynamic learning approach to which involves exploring, asking, making discoveries, and testing discoveries to seek new insights. Hands-on science learning such as inquiry can help students think critically and give students confidence in their skills to solve a problem [17]. Inquiry learning is student center learning [18] to build critical thinking skills, reasoning and subsequent creativity [19]. [20] explaining that inquiry-based learning techniques can encourage students' critical thinking skills. [21] inquiry learning is a dynamic learning approach to which involves exploring, asking, making discoveries, and testing discoveries to seek new insights. The focus of this research is to describe the critical thinking skills in the topic of the properties of light by using learning inquiry mend the skills of critical thinking in students.

2. Methods

The research method used in this research is quasi experiment or Weak Experimental Design, with the basic pattern "The One Group Pretest - Posttest Design" and descriptive method. Subjects in this study are the students of grade V of Kemangsen primary school in the academic year 2014-2015. The subject of the research are 22 students, with 12 male students and 10 female students. Critical thinking skills trained using inquiry learning in light topic. This research was conducted on light material in science teaching by using inquiry model. The scoring criteria used are the modified rubric score from Facione [24] (the solo taxonomic rubric as Table 1 follows).

Table 1. Indicator of Critical Thinking Skill of Student Science through Inquiry Learning.

Indicator	Information	Score
Interpretations	Students can't make interpretations the data provided	0
	Students make interpretations of the data provided but wrong and not in accordance with the context of the problem	1
	Students interpret the data given incorrectly but in accordance with the context of the problem	2
	Students interpret the data given correctly and according to the context of the problem but not complete	3
	Students interpret the data correctly, according to the context of the problem and complete	4
Inference	Did not make any conclusions.	0
	Make inaccurate conclusions and not in accordance with the context of the	1
	question.	2
	Make inaccurate conclusions, though tailored to the context of the question.	3
	Make conclusions appropriately, according to context but not complete.	4
	Make conclusions appropriately, contextually and completely	
Explanation	Did not make any explanation.	0
	Make an inappropriate explanation and not in accordance with the context of the question.	1

Indicator	Information	Score
	Make an inappropriate explanation even if adapted to the context of the question.	2
	Make annotations appropriately, in context but not complete.	3
	Make Explanations appropriately, contextually and exhaustively	4
Evaluate arguments	Do not evaluate arguments from statements given	0
	Provide an evaluation of the arguments of the statements given but are not appropriate and not in accordance with the context of the problem.	1
	Provide an evaluation of the arguments of the statements given incorrectly, even though they are adapted to the context of the problem.	
	Provide an evaluation of the arguments from the statements given correctly, logically but incompletely.	2
	Provide an evaluation of the arguments from the statements given correctly, logically and completely.	3
		4

Data analysis is divided into 2 parts, namely analysis of each indicator of critical thinking skills and analysis of critical thinking skills as a whole. The percentage of critical thinking skills obtained from the calculation is then categorized according to Table 2 below.

Table 2. Category Percentage of Critical Thinking Skills [25].

Interpretations	**Category**
81,25 < X ≤ 100	Very high
71,5 < X . 81,25	High
62,5 < X ≤ 71,5	Medium
43,75 < X . 62,5	Low
0 < X ≤ 43,75	Very Low

Analysis of critical thinking skills as a whole is done by using hypothesis testing using t-test. Before, performing the t-test it is necessary to test the normality to test that the sample data comes from normally distributed populations or not. Normality test used in this study using the Kolmogrov-Smirnov normality test. If the normality test obtained normal distributed data then the data calculation using parametric statistical method that is using test-t dependent sample or Paired t-Test.

3. Results and Discussion

3.1. The result of critical thinking skills of each indicator.

The result of students' critical thinking skills of class V can be known from the evaluation result done at a pretest and a posttest. The result of evaluation of students 'critical thinking skills for each indicator in Table 3 below is done by using descriptive method measured based on the scoring guidance of students' critical thinking skills.

Table 3. Frequency Distribution of Critical Thinking Skill in a Pretest

		Indicator							
Indicator	Category	Interpretations		Inference		Explanation		Evaluate arguments	
		F	%	F	%	f	%	F	%
81,25 < X ≤ 100	Very high	0	0	0	0	0	0	0	0
71,5 < X . 81,25	High	0	0	0	0	0	0	0	0
62,5 < X ≤ 71,5	Medium	0	0	0	0	0	0	0	0
43,75 < X . 62,5	Low	3	13,63	5	28,73	4	18.18	5	28,73
0 < X ≤ 43,75	Very Low	19	86,37	17	72,27	18	81,82	17	72,27
		22	100	22	100	22	100	22	100

Based on Table 3, critical thinking skills each indicator on the pretest of each indicator has a very low and low category. This result indicates that the capacity critical thinking skill student is very low. This was not attributed to the role of teachers who had not yet maximally and sustainable trying to involve the students in the activities of a critical thinking skill can even standard intellectual a basis for the capacity critical thinking skill usually not to be taught in schools. The Frequency Distribution of Critical Thinking Skill in a Posttest is shown in Table 4 below.

Table 4. Frequency Distribution of Critical Thinking Skill in a Posttest

Indicator	Category	Indicator							
		Interpretations		Inference		Explanation		Evaluate arguments	
		F	%	F	%	F	%	F	%
81,25 < X ≤ 100	Very high	16	72,73	17	77,27	18	81,82	14	63,63
71,5 < X . 81,25	High	6	27,27	5	22,73	2	9,10	3	13,63
62,5 < X ≤ 71,5	Medium	0	0	0	0	1	4,54	2	9,10
43,75 < X . 62,5	Low	0	0	0	0	1	4,54	2	9,10
0 < X ≤ 43,75	Very Low	0	0	0	0	0	0	1	4,54
		22	100	22	100	22	100	22	100

Table 4 shows that learning inquiry having a fairly significant effect to train critical thinking skill. One pf the four indicators, indicators evaluate argument having the lower in of appeals that other. Analytical capability this requires a exercise seamless and intellectual level adequate. Learning inquiry to boost the ability of investigation requires students to connect the investigation by scientific knowledge causing the students to uses reasoning scientific and critical thinking in developing concepts on the science. The data of critical thinking skills of each indicator presented in the following table 5.

3.1.1. Indicator 1: Interpretation.

Based on Table 5, it can be seen that there is a significant increase in interpretation skills after inquiry learning. In the phase of collecting and processing student data in demand to be able to do 2 aspects of interpretation are: 1) linking the results of observations and 2) finding the pattern of data relationships in observation. This is in line with the opinion of Carlson [26] which states that inquiry learning applied in the learning process can improve students' skills in making observations and put forward answers to a problem through data interpretation to obtain conclusion. Inquiry learning not only requires students to be able to conduct the investigative process independently, but also requires students to be able to understand the implications of an experimental result. The Percentage of critical thinking skills of each indicator is shown in Table 5 below.

Table 5. Percentage of critical thinking skills of each indicator

Indicator critical thinking skills	Pretest		Posttest	
	Percentage (%)	Category	Percentage (%)	Category
Interpretations	36,36	Very Low	85,27	Very high
Inference	38,59	Very Low	86,37	Very high
Explanation	39,02	Very Low	84,06	Very high
Evaluate arguments	34,45	Very Low	78,02	High

3.1.2. Indicator 2: Inference.

Based on these data it can be concluded that students' critical thinking skills of inference increases significantly after inquiry learning is done on the students. Inference is one of the activities needed in inquiry learning especially in phase IV collecting and processing data. [27] which states that inquiry learning requires direct experience of inquiry and practice on an ongoing basis so that students are not limited to understanding questions by learning words like "hypothesis" and "inference" or with memorize procedures but students gain a deep understanding of the characteristics of the steps of the scientific method.

3.1.3. Indicator 3: Explain.

Based on the data it can be concluded that the skills Student's critical thinking indicator improved significantly after inquiry learning was conducted on the students. The high critical thinking skills of students in this indicator· is not separated from the role of inquiry learning in the fifth phase of presenting the results of the investigation. In this phase students are given the opportunity to explain the results of their inquiry directly by presenting in front of the class or indirectly by using the question, given the opportunity to obtain a meaningful understanding of the concept [27] reveals that the inquiry learning model develops understanding skills.

3.1.4. Indicator 4: Evaluate the argument.

The indicator evaluating the argument is an indicator of critical thinking skills that obtains the lowest percentage of skills compared to other indicators trained in both pretest and posttest. Based on these data it can be concluded that students' critical thinking skills to evaluate arguments increased significantly after inquiry learning was done on the students. The teacher provides some statements or arguments related to the investigation that have been done then the students are instructed to be able to evaluate the given argument whether the argument whether the argument is the 2000 National Research Council (NRC) in Olson and Horsley [27]. Inquiry learning model can train students to build answers and think smartly in finding various alternative solutions to the problems posed by teachers.

3.2 Outcomes of overall critical thinking skills

Overall critical thinking skills were analyzed using N-gain and t-test. In table 6 can be concluded there is improvement of critical thinking skills of student before and after conducting inquiry study. This finding is in line with Estes [28] explaining that inquiry learning is capable of developing critical thinking skills and abilities in science literacy. Martin [29] which states that students who are taught science by using inquiry in primary school can develop literacy skills in science, independent thinking, critical thinking, and problem solving. In improving students' critical thinking skills. The Data Paired Samples Test critical thinking Skill is shown in Table 6 below.

Table 6. Data Paired Samples Test critical thinking Skill.

Paired Samples Test

	Paired Differences							
				95% Confidence Interval of the Difference				
	Mean	Std. Deviation	Std. Error Mean	Lower	Upper	T	df	Sig. (2-tailed)
Pair 1 pretest – posttest	-46,30591	7,95090	1,69514	-49,83114	-42,78068	-27,317	21	,000

4. Conclusion

Inquiry learning can be used to train students' critical thinking skills. Inquiry learning students will be actively involved in learning both physically and mentally (hands on activity) through experimenting activities, observing, asking, analyzing data, and making conclusions. Student centered inquiry learning has been shown to improve students' critical thinking skills. Critical thinking skills of the students to continue to develop should be to trained critical thinking skill continuously adjusted with the characteristics of learning that one of them using inquiry learning.

Reference

[1] Corich SP. 2011. *Automating the Measurement of Critical Thinking in Discussion Forum.* Disertasi. Palmerston North. Massey Universities. Available from:

[2] Hashemi SA. 2011. *The Use of Critical Thinking in Social Science Textbooks of High School: A field Study of Fars Province in Iran*. International Journal of Instruction. Vol. **4**, No. 1, 63-77. Available from: https://eric.ed.gov/?id=ED522910

[3] Hashemi, SA. 2011. *The Use of Critical Thinking in Social Science Textbooks of High School: A field Study of Fars Province in Iran*. International Journal of Instruction. Vol. **4**, No. 1, 63-77. Available from: https://eric.ed.gov/?id=ED522910

[4] Starkey L, *Critical Thinking Skills Succes in 10 Minutes A Day*. Second edition. New York: Express Learning. 2010.

[5] Molan M, *Logika Ilmu dan Seni Berpikir Kritis*. Jakarta: PT. Indeks. 2012.

[6] Corich SP, *Automating the Measurement of Critical Thinking in Discussion Forum*. Disertasi. Palmerston North. Massey Universities. 2011. Available from:

[7] Rahma AN, *Pengembangan Perangkat Pembelajaran Model Inkuiri Berpendekatan SETS Meteri Kelarutan dan Hasil Kali Kelarutan untuk Menumbuhkan Keterampilan Berpikir Kritis dan Empati Siswa Terhadap Lingkungan*. Vol. **JERE** 1, No. 2, 133-138. Journal of Educational Research and Evaluation. 2012

[8] Fahim M, *Manipulating Critical Thinking Skills in the Test Taking*. Vol. **4**, No. 1, 153-160. International Journal of Education. 2012.

[9] Ruggiero RV, *Becoming a Critical Thinker: A Master Student Text*. A Sixth edition. New York: Houhton Mifflin Company. Choy, S. C. dan San Oo, P. 2012. *Reflective Thinking and Precursor for Incorporating Critical Thinking into the Classroom?*. Vol. **2**, (1), 167-182. International Journal of Instruction. 2009.

[10] Fahim M, *Manipulating Critical Thinking Skills in the Test Taking*. Vol. **4**, (1), 153-160. International Journal of Education. 2012.

[11] Choy SC and San OoP. *Reflective Thinking and Precursor for Incorporating Critical Thinking into the Classroom?*. Vol. 2, (1), 167-182. International Journal of Instruction. 2012.

[12] King A, *Designing The Instructional Process To Enhance Critical Thingking Across The Curiculum. Inquiryng Minds Really Do Want to Know : Using Questioning to Teach Critical Thinking*. Vol.**22**, (1), 13-17. Society For The Teaching Of Psicology. 1995.

[13] Adams JW, *A Case Study: Using Lesson Study to Understand Factors That Affect Teaching Creative and Critical Thinking in the Elementary Classroom*. Disertasi. Philadelphia: Drexel University. 2013. Available from: http://idea.library.drexel.edu/bitstream/1860/4131/1/Adams_JanetEdD.pdf

[14] BNSP, *Standar isi 2006 mata pelajaran IPA SD*. Jakarta. Depdiknas. P.484. 2006.

[15] Filsaime and Dennis K, *Menguak Rahasia Berpikir Kritis dan Kreatif.* Jakarta: Prestasi Pustakaraya. 2008.

[16] Bailin S, *Critical Thinking and Science Education. Journal Science and Education*. Vol **11**, 361-375. 2002.

[17] Molan M, *Logika Ilmu dan Seni Berpikir Kritis*. Jakarta: PT. Indeks. 2012.

[18] Spencer TL and Walker TM, Creating a Love for Science for Elementary Students through Inquiry-based Learning. Vol. **4**, (2), 18-24. *Journal of Virginia Science Education.* 2011.

[19] Paulu N and Martin M. 1992. *Helping Your Child Learn Science*. New York: U.S. Departement of Education Office of Educational Reasearch and Improvment. 1992.

[20] Spencer TL and Walker TM, Creating a Love for Science for Elementary Students through Inquiry-based Learning. Vol. **4**, (2), 18-24. *Journal of Virginia Science Education.* 2011.

[21] Ibrahim M Dkk, *Dasar Proses Belajar Mengajar : Hakikat IPA, Kurikulum Tingkat Satuan Pendidikan, Media Pembelajaran, Perancangan Pembelajaran sampai kepada ciri Guru Profesional*. Surabaya : Unesa Press. 2010.

[22] Lampert N 2011 A study of an after-school art programme and critical thinking *International Journal of Education through Art* **7**(1) pp 55-67

[23] Spencer TL and Walker TM 2011 Creating a Love for Science for Elementary Students through Inquiry-based Learning *Journal of Virginia Science Education* **4**(2) 18-24.

[24] Facione PA 1990 *Critical Thinking: A Statement of Expert Consensus for Purposes of Educational Assessment and Instruction* (The Delpy Report. California: California Academic Press)
[25] Studijos K 2006 Developing Critical Thinking through Cooperative Learning *Studies About Languages* **NR**(9) 77-84
[26] Lampert N 2011 A study of an after-school art programme and critical thinking *International Journal of Education through Art* **7**(1) pp 55-67.
[27] Parr B and Edwards MC 2004 Inquiry Based Instruction in Secondary Agricultural Education: Problem Solving an Old Friend Revisited *Journal of Agricultural Education* **45**(4) pp. 106-117
[28] Setyowati A 2011 Implementasi Pendekatan Konflik Kognitif dalam Pembelajaran Fisika untuk Menumbuhkan Kemampuan Berpikir Kritis Siswa Kelas VIII Jurnal Pendidikan Fisika **7** 89-96
[29] Sanjaya W 2010 *Strategi Pembelajaran Berorientasi Standar Proses Pendidikan* (Jakarta: Kencana Prenada Media Group)

IOP Conf. Series: Journal of Physics: Conf. Series **1108** (2018) 012129 doi:10.1088/1742-6596/1108/1/012129

The Use of Web Technologies for Practice Based Learning Model in Vocational Education: A Literature Review

Kustori[1,*], Rusijono[2] and A Mariono[3]

[1,2,3] Educational Technology Department, Universitas Negeri Surabaya, INDONESIA
[*]Corresponding email: kustorikustori@mhs.unesa.ac.id

Abstract. Practice-based learning model is about to making the learning environment as realistic as possible and requires students to demonstrate what they know to link theory and practice. The nature of web technologies such as user centered, flexibility, mobility and multimedia capability are the key features that make web technology as a potential match for practice based learning. Our objectives were to identify the use of web technologies for practice based learning model in vocational education and to summarize strengths and limitations identified in the literature. The research methodology for this study was to review published studies and research on practice-based learning and web technologies in vocational education, the range of which included literature reviews and empirical research. We searched for the term "web technologies" in combination with "practice based learning" and "vocational education" in various literatures databases, and looked for additional studies by key authors. Searches finds that web technologies has been widely used for practice based learning model in vocational education in developed country in America, Europe, and Australia. This comprehensive literature review concludes with some of the advantages, drawbacks, and best practice of web technologies for practice-based learning model.

1. Introduction

This article explores the use of web technologies for practice based learning model in vocational education. Practice-based learning model is about making the learning environment as realistic as possible and requires students to demonstrate what they know to linking theory and practice [1]. It combines the current and innovative educational curriculum of the school's academic degree programs with real world experiences [2]. It is explicitly designed to relate to professional practice standards [3]. Many vocational education institution have been adopted it as part of their curriculum. But the requirement of practice-based learning to linking theory and practice is not easy to implement. The institution should put time, cost, personal, and many other things into consideration.

Web technologies may help vocational education institution to address the issue. The nature of web technologies such as user-centered, flexibility, mobility and multimedia capability are the key features that make web technology as a potential match for practice based learning. Although web technologies has been used widely used in many education and work field, It requires enhanced skills to engage learners in meaningful interaction and to overcome the transactional distance of online learning[4]. This comprehensive literature review concludes with some of the advantages, drawbacks, and best practice of web technologies for practice-based learning model.

2. Methods

The research methodology for this study was to review published studies and research on practice-based learning and web technologies in vocational education, the range of which included literature reviews and empirical research. We searched for the term “web technologies” in combination with “practice based learning” and “vocational education” in various literatures databases, and looked for additional studies by key authors.

The primary literature sources were journal articles and full texts. The major sources we used for data collection were online databases including ERIC, EBSCO, PsycINFO, ContentFirst, Google Scholar, SAGE Online, Project Muse, Education Full Text, and Academic Search Premier.

3. Results and Discussion

3.1. The essential principle of practice-based learning

There is no single universally agreed definition of practice-based learning; it is interconnected with and interdependent on other professional activities such as service development and regulation, as well as educational activities such as assessment and accreditation. In order to understand the practice-based learning, it is important to know its essential principles[5]. The essential principles of practice-based-learning are adaptive curriculum, workplace experience, and stakeholder collaboration.

3.1.1. Adaptive Curriculum

In order to overcome skills mismatch, the curriculum of practice-based learning must be adapted to the needs of workplace. This principle has been indirectly used in many countries such as Australia, Ireland, Finland, etc [6]. It is clear that the need of workplace keep changing overtime and the curriculum need to adapt to that situation.

Australia has been moving from a strategic planning model to a student demand driven system where Each state and territory has adapted different skills forecasting methods to meet local needs. National forecasts by industry were converted into regional forecasts, broken down into the 341 occupational unit groups of the Australian Standard Classification of Occupations. These were then used to determine the employment outlook for workers by age, sex, qualifications and hours worked per week.

Ireland use two mechanisms to anticipating skills needs. First, The ‘Expert Group on Future Skills Needs’ includes representatives of social partners, government departments, industrial development organizations and education and training bodies. Its objectives include identifying skills needs, developing techniques that will assist skills forecasts and advising on decisions related to training policies. It produces long-term forecasts, as well as projections of future demand by occupational groups under different growth. Two, The ‘FÁS/ESRI Manpower Forecasting Programme’ aims to provide information on the changing pattern of occupations and to identify skills needs in broad occupational fields.

Finland Anticipating Educational Needsby using labour force projections, it aims to forecast demand for new recruits by occupational groups and the supply of new job-seekers. Based on these forecasts, the anticipated educational need is determined by occupational field and level of education.

3.1.2. Workplace Experience

The employees valued work experience as the main source of their competence. They also developed their competence mainly through learning at work which highlight he importance accorded to experience in competence and in workplace learning [7]. As part of the curriculum, students often take part in work experience or work placement. Students benefit from the partnerships established between industry, enterprises, unions and schools that better prepare them for the world of work[8].

Students are placed with employers primarily to observe and learn – not to undertake activities which require extensive training or experience. Students need to participate in a variety of workplace learning activities while they are still at school to assist in their development of a realistic understanding of the world of work. These activities are an integral part of their learning and career development and help students to: develop employability skills; explore possible career options; understand employer expectations; and increase their self-understanding, maturity, independence and self-confidence[9].

3.1.3. Stakeholder collaboration
Successful practice demands the involvement of all stakeholders from vocational education field. Because Practice-based learning is interconnected with and interdependent on other professional activities such as service development and regulation, as well as educational activities such as assessment and accreditation. Partnerships involving all stakeholders and services should be established, formalized and co-ordinated. This includes establishing co-operation structures with local companies for practical training and/or jobs after graduation. Stakeholders are committed to quality assurance and improvement strategies.

3.2. Indicator of practice-based learning
Practice-based learning has several indicator [10], including but not limited to:

- Students are learning in the workplace in either paid or unpaid positions.
- Students are learning through participation in industry-based tasks and/or projects as part of the curriculum.
- Students are learning about skills for career management as part of their curriculum.
- Students are learning through activities in class time that emulate activities undertaken by a professional (e.g. laboratories have appropriate standards of dress, safety procedures, standard operating procedures, quality control).
- Students are learning skills in using software tools that are identical to that used by practicing professionals.
- Students are learning through simulations of workplace environments as part of their curriculum.
- Students are assessed and given feedback on Graduate Qualities indicators that have been mapped on to professional competencies.
- Academic staff teaching in the course is practicing professionals with opportunities to share their experiences during classes students are able to connect with their professional body - and their ethics; code of conduct - as part of curriculum activities.

3.3. Web technologies for practice-based learning
The higher education is in the process of moving from face-to-face courses using objectivist, teacher-centered pedagogy and offered by tens of thousands of local, regional, and national universities; to online and hybrid courses using digital technologies to support constructivist, collaborative, student-centered pedagogy, offered by a few hundred "mega-universities" that operate on a global scale [11].

3.3.1. Underlying technologies
Underlying these methodologies is digital support for the techniques of voting, scaling, hypertext, visualization, and the structuring of collaborative communication protocols and the structuring, filtering, and organization of collaborative discourse content. These technologies may be used in the following manner:

- Voting to direct or focus the discussion on areas of group differences and to allow for dynamic (ongoing) changes in evaluation of contributed material;
- Scaling to promote collective understanding of the group's views, degrees of agreement, and shared meanings;

- Hypertext (the two-way linking and typing of both links and nodes) to allow the construction and expression of complex relationship structures and individual and collective cognitive maps;
- Visualization to develop the functional equiva- lent of the periodic table of the elements for all other fields of human endeavors;
- Communication protocol structuring to allow for equality of participation by type of communications structuring; and
- Content structuring to allow asynchronous con- tributions to be automatically categorized and organized and to facilitate individual problem solving within a group process.

3.3.2. Advantages

There are benefits to the students, the organizations, and to the society, as well as more direct cost-benefit factors. The major driving forces for digital substitution processes include:

- The value to the student is the flexibility of being able to integrate education with the demands of work and family.
- Learning effectiveness in online (ALN) or blended courses is equal to or better than in entirely face-to-face courses.
- The value to the instructor is being able to treat all students equally, and to prepare and deliver the materials of the course as a single entity.
- The value to the organization is not having to duplicate any administrative or support function as a separate entity for distance learning.
- The growing competitive environment in higher education and the need to provide quality online instruction as a matter of long-term survival.

3.3.3. Drawbacks

Despite the claims that web technology can improve the education quality, in the context of practice-based learning, it still has drawbacks.

- Vocational education requires special attention when offered in an online medium when compared to other fields of education[11]. This study was done in 2005 and makes a strong argument for the use of online instructional technologies to teach engineering courses with the improvements in technology.
- The absence of vital personal interactions, not only between learners and instructors, but also among colleague learners [12].
- Not all fields or discipline can employ the e-learning technique in education. For instance the purely scientific fields that include practical cannot be properly studies through e-learning. Researches have argued that e-learning is more appropriate in social science and humanities than the fields such as medical science and pharmacy, where there is the need to develop practical skills.
- E-learning may also probably be misled to piracy and plagiarism, predisposed by inadequate selection skills, as well as the ease of copy and paste.

3.3.4. Discussion

Despite some challenges discussed, its adoption in some institutions has increased faculty and learner's access to information and has provided a rich environment for collaboration among students which have improved academic standards [13]. By 2004 at least two million higher-education students in the U.S. were engaged in distance education utilizing various ALN technologies where whole classes can engage in a continuous discourse and group project work independent of time, place, and synchronous constraints of participation [14]. Over the past decade, the number of online courses and programs has grown tremendously [15, 16, 17]. With increased demand for online learning as well as more institutions of higher learning striving to provide diverse educational opportunities, it indicates that web technologies continues to grow as a viable means of providing increased access to a greater number of students [15, 18]. Although online learning continues to grow rapidly, developers and deliverers of online learning need more understanding of how students perceive and react to elements of online learning along with how to apply these approaches most effectively to enhance learning [19].

4. Conclusion

This literature review concludes that web technologies can be used for practice based learning with advantages and drawbacks. There are many challenges to overcome in the implementation, but the benefit of web technologies are greater than the benefits of traditional learning if it is used and applied in proper ways. Futhermore, Future research incorporating the implementation of web technologies, ideally in combination with other technologies for practice based learning model in vocational education, would be of value.

References

[1] University of South Australia 2016 *Practice based learning*. Retrieved from http://w3.unisa.edu.au/academicdevelopment/engagement/practice.asp.
[2] Colorado School of Public Health 2017 *Practice-Based Learning*. Retrieved June 10, 2017, from publichealth.colostate.edu: http://publichealth.colostate.edu/practice-based-learning/
[3] QAA Scotland 2011 *Key aspects of practice-based learning in teaching, nursing and social work in Scotland* (Glasgow: The Quality Assurance Agency Scotland) Retrieved from http://www.qaa.ac.uk/en/Publications/Documents/practice-based-learning-guide.pdf
[4] Sargeant J, Curran V and Allen M 2006 Facilitating Interpersonal Interaction and Learning Online: Linking Theory and Practice *The Journal of Continuing Education in the Health Professions* **26**(1) 128-136
[5] Kustori 2018 The Essential Principles of Practice-Based Learning *Advances in Social Science, Education and Humanities Research* **173**(1) 170-172
[6] OECD 2009 Learning For Jobs: OECD Policy Review Of Vocational Education And Training (Paris: OECD) Retrieved from http://www.oecd.org/edu/skills-beyond-school/43898253.pdf
[7] Paloniemi S 2006 Experience, competence and workplace learning *Journal of Workplace Learning* **18**(7) 439-450
[8] Goverment of South Australia 2017 *Workplace learning for students*. Retrieved from Department for Education and Child Development: https://www.decd.sa.gov.au/teaching/curriculum-and-teaching/numeracy-and-literacy/workplace-learning-students. (June 10, 2017)
[9] Victoria State Government 2017 *Work Experience and Structured Workplace Learning*. Retrieved from Victoria State Goverment. (July 07, 2017)
[10] University of South Australia 2016 *Practice based learning*. Retrieved from http://w3.unisa.edu.au/academicdevelopment/engagement/practice.asp.
[11] Bourne J, Harris D and Mayadas F 2005 Online engineering education: Learning anywhere, anytime *Journal of Engineering Education* **94**(1) 131-146.
[12] Young J R 1997 Rethinking the Role of the Professor in an Age of High-Tech Tools *The Chronicle of Higher Education* **44**(6)
[13] Arkorful V and Abaidoo N 2014 The role of e-learning, the advantages and disadvantages of its adoption in Higher Education *International Journal of Education and Research* **12**(2)
[14] Allen E and Seaman J 2004 *Entering the Mainstream: The Quality and Extent of Online Education in the United States, 2003 and 2004*. Sloan Center for Online Education at Olin and (Babson Colleges: Needham, MA.)
[15] Allen I E and Seaman J 2008 Staying *the Course: Online Education in the United States*. (Needham MA: Sloan Center for Online Education)
[16] Sugar W, Martindale T and Crawly F 2007 One professor's face-to-face teaching strategies while becoming an online instructor *The Quarterly Review of Distance Education* **8**(4) 365-385
[17] Waits T and Lewis L 2003 *Distance education at degree granting post secondary institutions: 2000-2001*. U.S. Department of Education, National Center for Educational Statistics, NCES 20003-017 (Project Officer: Bernard Greene. Washington, DC.)

[18] Saba F 2005 Critical issues in distance education: A report from the United States *Distance Education* **26**(2) 255-272
[19] Koohang A and Durante A 2003 Learners' perceptions toward the web-based distance learning activities/assignments portion of an undergraduate hybrid instructional model *Journal of Informational Technology Education* **2** 105-113

Institute of Physics
Dirac House, Temple Back
Bristol BS1 6BE UK

ISSN: 1742-6588
ISBN 978-1-5108-7714-6

CURRAN ASSOCIATES INC.
proceedings.com

9th Russian National Conference on Irreversible Processes in Nature and Technics 2017 (9RNC-IPNT)

Journal of Physics: Conference Series Volume 918

Moscow, Russia
25 – 28 January 2017

IOP
Institute of Physics